HANDBOOK OF SEA-LEVEL RESEARCH

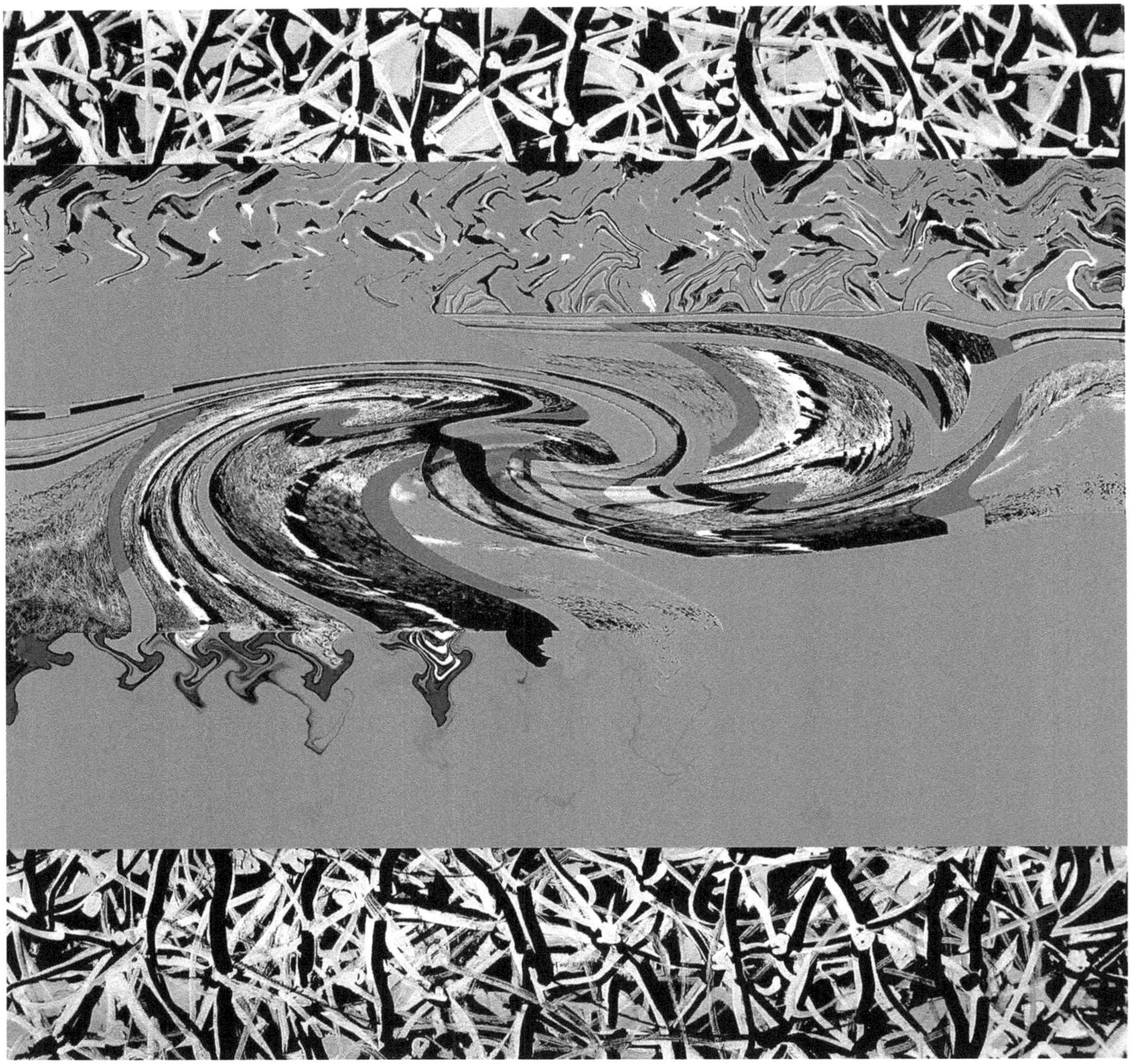

"not in spatial chaos but by time"

Handbook of Sea-Level Research

EDITED BY

Ian Shennan, Antony J. Long, and Benjamin P. Horton

AGU American Geophysical Union WILEY

This edition first published 2015 © 2015 by John Wiley & Sons, Ltd

Registered Office
John Wiley & Sons, Ltd, The Atrium, Southern Gate, Chichester, West Sussex, PO19 8SQ, UK

Editorial Offices
9600 Garsington Road, Oxford, OX4 2DQ, UK
The Atrium, Southern Gate, Chichester, West Sussex, PO19 8SQ, UK
111 River Street, Hoboken, NJ 07030-5774, USA

For details of our global editorial offices, for customer services and for information about how to apply for permission to reuse the copyright material in this book please see our website at www.wiley.com/wiley-blackwell.

Library of Congress Cataloging-in-Publication data has been applied for.

A catalogue record for this book is available from the British Library.

Wiley also publishes its books in a variety of electronic formats. Some content that appears in print may not be available in electronic books.

Cover Image: Girdwood marsh and ghost forest, Turnagain Arm, Alaska. The Mw 9.2 earthquake of March 27, 1964 caused ~1.5 m subsidence, death of the trees and the onset of tidal flat sedimentation on the top of the forest peat soil. Land uplift and sedimentation since 1964 aided recolonization of the tidal flat sediment by marsh plant communities in less than 50 years. Sediments from cores through the marsh sediments reveal evidence of six previous great earthquakes in the last 4000 years. Photograph by Ian Shennan, September 2013.

Set in 10/12pt Melior by SPi Publisher Services, Pondicherry, India

1 2015

The frontispiece "not in spatial chaos but by time" is a composition by Eleonora Tammes based on one of her paintings – title: "embrace within I" 45 × 55 cm (2012) acrylic, paintmarker, graphite on paper (from the Break Away series), and a set of cross section photographs (Pattagansett River Marsh & Menunketesuck River Marsh, CT) created following Orson's typical coring method ("Orsonian method").
Source: Reproduced with permission of Eleonora Tammes

Contents

List of contributors vii
Preface xi
About the companion website xiii

1. Introduction 1
   Ian Shennan, Antony J. Long, and Benjamin P. Horton

2. Handbook of sea-level research: framing research questions 3
   Ian Shennan

PART 1: FIELD TECHNIQUES FOR SEA-LEVEL RECONSTRUCTION

3. Pre-fieldwork surveys 29
   Robert C. Witter

4. Coastal sediments 47
   Alan R. Nelson

5. Geomorphological indicators of past sea levels 66
   Harvey M. Kelsey

6. Coastal caves and sinkholes 83
   Peter J. van Hengstum, David A. Richards, Bogdan P. Onac, and Jeffrey A. Dorale

7. Coral reefs 104
   Yusuke Yokoyama and Tezer M. Esat

8. Coral microatolls 125
   Aron J. Meltzner and Colin D. Woodroffe

9. Archeological and biological relative sea-level indicators 146
   Christophe Morhange and Nick Marriner

10. GPS and surveying 157
    James Foster

11. Reference water level and tidal datum 171
    Sarah A. Woodroffe and Natasha L. M. Barlow

PART 2: LABORATORY TECHNIQUES

12. Techniques and applications of plant macrofossil analysis in sea-level studies 183
    Martyn Waller

13. Foraminifera 191
    Robin Edwards and Alex Wright

14. Pollen and spores of terrestrial plants 218
    Christopher E. Bernhardt and Debra A. Willard

15. Diatoms 233
    Yongqiang Zong and Yuki Sawai

16. Ostracods and sea level 249
    Thomas M. Cronin

17. Mollusca 258
    Jessica E. Pilarczyk and Donald C. Barber

18. Fixed biological indicators 268
    Alessio Rovere, Fabrizio Antonioli, and Carlo Nike Bianchi

19. Testate amoebae 281
    Dan J. Charman

20. Stable carbon isotope and C/N geochemistry of coastal wetland sediments as a sea-level indicator 295
    Nicole S. Khan, Christopher H. Vane, and Benjamin P. Horton

21. Loss on ignition and organic content 312
    Andrew J. Plater, Jason R. Kirby, John F. Boyle, Timothy Shaw, and Hayley Mills

22. Grain size analysis 331
Adam D. Switzer and Jeremy Pile

PART 3: DATING METHODS

23. Radiocarbon dating and calibration 349
Torbjörn E. Törnqvist, Brad E. Rosenheim, Ping Hu, and Alvaro B. Fernandez

24. ^{210}Lead and ^{137}Cesium: establishing a chronology for the last century 361
D. Reide Corbett and J.P. Walsh

25. Chronohorizons: indirect and unique event dating methods for sea-level reconstructions 373
Wil Marshall

26. Uranium-thorium dating 386
Andrea Dutton

27. The application of luminescence dating in sea-level studies 404
Mark D. Bateman

PART 4: MODELING

28. Glacial isostatic adjustment 421
Glenn A. Milne

29. Tidal modeling 438
Stephen D. Griffiths and David F. Hill

30. Compaction 452
Matthew J. Brain

31. Transfer functions 470
Andrew C. Kemp and Richard J. Telford

32. Using chronological models in late Holocene sea-level reconstructions from saltmarsh sediments 500
Andrew C. Parnell and W. Roland Gehrels

33. Paleogeography 514
Geert-Jan Vis, Kim M. Cohen, Wim E. Westerhoff, Johan H. Ten Veen, Marc P. Hijma, Ad J.F. van der Spek, and Peter C. Vos

34. A protocol for a geological sea-level database 536
Marc P. Hijma, Simon E. Engelhart, Torbjörn E. Törnqvist, Benjamin P. Horton, Ping Hu, and David F. Hill

PART 5: DIRECT MEASUREMENTS

35. Sea-level measurements from tide gauges 557
Philip L. Woodworth, David T. Pugh, and Andrew J. Plater

Index 575

List of contributors

Fabrizio Antonioli
ENEA-UTMEA-TER, Roma, Italy
fabrizio.antonioli@enea.it

Donald C. Barber
Department of Geology, Bryn Mawr College, Bryn Mawr, PA, USA
dbarber@brynmawr.edu

Natasha L. M. Barlow
Sea Level Research Unit, Department of Geography, Durham University, Science Laboratories, Durham, UK
n.l.m.barlow@durham.ac.uk

Mark D. Bateman
Geography Department, University of Sheffield, Sheffield, UK
M.D.Bateman@sheffield.ac.uk

Christopher E. Bernhardt
US Geological Survey, Eastern Geology and Paleoclimate Science Center, Reston, VA, USA
cbernhardt@usgs.gov

Carlo Nike Bianchi
DISTAV, Università degli Studi di Genova, Genoa, Italy
nbianchi@dipteris.unige.it

John F. Boyle
Department of Geography and Planning, School of Environmental Sciences, University of Liverpool, Liverpool, UK
jfb@liverpool.ac.uk

Matthew J. Brain
Sea Level Research Unit, Department of Geography, Durham University, Science Laboratories, Durham, UK
matthew.brain@durham.ac.uk

Dan J. Charman
Department of Geography, College of Life and Environmental Sciences, University of Exeter, Exeter, UK
d.j.charman@exeter.ac.uk

Kim M. Cohen
TNO, Geological Survey of the Netherlands, Utrecht, The Netherlands
Deltares, Applied Geology and Geophysics, Utrecht, The Netherlands
Department of Physical Geography, Utrecht University, Utrecht, The Netherlands
k.m.cohen@uu.nl

D. Reide Corbett
East Carolina University, Greenville, NC, USA
UNC Coastal Studies Institute, Wanchese, NC, USA
corbettd@ecu.edu

Thomas M. Cronin
US Geological Survey, Reston, VA, USA
tcronin@usgs.gov

Jeffrey A. Dorale
Department of Earth and Environmental Science, University of Iowa, Iowa City, IA, USA
jeffrey-dorale@uiowa.edu

Andrea Dutton
Department of Geological Sciences, University of Florida, Gainesville, FL, USA
adutton@ufl.edu

Robin Edwards
School of Natural Sciences, Museum Building, Trinity College Dublin, Dublin, Ireland
robin.edwards@tcd.ie

Simon E. Engelhart
Department of Geosciences, University of Rhode Island, Kingston, RI, USA
engelhart@uri.edu

Tezer M. Esat
Australian Nuclear Science and Technology Organization, Institute for Environmental Research, Kirrawee, Australia
Research School of Earth Sciences and Research School of Physical Sciences and Engineering, The Australian National University, Canberra, Australia
tezer.esat@anu.edu.au

Alvaro B. Fernandez
Department of Earth and Environmental Sciences, Tulane University, New Orleans, LA, USA
afernan4@tulane.edu

James Foster
University of Hawai'i, Honolulu, HI, USA
jfoster@soest.hawaii.edu

W. Roland Gehrels
Environment Department, University of York, Heslington, York, UK
roland.gehrels@york.ac.uk

Stephen D. Griffiths
Department of Applied Mathematics, University of Leeds, Leeds, UK
sdg@maths.leeds.ac.uk

Peter J. van Hengstum
Department of Marine Sciences, Texas A&M University at Galveston, Galveston, TX, USA
vanhenp@tamug.edu

Marc P. Hijma
Department of Earth and Environmental Sciences, Tulane University, New Orleans, LA, USA
Deltares, Applied Geology and Geophysics, Utrecht, The Netherlands
marc.hijma@deltares.nl

David F. Hill
School of Civil and Construction Engineering, Oregon State University, Corvallis, OR, USA
david.hill@oregonstate.edu

Benjamin P. Horton
Sea Level Research, Department of Marine and Coastal Sciences, Rutgers University, USA
Earth Observatory of Singapore and Division of Earth Sciences, Nanyang Technological University, Singapore
bphorton@marine.rutgers.edu

Ping Hu
Department of Earth and Environmental Sciences, Tulane University, New Orleans, LA, USA
phu@tulane.edu

Harvey M. Kelsey
Department of Geology, Humboldt State University, Arcata, CA, USA
harvey.kelsey@humboldt.edu

Andrew C. Kemp
Department of Earth and Ocean Sciences, Tufts University, Medford, MA, USA
Andrew.Kemp@tufts.edu

Nicole S. Khan
Sea Level Research, Department of Earth and Environmental Science, University of Pennsylvania, Philadelphia, PA, USA
Department of Marine and Coastal Sciences, School of Environmental and Biological Sciences, Rutgers University, New Brunswick, NJ, USA
khann@sas.upenn.edu

Jason R. Kirby
School of Natural Sciences and Psychology, Department of Geography, Liverpool John Moores University, Liverpool, UK
J.R.Kirby@ljmu.ac.uk

Antony J. Long
Sea Level Research Unit, Department of Geography, Durham University, Durham, UK
a.j.long@durham.ac.uk

Nick Marriner
Laboratoire Chrono-Environnement, UMR 6249 CNRS, Université de Franche-Comté, UFR ST, Besançon, France
marriner@cerege.fr

Wil Marshall
School of Geography, Earth and Environmental Sciences, University of Plymouth, Plymouth, UK
wmarshall@plymouth.ac.uk

Aron J. Meltzner
Earth Observatory of Singapore, Nanyang Technological University, Singapore
meltzner@ntu.edu.sg

Hayley Mills
British Oceanographic Data Centre, National Oceanography Centre, Liverpool, UK
haymil@bodc.ac.uk

Preface

About 15 years ago Orson van de Plassche and we discussed putting together a successor to the extremely successful reference book *Sea-level Research: A Manual for the Collection and Evaluation of Data* that he edited. It was published in 1986 but never re-printed. For at least 15 years it was a widely used reference and highlighted the need for a handbook approach for research that aimed to investigate sea-level changes beyond the range of modern instrumental records. It was especially useful in transferring approaches and skills beyond the research heartlands of NW Europe and North America. By 2000, we had an outline and discussed the complications of organizing topics

Pattagansett River Marsh, East Lyme, CT; one of the sites used in sea-level research, illustrating the resilience of marsh environments to extreme events (van de Plassche et al., 2004, 2006). The upper photograph is from August 2010 and the lower one from March 2013, five months after Hurricane Sandy and the accompanying surge which caused significant damage along the east coast of the USA. The marsh survived relatively intact, with a thin veneer of fine-grained sediment unevenly distributed across only some parts of the marsh, with localized erosion, including small marsh cliffs and the foundation of an access road. The sign in the foreground of the 2010 photograph was also destroyed by the storm surge. Photographs courtesy of Andy Kemp.

Pattagansett River Marsh, East Lyme, CT. July 2013. Photography by Eleonora Tammes. The inscription reads "Dr Orson van de Plassche. He was a leading light in sea level and hurricane research, a meticulous and energetic field scientist and one of the best coastal geologists and stratigraphers of his generation. The wetlands will benefit us all into the future, and we thank him. East Lyme/Niantic Land Trust (2009)."

into chapters. With sea-level research covering such a wide range of disciplines, approaches, timescales, spatial scales, environments, laboratory methods, and analytical procedures, we found it challenging to decide on a framework of chapters that would be practical yet minimize repetition. In the end, we outlined sections and chapters that reflected the actual sequence of activities that many, but not all, researchers followed at that time.

Unfortunately, we never got it off the ground before Orson's untimely passing in 2009.

Orson spent much of his career seeking answers to research questions about the overprinting and interlinking of millennial-scale sea-level change with shorter-term variations, changes over the last 150 years, hurricanes, and storm surges. Having defined the question, his research typically encompassed meticulous fieldwork in marsh environments (photographs page xi), sample collection, innovative methods of dating, multi-proxy evidence of environmental reconstruction, and rigorous data analysis, incorporating modeling procedures where relevant.

Having received widespread approval and support at the final meeting of IGCP 495 in 2009 and then from IGCP 588, the idea was revived and we started inviting authors from across the research community.

The organization of the book broadly follows the structure we had outlined with Orson around 2000. Chapters 1 and 2 introduce the aims and context of the *Handbook*, what it does and does not aim to cover, approaches to sea-level research, common terms, and some suggestions for promoting clearer understanding, both within and beyond the sea-level research community. We have organized the remaining chapters into parts that broadly reflect different activities in research. Part 1 comprises nine chapters on field techniques for reconstructing sea level, and Part 2 provides 11 chapters on the main laboratory techniques. Five chapters on dating methods form Part 3, followed by 7 chapters on different modeling approaches in Part 4. The final chapter, in Part 5, focuses on the instrumental record of sea-level change.

For all who have had the privilege to work with Orson or be inspired by his work we hope that this Handbook will be a suitable successor to the 1986 Manual. Just as the marsh was resilient to Hurricane Sandy, so the plaque was replaced as the sentiments expressed on it are equally resilient. Similarly, this Handbook follows the Manual as an illustration of how the pursuit of excellence in sea-level reconstruction continues.

Ian Shennan, Antony Long and Ben Horton
Durham and New Brunswick January 2015

The Handbook is a contribution to IGCP 588 *Preparing for coastal change: A detailed process-response framework for coastal change at different timescales* and PALSEA2.

REFERENCES

van de Plassche, O., Wright, A.J., van der Borg, K., and de Jong, A. F. M. (2004) On the erosive trail of a 14th and 15th century hurricane in Connecticut (USA) salt marshes. Radiocarbon, 46, 775–784.

van de Plassche, O., Erkens, G., van Vliet, F., Brandsma, J., van der Borg, K., and de Jong, A. F. M. (2006) Salt-marsh erosion associated with hurricane landfall in southern New England in the fifteenth and seventeenth centuries. Geology, 34(10), 829–832.

About the companion website

This book is accompanied by a companion website:

www.wiley.com/go/shennan/sealevel

The website includes:
Pdfs of all figures from the book for downloading
Powerpoints of all tables from the book for downloading
High-resolution images of microfossils
Modeling code
Database materials

Chapter 1
Introduction

IAN SHENNAN[1], ANTONY J. LONG[1], AND BENJAMIN P. HORTON[2,3]

[1]*Sea Level Research Unit, Department of Geography, Durham University, Durham, UK*
[2]*Sea Level Research, Department of Marine and Coastal Sciences, Rutgers University, USA*
[3]*Earth Observatory of Singapore and Division of Earth Sciences, Nanyang Technological University, Singapore*

1.1 AIMS OF THE HANDBOOK

Every year, countless articles, books, social media, and TV programs debate the importance (or not) of sea-level change; the change debated is mainly sea-level rise, but sea-level fall is also possible and important. Millions of people live along the coast, estuaries, and adjacent coastal lowlands and many will concur with the view that "Due to sea level rise projected throughout the 21st century and beyond, coastal systems and low-lying areas will increasingly experience adverse impacts such as submergence, coastal flooding, and coastal erosion." (IPCC 2014, page 17). Significant debate over some of the causes and consequences of recent and future sea-level change remains within and between many communities including science, the media, coastal residents, industry, politics, and governments. Environments, communities, livelihoods, real-estate, and cash are all at stake. While improved understanding of the causes, impacts, and responses to sea-level change are of undeniable importance in many disciplines, these are not the subject of this book. With the range of technology available today, we can measure changes every few minutes with tide gauges and across entire oceans using satellites. Yet these offer only a small part of the wider picture. How does the variability we observe using these technologies fit with longer-term trends? Do our decades of instrumental observations adequately cover the natural variability and extreme events of the past and those we are likely to experience in the future? Most researchers would probably give a very guarded answer; some would give a forthright "No". We need to bring together the evidence across a greater range of timescales, from seconds to millennia or even longer, for a complete analysis.

The aims of this book are therefore to entice and guide the reader beyond their initial interest and discipline, to enable them to tackle new questions. This will hopefully also lead to the reader asking new questions and, ultimately, proposing new answers based on carefully collected observations, analyses, and models developed in the field and the laboratory from sites all over the world.

1.2 SEA-LEVEL RESEARCHERS

Sea-level research is primarily an observational science and we must realize what imperfect observers we are. Unlike experimental science where observations can usually be replicated and verified by others using the same or equivalent methods, we frequently deal with observations that have incomplete distributions through time and/or space. When looked at by another researcher, objectively and dispassionately, how many interpretations deteriorate into a collection of inferences, guesses, or hunches based on too little data, much of which is inconclusive or influenced by decisions made by the original author? With this in mind, it is easy to state that one clear recommendation of this Handbook is to encourage all researchers to make available their data for others to use as the basis for alternative analyses and interpretation. If we are fortunate to act as a reviewer or an editor for a peer-reviewed publication, we must ask authors to include the raw data either in the paper or an online repository linked to the article.

While our interpretations may remain unchallenged or un-falsified for only a few years, we should aim for our data to stand the test of time and be readily available. After all, digital media, international data repositories, and scientific journals encouraging online supplementary information files make this much more feasible than it has ever

Handbook of Sea-Level Research, First Edition. Edited by Ian Shennan, Antony J. Long, and Benjamin P. Horton.

been. But the reality is that we are human beings, living in different socio-economic and political environments where different pressures may work against this aspiration of openness. Career progression and demands from employers, research funding bodies, government, peers, students, and the media may all influence a researcher at different times throughout their career; we cannot hide from this fact. Similarly, our educational background, training, and experiences will influence the approach we take.

A review of sea-level research since the publication of Orson's original Manual (van de Plassche, 1986) will quickly reveal examples of how theories we thought we had right were in fact wrong. As a consequence, this Handbook does not set out to promote a single paradigm for sea-level research or a single "right" way of doing things. Rather, it aims to illustrate approaches and methods that have produced observations, analyses, and interpretations which have then stood the test of scrutiny by peers, mainly through the review process of scientific journal publication, but also at conferences and field discussion meetings. For many sea-level researchers we should acknowledge that defending one's work in the field or at a small workshop may well be more intimidating and rigorous than at an international meeting or through the journal review process. Despite the need to publish for career progression, we should not underestimate the value of field meetings and workshops, through international organizations such as the International Geoscience Programme (IGCP) and the International Union for Quaternary Science (INQUA), to generate open debate and different perspectives on the way we make observations. Such debates and perspectives can provide the catalyst for new ideas and the development of new methods and techniques for data collection, analysis, and hypothesis testing. Attendance at such meetings can however be difficult due to their location, their cost, or other reasons. This is where we hope that this handbook will serve a real purpose by making available to readers many of the approaches and methods of sea-level research developed at such events in a single volume. If we come close to achieving this aim, we will have achieved one of the prime motivations of the first Manual and produced something that is fitting testimony to Orson's original vision.

REFERENCES

IPCC, 2014: Summary for policymakers. In: Climate Change 2014: Impacts, Adaptation, and Vulnerability. Part A: Global and Sectoral Aspects. Contribution of Working Group II to the Fifth Assessment Report of the Intergovernmental Panel on Climate Change [Field, C.B., Barros, V.R., Dokken, D.J., Mach, K.J., Mastrandrea, M.D., Bilir, T.E., Chatterjee, M., Ebi, K.L., Estrada, Y.O., Genova, R.C, Girma, B., Kissel, E.S., Levy, A.N., MacCracken, S., Mastrandrea, P.R., and White, L.L. (eds)]. Cambridge University Press, Cambridge, United Kingdom and New York, NY, USA, pp. 1–32.

van de Plassche, O. (1986) *Sea-Level Research: A Manual for the Collection and Evaluation of Data*. GeoBooks, Norwich.

Chapter 2
Handbook of sea-level research: framing research questions

IAN SHENNAN
Sea Level Research Unit, Department of Geography, Durham University, Durham, UK

2.1 INTRODUCTION

Most sea-level researchers tend to be pragmatists, more interested in results than theory and methodology. While a simple question such as "Was sea level at time *t* different in estuary A compared to estuary B?" seems perfectly suitable for statistical analysis using an analysis of variance approach, we rarely see hypotheses rigorously tested using measures of probability and statistical significance. Implicitly however, we would attempt to assess within-site variability, in both estuaries, before assessing the difference between the two. Our aim here is not to present a detailed dissection of the scientific method, hypothesis formulation and testing, or the range of statistical methods available, as numerous other texts more than adequately achieve these (e.g., Davis, 2002; McKillup and Dyar, 2010). We do, however, aim to point out some of the essential steps to help avoid the pitfalls of poorly derived interpretations, such as those outlined in Chapter 1. We shall explore the range of ways in which research questions can be framed as starting points for field and laboratory analysis. Once completed, our research only becomes valuable if the results can be transmitted to other researchers and the wider community. That depends on the methods being reliable and understandable. Ideally, it also requires a clear distinction between observations and interpretations, while realizing that in some instances the distinction between each may not be clear.

2.2 SOME NECESSARY METHODOLOGY: IDEAS AND TESTABLE HYPOTHESES

Many studies of sea-level change are implicitly inductive, depending on inferences and reasoning, where extrapolations may be made to develop generalizations. An inductive approach could typically follow a line of reasoning illustrated in Fig. 2.1: reasoning from experience, the argument following a route from describing the effect and leading to an account of the cause. The reader is left to decide on the validity of the interpretation or conclusion by their assessment of the strength or weakness of the argument and its cogency.

Many good studies take this approach, producing results that have proved reliable and stood the test of time. Nevertheless, there is a danger that an inductive approach can lead to the possible result of a theory into which all subsequent observations are fitted or, as Chamberlin (1897, p. 840) put it more eloquently:

> *"The moment one has offered an original explanation for a phenomenon which seems satisfactory, that moment affection for [your] intellectual child springs into existence, and as the explanation grows into a definite theory [your] parental affections cluster about [your] offspring and it grows more and more dear to [you]. While [you] persuade [yourself] that [you] hold it still as tentative, it is none the less lovingly tentative and not impartially and indifferently tentative. So soon as this parental affection takes possession of the mind, there is apt to be a rapid passage to the unreserved adoption of the theory. There is then imminent danger of an unconscious selection and of a magnifying of phenomena that fall into harmony with the theory and support it and an unconscious neglect of phenomena that fail of coincidence. The mind lingers with pleasure upon the facts that fall happily into the embrace of the theory, and feels a natural coldness toward those that assume a refractory attitude.*

Handbook of Sea-Level Research, First Edition. Edited by Ian Shennan, Antony J. Long, and Benjamin P. Horton.

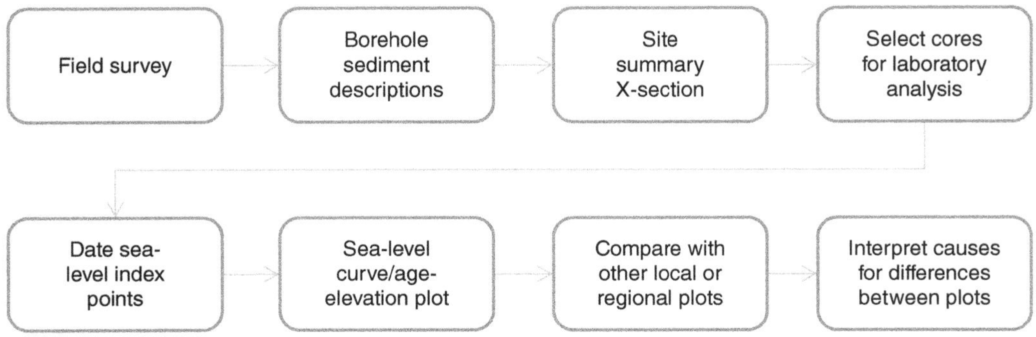

Fig. 2.1. One route to the analysis and explanation of relative sea-level changes.

Instinctively there is a special searching-out of phenomena that support it, for the mind is led by its desires. There springs up also unwittingly a pressing of the theory to make it fit the facts and a pressing of the facts to make them fit the theory."

Alternatively, and particularly in the experimental disciplines, there is a conventional view of the scientific method that uses the process of hypothesis formulation, testing and falsification. This is less easy to apply in practice within field-based science as the complexities of interactions of processes within nature make it too simplistic to consider a single cause and effect from the outset. The approach of multiple working hypotheses therefore has many advantages (Chamberlin, 1897). This is as much an approach to thinking and enquiring about complex interactions as well as a way of explaining to others the process of analysis that leads to explanation. As a method of complex and parallel thinking, it does not easily transcribe into linear sentences or flowcharts. Figure 2.2 is simply one attempt to visualize the process. By testing several hypotheses (in the broadest sense, these include theories, ideas, hunches, and explanations proposed by others), this approach avoids the potential pitfall of collecting only a narrow range of data that are unconsciously supporting a single idea. Some researchers may feel uncomfortable with the outcome that more than one explanation remains at the end of a long period of effort, but this must be balanced by the strengths of increased objectivity and increased ability to recognize your own errors and ignorance.

A key stage in the approach is the formulation of each potential idea, hunch, and hypothesis in a testable form (Fig. 2.2). If you cannot determine the means to present a hypothesis in a form that can be tested using observed phenomena that can be measured, then it should remain as speculation and stated as such.

Two examples illustrate these contrasting approaches. In both examples the research topics have important consequences for society to consider, if the results are deemed reliable. The question of the threat of future sea-level rise to the Maldives is well known in both the scientific and popular media. A field-based research program led to a conclusion that the Maldives are not experiencing relative sea-level rise at present and are unlikely to suffer flooding in the near future (Mörner et al., 2004). They adopted an approach of data evaluation and reasoning comparable to Fig. 2.1, with their conclusions dependent upon how the reader accepts their logic, evidence, interpretations, and strength of argument. Numerous authors reject their arguments and interpretations, suggesting that Mörner et al. (2004) present too few details, misinterpret data, do not consider alternative explanations, and suggest mechanisms for change that are implausible and unsupported by evidence (Kench et al., 2005; Woodroffe, 2005; Woodworth, 2005).

In contrast, Atwater (1987) takes an approach comparable to Figure 2.2 to test the hypothesis of great Holocene earthquakes occurring along the Cascadia subduction zone, despite there being no evidence for them documented in historical records. He derives a hypothesis by comparisons with the submergence of marshes during great earthquakes in Chile in 1960 and Alaska in 1964. After presenting the field data, he considers the alternative explanations for rapid marsh submergence, i.e. other working hypotheses, and explains the criteria for rejecting processes such as storms or floods. He ends by outlining the hypothesis left standing and how it may be developed further and tested with new data. This paper has

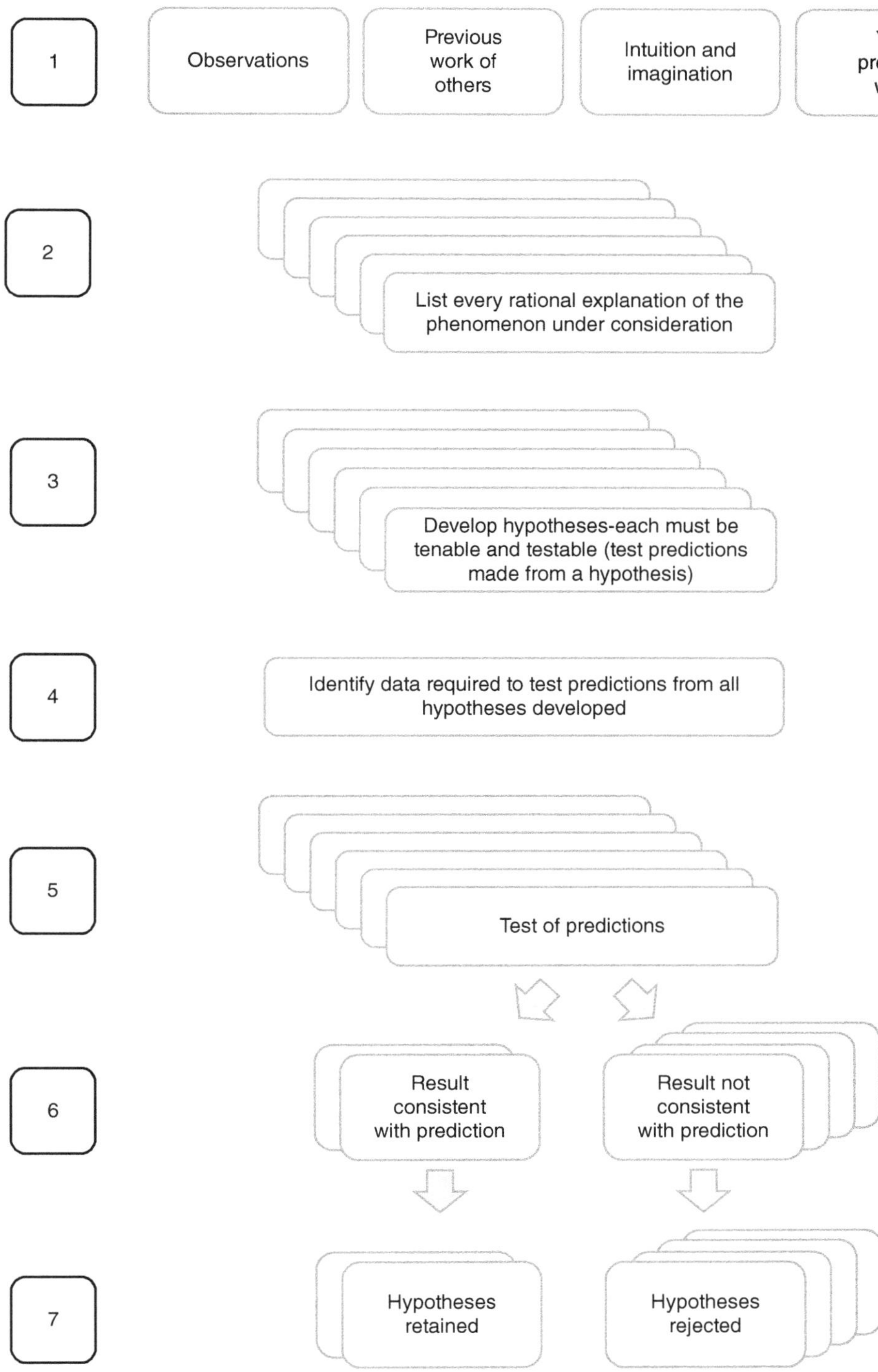

Fig. 2.2. Cartoon depicting seven stages in formulating and testing multiple working hypotheses, in this case leading to two of the original six hypotheses remaining as possible explanations.

stood the test of time, the principles adopted, expanded, and modified to study great earthquakes in other subduction zones around the world. Following the publication of the paper in 1987, studies in adjacent coasts quickly highlighted the question that, as the field location becomes more distant from the rupture zone, at what point does earthquake-induced submergence become unrecognizable from other non-seismic causes. This led to the proposal of specific criteria for assessing evidence of coseismic submergence (Nelson et al., 1996), criteria which still form the basis for working hypotheses in recent studies (Grand Pre et al., 2012).

In practice, very few of the studies referred to in this Handbook explicitly adopt a deductive, hypothesis testing approach as outlined above. Given the potential benefits of a multiple working hypothesis approach, why is this? There could be many explanations, but this would require a lengthy review and appraisal that is beyond the scope of the Handbook. We could suggest that one of the reasons is the fact that the studies cited are mainly from the refereed peer-reviewed academic literature. This aims to eliminate bad science at many stages, including in-house peer review, review of grant proposals and papers, and defense of conference presentations. There are relatively few examples of poor science in the sea-level literature leading to extreme, published criticism and refutation. As noted above many good studies appear to follow an essentially inductive approach, but there may be a difference between the details of the methodological approach explicitly presented in a scientific article and those presented in the research grant proposal that led to the research in the first place. To oversimplify, overly detailed and potentially repetitive descriptions of hypotheses and null-hypotheses rarely lead to easily read research articles, yet they are frequently set out in grant proposals. Authors may well have considered multiple working hypotheses and followed rigorous procedures, including those outlined in this chapter, but they do not always need repeating in every paper. If weaknesses in project design, execution, and presentation exist, then they are likely to be exposed at a number of stages before reaching publication in peer-reviewed scientific media.

2.3 OPERATIONAL DEFINITIONS

As interdisciplinary research develops, historical differences in terminology often appear. As our analyses depend upon numerical measures we require not only an accepted use of a term, but a known and hopefully agreed method of measuring it. An operational definition defines a variable or object by means of the specific process used to determine its presence and quantity, that is, not just what it is but how you measure it.

The concepts of eustatic sea level and relative sea level emerged following the 19th century debates about the relationships between shoreline features and the advance of retreat of ice sheets (Jamieson, 1865) and led to the discovery of the process we now know as glacial isostatic adjustment (GIA), summarized as Figure 2.3a. Subsequent analyses of complex linkages and time lags within the ice–Earth–ocean system led to new or revised terms. At different times over recent decades we have seen terms such as real sea level, apparent sea level, relative sea level, eustatic sea level, effective eustatic value, and ice-melt-equivalent sea level. Our dilemma is how to use terms familiar to a broad audience, but used slightly differently by authors across various scientific disciplines. Through this Handbook we adopt the following definitions.

2.3.1 Eustasy and eustatic sea-level change

Eustatic sea level is the sea level that would result from distributing water evenly across a rigid, non-rotating planet and neglecting self-gravitation in the surface load (Mitrovica and Milne, 2003). Because we know the planet is not rigid nor non-rotating and that there is a process of self-gravitation, we simply have to stress that it is highly unlikely that any location on Earth will actually record eustatic sea-level change through time, unless the contributions of non-eustatic processes cancel out (Milne and Mitrovica, 2008; Fig. 2.3b).

2.3.2 Sea level, relative sea level, and relative sea-level change

The terms sea level, relative sea level and relative sea-level change have also had different usages. In scientific literature sea level is relative in two respects: elevation relative to the surface of the Earth, and elevation relative to the present. This has led to some ambiguity in the use of terms. The GIA modeling community commonly presents explicit definitions using mathematical notation (Mitrovica and Milne, 2003). In contrast, the geological community appears less precise in their use of different terms (van de Plassche, 1986a; Devoy, 1987). GIA researchers define sea level as the elevation of the geoid (mean height of the sea surface averaged over several decades) in relation to the solid surface of the Earth (Fig. 2.4), whereas a common geological use is to call this relative sea level. In contrast, other geological studies use relative sea-level change in referring to change as a process rather than a measurement, not attributed to any specific cause such as land uplift or eustatic change. It is also used to

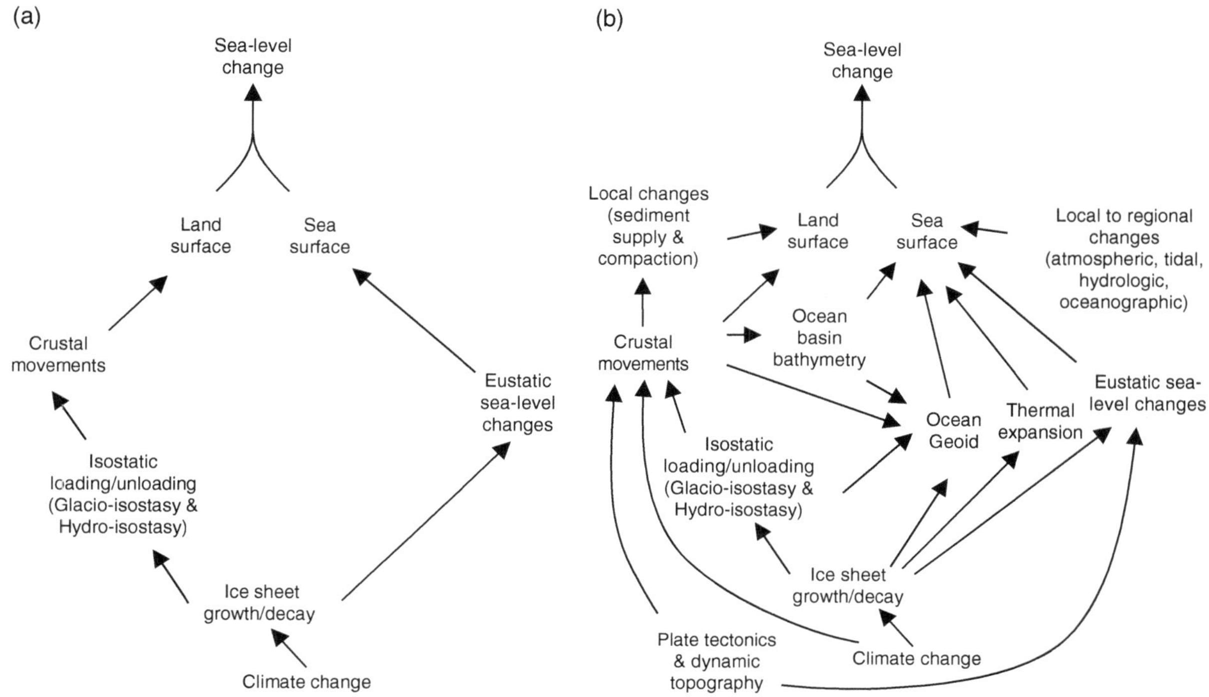

Fig. 2.3. Depiction of global- to local-scale processes that contribute to relative sea-level changes: (a) simple model of glacio-isostasy and eustasy; and (b) model to help frame multiple working hypotheses of sea-level change over different timescales; knowledge of these processes and their interactions served as a framework for researching late Quaternary relative sea-level changes that was common in the geological sea-level community in the late 20th century. Note that it does not include rotation and the influence of sediment erosion and deposition on the geoid and solid Earth.

describe other processes, for example in discussion of the relationships between sea level, intertidal sedimentation, and coastline advance or retreat, whether at present or in the fossil record.

In this Handbook we propose adoption of the terms as defined by Mitrovica and Milne (2003). For each geographical location φ and time t, sea level (SL) is the difference between the geoid G and the solid rock or sediment surface of the Earth R, both measured with reference to the centre of the Earth (Fig. 2.4a). With this definition, sea level is the same as the common geological usage of the term relative sea level:

$$\mathrm{SL}(\varphi,\tau) = G(\varphi,\tau) - R(\varphi,\tau) \quad (2.1)$$

It follows from Equation (2.1) and Figure 2.4a that topography T is simply the negative of sea level:

$$T(\phi,t)=R(\phi,t)-G(\phi,t)=-SL(\phi,t) \quad (2.2)$$

A change in sea level is given by the change in sea surface height minus the change in solid surface height over the period of interest (Fig. 2.4b):

$$\Delta\mathrm{SL}(\phi,t)=\Delta G(\phi,t)-\Delta R(\phi,t) \quad (2.3)$$

in which

$$\Delta G(\phi,t)=G(\phi,t)-G(\phi,t_0) \quad (2.4a)$$

$$\Delta R(\phi,t)=R(\phi,t)-R(\phi,t_0) \quad (2.4b)$$

and t_0 is a chosen reference time.

For Quaternary and Holocene timescales a common convention is to define a change in relative sea level (RSL) as a change relative to present. For any time in the past, where t_{past} is time before present, we therefore have:

$$\mathrm{RSL}(\phi,t_{past})=\mathrm{SL}(\phi,t_{past})-\mathrm{SL}(\phi,t_{present}) \quad (2.5)$$

Many studies use the convention of years Before Present (BP), where 0 BP is AD 1950, taken from the calibration of radiocarbon ages (Chapter 23). Whether adopting this convention or not, authors should state clearly their zero age datum.

The rate of relative sea-level change at $t=0$ according to Equation (2.3) is the same as describing the relative vertical motion at the shoreline

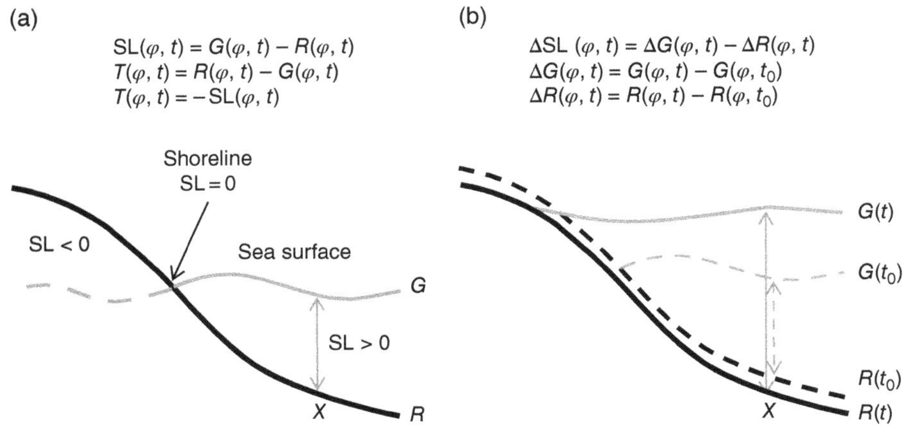

Fig. 2.4. Definition of sea-level change and associated variables: (a) Sea level (SL), topography T, sea surface or geoid G (the time-averaged sea surface, over several decades, approximates the ocean geoid), and solid or rock surface R. G and R are measured relative to the centre of the Earth for geographical location φ and time t. (b) Sea-level change (ΔSL) for example at location X is the length of the solid arrow minus the length of the dashed arrow and t_0 is a chosen reference time. *Source*: Shennan et al. (2012). Reproduced with permission of John Wiley & Sons.

between the solid Earth surface and the sea surface (ocean geoid). The rate of present relative sea-level change is the negative of the rate of present relative land-level (topography) change (Equation (2.2)). Importantly, this is very different to the change in the elevation of the rock surface relative to the centre of the Earth, $\Delta R(\varphi, t)$, commonly termed vertical land motion.

This is an important point. The total isostatic effect of the glacial rebound process includes the redistribution of ice and water loads on both the solid Earth and the mean sea surface (over several decades this approximates the ocean geoid), and rotational contributions to the redistribution of ocean mass. The gravitational and rotational effects mean that the present rate of relative land-level change is not the same as the present rate of vertical land motion, measured by instruments such as GPS. The difference varies spatially by +2.5 to −1.5 mm a^{-1} globally and with regional trends, for example from ~−0.1 to −0.3 mm a^{-1} across the British Isles (Shennan et al., 2012).

2.3.3 Sea-level indicator and sea-level index point

International Geological Correlation Programme (IGCP) Projects 61 and 200 (van de Plassche, 1986b) promoted a universally applicable approach to the reconstruction of relative sea-level histories for different locations and environments. Reliable estimates of past changes in sea level relative to present, with quantified uncertainty terms, come from *in situ* sediments, fossil organisms, morphological, and archeological features whose origin was controlled by the paleo sea level. Each of these features is an example of a sea-level indicator. Chapters 3–11 in Part I of the Handbook describe the major types used over Quaternary to present timescales. Where such indicators survive they can be used as sea-level index points (often abbreviated to SLIs or SLIPs) by defining four attributes: location, age, elevation, and tendency.

In many studies we see sea-level index points described as observations, often to separate them from estimates or predictions from GIA models. The term observation is technically incorrect as the derivation of one or more attributes of a sea-level index point frequently involves some sort of model, for example the age model and the dating method, the atmospheric correction and reference frame correction for GPS measurements, or the statistical model in assigning the paleo-sea level from the sea-level indicator. These form the basis of chapters in Parts II, III and IV of the Handbook. Rather than observation, proxy reconstruction is a more accurate description.

2.3.4 Location attribute of a sea-level index point

Once it is established that the indicator has not been transported from its original location, this attribute consists simply of its geographical coordinates, measured to a national or international convention (typically national grid reference and/or latitude and longitude). For studies that use regional collections of index points, for example to test GIA models or mapping coseismic subsidence, the geographical coordinates of some early published data may not be known sufficiently precisely and the sample is rejected. For the majority of index points within recent databases, the location is known to 10 m or less.

2.3.5 Age attribute of a sea-level index point

While the majority of index points used for studies of sea-level changes since the Last Glacial Maximum (LGM) have their ages measured by radiocarbon techniques, and most modern studies use calibrated ages expressed with 95% confidence limits, there is an increasing range of techniques available to cover Quaternary–present timescales (Chapters 23–27). Morphological expressions of past sea level, including beaches, notches, cliffs, and shorelines (Chapter 5), have the great advantage that they provide a spatial expression of the consistency or variation in a past sea level, but their use is severely limited if they cannot be dated reliably. Organic material incorporated in raised shorelines for example is not always *in situ*, so only provides a limiting age of the feature. Although not always possible, the most reliable sea-level index points should have at least one type of corroborating evidence to support the age estimate. Perhaps more importantly, there should be an analysis and evaluation of the continuity, or lack, of sedimentation or formation between the sea-level indicator and paleo sea level. A properly devised and completed field program (Chapter 4) should minimize the likelihood of sampling eroded sequences or those with a long or indeterminable hiatus, but studies that explicitly address rapid changes in sea level (e.g., storms, tsunami, or earthquakes) may have to accept that certain types of index points will give either maximum or minimum ages on an event and the data analysis procedures must allow for this (Lohne et al., 2007; Carver and Plafker, 2008).

2.3.6 Elevation attribute of a sea-level index point: indicative meaning, reference water level, and indicative range

While few sea-level index points formed exactly at paleo mean sea level, we do not need to measure them in relation to mean sea level. It is reasonable to ask how this can be the case, when our aim is to produce a graph of mean sea level over the last 10,000 years. The key is to remember Equation (2.5) that defines relative sea-level change, and determine the operational definition of elevation for your particular sea-level indicator.

Let us consider the examples of a line of dead barnacles exposed above present high tide (Fig. 2.5). This is a very simple index point. In this example, the fossil indicator (the dead barnacles) lies approximately 1.7 m directly above the modern equivalent, so the relative sea-level change is ~−1.7 m between the age of the fossil sample and the present. To reach this conclusion we do not need to know either the position of mean sea level at present or at the time when the fossil barnacles were living, although we do assume that the environmental conditions have not changed and that the relationship of barnacle growth to inundation frequency is therefore the same. Such indicators were widely used immediately after the M_w 9.2 earthquake in Alaska in 1964 to map the extent and magnitude of coseismic crustal deformation across hundreds of kilometers of coastline (Plafker, 1969). These types of indicator remain an important method of measuring relative land and sea-level change, with dead coralline algae (*Lithothamnium*) raised above the lower intertidal zone providing the first estimates of the spatial pattern of land-level changes produced by the M_w 8.8 2010 Chilean Earthquake (Farías et al., 2010).

Unfortunately, for most studies of sea-level change we do not have the luxury of the fossil indicator lying directly above or even in line of sight of its modern equivalent. Some index points are more than 100 km from the present coast or hundreds of meters above it. We therefore require a means of measuring the modern and fossil samples to a common datum. This can be anything ranging from, at a local scale, a mark the researcher makes in a rock and then instrumentally levels both to a national leveling datum, tide gauge, or GPS vertical reference frame (Chapter 10).

The measurement we require for elevation, our operational definition, is the "indicative meaning" (van de Plassche, 1986a). Indicative meaning comprises two parameters: the reference water level and the indicative range (Chapter 11). Reference water level is the modern equivalent elevation at which the indicator occurs, defined with reference to a tide level. In our example of the upper limit of barnacles this is around mean high tide (Kaye, 1964). Local conditions, such as

Fig. 2.5. Line of the upper limit of dead barnacles *Balanus balanoides* seen above head-height on a sheltered cliff near Cordova, Alaska. The equivalent upper limit of live forms is found at the base of the cliff, next to the person's feet. The difference in elevation between upper limits of the dead and live barnacles indicates the net effect of coseismic uplift in 1964 and post-seismic recovery 1964–2009. *Source*: Photograph by Ian Shennan.

exposure to waves, water salinity within an estuary or fjord, and tidal range, cause this value to vary, which we quantify as the indicative range, that is, the vertical range over which the indicator occurs at present. To allow for the present variability of the upper limit of barnacles in south-central Alaska, Plafker (1969) suggests ±0.30 m. Defining the reference water level and indicative range for every type of sea-level indicator, and the spatial scale over which a set of modern observations can be applied, are ongoing research challenges and central to many of the following chapters.

Few sea-level index points formed exactly at paleo mean sea level, and many more represent environments within the upper part of the tidal range. In total they cover the full tidal range, the shallow sub-tidal zone, and, for limiting dates, beyond these upper and lower limits. Limiting dates come from either samples from freshwater environments inland of the paleo coastline and at or above the past high-tide level, or fully marine environments for which only a minimum water depth can be given. In reconstructing relative sea-level change, on a plot of elevation against age the dated freshwater sample must lie at or above data points such as those sea-level index points whose origin was directly controlled by the paleo-sea level. The same argument applies to dated samples that represent marine environments; they must lie at or below the other index points on the altitude versus age plot.

Because different sea-level indicators form at different elevations with respect to present tide levels, researchers must be familiar with the definitions used for different tide levels at all locations their project refers to as they can change between regions and nations. Semidiurnal

Table 2.1. The two main systems for describing average tide levels. The time period used for calculating the average is typically of the order 19 years or more in order to average out 18.6 year cycles in the tidal amplitudes and phases (Pugh, 1996; http://woce.nodc.noaa.gov; http://tidesandcurrents.noaa.gov).

Diurnal and mixed tidal pattern		Semidiurnal tidal pattern	
HAT Highest Astronomical Tide	Elevation of the highest predicted astronomical tide	HAT Highest Astronomical Tide	Elevation of the highest predicted astronomical tide
MHHW Mean Higher High Water	Average of the higher high water height of each tidal day	MHWS Mean High Water Springs	Average of high water heights occurring at the time of spring tides
MHW Mean High Water	Average of all the high water heights	MHW Mean High Water	Average of all the high water heights
MLHW Mean Lower High Water	Average of the lower high water height of each tidal day	MHWN Mean High Water Neaps	Average of high water heights occurring at the time of neap tides
MTL Mean Tide Level	Arithmetic mean of mean high water and mean low water	MTL Mean Tide Level	Arithmetic mean of mean high water and mean low water
MSL Mean Sea Level	Arithmetic mean of hourly heights observed	MSL Mean Sea Level	Arithmetic mean of hourly heights observed
MHLW Mean Lower Low Water	Average of the higher low water height of each tidal day	MLWN Mean Low Water Neaps	Average of low water heights occurring at the time of neap tides
MLW Mean Low Water	Average of all the low water heights	MLW Mean Low Water	Average of all the low water heights
MLLW Mean Lower Low Water	Average of the lower low water height of each tidal day	MLWS Mean Low Water Springs	Average of low water heights occurring at the time of spring tides
LAT Lowest Astronomical Tide	Elevation of the lowest astronomical predicted tide	LAT Lowest Astronomical Tide	Elevation of the lowest astronomical predicted tide

oscillations dominate in most shelf seas and there are two main systems for describing tide levels (Table 2.1). One system separates the spring tides, the semidiurnal tides of increased range occurring approximately twice a month as the result of the moon being either new or full, from the intervening neap tides, where the range is smaller and which occur near the first and last lunar quarters. Where the tide is mixed, with diurnal and semidiurnal patterns both evident, the spring and neap system is less appropriate. In these situations there are different averages for two high waters and two low waters per tidal cycle. Mixing the two systems (Table 2.1) in a research project without recomputing elevations to a common system will lead to error.

To illustrate the calculation of relative sea-level change we develop the example of the barnacles (Fig. 2.5) and apply the same principles to a marsh sedimentary sequence (Figs 2.6 and 2.7).

The procedure for defining the indicative meaning for our example is as follows. First we take a number of measurements of the elevation of the upper limit of living barnacles, using a random sampling strategy and a total number of samples n. There is no fixed value for n, as many statistical tests depend upon our variable demonstrating a normal distribution, so we should check this while still in the field and collect further observations if required. Except where a distribution is grossly different from a normal distribution, such as a bimodal or multi-modal distribution, a sample size of 25 or more is a reasonable target to aim for (McKillup and Dyar, 2010).

The reference water level RWL defined from our set of observations is simply the sample mean:

$$\mathrm{RWL} = \bar{X} = \frac{1}{n}\sum_{i=1}^{n} X_i \qquad (2.6)$$

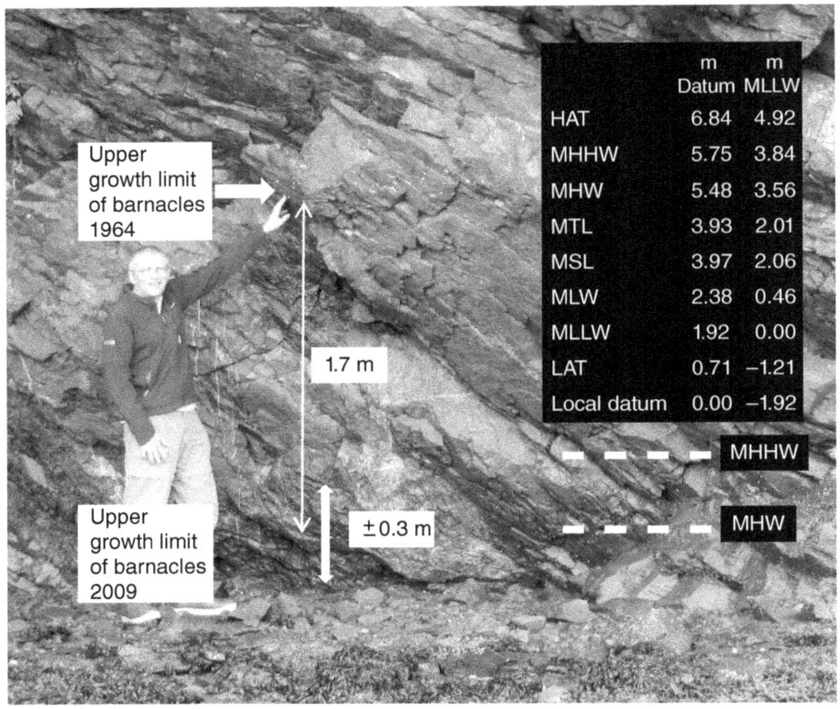

Fig. 2.6. Line of the upper limit of dead barnacles *Balanus balanoides* ~1.7 m above the equivalent upper limit of live forms, and the tidal information for the tide gauge at Cordova. To allow for the present variability of the upper limit of barnacles in south-central Alaska, Plafker (1969) suggests ±0.30 m. *Source*: Photograph Ian Shennan.

Fig. 2.7. Marsh sequence, 10 km east of Cordova. Two buried peat layers, dated at ~850 and 1500 cal a BP separated by intertidal silt. A temporary bench mark (TBM) was surveyed to 4.10 m above local zero datum, and the lower peat is 0.30 m below the TBM (values for illustration only). No modern equivalent peat forming at the site, but studies elsewhere indicate the indicative meaning for the formation of peat adjacent to tidal flat is a reference water level of midway between MHHW and HAT and an indicative range of ±0.25 m. *Source*: Photograph by Ian Shennan. For color details, please see Plate 1.

To determine the indicative range, we first calculate the sample variance s^2 and the sample standard deviation s, where

$$s^2 = \frac{1}{n-1}\sum_{i=1}^{n}\left(X_i - \bar{X}\right)^2 \tag{2.7}$$

We then calculate the indicative range IR using the 95% confidence interval, as ±1.96 times the sample standard deviation, that is,

$$\mathrm{IR} = \pm 1.96s \tag{2.8}$$

In this example of the upper limit of barnacles, we would report the indicative meaning as Mean High Water ±0.30 m, where Mean High Water is the reference water level and ±0.30 m is the indicative range.

As noted above where the fossil indicator is measured direct to the modern equivalent, in our example relative sea-level change is simply 1.7±0.30 m. We should add an additional uncertainty by measuring the variation in the upper limit of the dead barnacles, say ±0.20 m, so we sum the uncertainty terms to obtain the total elevation error estimate e_e using the general equation:

$$e_e = \sqrt{e_1^2 + e_2^2 + \ldots + e_n^2} \tag{2.9}$$

where $e_1, \ldots, e_n$ are the separate sources of uncertainty, that is, $\sqrt{0.20^2 + 0.30^2} = 0.36$ m.

But in most cases we need to take the intermediate stage of relating the elevation of the fossil indicator to another datum, typically via surveying to benchmarks, GPS, or local temporary tide measures (Chapters 10 and 11). If the live forms were not seen in Figure 2.6, we would have surveyed from the fossil barnacles to a benchmark nearby and recorded the elevation with respect to the local datum, obtaining a value of 7.18 m. Using tidal information from a local tide station, we need to know the value of the reference water level, in this case MHW, with respect to the local datum, i.e. 5.48 m. To obtain the value for relative sea-level change we use Equation (2.5) and substitute the elevation of the fossil sample E_s and the reference water level for the sea-level indicator $\mathrm{RWL_i}$, both measured to the same datum. This yields a general equation to apply to every index point:

$$\mathrm{RSL}(\varphi, t_{\mathrm{past}}) = E_s - \mathrm{RWL_i} \tag{2.10}$$

For the more likely scenario of a field site distant from both a tide gauge and the modern example of the sea-level indicator we must measure to a common datum, frequently via a temporary datum set-up in the field (Fig. 2.7). Using the values in Figure 2.7, we derive relative sea level at c. 1500 cal a BP using Equation (2.10). First we measure the elevation of the index point (the base of the peat layer) relative to our temporary benchmark (TBM) in the field, in this case –0.30 m. We then measure the elevation of our TBM relative to local zero datum using a benchmark or other methods described in Chapters 10 and 11. In our example, this is 4.10 m above local datum so E_s = 4.10 – 0.30 m. We now require the reference water level for the base of peat overlying intertidal silt. We may obtain a value from previous literature, our own measurements at another site, or one based on microfossil content and transfer functions (Chapters 12–17, 31). Let us suppose these give us a value that is midway between Mean Higher High Water (MHHW) and Highest Astronomical Tide (HAT), with a 95% confidence interval of ±0.25 m. So $\mathrm{RWL_i}$ = (6.84 – 5.75)/2 m and relative sea-level change in approximately 1500 years is therefore (4.10 – 0.30) – [(6.84 – 5.75)/2] ± 0.25 m. Note that this error term is only the indicative range and excludes further uncertainties that must be quantified (as noted above and in Chapter 37). These should also include surveying errors linking the temporary benchmark to the local datum.

These example calculations all assume that the tidal range parameters have not changed over time, but this is rather unlikely; numerous modeling studies over the last 30 years suggest changes that range from large-scale open-ocean tides through to changes within an estuary or barrier system (Chapter 29). With reference to measuring the indicative meaning of sea-level indicators, this has a number of important implications. We can illustrate this with reference to a typical situation where tidal range presently varies spatially within an estuary (Fig. 2.8).

The Forth Estuary (Fig. 2.8) is typical of numerous estuaries and delta areas, where the high tides increase up-estuary while lower tides may be relatively constant then rapidly diminish, thus reducing the tidal range. Depending on the local geography, increased storage areas in the up-estuary region also mean the height of high tides also decreases; this is called the flood basin effect. These tidal variations must be accommodated when assigning indicative meanings of a sea-level

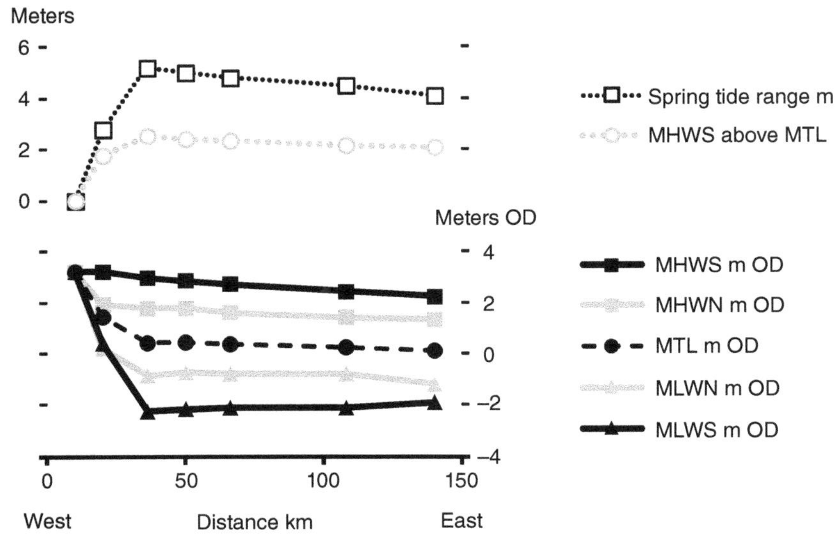

Fig. 2.8. Tidal data for the Forth Estuary, Scotland (Admiralty tide tables). OD: Ordnance Datum, the national datum for the UK.

indicator to the national datum (Chapter 11) if sea-level index points are also surveyed to the national datum. For example, if we were to survey along the Forth estuary the height of a fossil landform that we interpret as indicating a past high tide, measure its height relative to the national datum, and not allow for the trend from west to east, we would obtain an incorrect estimate of any fossil shoreline gradient. We must also consider the paleogeography of the estuary through time, whether the tidal limit was further to the east or to the west at any point. The abrupt decrease in tidal range would have an effect on the interpretation of some types of indicators. Most sea-level reconstructions deal with close-to-high-tide indicators as they are generally the most frequently preserved; for the example in Figure 2.8, these would be affected very little. The exception is where their indicative meaning is calculated based on standardized water levels using the tidal range or half-tidal range (the two upper lines in Fig. 2.8) as the standardizing measure (Chapters 13, 15, and 31).

These all point to the need for more research on modeling past tidal change (Chapter 29) and in particular the validation of tidal model predictions (Scourse, 2013), using different types of sea-level index points that have different reference water levels.

The importance of understanding the principles and definitions of indicative meaning, reference water level, and indicative range cannot be understated. They allow us to reconstruct relative sea-level change over many timescales, centennial to previous interglacials, using multiple lines of evidence derived from all the types of indicators described in later chapters; they also allow us to estimate changes in tidal range. In order for the research community to perform this as objectively as possible, it requires all our archived data to be publically available in a raw form. A simple data table of age and relative sea level with an error term is too little. Archives and databases (Chapter 34) must have all elevation data, such as tidal and national datums relevant to each index point, clearly tabulated. While the original paper may use the data for one purpose, later projects may use the same data to help investigate additional variables, for example, to test the hypothesis of no change in tidal range through the period of time under consideration. As noted earlier, with the opportunities for long-term online storage of data through data centers and journal supplementary files, we have no excuse not to.

2.3.7 Tendency attribute of a sea-level index point

The tendency of a sea-level index point describes the increase (positive sea-level tendency) or decrease (negative sea-level tendency) in marine influence recorded by the index point. The study of sea-level tendencies was developed to analyze sea-level data at a variety of spatial scales, to consider the spatial relationships between different types of sea-level indicator and how they may respond differently to sea-level change, and as an objective means of correlating sea-level chronologies (Shennan, 1982; Shennan et al., 1983). Most analyses of the pattern of sea-level tendencies consider changes at the local scale, either within or between estuaries, such as relationships between horizontal shifts in the coastal sedimentary environments, sediment accumulation, consolidation, and the potential driving mechanism of relatively small (decimeter-scale) vertical changes in sea level (Allen, 1997). In the analysis of larger-scale processes, such as testing relative sea-level predictions from isostatic adjustment models, observations of a consistent change in sign of the tendency across a region are useful in determining the turning points in relative sea-level chronologies, whether from the near-field or far-field (Shennan et al., 2005; Woodroffe, 2009). Furthermore, sea-level tendencies show whether relative sea level was slightly above or below present in the last few thousand years, thus identifying the geographic limits of the transition from mid-Holocene relative land uplift to relative submergence. Identification of sea-level tendencies was central to the development of some microfossil-based approaches to the reconstruction of relative sea-level changes throughout multiple earthquake deformation cycles (Long and Shennan, 1994) and the subsequent proposal of specific criteria for assessing evidence of coseismic submergence (Nelson et al., 1996). Overall, there has been much less focus on the identification and analysis of sea-level tendencies compared to the age and elevation attributes. In Section 2.4 we consider further applications of the sea-level tendency method that may enhance future research efforts.

2.3.8 Dissemination and display of sea-level data

Many articles published since the original Manual (van de Plassche, 1986b) demonstrate the values of one of the tasks set by the Manual: to encourage different disciplines to discuss and share knowledge and experience of the wide variety of evidence available. While data tables, online repositories, and databases go a long way to aiding dissemination of evidence there remains room for improvement. The time and expertise required for database compilation and validation should not be underestimated (Chapter 34), and partly explain the absence of any integrated global database of sea-level index points that includes all the necessary information. Although an age–elevation plot of sea-level index points and accompanying data table of age and relative sea level with error terms are effective summaries, and modern digital archives are easily transferable, they frequently omit information that could be used by subsequent users. This missing information may be as important as the index points on the graph and in the table, but to extract it requires detailed reading of the rest of the paper as well as summarizing in an equally understandable and accessible format. The early databases focused on radiocarbon-dated index points, and data storage, manipulation, and validation were limited by computing software and programming skills. Modern hardware and software provide the potential for much more sophisticated approaches.

The following examples (Fig. 2.9) illustrate a few of the possible enhancements that should be considered in future data repositories and databases: (1) presentation and enumeration of relative sea-level change before or after a dated sample or between dated samples in the same core or section; (2) incorporation of limiting data (defined in Section 2.3.6) in both hypothesis testing and quantitative measure of best fit; and (3) display of uncertainty and in the measurement of goodness of fit.

Figure 2.9a shows the data from raised tidal marshes and isolation basins in Knapdale, Scotland (Shennan et al., 2012). Using just the sea-level index points (open circles), we may favor one model (solid line) over the other based either on visual fit or a least-squares measure, taking into account the distribution of residuals and clustering of ages. If we add in the freshwater limiting dates (shaded circles), we may revise our conclusion and reject the solid line model as it lies well above the oldest sample. Limiting dates are less straightforward to apply in a least-squares calculation, as a model prediction that is for example 0.1 m below a freshwater limiting date is no more valid than one 10 m below but it would be erroneous to use the amount of disagreement, for example distance above such a limiting date, as a quantitative measure of fit. In fact the sedimentary data behind each index point include even more information that we may use to test the finer details of sea-level changes. The RSL model predictions in Figure 2.9a and b reveal the predicted local expression of meltwater pulse 1a, an oscillation at ~14 cal ka BP, and the magnitude of the mid-Holocene highstand that constrain different

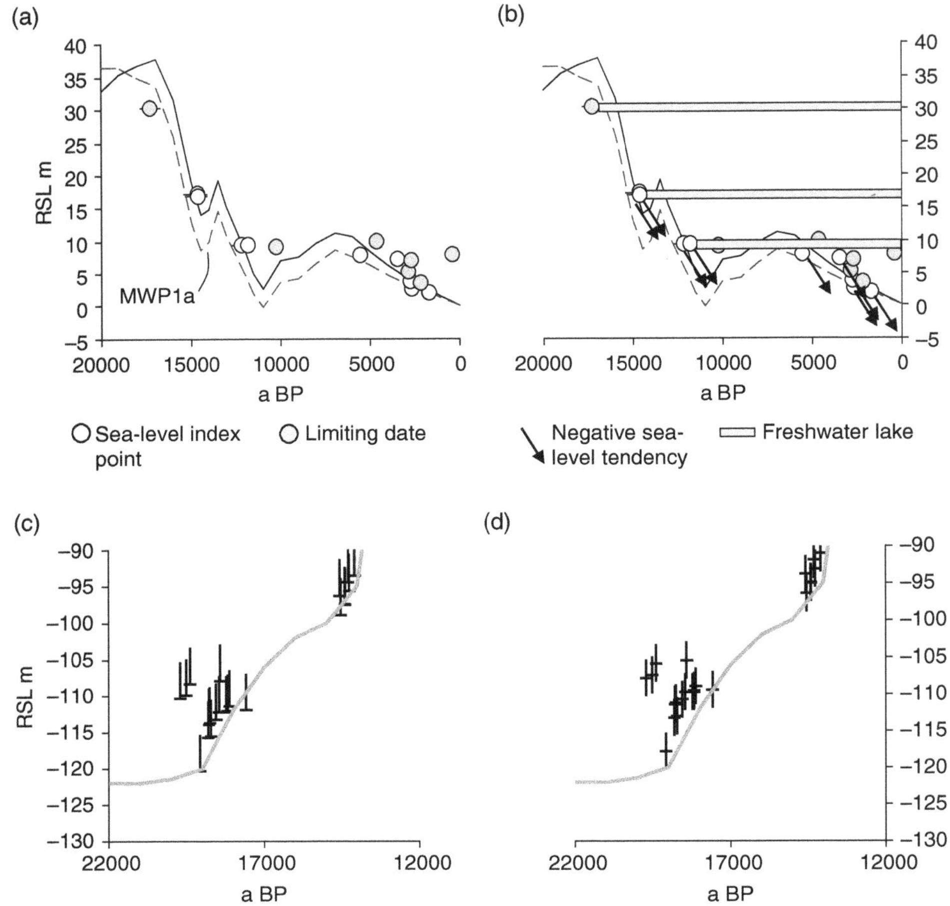

Fig. 2.9. Visualization of sea-level data. (a) Sea-level index points, terrestrial limiting dates and two GIA model predictions of RSL (Shennan et al., 2012). Error terms (95% confidence limits) are visible where they are greater than the size of the symbol. (b) Same data with additional information based on the sedimentary record of each index point. *Source*: Data for (a) and (b) from Shennan et al. (2012). (c) Part of the Barbados record sea-level index points using only samples from *Acropora palmata* (Peltier and Fairbanks, 2006) and a model prediction of RSL, horizontal bars indicate sample depth. (d) Same Barbados data, except that the horizontal bars indicate the mid-point of the indicative range. *Source*: Data for (c) and (d) from Peltier and Fairbanks (2006).

models of global meltwater discharge. We may therefore wish to utilize the information on the tendency of sea-level movement to see how these fit with the direction of change in the model predictions (Fig. 2.9b). We can also use the fact that the three lakes were always freshwater from the time of the dated horizon to the present. For the lake at the middle elevation this information would lead us to reject the oscillation at ~14 cal ka BP in the solid line model; the lower elevation lake is freshwater in the mid-Holocene, which would also lead us to reject the solid line model.

The next generation of databases should aim to incorporate such data in a form that wider communities can access without recourse to the detail of individual site descriptions, microfossil diagrams, or the results of transfer function analyses. It will mean a different structure, allowing for interpretations of both elevation and age parameters. This may seem a step too far removed from "observations" of sea level, but this is not really the case. As noted before, all reconstructions of past sea level involve some sort of judgment, interpretation, or modeling. After all, every sea-level database so far created has had to be recalibrated each time a new radiocarbon calibration dataset comes out. These new databases will not only have the dated samples themselves, but also for those parts of the record younger or older than the sample for which we can obtain viable age and elevation estimates. This allows us to interpolate relative sea-level changes at time-steps that depend on the resolution of the age model derived from the sequence, with examples using these approaches ranging from sea-level changes at the Late Glacial Maximum to the last few centuries (Yokoyama et al., 2001, 2003; Shennan and Milne, 2003; Kemp et al., 2011; Barlow et al., 2012).

We noted above the complication of making data available in one form that will both summarize the data for ease of interpretation and make the data usable for different analytical approaches. We can see from past articles that some field-based scientists take exception to the apparent lack of understanding of landforms and sediments in the way others subsequently use their observations in numerical analyses of large datasets (McCabe, 2008). We hope that by adopting the approaches and definitions outlined above and in the other chapters in the Handbook we can go some way to minimizing any misunderstandings. How we plot and tabulate our data can either help or unintentionally hinder future users of that information. Consider the portrayal of the sea-level data in Figures 2.9c and d which illustrate a complication that may arise in the translation of field data to data tables and subsequent numerical analysis. The Barbados sea-level record is one of the most used datasets. While there remains debate on the indicative meaning for corals (Chapter 7), let us take the conservative estimate that those on the species *Acropora palmata* represent environments that form in up to 5 m water depth. Figure 2.9c and d replicate two approaches that have been used to plot the same data with the same correction for long-term uplift. In Figure 2.9c we plot the horizontal bar at the elevation of the sample, so all of the 5 m vertical uncertainty plots above (Peltier and Fairbanks, 2006). In contrast, we plot the horizontal line at the mid-point of the elevation uncertainty in Figure 2.9d (Peltier et al., 2002).

The plots show the same vertical range for each index point but they raise a series of interesting questions. First, does one method help more in visualizing and explaining the data? Is it easier to see from Figure 2.9c that the horizontal line represents the past sea level inferred from the upper limit of coral growth and the inverted "T" is a good visual representation of the preferred living depth of *Acropora palmata*? Or does the graphical method unintentionally imply a bias to that elevation for which there is no supporting evidence? This is out of the control of the writer, but different readers may make different inferences from that intended. Consider too the evaluation of fit to a model prediction. Does the position of the horizontal line influence the reader's assessment of goodness-of-fit between the model and the data? Most likely, if we asked 100 readers we would get a range of answers across all three possibilities in favor of 9c, 9d, or no difference. If we take this to the next stage removed from the original article and use the age and elevation data in a numerical analysis of goodness-of-fit, how do we account for the uncertainty error? This is an active area of research (see Chapter 32), but suffice to say that if the goodness-of-fit uses only the intersection of the horizontal and vertical lines for each data point then the method of portraying the data in the original paper would give a systematic offset equivalent (in this example) to half of the indicative range.

2.4 FRAMING A RESEARCH QUESTION

Many of the challenges of reconstructing past sea level are not new. Godwin (1940) plotted 34 index points, with error ranges, on an age–elevation graph to quantify mid and late Holocene relative sea-level change in the Fenland of eastern England. The paper presented many of the issues that still confront sea-level researchers. Godwin highlighted the spatial variations within the Fenland at which peat and minerogenic deposits could form in relation to tidal cycles and the importance of considering vegetation succession in interpreting relative sea-level data. He also highlighted the need to sometimes look beyond the immediate study area in order to seek a modern equivalent of a fossil environment, and also the importance of sediment compaction. He understood how scale could be used to help discriminate between local and regional events. The discovery of the radiocarbon dating technique in the 1950s provided the stimulus for many new studies around the world, and a lively debate ensued regarding the nature of eustatic or global sea-level change since the LGM. Since the 1960s we have seen a recurring debate on the general pattern of eustatic sea-level change since the LGM; was it a generally smooth function over the last 10,000 years, or characterized by a series of sub-millennial oscillations of decimeter–meter scale superimposed on the broader scale pattern of change? Are these climatically induced changes, or noise in the record and uncertainties in the age and elevation estimates of sea-level points (Curray, 1961; Fairbridge, 1961; Jelgersma, 1961; Bloom, 1964; Mörner, 1969; Tooley, 1974)?

Despite better age controls and improved sampling techniques, these complexities continue to confound interpretations and key research questions about the vertical and temporal resolutions with which we can quantify past sea-level change. Much of the difference of opinion may be resolved by a more explicit consideration of scales, both spatial and temporal, within a hypothesis testing framework (Fig. 2.2). Not all studies address these issues explicitly, with the danger that factors important at one scale may not be adequately resolved for a study of processes assumed dominant at another scale. We need to consider the relevant temporal and spatial scales for each of the parameters in Figure 2.3. For each hypothesis we should question what the dependent variables are and what the independent variables are. For example, when studying the spatial pattern of isostatic deformation of the crust, we could either assume that local factors that control each sea-level index point used have been quantified, taken into account, or removed from subsequent analyses; alternatively, we take an estimate of isostatic deformation derived from other analyses and use the dispersion from this estimate as the basis for analysis of local scale factors.

We can develop this scale parameterization by thinking about the linkages of the processes (Fig. 2.3) as a sea-level equation. For each geographical location φ the change in relative sea level (ΔRSL) at time t, where t is the time relative to present (Shennan et al., 2012), is defined:

$$\Delta \mathrm{RSL}(\varphi,t) = \Delta \mathrm{EUS}(t) + \Delta \mathrm{ISO}(\varphi,t) + \Delta \mathrm{TECT}(\varphi,t) + \Delta \mathrm{LOCAL}(\varphi,t) + \Delta \mathrm{UNSP}(\varphi,t) \tag{2.11}$$

$\Delta\mathrm{EUS}(t)$ is the time-dependent eustatic sea level, derived from the model of ice history, that would result by distributing any meltwater evenly across a rigid, non-rotating planet and neglecting self-gravitation in the surface load. $\Delta\mathrm{ISO}(\varphi, t)$ is the total isostatic effect of the glacial rebound process including the ice load (glacio-isostatic), water load (hydro-isostatic), and rotational contributions to the redistribution of ocean mass. $\Delta\mathrm{TECT}(\varphi, t)$ is the tectonic effect, which includes processes operating over long timescales such as plate motions, mountain building, and dynamic topography, and short-term such as uplift and subsidence during great earthquakes. $\Delta\mathrm{LOCAL}(\varphi, t)$ is the total effect of local processes within the coastal system. Since we use observations from the geological record to reconstruct relative sea-level elevation, we can express these as the sum of $\Delta\mathrm{TIDE}(\varphi, t) + \Delta\mathrm{SED}(\varphi, t)$. $\Delta\mathrm{TIDE}(\varphi, t)$ is the total effect of tidal regime changes and any other influences, such as dynamic oceanographic and atmospheric effects, that may change the reference water level of a sea-level index point. $\Delta\mathrm{SED}(\varphi, t)$ is the total effect of sediment consolidation since the time of deposition. Numerous studies show that this can be both a major process in coastal evolution and a key variable in reconstructing relative sea-level change from index points taken from Holocene sediments. $\Delta\mathrm{UNSP}(\varphi, t)$ is the sum of other unspecified factors, either not quantified or thought of. Implicitly, most studies assume their total effect close to zero and random.

When we plot reconstructions of relative sea-level change using geological sea-level index points and find a scatter of data points that do not overlap, this indicates one or more of the following: (1) underestimation of age and/or elevation error terms; (2) plotting data from a too-large spatial extent, so incorporating differential GIA; (3) underestimation of $\Delta\mathrm{LOCAL}(\varphi, t)$; and (4) evidence of $\Delta\mathrm{UNSP}(\varphi, t)$.

Depending upon our working hypothesis, one or more of the factors on the right- hand side of Equation (2.11) are assumed constant, known, or negligible depending upon the appropriate temporal or spatial scale. If these are not considered, then the study is in danger of becoming just a description of sea-level data observed. As the timespan considered becomes smaller, then the spatial scale of dependent variables to be considered also decreases.

Let us develop the large question raised earlier. Is Holocene eustatic sea level, representing the global-scale change in mass of meltwater transferred from the continents to the oceans, generally a smooth function or characterized by a series of sub-millennial oscillations of decimeter–meter scale superimposed on the broader scale pattern of change? This is a different question from looking at oscillations that result from dynamic changes in sea level, that is, $\Delta\mathrm{TIDE}(\varphi, t)$ compared to $\Delta\mathrm{EUS}(t)$, with the former varying through space and time. Even though $\Delta\mathrm{EUS}(t)$ is only time-dependent, we already know through our understanding of the relationships in Figure 2.3b that RSL no single location will record eustatic sea-level change since the LGM, so we require data

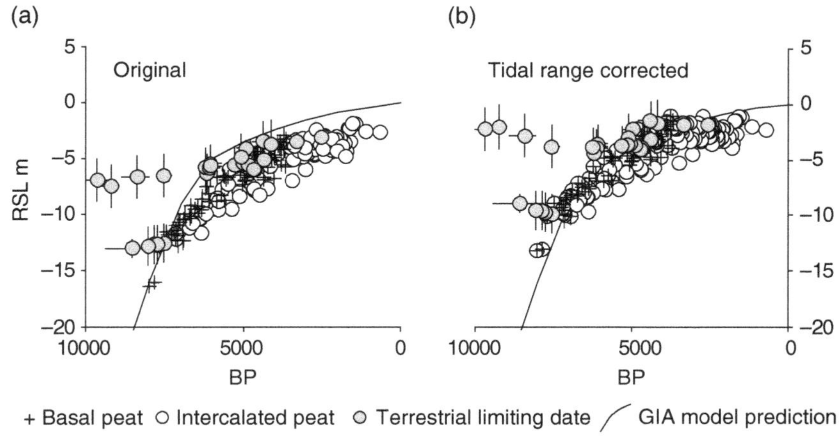

Fig. 2.10. Plots of Holocene sea-level index points from the Fenland Estuary: (a) original data; and (b) elevations corrected for modeled changes in tidal range. *Source*: Data and model predictions from Shennan et al. (2012). Error terms (95% confidence limits) are visible where they are greater than the size of the symbol.

from different regions in order to test the sensitivity of different models of eustasy (Chapter 28), but there is a tradeoff with the other unknowns in Equation (2.11).

We therefore undertake a series of tests, keeping some variables unchanged while varying others systematically. Our first test may be that during the Holocene $\Delta EUS(t)$ is a generally smooth function. Only after testing this, the simplest hypothesis, do we introduce more complicated hypotheses. We should also consider whether data already exist, including those in large databases (Chapter 34), or do we require new data to perform our analyses.

Let us consider this first test using a large number of sea-level index points from the Fenland estuary on the east coast of the UK (Fig. 2.10). The plot of all our sea-level index points shows a scatter of data points that do not overlap (Fig. 2.10a), but on what criteria can we propose to accept or reject our GIA model prediction? We could perform a least-squares fit to determine whether this GIA model performs better than another model. Alternatively, we could use a series of further questions to gain a better understanding. In this example the scatter could indicate one or more of the following: (1) misidentification or limited interpretation of some of the data points; (2) underestimation of age and/or elevation error terms; (3) plotting data from too large a spatial extent, so incorporating differential GIA; (4) underestimation of $\Delta LOCAL(\varphi, t)$; and (5) evidence of $\Delta UNSP(\varphi, t)$.

In addressing the first of these points, knowledge of the different types of index points clarifies what we should expect as a good fit. All the index points that are terrestrial limiting dates should lie on or above the GIA model prediction of relative sea level. Basal data points come from samples at the base of the Holocene sequence, likely to have no or minimal effects of consolidation, whereas intercalated data points from within the Holocene sediment sequence are likely to be at a lower elevation than at the time of deposition due to sediment consolidation. We could aim to model these effects (Chapter 30), or define best fit, by a solution that plots through the scatter of index points from basal peat and within or above the scatter of index points from intercalated peats. On this basis we would reject the GIA model prediction (Fig. 2.10a). We must similarly formulate methods of testing the other points above. For example, modeling tidal range changes (Chapter 29) in combination with using our classification of the different types of index points explicitly tests the issues of underestimating $\Delta LOCAL(\varphi, t)$. It provides a good fit with the GIA model prediction of sea-level change for this region (Fig. 2.10b), suggesting no systematic underestimation of age or elevation error terms or systematic influence of $\Delta LOCAL(\varphi, t)$. In theory the alternative hypothesis remains that these two exist but simply cancel each other out by chance. In the absence of further tests, the pragmatic approach is to accept that they have been adequately accounted for in our analysis. We would then repeat the analysis with the same GIA and ice model parameters to other locations or estuaries to assess the global solution.

But this only goes part of the way to addressing our earlier question of the nature of eustasy and dynamic climatic, oceanographic, and atmospheric influences on regional and local sea-level change. Decimeter to meter scale oscillations could not be rejected from the data in Figure 2.10. So how can we test for changes at this scale? For this question, the basal peat dates actually constrict our analysis since basal peat forms by waterlogging and plant decomposition and this process will not readily record a fall in sea level (unless we can identify a period of drying out and a potential hiatus). We require a better, unbiased sampling design in this situation. In contrast, intercalated peat layers may indicate oscillations in sea level. Field surveys can trace the extent of transgressive and regressive overlaps over many tens of kilometers, but separating the spatial expression of changing sedimentary environments and any vertical change in sea level has been a long-debated topic; there are numerous examples of strongly argued views on both sides (Tooley, 1974; Kidson, 1982). Can we build on these different views to test a hypothesis less focused on strength and persuasiveness of written argument?

We note above that the tendency of a sea-level indicator describes the increase (positive sea-level tendency) or decrease (negative sea-level tendency) in marine influence recorded by the indicator. This provides a framework to bring together the contrasting views regarding sea-level oscillations over centuries and millennia. To identify sub-millenial-scale changes we would expect to record changes at a number of sites within the same area, say an estuary. The expression of the change in vegetation, stratigraphy, morphology, or microfossils will be site specific, but a change in sea level of more than local significance should be recorded over the wider estuary. This change in sea level should be reflected in the mix of tendencies recorded in the system, reflecting whether or not sea-level change is the driving force for coastal evolution at that site at that time. While age–elevation plots of sea-level index points (with their associated errors) from multiple locations in the estuary are unlikely to reveal oscillations of sea level, analysis of the tendency for each index point, for example the lithostratigraphic and microfossil changes above and below each dated sample, could identify trends through time (Shennan, 1982; Tooley, 1982; Shennan et al., 1983). These initial attempts foundered due to insufficient numbers of index points and the precision of many radiocarbon ages. With many more data now available, and increased precision through technical developments in dating indicators and the development of methods to reconstruct relative sea-level change between dated levels (Edwards and Horton, 2000), we can revisit this method of tendency analysis. The combination of quantitative estimates of elevation change (e.g., Chapter 31) and age modeling (Chapter 32) provides a higher-resolution approach comparable to tendency analysis.

Let us return to the question of sub-millennial sea-level oscillations of decimeter–meter scale. Consider three estuaries sufficiently far apart to register different rates of isostatic movement (Fig. 2.11). At the central site the model shows an oscillation as a sine wave, wavelength (λ) 1000 years, and amplitude (A) 0.5 m, a second site where there is a GIA signal of relative uplift of 1 mm a^{-1}, and a third site with a GIA signal of subsidence, shown as −1 mm a^{-1}. These GIA differences modify the relative sea-level curve at each site as shown. We can use these modifications to derive testable working hypotheses. For example, at the central site, sea-level rise occurs for 500 years, from 750 to 250 BP. But at the site with GIA of +1 mm a^{-1} this period is shorter, as it starts later and finishes earlier, and we see the converse at the subsiding site: a longer period of sea-level rise starting earlier and finishing later.

We can utilize this simple relationship to develop a testable hypothesis. A number of recent studies show that depositional age models (Chapter 32) with a sufficient density of high-precision age controls through a sediment profile can provide sub-centennial age resolution with 95% confidence intervals perhaps as precise as ±20 years (Kemp et al., 2011; Barlow et al., 2012; Gehrels et al., 2012). By identifying the age of the turning points in relative sea level at each site we can test which models of sea-level oscillations, defined by their wavelength and amplitude, are compatible or rejected by the age difference and GIA parameters (Fig. 2.11b and c).

It is not uncommon (although potentially frustrating for the authors) for studies to show changes in sediment stratigraphy and microfossil content through profiles that suggest relative sea-level change, but possible oscillations lie within the error terms of reconstructions. Figure 2.12 shows one such example. Here we can hypothesize an oscillation in sea level, supported by changes in individual diatom species, summary assemblages classified by salinity preference, transfer-function-derived reconstructions of sea level, and periods of time dominated by positive and negative sea-level tendencies. We can use the age profile to define the duration of each period and the turning points. Should we have comparable data from another location, we could use the relationships in Figure 2.11 to test hypotheses about the magnitudes of GIA land motions, sea-level change at both sites, and the amplitude and wavelength of the oscillations at both sites. This requires careful consideration of which variables we assume are known and which are unknown. One hypothesis would be that our two sites are still undergoing GIA motions driven by ice melt since the LGM and we have

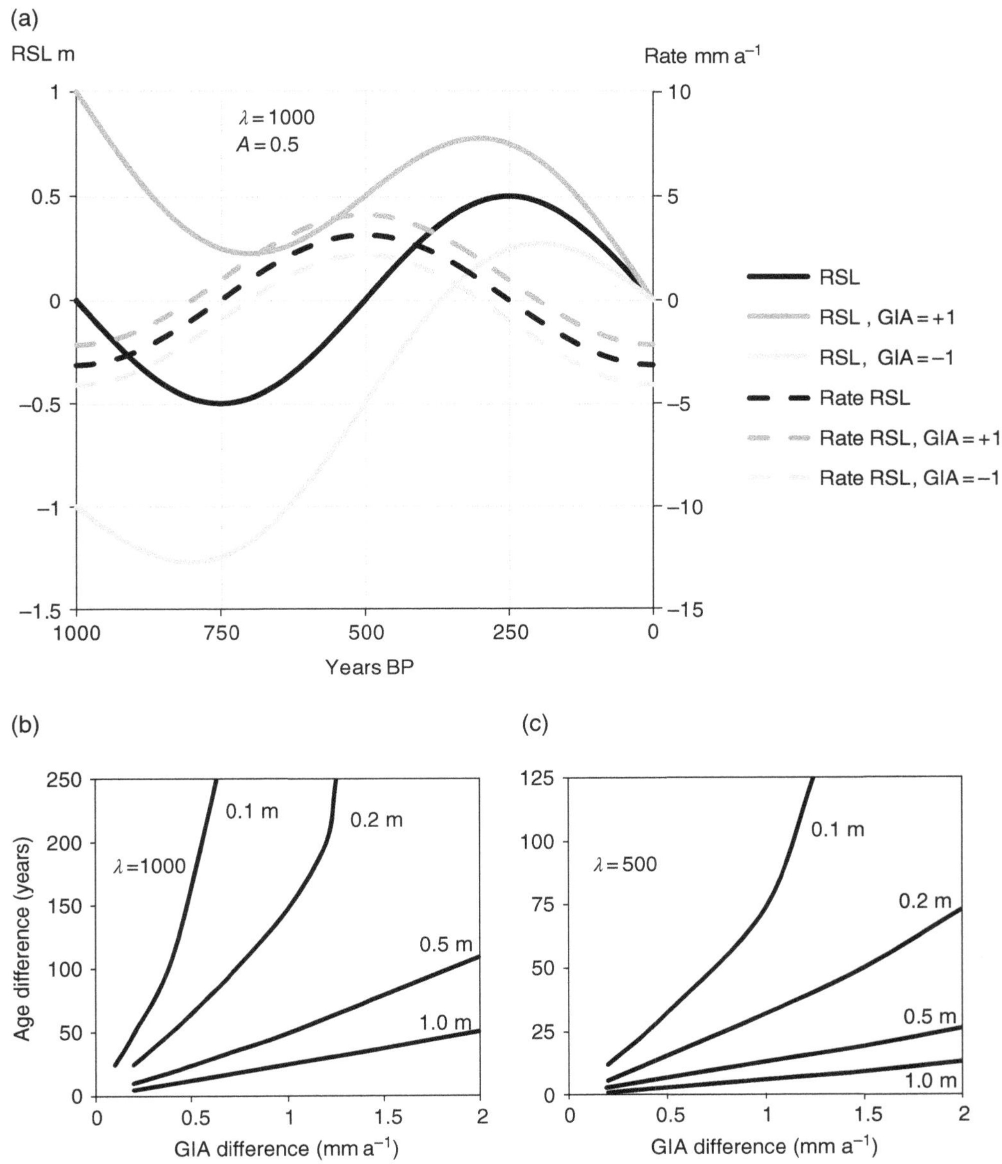

Fig. 2.11. Model for testing decimeter-scale changes in sea level. (a) Three estuaries with different rates of GIA motion. The central site, zero GIA motion, has an oscillation as a sine wave of wavelength λ 1000 years, amplitude A 0.5 m. A second site, where there is a GIA signal of relative uplift of 1 mm a^{-1} and a third site with a GIA signal of subsidence, shown as −1 mm a^{-1}. We can use these modifications to derive testable working hypotheses. For example, at the central site, sea-level rise occurs for 500 years from 750 to 250 BP. At the site with a GIA of +1 mm a^{-1} this period is shorter as it starts later and finishes earlier; the converse is true at the subsiding site. If we know the difference in GIA rate between sites and the difference in age of the change in sign of RSL change, then we estimate the possible amplitude of an oscillation within the range 0.1–1.0 m for: (b) λ = 1000 a; and (c) λ = 500 a.

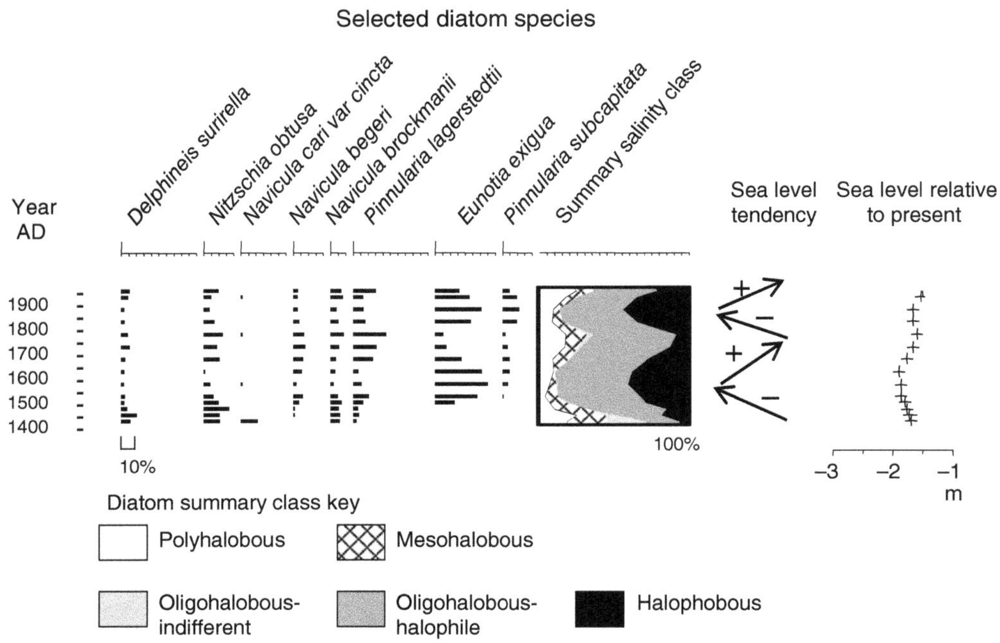

Fig. 2.12. Example of microfossil data indicating oscillations of sea level at Girdwood, Alaska. *Source*: Diatom data and sea-level reconstruction from Shennan and Hamilton (2006) and age model from Barlow et al. (2012). Relative sea-level rise since AD1900 is dominated by ~1.5 m land subsidence during the M_w 9.2 megathrust earthquake in AD 1964.

model predictions of these GIA rates. Our analysis therefore tests the amplitude and wavelength of a sea-level oscillation constrained by the age differences and we consider the sea-level oscillation caused by regional to global processes, such as ice-cap melting or ocean-scale oceanographic effects. Alternatively, in locations close to plate boundaries (e.g., as shown in Fig. 2.12), the GIA signal driven by ice melt since the LGM diminishes much more rapidly; we may therefore hypothesize a common sea-level signal at our two sites, with the sea-level oscillation a test of the spatial and temporal expression of a more local signal. For example, we can frame the hypotheses to test both Earth model parameters and the ice model parameters for events such as Little Ice Age fluctuations (Barlow et al., 2012).

2.5 CONCLUSIONS

To communicate the results of our sea-level research with we must avoid possible confusion by using consistent terminology. The seemingly simple concept of relative sea-level change is a case in point. To avoid confusion, we recommend the use of terms defined in Equations (2.1) to (2.5). Relative sea-level change always includes both time, relative to present or another clearly defined zero, and elevation, between the sea surface and land surface. At any location the present rate of relative land-level change is therefore the negative of relative sea-level change. We emphasize that this is not the same as the rate of vertical land motion. Other common terms that would enhance communication of sea-level research are sea-level indicator, sea-level index point, indicative meaning, reference water level, indicative range, and sea-level tendency. These can be applied across all timescales, from glacial/interglacial cycle to present.

Consider more explicit use of multiple working hypotheses, so that the reader can more easily judge the interpretations with less emphasis on the persuasiveness of the written argument that may reflect too much personal expression. Repeated citations of previous work do not make the interpretation more reliable.

It may be helpful to frame research questions by completing a mental or physical checklist of the likely variables, their status, and your knowledge of them. This could include use of a tabular form

Table 2.2. Checklist for sea-level studies

Scale of my study			
What do I mean by : Site/local/regional/ oceanic/global			
Possible variables	Status*	Source of data	Value or range
Tidal range			
Storms			
River effects			
Oceanographic effects			
Sediment supply			
Compaction			
Site morphology			
Exposure			
Crustal motions			
GIA (total effect)			
Tectonics			
Ice load			
Climate change			
Eustasy			

*Think of a multivariate equation of the form $y = a + bx + cy^t$ to classify variables as dependent, independent, constant, indeterminable, or not relevant at the perceived spatial and temporal scale of study

of Figure 2.3b, leaving extra space for your own additions (Table 2.2).

Other examples of standard practice are explicitly defining your operational definitions, seeking replication of analyses wherever possible, and demonstrating sampling and experimental errors. Surely it should be standard to present the final error terms not as 1σ, but rather as 95% confidence estimates, 1.96 × the standard deviation, or standard error? Total uncertainty in one dimension is calculated as $\sqrt{e_1^2 + e_2^2 + \ldots + e_n^2}$, where $e_1, \ldots, e_n$ are the separate sources of uncertainty, expressed as 95% confidence limits. Do not strive for over-precision. We must accept that the coast is a dynamic environment and we are summarizing some very complex interactions. When we reach a summary measure or diagram and see a scatter that scatter may indicate, for example: misidentification or misinterpretation of some of the data points; underestimation of age and/or elevation error terms; selecting data from an area which is too large so that one of the variables we took as constant is not; underestimating very local, small-scale processes; or the result of a process we had not even considered or thought of.

Finally, we cannot underestimate the value of replication. While our research does not allow for the controlled conditions of experimental science, we can at least strive to replicate our measurements and analyses, and use replicates to test our conclusions. If we conclude relative sea-level change of a certain magnitude and age at one location, is that interpretation compatible with every observation we have made at that site and with other cores or landforms or archeology at the site, within the estuary, and 10 km or 100 km away? If the answer is no, should we reject our interpretation? Or maybe there is more than one explanation.

ACKNOWLEDGEMENTS

Comments by Antony Long, Ben Horton and Ed Garrett improved this contribution.

REFERENCES

Allen, J.R.L. (1997) Simulation models of salt-marsh morphodynamics: some implications for high-intertidal sediment couplets related to sea-level change. Sedimentary Geology, 113, 211–233.

Atwater, B.F. (1987) Evidence for great Holocene earthquakes along the outer coast of Washington State. Science, 236, 942–944.

Barlow, N.L.M., Shennan, I., and Long, A.J. (2012) Relative sea-level response to Little Ice Age ice mass change in south central Alaska: Reconciling model predictions and geological evidence. Earth and Planetary Science Letters, 315–316, 62–75.

Bloom, A.L. (1964) Peat accumulation and compaction in a Connecticut coastal marsh. Journal of Sedimentary Petrology, 34, 599–603.

Carver, G., and Plafker, G. (2008) Paleoseismicity and neotectonics of the Aleutian Subduction Zone: an overview. In *Active Tectonics and Seismic Potential of Alaska* (eds Freymueller, J. T., Haeussler, P. J., Wesson, R., and Ekstrom, G.) American Geophysical Union, Washington, pp. 43–63.

Chamberlin, T.C. (1897) Studies for students: the method of multiple working hypotheses. The Journal of Geology, 5, 837–848.

Curray, J.R. (1961) Late Quaternary sea level: a discussion. Geological Society of America Bulletin, 72, 1707–1712.

Davis, J. (2002) *Statistics and Data Analysis in Geology*, 3rd edition. Wiley, Hoboken.

Devoy, R.J.N. (1987) Introduction: first principles and the scope of sea-surface studies. In *Sea Surface Studies: A Global View* (ed. Devoy, R. J. N.). Croom Helm, Beckenham, pp. 1–30.

Edwards, R.J., and Horton, B.P. (2000) Reconstructing relative sea-level change using UK salt-marsh foraminifera. Marine Geology, 169, 41–56.

Fairbridge, R.W. (1961) Eustatic changes in sea level. Physical and Chemical Earth, 4, 99–185.

Farías, M., Vargas, G., Tassara, A., Carretier, S., Baize, S., Melnick, D., and Bataille, K. (2010) Land-level changes produced by the Mw 8.8 2010 Chilean earthquake. Science, 329, 916.

Gehrels, W.R., Callard, S.L., Moss, P.T., Marshall, W.A., Blaauw, M., Hunter, J., Milton, J.A., and Garnett, M.H. (2012) Nineteenth and twentieth century sea-level changes in Tasmania and New Zealand. Earth and Planetary Science Letters, 315–316, 94–102.

Godwin, H. (1940) Studies of the Post-Glacial history of the British vegetation. III. Fenland pollen diagrams. IV. Post-Glacial changes of relative land and sea-level in the English Fenland. Philosophical Transactions of the Royal Society of London, B, 230, 239–303.

Grand Pre, C.A., Horton, B.P., Kelsey, H.M., Rubin, C.M., Hawkes, A.D., Daryono, M.R., Rosenberg, G., and Culver, S.J. (2012) Stratigraphic evidence for an early Holocene earthquake in Aceh, Indonesia. Quaternary Science Reviews, 54, 142–151.

Jamieson, T.F. (1865) On the history of the last geological changes in Scotland. Quarterly Journal of the Geological Society, 21, 161–204.

Jelgersma, S. (1961) Holocene sea-level changes in the Netherlands. Mededelingen Geologische Stichting, C-IV, 7, 1–100.

Kaye, C.A. (1964) The upper limit of barnacles as an index of sea-level change on the New England Coast during the past 100 years. The Journal of Geology, 72, 580–600.

Kemp, A.C., Horton, B.P., Donnelly, J.P., Mann, M.E., Vermeer, M., and Rahmstorf, S. (2011) Climate related sea-level variations over the past two millennia. Proceedings of the National Academy of Sciences, 108, 11017–11022.

Kench, P.S., Nichol, S.L., and McLean, R.F. (2005) Comment on "New perspectives for the future of the Maldives" by Mörner, N.A., et al. [Global Planetary Change 40 (2004), 177–182]. Global and Planetary Change, 47, 67–69.

Kidson, C. (1982) Sea-level changes in the Holocene. Quaternary Science Reviews, 1, 121–151.

Lohne, Ø.S., Bondevik, S., Mangerud, J., and Svendsen, J.I. (2007) Sea-level fluctuations imply that the Younger Dryas ice-sheet expansion in western Norway commenced during the Allerød. Quaternary Science Reviews, 26, 2128–2151.

Long, A.J., and Shennan, I. (1994) Sea level changes in Washington and Oregon and the "Earthquake deformation cycle". Journal of Coastal Research, 10, 825–838.

McCabe, A.M. (2008) Comment: Postglacial relative sea-level observations from Ireland and their role in glacial rebound modelling (A. J. Brooks, S. L. Bradley, R. J. Edwards, G. A. Milne, B. Horton and I. Shennan (2008) Journal of Quaternary Science 23: 175–192). Journal of Quaternary Science, 23, 817–820.

McKillup, S., and Dyar, M.D. (2010) *Geostatistics Explained: An Introductory Guide for Earth Scientists*, 1st edition. Cambridge University Press, Cambridge.

Milne, G.A., and Mitrovica, J.X. (2008) Searching for eustasy in deglacial sea-level histories. Quaternary Science Reviews, 27, 2292–2302.

Mitrovica, J.X., and Milne, G.A. (2003) On post-glacial sea level: I. General theory. Geophysical Journal International, 154, 253–267.

Mörner, N.A. (1969) The Late Quaternary history of the Kattegat Sea and the Swedish west coast; deglaciation, shore-level displacement, chronology, isostasy and eustasy. Sveriges Geoliska Undersökning, C-640, 1–487.

Mörner, N.A., Tooley, M.J., and Possnert, G. (2004) New perspectives for the future of the Maldives. Global and Planetary Change, 40, 177–182.

Nelson, A.R., Shennan, I., and Long, A.J. (1996) Identifying coseismic subsidence in tidal-wetland stratigraphic sequences at the Cascadia subduction zone of western North America. Journal of Geophysical Research, 101, 6115–6135.

Peltier, W.R., and Fairbanks, R.G. (2006) Global glacial ice volume and Last Glacial Maximum duration from an extended Barbados sea level record. Quaternary Science Reviews, 25, 3322–3337.

Peltier, W.R., Shennan, I., Drummond, R., and Horton, B. (2002) On the postglacial isostatic adjustment of the British Isles and the shallow viscoelastic structure of the Earth. Geophysical Journal International, 148, 443–475.

Plafker, G. (1969) Tectonics of the March 27, 1964, Alaska earthquake. US Geological Survey Professional Paper, 543-I, 74.

Pugh, D.T. (1996) *Tides, Surges and Mean Sea-Level*, 1st edition. John Wiley & Sons, Chichester.

Scourse, J. (2013) Quaternary sea-level and palaeotidal changes: a review of impacts on, and responses of, the marine biosphere. Oceanography and Marine Biology: An Annual Review, 51, 1–70.

Shennan, I. (1982) Interpretation of Flandrian sea-level data from the Fenland, England. Proceedings of the Geologists' Association, 93, 53–63.

Shennan, I., and Milne, G. (2003) Sea-level observations around the Last Glacial Maximum from the Bonaparte Gulf, NW Australia. Quaternary Science Reviews, 22, 1543–1547.

Shennan, I., and Hamilton, S. (2006) Coseismic and pre-seismic subsidence associated with great earthquakes in Alaska. Quaternary Science Reviews, 25, 1–8.

Shennan, I., Tooley, M.J., Davis, M.J., and Haggart, B.A. (1983) Analysis and interpretation of Holocene sea level data. Nature, 302, 404–406.

Shennan, I., Hamilton, S., Hillier, C., and Woodroffe, S. (2005) A 16 000-year record of near-field relative sea-level changes, northwest Scotland, United Kingdom. Quaternary International, 133–134, 95–106.

Shennan, I., Milne, G., and Bradley, S. (2012) Late Holocene vertical land motion and relative sea-level changes: lessons from the British Isles. Journal of Quaternary Science, 27, 64–70.

Tooley, M.J. (1974) Sea-level changes during the last 9000 years in northwest England. The Geographical Journal, 140, 18–42.

Tooley, M.J. (1982) Sea-level changes in northern England. Proceedings of the Geologists' Association, 93, 43–51.

van de Plassche, O. (ed.) (1986a) Introduction. In *Sea-level Research: A Manual for the Collection and Evaluation of Data*, Geo Books, Norwich, pp. 1–26.

van de Plassche, O. (ed.) (1986b) *Sea-Level Research: A Manual for the Collection and Evaluation of Data.* GeoBooks, Norwich.

Woodroffe, C.D. (2005) Late Quaternary sea-level high-stands in the central and eastern Indian Ocean: A review. Global and Planetary Change, 49, 121–138.

Woodroffe, S.A. (2009) Testing models of mid to late Holocene sea-level change, North Queensland, Australia. Quaternary Science Reviews, 28, 2474–2488.

Woodworth, P.L. (2005) Have there been large recent sea level changes in the Maldive Islands? Global and Planetary Change, 49, 1–18.

Yokoyama, Y., De Deckker, P., Lambeck, K., Johnston, P., and Fifield, L.K. (2001) Sea-level at the Last Glacial Maximum: evidence from northwestern Australia to constrain ice volumes for oxygen isotope stage 2. Palaeogeography, Palaeoclimatology, Palaeoecology, 165, 281–297.

Yokoyama, Y., Deckker, P.D., and Lambeck, K. (2003) Reply to "Sea-level observations around the Last Glacial Maximum from the Bonaparte Gulf, NW Australia" by I. Shennan and G. Milne. Quaternary Science Reviews, 22, 1549–1550.

Chapter 3
Pre-fieldwork surveys

ROBERT C. WITTER
US Geological Survey, Alaska Science Center, Anchorage, AK, USA

3.1 INTRODUCTION

In sea-level studies, initial surveys at the office or library can increase a project's likelihood of success. Pre-fieldwork surveys should begin with a thorough review of prior research literature that appraises available data, identifies data gaps, and places the project objectives into a broader scientific context. Whereas peer-reviewed journal articles may contain a wealth of research findings, often the most useful maps, historical documents, images, and other data critical for sea-level research are discovered by searching government files, libraries, museums, unpublished reports, or, increasingly, online digital data collections. The following section includes examples of online sources of data, with an emphasis on available resources in the United States (comparable online resources are available in many other countries).

The common datasets used in sea-level studies, including topographic data, bathymetric data, remote sensing data, tide gauge data, and tidal predictions, are reviewed in this chapter. It is then discussed how data uncovered during pre-fieldwork surveys can be used to assess the geomorphology of a study area, which may focus research questions and help formulate hypotheses about sea-level history. The chapter concludes with a summary of how historical charts, geomorphological interpretations, and written accounts from AD 1700 of a tsunami in Japan laid the foundation for subsequent research and, in the latter example, led to a scientific breakthrough.

After reviewing diverse data pertinent to the research topic, the next step in a pre-fieldwork survey is to assess the geomorphology of the study area. An important initial objective is to map the geomorphology of the study site and surrounding area. A geomorphological map depicts shorelines and other coastal landforms that indicate past sea levels and identify sites where evidence of sea-level changes might be examined in the field. Most field-based sea-level studies begin with interpretation of coastal landforms that indicate past sea level, which are described at greater length in Chapter 5. The increasing availability of high-resolution imagery and geographical information system (GIS) data of much of the Earth simplifies this task, which is now routine with the advent of digital globes such as Google Earth (Fig. 3.1). Preliminary geomorphological mapping of surficial deposits may show the spatial distribution of coastal sediment. Methods to sample, map, describe, and interpret coastal sediment deposited in quiet-water environments are summarized in Chapter 4. Vertical profiles across coastal landforms may also illustrate the magnitude of sea-level change. Finally, shoreline mapping techniques used to assess historical shoreline change are reviewed. Assessing and mapping the geomorphology of the study area before going into the field has several advantages, including: (1) promoting effective hypothesis testing; (2) aiding site selection and evaluation; (3) highlighting the main processes that have shaped the local landscape; and (4) providing baseline data on past sea levels.

Most pre-fieldwork surveys can be accomplished using standard interpretive methods in geomorphology with simple tools and access to resources for literature review, data compilation, and image analysis. Standard geomorphological mapping techniques are described by Smith et al. (2011). Coastal processes and geomorphology are addressed in a number of textbooks (e.g., Curray, 1964; Masselink and Hughes, 2003; Bird, 2008; Anthony, 2009). The tools required for the exercises discussed here include a stereoscope for viewing stereo-paired images, cartographic tools, and digital mapping (GIS) software (e.g., Google

Handbook of Sea-Level Research, First Edition. Edited by Ian Shennan, Antony J. Long, and Benjamin P. Horton.

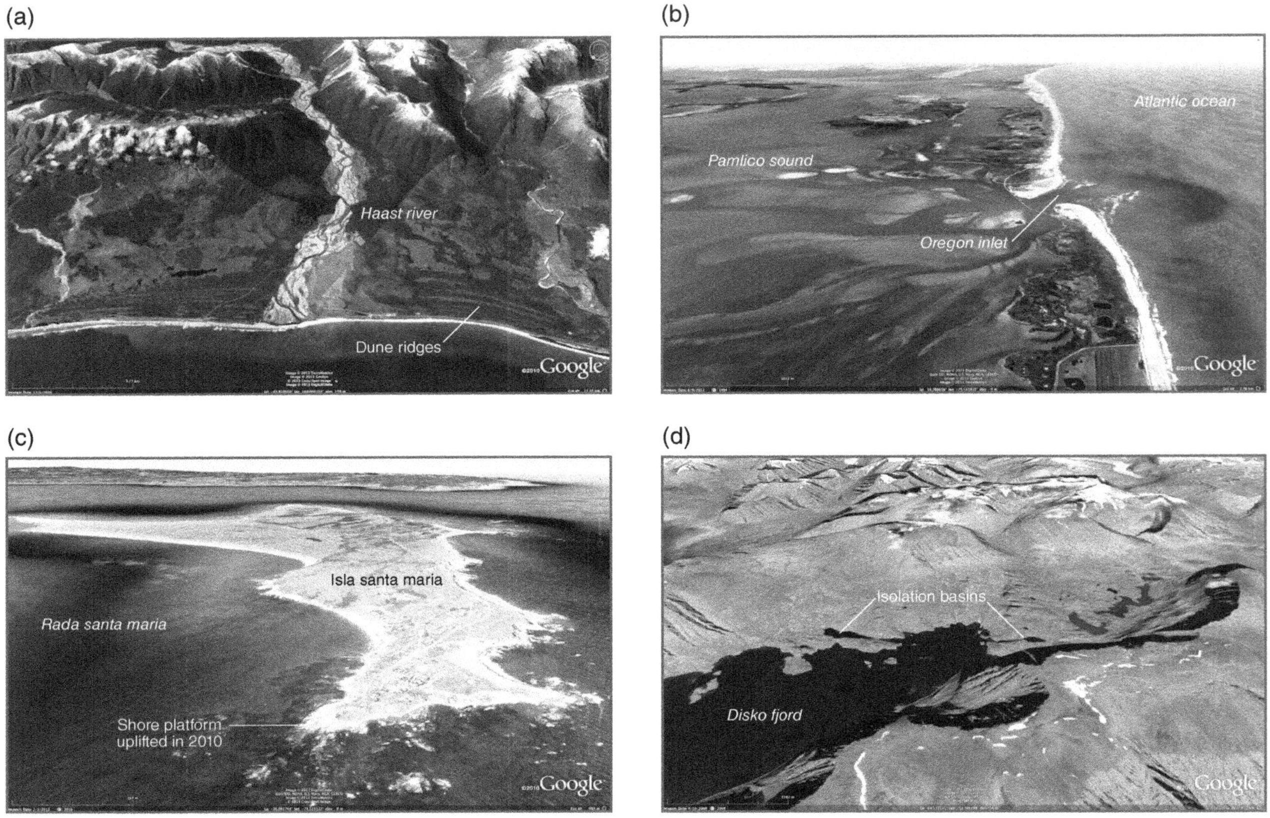

Fig. 3.1. Examples of coastal landscapes viewed with Google Earth. (a) Prograding coastal dune ridges in South Westland, New Zealand mark progressive seaward shifts of the shoreline in response to increased sediment loads in the Haast River following earthquakes on the Alpine Fault (Wells and Goff, 2007). (b) Late Holocene sea-level rise controls the configuration of Oregon Inlet and the barrier islands of the Outer Banks, North Carolina, USA. (c) Sudden tectonic uplift during the 2010 Maule earthquake in central Chile exposed shore platforms surrounding Isla Santa Maria near Concepción. (d) Sediments of isolation-basin lakes around the shores of Disko Fjord in western Greenland record relative sea-level changes related to glacial isostatic adjustment caused by changes in mass of the Greenland ice sheet (Long et al., 2011). *Source*: Google Earth images (© 2010 Google) include data from (a) TerraMetrics, GeoEye, Cnes/Spot Image, and DigitalGlobe (b) DigitalGlobe, GeoEye, and TerraMetrics; (c) DigitalGlobe, NOAA, US Navy, NGA, GEBCO, TerraMetrics, and Cnes/Spot Image; and (d) DigitalGlobe, GeoEye, US Geological Survey, and IBCAO. For color details, please see Plate 2.

Earth, GRASS GIS, ArcGIS). Methods involving geomorphological analysis with sophisticated GIS software or computer modeling are beyond the scope of this chapter.

Thorough surveys in advance of fieldwork often include communication with local officials, landowners or residents, and, most importantly, a site visit. Some researchers may view these initial efforts at assessing and exploring a site as too costly or unnecessary. However, in my experience, a few phone calls and a site visit can optimize site selection, streamline logistical planning, facilitate land access and permitting, and help assess the practical needs to successfully carry out a demanding field campaign. For example, thorough pre-fieldwork preparation can discover important information about tides, river discharge, wildlife, and sea-ice and other weather-related conditions that may influence sampling or access to the site. Time invested in establishing local contacts and becoming familiar with the site may pay dividends in the long term by minimizing delays and reducing unanticipated costs. The value of seeking local knowledge of shoreline history should not be underestimated. Residents and others familiar with the coastal landscape may offer personal accounts or leads on information crucial to the study, which can be followed up with research and review. Making local contacts often has the added benefit of facilitating transfer of scientific findings to the public.

Along the Pacific Northwest coast of the US, fruitful interactions with a local contact helped educate the public about tsunami hazards. Al Aya, a resident of Cannon Beach, Oregon, provided crucial local knowledge about the impacts of the 1964 Alaska tsunami and helped use science findings (Witter, 2008) to increase tsunami preparedness among community members and local stakeholders. Key ties to the Cannon Beach Historical Society helped unearth historical photographs of buildings damaged by the tsunami and the destruction of a bridge transported by the tsunami, providing important validation for tsunami simulations (Witter, 2008). In addition, his position as president of the Cannon Beach Rural Fire Protection District helped communicate the results of the tsunami study to emergency response personnel responsible for tsunami hazard planning.

3.2 DATA COMPILATION

Collecting and assembling data during pre-fieldwork surveys can be carried out more efficiently and cost-effectively if the researcher knows the types of data that will most likely aid the research, where to acquire the data, and how the data will be interrogated. This section briefly reviews common forms of topographic, bathymetric, remote sensing, and other data relevant to sea-level studies. Usually, the type of data desired will determine what data sources to explore and the appropriate tools needed for geomorphological interpretation and analyses. Old charts and drawings, topographic maps, aerial photography, microfiche, and research papers can be found at university libraries, museums, and government data centers. Increasingly, these often-rare data are scanned or digitized, and made available electronically.

Digital maps, imagery, and other data are becoming more accessible through public and commercial websites (e.g., Fig. 3.1). Compiling digital data is most useful when it can be rendered with computer applications (e.g., Google Earth, ArcGIS, NASA's World Wind) that allow images and other data to be compared, georeferenced, interpreted, and analyzed. Greater volumes of data are released into the public domain every year, including the near-global coverage of Shuttle Radar Topography Mission (SRTM) data released in 2004. There are many online sources for digital aerial photography, topographic data, satellite imagery, elevation models, and thematic layers. The following institutions, which offer free downloads of global-scale data, provide an initial starting point: Center for Southeast Asian Studies Kyoto University (http://www.cseas.kyoto-u.ac.jp/database/database _en .html); CGIAR Consortium for Spatial Information (http://srtm.csi.cgiar.org/); the National Oceanic and Atmospheric Administration's (NOAA) National Geophysical Data Center (NGDC, http://www.ngdc.noaa.gov/mgg/topo/); and US Geological Survey's Earth Resource Observation and Science (EROS) Center (http://eros.usgs.gov/). Also look into data available from local governments and academic institutions in the region being studied.

When compiling data prior to fieldwork, consider the field objectives and scope of the research plan to ensure that the data collected will be useful and of sufficient quality to meet the research goals. Before starting a search, specify the time frame of relevant data, the spatial resolution required, a schedule, and the budget available for data acquisition. Determine what tools and computer software are necessary to view, interpret, and analyze the data. Thoughtful planning will promote a more efficient, cost-effective pre-fieldwork survey. For example, a preliminary survey of a site only a few tens of square kilometers in area may be completed quickly with access to maps, digital data, and high-resolution satellite images (e.g., Pleiades, GeoEye, IKONOS sensors at 0.4–0.8 m resolution; Table 3.1) available at public institutions and online sources such as Google Earth. However, if the survey spans hundreds of square kilometers, the costs of data compilation and acquisition will likely increase. In these cases, smaller-scale images acquired by medium-resolution sensors (e.g., SPOT and Landsat at 20–30 m resolution) may be more cost-effective.

3.2.1 Topographic and bathymetric data

Topographic data depict natural and cultural features on the Earth's surface and their relief. Such data include early drawings surveyed with plane table and alidade, and topographical maps derived by photogrammetry techniques. In sea-level applications, topographical maps are indispensable because they delineate lines of equal elevation (contours) which relate landforms to their vertical position above sea level. National Ocean Service (NOS) T-sheets (e.g., Fig. 3.2b), one type of early US topographical map dating from

Table 3.1. High- to moderate-resolution satellite sensors for remote sensing applications

Satellite sensor	Operator	Period of operation	Resolution (m)	Spectral range	Mapping applications
GeoEye-1	DigitalGlobe	2008–present	0.4 1.7	Panchromatic, visible-NIR	Interpretation, spectral analysis, DEM
IKONOS	DigitalGlobe	1999–present	0.8 3.2	Panchromatic, visible-NIR	Interpretation, spectral analysis, DEM
Pleiades-1A, Pleiades-1B	Astrium	2011–present	0.5 0.5	Panchromatic, visible-NIR	Interpretation, spectral analysis, DEM
Worldview-1, Worldview-2	DigitalGlobe	2007–present	0.5 2.0	Panchromatic, Visible-NIR	Interpretation, spectral analysis, DEM
QuickBird	DigitalGlobe	2001–present	0.6 2.4	Panchromatic, visible-NIR	Interpretation, spectral analysis, DEM
SPOT-5, SPOT-6	Astrium	2002	1.5–2.5 5.0–6.0	Panchromatic, visible-NIR	Interpretation, spectral analysis, DEM
ALOS	NASDA	2006–present	2.5 10 10–100	Panchromatic, nisible-NIR, microwave	Interpretation, spectral analysis, DEM
RapidEye	RapidEye	2008–present	6.5	Visible-NIR	Interpretation, spectral analysis
Corona	CIA	1960–1972	1.8–7.5	Panchromatic	Interpretation
Terra ASTER	NASA	1999–present	15 30 90	Visible-NIR, shortwave-IR thermal-IR	Interpretation, spectral analysis, DEM
Landsat 5TM, Landsat 7ETM[2]*	NASA	1984–present	15 30 60–120	Panchromatic, visible-NIR thermal-IR	Interpretation, spectral analysis, DEM
TerraSAR-X	Astrium	2008present	1, 3, 18	Microwave	DEM
Envisat	ESA	2002–2012	30–150	Microwave	DEM
ERS-1, ERS-2	ESA	1991–2011	25	Microwave	DEM
SRTM	NASA	2000	30–90	Microwave	DEM

CIA: US Central Intelligence Agency; DEM: digital elevation model; ERS: European Remote-Sensing satellite; ESA: European Space Agency; IR: infrared; NASA: National Aeronautics and Space Administration, USA; NIR: near-infrared; NASDA: National Space Development Agency of Japan; SRTM: Shuttle Radar Topography Mission.
*Landsat-1 was launched in 1972 and provided some of the earliest satellite-based remote sensing imagery.

the 1830s, depict the boundary between land and sea as the high water line (HWL). The HWL is a proxy-based shoreline, typically the wet-dry sand line on the beach surveyed after high tide (Shalowitz, 1964). The National Oceanic and Atmospheric Administration (NOAA) has digitized and georeferenced NOS T-sheets and posted the entire collection for download and viewing via Google Earth (http://specialprojects.nos.noaa.gov/tools/shorelinesurvey.html). Other forms of digital topographic data include digital line graphs (DRG) consisting of point, line, and polygon vector data, digital raster graphics (DRG; usually scanned map images), and digital elevation models (DEMs). Digital elevation models represent the surface of the Earth (and sea floor) in the form of a data grid that assigns an elevation to each point or pixel. DEMs can be derived from photogrammetry, interferometry using synthetic aperture radar (SAR) data, and light detection and ranging (LiDAR) point clouds.

Bathymetric data represent the morphology and relief of the seafloor, the underwater equivalent of topography. For sea-level research, early bathymetric data from soundings, nautical charts, and hydrographic surveys may document historical sea-level changes (e.g., Fig. 3.2a). Digital bathymetric data are available for many of the Earth's oceans and include trackline data, hydrographic surveys, sidescan sonar, multibeam bathymetry, and LiDAR bathymetry. The NGDC has been archiving and distributing digital bathymetric data collected by international marine cruises since the early 1950s. Worldwide digital bathymetric data are freely available from NGDC and the International Hydrographic Organization.

Historical navigation charts and maps of Oregon Inlet (Fig. 3.2), a passage through the northern

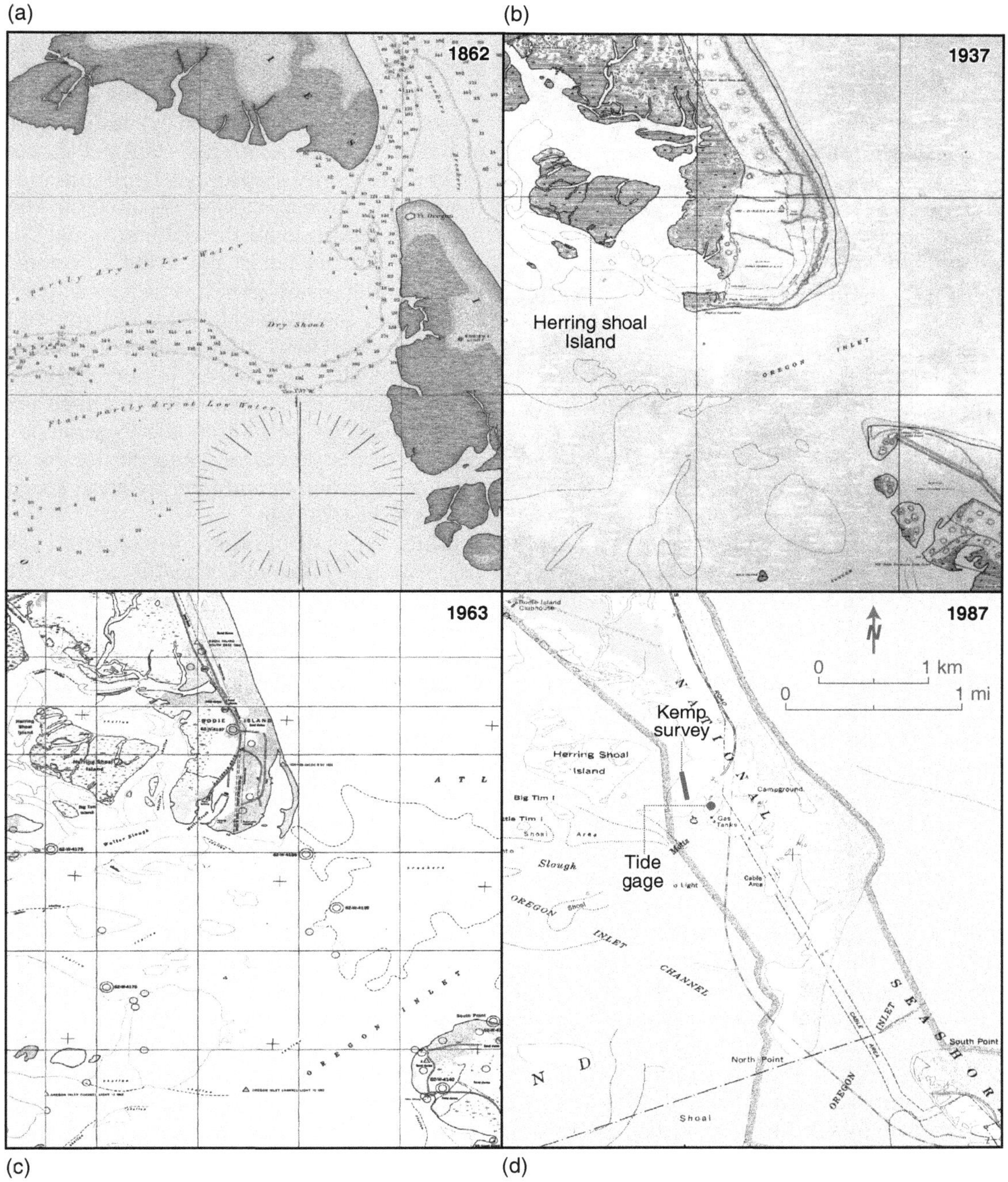

Fig. 3.2. Historical charts and topographic maps (at same scale) track the southward migration of Oregon Inlet over 125 years on the Outer Banks, North Carolina. (a) Herring Shoal Island is absent from the 1862 US Coast Survey chart, but prominent on (b) the 1937 NOS T-sheet (#5580). Both the 1937 T-sheet and (c) the 1963 chart depict the development of a broad saltmarsh east of Herring Shoal Island and north of Motts Creek. (d) The 1987 USGS topographic map shows the location of a tide gauge and field survey of modern foraminifera used in a sea-level study by Kemp et al. (2009).

barrier islands of the Outer Banks, North Carolina, USA, offer an example of how topographic data may aid in selection of a study site and addressing research questions. Oregon Inlet, which connects Pamlico Sound with the Atlantic Ocean (Fig. 3.1b), opened during a hurricane in 1846; by 1989 the inlet had migrated over 3.9 km south to near its present location (Pilkey and Fraser, 2003). The natural instability of the transgressive, overwash-dominated barrier island system presents many challenges to Outer Banks communities and the local economy. Since the early 1950s, construction and engineering projects have concentrated on stabilizing the inlet and adjacent islands, including construction of the Oregon Inlet Bridge (1962–1963; Fig. 3.2c), dredging the inlet channel, erecting rock jetties, armoring the backshore with revetments and sand ridges, and extensive campaigns to renourish beaches (Riggs et al., 2009). The map series in Figure 3.2 chronicles not only the southward migration of Oregon Inlet over a period of 150 years, but also the birth of Herring Shoal Island and the inception of nearby back-barrier saltmarshes (Fig. 3.2b) that archive microfossil evidence of historical sea-level rise.

A study of historical sea-level change at the Outer Banks by Kemp et al. (2009) illustrates the benefits of carefully examining historical topographic data to optimize site selection and help interpret the fossil record. To reconstruct rates of sea-level change over the past two centuries, Kemp and co-workers first surveyed contemporary distributions of foraminifera and diatoms in back-barrier marshes at three Outer Banks sites. One of the modern surveys was located at a site ~1 km north of Oregon Inlet (Fig. 3.2d) near a tide gauge used to validate their microfossil-based sea-level curve. Next, they identified fossil assemblages in saltmarsh sediment from a core at one of the sites. However, Kemp and others would likely have been frustrated had they cored the emerging saltmarsh north of Oregon Inlet. Historical charts and maps document that the marsh near Oregon Inlet was likely too young (<150 years old) to contain a 200-year-old sea-level record (Fig. 3.2). In fact, few back barrier marshes along the Outer Banks exceed 300 years in age. Instead, Kemp et al. (2009) cored an older saltmarsh near Salvo, ~30 km south of Oregon Inlet, far removed from the effects of inlet migration. The historical maps also may help interpret the possible impacts on the fossil record of changes in wave climate, tidal range, and salinity gradients caused by inlet migration. Kemp et al. (2009) showed that the rate of sea-level rise increased from 2.6 mm a^{-1} since 1800 to 4.6 mm a^{-1} in the 20th century.

3.2.2 Remote sensing data

Remote sensing data encompass many types of data acquired through a variety of airborne, space-borne, or fixed sensors. Early remote sensing involved aerial photography, available as early as 1937 in some areas of the US, taken as black and white, color, and color infrared images. These data have proven extremely important as time-series components to detect coastal changes caused by erosion, deposition, and shoreline migration (e.g., Hapke et al., 2009) and for providing early snapshots of landforms untouched by urban development. Stereo-paired air photos are extremely effective for three-dimensional (3D) visualizations of the landscape using a simple stereoscope. Other useful aerial images include digital orthophoto quadrangles (DOQs) and high-resolution orthoimagery. These orthorectified images have been planimetrically corrected by removing the effects of terrain and sensor (camera) tilt to produce a constant scale across the image that allows accurate feature measurements. Practical image scales for geomorphological mapping and interpretation are typically greater than 1:24,000.

Other types of photography targeted at detecting changes through coastal monitoring are oblique aerial photographs and video surveillance of coastlines. Several organizations now post oblique photography online, including the California Coastal Records Project (http://www.californiacoastline.org/), the Alaska ShoreZone Coastal Inventory and Mapping Project (http://alaskafisheries.noaa.gov/shorezone/default.htm), and the University of Hawai'i Coastal Geology Group (http://www.soest.hawaii.edu/coasts/). Video cameras have been used as fixed sensors to investigate shoreline change, wave dynamics, and nearshore bathymetry (Pajak and Leatherman, 2002; Holman and Haller, 2013). For example, McBride et al. (1992) used video recordings to monitor shoreline change and hurricane impacts on barrier islands in Louisiana. In 2011, digital videos were used to estimate the velocity of the Tohoku-Oki tsunami that devastated Japan (Goto et al., 2011; Fritz et al., 2012) and document its impact on beaches and nearshore environments (Tappin et al., 2012). Several review articles summarize how remote sensing techniques have been applied to interpret coastal geomorphology and aid sea-level research (Smith and Pain, 2009; Klemas, 2011; Holman and Haller, 2013).

LiDAR, which includes airborne or terrestrial laser scanning, yields extremely dense and accurate elevation measurements on land and water depths in nearshore surveys. The advantages of LiDAR over other methods include high-resolution point clouds (multiple data points per meter), centimeter accuracies, and the ability to collect ground measurements in heavily forested terrain (Fig. 3.3). LiDAR surveys can typically cover tens of square kilometers in a few days. Despite a high demand for LiDAR data and derivative products such as bare-earth DEMs that remove buildings and vegetation, high data acquisition and processing costs result in limited coverage compared to other data. However, LiDAR shoreline and bathymetric data to depths of 70 m in clear water are available for the entire US coast from NOAA's Coastal Services Center (http://www.csc.noaa.gov/digitalcoast/data/coastallidar).

Satellite sensors acquire images across a broad range of the electromagnetic spectrum (from visible wavelengths to microwaves) and offer medium (30–90 m) to high (<10 m) resolution data for geomorphological interpretation and digital elevation models (Table 3.1). NASA's launch of Landsat-1 in 1972 marked the beginning of routine, repeated satellite-based observations for the purposes of studying and monitoring Earth's surface. The multispectral scanner on Landsat-1 captured over 300,000 images and, along with its sister satellites, has provided over 40 years of repeated coverage of much of the world (Landsat-8 was launched on 11 February 2013). The Shuttle Radar Topography Mission (SRTM) collected 30–90 m resolution elevation data over 11 days in February 2000 with near-global coverage. The data obtained have provided the most complete publicly available digital elevation model of Earth. Today, a growing constellation of commercial satellites are collecting higher-resolution multispectral images, obtaining SAR data for high-resolution DEMs and revisiting targets as frequently as every few days (e.g., GeoEye, IKONOS, WorldView satellites; Table 3.1). Satellite images were very effective for determining relative sea-level changes caused by the 2004 Sumatra–Andaman Islands earthquake and the geomorphological impacts of the Indian Ocean tsunami (Meltzner et al., 2006; Smet et al., 2008; Choowong et al., 2009).

Following the great 2004 Sumatra–Andaman Islands earthquake, Meltzner et al. (2006) used satellite imagery and a tidal model to place minimum constraints on sea-level changes induced by rupture of the underlying subduction-zone megathrust (Fig. 3.4). They identified coastal elevation changes relative to sea level using remote sensing, in advance of fieldwork and for locations too remote or costly to investigate in the field. Meltzner et al. (2006) compared satellite images (ASTER, SPOT, IKONOS, QuickBird, and Landsat; Table 3.1) acquired before and after the earthquake to identify areas with different extents of coral reef and beach exposure. In false-color satellite images, such as the ASTER images in Figure 3.4, coral reefs vary in color from deep blue where deeply submerged, to lighter, brighter blue where submerged in shallow water, to pink or white where exposed above water. Red reefs on false-color images indicate algae growing on coral heads that were raised above sea level and died. To identify evidence for coastal uplift, Meltzner et al. (2006) selected post-earthquake images with greater reef exposure than a pre-earthquake image taken at a lower tide. A regional tidal model was used to determine the elevation of tide at the location and exact acquisition time of each image. Differences in the predicted tides for the paired images provided minimum estimates of tectonic displacement with 2σ errors of 0.24 m (Meltzner, 2010). Identifying coseismic subsidence followed a similar approach, but in this case pre-earthquake images were selected with more reef exposure than a post-earthquake image at a lower tide. Meltzner et al. (2006) applied the regional tidal model, TPXO (Egbert and Erofeeva, 2002), to estimate minimum coseismic uplift and subsidence, which enabled them to more accurately delineate the lateral extent and downdip limit of the 2004 earthquake rupture.

3.2.3 Tidal models and historical tide gauge data

Tidal models and historical tide gauge data offer sea-level predictions and measurements that can direct fieldwork and help frame and test research hypotheses. For example, referring to tide tables and web-based tide predictions (e.g., NOAA Tide Predictions; see http://tidesandcurrents.noaa.gov/tide_predictions.shtml) prior to fieldwork can identify windows of time when tide levels will be most conducive to accomplish field objectives. Also helpful are regional tidal models that predict sea level for remote sites far from tide gauges at specific dates and times (e.g., TPXO; Egbert and Erofeeva, 2002). Such predictions have been used to estimate the level of tide at the exact time a satellite image was acquired, as described above (e.g., Meltzner et al., 2006). Models have also been employed to relate tidal measurements in the field to mean sea level and other tidal datum (e.g., Hawkes et al., 2010). Establishing reference water levels

Fig. 3.3. (a) Uplifted marine terraces of the Yakataga coast, near Icy Bay, Alaska revealed by a digital elevation model (DEM) derived from LiDAR topographic data. Post-processing the LiDAR data removed first returns from vegetation to produce a "bare-earth" DEM with 1 m horizontal and 0.2 m vertical resolution. Thin black lines depict 4 m elevation contours. Plafker et al. (1982) identified four marine terraces T1–T4 (from oldest to youngest) that record tectonic uplift related to collision and subduction of the Yakatat microplate in southern Alaska. The LiDAR image reveals details of the marine terrace geomorphology, including eroded seacliffs along abandoned terrace backedges, and alluvial fans deposited on terrace surfaces that have been subsequently dissected by stream and river channels as a result of progressive uplift. *Source*: Derived from LiDAR topographic data. Reproduced with permission of Ron Bruhn. (b) Photograph of marine terraces flanking the channel and alluvial fan of Little River, near Icy Bay, Alaska. *Source*: Photograph by I. Shennan. For color details, please see Plate 3.

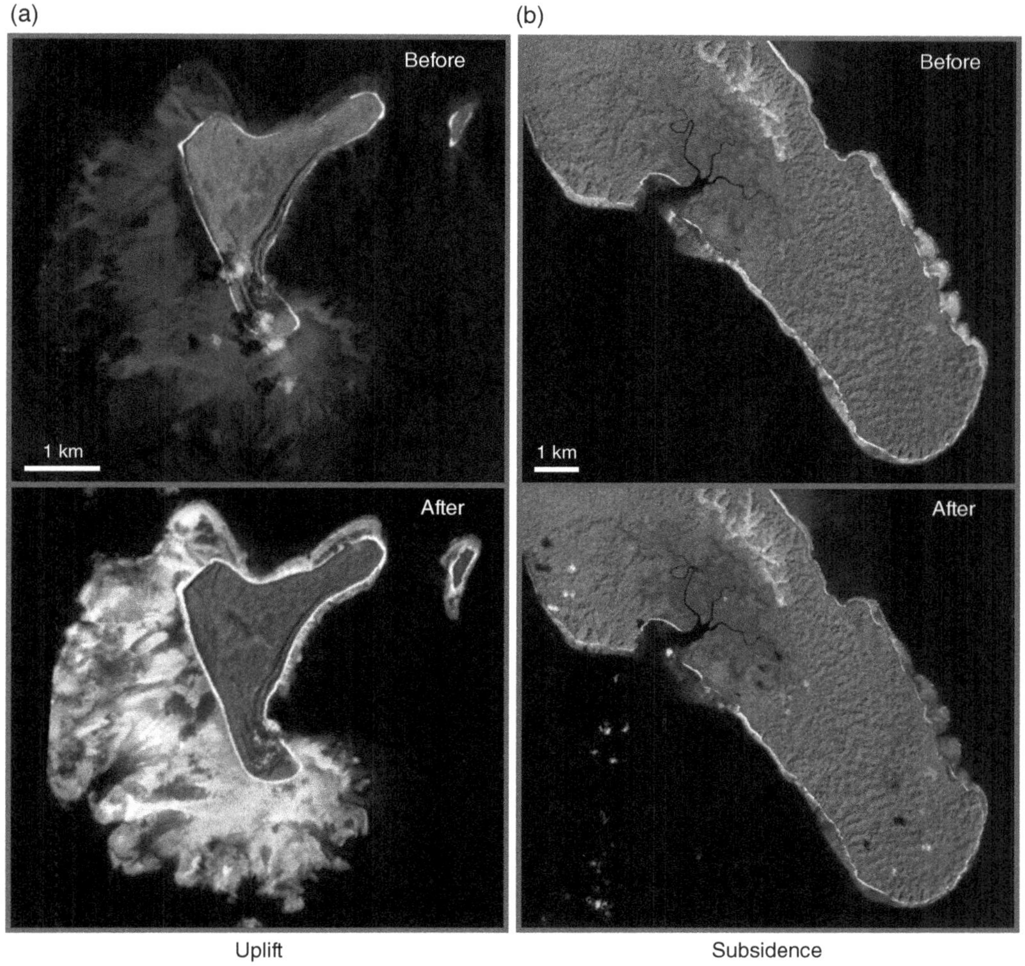

Fig. 3.4. Advanced Spaceborne Thermal Emission and Reflection Radiometer (ASTER) images of North Reef Island (a) before (top) and after (bottom) the great 2004 Sumatra–Andaman Islands earthquake. Submerged coral reefs fringe the island before the earthquake. Emergent reefs, uplifted during the earthquake, surround the island in post-earthquake images. (b) ASTER images of the southeast tip of Havelock Island show shallow coral reefs and beaches exposed before (top) the earthquake and the same shoreline features submerged by deeper water after (bottom) the earthquake as a result of tectonic subsidence. *Source*: Reproduced with permission of Aron Meltzner. For color details, please see Plate 4.

and defining tidal datums are addressed more fully in Chapter 10, while Chapter 29 focuses on the use of tidal modeling in sea-level research.

Historical tide gauge data has been relied on to validate historical sea-level curves reconstructed from geological data (e.g., Donnelly et al., 2004; Kemp et al., 2009), and identify sea-level trends and historical sea-level changes related to tectonic processes (Burgette et al., 2009), glacial isostatic adjustment (Larsen et al., 2005), climate cycles (e.g., ENSO; Allan and Komar, 2006), or a combination of these processes (Komar et al., 2011). Examples of tectonic applications include: (1) tide gauge data along the US west coast (Oregon and Washington) used to extend geodetic estimates of contemporary strain rates during interseismic periods (e.g., Burgette et al., 2009) and (2) data that revealed periodic pulses of crustal deformation related to slow-slip earthquakes (Schmidt et al., 2009; Krogstad et al., 2011). Hydrographs in coastal rivers and estuaries and water-well levels from coastal regions may also track historical sea-level change.

Along the Alaska–Aleutian subduction zone, tide gauges have recorded short-term trends of gradual sea-level change over decades as well as sudden changes in relative sea level induced by earthquake-related surface deformation (Fig. 3.5). Tide gauge data in this region also likely reflect sea-level change

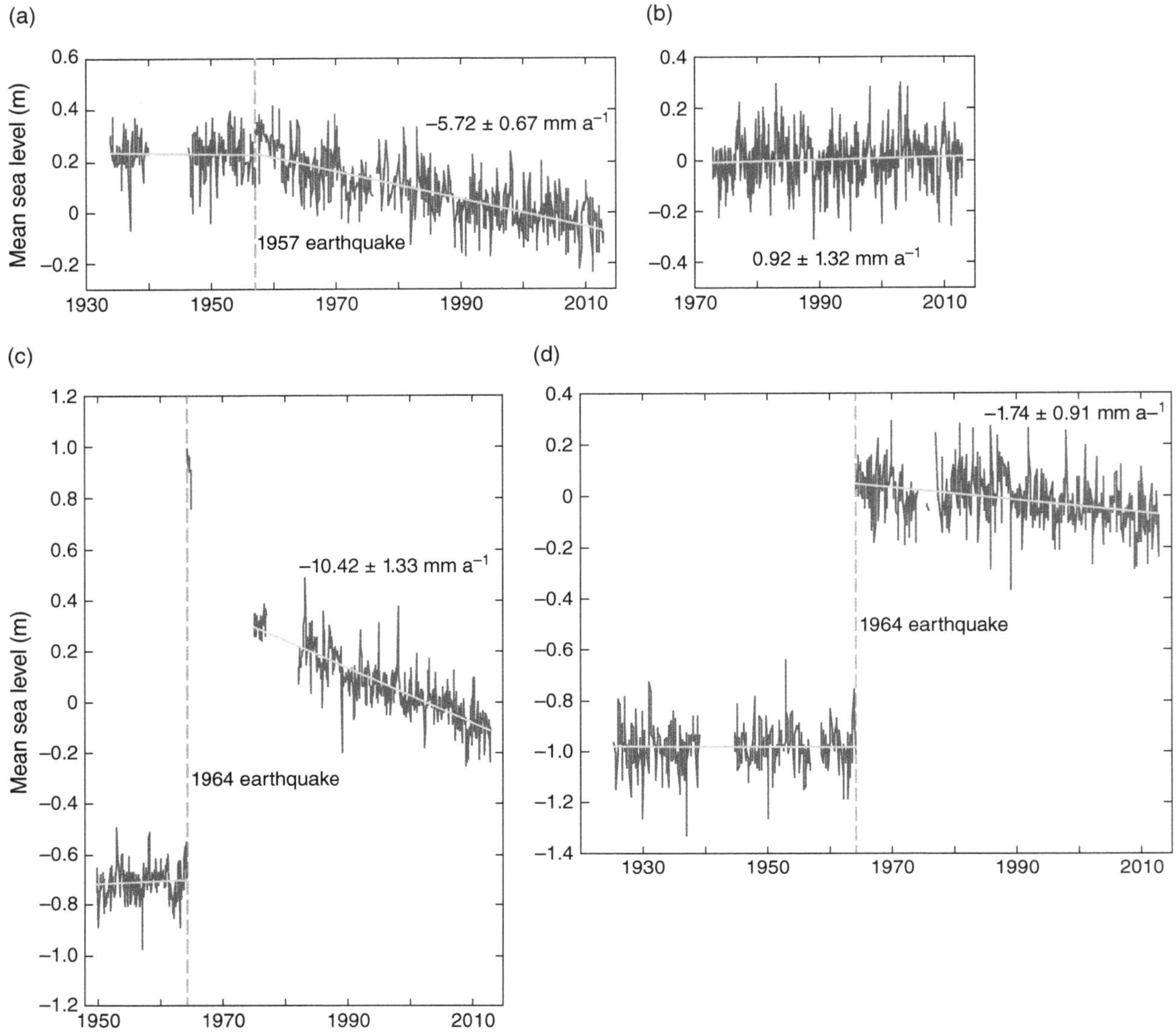

Fig. 3.5. Historical linear trends of mean sea level at four sites along the Alaska–Aleutian subduction zone (NOAA data available at http://tidesandcurrents.noaa.gov/index.html). Vertical lines on 3 of the 4 plots mark times of significant historical earthquakes near the tide gauges. (a) At Dutch Harbor on Unalaska Island, mean sea level showed little change prior to the 1957 Andreanof Islands earthquake (magnitude 8.6). After the earthquake, mean sea level fell at a rate of −5 to −6 mm a^{-1}, possibly as a result of aseismic slip on a deeper part of the plate interface below the area that ruptured in 1957. (b) At Sand Point, located in a region that has not experienced a historical great earthquake larger than magnitude 8, mean sea level has risen since 1970 at a rate of ~1 mm a^{-1}. (c) At Kodiak Island, the slow rise of mean sea level was interrupted by over 1 m of sudden tectonic subsidence during the great (magnitude 9.2) 1964 Prince William Sound earthquake. Since 1964, mean sea level at Kodiak has fallen at a linear rate of −9 to −11 mm a^{-1}. (d) At Seward, the trend of mean sea level was flat prior to about 1 m of subsidence during the 1964 Prince William Sound earthquake, and since the earthquake mean sea level has fallen at a rate of −1 to −2 mm a^{-1}.

related to glacial isostatic adjustment and volcanic processes. Four gauges in Alaska – Unalaska, Sand Point, Kodiak Island, and Seward – record 50–80 years of data that exhibit variations in sea-level trends along this tectonic coast, reversals in trend, and episodes of instantaneous sea-level rise of a meter or more corresponding to historical earthquakes. At Unalaska (Fig. 3.5a), tide gauge data show a stable sea level or a slight negative tendency before the great 1957 Andreonof Islands earthquake. The data do not indicate a clear displacement of the tide gauge at the time of the earthquake. However, after the earthquake, tidal data show a trend of sea-level fall at a rate of $-5.7\,mm\,a^{-1}$. The sudden increase in rate of sea-level fall suggests that aseismic after-slip on the part of the plate interface downdip of the 1957 earthquake rupture has uplifted Unalaska. In the Shumagin Islands, the tide gauge at Sand Point (Fig. 3.5b) indicates a trend of sea-level rise over the past 40 years at a rate of $0.9mm\ a^{-1}$. No historical earthquakes greater than magnitude 8 have occurred on this part of the megathrust, which is thought to be poorly coupled based on geodetic data (Freymueller et al., 2008).

Before the great (magnitude 9.2) 1964 Prince William Sound earthquake, the Kodiak Island tide gauge indicated slowly rising sea level that abruptly ceased when the site subsided by more than a meter during the earthquake (Fig. 3.5c). The post-1964 sea-level trend is a rapid sea level fall at $-10.4\,mm\ a^{-1}$. Sea-level trends for the Seward tide gauge are similar to Kodiak Island, except that sea level has fallen at a slower rate of $-1.7\,mm\ a^{-1}$ since the 1964 earthquake (Fig. 3.5d). These data have guided paleogeodesy and paleotsunami studies in the eastern Aleutians by revealing spatial variations in historical changes of relative sea level over coseismic, post-seismic and interseismic phases of the earthquake deformation cycle (Engelhart et al., 2012; Nelson et al., 2012; Witter et al., 2014).

3.3 ASSESSING GEOMORPHOLOGY

Geomorphological maps illustrate relationships between landforms, prior geological structures, and ongoing processes (tectonic, volcanic, and ice loading) at the Earth's surface. Smith et al. (2011) review geomorphological mapping techniques and analyses in a variety of environments. For sea-level research, geomorphological maps are used to describe coastal landforms (e.g., terraces, estuaries, fjords, barrier spits and islands, beach ridges, strandlines, and dunes; Fig. 3.6) including their shape and dimensions, relief, relative ages, and the depositional environments wherein they formed. Maps also delineate topographic, tonal and vegetation lineaments, slope breaks, troughs or depressions, side hill benches, and other landforms that may partly reflect the position of past sea level. There is a diversity of geomorphological maps of coastal environments, including those depicting the chenier-plain of the Mississippi River delta (Penland and Suter, 1989; McBride et al., 2007), coastal dunes and ancient shorelines of Oregon and Washington (Cooper, 1958), and glacial landforms along the west coast of New Zealand's South Island (see interactive map at http://maps.gns.cri.nz/website/csigg/).

Geomorphological maps are enhanced by maps of surficial deposits and topographic profiles that show landform topography in greater detail. Maps of surficial deposits are constructed from interpretations of the inferred depositional environments depicted on a geomorphological map as well as a variety of other information including topographic data, air photos and satellite images, geologic data, and borehole and well logs. Topographic profiles across terrace backedges, strandlines, beach ridges, notches, and other features define the elevations of possible shoreline indicators and the spatial relationships among landforms. If alternative interpretations arise or new questions emerge during landform mapping, flag the feature or area for further evaluation in the field. Geomorphological maps are not complete until they have been validated in the field, as the following case illustrates.

Preliminary interpretations of coastal geomorphology on Simeonof Island, part of the Shumagin Island archipelago near Sand Point, Alaska, were shown to be incorrect by fieldwork in 2011. After literature review and drafting a preliminary geomorphological map from interpretation of satellite imagery (Fig. 3.6), Witter et al. (2014) visited Simeonof Island to reconstruct post-glacial relative sea-level changes and investigate evidence for megathrust earthquakes and tsunamis. In addition, subtle scarps, beach ridges, tonal lineaments, and discontinuous slope breaks identified on satellite imagery (Fig. 3.6) suggested shoreline indicators might be present, which would be consistent with earlier reports of emergent marine shorelines associated with archeological sites (Johnson and Winslow, 1991; Winslow, 1991).

Although field investigations confirmed that topographic scarps ringing a small intertidal lagoon were former shorelines, the shorelines were not marine. Instead, field investigations discovered 10,000-year-old freshwater peat, dated with radiocarbon, exposed in coastal bluffs within a few meters of sea level. Witter et al. (2014) concluded that the shorelines marked past high lake-water levels in an interconnected network of freshwater

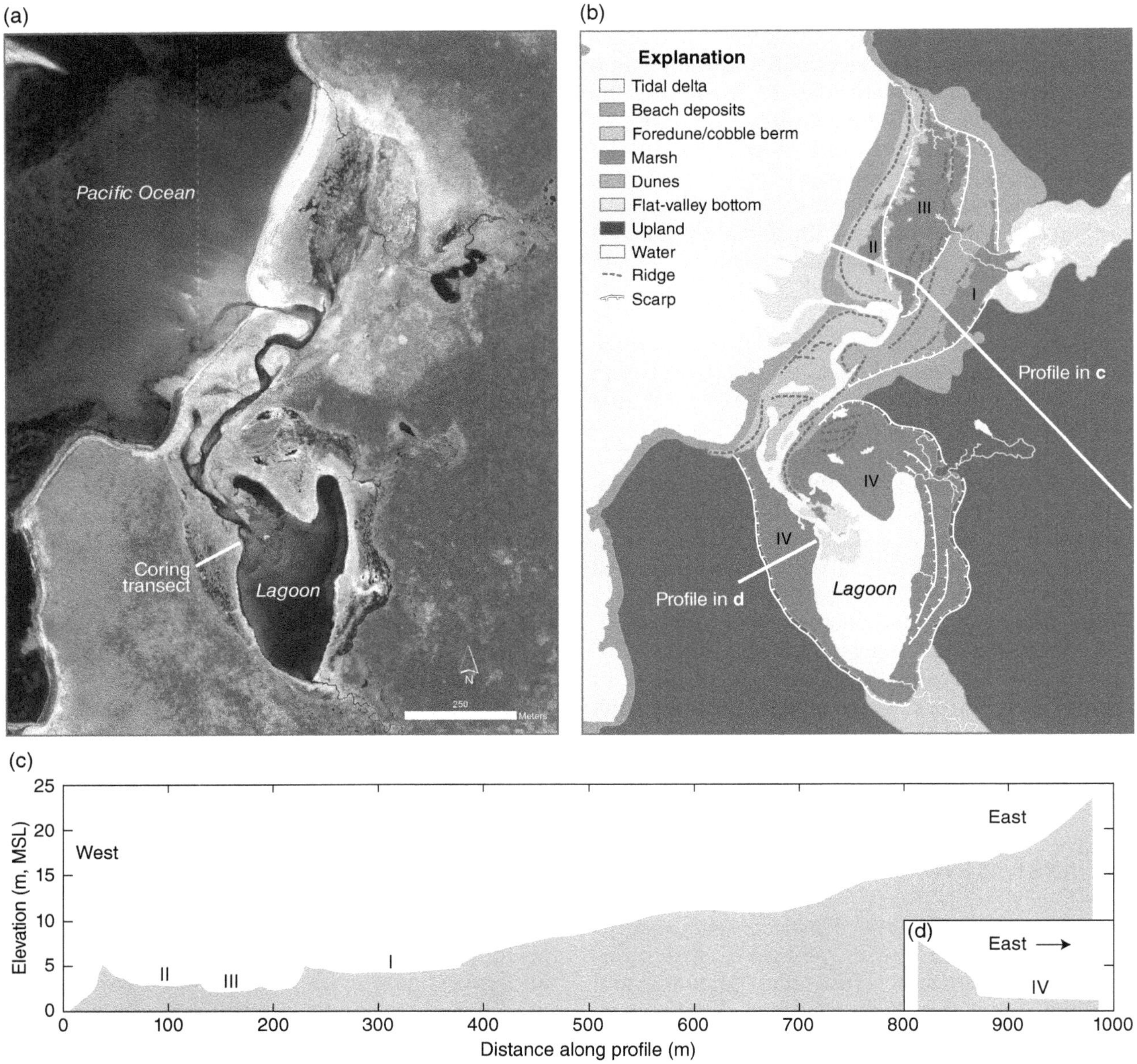

Fig. 3.6. Geomorphological map of an intertidal lagoon on Simeonof Island near Sand Point, Alaska. (a) 2009 satellite image of study area. Cores along the transect near the lagoon indicated that the basin is filled with 1–1.5 m of freshwater peat. (b) Interpreted coastal geomorphology. *Source*: GeoEye image of Simeonof Island from May 2009. © 2009 DigitalGlobe, NextView. (c) Topographic profile corresponding to transect line in image and map above. The map and profile delineate scarps that may reflect recessional shorelines related to progressive draining of coastal lakes rather than to a series of emergent marine strandlines. Slowly rising sea level over the past 3500 years or stable sea level with severe coastal erosion allowed the sea to invade the lakes (Witter et al., 2012). For color details, please see Plate 5.

lakes. At one site (Fig. 3.6), an intertidal lagoon formed when slowly rising late Holocene sea level had sufficiently eroded the coast and breached the lake system, invading a former freshwater basin. This example demonstrates the importance of verifying mapped geomorphological interpretations in the field.

Greater attention to coastal change over the past three decades has motivated coastal mapping to quantify rates of shoreline retreat, barrier island migration, and dune erosion (e.g., Leatherman, 1983; Crowell et al., 1991; Hapke et al., 2009). Until the late 1980s, techniques defined the shoreline as the high water line (HWL) mapped on NOS T-sheets

Fig. 3.7. Comparison of shoreline data from 1927 to 2009 at Neskowin, Oregon, USA. (a) 1927 NOS T-sheet (#4337) depicting the high water line (HWL) as a bold shoreline backing a stippled beach. *Source*: LiDAR data reproduced with permission of Jon Allan. (b) 1953 black and white aerial photography acquired by the US Geological Survey showing the HWL coincident with the boundary between wet and dry beach sand. *Source*: Photographs by Jon Allan. Reproduced with permission. (c) Shaded relief map derived from LiDAR data acquired by the State of Oregon in 2009. The blue line traces the mean shoreline position for the period 2007–2013 measured by GPS survey. Dashed line marks one standard deviation about the mean (unpublished data by Jon Allan). *Source*: LiDAR data reproduced with permission of Jon Allan. (d) 2009 digital ortho-imagery acquired by the US Department of Agriculture's National Agriculture Imagery Program (NAIP). (e) Late Holocene tree stumps exhumed in the surf zone near Proposal Rock. *Source*: Photograph by Jon Allan. Reproduced with permission. For color details, please see Plate 6.

(Shalowitz, 1964) and delineated on aerial photography by tonal contrasts between wet and dry beach sand. Mapping proxies such as HWL as the shoreline came with many potential errors related to the source of information (e.g., paper maps shrink, cameras tilt off vertical, and camera lenses impart radial distortion), seasonal and tidal variations of the HWL, and errors in interpretation and delineation of the HWL (Moore, 2000; Pajak and Leatherman, 2002). Over the past decade, orthorectified digital imagery has been used to define the modern shoreline as a specific tidal datum (e.g., mean high water, MHW), which can be determined from GPS or LiDAR data referenced to a tide gauge (Stockdon et al., 2002; Sallenger et al., 2003). Calculating shoreline change rates over decades entails differencing the shoreline position on old maps and vintage aerial photography against more recent orthorectified imagery. Care must be taken when comparing proxy-based shorelines from older data sources to modern datum-based shorelines to avoid introducing additional errors related to biases in shoreline proxies (Ruggiero and List, 2009).

Neskowin, Oregon, a village of 170 inhabitants and a popular vacation destination on the Pacific coast of the western US (Fig. 3.7), has struggled to cope with the impacts of sea-level rise and a 25-year trend of increasing wave heights attributed to climate change (Allan and Komar, 2006). The climbing trend of wave heights superimposed onto rising sea level has increased the frequency of property losses due to ocean flooding and coastal erosion, particularly during the severe winter storms of El Niño years (Komar et al., 2011). Once protected by a broad beach and active foredune (Fig. 3.7a), the community now faces a much narrower beach encroaching on an eroded remnant of the foredune (Fig. 3.7d). To combat the hazard, residents have hardened the shoreline by placing rock revetments (riprap) along the entire length of the community exposed to the sea, much of it installed since 1999. Ongoing analysis and regular monitoring of shoreline change at Neskowin takes advantage of early NOS T-sheets, time-series aerial images from the 1930s onward, and coastal LiDAR (Fig. 3.7). Residents are cognizant of the warning served by a fossil forest partially exhumed on the beach at Neskowin (Fig. 3.7e; Hart and Peterson, 2007), which provides graphic evidence of sea-level rise over the past few thousand years. The overall goals of shoreline change studies at Neskowin are to advise the community on its options to mitigate increasing erosion hazards based on projections of future sea-level rise and elevated storm waves.

3.4 EXAMPLES OF APPLYING HISTORICAL DATA TO ENHANCE SEA-LEVEL RESEARCH

Three studies that used historical sea-level data to uncover unanticipated or rare information that enhanced the project outcomes are described in this section. The first case used historical nautical charts surveyed during the 1835 voyage of the HMS *Beagle* to estimate the amount of relative sea-level rise in the interseismic period following the 1835 Concepción, Chile earthquake. The second case used pre-and post-earthquake images to estimate coseismic subsidence near Iloca, Chile caused by the 2010 Maule earthquake (magnitude 8.8), a near repeat of the 1835 Concepción earthquake. The third study relied on historical records of a tsunami that hit Japan in AD 1700 to identify the tsunami's source earthquake that struck the Cascadia subduction zone along western North America.

Along subduction zone coasts, the earthquake deformation cycle has been invoked to explain gradual (interseismic) relative sea-level change between great earthquakes and sudden (coseismic) relative sea-level change during megathrust earthquake ruptures. These changes in relative sea level are consistent with theoretical models of earthquake deformation where vertical displacement of the coast occurs slowly during the interseismic period and is elastically restored suddenly during an earthquake by the opposite sense of vertical displacement. In a few fortuitous cases, evidence of relative sea-level changes related to the earthquake deformation cycle and its impact on coastal landforms has been recorded by historical nautical charts and bathymetric data (Malloy, 1965; Wesson, 2008). For example, comparisons between soundings taken in 1927 and hydrographic and geophysical seismic surveys completed after the 1964 Alaska earthquake provided estimates of fault displacements offshore Montague Island (Malloy, 1965), and these estimates compared well with measured displacements on land (Plafker, 1967). Wesson et al. (2010) used a similar approach to quantify interseismic subsidence at Santa Maria Island, off the coast of south-central Chile (Fig. 3.1c). In March of 1835, Robert FitzRoy, captain of the HMS *Beagle*, visited the island a few weeks after a magnitude ~8.5 earthquake destroyed the city of Concepción. FitzRoy reported dead intertidal organisms on shore platforms along Santa Maria Island that had been elevated by 2.4–3 m as a result of earthquake uplift. Wesson et al. (2010) compared 1835 navigation charts (surveyed by officers of the *Beagle*) that documented the uplifted shore platforms with Chilean charts from 1886 that showed narrower shore platforms. The ensuing post-earthquake subsidence of the island implied by the difference between historical charts was too rapid to be explained by coastal erosion. Continuing subsidence into the present was corroborated by accounts of long-time residents, GPS measurements, and bathymetric surveys of the adjacent bay (Wesson et al., 2010). Shortly after Wesson's study, Santa Maria Island was uplifted again by 1.5–2.1 m during the 2010 Chile earthquake (Melnick et al., 2012), completing an earthquake cycle.

Although uplift characterized most of the coastal deformation above the 500-km-long rupture that produced the February 2010 Maule earthquake, some areas of the Chilean coast subsided. Earthquake subsidence at one site, the mouth of the Mataquito River near the town of Iloca, was discernible from before and after images posted on Google Earth and later verified by field reconnaissance (Fig. 3.8). Prior to the 2010 earthquake, Google Earth images from 2007 and 2009 show the Mataquito River

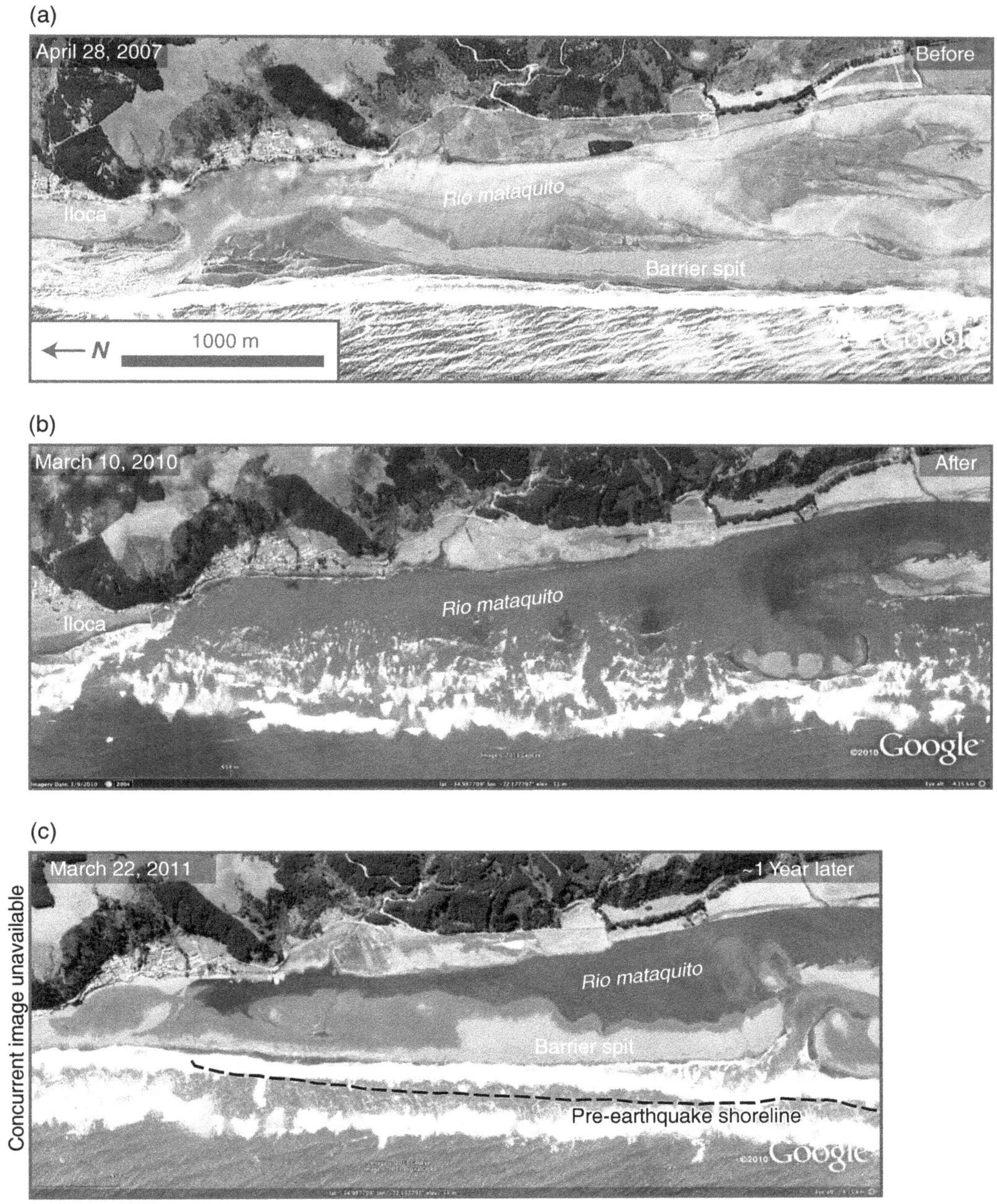

Fig. 3.8. Changes in the configuration of the barrier spit at the mouth of Rio Mataquito following the 27 February 2010, *M* 8.8 Maule, Chile earthquake shown by Google Earth images (© 2010 Google). (a) The long, unbroken barrier spit in April 2007 diverted the mouth of Rio Mataquito to the north near the village of Iloca. Google Earth images from September 2009 show a similar spit configuration with little shoreline change prior to the 2010 earthquake. (b) Post-earthquake images from March 2010 show the disappearance of the spit due to wave erosion. (c) In March 2011, the reconstructed barrier placed the river mouth about 4 km south of Iloca with the shoreline 150 m landward of its 2009 position (dashed shoreline). For color details, please see Plate 7.

flowing northward toward its mouth at Iloca, protected by a narrow barrier sand spit (Fig. 3.8a). Immediately after the earthquake, the barrier spit vanished as shown on images from March 2010 (Fig. 3.8b). Post-earthquake field reconnaissance (GEER, 2010) attributed the disappearance of the spit to erosion caused by tsunami inundation and ocean waves strengthened by ~0.5 m of coseismic subsidence (relative sea-level rise). Subsequent studies agreed with initial estimates of subsidence at the river mouth near Iloca, based on ocean flooding in a former upland field and the disappearance of offshore sand bars (Farías et al., 2010; Kelson et al., 2012). Additional areas of coastal subsidence were identified at Bucalemu, 33 km north of Iloca, and along the Bío-Bío River near Concepción. Over one year later, Google Earth images from March 2011 showed a new barrier spit shifted landward by over 100 m and the relocation of the river mouth 4 km south of Iloca. Careful examination of the before and after images reveals erosion along the landward shoreline of the estuary that betrays the lasting impacts of subsidence (Fig. 3.8c). Villagran et al. (2013) used airborne and satellite images, along with field surveys, to monitor post-earthquake coastal erosion and to examine how wave climate, longshore currents, and river hydrology influenced the rapid recovery of the spit.

In Japan, written documents register damage caused by tsunamis that inundated Honshu's Pacific coast as early as 1586 (Watanabe, 1998), including a tsunami with an unknown, non-Japanese source in January 1700 (Satake et al., 1996). After precluding that earthquakes in Kamchatka, the Aleutians, and South America generated the 1700 tsunami, Satake et al. (1996) used a hydrodynamic model to infer that the source was an earthquake on the Cascadia subduction zone, estimated the magnitude at 9, and determined its occurrence as probably about 21:00 local time on 26 January 1700. In a follow-on study, Satake et al. (2003) scrutinized descriptions of the 1700 tsunami, resultant damages, tidal elevation at the time of the tsunami, and subsequent relative sea-level change to explore the uncertainty of the estimated tsunami heights. Their analyses led to minimum, maximum, and preferred estimates of tsunami height which, when tested against computed tsunami heights, confirmed the giant size of the earthquake and provided estimates of the amount of slip on the Cascadia megathrust and the amount of seismic energy released during the earthquake. Building on the breakthrough of Satake et al. (1996, 2003), Atwater et al. (2005) linked geological evidence of giant earthquakes and tsunamis in Cascadia to the old Japanese descriptions of tsunami wave impacts in 1700, and described how the new findings apply to seismic hazards in the western US and Canada. In this case, historical documents regarding a tsunami in Japan that lacked a Japanese source earthquake led to enormous breakthroughs in understanding Cascadia's earthquake potential on the other side of the Pacific Ocean.

These summaries illustrate some of the many potential benefits of thorough pre-fieldwork surveys. Whereas surveys that supplement fieldwork may not always precede field investigations, when they do, thorough data compilation and geomorphological review may lead to discoveries that aid the research or redirect planned fieldwork. Time spent assembling historical and other geographic data on the study site may offer new hypotheses to address, aid in site selection, optimize the approach and methods used in the field, and reduce costs by streamlining field logistics. In particular, constructing preliminary maps of coastal geomorphology and surficial deposits may provide insights about historical shoreline change and related surficial processes, as well as identify sites to examine closely in the field.

REFERENCES

Allan, J.C., and Komar, P.D. (2006) Climate controls on U.S. West Coast erosion. Journal of Coastal Research, 22, 511–529.

Anthony, E.J. (2009) *Shore Processes and their Palaeoenvironmental Applications*. Elsevier, Amsterdam.

Atwater, B.F., Satoko, M.-R., Kenji, S., Yoshinobu, T., Kazue, U., and Yamaguchi, A.D.K. (2005) Orphan Tsunami of 1700. USGS Professional Paper 1707.

Bird, E.C.F. (2008) *Coastal Geomorphology: An Introduction*. John Wiley and Sons Ltd, West Sussex.

Burgette, R. J., Weldon, R. J., II, and Schmidt, D. A. (2009) Interseismic uplift rates for western Oregon and along-strike variation in locking on the Cascadia subduction zone. Journal of Geophysical Research, 114, doi: 10.1029/2008JB005679.

Choowong, M., Phantuwongraj, S., Charoentitirat, T., Chutakositkanon, V., Yumuang, S., and Charusiri, P. (2009) Beach recovery after 2004 Indian Ocean tsunami from Phang-nga, Thailand. Geomorphology, 104, 134–142.

Cooper, W. S. (1958) *Coastal Sand Dunes of Oregon and Washington*. Geological Society of America, Memoir 72.

Crowell, M., Leatherman, S.P., and Buckley, M.K. (1991) Historical shoreline change: error analysis and mapping accuracy. Journal of Coastal Research, 7, 839–852.

Curray, J.R. (1964) Transgressions and regressions. In: *Papers in Marine Geology* (ed. Miller, R.L.). Macmillan, New York, Shepard Commemorative Volume, pp. 175–203.

Donnelly, J.P., Cleary, P., Newby, P., and Ettinger, R. (2004) Coupling instrumental and geological records of sea-level change: evidence from southern New England of an increase in the rate of sea-level rise in the late 19th

century. Geophysical Research Letters, 31, L05203, doi: 10.1029/2003GL018933.
Egbert, G.D., and Erofeeva, S.Y. (2002) Efficient inverse modeling of barotropic ocean tides. Journal of Atmospheric and Oceanic Technology, 19, 183–204.
Engelhart, S.E., Kemp, A.C., Nelson, A.R., Briggs, R.W., Haeussler, P.J., Culver, S.J., Angster, S.J., and Bradley, L.–A. (2012) Application of salt-marsh and tidal-flat foraminifera to identify late Holocene land-level changes caused by megathrust earthquakes at Sitkinak Island, Alaska. Geological Society of America, Abstracts with Programs, 44, 301.
Farías, M., Vargas, G., Tassara, A., Carretier, S., Baize, S., Melnick, D., and Bataille, K. (2010) Land-level changes produced by the M_w 8.8 2010 Chilean earthquake. Science, 329, 916.
Freymueller, J. T., Cohen, S. C., Cross, R., Elliott, J., Fletcher, H., Larsen, C., Hreinsdóttir, S., and Zweck, C. (2008) Active deformation processes in Alaska, based on 15 years of GPS measurements. In: *Active Tectonics and Seismic Potential of Alaska* (eds Freymueller, J.T. et al.). American Geophysical Union, Washington, DC, pp. 1–42.
Fritz, H.M., Phillips, D.A., Okayasu, A., Shimozono, T., Liu, H., Mohammed, F., Skanavis, V., Synolakis, C.E., and Takahashi, T. (2012) The 2011 Japan tsunami current velocity measurements from survivor videos at Kesennuma Bay using LiDAR. Geophysical Research Letters, 39, L00G23, doi: 10.1029/2011GL050686.
Geoengineering Extreme Events Reconnaissance (GEER) Team (2010) Geo-Engineering Reconnaissance of the 27 February 2010 Maule, Chile Earthquake. Available at http://www.geerassociation.org/GEER_PostEQReports/Maule_Chile_2010/Cover_Chile_2010.html (accessed 14 July 2014).
Goto, K., Chagué-Goff, C., Fujino, S., Goff, J., Jaffe, B., Nishimura, Y., Richmond, B., Sugawara, D., Szczucinski, W., Tappin, D.R., Witter, R.C., and Yulianto, E. (2011) New insights of tsunami hazard from the 2011 Tohoku-oki event. Marine Geology, 290, 46–50, doi: 10.1016/j.margeo.2011.10.004.
Hapke, C.J., Reid, D., and Richmond, B. (2009) Rates and trends of coastal change in California and the regional behavior of the beach and cliff system. Journal of Coastal Research, 25, 603–615.
Hart, R., and Peterson, C. (2007) Late-Holocene buried forests on the Oregon Coast. Earth Surface Processes and Landforms, 32, 210–229.
Hawkes, A. D., Horton, B. P., Nelson, A. R., and Hill, D. F. (2010) The application of intertidal foraminifera to reconstruct coastal subsidence during the giant Cascadia earthquake of AD 1700 in Oregon, USA. Quaternary International, 221, 116–140, doi: 10.1016/j.quaint.2009.09.019.
Holman, R., and Haller, M.C. (2013) Remote sensing of the nearshore. Annual Review of Marine Science, 5, 95–113.
Johnson, L.L., and Winslow, M.A. (1991) Paleoshorelines and prehistoric settlement on Simeonof and Chernabura Islands, outer Shumagin Islands, Alaska, In: *Paleoshorelines and Prehistory: An Investigation of Method* (eds Johnson, L.L., and Stright, M.). CRC Press, Boca Raton, Florida, pp. 171–186.
Kelson, K., Witter, R.C., Tassara, A., and Ryder, I. (2012) Coseismic tectonic surface deformation during the 2010 Maule, Chile, M_w 8.8 earthquake. Earthquake Spectra, 28, doi: 10.1193/1.400042.
Kemp, A.C., Horton, B.P., Corbett, D.R., Culver, S.J., Edwards, R.J., and van de Plassche, O. (2009) The relative utility of foraminifera and diatoms for reconstructing late Holocene sea-level change in North Carolina, USA. Quaternary Research, 71, 9–21.
Klemas, V. (2011) Remote sensing techniques for studying coastal ecosystems. Journal of Coastal Research, 27, 2–17.
Komar, P.D., Allan, J.C., and Ruggiero, P. (2011) Sea level variations along the U.S. Pacific Northwest Coast: tectonic and climate controls. Journal of Coastal Research, 276, 808–823.
Krogstad, R., Schmidt, D.A., Weldon, R.J., and Burgette, R.J. (2011) Net strain accumulation in the slow slip zone along the Cascadia Subduction Zone constrained by leveling and tide gauge data. American Geophysical Union, Fall Meeting 2011, abstract #S23B-2283.
Larsen, C.F., Motyka, R.J., Freymueller, J.T., Echelmeyer, K.A., and Ivins, E.R. (2005) Rapid viscoelastic uplift in southeast Alaska caused by post–Little Ice Age glacial retreat. Earth and Planetary Science Letters, 237, 548–560.
Leatherman, S.P. (1983) Shoreline mapping; a comparison of techniques. Shore and Beach, 51, 28–33.
Long, A.J., Woodroffe, S.A., Roberts, D.H., and Dawson, S. (2011) Isolation basins, sea-level changes and the Holocene history of the Greenland Ice Sheet. Quaternary Science Reviews, 30, 3748–3768.
Malloy, R.J. (1965) Crustal uplift southwest of Montague Island, Alaska. Science, 146, 1048–1049.
Masselink, G., and Hughes, M.G. (2003) *Introduction to Coastal Processes and Geomorphology*. Arnold, London.
McBride, R.A., Penland, S., Hiland, M.W., Williams, S.J., Westphal, K.A., Jaffe, B.E., and Sallenger, A.H. (1992) Analysis of barrier shoreline change in Louisiana from 1853 to 1989. In: *Louisiana Barrier Island Erosion Study: Atlas of Shoreline Changes in Louisiana from 1853 to 1989*. US Geological Survey, Miscellaneous Investigation Series I-2150-A, p. 36–97.
McBride, R.A., Taylor, M.J., and Byrnes, M.R. (2007) Coastal morphodynamics and Chenier-Plain evolution in southwestern Louisiana, USA: A geomorphic model. Geomorphology, 88, 367–422.
Melnick, D., Cisternas, M., Moreno, M., and Norambuena, R. (2012) Estimating coseismic coastal uplift with an intertidal mussel: calibration for the 2010 Maule Chile earthquake (M_w 8.8). Quaternary Science Reviews, 42, 29–42.
Meltzner, A.J. (2010) *Earthquake Recurrence, Clustering, and Persistent Segmentation near the Southern End of the 2004 Sunda Megathrust Rupture*. PhD thesis, California Institute of Technology, Pasadena, California.
Meltzner, A.J., Sieh, K., Abrams, M., Agnew, D.C., Hudnut, K.W., Avouac, J.-P., and Natawidjaja, D.H. (2006) Uplift and subsidence associated with the great Aceh-Andaman earthquake of 2004. Journal of Geophysical Research, 111, B02407, doi: 10.1029/2005JB003891.
Moore, L.J. (2000) Shoreline mapping techniques. Journal of Coastal Research, 16, 111–124.

Nelson, A.R., Briggs, R.W., Engelhart, S.E., Gelfenbaum, G., Dura, T., Bradley, L.–A., and Vane, C.H. (2012) Holocene tsunami recurrence along the eastern Aleutian-Alaskan megathrust, Chirikof Island, Alaska. Geological Society of America, Abstracts with Programs, 44, 301.

Pajak, M.J., and Leatherman, S. (2002) The high water line as shoreline indicator. Journal of Coastal Research, 18, 329–337.

Penland, S., and Suter, J.R. (1989) The geomorphology of the Mississippi River chenier plain. Marine Geology, 90, 231–258.

Pilkey, O.H., and Fraser, M.E. (2003) *A Celebration of the World's Barrier Islands*. Columbia University Press, New York.

Plafker, G. (1967) Surface faults on Montague Island associated with the 1964 Alaska earthquake. The Alaska Earthquake, March 27, 1964: Regional Effects. US Geological Survey Professional Paper 543-G.

Plafker, G., Hudson, T., Rubin, M., and Dixon, K.L. (1982) Holocene marine terraces and uplift history in the Yakataga seismic gap near Icy Cape, Alaska. US Geological Survey Circular 844, 111–115.

Riggs, S.R., Ames, D.V., Culver, S.J., Mallinson, D.J., Corbett, D.R., and Walsh, J.P. (2009) Eye of a human hurricane: Pea Island, Oregon Inlet, and Bodie Island, northern Outer Banks, North Carolina. In: *America's Most Vulnerable Coastal Communities* (eds Kelley, J.T., Pilkey, O.H., and Cooper, J.A.G.) Geological Society of America, Special Paper 460, pp. 43–72.

Ruggiero, P., and List, J.H. (2009) Improving accuracy and statistical reliability of shoreline position and change rate estimates. Journal of Coastal Research, 255, 1069–1081.

Sallenger, A.H., Jr., Krabiii, W., Swift, R.N., Brock, J., List, J.H., Hansen, M., Holman, R.A., Manizade, S., Sontag, J., Meredith, A., Morgan, K., Yunkel, J.K., Frederick, E.B., and Stockdon, H.F.(2003) Evaluation of airborne topographic LiDAR for quantifying beach changes. Journal of Coastal Research, 19, 125–133.

Satake, K., Shimazaki, K., Tsuji, Y., and Ueda, K. (1996) Time and size of a giant earthquake in Cascadia inferred from Japanese tsunami records of January 1700. Nature, 379, 246–249.

Satake, K., Wang, K., and Atwater, B.F. (2003) Fault slip and seismic moment of the 1700 Cascadia earthquake inferred from Japanese tsunami descriptions. Journal of Geophysical Research, 108, doi: 10.1029/2003JB002521.

Schmidt, D.A., Weldon, R.J., and Gao, H. (2009) Combining slow slip events, historical uplift, and contemporary geodesy to map interseismic strain along the central Cascadia Subduction Zone. American Geophysical Union, Fall Meeting 2009, abstract #U52A-02.

Shalowitz, A. (1964) *Shore and Sea Boundaries*. US Government Printing Office, Washington, DC.

Smet, S., Michel, R., and Bollinger, L. (2008) Uplift of the 2004 Sumatra-Andaman earthquake measured from differential hyperspectral imagery of coastal waters. Journal of Geophysical Research, 113, B09403, doi: 10.1029/2007JB005317.

Smith, M.J., and Pain, C.F. (2009) Applications of remote sensing in geomorphology. Progress in Physical Geography, 33, 568–582.

Smith, M.J., Paron, P., and Griffiths, J.S. (eds.) (2011) *Geomorphological Mapping: Methods and Applications*. Elsevier, London, Developments in Earth Surface Processes 15.

Stockdon, H.F., Sallenger, A.H., Jr., List, J.H., and Holman, R.A. (2002) Estimation of shoreline position and change using airborne topographic LiDAR data. Journal of Coastal Research, 18, 502–513.

Tappin, D.R., Evans, H.M., Jordan, C.J., Richmond, B., Sugawara, D., and Goto, K. (2012) Coastal changes in the Sendai area from the impact of the 2011 Tōhoku-oki tsunami: Interpretations of time series satellite images, helicopter-borne video footage and field observations. Sedimentary Geology, 282, 151–174.

Villagran, M., Cienfuegos, R., Catalán, P. A., and Almar, R. (2013) Morphological response of central Chile sandy beaches to the 8.8 Mw 2010 earthquake and tsunami. Presented at the Coastal Dynamics 2013, Bordeaux, France, pp. 1823–1834.

Watanabe, H. (1998) *Comprehensive List of Tsunamis to Hit the Japanese Islands*, 2nd edition. University of Tokyo Press, Tokyo (in Japanese).

Wells, A., and Goff, J. (2007) Coastal dunes in Westland, New Zealand, provide a record of paleoseismic activity on the Alpine fault. Geology, 35, 731–734.

Wesson, R.L. (2008) Historical nautical charts and hydrographic surveys as recorders of vertical displacement above shallow portions of subduction zones. American Geophysical Union, Fall Meeting 2008, abstract #T23B-2034.

Wesson, R., Melnick, D., Cisternas, M., Ely, L., and Moreno, M. (2010) Long-term inter-seismic subsidence of Santa Maria Island, Chile (37°S) supported by resurvey of 1835 HMS Beagle soundings. American Geophysical Union, Chapman Conference, Viña del Mar, Chile.

Winslow, M.A. (1991) Modeling paleoshorelines in geologically active regions: Applications to the Shumagin Islands, southwest Alaska. In: *Paleoshorelines and Prehistory: An Investigation of Method* (eds Johnson, L.L., and Stright, M.). CRC Press, Boca Raton, Florida, pp. 151–169.

Witter, R. (2008) Prehistoric Cascadia tsunami inundation and runup at Cannon Beach, Clatsop County, Oregon. Oregon Department of Geology and Mineral Industries, Open File Report O-08-12, 46 p.

Witter, R.C., Briggs, R.W., Engelhart, S.E., Gelfenbaum, G., Koehler, R.D., and Barnhart, W.D. (2014) Little late Holocene strain accumulation and release on the Aleutian megathrust below the Shumagin Islands, Alaska. Geophysical Research Letters, 41(7), 2359–2367.

Part 1

Field Techniques for Sea-Level Reconstruction

Chapter 4
Coastal sediments

ALAN R. NELSON
Geologic Hazards Science Center, US Geological Survey, Golden, CO, USA

4.1 INTRODUCTION

Most sea-level histories have been reconstructed with stratigraphic evidence from sequences of coastal sediment. The timespans of sequences used in reconstructing sea-level history range from decades for studies of the rate of the 20th century sea-level change to tens of millions of years in studies of the global transgessions and regressions recorded in Cretaceous sedimentary basins. Rather than review the past use of coastal sediment in reconstructing sea-level history, the focus in this chapter is on methods of gathering field data on coastal sediment used in studies of relative sea-level change during the Holocene, especially the past 6000 years when rates of change had slowed relative to earlier Holocene rates. Some laboratory methods (not covered elsewhere in this volume) that enhance the field characterization of coastal sediment are also mentioned. Equally important aspects of sea level field investigations, such as coastal geomorphology or surveying (for example, sampling locations and elevations), are addressed in Chapters 9 and 10. Research approaches used in sea-level reconstruction, including key concepts such as relative sea level, indicative meaning, sea-level index point, and sea-level tendency, are explained in Chapters 1 and 2.

Thorough familiarity with the basic concepts and methods of sedimentology and stratigraphy, particularly at the few-kilometers scale of many Holocene sea-level study sites, is required to interpret sequences of coastal sediment as a record of sea-level change. Key stratigraphic concepts, such as unconformity, disconformity, transgression, regression, progradation, accommodation space, onlap, offlap, toplap, downlap, uplap, and many others, may be as critical in interpreting sea-level stratigraphy at sites only a few hundred meters across as they are in interpreting the stratigraphy of basins hundreds of kilometers long (e.g., Allen, 2000; van de Plassche et al., 2006). For basics, see Stow (2005) and Tucker (2011) for field methods, and Miall (1999, 2010), Coe et al. (2003), Prothero and Schwab (2003), and Boggs (2011) for concepts.

Sediment mapping in the field ideally follows two different but complementary approaches: traditional lithostratigraphy and process-based lithofacies analysis (Miall, 1999; Stow, 2005). The focus of lithostratigraphy on the architecture and chronology of coastal sediment sequences provides a framework in which to interpret the history of shoreline change over hundreds to thousands of years, whereas lithofacies analysis leads to models that help explain the site- to regional-scale processes that led to changes (Fig. 4.1). Lithostratigraphy is primarily descriptive and focuses on the spatial relations among bodies of sediment (lithostratigraphic units) or biofacies (biostratigraphic units), including the characteristics of stratigraphic contacts between adjacent units, the age or time span of units, and on the three-dimensional (3D) architecture of lithostratigraphic sequences. Much of Part II (Chapters 12–22) is devoted to biofacies analysis because of its importance in reconstructing Holocene sea-level history. In lithofacies analysis, geologists describe the characteristics of bodies of sediment (lithofacies) and the depositional, erosional, deformational, diagenic, or weathering features within or between the bodies, summarize their lateral and vertical relations, and then construct models that explain the processes that led to lithofacies deposition and modification. Especially at sea-level study sites, sedimentary process–response models are key to understanding the interplay among site geomorphology, climate variability, sediment influx and erosion, sea-level rise or fall, coastal subsidence or uplift, and sediment diagenesis and compaction (e.g., Reading, 1986; Allen, 2000; Long et al., 2006; Baeteman et al., 2011).

Handbook of Sea-Level Research, First Edition. Edited by Ian Shennan, Antony J. Long, and Benjamin P. Horton.

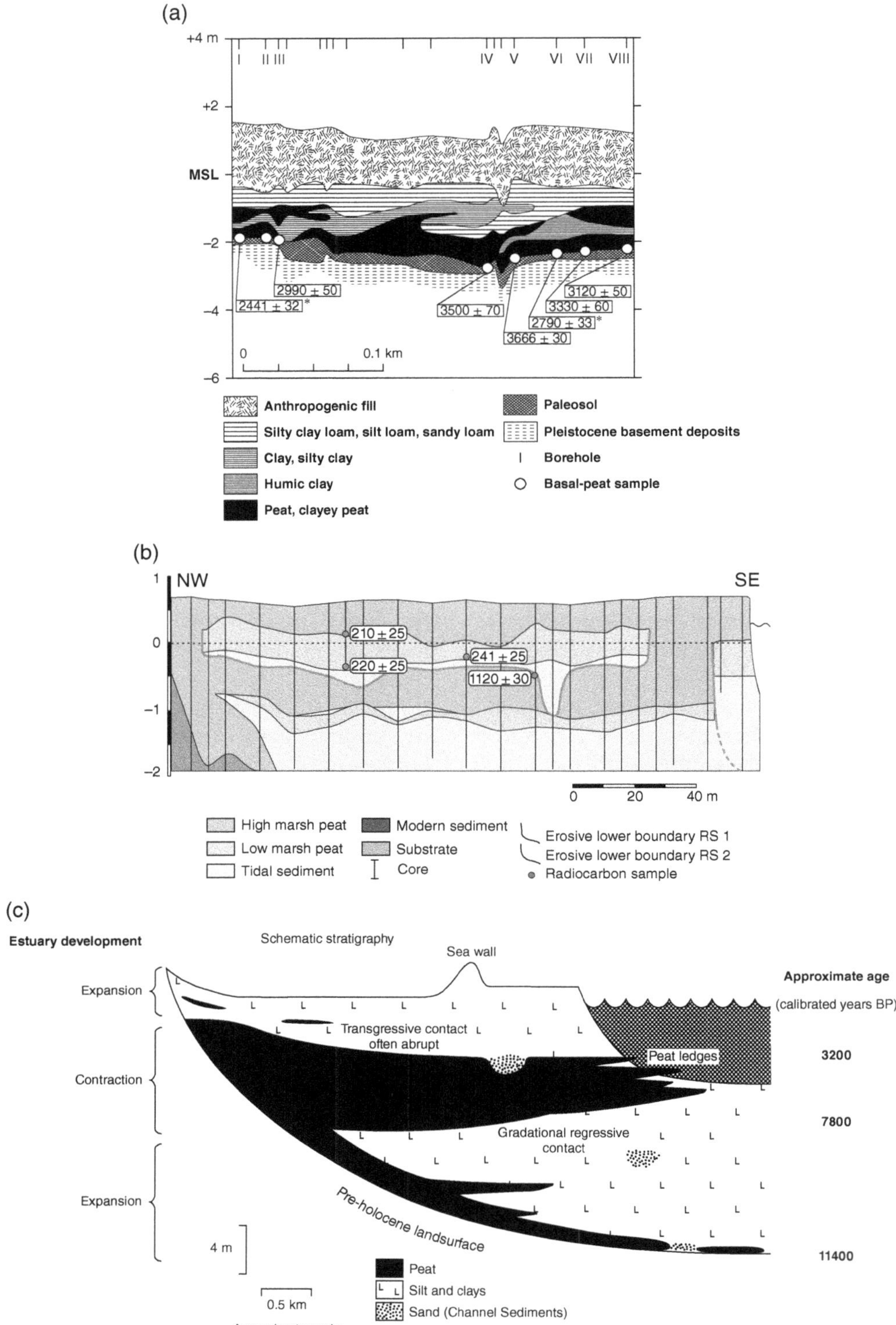

Fig. 4.1. Examples of figures from three studies of the history of Holocene sea-level that used traditional lithostratigraphy and the development of lithofacies models to differing degrees, depending on the objectives of the study. Lithostratigraphy in many tens to hundreds of cores was the data framework in all three studies. Lithofacies used to make paleoenvironmental interpretations were also carefully described in each study. (a) Törnqvist et al. (2004) used an implied lithofacies model of paleosol submergence and burial to select radiocarbon samples that were used to document the pattern of late Holocene sea-level rise in the Mississippi delta. (b) Detailed lateral relations of saltmarsh lithofacies were the focus for developing a lithofacies model of two regressive sequences overlying erosional contacts, attributed to erosion of a Connecticut saltmarsh during hurricanes. *Source*: van de Plassche et al., 2006. (c) Detailed microfossil analyses allowed Long et al. (2000) to emphasize biofacies analysis as much as lithofacies characteristics in developing a lithofacies model for the development of the Southampton Water estuary. These authors then went on to compare their model with the stratigraphy in two other estuaries, and then to explicitly synthesize all data into a lithofacies model (shown in (c)) that explains the Holocene history of estuaries in southern England.

Facies models are 2D or 3D frameworks depicting the stratigraphic relations among facies with an emphasis on the processes that produced them (Reading, 1986; Walker, 1992; Miall, 1999; Stow, 2005). Models might take the form of: correlated core sections perpendicular to a former prograding shoreline; block diagrams showing complex relations among fluvial, littoral, and tidal marsh lithofacies near the mouth of a spit-restricted estuary; or a series of paleogeographic maps depicting the stratigraphic and geomorphic evolution of an embayment over thousands of years (e.g., Dalrymple et al., 1992; Fig. 4.1c). Facies analysis begins with the detailed description of sediment properties (lithofacies), fossils (biofacies), or less commonly used properties, such as magnetic characteristics or distinctive signatures on geophysical reflection profiles. Description is followed by interpretation of lithofacies and their lateral and vertical relations in terms of sedimentary and other site processes, and finally by synthesis into a model that explains the history of sedimentary environmental change at a site. Use of specialist terms, such as "lithofacies association," "facies assemblage," "architectural element," or "bounding surface," is not required in studies of Holocene sea-level change, but an understanding of the concepts such terms express is needed to develop lithofacies models. In comparison with pre-Holocene studies, a major advantage in the application of lithofacies models to reconstruct Holocene sea-level history is that many coastal landscapes are similar to (or, in the case of late Holocene environments, little changed from) landscapes of the times when sediment sequences were deposited. Exceptions include extensively modified coastlines in highly developed regions, such as parts of eastern North America, Europe, and Japan.

Lithofacies models are of prime importance in studies of Holocene sea-level change at several scales. To a greater degree than in most other studies of sedimentary history or environments, the goal of Holocene sea-level reconstruction is to identify, date, and evaluate sea-level index points. Sea-level researchers in temperate latitudes have commonly focused on the quiet-water environments, particularly tidal wetlands, of estuaries and embayments where index points with precise indicative meanings are most likely to be preserved. For this reason, large-scale lithofacies models, such as those for estuarine depositional systems (e.g., Kosters 1989; Dalrymple et al., 1992; Long et al., 2000; Abrahim et al., 2008), are the starting point for site selection and fieldwork. At this scale, sea-level studies commonly apply existing lithofacies models rather than develop new ones. Lithofacies analysis is even more important at the scale of sea-level study sites because existing lithofacies models are rarely detailed enough to provide an adequate process–response framework within which to interpret lithofacies at the shallow depths typical of most sea-level studies (<10 m; e.g., Long et al., 2006). For example, both lithologic and biologic indicators of sea-level change will be more distinct where a high marsh is submerged by 0.5 m of sea-level rise than where the same rise submerges a low marsh. Further, because tidal environments shift laterally over time, a site with the most precise sea-level indicator during one time period may poorly record changes at a later time. Many studies of Holocene sea-level history could be improved by a more explicit description of the large-scale and site-scale lithofacies models used to infer the history of sea-level change.

4.2 EXPOSING COASTAL SEDIMENT

Safety in the field comes first. Field safety resources range from minimal recommendations, such as those in Coe et al. (2003) and Stow (2005), to comprehensive safety guides (e.g., O'Reilly, 1992) or the safety plans for specific types of fieldwork available on the web. Everyone needs adequate clothing for all possible weather conditions and training in first aid for situations far from emergency medical help. Checking your list of field and safety equipment again, just before going out, is more likely to further the goals of your field research than a few extra minutes of field time. Always carry a first aid kit adequate for yourself and 1–2 companions, whose components you have personally selected (commercial first aid kits are rarely adequate). For especially dangerous activities such as sampling steep exposures of unconsolidated sediment, taking vibracores or geoslicer slabs, describing stratigraphy in excavations >1.5 m deep, sampling in surf or along fast-flowing rivers, or taking sediment peels with toxic chemicals, even very experienced investigators should have a safety plan and brief everyone on it. Keep an eye on your field companions; if they are not adequately equipped or taking sensible precautions in dangerous situations, it may severely impact your ability to complete the field study.

4.2.1 Exposures

Coastal stratigraphy is most easily mapped, fully sampled, and accurately interpreted in near-vertical exposures (Tröels-Smith, 1955; Long et al., 1999; e.g., Atwater and Hemphill-Haley, 1997). Exposures typically provide views of coastal stratigraphy that are orders of magnitude more extensive than those in tens of small-diameter cores. Natural exposures are usually stable, and differential erosion by water frequently makes subtle differences in lithology more distinct. Unfortunately, natural exposures more than a meter high are quite rare at coastal sites likely to host sea-level indicators with precise indicative meanings. Pre-existing artificial exposures, such as along recently cleaned drainage canals or excavations for building foundations, are rarely open for more than days, require permission for access, and may be unstable.

Many sea-level studies would benefit from greater use of excavations tailored to specific site objectives. Relative to the value of stratigraphic information obtained, size of potential samples, and field time saved, excavations by rubber-tired backhoe loaders and track-mounted hydraulic excavators can be very cost-effective even at sites far from population centers (McCalpin, 2009, p. 64–76). Soft, wet ground that will not support heavy equipment, sandy or clast-rich sediment that will not maintain exposed walls, and permission for access (with probable damage to vegetation), rather than cost or availability, are the main limitations in using heavy construction equipment in most sea-level studies. Shovels are an inexpensive, minimally destructive, and underutilized alternative to powered equipment for exposing coastal sediment to depths of 1–2 m. Sediment that is difficult to penetrate with hand-operated corers may yield easily to picks and shovels.

Excavation design balances optimal exposure of stratigraphy against access and safety. Excavations of all sizes usually combine near-vertical (>80°), sloping, or benched walls McCalpin, 2009, p. 66–76). Sloped or benched walls are more stable (much safer) and more accessible than vertical walls, but involve moving large amounts of sediment. Visualizing 3D stratigraphy on sloping or benched walls is more difficult than on vertical walls. Once walls are photographed and mapped digitally however, vertical sections are easily constructed with computer software. For excavation with heavy equipment, investigators should consult the extensive literature on excavation design and safety and be familiar with local to national safety regulations (e.g., Occupational, Safety and Health Administration, http://www.osha.gov/SLTC/trenchingexcavation/). Excavation walls should be oriented and sloped for optimal lighting (consider the position of the sun and safe placement of spoil from the excavation) and carefully cleaned before mapping; key tools include flat-bladed shovels, trowels, and small Japanese garden hoes (Fig. 4.2c). At wet sites, either pumps or bailing buckets of appropriate sizes should be ready for use before excavation begins. Do not allow water to accumulate in the excavation, as repeated wetting of walls destabilizes them. As soon as practical after study, excavations should be filled following local construction codes or with the original sediment mounded over the hole so that later settling will approximately restore the former ground surface. Blocks of turf and grass seed placed last speed site restoration.

Exposures can be mapped in detail if a reference grid in a vertical plane is installed, a series of close-up photographs of exposure walls taken, and a mosaic of photographs constructed prior to detailed description and interpretation of lithostratigraphy (e.g., Lienkaemper et al., 2002; DuRoss et al., 2012). These and other methods of wall mapping are described by McCalpin (2009, p. 92–97). Even large outcrops exposed only during the lowest tides can be cleaned, markers or a horizontal grid installed, and photographs taken before the tide rises (Long et al., 1999; e.g., Atwater and Hemphill-Haley, 1997).

High, full cloud cover thick enough to hide the sun gives optimal lighting for photographs. Photographs are rectified (undistorted) and placed on a scaled photomosaic grid (commonly 1 × 1 m) with raster-based software. Photo-stitching software also merges photographs well, but markers and a reference grid are still needed to remove photograph distortion and to place them into a vertical reference grid. We print mosaics on 28 × 43 cm paper (at scales of 1:4 to 1:20) and clip them to lightweight (foam or plastic) mapping boards under an overlay of clear

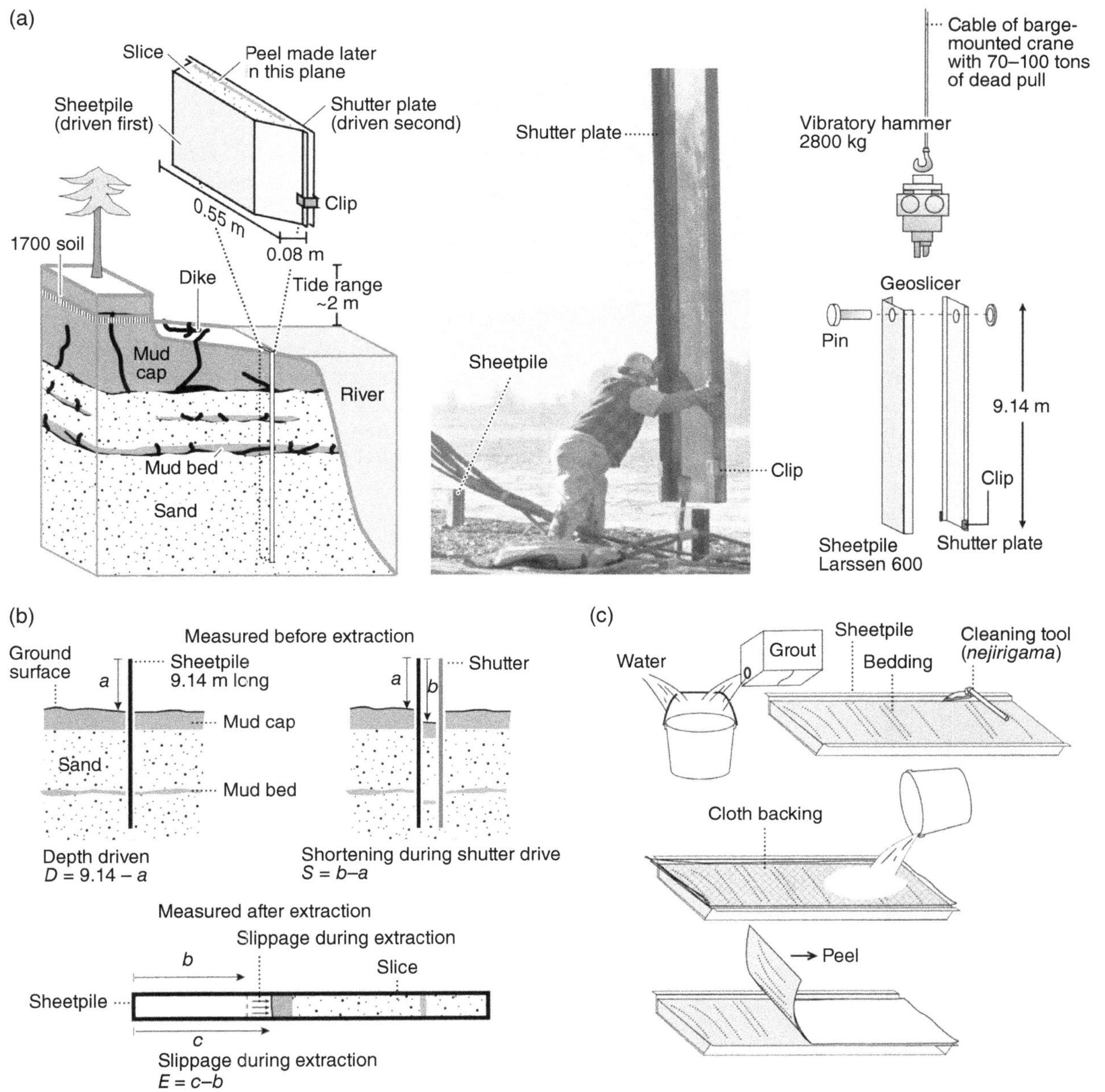

Fig. 4.2. Methods of geoslicing and peeling used at tidal sites along the Columbia River, Washington (from Takada and Atwater, 2004). (a, b) 9-m-long pilings were vibrated into river bank sediment, a shutter plate was hydraulically pushed down to seal off the geoslice, and the geoslicer was pulled up using a large crane on a barge. (c) Peels were made by pouring hydrophilic grout onto a geoslice (shutter plate removed) that had been cleaned (smoothed) with a nejirigama (handheld Japanese garden hoe). *Source*: Takada and Atwater (2004). © Seismological Society of America.

polyester. Stratigraphic contacts, sample locations, and structures are quickly drawn on the overlay during the next low tide. As long as color mosaics can be printed before fieldwork at the outcrop recommences, photomosaic methods are faster and more accurate than traditional methods of drawing contacts on graph paper or surveying in numerous control points with a total station, especially at tidal sites with difficult access. Tablet computers (Section 4.3.6) eliminate the need for mapping boards and printed mosaics, but not for mosaics constructed with rectified photographs. Photos of shovel pits or low outcrops (with a labeled tile and scales extending to the base of the exposure) or cores (Fig. 4.3b) allow them to be rectified into a vertical plane or reference grid (measure pit wall slope) and later correlated with photomosaics of other exposures or cores (see Section 4.4).

When sampling exposures, rectangular columns of sediment termed "monoliths" cut from exposure walls provide large samples that can be thoroughly studied in the laboratory (Aaby and Digerfeldt, 1986; de Vleeschouwer et al., 2010). Monoliths may consist of blocks or stacks of blocks, cut with hand tools from the smoothed vertical wall of an exposure, and wrapped with sufficient plastic film and aluminum foil to provide stability for transport. More commonly, rectangular plastic or lightweight metal trays, open on one side, are oriented vertically and pushed into the smoothed wall of the exposure to create a vertical column of sediment protected on three sides by the tray. Once the stratigraphic position of the monolith has been measured, photographed, and its orientation labeled, it is cut from the wall and trimmed and wrapped for transport. Exposures many meters high can be sampled with a series of vertically overlapping monoliths (each <0.5 m long due to sediment weight).

4.2.2 Corers

The most common means of accessing coastal sediment that cannot be viewed in exposures is with hand-operated coring devices such as gouge corers, chamber corers, and small piston corers. The ease of carrying corers in wet, rugged terrain, their ability to sample hundreds to many thousands of years of sediment, and their low cost explains their popularity in sea-level studies worldwide. Here we focus on the most common types of corers used to take <10-m-long cores primarily on land; see Glew et al. (2001) for the wide range of corers designed for lakes or lagoons.

Gouge corers consist of tubes of hardened steel alloy with about 40% of the side of the tube removed (Fig. 4.3a). Extensions (1 m long; also termed "rods") are attached and the gouge is pushed to the sampling depth, rotated a few times to cut a segment of core sediment, pulled to the surface, and then trimmed with a knife to

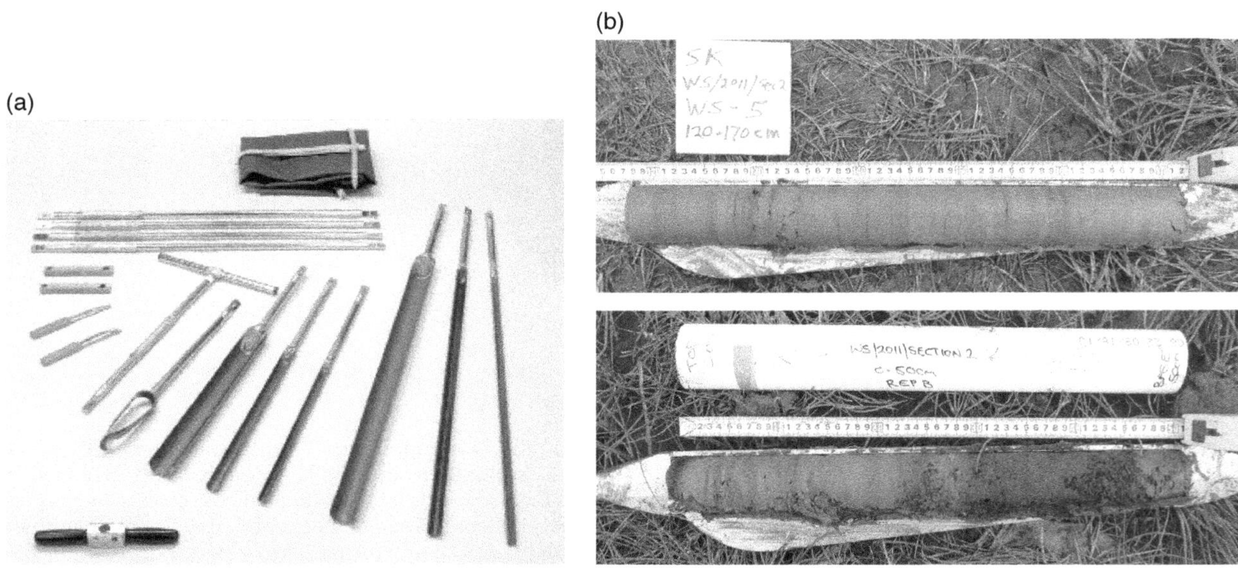

Fig. 4.3. (a) Gouge corers manufactured by Eijkelkamp Company, The Netherlands. Similar soil and soft sediment sampling equipment is manufactured in the United States, Australia, and Japan. Long gouge corers collect 1-m-long sections of sediment, short corers 0.5 m. Extensions with bayonet connections are shown, but threaded screw connections are much stronger. If a little oil is dripped onto screw connections each time the equipment is assembled, the connections always come apart easily. (b) Two segments of a Russian core from a tidal lagoon on Sitkinak Island, Alaska. In the lower photo, the PVC plastic pipe tray is about to be placed on the core segment, the corer turned over, the core transferred to the tray, photographed again, labeled, and then wrapped in plastic wrap for transport. Scales and labels as shown are required if rectified (undistorted) photomosaics will be made of the core. Using the standard convention of up towards the left, with the scale on the right side of the core, saves considerable time when constructing photomosaics. *Source*: Photographs by Alan Nelson.

expose the core segment. The most popular gouge corers come in lengths of 0.5 m and 1.0 m and diameters of 20–60 mm. All gouge corers compact soft or peaty sediment to varying degrees, and distorted sediment, missing core section, and depth measurement errors due to non-vertical coring are common (Long et al., 1999; Törnqvist et al., 2004). Narrow gouge corers (<30 mm diameter) are best suited to site exploration and for mapping marker beds in many tens of cores over hundreds of meters. Narrow gouge corers routinely recover sediment from depths of 6–8 m; recovery of peat from >20 m depth has been reported using lightweight aluminum extensions. The larger-diameter gouge corers (≥50 mm diameter), or "fat gouges," sample 4–6 times the sediment volume of narrow gouge corers, making them suitable for obtaining samples for isotopic dating or microfossil analyses. However, because the gouge corer surface area to sediment volume ratio is lower in fat gouge corers than in narrow gouge corers, fat gouge corers do not retain sandy or peaty sediment well unless dense mud or sand near the base of the core segment prevents sediment from sliding out during segment recovery. Gouge corers are designed primarily for collecting cores to be described in the field rather than for returning cores to the laboratory for detailed study.

Chamber corers were designed to reduce the sample disturbance typical of gouge corers. The Hiller chamber corer, first described in 1868, scrapes sediment into a 40-mm-diameter chamber, over which a vertical fin slides when the corer is rotated at the sampling depth prior to recovery. The Hiller corer can be driven to greater depths than other chamber corers, but the sediment core in the chamber is easily contaminated and cannot be removed for transport to the lab (Aaby and Digerfeldt, 1986). The similar Vrijwit auger employs a vertical plate that slides down the open wall of a rectangular chamber to cut a slice of sediment 40 mm wide and 0.5 m long. The most widely used type of chamber corer is the "Russian" corer described by Belekopytov and Beresnevich (1955); see also Aaby and Digerfeldt (1986) and Glew et al. (2001). The most popular model consists of a heavy steel or aluminum shaft with a 0.5-m-long, 50-mm-diameter chamber, a downward tapering fin, and a horizontally rotating blade that protects the sample in the chamber (Fig. 4.3b). After the Russian corer is pushed down to the sampling depth, the handle is rotated 180° clockwise to fill the chamber with a half-cylinder of sediment and allow the blade to cover the sample as the corer is pulled up. At the surface, the corer is rotated back 180° to expose the sample on the flat surface of the blade (Fig. 4.3b). Because it cuts horizontally into sediment, the Russian corer extracts uncompacted, relatively undisturbed samples of about three times the volume of sediment obtained with narrow gouge corers. To ensure complete recovery of the entire section, core segments are taken with 0.1 m depth overlaps with the segments above and below them (de Vleeschouwer et al., 2010). Cores that consist of a series of Russian core segments are usually sufficient for microfossil analyses; replicate cores should be collected for dating (^{14}C, ^{137}Cs, ^{210}Pb), geochemical, and other analyses (e.g., Last and Smol, 2001b). Principal disadvantages of Russian corers include: they are occasionally clogged by wood or pebbles; they cannot recover undisturbed samples of undecayed, root-bound peat; they cannot penetrate stiff, muddy, or sandy sediment as easily as gouge corers; and the height of the corer's nose cone prevents the lowest 0.1 m of sediment penetrated by the corer from being sampled. Large Russian corers (1.5–2 m long and 120 mm diameter) have been used for specialized sampling (Aaby and Digerfeldt, 1986; de Vleeschouwer et al., 2010).

Piston corers are more complex and require greater skill to operate than gouge or chamber corers; for a particular application, specialized experience is needed to select from among the variety of designs and diameters available (Wright et al., 1965; Aaby and Digerfeldt, 1986; Wright, 1991; Glew et al., 2001). Piston corers attached to successive extension rods typically sample coastal sediment beneath floating platforms or ice in shallow lakes and lagoons, or in thick sections of peat in bogs, where the sediment is soft enough to push them down to sampling depths. In water depths of more than a few meters, piston coring is most successful inside casing (100 mm diameter, thin-walled, steel pipe) extending from the coring platform to the bottom of the lake. A major advantage of piston corers is that the fixed position of the piston with each drive of the core barrel into unsampled sediment creates a seal that prevents significant shortening through deformation of the sediment in the barrel (Glew et al., 2001). Some piston corers use plastic liners inside the core barrel; liners help protect fragile sediment during transport, but prevent viewing of the stratigraphy until the liner is cut open. The most commonly used piston corer is the 1-m-long, 50-mm-diameter Livingston corer (Livingston, 1955; Wright et al., 1965), which does not use liners. Instead, 1-m-long core segments are extruded into plastic trays in the field by using the piston to force the sediment out the end of the core barrel (commonly with a hand-operated winch), a process that may compact or distort the core (Wright, 1991).

For taking large-diameter cores or cores in stiff, muddy, or sandy sediment vibracores are widely used, traditionally with a 76-mm-diameter aluminum irrigation pipe in North America and more recently with rectangular slab corers, termed "geoslicers," of various dimensions in Japan. Safe and successful vibracoring or geoslicing are specialized sampling methods that require knowledge of the literature on the equipment and procedures as well as practice with experienced personnel. Pipe vibracorers employ a custom-made steel or aluminum clamp that holds a commercial cement vibrator wand against the top of the pipe (Lanesky et al., 1979; Smith et al., 1984; Thompson et al., 1991; Fig. 4.4a). A gasoline engine rotates a shaft inside a cable attached to the wand producing strong vibrations that are translated to the sharpened, bottom end of the pipe. The vibrations liquefy saturated sediment at the end of the pipe, allowing it to slowly descend to depths of 1–15 m (Glew et al., 2001). After the aluminum pipe stops

Fig. 4.4. Methods of vibracoring in tidal marshes in Oregon, USA. (a) Vibracoring in a diked former tidal marsh along the Coquille River (Sevenmile site of Witter et al., 2003). (b) Harvey Kelsey and Rob Witter compare five vibracores from a tidal marsh at Siletz Bay. The dark beds are peat-rich O horizons of former high marshes that suddenly subsided during great subduction-zone earthquakes (e.g., Witter et al., 2003). (c) Light tent that provides uniform, diffuse lighting for photographing vibracores shortly after they are split. Tent is easily disassembled for transport.

descending, amounts of sediment recovery and compaction are measured (e.g., Fig. 4.2b), and the pipe is extracted (carefully, as 1 m of filled vibracore pipe weighs 14 kg) with a hand-operated winch attached to a strong ladder (Fig. 4.4a) or custom-built tripod (Thompson et al., 1991). Handmade (sheet aluminum) or machined-steel core catchers greatly improve recovery, but slow vibracorer descent and may distort soft sediment. Better penetration and recovery is reported with large-diameter (76 mm) versus small-diameter (50 mm) pipe. Core shortening and compaction in vibracorers may approach 30–50% in peat-rich sediment; the addition of a piston in the pipe may reduce compaction. Cores are split with electric metal shears and a serrated knife.

Geoslicers retrieve rectangular slabs of moist sediment many times wider than vibracores, making them ideal for study of sedimentary structures and complex stratigraphy, and for extensive sampling of sediment. In geoslicing, a rectangular tray, open on one side, is driven into the ground with a weighted vibrator; a shutter plate is then vibrated down to seal off the open side of the tray (Nakata and Shimazaki, 1997; Haraguchi et al., 1998; Fig. 4.2). Geoslicers range in size from hand-pounded models that take slabs 0.15 m wide and 1 m long to custom-made equipment that takes slabs 12 m wide and 11 m long (McCalpin, 2009, p. 100). Large geoslicers require truck-mounted cranes, but smaller models can be transported off road by 3–4 investigators. A trained crew is needed, as well as equipment for rapidly describing and sampling slabs, or for transporting and storing them for later study. Geoslicing is particularly applicable at sites with shallow groundwater tables (e.g., Kaneda et al., 2008) where ground disturbance must be minimized, or where the site accessible for coring is only a few meters wide. The costs of equipment, trained personnel, and slab storage are the primary deterrents to the more widespread use of geoslicers in sea-level studies.

A wide variety of other powered coring systems can be used where site conditions and budgets permit. They include augers, hollow-stem augers, direct-push hydraulic systems using either tubes (geoprobes) or rectangular samplers, and drilling equipment of many types and sizes. Most are truck- or trailer-mounted (McCalpin, 2009, p. 97–100), but a few small vibrating systems can be carried to the site by 1–3 people. Powered corers allow recovery of sediment too hard for hand-operated corers, or where many large-diameter cores are needed. Some systems allow core samples to be returned to the lab in plastic liners and can take continuous cores to depths of many tens of meters. Such methods are expensive and require a site capable of supporting the equipment.

4.3 DESCRIBING COASTAL SEDIMENT

Coastal lithofacies are described with reference to their lateral and vertical architecture, including lithology (composition); contacts (boundaries) that separate lithofacies; depositional (including bedding) or post-depositional structures; and weathering features within lithofacies. Well-organized, carefully annotated field sketches and core logs (e.g., Stow, 2005, p. 19–22), with interpretative notes added before leaving the site, are the foundation of many widely cited sea-level publications. We supplement the basics of sediment description in the field, covered well by many stratigraphy and sedimentology texts (e.g., Stow, 2005; Tucker, 2011), with comments directed at sediment description in the quiet-water sedimentary environments favored by Holocene sea-level researchers.

4.3.1 Lithology

The lithology, or texture (grain-size distribution) and composition, of sediment is described in the field using standardized qualitative to semi-quantitative classification systems. Geologic descriptions of sediment containing small amounts of organic material usually employ two or three main components (e.g., silty sand, sandy gravel; Last, 2001; Farrell et al., 2012), commonly with additional adjectives. Soil scientists estimate three clastic matrix components (sand, silt, clay) and relative proportions of decomposed humus in the surface horizons of soils. With experienced fingers and percentage charts, major sediment components (sand, silt, clay, pebbles, cobbles, boulders) can be described in the field to within a few percentage points of their laboratory-measured values (e.g., Schoeneberger et al., 2002). Studies of Holocene sea-level history commonly focus on sediment sequences with much higher percentages of organic material than most geologic studies. The most useful field system for describing sediment with significant percentages of organic material is the system of Tröels-Smith (1955; Fig. 4.5).

Tröels-Smith (1955) proposed a descriptive system based on the earlier genetic system of von

Elements (components) of each lithofacies

	Sh	Substantia humosa	Humous substance, highly decomposed, homogeneous microscopic structure
Turfa	Tb^{0-4}	T. bryophytica	Rooted moss peat*
(rooted peat)	Tl^{0-4}	T. lignosa	Rooted woody peat - stumps, roots, rootlets, branches, trunks, stems, of woody plants*
	Th^{0-4}	T. herbacea	Rooted herbaceous peat - roots, rootlets, rhizomes, stems, leaves of herbaceous plants*
Detritus	Dl	D. lignosus	Unrooted fragments of woody plants >2 mm
(detrital)	Dh	D. herbosus	Unrooted fragments of herbaceous plants >2 mm
	Dg	D. granosus	Fragments of woody and herbaceous plants and animal fossils (not molluscs) < 2mm and >0.1 mm
Limus	Ld^{0-4}	L. detrituous	Plants and animals or fragments of them <0.1 mm*, except those below
(gyttja)	Lso	L. siliceus organo.	Diatoms and other siliceous fossils or their fragments <0.1 mm
	Lc	L. calcareus	Marl, not hardened like calcareous tufa, etc. < 0.1 mm
	Lf	L. ferrugineus	Iron-oxide, not hardened <0.1 mm
Argilla	As	A. steatodes	Particles of clay <0.002 mm
(mud)	Ag	A. granosa	Particles of silt 0.002–0.06 mm
Grana	Ga	G. arenosa (-f, -m, -co)	Particles of sand 0.06–2 mm; fine 0.06–0.2, medium 0.2–0.6, coarse 0.6–2.0 mm
(sand and gravel)	Gg	G. glareosa (-f, -m, -co)	Particles of gravel 2–60 mm; fine 2–6, medium 6–20, coarse 20–60 mm

* includes minor amounts of humous substance
Each element proportion (1 = 12–25%, 2 = 25–50%, 3 = 50–75%, 4 = >75%) is listed with largest proportion first. Proportions sum to four. " +" indicates less important components (<12%; e.g., "Ld^{+}")." +" and "-" may be used to modify element proportions (e.g.,"$Th2^{-}$" means ~ 25–30% Th; "$Ag3^{+}$" means ~70–75% Ag); "tr" indicates a trace (<2%) of an element ("Ga tr"). The humicity of turfa and Ld may be estimated with the 0–4 scale below. The proportion of "foreign" elements (e.g., charcoal, macrofossils) are estimated separately from the primary elements and are shown within []. Irregularly distributed mixtures of two lithologies are separated by /. Interbedded lithologies are separated by —.

Humicity

Super-script	Preservation of plant structure	Color and consistency on squeezing	Percent Sh which passes fingers on squeezing
0	fresh and well preserved	clear, colorless water; no Sh	0 Sh
1	well preserved	dark-colored water; mino r Sh	<25 Sh
2	partly decayed but distinct	dark water and Sh (opaque)	<50 Sh
3	decayed and indistinct	mostly Sh (opaque), little water	<75 Sh
4	absent or barely discernable	no water, all Sh (opaque)	100 Sh

Lithofacies properties

Color – Use Munsel color chart (5YR, 7.5YR, 10YR, 2.5Y, 5Y, 7.5Y, and 10Y are most commonly used pages).

Stratification – Use Tröels-Smith Smith 0–5 scale, or geologic terms (weak, indistinct, strong, non-parallel, wavy, etc.); show type of stratification by reference to keys (e.g., Stow, 2005) or with sketch.

Elasticity – Estimate consistence and plasticity (see Troels-Smith 1955, p. 53) following soil description methods of Schoeneberger et al. (2003, p. 2-46,50), or use Tröels-Smith 0–5 scale for elasticity.

Dryness – Use Tröels-Smith Smith 0–5 scale or soil description terms (dry, moist, wet, saturated).

Structure – Estimate type (massive, homogeneous, fibrous, felted, granular, angular blocky, subangular blocky, platy, wedge, prismatic), size (list range in mm), and strength (0–4 scale; see Schoeneberger et al., 2003, p. 2–38).

Lower contact (boundary) - Describe lower contact thickness (in mm), topography (smooth, planar, irregular, wavy, broken; or use a sketch), and distinctness (use 0–5 scale).

Fig. 4.5. The Tröels-Smith (1955) system of sediment description, as modified by Aaby and Berglund (1986) and Long et al. (1999) with some additional recommendations.

Post and Granlund (1926); see also Birks and Birks (1980); Aaby and Berglund (1986); Kershaw (1997); and Long et al. (1999). Tröels-Smith descriptions consist of estimates of lithologic components ("elements" of Tröels-Smith, 1955) that sum to four, physical properties (the degree of darkness, the degree of stratification, the degree of elasticity, the degree of dryness, and boundary), and the humicity or degree of decomposition (humification) of peat-rich components. Tröels-Smith (1955) used Latin names for components and properties in order to make the system more universal and to avoid confusion with previously used terms in several northern European languages. Other systems for describing organic-rich sediment have been proposed (e.g., Kearns and Davison, 1983; Nelson et al., 1996; Schnurrenberger et al., 2003), but are less useful because they were designed for restricted environments (e.g., lakes, clastic sediments, soils developed on terrestrial deposits) or use only two or three primary components. Some strengths of the Tröels-Smith system are its non-genetic focus, the variety of organic materials that can be described, and its use of four components (an effective compromise of detail versus the number of components that can be identified in the field).

Each of the four primary Tröels-Smith components comprise about 5–25% (by volume) of a lithofacies (Fig. 4.5). Less-important components (<12%) are designated with a plus ("+"); some investigators use trace ("tr") for components <2%. Many investigators also modify the primary component number with a plus or minus superscript if they estimate that a primary component is close to the upper or lower percentage range for that component. For example, a silty peat thought to contain 70–75% peat derived from rooted herbaceous plants would be designated "$Th3^{+}$, Ag1." More detailed use of plus-minus superscripts (for example, "$Th2^{--}$" or "$Ga1^{++}$") is discouraged because consistent description of such subtle compositional differences is difficult. Humicity superscripts may be used to show the degree of decomposition of "Th" or "Ld" (Fig. 4.5). Thus, humicity could be shown for turfa components (e.g., $Th^{2}1$ or $Th^{3}2$) where the relative degree of peat decomposition was an important distinguishing characteristic. The alternative (recommended) way of recording the degree of decomposition is to use "Sh" to record the proportion of completely decomposed organic material; "Th" or "Ld" with a humicity of 4 is equivalent to "Sh" (Fig. 4.5; Tröels-Smith, 1955, p. 47). Long et al. (1999) explain related inconsistencies in how Tröels-Smith (1955) used the "Ld" component. Amounts of other materials, such as charcoal or macrofossils, are estimated separately from the four primary components. Some components, such as "Th," are most accurately estimated by breaking sediment into fragments along horizontal and vertical planes and examining them with a hand lens. Lithologic properties that do not fit easily into the Tröels-Smith system (e.g., soil structure, weathering features, or burrows) are described with words.

Several minor modifications have been made to Tröels-Smith's (1955) original system and I recommend additional changes (Fig. 4.5). Aaby and Berglund (1986) combined several of the original *Grana* catagories into *Grana minora* and *Grana majora* to match modern geologic classification systems for sand and clasts. Color (hue, value, and chroma; rather than only darkness) is a basic property of sediment and should be described with reference to the Munsell color charts used by soil scientists (Aaby and Berglund, 1986; Long et al., 1999). Fieldworkers can measure their ability to distinguish color on the web (e.g., http://www.xrite.com/online-color-test-challenge). Description of colors, shapes, and dimensions of mottles (e.g., Schoeneberger et al., 2002) may be useful in inferring changes in water levels. Stratification, elasticity, and dryness can be described with 5-point scales as advocated by Tröels-Smith (1955), but elasticity is more accurately described as "consistence" and "plasticity" if investigators are familiar with methods for describing geologic soil profiles (e.g., Birkeland, 1999; Schoeneberger et al., 2002; Schaetzl and Anderson, 2005). General terms may be used for dryness, but I recommend describing structure with specific soil description terms. Many lithofacies have no structure (massive), but for those that do it may be the most important property for inferring genesis.

A widely recognized but little discussed aspect of describing sediment with systems such as Tröels-Smith is property descriptor drift (e.g., Birkeland, 1999; Long et al., 1999). Drift afflicts all field investigators, but experienced investigators recognize it early and make corrections or redescribe sediment. For example, if Tröels-Smith lithology in tens of cores ranges only between "Th2, Sh2, Ag^{+}," "Th2, Sh1, Ag1," and "Sh2, Th1,

Ag1," after tens of hours of core description investigators' fingers will become more and more sensitive to slight lithologic changes. Unless an investigator is vigilant, "Th2, Sh1, Ag1" will tend to be described as "Ag2, Th1, Sh1" or "Sh2, Th1, Ag1" as "Sh3, Ag1, Th^{+}" or even "Sh4, Ag^{+}, Th^{+}." Greater drift is common. All investigators benefit from occasional recalibration of their field description skills against laboratory-measured samples and among co-investigators in the field (Long et al., 1999).

Microscopic study of sediment confirms field descriptions and allows sediment components to be studied in greater detail (Birks and Birks, 1980; Aaby and Berglund, 1986; Last and Smol, 2001a, 2001b). For fine-grained sediment, smear slides are a widely under-utilized, quick, low-cost method for identifying potentially important components such as tephra, spicules, diatoms, foraminifera, ostracods, charcoal, clay, carbonate coatings of various types, iron-oxide, vivianite, pryrite, and other authogenic minerals. Modern adhesives allow researchers to cure many high-quality slides in minutes under ultraviolet light. Only a pencil point of sediment is needed. Ideally, examination of smear slides leads to other laboratory analyses, back to more detailed descriptions of field samples or cores, and then to examining more smear slides until sediment genesis is thoroughly understood. The best resource is LacCore's (University of Minnesota) website (http://lrc.geo.umn.edu/laccore).

4.3.2 Bedding and stratigraphic unit contacts

Changes in lithology and bedding are the primary attributes used to group packages of sediment into lithofacies with mappable contacts. Grouped together by Tröels-Smith (1955) as stratification, bedding is defined as ≥10 mm thick, whereas lamination (<10 mm thick) is considered with depositional structures. Both are described with measurements of thickness (in millimeters), references to sketches or keys of different types of bedding (e.g., Stow, 2005, p. 30, 42, 46, 55), and photographs.

Contacts between lithofacies, termed "boundaries" by Tröels-Smith (1955), require a description of their thickness (in millimeters), topography, and overall distinctness (use soil science terms; Fig. 4.5). Where these characteristics vary, a range should be described (e.g., "irregular to broken" or "very indistinct to distinct"). Include a sketch of the topography and variability of important contacts. Particularly in sea-level studies where the direction and rate of environmental change is paramount, contacts should be scrutinized to distinguish among erosional unconformities, disconformities, and post-depositional types of contacts, such as soil horizon boundaries (Section 4.3.4). Breaking a piece of sediment spanning a contact along a vertical plane and examining the contact with a hand lens gives a fresh view.

4.3.3 Depositional structure

A huge range of depositional structures found in coastal sediment are described in the literature and summarized and illustrated in textbooks and specialist volumes (e.g., Reading, 1986; Kemp et al., 2001; Stow, 2005; Davis and Dalrymple, 2012). In quiet-water tidal environments, structures produced by plant roots or burrowing animals are most likely to be preserved. Knowledge of plant macrofossils (Chapter 12), particularly rhizomes and roots, can be helpful in interpreting indistinct or fragmentary structures seen in cores and exposures of muddy or peaty sediment. Although decades of decay may severely alter macrofossil morphology, study of living plants and animals in the field is the best guide to interpreting sedimentary structures at many sea-level sites.

Sediment peels have a long history in soil science and sedimentology (McCalpin, 2009, p. 100–101; e.g., Kidwell et al., 1985; Barrett-Mold et al., 2009) and are of special value in preserving sea-level indicator stratigraphy. Even peels of vertical stratigraphy more than a meter square can be returned to the laboratory for later detailed study and permanent documentation (Fig. 4.2c; e.g., Kaneda et al., 2008). Many materials and methods have been used including glue, latex, epoxy resin, polyester resin, acrylic resin, cement, hydrophilic grout, and polyurethane foam. Because peel compounds must soak into sediment, better results are obtained on fine-grained sediment if a vertical slab can be cut from an exposure or core and laid flat before applying the compound. Some peeling methods that produce strong, flexible peels of wet, muddy-to-peaty sediment employ toxic compounds that require the use of goggles, masks, and special safety procedures (e.g., Takada and Atwater, 2004).

In sea-level studies, greater confidence in field interpretations may be obtained through laboratory study of micro-scale sedimentary structures. Such analyses range from microscope examination of field-collected peels in the laboratory to detailed study of sediment physical properties, chemistry, and micromorphology (e.g., Last and Smol 2001a, 2001b; Francus 2004; Schaetzl and Anderson, 2005). In some coral reef environments, slabs of coral-head limestone are examined under a microscope to identify sea-level index points with precisions of a few centimeters over a few years (Chapter 8). A low-cost compromise between field descriptions and high-resolution laboratory methods is to describe oriented blocks of sediment under a low-power (10–50×) binocular microscope. Vertical slabs 3–20 mm thick cut from blocks or cores of appropriate size commonly reveal lithologic components and structures not noticed in the field. Additional structural details may be apparent after the slabs air dry.

A development of the past two decades of considerable potential for sea-level studies is the non-destructive scanning of cores (whole or split) or slabs with multi-sensor core loggers, acoustic logging systems, and elemental scanners. A variety of sediment properties can be determined or inferred using high-quality incremental measurements of gamma-ray attenuation bulk density, P-wave velocity, magnetic susceptibility, digital color imaging, X-ray fluorescence, and gamma spectroscopy (Last and Smol, 2001b; Zolitschka et al., 2001; Francus, 2004; Rothwell and Rack, 2006). Most of these methods provide a rapid means of identifying changes in sediment properties or marker beds in long sections of massive, indistinct sediment. Methods for imaging core sediment with high-resolution, visible-light or infrared cameras, 3D X-radiography (X-ray computed tomography or CT scanning), magnetic resonance imaging, confocal laser microscopy, and wavelength-specific imaging may be particularly useful for identifying subtle structures diagnostic of sediment disturbance, degree of decomposition, or depositional processes. Considerable progress has also been made in measuring sediment grain-size distributions directly from photographs (e.g., Buscome and Rubin, 2012). The increasing variety of scanning methods available, their improving resolution, and the wider availability of scanning instruments and software for data analysis (with resulting lower costs) ensure that sediment scanning will become increasingly important in sea-level studies.

4.3.4 Soils and post-depositional structure

Because few shoreline sequences record continuous deposition, recognition of periods of subaerial exposure of sediment is critical in reconstructing sea-level change. In some environments, distinctive post-depositional structures such as mud cracks may show intermittent exposure (Davis and Dalrymple, 2012). Most often however, years or more of exposure are recognized by the type and degree of soil development on coastal sediment. For this reason, an understanding of how soils develop on differing lithologies in different geomorphic and climatic settings, and familiarity with characteristics of common soil horizons are required to identify and interpret modern and fossil soils, particularly those of wetland environments with abundant plant remains (Schaetzl and Anderson, 2005). Although most soils in sequences of Holocene coastal sediment are weakly developed, some characteristics (e.g., thickness and types of horizons, color, soil structure, clay or iron accumulation, degree of decomposition) may be time-dependent and may therefore be used to estimate relative differences in periods of exposure (e.g., Birkeland, 1999). Given sufficient time for its development, soil structure is commonly a diagnostic characteristic that outlasts other horizon characteristics following submergence and burial of former soils during sea-level rise. Correct names and suffixes should be used for identified soil horizons; for example, O, AO, AB, and AC horizons are common on young soils, whereas older soils may show thin B, BA, EA, Bw, Bh, and BC horizons of the US soil classification system (Schoeneberger et al., 2002). Canadian and UK soil classification systems are similar.

In studies of sequences of tidal sediment there is sometimes confusion as to whether an organic-rich layer is a deposit (bed) or a soil horizon. Soil horizons develop in or on sediment (parent material) at or near the surface, at least partly as a result of plant growth (Birkeland, 1999; Schaetzl and Anderson, 2005), whereas deposits involve sediment transport and usually require a water column in which to form. For example, a high marsh peat tens of centimeters thick that accumulated gradually over many hundreds of years is a soil horizon (O horizon of the US classification system). If the peat was later buried and compacted by 2 m of overlying tidal mud, however, it might be mistakenly mapped as a deposit. A Tröels-Smith description of lithology with emphasis on soil structural characteristics, followed by low-power microscope examination, will, in almost all cases, distinguish a soil horizon from a deposit.

4.3.5 Final comment on description

How many lithofacies should be described in each core or exposure? Where a change in one or more sediment properties can be identified, a contact is mapped and separate lithofacies described. I have described between 1 and 25 lithofacies in 1-m-long sections of gouge cores. Very detailed lithofacies descriptions can easily be condensed into simpler sections, but not the reverse. As figures in publications are simplified summaries of field data, investigators who want to indicate the level of detail of field descriptions can state the number (mean and range) of lithofacies described per meter of core or exposure.

Before leaving the field, record your best guess as to the genesis of each lithofacies described (e.g., "pond deposit," "A horizon of forest soil," "high-marsh peat"). Include notes about the sediment properties that led to your inference about genesis. Review these inferences as fieldwork progresses.

4.3.6 Recording field data

Waterproof field notebooks or clipboard forms remain the most effective, low-cost method for recording field data in most investigations of coastal sediment. Wide notebooks (180 × 220 mm or 200 × 250 mm) allow more detailed sketches but do not fit into field vest pockets as narrow (125 × 200 mm) notebooks do. Field description forms designed by the investigator for a specific investigation (e.g., Fig. 4.6) save time and ensure systematic recording of data. They may however be hard to manipulate in rain and wind and forms are more easily misplaced than notebook pages. I use a small audio recorder to supplement my primary written field notes, make interpretative comments, or to record notes while driving or in extremely unpleasant weather.

Photographs are an integral part of all aspects of data collection in the field. Because consumer-grade cameras take high-resolution photographs at insignificant per-photo cost, all cores, exposures, sampling locations, and most sediment features should be photographed as they are described. With appropriate labels and scales in the photograph (Fig. 4.3b), even small sedimentary structures can be documented in great detail; sizes of grains, clasts, and structures can later be measured directly from photographs (e.g., Buscombe and Rubin, 2012). Software may enhance exposure, contrast, color, and other aspects of photographs to reveal subtle bedding or other structures not visible in original photographs. Color is most accurately described using Munsell charts in the field, but good results can be obtained later from photographs if a color card is placed in the photo. In fact, if a color card is placed in a high-resolution photograph for calibration, color scanning instruments can measure the spectral characteristics of sediment in cores or exposures long after the field season.

Tablet-sized computers are rapidly replacing paper-based methods in field research. To a much greater degree than just a few years ago, interpretation and synthesis of field data may now be part of the data collection process (Pavlis et al., 2010; Whitmeyer et al., 2010). Integrated software for lightweight (<1.5 kg) devices greatly extends traditional notebook-and-camera data collection: notes (transcribed oral or written) can be organized in multiple formats; photographs annotated; sketches and samples keyed to stratigraphic positions within cores or exposures; and all data referenced to GPS positions on maps or imagery, manipulated in the field with simplified GIS software and later transferred to full-featured office software (De Donatis et al., 2008; Pavlis et al., 2010). Advanced tablet systems provide 3D summaries of data as it is collected which helps guide fieldwork, highlight incorrect or missing data, and vastly reduces the time needed to compile and interpret data in the office. The more expensive tablets are durable enough for routine fieldwork (Whitmeyer et al., 2010), although they might not survive repeated exposure to mud, days of steady rain, or a dunking in the sea as well as a waterproof notebook. Other advantages include: (1) the amount of data (maps, cross-sections, imagery, publications, field guides, your field companions' notes and data from the previous day) that can be accessed in the field; (2) wireless connection of peripheral devices such as cameras and precise

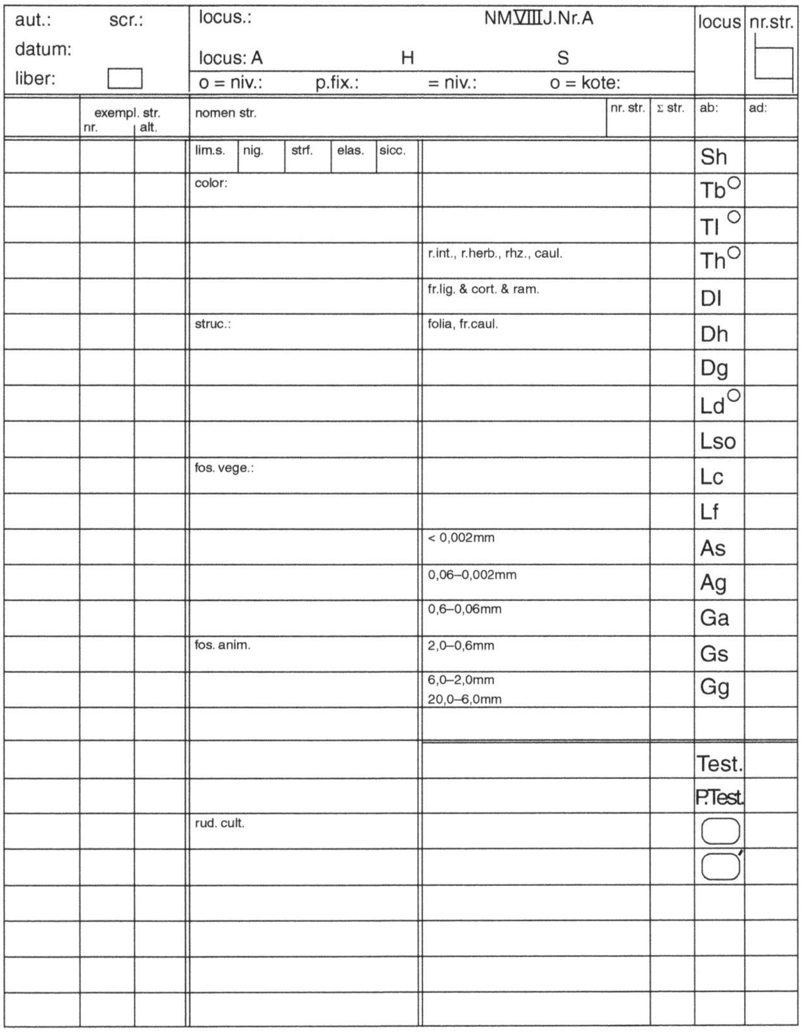

Fig. 4.6. A field description form designed and used by J. Tröels-Smith (form provided by Michael Tooley, 1993). Tröels-Smith used a separate form for each lithofacies. Tablet computers are rapidly replacing the use of such paper forms in the field.

GPS equipment; and (3) back-ups of all data at least daily.

The major disadvantage of integrated tablet systems rugged enough for continuous field use is high cost. Other significant disadvantages include the amount of time needed to learn system software, to modify field procedures to conform to the requirements of the software, or to modify the software to fit the needs of data collection. Data entry on tablet systems typically takes longer than with pen and notebook and displays remain difficult to read in bright sunlight (Pavlis et al., 2010). Because even the most expensive systems are not failsafe, many digital field geologists still back up important data in a field notebook (Whitmeyer et al., 2010).

Until costs and learning curves for fully integrated digital mapping systems decrease, I recommend using lower-cost, less-well-integrated devices and software to supplement data collection with a notebook and camera. For example, transcribing oral notes, annotating photographs with notes and GPS locations, filing out field description forms, and emailing notes or photographs home can now be accomplished with mass-market tablet software and inexpensive cell phone applications. As all systems become obsolete within a few years, such less-integrated but much-less-expensive systems are the most cost-effective. Nevertheless, sea-level researchers should be investigating which integrated systems will best meet their needs in the near future.

4.4 CORRELATING AND SYNTHESIZING SEDIMENT SEQUENCES

Whether the goal is to evaluate a single sea-level index point or to develop a detailed history of sea-level change at a series of sites, field data need to be compiled, compared, and synthesized into a lithofacies model that summarizes the distribution of sedimentary environments and explains how they have changed over time. For small-scale studies at a single site where a few distinct lithofacies have been identified in <20 cores, data can be correlated and effectively summarized on 2D diagrams on paper or with illustration software (e.g., Fig. 4.1). For sites with more complex lithofacies sequences, very detailed stratigraphic data, or data from tens to hundreds of cores, correlation and synthesis proceed in stages.

The first stage involves comparing data from cores and exposures, usually along transects perpendicular to former shorelines or environmental gradients at a site, to place all cores and exposures within a common stratigraphic framework. The chief problem is how to simultaneously compare all data among cores and exposures. An early approach to the problem is the system developed by Tröels-Smith (1955) for graphically portraying lithologic descriptions of exposures and cores. This complex system requires considerable study and experience to understand, even in simplified form (Kershaw, 1997; Long et al., 1999).

A more effective method is to use photographic (raster-based) and illustration (vector-based) software to build mosaics of rectified photographs of cores and exposures and then add lithology and other data alongside the photomosaic. Each photo-data-mosaic is then compared (vertically and laterally) with nearby mosaics until correlation points are identified and lithofacies changes mapped. Formerly, we made mosaics by taping color prints of cores to 0.5-m-long strips of cardboard and writing the data alongside. High-resolution photos, photographic and illustration software, and large computer monitors now make building mosaics, expanding and condensing stratigraphic sequences, finding correlation points among cores, and extending correlations across a site much faster and more accurate than with paper-based methods. Two shortcomings of mosaic and illustration software methods are that: (1) expanding or condensing sediment sections to line up correlation points is not automated; and (2) lithologic, fossil, and other data on mosaics or in spreadsheets are not linked directly to correlation diagrams. These and many other time-saving features are available in software developed for petroleum geologists, which is very expensive. Investigators might consider less-expensive data analysis software that constructs 3D stratigraphic sections or maps, drapes them on digital-globe or other imagery, and provides views from any angle (e.g., Pavlis et al., 2010; Whitmeyer et al., 2010). For sea-level studies with modest budgets, especially where data for key cores are quite detailed (methods described in Part II) or continuous (X-ray density, magnetic susceptibility), researchers should explore the open-source software developed for making on-screen comparisons, annotations, and analyses of photographic, instrumental, and laboratory data from marine and lacustrine cores (Last and Smol, 2001b; Francus, 2004; Rothwell and Rack, 2006, available at http://lrc.geo.umn.edu/laccore).

In the final stage, correlated lithofacies data are summarized, synthesized with paleoecologic or biostratigraphic data, and interpreted in terms of process-based models as summarized in Section 4.1. For studies with areally extensive or detailed data, synthesizing correlations and models in several progressively more simplified diagrams will commonly result in more accurate interpretations of sea-level history than going directly from detailed field and laboratory data to publication-scale summary figures.

Before completing the final stage of the study, investigators need to step back from correlation and synthesis and ask themselves some broader questions. Have sufficient field data been gathered to support site-scale lithofacies models and the inferred sea-level history? A few tens of cores may be sufficient in a shallow lake or small lagoon, but many tens to hundreds of cores (or outcrop sections) spaced along multiple transects every 5–10 m, perhaps with closer spacing where environmental gradients are steep, may be required to sufficiently map lithofacies changes to confidently interpret sea-level index points and reconstruct sea-level history at complex sites (e.g., Atwater and Hemphill-Haley, 1997; Witter et al., 2003; van de Plassche, 2006). Such thorough field investigations give investigators the confidence to ask questions with regional implications. How does the inferred sea-level history compare with other histories and with geophysical models of sea-level change for the region? What factors best explain the differences among the region's sea-level histories at different sites?

ACKNOWLEDGEMENTS

My research in coastal environments has been supported by the Earthquake Hazards Program of the US Geological Survey. Much of this work has benefited from collaboration with colleagues supported by the US National Science Foundation. Discussions with Ian Shennan, Brian Sherrod, Lee-Ann Bradley, Rob Witter, Harvey Kelsey, Simon Engelhart, Ben Horton, Brian Atwater, Andrea Hawkes, Chris Goldfinger, Ann Morey, Bobbi Conard, Rich Briggs, Andrew Kemp, and many others have been especially helpful. Comments by the editors, Rob Witter, and two anonymous reviewers improved this summary.

REFERENCES

Aaby, B., and Berglund, B.E. (1986) Characterization of peat and lakes deposits. In: *Handbook of Holocene Palaeoecology and Palaeohydrology* (ed. Berglund, B.E.), John Wiley and Sons, Chichester, pp. 231–246.

Aaby, B., and Digerfeldt, G. (1986) Sampling techniques for lakes and bogs. In: *Handbook of HolocenePpalaeoecology and Palaeohydrology* (ed. Berglund, B.E.), John Wiley and Sons, Chichester, pp. 181–194.

Abrahim, G.M.S., Nichol, S.L., Parker, R.J., and Gregory, M.R. (2008) Facies depositional setting, mineral maturity and sequence stratigraphy of a Holocene drowned valley, Tamaki estuary, New Zealand. Estuarine, Coastal and Shelf Science, 79, 133–142.

Allen, J.R.L. (2000) Morphodynamics of Holocene salt marshes: a review sketch from the Atlantic and southern North Sea coasts of Europe. Quaternary Science Reviews, 19, 1155–1231.

Atwater B.F., and Hemphill-Haley E. (1997) Recurrence intervals for great earthquakes of the past 3,500 years at northeastern Willapa Bay, Washington. US Geological Survey Professional Paper 1576.

Baeteman, C., Waller, M., and Kiden, P. (2011) Reconstructing middle to late Holocene sea-level change: a methodological review with particular reference to "A new Holocene sea-level curve for the southern North Sea" presented by K.-E. Behre. Boreas, 40, 557–572.

Barrett-Mold, C., Burningham, H., and French, J.R. (2009) Stratigraphic insights from sedimentary peels of littoral estuarine depositional systems. Journal of Coastal Research, 56, 584–588.

Belokopytov, I.E., and Beresnevich, V.V. (1955) Giktrof's peat borers. Torfyanaya Promyshlennost', 8, 9–10.

Birkeland, P.W. (1999) *Soils and Geomorphology*, 3rd edition. Oxford University Press, Oxford.

Birks, H. J. B., and Birks, H. H. (1980) *Quaternary Palaeoecology*. University Park Press, Baltimore, Maryland.

Boggs, S., Jr. (2011) *Principals of Sedimentology and Stratigraphy*, 5th edition. Prentice Hall.

Buscombe, D., and Rubin, D.M. (2012) Advances in the simulation and automated measurement of well-sorted granular material: 2. Direct measures of particle properties. Journal of Geophysical Research, 117, F02002, doi: 10.1029/2011JF001975.

Coe, A.L., Bosence, D.W.J., Church, K.D., Flint, S.S., Howell, J.A., and Wilson, R.C.L. (2003) *The Sedimentary Record of Sea-Level Change*. Cambridge University Press, Cambridge.

Dalrymple, R.W., Zaitlin, B.A., and Boyd, R. (1992) Estuarine facies models: conceptual basis and stratigraphic implications. Journal of Sedimentary Petrology, 62, 1130–1146.

Davis, R.A., Jr., and Dalrymple, R.W. (eds) (2012) *Principles of Tidal Sedimentology*. Springer, New York, Heidelberg.

De Donatis, M., Susini, S., and Delmonaco, G. (2008) Digital geologic mapping methods: from field to 3D model. International Journal of Geology, 2, 47–52.

De Vleeschouwer, F., Chambers, F.M., and Swindles, G.T. (2010) Coring and sub-sampling of peatlands for paleoenvironmental research. Mires and Peat, 7, 1–10.

DuRoss, C.B., Personius, S.F., Crone, A.J., McDonald, G.N., and Lidke, D.J. (2012) Paleoseismic investigation of the northern Weber segment of the Wasatch fault zone at the Rice Creek trench site, North Ogden, Utah. Utah Geological Survey Special Study 130, Paleoseismology of Utah 18.

Farrell, K.M., Harris, W.B., Mallinson, D.J., Culver, S.J., Riggs, S.R., Pierson, J., Self-Trail, J.M., Lautier, J.C. (2012) Standardizing texture and facies codes for a process-based classification of clastic sediment and rock. Journal of Sedimentary Research, 82, 364–378.

Francus, P. (ed.) (2004) *Tracking Environmental Change Using Lake Sediments:* Volume 7, Image Analysis, Sediments and Paleoenvironments. Kluwer Academic Publishers, Dordrecht, Netherlands.

Glew, J.R., Smol, J.P., and Last, W.M. (2001) Sediment core collection and extrusion. In: *Tracking Environmental Change Using Lake Sediments: Volume 1, Basin Analysis, Coring, and Chronological Techniques* (eds Last, W. M., and Smol, J. P.), Kluwer Academic Publishers, Dordrecht, Netherlands, pp. 73–105.

Haraguchi, T., Nakata, T., Shimazaki, K., Imaizumi, T., Kojima, K., and Ishimaru, T. (1998) A new sampling method of unconsolidated sediments by long geo-slicer, a pile-type soil sampler. Journal of Japan Society Engineering, 39, 306–314 (in Japanese).

Kaneda, H., Kinoshita, H., and Komatsubara, T. (2008) An 18,000-year record of recurrent folding inferred from sediment slices and cores across a blind segment of the Biwako-seigan fault zone, central Japan. Journal of Geophysical Research, 113, B05401.

Kearns, F.L., and Davison, A.T. (1983) Field classification system of organic-rich sediments. In: *Mineral Matter in Peat: Its Occurrence, Form, and Distribution* (eds Raymond, R., Jr. and Andrejko, M.J.), Los Alamos National Laboratory, Los Alamos, New Mexico, pp. 147–157.

Kemp, A.E.S., Dean, J., Pearce, R.B., and Pike, J. (2001) Recognition and analysis of bedding and sediment fabric features. In: *Tracking Environmental Change Using Lake Sediments:* Volume 2, Physical and Geochemical Methods (eds Last, W. M., and Smol, J. P.), Kluwer Academic Publishers, Dordrecht, Netherlands, pp. 7–22.

Kershaw A.P. (1997) A modification of the Tröels-Smith system of sediment description and portrayal. Quaternary Australasia, 15, 63–68.

Kidwell, S.M., Moore, J.A., and Moore, J.R. (1985) Inexpensive field technique for polyester resin peels of structures in unconsolidated sediments. Marine Geology, 64, 351–359.

Kosters, E.C. (1989) Organic-clastic facies relationships and chronolostratigraphy of the Barataria interlobe basin, Mississippi delta plain. Journal of Sedimentary Petrology, 59, 98–113.

Lanesky, D.E., Logan, B.W., Brown, R.G., and Hine, A.C. (1979) A new approach to portable vibracoring underwater and on land. Journal of Sedimentary Petrology, 49, 654–657.

Last, W.M. (2001) Textural analysis of lake sediments. In: *Tracking Environmental Change Using Lake Sediments:* Volume 2, Physical and Geochemical Methods (eds Last, W. M., and Smol, J. P.), Kluwer Academic Publishers, Dordrecht, Netherlands, pp. 41–81.

Last, W.M., and Smol, J.P. (eds) (2001a) *Tracking Environmental Change Using Lake Sediments:* Volume 1, Basin Analysis, Coring, and Chronological Techniques. Kluwer Academic Publishers, Dordrecht, Netherlands.

Last, W.M. and Smol, J.P. (eds) (2001b) *Tracking Environmental Change Using Lake Sediments:* Volume 2, Physical and Geochemical Methods. Kluwer Academic Publishers, Dordrecht, Netherlands.

Lienkaemper, J., Dawson, T., Personius, S., Seitz, G., Reidy, L., and Schwartz, D. (2002) Logs and data from trenches across the Hayward Fault at Tyson's Lagoon (Tule Pond), Fremont, Alameda County, California. US Geological Survey Miscellaneous Field Studies Map MF-2386.

Livingstone, D.A. (1955) A lightweight piston sampler for lake deposits. Ecology, 36, 137–139.

Long, A.J., Innes, J.B., Shennan, I., and Tooley, M.J. (1999) Coastal stratigraphy a case study from Johns River, Washington, USA. In: *The Description and Analysis of Quaternary Stratigraphic Field Sections: Technical Guide 7* (eds Jones, A.P., Tucker, M.E., and Hola, J.K.), Quaternary Research Association, London, UK, pp. 267–286.

Long, A.J., Scaife, R.G., and Edwards, R.J. (2000) Stratigraphic architecture, relative sea level, and models of estuary development in southern England: new data from Southampton Water. In: *Coastal and Estuarine Environments: Sedimentology, Geomorphology and Geoarchaeology* (eds Pye, K., and Allen, J.R.L.), Geological Society of London, Special Publication 175, pp. 253–279.

Long, A.J., Waller, M.P., and Stupples, P. (2006) Driving mechanisms of coastal change: Peat compaction and the destruction of late Holocene coastal wetlands. Marine Geology, 225, 63–84.

McCalpin J.P. (ed.) (2009) *Paleoseismology.* Academic Press, San Diego.

Miall, A.D. (1999) *Principals of Sedimentary Basin Analysis*, 3rd edition. Springer-Verlag, New York.

Miall, A.D. (2010) *The Geology of Stratigraphic Sequences.* Springer, New York.

Nakata, T., and Shimazaki, K. (1997) Geo-slicer, a newly invented soil sampler for high-resolution active fault studies. Journal Geography (Chigaku Zasshi), 106(1), 59.

Nelson, A.R., Jennings, A.E., and Kashima, K. (1996) An earthquake history derived from stratigraphic and microfossil evidence of relative sea-level change at Coos Bay, southern Oregon coast. Geological Society of America Bulletin, 108, 141–154.

O'Reilly, G. (1992) *Planning for Field Safety.* American Geological Institute, Washington, DC.

Pavlis, T.L., Langford, R., Hurtado, J., and Serpa, L. (2010) Computer-based data acquisition and visualization systems in field geology: Results from 12 years of experimentation and future potential. Geosphere, 6, 275–294.

Prothero, D.R., and Schwab, F. (2003) *Sedimentary Geology: An Introduction to Sedimentary Rocks and Stratigraphy*, 2nd edition. W.H. Freeman.

Reading, H.G. (1986) *Sedimentary Environments and Facies*, 2nd edition. Blackwell, Oxford.

Rothwell, R.G., and Rack, F.R. (2006) New techniques in sediment core analysis: an introduction. In: *New Techniques in Sediment Core Analysis* (ed. Rothwell, R.G.), Geological Society of London, Special Publication, 267, pp. 1–29.

Schaetzl, R., and Anderson, S. (2005) *Soils: Genesis and Geomorphology.* Cambridge University Press, Cambridge.

Schnurrenberger, D.S., Russell, J.M., and Kelts, K.R. (2003) Classification of lacustrine sediments based on sedimentary components. Journal of Paleolimnology, 29, 141–154.

Schoeneberger, P.J., Wysocki, D.A., Benham, E.C., and Broderson, W.D. (eds) (2002) *Field Book for Describing and Sampling Soils*, Version 2.0. Natural Resources Conservation Service, US National Soil Survey Center, Lincoln, Nebraska.

Smith, D. G. (1984) Vibracoring fluvial and deltic sediments: tips on improving penetration recovery. Journal of Sedimentary Petrology, 54(2), 660–663.

Stow, D.A.V. (2005) *Sedimentary Rocks in the Field: A Color Guide.* Academic Press, London.

Takada, K., and Atwater, B.F. (2004) Evidence for liquefaction identified in peeled slices of Holocene deposits along the lower Columbia River, Washington. Bulletin of the Seismological Society of America, 94, 550–575.

Thompson, T.A., Miller, C.S., Doss, P.K., Thompson, L.D.P., and Baedke, S.J. (1991) Land-based vibracoring and vibracores analysis: Tips, tricks, and traps. Geological Survey of Indiana, Occasional Paper 58, Bloomington, Indiana.

Törnqvist, T.E., Newsom, L.A., van der Borg, K., de Jong, A.F.M., and Kurnik, C.W. (2004) Deciphering Holocene sea-level history on the U.S. Gulf Coast: A high-resolution record from the Mississippi delta. Geological Society of America Bulletin, 116, 1026–1039.

Tröels-Smith, J. (1955) Characterization of unconsolidated sediments. Geological Survey of Denmark, Series IV, No. 10.

Tucker, M.E. (2011) *Sedimentary Rocks in the Field: A Practical Guide.* Wiley-Blackwell, Chichester.

van de Plassche, O., Erkens, G., van Vliet, F., Brandsma, J., van der Borg, K., and de Jong, A.F.M. (2006) Salt-marsh erosion associated with hurricane landfall in southern New England in the fifteenth and seventeenth centuries. Geology, 34, 829–832.

von Post, L., and Granlund, E. (1926) Södra Sveriges torvtillgängar I. Sveriges Geologiska Undersökning C, 335, 1–127.

Walker, R.G. (1992) Facies, facies models, and modern stratigraphic concepts. In: *Facies Models: Response to*

Sea-Level Change (eds Walker, R.G., and James, N.P.), Geological Association of Canada, pp. 1–14.

Whitmeyer, S.J., Nicoletti, J., and De Paor, D.G. (2010) The digital revolution in geologic mapping. GSA Today, 20(4/5), 4–10.

Witter, R.C., Kelsey, H.M., and Hemphill-Haley, E. (2003) Great Cascadia earthquakes and tsunamis of the past 6700 years, Coquille River estuary, southern coastal Oregon. Geological Society of America Bulletin, 115, 1289–1306.

Wright, H.E. (1991) Coring tips. Journal of Paleolimnology, 6, 37–49.

Wright, H.E., Jr., Livingston, D.A., and Cushing, E.J., 1965. Coring devices for lake sediments. In: *Handbook of Palaeontological Techniques* (eds Kummel, B., and Raup, M.), Freeman, San Francisco, pp. 494–520.

Zolitschka, B., Mingram, J., van der Gaast, S., Jansen, J.H.F., and Naumann, R. (2001) Sediment logging techniques. In: *Tracking Environmental Change Using Lake Sediments:* Volume 1, Basin Analysis, Coring, and Chronological Techniques (eds Last, W. M. and Smol, J. P.), Kluwer Academic Publishers, Dordrecht, Netherlands, pp. 137–153.

Chapter 5
Geomorphological indicators of past sea levels

HARVEY M. KELSEY
Department of Geology, Humboldt State University, Arcata, CA, USA

5.1 INTRODUCTION

Geomorphic indicators of past sea levels represent a continuum of processes and landforms associated with sea level and sea-level change. In an attempt to honor this continuum, while at the same time discussing geomorphic features separately, this chapter discusses geomorphic indicators of past sea level in the following sequence. First, coral as a geomorphic indicator of past sea level is discussed. The recognition of late Pleistocene uplifted coral platforms as indicators of past sea level was one of the pioneering efforts in sea-level research. Decades later, coral was again the focus of pioneering work in late Holocene sea-level research when coral microatolls were used to reconstruct sea-level history to a much higher resolution of centimeters of sea-level change over the scale of years. Erosional landforms that indicate former sea levels are then discussed, starting with late Pleistocene geomorphic features, marine terraces, and shoreline angles, before describing the more highly resolved sea-level records that can be gleaned from similar features in the late Holocene through the use of the seacliff–shore platform junction. Tidal notches form another set of erosional indicators that are discussed. Finally, the suite of geomorphic landform indicators that are deposits are discussed. Coastal deposits were initially laid down near sea level and such deposits, if subsequently vertically or horizontally separated from the shoreline, may indicate past sea levels. The discussion on deposits focuses on beach ridges.

For each set of geomorphic indicators, the accuracy with which the geomorphic feature indicates a past sea level is discussed. "Indicative meaning" defines the relationship of the geomorphic feature to tidal range and thereby allows the relative sea-level change to be measured by differencing the elevation of the geomorphic feature from a specific reference water level (e.g., mean high tide or some other tidal datum) that formed the feature initially (van de Plassche, 1986). For geomorphic features that indicate a past sea level, knowledge of the tidal datum that the geomorphic feature represents (e.g., mean high tide) and the resolution at which the geomorphic indicator forms at that elevation is required. As an example, consider a survey that shows that a shore platform–seacliff junction (a sea-level indicator) eroded into bedrock forms at about mean high spring to mean high neap tide level. Such an indicative meaning, ranging over meters, is not precise compared to resolving past sea level using biostratigraphic data of sediments, but relative lack of precision is usually the case with geomorphic indicators. The exception is the instance where the geomorphic indicator is an organism (e.g., a coral microatoll) rather than an eroded landform.

Geomorphic indicators of former sea level are overwhelmingly concentrated on coasts that have undergone relative sea-level fall over a time span ranging from the last few thousand years to over hundreds of thousands of years. The reason for the bias is that relative sea-level rise will drown geomorphic features created by former sea levels, and these indicators become submerged, buried by sediment, and/or eroded. In some instances, submerged indicators may remain preserved; an example is a submerged tidal notch. Coastal sediments that aggrade within embayments are employed for sea-level reconstruction for the case of rising sea level in the last few thousand years, and the key role provided by coastal sediments provides good justification for Chapter 4 being devoted entirely to these deposits.

Preservation of geomorphic indicators of former sea level is usually the result of a period of

Handbook of Sea-Level Research, First Edition. Edited by Ian Shennan, Antony J. Long, and Benjamin P. Horton.

slowly changing or static relative sea level, when deposits or erosional landforms are created, followed by a relatively rapid relative sea-level lowering that subsequently subaerially preserves these deposits or erosional landforms. An example would be a shore platform formed in the tidal zone over hundreds–thousands of years that is subsequently coseismically uplifted 1–2 m during an earthquake. Other examples are provided in Sections 5.5–5.7. However, deposits recording static or slowly changing sea level over time can be preserved in cases of coastal progradation along sediment-rich coasts.

Rate of change of relative sea level affects the robustness of geomorphological features indicative of former sea level. In the case of eroding (non-prograding) coasts, a static or very slowly falling sea level will result in the concentration of wave energy and weathering at the same shoreline elevation over geomorphologically significant durations of time (at least thousands of years) such that shoreline features can form and then be preserved if relative sea level falls. In contrast, in the case of ongoing, rapid-falling relative sea level (≥50 mm a^{-1}), preservation of geomorphic indicators of former relative sea level may be unlikely. In high-latitude regions (e.g., Norway, Sweden, Spitzbergen, or the Hudson Bay region) however, beach ridges or shore platforms may form and survive under rapidly falling relative sea level; see Section 5.6.3 on preservation of geomorphic indicators of past sea levels at high latitudes.

5.2 MODEL FOR FORMATION OF LATE PLEISTOCENE EMERGENT SHORE PLATFORMS

Shore platforms form at times when the rate of glacioeustatic sea-level change is equal to the rate of local surface uplift (Mesolella et al., 1969; Bradley and Griggs, 1976). From the definition provided by Milne and Mitrovica (2008), the expression "glacioeustatic sea-level curve" refers to the time-dependent "spatially uniform height shift of the ocean surface to accommodate any mass gained/lost from grounded ice." The relative sea-level change at any specific coastal site is also affected by glacial isostatic adjustment (GIA) that causes the observable sea-level change to be different from the eustatic curve (Milne and Mitrovica, 2008). At such time that the rate of glacioeustatic sea-level change is equal to the rate of local surface uplift (and the rate of sea level change is modified to some extent by GIA), there is a relative sea-level stasis and wide platforms develop.

The details of platform formation versus the tempo of glacioeustatic sea-level rise and subsequent peaking at a highstand are slightly different for erosional marine terrace formation versus constructional reef terrace formation (see fig. 2 in Muhs et al., 2002), but both shore platform types form at, or temporally near, highstands. Upon successive sea-level fall brought on by glacioeustatic sea-level change, the platforms emerge and continue to rise tectonically. By the time of formation of a succeeding glacioeustatic sea-level highstand, the platforms are well above the reach of sea level. In such a manner, topographic "stair steps" of emergent marine platforms develop on both constructional (i.e., coralline; Mesolella et al., 1969; Chappell, 1974) and erosional (i.e., mid latitude; Bradley and Griggs, 1976; Lajoie, 1986) coasts (Fig. 5.1a, b). In the case of the coralline terraced landscape of the Huon Peninsula of New Guinea (Fig. 5.1a), the stair-stepped terraces have been dated and a sea-level curve for the highstands constructed (the "New Guinea sea-level curve") using known ages, highstand elevations, and a constant uplift rate (Chappell and Shackleton, 1986).

For raised platforms the emergent stair step consists of a tread, which is the paleo-shore platform, and a riser landward of the tread, which is the paleo-seacliff; "shoreline angle" or "platform inner edge" are both terms for the junction of the paleo-seacliff and the paleo-platform. In the case of raised late Pleistocene shore platforms, this junction is invariably covered by marine deposits and by colluvium shed off the adjoining seacliff (in Fig. 5.1b note that the platform inner edges in all cases are buried by marine deposits and colluvium). The elevation of the raised junction therefore cannot usually be measured directly but must be inferred by subtracting an assumed, or known, thickness of cover sediment. By contrast, junctions are well exposed in modern coastal settings and for some raised late Holocene platforms. These cases are discussed in Section 5.6.2.

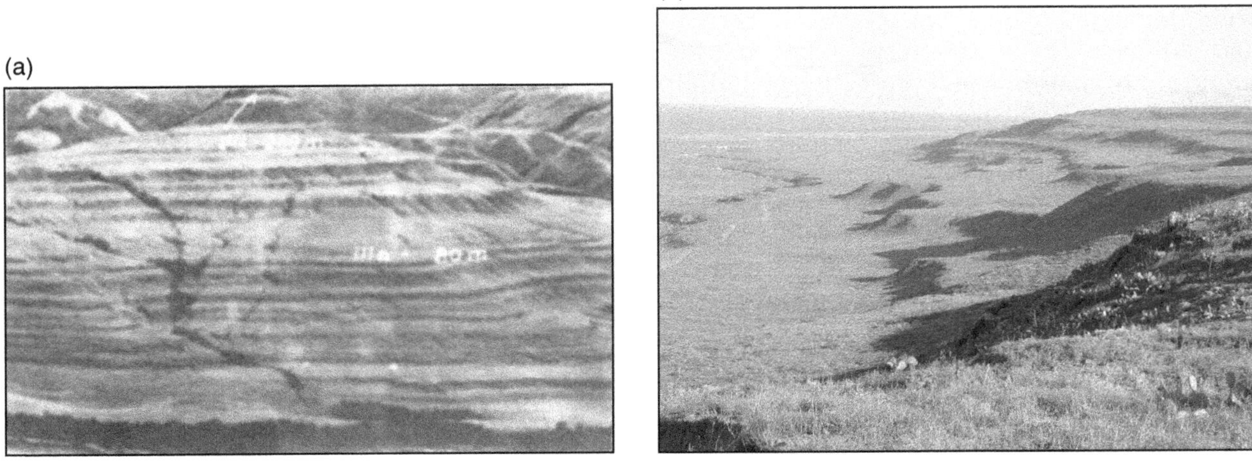

Fig. 5.1. (a) Coralline terraced landscape of Huon Peninsula, New Guinea (6°05' N, 147°34' E). *Source*: Reproduced from Chappell (1974, fig. 2). The white lettering on the image is the original annotation of Chappell (1974) and refers to his terrace identifying number and the elevation of the terrace in meters. *Source*: Reproduced with permission of the Geological Society of America. (b) Erosional marine terraced landscape exemplified by marine terraces on the west coast of San Clemente Island, California (32°53' N, 118°31' W) looking northward (Muhs et al., 2002). The most extensive terrace, in the foreground and middle ground on the left of the photo, is the c. 120 ka marine terrace. The youngest and lowest terrace, limited in extent and minimally exposed along the embayed coastline in the middle ground, is the 80–100 ka marine terrace. The higher and older terraces, in the center and right on the photograph, are mid-Pleistocene marine terraces (200 ka and older). *Source*: Photograph by Daniel Muhs. Reproduced with permission. For color details, please see Plate 8.

5.3 GEOMORPHIC INDICATORS OF PAST SEA LEVEL AND GLACIAL ISOSTATIC ADJUSTMENT IN THE NEAR AND FAR FIELD

Raised platform inner edges on stair-stepped terraced coastlines form a unique dataset comprising a set of elevations for inner edges, where each inner edge corresponds to a specific occurrence of a global sea-level highstand. Assuming that the tectonic uplift rate responsible for raising the platforms is constant on a given coast and that the New Guinea sea-level curve (Chappell and Shackleton, 1986) is a global sea-level curve that provides relative elevation of these highstands, then for any given stair-stepped marine terrace sequence, a set elevations of platform inner edges uniquely corresponds, one for one, to a set of global eustatic sea-level highstands (e.g., Lajoie, 1986). Following this reasoning, Bull (1985) proposed that by knowing the elevation of shoreline angles of a particular marine terrace flight, one could graphically match a set of terraces one-for-one to a set of specific-aged highstands using an assumed constant uplift rate. The flaw to this reasoning is most elegantly articulated by Lambeck and Nakada (1992) and Muhs et al. (2012, p. 20), who argue against the concept of a global sea-level curve. The flaw entails not including another variable: glacial isostatic adjustment (GIA).

Although a review of GIA is not within the scope of this chapter (and is thoroughly covered elsewhere; e.g., Nakada and Lambeck, 1989; Mitrovica and Peltier, 1991; Milne and Mitrovica, 2008), GIA is a response to load transfer of glacier ice to ocean basins through melting that perturbs the mantle both in the near field, which is proximal to the ice sheets, and in the far field, which includes the equatorial regions that are removed for the immediate effects of ice mass redistribution. In addition, load transfer of glacier ice to ocean basins perturbs the mantle differently beneath coastlines of major continental masses as opposed to beneath coastlines of oceanic islands. The net result is that the factor of global isostatic adjustment varies in space and in time. Although uplifted shoreline angles are geomorphic relics of former sea-level highstands of glacioeustatic sea-level fluctuations, two identically aged shoreline angles at different parts of the globe will have elevations dependent both on local tectonic uplift rate and local glacial isostatic adjustment effects on the immediate coastline. Changes in ocean water temperature and salinity also lead to departures from a global sea-level curve, but these departures are relatively minor compared to GIA (Milne and Mitrovica, 2008).

5.4 GEOMORPHIC INDICATORS OF PAST SEA LEVEL AND THE ISSUE OF INHERITANCE AND REOCCUPATION

Geomorphic indicators are here separated into late Pleistocene and late Holocene as the latter should have more resolved indicative meaning. The late Holocene indicator was formed within a few thousand years of the present time and the well-exposed modern shoreline is an analog for the genesis and formation of the late Holocene indicator landform. However, is the geomorphic indicator late Holocene or late Pleistocene in age? The nature of glacioeustatic sea levels is that, in general, each glacioeustatic sea-level highstand (marine oxygen isotope stages 5, 7, 9, etc.) is about the same height as the previous. For example, on the tectonically stable Florida Keys, reefs dating to marine oxygen isotope stage 5 (~120 ka) and 7 (~200 ka) apparently had paleo-sea levels that were only slightly different from each other, although both were slightly higher than present (Muhs et al., 2011). On coasts with little tectonic uplift, late Holocene sea-level highstands could therefore reoccupy (inherit) the same elevation as the late Pleistocene sea-level highstand, and investigators should be aware of this possible scenario for modern shore platforms (Fig. 5.2).

As a consequence, geomorphic features created in the late Pleistocene can be reoccupied and modified in the late Holocene by sea levels that reach the same elevations (Phillips, 1970; Trenhaile, 1972; Kelsey, 1990; Kelsey and Bockheim, 1994). Notches and shore platform–seacliff junctions can therefore both be polygenetic in that a Holocene relative sea level may occupy a landform (platform or notch) that is actually inherited from the late Pleistocene. Field sea-level researchers need to be aware of this potential, and the coastal site illustrated in Figure 5.2 is a potential case in point.

Fig. 5.2. The seacliff–shore platform junction at Drake's Beach, Point Reyes Peninsula, California at low tide (38°01.5' N, 122°58' W). The platform is prominently displayed, as are the tide drainage channels cutting the platform. Platform and seacliff eroded in siltstone and sandstone of the Miocene–lower Pliocene Purisima Formation. The wide, prominent modern platform, the lack of any older uplifted platforms in the landscape, and the low uplift rates along this section of the coast adjacent to the San Andreas fault all allow for the possibility that the platform may have been originally formed in the late Pleistocene and then inherited and reoccupied in the late Holocene. *Source*: Photograph by S. M. Cashman. Reproduced with permission. For color details, please see Plate 9.

5.5 CORAL AS A GEOMORPHIC INDICATOR OF PAST SEA LEVEL

5.5.1 Constructional (low latitude, coralline) late Pleistocene marine shore platforms

Growth of a coralline fringing reef or a shallow lagoon with a coralline outer barrier ridge produces constructional shoreline platforms. Such platforms are recognized in many low-latitude coastal environments where corals thrive.

In cases where tectonic uplift is ongoing, coralline shore platforms are preserved as stair-stepped coralline terraced landscapes (Fig. 5.1a). Using radiometric dating techniques for coral, such terraced landscapes have been used to date former glacioeustatic highstands of sea level.

Two of the earliest pioneering and well-documented examples of paleo-sea-level histories using terraced coralline landscapes are Barbados and Papua New Guinea. Barbados is a small island (431 km^2) on the eastern margin of the Caribbean Sea with a terraced coral cap that overlies deformed Tertiary strata (Mesolella et al., 1969; Radtke and Schellmann, 2004). Each individual terrace represents an elevated reef tract. The raised terraced reef tracts were assumed to represent major eustatic sea-level highstands and radiometric ages define the times of these highstands (Mesolella et al., 1969, Radtke and Schellmann, 2004). On the Huon Peninsula of New Guinea (Fig. 5.1a), the stair-stepped, terraced landscape consists of an off-lapping sequence of more than 20 coral reefs, and dating of the reefs enabled inference of a eustatic sea-level curve for the late Pleistocene (Bloom et al., 1974; Chappell, 1974; Chappell and Shackleton, 1986).

The New Guinea record provides a good example of the level of accuracy of sea-level reconstruction possible for late Pleistocene sea levels. On the Huon Peninsula, the outer reef crest assemblage of flat, narrow (tens of meters wide) fringing reefs, or the barrier ridge on the outer margin of uplifted shallow lagoons, was considered the most reliable indicator of former sea level. Of course, corals can only grow below sea level and these late Pleistocene sea-level indicators must be lowest tide markers. However, growth up to the proximity of the lowest tide was a sufficiently resolved former sea level in the case where workers were trying to assess late Pleistocene sea-level highstands within the context of 100-meter-plus climatically driven oscillations in sea level. Finally, there is a vast literature on the accumulation of modern carbonate reefs relative to sea level (well summarized in Kim et al., 2012), and workers extensively using coralline platforms for sea-level reconstruction should be familiar with these processes.

5.5.2 Coralline indicators of late Holocene relative sea-level change: microatolls

An exceptionally informative geomorphic indicator of past sea levels in the late Holocene are coral microatolls (Fig. 5.3). Microatolls are so named because, although on the order of 0.5 m to a few meters in diameter, many specimens have raised rims and low centers (Fig. 5.3a) that much resemble the atoll islands first well publicized by Charles Darwin (Darwin, 1842); however, other microatoll specimens have different morphologies (Fig. 5.3b). Although Chapter 8 is entirely devoted to microatolls as a sea-level indicator, it is appropriate to put microatolls in perspective with other geomorphic indicators within this chapter.

Microatolls, consisting of several species of the coral genera *Porites* and *Goniastrea*, grow upward to an extent limited by the lowest tides, and the uppermost limit of growth is called the highest level of survival (HLS). Taylor et al. (1987) initially introduced the HLS as an indicator of late Holocene sea-level change. The form and morphology of microatolls can therefore be interpreted in terms of gradual and abrupt changes, both positive and negative, in relative sea level (Fig. 5.3; Taylor et al., 1987, Sieh et al., 1999; Zachariasen et al., 1999, 2000).

Using coastal field sites in Vanuatu and Sumatra, Taylor et al. (1987) and Sieh et al. (1999), respectively, both suggested that microatolls are valuable as geodetic markers of earthquake-induced changes in relative sea level. Corals have color banding that can be used to discern annual growth (Taylor et al., 1987), and developments in U-series dating (Edwards et al., 1988) allow age determination uncertainty for these corals to be on the order of a few years to a few decades. The annual growth record combined with well-resolved ages for selected annular rings means that coral microatolls are not only geomorphically distinct but potentially precise paleogeodetic markers. Some limitations to the use of microatolls as paleogeodetic markers include the tendency for them to live for a limited period of time, so that data often have to be combined from several areas to get a late Holocene relative sea-level record, and the observation that a microstoll's highest level of survival may be influenced by ponding on a reef platform.

Paloegeodesy from microatolls can be as precise as 20 mm of change over a few years to a few decades (Sieh et al., 1999; Zachariasen et al., 2000). Abrupt coral death can be inferred to be a result of an abrupt coseismic elevation shift of the coral to an elevation above the highest level of survival (e.g., Fig. 5.3c; Sieh et al., 2008). Microatoll paloegeodetic records have also been interpreted in terms of long-term paleoseismic behavior of the Sunda megathrust (Sieh et al., 2008) and in terms of a combination of earthquakes and non-tectonic "sea-level anomalies," which are aperiodic oscillations of

Fig. 5.3. Microatolls that were killed, wholly or in part, due to coseismic uplift during the December 2004 Aceh Sumatra earthquake. The three sites, located in the vicinity of 2°50' N, 95°49' E, are described in detail in Meltzner (2010). (a) Microatoll with cup-shaped geometry that implies that rapid submergence was occurring before the abrupt coseismic uplift event that emerged, and killed, the microatoll. Langi site, northern Simeulue Island, Sumatra. *Source*: Photograph by D. H. Natawidjaja. Reproduced with permission of Aron Meltzner. (b) Uneroded concentric ridges and swales characterize this microatoll and attest to annual growth outward once the coral reached the highest level of survival (HLS) dictated by low tide. Outward growth terminated when the coral was killed by coseismic uplift. This specimen is from the Lewak site, northern Simeulue, but similar specimens can be found in the circum-Pacific in tropical regions. *Source*: Photograph by D. H. Natawidjaja. Reproduced with permission of Aron Meltzner. (c) Most of this microatoll was abruptly killed by abrupt relative sea-level fall, although the lowest 5 cm of the colony is still alive in this photo. The fall in relative sea level is measured as the elevation difference between the microatoll's dead flat top (the former HLS) and the highest living corallites today (the new HLS, a few centimeters above water level in this photo). Photo was taken by J. Galetzka at the Lewak site, northern Simeulue, on 5 February 2005. *Source*: Photograph by J. Galetzka. Reproduced with permission of Aron Meltzner. For color details, please see Plate 10.

sea-surface temperatures (Meltzner et al., 2010). By utilizing specimens that can survive vertical crustal displacements yet stay with the zone of survival and grow for long periods of time, a time series of sea-level relative elevation change can be constructed for up to 300 years (Meltzner et al., 2012). Such time series have produced exceptionally detailed (millimeter-scale over a single year) sea-level records (Meltzner et al., 2012; Philibosian et al., 2012) and have produced records for periods back in time as far as 1200–1350 AD (Philibosian et al., 2012).

5.6 EROSIONAL LANDFORMS THAT INDICATE FORMER SEA LEVELS

5.6.1 Late Pleistocene emergent shoreline platforms

In mid-latitude non-tropical coastal areas, shoreline platforms are an erosional response to selective chemical and bioerosional weathering and/or wave-induced abrasion in the tidal zone (for an excellent discussion, see chapter 9 in Trenhaile, 1987). For tectonically uplifting coasts, successions of uplifted shore platforms may be preserved as stair-stepped flights of marine terraces (Fig. 5.1b). Similar to the case for uplifted terrace fights consisting of former carbonate fringing reefs, the erosional stair-stepped marine terraces can be related to a succession of sea-level highstands (Bradley and Griggs, 1976; Pillans, 1983).

The inner edge, or shoreline angle, of an uplifted shore platform is a paleo horizontal contour line (within an error margin of about 1–4 m dictated by lithology-driven and coastal-morphology-driven variation in inner edge position). Variation in elevation of the inner edge of uplifted platforms in a shore-parallel sense is an indication of shore-parallel variation in surface uplift rates. Tracking the elevation variation of uplifted inner edges of marine terraces is an effective means to detect faults, regional tectonic tilts, and long-wavelength (multiple kilometer) folds active in the Quaternary (e.g., Bradley and Griggs, 1976; Pillans, 1983; Kelsey, 1990; Kelsey and Bockheim, 1994).

Elevated late Pleistocene emergent shoreline platforms are therefore useful for calculating long-term uplift rates over tens of thousands of years, and useful for determining rates of regional tilt over the same timescale. Inner edge elevations of uplifted buried platforms are often difficult to measure directly however, except in specific cases of fortuitous road cuttings or borehole data. The inability to directly measure variation or complexity in inner edge elevations along a flight of uplifted shore platforms, as well as the inability to determine timing of inner edge formation, precludes the use of late Pleistocene emergent shoreline platforms for determining vertical elevation changes caused by coseismic or interseismic tectonic movements that occur on the timescale of hours to hundreds of years. It is for this reason that late Holocene erosional platform–cliff junctions are a focus of applied sea-level research.

5.6.2 Late Holocene platform–seacliff junctions

Compared to late Pleistocene shore platforms, the resolution of sea-level changes in the late Holocene is better because the shore platform–seacliff junction is usually exposed (e.g., Fig. 5.4) and amenable to surveying; the junction also may be amenable to dating through cover sediment packages or in-growth-position organisms that bore into the platform. On bedrock coasts, platform junctions tend to be close to the high tidal level, specifically between the elevation of mean high neap to mean high spring tides (Fig. 5.4; Trenhaile, 1972). Nonetheless, the elevation variation of this junction in a shore-parallel direction can be as much as 4 m or more. This variation in height of the cliff–platform junction is partly related to the seasonal variability of the height of spring and neap tides but mostly related to the variation in lithologic resistance to erosion of the platform (junctions tend to be higher in areas of hard than in soft rocks) and to the location of the specific site on a headland versus an embayment (Wright, 1970; Trenhaile, 1972). Causes of this variation are reinforced in some cases and cancel each other out in other cases, that is, junctions are higher than in the adjacent bays on most headlands but the reverse may be true in some cases as higher wave exposure and weathering on headlands preferentially lowers junctions (Trenhaile, 1987).

The most productive application of the platform–cliff junction in sea-level research is to reconstruct sea-level change in the late Holocene. Such studies have focused on the combined problem of sea-level change caused by abrupt vertical tectonic movements and sea-level change, more gradual in nature, as a consequence of glacial isostatic adjustment in the far field away from former ice sheets. For both applications, the inner edge of the late Holocene platform (or the shoreline angle) is the optimum elevation marker of a raised Holocene platform. As discussed above, however, this inner edge forms at a variable elevation relative to the tidal cycle and to the position along the coast.

In reconstructing former late Holocene sea levels, the inherent elevation variability for a platform junction can be avoided by surveying the modern junction relative to the relict, uplifted late Holocene junction at the same coastal site and measuring the difference between the two elevations. By doing so, Hull (1987) was able to constrain the maximum amount of coseismic tectonic uplift (6 m) for an

(a) (b) (c)

Fig. 5.4. Modern examples of the shore platform inner edge, which is also called the platform–seacliff junction or the shoreline angle. (a) Platform inner edge eroded into highly erosive, interbedded (3–6 cm thick) middle Miocene sandstone and argillite, northern California, USA. People are astride the inner edge. Note that the well-rounded beach cobbles, which partially cover the platform, are a distinctly different, much more resistant massive sandstone (40°06' N, 124°06.5 W). (b) Close-up view of modern platform inner edge, La Paz Peninsula, Mexico. Platform and seacliff consist of andesitic lahar deposits (24°19' N, 110°12' W). (c) Modern shore platform and platform inner edge at Arisaig, northern coast of Nova Scotia (45°46' N, 62°09' W). The platform is cut in landward-dipping Silurian sandstone and siltstone (Porter et al., 2010). *Source*: Photograph (a) by Harvey Kelsey and photographs (b) and (c) by A. S. Trenhaile. Reproduced with permission. For color details, please see Plate 11.

earthquake about 2300–2750 ^{14}C a BP. The age determination was on shells from intertidal organisms within a 1–6-m-thick marine cover sediment sequence that overlay the late Holocene shore platform. Using similar strategies, emergent late Holocene platforms, often dated by means of associated cover sediment, have been employed to document a stand of sea level higher than present in the latter part of the Holocene in the eastern Pacific and Indian ocean regions (e.g., Woodroffe et al., 1995).

Another useful strategy for assessing changes in relative sea level is to measure the difference between the inner edge elevation of two uplifted platforms at the same site. Because they are taken at the same site, these differences avoid the problem of inherent variability of the inner edge elevation in a shore-parallel direction along the coast. Using such a differencing strategy, various workers in regions as diverse as New Zealand (Wilson et al., 2006) and the Philippines (Ramos and Tsutsumi, 2010) have determined specific relative sea-level drops, interpreted to be related to coseismic uplifts, to an elevation resolution of less than a meter. In another example, Merritts (1996) inferred the amount of coseismic uplift for each of the multiple earthquakes on the coast in the vicinity of Cape Mendocino, California by measuring, and differencing, along-coast elevation of platform–cliff junctions of late Holocene emergent platforms. However, even for platforms with ages younger than a few thousand years, the uplifted platforms were partly buried by sediment or removed by coastal erosion, and elevation of uplifted platform–cliff junctions had to be extrapolated from limited platform exposures. The ^{14}C ages of the late Holocene coseismically uplifted platforms were determined from *pholads* (mollusks that bore into rock platforms) or from platform cover sediment that either consisted of marine sediment containing shells or colluvial sediment containing wood.

5.6.3 Erosional indicators of past sea levels at high latitudes

Processes that operate on Holocene shore platforms at high latitudes are the same as those that operate at mid–low latitudes, that is, mechanical wave action, chemical weathering, and bioerosional weathering (Trenhaile, 1983). At high latitudes, freeze–thaw cycles are a significant mechanism for triggering recession of sea cliffs. Although the process of freeze-and-thaw may accelerate cliff retreat, the process is not the primary mechanism responsible for planation of shore platforms (Trenhaile, 1983). The indicative meaning of the shoreline angle is therefore unaffected by the importance or lack of importance of frost action.

In areas of isostatic rebound where sea-level fall can be rapid (1 to >10 mm a^{-1}), modern platforms appear to reflect modern sea level (although documentation is mostly qualitative observation rather than surveys; Trenhaile, 1983). Nonetheless, it appears that shore platforms can both form and adjust rapidly to modern sea level along coasts subject to isostatic rebound.

5.6.4 Notches as sea-level indicators

Notches are geomorphically distinct undercuttings, centimeters to a few meters deep, formed in bedrock cliffs on coasts (Fig. 5.5), and are potential indicators of former sea levels. Notches have been extensively investigated for sea-level reconstruction and paleoseismic applications in the Mediterranean region (Pirazzoli, 2005; Evelpidou et al., 2012) where they were first publicized (Higgins, 1980; Pirazzoli, 1980), but notches have also been used for sea-level reconstruction in other parts of the world (Liew et al., 1993).

Similar to the formation of shore platforms, the formation of notches requires periods of stable relative sea level that are longer than the effective period of notch formation. Notches can then be preserved as indicators of former sea level if relative sea level then abruptly rises or falls.

The indicative meaning of notches, that is, the relationship of the formation of notches to the tidal regime, was first critically addressed by Pirazzoli (1986). He proposed that tidal notches can have a well-defined indicative meaning because the notch apex (point of maximum inward growth) is at mean tide level and the upper and lower elevation limits of the notch indicate the range of the tides plus wave splash. Tidal notch formation mechanisms include some combination of bioerosion and dissolution (Pirazzoli, 1986). However, Pirazzoli noted that notches can be formed at or near sea level by other mechanisms including abrasion by clast-laden wave action and enlargement by abrasion erosion along horizontal zones of structural weakness. According to Pirazzoli (1986), such features, although notch-like in morphology, do not have

(a)

(b)

Fig. 5.5. (a) Modern tidal notch in tidal zone, La Paz Peninsula, Mexico; notch is eroded into andesitic lahar deposits (24°19' N, 110°12' W). (b) Uplifted paleo-tidal notch, Barbados Island, western Caribbean; notch is eroded into coralline limestone (13°10' N, 59°33' W). *Source*: Photograph (a) by A. S. Trenhaile and photograph (b) by Anthony Long. Reproduced with permission. For color details, please see Plate 12.

the clearly defined indicative meaning that is a characteristic of tidal notches.

As is the case with shore platforms, relative sea-level reoccupation can occur during notch genesis. Late Holocene tidal (or non-tidal) notches can potentially reoccupy late Pleistocene notches because glacioeustatic sea-level highstands rise to approximately the same levels on stable coasts or coasts with low tectonic uplift rate ($<0.1\,\text{mm a}^{-1}$) and/or negligible GIA-induced net relative sea-level change from one highstand to the next.

Late Holocene notches are potentially more promising for indicative meaning than late Pleistocene counterparts, because late Holocene notch positions can be directly tied to modern sea levels (compare Fig. 5.5a and b). Recent studies have used this attribute to assess the magnitude of coseismic uplift events (earthquakes) from raised Holocene notches (Liew et al., 1993; Stiros et al., 2000).

Finally, as is the case with all geomorphic indicators of former sea level, the indicator is of little use for sea-level studies if the indicator cannot be assigned an age. In the case of paleo notches, they can be dated either directly or indirectly. The direct approach is to find remnants of dead in-growth-position organisms that bioeroded the notch, for example barnacles or borings that contain the organism that created the bore. The indirect approach is to associate a submerged late Holocene notch with an adjacent coastal salt-marsh that has undergone accretion over the time of notch formation and submergence. In such a case, marsh accretion rates (determined by radiocarbon age determinations) can be related to submerged notches such that the notch provides the inference that the submergence was rapid and the ages provide the chronology of the submergence (e.g., Nixon et al., 2009). In many cases however, relict tidal notches are barren of in-growth-position organisms or are not associated with nearby sediment packages that record the same relative sea-level change.

In summary, the applicability of uplifted, or submerged, notches for sea-level reconstruction relates directly to the accuracy at which these relict notches can be assigned to a paleo-tidal regime operating at a specific time in the past. For investigators who encounter relict notches (paleo-notches) in areas geographically removed from well-studied regions, it is important to assess whether modern notches are forming in the tidal range and determine how the notch geometry relates to the tidal range (i.e., what is the indicative meaning of the notch?). For example, the notch depicted in Figure 5.5a developed in andesitic lahar deposits along a coastal segment of the La Paz Peninsula, Mexico. Elsewhere along this same coast, the tidal zone-upland transition is not a notch but is an inner edge of a platform; that is, there is a continuum of the notch geometry to the platform inner edge geometry. Because platform inner edges on average form at high tide levels, does the notch in this Mexico coast example form at high tide or at mean tide, as suggested in the literature? Only local surveying (in this example, on the La Paz Peninsula coast) can address such uncertainty.

5.7 EMERGENT HOLOCENE SHORELINE DEPOSITS: BEACH RIDGES

Emergent coastal deposits are geomorphic indicators of past sea levels; because such deposits are ubiquitous globally, they are potentially one of the richest sources of information on former relative sea levels. In a broad sense, all coastal deposits are associated with shorelines and thus associated with present or past sea levels, but deposits most often ascribed to have indicative meaning (that is, indicate a past sea level) are variously referred to as beach ridges, storm berms, and strandlines.

But confounding any discussion of beach ridges, storm berms, and strandlines as indicators of past sea level is terminology. There is much well-intentioned, but cumulatively confusing, literature that attempts to orchestrate a common language for beach ridges, storm berms, and strandlines (e.g., Roy et al., 1994; Otvos, 2000; Hesp et al., 2005). Because of the ongoing debate among experts as to what terms refer to what coastal features, beach ridges and related terms are defined here before discussing the extent to which these geomorphic features indicate former sea levels.

Beach ridges are shore-parallel ridges that consist of marine sediment and are formed by wave swash. Berms are features that can be generated by tides and are ephemeral. Storm berms, in contrast, are not ephemeral and are generated by wave swash during storms and high tide with higher elevation of deposition, possibly enhanced by storm surge. Storm berms are therefore one end-member of beach ridges formed by wave swash, and I refer to storm berms as beach ridges. Rate of beach ridge addition is a function of sediment supply, sea-level trends, and storm climate (Roy et al., 1994). In contrast, foredunes are dunes constructed by aeolian processes; if excavated, foredunes are distinct sedimentologically and do not have foreset beds characteristic of wave swash. Beach ridges that are no longer active can be capped by aeolian deposits.

Strandplains are broad accumulations of sediment formed in parallel or semi-parallel ridges oriented approximately parallel to the coastline (Roy et al., 1994; Hein et al., 2012). Strands are linear geomorphic features parallel to the modern shoreline that define former shorelines. Strandplains may represent a composite of wave-swash-created beach ridges, foredunes or even emergent offshore bar features. The progradational nature of strandplains allows them to serve as recorders of paleoenvironmental change (Wells, 1996; Goy et al., 2003; Hein et al., 2012). For instance, regressive strandplain sequences can chronicle sea-level fall in the late Holocene. At Pinheira, southern Brazil, falling sea level led to forced coastal progradation. The result was a 5-km-wide strandplain that prograded as offshore sediment was reworked and supplied. The plain consists of a succession of closely spaced arcuate, low, 1–4-m-high beach and dune ridges (Hein et al., 2012). In order to investigate the deposit geometry that accompanies such a progradation, Hein et al. (2012) demonstrate the utility of the active-source imaging technique of ground-penetrating radar (GPR) for investigating subsurface stratigraphy and sea-level history on coastal plains.

Given the above definitions, then beach ridges (including "storm berms") are wave-swash-created features that, if separated from the active beach by relative sea-level fall or by coastal progradation, can be sea-level indicators. Beach ridges cannot form above spring high tides or above storm wave conditions; beach ridges can therefore document relative sea-level fall if the features are meters above spring high tide level or storm swash levels. Wells (1996), following Badukov (1986), suggests determination of mean sea level for beach ridges can be derived from the location of the swash zone on the beach face of a beach ridge. Tide range determination involved excavation of the lower slope of the beach ridge at the base of the former beach face to identify the foreset beds of the swash zone (Wells, 1996). If the reference water level for beach ridges is within the lower slope of the beach ridge, then the indicative range is large because such a reference water level could be the upper swash zone of spring high tides or the relatively higher upper swash zone of less frequent storms. A beach ridge may therefore indicate a sea level datum only over a broad vertical range, and that range may be greater than the height difference of adjacent beach ridges.

The elevations of two adjacent paleo-beach ridges can be differenced to assess relative sea-level fall between two beach ridges. Elevation differencing of the base of beach ridges and not the crests is however desirable, because paleo-beach ridge elevation can be affected by

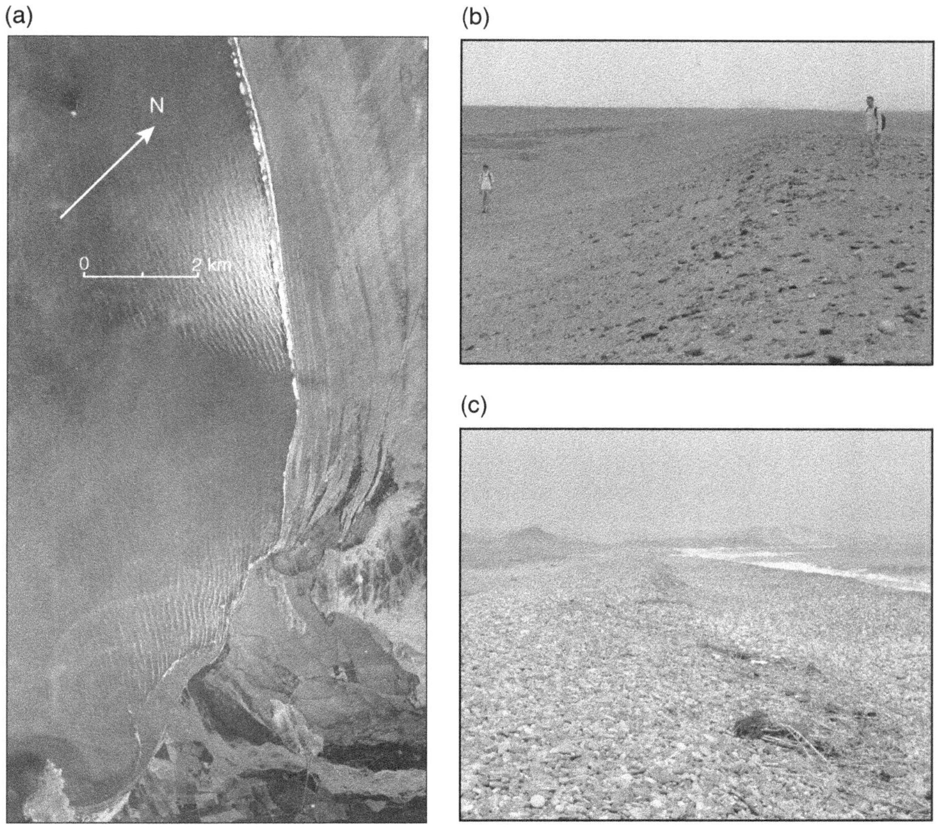

Fig. 5.6. (a) Mosaic of three vertical aerial photographs depicting beach ridges on the progradational beach-ridge plain of the Santa beach ridge complex, latitude 8° S, Peru. Santa River empties to Pacific Ocean to the southeast of the beach-ridge plain. Note the plume of sediment at the river mouth that is being transported north. Source of photograph: Servicio Aerofotográfico Nacional (Peru), Project 147-66, 20 March 1967. *Source*: Photographs mosaicked and contributed by D. Sandweiss. Reproduced with permission. (b) Abandoned inactive beach ridge, Santa beach ridge complex, latitude 8° S, Peru. View to north-northwest. (c) Modern beach ridge, seaward edge of Santa River progradational plain, latitude 8° S, Peru. View to south-southeast. *Source*: Photographs (b) and (c) by David Reid. Reproduced with permission. For color details, please see Plate 13.

sediment supply, storminess, and aeolian caps (Orford et al., 2003).

On progradational plains, beach ridges often occur as multiple parallel ridges with swales between them. There is not any one cause for the isolation of a beach ridge and the deposition of another in a more seaward position. Various case studies and review articles (Sandweiss, 1986; Tanner, 1995; Wells, 1996; Goy et al., 2003; Orford et al., 2003) argue that beach ridges can become isolated by a storm as it may make the ridge less accessible to wave swash, by sea-level change of the order 1–1.5 m, or by a climate-induced change in sediment supply. Each of these factors can reinforce the others, leading to the isolation of a beach ridge.

The Santa beach ridge complex in coastal Peru, at latitude 8°S, consisting of as much as 4 km of coastal progradational plain and as many as 15 individual beach ridges (Fig. 5.6), is a good example of the complexity of identifying the cause of abandonment of one beach ridge and the establishment of the next seaward ridge. Although Wells (1996) justifiably argues from sedimentologic evidence that the ridges are marine, establishing a sea-level record from a set of coast-parallel beach ridges is problematic because: (1) the ridges are difficult to date (Sandweiss et al., 1998); and (2) in Peru the combination of earthquakes and El-Niño-triggered storms can mobilize sediment from coastal drainages that result in episodes of high sediment delivery to the coast (Sandweiss, 1986). But each sediment-influx-related El Niño event does not engender a new ridge (Rogers et al., 2004). And sea level in the last 4000 years has been static or has dropped by 1–2 m or less (Sandweiss, 1986; Wells, 1996). Therefore, using beach ridge elevation to chronicle sea-level change in coastal Peru

Fig. 5.7. (a) Uplifted beach ridges at Cape Turakirae, South Island, New Zealand (41°26' S, 174°55' E). Uplift of all ridges is inferred to be associated with coseismic displacement on the nearby Wairarapa fault, and the most recent, historic beach ridge uplift occurred during coseismic displacement on this fault in 1855. Individual uplifts range in magnitude from 3.5 to 7.1 m (McSaveney et al., 2006; Little et al., 2009). The two stair-stepped treads of marine terraces in the middle ground are oxygen isotope stage 5 (80–125 ka) and stage 7 (~200 ka) uplifted shore platforms (McSaveney et al., 2006). *Source*: Photograph by Lloyd Homer. Reproduced with permission of GNS Science, New Zealand. (b) Late Holocene uplifted beach ridges along the central-east coast of Isla Mocha, which is an island at latitude 38° S off the coast of Chile (see Nelson and Manley, 1992). Isla Mocha sits above the Chilean subduction zone. *Source*: Photograph by Alan R. Nelson. Reproduced with permission. For color details, please see Plate 14.

is imprecise because the uncertainty in the indicative meaning of the base elevation of a beach ridge is at least as large or larger than the multi-thousand-year period of relative sea-level change. Relative sea-level change alone is therefore not a convincing driver for ridge isolation. Isolation of individual ridges of the Santa beach ridge complex and establishment of the next landward ridge may well be triggered by climate-induced change in sediment delivery to the coastal zone; isolation probably occurs in conjunction with other factors, which could include vertical tectonic displacements, increased storminess and/or minor (of the order decimeters over hundreds of years) GIA-related relative sea-level fall.

On tectonically active coasts where there is minimal or no progradational beach plain, beach ridges can become isolated by tectonic, usually coseismic, uplift (Wellman, 1969; Nelson and Manley, 1992; McSaveney et al., 2006). Workers can identify the mechanism of beach ridge isolation as tectonic in these cases because storms and high tides, while creating the beach ridges in the first place, are not a mechanism to isolate them. These coasts are not progradational plains, and therefore a change in sediment supply is not a potential factor leading to beach ridge isolation. For instance, Turakirae Head on the south end of the North Island of New Zealand prominently features four beach ridges that are vertically separated from each other by 3.4–7.1 m (Fig. 5.7a; Aston, 1912; Wellman, 1969; McSaveney et al., 2006). The crest of the modern beach ridge is located 2–7 m above mean sea level and presumably marks the run-up limit of present-day storm waves. The elevation difference of adjacent raised ridges is a measure of coseismic uplift that caused the stranding of each higher elevation ridge (Fig. 5.7a; Little et al., 2009).

In a similar case study, Nelson and Manley (1992) document beach ridges on the coast of Isla Mocha offshore of mainland Chile and above the Chilean subduction zone megathrust. A set of beach ridges was coseismically uplifted in the 1960 Chilean subduction zone earthquake, and there are older, higher beach ridges that may have been isolated from sea level by the same mechanism (Fig. 5.7b).

Although shore-parallel lines of raised beach ridges on a tectonically active coast can be a signature of abrupt relative sea-level fall brought on by coseismic uplift of the land, aseismic uplift, if fast enough, can elevate beach ridges formed by infrequent storm events and can therefore leave a record of relative sea-level fall. Nelson and Manley (1992) also identify a beach ridge that has been isolated above sea level along the same coast since the 1960 subduction zone earthquake. They infer that this beach ridge was deposited by storms and isolated by unusually rapid aseismic tectonic uplift related to strain on upper plate faults within the subduction zone margin.

In addition, beach ridges can also be stranded above highest tide levels by GIA-induced relative sea-level fall. Such raised beaches are common in equatorial regions (e.g., Fletcher and Jones, 1992, Woodroffe et al., 1995; Grossman and Fletcher, 1998) where ocean syphoning (Mitrovica and Peltier, 1991; Mitrovica and Milne, 2002) has occurred.

Beach ridge formation in high latitudes, with examples from the Arctic, deserves separate mention because raised beach ridges in high latitudes are ubiquitous and are the most common coastal landform (Sanjaume and Tolgensbakk, 2009). A prominent reason for the ubiquity of beach ridges as evidence of former shorelines in high latitudes is that retreat of ice resulted in isostatic rebound and relative sea-level fall. In Spitzbergen (the main island of Svalbard, latitude 79°N), more than 100 beach ridges formed in this manner; however, in addition to fall in relative sea level, requirements for beach ridge formation include adequate exposure of the shore to wind fetch and seasonally ice-free conditions so that waves can form ridges (Brückner et al., 2002). Although beach ridges are useful as an indicator of sea-level change in the Arctic (Long et al., 2012), the number of preserved beach ridges on neighboring sections of coast may vary due to intrinsic factors. For instance, the Varanger Peninsula of northern Norway (70°40' N) is an example of a region where, even with isostatic rebound being the same among neighboring bays, the number of raised beach ridges varies from one neighboring bay to the next because local ridge formation is controlled more by local sediment supply, offshore gradient, and site-specific coastal evolution than by more regional factors such as storminess and rates of regional sediment supply (Sanjaume and Tolgensbakk, 2009).

Because beach ridge formation in high latitudes has persisted throughout the Holocene, that is, persisted into time periods younger than the period of major ice melting and consequent isotatic adjustments, other factors in addition to isostatic uplift play a role in the formation and preservation of high-latitude beach ridges. Beach ridges can form in the latter part of the Holocene on progradational plains, and the main factors determining whether ridge progradation takes place or not are wave climate and/or high sediment input to the beach (Mason and Jordon, 1993; St-Hilaire-Gravel et al., 2010). And varying summer ice extent may be a main determinant of beach ridge morphology and preservation. Reduced summer sea ice results in more open water that in turn leads to higher wave energy for prolonged periods; reduced ice, more fetch, and higher wave energy therefore leads to better-developed beach ridges (Lowther Island, Canadian Arctic, 74°30'N; St-Hilaire-Gravel et al., 2010). If lack of ice cover, increased fetch and consequent higher wave energy can produce beach ridges on progradational plains, then the main factors influencing Holocene beach ridge isolation and character may be factors other than relative sea-level fall.

Finally, for those cases in the Arctic where beach ridges can be used as an indicator of regional sea-level change, beach ridge chronology is essential to track history of relative sea-level change. For northern Spitzbergen in the high Arctic (79°N), age determinations from juvenile articulated shells incorporated into beach ridges during storms provide reliable sample age determinations that compare well with other chronologies derived from driftwood or whalebone (Long et al., 2012).

As a final note on emergent Holocene shoreline deposits, transgressive barrier island systems have received much attention; on the east coast of the US, these depositional systems form in the context of rising sea level throughout the Holocene. For transgressive barrier island systems however, rates of sea-level rise are most effectively documented through saltmarsh stratigraphy, which is discussed in Chapter 5.

ACKNOWLEDGEMENTS

Support over the last three decades from both the US National Science Foundation and the US Geological Survey's National Earthquake Hazard Reduction Program has nourished my research in coastal geomorphic and neotectonic processes. This chapter benefited from thoughtful reviews by two anonymous reviewers and from collegial discussion with Aron Meltzner, Dan Muhs, David Reid, Dan Sandweiss and Alan Trenhaile, all of whom generously contributed photographs from their personal collections. Others who either contributed their photographs or assisted in permissions to use photographs include Alan Nelson, Antony Long, Gina Harlow and Margaret Low. Matt Strzelecki and Antony Long helped turn my attention to beach ridges and shore platforms in high-latitude settings, and Ian Shennan provided feedback on several occasions during the review process. This chapter was prepared while Kelsey was a Visiting Scholar in the Department of Earth and Space Sciences at the University of Washington.

REFERENCES

Aston, B.C. (1912) The raised beaches of Cape Turakirae. Transactions and Proceedings of the New Zealand Institute, 44, 208–213.

Badukov, D. D. (1986) Ancient shorelines as indicators of sea level. Journal of Coastal Research, 2, 147–157.

Bloom, A.L., Broecker, W.S., Chappell, J.M.A., Matthews, R.K., and Mesolella, K.J. (1974) Quaternary sea level fluctuations on a tectonic coast: new 230Th/234U dates from the Huon Peninsula, New Guinea. Quaternary Research, 4, 185–205.

Bradley, W.C., and Griggs, G.B. (1976) Form, genesis and deformation of central California wave-cut platforms. Geological Society of America Bulletin, 87, 433–449.

Brückner, H., Schellmann, G., and van der Borg, K. (2002) Uplifted beach ridges in northern Spitsbergen as indicators for glacio-isostasy and palaeo-oceanography. Zeitschrift für Geomorphologie, 46, 309–336.

Bull, W.B. (1985) Correlation of flights of global marine terraces. In: *Tectonic Geomorphology* (eds Morisawa, M., and Hack, J.T.), Allen & Unwin, Boston, pp. 129–152.

Chappell, J.M.A. (1974) Geology of coral terraces, Huon Peninsula, New Guinea: A study of Quaternary tectonic movements and sea level changes. Geological Society of America Bulletin, 85, 553–570.

Chappell, J., and Shackleton, N.J. (1986) Oxygen isotopes and sea level. Nature, 324, 137–140.

Darwin, C. (1842) *The Structure and Distribution of Coral Reefs*. Smith, Elder and Co., London.

Edwards, R.L., Taylor, F., and Wasserburg, G. (1988) Dating earthquakes with high-precision thorium-230 ages of very young corals. Earth and Planetary Science Letters, 90, 371–381.

Evelpidou, N., Kampolis, I, Pirazzoli, P.A., and Vassilopouos, A. (2012) Global sea-level rise and the disappearance of tidal notches. Global and Planetary Change, 92–93, 248–256.

Fletcher, C.H., and Jones, A.T. (1996) Sea-level highstand recorded in Holocene shoreline deposits on Oahu, Hawaii. Journal of Sedimentary Research, 66, 632–641.

Goy, J.L., Zazo, C., and Dabrio, C.J. (2003) A beach-ridge progradational complex reflecting periodical sea-level and climate variability during the Holocene, Gulf of Almeria, western Mediterranean. Geomorphology, 50, 251–268.

Grossman, E.E., and Fletcher, C.H. (1998) Sea level higher than present 3500 years ago on the northern main Hawaiian Islands. Geology, 26, 363–366.

Hein, C.J., Fitzgerald, D.M., Cleary, W.J., Albernaz, M.B., De Menezes, J., and Klein, A. (2012) Evidence for a transgressive barrier within a regressive strandplain system: Implications for complex coastal response to environmental change. Sedimentology, 60(2), 469–502.

Hesp, P.A., Dillenburg, S.R., Barboza, E.G., Tomazelli, L.J., Ayup-Zouain, R.N., Esteves, L.S., Gruber, N.L.S., Toldo, E.E., Tabajara, L., and Clerot, L. (2005) Beach ridges, foredunes or transgressive dunefields? Definitions and an examination of the Torres to Tramandaí barrier system, Southern Brazil. Annals of the Brazilian Academy of Sciences, 77, 493–508.

Higgins, C.H. (1980) Nips, notches, and solution of coastal limestone: an overview of the problem with examples from Greece. Estuarine and Coastal Marine Science, 10, 15–30.

Hull, A.G. (1987) A late Holocene marine terrace on the Kidnappers coast, North Island, New Zealand: Some implications for shore platform development processes and uplift mechanism. Quaternary Research, 28, 183–195.

Kelsey, H.M. (1990) Late Quaternary deformation of marine terraces on the Cascadia subduction zone near Cape Blanco, Oregon. Tectonics, 9, 983–1014.

Kelsey, H.M., and Bockheim, J.G. (1994) Coastal landscape evolution as a function of eustasy and surface uplift rate, Cascadia margin, southern Oregon. Geological Society of America Bulletin, 106, 840–854.

Kim, K., Fouke, B., Petter, A., Quinn, T., Kerans, C., and Taylor, F. (2012) Sea-level rise, depth-dependent carbonate sedimentation and the paradox of drowned platforms. Sedimentology, 59, 1677–1694.

Lajoie, K.R. (1986) Coastal Tectonics. In: *Active Tectonics* (ed. Wallace, R.E.), National Academic Press, Washington, DC, pp. 95–124.

Lambeck, K., and Nakada, M. (1992) Constraints on the age and duration of the last interglacial period and on sea-level variations. Nature, 357, 125–128.

Liew, P.M., Pirazzoli, P.A., Hsieh, M.L., Arnold, M., Barusseau, J.P., Fontugne, M., and Giresse, P. (1993) Holocene tectonic uplift deduced from elevated shorelines, eastern Coastal Range of Taiwan. Tectonophysics, 222, 55–68.

Little, T.A., Van Dissen, R., Schermer, E., and Carne, R. (2009) Late Holocene surface ruptures on the southern Wairarapa fault, New Zealand: Link between earthquakes and the uplifting of beach ridges on a rocky coast. Lithosphere, 1, 4–28.

Long, A.J., Strzelecki, M.C., Lloyd, J.M., and Bryant C. (2012) Dating High Arctic Holocene relative sea level changes using juvenile articulated marine shells in raised beaches. Quaternary Science Reviews, 48, 61–66.

Mason, O., and Jordan, J. (1993) Heightened North Pacific storminess during synchronous late Holocene erosion of northwest Alaska beach ridges. Quaternary Research, 40, 55–69.

McSaveney, M.J., Graham, I.J., Begg, J.G., Beau, A.G., Hull, A.G., Kim, K., and Zondervan, A. (2006) Late Holocene uplift of beach ridges at Turakirae Head, south Wellington coast, New Zealand. New Zealand Journal of Geology and Geophysics, 49, 337–358.

Meltzner, A.J. (2010) Earthquake recurrence, clustering, and persistent segmentation near the southern end of the 2004 Sunda megathrust rupture. PhD thesis, Californian Institute of Technology, Pasadena.

Meltzner, A.J., Sieh, K., Chiang, H.-W., Shen, C.-C., Suwargadi, B.W., Natawidjaja, D.H., Philibosian, B.E., Briggs, R.W., and Galetzka, J. (2010) Coral evidence for earthquake recurrence and an A.D. 1390-1455 cluster at the south end of the 2004 Aceh-Andaman rupture. Journal of Geophysical Research, 115, B10402, doi: 10.1029/2010JB007499.

Meltzner, A.J., Sieh, K., Chiang, H.-W., Shen, C.-C., Suwargadi, B.W., Natawidjaja, D.H., Philibosian, B. and Briggs, R.W. (2012) Persistent termini of 2004- and 2005-like ruptures of the Sunda megathrust. Journal of Geophysical Research, 117, B04405, doi: 10.1029/2011JB008888.

Merritts, D.J. (1996) The Mendocino triple junction: Active faults, episodic coastal emergence, and rapid uplift. Journal of Geophysical Research, 101, 6051–6070.

Mesolella, K.J., Matthews, R.K., Broecker, W.S., and Thurber, D.L. (1969) The astronomical theory of climatic change, Barbados data. Journal of Geology, 77, 250–274.

Milne, G.A., and Mitrovica, J.X. (2008) Searching for eustasy in deglacial sea-level histories. Quaternary Science Reviews, 27, 2292–2302.

Mitrovica, J.X., and Peltier, W.R. (1991) On postglacial geoid subsidence over the equatorial oceans. Journal of Geophysical Research, 96, 20053–20071.

Mitrovica, J.X., and Milne, G.A. (2002) On the origin of late Holocene sea-level highstands within equatorial ocean basins. Quaternary Science Reviews, 21, 2179–2190.

Muhs, D.R., Simmons, K.R., Kennedy, G.L., and Rockwell, T.K. (2002) The last interglacial period on the Pacific Coast of North America: Timing and paleoclimate. Geological Society of America Bulletin, 114, 569–592.

Muhs, D.R., Simmons, K.R., Schumann, R. and Halley, R.B. (2011) Sea-level history of the past two interglacial periods: new evidence from U-series dating of reef corals from south Florida. Quaternary Science Reviews, 30, 570–590.

Muhs, D.R., Simmons, K.R., Schumann, R.R., Groves, L.T., Mitrovica, J.X., and Laurel, D. (2012) Sea-level history during the Last Interglacial complex on San Nicolas Island, California: implications for glacial isostatic adjustment processes, paleozoogeography and tectonics. Quaternary Science Reviews, 37, 1–25.

Nakada, M., and Lambeck, K. (1989) Late Pleistocene and Holocene sea-level change in the Australian region and mantle rheology. Geophysical Journal International, 96, 497–517.

Nelson, A.R., and Manley, W.F. (1992) Holocene coseismic and aseismic uplift of Isla Mocha, south-central Chile. Quaternary International, 15/16, 61–76.

Nixon, F.C., Reinhardt, E.G., and Rothaus, R. (2009) Foraminifera and tidal notches; dating neotectonic events at Korphos, Greece. Marine Geology, 257, 41–53.

Orford, J.D., Murdy, J.M., and Wintle, A.J. (2003) Prograded Holocene beach ridges with superimposed dunes in north-east Ireland: mechanisms and timescales of fine and coarse beach sediment decoupling and deposition. Marine Geology, 194, 47–64.

Otvos, E.G. (2000) Beach ridges – definitions and significance. Geomorphology, 32, 83–108.

Philibosian, B., Sieh, K., Natawidjaja, D.H., Chiang, H.-W., Shen, C.-C., Suwargadi, B.W., Hill, E.M., and Edwards R.L. (2012) An ancient shallow slip event on the Mentawai segment of the Sunda megathrust, Sumatra. Journal Geophysical Research, 117, B05401, doi: 10.1029/2011JB009075.

Phillips, B.A.M. (1970) The significance of inheritance in the interpretation of marine and lacustrine coastal histories. Lakehead University Review, 3, 36–45.

Pillans, B. (1983) Upper Quaternary marine terrace chronology and deformation, South Taranaki, New Zealand. Geology, 11, 292–297.

Pirazzoli, P.A. (1980) Formes de corrosion marine et vestiges archéologiques submerges: interpretation néotectonique de quelques exemples en Grèce et en Yougoslavie. Annales de l'Institut Océanographique, 56, 101–111.

Pirazzoli, P.A. (1986) Marine notches in sea-level research. In: *A Manual for the Collection and Evaluation of Data* (ed. van de Plassche, O.), Geo Books, Norwich, pp. 361–400.

Pirazzoli, P.A. (2005) Marine erosion features and bioconstructions as indicators of tectonic movements, with special attention to the eastern Mediterranean area. Zeitschrift für Geomorphologie, Supplement 137, 71–77.

Porter, N.J., Trenhaile, A.S., Prestanski, K., and Kanyaya, J.I. (2010) Patterns of surface downwearing on shore platforms in eastern Canada. Earth Surface Processes and Landforms, 35, 1793–1810.

Radtke, U., and Schellmann, G. (2004) *The Marine Quaternary of Barbados*, Kolner Geographische Arbeiten, Heft 81, 137 pp., Geographisches Institut der Universitat zu Koln, Germany.

Ramos, N.T., and Tsutsumi, H. (2012) Evidence of large prehistoric offshore earthquakes deduced from uplift Holocene marine terraces in Pangasinan Province, Luzon Island, Philippines. Tectonophysics, 495, 145–158.

Rogers, S.S., Sandweiss, D.H., Maascht, K.A., Belknapt, D.F., and Agouris, P. (2004) Coastal change and beach ridges along the northwest coast of Peru: Image and GIS analysis of the Chira, Piura, and Colan beach-ridge plains. Journal of Coastal Research, 20, 1102–1125.

Roy, P.S., Cowell, P.J., Ferland, M.A., and Thom, B.G. (1994) Wave-dominated coasts. In: *Coastal Evolution, Late Quaternary Shoreline Morphodynamics* (eds Carter, R.W.G., and Woodroffe, C. D.), Cambridge University Press, Cambridge, pp. 121–186.

Sandweiss, D.H. (1986) The beach ridges at Santa, Peru: El Nino, uplift, and prehistory. Geoarchaeology: An International Journal, 1, 17–28.

Sandweiss, D.H., Maasch, K.A., Belknap, D.F., Richardson, J.B., and Rollins, H.B. (1998) Discussion of: Lisa E. Wells, 1996. The Santa Beach Ridge Complex, Journal of Coastal Research, 12, 1–17. Journal of Coastal Research, 14, 367–373.

Sanjaume, E. and Tolgensbakk, J. (2009) Beach ridges from the Varanger Peninsula (Arctic Norwegian coast): Characteristics and significance. Geomorphology, 104, 82–92.

Sieh, K., Ward, S., Natawidjaja, D., and Suwargadi, B. (1999) Crustal deformation at the Sumatran subduction zone revealed by coral rings. Geophysical Research Letters, 26, 3141–3144.

Sieh, K., Natawidjaja, D.H., Meltzner, A.J., Shen, C.-C., Cheng, H., Li, K.-S., Suwargadi, B.W., Galetzka, J., Philibosian, B., and Edwards, R.L. (2008) Earthquake supercycles inferred from sea-level changes recorded in the corals of West Sumatra. Science, 322, 1674–1678, doi: 10.1126/science.1163589.

Stiros, S.C., Laborel, J., Laborel-Deguen, F., Papageorgiou, S., Evin, J., and Pirazzoli, P.A. (2000) Seismic coastal uplift in a region of subsidence: Holocene raised shorelines of Samos Island, Aegean Sea, Greece. Marine Geology, 170, 41–58.

St-Hilaire-Gravel, D., Bell, T.J., and Forbes D.L. (2010) Raised gravel beaches as proxy indicators of past sea-ice and wave conditions, Lowther Island, Canadian Arctic Archipelago. Arctic 63, 213–226.

Tanner, W.F. (1995) Origin of beach ridges and swales. Marine Geology, 129, 149–161.

Taylor, F.W., Frohlich, C., Lecolle, J., and Strecker, M. (1987) Analysis of partially emerged corals and reef terraces in the central Vanuatu arc: Comparison of contemporary coseismic and nonseismic with Quaternary vertical movements. Journal of Geophysical Research, 92, 4905–4933.

Trenhaile, A.S. (1972) The shore platforms of the Vale of Glamorgan, Wales. Transactions of the Institute of British Geographers, 56, 127–144.

Trenhaile, A.S. (1983) The development of shore platforms in high latitudes. In: *Shorelines and Isostasy* (eds Smith, D.E., and Dawson, A.G.), Academic Press, London, Institute of British Geographers Special Publication 16, pp. 77–93.

Trenhaile, A.S. (1987) *The Geomorphology of Rock Coasts.* Oxford University Press, Clarendon, Oxford, UK.

van de Plassche, O. (ed.) (1986) *Sea-Level Research: A Manual for the Collection and Evaluation of Data*, Geo Books, Norwich.

Wellman, H.W. (1969) Tilted marine beach ridges at Cape Turakirae, New Zealand. Tuatara, 17, 82–93.

Wells, L.E. (1996) The Santa Beach ridge complex: sea-level and progradational history of an open gravel coast in central Peru. Journal of Coastal Research, 12, 1–17.

Wilson, K., Berryman, K.B., Litchfield, N., and Little, T. (2006) A revision of mid–late Holocene marine terrace distribution and chronology at the Pakarae River mouth, North Island, New Zealand. New Zealand Journal of Geology & Geophysics, 49, 477–489.

Woodroffe, C.D., Murray-Wallace C.V., Bryant, E.A., Brooke, B., Heijnis, H., and Price, D.M. (1995) Late Quaternary sea-level highstands in the Tasman Sea: evidence from Lord Howe Island. Marine Geology, 125, 61–72.

Wright, L.W. (1970) Variation in the level of the cliff/shore platform junction along the south coast of Great Britain. Marine Geology, 9, 347–353.

Zachariasen, J., Sieh, K., Taylor, F., Edwards, R., and Hantoro, W. (1999) Submergence and uplift associated with the giant 1833 Sumatran subduction earthquake: Evidence from coral microatolls. Journal of Geophysical Research, 104, 895–919.

Zachariasen, J., Sieh, K., Taylor, F., and Hantoro, W. (2000) Modern vertical deformation at the Sumatran subduction zone: Paleogeodetic insights from coral microatolls. Bulletin Seismological Society of America, 90, 897–913.

Chapter 6
Coastal caves and sinkholes

PETER J. VAN HENGSTUM[1], DAVID A. RICHARDS[2], BOGDAN P. ONAC[3], AND JEFFREY A. DORALE[4]

[1]*Department of Marine Sciences, Texas A&M University at Galveston, Galveston, TX, USA*
[2]*School of Geographical Sciences, University of Bristol, Bristol, UK*
[3]*School of Geosciences, University of South Florida, Tampa, FL, USA*
[4]*Department of Earth and Environmental Science, University of Iowa, Iowa City, IA, USA*

6.1 INTRODUCTION

For over a century, karst scientists have appreciated that the vertical migration of groundwater strongly impacts the subterranean realm by regulating cave and sinkhole formation, speleothem precipitation, and subterranean karst ecosystems (Grund, 1903; Cvijić, 1918; Swinnerton, 1929; Davis, 1931; Iliffe et al., 1983). Grund (1903) first proposed that sea level controls groundwater elevation in both inland and coastal karst, but further analysis by Cvijić (1918) demonstrated that the water table elevation on inland karst terrain is instead linked to multiple variables including topography, structural geology, and climate (Watts, 1969; Bates, 1995). In the coastal zone, however, Grund's assertion holds true where groundwater within coastal carbonate terrain is governed by defined hydrogeologic principles (Reilly and Goodman, 1985; Vacher, 1988; Vacher and Wallis, 1992) and it concomitantly migrates during sea-level oscillations (Gascoyne et al., 1979; Harmon et al., 1981; Richards et al., 1994). Even early naturalists speculated that modern bluehole environments were intimately linked to sea-level change (Northrop, 1890; Agassiz, 1894). The near-synchronous oscillation of coastal groundwater and sea level during glacioeustatic cycles is the basis of all sea-level research in caves and sinkholes.

Despite the simplicity of this basic tenet, several common problems promote confusion for non-karst researchers on how sea level influences environmental change in karst environments, and how this influence is subsequently recorded in the karst geological record. First, diverse nomenclature is used to address karst environments including sinkholes, anchialine caves, cenotes, submarine caves, and blueholes. More importantly, the different areas of karst research remain considerably compartmentalized. For example, investigations on: (1) sea-level change as recorded by speleothems (e.g., Harmon et al., 1978; Dorale et al., 2010); (2) analysis of subterranean environments and ecosystems (Iliffe, 1987; Alverez Zarikian et al., 2005; Gabriel et al., 2009); and (3) cave and sinkhole sedimentology occur almost independently (e.g., Alverez Zarikian et al., 2005; Fornós et al., 2009; Lane et al., 2011). However, all these lines of inquiry must first address how sea-level change impacts these environments.

Here we present a simple model for environmental succession in coastal karst basins under sea-level forcing, and use this model as a framework for reviewing sea-level indicators preserved in caves and sinkholes. This model explains how common karst sea-level indicators develop (flank margin caves, speleothems, sediments), and why some of these indicators are sea-level index points (SLIPs) with a clearly defined indicative meaning, whereas others can only provide maximum or minimum constraints on sea level. We also provide the indicative meaning of these sea-level indicators, where appropriate, which is a clearly defined relationship between the indicator and a contemporaneous reference tide level (van de Plassche, 1986). The model presented here dovetails geological, environmental, biological, and ecosystem research, and it provides a simple framework for analyzing the sea-level proxies that

Handbook of Sea-Level Research, First Edition. Edited by Ian Shennan, Antony J. Long, and Benjamin P. Horton.

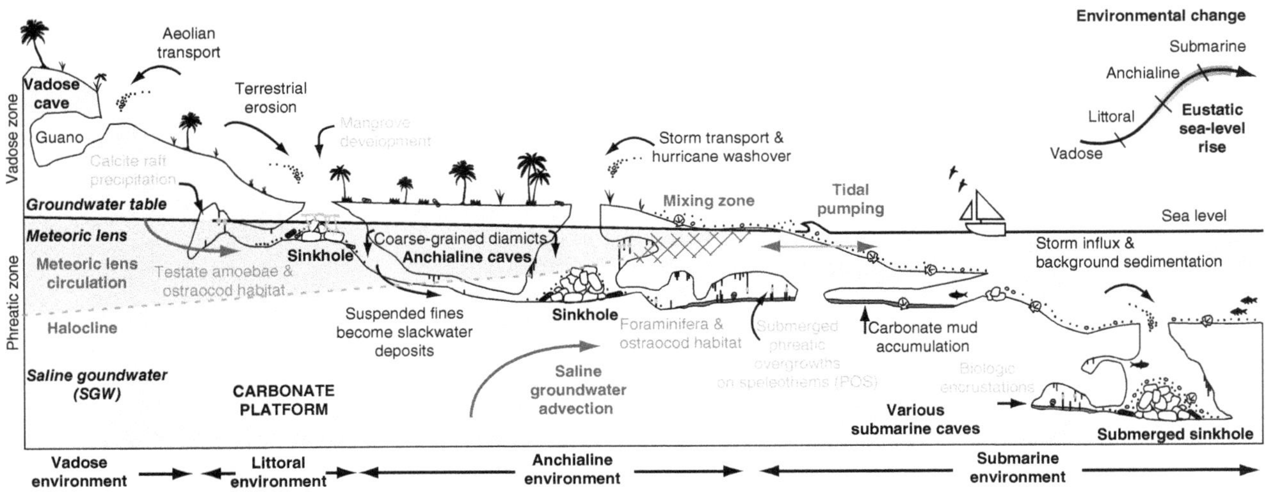

Fig. 6.1. Coastal karst basins (CKBs) provide accommodation space for unique environments, sedimentation patterns, and ecosystems that are distinct from all other coastal environments. All four environmental categories can be observed with increasing distance from the coastline, and sea-level rise causes predictable environmental succession in CKBs (top right corner). The distinctive environmental change that occurs during sea-level oscillations is preserved in the geologic record through speleothems and sediments in CKBs. Green: microfossil remains; red: sea-level indicators; blue: groundwater hydrography and flow; black: common sedimentary processes. For color details, please see Plate 15.

are preserved in coastal karst basins (CKBs, described in the following section).

Before proceeding further, it should be noted that geologic sampling in many CKBs can only be accomplished by individuals with specialized training in technical cave diving. The American Academy of Underwater Sciences (AAUS) does outline minimal procedures for cave diving, but prudent divers expand considerably on these protocols and seek extensive individual training and experience before scientific objectives are ever attempted in underwater caves. Even after such training, most scientists still consult regional experts to guarantee team safety, ensure minimal damage and impact to caves, and ensure the inherent aesthetic beauty of these sensitive environments remains preserved.

6.2 COASTAL KARST BASINS: ENVIRONMENTS AND SUCCESSION

Before cave and sinkhole sea-level indicators can be evaluated in the temporal dimension, we must first establish how different subterranean karst environments are spatially related. Some researchers have inferred sea-level information from inland karst landscapes, but inland groundwater-level changes are more complex and governed by regional geology and climate (Florea et al., 2007; Gulley et al., 2013). In contrast, coastal karst is narrowly defined as the landward limit of marine tidal influences on the coastal aquifer in modern or ancient settings (Guilcher, 1988), and this zone is where the vast majority of sea-level research occurs. With the exception of flank margin caves (Section 6.3), sea-level indicators from inside caves and sinkholes form in very specific and predictable subterranean environments.

Recently, the term "coastal karst basins" (CKBs) was introduced by van Hengstum et al. (2011) to collectively refer to the myriad of geomorphological features in coastal carbonate terrains that formed from the long-term dissolution and modification (speleogenesis) of limestone (Fig. 6.1). CKBs include caves and sinkholes that may be positioned in the present vadose (unsaturated) zone, the phreatic (saturated) zone, or both. This term is purely a geomorphologic descriptor for all the karst basins that can potentially preserve sea-level proxies, as opposed to using many different terms that relate to a physical process (e.g., phreatic cave), a physical subaerial entrance (cave vs. sinkhole), or geographically-specific labels (e.g., cenotes, blueholes).

In order to use karst sea-level proxies, researchers must also understand the hydrography and stratification of coastal aquifers. The subsurface beneath coastal carbonate platforms can be broadly divided into the vadose zone above the groundwater table, and the phreatic zone below the groundwater table (Fig. 6.1). The phreatic zone can then be further subdivided into an upper meteoric lens of variable salinity that rests buoyantly above the saline groundwater mass that has variable dissolved oxygen concentration. The upper meteoric lens is typically oxic, but its salinity generally increases towards the coast via mixing with seawater (Beddows et al., 2007). The boundary between the meteoric lens and saline groundwater is demarcated by a sharp halocline further inland (Reilly and Goodman, 1985), but groundwater mixing proximal to the coastline causes this sharp boundary to destabilize into a mixing zone with several haloclines. In contrast, the lower saline groundwater is of marine salinity, but the dissolved oxygen varies from anoxic (e.g., Sawmill Sink, Bahamas) to completely oxic (e.g., Green Bay Cave, Bermuda), depending upon how circulated the saline groundwater in the aquifer is with the ocean. The circulation of water from the ocean into carbonate platforms commonly occurs through two processes. The first process is tidal pumping, which causes the horizontal movement of seawater from the ocean into the carbonate platform through the limestone bedrock (Beddows et al., 2007; Martin et al., 2012) or through submarine caves and conduits (Cate, 2009). This process is often observed as vertical oscillation of the groundwater level in caves and sinkholes (Martin et al., 2012), which may or may not be synchronous with local tides (Cate, 2009). The second process is the incorporation of ocean water into broad subterranean circulation cells driven by geothermal heating (Kohout, 1965; Moore et al., 1992). The hydrography of coastal groundwater forms a critical environmental and physico-chemical backdrop for sea-level research in CKBs.

The most important principle of karst sea-level research is that coastal groundwater and sea-level oscillate in near synchrony during glacioeustatic cycles. As such, sea-level proxies from CKBs must delimit the former elevation of the groundwater table and its relative shifts through time and, by corollary, relative sea-level change. One caveat to this principle is that the groundwater table is not a perfectly horizontal planar surface. The hydraulic head refers to the groundwater surface elevation at a specific point, such as at the location of a sea-level indicator. The vertical offset between the water table at this location and sea-level over a geographic distance is the hydraulic gradient, and this offset is one potential source of vertical uncertainty in all karst-based sea-level reconstructions. In general however, coastal eogenetic karst has such high porosity that the hydraulic gradient is generally low and considered negligible in most studies. For example, the hydraulic gradient in the northwestern Yucatan Peninsula (Mexico) is 5–10 $mm\,km^{-1}$ (Marin and Perry, 1994). In cases where a CKB is hosted by carbonate lithology with low porosity (e.g., telogenetic karst, Vacher and Mylroie, 2002), it would be prudent to measure hydraulic gradients to ensure this assumption is reasonable or make appropriate corrections.

Over the last 30 years, advances in technical scuba diving have enabled biologists to investigate the unique subterranean ecosystems that develop in karst landscapes (Holthuis, 1973; Sket and Iliffe, 1980; Iliffe et al., 1983; Pohlman et al., 1997). Their research constrains four different subterranean environments in CKBs (vadose, littoral, anchialine, and submarine) that are defined by their relationship to coastal groundwater, biology and multiple physico-chemical processes (e.g., sedimentation, groundwater dynamics, terrestrial chemical flux).

- Vadose environments are the most commonly explored environments because they are positioned in the unsaturated zone above the water table, and they are where most speleothems form (Fig. 6.1). Sea-level proxies in vadose environments are maximum sea-level indicators only (e.g., speleothems).
- Littoral environments occur when the groundwater table is predominantly positioned in a CKB, and they may or may not be open to the ocean. Littoral environments occur at sea level, within constraints of the local hydraulic gradient, because the absolute elevation of groundwater table and sea level are very similar on coastal carbonate terrain. Sea-level index points with clear indicative meaning form in littoral environments (e.g., calcite rafts, phreatic overgrowths on speleothems).
- Anchialine environments are the most physically and chemically complex of karst environments and their definition has been expanding over time: (1) they have restricted subaerial access (Stock et al., 1986) and no subaerial connection to the ocean (Holthuis, 1973); (2) have noticeable marine and terrestrial influences (Stock et al., 1986), but terrestrial processes dominate (van Hengstum and Scott, 2011); (3) they have hydrographic connection to the

ocean through the subterranean circulation of groundwater through porous bedrock (Stock et al., 1986); (4) their tides are observable (Holthuis, 1973) by either vertical migration of the groundwater table or horizontal pumping of seawater into the CKB; and (5) they can be flooded by either saline groundwater (SGW) or the meteoric lens (ML). These environments produce minimum sea-level indicators only. For example, speleothem dissolution by a paleohalocline must have occurred below a contemporaneous sea level, but the vertical elevation of that paleohalocline will be related to local hydrogeology and likely not the contemporaneous sea level.

- Submarine environments: (1) are completely flooded by only saline groundwater and typically have physical entrances that are flooded by sea level; (2) are dominated by marine processes (van Hengstum and Scott, 2011); (3) their saline groundwater often circulates with the ocean by tidal pumping (Beddows et al., 2007; Martin et al., 2012); and (4) they can be oxygen stratified if promoted by local oceanography and basin geometry (e.g., Jewfish Sink, Florida; Graman and Garey, 2005). These environments produce minimum sea-level indicators only, such as serpulid worm encrustations on stalagmites.

In summary, the long-term dissolution and modification of carbonate creates karst landscapes with many types of CKBs. Once formed however, four specific environments can develop in CKBs that host unique ecosystems, discrete hydrographic characteristics, sedimentary processes, and promote specific speleothem processes. These four environments can be co-located in large cave systems (e.g., Walsingham Cave, Bermuda and Ox Bel Ha Cave, Mexico), and align karst environments with other coastal environments that can be spatially defined by their relationship to sea level (e.g., saltmarshes versus lagoons). During inundation of a carbonate platform from sea-level rise or coastal erosion, CKBs can progress from first being vadose environments, followed by littoral environments, then anchialine environments, and finally into submarine environments. This pattern of succession influences the ecosystems and sedimentation within CKBs, and encompasses all aspects of sea-level research in coastal caves and sinkholes.

6.3 FLANK MARGIN CAVES AS SEA-LEVEL INDICATORS

Cave formation in the coastal zone is complex and often occurs during multiple sea-level changes, so relating cave and sinkhole elevations to a specific sea level is a challenging exercise at best. However, flank margin caves (Fig. 6.2) and zones of cavernous porosity that develop in former mixing zones are the one exception that can be used as sea-level indicators (Mylroie and Carew, 1988; Smart et al., 2008). Flank margin caves develop at sea level when groundwaters with different salinity and carbonate chemistry mix with seawater in the coastal zone (Mylroie and Carew, 1990). The saturation curve for calcite is not linear, and when two solutions with different pCO_2 mix the resulting solution may be undersaturated with respect to calcite and become acidic (Fig. 6.2; Plummer, 1975; Wigley and Plummer, 1976; Back et al., 1986). This chemical process occurs in both modern littoral and anchialine environments, and the archetypal location for this process is in The Bahamas (Smart et al., 1988). This groundwater mixing causes flank margin caves to develop along the periphery of carbonate platforms worldwide (Mylroie and Carew, 1990; Mylroie and Mylroie, 2007). Recent numerical modeling results now show that this physical process does indeed create the exact physical geometry of the flank margin caves that can be observed in outcrop (Labourdette et al., 2007). One helpful tool for identifying flank margin caves in outcrop is meter-scale curvilinear features known as "wall cusps" (Frank et al., 1998). The elevation of flank margin caves is a sea-level indicator because the mixing of groundwater and seawater is directly linked to the position of sea level (Mylroie and Carew, 1988; Mylroie and Mylroie, 2009).

A great example of flank margin caves is in the Torbay region of southern England (Proctor, 1988; Proctor and Smart, 1991). Based on the gross cave morphology, Proctor (1988) concluded that these caves developed in a phreatic environment from the mixing of meteoric and marine water, similarly to flank margin caves in The Bahamas (Fig. 6.3). The Torbay caves were likely to have formed in anchialine and littoral settings during late Pleistocene sea-level highstands (Proctor and Smart, 1991).

Perhaps the greatest limitation of using flank margin caves in sea-level research is the difficulty of dating the formation event of flank margin caves. Even in situations where several speleothems

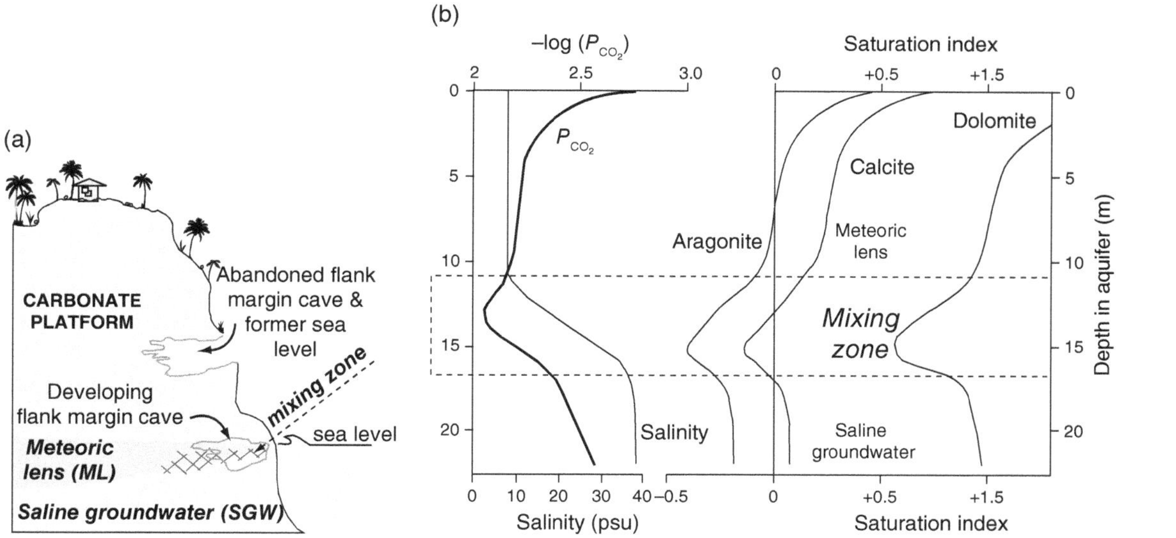

Fig. 6.2. (a) Model of flank margin cave development on a carbonate platform *Source*: adapted from Mylroie and Carew, 1995. The hatched pattern near the platform margin indicates the position of the mixing zone. (b) The saturation index of aragonite and calcite in a modern mixing zone in a Bahamian bluehole *Source*: data from Smart et al., 1988, indicating that limestone dissolution will occur when water from the meteoric lens mixes with saline groundwater in coastal karst landscapes. This process generates flank margin caves on the periphery of global carbonate platforms.

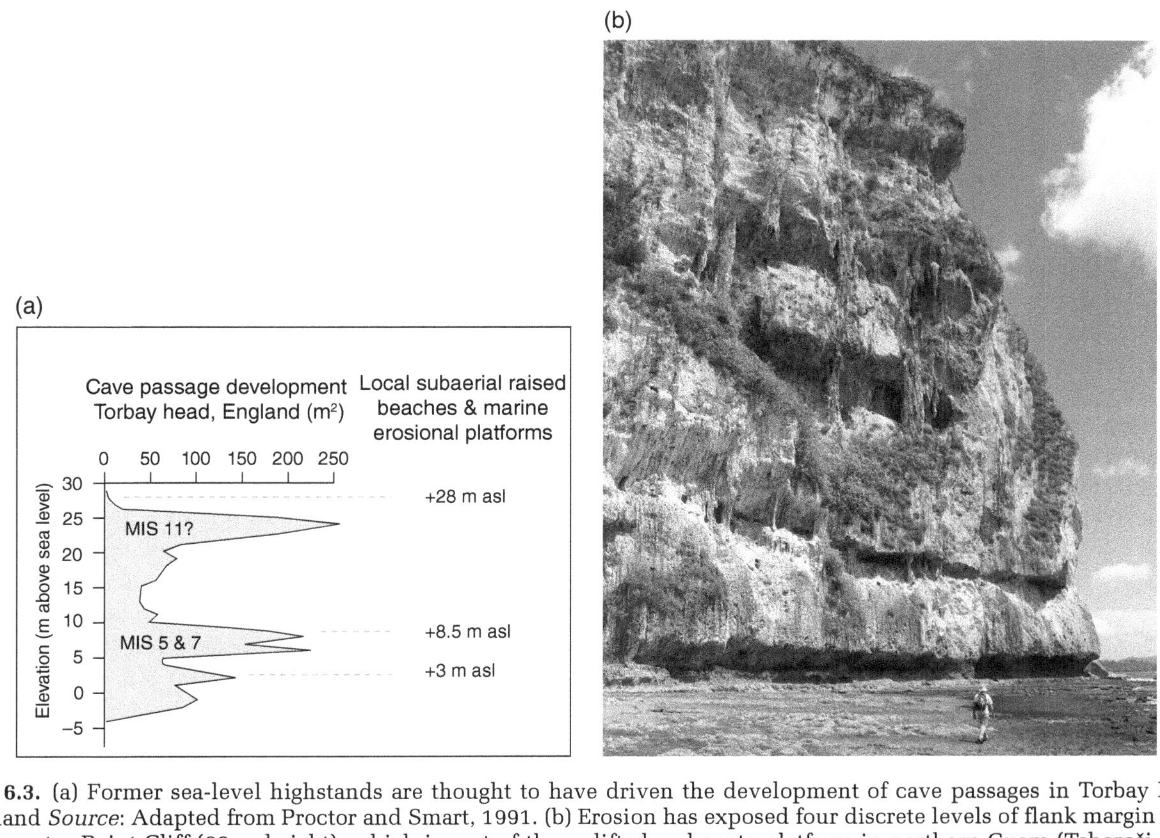

Fig. 6.3. (a) Former sea-level highstands are thought to have driven the development of cave passages in Torbay Head, England *Source*: Adapted from Proctor and Smart, 1991. (b) Erosion has exposed four discrete levels of flank margin caves at Amantes Point Cliff (80 m height), which is part of the uplifted carbonate platform in northern Guam (Taboroši et al., 2003). The low tide also reveals the modern marine tidal notch. MIS: Marine Isotope Stage. *Source*: Photograph by John Mylroie. Reproduced with permission.

from a single cave can be dated by U-series methods, this provides only a minimum age for the cave. Only in a few rare circumstances have cave formation events been dated (Polyak et al., 1998). Tectonic or isostatic movements of a carbonate platform may further confound simple interpretation of sea-level positions from the elevation of flank margin caves. A final limitation is that mixing corrosion may develop anywhere between the top of the saline groundwater and the top of the meteoric lens, which adds significant vertical uncertainty (±10m) for relating flank margin cave to an exact sea level position (Mylroie and Carew, 1988). As a result, flank margin caves are predominantly used as supporting evidence for previously known sea levels (Proctor and Smart, 1991). Nevertheless, flank margin caves can provide strong evidence for Quaternary sea levels in conjunction with other proxies.

6.4 SPELEOTHEMS AS SEA-LEVEL INDICATORS

The potential of using speleothems to reconstruct former sea levels was recognized in the early 1970s along two research tracks, both of which yield sea-level index points. The first track used speleothem growth hiatuses to provide evidence for previous inundation of a CKB by sea-level rise, and the second track used littoral speleothems as direct sea-level indicators. For example, Spalding and Matthews (1972) used ^{14}C and U–Th methods to date a submerged stalagmite from Ben's Hole (Grand Bahama Island), and Ginés and Ginés (1974) described the encrustations from Cova de sa Bassa Blanca (Mallorca) and their relationship with the Pleistocene sea-level changes in the Mediterranean basin. Shortly thereafter, Harmon et al. (1978) used U-series methods to provide age estimates for submerged stalagmites from Bermuda to constrain relative sea level in the western Atlantic during the last interglacial. The advent of mass spectrometric methods for measuring U-series isotopes greatly improved the dating of speleothems (Edwards et al., 1987; Richards and Dorale, 2003), allowing Li et al. (1989) and Richards et al. (1994) to pinpoint maximum sea levels during the last glacial period. Speleothem sea-level proxies have since made prominent contributions to our understanding of Quaternary climate and ice sheet interactions because sea-level data can be recovered from tectonically stable regions (e.g., Bahamas, part of the Mediterranean Basin) where other evidence may be lacking (e.g., marine or reef terraces).

6.4.1 Speleothems that form in vadose environments

Speleothems primarily form in air-filled subterranean void spaces that are vadose environments. In karst regions, carbonic acid generated by the dissolution of carbon dioxide in the atmosphere and soil produce acidified waters that then cause the dissolution of carbonate terrains. During dissolution, these waters become supersaturated with respect to calcite or aragonite and, upon entering a subterranean void space such as a cave with a lower partial pressure of CO_2, the gravity-fed drips degas and precipitate of a variety of $CaCO_3$ mineral forms and phases, but predominantly calcite (Fig. 6.4). Where speleothems are formed from continuous periods of crystal growth, they become ideal archives for past climate and environmental conditions (see extensive review by Fairchild and Baker, 2012). Stalagmites are the most commonly used speleothems to provide unambiguous maximum sea-level constraints. However, other speleothems can also prove useful when stalagmite availability is limited (e.g., shelfstones, stalactites, and flowstones).

6.4.2 The switch from vadose to phreatic environments (littoral, anchialine, submarine)

Early researchers immediately recognized that speleothems stop growing when sea level inundates a coastal cave (Gascoyne et al., 1979; Li et al., 1989). These growth hiatuses signal the switch between vadose and phreatic environmental conditions in this particular location of a CKB, and so speleothems can potentially preserve multiple sea-level oscillations at a specific geographic locale (Fig. 6.5). During the inundation when the speleothem is submerged, a characteristic mineralogical signature may form on the speleothem surface from chemical or biological processes, such as: (1) the presence of corroded layers caused by dissolution at the halocline (Li et al., 1989; Surić et al., 2009); (2) layers of biogenic encrustations within stalagmites or only at their surface (Mylroie and Carew, 1988; Bard et al., 2002; Surić et al., 2005); (3) traces of marine boring organisms (Antonioli and Oliverio, 1996); (4) re-crystallized lamina; and (5) deposition of various trace elements or minerals from either oxic or anoxic

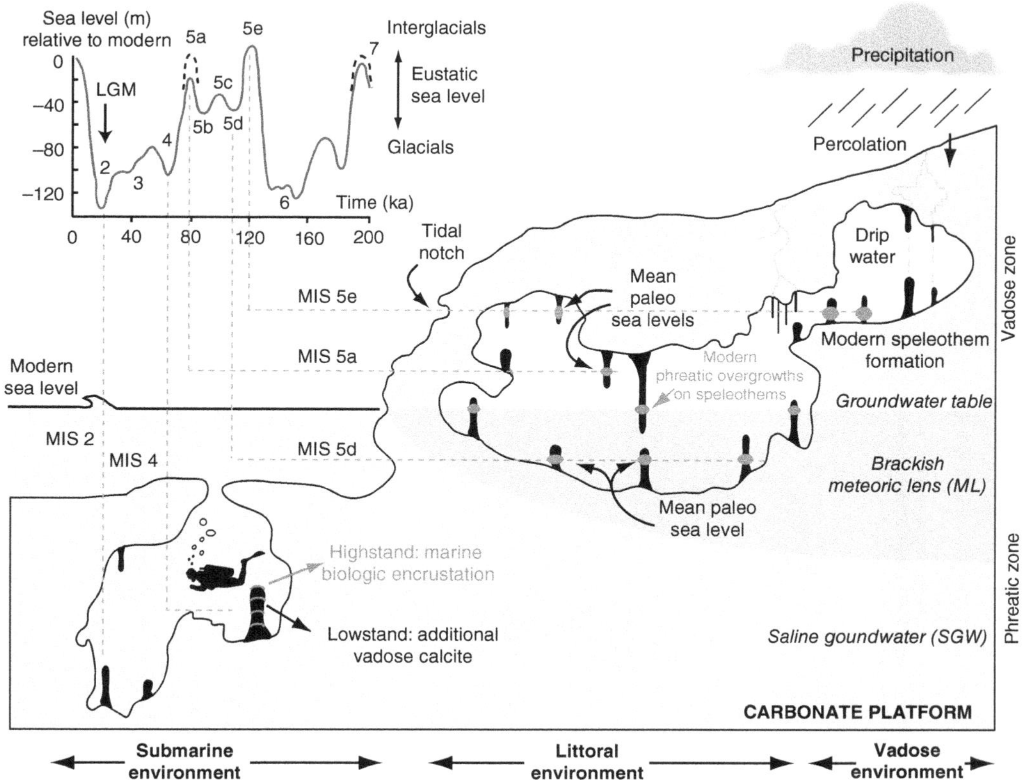

Fig. 6.4. The various speleothem sea-level indicators and their development in CKBs. All currently submerged speleothems have stopped growing, providing a maximum sea-level indicator at the time they stopped growing (Harmon et al., 1978; Richards et al., 1994), whereas marine biologic encrustations provide a minimum sea-level indicator (Dutton et al., 2009; Surić et al., 2009). In contrast, phreatic overgrowths on speleothems (POS) provide a sea-level index point, with the overgrowth itself providing a tide-range indicator during growth (Fornós et al., 2002). The environmental designations at the base of this conceptual model are only relative to the elevation of modern sea level as drawn. Note that the meteoric lens in this case is most likely of brackish salinity.

groundwater, such as halite, gypsum, hematite, goethite, lepidocrocite (Onac et al., 2001; Antonioli et al., 2004; Surić et al., 2009). These secondary features provide visual clues for detecting growth hiatuses in speleothems, which in turn can be used to decipher the sea-level history at a specific elevation in a CKB.

A limitation of using speleothem growth cessation as a sea-level indicator is that other climatic and hydrologic factors can stop speleothem growth. Examples of these factors include prolonged drought, permafrost/ice cover, or percolation of water into a cave that is undersaturated with respect to bicarbonate. Detailed petrographic study can help resolve the mechanism responsible for the growth hiatus, which can include changes in fabric, mineralogy, or sharp color changes from thin layers (millimeter to sub-millimeter) of detrital impurities.

Dating the initiation and cessation of speleothem growth and thereby delimiting the hiatus can be achieved by subsampling and dating the carbonate above and below the growth hiatus (Li et al., 1989; Lundberg and Ford, 1994; Hodge et al., 2008; Surić et al., 2009). The age below the hiatus will provide a maximum age estimate for when this location in the CKB was air-filled when sea level was positioned below the speleothem elevation, followed by a time gap from when sea level flooded the CKB with groundwater or seawater and the speleothem drowned (Fig. 6.5). When dating such material, numerous subsamples below the surface should be analyzed because the outermost material is most likely altered from being submerged for multiple millennia. The age after the hiatus marks speleothem re-growth during a subsequent sea-level fall (Li et al., 1989; Richards et al., 1994; Moseley et al., 2013), but this age must be considered a minimum age constraint for any sea-level fall because speleothems may not immediately resume growth after

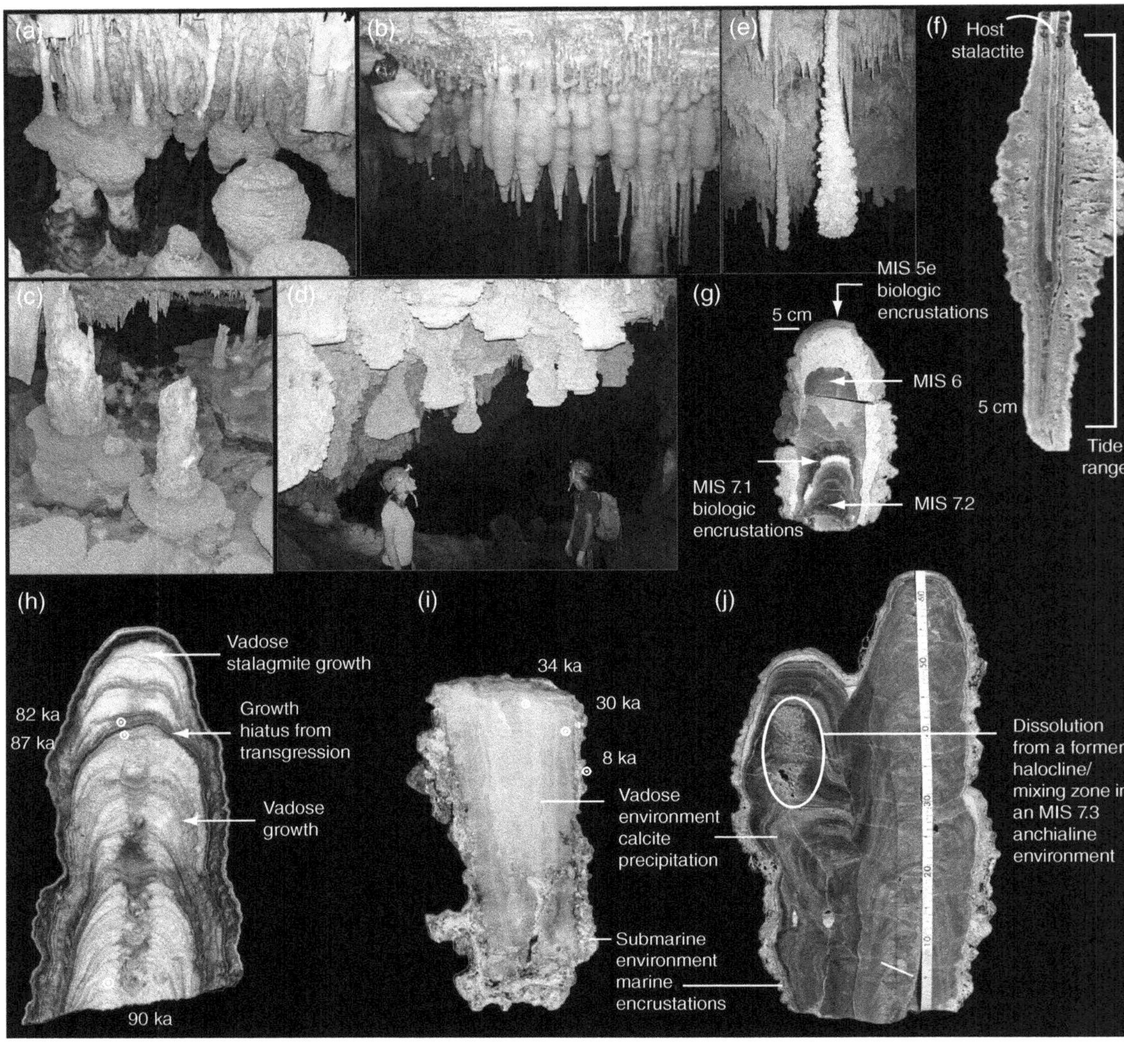

Fig. 6.5. Speleothem sea-level indicators. (a–c) Recent phreatic overgrowths on speleothems (POS) from Mallorcan littoral cave environments. (d, e) Pleistocene POS in Mallorca indicative of prior sea-level highstands. (f) POS developed on a stalactite where thickest point represents mean tide level. (g) Stalagmite ASI from Argenterola Cave (Italy), indicating two cycles of drowning by sea-level rise when the cave became a submarine environment suitable for encrusting marine invertebrates (Bard et al., 2002); *Source*: Photograph by Fabrizio Antonioli. Reproduced with permission. (h) Stalagmite K14 from U Vode Pit (Croatia) depicting a classic growth hiatus from a drowning event *Source*: Surić et al., 2009. Reproduced with permission of Elsevier. (i) Stalactite P-23 from Tilhovac Bay Cave (Croatia) depicting biologic encrustations as cave flooded by Holocene sea-level rise *Source*: Surić et al., 2005. Reproduced with permission of Elsevier. (j) Stalagmite ASN from Argentarola Cave (Italy) depicting typical vadose environment calcite precipitation, calcite dissolution from a halocline in an anchialine environment during MIS 7.3, and biologic encrustations developed in a subsequent submarine environment (Dutton et al., 2009). *Source*: Photograph by and reproduced with permission of Fabrizio Antonioli. MIS: Marine Isotope Stage. For color details, please see Plate 16.

the location becomes air-filled. It may take decades or millennia before the prerequisite climatic, land surface, and cave hydrogeological conditions for speleothems to resume growth, during which time local sea level may have fallen several decimeters or even meters. Dating the initiation of growth after a hiatus is challenging, particularly in slower-growing flowstones. Ideally, where researchers can take advantage of the sub-millimeter resolution afforded by MC-ICPMS methods (see Chapter 26), extrapolation to an initiation age can be closely estimated. Researchers should also be cautious of initial Th contamination when sampling close to hiatuses and consider isochron techniques that rely on multiple age estimates from a single layer of calcite growth exhibiting a range of Th concentrations.

In summary, speleothem growth hiatuses only indicate the timing of when a CKB transitions between vadose environments (speleothem growth) and phreatic environments (hiatus). A precise sea-level elevation cannot be resolved by a single speleothem alone because speleothem growth hiatuses do not have indicative meaning. As such, speleothem growth is a maximum sea-level indicator only, and their growth hiatuses are a minimum sea-level proxy. Nevertheless, multiple speleothems can be used together to provide tight vertical constraints on the absolute elevation of former sea levels (Richards et al., 1994).

6.4.3 Phreatic overgrowths on speleothems in littoral cave environments

A sea-level index point (SLIP) requires spatial geographic positioning, an elevation, an absolute age, and an indicative meaning (van de Plassche, 1986). Occasionally, the interaction between fresh groundwater and seawater can produce aqueous geochemical conditions that are favorable for speleothems to develop into SLIPs in littoral cave environments (Mylroie and Carew, 1990; Tuccimei et al., 2007). The geochemistry of these conditions is still not fully understood, but they cause a carbonate encrustation to precipitate at the water table around pre-existing structures such as stalagmites or stalactites that are partly immersed under water (Fig. 6.4). Structurally, the host speleothem (stalactite, stalagmite) is a primary depositional structure, whereas the encrusting carbonate is a secondary depositional structure; the latter are known collectively as phreatic overgrowths on speleothems (POS). We suggest avoiding the expression "marine overgrowths on speleothems" (Ford and Williams, 2007; Fairchild and Baker, 2012) because biogenic encrustations on speleothems occur when the CKB is a submarine environment, whereas POS are carbonates that are inorganically precipitated in brackish water lenses flooding a littoral cave environment.

The final morphology of POS depends on the size, shape, and extent of immersion of the host speleothem by the brackish meteoric lens when the POS formed (Fig. 6.5). For example, if only the tip of the stalactite is immersed by groundwater, this stalactite will become covered by an asymmetric POS that is a knob-like carbonate encrustation that develops at mean sea level and narrows upward to the highest tide range. For partially submerged stalagmites, the reverse holds with thinning towards the base. The most optimal POS form on speleothems that stay immersed throughout the full vertical tidal range, producing a spindle or fusiform-shaped POS, where again the widest section of the POS corresponds to the mean sea-level position and the tapered ends mark the tidal range.

In addition to POS formation, carbonate protrusions up to 25 cm in width may form continuous bands at the same elevation as sea level and POS (Fig. 6.5c). These appear as halved POS when sectioned along their growth axis. It is important that POS are not misidentified as the more ubiquitous speleothem known as shelfstone. Shelfstone forms laterally extensive mineral deposits several centimeters in thickness along cave walls, often in pools of water formed by the percolation of drip water (Hill and Forti, 1997). Indeed, shelfstones can record the water level of perched aquifers or vadose cave pools, but shelfstones are morphologically distinct from the POS.

Using modern analogs in Mallorcan littoral cave environments with a tidal range less than 50 cm, both types of overgrowths (POS and carbonate wall protrusion) were found precipitating in a brackish meteoric lens at, above, and a few centimeters below present sea level (Dorale et al., 2010). Considering that the caves hosting POS are proximal to the modern coastline, the water table of the caves is, and was in the past, coincident with sea level because of negligible hydraulic gradient between the cave site and the ocean. These results conclusively indicate that POS are sea-level index points (SLIPs) because they have indicative meaning: a quantified relationship with tides, a geographic position, and they can be dated by U-series methods.

The coastal caves of Mallorca provide the archetype of these SLIPs (Vesica et al., 2000; Dorale et al., 2010; Tuccimei et al., 2010), but they have also been found in Bermuda and elsewhere (Harmon et al., 1978; Grimes, 2001; Urushibara-Yoshino, 2003; Tuccimei et al., 2007). POS also exist at different depths within the flooded passages of Mallorca's coastal caves, thus providing evidence for prior sea-level highstands and lowstands with unprecedented vertical and temporal resolution (Dorale et al., 2010). The advantages of using POS for reconstructing sea level are straightforward and powerful: within a single cavity several sea-level stands can be documented from multiple SLIPs. POS can also provide a test of the tectonic stability of the area because POS ages and elevations can be compared to other geomorphologic and sedimentologic evidence for sea-level change. Perhaps one of the greatest advantages of POS is that they exist in a protected cave environment that is far removed from the processes that disrupt or remove other sea-level indicators from the subaerial geologic record.

6.4.4 The effect of anchialine environments on speleothems

The imprint of anchialine environmental conditions on speleothems is understood because modern analog studies on speleothems are lacking. However, the one diagnostic environmental feature of anchialine caves is the presence of a halocline or mixing zone in the CKB. As previously discussed, the mixing of groundwater creates caustic conditions to carbonate minerals, which can also impact speleothems. In a sample from Argentarola Cave in Italy, Dutton et al. (2009) documented how a paleohalocline or mixing zone partly dissolved a speleothem (Fig. 6.5). This sample was dated to Marine Isotope Stage 7.3, and precisely documents prior anchialine environmental conditions present in Argentarola Cave.

Petrographic analysis may provide further evidence of anchialine conditions in speleothems. In geologic samples from Evelyn's Bluehole in the Bahamas (not speleothems), Smart et al. (1988) documented the precipitation of FeS_2 at the base of the mixing zone, most likely from combined hydrogeologic and bacterial process. Most speleothems in Bermudian underwater caves are heavily coated in a dark-mineral precipitates, similar to the Fe and Mn oxide layers observed in speleothems from Argentarola Cave (Antonioli et al., 2004). Perhaps these oxide layers developed in favorable geochemical conditions in prior anchialine cave environments, but further research is needed to be definitive. However, evidence for anchialine environmental conditions in speleothems, whether mineral precipitates or dissolution in a halocline, will provide a minimum sea-level indicator only. Because sea-level must have been positioned above the halocline or mixing zone below a paleo-meteoric lens (Fig. 6.1), the preserved speleothem evidence will not have indicative meaning.

6.4.5 Biologic encrustations on speleothems in submarine cave environments

Submarine environments are created in CKBs when sea-level rise completely inundates a carbonate landscape and floods CKBs with seawater. When the physical geometry of a CKB allows open connectivity between the CKB and ocean, sessile marine organisms colonize the submerged hardgrounds and speleothems. Submerged speleothems from submarine cave environments often contain biogenic overgrowth crusts, such as serpulid worms and various boring organisms. The first to exploit this attribute for sea-level research was a team lead by Gascoyne, who used five stalagmites from a Bahamian bluehole to constrain a lowstand to at least 42 m bsl between 160 and 139 ka (Gascoyne et al., 1979). Later, Alessio et al. (1992) generated a sea-level curve over the past 40 ka based on ^{14}C ages of marine encrusting organisms on submerged speleothems from Italy. To date, the most detailed studies on stalagmites with sequences of calcite precipitation during lower sea levels interspaced with marine biogenic overgrowths corresponding to periods of higher sea levels were carried out in Argentarola Cave by Bard et al. (2002) and Antonioli et al. (2004). Several speleothems have been recovered from this cave between depths of 3.5 and 21.7 m bsl (Fig. 6.5). Particularly important was a stalagmite that contained five marine layers and four vadose calcite layers that preserved a 214 ka record of sea-level change. With two additional submerged speleothems, Dutton et al. (2009) documented three sea-level highstands between 245 and 190 ka. It must be remembered, however, that marine-encrusted layers in speleothems can only be used as a minimum sea-level indicator, unless indicative meaning can be established.

In another noteworthy example, Antonioli and Oliverio (1996) recovered a stalagmite from a depth of 48 m bsl in the Scaletta-Punta Lacco cave system (Capo Palinuro, Italy), where the marine encrusted layers provided some indicative meaning from a prior submarine cave environment. The recovered stalagmite was bored by the bivalve *Lithophaga lithophaga* (see Plate 31), and at several locations the shells were actually sealed-in by calcite when the cave returned to a vadose environment. *Lithophaga lithophaga* is considered an early colonizer of bare limestone mostly in the midlittoral tide (see section 18.4.2), indicating when the cave was inundated and submerged. Precise ^{14}C or U-series dating on such shells may provide supplementary data to help reconstruct local sea-level changes. In summary, the biologic encrustations on speleothems that form in submarine environments provide a minimum sea-level indicator only, but there may be some opportunity to derive indicative meaning for the speleothem if biologic sea-level indicators are also preserved.

6.5 CAVE SEDIMENTARY DEPOSITS AS SEA-LEVEL INDICATORS

Considerable advances have occurred in the last ten years on cave and sinkhole sedimentology, motivated by evidence that flooded caves and sinkholes contain significant records of paleohydrogeology (Alverez Zarikian et al., 2005; van Hengstum et al., 2010; van Hengstum and Scott, 2012), hurricane frequency (Gischler et al., 2008; Lane et al., 2011; van Hengstum et al., 2014; Denomee et al., 2014), oceanography (Kitamura et al., 2007; Yamamoto et al., 2010), and sea-level change (Gabriel et al., 2009; van Hengstum et al., 2011). Sediment records from sinkholes can even preserve evidence of terrestrial climate variability that can be correlated to deep-sea oceanographic climate records (Grimm et al., 1993).

The diversity of sediment records from CKBs currently available in the literature now allows for an initial attempt at a broader facies analysis in caves and sinkholes. These sediment records, when considered collectively, indicate that CKBs follow Walther's Law of Correlation of Facies, as translated by Middleton (1973): "...various deposits of the same facies areas are formed beside each other in [modern] space, though in cross section we see them lying on top of each other". Recently, Springer associated cave sedimentary deposits with the hydrography of the original cave depositional environments: vadose versus phreatic environments (Springer and Kite, 1997; Springer et al., 1997). This is in contrast to the more traditional approach of interpreting cave sediments based on their architectural and lithologic features alone. The key concept underpinning Springer's innovative facies analysis is that water levels intrinsically control cave sedimentary environments and their resultant facies.

This conceptual framework of phreatic versus vadose environments can be expanded in the coastal zone. This is because all coastal subterranean environments established by biologists (vadose, littoral, anchialine, submarine) are now vertically represented in stratigraphic succession, as per Walther's Law of Correlation of Facies (van Hengstum and Scott, 2011; van Hengstum et al., 2011). This is important for sea-level research because the sedimentary units in CKBs are not haphazardly organized. In contrast, they can be organized into discrete facies corresponding to specific depositional environments that are ultimately controlled by sea-level change. These four primary facies can be identified worldwide, and are described in further detail in the following sections.

6.5.1 Vadose facies

The vadose facies is often spatially discontinuous and heterolithic. This is because it is dominated by allochthonous sediment transported into the CKB, although some autochthonous sediment is also present. CKBs often have openings to the subaerial surface, sometimes referred to as karst windows, that allow wind, rivers, ice, gravity, or precipitation to transport terrestrial materials such as organics, animal remains, soils, and sediments into the CKB. In contrast, the autochthonous sediment remains that accumulate in CKBs are often related to cave weathering by-products or biogenic remains, such as clastic accumulations of limestone residuum (Springer et al., 1997) or bat guano (Wurster et al., 2008), respectively. These spatially diverse sediment sources promote discontinuous sedimentary units that can be challenging to interpret. However, all of these discontinuous units can be linked under a broad vadose facies because they are deposited when the CKB was a vadose environment.

Three of the most common sedimentary architectures of the vadose facies include: (1) sedimentary features and beds that are analogous to fluvial systems; (2) tallus cones or debris piles; and (3) unconformities related to non-deposition. Fluvial-style sedimentary architectures are created as meteoric water migrates through caves and sinkholes in order to reach the phreatic zone, thus creating the physical process to transport and rework sediments in the subsurface. The resultant sedimentary units often contain fluvial-like sedimentary beds such as thalweg and channel units (White, 2007). These fluvial-style units often preserve primary sedimentary features such as laminations or ripples, and their particle size distributions can reflect hydrodynamic conditions during sedimentation (Springer et al., 1997; Bosch and White, 2007). One important limitation is that these subterranean rivers are constrained by the cave itself, so older sedimentary deposits are often cannibalized to form newer deposits. Another common sedimentary architecture of the vadose facies is the vertical accumulation of sediment into a talus cone or debris pile. These features commonly develop from allochthonous sediment that is transported through a karst window and deposited under gravitational forcing in a subterranean void space. Slumping and poorly sorted sediments are common in these features. However, talus cones are known to provide significant paleoenvironmental information on karst landscapes because terrestrial plant and animal remains become systematically isolated from subaerial weathering and decay. Talus cones may even become interspaced by speleothem deposits that provide sufficient material for U-series radiometric dating (Hearty et al., 2004). Lastly, the lack of sedimentation in a CKB will inevitably cause an unconformity to develop in the stratigraphic record, which is of little value to any paleoenvironmental reconstruction. It remains important, however, to obtain diagnostic sedimentary structures or biologic remains to confirm an interpretation that a sedimentary deposit is indeed part of a vadose facies, such as mud cracks in a currently submerged Mallorcan cave (Fornós et al., 2009; Fig. 6.6a), or a Pleistocene owl roost in a modern Bahamian anchialine cave (Steadman et al., 2007). From a sea-level perspective, these sedimentary deposits do not have indicative meaning because they do not have a definitive relationship with sea level; they can only be used as a maximum sea-level indicator.

6.5.2 Littoral facies

The most important facies for sea-level research is the littoral facies. As sea-level rise causes the concomitant vertical migration of groundwater, CKBs become partially flooded and the new littoral conditions initiate distinct sedimentary processes and permanent aquatic ecosystems. Sediments comprising the littoral facies may also be allochthonous or autochthonous because material from the adjacent subaerial surface or coastal environments can still become transported into the CKB. The most important environmental difference is that CKBs with standing water have different opportunities for *in situ* sedimentary processes that can potentially be informative for sea level. Based on current research, it is best to differentiate the littoral facies that can accumulate in caves versus subaerial sinkholes because sinkholes can provide a habitat for coastal vegetation.

Freshwater marshes and brackish mangrove swamps can develop when littoral environmental conditions are present in sinkholes, at both inland (Watts, 1969; Bates, 1995) and coastal settings (Gabriel et al., 2009; Lane et al., 2011). As sea-level and groundwater rise, brackish water can begin to flood coastal sinkholes and create ideal habitat for mangroves such as *Rhizophora* and *Avicennia* (Fig. 6.6b). In a modern analog setting, mangroves can be observed in littoral conditions around the periphery of Whiskey Sinkhole in Bermuda or Yax Chen Cenote in the Yucatan Peninsula, Mexico. It is therefore not surprising that mangrove peat can be recovered from the successions in flooded sinkholes on carbonate terrain (Gabriel et al., 2009). When these basal brackish peats from sinkholes are directly above a carbonate hardground, such as the sinkhole bottom, they can be used as effective relative sea-level indicators because they had to develop within the tidal limits of the mangrove species giving rise to the peat deposit (Milne and Peros, 2013).

However, the simple act of flooding a sinkhole by sea-level rise does not necessarily produce a well-developed littoral facies in the stratigraphic record. For example, Holocene successions recovered from Little Salt Spring, Florida (11 m length, dated to 12.2 ka), were recovered from a depth of 70 m below groundwater level. Assuming 60–70 m of sea-level lowering prior to 12.2 ka (Siddall et al., 2003), transient littoral conditions which were part of the Holocene prehistory of Little Salt Spring could be anticipated, yet no littoral facies is preserved. It is also prudent to take a multiproxy approach and use microfossils (e.g., benthic foraminifera, pollen) or sedimentary geochemistry (e.g., $\delta^{13}C_{org}$) to help provide diagnostic evidence of brackish or marine conditions (see Chapters 13 and 19). For example, Mullet Pond is another sinkhole basin in Florida where Holocene infilling successions have been recovered (Lane et al., 2011), but microfossil analysis indicated that the basal littoral facies was freshwater peat. Lastly, the organic matter comprising the basal peat deposit in Runway Sinkhole (The Bahamas) was most likely derived from an adjacent mangrove swamp, not necessarily *in situ* (Kovacs et al., 2013). Although the environmental changes in Runway Sinkhole appear broadly related to regional sea-level deceleration in the late Holocene, the basal peats themselves could not be used as strict

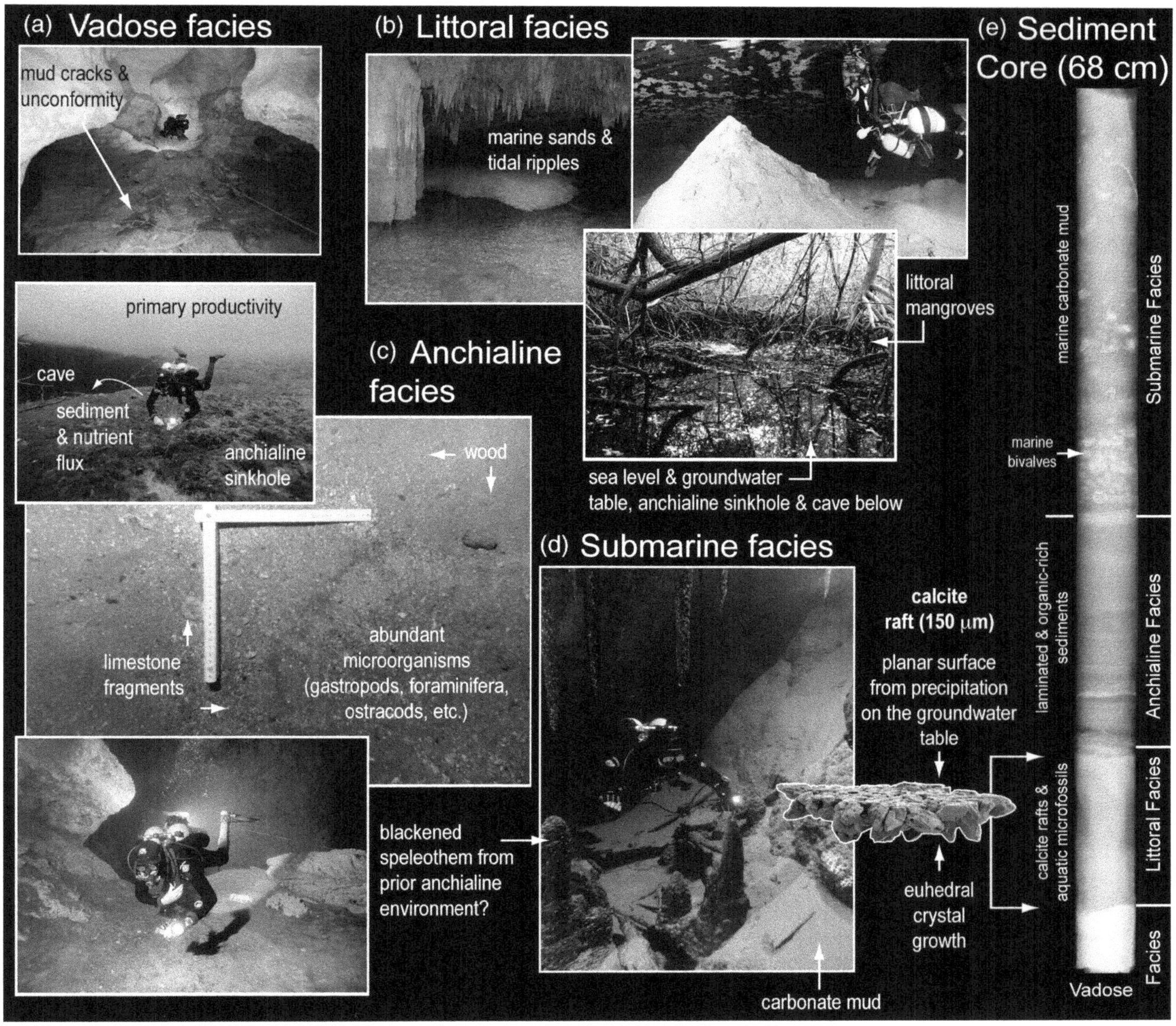

Fig. 6.6. Sediments and facies in CKBs. (a) Mud cracks on sediments that were originally deposited in an underwater cave during a previous vadose environment, and a subsequent unconformity from non-deposition (Mallorca, photograph by Joan Fornos, used with permission). (b) Sediment deposited in littoral caves that are open to the ocean can be primarily marine in character and can even have tidal indicators such as ripples (Bermuda, photograph by Peter van Hengstum), but calcite rafts can precipitate at the water table in isolated littoral cave environments that are not physically connected to the ocean (Mallorca, photograph by Joan Fornos, used with permission). In sinkholes, mangroves commonly grow at sea level (Whiskey Sinkhole, Bermuda). (c) Anchialine facies often comprise both allochthonous sediments eroding into sinkholes and caves, as well as autochthonous sediments from *in situ* cave processes (Mexico and Bermuda, photographs by Daniel Riordan and Peter van Hengstum, used with permission). (d) Carbonate mud and marine invertebrates are common in submarine environments that are well-circulated with the ocean. (e) A sediment core documenting the succession of Green Bay Cave (GBC5) through all four CKB environments in response to Holocene sea-level rise (van Hengstum et al., 2011). *Source*: Photographs by Peter van Hengstum. For color details, please see Plate 17.

sea-level indicators. These examples illustrate that basal brackish peats with indicative meaning can be preserved in sinkholes, but prudence is required in ensuring it is not basal freshwater peat (note that non-deposition in the sinkhole may equally preclude any peat preservation).

In contrast to sinkholes, the isolation of caves promotes different sedimentary styles that create different littoral facies. Two sedimentary units are currently known to comprise the littoral facies: calcite rafts and micrite. In partly flooded caves, calcite rafts are a specific speleothem that form at the air–water interface from the off-gassing of CO_2 from $CaCO_3$-saturated groundwater (Jones, 1989; Taylor and Chafetz, 2004; Figs 6.6b and 6.7). The resultant mineral morphology is distinct: a planar surface characterizes the mineral face at the air–water interface while the opposite side has developed euhedral crystal growth (Jones, 1989; van Hengstum et al., 2011). When the force of gravity surpasses the buoyant force supporting the calcite raft, the mineral sinks to the bottom of the water flooding the cave and becomes part of the sediment record (Taylor and Chafetz, 2004; Taylor et al., 2004). One of the first examples of using calcite rafts as a groundwater level proxy was by Auler and Smart (2001) in an inland Brazilian cave system. However, calcite rafts can also form extensive sedimentary deposits in modern littoral cave systems (Fornós et al., 2009). In the stratigraphic record, van Hengstum et al. (2011) used calcite rafts as a sea-level indicator from the successions in Green Bay Cave, Bermuda, which provided previously unavailable sea-level information from the middle Holocene in Bermuda. The indicative meaning is that calcite rafts will form anywhere within the full tide range of a given region. However, they can also continue to form as water level rises in the cave far above a reference level. In order to be used as a sea-level indicator, sediments containing calcite rafts must therefore be accumulating on a non-compactable surface with associated evidence that is diagnostic of brackish conditions (e.g., microfossils).

The other cave sedimentary unit associated with sea level is the deposition of fine-grained carbonate muds known as micrite (grain sizes <8 μm). The mixing of fresh and saline groundwater can cause micrite to precipitate directly from the water column (Jones, 1989; Lin et al., 2012). This process can be currently observed in the littoral zone of Cow Cave, Bermuda. In a late Pleistocene example, lithified micrite can be observed at 34 m bsl in Dan's Cave, The Bahamas, most likely precipitated in a paleo-mixing zone. In Green Bay Cave, Bermuda, van Hengstum et al. (2011) documented a spatially extensive deposit of calcite rafts in a sedimentary matrix of micrite. However, if micrite units are not preserved with other sedimentary evidence, they can only be used as a minimum sea-level indicator because the mixing zone can be located anywhere below sea level. As with other sedimentary sea-level proxies, micrite directly

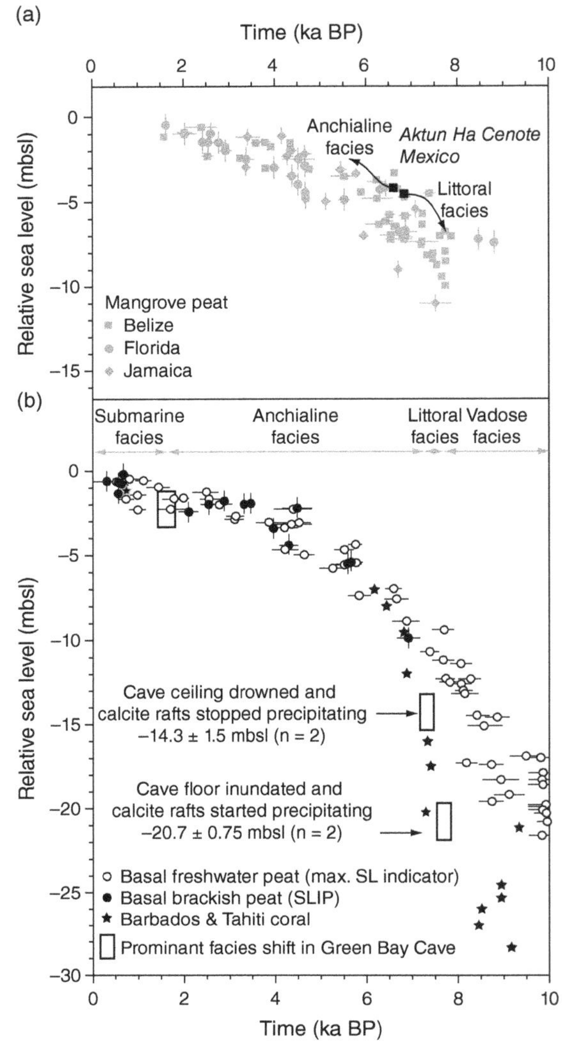

Fig. 6.7. A comparison of Holocene sea-level trends in the Caribbean and Atlantic regions and facies changes in CKBs using data which may need further correcting (<2 m) in response to local glacioeustatic adjustment (GIA). (a) Caribbean basal brackish peat data *Source*: data from Toscano and McIntyre, 2003. compared to brackish peat from a littoral facies preserved in Aktun Ha Cenote, Mexico. However, the brackish peat from Aktun Ha is only a maximum sea-level indicator because it was not recovered from directly above a hard ground. (b) Holocene sea-level data in Bermuda from basal freshwater peat and basal brackish peat compared to facies shifts in Green Bay Cave (see GBC5 in Fig. 6.6), Bermuda. Environmental conditions in the cave naturally progress as sea level inundates the cave basin. The first initiation of calcite rafts and micrite from the littoral facies at 7.7 ka provides a sea-level indicator at 20.7 m bsl (van Hengstum et al., 2011). The data from Bermuda are compared to coral sea-level index points (±5 vertical uncertainty) from Barbados and Tahiti (Bard et al., 1996; Peltier and Fairbanks, 2006).

deposited above a lithologic hardground will provide the most accurate minimum sea-level information. With both calcite rafts and micrite muds, it would be prudent to obtain multiple ages on a deposit from the same cave by radiocarbon dating material (e.g., shells, twigs) deposited contemporaneously with the micrite (van Hengstum et al., 2011). This is because the inherent assumption in the technique is that the earliest appearance of calcite rafts or micrite are indicative of when the tidal range immediately flooded the sample locale, an assumption that should be rigorously tested with multiple data points.

6.5.3 Anchialine facies

Anchialine environments develop when littoral environmental conditions are no longer the primary environmental forcing in a CKB, which typically result from groundwater fully submerging the CKB (Fig. 6.1). The anchialine facies can only ever be used as a minimum sea-level indicator because the position of sea level must be located above the elevation of the accumulating stratigraphy. Heterolithic sedimentary units also characterize the anchialine facies in the same way as the littoral facies, which both contain terrestrial and aquatic sedimentary constituents. Anchialine environments are often characterized by pond-like sinkholes interconnected by flowing groundwater to submerged anchialine caves (Fig. 6.6c). This creates a strong source-effect because sediments, organic matter, and nutrients from pond-like sinkholes are transported into adjacent cave passages. The result is that the anchialine facies often laterally attenuates in both volume and particle size distribution with increasing distance away from sediment sources. For example, coarse-grained sediments such as terrestrial organics enter the CKB through a point source (e.g., fissure or sinkhole) and create debris piles of poorly sorted diamict. However, flowing groundwater causes these diamict units to laterally intragrade into fine-grained slackwater units more distally into the cave (van Hengstum et al., 2011). Laminated or massive deposits of fine silt are common sedimentary deposits far removed from the point source, where the sediment has been laterally transported by subterranean hydrographic currents until eventually settling out under gravitational forcing.

Sedimentary material from aquatic invertebrates are also common in the anchialine facies such as bivalves, crustaceans (e.g., ostracods), biogenic silica from diatoms or sponges, and protists (e.g., foraminifera, testate amoeba). These biologic remains provide critical information for the original environmental conditions during deposition of the anchialine facies because the abundance, distribution, and biodiversity of aquatic invertebrates are primarily controlled by groundwater salinity and oxygenation. For example, foraminifera and testate amoebae have been found to populate CKBs according to similar salinity gradients that are found at other coastal environments (van Hengstum et al., 2008). Nutrient availability can also control aquatic invertebrates in anchialine environments, because the lack of sunlight in underwater caves hampers primary productivity and, indeed, the entire ecosystem. The most useful features of the aquatic invertebrates for sea-level research is that they can provide evidence for the salinity regime under which the sedimentary unit was deposited and they can also be radiocarbon dated. The benthic foraminfer *Physalida simplex* has been found in many North Atlantic anchialine caves and sinkholes in both modern settings and stratigraphic succession, a taxon tolerant to wide salinity ranges and dissolved oxygen concentrations (van Hengstum and Scott, 2012). The anchialine facies can then be used as a minimum sea-level indicator, because the position of sea level must have been above the cave itself. In two late Pleistocene examples, sediments containing benthic foraminifera attributable to a prehistoric anchialine cave environment were used as minimum sea-level indicators in Mexico (van Hengstum et al., 2009) and southwest England (Proctor and Smart, 1991).

6.5.4 Submarine facies

Submarine environments in CKBs develop when sea level completely inundates a carbonate platform and marine processes become the primary environmental forcing within the CKB itself. This process has been occurring throughout geologic time, which is evidenced by the discovery of Jurassic-aged (201–145 Ma) submarine cave deposits (Taylor and Palmer, 1994; Aubrecht and Schlögl, 2011). The submarine facies are both spatially extensive and laterally diverse sedimentary units, and they predominantly comprise carbonate sediments. If a CKB is located offshore, the submarine facies may be completely devoid of terrestrial material. Marine fauna can be abundant in submarine environments of CKBs, but their distribution and abundance is typically linked to the nutrients transported into CKBs by tidal currents or sunlight penetration (Zabala et al., 1989; Fichez, 1990). Biologic sedimentary constituents in the submarine facies can include bivalves, sponge spicules, foraminifera, ostracods, or any other biogenic sediments transported into the cave from adjacent marine environments (Kitamura et al., 2007; van Hengstum et al., 2011). Both allochthonous and autochthonous sediments of primarily marine origin therefore dominate the submarine facies, but some terrestrially derived sediments may be present if a CKB is proximal to the coastline.

Regional oceanography (e.g., tides, currents, seawater geochemistry) also influences the sedimentary architecture and structures observed in the submarine facies. For example, ooids can be abundant in CKBs that are dominated by tidal currents and seawater supersaturated in aragonite (e.g., Reel Breaker Cave, Abaco Island, The Bahamas). Deep offshore sinkholes, also known as blueholes, are typically oxygen-stratified with anoxic seawater flooding the sediment–water interface. In such settings, long-term quiescent intervals can be punctuated by storm activity along the coastline to create alternating laminations of coarse-grained allochthonous sediment and fined-grained non-bioturbated autochthonous sediment (Gischler et al., 2008; Denomee et al., 2013; van Hengstum et al., 2014). Tidal currents at the entrance to submarine caves also cause sediment sorting, where fines are transported deeper into the cave or elsewhere. For example, in a Japanese submarine cave, winnowing by tidal currents caused coarse-grained deposits to accumulate at the cave entrance, yet only fine-grained carbonate sediment accumulated more distally into the system (Omori et al., 2010). Fine-grained carbonate sedimentation is also widespread throughout Bermuda's submarine cave systems (Fig. 6.6d). Importantly, these sedimentary process and particle size distributions have been observed both in the modern setting and distinguished in the stratigraphic record in the same system (van Hengstum and Scott, 2011; van Hengstum et al., 2011). In some cases, oscillating seawater under tidal forcing can even create tidal ripples at the sediment–water interface. Similarly to the anchialine facies, dating basal sediments attributable to a submarine facies can only provide a minimum sea-level indicator because sea level was positioned above the cave level during its deposition.

6.5.5 Limitations

Perhaps the greatest limitation for using CKB sedimentary facies as sea-level indicators is that these environments naturally promote spatially discontinuous sedimentary units. In contrast to other coastal environments, CKB sediments are predominantly allochthonous and enter the system through point sources. This "point source effect" naturally promotes areas with very low to nonexistent sedimentation rates, in turn causing time-averaging and unconformities in the stratigraphic record. The rate and magnitude of groundwater flow further complicates sedimentation, which causes lateral sediment sorting and possible reworking. These lateral depositional characteristics mean that not all CKBs have sediment records suitable for reconstructing environmental change (Osborne, 1984). Anoxic groundwater conditions can further hamper efforts to confidently interpret CKB facies by reducing the abundance of paleobiological proxies in the sedimentary record. These limitations indicate that several sediment cores should be investigated with a multi-proxy approach in order to confidently reconstruct the environmental history of a CKB.

6.6 SUMMARY

In conclusion, sea-level change drives environmental change in CKBs worldwide, which in turn forces CKBs to transition through predictable environmental states (vadose, littoral, anchialine, and submarine). Figure 6.8 summarizes the different environments that develop in CKBs, and organizes their sea-level indicators according to the different environments in which they develop. The primary issue for researchers is distinguishing between the speleothems or sedimentary facies that constrain maximum or minimum sea-level elevations, versus those that have a more defined indicative meaning. Researchers working with important speleothems and cores samples should take care to archive and preserve such samples for possible reanalysis in the future by more advanced methodologies, as there is little doubt CKBs will continue to refine our understanding of Quaternary sea levels.

Coastal karst basin environments	Groundwater conditions	Cave SL indicators	Sedimentary deposits	Speleothems & effects	Proxy value
Vadose	Vadose zone • No groundwater		Vadose facies • No aquatic microfossils • Subaerial sedimentary structures	Speleothem precipitation (e.g., stalagmite, stalactite) ↓	***Maximum SL indicator***
Littoral	Intertidal • Meteoric lens • Mixing zone	Flank margin caves	Littoral facies • Calcite rafts, micrite • Cave fauna	Calcite raft precipitation littoral phreatic overgrowths on speleothems (POS)	***SL index point***
Anchialine	Phreatic zone • Meteoric lens • Mixing zone, or halocline • Saline groundwater	Flank margin caves	Anchialine facies • Organic-rich sediments • Brackish or O_2-stressed fauna	Dissolution, corrosion non-calcite mineral precipitates ↑	***Minimum SL indicator***
Submarine	Phreatic zone • SGW only		Submarine facies • Marine sedimentation • Marine fauna	Biogenic overgrowths (e.g., serpulid worm tubes) ↑	

Fig. 6.8. The transition of environments in coastal karst basins (CKBs) is driven by sea-level forcing, which in turn alters their internal hydrography, sedimentary processes, and potential for speleothems to precipitate or change. The vertical uncertainty on sea-level index points (SLIPs) from littoral environments is controlled by the local tidal ranges and local hydraulic gradients, and their horizontal uncertainty is associated with radiometric dating techniques.

ACKNOWLEDGEMENTS

The authors gratefully acknowledge technical cave divers worldwide whose primary observations and curiosity have helped provide academics with critical geological samples and information for over 30 years. Furthermore, many of the ideas presented here are the amalgamation of conceptual threads that both speleothem, stratigraphic, and biological researchers have been converging upon for years, and we have benefited from communicating with many global karst researchers. We thank Fabrizio Antonioli, Edouard Bard, Andrea Dutton, Joan Fornós, John Mylroie, Eduard Reinhardt, Danny Riordan, Maša Surić, and Bruce Williams for providing images that significantly enhanced the illustrations.

REFERENCES

Agassiz, A. (1894) A reconnaissance of the Bahamas and elevated reefs of Cuba in the steam yacht "Wild Duck", January to April 1893. Bulletin of the Museum of Comparative Zoology at Harvard College, 26, 1–203.

Alessio, M., Allegri, L., Antonioli, F., Belluomini, G., Ferranti, L., Improta, S., Manfra, L., and Proposito, A. (1992) Risultati preliminari relativi alla datazione di speleotemi sommersi nelle fasce costiere del Tirreno centrale. Giornale di Geologia, 54, 165–194.

Alverez Zarikian, C.A., Swart, P.K., Gifford, J.A., and Blackwelder, P.L. (2005) Holocene paleohydrology of Little Salt Spring, Florida, based on ostracod assemblages and stable isotopes. Palaeogeography, Palaeoclimatology, Palaeoecology, 225, 134–156.

Antonioli, F., and Oliverio, M. (1996) Holocene sea-level rise recorded by fossil mussel shells in submerged speleothem in the Mediterranean sea. Quaternary Research, 45, 241–244.

Antonioli, F., Bard, E., Silenzi, S., Potter, E.K., and Improta, S. (2004) 215-kyr history of sea-level oscillations from marine and continental layers in Argenterola Cave speleothems. Global and Planetary Change, 43, 57–68.

Aubrecht, R., and Schlögl, J. (2011) Jurassic submarine troglobites: is there any link to the recent submarine cave fauna? Hydrobiologia, 677, 3–11.

Auler, A.S., and Smart, P.L. (2001) Late Quaternary paleoclimate in semiarid northeastern Brazil from U-series dating of travertine and water-table speleothems. Quaternary Research, 55, 159–167.

Back, W., Hanshaw, B.B., Herman, J.S., and Van Driel, J.N. (1986) Differential dissolution of a Pleistocene reef in the ground-water mixing zone of a coastal Yucatan, Mexico. Geology, 14, 137–140.

Bard, E., Hamelin, B., Arnold, M., Montaggioni, L., Cabioch, G., Faure, G., and Rougerie, F. (1996) Deglacial sea-level record from Tahiti corals and the timing of global meltwater discharge. Nature, 382, 241–244.

Bard, E., Antonioli, F., and Silenzi, S. (2002) Sea-level during the penultimate interglacial period based on submerged stalagmite from Argentarola Cave (Italy). Earth and Planetary Science Letters, 196, 135–146.

Bates, A.L., Spiker, E.C., Hatcher, P.G., Stout, S.A., and Weintraub, V.C. (1995) Sulfer geochemistry of organic

rich sediments from Mud Lake, Florida, USA. Chemical Geology, 121, 245–262.
Beddows, P.A., Smart, P.L., Whitaker, F.F., and Smith, S.L. (2002) Density stratified groundwater circulation on the Caribbean coast of Yucatan Peninsula, Mexico. In *Hydrogeology and Biology of Post-Paleozoic Carbonate Aquifers* (eds Martin, J.B., Wicks, C.M., and Sasowsky, I.D.), Karst Waters Institute, Leesburg, Virginia, pp. 129–134.
Beddows, P.A., Smart, P.L., Whitaker, F.F., and Smith, S.L. (2007) Decoupled fresh-saline groundwater circulation of a coastal carbonate aquifer: spatial patterns of temperature and specific electrical conductivity. Journal of Hydrology, 346, 18–32.
Bosch, R.F., and White, B. (2007) Lithofacies and transport of clastic sediments in karstic aquifers. In *Studies of Cave Sediments* (eds Sasowsky, I.D., and Mylroie, J.), Kluwer Academic Publishers, pp. 1–22.
Cate, J.R. (2009) Assessing the impact of groundwater pollution from marine caves on nearshore seagrass beds in Bermuda. MSc thesis, Texas A&M, Galveston.
Cvijić, J. (1918) Hydrographie souterraine et evolution morphologique du karst. Recueil des Travaux de l'Institute de Graphie Alpine, 4, 375–426.
Davis, W.M. (1931) The origin of limestone caverns. Science, 73, 327–331.
Denomee, S.J., Bentley, S.J., and Droxler, A.W. (2014) Climatic controls on hurricane patterns: a 1200-y near-annual record from Lighthouse Reef, Belize. Scientific Reports, 4, article no. 3876, doi: 10.1038/srep03876.
Dorale, J.A., Onac, B.P., Fornós, J.J., Ginés, J., Ginés, A., Tuccimei, P., and Peate, D.W. (2010) Sea-level highstand 81,000 years ago in Mallorca. Science, 327, 860–863.
Dutton, A., Bard, E., Antonioli, F., Esat, T.M., Lambeck, K., and McCulloch, M.T. (2009) Phasing and amplitude of sea-level and climate change during the penultimate interglacial. Nature Geoscience, 2, 355–359.
Edwards, R.L., Chen, J.H., and Wasserburg, G.J. (1987) ^{238}U-^{234}U-^{230}Th-^{232}Th systematics and the precise measurement of time over the past 500,000 years. Earth and Planetary Science Letters, 81, 175–192.
Fairchild, I.J., and Baker, A. (2012) *Speleothem Science: From Processes to Past Environments.* John Wiley & Sons.
Fichez, R. (1990) Decrease in allochthonous organic inputs in dark submarine caves, connection with lowering in benthic community richness. Hydrobiologia, 207, 61–69.
Florea, L.J., Vacher, H.L., Donahue, B., and Naar, D. (2007) Quaternary cave levels in peninsular Florida. Quaternary Science Reviews, 26, 1344–1361.
Ford, D.C., and Williams, P.W. (2007) *Karst Hydrogeology and Geomorphology.* John Wiley & Sons.
Fornós, J.J., Gelabert, B., Ginés, A., Ginés, J., Tuccimei, P., and Vesica, P. (2002) Phreatic overgrowths on speleothems: a useful tool in structural geology in littoral karstic landscapes. The example of eastern Mallorca (Balearic Islands). Geodinamica Acta, 15, 113–125.
Fornós, J.J., Ginés, J., and Gràcia, F. (2009) Present-day sedimentary facies in the coastal karst caves of Mallorca island (western Mediterranean). Journal of Cave and Karst Studies. 71, 86–99.
Frank, E., Mylroie, J., Troester, J., Alexander, E.C.J., and Carew, J. (1998) Karst development and speleogenesis, Isla de Mona, Puerto Rico. Journal of Cave and Karst Studies, 60, 73–83.
Gabriel, J.J., Reinhardt, E.G., Peros, M.C., Davidson, D.E., van Hengstum, P.J., and Beddows, P.A. (2009) Palaeoenvironmental evolution of Cenote Aktun Ha (Carwash) on the Yucatan Peninsula, Mexico and its response to Holocene sea-level rise. Journal of Paleolimnology, 42, 199–213.
Gascoyne, M., Benjamin, G.J., Schwarcz, H.P., and Ford, D.C. (1979) Sea-level lowering during the Illinoian Glaciation: evidence from a Bahama "Blue Hole". Science, 205, 806–808.
Ginés, A., and Ginés, J. (1974) Consideraciones sobre los mecanismos de fosilización de la Cova de sa Bassa Blanca y su paralelismo con las formaciones marinas del Cuaternario. Boletin de la Sociedad de Historia Natural de las Baleares, 19, 11–28.
Gischler, E., Shinn, E.A., Oschmann, W., Fiebig, J., and Buster, N.A. (2008) A 1500-year Holocene Caribbean climate archive from the Blue Hole, Lighthouse Reef, Belize. Journal of Coastal Resarch, 24, 1495–1505.
Graman, K.M., and Garey, J.R. (2005) The transition of a freshwater karst aquifer to an anoxic marine system. Estuaries, 28, 686–693.
Grimes, K.G. (2001) Karst features of Christmas Island (Indian Ocean). Helictite, 37, 41–58.
Grimm, E.C., Jacobson, G.L., Watts, W.A., Hansen, B.C.S., and Maasch, K.A. (1993) A 50,000-year record of climate oscillations from Florida and its temporal correlation with the Heinrich events. Science, 261, 198–200.
Grund, A. (1903) Die Karsthydrographie: Studien aus Westbosnien. Geographische Abhandungen, 7, 1–200.
Guilcher, A. (1988) *Coral Reef Geomorphology.* John Wiley & Sons.
Gulley, J.D., Martin, J.B., Moore, P.J., and Murphy, J. (2013) Formation of phreatic caves in an eogenetic karst aquifer by CO_2 enrichment at lower water tables and subsequent flooding by sea-level rise. Earth Surface Processes and Landforms, 38, 1210–1224.
Harmon, R.S., Schwarcz, H.P., and Ford, D.C. (1978) Late Pleistocene sea level history of Bermuda. Quaternary Research, 9, 205–218.
Harmon, R.S., Land, L.S., Mitterer, R.M., Garett, P., Schwarcz, H.P., and Larson, G.J. (1981) Bermuda sea level during the last interglacial. Nature, 289, 481–483.
Hearty, P.J., Olson, S.L., Kaufman, D.S., Edwards, R.L., and Cheng, H. (2004) Stratigraphy and geochronology of pitfall accumulations in caves and fissures, Bermuda. Quaternary Science Reviews, 23, 1151–1171.
Hill, C.A., and Forti, P. (1997) *Cave Minerals of the World,* 2nd edition. National Speleological Society, Huntsville, Alabama.
Hodge, E.J., Richards, D.A., Smart, P.L., Ginés, A., and Mattey, D.P. (2008) Sub-millennial climate shifts in the western Mediterranean during the last glacial period recorded in a speleothem from Mallorca, Spain. Journal of Quaternary Science, 23, 713–718.
Holthuis, L.B. (1973) Caridean shrimps found in land-locked saltwater pools at four Indo-west Pacific localities (Sinai peninsula, Funafuti atoll, Maui and Hawaii

islands), with the description of the new genus and four new species. Zoologishe Verhandelingen, 128, 1–48.

Iliffe, T.M. (1987) Observations on the biology and geology of anchialine caves. In *Proceedings of the Third Symposium on the Geology of the Bahamas* (ed. Curan, H.A.), pp. 73–80, CCFL Bahamian Field Station.

Iliffe, T.M., Hart, C.W.J., and Manning, R.B. (1983) Biogeography and the caves of Bermuda. Nature, 302, 141–142.

Jones, B. (1989) Calcite rafts, peloids, and micrite in cave deposits from Cayman Brae, British West Indies. Canadian Journal of Earth Sciences, 26, 654–664.

Kitamura, A., Yamamoto, N., Kase, T., Ohashi, S., Hiramoto, M., Fukusawa, H., Watanabe, T., Irino, T., Kojitani, H., Shimamura, M., and Kawakami, I. (2007) Potential of submarine-cave sediments and oxygen isotope composition of cavernicolous micro-bivalve as a late Holocene paleoenvironmental record. Global Planetary Change, 55, 301–316.

Kohout, F.A. (1965) A hypothesis concerning cyclic flow of salt water related to geothermic heating in the Floridian aquifer. Transactions of the New York Academy of Sciences, 2, 249–271.

Kovacs, S.E., van Hengstum, P.J., Reinhard, E.G., Donnelly, J.P., and Albury, N.A. (2013) Late Holocene sedimentation and hydrologic development in a shallow coastal sinkhole on Great Abaco Island, The Bahamas. Quaternary International, 317, 118–132.

Labourdette, R., Lascu, I., Mylroie, J., and Roth, M. (2007) Process-like modeling of flank-margin caves: from genesis to burial evolution. Journal of Sedimentary Research, 77, 965–979.

Lane, P., Donnelly, J.P., Woodruff, J.D., and Hawkes, A.D. (2011) A decadally-resolved paleohurricane record archived in the late Holocene sediments of a Florida sinkhole. Marine Geology, 287, 14–30.

Li, W.X., Lundberg, J., Dickin, A.P., Ford, D., Schwarcz, H.P., McNutt, R., and Williams, D. (1989) High-precision mass-spectrometric uranium-series dating of cave deposits and implications for palaeoclimate studies. Nature, 339, 534–536.

Lin, C.Y., Musta, B., and Abdullah, M.H. (2012) Geochemical processes, evidence and thermodynamic behavior of dissolved and precipitated carbonate minerals in a modern seawater/freshwater mixing zone of a small tropical island. Applied Geochemistry, 29, 13–31.

Lundberg, J., and Ford, D.C. (1994) Late Pleistocene sea level change in the Bahamas from mass spectrometric U-series dating of submerged speleothem. Quaternary Science Reviews, 13, 1–14.

Marin, L.E., and Perry, E.C. (1994) The hydrogeology and contaminant potential of northwestern Yucatan, Mexico. Geofis International, 33, 619–623.

Martin, J.B., Gulley, J., and Spellman, P. (2012) Tidal pumping of water between Bahamian blue holes, aquifers, and the ocean. Journal of Hydrology, 416–417, 28–38.

Middleton, G.V. (1973) Johanne Walther's law of the correlation of facies. Geological Society of America Bulletin, 84, 979–988.

Milne, G.A., and Peros, M. (2013) Data-model comparison of Holocene sea-level change in the circum-Caribbean region. Global Planetary Change, 107, 119–131.

Moore, Y.H., Sotessell, R.K., and Easley, D.H. (1992) Freshwater/sea-water relationship within a ground-water flow system, northeastern coast of the Yucatan Penisula. Ground Water, 30, 343–350.

Moseley, G.E., Smart, P.L., Richards, D.A., and Hoffmann, D.L. (2013) Speleothem constraints on marine isotope stage (MIS) 5 sea-levels, Yucatan Peninsula, Mexico. Journal of Quaternary Science, 28, 293–300.

Mylroie, J.E., and Carew, J.L. (1988) Solution conduits as indicators of late Quaternary sea level position. Quaternary Science Reviews, 7, 55–64.

Mylroie, J.E., and Carew, J.L. (1990) The flank margin model for dissolution cave development in carbonate platforms. Earth Surface Process Landforms, 15, 413–424.

Mylroie, J.E., and Carew, J.L. (1995) Geology and karst geomorphology on San Salvador Island, Bahamas. Carbonates and Evaporites, 10, 193–206.

Mylroie, J.R., and Mylroie, J.E. (2007) Development of the carbonate island karst model. Journal of Cave and Karst Studies, 69, 59–75.

Mylroie, J.E., and Mylroie, J.R. (2009) Caves as sea level and uplift indicators, Kangaroo Island, South Australia. Journal of Cave and Karst Studies, 71, 32–47.

Northrop, J.I. (1890) Notes on the geology of the Bahamas. Transactions of the New York Academy of Sciences, 10, 4–23.

Omori, A., Kitamura, A., Fujita, K., Honda, K., and Yamamoto, N. (2010) Reconstruction of light conditions within a submarine cave during the past 7000 years based on the temporal and spatial distribution of algal symbiont-bearing large benthic foraminifers. Palaeogeography, Palaeoclimatology, Palaeoecology, 292, 443–452.

Onac, B.P., Mylroie, J.E., and White, W.B. (2001) Mineralogy of cave deposits on San Salvador Island, Bahamas. Carbonates and Evaporites, 16, 8–16.

Osborne, R.A.L. (1984) Lateral facies changes, unconformities and stratigraphic reversals: their significance for cave sediment stratigraphy. Cave Science, 11, 175–184.

Peltier, W.R., and Fairbanks, R.G. (2006) Global glacial ice volume and Last Glacial Maximum duration from an extended Barbados sea level record. Quaternary Science Reviews, 25, 3322–3337.

Plummer, L.N. (1975) Mixing of sea water with calcium carbonate groundwater. In *Quantitative Studies in the Geological Sciences* (ed. Whillen, E.H.T.), Geological Society of America, pp. 219–236.

Pohlman, J.W., Iliffe, T.M., and Cifuentes, L.A. (1997) A stable isotope study of organic cycling and the ecology of an anchialine cave ecosystem. Marine Ecology Progress Series, 155, 17–27.

Polyak, V.J., McIntosh, W.C., Guven, N., and Provencio, P. (1998) Age and origin of Carlsbad Cavern and related caves from $^{40}Ar/^{39}Ar$ of alunite. Science, 279, 1919–1922.

Proctor, C.J. (1988) Sea-level related caves on Berry Head, South Devon. Cave Science, 15, 39–49.

Proctor, C.J., and Smart, P.L. (1991) A dated sediment record of Pleistocene transgressions on Berry Head, Southwest England. Journal of Quaternary Science, 6, 233–244.

Reilly, T.E., and Goodman, A.S. (1985) Quantative analysis of saltwater-freshwater relationships in groundwater systems: a historical perspective. Journal of Hydrology, 80, 125–160.

Richards, D.A., and Dorale, J.A. (2003) Uranium-series chronology and environmental applications of speleothems. In *Uranium-Series Geochemistry* (eds Bourdon, B., Henderson, G.M., and Lundstrom, C.C.), Mineralogical Society of America, Washington, DC, pp. 407–460.

Richards, D.A., Smart, P.L., and Edwards, R.L. (1994) Maximum sea levels for the last glacial period from U-series ages of submerged speleothems. Nature, 367, 357–360.

Siddall, M., Rohling, E.J., Almogi-Labin, A., Hemleben, C., Meischner, D., Schmelzer, I., and Smeed, D.A. (2003) Sea-level fluctuations during the last glacial cycle. Nature, 423, 853–858.

Sket, B., and Iliffe, T.M. (1980) Cave fauna of Bermuda. Internationale Revue der Gesmten Hydrobiologie, 65, 871–882.

Smart, P.L., Dawans, J.M., and Whitaker, F. (1988) Carbonate dissolution in a modern mixing zone. Nature, 355, 811–813.

Smart, P.L., Moseley, G.E., Richards, D.A., and Whitaker, F.F. (2008) Past high sea-stands and platform stablity: evidence from Conch Bar Cave, Middle Caicos. In *Developing Models and Analogs for Isolated Carbonate Platforms, Holocene and Pleistocene Carbonates of Caicos Platforms, British West Indies* (eds Morgan, W., and Harris, P.), Society of Economic Paleontologists and Mineralogists, Tulsa, Oklahoma, pp. 203–210.

Spalding, R.F., and Matthews, T.D. (1972) Submerged stalagmites from caves in the Bahamas: indicators of low sea level stand. Quaternary Research, 2, 470–472.

Springer, G.S., and Kite, J.S. (1997) River derived slackwater sediments in caves along Cheat River, West Virgina. Geomorphology, 18, 91–100.

Springer, G.S., Kite, J.S., and Schmidt, V.A. (1997) Cave sedimentation, genesis, and erosional history in the Cheat River Canyon, West Virginia. Geological Society of America Bulletin, 109, 542–532.

Steadman, D.W., Franz, R., Morgan, G.S., Albury, N.A., Kakuk, B., Broad, K., Franz, S.E., Tinker, K., Pateman, M.P., Lott, T.A., Jarzen, D.M., and Dilcher, D.L. (2007) Exceptionally well preserved late Quaternary plant and vertebrate fossils from a blue hole on Abaco, The Bahamas. Proceedings of the National Academy of Sciences, 104, 19897–19902.

Stock, J.H., Iliffe, T.M., and Williams, D. (1986) The concept "anchialine" reconsidered. Stygologia, 2, 90–92.

Surić, M., Juračić, M., Horvatinčić, N., and Bronić, I.K. (2005) Late-Pleistocene-Holocene sea-level rise and the pattern of coastal karst inundataion: records from submerged speleothems along the Eastern Adriatic Coast (Croatia). Marine Geology, 214, 163–175.

Surić, M., Richards, D.A., Hoffmann, D.L., Tibljaš, D., and Juračić, M. (2009) Sea-level change during MIS 5a based on submerged speleothems from the eastern Adriatic Sea (Croatia). Marine Geology, 262, 62–67.

Swinnerton, A.C. (1929) Changes in base-level indicated by caves in Kentucky and Bermuda. Geological Society of America Bulletin, 40, 194.

Taboroši, D., Jenson, J.W., and Mylroie, J.E. (2003) Zones of enhanced dissolution and associated cave morphology in an uplifted carbonate island aquifer, north Guam, Mariana Islands. Speleogenesis and Evolution of Karst Aquifers, 1, 16 p.

Taylor, M.P., Drysdale, R.N., and Carthew, K.D. (2004) The formation and environmental significance of calcite rafts in tropical tufa-depositing rivers in northern Australia. Sedimentology, 51, 1089–1101.

Taylor, P., and Palmer, T.J. (1994) Submarine caves in a Jurassic Reef (La Rochelle, France) and the evolution of cave biotas. Naturwissenchaften, 81, 357–360.

Taylor, P., and Chafetz, H.S. (2004) Floating rafts of calcite crystals in cave pools, central Texas, U.S.A.: crystal habit vs. saturation state. Journal of Sedimentary Research, 74, 328–341.

Toscano, M.A., and McIntyre, I.G. (2003) Corrected western Atlantic sea-level curve for the last 11,000 years based on calibrated ^{14}C dates from Acropora palmata framework and intertidal mangrove peat. Coral Reefs, 22, 257–270.

Tuccimei, P., Fornós, J.J., Ginés, A., Ginés, J., Gràcia, F., and Mucedda, M. (2007) Sea level change at Capo Caccia (NW Sardinia) and Mallorca (Balearic Islands) during oxygen isotope substage 5e, based on Th/U datings of phreatic overgrowths on speleothems. In *Monografies de la Societat d'Historia Natural de les Balears* (eds Pons, G.X., and Vicens, D.), Societat D'Història Natural de Les Balears, Palma de Mallorca, pp. 121–135.

Tuccimei, P., Soligo, M., Ginés, J., Ginés, A., Fornós, J., Kramers, J., and Villa, I.M. (2010) Constraining Holocene sea levels using U-Th ages of phreatic overgrowths on speleothems from coastal caves in Mallorca (Western Mediterranean). Earth Surface Processes and Landforms, 35, 783–790.

Urushibara-Yoshino, K. (2003) Karst terrain of raised coral islands, Minamidaito and Kikai in the Nansei Islands of Japan. Zeitschrift für Geomorphologie, N.F. Supplement, 131, 17–31.

Vacher, H.L. (1988) Dupuit-Ghyben-Herzberg analysis of strip island lenses. Geological Society of America Bulletin, 100, 580–591.

Vacher, H.L., and Wallis, T.N. (1992) Comparative hydrogeology of fresh-water lenses of Bermuda and Great Exuma Island, Bahamas. Ground Water, 30, 15–20.

Vacher, H.L., and Mylroie, J.E. (2002) Eogenetic karst from the perspective of an equivalent porous medium. Carbonates and Evaporites, 17, 182–196.

van de Plassche, O. (1986) *Sea-level Research: a Manual for the Collection and Evaluation of Data.* Geo Books, Norwich.

van Hengstum, P.J., and Scott, D.B. (2011) Ecology of foraminifera and habitat variability in an underwater cave: distinguishing anchialine versus submarine cave environments. Journal of Foraminiferal Research 41, 201–229.

van Hengstum, P.J., and Scott, D.B. (2012) Sea-level rise and coastal circulation controlled Holocene groundwater development and caused a meteoric lens to collapse 1600 years ago in Bermuda. Marine Micropaleonology, 90–91, 29–43.

van Hengstum, P.J., Reinhardt, E.G., Beddows, P.A., Huang, R.J., and Gabriel, J.J. (2008) Thecamoebians (testate amoebae) and foraminifera from three anchialine cenotes in Mexico: Low salinity (1.5 - 4.5 psu) faunal transitions. Journal of Foraminiferal Research, 38, 305–317.

van Hengstum, P.J., Reinhardt, E.G., Beddows, P.A., Schwarcz, H.P., and Garbriel, J.J. (2009) Foraminifera and testate amoebae (thecamoebians) in an anchialine cave: surface distributions from Aktun Ha (Carwash) cave system, Mexico. Limnology and Oceanography, 54, 391–396.

van Hengstum, P.J., Reinhardt, E.G., Beddows, P.A., and Gabriel, J.J. (2010) Investigating linkages between Holocene paleoclimate and paleohydrogeology preserved in Mexican underwater cave sediments. Quaternary Science Reviews, 29, 2788–2798.

van Hengstum, P.J., Scott, D.B., Gröcke, D.R., and Charette, M.A. (2011) Sea level controls sedimentation and environments in coastal caves and sinkholes. Marine Geology, 286, 35–50.

van Hengstum, P.J., Donnelly, J.P., Toomey, M.R., Albury, N.A., Lane, P., and Kakuk, B. (2014) Heightened hurricane activity on the Little Bahama Bank from 1350 to 1650 AD. Continental Shelf Research, 86, 103–115, doi: 10.1016/j.csr.2013.04.032.

Vesica, P., Tuccimei, P., Turi, B., Fornós, J.J., Ginés, J., and Ginés, A. (2000) Late Pleistocene paleoclimates and sea-level change in the mediterranean as inferred from stable isotope and U-series studies of over-growths on speleothems, Mallorca (Spain). Quaternary Science Reviews, 19, 865–879.

Watts, W.A. (1969) A pollen diagram from Mud Lake, Marion County, North-Central Florida. Geological Society of America Bulletin, 80, 631–642.

White, W.B. (2007) Cave sediments and paleoclimate. Journal of Cave and Karst Studies 69, 76–93.

Wigley, T.M.L., and Plummer, L.N. (1976) Mixing of carbonate waters. Geochimica et Cosmochimica Acta, 40, 989–995.

Wurster, C.M., Patterson, W.P., McFarlane, D.A., Wassenaar, L.I., Hobson, K.A., Beavan-Athfield, N., and Bird, M.I. (2008) Stable carbon and hydrogen isotopes from bat guano in the Grand Canyon, USA, reveal Younger Dryas and 8.2 ka events. Geology, 36, 683–686.

Yamamoto, N., Kitamura, A., Irino, T., Kase, T., and Ohashi, S. (2010) Climatic and hydrologic variability in the East China Sea during the last 7000 years based on oxygen isotopic records of the submarine caverniculous microbivalve Carditella iejimensis. Global Planetary Change, 72, 131–140.

Zabala, M., Riera, T., Gili, J.M., Barange, M., Lobo, A., and Penuelas, J. (1989) Water flow, trophic depletion, and benthic macrofauna impoverishment in a submarine cave from the western Mediterranean. Marine Ecology, 10, 271–287.

Chapter 7
Coral reefs

YUSUKE YOKOYAMA[1,2,3] AND TEZER M. ESAT[4,5]

[1]*Atmosphere and Ocean Research Institute, University of Tokyo, Chiba, Japan*
[2]*Department of Earth and Planetary Science, University of Tokyo, Tokyo, Japan*
[3]*Department of Biogeosciences, Japan Agency for Marine-Earth Science and Technology Organization, Yokosuka, Japan*
[4]*Australian Nuclear Science and Technology Organization, Institute for Environmental Research, Kirrawee, Australia*
[5]*Research School of Earth Sciences and Research School of Physical Sciences and Engineering, The Australian National University, Canberra, Australia*

7.1 INTRODUCTION

Variations in climate can directly influence global sea levels. However, changes in sea level are also dependent on other factors such as tectonics, isostasy, and the effects of gravity fields of massive transient ice sheets (Fig. 7.1; Yokoyama and Esat, 2011). Locally determined sea levels are referred to as "relative sea level" and can be derived from direct observation or through various geological, geomorphic, biological, and geochemical proxies (e.g., Yokoyama et al., 2012). Global or "absolute" sea levels can then be estimated through appropriate Earth and ice-sheet models that account for the other factors. From approximately over a million years ago, global climate extremes intensified and global sea levels fluctuated between two radical states corresponding to the presence or absence of huge ice sheets largely concentrated on Northern Hemisphere land masses (Fig. 7.2). These changes occurred with a period of about 100,000 years such that, for over 90% of the time, the planet was in the grip of an ice age with approximately 100 m lower sea levels. Warmer temperatures and high sea levels, similar to the present conditions, occurred for less than 10% of each cycle (Milankovitch, 1930). In this sense, "ice volume" is synonymous with climate and global sea levels.

Oxygen isotope measurements in foraminiferal tests from deep-sea sediment cores can provide a continuous record of past climate and sea-level changes (e.g., Lisiecki and Raymo, 2005). However, it is not possible to directly date these records to beyond the range of the radiocarbon method which extends back to about 50,000 years ago (Fig. 7.2). There are additional complications such as reservoir effects and up to 1000 years plus time lags due to variable global ocean circulation (e.g., Skinner and Shackleton, 2005, 2006; Esat and Yokoyama, 2008; Matsumoto and Yokoyama, 2013). During the formation of foraminiferal carbonate shells, additional oxygen isotope fractionation occurs depending on ambient temperatures. It can be as large as 20% of the full ice volume effect and further complicates the oxygen-isotope–sea-level correspondence.

At recent times the importance of directly datable sea-level records has come to the fore following new discoveries of widespread rapid fluctuations in climate and sea-level (Fig. 7.2). These are beyond the scope of the astronomical theory of insolation variations at the Earth's surface due to orbital fluctuations. Orbital timescales operate over many thousands of years, whereas the rapid climate and sea-level fluctuations are multidecadal (Yokoyama and Esat, 2011). The rapid climate and hence temperature fluctuations raises questions regarding the accuracy of temperature corrections to oxygen isotope records. The best candidates for the task of directly datable sea-level indicators are currently reef-building corals that concentrate seawater uranium in their skeletons by a factor of about 1000 and can therefore be dated by the Uranium-series method.

Attempts at converting foram oxygen isotope data directly to sea level have a long history. Gross sea-level variations have been derived over many glacial–interglacial intervals by comparison with the predictions of astronomical theory. Another

Handbook of Sea-Level Research, First Edition. Edited by Ian Shennan, Antony J. Long, and Benjamin P. Horton.

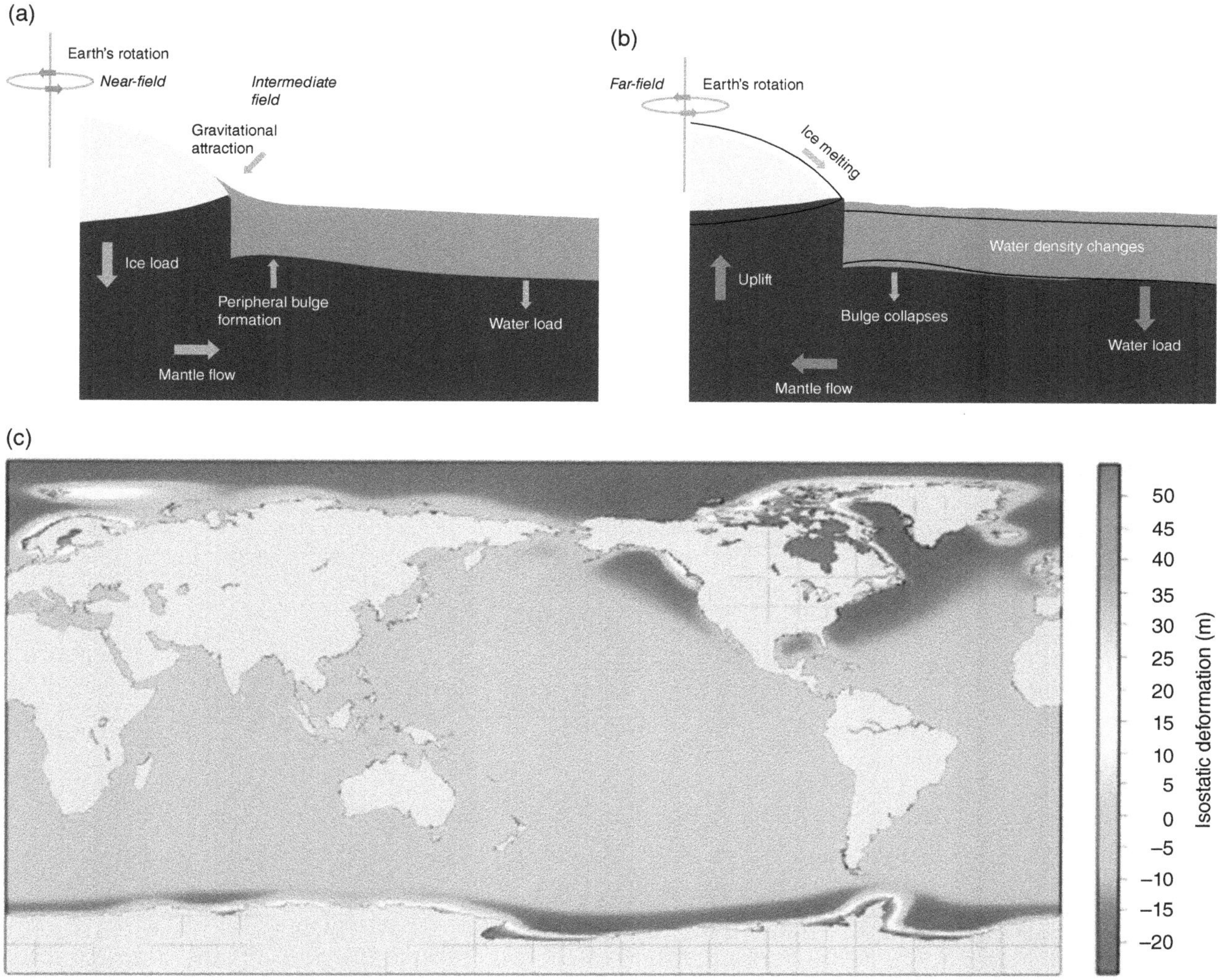

Fig. 7.1. Solid Earth deformation due to growth and decay of large ice sheets. (a) Near the large ice sheet during the ice growth phase, loading of the lithosphere deforms the mantle causing flow in mantle materials away from the center of the ice sheet. (b) Melting of the large ice mass produces extra water in the oceans causing a rise in sea-level. At far-field locations, where most of coral reefs are situated, the additional water loading in ocean basins tends to cause uplift of land masses. (Yokoyama & Esat, 2011. Reproduced with permission of The Oceanography Society). (c) Calculated sea-level change during the last ice age with the eustatic component removed (Yokoyama et al., 2012. Reproduced with permission of John Wiley & Sons). Low-latitude far-field sites are relatively unaffected by deformations induced by former ice sheets. Coral reefs, which respond to all of these changes, therefore provide a unique archive of paleo-sea levels. For color details, please see Plate 18.

method involves minimizing the dependence of oxygen isotopes on temperature by using benthic foraminifera on the basis that deep-sea temperatures tend to be stable (Fig. 7.2).

An interesting comparison between fossil-coral-derived sea levels and benthic oxygen-isotope sea levels was made using data from the uplifted coral terraces at Huon Peninsula, Papua New Guinea (PNG). The two sea-level curves appeared to deviate by as much as 40 m over the last glacial period from 60–30 ka years ago. It pointed to additional unknown complications in the derivation of oxygen-isotope-based ice volumes, such as the deep-sea temperature estimations, or equally to problems with U-series dates or to uncertainties in terrace uplift rates (Chappell et al., 1996). Later, detailed surveys of terrace heights and high-resolution Uranium-series dates resolved the conflict between the coral- and oxygen-isotope-based sea-level estimates and the records were reconciled. However, the utility of the oxygen isotope method over short timescales that also involve rapid ocean circulation changes is not proven. Directly dated sea-level indicators are therefore essential to better

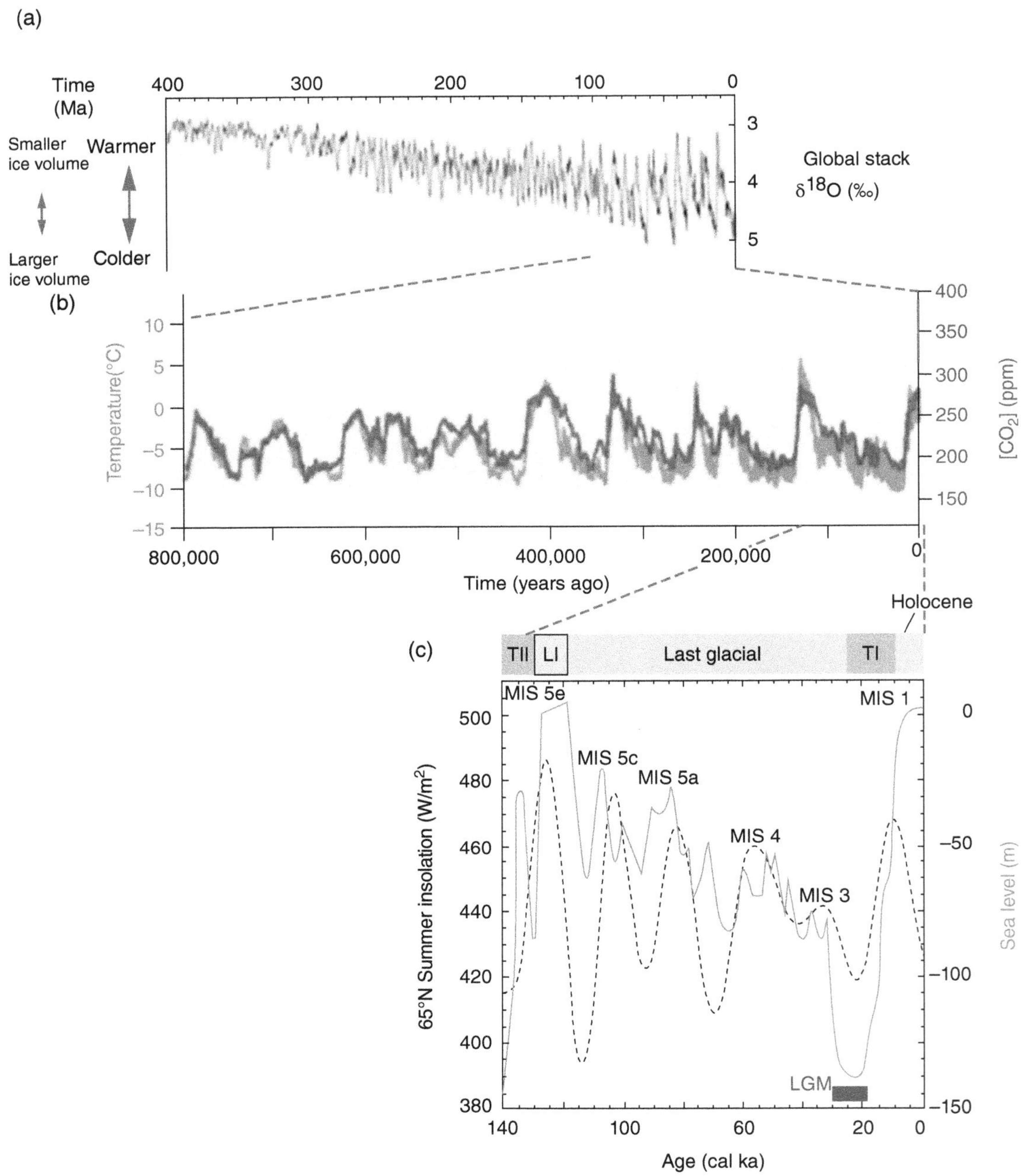

Fig. 7.2. (a) A 4 Ma record of paleo-sea-level reconstructions using stacked deep-sea sediment archives (data from Lisiecki and Raymo, 2005); (b) comparisons between Antarctic surface temperatures and atmospheric CO_2 records from Vostok (data from EPICA Community Members, 2004) and (c) Northern Hemisphere high-latitude insolation curve with mainly coral-based sea-level reconstructions (data from Yokoyama and Esat, 2011). Several distinct climate time windows are identified such as TII (termination II, namely penultimate deglaciation, 129–150 ka); TI (termination I, the last deglaciation, 10.4–24 ka); and LGM (19–26 ka). For color details, please see Plate 19.

understand the relationships between sea level and climate. It highlights the importance of determining accurate paleo-sea levels using the most reliable dating methods, that is, U-series dating with high-precision analytical methods (Edwards et al., 1987; Esat, 1995; Stirling et al., 1995).

In this chapter, we review methods of coral-based sea-level derivations. Technical advances over the past two decades in sampling and in analytical methods have resulted in a better understanding of past sea levels. The high-precision data have highlighted the close relationship between rapid climate variability and equally rapid changes in global ice volumes and sea levels (Yokoyama and Esat, 2011).

7.2 NATURE OF CORAL REEF SYSTEMS

Coral reefs can be classified relative to the range of sea-level depths of their natural habitat. Zooxanthellae, the symbiotic algae living with corals, requires light and warm temperatures which limits the depth range of these corals; this type of coral reef is therefore a good indicator of sea level. From their observations at Caribbean coasts, Neumann and Macintyre (1985) classified coral reefs into three broad categories of: keep-up; catch-up; and give-up types. Keep-up reefs are mainly of *Acropora palmata* with an estimated water depth range of less than 5 m, and their reef crests always extend close to the sea surface (Table 7.1). This classification is based on depth distributions in the Caribbean (Lighty et al., 1982). On occasion however, *A. palmata* has been reported living at greater depths down to 17 m (Goreau and Wells, 1967; Hubbard, 2009). There is broader coral diversity within Pacific coral reefs compared with those in the Atlantic, so that a mono-specific classification for the purposes of constraining habitat depths and hence sea level is difficult (Davis and Montaggioni, 1985; Hopley, 1986). An alternative formulation of using an assemblage of more than two species to constrain shallow water facies was proposed. Bard et al. (1996) and Montaggioni et al. (1997) reported that *Acropora*, *Porites* and *Hidrolithon (Porolithon) onkodes* combination can only be found at water depths shallower than 5 m. This proposal was originally challenged by Blanchon (1998), but cumulative observations from Indo-Pacific support the original arguments (Montaggioni and Bard, 1998).

Coatings of coralline algae provide additional opportunities to constrain the habitat depth of Pacific coral reefs. Algae growth is more sensitive to light levels and can reinforce coral-based paleowater depth reconstructions (Braga and Aguirre, 1995; Webster et al., 2004a). Seard et al. (2011) and Camoin et al. (2012) refined habitat depth estimates of Indo-Pacific coral reefs by including constraints from coral and algal assemblages.

The second category in Macintyre's classification is the catch-up reef. Here, the initial rise in sea level is faster than the reef accretion rate.

Table 7.1. Dominant coral species found in Atlantic in terms of water depth

Depth (m)	Dominant species
<5	*Acropora palmata*
5–15	*Acropora cervicornis*
>20	*Montastrea annularis*

However, reef growth eventually catches up with sea level by filling in the space created by the earlier sea-level rise. In the Atlantic, *Acropora cervicornis* has been observed to fulfill this role down to water depths of 15 m.

During the course of the last deglaciation several rapid rises in sea level occurred. In these circumstances existing reefs were stressed as, in addition to the rapid rises in sea level, high turbidity and low temperatures impeded and terminated the vertical growth of the reef (Webster et al., 2004b). These give-up-type reefs have also been referred to as "drowned reefs".

An additional reef classification was proposed from observations of uplifted coral terraces at Huon Peninsula, PNG (Esat and Yokoyama, 2006a). The last ice age climate was dominated by rapid shifts in climate in a well-defined pattern of over 6–7 ka cycles called Heinrich and Dansgard-Oeschger events from evidence of ice-rafted debris layers in North Atlantic deep-sea cores. It was also established that the start of each cycle was initiated by discharge of armadas of icebergs from the Laurentide ice sheet into the North Atlantic, which precipitated a particularly cold period by interrupting the Gulf Stream flow. The ice discharge amounted to over 20% of the mass of the ice sheet and caused rapid rises in sea level by as much as 10–35 m (Yokoyama et al., 2001a, b; Chappell, 2002; Siddall et al., 2003). By some estimates, the sea level rises occurred over decadal timescales (Yokoyama et al., 2001a; Siddall et al., 2003). The ice age coral terraces at Huon Peninsula were constructed in response to the sudden creation of habitat space over a substrate of older coral deposits that were suitable for re-colonization. Additional coral growth occurred during the subsequent slower sea-level fall lasting for over 1000 years. The net result is a terrace structure of a stack of individual corals that are not in stratigraphic age order and labeled as "pack-up" reefs (Esat and Yokoyama, 2006a). The phenomena is unique for the last ice age and contrasts with times of glacial–interglacial transition when the rate of sea-level rise is sufficiently slow for reefs to keep-up or

Fig. 7.3. Images of corals from Huon Penninsula, Papua New Guinea. (a) Present-day nearshore seafloor around 3 m water depth shows shallow-water-living corals. (b) A wall of Holocene uplifted reef includes robust branching Acropora corals as well as platy and encrusting Acropora. (c) A photo taken offshore from Bobongara section showing a full section ranging from the Holocene reef to the last interglacial reef. (d) Kanzarua section looking up from 30 ka reef to MIS 4 and MIS 5 terraces. *Source*: Photographs by Yusuke Yokoyama. For color details, please see Plate 20.

catch-up with sea level. Reconstructing past sea levels from this type of setting is challenging but can be achieved by selecting a large number of samples for dating from the whole section as well as undertaking coral and coralline algae facies analyses (Esat and Yokoyama, 2006a).

It is believed that Indo-Pacific reefs have common coral and coralline algal (coralgal) assemblages. Veron and Kelley (1988) studied reef corals of southern Papua New Guinea (PNG) and the Indo-Pacific and found almost identical coralgal assemblages as is found in the Great Barrier Reef. The southern reefs of PNG had been thought of as the northward extension of the Great Barrier Reef; the finding is therefore in accord with this observation. Huon Peninsula in PNG has been investigated extensively as a prime location for sea-level change studies (e.g., Bloom et al., 1974; Chappell, 1974; Chappell et al., 1996; Stein et al., 1993; Esat et al., 1999; Yokoyama et al., 2001a, b). Chappell (1974) briefly described coral assemblages of the Huon Peninsula (Fig. 7.3). This was followed by detailed studies undertaken by other research groups at Sialum and Hubegong coasts (Nakamori et al., 1995) and Madang Lagoon (Pandolfi and Minchin, 1995). Nakamori et al. (1995) categorized coral species into four groups, namely: lagoon (L: 0–10 m); shallow reef slope (S: 0–3 m); middle reef slope (M: 3–5 m); and deep reef slope assemblages (D: 5–30 m). Dominant species for each assemblages are: *Acropora formosa*, *A. pulchra*, *Montipora digitata*, *Porites cylindrical*, and *P.* (*Synaraea*) *rus* for lagoon assemblage; *Acropora hyacinthus*, *A. monticulosa*, *A. humilis*, and *A. robusta* for shallow lagoon assemblage; *Acropora palifera* for middle reef slope assemblage; and *Platygyra* spp. *Mycedium elephantotus*, *Oxypora* sp., and *Leptoseris scabra* for deep reef slope assemblage (Table 7.2). Comparisons between modern and Pleistocene assemblages show some discrepancies in species richness but strong similarities are present in zonation patterns with depth (Pandolfi, 1996) both in the current and fossil reefs (Fig. 7.3). The behavior of modern analogs of fossil corals can therefore be used as examples to better understand the fossil species paleo-water depth distribution.

Table 7.2. Hermatypic assemblages found at Huon Peninsula

	Depth (m)	Assemblage	Dominant species	Associated species
Lagoon	0–10	Lagoon	*Acropora formosa* A. *pulchra* *Montipora digitata* *Porites cylindrical* *P. (Synaraea) rus*	*Acropora microphthalma* *Montipora foliosa* *Porites* spp. (massive)
Reef slope	0–3	Shallow reef slope	*Acropora hyacinthus* A. *monticulosa* *A. humilis* *A.robusta*	*Pocillopora verrucosa* *Acropora danai* A. *florida* *A. gemmifera*
	3–5	Middle reef slope	*Acropora palifera*	*Montipora* spp. *Pocillopora verrucosa*
	5–30	Deep reef slope	*Platygyra* spp. *Mycedium elephantotus* *Oxypora* sp. *Leptoseris scabra*	*Porites* spp. (massive) Faviidae corals

Source: Nakamori et al. (1995). Reproduced with permission of Taylor & Francis.

Table 7.3. Modern and fossil Hermatypic coral assemblages found in the Ryukyu area

Habitat	Water depth (m)	Dominant species	Associate species
Moat to inner reef flat	0–5	*Acropora formosa, A. aspera, Porites cylindrica, Stylophora pistillata*	*Seriatopora* spp., *Porites* spp. (massive), *Acrheria horrescens*
Reef crest to upper reef slope	0–5	*Acropora hyacinthus, A. montiulosa, A. donai, Pocillopora verrucosa, Goniastrea retiformis*	*Acropora palifera*
Upper reef slope	5–20	*Acropora palifera, Favia stelligera, Platygyra sinensis*	Faviid corals
Middle reef slope	20–30	*Oxypora* spp., *Pectinia* spp., *Mysedium* spp.	*Echinophyllia* spp., Encrusting corals
Lower reef slope	30–50	*Leptoseris yabei, L. papyacea*	*Cycloseris* spp., *Pachyseris* spp., *Diaseris* spp.

Source: Sagawa et al. (2001). Reproduced with permission of Elsevier.

Reefs around the Ryukyu islands of Japan are located in the north-western Pacific. Several authors have studied the distribution of these coral assemblages in relation to their habitat and water depth. Sagawa et al. (2001) studied modern reefs and divided them into five distinct coral assemblages (Table 7.3) from 0 m to 50 m water depth by extending the original criteria proposed by Nakamori (1986). Their five sub-groups are: moat to inner reef flat assemblage (A: 0–5 m); reef crest to upper reef slope assemblage (B: 0–5 m); upper reef slope assemblage (C: 5–20 m); middle reef slope assemblage (D: 20–30 m); and lower reef slope assemblage (E: 30–50 m) (Fig. 7.4). The dominant and associated species are listed in Table 7.2. They found a close correspondence with Papua New Guinea modern coral assemblages observed by Nakamori et al. (1995), indicating that the Western Pacific coral zonation pattern with depth is ubiquitous. Two major assemblages characterized by Webster et al. (2004a) also are in accord with these earlier studies. They reported shallow reef facies (<5 m) dominated by robust branches or ridges of the *Acropora palifera*, and *A. humilis* group and the tabulate *A. hyacinthus* group (Fig. 7.3). These species are associated with encrusting *Montipora* sp. (e.g., *M. tuberculosa, M. informis*) and submassive to massive *Porites* sp. (e.g., *P. horizontalata, P. lobata*) and minor encrusting *Siderastrea savignyana, Psammocora superficialis*, and faviids (e.g., *Favia laxa, Montastrea mulitipunctata*). For depth ranges somewhat wider than considered above (<10 m), the assemblages include dominant encrusting *Montipora* sp. (e.g.,

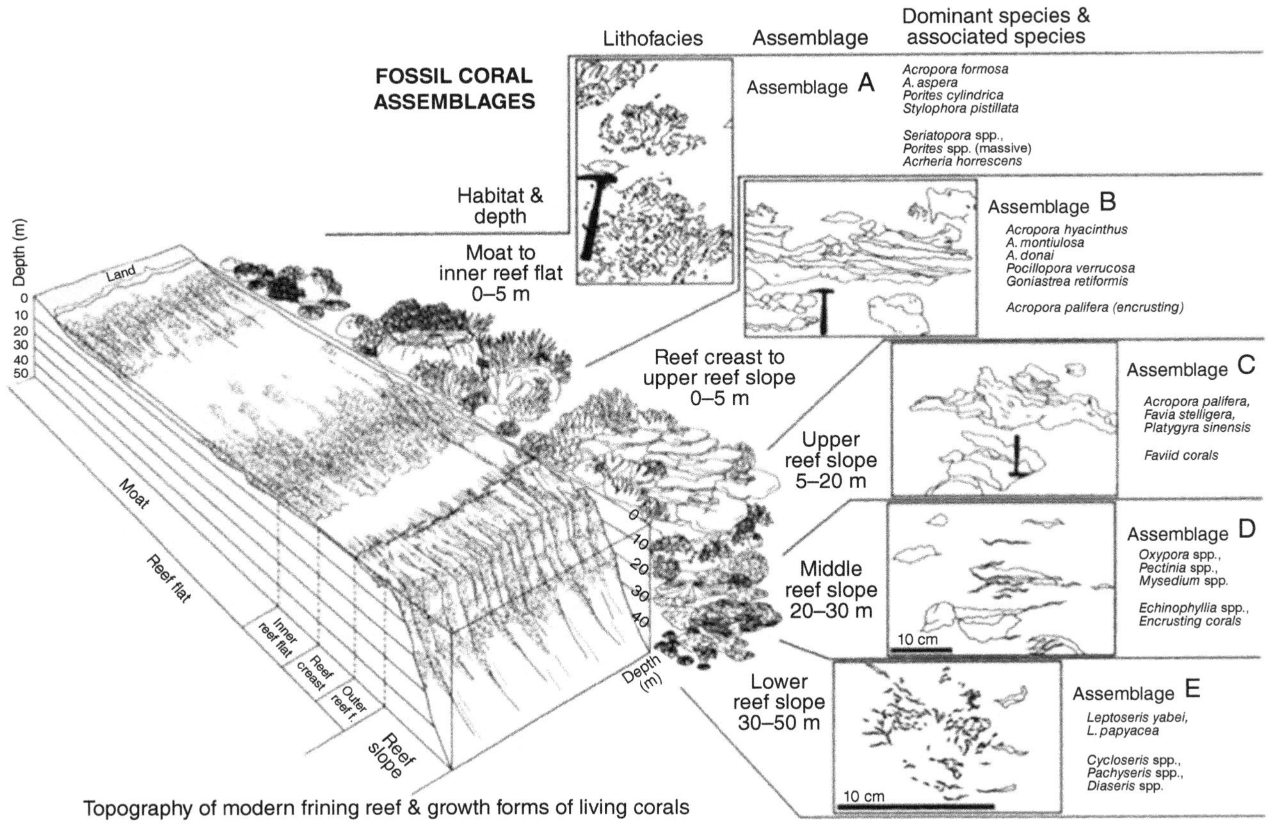

Fig. 7.4. Coral assemblages at Ryukyu Island (Japan) described by Sagawa et al. (2001). Reproduced with permission of Elsevier.

M. monasteriata, *M. corbettensis*) and Porites sp. (eg. *P. horizontalata*) with associated faviids and agariciids (*Siderastriidea savignyama*, *Pseudosiderastrea tayamai*, and *Psammocora* sp.).

West Indian Ocean coral reefs have also been found to have similarly zoned assemblages (Camoin et al., 1997; Montaggioni and Faure, 1997). Abundant *Acropora* is observed as a main reef frame component and at Mayotte (west of Madagascar Island, Mozambique Channel) they seem to have grown as keep-up reefs over the last 10,000 years (Camoin et al., 1997). The differences between Pacific and West Indian Ocean reefs can probably be attributed to different wave energy environments. Comparisons between Pacific and Indian Ocean coral reef assemblages were systematically investigated by Cabioch et al. (1999). They concluded that coral assemblages within depths of 6 m are *Acropora danai/robusta*, *A. humilis*, and *A. digitifera* associated with vermetid gastropods and thick coralline crusts of *Hydrolithon* cf. *onkodes* (Fig. 7.5) and *Neogoniolithon* cf. *fosliei*. In summary, further studies are required to better understand and constrain the depth distribution of coralgal assemblages including microborer ichnocoenoses in corals (e.g., Rémy et al., 2011). However, the depth zonations of coralgal assemblages in the Indo-Pacific oceans appear to have a common basis which can be usefully employed to constrain and reconstruct paleo-sea levels.

7.3 DATING

Corals in the oceans can be quite selective as to which major or trace elements they take up in their $CaCO_3$ matrix. For some elements, the uptake is temperature dependent and has been exploited as a sensitive proxy for ocean temperatures (Beck et al., 1997). Atmospheric radiocarbon rapidly equilibrates with sea-surface water and is taken up and preserved in the coral skeleton. Radiocarbon abundance measurements can then date the timing of past sea-level changes (Yokoyama et al., 2000b; Yokoyama and Esat, 2004). Originally, conventional radiocarbon dating using scintillation

Fig. 7.5. Present-day example of (left) shallow-living coralgal algae (*Hidrolithon* (*Porolithon*) *onkodes*) and (right) *Acropora* at Heron Island, Great Barrier Reef. *Source*: Photographs by Yusuke Yokoyama. For color details, please see Plate 21.

counters required large quantities of coral sample (~10 g) and long measurement times (>3–4 days/sample, not including the preparation time). Currently, accelerator mass spectrometer (AMS) dating techniques require lesser quantities of sample (10 mg or less), are much quicker (minutes/sample), and provide significantly higher precision (see Yokoyama and Esat, 2004; Yokoyama et al., 2007). However, the range limit of radiocarbon dating method (which is restricted to less than half of a glacial–interglacial cycle at 50,000 years), ocean reservoir effects, and the dynamic nature of ocean circulation currents can limit the usefulness of the method. Recent advances in Quaternary paleoclimate reconstructions, such as those from Greenland ice cores, show rapid, millennial-scale climate oscillations that point to changes in Greenland surface temperatures of more than 15° C repeatedly during the last glacial period (Dansgaard et al., 1993; Alley, 2000). These extreme climate events appear to be coupled to equally rapid sea-level changes where the timescale is exclusively derived from coral dating. In these circumstances, radiocarbon reservoir age and calendar age calibration uncertainties have become obstacles in cross-correlating rapid changes in climate and sea level for a better understanding of the climate system. For a more precise and accurate verification of the timing of sea-level changes, a dating system independent of the carbon cycle is essential and can be provided by U-series disequilibrium dating (Stirling and Anderson, 2009). In turn, corals appear to be the most suitable candidates for U-series dating and attempts at dating other carbonate materials such as mollusks or carbonate sands have not been as successful. Cave deposits and speleothems can also be successfully dated. However, they do not have a direct connection to sea level except in rare cases when the caves are repeatedly submerged and exposed due to sea-level changes (Richard et al., 1994). When combined with carbon or oxygen isotopes, speleothem ages serve as a very important climate record due to the sensitivity of oxygen isotopes to climate variability such as the monsoon cycles of South East Asia (Wang et al., 2001; Cheng et al., 2009).

In natural environments uranium can exist in several oxidation states. In its +VI oxidation state, uranium is soluble whereas it is particle reactive and insoluble in its +IV state. In oxygenated seawater environments, uranium is highly soluble and forms uranyl (U^{6+}) carbonate complexes. The daughter isotope of U used in U-series dating is ^{230}Th and has a single +IV oxidation state. Th is highly particle reactive and insoluble. It is readily removed from freshwater and seawater attached to colloids and to particulates. Pristine live corals essentially contain no Th and, over time, ^{230}Th

builds up in the coral skeleton from the decay of ^{234}U. The extreme fractionation of the parent U from the daughter Th forms the basis for U-series dating as the system subsequently approaches secular equilibrium where the activities of all the nuclides in the decay chain are equal.

7.4 OPEN-SYSTEM PROBLEM

In any radioactive decay dating scheme, the parent and daughter nuclides have to remain isolated from outside influences so that either loss from the system or external addition can be verified. A major preoccupation in coral U-series dating is the notion of open-system behavior, which is the main source of uncertainty and argument as to the veracity of the dating results. Corals are susceptible to diagenesis; this term can cover a multitude of processes, one of which is a mineral transormation from aragonite to calcite. Numerous studies have been undertaken to understand open-system behavior in corals as it relates to uranium and thorium nuclides (e.g., Bar-Matthews et al., 1993; Stein et al., 1993; Gallup et al., 1994). None of these have been successful in establishing a clearly defined mechanism or a correlation that has gained general acceptance. Clearly the basic processes have not yet been fully recognized. However, a set of selection criteria have been formulated to help recognize open-system behavior, although no one particular test is likely to be either a necessary or a sufficient condition.

7.4.1 Microscopic analysis

Optical and electron microscopes are often used to screen for diagenically altered samples. Original textures of coral aragonitic skeletons are visible at high magnification; the degree of contamination in samples that have coatings of secondary calcite and secondary aragonite can be estimated and, in some cases, they can be mechanically removed. Both longitudinal and latitudinal thin-sections are usually made and checked for secondary void filling and alteration. Evidence for alteration is often found around delicate lattice-work or septa in between solid wall sections in *Faviid* corals. Where possible, these structures are routinely removed with mechanical cleaning using various dental drills. Any samples that contain extensive alteration features are rejected. The simple test of looking for signs of "glitter" in sunlight is almost as effective in recognizing calcite, as tests with X-ray diffraction confirm that 2–3% calcite contamination can be visually recognized.

7.4.2 XRD analysis

Original aragonitic skeleton can be replaced by secondary calcite since calcite is more stable than aragonite in such environments once corals are exposed to freshwater environments (Deer et al., 1967). The presence of calcite may be indicative of U and Th exchange after deposition. This can be detected using X-ray diffraction analysis (XRD). However, the sensitivity of XRD is usually limited to c. 1–2% calcite.

7.4.3 Trace elements and stable isotopes

A number of studies have attempted to establish a correlation between trace elements, stable isotopes and U–Th ages (e.g., Chen et al., 1991; Bar-Matthews et al., 1993; Stein et al., 1993). For examples, Na, Mg, Sr, $\delta^{13}C$ and $\delta^{18}O$ were studied by Stein et al. (1993) in corals from Huon Peninsula. It was concluded that neither trace element nor stable isotope analyses were sensitive enough to recognize low levels of alteration. Chen et al. (1991) and Bar-Mattews et al. (1993) have reported generally lower concentrations of Na, Mg, and Sr in fossil corals from the Huon Peninsula and the Bahamas compared with modern corals, which they attributed to secondary calcite formation. However, quantitative relationships between U-series age of the corals and trace element concentrations have not yet been established.

7.4.4 $\delta^{234}U(T)$

It is generally assumed that the seawater $^{234}U/^{238}U$ ratio has remained constant over the time range covered by U-series (<700 ka) and hence $\delta^{234}U(T)$ (the fractional difference between seawater and secular equilibrium values) in fossil corals should be similar to that in modern corals. The value of modern $\delta^{234}U(T) = 149 \pm 1‰$ (Gallup et al., 1994; Stirling et al., 1995) and $144 \pm 7‰$ (Chen et al., 1986), respectively. The acceptance criteria for $\delta^{234}U(T)$ adopted for Huon Peninsula PNG coral samples were 140–158‰, whereas a narrower range of 4‰ was employed by Stirling et al. (1995) for the last interglacial corals from Western Australia. Recent revisions to the decay constants for ^{234}U and ^{230}Th reduce $\delta^{234}U$ in general by about 2‰ without affecting the calculated ages. The seawater value reduces to $\delta^{234}U(T) = 147 \pm 1‰$ (Cheng et al., 2000). The $\delta^{234}U$ criterion is the best quantitative measure available as an indicator of uranium and thorium open-system behavior. Gallup et al. (1994) have shown a direct correlation between $\delta^{234}U(T)$ values and U-series ages. The increase in $\delta^{234}U(T)$ beyond the present-day value (147‰) appeared to be approximately correlated with older apparent ages in all cases. Similar trends can be seen in last interglacial Huon corals (Stein et al., 1993), Barbados corals (Potter et al., 2004), and in Western Australian corals (Stirling et al., 1998). A demonstrable explanation for such a correlation remains elusive, although Thompson et al. (2003) have recently proposed a plausible mechanism to better explain U-series nuclides behavior. However, Scholz and Mangini (2007) quantitatively evaluated open-system model ages and concluded that the Thompson model could only be valid in very limited circumstances (Andersen et al., 2008). Open-system model ages continue to be heavily debated (Esat and Yokoyama, 2006b, 2010).

7.4.5 Total U concentration

Modern corals contain about 3 ppm of U within the approximate range of 2.5–3.5 ppm. Within these limits, the uranium concentration in corals can also be used for checking the quality of corals for uranium series dating. There is some indication in the available data that branching *Acropora*-type corals have systematically higher U contents (Stirling et al., 1998).

7.4.6 ^{232}Th concentration

Pristine corals from oceanic islands normally contain less than 0.5 ppb of ^{232}Th (Edwards et al., 1987; Chen et al., 1991). In continental rocks ^{232}Th is a common isotope (>1 ppm concentration). Excess ^{232}Th corals is indicative of external U and ^{230}Th addition and results in older measured ages. The measured concentration of ^{232}Th and the expected $^{238}U/^{232}Th$ ratio can therefore be used to correct for the contribution of detrital materials. In most studies however, selected corals have negligible ^{232}Th abundances.

7.5 $^{234}U/^{238}U$ IN OCEANS

Uranium in the oceans is expected to have a relatively long residence time; the seawater isotopic composition of Uranium ($^{234}U/^{238}U$) should therefore have been invariant for over the past 300–600 ka. This was supported by modeling (Richter and Turekian, 1993; Henderson and Anderson, 2003) as well as from dating of pristine corals mainly from the last interglacial period (e.g., Stirling et al., 1995, 1998). However, recent studies using last glacial corals (Yokoyama and Esat, 2004; Esat and Yokoyama, 2006b) as well as measurements in rivers (Robinson et al., 2004) suggest possible isotopic variations over glacial–interglacial timescales as well as over much shorter periods of millennia. During warm interglacials, redox-sensitive uranium nuclides can be deposited at coastal and continental shelf sediments by rivers under reducing conditions. The deposits are enriched with ^{234}U because of preferential rainwater leaching from weakened lattice sites in minerals due to α-recoil following ^{238}U decay. When sea level falls during glacial periods, the continental shelf sediments are exposed and oxidized. During subsequent sea-level rise the ^{234}U-enriched U is released close to shorelines where corals are actively growing. As the sea-level slowly rises and then stabilizes, there is a ^{234}U gradient in seawater between coastal regions and the open ocean which can last for several thousand years. In principle, this mechanism is able to account for the observed correlation in corals between ^{234}U and age. The $\delta^{234}U$ closed-system indicator should therefore be applied with caution in circumstances where there is an apparent correlation between $\delta^{234}U$ and age (Andersen et al., 2008).

7.6 SEA LEVELS RECONSTRUCTED FROM UPLIFTED REEFS

A number of coral-based sea-level studies have been published from tectonically active regions. At stable margins, reefs that would have grown at different times but during similar global sea levels overlap, and those that grew during cycles of low sea level are now below present sea level. At sites with continuous uplift, these coral reefs were uplifted and became vertically separated while those that would have been below current sea level are now accessible for easy sampling (Bloom et al., 1974). For example, at stable sites coral reefs of the last glacial period are currently underwater and only accessible with drilling. For these reasons, tectonically uplifting sites such as Barbados and Huon Peninsula, PNG have been the focus of attention over many years as they also appear to have relatively constant uplift rates. At Barbados (uplift rate approximately 0.25 mm a^{-1}), marine isotope stage (MIS) 5a and 5c are currently at present sea level (Schellmann and Radtke, 2004), whereas at Huon Peninsula (uplift rate up to 3 mm a^{-1}) a number of MIS 3 coral terraces are exposed (Lambeck et al., 2002).

Long-term uplift rate estimates are a key parameter for constructing sea-level histories from these tectonically active locations. The height and age of the last interglacial terrace is often used together with global sea-level estimates to calculate uplift rates. Global last-interglacial sea-level values were believed to have been about 4 m higher than the present (Stirling et al., 1998); however, recent estimates point to higher values of 5.5–9 m (Dutton and Lambeck, 2012). If confirmed, the new values would only slightly affect (less than 1%) the previous uplift estimates. At Huon Peninsula uplift rates derived from both last interglacial and Holocene terrace heights and ages (Ota et al., 1993) are internally consistent and can be used with confidence to extract precise sea-level histories.

Barbados is located at a tectonically active fore-arc. It hosts exposed Pleistocene coral terraces that range in age from 640 ka to 60 ka, consisting of 15 coral terraces (Broecker et al., 1968; Mesolella et al., 1969; James et al., 1971; Matthews, 1973; Bender et al., 1979; Edwards et al., 1987; Gallup et al., 2002; Schellmann and Radtke, 2004). The ubiquitous presence of shallow living corals *Acropora palmata* are of significant help in reconstructing sea-level histories although, in some instances, they have been observed at considerable depths (17 m; Goreau and Wells, 1967). Uplift rates around the island vary due to the presence of several active faults (Taylor and Mann, 1991) and regional uplift rates have to be known to provide an accurate sea-level history. Barbados is sometimes referred to as the best site to "tune" glacio-hydro-isostatic effects (Peltier, 1994) as it is still under the influence of mantle flow from the redistribution of surface loading due to the melting of the Laurentide ice sheet (Yokoyama and Esat, 2011). It is characterized as an intermediate field site (Lambeck et al., 2002; Mitrovica and Milne, 2003; Potter and Lambeck, 2003) and therefore caution is needed when using Barbados relative sea-level data to "tune" modeling parameters (Lambeck et al., 2003).

Huon Peninsula, Papua New Guinea has numerous uplifted coral terraces extending back in time for at least 340 ka (Veeh and Chappell, 1970; Chappell 1974). The Huon coral terraces extend for over 80 km along the coastline (Fairbridge, 1960; Chappell, 1974) where the tectonic uplift rate varies from 0.5 m ka^{-1} to almost 4 m ka^{-1} from northwest to southeast (Chappell, 1974; Ota et al., 1993). A number of studies of Huon coral terraces have focused on the Quaternary sea levels of the last glacial–interglacial cycle (Veeh and Chappell, 1970; Bloom et al., 1974; Chappell, 1974, 1983; Bloom and Yonekura, 1985; Chappell and Shackleton, 1986; Edwards et al., 1993; Stein et al., 1993; Omura et al., 1995). The preservation of corals older than the last interglacial is not as good and has limited the investigation of older time periods. However, Huon Peninsula is a prime site for studying the sea levels of the last ice age when they were 60–100 m lower than the present. More recently, it has played a significant role in identifying rapid sea-level changes mainly because of the "spacing-out" effect associated with high uplift rates. At other (stable or slower uplifting) sites, corals of slightly different ages would have been piled up together, smearing the evidence for millennial-scale sea-level and climate variations (Chappell et al., 1996; Esat et al., 1999; Yokoyama et al., 2000a; 2001a, b; Chappell, 2002; Cutler et al., 2003; Potter et al., 2004). Repeated disintegration of Northern Hemisphere ice sheets, called Heinrich events (Heinrich, 1988; Broecker, 1998), have impacted the global climate due to the weakening of the Atlantic Meridional Overturning Circulation (AMOC) that cooled the North Atlantic regions as well as affected global climate during the last ice age (Clark et al., 2009). The amount of ice discharge was originally thought to be no more than the equivalent of a few meters of sea level (MacAyal, 1993). However, detailed U-series-dated coral analyses revealed 10–15 m sea-level rise for each episode (Yokoyama et al., 2001a, b). This finding was subsequently confirmed (Chappell, 2002; Siddal et al., 2004) and corresponds to about 20% of the Laurentide ice sheet volume. Estimates from subsequent studies of sediment cores from the Red Sea indicated even larger sea-level rises up to about 30 m (Siddall et al., 2004).

One of the important outcomes from Huon Penninsula relates to the ages (c. 32 ka) and elevations (70–80 m below current sea level) of terraces present at a time just prior to the sea-level fall into the Last Glacial Maximum (LGM). The rapid growth of global ice sheets within a few thousand years causing draw-down of global sea level to 125 m or more attests to the severity of this event.

Vanuatu is located at the boundary between the Australian and the Pacific plates and has a high (3 mm a^{-1}) uplift rate (Cabioch et al., 2003), confirmed by geodetic observations (Larson et al., 1997). Some deglacial reef sequences are present above sea level. Cabioch et al. (2003) have reported changes from catch-up reef assemblages to keep-up-type reefs before and after 11.3 ka. This is an interesting time period that corresponds to the resumption of deglaciation following the Younger Dryas cold event from about 12.6 ka to 11.3 ka. Similar results have been reported from Mayotte Island from the West Indian Ocean (c. 11.6 ka: Zinke et al., 2003) and PNG (Edwards et al., 1993). However, there are too few U-series ages to reach any firm conclusions.

Uplifted coral terraces at Kikai Island were reported from central Ryukyus, south-western Japan (Omura, 1988; Sasaki et al., 2004). Based on the height and age of the last interglacial reef, the uplift rate is about 1.7 mm a^{-1} and some ice age reefs are exposed. Five discrete coral reef terraces were identified at approximately 66 ka, 64 ka, 62 ka, 55 ka and at 52 ka. They appear to belong to the MIS 5a and 5c sequence of sea-level changes (Potter et al., 2004) and the youngest two that date to around 55–52 ka may be associated with Heinrich event 5a.

Because of the limited occurrence of rapidly uplifting coastal margins with coral growth, coral-based sea-level reconstructions are still scarce. The problem is greater for periods before the last interglacial due to a lack of preservation of good-quality corals; there are only a few U-series ages for this period of time (e.g., Bard et al., 1996; Galewsky et al., 1996; Camoin et al., 2001; Stirling et al., 2001; Ayling et al., 2006; Scholz et al., 2006). Henderson Island in the South Pacific is one such location where well-developed uplifted coral terraces are available (Stirling et al., 2001). Paleo-reef crests are clearly preserved and high-precision U-series dating identified clusters of ages 320 ka, 307 ka, 240 ka, and 234 ka (Stirling et al., 2001; Andersen et al., 2008). These ages agree well with the timing of climate cycles predicted by astronomical theory (Milankovitch, 1930).

7.7 SEA LEVELS RECONSTRUCTIONED FROM SUBMERGED REEFS

A major study on coral drill core records offshore from Barbados was published in 1989. Fairbanks (1989) documented radiocarbon age and depth relations of *Acropora palmata* corals from a series of drowned reefs around the island extending over the past 17 ka. The same samples were also dated using U-series methods (Bard et al., 1990; Fairbanks et al., 2005). The maximum local elevation of sea level during the LGM was determined to be 121±5 m and was later confirmed by sea-level studies from Sunda Shelf (Hanebuth et al., 2000) and Bonaparte Gulf (Yokoyama et al., 2001c, d; DeDeckker and Yokoyama, 2009). The local relative sea-levels were corrected for glacial isostatic adjustments (GIA) (Yokoyama et al., 2000b; Lambeck et al., 2002). The record showed two periods of accelerated sea-level rise called meltwater pulses Mwp-1A and Mwp-1B at c. 13.61–14.08 ka and 11.4–11.1 ka respectively. These periods correspond to enhanced Northern Hemisphere high-latitude insolation and approximately bracket the Younger Dryas cold episode (YDE: Broecker, 1998). However, subsequent studies of coral reefs from Tahiti place Mw-1A further back in time to the start of the Bølling warming at c. 14.7 ka, whereas the presence of Mwp-1B could not be confirmed (Deschamps et al., 2012).

Tahiti is a volcanic island located at the South Pacific which is experiencing slow subsidence. This is in contrast to Barbados or Vanuatu where the islands are being uplifted in an active margin. Subsidence here is due to the relaxation of the volcanic load stresses in the lithosphere; the relaxation time in this region is estimated to be of the order of c. 10^7 years (Lambeck, 1981). The subsidence rate is estimated to be between about 0.25 mm a^{-1} (Bard et al., 1996) and 0.1 mm a^{-1} (Lambeck et al., 2002), depending on considerations of geodynamic components (Nakada and Lambeck, 1987). The site is better suited to reconstructing global ice-volume changes as it is less influenced by direct effects of glacial isostasy (Bard et al., 1996; Lambeck et al., 2002; Yokoyama and Esat, 2011).

Earlier, shore-based drilling recovered corals corresponding to the latter half of the last deglacial sequence that had grown continuously over the past 14 ka (Bard et al., 1996; Montaggioni et al., 1997). Additionally, a detailed study of six boreholes at the Tahitian reef near Papeete revealed a three-dimensional structure of reefs constructed during the last deglacial sea-level rise (Cabioch et al., 1999). However, none of the drill holes were able to reach the sequences corresponding to Mwp1A (Bard et al., 2010). Additional bore holes were purposely drilled at an angle to reach deeper sequences. However, these could not extend the record to periods older than 14 ka; the results raised questions regarding the existence of Mwp1B, previously reported in Barbados cores (Fairbanks 1989; Blanchon and Shaw, 1995; Bard et al., 2010). The data instead hinted at the possibility of a 5 m sea-level rise just before the Younger Dryas.

The technical limitations of shore-based drilling motivated a large-scale offshore operation for the express purpose of recovering evidence for Mwp-1A as well as for the timing and sea-level elevation of the LGM. In 2004, the Integrated Ocean Drilling Project (IODP) Expedition 310 was undertaken offshore from Tahiti (Camoin et al., 2007). Underwater recovery of drill-core corals extended over the last deglacial sequence as far back as 16 ka and covered the entire section of the expected location of Mwp-1A. Three coring sites in Tiarei, Maraa, and Faaa provided numerous coral samples to define Mwp-1A. Analysis techniques included high-resolution U-series dating by thermal ionization mass spectrometry (TIMS) and by multi-collector inductively coupled plasma mass spectrometry (MC-ICPMS). The coral assemblages were divided according to six criteria, where the depths quoted in brackets represent the habitat depth range of the named corals: (1) PM assemblage which consisted of robust branching *Pocillopora*/massive *Montipora* (0–10 m); (2) mP assemblage of massive *Porites* (0–25 m); (3) tA assemblage indicating tabular *Acropora* (5–15 m); (4) PP assemblage that consisted of branching *Porites*/*Pocilopora* (PP) (5–15 m); (5) PPM assemblage of branching *Porites*/encrusting *Porites* and *Montipora* (5–25 m); and (6) AFM assemblage typically found in deeper water (>20 m) consisting of *Agaricid* and *Faviid* (Deschamps et al., 2012). The occurrence of Vermetid gastropods can also help to identify paleo-water depths and typically indicate shallower-water (<5–6 m) environments (Blanchon and Blakeway, 2003). The above divisions were developed to identify paleo-water depths as similar criteria used for nearshore facies interpretation could not be applied to these offshore sites. For example, the occurrence of the robust branching *Acropora* gr. *robusta-danai* prevented the use of APH (*Acropora* gr. *robusta-danai*, *Pocillopora cf. verrucosa*, *Hydrolithon* (*Porolithon*) *onkodes*) assemblages discussed in earlier sections (Bard et al., 1996; Montaggioni et al., 1997). It is striking that there are no robust branching *Acropora* in the Tiarei sequence and they are scarce prior to 13.9 ka in Faaa and Maraa reef (Deschamps et al., 2012). This indicates drastic changes in environmental conditions before and after 13.9 ka, also reflected in habitat variations in response to sea-level-induced environmental changes (Abbey et al., 2011; Camoin et al., 2012). The extensive presence of microbialites is also an important signature of Tahitian corals which provides better-consolidated reef framework together with other components such as corals and coralline algae (Seard et al., 2011).

The new offshore record from Tahiti shows a 14–18 m sea-level rise that started at about 14.65 ka and ended at 14.31 ka. If this is the same event (Mwp1A) as defined in the Barabados record, its timing at Tahiti is inconsistent with its timing at Barbados. The rapid rate of sea-level rise during Mwp1A is as much as 40 mm a^{-1}. Geophysical modeling was applied to monitor the changes in heights of sea surface globally due to redistribution of surface mass as well as variation in gravitational attraction from large ice sheets. The process is known as fingerprint modeling (Nakada and Lambeck, 1987; Mitrovica et al., 2001, 2011). In the case of Tahiti Mwp1A, the models suggest a large contribution from Antarctic melting, which may possibly reconcile the conflicting results from Barbados and Tahiti. However, geological evidence from Antarctica based on cosmogenic exposure dating is in conflict with any significant volume of ice melting at this time (e.g., Bentley et al., 2010; Mackintosh et al., 2011; Whitehouse et al., 2012). The timing and provenance of Mwp1A therefore remains uncertain at present.

Corals recovered offshore from Tahiti were also useful for constraining the penultimate deglaciation sea level. The sea-level data for this period are scarce because of limited availability and the state of preservation of coral samples from this age (Yokoyama and Esat, 2011). The coral record from Huon Peninsula and Barbados showed rapid fluctuations in sea level during this period (Esat et al. 1999; Gallup et al., 2002). Initially, the sea-level rose with insolation up to about 10 m of present sea level and then plunged by as much as 60 m or more before rapidly returning to last-interglacial levels. At Tahiti, coral samples were identified

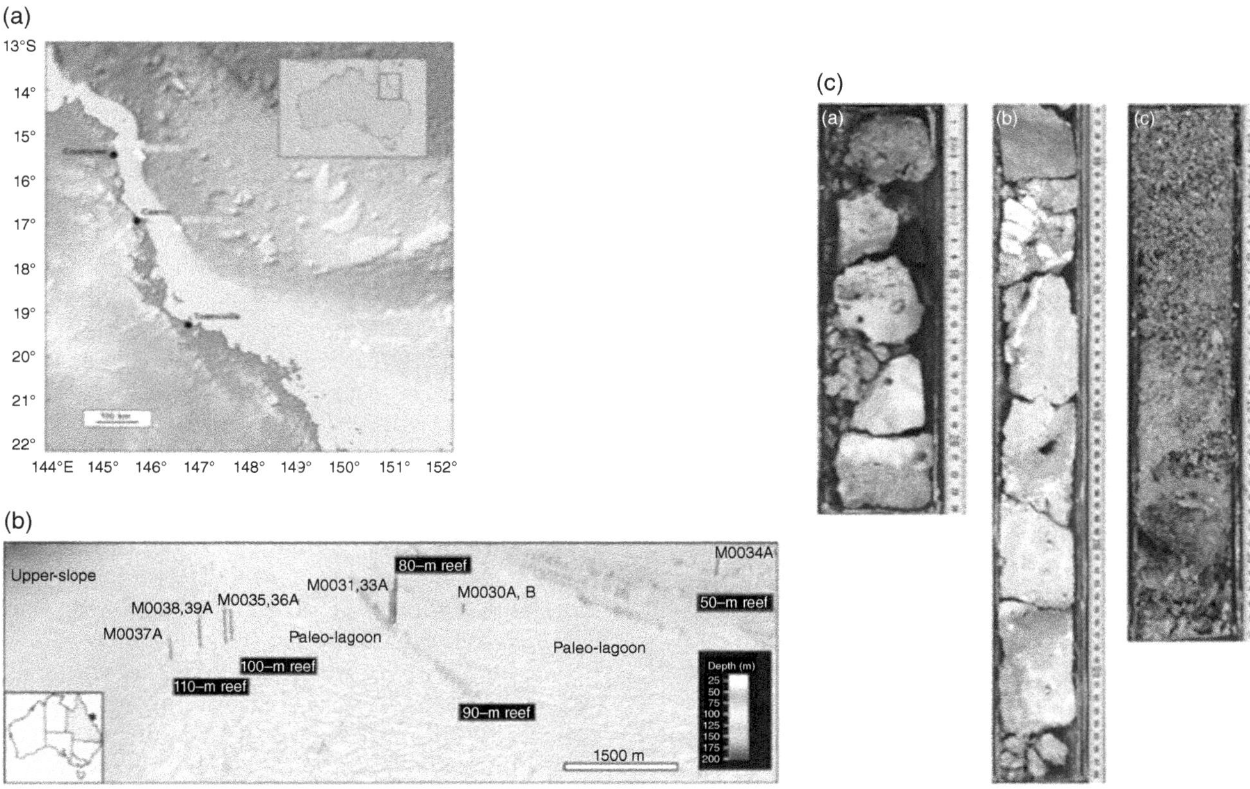

Fig. 7.6. A study of Quaternary sea levels with corals drilled from the Great Barrier Reef, Eastern Australia. *Source*: Yokoyama et al., 2011. (a) Location of drilling sites of IODP Expedition 325; (b) a high-resolution multi-beam image of Hydrographer's Passage transect HYD-0C; and (c) representative lithology in cores. Section A is a coralgal boundstones typically found at the top of the reef structure. Coralgal-microbialite boundstones and unlithfied sediments can be seen in sections B and C respectively. For color details, please see Plate 22.

defining the low sea-level points before and after the 60 m peak in sea level. Both coral as well as foraminifera assemblages supported these fluctuations (Thomas et al., 2009; Fujita et al., 2010).

Tahiti is a volcanic island and the rock and sediment substrate around the island flanks is steep. IODP Expedition 310 could not recover the last deglacial sequence entirely and corals older than 16 ka were not sampled. The pre-Mwp1A period, the time period of the LGM (Yokoyama et al., 2000b; Clark et al., 2009), is yet to be recovered (Camoin et al., 2012).

The Great Barrier Reef (GBR) has been extensively studied to determine Holocene sea levels but there is very little evidence for last deglacial sequences (see Veeh and Veevers, 1970; Marshall and Thom, 1976; Hopley, 2006; Yokoyama et al., 2006). Offshore coral drilling was recently attempted (Fig. 7.6) in 2010 by the IODP Expedition 325 (Yokoyama et al., 2011). Australia's GBR is the largest reef in the world and extends out 2000 km, located along a tectonically stable continental margin (Fig. 7.6). It is an ideal site for sea-level studies because of its distance from former continental ice sheets in a similar fashion to Tahiti (Yokoyama and Esat, 2011). Eustatic, global sea levels can be derived from this location with much less uncertainty after correcting for GIA adjustments due to loading/unloading of ocean basins to account for global ice-volume variations (Yokoyama et al., 2011).

The presence of submerged reefs over a distance of more than 100 km from north to south was clearly evident in seismic survey data (Abbey et al., 2011; Bridge et al., 2011) at depths in the range 40–130 m. For example, one of the four transects drilled during the expedition labeled HYD-01C is located on the Mackay Shelf in Hydrographer's Passage. It consists of a double-fronted barrier reef at 90 m and at 80 m followed by a 2-km-wide paleo-lagoon at 70 m depth. A similar paleo-lagoon exists behind another double reef system at 100 m and at 110 m from the edge of the continental shelf. A complex reef system comprising a series of vertically stacked coral structures called pinnacles occur at a depth of 50 m.

Offshore sampling by drilling was undertaken in 2010 for 2 months but the recovery of material was lower than that for the Tahiti IODP Expedition 310 (57.5% for Tahiti, 27.2% for GBR; Yokoyama et al., 2011). The drill cores were sampled for various types of analyses including sedimentological and radiometric dating. Nine lithological types were recognized in the cores (Fig. 7.6). The uppermost section of many holes are covered by modern or sub-recent materials. They range from lime sand to pebbles, occasionally filled by muddy matrix. Pebbles are made up of coralgal boundstone, coralline algae, serpulid worm tubes, mollusk shells, and bryozoans. *Halimeda* have been identified, as well as larger benthic foraminifera (abundant and well-preserved or stained specimens of *Alveolinella*, *Amphistegina*, *Cycloclypeus*, *Elphidiidae*, *Heterostegina*, *Operculina*, and *Sphaerogypsina*). Most of the sections recovered are dominated by coralgal-microbialite and coralgal boundstone. A thickness of 10–30 m of the coralgal-microbialite boundstones are composed of corals partially coated with coralline algae and vermetids and encrusted in microbialites. Volumetrically, these are the major components.

Coral assemblages are dominated by massive *Isopora*, branching *Acropora*, and *Seriatopora*, but massive *Porites* and *Faviidae* are locally abundant (Fig. 7.6). *Hydrolithon onkodes* (Fig. 7.5) is the most abundant coralline alga, together with *Lithophyllum prototypum* and *Neogoniolithon fosliei*. Stratigraphically, the coralgal boundstones consistently overlie coralgal-microbialite units. Skeletal packstone to grainstone, up to 13 m thick, underlies the unconsolidated sediment or the coralgal/coralgal-microbialite boundstones in most holes. These lithologies are variably cemented and composed of fragments of shells, coral, coralline algae, *Halimeda*, and abundant larger benthic foraminifera. The top of the cemented lithologies is commonly bored by worms and sponges. Features indicating subaerial exposure, such as calcrete deposits, brownish staining, and rhizoliths, appear at the top of grainstones in the shallowest holes on transect NOG-01B. In Hole M0036A, a dark-colored boundstone of c. 1.5 m thickness, overlies the packstone. This blackened boundstone comprises encrusting coral and thin coralline algae in its upper part, and a boundstone of coral, thin foliose coralline algae, and worm tubes in the lowest 10 cm. The dominant coral is massive *Goniopora*, with fragments of massive *Faviidae* and fine-branching *Seriatopora*.

Detailed radiometric dating is currently in progress; however, preliminary dates have been obtained in subsamples of several sections in different cores. Additional diagenetic screening of these samples using XRD and SEM is ongoing. All sample ages in the cores appear to be in stratigraphic order. This provides confidence in the accuracy of the dates and that the cores have been sampled in *in situ* reef framework; the ages are also consistent with sedimentologic observations. The majority of the deeper holes have U–Th dates near their bases, indicating that the LGM was sampled (age range 25–20 ka; Fig. 7.7). Shallower samples from these cores (dated by radiocarbon) record dates as recent as 13–14 ka, which suggests that the early portion of the deglacial has also been captured (Yokoyama et al., 2011). Holes drilled in shallower water have ages as young as 10 cal ka BP (Fig. 7.7), indicating that early Holocene coral reef samples have been collected. There appears to be a sharp decline in the number of samples post-dating 10 cal ka BP and may reflect a possible reef drowning event at this time.

7.8 CONCLUSIONS

Coral reefs are ideally suited to directly determine paleo-sea levels as reef development is sensitive to water depth due to the association of symbiotic algae, zooxanthellae, cohabiting with corals. Corals are essentially the only marine organisms for which the "closed-system" requirement can be satisfied for accurate radiometric dating through careful sampling. A number of specific tests ensure closed-system behavior for uranium and thorium to facilitate U-series dating using TIMS and ICP-MS over the last 600 ka (Yokoyama and Esat, 2004; Stirling and Andersen, 2009). Due to technological advances in offshore underwater sampling as well as in analytical dating techniques, past sea levels can be reconstructed at high temporal resolution. At this level it becomes possible to usefully compare coral data with other climate proxies, such as ice core records (Yokoyama and Esat, 2011). A further advantage of using coral reefs is that they are located at sites far from former ice sheets, meaning that the ice loading component of GIA is negligible so that the

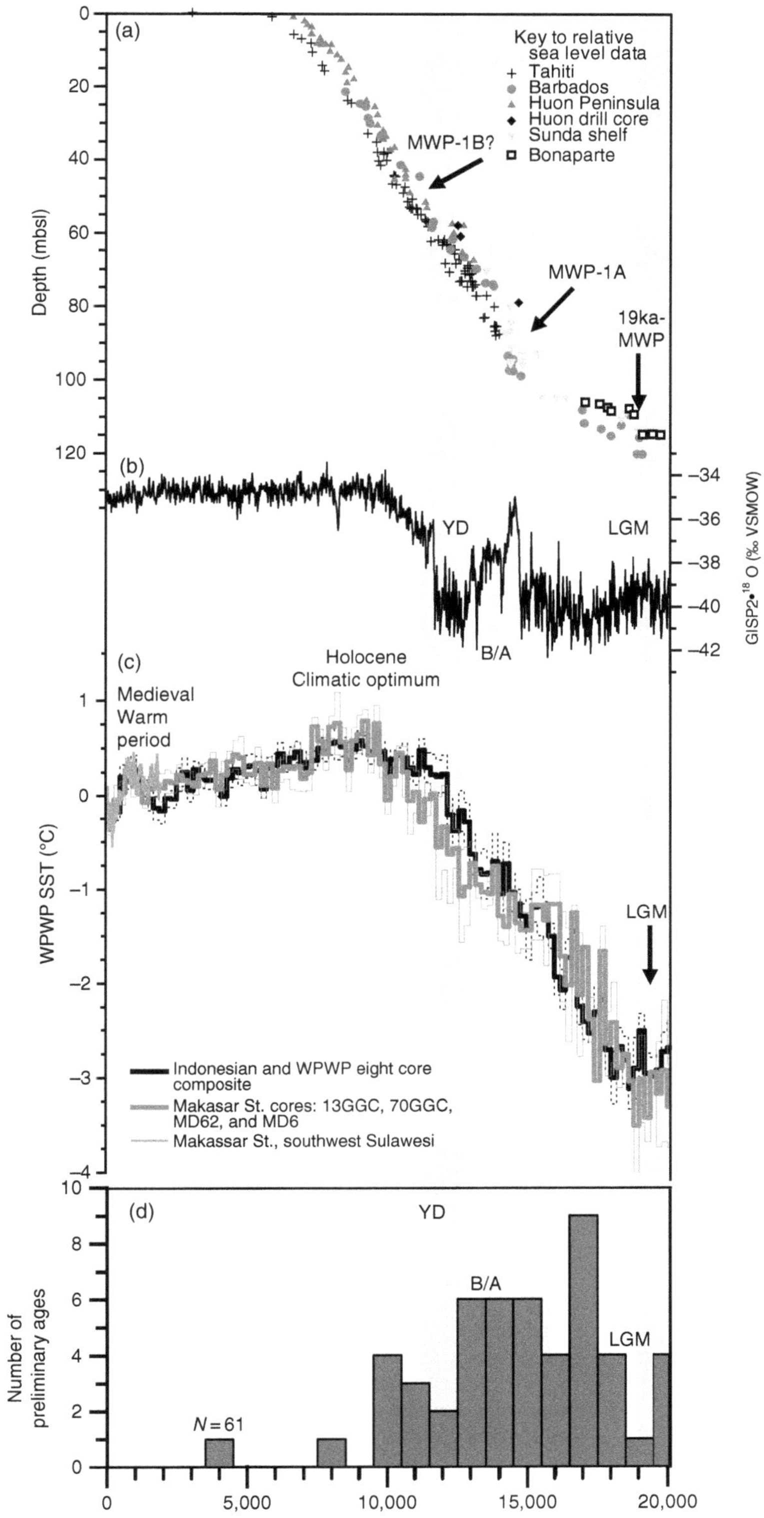

Fig. 7.7. (a) Comparisons between previously reported sea levels during the past 20 ka; (b) Greenland ice core $\delta^{18}O$; (c) sea-surface temperature history of the Western Pacific Warm Pool; and (d) preliminary dating results of GBR IODP Expedition 325 samples. Reef materials obtained from the GBR nicely capture the key intervals of the past climate events including several meltwater pulses, Younger Dryas cold event and the LGM. *Source*: Yokoyama et al., 2011. For color details, please see Plate 23.

relative sea-level observation obtained from corals are closely related to the global sea level (Yokoyama and Esat, 2011). Extensive spatial and temporal archives of coral-based sea-level records have provided pivotal paleoclimate data, and have been eagerly anticipated by the scientific community. Recognition of rapid Quaternary climate changes in the Earth's past has been made possible by coral-reef-based sea-level records. The same records also provide baseline examples to gauge the veracity of extrapolations for future climate change predictions.

ACKNOWLEDGEMENTS

Part of this research is supported by a grant from the Japan Society for the Promotion of Science (JSPS) through the funding program for "Next Generation World-Leading Researchers (NEXT Program GR031)", initiated by the Council for Science and Technology Policy (CSTP) as well as KAKENHI 26247085. This paper is a contribution to INQUA commission on Coastal and Marine Processes and PAGES PALSEA program.

REFERENCES

Abbey, E.A., Webster, J. M., Braga J.C., Sugihara, K., Wallace, C., Iryu, Y., Potts, D., Done T., Camoin, G., and Seard, C. (2011) Variation in postglacial coralgal assemblages and their paleoenvironmental significance, IODP Expedition 310, Tahiti Sea Level. Global and Planetary Change, 76, 1–15.

Alley, R.B. (2000) Ice core evidence of abruput climate changes. Proceedings of the National Academy of Sciences of the United States of America, 97(4), 1331–1334.

Andersen, M.B., Stirling, C.H., Potter, E.-K., Halliday, A.N., Blake, S.G., McCulloch, M.T., Ayling, B.F., and O'Leary, M. (2008) High-precision U-series measurements of more than 500,000 year old fossil corals. Earth and Planetary Science Letters, 265, 229–245.

Ayling, B.F., McCulloch, M.T., Gagan, M.K., Stirling, C.H., Andersen, M.B., and Blake, S.G. (2006) Sr/Ca and d18O seasonality in a Porites coral from the MIS 9 (339–303 ka) interglacial. Earth and Planetary Science Letters, 248, 462–475.

Bar-Mathews, M., Wasserberg, G.J., and Chen, J.H. (1993) Diagenesis of fossil coral skeletones, Correlation between trace elements, textures, and 234U/238U. . Geochimica et Cosmochimica Acta, 57, 257–276.

Bard, E., Hamelin, B., and Fairbanks, R.G. (1990) U–Th ages obtained by mass spectrometry in corals from Barbados, sea level during the past 130,000 years. Nature, 346, 456–458.

Bard, E., Hamelin, B., Arnold, M., Montaggioni, L.F., Cabioch, G., Faure, G., and Rougerîe, F. (1996) Deglacial sea-level record from Tahiti corals and the timing of global meltwater discharge. Nature, 382, 241–244.

Bard, E., Hamelin, B., and Delanghe-Sabatier, D. (2010) Deglacial Meltwater Pulse 1B and Younger Dryas Sea Levels Revisited with Boreholes at Tathiti. Nature, 327, 1235–1237.

Beck, J.W., Récy, J., Taylor, F., Edwards, R.L., and Cabioch, G. (1997) Abrupt changes in early Holocene tropical sea surface temperature derived from coral records. Nature, 385, 705–707.

Bender, M., Fairbanks, R.G., Taylor, F.W., Matthews, R.K., Goddard, J.G., and Broecker, W.S. (1979) Uranium-series dating of the Pleistocene reef tracts of Barbados, West Indies. Geological Society of America Bulletin, 90(6), 577–594.

Bentley, M.J., Fogwill, C.J., Le Brocq, A.M., Hubbard, A.L., Sugden, D.E., Dunai, T.J., and Freeman, S.P.H.T. (2010) Deglacial history of the West Antarctic Ice Sheet in the Weddell Sea embayment, Constraints on past ice volume change. Geology, 38, 411–414.

Blanchon, P. (1998) Continuous record of reef growth over the past 14 k.y. on the mid-Pacific island of Tahiti, Comment. Geology, 26(5), 479–480.

Blanchon, P., and Shaw, J. (1995) Reef drowning during the last deglaciation, Evidence for catastrophic sea-level rise and ice-sheet collapse. Geology, 23, 4.

Blanchon, P., and Blakeway, D. (2003) Are catch-up reefs an artefact of coring. Sedimentology, 50, 1271–1282.

Bloom, A.L., and Yonekura, N. (1985) Coastal terraces generated by sea-level change and tectonic uplift. In *Models in Geomorphology* (ed. M.J. Woldenberg), Allen and Unwin Inc., Winchester, Massachusetts, pp. 139–153.

Bloom, A.L., Broecker, W.S., Chappell, J., Matthews, R.K., and Mesolella, K.J. (1974) Quaternary sea level fluctuations on a tectonic coast, new 230Th/234U dates from the Huon Peninsula, New Guinea. Quaternary Research, 4, 185–205.

Braga, J.C.A., J. (1995) Taxonomy of fossil coralline algal species, Neogene Lithophylloideae (Rhodophyta, Corallinaceae) from southern Spain. Review of Palaeobotany and Palynology, 86, 265–285.

Bridge, T.C.L., Done, T.J., Beaman, R.J., Friedman, A., Williams, S.B., Pizarro, O., and Webster, J.M. (2011) Topography, substratum and benthic macrofaunal relationships on a tropical mesophotic shelf margin, central Great Barrier Reef, Australia. Coral Reefs, 30, 143–153.

Broecker, W.S. (1998) Paleocean circulation during the last deglaciation: a bipolar seesaw? Paleoceanography, 13(2), 119–121.

Broecker, W.S., Thurber, D.L., Goddard, J., Ku, T.-L., Matthews, R.K., and Mesolella, K.J. (1968) Milankovitch hypothesis supported by precise dating of coral reefs and deep-sea sediments. Science, 159(3812), 297–300.

Cabioch, G., Camoin, G.F., and Montaggioni, L.F. (1999) Postglacial growth history of a French Polynesian barrier reef tract, Tahiti, central Pacific. Sedimentology, 46, 985–100.

Cabioch, G., Banks-Cutler, K.A., Beck, W.J., Burr, G.S. Corrége, T., Edwards, R.L., and Taylor, F.W. (2003) Continuous reef growth during the last 23 cal kyr BP in a tectonically active zone (Vanuatu, SouthWest Pacific). Quaternary Science Reviews, 22, 1771–1786.

Camoin, G.F., Colonna, M., Montaggioni, L.F., Casanova, J., Faure, G., and Thomassin, B.A. (1997) Holocene sea level changes and reef development in the southwestern Indian Ocean. Coral Reefs, 16, 247–259.

Camoin, G., Ebren, P., Eisenhauer, Bard, E., and Faure, G. (2001) A 300,000-yr coral reef record of sea level changes, Mururoa atoll (Tuamotu archipelago, French Polynesia). Palaeogeography, Palaeoclimatology, Palaeoecology, 175, 325–341.

Camoin, G.F., Iryu, Y., McInroy, D.B. and the IODP Expedition 310 Scientists (2007) IODP Expedition 310 reconstructs sea level, climatic, and environmental changes in the South Pacific during the last deglaciation. Scientific Drilling, 5, 4–12.

Camoin, G.F., Seard, C., Deschamps, P. et al. (2012) Reef response to sea-level and environmental changes during the last deglaciation. IODP Expedition 310 'Tahiti Sea Level'. Geology, 40, 643–646.

Chappell, J. (1974) Geology of coral terraces, Huon Peninsula, Papua New Guinea: A study of Quaternary tectonic movements and sea-level changes. Bulletin of Geological Society of America, 85, 553–570.

Chappell, J. (1983) Evidence for smoothly falling sea levels relative to north Queensland, Australia, during the past 6000 years. Nature, 302, 406–408.

Chappell, J. (2002) Sea level changes forced ice breakouts in the Last Glacial Cycle, New results from coral terraces. Quaternary Science Reviews, 21(10), 1229–1240.

Chappell, J., and Shackleton, N.J. (1986) Oxygen isotopes and sea level. Nature, 324, 137–140.

Chappell, J., Omura, A., Esat, T.M., McCulloch, M.T., Pandolfi, J., Ota, Y., and Pillans, B. (1996) Reconciliation of late Quaternary sea levels derived from coral terraces at Huon Penninsula with deep sea oxygen isotope records. Earth and Planetary Science Letters, 141, 227–236.

Chen, J.H., Edwards, R.L., and Wasserburg, G.J. (1986) 238U, 234U, 232Th in sea water. Earth and Planetary Science Letters, 80, 241–251.

Chen, J.H., Curran, H.A., White, B., and Wasserburg, G.J. (1991) Precise chronology of the last interglacial period, 234U–230Th data from fossil coral reefs in the Bahamas. Geological Society of America Bulletin, 103, 82–97.

Cheng, H., Edwards, R.L., Hoff, J., Gallup, C.D., Richards, D.A., and Asmerom, Y. (2000) The half-lives of uranium-234 and thrium-230. Chemical Geology, 169(1–2), 17–33.

Cheng, H., Edwards, R.L., Broecker, W.S., Denton, G.H., Kong, X., Wang, Y., Zhang, R., and Wang, X. (2009) Ice age terminations. Science, 326, 248–252.

Clark, P.U., Dyke, A.S., Shakun, J.D., Carlson, A.E., Clark, J., Wohlfarth, B., Mitrovica, J.X., Hostetler, S.W., and McCabe, A.M. (2009) The Last Glacial Maximum. Science, 325, 710–714.

Cutler, K.B., Edwards, R.L., Taylor, F.W., Cheng, H., Adkins, J., Gallup, C.D., Cutler, P.M., Burr, G.S., and Bloom, A.L. (2003) Rapid sea-level fall and deep ocean temperature change since the last interglacial period. Earth and Planetary Science Letters, 206(3–4), 253–271.

Dansgaard, W., Johnsen, S.J., Clausen, H.B., Dahl-Jensen, D., Gundestrup, N.S., Hammer, C.U., Hvidberg, C.S., Steffensen, J.P., Sveinbjörnsdottir, A.E., Jouzel, J., and Bond, G. (1993) Evidence for general instability of past climate from a 250-kyr ice-core record. Nature, 364, 218–220.

Davis, P., and Montaggioni, L. (1985) Reef growth and sea-level change, the environmental signature. Proceedings of the Fifth International Coral Reef Symposium, Tahiti, 477–515.

DeDeckker, P., and Yokoyama, Y. (2009) Micropalaeontological evidence for Late Quaternary sea-level changes in Bonaparte Gulf, Australia. Global and Planetary Change, 66, 85–92.

Deer, W.A., Howie, R.A., and Zussman, J. (1967) *Rock Forming Minerals*. Longmans and Green Co. Ltd., London.

Deschamps, P., Durand, N., Bard, E., Hamelin, B., Camoin, G., Thomas, A.L., Henderson, G.M., Okuno, J., and Yokoyama, Y. (2012) Ice-sheet collapse and sea-level rise at the Bølling warming 14,600 years ago. Nature, 483(29), 559–564.

Dutton, A., and Lambeck, K. (2012) Ice volume and sea level during the last interglacial. Science, 337, 216–219.

Edwards, R.L., Chen, J.H., and Wasserburg, G.J. (1987) 238U–234U–230Th–232Th systematics and the precise measurement of time over the past 500,000 years. Earth and Planetary Science Letters, 81, 175–192.

Edwards, R.L., Beck, J.W., Burr, G.S., Donahue, D. J., Chappell, J.M.A., Bloom, A. L., Druffel, E.R.M., and Taylor, F.W. (1993) A large drop in atmospheric 14C/12C and reduced melting in the Younger Dryas, documented with 230Th ages of corals. Science, 260, 962–968.

EPICA Community Members (2004) Eight glacial cycles from Antarctic ice core. Science, 429, 623–628.

Esat, T.M. (1995) Charge collection thermal ion mass spectrometry of thorium. International Journal of Mass Spectrometry and Ion Processes, 148, 159–171.

Esat, T.M., and Yokoyama, Y. (2006a) Growth patterns of the last ice age coral terraces at Huon Peninsula. Global and Planetary Change, 54, 216–224.

Esat, T.M., and Yokoyama, Y. (2006b) Variability in the Uranium isotopic composition of the oceans over glacial–interglacial timescale. Geochimica et Cosmochimica Acta, 70, 4140–4150.

Esat, T.M., and Yokoyama, Y. (2008) Issues in radiocarbon and U-series dating of corals from the last glacial period. Quaternary Geochronology, 3, 244–252.

Esat, T.M., and Yokoyama, Y. (2010) Coupled uranium isotope and sea-level variations in the oceans. Geochimica et Cosmochimica Acta, 74, 7008–7020.

Esat, T.M., McCulloch, M.T., Chappell, J., Pillans, B., and Omura, A. (1999) Rapid fluctuations in sea level recorded at Huon Peninsula during the Penultimate deglaciation. Science, 283, 197–201.

Fairbanks, R.G. (1989) A 17,000-year glacio-eustatic sea level record, influence of glacial melting dates on Younger Dryas event and deep ocean circulation. Nature, 342, 637–642.

Fairbanks, R.G., Mortlock, R.A., Chiu, T.-C., Cao, L., Kaplan, A., Guilderson, T.P., Fairbanks, T.W., Bloom, A.L., Grootes, P.M., and Nadeau, M.-J. (2005) Radiocarbon calibration curve spanning 0 to 50,000 years BP based on paired 230Th/234U/238U and 14C dates on pristine corals. Quaternary Science Reviews, 24, 1781–1796.

Fairbridge, R.W. (1960) The changing level of the sea. Scientific American, 202, 7079–7084.
Fujita, K., Omori, A., Yokoyama, Y., Sakai, S., and Iryu, Y. (2010) Sea-level rise during Termination II inferred from large benthic foraminifers, IODP Expedition 310, Tahiti Sea Level. Marine Geology, 271, 149–155.
Galewsky, J., Silver, E.A., Gallup, C.D., Edwards, R.L., and Potts, D.C. (1996) Foredeep tectonics and carbonate platform dynamics in the Huon Gulf, Papua New Guinea. Geology, 24, 819–822.
Gallup, C.D., Edwards, R.L., and Johnson, R.G. (1994) The timing of High Sea Levels over the past 200,000 years. Science, 263, 796–800.
Gallup, C.D., Cheng, H., Taylor, F.W., and Edwards, R.L. (2002) Direct determination of the timing of sea level change during termination II. Science, 295, 310–313.
Goreau, T.F., and Wells, J.W. (1967) The shallow-water scleractinia of Jamaica: Revised list of species and their vertical distribution range. Bulletin of Marine Science, 17, 442–453.
Hanebuth, T., Stattegger, K., and Grootes, P.M. (2000) Rapid flooding of the Sunda Shelf, a late-glacial sea-level record. Science, 288, 1033–1035.
Heinrich, H. (1988) Origin and consequences of cyclic ice rafting in the Northeast Atlantic Ocean during the past 130,000 years. Quaternary Research, 29(2), 142–152.
Henderson, G.M., and Anderson, R.F. (2003) The U-series toolbox for paleoceanography. Reviews in Mineralogy and Geochemistry, 52(1), 493–531.
Hopley, D. (1986) Corals and reefs as indicators of paleo-sea levels with special reference to the Great Barrier Reef. In *Sea-Level Research, A Manual for the Collection and Evaluation of Data* (ed. O. van de Plassche), Geo Books, Norwich, pp. 195–228.
Hopley, D. (2006) Coral reef growth on the shelf margin of the Great Barrier Reef with special reference to the Pompey Complex. Journal of Coastal Research, 22(1), 150–158.
Hubbard, D.K. (2009) Depth-related and species-related patterns of Holocene reef accretion in the Caribbean and western Atlantic: a critical assessment of existing models. International Association of Sedimentologists. Special Publication, 41, 1–18.
James, N.P., Mountjoy, E.W., and Omura, A. (1971) An early Wisconsin reef terrace at Barbados, West Indies, and its climatic implications. Geological Society of America Bulletin, 82, 2011–2018.
Lambeck, K. (1981) Lithospheric response to volcanic loading in the Southern Cook Islands. Earth and Planetary Science Letters, 55, 482–496.
Lambeck, K., Yokoyama, Y., and Purcel, A. (2002) Into and out of the Last Glacial Maximum, sea-level change during oxygen isotope stages 3 and 2. Quaternary Science Reviews, 21, 343–360.
Lambeck, K., Purcell, A., Johnston, P., Nakada, M., and Yokoyama, Y. (2003) Water-load definition in the glacio-hydro-isostatic sea-level equation. Quaternary Science Reviews, 22, 309–318.
Larson, K.M., Freymueller, J., and Philipsen, S. (1997) Global plate velocities from the Global Positioning System. Journal of Geophysical Research, 102, 9961–9981.
Lighty, R.G., Macintyre, I.G., and Stuckenrath, R. (1982) Acrpora palmate reef Framework, a reliable indicator of sea level in the western Atlantic for the past 10,000 years. Coral Reefs, 1, 125–130.
Lisiecki, L.E., and Raymo, M.E. (2005) A Pliocene–Pleistocene stack of 57 globally distributied benthic d18O records. Paleoceanography, 20, PA1003.
MacAyeal, D.R. (1993) Binge/purge oscillations of the Laurentaide ice sheet as a cause of the North Atlantic's Heinrich events. Paleoceanography, 8, 775–784.
Mackintosh, A., Golledge, N., Domack, E., Dunbar, R., Leventer, A., White, D., Pollard, D., DeConto, R., Fink, D., Zwartz, D., Gore, D., and Lavoie, C. (2011) Retreat of the East Antarctic ice sheet during the last glacial termination. Nature Geoscience, 4, 195–202.
Marshall, J.F., and Thom, B.G. (1976) The sea-level in the last interglacial. Nature, 263, 120–121.
Matsumoto, K., and Yokoyama, Y. (2013) Reduction in atmospheric Δ14C in simulations of Atlantic overturning circulation shutdown. Global Biogeochemical Cycles, 27, 1–9.
Matthews, R.K. (1973) Relative elevation of late Pleistocene high sea level stands, Barbados, West Indies. Science, 156, 638–640.
Mesolella, K.J., Matthews, R.K., Broecker, W.S., and Thurber, D.L. (1969) The astronomical theory of climatic change: Barbados data. Journal of Geology, 77, 250–274.
Milankovitch, M. (1930) Mathematische kilmalehre und astronomische theorie der klimaschwankungen. In *Handbuch der Klimatologie* (eds Köppen, W., and Geiger, R.), Band1(TeilA), 1–176.
Mitrovica, J.X., and Milne, G.A. (2003) On the postglacial sea level, I General Theory. Geophysical Journal International, 154, 253–267.
Mitrovica, J.X., Tamisiea, M.E., Davis, J.L., and Milne, G.A. (2001) Recent mass balance of polar ice sheets inferred from patterns of global sea-level change. Nature, 409, 1026–1029.
Mitrovica, J.X., Gomez, N., Morrow, E., Hay, C., Latychev, K., and Tamisiea, M.E. (2011) On the robustness of predictions of sea level fingerprints. Geophysical Journal International, 187, 729–742.
Montaggioni, L.F., and Faure, G. (1997) Response of reef coral communities to sea-level rise, A Holocene model from Mauritius (Indian Ocean). Sedimentology, 44, 1053–1070.
Montaggioni, L.F., and Bard, E. (1998) Continuous record of reef growth over the past 14 ky on the mid-Pacific island of Tahiti, Reply. Geology, 26(5), 479–480.
Montaggioni, L.F., Cabiochi, G., Camoin, G., Bard, E., Ribaud-Laurenti, A., Faure, G., Dejardin, P., and Récy, J. (1997) Continous record of reef growth over the past 14 k.y. on the mid-Pacific island of Tahiti. Geology, 25, 555–558.
Nakada, M., and Lambeck, K. (1987) Glacial rebound and relative sea-level variations, a new appraisal. Geophysical Journal of Royal Astronomical Society, 90, 171–224.
Nakamori, T. (1986) Community structures of Recent and Pleistocene hermatypic corals in the Ryukyu islands, Japan. Science Report of Tohoku University, 2nd series (Geology), 56, 71–133.
Nakamori, T., Campbell, C.R., and Wallensky, E. (1995) Living Hermatypic coral assemblages at Huon Penninsula Papua New Guinea. Journal of Geography, 104(5) 743–757.

Neuman, A.C., and Macintyre, I. (1985) Reef response to sea level rise, keep-up, catch-up or give-up. Proceedings of the Fifth International Coral Reef Symposium, 3, 105–110.

Omura, A. (1988) Geologic history of the Kikai Island, Central Ryukyus, Japan, summary of uranium-series dating of fossil corals from the Ryukyu Limestone. Memoirs of the Geological Society of Japan, 29, 253–268.

Omura, A., Chappell, J., Bloom, A.L., McCulloch, M., Esat, T.M., Pillans, B.J., Ota, Y., Sasaki, K., Nakamori, R., Matsuda, S., and Kawada, Y. (1995) Re-evaluation of a-spectrometric 230Th/234U ages for late Pleistocene coral reef terraces of Huon Peninsula, Papua New Guinea. Journal of Geography, 104(5), 758–776.

Ota, Y., Chappell, J., Kelley, R., Yonekura, N., Matsumoto, E., Nishimura, T., and Head, J. (1993) Holocene coral reef terraces and coseismic uplift of Huon Peninsula, Papua New Guinea. Quaternary Research, 40, 177–188.

Pandolfi, J.M. (1996) Limited membership in Pleistocene reef coral assemblages from the Huon Peninsula, Papua New Guinea, constancy during global change. Paleobiology, 22(2), 152–176.

Pandolfi, J.M., and Minchin, P.R. (1995) A comparison of taxonomic composition and diversity between reef coral life and death assemblages in Madang Lagoon, Papua New Guinea. Palaeogeography, Palaeoclimatology, Palaeoecology, 119, 321–341.

Peltier, W.R. (1994) Ice age paleotopography. Science, 265, 195–201.

Potter, E.-K., and Lambeck, K. (2003) Reconciliation of sea-level observations in the Western North Atlantic during the last glacial cycle. Earth and Planetary Science Letters, 217, 171–181.

Potter, E.-K., Esat, T.M., Schellmann, G., Radtke, U., Lambeck, K., and McCulloch, M.T. (2004) Suborbital-period sea-level oscillations during marine isotope substages 5a and 5c. Earth and Planetary Science Letters, 225, 191–204.

Rémy, R., Véronique, C., Guy, C., Norbert, F., and Burr, S.B. (2011) Microborer ichnocoenoses in Quaternary corals from New Caledonia: reconstructions of paleo-water depths and reef growth strategies in relation to environmental changes. Quaternary Science Reviews, 30, 2827–2838.

Richard, D.A., Smart, P., and Edwards, R.L. (1994) Maximum sea levels for the last glacial period from U-series ages of submerged speleothems. Nature, 367, 357–360.

Richter, F.M., and Turekian, K.K. (1993) Simple models for the geochemical response of the ocean to climatic tectonic forcing. Earth and Planetary Science Letters, 119, 121–131.

Robinson, L.F., Henderson, G.M., Hall, L., and Matthews, I. (2004) Climatic control of riverine and seawater uranium-isotope ratios. Science, 305, 851–854.

Sagawa, N., Nakamori, T., and Iryu, Y. (2001) Pleistocene reef development in the southwest Ryukyu Islands, Japan. Palaeogeography, Palaeoclimatology, Palaeoecology, 175, 303–323.

Sasaki, K., Omura, A., Murakami, K., Sagawa, N., and Nakamori, T. (2004) Interstadial coral reef terraces and relative sea-level changes during marine oxygen isotope stages 3–4, Kikai Island, central Ryukyus, Japan. Quaternary International, 120, 51–64.

Schellmann, G., and Radtke, U. (2004) A revised morpho- and chronostratigraphy of the Late and Middle Pleistocene coral reef terraces on Southern Barbados (West Indies). Earth Science Reviews, 64, 157–187.

Scholz, D., and Mangini, A. (2007) How precise are U-series coral ages? Geochimica et Cosmochimica Acta, 71(8), 1935–1948.

Scholz, D., Mangini, A., and Meischner, D. (2006) U-redistribution in fossil reef corals from Barbados, West Indies, and sea level reconstruction for MIS 6.5. In *The Climate of Past Interglacials* (eds Sirocko, F., Claussen, M., Sanchez-Goni, M.F., and Litt, T.), Elsevier, Amsterdam, pp. 119–140.

Seard, C., Camoin, G., Yokoyama, Y., Matsuzaki, H., Durand, N., Bard, E., Sepulcre, S., and Deschamps, P. (2011) Microbialite development patterns in the last deglacial reefs from Tahiti (French Polynesia; IODP Expedition #310), implications on reef framework architecture. Marine Geology, 279, 63–86.

Siddall, M., Rohling, E. J., Almogi-Labin, A., Hemleben, C., Meischner, D., Schmelzer, I., and Smeed, D. A. (2003) Sea-level fluctuations during the last glacial cycle. Nature, 423, 853–858.

Siddall, M., Smeed, D.A., Hemleben, C., Rohling, E.J., Schmelzer, I., and Peltier, W.R. (2004) Understanding the Red Sea response to sea level. Earth and Planetary Science Letters, 225, 421–434.

Skinner, L.C., and Shackleton, N.J. (2005) An Atlantic lead over Pacific deep-water change across Termination I, implications for the application of the marine isotope stage stratigraphy. Quaternary Science Reviews, 24, 571–580.

Skinner, L.C., and Shackleton, N.J. (2006) Deconstructing Terminations I and II, revisiting the glacioeustatic paradigm based on deep-water temperature estimates. Quaternary Science Reviews, 25, 3312–3321.

Stein, M., Wasserburg, G.J., Aharon, P., Chen, J.H., Zhu, Z.R., Bloom, A., and Chappell, J.M.A. (1993) TIMS U-series dating and stable isotopes of the last interglacial event in Papua New Guinea. Geochimica et Cosmochimica Acta, 57, 2541–2554.

Stirling, C.H., and Andersen, M.B. (2009) Uranium-series dating of fossil coral reefs, extending the sea-level record beyond the last glacial cycle. Earth and Planetary Science Letters, 284, 269–283.

Stirling, C.H., Esat, T.M., McCulloch, M.T., and Lambeck, K. (1995) High-precision U-series dating of corals from Western Australia and implications for the timing and duration of the Last Interglacial. Earth and Planetary Science Letters, 135, 115–130.

Stirling, C.H., Esat, T.M., Lambeck, K., and McCulloch, M.T. (1998) Timing and duration of the last interglacial; evidence for a restricted interval of widespread coral reef growth. Earth and Planetary Science Letters, 160, 745–762.

Stirling, C.H., Esat, T.M., Lambeck, K., McCulloch, M.T., Blake, S.G., Lee, D.-C., and Halliday, A.N. (2001) Orbital forcing of the marine isotope stage 9 interglacial. Science, 291, 290–293.

Taylor, F.W., and Mann, P. (1991) Late Quaternary folding of coral reef terraces, Barbados. Geology, 19, 103–106.

Thomas, A.L., Henderson, G., Deschamps, P., Yokoyama, Y., Mason, A.J., Bard, E., Hamelin, B., Durand, N., and Camoin, G. (2009) Penultimate deglacial sea level timing from uranium/thorium dating of Tahitian corals. Science, 324, 1186–1189.

Thompson, W.G., Spiegelman, M.W., Goldstein, S.L., and Speed, R.C. (2003) An open-system model for the U-series age determinations of fossil corals. Earth and Planetary Science Letters, 210, 365–381.

Veeh, H.H., and Chappell, J. (1970) Astronomical theory of climate change, support from New Guinea. Science, 167(3919), 862–865.

Veeh, H.H., and Veevers, J.J. (1970) Sea-level at 175m off the Great Barrier Reef 13,600 to 17,000 years ago. Nature, 226, 536–537.

Veron, J.E.N., and Kelly, R. (1988) Species stability in hermatypic corals of Papua New Guinea and the Indo-Pacific. Memoirs of the Association of Australasian Palaeontologists, 6, 1–69.

Wang, Y.J., Cheng, H., Edwards, R.L., An, Z.S., Wu, J.Y., Shen, C.-C., and Dorale, J.A. (2001) A high-resolution absolute-dated late Pleistocene monsoon record from Hulu Cave, China. Science, 294, 2345–2348.

Webster, J.M., Clague, D.A., Riker-Coleman, K. et al. (2004a) Drowning of the –150 m reef off Hawaii: A casualty of global meltwater pulse 1A? Geology 32, 249–252.

Webster, J.M., Wallace, L., Silver, E., Potts, D., Braga, J.C., Renema, W., Riker-Coleman, K., and Gallup, C. (2004b) Coralgal composition of drowned carbonate platforms in the Huon Gulf, Papua New Guinea; implications for lowstand reef development and drowning. Marine Geology, 204, 59–89.

Whitehouse, P.L., Bentley, M.J., and Le Brocq, A.M. (2012) A deglacial model for Antarctica, geological constraints and glaciological modelling as a basis for a new model of Antarctic glacial isostatic adjustment. Quaternary Science Reviews, 32, 1–24.

Yokoyama, Y., and Esat, T.M. (2004) Long term variations of uranium isotopes and radiocarbon in surface seawater as recorded in corals. Global Environmental Change in the Ocean and on Land, 1, 279–309.

Yokoyama, Y., and Esat, T.M. (2011) Global climate and sea level-enduring variability and rapid fluctuations over the past 150,000 years. Oceanography, 24(2), 54–69.

Yokoyama, Y., Esat, T.M., Lambeck, K., and Fifield, L.K. (2000a) Last ice age millennial scale climate changes recorded in Huon Peninsula corals. Radiocarbon, 42, 383–401.

Yokoyama, Y., Lambeck, K., DeDeckker, P., Johnston, P., and Fifield, L.K. (2000b) Timing of the Last Glacial Maximum from observed sea-level minima. Nature, 406, 713–716.

Yokoyama, Y., Esat, T.M., and Lambeck, K. (2001a) Last Glacial sea-level change deduced from uplifted coral terraces of Huon Penninsula, Papua New Guinea. Quaternary International, 83–85, 275–283.

Yokoyama, Y., Esat, T.M., and Lambeck, K. (2001b) Coupled climate and sea-level changes deduced from Huon Peninsula coral terraces of the last ice age. Earth and Planetary Science Letters, 193, 579–587.

Yokoyama, Y., Purcell, A., Lambeck, K., and Johnston, P. (2001c) Shore-line reconstruction around Australia during the Last Glacial Maximum and Late Glacial Stage. Quaternary International, 83–85, 9–18.

Yokoyama, Y., DeDeckker, P., Lambeck, K., Johnston, P., and Fifield, L.K. (2001d) Sea-level at the Last Glacial Maximum: evidence from northwestern Australia to constrain ice volume for oxygen isotope stage 2. Palaeogeography, Palaeoclimatology, Palaeoecology, 165, 281–297.

Yokoyama, Y., Purcell, A., Marshall, J.F., and Lambeck, K. (2006) Sea-level during the early deglaciation period in the Great Barrier Reef, Australia. Global and Planetary Change, 53, 147–153.

Yokoyama, Y., Miyairi, Y., Matsuzaki, H., and Tsunomori, F. (2007) Relation between acid dissolution time in the vacuum test tube and time required for graphitization for AMS target preparation. Nuclear Instruments and Methods in Physics Research Section B, 259, 330–334.

Yokoyama, Y., Webster, J.M., Cotterill, C., Braga, J.C., Jovane, L., Mills, H., Morgan, S., Suzuki, A., and the IODP 325 Scientists (2011) IODP Expedition 325: The Great Barrier Reef reveals past sea-level, climate and environmental changes since the Last Ice Age. Scientific Drilling, 12, 32–45.

Yokoyama, Y., Okuno, J., Miyairi, Y., Obrochta, S., Demboya, N., Makino, Y., and Kawahata, H. (2012) Holocene sea-level change and Antarctic melting history derived from geological observations and geophysical modeling along the Shimokita Peninsula, northern Japan. Geophysical Research Letters, 39, L13502.

Zinke, J., Reijmer, J.J.G., Thomassin, B.A., Dullo, W.-Chr., Grootes, P.M., and Erlenkeuser, H. (2003) Postglacial flooding history of Mayotte Lagoon (Comoro Archipelago, southwest Indian Ocean). Marine Geology, 194, 181–196.

Chapter 8
Coral microatolls

ARON J. MELTZNER[1] AND COLIN D. WOODROFFE[2]
[1]*Earth Observatory of Singapore, Nanyang Technological University, Singapore*
[2]*School of Earth and Environmental Sciences, University of Wollongong, Wollongong, NSW, Australia*

8.1 INTRODUCTION

Microatolls are coral colonies with living outer margins but with flat, dead upper surfaces. They are commonly found growing in the lower intertidal zone on reef flats and in shallow-water lagoonal environments. Microatoll morphology is most often developed in massive corals; however, branching and foliaceous corals may also develop a similar form when their upward growth is constrained (Fig. 8.1).

The term "microatoll" was adopted to describe large, flat-topped corals with a subcircular ring of living polyps around the margin; their name derives from the fact that their shape is reminiscent of the much larger, true atolls they are sometimes found upon. Large colonies of coral with this growth form were noted on the Cocos (Keeling) Islands in the Indian Ocean by Darwin (1842). Dana (1849, p. 39) correctly deduced that they formed when sea level restricted their further upward growth, describing how "corals, when growing beneath the water, form solid hemispheres, or rounded hillocks; but on reaching the surface, the top dies, and enlargement takes place only on the sides. In this manner the hemisphere is finally changed to a broad cylinder with a flat top." They were referred to as "miniature atolls" by Guppy (1886), and were first called "microatolls" by Krempf (1927, pp. 13–19). Kuenen (1933, pp. 64–65) defined a microatoll as "a colony of corals [with] a raised rim, more or less completely surrounding a lower, dead surface."

Although some authors have ascribed the formation of microatolls to other causes such as the accumulation of sediment on their interior upper surfaces and the consequent death of the colony's upper corallites (e.g., Wood-Jones, 1910, p. 107), it is now clear that it is exposure at lowest tides that prevents further upward growth; microatolls therefore have value as sea-level indicators (Smithers, 2011). The term is now used to describe single coral colonies that are usually, but not always, circular in planform with living vertical sides, and subhorizontal dead upper surfaces. "Fossil" coral microatolls are those in which no living polyps survive – the colony died completely at some time in the past – but the morphology is preserved.

The upper surface of a microatoll has been described as the microatoll plane, and there are commonly a series of concentric ridges, or annuli, that can be seen across this plane. Microatolls growing in close proximity tend to exhibit very similar form, suggesting that they are responding to the same environmental cues. Hopley (1982, p. 113) reviewed their usefulness as indicators of relative sea-level tendency, recognizing four common microatoll forms including: (1) the "classical" coral microatoll form with a relatively flat upper surface; (2) a "top hat" morphology with an elevated center and lower rim; (3) an "upgrown" microatoll form with outer rims substantially elevated above the interior microatoll plane; and (4) "multiple ringed" morphologies in which the upper limit to coral growth has oscillated over time (Fig. 8.2) (Flora and Ely, 2003). While upward growth of a microatoll is restricted by the water/air interface, downward growth is restricted by the sand/water interface (Scoffin and Stoddart, 1978).

Microatolls of *Porites lutea* and *Porites lobata* grow at rates of 5–25 mm a^{-1} and can reach diameters of 9 m. Scoffin and Stoddart (1978) determined that for the northern Great Barrier Reef, where microatolls grow in "open water" settings, they tend to have living rims at an elevation around mean low water springs (MLWS); this relationship has been commonly assumed to be

Handbook of Sea-Level Research, First Edition. Edited by Ian Shennan, Antony J. Long, and Benjamin P. Horton.

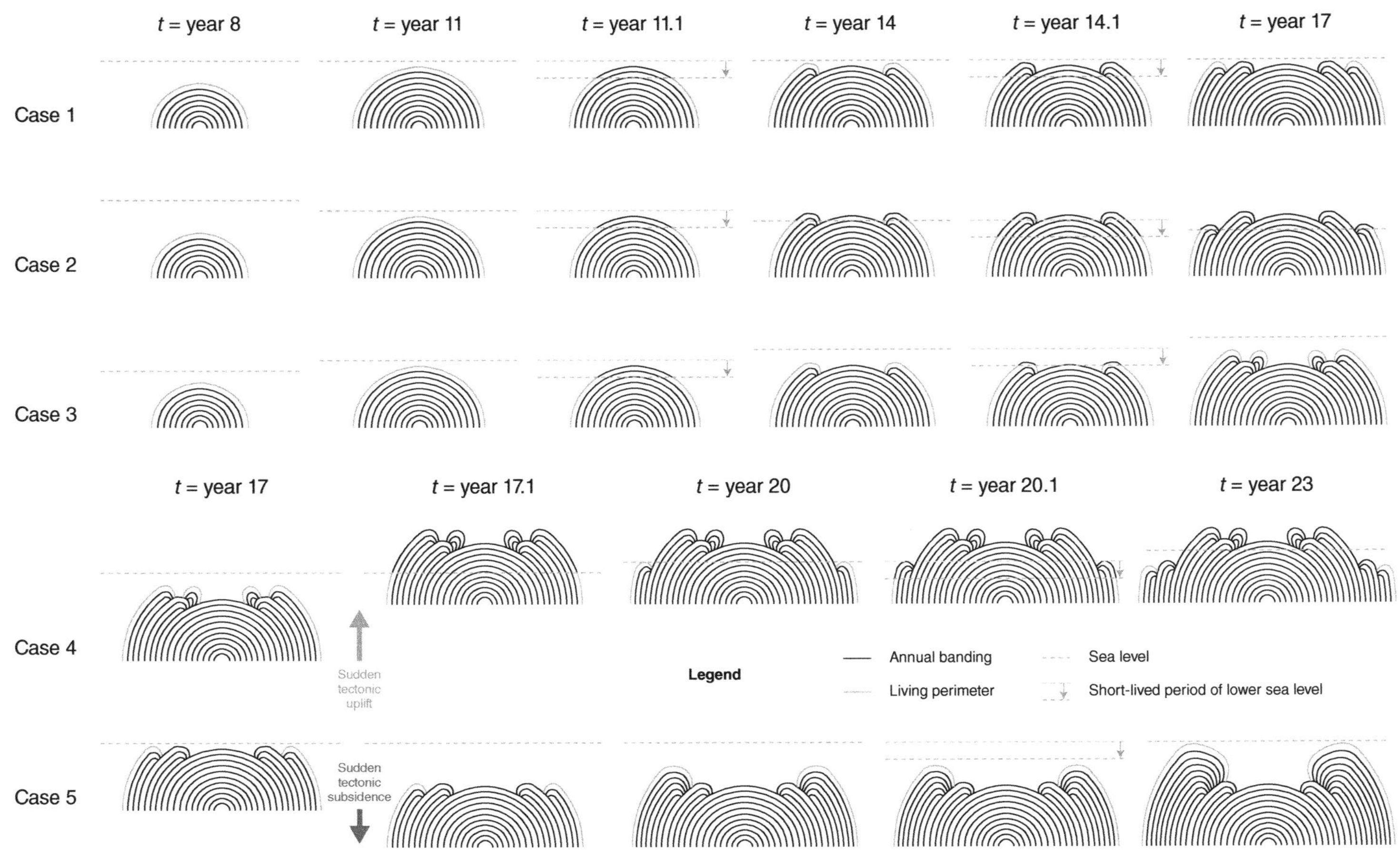

Fig. 8.1. Upper: microatoll development under different relative sea-level history scenarios: under stable sea-level conditions (Case 1); under gradually falling sea level (Case 2); and under gradually rising sea level (Case 3). In all three cases, we superimpose on the long-term trend a realistic interannual variability: at year 11.1 and 14.1, we simulate temporary local sea-level lowerings such as those that tend to occur during La Niña events in the central Pacific, during El Niño events in the western Pacific, or during positive IOD events in the eastern Indian Ocean. Note that concentric annuli form in open-water (unmoated) environments simply from year-to-year fluctuations in the low water level and highest level of survival. Also note the "out-of-sequence" interior annuli that form where relative sea level is rising rapidly enough; this overgrowth protects the underlying inner surface of the microatoll from erosion. Lower: microatoll development affected by sudden (tectonic) changes in land level. Case 4 illustrates the microatoll from Case 3, followed by coseismic uplift at year 17.1; Case 5 illustrates the microatoll from Case 1, followed by coseismic subsidence at year 17.1. In each case, the long-term trend is superimposed on a typical ENSO or IOD cycle, with an additional short-lived period of lower local sea level at year 20.1. In Case 4, the uplift must have been sudden at year 17.1, but if we found the microatoll in Case 5, we could not distinguish between sudden coseismic subsidence at year 17.1 and rapid interseismic subsidence (at an average rate exceeding the coral's growth rate) beginning at year 17.1. For color details, please see Plate 24.

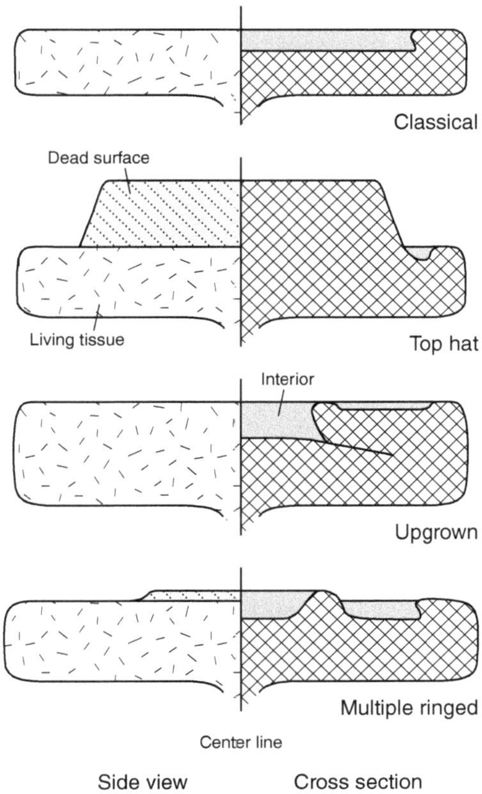

Fig. 8.2. Four common microatoll forms. *Source*: Adapted from Hopley (1982, figure 4.6, p. 113). Reproduced with permission of John Wiley & Sons.

applicable elsewhere, but neither a precise nor a universal relationship has been demonstrated. When microatolls grow in areas not freely connected to the open ocean, such as behind storm ridges, they may grow to higher elevations because the ebb tide is moated (Hopley, 1982, p. 109).

Many massive corals, including *Porites*, are annually banded. The banding results from seasonal contrasts in density and fluorescence. Although the exact causes of these contrasts in *Porites* corals are not well understood (Scoffin et al., 1989), Juillet-Leclerc and Reynaud (2010) studied the growth of the branching coral *Acropora* under different light conditions, identifying relationships between growth rates, microstructures, and photosynthetic activity. In particular, they recognized two distinct growth modes: skeleton thickening is light-enhanced, whereas extension growth is light-suppressed. Coral formed during thickening is denser than that formed during extension.

Microatolls provide information in their growth patterns about present low water level and past sea-level change. They are particularly valuable paleo-sea-level indicators not only because of their vertical precision, but also because individual coral colonies grow and respond to sea level for decades to centuries; unlike shorter-lived organisms, they contain a filtered record of sea level over a sustained period. Individual microatolls may record relative sea level over a period of decades to more than a century and, by combining records from microatolls of different ages, coral microatolls can inform multi-century to millennial-scale relative sea-level trends. In tectonically active settings, microatolls can record both gradual (interseismic and post-seismic) and sudden (coseismic) vertical changes.

This chapter provides an overview of how to use coral microatolls to reconstruct a history of relative sea level for a site. We discuss the utility and limitations of microatolls; considerations for site and sample selection; field techniques; documentation strategies; slab processing, analysis, and interpretation; dating; and vertical accuracy, sources of error, and indicative meaning. We also include a note on permit considerations.

8.2 RELATIVE SEA LEVEL: UNDERSTANDING WHAT MICROATOLLS DO AND DO NOT RECORD

Microatolls start their lives as hemispherical corals below the intertidal zone and for a period of time grow radially outward and upward, unconstrained by low tide or the effects of exposure. They begin to take their microatoll form when they grow high enough, or relative sea level falls low enough, that a coral colony's upper surface is exposed (if only briefly) and those polyps that are desiccated die. The death of the uppermost part of a coral colony, due to exposure at extreme low water, is called a diedown. These coral colonies continue to live below their highest level of survival, or HLS (Taylor et al., 1987), and they continue to grow outward and upward from below that highest level of survival. Because of fluctuations in sea level, a site's highest level of survival can vary by several centimeters or more from one year to the next, an amplitude that exceeds the annual growth rate of the coral (commonly ~10 mm a^{-1}, with a demonstrated range of 5–25 mm a^{-1} for *Porites*; Isdale, 1984; Scoffin et al., 1989). When a microatoll experiences a diedown, its highest level of survival is controlled by low water levels; subsequently, the height to which a coral can grow is limited instead by its growth rate. We will use the expression "highest level of survival" to refer to the elevation down to which a coral dies following an uplift or a period of particularly low sea level; the rest of the time, we will use the expression "highest level of growth" (or HLG) to emphasize that a coral's elevation is limited only by the rate at which it can grow back up to the theoretical highest level at which it could then survive. Some authors have referred to these collectively as the height of living coral or HLC.

If the lowest water level of the year at a site was the same from one year to the next, if the lowest water level always occurred at the same time of day and in the same season, and if the air temperature, cloud cover, wind, and waves were always the same at those times of maximum exposure, then, in the absence of any land-level changes affecting the site, we might expect a coral microatoll to experience a small diedown each and every year. In this hypothetical situation, any small changes in elevation from one year to the next could be attributed to longer-term trends in relative sea level. Yet none of those required conditions are true. Even if weather effects are ignored, the lowest astronomical tide of the year varies from one year to the next, and that year-to-year variation tends to be bigger where the tidal range is larger. In regions where the El Niño/Southern Oscillation (ENSO) or Indian Ocean Dipole (IOD) influences sea level, markedly lower or higher water levels can persist for weeks to months during strong ENSO or IOD events; these sea-level anomalies can have amplitudes of plus or minus tens of centimeters (Woodroffe and McLean, 1990; Meltzner et al., 2010). Weather and waves can vary dramatically from one hour to the next, let alone from one year to the next. Storm surges can dramatically affect sea level, as can monsoonal wind blowing across broad, shallow continental shelves. Distant storms on the open ocean can drive swells that, in turn, modulate wave heights and affect wave splash. Finally, all else being equal, an exposed coral is more likely to die down at noon on a scorching, sunny day than overnight or on a cool, rainy day. For any and all of these reasons, we might expect microatolls to have bigger diedowns some years than others, and perhaps no diedowns at all in other years. Microatolls are remarkably useful proxies for former sea level, but we must take care to understand what microatolls do and do not record.

In Figure 8.1, we illustrate microatoll development under several different relative sea-level history scenarios. The first three cases show how a microatoll might develop under stable sea-level conditions, under gradually falling sea level, and under gradually rising sea level. In all three cases, we superimpose on the long-term trend a realistic interannual variability: at 11.1 and 14.1 years, we simulate temporary local oceanographic lowerings such as those that tend to happen during La Niña events at Kiritimati (Christmas) Island in the central Pacific or during positive IOD events in Sumatra. Several conclusions to draw from these simple cartoons include: (1) diedowns do not occur every year but instead respond to local minima in the relative sea-level time series; (2) corals can sometimes take several years or more to grow back up to their theoretical highest level of survival following a diedown; and (3) the relative elevations of the concentric annuli can inform us about the overall trend of sea level during the coral's lifetime, but we must take care to interpret each annulus correctly. Regarding the second conclusion, we note that a coral might die down further during a severe ENSO or IOD event than during a mild event, and hence it would take a coral longer to grow back up to its theoretical highest level of survival following a severe event than following a mild event; that said, a coral with a fast growth rate will grow back up to its theoretical highest level of survival more quickly than a coral with a slow growth rate. Regarding the third conclusion, we note that in the case of stable sea level (Case 1), the two annuli are at the same height; in the case of falling sea level (Case 2), the outer annulus is lower; and in the case of rising sea level (Case 3), the outer annulus is higher, but there is an extra ("out-of-sequence") interior annulus, equivalent in age to the outer annulus; this extra annulus grew because of the peculiarities of this coral's sea-level history. Although this out-of-sequence interior annulus is shown in Figure 8.1 at exactly the same elevation as the outer annulus, we find that the interior growth can be higher than the outer contemporary annulus (by several centimeters or more) perhaps because water can sometimes pool on the interior upper surface of a microatoll at low tide and allow for locally higher growth.

Such out-of-sequence interior annuli are common in areas of rising sea level, and they actually protect the underlying inner surface of the microatoll from erosion. That said, one must be careful not to misinterpret a coral's relative sea-level history based on out-of-sequence annuli. Furthermore, we ignore the higher interior annuli when calculating a coral's highest level of growth (or survival) for a given year.

At least two studies have shown that a microatoll's diedowns, which can be separated by less than a year to more than a decade, track low water levels. Woodroffe and McLean (1990) compared a representative microatoll from Kiribati with a 3-month running mean of monthly sea level recorded on a tide gauge 140 km away, and showed that the diedowns on the microatoll coincided with the lowest water levels at the tide gauge. More recently, Meltzner et al. (2010) compared satellite altimetry data with Sumatran microatolls and concluded that every diedown on the microatolls since the beginning of satellite altimetry in 1992 could be explained by coseismic uplift or by periods of anomalously low sea surface height. All in all, microatolls preserve a record of the water level that limits upward coral growth (not mean sea level), but microatoll response to rapid sea-level rise is limited by coral growth rate. Still, the exact causes of microatoll diedowns, and the exact relationship between a coral's highest level of survival and the tidal cycle, are not well understood. We explore this relationship in Section 8.9.

8.3 SELECTING A SITE: WHAT MAKES A GOOD SITE?

Most field geologists come to realize, usually sooner rather than later, that the geologist is at the mercy of the geology, and that a perfect field site is difficult (often impossible) to come by. There are situations in which any site with more than a handful of microatolls is a good site. But if the geologist has some flexibility in choosing a field area or has several decent potential sites to choose from within the field area, then there are certain factors he or she should consider.

At bare minimum, a good site will contain one or more populations of coeval microatolls, not simply a few isolated dissimilar coral colonies; each unique population should contain several (≥4–10) morphologically similar microatolls at the same elevation. If the researcher is interested in studying only living microatolls, there should be a sizable selection of living microatolls scattered over the site. If the researcher is interested in reconstructing past sea level with fossil microatolls, there should be a population of living microatolls plus populations from one or more distinct generations of fossil microatolls; the more unique generations at a site, the better the site.

For studies involving fossil microatolls, the researcher should attempt to identify and differentiate different generations of fossil microatolls, although this may not always be possible by field inspection alone and may require range-finder dates to confirm approximate ages of different colonies (if practical). If microatolls at a site are all responding to changing sea level over their lifetimes, then all microatolls alive at a site at a particular time should have grown with a similar morphology. Some microatolls might have reached their highest level of survival earlier than others, so some microatolls of a particular generation might have more annuli than others, but the outer annuli should all be similar if the microatolls died synchronously (Fig. 8.3). Microatolls growing at a different time would have responded to a different sea-level history, so a different generation should have a different morphology and might be at a different elevation. To illustrate this point we consider the fossil microatolls found at a site on northern Simeulue Island, off the west coast of Sumatra (Meltzner et al., 2010). At least eight fossil microatolls with a low interior and high outer double rim (Fig. 8.4a, b), all of which had outer-rim elevations within a narrow range of ~10 cm, were found scattered over ~200 m at the site; interspersed among these were at least seven fossil microatolls of another generation with a low interior, high middle annulus, and intermediate outer annulus (Fig. 8.4c, d), consistently ~20 cm lower than the first generation. By recognizing in the field that this site had multiple generations of fossil microatolls – in addition to a generation of recently living microatolls that died from uplift in 2004 – they anticipated this would be a good site. They invested substantial time and effort into developing and documenting this site, and it turned out to be central to uncovering the 14th–15th century predecessors of the 2004 earthquake.

Because microatolls' reference water level can be complicated to determine (an issue we discuss in Section 8.9) finding living microatolls is

Fig. 8.3. Potential variability in contemporaneous microatolls. Microatolls (a–c) all experienced the relative sea-level history depicted in Cases 3 and 4 of Figure 8.1, having formed under gradually rising sea level with a modest uplift event at year 17.1. Microatoll (b) is the specimen from Case 4 in Figure 8.1. Microatoll (a) started growing at the same time in a depression but at the same basal elevation as microatoll (b), but irregular topography resulted in less accommodation space. The substrate to the left of the coral is sufficiently high that the microatoll's left side did not survive the diedown at year 17.1. The substrate to the right of the coral is slightly lower, which allowed the right side to survive the year-17.1 diedown, but not the lower diedown at year 20.1. Microatoll (c) grew from the same basal elevation as microatoll (b) but started growing only in year 4; its highest level of growth was therefore lower at year 11.1 and it did not die down then. After year 14.1, microatoll (c) recorded a similar history as microatoll (b). White horizontal lines denote elevations of annuli that could be surveyed in the field. For color details, please see Plate 25.

Fig. 8.4. Photos of microatolls all within ~200 m of one another from LKP-B site of Meltzner et al. (2010), northern Simeulue, Sumatra, Indonesia. (a, b) Examples from a population of fossil microatolls characterized by a low interior and high outer double rim, all of which had outer-rim elevations within a narrow range of ~10 cm. (c, d) Examples from a second population interspersed among the first; these are characterized by a low interior, high middle annulus, and intermediate outer annulus, consistently ~20 cm lower than the first population. The authors initially anticipated these two populations represented different generations of microatolls; dating confirmed that the first population (a, b) died around AD 1394, and the second population (c, d) died ~56 years later. *Source*: Photographs from Aron Meltzner (2010).

important for reconstructing past relative sea level, even for studies focused on fossil microatolls. At sites where both fossil and living microatolls are found, constructing a relative sea-level history for the site is a process of directly comparing the elevations of the fossil microatolls with those of their living counterparts. It is not necessary to know the exact relationship between the highest level of survival and the tidal cycle; we simply assume that that relationship, at that site, has remained the same over the lifetimes of all the microatolls at the site, and the variation in the elevations of both the living and fossil populations at the site must be known. If living microatolls cannot be found however, it is necessary to find nearby sites with living microatolls and to carefully determine the relationship between the highest level of survival and the tides, that is, the indicative meaning of the microatolls.

8.4 SURVEYING AND DOCUMENTATION STRATEGIES

Recognizing the highest living corallites on a living coral is normally easy to do, once you know what to look for (Fig. 8.5). Living coral should be brightly pigmented although, if it has recently bleached, it may appear white. In contrast, once corallites die they tend to be quickly covered by algae if they remain below approximate mean tide. If a microatoll has experienced a diedown within the past year or so, its highest level of survival should appear as a smooth, nearly horizontal level separating brightly pigmented coral below from dark algae-covered coral above. If the boundary between living and dead coral is irregular or not roughly horizontal, this suggests that parts of the coral may have died for reasons other than reaching the highest level of survival; in cases such as these, the highest living coral on the colony provides only a lower bound (minimum estimate) of the highest level of survival. As the coral grows upward and outward from below its highest level of survival, a lip will gradually form around the perimeter of the microatoll at its new highest level of growth; if the microatoll goes several years or more without an additional diedown, the outer lip will grow into a full-fledged outer ring or outer rim. After 3–5 years from a microatoll's most recent diedown, it should have an outer rim that is brightly pigmented. Inside of the outer rim however, the top of the microatoll should be dark and algae-covered. The highest living coral on the outer rim is, by definition, the highest level of growth. On a fossil microatoll, the highest level of growth may be defined either as the highest preserved coral on the outer rim, or as the highest preserved coral anywhere on the colony. This can be ambiguous for a microatoll whose center is higher than its perimeter, so authors should specify their intended meaning of the highest level of growth.

We have already discussed the importance of surveying the highest level of growth on multiple coeval microatolls, preferably distributed over the footprint of the site, for each unique generation of microatolls at a site. Such an exercise in the field will subsequently facilitate the arguments: (1) that any sampled (slabbed) colony is not an anomaly and instead represents an entire population of morphologically similar microatolls; (2) that the sampled microatoll's elevation is representative of the entire population's elevation; and (3) that no part of the site slumped or settled with respect to another part of the site. This will also allow the researcher to quantify the range of the highest level of growth or highest level of survival at the site, and to reduce vertical uncertainties.

In addition to surveying the highest level of growth on multiple microatolls, it is important to survey multiple highest level of growth points on each microatoll to establish consistency or to quantify variability. Smithers and Woodroffe (2000) established the practice of measuring the highest level of survival or highest level of growth at four cardinal and four intermediate compass points (N, NE, E, SE, S, SW, W, NW) around each microatoll perimeter, but they skipped measurements at compass points lacking living coral. They noted that microatolls with incomplete or locally depressed living rims are common, and can be caused by partial burial, bioerosion or physical damage.

Where feasible, it is also wise to survey multiple points on multiple concentric annuli of a microatoll, to demonstrate that a microatoll has not experienced tilting either during its life or after its death. Additionally, in some cases, one or several microatolls at a site will die while most others survive; this may be due to a biological response on an individual coral, or due to the fact that microatolls living in deeper water (growing up to the highest level of survival but from a

Fig. 8.5. Recognizing the highest living corallites (highest level of survival or highest level of growth) on a microatoll is normally easy to do, if you know what to look for. (a) Close-up and (b) wider view of highest level of survival ~3 cm above the water line on a coral at the PPY-A site of Meltzner et al. (2012), northeastern Simeulue, Sumatra, Indonesia. This coral experienced a diedown of ~40 cm due to uplift in the December 2004 and March 2005 earthquakes, but the coral survived around its base and continued to grow outward and upward from below its highest level of survival. Three years later, in an earthquake in February 2008, the microatoll experienced another diedown of ~2 cm. Above the 20 cm mark on the ruler, the coral is dead and covered by 4 years' accumulation of algae. At the 20 cm mark ~5 cm above the water line, a lip circumscribes the microatoll, marking the coral's highest level of growth from early 2005 to early 2008. The uppermost ~2 cm of this lip, from ~20 cm to ~18 cm, is thinly coated by algae; these are the polyps that died in 2008. Below ~18 cm, the coral is alive and polyps are not coated by algae. An incipient outer lip is forming ~2 cm below the previous lip, although this is not obvious in the photographs (taken in February 2009). When considering the scale, note the optical illusion that only the upper ~5 cm, and the reflection of those ~5 cm, are visible due to the bright reflection of the metallic ruler. (c) Stark contrast between dull, algae-covered, dead polyps above the highest level of survival, and bright, living polyps below on recently uplifted microatolls along the coast of northern Tetepare Island, Western Province, Solomon Islands (June 2012). (d) Field of living microatolls in a moated pool, Mapur Island, Riau Islands, Indonesia (April 2012). Again, note the contrast between the bright living corallites on the rim and perimeter of the colony, and the dull, fine sediment-covered inner upper surface. *Source*: Photographs by Aron Meltzner. For color details, please see Plate 26.

deeper substrate) might survive a low diedown that would kill shallower microatolls. In such cases, it is helpful to have elevation data on multiple annuli on each microatoll, to allow the researcher to correlate parts of one microatoll to another (Fig. 8.3) without extracting a cross-section from each colony.

Lastly, we point out the importance of documenting a microatoll's radial symmetry in the field with photographs, careful sketches, notes,

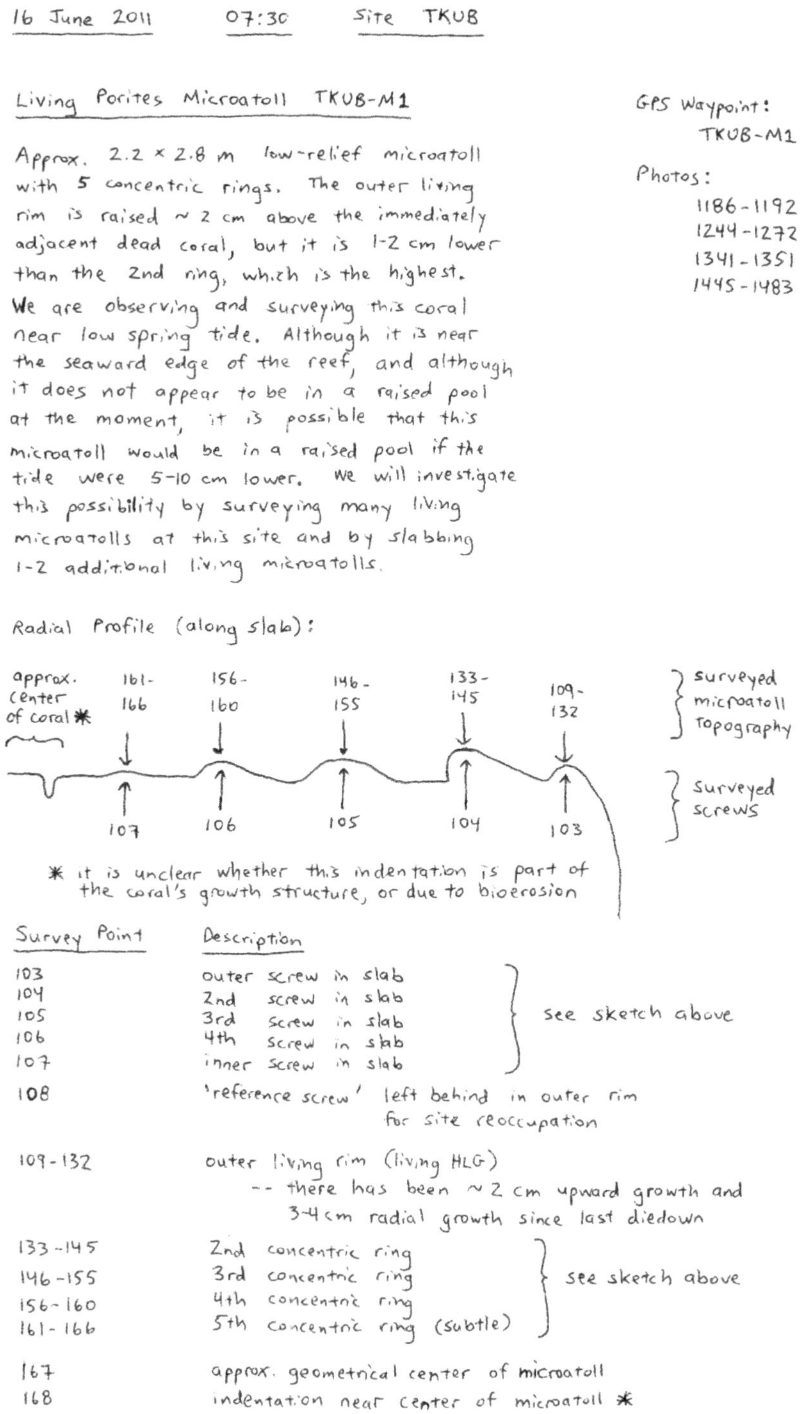
16 June 2011 07:30 Site TKUB

Living Porites Microatoll TKUB-M1

GPS Waypoint: TKUB-M1

Photos:
1186-1192
1244-1272
1341-1351
1445-1483

Approx. 2.2 × 2.8 m low-relief microatoll with 5 concentric rings. The outer living rim is raised ~2 cm above the immediately adjacent dead coral, but it is 1-2 cm lower than the 2nd ring, which is the highest. We are observing and surveying this coral near low spring tide. Although it is near the seaward edge of the reef, and although it does not appear to be in a raised pool at the moment, it is possible that this microatoll would be in a raised pool if the tide were 5-10 cm lower. We will investigate this possibility by surveying many living microatolls at this site and by slabbing 1-2 additional living microatolls.

Radial Profile (along slab):

Survey Point	Description	
103	outer screw in slab	see sketch above
104	2nd screw in slab	
105	3rd screw in slab	
106	4th screw in slab	
107	inner screw in slab	
108	'reference screw' left behind in outer rim for site reoccupation	
109-132	outer living rim (living HLG) -- there has been ~2 cm upward growth and 3-4 cm radial growth since last diedown	
133-145	2nd concentric ring	see sketch above
146-155	3rd concentric ring	
156-160	4th concentric ring	
161-166	5th concentric ring (subtle)	
167	approx. geometrical center of microatoll	
168	indentation near center of microatoll *	

Fig. 8.6. Example field notes showing *in situ* documentation of key observations, including: a general description of the microatoll to be slabbed and its surroundings; the GPS location; photographs; a radial profile along the slab showing surveyed features, locations of surveyed screws, and corresponding survey points. This reflects the minimum information necessary for every slabbed microatoll; more information is often needed in complicated or irregular situations.

and a careful survey (Fig. 8.6). If there has been moderate erosion some diedowns may not be preserved in a particular cross-section, even if a concentric annulus has formed on the surface. Nonetheless, a radially symmetric annulus generally forms in the years following an even, radially symmetric diedown, so the presence of an annulus on the microatoll's upper surface, even if subtle, suggests that a diedown had occurred and that a coral had reached its highest level of survival at that point in the coral's lifetime.

8.5 SAMPLING STRATEGIES: METHODS AND SELECTION OF MICROATOLLS TO SAMPLE

Options for sampling a microatoll include slabbing (extracting a radial or diametrical cross-sectional slab from the microatoll, using a saw) and coring (extracting a core from the microatoll, using a core drill). We begin this section with a discussion of the advantages of slabbing over coring. With only a cylindrical core, only the coral's surface morphology can be used to infer diedowns, an often reliable but nonetheless indirect method. With just a core, the exact timing or the exact elevation of each diedown cannot be determined. With a slab however, the coral's banding pattern can be examined to confirm that the coral's concentric annuli result from diedowns and that the growth is continuous throughout the microatoll. (In rare cases, a coral will die entirely and be subsequently recolonized; this process may impart on the coral a morphology that resembles a microatoll.) The number of annual bands between successive diedowns can also be precisely counted, and the coral's highest level of survival or highest level of growth can be determined each year or as frequently as desired.

As a rule of thumb a researcher should slab at least one microatoll per unique generation, but a second slab (from a second microatoll) may be considered: (1) to assess any vertical error in the highest level of survival; (2) in the situation where one microatoll from a particular generation may be better preserved earlier in its history and another microatoll may be better preserved later in its history; or (3) if there is uncertainty or reason to believe that the microatolls may be from different generations or may have died at different times. Additionally, dating issues should be considered at the slabbing stage: *Porites* sp. corals tend to provide the best slabs morphologically, yet Zachariasen (1998, §3.4b), Natawidjaja et al. (2006), Sieh et al. (2008), and Meltzner et al. (2010) found that, in comparison to *Porites* corals, *Goniastrea* corals tend to have far lower thorium concentrations (i.e., they are "cleaner") and hence provide a more precise estimate of the coral's age by U–Th dating techniques, often to better than ±5 years. For this reason, if a population of fossil microatolls includes both genera of corals, it may be advantageous to take a slab from a *Porites* microatoll to provide the best relative sea-level history, and an additional slab or sample from a *Goniastrea* microatoll to provide the best estimate of the generation's age. If the corals are from the same generation, they should yield replicate U–Th ages but the *Goniastrea* sample will commonly yield a much more precise age, because it was so much "cleaner" than the *Porites* sample. We note however that comparisons of highest level of survival from different species should be made with caution, as systematic differences in highest level of survival exist between different genera and species (Scoffin et al., 1997; Natawidjaja et al., 2006).

Lastly, we discuss considerations for choosing the best microatolls to slab. An important point is that the biggest colonies of a generation are not necessarily the best, particularly if they have few annuli or are more eroded. A large colony with few concentric annuli may have grown for a long time, but it would have been below its highest level of survival for most of its life; its record of relative sea level is therefore limited to the short period spanned by its annuli. Instead, a researcher should search for the microatoll from each generation with the most concentric annuli, regardless of the size of the microatoll itself. Sometimes, the inner annuli of a microatoll will be overgrown (e.g., Fig. 8.7), and the overgrowth ("out-of-sequence" interior annuli, as discussed in Section 8.2) will protect the inner annuli and inner diedowns from subsequent erosion. These microatolls can be the best microatolls to slab because of their superb preservation, but recognizing which of these microatolls will have a good set of hidden, protected inner annuli is challenging and requires a combination of experience, skill, and luck on the part of the researcher.

8.6 SLABBING TECHNIQUES AND SUBSEQUENT SLAB PROCESSING

Once a microatoll is selected for slabbing, the best radius should be chosen for the actual slab. Considerations for choosing a radius include: finding the radius where the most diedowns tend to be best preserved and least eroded; avoiding cracks, irregularities, or areas of substantial bioerosion; and aligning the slab with the coral's growth direction as closely as possible. In some cases, cutting an entire diameter or

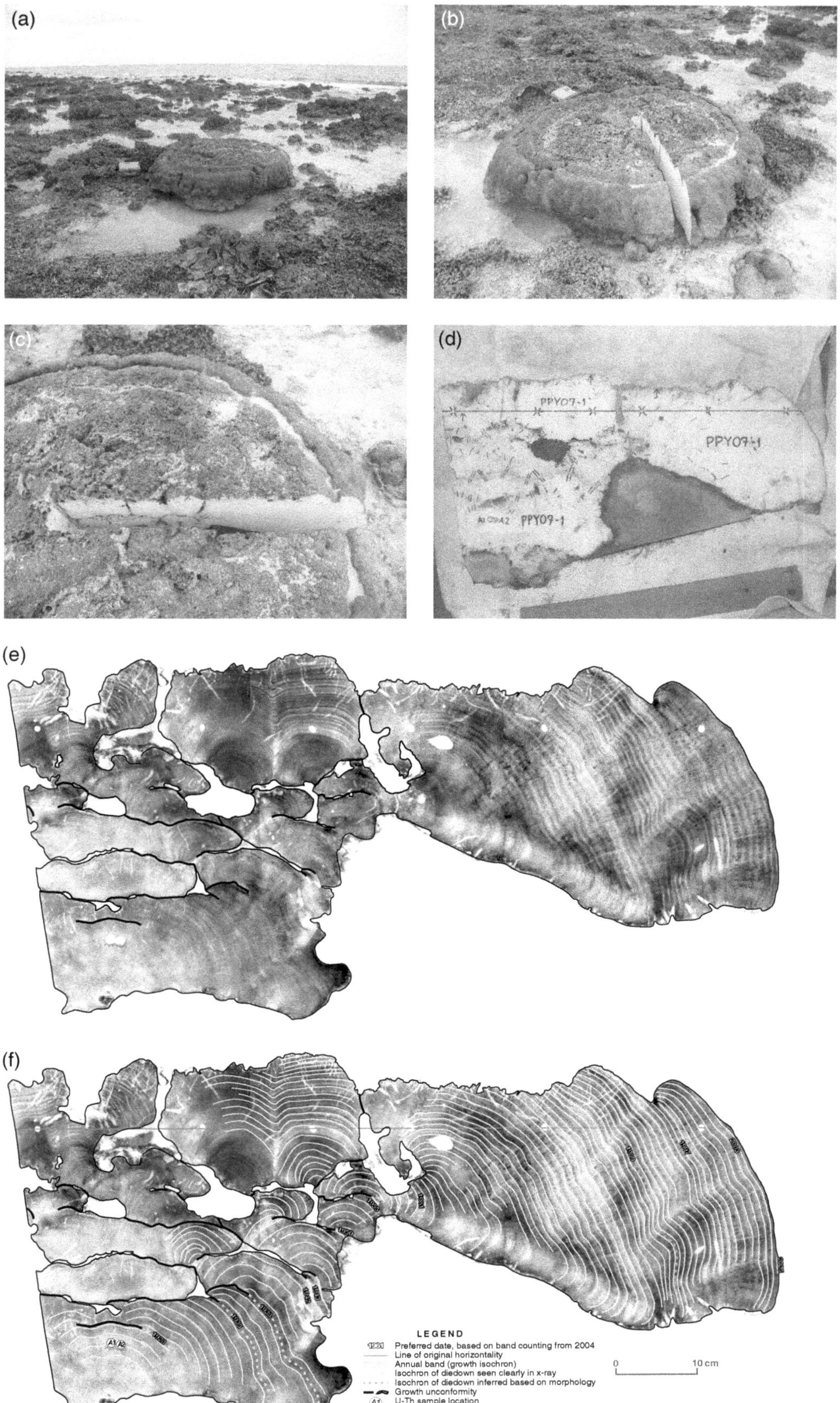

Fig. 8.7. PPY-1 microatoll and slab. (a–c) Views of the microatoll in the field. Removal of the slab allowed us to view below the inner surface of the microatoll, but prior to the slab's removal, only the overgrowth (the flat inner surface of the microatoll) was visible. (d) Photograph of the slab after removal and initial processing. The slab was subsequently X-rayed, and the X-rays were (e) mosaicked and (f) annotated. Although some erosion of the low interior annuli occurred and is evident on the X-rays, the erosion was minor. Once the low interior annuli were overgrown, they were protected from additional erosion. *Source*: Photographs by Aron Meltzner. For color details, please see Plate 27.

more than one radial slab might be advantageous. Once the location of the slab has been chosen, it is helpful to mark the location where the cuts are to be made; a string or other straightedge can be useful for ensuring that the cuts are not crooked.

It will also be important to tie the slab's elevation into the overall site survey, and to retain information about the slab's original horizontality. Screws should be placed at intervals along the radius or diameter of what will become the slab. We use an "eggbeater"-type hand drill to drill small holes into the top of the coral where the screws will be located; we then insert plastic anchors and screw in the screws. We recommend at least four–five screws per slab. As a minimum, two screws are needed to reconstruct the horizontal, but a third screw is important for verification and redundancy; a fourth (and fifth) screw is recommended because sometimes a piece of the slab breaks off or a screw can fall out, and in the end it is important to have at least three useable screws. For slabs longer than ~1.75 m, six or more screws should be considered. The screws should be surveyed carefully before the slab is cut. In addition to the screws in the slab, we recommend placing an additional "reference" screw elsewhere in the microatoll which will be left behind after work at the site is completed; this can serve as a useful survey benchmark in case the site ever needs to be reoccupied and resurveyed.

For slabbing microatolls, several types of saws are available. In the Cocos (Keeling) Islands, where microatolls tend to be pancake-shaped (thin and wide), C.D.W. and colleagues initially mounted a diamond rock-saw blade on a modified brush cutter (string trimmer), but later used a modified circular saw to cut slabs. At Christmas (Kiritimati) Island, they used a petrol-powered concrete cutting saw. Chappell and others have used a cross-cut saw. In Sumatra, where microatolls tend to be taller because of interseismic subsidence, A.J.M. and colleagues have used a hydraulic chainsaw. ICS® (http://icsbestway.com/) manufactures diamond-coated hydraulic chainsaws designed for cutting concrete underwater, with guide bars up to 63 cm long; this allows for cutting microatoll slabs nearly 63 cm deep, both above and below water level (Fig. 8.8a). Zachariasen (1998, §2.2c) described in detail the method for extracting a slab with such a chainsaw. However, living or recently dead coral can be cut without too much difficulty using only a hand saw.

After sawing and extracting an approximately 8–10-cm-thick slab along a chosen radius or diameter of the microatoll, samples must be extracted for dating and the slab must be cut into slices thin enough to be X-rayed. Several important steps are involved in processing the slab (e.g., Fig. 8.8).

(1) A line of original horizontality should be constructed and drawn (with permanent marker) on one face of the slab. This can be done by using the surveyed elevation of each screw in the slab and projecting down from each screw to an arbitrary zero elevation; a line through the set of projected points should indicate what was horizontal when the coral (and slab) was observed *in situ*. The elevation of this horizontal line relative to one of the screws (or to the site survey) should be noted.
(2) A set of small holes should be drilled along the horizontal line, through the entire slab; these small holes will collectively define a plane of original horizontality and will allow the horizontal to be identified on an X-ray of any slice of the slab.
(3) Samples should be extracted for radiometric dating using a small (~1 cm diameter) core drill, and the samples should be labeled on the face of the slab. If the samples for dating do not extend through the entire thickness of the slab, it is important to use a drill to extend the hole from each sample entirely through the slab; again, this is so that the samples can be located on an X-ray of any slice of the slab.
(4) If the slab has broken into multiple pieces, these pieces should all be glued back together. Special epoxies exists, created by mixing an epoxy resin and a hardener, that are well suited for this purpose. If there are small, irregular pieces of coral protruding from the microatoll's upper surface, they should be "anchored" to the rest of the coral with generous amounts of epoxy or concrete so that they do not break off when the slab is cut into thin slices.
(5) Photographs should be taken of the face of the slab, with the horizontal line and holes marked and dating samples labeled.
(6) Wooden boxes may be built to hold each slab upright.

Fig. 8.8. Steps involved in obtaining and processing the slab. (a) Cutting a slab with an ICS® diamond-coated hydraulic chainsaw and 63-cm guide bar and chain. This is not the microatoll from which the slab in (b) and (c) came. (b) The pieces of a slab (one radius of a microatoll) have been fitted together, and a line of original horizontality has been drawn on the slab by measuring down an appropriate distance from each surveyed screw; the locations of the screws are indicated by upward pointing arrows. The elevation of the horizontal line relative to the outer screw (the left-most screw in this photo) is indicated and, because the outer screw was surveyed, the elevation of the line can be determined relative to the entire site survey. (c) A set of small holes has been drilled along the horizontal line, collectively defining a plane of original horizontality; samples (labeled A1, A2, B1, B2, etc.) have been extracted for radiometric dating; the various pieces of the slab have been glued together. A "hole" in the face of the slab near the upper right has been filled with epoxy, so that the upper right piece of the slab does not break off during slicing. (d) Slabs being set upright in concrete in wooden boxes. (e) A marble saw is used to cut each slab into two or three parallel slices. *Source*: Photographs by Aron Meltzner.

(7) The slabs should be positioned upright in the boxes, and concrete could be poured around the base of each slab to lower the center of gravity and prevent the slab from falling over. Care should be taken to ensure that the slab is positioned as close to vertical as possible; carpenter's levels are handy for this.

(8) Once the concrete is dry (usually after 2 days, depending upon atmospheric conditions), the slabs can be transported to a location to be sliced into thin 5–8-mm-thick "bread slices". In Indonesia, A.J.M. and colleagues transport the upright slabs by pick-up truck to a nearby marble tile factory. At the factory, a marble saw cuts each slab into two or three parallel

slices. There is a tradeoff in the thickness of the slices: because X-rays show an average density through the slice rather than preferentially sampling one depth or surface, bands are more likely to show up clearly in thinner slices (Barnes et al., 1989); however, a thinner slice is fragile and prone to fragmenting. We find a thickness of 6–7 mm to be a good balance for the coral slices.

(9) Finally, the preferred slice from each slab should be X-rayed. Increasingly, digital X-rays of sufficient resolution (~250 pixels per inch or better) are available; otherwise, traditional X-ray film negatives should be obtained and scanned at sufficient resolution.

We recommend that the first step (drawing a horizontal line) be completed before the slab is transported far, to minimize slab break-up or the loss of any screws; however, steps (4) and onward are best completed in a laboratory setting (a workshop or other sheltered location) once fieldwork is complete.

8.7 DATING STRATEGIES

For living coral slabs and for dead coral slabs whose date of death is known precisely, annual density bands can be counted backwards from the outer edge to determine the true age of any part of the coral. The validity and accuracy of band counting is supported by Smithers and Woodroffe (2001), who demonstrated synchronous undulations, spanning most of the 20th century, over the upper surface of two microatolls collected live from the eastern and southern reef flats of the Cocos (Keeling) Islands. Additionally, they verified an annual periodicity for the fluorescent bands observed in these samples by comparison with density bands, seasonal trends in stable carbon isotopes, alizarin staining, and reference to a skeletal stress band that formed during a severe coral stress event in March 1983. On northern Simeulue Island, Indonesia, Meltzner et al. (2010) demonstrated synchronous diedowns from 1961 to 2004 based on band counting on all slabbed microatolls collected following coseismic uplift in 2004; in particular, band counting indicates that the largest non-tectonic diedowns in that period all align with known positive IOD events in 1961, 1982, and 1997 (Rao et al., 2002), when sea surface heights in the eastern tropical Indian Ocean would likely have been lowest. These correlations, which persist over four decades, support the accuracy of band counting.

For fossil corals whose age is not independently known, small samples can be drilled out of the slab (during the slab preparation stage described in the previous section) for dating. Commonly applied dating methods include radiocarbon (^{14}C) dating and uranium–thorium (U–Th) techniques, although each method has its own pitfalls. For ^{14}C analysis, the marine reservoir effect (ΔR) is often poorly known and must be assumed or determined independently. For U–Th analysis, the system must be assumed to have been closed to both uranium and thorium, and the initial $^{230}Th/^{232}Th$ ratio is assumed or determined by isochron (Shen et al., 2008). In addition, McGregor et al. (2011) demonstrated the utility of uranium-series age dating using laser ablation multi-collector inductively coupled plasma mass spectrometry (LA MC-ICPMS) to rapidly and inexpensively find low-precision ages on a large suite of fossil corals.

Whether ^{14}C or U–Th dating techniques are used, increasingly precise estimates of a microatoll's age can be obtained by considering weighted averages of dates from multiple samples from a single slab. For U–Th analyses, consider a sample from a band $n \pm \varepsilon$ years prior to the outer preserved edge of the slab (determined by counting annual bands; with clear banding, ε is effectively zero) that gives a corrected U–Th age of $t_{\text{sample}} \pm 2\sigma_{\text{sample}}$ years before "present" ($2\sigma_{\text{sample}}$ is the analytical 2σ error and, for the sake of simplicity, ε is also treated as a 2σ error); then the age of the outer preserved edge of the slab is given by:

$$t_{\text{edge}} = t_{\text{sample}} - n$$

and the error of that age estimate is the sum of the two errors added in quadrature:

$$2\sigma_{\text{edge}} = \left[(2\sigma_{\text{sample}})^2 + \varepsilon^2\right]^{1/2}$$

If there are multiple samples from various bands on a single slab, each $n_i \pm \varepsilon_i$ years prior to the outer preserved edge of the slab, giving respective ages of $t_i \pm 2\sigma_i$ years before "present", then the age of the outer preserved edge of the slab is given by:

$$\frac{\sum_i \left[(t_i - n_i) \big/ \left((2\sigma_i)^2 + \varepsilon_i^2 \right) \right]}{\sum_i \left[1 \big/ \left((2\sigma_i)^2 + \varepsilon_i^2 \right) \right]}$$

and, unless overdispersion has occurred, the error of that age estimate is:

$$\sqrt{\frac{1}{\sum_i \left[1 \big/ \left((2\sigma_i)^2 + \varepsilon_i^2 \right) \right]}}$$

The analogous procedure for radiocarbon dating – taking a weighted average of ^{14}C dates from multiple samples from various bands on a slab – is more complicated, but it can be handled explicitly in OxCal (available at http://c14.arch.ox.ac.uk/embed.php?File=oxcal.html) using the D_Sequence() or Combine() function (Bronk Ramsey, 2008).

If a compelling argument can be made that two microatolls at a site lived coevally, their dates can be combined in a manner similar to that just described. Meltzner et al. (2010, 2012) considered several cases in which multiple microatolls from a site yielded similar but insufficiently precise radiometric ages. By carefully counting and comparing the intervals between successive diedowns on each microatoll and noting the highest level of survival following each diedown, they identified slabs with matching diedown histories, that is, places where the slabs recorded similar elevation changes and identical intervals between the respective diedowns. Using this "diedown matching" technique (which is in some ways akin to matching the thickness of successive magnetic stripes on the seafloor or of successive magnetostratigraphic units in sedimentary sequences, and in other ways akin to matching annual growth ring histories in trees) they were able to identify precisely the portions of the microatolls that corresponded to one another, even if different growth rates or other irregularities initially masked the temporal overlap. The ability to correlate individual annual bands on the various slabs allowed all the dates to be combined into a single, far more precise weighted mean than would otherwise have been possible.

Chapters 23 and 26 discuss the methodologies of ^{14}C and U–Th dating, respectively, in greater detail. These chapters explore potential sources of error associated with each technique and potential issues regarding sample integrity, including alteration and open-system diagenesis. We refer readers to these chapters for further valuable information.

8.8 VERTICAL ACCURACY AND SOURCES OF ERROR

In any year, the height to which a microatoll's living coral rim grows (the highest level of growth) may lag beneath the level at which it may potentially survive (the highest level of survival) if the rate of relative sea-level rise exceeds the coral growth rate. Where the limiting water level rises at a rate greater than the coral growth rate, including in cases of coseismic or rapid post-seismic tectonic subsidence, the upper level of coral growth will decouple from the water level and the living coral rim may be growing up to a still rising, stable, or even falling water level. Under these circumstances a microatoll's upper surface morphology may inaccurately reflect the rate and even direction of relative sea-level fluctuations over timescales of months to years. For these reasons, we remind the reader that although the diedowns track low water levels and overall microatoll morphology provides valuable information on multi-decadal trends in relative sea level, care must be taken to understand the limitations of a microatoll record. For example, Spencer et al. (1997) determined that it would take two decades for the living coral rim around a microatoll on Tongareva, Cook Islands, to re-grow to near its limiting water level following a marked (>20 cm) fall in rim height associated with the 1982–1983 ENSO event. Nevertheless, the sea-level variations of tens of centimeters that accompany severe ENSO episodes in the central Pacific or IOD events in the eastern Indian Ocean tend to be recorded as minor annuli of a few centimeters on the tops of microatolls on several Pacific island reef flats (Woodroffe and McLean, 1990; Spencer et al., 1997) or as annuli with up to 10–20 cm relief on the reef flats off northern Sumatra (Meltzner et al., 2010).

The utility of microatolls as paleo-sea-level indicators notwithstanding, there can be geographical variation across a reef in the elevation that limits coral upward growth (Table 8.1). Chappell et al. (1983) found the tops of microatolls within any single field on the inner Great

Table 8.1. Documented highest level of growth (HLG) variability on living open-ocean *Porites* microatolls

Location	Tidal range	HLG variation per microatoll	HLG variation per site	Reference
Inner Great Barrier Reef, Australia	c. 3.0 m		Tops of microatolls within a field usually lie within a vertical range of 10 cm	Chappell et al. (1983)
Mentawai Islands, Sumatra, Indonesia	c. 1.1 m	< ±3 cm	c. 5–15 cm	Zachariasen et al. (2000)
Cocos (Keeling) Islands, Australia	c. 1.2 m	2–3 cm when outliers obviously not constrained by water level are omitted from analysis	>10 cm at 9 of 19 sites; >20 cm at 3 of 19 sites; 282 living microatolls at a total of 19 sites demonstrated overall variation of 40 cm	Smithers and Woodroffe (2000)

Barrier Reef, where the tidal range is ~3 m, to usually lie within a vertical range of 10 cm. Likewise, in the Mentawai Islands off the west coast of Sumatra, Indonesia where the tidal range is ~1.1 m and reefs tend to be narrow, Zachariasen et al. (2000) reported the typical variation of highest level of survival on a single *Porites* colony to be less than ±3 cm, and the variation in the average highest level of survival between microatolls at a single site to be about 5–15 cm. On the Cocos (Keeling) Islands, with a tidal range of ~1.2 m, Smithers and Woodroffe (2000) similarly observed that the highest level of survival around most individual microatolls fluctuates within a 2–3 cm range when outliers obviously not constrained by water level are omitted from the analysis. In addition, they reported that the tops of open-water microatolls are typically elevated between MLWS and MLWN (mean low water neaps) around the atoll as a whole. Yet accurate surveying of 282 living *Porites* microatolls at a total of 19 sites – although showing consistency in the heights of adjacent microatolls in the same field – demonstrated an overall variation of up to 40 cm. Irrespective of whether the microatolls were in an open reef flat, interisland passage, or lagoonal habitat, they found that the highest level of growth (or highest level of survival) on the various surveyed microatolls spanned more than 10 cm at 9 of the 19 sites, and more than 20 cm at 3 sites. Although some of that variability may be related to subtle hydrodynamic conditions such as wave set-up, wave splash, and local tidal conditions, Smithers and Woodroffe (2000) proposed that considerable variability may arise from subtle moating (or ponding), recognizing that "microatolls at many sites clearly survive at elevations substantially above contemporaries lower in the tidal range … usually in small shallow ponds held in subtle depressions in the reef surface as the tide ebbs." Smithers and Woodroffe (2000) further noted that "although individual pools are shallow and may raise water levels only slightly, the cumulative impact of numerous shallow ponds at progressively higher elevations can produce the large depth ranges occupied by microatolls at many sites."

Moating or ponding can play an important role in the upper limit to coral growth, even in cases where the presence of elevated pools is not obvious. Although one pool may raise the water level at extreme low tide by only a centimeter or two over the level in an adjacent pool immediately seaward, and although this may be observable only at extreme low tide, the cumulative effect of multiple subtle ponds at progressively higher elevations can be to raise the highest level of survival in the highest pools by >10 cm on the wider and more physiographically complex reefs (Smithers and Woodroffe, 2000). Ironically, in high wave-energy environments, those highest pools tend to be the most protected and the most conducive to large microatoll growth, so the best microatolls at a site may all have a raised highest level of survival due to the effects of moating. It is therefore imperative that in any microatoll study the researcher considers the reef morphology, both at present and how it might have appeared in the past. The highest level of survival history on a microatoll affected by moating tends to reflect changes in the elevation of the moat sill, rather than regional changes in relative sea level (Smithers, 2011). If the researcher can recover and analyze contemporaneous microatolls that grew hundreds of meters or even kilometers apart, and if these microatolls yield consistent histories with diedowns in the same years at approximately the same elevations, then it is unlikely that either microatoll was affected by moating.

Sediment compaction, settling, or slumping of the substrate can also affect the elevation of microatolls. This can be a problem particularly in tectonically active settings where sites may be frequently affected by shaking, although muddy substrates can be problematic anywhere. At one site off the west coast of Sumatra, Sieh et al. (2008) documented a population of microatolls that was uplifted and died around AD 1380; the highest level of survival of these microatolls spanned 40 cm. Although this range is identical to that documented in the Cocos (Keeling) Islands by Smithers and Woodroffe (2000), the narrowness of the reef off Sumatra, the lack of moating apparent in contours of substrate elevation, and the site's proximity to sources of strong shaking collectively suggest that shaking-induced settling or slumping is a more reasonable explanation at the Sumatra site for the range in the highest level of survival. The microatolls were systematically higher landward and lower seaward, suggesting that progressively more settling had occurred closer to the reef edge. The settling probably occurred during an earthquake at about AD 1380, because the highest level of survival variation for younger microatolls is far less. At another site further north along the outer-arc chain of islands, Meltzner et al. (2010) identified a cluster of microatolls near the reef edge that was 30 cm lower than the rest of the contemporaneous population of microatolls. Again, the problem was localized; unrecognized subtle moating on this low-relief reef cannot easily explain such an abrupt and large variation in the highest level of survival elevations, leaving localized slumping as the most likely explanation.

Whatever the causes for the variability in the highest level of survival, it is useful to document not only the elevation of the highest level of survival on a single (slabbed) microatoll from a particular generation, but also the variability of that highest level of survival. Had only a single microatoll or a single cluster of adjacent microatolls been surveyed at a site with large highest level of survival variation, it would be difficult to determine how well the surveyed highest level of survival represented relative sea level. Surveying coeval highest level of survival on multiple colonies spread over a site gives a sense of the reliability and accuracy of those elevations. Although there is no perfect way to do this for fossil microatolls, we suggest surveying the highest level of survival on a suite of morphologically similar microatolls. If all such microatolls are at nearly identical elevations, an argument can be made that the slab's highest level of survival accurately represents the highest level of survival for that generation; as Chappell et al. (1983) described of their work on the northern Great Barrier Reef, "only those groups of dead microatolls that were extensive and consistent in height were sampled for dating." If there is large variability in the surveyed elevations however, the researcher should decide whether to investigate further or to use the range of elevations to estimate the vertical error. Alternatively, if a population is not extensive, the researcher should consider assigning vertical errors that match or exceed the largest documented vertical spread of any population at the site.

It is also important to recognize whether a microatoll, when found by a researcher, is in its original growth position. Large microatolls may be secured to the substrate beneath them by only a small pedestal from which they can be detached or, in some cases, microatolls may rest on a sandy substrate their entire lives. Occasionally corals will be transported during large storms or tsunami. If a transported coral is deposited right-side-up, it could be interpreted as *in situ* when it is not (McLean et al., 1978). This is another reason for demonstrating that the sampled microatoll is part of a population of morphologically similar microatolls all at the same elevation; it is unlikely that multiple microatolls would all be transported and deposited at the same level.

McLean et al. (1978) highlighted a number of assumptions that are inherent in sea-level studies using microatolls: (1) fossil features equivalent in form and composition, but at higher or lower levels than their contemporary counterparts, developed with a sea level that was higher or lower than present, the magnitude of sea-level change being the difference between the two levels; (2) features inferred to be related to present sea level are indeed so related; (3) ranges in elevation between equivalent contemporary features for all reefs surveyed are similar to the ranges of relict features; and (4) there have been no significant changes over time in the tidal range or wave climate at a particular site.

In general, it is best to relate the elevations of fossil microatolls to elevations of living counterparts nearby, rather than attempting to relate fossil microatoll elevations to low water level or even to living microatoll elevations tens of kilometers away. A key problem of relating the highest level of survival to low water level is that the relationship is poorly defined and could vary considerably from one environment or region to another; we discuss this problem further in Section 8.9.

On Christmas (Kiritimati) Island in the Line Islands, Woodroffe et al. (2012) inadvertently discovered the effect of a strong geoidal gradient that has implications for surveying over long distances in any sea-level study. Using real-time kinematic (RTK) GPS, they surveyed the elevations of an extensive suite of more than 100 fossil microatolls spanning ~40 km across the island and found disparities from one end of the island to the other that reveal a significant gradient of the geoid. Although the elevations of all fossil microatolls on the island are within a remarkably narrow range of 0.1 m of adjacent living microatolls, the living microatolls themselves, and contemporaneous sets of fossil microatolls, showed a variation of ~1 m in the RTK GPS survey, a range comparable to the variation around the island in the height of the geoid itself (Woodroffe et al., 2012). This should serve as a note of caution to any researcher surveying and directly comparing the elevations of sea-level markers over distances of more than a few kilometers: variations in the height of the geoid matter!

8.9 INDICATIVE MEANING

The upper limit to coral growth (a coral's highest level of survival) is generally thought to be constrained near to the elevation of a critical low water level by prolonged subaerial exposure, possibly due to the influence of desiccation, temperature stress, UV exposure, or some combination. However, the exact nature of these controls is poorly understood: a coral's highest level of survival may be influenced by the duration of exposure, intensity of solar radiation, air and water temperature, frequency of wave splash, capillary action on the coral itself, or other factors. It is not yet clear how best to define a reference water level for coral microatolls' highest level of survival. The reference water level certainly varies by genus (and perhaps by species) and is setting-dependent; ultimately, it may well prove to be a parameter that is site-specific. Even considering only *Porites* microatolls in open-water settings, and ignoring moated colonies, does not resolve the problem.

Different studies at different sites have led to different estimates of reference water level. On the northern Great Barrier Reef where the tidal range is ~3 m, Scoffin and Stoddart (1978) found open-water microatolls' highest level of survival to occur around MLWS. In the Cocos (Keeling) Islands where the tidal range is ~1.2 m, Smithers and Woodroffe (2000) reported that the tops of open-water *Porites* microatolls are typically elevated midway between MLWS and MLWN. Off the west coast of Sumatra where the overall tidal range varies from 0.8 to 1.0 m, Meltzner et al. (2010) took a different approach, arguing that the highest level of survival for *Porites* more closely tracks the annual extreme low water (for years in which there are diedowns), lying 19 ± 8 cm above the lowest water level recorded at a site in a given year. Considering all the factors that may influence a coral's highest level of survival, however, it would be surprising if the relationship determined for the west coast of Sumatra applied generally.

We expect the relationship between the highest level of survival and extreme low water, MLWS, MLWN, or any other part of the tidal cycle to depend in part upon the shape of the tidal curve at a site, the site's typical wave environment, and perhaps even the morphology of the reef. A reference water level determined in a microtidal region characterized by narrow reefs with high wave energy and semidiurnal tides is not expected to apply in a mesotidal environment, a region with low wave energy or diurnal tides (if the duration of exposure is important), or on broad reefs. Similarly, the indicative range is expected to be larger on wider and more physiographically complex reefs. If knowledge of the indicative meaning of coral microatoll's highest level of survival is desired, a researcher must collect sufficient data to determine the relationship for his or her field area, and cannot simply rely on values determined elsewhere.

8.10 RECONSTRUCTING RELATIVE SEA-LEVEL TIME SERIES

After microatoll slices have been X-rayed, the X-rays should be mosaicked together in a raster graphics editor such as Adobe® Photoshop®. At this stage, any deformation of the slice caused by cracks can be corrected for by fitting individual fragments together, and the contrast of the X-rays can be improved so that banding is clearer, as long as care is taken to ensure that any adjustments merely emphasize the banding and growth patterns visible in the original X-rays and that no artifacts (such as false bands) are introduced. Subsequently, the mosaicked X-ray image can be imported into a vector graphics editor such as Adobe® Illustrator®, wherein annual bands can be traced. As a note of caution, it should be checked that the scale on the final image represents the actual scale of the coral slice and that the horizontal line, indicated by a line of holes in the slab X-rays (if the procedures recommended in Section 8.6 were followed) is indeed horizontal.

Relative sea-level histories can be reconstructed by accurately measuring the highest level of growth or highest level of survival at regular intervals on each annotated slab X-ray mosaic. Smithers and Woodroffe (2001) measured the highest living coral at two points per annual growth band (for semiannual resolution); Meltzner et al. (2010) measured only one point per year in most years, but for every diedown they measured both the highest level of growth immediately before each diedown and the highest level of survival immediately thereafter. When plotting these time series, Meltzner et al. (2010) were careful to distinguish four different types of data obtained from the coral slice: (1) highest level of survival elevations immediately after a diedown; (2) highest level of growth elevations immediately before a diedown; (3) uneroded highest level of growth elevations where no diedown occurred (a minimum bound on the highest level of survival for that year); and (4) eroded highest level of growth or highest level of survival elevations, where the upper part of the growth band has been removed by erosion (a minimum bound on both the highest level of growth and highest level of survival for that year). Methods to determine rates of relative sea-level change, the significance of those rates, and appropriate errors on those rates have been extensively discussed by Meltzner et al. (2010, 2012). The time series records from individual microatolls at a site can be combined into a single plot for the site, and all time series can be plotted relative to the highest level of growth at present or at any arbitrarily chosen point in time.

8.11 DIFFERENTIATING TECTONIC FROM NON-TECTONIC DIEDOWNS

Criteria for distinguishing oceanographically induced diedowns (those caused by sea-level fluctuations) from tectonically induced diedowns (those caused by changes in land level) have been discussed by Taylor et al. (1987), Zachariasen (1998, §1.5c), and Meltzner et al. (2010). These authors recommend considering the microatoll morphology, contemporaneous microatoll records from other sites regionally, and any historical information available independently. First, the amplitude of the diedown matters: diedowns larger than 10 cm are more likely to be tectonic (Taylor et al., 1987), although ENSO- or IOD-related diedowns of ~15 cm have been documented (Meltzner et al., 2010). Second, the geographic distribution of the diedown and the spatial variability of the amplitude are important: the amplitude of tectonic uplifts tends to vary markedly over short distances, whereas diedowns caused by temporarily lower sea level (as can occur during ENSO or IOD events) should be similar over distances of tens to hundreds of kilometers and should occur simultaneously at sites in tectonically active areas and in adjacent stable regions. Third, the duration of the sea-level change is telling: tectonically induced relative sea-level changes are more likely to persist through time, as a step function, whereas oceanographic fluctuations should be short-lived, as a delta function (Zachariasen, 1998, §1.5c). Meltzner et al. (2010) suggested that, following a large non-tectonic diedown, the coral should have unrestricted upward growth without additional diedowns until that upward growth approaches the coral's previous highest level of growth. In contrast, a tectonic uplift should not be followed by unusual unrestricted upward growth and subsequent diedowns should all be lower than diedowns preceding the uplift (Fig. 8.1, Case 4). Lastly, coincidence in time and location between known earthquakes and changes in relative sea level favors a tectonic cause.

8.12 ENVIRONMENTAL AND ADMINISTRATIVE ISSUES

We conclude with a note about environmental and administrative considerations. It is advisable to ensure that the necessary permits have been researched and obtained well before field sampling. If living corals are to be sampled from shallow water then a permit will likely be needed, particularly in marine protected areas, and it will be necessary to check zoning plans to ensure that research can be undertaken and the necessary slabs or cores of coral can be collected. There is little consistency as to which organizations will need to be approached; it may be environmental agencies, natural resources divisions, or land and wildlife agencies and, in many countries, clearance may be required from fisheries authorities, either to collect or to export coral samples. It will also be important to consider whether the slabs of coral collected can be freighted between countries under the provisions of the Convention on International Trade in Endangered Species of Wild Fauna and Flora (CITES). Coral is listed under CITES, and regulations are becoming increasingly stringent. Live corals, dead corals, and "live rock" require CITES permits for export from and import into most countries, often with specifications to genus, or even species level. Fossil coral, and coral that is dead but may have been alive when collected and in which corallites remain distinct, remain contentious; some countries require a CITES permit to import, regardless of whether an export permit is required by the laws of the country of origin. Coral samples collected from on land may be subject to quarantine restrictions, if only to ensure there are no traces of soil, plants, or insects. It is prudent to have applied for the necessary quarantine clearance or permits before undertaking fieldwork.

ACKNOWLEDGEMENTS

We thank J. Majewski, two anonymous reviewers, and the editors for constructive feedback that led to substantial improvements. We also appreciate our colleagues who worked with us on coral microatoll research over the years; their questions and observations inspired considerable material in this chapter. This is Earth Observatory of Singapore contribution 55.

REFERENCES

Barnes, D.J., Lough, J.M., and Tobin, B.J. (1989) Density measurements and the interpretation of X-radiographic images of slices of skeleton from the colonial hard coral *Porites*. Journal of Experimental Marine Biology and Ecology, 131, 45–60.

Bronk Ramsey, C. (2008) Deposition models for chronological records. Quaternary Science Reviews, 27, 42–60.

Chappell, J., Chivas, A., Wallensky, E., Polach, H.A., and Aharon, P. (1983) Holocene palaeo-environmental changes, central to north Great Barrier Reef inner zone.BMR Journal of Australian Geology and Geophysics, 8, 223–235.

Dana, J.D. (1849) *United States Exploring Expedition, During the Years 1838, 1839, 1840, 1841, 1842, Under the Command of Charles Wilkes, U.S.N., Vol. X: Geology*. C. Sherman, Philadelphia. Available at http://archive.org/details/unitedstatesexpl10unitrich (accessed 18 July 2014).

Darwin, C. (1842) *The Structure and Distribution of Coral Reefs: Being the First Part of the Geology of the Voyage of the Beagle, under the Command of Capt. Fitzroy, R.N.during the Years 1832 to 1836*. Smith Elder and Co., London. Available at http://darwin-online.org.uk/content/frameset?itemID=F271&pageseq=1&viewtype=text (accessed 18 July 2014).

Flora, C.J., and Ely, P.S. (2003) Surface growth rings of *Porites lutea* microatolls accurately track their annual growth. Northwest Science, 77, 237–245. Available at http://www.vetmed.wsu.edu/org_nws/NWSci%20journal%20articles/2003%20files/Issue%203/v77%20p237%20Flora%20and%20Ely.PDF.

Guppy, H.B. (1886) Notes on the characters and mode of formation of the coral reefs of the Solomon Islands, being the results of observations made in 1882–84, during the surveying cruise of H.M.S. "Lark". Proceedings of the Royal Society, Edinburgh, 13, 857–904. Available at http://books.google.com/books?id=3tgfAQAAIAAJ&pg=PA857.

Hopley, D. (1982) *The Geomorphology of the Great Barrier Reef: Quaternary Development of Coral Reefs*. John Wiley, and Sons Inc., New York.

Isdale, P. (1984) Fluorescent bands in massive corals record centuries of coastal rainfall. Nature, 310, 578–579.

Juillet-Leclerc, A., and Reynaud, S. (2010) Light effects on the isotopic fractionation of skeletal oxygen and carbon in the cultured zooxanthellate coral, *Acropora*: implications for coral-growth rates. Biogeosciences, 7, 893–906.

Krempf, A. (1927) *Travaux du Service Océanographique des Pêches de l'Indochine ..., 2e Mémoire: La Forme des Récifs Coralliens et le Régime des Vents Alternants*. Gouvernement Général de l'Indochine, Saigon.

Kuenen, P.H. (1933) *Wetenschappelijke uitkomsten der Snellius-expeditie onder leiding van P.M. van Riel (The Snellius-Expedition in the Eastern Part of the Netherlands East-Indies 1929–1930, under Leadership of P.M. Van Riel), vol. 5 (Geological Results), Part 2: Geology of Coral Reefs*. E. J. Brill, Leiden.

McGregor, H.V., Hellstrom, J., Fink, D., Hua, Q., and Woodroffe, C.D. (2011) Rapid U-series dating of young fossil corals by laser ablation MC-ICPMS. Quaternary Geochronology, 6, 195–206.

McLean, R.F., Stoddart, D.R., Hopley, D., and Polach, H. (1978) Sea level change in the Holocene on the northern Great Barrier Reef. Philosophical Transactions of the Royal Society, London Series A, 291, 167–186.

Meltzner, A.J. (2010) Earthquake recurrence, clustering, and persistent segmentation near the southern end of the 2004 Sunda megathrust rupture. PhD thesis, California

Institute of Technology, Pasadena, CA. Available at http://resolver.caltech.edu/CaltechTHESIS:06012010-082222484.

Meltzner, A.J., Sieh, K., Chiang, H.-W., Shen, C.-C., Suwargadi, B.W., Natawidjaja, D.H., Philibosian, B.E., Briggs, R.W., and Galetzka, J. (2010) Coral evidence for earthquake recurrence and an A.D. 1390–1455 cluster at the south end of the 2004 Aceh–Andaman rupture. Journal of Geophysical Research, 115, B10402.

Meltzner, A.J., Sieh, K., Chiang, H.-W., Shen, C.-C., Suwargadi, B.W., Natawidjaja, D.H., Philibosian, B., and Briggs, R.W. (2012) Persistent termini of 2004- and 2005-like ruptures of the Sunda megathrust. Journal of Geophysical Research, 117, B04405.

Natawidjaja, D.H., Sieh, K., Chlieh, M., Galetzka, J., Suwargadi, B.W., Cheng, H., Edwards, R.L., Avouac, J.-P., and Ward, S.N. (2006) Source parameters of the great Sumatran megathrust earthquakes of 1797 and 1833 inferred from coral microatolls. Journal of Geophysical Research, 111, B06403.

Rao, S.A., Behera, S.K., Masumoto, Y., and Yamagata, T. (2002) Interannual subsurface variability in the tropical Indian Ocean with a special emphasis on the Indian Ocean Dipole. Deep-Sea Research II, 49, 1549–1572.

Scoffin, T.P., and Stoddart, D.R. (1978) The nature and significance of microatolls, Philosophical Transactions of the Royal Society, London, Series B, 284, 99–122.

Scoffin, T.P., Tudhope, A.W., and Brown, B.E. (1989) Fluorescent and skeletal density banding in *Porites lutea* from Papua New Guinea and Indonesia. Coral Reefs, 7, 169–178.

Scoffin, T.P., Brown, B.E., Dunne, R.P., and LeTissier, M.D.A. (1997) The controls on growth form of intertidal massive corals, Phuket, south Thailand. Palaios, 12, 237–248.

Shen, C.-C., Li, K.-S., Sieh, K., Natawidjaja, D., Cheng, H., Wang, X., Edwards, R.L., Lam, D.D., Hsieh, Y.-T., Fan, T.-Y., Meltzner, A.J., Taylor, F.W., Quinn, T.M., Chiang, H.-W., and Kilbourne, K.H. (2008) Variation of initial $^{230}Th/^{232}Th$ and limits of high precision U-Th dating of shallow-water corals. Geochimica et Cosmochimica Acta, 72, 4201–4223.

Sieh, K., Natawidjaja, D.H., Meltzner, A.J., Shen, C.-C., Cheng, H., Li, K.-S., Suwargadi, B.W., Galetzka, J., Philibosian, B., and Edwards, R.L. (2008) Earthquake supercycles inferred from sea-level changes recorded in the corals of West Sumatra. Science, 322, 1674–1678.

Smithers, S. (2011) Microatoll. In *Encyclopedia of Modern Coral Reefs: Structure, Form and Process* (ed. Hopley, D.), Springer, Dordrecht, The Netherlands, Encyclopedia of Earth Sciences Series, pp. 691–696.

Smithers, S.G., and Woodroffe, C.D. (2000) Microatolls as sea-level indicators on a mid-ocean atoll. Marine Geology, 168, 61–78.

Smithers, S.G., and Woodroffe, C.D. (2001) Coral microatolls and 20th century sea level in the eastern Indian Ocean. Earth and Planetary Science Letters, 191, 173–184.

Spencer, T., Tudhope, A.W., French, J.R., Scoffin, T.P., and Utanga, A. (1997) Reconstructing sealevel change from coral microatolls, Tongareva (Penrhyn) Atoll, northern Cook Islands. In *Proceedings of the 8th International Coral Reef Symposium* (eds Lessios, H. A., and Macintyre, I. G.), Smithsonian Tropical Research Institute, Panama, vol. 1, 489–494. Available at http://www.reefbase.org/download/download.aspx?type=1&docid=7727.

Taylor, F.W., Frohlich, C., Lecolle, J., and Strecker, M. (1987) Analysis of partially emerged corals and reef terraces in the central Vanuatu Arc: comparison of contemporary coseismic and nonseismic with Quaternary vertical movements. Journal of Geophysical Research, 92, 4905–4933.

Wood-Jones, F. (1910) *Coral and Atolls: Their History, Description, Theories of their Origin both before and since that of Darwin, the Influence of Winds, Tides and Ocean Currents on their Formation and Transformations, their Present Condition, Products, Fauna and Flora*. Lovell Reeve, and Co., Ltd., London. Available at http://ocp.hul.harvard.edu/dl/expeditions/005186246 (accessed 18 July 2014).

Woodroffe, C., and McLean, R. (1990) Microatolls and recent sea level change on coral atolls. Nature, 344, 531–534.

Woodroffe, C.D., McGregor, H.V., Lambeck, K., Smithers, S.G., and Fink, D. (2012) Mid-Pacific microatolls record sea-level stability over the past 5000 yr. Geology, 40, 951–954.

Zachariasen, J. (1998) Paleoseismology and paleogeodesy of the Sumatran subduction zone: a study of vertical deformation using coral microatolls. PhD thesis, California Institute of Technology, Pasadena, CA. Available at http://resolver.caltech.edu/CaltechETD:etd-04112005-132058.

Zachariasen, J., Sieh, K., Taylor, F.W., and Hantoro, W.S. (2000) Modern vertical deformation above the Sumatran subduction zone: paleogeodetic insights from coral microatolls. Bulletin of the Seismological Society of America, 90, 897–913.

Chapter 9
Archeological and biological relative sea-level indicators

CHRISTOPHE MORHANGE[1] AND NICK MARRINER[2]
[1]*Université Aix-Marseille, IUF, CEREGE UMR 7330, Europôle de l'Arbois, Aix-en-Provence, France*
[2]*Laboratoire Chrono-Environnement, UMR 6249 CNRS, Université de Franche-Comté, UFR ST, Besançon, France*

9.1 INTRODUCTION

The great antiquity of human occupation in the Mediterranean has left rich archeological evidence along its coastlines, including harbors and fish tanks. Within this context, it has long been recognized that certain archeological structures can provide interesting insights into the direction and amplitude of relative sea-level changes since Antiquity. Over the past century, a number of authors have investigated the use of archeological markers to probe relative sea-level changes (e.g., Negris, 1903, 1904, 1921; Cayeux, 1907, 1914; Flemming 1969, 1979–80; Schmiedt, 1972; Pirazzoli, 1976a, b; Blackman, 1973, 1982a, b; Galili et al., 1988; Galili and Nir, 1993; Antonioli and Leoni, 1998; Stiros, 1998; Morhange et al., 2001; Sivan et al., 2001, 2004; Lambeck et al., 2004; Auriemma and Solinas, 2009; Faivre et al., 2010; Anzidei et al., 2011, 2013; Evelpidou et al., 2012; Mourtzas, 2012, among many others). Relative sea level (RSL) archeological evidence is particularly rich in the ancient worlds, including Atlantic Europe with the pioneering development of waterfront archeology (e.g., Milne and Hobley, 1981; Van de Noort and O'Sullivan, 2006), the Mediterranean and the Near East (e.g., Marriner, 2009; Carayon et al., 2011; Hein et al., 2011), India (e.g., Rao, 1988; Gaur, 2006) and China (references in Chinese). In a general context of RSL stability since 6000 cal. BP (van Andel, 1989; Lambeck and Bard, 2000), archeological heritage provides a unique opportunity to refine RSL variations in the highly diversified crustal context of the Mediterranean (Stewart and Morhange, 2009). In this chapter, we emphasize the use of fixed bioindicators in archeological contexts to further the precision of relative sea-level variations and trends during the mid–late Holocene. Archeological indicators on their own may be problematic in referencing RSL, but when combined with fixed bioindicators their value can be greatly increased.

One of the advantages of archeological structures relates to their great antiquity. For instance, the oldest harbor structures have been dated to the Old Kingdom c. 2600–2300 BC at Ayn Soukhna in the Red Sea (Tallet, 2009). By contrast, the oldest maritime installations in the Mediterranean seem to be more recent and have been attributed to the Iron Age. For example, radiometric dating has constrained the Phoenician mole at Athlit to the 9th century BC (Haggi and Artzy, 2007). A similar example is also known from the Syrian coast at Tabbat el-Hammam, where the archeological evidence supports a 9th/8th century BC age (Braidwood, 1940). For the Mediterranean, this restricts the use of archeological remains to the last c. 3000 years although drowned coastal sites dating from earlier periods can provide insights into broad sea-level tendencies during the early–mid Holocene (Galili et al., 1988; Sartoretto et al., 1995).

Analysis of harbor works and the fixed and boring marine organisms attached to waterfront structures has long been recognized as a potential source of sea-level data, for example the Roman columns of the market of Pozzuoli in southern Italy (Lyell, 1830; Fig. 9.1) or the Roman harbor of Marseille in France (Pirazzoli and Thommeret, 1973). Such data are fundamental to understanding the vertical distribution of coastal remains. Where precise vertical relationships can be established between archeological structures and past biological sea levels, it has been possible to

Handbook of Sea-Level Research, First Edition. Edited by Ian Shennan, Antony J. Long, and Benjamin P. Horton.

Fig. 9.1. Frontispiece of the three pillars of the Roman market at Pozzuoli (southern Italy) which have become an icon of uniformitarianism since their publication in Lyell's Principles of Geology (1830). Lyell argued that the rise and fall of these coastal archeological remains showed that the land had undergone significant vertical movements since Antiquity.

accurately reconstruct relative sea-level trends since Antiquity at a number of Mediterranean sites (Marriner and Morhange, 2007). Paradoxically, this simple methodology is rarely used.

9.2 HISTORICAL RESEARCH CONTEXT IN THE MEDITERRANEAN

Since the early 20th century, a number of scholars have undertaken systematic surveys of submerged port structures (e.g., in Egypt, Jondet 1916; in Greece, Paris 1915, 1916; and in Dalmatia, Degrassi 1955). At the beginning of the 20th century – in a scientific context dominated by the dogma of a stable sea level during historical times as advocated by Suess (1885–1908) – a debate opposed the Greek Negris (1903, 1904, 1921) and the influential French geologist Cayeux (1907, 1914) regarding the position of submerged archeological remains around the coasts of the Mediterranean (e.g., Delos, Lefkada, and Aegina). For Negris, the observed meter-scale submergence was not to be linked with localized phenomena but was rather a ubiquitous basin-wide rise in sea level. By contrast, Cayeux, who was very close to Suess, suggested that the general level of the Mediterranean had not varied significantly since Antiquity. At Delos, Cayeux identified archeological examples of sea-level changes that he interpreted as being the result of sediment compaction. He ignored the well-documented example of Phalasarna's uplifted harbor (western Crete), refusing to concede the possibility of regional sea-level changes since Antiquity. Around the same time, the seminal archeological synthesis of Lehmann-Hartleben (1923) furnished one of the most comprehensive and authoritative early studies on Mediterranean port infrastructures, but with the noticeable absence of any discussion of relative sea-level changes.

Much of this RSL debate ended from the 1950s onwards, when high-resolution archeological excavations and the advent of radiometric dating techniques ushered in much greater temporal control. During the 1970s and 1980s, scholars such as Flemming (1969), Pirazzoli (1976b), Blackman (1982a, b), and Raban (1985) resuscitated sea-level geoarcheological research as an aid to understanding ancient sites. In a similar vein, Schmiedt (1972) used Roman fish tanks to precisely reconstruct sea-level changes during the past 2000 years along the western coast of Italy.

9.3 ARCHEOLOGICAL ZONATION AND FUNCTIONAL HEIGHTS

Auriemma and Solinas (2009) recently presented a very complete synthesis of archeological sea-level proxies. Many different archeological structures that were originally emerged, or in contact with seawater, today lie below mean sea level and therefore attest to a relative change in the position of the sea surface and the structure. Following the work of Flemming (1969, 1979–80) and Flemming and Webb (1986), the methodology consists of finding the original and functional position of the analyzed remains and their relationship to sea level. *Sensu stricto*, in the absence of fixed biological fauna archeological structures can rarely be used as precise index points; rather, they are employed to generate an indicative meaning.

The "functional height" of an archeological benchmark corresponds to the elevation of

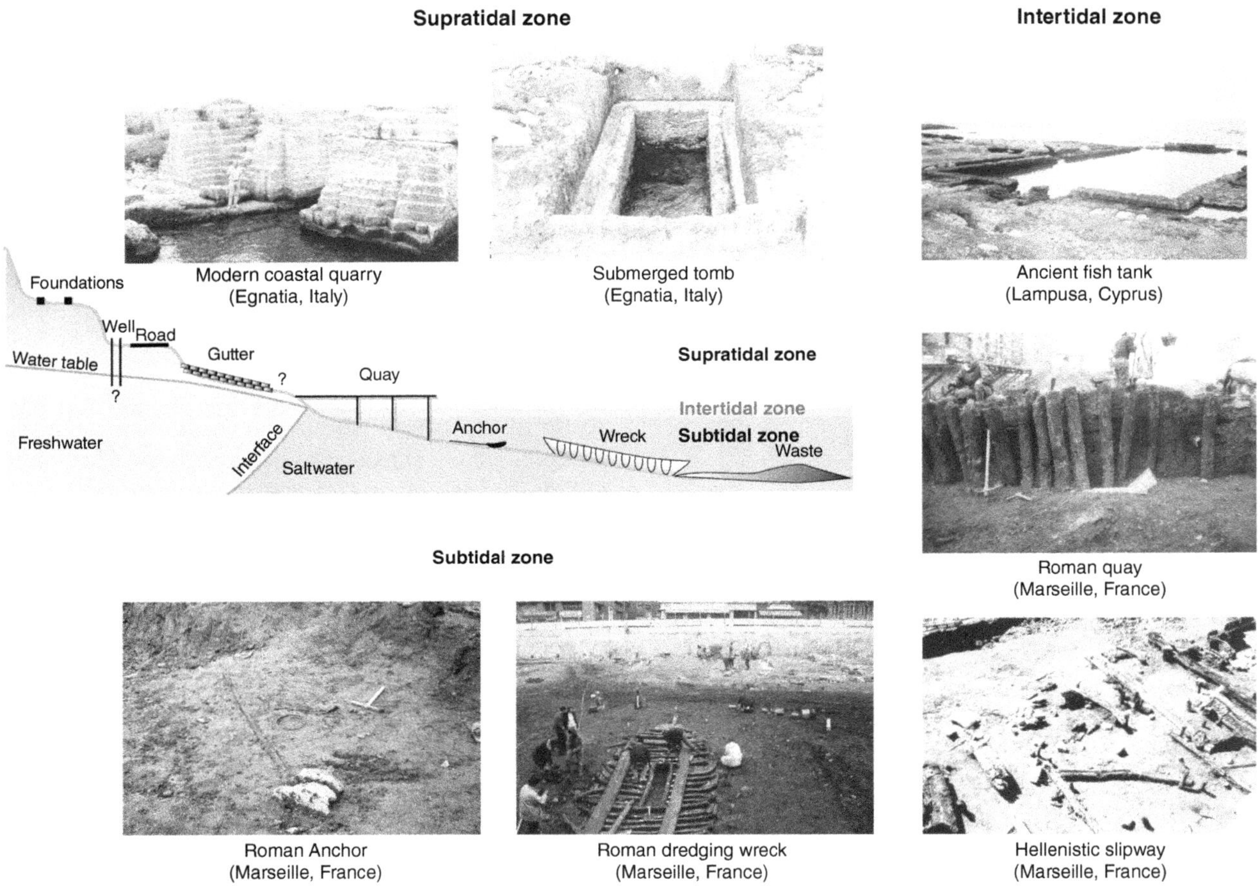

Fig. 9.2. Archeological RSL indicators adapted from Flemming and Webb (1986). *Source*: Photographs by the authors organized into three categories from base to top: (1) submerged archeological zone including wrecks and harbor foundations (subtidal); (2) interface structures zone comprising harbor installations (quays, piers, breakwaters, slipways, etc.) and fish tanks (intertidal); and (3) emerged structures zone including buildings, tombs, and quarries (supratidal).

specific architectural parts with respect to an averaged sea-level position at the time of their construction. Functional elevations define the minimum elevation of the structure above the highest local tides (Lambeck et al., 2004; Antonioli et al., 2007). In practical terms this is very difficult to achieve because the functional heights and the error bars are estimated on the basis of present analogs, which are not always related to past archeological structures. A good example of this approach is provided by recent work on Lechaion harbor in the Corinthian Gulf (Mourtzas et al., 2014) and Delos harbor in the Aegean Sea (Mourtzas, 2012).

A variety of archeological remains can be used to reconstruct sea-level changes with varying degrees of precision. An archeological zoning exists that can be organized into three categories from base to top (Fig. 9.2):

(1) Submerged structures including harbor foundations and wrecks. These remains give an indication of the direction of sea-level change but are generally low-precision indicators for the amplitude of movement.
(2) Interface structures constitute harbor installations proper (quays, piers, breakwaters, equipped banks, slipways, etc.) and fish tanks (Higginbotham, 1997; Evelpidou et al., 2012). For instance, excavations of the Roman and Medieval harbors of London unearthed a wide variety of this type of proxy in a meso-tidal context (Milne, 1985, 2003). This type of remain tends to yield quite precise RSL data because their function is directly related to sea level. The vertical error is usually speculated from present-day analogs.
(3) Emerged structures comprise residential units: villae maritimae (Lafon, 2001), buildings, or

town quarters (flooring, roads, and pavements, etc.), tombs, and quarries. Again, this type of indicator can provide information on the direction of sea-level changes. For instance, Late Roman tectonic movements in south Lebanon have led to the drowning of town quarters on the southern portion of the paleo-island of Tyre (Marriner et al., 2008).

Some research has also used more indirect proxies such as well bottoms (e.g., Sivan et al., 2004 in Caesarea, Israel), sewage outlets (e.g., Toker et al., 2012 in Akko, Israel) and flooring in churches (e.g., St Nicholas Basilica, Bari, Italy; Pagliarulo et al., 2013) to reconstruct water table changes in coastal areas. The RSL measurements are always indirect and imprecise because they are linked to the mobility of the coastal aquifer, which is affected by climate variability, groundwater extraction, and sea-level changes.

In order to use these indicators, the archeological interpretation must ensure the "maritime" function of the interface structure, and clarify the typology. Further considerations include the building techniques, which are important markers of height or depth at the time of construction (foundation versus elevation, e.g., Papageorgiou et al., 1993 for the elevated harbor of Aigeira on the uplifted coast of fault-controlled North Peloponnesia), as well as the "functional" elements, namely the relationship between the emerged part of the archeological remains compared to past and present mean sea level (Auriemma and Solinas, 2009). It is therefore important to determine the time of construction, its period of use, and the dynamics of its abandonment or destruction (Marriner and Morhange, 2006). These can be established using archeological excavation of the study area and high-resolution geoarcheological investigations. The emphasis is on multidisciplinary work to establish the precise functional depth in relation to present mean sea level and strong chronological brackets using dates based on both the typology of pottery and archeological structures (Fig. 9.2).

9.4 METHODOLOGICAL CONSIDERATIONS

The use of RSL archeological indicators is at the origin of two main uncertainties that can bias the precision of sea-level index points.

9.4.1 Difficulties in establishing the former functional height

Establishing the functional heights of archeological indicators is key to estimating local sea-level change. This parameter is defined as the elevation of specific architectural parts of an archeological structure with respect to an estimated mean sea level at the time of its construction by comparison to present contexts. The assumed functional height is dependent on the type of structure, its use, and the geomorphological and coastal hydrodynamic contexts (exposure, tidal amplitude, river discharge, etc.). As outlined in the previous section, functional heights define the minimum elevation of the structure above the mean high water mark (Fig. 9.2). Nonetheless, there is a great diversity of ancient coastal remains and many of these have undergone significant erosion due to wave dynamics in the intertidal zone. It is often the estimation of the functional height that can pose a problem of precision. The error bar in tens of centimeters is often greater than the absolute measurement.

9.4.2 Difficulties in estimating the amount of submersion

Many archeological structures are poorly preserved due to tidal wave action and subtidal bioerosion. Some examples of archeological features include (Fig. 9.2) fish tanks and harbor structures, as described in the following sections.

9.4.2.1 Fish tanks

Fish tanks are assumed to be the most reliable types of archeological indicators because they have a relatively precise relationship with sea level at the time of construction between the 1st century BC and the 1st century AD (Higginbotham, 1997). For instance, fish tanks have been widely used to reconstruct sea-level variations on the Tyrrhenian coast of Italy by Schmiedt (1972), Pirazzoli (1976a, b), Lambeck et al. (2004), and Evelpidou et al. (2012).

Fish tank remains can yield information on past sea-level positions. The external perimeter of fishpond walls cannot provide precise data on the ancient sea level because: (1) its summit is not directly linked to mean sea level; and (2) there is a great plethora of architectural types (Carre et al., 2011). By contrast, analysis is usually more precise when confined to the reference heights gathered from walkways, canals, and intertidal closing gates.

(1) Walkways are narrow paths running along the inner basins. Originally they were used for maintenance purposes and are therefore often considered to lie above mean sea level. Unfortunately, these structures are not very common and only indicate the direction of RSL movement. In some cases, such as the Lucullus fish tank (Circeo National Park, Italy), lower foot-walks were built below the openings for water arrival (Chiappella, 1965; Pirazzoli, 1976a).
(2) Canals were used to refill and empty the basins with water. They can correspond to mean sea level when they function as sluice gates, but can also be immersed fixed gates such as at Fréjus (Morhange et al., 2013a) and should therefore be studied with great care.
(3) *In situ* intertidal closing gates, which are precise indicators of RSL change, are exceptionally rare due to their original location in the wave-breaking zone.

In conclusion, archeological RSL proxies must be used with great care in fish tank contexts. Most publications are overconfident with regards to the precision of these structures, often quoted as being ±5 cm (Lambeck et al., 2004). Corrections for present tides and pressure do not overcome these uncertainties.

9.4.2.2 Harbor structures

Harbor contexts are interesting due to the diversity of their waterfront interface structures. In most cases however, they present an important margin of vertical error due to the poor state of preservation of most remains (intertidal dismantlement by natural and human processes over several millennia) and uncertainties with regards to functional heights in relation to former sea level. These sea-level markers are pier and quay surfaces with three important elements to determine: (1) the draught of ancient ships, which was smaller than present-day vessels (Boetto, 2010); (2) the tidal range, which varies depending on the study area; and (3) harbor function and their hierarchy (Auriemma and Solinas, 2009). In most circumstances, the original work surfaces are not preserved due to long-term wave action. By chance, in silted areas excavations can unearth well-preserved interface structures such as at Portus (Rome's harbor), Marseille, or Naples. As we demonstrate in the following section, archeological approximations can only be resolved using a multidisciplinary approach that integrates the use of biological indicators (Morhange et al., 2013b, c; also see Chapter 18).

9.5 PRINCIPLES OF BIOLOGICAL ZONATION OF BENTHOS ON ARCHEOLOGICAL REMAINS

The archeological indicators described in the previous section are by no means independent. For example, ancient harbors are important stratigraphical archives (e.g., Marriner and Morhange, 2007) and many Mediterranean sea-level studies typically combine sedimentological, geomorphological, and archeological indicators. However, few of these indicators are valuable without associated biological proxies. Traditionally, biological proxies provide dateable radiocarbon material from which to establish sea-level histories, but it is notably their precision as reference markers for former sea levels that are of particular interest. Over the last two decades or so, the use of biological sea-level indicators in the study of Mediterranean sea-level changes has gradually evolved from a descriptive to a multidisciplinary approach integrating many of the proxies outlined above (Laborel and Laborel-Deguen 1994). It is an approach based on the recognition that the vertical distribution of the fauna and flora of rocky shores shows a pattern of juxtaposed ecological belts, known as biological zonation (Stephenson and Stephenson, 1972; Péres 1982; Kelletat, 1988; and Chapter 18).

RSL biological proxies are mediated by physical factors. According to biological zonation, marine benthic animals and plants are finely adapted to very precise ecological conditions such as light intensity, turbidity, water salinity, temperature, tidal, and surf exposure. However, biological interactions can be important. Littoral flora and fauna are organized in three subhorizontal belts (Péres and Picard, 1964; Stephenson and Stephenson, 1972; Laborel and Laborel-Deguen, 1994; Stewart and Morhange, 2009). Marine biological studies have shown that on archeological structures (including harbor quays) there exists a precise biological zonation. Consequently, changes in local ecological conditions such as

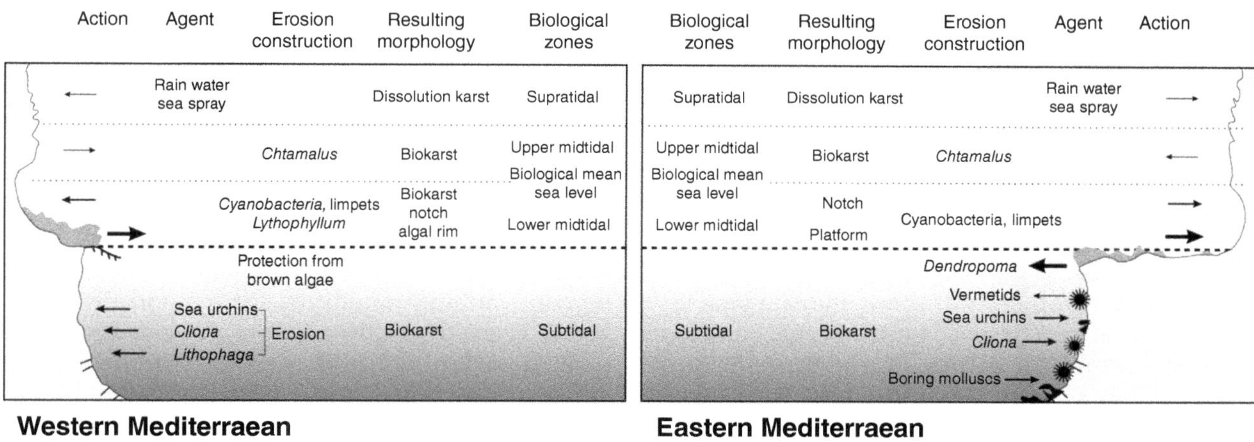

Fig. 9.3. Coastal profile showing the main characteristics of bioconstruction and biodestruction on limestone coasts in (a) the western Mediterranean; and (b) the eastern Mediterranean. *Source*: Adapted from Laborel and Laborel-Deguen, 1994.

relative sea-level change are followed by a concomitant quantitative and qualitative modification of the organisms with replacement by more tolerant forms. Laborel (1986, 1987) has discussed this biological zonation in detail, and demonstrated its scope in measuring past sea levels. Several parallel zones can be recognized (Fig. 9.3) and these are outlined as follows.

(1) A supratidal zone, wetted by surf but never or rarely submerged, in which the biomass is very low and mainly represented by boring endolithic cyanobacteria and grazing gastropods.
(2) An intertidal zone submerged by tides and waves on a regular basis, which displays a pattern of parallel algal and faunal belts, with biomass and species diversity increasing downwards. Cyanobacteria, limpets (*Patella* spp.), and Chitons are the main bio-eroders in this zone. Constructional elements such as the rim-building coralline rhodophyte *Lithophyllum byssoides* may develop in the northwest Mediterranean. A submerged intertidal notch can be carved into archeological structures such as in the ancient harbor of Aegina in the Saronic Gulf (N.D. Mourtzas, pers. comm., 2013). This erosional form is very useful in accurately determining a past sea level but very difficult to date.
(3) A subtidal zone whose upper limit is marked by a sudden increase in biodiversity, thus defining a biological sea level that ranges down to the lower limit of marine phanerogams (*Posidonia oceanica*) and photophilous algae, that is, to a mean depth of about 35 m. The upper part of this subtidal zone is densely populated by brown algae (*Cystoseira*), Coralline Rhodophytes, fixed vermetid gastropod molluscs (such as *Dendropoma* sp.), and cirrhipeds, for example *Balanus* spp. Active erosive agents, such as clionid boring sponges, sea-urchins, and rock-boring mussels (*Lithophaga*, *Hyatella*, *Coralliophaga* spp.), are responsible for rapid underwater erosion of the limestone outcrop such as the Roman columns of the market of Pozzuoli (Morhange et al., 2006).

The limit between the intertidal and the subtidal zones corresponds to the "biological sea level" or mean sea level (Laborel, 1986). Biological MSL corresponds to the base of the tidal zone. The influence of local variations in coastal morphology upon surf exposure explains why this biological limit undulates locally, reflecting the level of energy. Aperiodic sea-level oscillations linked to atmospheric pressure or wind variations are included in the "average" biological signal, translated by a precise marine zoning of living organisms with a lifespan of more than one year. Biological zonation is the cumulative expression of all these parameters at different timescales. In a harbor context, it is particularly interesting to note that the environment is artificially protected and that, as a result, the biological zonation is very precise and not significantly affected by high-energy events such as storms. Although species corresponding to such zones may differ between the western and eastern Mediterranean, biological zonation has the potential to yield very precise RSL index points.

Biological markers can be grouped on the basis of their bathymetric relationship to mean sea level.

(1) Sea-level indicators proper: The most appropriate organisms for RSL studies are those with a very narrow vertical life range, near sea surface. For example, in the Mediterranean Sea, biological sea level is best characterized by the development of a few marine species with a very narrow depth range, located immediately above (e.g., *Lithophyllum* rim) or below (e.g., *Dendropoma*) the mean waterline.
(2) Biological indicators of submersion: (a) Boring species: boring mussels include *Lithophaga lithophaga* and several species of *Petricola* and *Coralliophaga.* (b) Subtidal builders; these building species have a wide ecological range.

Although they do not show a precise relationship with sea level, they can yield interesting information about paleobathymetry. The upper limit of biological perforations by *Cliona* and *Lithophaga* (Laborel and Laborel-Deguen, 1994) are excellent proxies with a centimeter-scale indicative range in the case of artificially protected coastal environments such as ancient harbors. Consequently, bioconstructions and the upper limits of bioerosive elements (marine burrows and perforations), and fixed invertebrates (oysters, barnacles, solitary vermetids) are commonly used as biological sea-level indicators on archeological structures (Fig. 9.4).

The long-term stability of biological belts results from the fact that the zones are defined by species living at least a few years; they therefore tend to be confined to horizontal belts permitting their long-term survival. Consequently, if no significant changes in the relative sea level occur (as well as in the currents, temperature, and other characteristics of seawater), the biological zoning remains stable.

There are some further important methodological points to note. (1) Generally, sampling should be avoided at sites of strong exposure to surf because of the upward displacement of species zones. This is not usually the case in archeological contexts, which are well sheltered. (2) Measurement should occur between relic and current species (e.g., the upper limit of fossil and living balanids). (3) The precision of the height measurements depends on whether there is a clearly distinguishable upper limit (Baker and Haworth, 2000). The excavation of silted harbors allows this type of preservation. Laborel and Laborel-Deguen (1994) believed that, in sheltered environments, it is possible to accurately determine paleo-sea level to within ±5–10 cm.

If sea-level changes due to crustal effects, such as coseismic movements, biological zoning will be concomitantly modified (Stiros and Pirazzoli, 2008). Coastal species adapt to the new mean sea level, abandoning bands of rocks or archeological structures on which they were previously living. In many instances, such as uplifted or silted harburs, these bands are fossilized and can be used as precise sea-level indicators (Pirazzoli, 1991; Morhange et al., 1998).

Depending on the type of species, comparison of active and fossil biological zoning permits accurate identification of former sea levels (Morhange et al., 2001). In some instances, estimates for the velocity of the movement can even be obtained (episodic or slow movement, e.g., Laborel and Laborel-Deguen, 1994; Pirazzoli et al., 1996). Marrying archeological and biological proxies gives the most precise insights into the real marine conditions and RSL changes.

9.6 CONCLUSION

Tectonic uplift (e.g., the port of Phalasarna in western Crete probably around 365 AD) and silting up of sedimentary basins such as ancient harbors is particularly conducive to the preservation of fixed or boring marine organisms on archeological remains and their subsequent use as precise sea-level index points. The precision of the measurements depends upon the definition of a reliable benchmark (e.g., present biological sea level). Recent geoarcheological work embracing bioindicators and archeological remains has allowed progress to be made in the measurement of relative sea-level changes at archeological sites such as Marseille (Morhange et al., 2001) and Fréjus in France (Devillers et al., 2007; Morhange et al., 2013a), Pozzuoli (Morhange et al., 2006) and Portus of Rome in Italy (Goiran et al., 2009), Vis in Croatia (Faivre et al., 2010), Seleucia Pieria in Turkey (Erol et Pirazzoli, 1992), and Alexandria in Egypt (Goiran, 2001), among others.

Present posts in Toulon (France)

Roman post in Marseille (France)

Fig. 9.4. Biological zones on stakes from the ancient harbor of Marseille and present-day Toulon.

In terms of understanding mid–late Holocene coastal environments and archeological contexts in the Mediterranean, RSL modifications are generally a minor agent of change when compared with the important role of sedimentary budgets at base level, especially at sites on or close to deltaic systems. Today, it is widely recognized that close interaction between archeologists, geomorphologists, and biologists is needed to obtain the most robust RSL results.

REFERENCES

Antonioli, F., and Leoni, G. (1998) Siti archeologici sommersi e loro utilizzazione quali indicatori per lo studio delle variazioni recenti del livello del mare. Il Quaternario 11(1), 53–66.

Antonioli, F., Anzidei, M., Lambeck, K., et al. (2007) Sea-level change during the Holocene in Sardinia and in the northeastern Adriatic (central Mediterranean Sea) from archaeological and geomorphological data. Quaternary Science Reviews, 26, 2463–2486.

Anzidei, M., Antonioli, F., Lambeck, K., Benini, A., Soussi, M., and Lakhdar, R. (2011) New insights on the relative sea-level change during Holocene along the coasts of Tunisia and western Libya from archaeological and geomorphological markers. Quaternary International, 232(1–2), 5–12.

Anzidei, M., Antonioli, F., Benini, A., Gervasi, A., and Guerra, I. (2013) Evidence of vertical tectonic uplift at Briatico (Calabria, Italy) inferred from Roman age maritime archaeological indicators. Quaternary International, 288, 158–167.

Auriemma, R., and Solinas, E. (2009) Archaeological remains as sea level change markers: A review. Quaternary International, 206, 134–146.

Baker, R.G.V., and Haworth, R.J. (2000) Smooth or oscillating late Holocene sea-level curve? Evidence from the palaeo-zoology of fixed biological indicators in east Australia and beyond. Marine Geology, 163, 367–386.

Blackman, D.J. (1973) Evidence of sea level change in ancient harbours and coastal installations. In: *Marine Archeology* (ed. Blackman, D.J.), Butterworth, London, Colston Papers, 23, pp. 115–139.

Blackman, D.J. (1982a) Ancient harbours in the Mediterranean, part 1. The International Journal of Nautical Archaeology and Underwater Exploration, 11(2), 79–104.

Blackman, D.J. (1982b) Ancient harbours in the Mediterranean, part 2. The International Journal of Nautical Archaeology and Underwater Exploration, 11(3), 185–211.

Boetto, G. (2010) Le port vu de la mer: l'apport de l'archéologie navale à l'étude des ports antiques. Bollettino di Archeologia online, 2010, 112–128.

Braidwood, R.J. (1940) Report on two sondages on the coast of Syria, south of Tartous. Syria, 21, 183–221.

Carayon, N., Marriner, N., and Morhange, C. (2011) Geoarchaeology of Byblos, Tyre, Sidon and Beirut. Rivista di Studi Fenici, 39, 55–66.

Carre, M.-B. , Kovacic, V., and Tassaux, F. (2011) *L'Istrie et la mer: La côte du Parentin dans l'Antiquité*. Ausonius, France.

Cayeux, L. (1907) Fixité du niveau de la Méditerranée à l'époque historique. Annales de Géographie, 16, 97–116.

Cayeux, L. (1914) Les déplacements de la mer à l'époque historique. Revue Scientifique, 52, 577–586.

Chiappella, V. G. (1965) Esplorazione della cosiddetta "Piscina di Lucullo" sul lago di Paola. Atti Accademia Nazionale dei Lincei. Notizie degli scavi di Antichità, Serie 8(19), 46–160.

Degrassi, A. (1955) *I Porti Romani dell'Istria*. Anthemon, Firenze, pp. 119–169.

Devillers, B., Excoffon, P., Morhange, C., Bonnet, S., and Bertoncello, F. (2007) Relative sea-level changes and coastal evolution at Forum Julii (Fréjus, Provence). C. R. Geosciences, 339 (5), 329–336.

Erol, O., and Pirazzoli P. A. (1992) Seleucia Pieria: an ancient harbour submitted to two successive uplifts. International Journal of Nautical Archaeology, 21(4), 317–327.

Evelpidou, N., Pirazzoli, P., Vassilopoulos, A., Spada, G., Ruggieri, G., and Tomasin, A. (2012) Late Holocene sea level reconstructions based on observations of Roman fish tanks, Tyrrhenian coast of Italy. Geoarchaeology: An International Journal, 27, 259–277.

Faivre, S., Bakran-Petricioli, T., and Horvatincic, N. (2010) Relative sea-level change during the Late Holocene on the island of Vis (Croatia) – Issa, harbour archaeological site. Geodinamica Acta, 23(5–6), 209–223.

Flemming, N.C. (1969) Archaeological evidence for eustatic change of sea level and earth movements in the Western Mediterranean in the last 2000 years. Geological Society of America, Special Paper 109, pp. 125.

Flemming, N.C. (1979–80) Archaeological indicators of sea level. Oceanis 5 (hors-série), 149–166.

Flemming, N.C., and Webb (1986) C.O. (1986) Tectonic and eustatic coastal changes during the last 10.000 years derived from archeological data, Zeitschrift für Geomorphologie NF, 62, 1–29.

Galili, E., and Nir, Y. (1993) The submerged Pre-Pottery Neolithic water well of Atlit-Yam, Northern Israel, and its paleoenvironmental implications. The Holocene, 3(3), 265–270.

Galili, E., Weinstein-Evron, M., and Ronen, A. (1988) Holocene sea level changes based on submerged archaeological sites off the Northern Carmel Coast in Israel. Quaternary Research, 29, 36–42.

Gaur, A. S. (2006) *Glimpses of Marine Archaeology in India*. India Society for Marine Archaeology, Goa.

Goiran, J.-P. (2001) *Recherches géomorphologiques dans la région littorale d'Alexandrie, Egypte*. PhD thesis, Aix-en-Provence, Université de Provence.

Goiran, J.-P., Tronchere, H., Collalelli, U., Salomon, F., and Djerbi H. (2009) Découverte d'un niveau marin biologique sur les quais du Portus: le port antique de Rome. Méditerranée, 112, 59–67.

Haggi, A., and Artzy, M. (2007) The Harbor of Atlit in Northern Canaanite/Phoenician Context. Near Eastern Archaeology, 70, 75–84.

Hein, C. J., FitzGerald, D. M., Milne, G. A., Bard, K., and Fattovich, R. (2011) Evolution of a Pharaonic harbor on the Red Sea: Implications for coastal response to changes in sea level and climate. Geology, 39, 687–690.

Higginbotham, J. (1997) *Piscinae: Artificial Fishponds in Roman Italy*. University of North Carolina Press, North Carolina.

Jondet, G. (1916) Les ports submergés de l'ancienne île de Pharos. Mémoires presentés à l'Institut Egyptien (9), Le Caire.

Kelletat, D. (1988) Zonality of modern coastal processes and sea-level indicators. Palaeogeography, Palaeoclimatology, Palaeoecology, 68, 219–230.

Laborel, J. (1986) Vermetid gastropods as sea-level indicators. In: *Sea-level Research: A Manual for the Collection and Evaluation of Data* (ed. van de Plassche O.), GeoBooks, Norwich.

Laborel, J. (1987) Marine biogenic constructions in the Mediterranean. Scientific Reports of Port-Cros National Park, 13, 97–126.

Laborel, J., and Laborel-Deguen, F. (1994) Biological indicators of relative sea-level variations and of co-seismic displacements in the Mediterranean region. Journal of Coastal Research, 10(2), 395–415.

Lafon, X. (2001) Villa maritima. Recherches sur les villas littorales de l'Italie romaine. Bibliothèque des Écoles françaises d'Athènes et de Rome, 307. Roma. École française de Rome, 528 p.

Lambeck, K., and Bard, E. (2000) Sea-level change along the French Mediterranean coast for the past 30 000 years. Earth and Planetary Science Letters, 175, 203–222.

Lambeck, K., Anzidei, M., Antonioli, F., Benini, A., and Esposito, A. (2004) Sea level in Roman time in the Central Mediterranean and implications for recent change. Earth and Planetary Science Letters, 224, 563–575.

Lehmann-Hartleben, K. (1923) *Die Antiken Hafenanlagen im Ostlichen Mittelmeer.* Dietrich, Leipzig.

Lyell, C. (1830) *Principles of Geology.* John Murray, London.

Marriner, N. (2009) *Geoarchaeology of Lebanon's Ancient Harbours.* Archaeopress, British Archaeological Reports, Oxford.

Marriner N., and Morhange C. (2006) The 'Ancient Harbour Parasequence' Anthropogenic forcing of the stratigraphic highstand record. Sedimentary Geology, 186, 13–17.

Marriner, N., and Morhange, C. (2007) Geoscience of ancient Mediterranean harbours. Earth Science Reviews, 80, 137–194.

Marriner, N., Morhange, C., and Carayon, N. (2008) Ancient Tyre and its harbours: 5000 years of human-environment interactions. Journal of Archaeological Science, 35, 1281–1310.

Milne, G. (1985) *The Port of Roman London.* Batsford, London.

Milne, G. (2003) *The Port of Medieval London.* Tempus Publishing.

Milne, G., and Hobley, B. (eds) (1981) *Waterfront Archaeology in Britain and Northern Europe* (CBA Research Reports). Council for British Archaeology.

Morhange, C., Laborel, J., and Laborel-Deguen, F. (1998) Précision des mesures de variations verticales du niveau marin à partir des indicateurs biologiques, le cas des soulèvements bradysismiques de Pouzzoles, Italie du Sud (1969–1972; 1982–1984). Zeitschrift für Geomorphologie, NF, 42(2), 143–157.

Morhange C., Laborel J., and Hesnard A. (2001) Changes of relative sea level during the past 5000 years in the ancient harbour of Marseilles, Southern France. Palaeogeography, Palaeoclimatology, Palaeooecology, 166, 319–329.

Morhange, C., Marriner, N., Laborel, J., Todesco, M., and Oberlin, C. (2006) Rapid sea-level movements and noneruptive crustal deformations in the Phlegrean Fields caldera, Italy. Geology, 43(2), 93–96.

Morhange, C., Marriner, N., Excoffon, P., Bonnet, S., El-amouri, M., and Zibrowius, H. (2013a) Relative sea level changes during Roman times in the NW Mediterranean. The 1st century AD fish tank of Forum Julii (Fréjus, France). Geoarchaeology: An International Journal, 28, 363–372.

Morhange, C., Bony, G., Flaux, C., and Shah-Hosseini, M. (2013b) Comment on sea level changes since the Middle Ages along the coast of the Adriatic Sea: The case of St. Nicholas Basilica, Bari, Southern Italy. Quaternary International, doi: 10.1016/j.quaint.2013.03.039.

Morhange, C., Flaux, C., Pirazzoli, P.A., and Carre, M.B. (2013c) Comment on "Holocene Sea Level Change in Malta". Quaternary International, 303, 233–234.

Mourtzas, N.D. (2012) A palaeogeographic reconstruction of the seafront of the ancient city of Delos in relation to Upper Holocene sea level changes in the central Cyclades. Quaternary International, 250, 3–18

Mourtzas, N. D., Kissas, C., and Kolaiti, E. (2014) Archaeological and geomorphological indicators of the historical sea level changes and the related palaeogeographical reconstruction of the ancient foreharbour of Lechaion, East Corinth Gulf (Greece). Quaternary International, 332, 151–171.

Negris, P. (1903) Observations concernant les variations du niveau de la mer depuis les temps historiques et préhistoriques. Comptes Rendus de l'Academie des Sciences, Paris, 137(2), 222–224.

Negris, P. (1904) Nouvelles observations sur la dernière transgression de la Méditerranée. Comptes Rendus de l'Academie des Sciences, Paris, 2, 379–381.

Negris, P. (1921) Les ports submergés de l'ancienne île de Pharos. Bulletin de la Société Géologique de France, 21, 161–164.

Pagliarulo, R., Antonioli, F., and Anzidei, M. (2013) Sea level changes since the Middle Ages along the coast of the Adriatic Sea: The case of St. Nicholas Basilica, Bari, Southern Italy. Quaternary International, 288, 139–145.

Papageorgiou, S., Arnold, M., Laborel J., and Stiros, S. C. (1993) Seismic uplift of ancient Aigeira, Central Greece. The International Journal of Nautical Archaeology, 22(3), 275–281.

Paris, J. (1915) Contributions à l'étude des ports antiques du monde grec, note sur Léchaion. Bulletin de Correspondance Hellénique, 29, 6–16.

Paris, J. (1916) Contributions à l'étude des ports antiques du monde grec, les établissements maritimes de Délos. Bulletin de Correspondance Hellénique, 40, 5–73.

Péres J.-M. (1982) Major benthic assemblages. In: *Marine Ecology* (ed. Kinne, O.), Wiley, Chichester, vol. 5, pp. 373–522.

Péres, J.-M., and Picard, J. (1964) Nouveau manuel de bionomie benthique de la mer Méditerranée. Recueil des Travaux de la Stations Marine d'Endoume, 31(47), 137.

Pirazzoli, P. (1976a) Les variations du niveau marin depuis 2000 ans. Mémoire du Laboratoire de Géomorphologie de l'Ecole Pratique des Hautes Etudes, 30, 421.

Pirazzoli, P. (1976b) Sea level variations in the Northwest Mediterranean during Roman times. Science, 194, 519–521.

Pirazzoli, P. A. (1991) *World Atlas of Holocene Sea-Level Changes.* Elsevier, Oceanography Series, 58, pp. 300.

Pirazzoli, P., and Thommeret, J. (1973) Une donnée nouvelle sur le niveau marin à Marseille à l'époque romaine. Comptes Rendus de l'Academie des Sciences, Paris, 277(D), 2125–2128.

Pirazzoli, P. A., Laborel, J., and Stiros, S. C. (1996) Coastal indicators of rapid uplift and subsidence: examples from Crete and other Eastern Mediterranean sites. Zeitschrift für Geomorphologie, N.F., Suppl. Bd. 102, 21–35.

Raban, A. (ed.) (1985) Harbour archeology. *Proceedings of the First International Workshop on Ancient Mediterranean Harbours*, 1983, BAR International series, 257.

Rao, S. R. (ed.) (1988) *Marine Archaeology of Indian Ocean Countries*. National Institute of Oceanography in Dona Paula, Goa, India.

Sartoretto, S., Collina-Girard, J., Laborel, J., and Morhange, C. (1995) Quand la Grotte Cosquer a-t-elle étéfermée par la montée des eaux? Méditerranée, 82, 21–24.

Schmiedt, G. (1972) *Il Livello Antico del Mar Mediterraneo*. Olschki, Florence.

Sivan, D., Wdowinski, S., Lambeck, K., Galili, E., and Raban, A. (2001) Holocene sea-level changes along the Mediterranean coast of Israel, based on archaeological observations and numerical model. Palaeogeography, Palaeoclimatology, Palaeoecology, 167, 101–117.

Sivan, D., Lambeck, K., Toueg, R., Raban, A., Porath, Y., and Shirman, B. (2004) Ancient coastal wells of Caesarea Maritima, Israel, an indicator for relative sea level changes during the last 2000 years. Earth and Planetary Science Letters, 222, 315–330.

Stephenson, T.A., and Stephenson, A. (1972). *Life Between Tidemarks on Rocky Shores*. Freeman, San Francisco.

Stewart, I.S., and Morhange, C. (2009) Coastal geomorphology and sea-level change. In: *The Physical Geography of the Mediterranean* (ed. Woodward J.), Oxford University Press, Oxford, pp. 385–414.

Stiros, S.C. (1998) Archaelogical evidence for unusually rapid Holocene uplift rates in an active normal faulting terrain: Roman Harbor of Aigeira, Gulf of Corinth, Greece. Geoarchaeology, 13(7), 731–741.

Stiros, S., and Pirazzoli, P.A. (2008) Direct determination of tidal levels for engineering applications based on biological observations. Coastal Engineering, 55(6), 459–467.

Suess, E. (1885–1908) *Das Antlitz der Erde*. F. Tempsky, Vienna.

Tallet, P. (2009) Les Égyptiens et le littoral de la mer Rouge à l'époque pharaonique. Comptes Rendus Académie Inscriptions Belles Lettres 2, 687–719.

Toker, E., Sivan, D., Stern, E., Shirman, B., Tsimplis, M., and Spada, G. (2012) Evidence for centennial scale sea level variability during the Medieval Climate Optimum (Crusader Period) in Israel, eastern Mediterranean. Earth and Planetary Science Letters, 315–316, 51–61.

van Andel, T. (1989) Late Quaternary sea-level and archaeology. Antiquity, 63, 733–745.

Van de Noort, R., and O'Sullivan, A. (2006) *Rethinking Wetland Archaeology*. Duckworth, London.

Chapter 10
GPS and surveying

JAMES FOSTER
University of Hawai'i, Honolulu, HI, USA

10.1 INTRODUCTION

Measuring sea level is fundamentally a geodetic problem, and geodetic theory, techniques, and infrastructure are central to the challenge of determining global sea-level change. The two geodetic techniques most intimately linked to the monitoring and positioning of tide gauges are the Global Navigation Satellite Systems (GNSS) constellations, of which the US Global Positioning System (GPS) is most commonly used, and traditional vertical surveying or "leveling". These two techniques fulfill complementary roles in the determination of tide gauge stability and the maintenance of local and global reference frames, so that measurements of sea level at one location can be compared with others elsewhere. Both techniques are required elements at tide gauges (Fig. 10.1) that form the core of the Global Sea-Level Observing System (GLOSS) network.

In this chapter we will first introduce the basic principles of geodesy and its reference frames, and their importance to sea-level research. We then discuss vertical surveying and its application to the measurement of sea level, constraining the long-term stability of tide gauges, and detecting local vertical motions. Turning to GPS, we explain its role within the GLOSS, and the technical requirements and considerations that should be taken into account when equipping a tide gauge with GPS. We describe some of the contributions of GPS and surveying to studies of sea-level change, and conclude with a brief survey of the more exploratory applications of GPS and other satellite geodetic techniques to sea-level research.

10.1.1 Geodesy

Modern geodesy is based on three pillars: (1) "geokinematics" is the study of the shape of the Earth and the change of that shape in space and time; (2) "Earth rotation" tracks the orientation of the planet with respect to celestial bodies; and (3) the "gravity field" measures variations of the Earth's gravity and geoid (the surface that sea level would follow if it was in static equilibrium). Each of these pillars changes in response to dynamic Earth processes, and together they are key to measuring, understanding, and predicting changes at the Earth's surface. Changes in mass distribution associated with sea-level change, for example from the melting of ice sheets, perturb the angular momentum and hence rotation of the Earth. The reduction in the load from the ice sheet causes the surface of the Earth to deform; this and the change in the distribution of the water changes the gravity field. As Blewitt et al. (2010) note: "sea level variation cannot be understood outside the context of geodesy."

10.1.2 Reference frame

The geodetic construct that allows us to measure and understand the processes affecting the three pillars is the terrestrial reference frame. The formation and maintenance of the terrestrial reference frame is one of the core efforts of the geodetic community. It underpins much of modern solid Earth science, and reference frame stability sets the limit for the absolute accuracy for any global measurements. Blewitt et al. (2010) explore the philosophies and strategies for realizing an accurate reference frame for global geodetic studies. Here we summarize some of the key elements.

A reference frame defines how positions should be computed for a particular time. The realization of a frame essentially involves defining the positions of fiducial points (key benchmarks) for a given time, and assigning a time-dependent model of how those positions change. For global geodetic studies, the International Terrestrial Reference

Handbook of Sea-Level Research, First Edition. Edited by Ian Shennan, Antony J. Long, and Benjamin P. Horton.

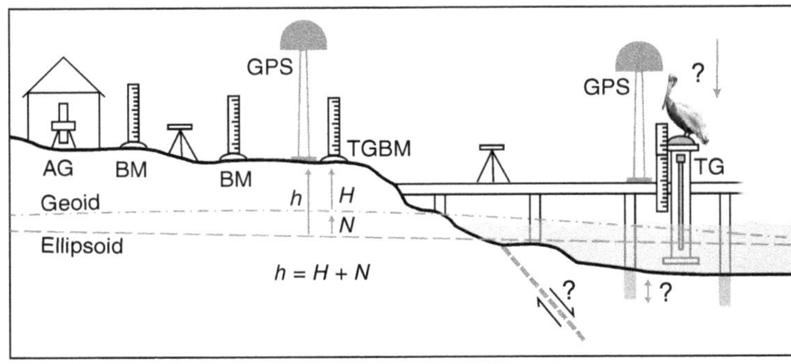

Fig. 10.1. An ideal continuous GPS tide gauge illustrating recommended instrumentation and some of the possible sources of vertical motions. TG: tide gauge; BM: benchmark; TGBM: tide gauge benchmark; AG: absolute gravimeter. Surveying staffs and levels are illustrated.

Frame (ITRF) provides that service. Datasets from several different observational techniques, including GPS, the French Doppler Orbitography and Radiopositioning Integrated by Satellite (DORIS) satellite positioning system, satellite laser ranging, satellite gravimetry, and very long baseline interferometry using radio telescopes all provide complementary constraints on the reference frame. They are analyzed together to determine the position and motions of a core set of fiducial stations. There have been several generations of ITRF, each significantly better defined and more stable than its predecessor, with efforts ongoing to continue to produce significant improvements in the future. Recent studies find that noise and errors in the recent releases of the ITRF now contribute only ~0.2 mm a^{-1} error to estimates of sea-level rise, with every reason to think that the stability will continue to improve. This level of stability that the latest version (ITRF2008) provides is good enough to allow GPS to play a crucial role tying tide gauges together into a single, accurately defined, global network, and constrain the contribution of vertical land motions. Knowing the absolute local land motion allows for the transformation of the relative sea-level change measured by tide gauges into accurate estimates of absolute sea-level change. It is also important to note that the space-based altimetry missions rely on the ITRF for positioning of the satellites and for ensuring that their measurements provide absolute values for sea levels.

10.2 SURVEYING

Prior to the advent of satellite geodesy and the ITRF, many different local reference frames were defined to serve the needs of a nation or region. These provided a common frame in which points and measurements could be compared. One of the techniques used to define the vertical component of most of these frames was the traditional leveling survey, as it is able to provide accurate ties to tide gauges. This is crucial, as measurements of mean sea level form the vertical reference datum for most national reference frames. Extensive discussions of the technical requirements and field procedures for leveling, and its application to tide gauge control, is available in Hicks et al. (1987) and Woodworth (2002). Here, we describe some of the more important issues relevant to monitoring sea level.

A primary purpose of the surveying effort at a tide gauge was to ensure continuity of the mean sea-level observations for the national reference frame. In the event that a gauge was destroyed or needed upgrading, its replacement was installed and connected to the previous measurement records using leveling. While this role is still as valid as it ever was, there is a new emphasis on the necessity for robust geodetic control of the area around each tide gauge. Interest in measuring sea level with increased accuracy, and especially in determining long-term trends, requires a stable local vertical reference. This reference can be used to detect any motions of the tide gauge or its support structure with respect to the local bedrock.

In order to acquire continuous, accurate measurements of the height of the sea surface, tide gauges are typically installed in deeper water where the depth is considerably more than the full tidal range and, if possible, the site is sheltered from wave action. In practice, for many locations this means the gauge is installed on a pier or some other man-made structure reaching out beyond the coast. This introduces the question of whether the structure on which the gauge is installed is itself stable. In addition, for various reasons, a gauge may have been installed in hard-to-access locations (e.g., inside a building or in a high-security zone) which presents a problem for physical access to the vertical reference level that it provides. These issues highlight the need for a stable point near the gauge which can be easily accessed and provide the tie to sea level.

10.2.1 Tide gauge benchmark control

To control for structure stability, provide access to the reference level, and enable the recovery of the vertical reference level in the event of a change in tide gauge, each gauge is accompanied by a local network of benchmarks (BMs). A BM is any physical point (e.g., disc, pipe, scribe-line) that can be used to help define a vertical or horizontal reference frame. Most commonly it is a metal disc mounted into rock or concrete, but many alternatives exist. Depending on the style of BM and its location (some are more robust and stable than others), each benchmark has a stability rating based on its expected, or monitored, stability. The recommendation is that there should be a network of ~10 BMs, with as many as possible installed in locations that are expected to be secure and stable (bedrock if possible). Historically, many locations had fewer BMs installed. There is typically attrition of the network over time as local development and construction destroys BMs, so many tide gauges have a less-robust control network than might be desired. For new installations it is recommended that the BMs are placed at a minimum spacing of 60 m, a maximum spacing of 1 km, within a 1.6 km radius of the gauge. BM locations should be chosen to ensure that no more than one could be damaged or destroyed by any predictable event.

One of these BMs is designated as the primary or tide gauge benchmark (TGBM), and it is this point to which the local sea level is referenced. It is selected on the basis of its stability, while being as close to the tide gauge and as accessible as possible. Leveling between the tide gauge (both the gauge itself as well as the staff installed next to it, if one exists) is performed to determine the height offsets between each component, and determine the site correction term. This is a constant which is added to the raw measured ranges/depths of sea level in order to reference them to the TGBM, and which defines the official site sea-level datum. Leveling is performed when a new gauge is installed or any changes are made to an installation; if necessary, the correction term is recalculated and recorded in the local data logger. A log of all correction term changes is maintained (but may not be easily retrievable); it is applied automatically to the data stream, ensuring transparency to the end-user.

Reoccupation of the benchmark network is recommended annually if possible, but at a minimum of every two years. Each leveling run must always include the gauge and TGBM, and as many of the other BMs as can be located and occupied. In this way the stability of the vertical reference is maintained in the event of damage to the gauge or TGBM, and any motion of the gauge can be identified and corrected in the sea-level record.

10.2.2 Technical issues for surveying

Despite being one of the older geodetic surveying techniques with the basic instrumentation largely unchanged for many decades, leveling is still unparalleled in its ability to measure vertical offsets over short distances. The principal is simple: an optical sighting instrument, the "level", is set up midway between the two points whose relative heights are to be measured, and adjusted to be aligned horizontally via spirit bubbles. A graduated staff is positioned on one benchmark, and a sub-millimeter-precision reading is made for the position of the level's cross-hair sight on the staff's scale. A similar reading to the other benchmark allows for a simple subtraction of the reading at one rod from the other to give the height offset. The set-up is then rotated, with the level relocated to be between one of the previous marks and a new one, and the process is repeated. Although simple in theory, in order to achieve the very highest order of accuracy in the field and avoid systematic errors, several issues need to be taken into consideration and corrections may need to be applied. Good field procedure can however mitigate the most common and significant problems. For tide gauge control, with the limited shot and network length the primary errors sources are: (1) collimation error; (2) the Earth's curvature; and (3) atmospheric refraction. A collimation error occurs when the optical sight is not perfectly horizontal when the instrument is leveled. In this case, there will be a vertical bias in the reading. By ensuring that the instrument is set-up midway between the two benchmarks, however,

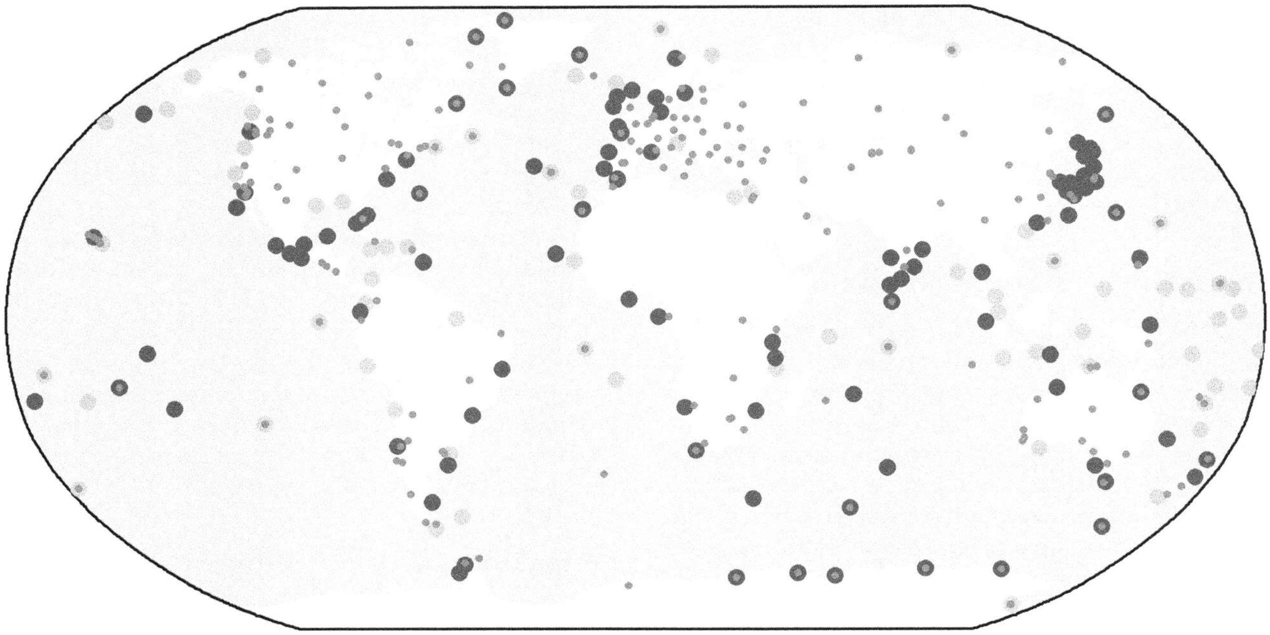

Fig. 10.2. Map of core IGS sites (232 dots), TIGA sites with GPS less than 10 km from tide gauge (224 light circles) and TIGA sites with GPS less than 1 km from the tide gauge (113 dark circles). For color details, please see Plate 28.

this bias cancels. This field procedure also minimizes the effect of the Earth's curvature which can introduce vertical biases in the readings. Similarly, changes in the air density due to temperature gradients change the way the light ray is refracted; again, ensuring equal sight lengths cancels most of this effect (although there may be some residual errors in highly variable atmospheric conditions). A well-performed leveling survey over a tide gauge network can expect formal accuracies of better than 1 mm in the height differences between BMs.

10.3 GPS

Global Navigation Satellite Systems (GNSS) are space geodetic systems that use signals broadcast from satellites to achieve precise positioning of receiving antennas. GPS is the best known of these, although the Russian GLONASS is also operational, and the European Galileo and Chinese Compass/BeiDou systems are expected to have fully functioning constellations of satellites over the next few years. The French system DORIS is worthy of a special mention here as it is one of the two systems, along with GPS, that are used to provide precise orbit determination for remote sensing satellites, including the altimetric missions that are responsible for monitoring sea level in the open oceans. We will focus here on the GPS system, as it is the one in current use for geodetic surveying, but most issues pertaining to it also apply (or will apply when they become operational) to the other GNSS constellations.

The GPS satellites broadcast radio signals at two (soon to be three) different frequencies. The signals contain information about the time and positions of the satellites. The importance of having at least two frequencies is that the refraction of the signal in the ionosphere can then be modeled and corrected. By recording signals from four or more satellites, a receiver can determine its position with meter-scale accuracy. This forms the basis of conventional GPS navigation and positioning. In order to determine sub-centimeter accuracy positions for GPS antennas, rather than just using the information encoded in the signals, the carrier radio waves themselves are tracked. The phase of the signals can be measured with sub-millimeter accuracy and, with care, these observations can be processed to achieve millimeter accuracies for the position of a GPS antenna with respect to a reference GPS site(s). In order to support global access to high-quality reference sites, the International Association of Geodesy formed the International GNSS Service (IGS), a voluntary federation of more than 200 worldwide agencies, to provide the framework for the archiving and processing of data from a global network of GPS sites. This core network (Fig. 10.2) constitutes the GPS contribution to the ITRF, and allows local users to tie their GPS measurements into the global reference frame. By including data from nearby IGS fiducial sites, data from a well-designed GPS site can be processed to produce estimates of absolute horizontal and vertical velocities for that location accurate to ~1 mm a^{-1} or better.

Each technique that contributes to the ITRF has improved its performance over time. For GPS the improvements have come from a variety of sources, including more accurate data processing. Processing of GPS data is performed in daily batches, with a single position estimated for each site for that 24 hour period. Included in the processing are models for all of the many components of the system that impact the exact position of the GPS antenna at that time, or that influence the radio signal that the antenna is receiving. These includes models for how the gravity fields of the Sun and Moon perturb the Earth's surface, how each antenna's measurement of the GPS signal changes depending on its orientation with respect to the satellite, and how the Earth's atmosphere refracts the GPS signal as it propagates through it to the surface. Improvements to the modeling and treatment of each of these elements have contributed to increasing reference frame stability. An additional contribution with a large impact has come from the huge increase in the number of continuous sites reporting data to the IGS over the last decade. This densification of the global network means that individual sites with high noise levels, or other problems, become less influential to the overall solution and the reference frame therefore becomes more robust.

A final major contributor to better GPS reference frame stability is from the increasingly long occupations at a growing number of stable GPS sites. With longer time windows, the impacts of both natural and anthropogenic processes that generate annual cycles and long-period perturbations in the vertical position of a site become better characterized. For example, sites near high mountain ranges may see an annual subsidence cycle due to the weight of the snow pack on the mountains in the winter. Similarly, changes in the amount of water in an aquifer underneath or near a GPS site can create vertical signals that may be approximately annual, but also have longer-term periodicities depending on pumping and climate cycles. For a GPS site that has only been active for a year or so, these signals can make it very difficult to define an accurate reference position and velocity. As the occupation time increases however, these long-period signals can be modeled more accurately and removed from the site position and velocity estimates. The other important impact of these noise sources is in the formal error estimates for the velocity of a GPS site. Most standard least-squares-based solutions for site positions and velocities assume that errors are normally distributed, that is, that the noise spectrum is "white." Long-period perturbations add a "red" component to the noise spectrum. Formal errors based on an assumed white noise spectrum will be optimistically small. In order to generate more realistic errors, the velocities should be recalculated using a more appropriate noise model. The most common approach for this is to analyze the position time series using a combined noise model that estimates a white component and a "flicker" component that describes the long-term correlations.

10.3.1 GPS tide gauge benchmark monitoring project

GPS and its demonstrated capability to accurately tie together global measurements was quickly recognized as an important development for the measurements of global sea level, and the incorporation of satellite-based systems such as GPS into tide gauge monitoring networks was recommended. In response to this need, and the growing number of GPS sites located near tide gauges, the IGS initiated the GPS Tide Gauge Benchmark Monitoring Pilot Project (TIGA-PP). Formed in 2001, the purpose of TIGA was to analyze GPS data from stations at, or near, tide gauges on a continuous basis, to promote the establishment of high-quality continuous GPS sites at tide gauges (Fig. 10.2), and to produce geocentric coordinates and time series of the TIGA station network. After 9 years of successful operation, this pilot project has recently been transitioned to become an IGS Working Group which will work to further improve the quality of its products, seek ties to other techniques that can help the determination of vertical motion, and promote the maintenance and expansion of GPS/GNSS sites at tide gauges globally. All new GPS sites installed at, or near, tide gauges should seek recognition by TIGA, and have their data stream added to the archives (www.sonel.org) and included in the global processing effort.

10.3.2 Installing GPS at tide gauges

The measurement of vertical motions using GPS is a difficult and exacting process, and many parameters need to be taken into account if a new GPS installation is to be useful. There are many issues that are common to any new geodetic grade GPS site so, while every site is unique and may require individual solutions, there are useful general rules. Furthermore, installations at tide gauges have additional and very specific constraints and complexities; it is therefore worthwhile to provide some discussion of the primary issues, and offer guidelines and recommendations for a general approach to equipping a tide gauge site with GPS. A detailed exploration of technical issues for installing GPS at tide gauge is described by Bevis et al. (2002); here we summarize some of the key points that may impact both initial site selection for new installations, as well as the degree of confidence which can be assigned to velocity solutions from existing sites.

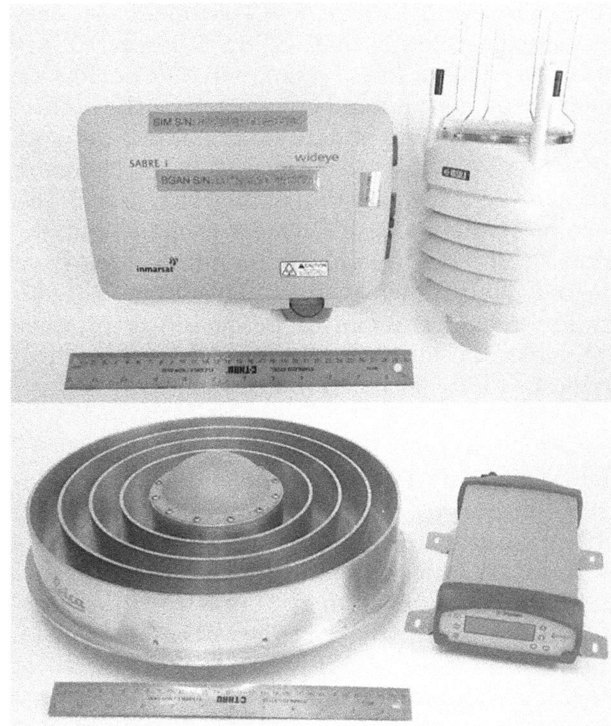

Fig. 10.3. Typical equipment for a continuous GPS at tide gauge installation. Top left: BGAN satellite communications terminal/antenna; top right: Vaisala WXT520 meteorological sensor; bottom left: Leica chokering antenna; and bottom right: Trimble NetR9 continuous GPS receiver.

10.3.2.1 GPS equipment

Many companies produce various models of GPS receivers and antennas. The application of GPS to produce accurate, long-term measurements of height is, however, extremely exacting, and demands high-quality equipment in order to achieve the desired accuracy. The basic requirement for a tide gauge GPS receiver is that it is designed for continuous geodetic monitoring, and is a dual-frequency system (Fig. 10.3). Many of these exist, and the choice is largely one of price or personal preference: for example, in order to maintain consistency with existing deployments, for preferred technical support, for ease of data flow, or perhaps low power requirements. For antennas however, the choice is smaller: due to the demanding nature of the task, a choke-ring antenna is strongly recommended. This design includes a broad collar, whose role is to destructively interfere signals arriving at low elevation angles in order to reduce the impact of reflected signals, "multipath," on the recorded observations. The addition of a radome is also recommended to protect the antenna. Radome materials and shapes are designed to minimize their effect on the GPS signal, and for the refractive effect that they do induce to be as predictable as possible so that it can be modeled during the processing. Seasonal issues, such as snow and ice, may dictate the particular shape and type of radome deployed, but the Southern California Integrated GPS Network (SCIGN) hemispherical type is widely used, and has a well-characterized model.

Although not required, the addition of a meteorological sensor (met-pack) is also potentially valuable (Figs 10.3 and 10.4). Several companies produce GPS-compatible units that plug into the receiver which then logs the met-pack's measurements of pressure and temperature, and often several other meteorological parameters, within the GPS data stream. The met-pack enables a site

Fig. 10.4. Continuous GPS site TGCV at tide gauge in Cape Verde islands. The instrument box houses both tide gauge and GPS equipment. The satellite communications antenna, solar panel, and meteorological sensor are mounted on it. The tide gauge sensor is mounted on the dock to the left of the GPS monument. *Source*: Photograph by Dana Caccamise.

to operate for GPS meteorology, providing measurement of water vapor valuable for weather forecasting. As detailed investigations of local sea level require corrections for winds and air pressure, the synergy produced by having a GPS meteorological sensor is attractive. In addition, including pressure measurements in the GPS processing has been found to improve the repeatability of the vertical position estimates. Maintaining calibration of the pressure sensor, which can drift over the time period of a couple of years, is important and can add extra costs. If a met-pack is included at a site, the vertical offset between the met-pack pressure sensor and the GPS antenna should be recorded. The met-pack should be installed in a location where its readings of pressure and temperature sense the free atmosphere as much as possible rather than, for example, re-radiated heat or the pressure shadow from a building.

10.3.2.2 Monumentation

Any GPS installation is only as good as the monument on which the antenna is installed. The monument needs to be a rigid and stable platform for the antenna, while not introducing an additional multipath by having broad flat surfaces near the antenna from which the GPS signal can be reflected. There are many solutions to monumentation, with the general approach largely dictated by the conditions into which it will be installed. If there is no bedrock or stable concrete into which bolts can be drilled to anchor a mast-type monument (e.g., Fig. 10.4), then a tripod-style braced monument may be required with legs pounded deep into the ground. In other conditions, site-specific solutions may be required. With that in mind however, as a general rule installing antennas on building roofs or bolted to building walls is not preferred (unless it is known that the building is rigid and stable with millimeter confidence, and well anchored to the ground). If this is the only available solution the monument should be anchored to the most stable structural element possible: a corner or a support beam, with strong connection to the foundation.

The design and construction material of the monument should be built to last as long as possible in the conditions in which it will be installed: at least 10 years should be the goal. Stainless steel or aluminum provides excellent corrosion resistance above ground, although aluminum can corrode quickly when buried. Consideration should be given to the possibility of electrochemical effects where dissimilar metals are in contact. If it is at all possible, it is best to never have to remove and replace an antenna as this almost inevitably introduces steps in the data, reducing the accuracy and confidence level of velocity estimates. The monument must however be designed so that if the antenna is replaced, the new antenna's reference point occupies exactly the same location in space as the previous one.

10.3.2.3 Communications

While the equipment and site selection issues have not changed significantly in the last decades, one issue that has progressed significantly is the availability and cost of global communications. The target for data delivery for continuous IGS GPS sites is to provide daily files of 30-second sampled data within a day to the central GPS data processing and archiving centers (although for TIGA-only sites this is not a fixed requirement). For many locations in developed countries there may be easy access to local wireless or hard-wired connections to the internet, for which most modern GPS receivers are enabled. This is often not the case for other locations globally. In areas with reliable cell phone networks, GSM cell modems provide a cheap possible solution for communication. Alternatively, if sufficient budget is available, some form of satellite communications may be the only viable option. Ongoing developments in this area have significantly reduced the pricing for data transmission and, with the exception of the polar regions where satellite coverage is very poor, systems such as Broadband Global Area Network (BGAN) (Fig. 10.3) provide reasonably cost-effective base stations and data plans.

10.3.2.4 Technical issues for GPS installation and operation

In order to achieve the highest accuracy of positioning for a GPS site, many conditions must be met; in many practical cases however, some of these may have to be compromised when finding the best-available solution to any given site's specific issues. As noted in Section 10.2, the location of tide gauges is often not optimal for ensuring their vertical stability. To provide the best possible correction for local vertical motion of the tide gauge, the best possible location for a GPS site would be immediately next to the gauge, sharing the same structural elements, and thus experiencing exactly the same motions. This situation is often an impractical option, however.

The single-most crucial condition for collecting good-quality GPS data is for the antenna to have an unobstructed view of the sky in all directions, preferably down to 10 degrees above the horizon but at least down to 15 degrees. With a partially obscured sky view, not only will the receiver record fewer observations but these observations will have a poor geometric distribution. This can lead to spurious apparent motions in the solution, depending on the exact distributions of the satellites over time. It is unfortunately the case that many tide gauges, especially those with the most valuable long-term series, were originally installed to monitor tides and water depths for shipping and so are located in busy harbor areas. The environment here is typically cluttered with warehouses, cranes, and transient piles of containers, not to mention the ships themselves which may tie up alongside the tide gauge.

A secondary but typically related issue is the problem of multipath, the effect caused by the GPS signals reflecting off nearby surfaces and interfering at the GPS antenna. Multipath not only increases the noise level, but can also cause temporally correlated excursions in the position solutions. In the worst cases multipath can render GPS observations worthless. To minimize its impact the antenna should not be located near to large, strongly reflecting surfaces, such as the corrugated iron warehouse sidings and roofs found at many ports.

Some other general site questions that should be considered are: is it possible to firmly anchor the GPS monument and be sure there is no monument instability? Is the site secure? Are communications and power available? Is the site protected from impacts, for example by trucks or ships? Finally, the history of the tide gauge and its leveling network should be considered. If the TGBM has a long-term history and demonstrated stability, it might be most advantageous, if necessary, to locate the GPS next to it rather than the gauge itself.

One possible solution to extreme site issues at the tide gauge location is to install the GPS on land within the leveling network, at a site that satisfies the primary conditions. An additional potential advantage to this approach is that, with good site selection, the GPS will now be measuring the local tectonic motions, rather than a combination of those and local structural stabilities. In this case, the leveling must be relied on to provide the necessary characterization of the structural stability of the tide gauge location (Fig. 10.5). This configuration weakens the ability of the GPS to provide absolute corrections for the sea-level measurements, but provides valuable information about broader regional tectonics that might have important implications for the local impact of sea-level change. It is however often true that the area along the leveling network is equally industrial as at the tide gauge, and finding a suitable location there may prove to be just as difficult. The best of both worlds would be at least two GPS sites, one at the tide gauge and the other installed in bedrock within the leveling network (Fig. 10.1). This is in fact the configuration recommended by some groups,although of course this doubles the site cost. Alternatively, with some luck a GPS site from another network may be close enough to be able to provide the regional tectonic control.

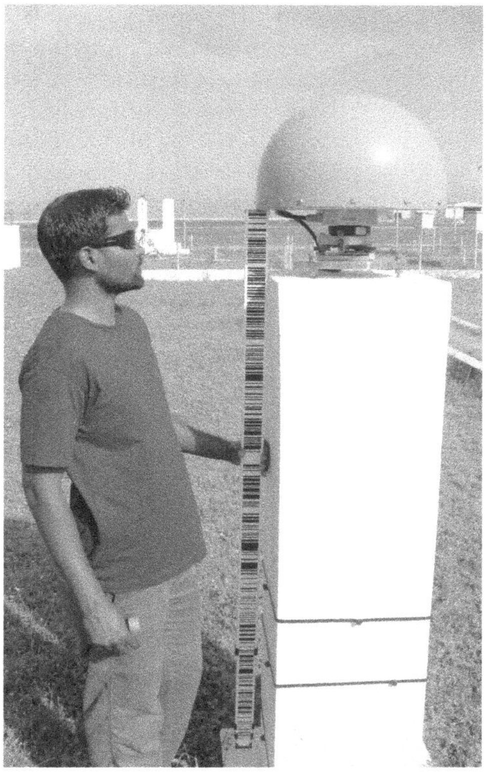

Fig. 10.5. Performing a leveling survey to a GPS antenna at the MAL1 continuous GPS site in the Maldives. Several measurements around the edge of the antenna should be made to confirm that the antenna is level, or to correct for any slight tilt. *Source*: Photograph by Jason Klem.

10.3.3 GPS data processing

It is entirely possible to install a GPS site at a tide gauge and analyze its position time series, yet never have to process the data. If the site is part of TIGA, especially if it is a regular IGS site, then its data will be processed by one or more of the designated analysis centers and the time series made publically available. If one or more sites are not part of these ongoing processing efforts then one of the processing packages that are used by the IGS analysis centers must be used to process the data in order to achieve sufficient accuracy. The three most used are GAMIT/GLOBK (GPS Analysis at MIT/Global Kalman filter) maintained by the Massachusetts Institute of Technology, Gipsy produced by Jet Propulsion Laboratory, and Bernese written by Astronomical Institute of the University of Berne. A fourth package Program for the Adjustment of GPS EphemerideS (PAGES) is supported by the US National Geodetic Survey. The institutes developing these processing packages have made an effort to make it as easy as possible to use them, and accurate relative positioning can be performed without an understanding of the internal workings of the software. The demanding nature of generating accurate global absolute vertical velocities requires more involvement, however. In order to generate useful solutions for this, a significant learning curve must be navigated, and is not for the faint of heart. The details are too technical for inclusion here, but a discussion of some of the relevant issues is provided by Steigenberger et al. (2006) and in the manuals for each of the processing packages.

As a general illustration of the steps involved, we summarize here how reasonably accurate positions for local GPS sites can be produced by connecting these sites to the ITRF. In order to tie a site successfully to the ITRF, data from four or more IGS sites must be included in the processing along with the precise solutions for the GPS satellite orbits provided by IGS. The processing software must be configured to use the same model tables and key parameter settings as the IGS analysis centers, in order to avoid introducing distortions between the locally run and the IGS published solutions. Each package includes routines to allow a local processing run to be incorporated into the published global solutions. This final step produces a final merged solution that includes the local GPS sites, properly tied to the global reference frame.

10.3.4 Absolute gravity

One final recommended element at GLOSS tide gauge sites to provide valuable additional constraints on vertical motions is absolute gravity (Fig. 10.1). The principles of measuring absolute gravity are straightforward applications of Newton's Laws: a weight is dropped, and the time it takes to fall a set distance measured. As gravity declines with distance from the center of mass, vertical motion of a benchmark also produces changes in the observed gravity. In practice, it is only in recent years that it has become practical to perform these measurements with an affordable and portable instrument of high-enough accuracy for it to be useful for measurements of tectonic motion. The commonly used Micro-g Lacoste FG5 provides a precision of better than 1 μgal and an accuracy of 2 μgal, which equates to an elevation change of ~1 cm. If the effects of tides are carefully corrected for, repeated measurements at a benchmark tied to the TGBM with leveling surveys can provide independent estimates of vertical motion, with accuracies similar to GPS.

10.4 GPS AND LEVELING SURVEYS AND THE VERTICAL DATUM

The role of GPS we have discussed up to this point is to act as a stable, long-term reference, to tie a tide gauge benchmark network into the global absolute reference frame. A common additional use of GPS is to survey points of interest, and determine their positions with respect to each other and with respect to the local reference datum. An example of this type of application would be mapping of proxy indicators for local sea level (see Chapter 11). When a continuous site exists locally to act as the reference site, preferably with the possibility of recording 1 Hz observations, survey GPS offers a low-cost, low-manpower technique for acquiring accurate relative positions for points across a wide area. If sub-centimeter accuracy is required, the points must be observed using the static survey approach with the GPS antenna mounted on a tripod, typically for several hours or more. For surveys with less-demanding requirements, a "stop-and-go" kinematic approach can be adopted with the survey GPS running continuously. The operator simply holds the antenna pole in position at each point of interest for a minute or more (assuming 1 Hz data collection) to collect enough data for a sufficiently accurate position solution.

10.4.1 Orthometric and ellipsoidal heights

While GPS has many advantages as a surveying tool over alternative techniques such as traditional leveling (or similar techniques such as total station surveys) it is important to recognize that the elevations these two techniques measure are fundamentally different. The elevation measured by GPS at a point is the ellipsoidal height (h in Fig. 10.1), whereas leveling measures orthometric heights, that is, the height above the geoid (H in Fig. 10.1). The geoid is a gravitational equipotential surface that is physically measureable (at least in principle) as it is defined by MSL (or where MSL would be if the ocean extended under land). In contrast, the reference ellipsoid is a mathematical construct describing the shape of the Earth's surface and its value at any point is given by a formula: it cannot be directly observed. Although the shapes of these surfaces are broadly similar they can differ locally by tens of meters and, as a consequence, leveling and GPS surveys produce different height values. For any given point, the ellipsoidal height $h = H + N$, where N is the geoid height (Fig. 10.1), that is, the height of the geoid above the ellipsoid at that location. Crucially, the ellipsoid and geoid can also have different local slopes, so measuring the height difference between two nearby points using GPS, and those same two points using leveling, may also generate different results. For the height difference between two points, a similar relationship holds: $dh = dH + dN$.

10.4.2 Geoid and the vertical datum

For sea-level work the relevant vertical datum for height measurements is MSL; when surveying proxy indicators for local sea level with GPS for example, some knowledge of the local geoid is required. In order to interpret the results for points surveyed using GPS, the height of the geoid at those points needs to be found and used to transform the GPS ellipsoidal heights into orthometric heights. In most cases it is either impossible or impractical to directly measure the geoid height at all points of interest. However, the integration of modern satellite-based data with traditional ground-based surveying is producing increasingly accurate maps of the geoid with ever-greater spatial resolution. The current best global geoid model is EGM2008 (Earth Gravity Model 2008). It is openly available on a 5 arc-minute grid and is found to agree with independent estimates of the geoid undulations to ~5–10 cm. In addition to the global model, many regions have even more accurate and detailed regional models that are also freely available to the public.

At most locations, over short kilometer-scale distances local differences in geoid height – which may be up to 10 cm or more – are represented in the geoid models with greater accuracy than the absolute height. Similarly, a GPS survey is much more accurate at establishing the ellipsoidal height difference between points rather than their absolute heights. For this reason, the best practical approach for transforming the GPS-surveyed elevations h of local points of interest into orthometric heights H requires a reference location where the orthometric and ellipsoidal heights are known simultaneously. A tide gauge with GPS fulfils this requirement (Fig. 10.1), and orthometric heights for points surveyed with GPS can be determined using the following steps: (1) use MSL, the reference orthometric height H_{TGBM} measured for the TGBM, and the GPS-determined ellipsoidal height of the TGBM to determine the geoid height N_{TGBM} at that point; (2) estimate the geoid height differences dN between the TGBM and the surveyed points by interpolating from the best-available geoid model for the area; and (3) using the GPS-determined ellipsoidal height differences dh between each point and the TGBM, the orthometric heights for the points can then be accurately estimated using $H = dh - dN + H_{TGBM}$.

10.5 APPLICATION OF GPS AND SURVEYING TO SEA-LEVEL RESEARCH

There are many ways in which geodetic techniques are employed to study sea level and sea-level change, and they can be categorized in different ways. As we focus here on GPS and surveying, we will summarize the state of current geodetic research by highlighting some of the ways in which these two techniques are being used to examine sea level and coastal change.

10.5.1 Leveling-based applications

Although the primary purpose of the benchmark control network is to maintain the vertical datum for the gauge, the data collected also offer the possibility for studying local stability. It is well known that vertical land motions are typically not predictably linear or absolutely periodic over time. This may result in systematic errors and biases when extrapolating measurements made within a limited time window over a much longer period. Although the aperture of the tide gauge leveling network is small, the time history of measurements may be many decades and can therefore be used to examine the long-term stability of the area around the tide gauge. This type of study can help identify BM instabilities, characterize local tectonic and structural motions, and revalidate the tide gauge correction terms.

The national leveling networks are an even more valuable resource. In many countries, the national network of benchmarks (or at least regional components of it) has been repeatedly measured. Although the increased baselines for these networks can introduce systematic errors, they also offer a unique and largely underexploited opportunity to examine broader-scale tectonic stability and vertical motions of a region. With long time windows and high relative accuracies, re-leveling campaigns can have very high signal to noise ratios. They have been used to identify local and regional subsidence and uplift patterns, and model the structural controls on these processes. Recognizing the magnitudes and spatial extent of sediment compaction, interseismic strain accumulation, and creep on faults for example, has the potential to greatly increase our understanding of the long-term vertical motions of the tide gauge and coastlines and the impacts of sea-level changes on local coastal stability. While leveling data alone provide only a relative measure of height changes, it is possible to take advantage of other observations to provide a reference frame for the leveling data: tide gauges provide the link to local sea level, while GPS or other systems can be used to connect the leveling data to the ITRF (Fig. 10.5).

10.5.2 GPS-based studies

GPS is employed in a range of ways in sea level studies: it can simply be used directly to correct tide gauge sea-level rise measurements, or it can be used to map regional tectonic motions in order to better understand and model the spatial and temporal variations in observed sea level. Some studies focus on how to improve the processing of GPS data and the reference frame in order to support ever more accurate measurements of sea-level change. Others use GPS as one of several techniques that together can tease out longer-term relative motions. Global-scale mapping of the motions produced by important large-scale sea-level-related processes, such as ice sheet melting and global isostatic adjustment (GIA), is crucial to obtain a better understanding of them and their impact.

Prior to the demonstrated stability of the recent iterations of the ITRF, sea-level studies using GPS had two options. One could explore relative vertical motions within smaller regional networks which remove most of the longer-wavelength reference-frame related errors. This approach provides very high-accuracy relative motions, and can be useful for local and regional interpretations of deformation fields. The alternative approach is to accept the increased error budget, while minimizing it as much as possible, and explore the contribution that the vertical velocities from GPS can make to estimating absolute vertical motions or sea-level change. As the number and length of time series of GPS positions at tide gauges has increased in concert with better reference frame stability, GPS has become increasingly central to the problem of understanding sea-level observations in a global context. The rapid proliferation of GPS at tide gauges has permitted a number of both regional and global investigations. An excellent example of this is provided by the Hawaiian Islands (Fig. 10.6). Here, the weight of the lava erupted by the growing volcanoes on the Island of Hawai'i causes the lithosphere to flex downwards in response, resulting in subsidence of the southeast end of the island chain. The subsidence rates can be determined from the GPS sites along the islands, and compared with the rates of sea-level rise estimated from tide gauges. The tectonic subsidence matches the higher observed sea-level rise rates

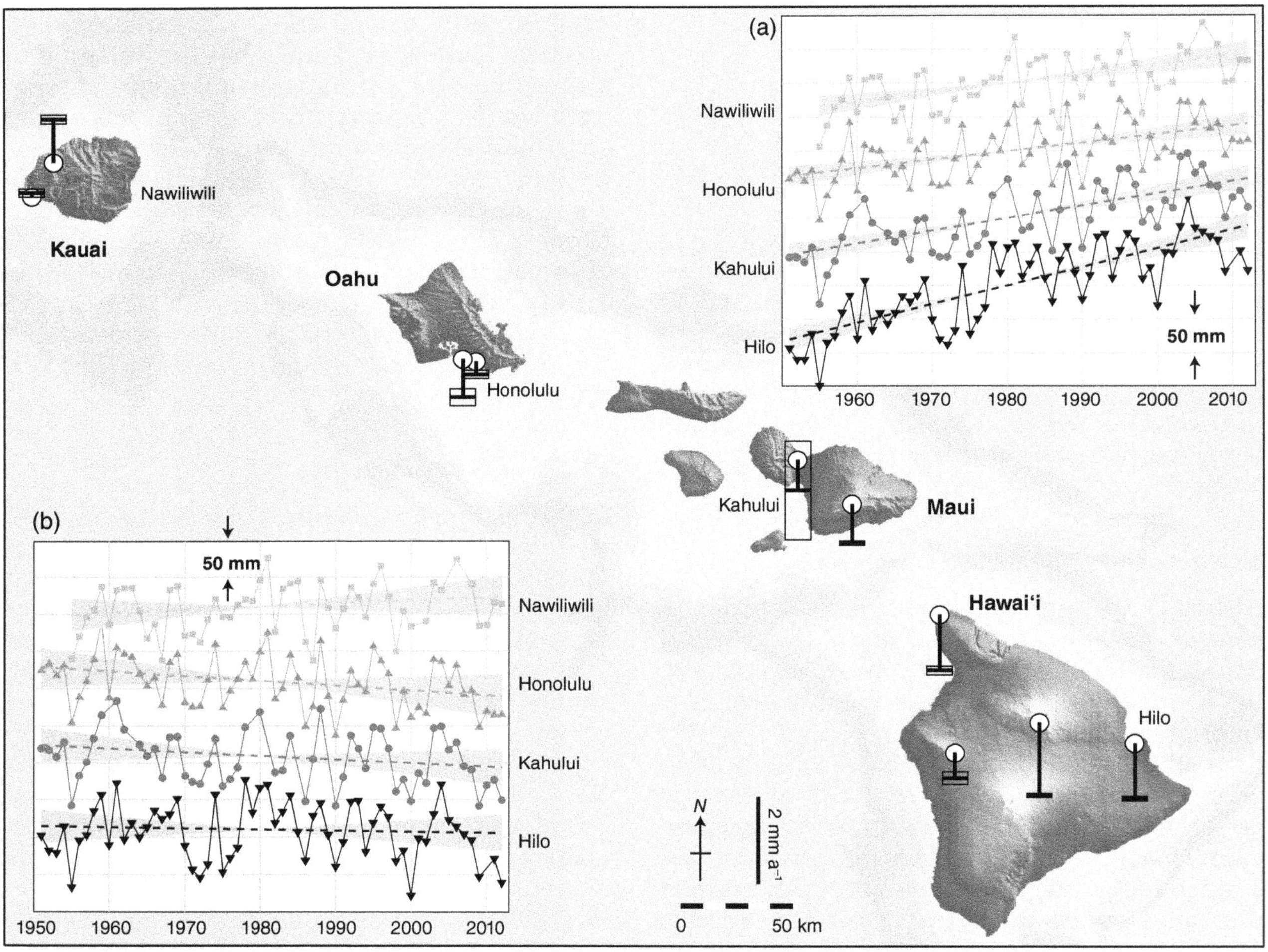

Fig. 10.6. Long-term vertical rates (vertical lines) and error bars (thin lines) for stable continuous GPS sites (circles) in Hawai'i. The increasing mass of the active volcanoes on Hawai'i Island are causing it and Maui to sink with respect to Oahu and Kauai, which are thought to be stable. (a) Annual sea-level changes for tide gauges in Hawai'i, with best-fit slopes (dashed lines) and 95% confidence limits (gray zones). Differences in the sea-level rise rates mirror the GPS vertical velocities. (b) Annual sea-level changes after the mean sea-level rise rate for Nawiliwili, Kauai, and Honolulu, Oahu, is removed, and corrections for the GPS vertical velocities applied. The residual slopes are not significantly different from zero. For color details, please see Plate 29.

on Maui and Hawai'i islands and, after correcting for this, there is no significant difference between the relative rates observed along the islands.

These GPS-based studies generally find that, as the quality of the reference frame and GPS velocity estimates improve, the absolute sea-level rise estimates at tide gauges tend to agree more closely with the rates estimated by satellite altimetry; occasional outlier sites are also clearly identifiable however, and can be investigated to determine the origin of the disagreement. For tide gauge installations that have had absolute gravity measurements, the extra and independent information it provides allows more confidence in estimating vertical motions. GPS and gravity measurements are used to examine the impact of ocean loading on vertical motions as well as for improving confidence in vertical motion resolution. Accurate models of GIA are important for predicting the long-term regional impacts of sea-level rise. As accurate measurements and accurate GIA models evolve together, recent results from GPS and gravity match the motions predicted by the recent models increasingly well.

The various datasets that are involved in measuring sea-level change have different strengths and weaknesses: tide gauges provide long, but rather noisy, time series of relative motions; leveling may span a long time interval but typically with only a small spatial footprint; whereas modern satellite altimetric measurements and GPS provide short, but absolute, time series. Some recent research efforts have focused on integrating many of these datasets, exploiting correlations between adjacent locations in an attempt to leverage each data type's strengths and offset their weaknesses. This approach has been found to significantly reduce the formal errors on the resulting rates, and produce improved consistencies between datasets.

10.5.3 Application of InSAR to mapping coastal vertical motions

While we have focused primarily on the role GPS can play in providing the reference frame and absolute velocities for sea-level change measurements, an important related topic is how local tectonic motions and their spatial variation impact relative sea-level rise. Although leveling and, in some places, sufficiently dense networks of GPS can provide the spatial coverage necessary to measure local variations in vertical motion, other techniques are especially suited to this type of application. Several satellites acquire interferometric synthetic aperture radar (InSAR) data, which is extremely sensitive to relative changes in the line of sight range to the ground across the observed area (typically ~50–100 km), and can produce dense maps of relative motion with horizontal resolutions of the order 20 m. Although this technique does not provide absolute motions, if there are GPS sites within the scene it is possible to use these to provide a tie to the global reference frame. There are different approaches to forming the tie but the simplest, and arguably most robust, is to define a radius around each GPS sites that is representative of the spatial scale of ground motions. Forming an average value from all the InSAR data within each circle, a mean translation and rotation can be defined that minimizes the differences between all the GPS and InSAR motion estimates. This transformation can then be applied to the entire InSAR scene to generate maps of absolute vertical motion. These maps provide a uniquely detailed image of the local vertical motions and can yield a greater understanding of the local impacts of sea-level rise.

10.6 NOVEL APPLICATIONS OF GPS TO SEA-LEVEL MEASUREMENTS

Although the current role of GPS in sea-level measurements is to provide a tie to a stable global reference frame and measure local vertical motions, there have been efforts to explore whether GPS can do more than that. Although it is unlikely that these techniques will ever be accurate and robust enough to replace tide gauge measurements of sea level, several groups have been developing techniques that use GPS directly as a sensor for measuring the height of the sea surface that may play useful roles in the future.

10.6.1 Reflectometry

GPS signals are reflected by the ocean surface, undergoing a polarization reversal in the process. With a suitably modified antenna oriented to point at the ocean, the reflected signal can be recorded and, through comparison with the directly received signal, the range to the sea surface calculated. Several systems have demonstrated their ability to determine the height of the water surface with an accuracy of a few centimetres from fixed platforms, or to the decimeter level from aircraft. Although of limited immediate application to the issue of measuring global sea level, this is a passive system that has been demonstrated on satellites. The low power requirements and relative simplicity increases the possibility of this type of system being deployed cheaply on constellations of satellites; this would provide a dense global coverage, albeit with somewhat lower accuracy than dedicated satellite altimetry missions. An alternative approach using the same reflected signal has been demonstrated using only a standard GPS installation. In this approach, the interference of the reflected signal with the direct path signal can be identified and modeled from the standard GPS data stream to give the height of the sea surface to the sub-decimeter level. Again, although no substitute for a real tide gauge, it opens up the possibility of densifying coverage by acquiring independent sea-level measurements from "installations of opportunity".

10.6.2 Ocean platforms

An alternative way of using GPS for sea-level observations is to make it part of the primary structure that measures the sea surface. One such approach is to equip ocean platforms with GPS and use them as floating tide gauges. This has been demonstrated with GPS buoys and GPS-equipped catamarans and ships; although the point accuracy of this type of measurement is lower than a fixed GPS tide gauge, it offers the possibility of high-density measurements in the open ocean and notably in the near-coastal zone where satellite altimetry may be unable to produce accurate measurements. It is therefore possible that increased instrumentation of this type of vehicle, including perhaps the global commercial shipping fleet, could complement and supplement satellite measurements.

10.7 CONCLUSION

The proliferation of continuous GPS sites at tide gauges, supported by an extremely stable geodetic reference frame and improved GPS processing, provides the opportunity to determine vertical motions of accuracy better than 1 mm a^{-1}. With robust, repeated, leveling ties to characterize any relative motion between GPS antenna and tide gauge, properly installed sites can expect to be able to resolve absolute sea-level changes with similar accuracy. Advanced data analysis approaches integrating all available geodetic data, perhaps also including models of regional oceanographic phenomena, show great promise for extracting the most information possible from all the diverse types of measurements, and may represent a very productive avenue for future efforts. As more sites are installed and more GNSS constellations are launched, the reference frame grows ever more stable. As new datasets from GPS reflectometry or from fleets of ocean-platform-based GPS sensors become available, the key role that geodesy plays in measuring and understanding sea-level change seems destined to increase in importance.

REFERENCES

Bevis, M., Scherer, W., and Merrifield, M. (2002) Technical issues and recommendations related to the installation of continuous GPS stations at tide gauges. Marine Geodesy, 25, 87–99.

Blewitt, G., Altamimi, Z., Davis, J., Gross, R., Kuo, C.-Y., Lemoine, F.G., Moore, A.W., Neilan, R.E., Plag, H.-P., Rothacher, M., Shum, C.K., Sideris, M.G., Schöne, T., Tregoning, P., and Zerbini, S. (2010) Geodetic observations and global reference frame contributions to understanding sea-level rise and variability. In *Understanding Sea-Level Rise and Variability* (eds Church, J. A., Woodworth, P. L., Aarup, T., and Wilson, W. S.), Blackwell Publishing Ltd, Chichester, UK.

Hicks, S.D., Morris, P.C., Lippincott, H.A., and O'Hargan, M.C. (1987) *User's Guide for the Installation of Bench Marks and Leveling Requirements for Water Levels.* NOS, NOAA, Rockville, MD.

Steigenberger, P., Rothacher, M., Dietrich, R., Fritsche, M., Rülke, A., and Vey, S. (2006) Reprocessing of a global GPS network. Journal of Geophysical Research: Solid Earth, 111, B05402.

Woodworth, P.L. (2002) *Manuals and Guides 14: Manual on Sea Level Measurement and Interpretation, Volume III - Reappraisals and Recommendations as of the Year 2000.* Intergovernmental Oceanographic Commission.

Chapter 11
Reference water level and tidal datum

SARAH A. WOODROFFE AND NATASHA L. M. BARLOW
Sea Level Research Unit, Department of Geography, Durham University, Science Laboratories, Durham, UK

11.1 INTRODUCTION

This chapter is concerned with relating the environment in which a proxy indicator accumulates to a reference water level. Reference water level is the elevation at which the indicator occurs, defined with reference to a tidal datum. A vertical datum is called a tidal datum when it is defined by a certain phase of the tide. Very few proxy indicators accumulate at mean sea level (MSL); when calculating how much relative sea-level (RSL) change has occurred in the past, it is therefore important to know both where and over what range in the tidal frame the indicator occurs and what correction is needed to ensure that different indicators relate to a common datum. This calculation is known as "indicative meaning" (Chapter 2) and is specific to each field site because the exact nature of the correction needed relates to the tidal range at the site. As the majority of precise sea-level indicators accumulate in the upper half of the intertidal zone, the most-common reference water level (RWL) used to calculate the indicative meaning is mean high water of spring tides (MHWST) or mean higher high water (MHHW; see Section 11.3 and Table 2.1 for definitions). This chapter introduces MHWST, MHHW, and other reference water levels and explains how these values are calculated and how they relate to other land survey datums used in RSL studies. We then go on to explain how a land survey datum may be calculated and used in sea-level research when in the field.

The chapter does not aim to provide details of tidal dynamics, but rather provide some context of tidal cycles in order to place reference water levels and tidal datums into perspective. For further reading about tidal dynamics see Pugh (1987, 2004). We have attempted to keep the discussion at the practical level necessary for field scientists undertaking RSL investigations, often in remote locations.

11.2 TIDAL CYCLES

For the purpose of this chapter, we focus on tides as periodic movements of local/regional sea-surface elevation which have coherent amplitude and phase relationships to some periodic geophysical force, normally the Moon and Sun. The response of the ocean to the forcing of the Moon and Sun is very complex and causes tides to vary around the world (see the Admiralty Manual of Tides NP120 for a detailed mathematical description of tidal theory). However, there are two common features of any tidal cycle: range and period. The range is the difference between two successive high and low levels (which is therefore equal to twice tidal amplitude), and the period is the time between one high (or one low) level and the next high (or low) level.

If Earth were a perfect sphere without large continents, everywhere would experience two equally proportioned high and low tides each lunar day. However, the presence of large land masses means tides are unable to move freely around the planet and therefore establishes complex tidal cycles within the ocean basins. Tidal cycles are semidiurnal at many locations, in that the tidal period is approximately half a lunar day. In other words, the time from one high tide to the next takes an average of 12 hours and 25 minutes so that two tidal cycles occur for each transit of the Moon (every 24 hours and 50 minutes; e.g., at Mombasa and Bermuda, Fig. 11.1). Some locations such as the Gulf of Mexico and Persian Gulf experience diurnal

Handbook of Sea-Level Research, First Edition. Edited by Ian Shennan, Antony J. Long, and Benjamin P. Horton.

tides with only one high and one low tide each day (e.g., at Karumba, Fig. 11.1). A mixed tidal regime occurs where both the diurnal and semi-diurnal components are important, resulting in a significant difference between the elevation of the two high waters and elevation of the two low waters (known as diurnal inequality; e.g., at San Francisco, Fig. 11.1). Mixed tides may also occur at the transition between areas of semi-diurnal and diurnal tides, or where there is strong distortion due to the local bathymetry (e.g., at Courtown, Fig. 11.1). All tidal cycles typically have ranges which increase and decrease in a cyclic manner over a 14-day period (Fig. 11.1). The maximum range (spring tides) occur a few days after the new and full moon, whereas minimum ranges (neap tides) occur shortly after the time of the first and third quarters of the moon.

The astronomical tidal spectrum consists of a number of harmonic constituents that are related to the mechanics of the Earth–Moon–Sun system. The two fundamental lunar (M2) and solar (S2) constituents, which have periodicity of 12 hours, 25 minutes, and 12 hours, respectively, are the primary drivers of the tidal cycle. They combine with a number of other constituents, and allow theoretical tides to be accurately predicted at any coastal location (see Chapter 29). Of particular note when calculating tidal datums is the nodal cycle of maximum lunar declination, which occurs over a period of 18.6 years. Maximum diurnal tides occur when lunar declination is greatest. In principle, calculating tidal datums therefore requires measurements over a 19-year period or longer, although this may not always be feasible.

The observed tidal cycle often differs from the predicted due to atmospheric processes and weather effects. In the first instance, there is an inverse relationship between sea level and atmospheric pressure where, in a theoretical world, a 1 millibar increase in atmospheric pressure results in a 1 cm decrease in sea level (Pugh, 1987). Dynamic processes such as wind stress and friction in shallow water means this exact inverted response is in reality uncommon. Tsunamis, hurricanes, and land-level change due to earthquakes can also cause non-tidal deviations of centimeters to several meters from the predicted cycle. Superimposed on the astronomical tidal cycle can also be natural oscillations called seiches, which have amplitudes of a few centimeters to several meters and periods of tens of seconds to several hours. These are due to oscillations caused by a local harbor, estuary, or bay and may be generated by a wind squall, surf beat, or submarine earthquake, for example (Rabinovich, 2009).

If there was no tidal forcing, no atmosphere and ocean currents, and no differences in water density, the sea surface would adjust to form an equipotential surface called the geoid, the shape of which depends on the distribution of mass in the Earth. The rotation of the Earth on its axis causes density anomalies which prevent the geoid from being an ellipsoid; instead, it takes the form of a rather bumpy uneven surface with local variations of several meters. In ideal conditions mean sea level (MSL) would follow the shape of the geoid exactly, but ocean density, ocean–atmosphere currents, and metrological processes cause it to deviate from the geoid usually by a few decimeters, but in some places by more than a meter.

Global positioning systems (GPS) use an ellipsoid coordinate system, which means GPS elevation is given relative to an ellipsoid vertical datum rather than elevation with respect to MSL (see Chapter 10). However, most modern GPS systems can make the computation to output orthometric (geoid) elevations that are then relative to MSL. A single GPS can achieve an accuracy of approximately 10 m. To achieve the accuracy desired for sea-level research (decimeter to centimeters), it is necessary to use differential GPS (dGPS). The underlying premise of dGPS is that that any two GPS receivers that are relatively close together will experience similar atmospheric errors. One or more GPS base stations with precisely known location calculate their position based upon satellite signals and compare this location to its known location. This base station location difference is then applied to data from a roving GPS receiver, either in real time in the field using radio signals or through post-processing after data capture using specialist processing software.

11.3 REFERENCE WATER LEVELS AND TIDAL DATUM: DEFINITION AND CALCULATION

Reference water levels and tidal datum are of specific importance when defining the elevation of proxy sea-level data and allow the analysis of long periods of sea-level variations. A certain

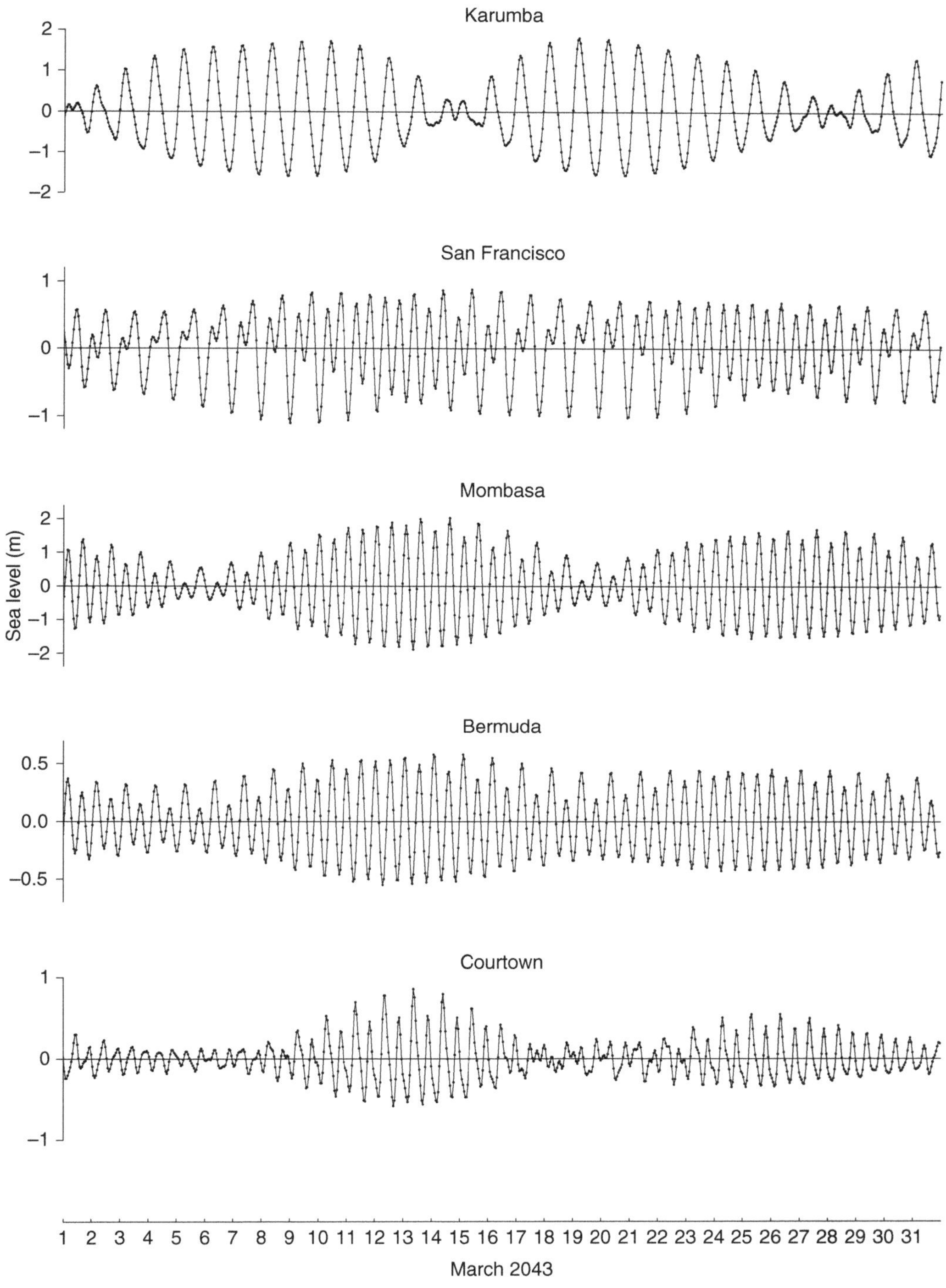

Fig. 11.1. Tidal predictions for March 2043 at five sites which have very different tidal regimes. At Karumba the tides are diurnal and at San Francisco they are mixed, whereas at both Mombasa and Bermuda semidiurnal tides are dominant. The tides at Courtown are strongly distorted by the influence of the shallow waters of the Irish Sea, and therefore have a complex mixed regime. *Source*: Pugh and Woodworth (2014). Reproduced with permission of Cambridge University Press.

phase of the tide is known as a reference water level. A tidal datum is a standard elevation of the stated reference water level, although the two terms are often used interchangeably. Reference water levels form an important part of defining the indicative meaning of a sea-level indicator (see Chapter 2). It is important to define the reference water level and establish its elevation when comparing data to provide a common reference frame.

Geodetic surveys often use MSL which is the arithmetic mean of hourly elevations measured over a specified period (commonly ≥19 years) at a given location. The Permanent Service for Mean Sea Level (PSMSL) collects and publishes the monthly and annual MSL values from the global network of tide gauges. In comparison, mean tide level (MTL) is the arithmetic average of all the elevations of mean high and mean low water at a given location. MTL often differs by millimeters to centimeters from MSL due to tidal distortions in shallow water.

Reference water levels and tidal datum other than MSL and MTL are also used to understand where in the tidal range sea-level indicators form (Fig. 11.2; Table 2.1). Mean high water springs (MHWST) and mean low water springs (MLWST) are the average spring tide high water level and spring tide low water level, respectively, averaged over a sufficiently long period in semidiurnal waters. Mean higher high water (MHHW) and mean lower low water (MLLW) are the average of the higher high or lower low water elevation of each tidal day. Mean high water (MHW) and mean low water (MLW) are the average of all high or low waters ideally observed over a 19-year period; these can also be applied in diurnal tidal settings where there is no discernible difference between the elevation of the daily high or low waters. The difference between MHW and MLW is the mean tidal range, and the difference between MHWST and MLWST is the spring tidal range. Macro-tidal regions are typically defined as having a mean tidal range of >4 m, a meso-tidal region has a tidal range of 2–4 m, and micro-tidal regions have a tidal range of <2 m.

The highest and lowest occurrence of tidal inundation can exceed the mean higher and lower tides. The highest astronomical tide (HAT) and lowest astronomical tide (LAT) is the highest and lowest tidal level that can be predicted to occur under any combination of astronomical conditions, and usually relates to the 18.6-year lunar declination cycle. LAT is often used to define Chart Datum. Chart Datum is the extreme elevation below which local or regional sea-level seldom falls, providing assurance to navigators of the minimum depth of water. It is defined in terms of local tidal characteristics and, although many authorities use LAT, some (e.g., USA National Oceanic and Atmospheric Administration or NOAA) use MLLW. This is important as many tide tables give elevations relative to Chart Datum, but for sea-level research purposes a theoretically horizontal geodetic datum relative to MSL is of greater value.

An issue with using MSL as a common datum to compare records over regional scales is that it varies spatially. There is therefore a need for a standardized vertical geodetic datum that has a surface of zero elevation to which elevations of other points can be referred, in order that all elevations and tidal datum be in a consistent format. Different nations and

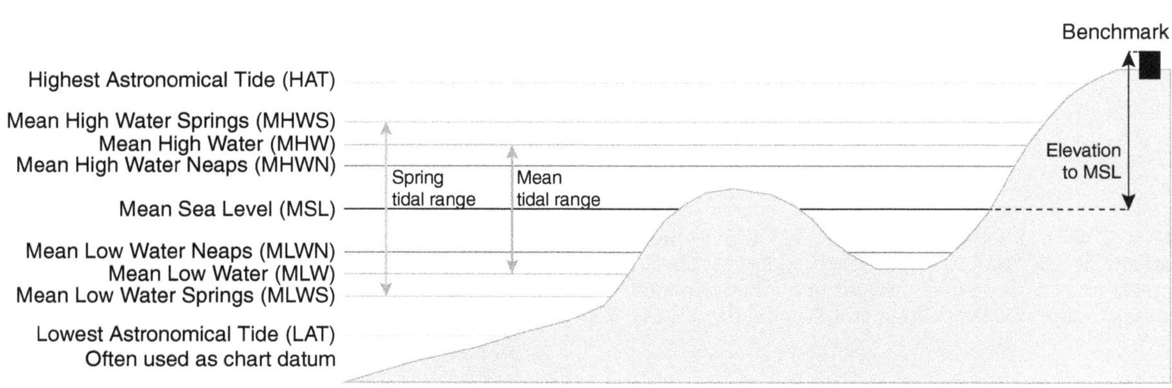

Fig. 11.2. Main reference water levels for a location that experiences semidiurnal tides. For definitions see Chapter 2, Table 2.1.

authorities define different land survey datums which are useful for regional leveling and collating sea-level data.

For example, in the United Kingdom all elevations are related to Ordnance Datum Newlyn (ODN) which is the mean sea level at Newlyn in southwest England, measured during 1915–1921. Following establishment of this level, it was applied around the UK by the Ordnance Survey (OS) who aimed to establish one benchmark of known elevation per square km in rural areas and erected ~500,000 third-order benchmarks giving vertical control for mapping and surveying around the UK. The last surveys of the benchmark network occurred in the 1980s however and, since 2000, the Ordnance Survey has only maintained a network of 190 fundamental benchmarks. In the USA, Canada, and Mexico, the North American Vertical Datum of 1988 (NAVD 88) and International Great Lakes Datum of 1985 (IGLD 85) leveling networks record MSL calculated from tide gauge measurements between 1982 and 1988 at the Father Point/Rimouski water level station in the St Lawrence River, Quebec, Canada (Zilkoski et al., 1992). In the Netherlands, the national vertical datum is the Normaal Amsterdam Peil (NAP) and in France it is Nivellement General de la France (NGF).

If using local tide predictions, it is important to note whether the elevations are given relative to Chart Datum or to a land survey datum. In most locations of known tidal values (e.g., tide gauge stations), resources such as Admiralty Tide Tables provide correction values to convert from Chart Datum to the regional/national leveling datum. To compare data from a range of locations using differing land survey datums requires specialist software or complicated computations; a geodetic transformation must be applied to convert elevations measured in one coordinate system to another. Where possible, a global reference frame such as GPS should be used, such as the latest of the International Reference Frame series (ITRF-2008; Altamimi et al., 2011).

Where there are no defined tidal datums, due to the absence of a local tide gauge or the presence of coastal geomorphology which causes deviation from tidal dynamics measured at the nearest tide gauge to the site of interest, it may be necessary to calculate the elevation of tidal datum by other means (discussed in the following section).

11.4 CALCULATING REFERENCE WATER LEVELS FOR PROXY INDICATORS AND FOSSIL DEPOSITS IN THE FIELD

There are a range of approaches used by researchers to relate proxy indicators to tidal datums and past sea levels. To enable plotting of RSL data from different studies in a region on a single graph, researchers should undertake one of the approaches described below to relate their data to tidal datums.

11.4.1 Comparing the elevation of modern and fossil counterparts

The easiest way of determining the amount of RSL change shown by an indicator is to directly compare the elevation of the fossil indicator to its modern counterpart. This is only possible if both modern and fossil indictors are found in the same location. If this is not the case, the indicators should be related to a land survey datum or tidal levels (see the following section). This approach is commonly used when reconstructing RSL from coral microatolls on tropical reef flats. Several studies have found both living and fossil microatolls at different elevations on the same reef flat (e.g., Mclean et al., 1978; Chappell et al., 1983). To calculate the amount of RSL change between the time when the fossil microatolls were living and the present, the elevation of living coral on a series of modern microatolls is leveled and the average elevation taken. This is then compared to the elevation of the fossil specimens to produce the amount of RSL change with an error (based on variability in the elevation of the modern counterparts; Fig. 11.3). This approach is very simple and eliminates the need to relate either modern or fossil microatolls to any tidal datums, but is only applicable if the sea-level record is to be compiled only from one form of proxy evidence at one location. This method has been widely used in far-field locations where a single sea-level indicator is used, RSL has been relatively stable during the middle–late Holocene (e.g., within c. 2–3 m of present), and the coastline morphology has not changed significantly during this time.

A RSL study using coral microatolls on Kiribati in the central Pacific takes this approach by measuring the uppermost elevation of living coral on 51 different modern microatolls around the atoll

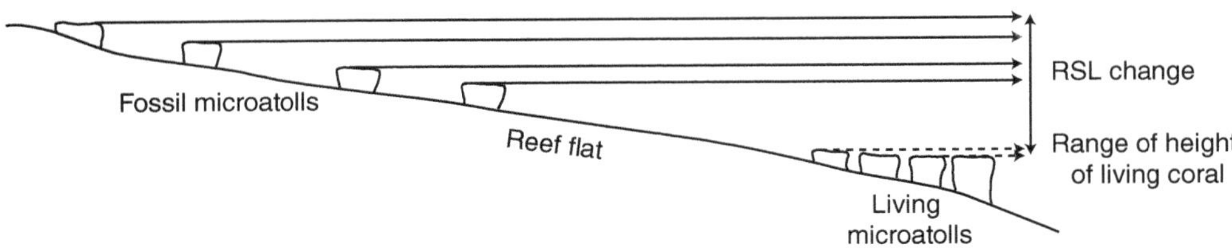

Fig. 11.3. Calculating RSL change from a series of living and fossil microatolls across a reef flat.

using dGPS, and relating these elevations to a permanent survey point on the island (with a known ellipsoidal elevation and the lowest observed tide level), before taking a mean value for each modern microatoll field and producing estimates of sea-level change from fossil examples relative to these means (Woodroffe et al., 2012). An important complication of using dGPS here is that there is a variation in the sea-surface elevation of up to 0.5–1 m around the atoll caused by a geoidal elevation gradient. This is a common issue around mid-ocean islands and seamounts and, unless identified and corrected for, introduces significant error into estimates of RSL change (Cazenave et al., 1980; Woodroffe et al., 2012).

11.4.2 Relating proxy indicators to land survey datum and tidal datum using benchmarks or dGPS

Many RSL studies from the UK and other northwestern European countries relate modern and fossil proxy indicators to land survey datums. This allows easy compilation of results from different locations, in particular if the coastline has moved a significant distance since the formation of the proxy indicators. When calculating the indicative meaning for a particular proxy indicator however, we also need to know the relationship between that indicator and its local constraining tidal level (e.g., MHWST or MTL). Information on the elevations of local tidal levels in land survey datums are often more difficult to ascertain.

To determine the elevation of a modern or fossil indicator in a land survey datum, fieldworkers often choose to physically occupy a benchmark site with survey equipment as part of the leveled survey of their field site (Fig. 11.4a). Isolation basin studies by Shennan and others in Arisaig, Scotland rely on local OS benchmarks with ODN elevations to relate basin sill elevations to the land survey datum (Shennan et al., 1993, 1994, 1999). In the UK where the lower-order OS benchmark network is no longer maintained, this method is becoming increasingly difficult. Existing benchmarks are being destroyed during construction and road maintenance, and the elevations of those that remain may become less reliable over time due to land movements.

A new solution in the UK to relating local measurements to ODN is to use survey-grade dGPS. The procedure involves setting up a tripod-mounted dGPS receiver over a suitable ground point at the field site, and collecting data from overhead GPS satellites for several hours (Fig. 11.4b). The user must then download data for the same time period from at least the three nearest permanent Ordnance Survey (OSNET) GPS receivers which are situated around the UK. These datasets can be post-processed together to convert the new GPS data into land survey datum (ODN in the UK) values. This provides a new temporary benchmark for the field site, which can then be transferred around the site using a level and staff. This method requires the availability of a survey-grade dGPS receiver for fieldwork and a network of permanent GPS receivers which record elevations in the relevant land survey datum. In the UK, the national mapping agency believes that maintaining 100 permanently recording OSNET GPS receivers and providing this data for post-processing is more cost-effective than re-surveying and maintaining the previously extensive benchmark network.

This new approach does away with the need to find a local benchmark, but does rely on "black box" modern technology and the availability of data from other local GPS satellites. Checking the reliability of a temporary benchmark created using dGPS can also be a problem. Where there is a local land survey datum benchmark available, you could set up the dGPS equipment over this existing benchmark and compare the elevation results (bearing in mind that some vertical land motion may have occurred at the location since the benchmark was last surveyed). As original

Fig. 11.4. Examples of field-based techniques to measure elevation: (a) using level and staff on a saltmarsh. *Source*: Photograph by Rod Combellick. Reproduced with permission; (b) a dGPS set up over a rock to establish a temporary benchmark elevation; and (c) a dGPS rover being used to measure the elevation of modern microatolls.

benchmarks still exist in many areas this is still possible, but will become increasingly difficult in the future in the UK as they are progressively destroyed by redevelopment or construction.

Unlike in the UK, in many other countries national networks of benchmarks are still maintained. In the United States and Canada there are several thousand coastal benchmarks, which are related to the NAVD88 or CGVD28 land survey datums. Although the United States and Canada still have a maintained benchmark network, it is unlikely that field sites used for RSL studies will be sufficiently close to one of these maintained benchmarks. The GPS-based method as described above for the UK, with post-processing to an appropriate datum, is therefore the best way to relate proxy indicators in North America (see Engelhart and Horton, 2012).

To understand how GPS-derived elevations relate to tidal datums, software such as VDatum in the United States converts between ellipsoidal and orthometric land survey datums and tidal datums. Alternatively, tidal prediction publications often include calculated tidal levels for ports where tidal readings have been taken in the past relative to a land survey datum. The reliability of these tidal levels depends on the length of dataset used to calculate them. In the UK this is normally at least one month of data; be aware however that at some locations this could be as little as a few scattered observations of high and low water.

11.4.3 Relating proxy indicators to MSL or MTL

There are many locations in the world where no land survey datum is available; users must therefore find an alternative common datum to which proxy indicators should be related. The most widely used datums are local MTL or MSL. In Greenland for example, there are very few currently operational tide gauges; although there is a GPS-maintained national reference system (REFGR), there are large distances between GPS receivers. Post-processed GPS elevation data is therefore not necessarily as precise (~0.5 m) as locally derived tidal data (~0.1 m or less). In this situation we must rely on tidal predictions to relate indicators to reference tidal levels. Predictions are produced for ports around Greenland and reported by the Danish Meterological Institute.

Tidal prediction software requires a period of tidal measurements to calculate harmonic constants used in the prediction process. That is why it is not possible to acquire tidal predictions for every coastal location on the map. There are many tidal predictors available online, including free tidal calculators. Often these free calculators will only give high and low tide and not 15-minute or hourly predictions, but these values are often available from the national mapping agency or hydrographic office of the country in question.

11.4.4 Timed water levels and published tidal predictions or tide gauge data

A single timed water level can transfer the tidal elevation from a prediction port or tide gauge location to a field site. A water level can be taken at any point in the tide, but it is better to measure a water level during slack water (at the point when the tide is changing at its high or low point) as the tide is likely to be closest to its predicted level during this period. This method can however be problematic because of the effects of tidal amplification or dampening depending on the coastal geometry of the area, the effect of waves on the ability to measure water level, and any effects caused by high or low barometric pressure in the area at the time of fieldwork compared to the average. This approach should only be taken where the distance between the prediction port and the field site is less than c. 2 km, the sea state is calm, and there are no differences in coastal geometry between the two locations (e.g., distance up estuary). If there is insufficient time to wait for the next high or low tide, the next best approach is to take measurements every 15 minutes for 2–3 hours and compare them to the nearest tidal predictions or tide gauge readings.

A much more robust approach is to repeatedly measure high or low tides at the field site. This is very important at locations where the coastal geometry changes between the prediction port or tide gauge and the field site. For example, the upper Cook Inlet in Alaska has a large tidal range and the funnel shape of Turnagain Arm results in significant spatial variations in tidal levels. Hamilton and Shennan (2005) overcame these problems by measuring local high tides of differing elevations and comparing them with the same high tides on the same day measured at the tide gauge at Anchorage, repeating the exercise on more than ten separate occasions at each field site. This allowed them to calculate field elevations in relation to MHHW, and determine patterns of tidal amplification or dampening in the estuary system.

11.4.5 Water level loggers

Using a portable water level logger in deep water at or close to a field site eliminates the need to measure the tidal cycle or high tides using leveling equipment. The logger should ideally be permanently submerged to record the whole of the tidal cycle, but this might not be possible in locations with a large tidal range. Depending on the size of memory in the logger, they can record tidal fluctuations at regular intervals (e.g., 10 min, 30 min, 1 hourly intervals) over several months, capturing more tidal variability than by repeat leveling. It is preferable to have two loggers, one in the water and a second on land to measure local atmospheric pressure. A tidal pole is also necessary to relate "tide gauge zero" datum to a temporary benchmark on land to which the proxy indicators are leveled. It is then easy to compare pressure-corrected tidal readings to tidal predictions for the nearest port. If they are close in timing and amplitude, then it is possible to use the published tidal predictions to relate the field measurements to MSL. If they are significantly different, it is better to calculate MSL and other tidal levels from the water level logger data instead. Ideally, regular tidal readings for at least a month and a matching atmospheric pressure dataset, collected locally or from available meteorological data, are needed for this.

11.4.6 Measuring water level offshore

It is sometimes necessary to take samples from a research vessel or small boat on tidal water, for instance when sampling sediments in a tidal pond, during sampling of subtidal surface sediments, or while taking cores offshore. The depth of water above the sampled site and time of measurement should always be noted, using either an echo-sounder or plumb line (bearing in mind that the accuracy of a hand-held echo-sounder may be less than using a plumb line held vertically in the water column, depending on the local water conditions when sampling). This can then be related to the local tide gauge, published tidal predictions or water level logger readings. If time allows, frequent depth soundings should be undertaken at the same location to produce a local tidal curve which can be compared to local tidal predictions or readings. There is more error in measuring tidal fluctuations offshore than with land-based methods because of difficulties in measuring water depth due to waves or soft sediments which may provide little resistance to the weight of the plumb line, or if the water level within the tidal pond is not always in equilibrium with the tides.

11.5 PROBLEMS WITH THESE APPROACHES AND EQUIPMENT REQUIRED

There are issues with all of the approaches described in the previous section. To minimize error, the best approach is to work very close to a functioning tide gauge or a maintained benchmark related to a land survey datum and to tidal levels. This is not often possible, but it is important to bear in mind that errors resulting from using other less-precise methods are unlikely to exceed the indicative range of the proxy indicators in use (apart from when measuring altitude offshore, where the errors may be substantially larger).

Most error is introduced when using timed water levels and relating them directly to published tidal predictions. Taking a single timed water level may introduce decimeter-scale error depending on the local atmospheric pressure during fieldwork, or because of differences in the timing of tidal fluctuations between the nearest port and the field location. However, these errors can often be reduced to the centimeter-scale by taking multiple readings over several days or weeks, comparing them to published tidal predictions or verified readings, and correcting the water level reading for atmospheric pressure and timing differences. Other problems with timed water levels include difficulty in measuring a quiet water level because of waves (centimeter- to decimeter-scale, depending on the location) and the availability of sufficiently detailed tidal predictions (hourly or more frequently, to the nearest centimeter is ideal to reduce this error). If there is a choice between using tidal predictions or tide gauge readings from the same port, tide gauge readings should always be used.

Most of the methods described above use standard surveying equipment with access to tidal data (predictions or readings). However, dGPS is increasingly being used in order to bring land survey datums into field sites (often with centimeter-scale error). This is specialized, heavy, and expensive equipment which may not be field-serviceable and requires frequent charging from mains or another source of power (e.g., solar, generator). If working in a remote area it makes sense to investigate the availability of tide gauges, tidal predictions, and benchmarks before deciding to use dGPS to establish the vertical datum.

11.6 SUMMARY

We have discussed the basic background information necessary for sea-level researchers to accurately relate proxy indicators to reference water levels and tidal datums. The methods used to calculate reference water levels have conventionally been by the use of benchmarks with known elevations to relate indicators to a national land survey datum. With the increasing number of studies in remote locations from the Arctic to the tropics, and the development of GPS technology, new approaches to calculating reference water levels using dGPS are becoming more common. The accuracy of these technologically advanced approaches in defining elevations in relation to MSL can be variable however, and the equipment may not be field-serviceable. When planning fieldwork in a remote location, using local tidal predictions and measuring the tide at the field site (to define any offsets) is probably still the most appropriate approach for field scientists to take.

REFERENCES

Altamimi, Z., Collilieux, X., and Métivier, L. (2011) ITRF2008: an improved solution of the international terrestrial reference frame. Journal of Geodesy, 85, 457–473.

Cazenave, A., Lago, B., Dominh, K., and Lambeck, K. (1980) On the response of the ocean lithosphere to sea-mount loads from Geos 3 satellite radar altimeter observations. Geophysical Journal of the Royal Astronomical Society, 63, 233–252.

Chappell, J., Chivas, A., Wallensky, E., Polach, H.A., and Aharon, P. (1983) Holocene palaeo-environmental changes, central to north Great Barrier Reef inner zone. BMR Journal of Australian Geology and Geophysics, 8, 223–235.

Engelhart, S., and Horton, B.P. (2012) Holocene sea level database for the Atlantic coast of the United States. Quaternary Science Review, 54, 12–25.

Hamilton, S., and Shennan, I. (2005) Late Holocene relative sea-level changes and the earthquake deformation cycle around upper Cook Inlet, Alaska. Quaternary Science Reviews, 24, 1479–1498.

Mclean, R.F., Stoddart, D.R., Hopley, D., and Polach, H. (1978) Sea-level change in Holocene on Northern Great Barrier Reef. Philosophical Transactions of the Royal Society, A, 291, 167–186.

Pugh, D.T. (1987) *Tides, Surges and Mean Sea Level*. John Wiley and Sons Ltd, Chichester, UK.

Pugh, D.T. (2004) *Changing Sea Levels: Effects of Tides, Weather and Climate*. Cambridge University Press, Cambridge.

Pugh, D.T., and Woodworth, P.L. (2014). *Sea-Level Science: Understanding Tides, Surges, Tsunamis and Mean Sea-Level Change*. Cambridge University Press.

Rabinovich, A.B. (2009) Seiches and harbor oscillations. In: *Handbook of Coastal and Ocean Engineering* (ed. Kim, Y.C.), World Scientific Publishing Company, Singapore, pp. 193–236.

Shennan, I., Innes, J.B., Long, A.J., and Zong, Y.Q. (1993) Late Devensian and Holocene relative sea-level changes at Rumach, near Arisaig, northwest Scotland. Norsk Geologisk Tidsskrift, 73, 161–174.

Shennan, I., Innes, J.B., Long, A.J., and Zong, Y. (1994) Late Devensian and Holocene relative sea-level changes at Loch nan Eala, near Arisaig, northwest Scotland. Journal of Quaternary Science, 9, 261–283.

Shennan, I., Tooley, M.J., Green, F.M.L., Innes, J.B., Kennington, K., Lloyd, J.M., and Rutherford, M.M. (1999). Sea level, climate change and coastal evolution in Morar, northwest Scotland. Geologie en Mijnbouw, 77, 247–262.

Woodroffe, C.D., McGregor, H.V., Lambeck, K., Smithers, S.G., and Fink, D. (2012) Mid-Pacific microatolls record sea-level stability over the past 5000 yr. Geology, doi: 10.1130/G33344.1.

Zilkoski, D.B., Richards, J.H., and Young, G.M. (1992) Results of the General Adjustment of the North American Vertical Datum of 1988. American Congress on Surveying and Mapping, Surveying and Land Information Systems, 52, 133–149.

Part 2

Laboratory Techniques

Chapter 12

Techniques and applications of plant macrofossil analysis in sea-level studies

MARTYN WALLER

Centre for Earth and Environmental Science Research, School of Geography, Geology and the Environment, Kingston University, Kingston upon Thames, Surrey, UK

12.1 INTRODUCTION

Plant macrofossils are defined as remains which can be seen with the naked eye, potentially including pieces of wood, propagules (seeds, fruits, and various other structures which through dispersal can give rise to new individuals), buds, leaves, epidermal tissues (which may be derived from stems, leaves, or roots/rhizomes), mosses, and any such remains when carbonized. The remains of the larger freshwater and marine algae are also generally included within this fossil category.

Macrofossils, in the form of the *in situ* remains of trees associated with terrestrial environments in the modern intertidal zone, provide some of the most compelling and accessible evidence available to demonstrate changes in relative sea-level (RSL) through time. It is unsurprising then that some of the earliest investigations of fossil floras from the Quaternary were made using coastal exposures. With an increased emphasis on establishing stratigraphic position, the study of plant macrofossils has developed alongside, and continues to make an important contribution to, sea-level studies. The technique can provide invaluable information on paleoenvironments and RSL movements. However, large volumes of sediment are required and the production and dispersal of plant macrofossils is highly variable. It is not therefore a technique which lends itself to detailed reconstructions of RSL through the production of transfer functions.

One advantage of using plant macrofossils over pollen analysis for vegetation reconstructions is that it extends the list of identifiable taxa, both in terms of the taxonomic precision which can be achieved and also to taxa/groups that produce very little in terms of pollen/spores, or the pollen of which is rarely preserved. Identification to species level considerably improves the detail of the environmental inferences which can be drawn. In addition, while a single propagule may travel some distance, most do not and macrofossil occurrence is generally regarded as indicating local presence. The propagules and leaves of terrestrial plants are ideal for accelerator mass spectrometry (AMS) radiocarbon dating avoiding potential errors such as reservoir effects. The analysis of plant macrofossils is not technically difficult and nor does it require substantial investments in equipment; however, it does require botanical expertise (in taxonomy and morphology) and knowledge of plant ecology.

This chapter outlines the methods used in the study of plant macrofossils and the issues which need to be considered when collecting material and interpreting macrofossil data in the context of sea-level studies. Three applications of plant macrofossil analysis in sea-level studies are then reviewed, namely: their use as part of stratigraphic studies which aim to reconstruct former environments in the coastal zone and infer RSL movements; studies of wood and associated remains in coastal deposits; and the application of macrofossil analysis to the investigation of isolation basins.

12.2 METHODS

As plant macrofossils are produced in relatively small quantities, large volumes of sediment are required for analysis. Large-diameter corers (e.g., 5-cm-diameter cylindrical samplers) are needed to collect material though, where this possible, it is best extracted directly from exposures using monolith boxes. Storage should be in the dark in a cold room at 3–4 °C. Any exposed surfaces should be trimmed before subsampling and, before each

Handbook of Sea-Level Research, First Edition. Edited by Ian Shennan, Antony J. Long, and Benjamin P. Horton.

sample is taken, instruments need to be cleaned to avoid contamination.

For quantitative analysis a known volume of sediment is required, typically $50\,cm^3$ or $100\,cm^3$ is measured by displacement in a large measuring cylinder. Some sediment types will disaggregate in water if left to soak for several hours, although peat often requires the addition of 5% KOH or gentle boiling in 5% KOH for a short period (e.g., 5 minutes). It is useful to separate out both the larger remains (e.g., wood) and the finer detritus (and smallest seeds) from the majority of the remains, which can be achieved using sieves of respectively 2 mm, 1 mm, 0.5 mm, and 0.2 mm mesh size.

Residue from the sieves is transferred into Petri dishes and examined using a low power (×10–50) stereomicroscope. Material can be moved from one end of the dish to the other with fossils picked out using a small artist's brush/tweezers. Epidermal tissues, small seeds, and moss leaves should to be examined further at higher magnifications (×100–400). Some small seeds, notably those belonging to the genus *Juncus* in studies of saltmarsh sediment, can occur in such large quantities that only subsamples of material from the finest sieve can sensibly be examined. If required for radiocarbon dating, samples should be stored in distilled water with the addition of a few drops of 5% HCl to prevent bacterial and fungal contamination. For long-term storage, a 1:1:1 mixture of glycerol, alcohol, and distilled water is recommended.

In northwest Europe it has long been the practice to record in the field, as part of the initial characterization of a deposit, the presence of a few common taxa which have a diagnostic macroscopic appearance. The tendency of peat to split naturally can be utilized, enabling even some of the more delicate macrofossils to be identified with the aid of a hand lens. These include the rhizomes of *Phragmites* (common reed), the subterranean axes of the sedges *Cladium* and *Eriophorum angustifolium*, the leaves of a few trees and shrubs (notably *Salix*), the leaves and axes of dwarf shrubs (e.g., *Calluna*), and the presence of Sphagna (bog mosses). Knowledge of the morphology of these remains has passed between generations of workers. However, the identification of most plant macrofossils is based on published descriptions and illustrations and the use of reference material.

Few publications cover the range of plant material likely to be encountered in macrofossil-based sea-level studies, although regional floras often provide useful descriptions and illustrations. Specialist publications are available, but tend to be geographically specific and may be difficult to access. For example, photographic atlases of the seeds of many European and North American plants have been published (Martin and Barkley, 1961; Cappers et al., 2006) and descriptions of the seeds of difficult groups may be available, as with the Cyperaceae (Berggren, 1969) and *Juncus* (Körber-Grohne, 1964) for parts of Europe. Epidermal tissue is much more difficult to identify. Few illustrated guides have been published (but see Grosse-Brauckmann, 1972, 1992 for species commonly recovered from peat deposits in northern Europe), and variation occurs between shoot and root tissue and between the inner and outer margins of leaves. Grosse-Brauckmann (1986) cautions that it is rarely possible to identify with certainty the roots or rootlets of herbaceous plants. Mosses are potentially among the easier macrofossils to identify, as determination of living material is often dependent on leaf anatomy and moss floras are available for many regions (see Dickson, 1986). However, aquatic and semi-aquatic moss taxa in particular exhibit a high degree of morphological and anatomical plasticity and may require confirmation of identification by a specialist.

Although it is tempting to use bark characteristics and color, microscopic examination, which calls for knowledge of wood anatomy (e.g., Wilson and White, 1986), is essential for the correct identification of wood and charcoal remains. For wood and charcoal, identification is achieved through the examination of anatomical features in section (wood) or on clean surfaces created by fracturing (charcoal) at × 100–400 magnification. Waterlogged wood can often be suitably sectioned using a razor blade, but may require hardening with alcohol. Very hard wood (and reference material) can be sectioned using a microtome. For charcoal, fractures should be created by pressure (not cutting) using a rigid blade (e.g., a scalpel). Fracture surfaces are examined microscopically under surface illumination while wood sections are viewed using transmitted (from beneath) light. Good photographic guides are available to aid wood identification for some regions (e.g., Schweingruber, 1978 for

Europe), and more widely for taxa of economic importance (e.g., Miles, 1978). As with other types of macrofossil, species determination is often, but not always, possible (e.g., in northern Europe, members of the Rosaceae are difficult to separate).

Individual macrofossils (seeds, epidermal tissue, and wood) exhibit variation within a species, which is not only environmentally and genetically determined but also influenced by processes related to deposition and fossilization. While publications are a useful aid, reference material is therefore often essential to confirm identification. Such specimens can be collected in the field (see Nesbitt, 1990) or be derived from fossil material identified by an experienced worker. Access to existing herbaria and specialist advice are important factors for any researcher considering adopting this technique.

The analysis of tree growth rings has a number of applications in sea-level studies (see Section 12.4). The techniques used are described in many texts (e.g., Schweingruber, 1988). Ideally samples (from sections or living trees) should be collected from stem bases and contain the full sequence from the central pith to the last ring ending in the bark. Deciduous trees, where the spring vessels are much larger than the summer wood, particularly oak (*Quercus robur*) and pine (*Pinus sylvestris*), have been considered the most suitable species. In recent years however the number of species of potential interest has been extended with, for example, growth layers (not necessarily of annual periodicity) recorded from some mangrove species (e.g., Verheyden et al., 2004).

With a variety of plant material generally distinguished, the results of macrofossil investigations are rarely expressed relatively as a percentage. Propagules can be presented as concentrations, the number of remains per unit of sediment volume or, if a sediment accumulation rate (age–depth model) is available, as "influx". Charcoal fragments can be counted and split into size classes (based on the length of the long axis). However, with most vegetative parts the number of fragments is likely to be a function of preservation and laboratory procedures and the data are best expressed on a relative scale of abundance using ordinal values. Standard packages for the plotting of biostratigraphic data, ideally those which can also undertake statistical analysis (e.g., stratigraphically constrained cluster analysis for zonation), are suitable for presenting the results of macrofossil analysis. An example of good practice is figure 5 in Yu et al. (2003) which demonstrates how a large dataset can be presented clearly by such means. In contrast to the results of pollen analysis, in sea-level studies macrofossil taxa have often been ordered by broad salinity preferences or into recurrent groups (taxa which show a similar pattern of stratigraphic occurrence); in the case of *in situ* macrofossils, these are likely to resemble in composition extant plant communities, thereby facilitating the identification of vegetation stages and changes in water level.

12.3 THE INTERPRETATION OF MACROFOSSIL DATA

The main prerequisite for any examination of plant macrofossil content is that material is well preserved. The oxidative destruction of plant remains in the coastal zone, due to fluctuating water levels, is often a limitation. Decomposition can occur both as the deposit forms and long after deposition has ceased. The degree of decomposition (or "humification"; the conversion of plant debris into humic acids) can be estimated in the field by squeezing small samples of organic material through the fingers. Fresh sediments are well structured and produce a colorless solution, whereas highly humified sediments produce a dark brown solution and sediment devoid of plant structure passes through the fingers. For detailed studies, macrofossils must also be abundant. For seeds, >50 per 100 cm^3 of sediment is a desirable number; fewer than 10 per 100 cm^3 would suggest further analysis is not worthwhile.

Interpretation of macrofossil data is most secure when investigating peat-forming plant communities where the presence of subterranean parts indicates that the remains grew *in situ*. Typically, much of the above-ground organic material produced by these communities decomposes and the resulting sediment largely consists of roots and rhizomes. Woodland communities may add significant quantities of detached (detrital) material derived from above-ground parts (leaves, branches, and trunks) which can accumulate in water-filled hollows between the tree trunks. However, McKee and Faulkner (2000) described a mangrove peat from Pelican Cays, Belize, where 80% of the sediment mass was composed of roots, noting the potential for surface litter to be exported in tidal systems.

The downward growth of roots may result in the formation of mixed macrofossil assemblages with elements derived from communities with different ecological tolerances (e.g., to salinity). This is most likely to occur close to obvious sediment boundaries and, if peat is overlain by sediment deposited in a marine/brackish environment, or followed by a period with little or no sedimentation, younger material may also be introduced by burrowing organisms. In these circumstances it is particularly important to be aware of the potential for intrusion and consider whether an assemblage is transitional or just mixed. The vertically penetrating rhizomes of *Phragmites* are a particularly common, but fortunately also an easy-to-recognize, feature of coastal sediments.

A number of additional interpretative problems arise when intertidal deposits are investigated for their macrofossil content with a view of reconstructing tidal constants from the presence of salt-tolerant (halophytic) plant communities. The plants and plant communities found in intertidal environments certainly occur in zones related to elevation, over which the frequency and duration of tidal inundation has a strong influence. However, there is considerable geographic variation in the composition of these communities; other factors such the nature of the substrate and land use are also influential, with particular complexities (e.g., due to the variable freshwater influence) in estuaries. Local information on the composition of the modern vegetation is therefore desirable if macrofossil data are to be used to reconstruct RSL movements. A second potential constraint is the dispersal of plant propagules by tidal flows. A number of ecological studies (e.g., Huiskes et al., 1995; Rand, 2000) have been undertaken into the dispersal of seeds of saltmarsh taxa with differing results as to the amount of exchange between vegetation zones. However, caution clearly needs to be exercised if only a small number of seeds (as is often the case) are recovered. For this reason, along with difficulties in identifying epidermal material, the reconstruction of intertidal plant communities may be more effectively achieved using stable carbon isotopes (e.g., Kemp et al., 2012). Comparisons of the utility of plant macrofossils with other proxy techniques in sea-level reconstructions, based on saltmarsh sediments, have been made by Freund et al. (2004) and Patterson et al. (2005).

12.4 THE APPLICATION OF PLANT MACROFOSSIL ANALYSIS IN SEA-LEVEL STUDIES

12.4.1 Characterization of coastal zone deposits

The primary application of plant macrofossil analysis in sea-level research has been as part of stratigraphic investigations which aim to characterize the deposits of the coastal zone. The scale of the macrofossil analysis undertaken has varied from identifying a few key indicator species in the field to the collection of material for detailed laboratory analysis. For example, Nelson et al. (2008) only specify the presence of the rhizomes of *Triglochin maritima* which are used to distinguish mudflat. In contrast, Allemeersch (1994) systematically examined subsamples recovered from a series of boreholes from the eastern part of the Belgian coastal plain. Both vegetative remains and propagules were identified enabling reconstructions to be based on comparison with extant plant communities. The amount of effort expended should sensibly reflect the importance of organic material in the sequences under investigation.

The utility of macrofossil analysis in such stratigraphic investigations lies in the potential to relate particular plant communities to reference water levels and, from the presence of communities indicative of successively higher or lower water levels, infer RSL movements. Behre (1986) defined progressive and retrogressive series. These are based on the concept of hydroseral succession, whereby a series of plant communities are able to successively colonize an increasingly shallow body of open water as a result of sediment accumulating during the preceding stage. The stages envisaged in natural coastal landscapes in northwest Europe are detailed in Figure 12.1. A progressive series indicates an increase in the height of the sediment surface relative to the groundwater level. It could be used to infer a lowering of the groundwater table and hence RSL fall. Organic material will also accumulate under conditions of environmental stability until terrestrialization occurs, however; accommodation space can be partially filled through the external input of sediment, which may also heavily influence sedimentation rates. Retrogressive series, with plant communities tolerant of successively wetter conditions recorded, are more certainly driven by rising water levels.

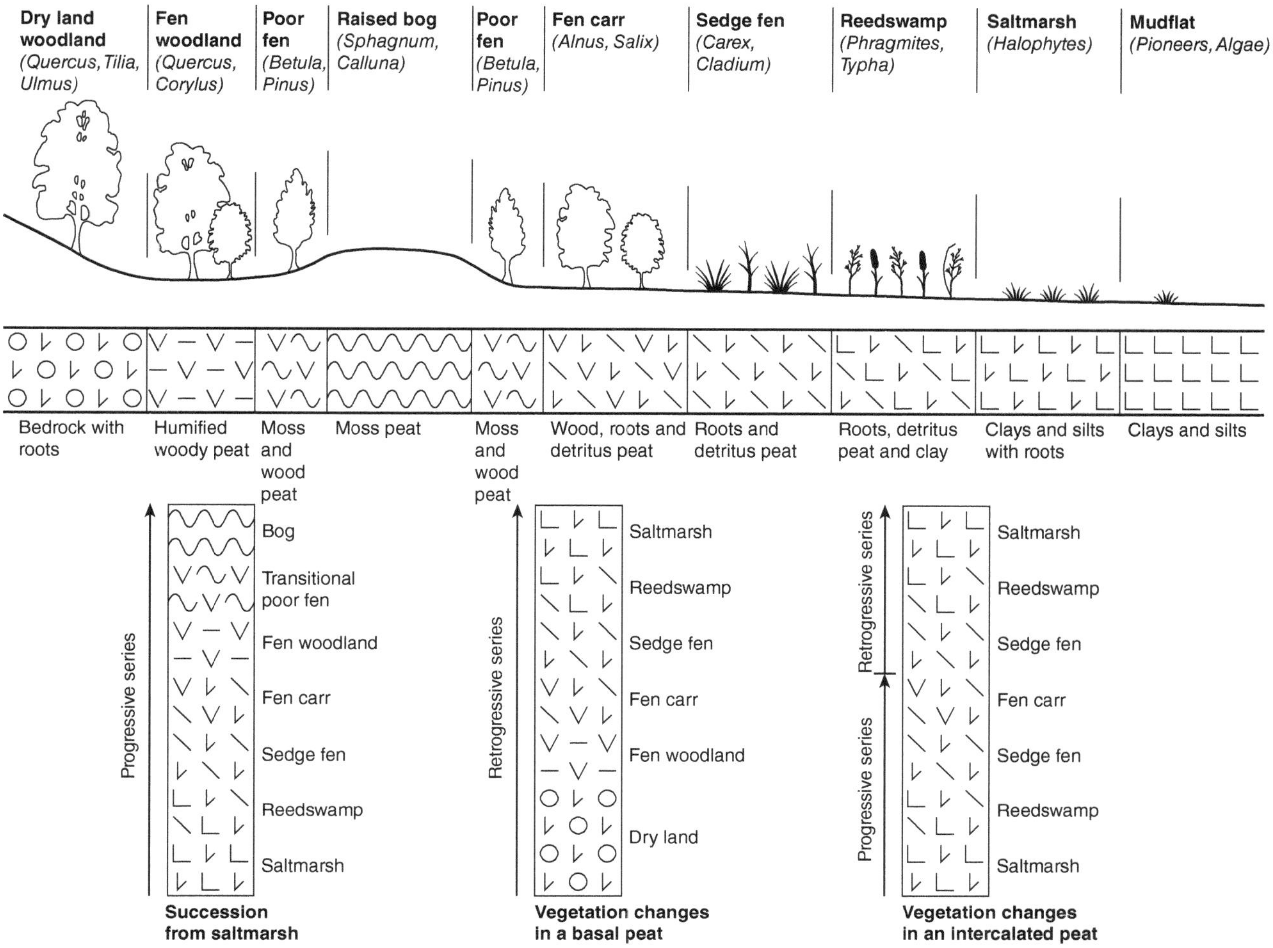

Fig. 12.1. Idealized natural coastal vegetation zonation for northwest Europe showing characteristic taxa, the nature of the sediments deposited and likely conformable stratigraphic relationships.

In a classic study from the Somerset Levels in southwest England, Clapham and Godwin (1948) describe a progressive series comprising sequentially of marine clay, *Phragmites* peat, wood peat, and *Sphagnum* peat. *Sphagnum* colonization indicates isolation from base-rich water, but not necessarily further vertical separation as a result of falling RSL. The structure of these mosses enables them to grow above and independently of the groundwater level. Samples containing Sphagna therefore cannot be used to make inferences as to the height of groundwater levels. Within the *Sphagnum* peat, Clapham and Godwin (1948) recorded layers dominated by the rhizomes of the sedge *Cladium*. These indicate the flooding of the peat surface with nutrient-enriched waters and re-establishment of groundwater control, but whether this was caused by coastal or catchment processes has been the subject of subsequent debate.

Rising water levels in coastal areas often result in the accumulation of peats over formerly dry surfaces, which exhibit retrogressive series. With the underlying substrata unlikely to compact, samples from such layers are particularly suitable as sea-level index points. However, organic material may initially accumulate in response to occasional flooding events. Knowledge of the retrogressive series can be used to ensure material taken for dating accumulated under the influence of the groundwater table. For example, Törnqvist et al. (1998) used macrofossil analysis to identify the first stage (fen woodland) indicative of permanently wet conditions.

The identification of such series is particularly important when selecting material for radiocarbon dating from sediment adjacent to transgressive and regressive contacts, as continuous sedimentation can be inferred from the presence of successive stages. In a study of St Lawrence Estuary, Québec, Bhiry et al. (2000) provide an example of the changes expected during the uninterrupted accumulation of an intercalated peat. At the onset of peat formation the presence of *Typha* and *Carex* indicate a semi-aquatic freshwater environment, which was followed by wood remains (*Pinus strobus* and *Tsuga canadensis*), indicating the development of a terrestrial environment. The reoccurrence of the semi-aquatic freshwater taxa occurs prior to the change back to minerogenic sedimentation. In contrast, a thin organic layer above the peat comprised organic material of a heterogeneous nature; rather than indicative of a short-lived return of freshwater conditions, this was interpreted as a drift deposit. The absence of stages is often indicative of a hiatus in deposition (either the erosion or non-deposition of sediment), but it can also be the result of abrupt change (particularly in regions subject to seismic activity).

12.4.2 Use of wood remains

Wood remains have been used for a variety purposes in sea-level studies. Distinction should be made between *in situ* finds of species associated with terrestrial environments and the remains of species which are adapted to grow in freshwater wetlands and intertidal environments. The presence of the former, exposed in the modern intertidal zone, gives rise to the phenomena known as "submerged forests". Foreshore exposures are useful in showing that trees with rooted stumps are *in situ*, can reveal that tree occurrence is confined to a particular stratigraphic level(s), and ease sample collection. However, remains are often poorly preserved and they have the obvious disadvantages of the absence of later deposits and the potential for contamination.

With establishment expected during periods of low RSL and demise potentially attributable to rising RSL, the occurrence of the remains of terrestrial trees in coastal deposits is likely to be of significance in sea-level studies. However, some caution is required in their interpretation. Lewis et al. (2003) indicate an entirely terrestrial origin for a deposit now fortuitously exposed in the coastal zone, partly as a result of the wood remains revealing the presence of wood decay fungi common to forest floors. The lowest altitudinal limit relative to sea level at which terrestrial species will grow is of particular importance, and this is often assumed to be above the highest tidal level. However, present-day relationships must be established. For example, Heyworth (1986) suggests that the lower limit of *Quercus robur* in the British Isles is mid-way between MHWST and HAT, regardless of the tidal range. The death of trees in coastal deposits cannot be assumed to be the result of inundation by saltwater. In conformable sequences, under conditions of rising RSL, terrestrial trees might be expected to be killed by freshwater ponding and, as is often the case in basal peats, be replaced by trees tolerant of waterlogging. Causes of death unrelated to water-level fluctuations are also possible. For example, Allen (1992) investigated prostrate trees exposed in the Severn Estuary, southwest England, and suggested, as they appeared to be both young and healthy (indicated by the accompanying leaf remains), that they were felled by strong winds. Finally, the predisposition to regard wood remains in coastal deposits as indicative of periods of relatively low sea-level in temperate areas must be set against the many records of thick organic sequences predominantly comprising the remains of wetland trees. The potential for such wet woodland communities to persist for thousands of years against rising RSL is clear from the sedimentary record (e.g., Waller, 1994). However, the degree to which this is dependent upon a protected setting and the importance of the external supply of sediment against internal factors (production and decomposition) has been difficult to establish due to the absence of modern analogs.

Mangroves occur predominantly between MSL and MHW (Ellison, 2008). Salinity tolerance varies between species with mangrove forests frequently exhibiting species zonations. The potential therefore exists to use mangrove remains to reconstruct both sea-level altitude and direction of movement. Distinction has been made between mangroves in estuarine and deltaic settings, where the external input of sediment is significant, and carbonate settings where mangroves may keep pace with rising RSL through the deposition of peat (Woodroffe, 1981). Pollen has been the tool most favored for paleoenvironmental reconstruction from mangrove peats, although macrofossils have also been studied. For example, in the McKee and Faulkner (2000) study from Pelican Cays, root fragments were identified and counted and revealed that the abundance of mangrove species varied with depth. Wooller et al. (2003) concentrated on leaf remains and used stable isotope signatures to distinguish between dwarf and large forms of *Rhizophora mangle*, associated respectively with the interior and edge of mangrove islands in Central America.

Tree ring analysis is generally seen as a method of deriving the absolute age of tree remains by statistically matching a sample against a master chronology. However, it can also be used to establish the age and health of trees at the time of death. In the context of sea-level studies the work of Munaut (1966) at Terneuzen, the Netherlands, is an early example. This showed that the establishment of *Pinus sylvestris* in a coastal bog occurred over a period of 125 years with the trees dying over a period of 40 years some 250 years later, which was attributed to a gradual rise in water level. In recent studies, Bhiry et al. (2000) use a reduced rate of ring growth in the last decades of growth to indicate deteriorating conditions, while Atwater and Yamaguchi (1991) use the absence of such a signal (the trees were healthy until the year of death) to indicate sudden submergence. The potential also exists to use ring analysis on living trees, for example ring counts can be made to determine the maximum age of a population of trees and thereby establish the minimum age of a barrier island.

12.4.3 Investigation of isolation basins

One relatively recently developed application of plant macrofossils analysis in sea-level studies has been in the investigation of isolation basins. Macrofossils have proved useful in determining, alongside dating evidence, the timing of lake isolation from, and connection to, brackish water in areas subject to shoreline displacement. In lakes with small catchment areas, macrofossil remains including propagules are likely to be indicative of a local presence. In addition to vascular plants, such studies have utilized macro marine brown algae (e.g., *Sphacelaria*, *Desmarestia*) and lacustrine green algae (e.g., *Chara*, *Nitella*).

An example of the use of plant macrofossils in such studies is that undertaken by Yu et al. (2003) from Lake Ryssjön, southeast Sweden. Diatoms indicate marine conditions were established in the lake from 7500 to 3000 cal a BP. In the macrofossil sequence this phase is represented by a decline in trees and shrubs (attributed to shoreline retreat) and increases in charcoal (attributed to soil erosion as a result of strong wave action), *Zannichellia palustris* (an aquatic plant found in both fresh and brackish water), *Ruppia maritime* (a seagrass), and *Chara* and *Nitella* (green algae). Modern observations from shallow bays in the Baltic suggest *Ruppia maritime* and *Zannichellia palustris* are replaced in deeper water by *Chara* and *Nitella*. Peaks in the latter taxa were therefore used to suggest minor local transgressions (within the marine phase), attributed to centennial-scale variations in wind pattern or coastal storminess.

One potential advantage of using plant macrofossils in such studies, given that other indicators of salinity are also likely to be preserved, is that they can be directly dated. Aquatic plants may derive carbon from dissolved inorganic carbonates however, and such macrofossils can therefore be influenced by the reservoir effect. If taxa tolerant of brackish conditions are utilized, then correction must be made for the marine reservoir age.

REFERENCES

Allemeersch, L. (1994) Peat in the Belgian Eastern coastal plain. Aardkundige Mededelingen, 6, 1–54.

Allen, J.R.L. (1992) Trees and their response to wind: mid Flandrian strong winds, Severn Estuary and inner Bristol Channel, southwest Britain. Philosophical Transactions of Royal Society, London, B, 338, 335–364.

Atwater, B.F., and Yamaguchi, D.K. (1991) Sudden, probably coseismic submerge of Holocene trees and grass in coastal Washington. Geology, 19(7), 706–709.

Behre, K.E. (1986) Analysis of botanical macro-remains. In: *Sea-level Research: A Manual for the Collection and Evaluation of Data* (ed. Van de Plassche, O.), Geo Books, Norwich, pp. 413–433.

Berggren, G. (1969) *Atlas of Seeds. 2: Cyperaceae.* Swedish Natural Science Research Council, Stockholm.

Bhiry, N., Garneau, M., and Filion, L. (2000) Macrofossil record of a middle Holocene drop in relative sea level at the St. Lawrence Estuary, Québec. Quaternary Research, 54(2), 228–237.

Cappers, R.T.J., Bekker, R.M., and Jans, J.E.A. (2006) *Digital Seed Atlas of the Netherlands.* Bakhuis, Groningen.

Clapham, A.R., and Godwin, H. (1948) Studies in the post-glacial history of British vegetation. VIII. Swamping surfaces in peats of the Somerset Levels. IX. Prehistoric trackways in the Somerset Levels. Philosophical Transactions of the Royal Society of London, B, 233, 233–273.

Dickson, J.H. (1986) Bryophyte analysis. In: *Handbook of Holocene Palaeoecology* (ed. Berglund, B. E.), John Wiley & Sons, Chichester, pp. 627–643.

Ellison, J.C. (2008) Long-term retrospection on mangrove development using sediment cores and pollen analysis: a review. Aquatic Botany, 89(2), 93–104.

Freund, H., Gerdes, G., Streif, H., Dellwig, O., and Watermann, F. (2004) The indicative meaning of diatoms, pollen and botanical macrofossils for the reconstruction of palaeoenvironments and sea-level fluctuations along the coast of

Lower Saxony; Germany. Quaternary International, 112(1), 71–87.
Grosse-Brauckmann, G. (1972) Über pflanzliche Makrofossilien mitteleuropäischer Torfe. I. Gewebereste krautiger Pflanzen und ihre Merkmale. Telma, 2, 19–55.
Grosse-Brauckmann, G. (1986) Analysis of vegetative plant macrofossils. In: *Handbook of Holocene Palaeoecology* (ed. Berglund, B. E.), John Wiley & Sons, Chichester, pp. 591–618.
Grosse-Brauckmann, G. (1992) Über pflanzliche Makrofossilien mitteleuropäischer Torfe. III. Früchte, Samen und einige Gewebe (Fotos von fossilen Pflanzenresten). Telma, 22, 53–102.
Heyworth, A. (1986) Submerged forests as sea-level indicators. In *Sea-level Research: A Manual for the Collection and Evaluation of Data* (ed. van de Plassche, O.), Geo Books, Norwich, pp. 401–411.
Huiskes, A.H.L., Koutstaal, B.P., Herman, P.M.J., Beeftink, W.G., Markusse, M.M., and Munck, W. De (1995) Seed dispersal of halophytes in tidal salt marshes. Journal of Ecology, 83(4), 559–567.
Kemp, A.C., Vane, C.H., Horton, B.P., Engelhart, S.E., and Nikitina, D. (2012) Application of stable carbon isotopes for reconstructing salt-marsh floral zones and relative sea level, New Jersey, USA. Journal of Quaternary Science, 27(4), 404–414.
Körber-Grohne, U. (1964) *Bestimmungsschlüssel für subfossile Juncus-Samen und Gramineen-Früchte, Schriftenreihe des Niedersächsischen Landesinstitutes für Marschen und Wurtenforschung*, Bd 7. W. Haarnagel, Hildesheim.
Lewis, B.A., Wrenn, J.H., Lewis, A.J., Alford, J.J., and Alford, D. (2003) Middle Wisconsinan and recent wet site mummied wood, humus, peat, and pollen, Santa Rosa Island, Florida. Review of Palaeobotany and Palynology, 126(3–4), 243–266.
Martin, A.C., and Barkley, W.D. (1961) *Seed Identification Manual*. University of California Press, Berkeley and Los Angeles.
McKee, K.L., and Faulkner, P. (2000) Mangrove peat analysis and reconstruction of vegetation history at the Pelican Cays, Belize. In *Atoll Research Bulletin* (eds Macintyre, I.G., and Rutzler, K.), National Museum of Natural History, Smithsonian Institution, Washington, DC, pp. 47–60.
Miles, A. (1978) *Photomicrographs of world woods*. Building Research Establishment Report. Department of the Environment HMSO, London.
Munaut, A.V. (1966) Reherches dendrochronologiques sur Pinus sylvestris. II. Première application des méthods dendrochronologiques á l'étude de pins sylvestres sub-fossiles (Terneuzen-Pays-Bas). Agricultura, 14(8), 361–389.
Nelson, A.R., Sawai, Y., Jennings, A.E., Bradley, L-A., Gerson, L., Sherrod, B.L., Sabean, J., and Horton, B.P. (2008) Great-earthquake palogeoesy and tsunamis of the past 2000 years at Alsea Bay, central Oregon coast, USA. Quaternary Science Review, 27(7–8), 747–768.
Nesbitt, M. (1990) Seed reference collections and archaeobotany. Circaea, 8(1), 21–39.
Patterson., R.T., Dalby, A.P., Roe, H.M., Guilbault, J-P., Hutchinson, I., and Clague, J.J. (2005) Relative utility of foraminifera, diatoms and macrophytes as high resolution indicators of paleo-sea level in coastal British Columbia, Canada. Quaternary Science Review, 24(18–19), 2002–2014.
Rand, T.A. (2000) Seed dispersal, habitat suitability and the distribution of halophytes across a salt marsh tidal gradient. Journal of Ecology, 88(4), 608–621.
Schweingruber, F.H. (1978) *Microscopic Wood Anatomy*. Swiss Federal Institute of Forestry Research, Birmensdorf.
Schweingruber, F.H. (1988) *Tree Rings, Basics and Applications of Dendrochronology*. Reidel, Dordrecht.
Törnqvist, T., van Ree, M.H.M., van 't Veer, R., and van Geel, B. (1998) Improving methodology for high-resolution reconstruction of sea-level rise and neotectonics by paleoecological analysis and AMS 14C dating of basal peats. Quaternary Research, 49(1), 72–85.
Verheyden, A., Kairo, J.G., Beeckman, H., and Koedam, N. (2004) Growth rings, growth ring formation and age determination in the mangrove Rhizophora mucronata. Annals of Botany, 94(1), 59–66.
Waller, M.P. (1994) Flandrian vegetational history of south-eastern England. Stratigraphy of the Brede valley and pollen data from Brede Bridge. New Phytologist, 126(2), 369–392.
Wilson, K., and White, D. (1986) *The Anatomy of Wood: Its Diversity and Variability*. Stobart, London.
Woodroffe, C.D. (1981) Mangrove swamp stratigraphy and Holocene transgression, Grand Cayman Island, West Indies. Marine Geology, 41(3–4), 271–294.
Wooller, M.J., Scharler, U., Smallwood, B., Jacobson, M., and Fogel, M. (2003) A taphonomic study of δ13C and δ15N values in Rhizophora mangle leaves for a multi-proxy approach to mangrove paleoecology. Organic Geochemistry, 34(9), 1259–1275.
Yu, S.-Y., Andrén, E., Barnekow, L., Berglund, B.E., and Sandgren, P. (2003) Holocene palaeoecology and shoreline displacement on the Biskopsmåla Peninsula, southeastern Sweden. Boreas 32(4), 578–589.

Chapter 13
Foraminifera

ROBIN EDWARDS[1] AND ALEX WRIGHT[2]

[1] *School of Natural Sciences, Museum Building, Trinity College Dublin, Dublin, Ireland*
[2] *Department of Earth and Life Sciences, VU University of Amsterdam, The Netherlands*

13.1 INTRODUCTION

Foraminifera are microscopic single-celled animals with hard shells, referred to as "tests" ("naked" forms without a test do not fossilize and so are not considered here), which inhabit a wide spectrum of marine environments. Their ubiquity, relatively high abundance, and preservation potential, coupled with their sensitivity to a range of environmental variables (e.g., temperature and salinity), has made them a favored tool of geoscientists seeking to reconstruct past ocean conditions. Pioneering work on marine oxygen isotopes during the 1950s and 1960s revealed that the carbonate tests of free-floating (planktonic) and bottom-dwelling (benthic) foraminifera acquired isotopic signatures that reflected the physico-chemical properties of the surrounding seawater. While the initial focus was on elucidating the relationship between oxygen isotope ratios and water temperature, it soon became evident that a significant component of the observed isotopic variation through time reflected the growth and decay of large terrestrial ice sheets. Since global ice volume is intimately linked to sea level over glacial–interglacial timescales, foraminiferal marine oxygen isotope data provide first-order information on global sea-level variations during the Quaternary (e.g., Waelbroeck et al., 2002; Rohling et al., 2009; Grant et al., 2012).

When considered over shorter (sub-millennial) timescales, spatial differences in sea level become increasingly significant (Chapter 2). It is therefore desirable to use complementary techniques that are capable of discriminating spatial patterns by recording local changes in relative sea level (RSL). Once again, foraminifera have proven to be extremely useful sea-level indicators, although for this application the focus of attention is on intertidal assemblage-based studies rather than deep-ocean isotopic data. While the use of intertidal foraminifera in sea-level research extends back to the 1970s, their popularity has increased from the late 1990s with the development of statistically based methods (transfer functions) capable of producing precise reconstructions (Chapter 31). Foraminifera now occupy a central place in studies evaluating the significance of recent changes in RSL, since they are capable of bridging the gap between short instrumental records and longer-term geological data (e.g., Kemp et al., 2011; Gehrels and Woodworth, 2012).

This chapter presents a brief introduction to the use of intertidal foraminifera in RSL reconstruction, with a particular focus on illustrations from temperate saltmarsh environments. After considering the theoretical basis for using foraminifera as sea-level indicators, the chapter concludes with a practical guide to sample collection, processing, and analysis.

13.2 USE OF FORAMINIFERA AS SEA-LEVEL INDICATORS

The use of intertidal foraminifera in RSL research stems from the need to determine how high in the intertidal zone a sediment sample accumulated (Chapter 2). Consequently, the effective application of foraminifera-based approaches has two principal requirements: (1) the vertical distributions of certain taxa correlate with elevation within the intertidal zone in quantifiable, time-invariant ways; and (2) the foraminifera contained within a sediment sample are autochthonous and the assemblages are unchanged from their initial incorporation into the sediment surface. Where these conditions are met, faunal data (i.e., the relative abundances of different foraminiferal species) can be used to infer sample height above mean tide

Handbook of Sea-Level Research, First Edition. Edited by Ian Shennan, Antony J. Long, and Benjamin P. Horton.

level (hereafter referred to as "tidal elevation") in accordance with uniformitarian principles.

13.2.1 Vertical zonation of intertidal foraminifera

Seminal papers on the use of intertidal benthic foraminifera as precise RSL indicators were published in the late 1970s and early 1980s by David Scott and colleagues (e.g., Scott and Medioli, 1978, 1980a, 1986; Scott et al., 1980, 1981). This early work was centered on the observation that distinct assemblages of saltmarsh foraminifera appeared to characterize different height intervals within the intertidal zone. While initially based on a limited number of saltmarshes, there is now abundant empirical evidence that distinct intertidal foraminiferal assemblages can be correlated with tidal elevation. This vertical zonation is a common feature of saltmarsh and mangrove environments and is recorded in regions with widely differing climate, vegetation, salinity, and tidal regimes (Fig. 13.1).

Many of the significant features initially highlighted by this pioneering work can be seen in the summary foraminiferal zonations presented in Figure 13.1. Firstly, the highest occurrence of saltmarsh foraminifera correlates closely with the upper limit of marine influence. Scott and Medioli (1978) suggest that the marked decrease in the number of foraminifera around the higher high water mark has the potential to relocate past RSL with a maximum precision of about ±5 cm (see Fig. 13.1, location 2).

A second feature of note is the general bipartite division between high marsh assemblages characterized by agglutinated foraminifera, and lower elevation assemblages dominated by taxa with calcareous tests (see Appendix). In most cases, the switch from calcareous to agglutinated assemblage is broadly coincident with the change from unvegetated tidal flat to vegetated saltmarsh platform. A transitional zone, often characterized by the agglutinated species *Miliammina fusca* in combination with several calcareous taxa, is commonly observed in the lowest marsh fringe. Similar general patterns are also reported from mangrove environments (e.g., Horton et al., 2003, 2005; Woodroffe et al., 2005; Berkeley et al., 2009), although these distributions are not universal (e.g., Barbosa et al., 2005).

Attention has traditionally been focused on the agglutinated saltmarsh taxa that, while widely distributed around the world, are vertically restricted to higher-elevation vegetated environments. Consequently, species such as *Jadammina macrescens* (also reported as *Trochammina macrescens* and *Jadammina polystoma*; see Appendix for discussion) or *Trochammina inflata* are generally considered diagnostic of these settings which are broadly coincident with the upper quarter of the tidal range. Within these high marsh environments, it is sometimes possible to identify more restricted assemblages that allow finer subzones to be delimited. However, as is evident from Figure 13.1, these subzones are more spatially variable in terms of composition, vertical position, and height range. This variability persists even after data are normalized to account for inter-site differences in tidal range (Chapter 31). Scott and Medioli (1978) suggested that such regional differences were probably due to the influence of environmental factors other than tidal elevation, and this is examined in more detail in Section 13.2.3.

13.2.2 Describing modern foraminiferal distributions

While Scott and Medioli (1978) noted that the abundance of foraminiferal tests within marsh deposits made them well suited to statistical analysis, much of the early work on vertical assemblage zones was based on visual assessment of the data. Vertical assemblage zones were described in terms of their dominant taxa (e.g., Fig. 13.1) and core material was interpreted by matching fossil assemblages to their modern equivalents (see Section 13.2.4). More recent work has tended to use statistical techniques to define and discriminate between assemblages and delimit vertical assemblage zones. The precise methods employed vary and a detailed consideration of their comparative merits is beyond the scope of this chapter. However, it is important to note that different approaches can subdivide a dataset in contrasting ways. Consequently, it is useful to consider the outputs from complementary forms of analysis when assigning assemblage zones.

For example, Figure 13.2 presents the surface distribution of saltmarsh foraminifera recorded at Hynes Brook in Newfoundland, Canada. Three vertical assemblage zones have been established on the basis of unconstrained cluster analysis and a complementary ordination technique called detrended correspondence analysis (DCA). Only common groupings identified by both forms of analysis are

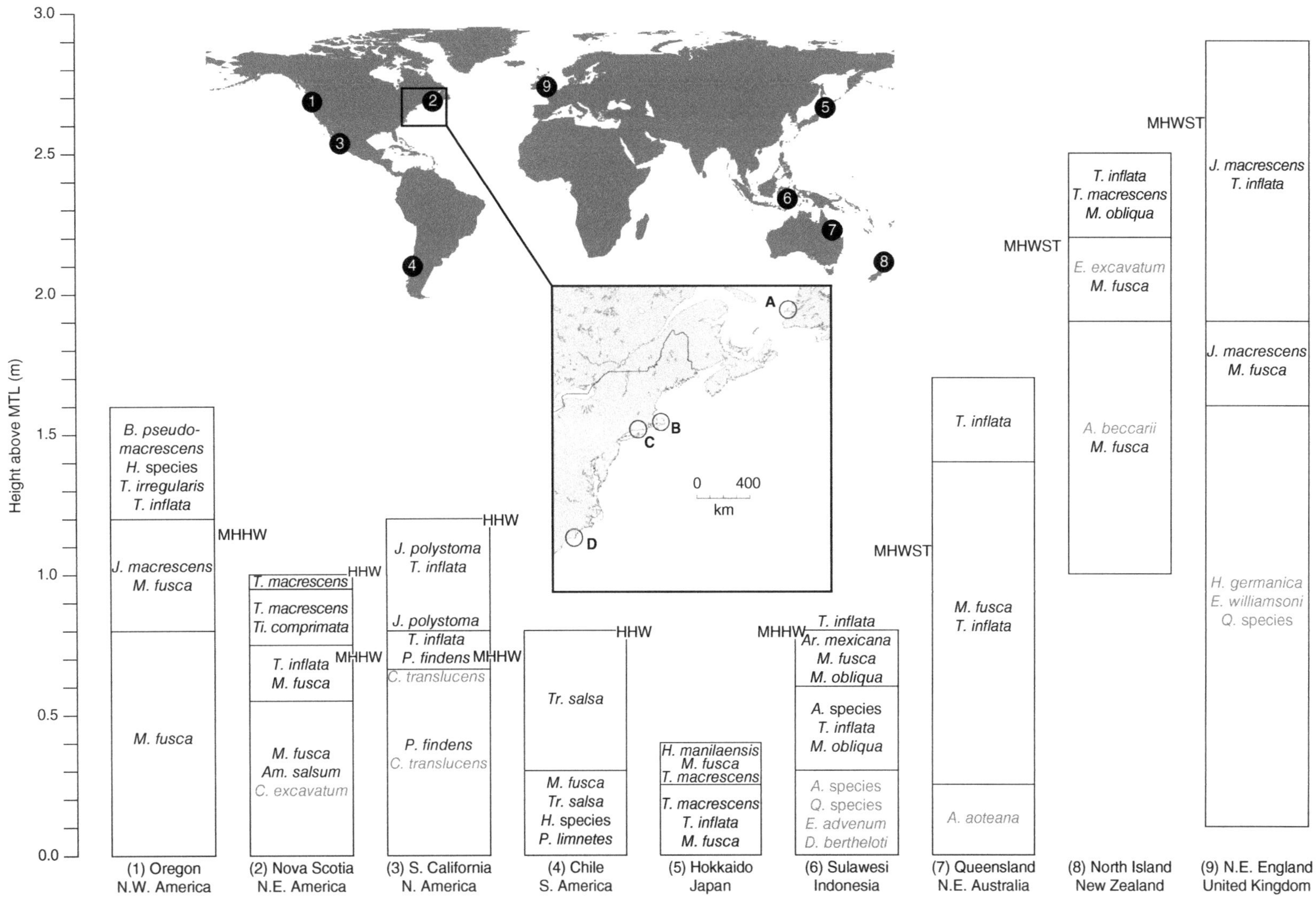

Fig. 13.1. Simplified vertical foraminiferal assemblage zones from selected saltmarsh and mangrove systems. Taxonomy follows source publications. Agglutinated taxa are in black, calcareous taxa are in gray. Key to genera: *A*: *Ammonia*; *Am*: *Ammotium*; *Ar*: *Arenoparrella*; *B*: *Balticammina*; *C*: *Cribroelphidium*; *D*: *Discorbinella*; *E*: *Elphidium*; *H*: *Haplophragmoides*; *J*: *Jadammina*; *M*: *Miliammina*; *P*: *Protoschista*; *Q*: *Quinqueloculina*; *T*: *Trochammina*; *Ti*: *Tiphotrocha*; *Tr*: *Trochamminita*. Tidal levels where available: HHW: higher high water; MHHW: mean higher high water; MHWST: mean high water spring tide; MTL: mean tide level. Inset map shows location of sites referred to in the text: A: Hynes Brook marsh, Newfoundland; B: Great Marshes, Massachusetts; C: Pattagansett River marsh, Connecticut; D: Elizabeth River marsh, North Carolina. *Source*: Data from 1. Hawkes et al. (2010); 2, 3. Scott and Medioli (1978); 4. Jennings et al. (1995); 5. Scott et al. (1996); 6. Horton et al. (2005); 7. Horton et al. (2003) and Woodroffe et al. (2005); 8. Hayward et al. (1999); 9. Horton (1999) and Horton and Edwards (2006).

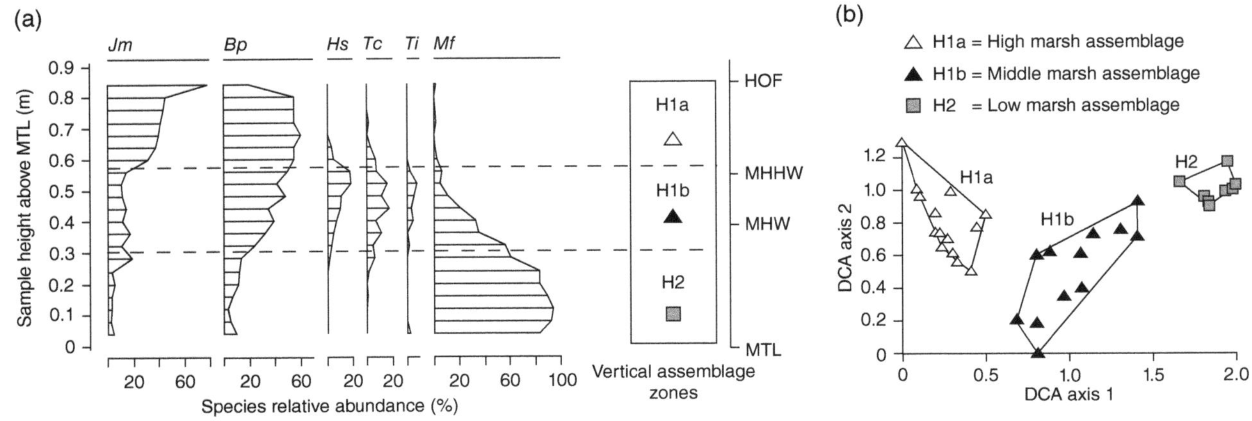

Fig. 13.2. Simplified surface distribution of foraminifera from Hynes Brook marsh, Newfoundland. (a) Relative abundances of foraminiferal species plotted with height above mean tide level. Cluster analysis discriminates a low marsh assemblage (H2) and a high marsh assemblage (H1) that can be further subdivided into upper (H1a) and middle (H1b) marsh assemblages on the basis of secondary taxa. Species abbreviations: *Jm*: *Jadammina macrescens*; *Bp*: *Balticammina pseudomacrescens*; *Hs*: *Haplophragmoides* species; *Tc*: *Tiphotrocha comprimata*; *Ti*: *Trochammina inflata*; Mf: *Miliammina fusca*. Tidal level abbreviations as in Figure 13.1 with the addition of MHW: mean high water, HOF: highest occurrence of foraminifera. (b) Detrended correspondence analysis of Hynes Brook data supports the three assemblage zones identified by cluster analysis. *Source*: Data from Wright et al. (2011).

used to delineate assemblage zones. Cluster analysis seeks to group samples together on the basis of how similar their foraminiferal assemblages are. It is therefore predicated on the notion that foraminiferal assemblages form distinct "natural" groups within the intertidal zone. In contrast, DCA distributes samples along a primary axis such that the most dissimilar foraminiferal assemblages are located at opposing ends. This axis represents a theoretical "environmental gradient" along which foraminiferal assemblage composition varies, and it is therefore capable of representing more continuous change across the intertidal zone. The process of distributing samples to maximize dissimilarity is repeated for axis 2 with the additional constraint that this secondary "environmental gradient" is uncorrelated with the first.

In the example of Hynes Brook, cluster analysis clearly separates a low marsh assemblage (H2), dominated by *M. fusca*, from the high marsh assemblages with abundant *Balticammina pseudomacrescens* and *J. macrescens*. This high marsh zone can be further subdivided into an upper (H1a) and middle (H1b) grouping on the basis of the relative abundances of secondary taxa (e.g., *Tiphotrocha comprimata* and *Haplophragmoides* species). Support for these groupings is given by the clear separation of samples on the DCA biplot, while dispersion along the secondary axis provides additional insights regarding intra-assemblage variability, particularly in the high marsh.

13.2.3 Spatial variability of foraminiferal distributions: causes and implications

While the foraminiferal assemblage zones presented in Figure 13.1 emphasize the global nature of intertidal vertical zonation, closer inspection of the details of assemblage composition reveals the spatial variability noted by Scott and Medioli (1978) in their original work. This variability ranges from differences in the relative abundances of the species present in particular parts of the intertidal zone to the complete absence of certain taxa and/or their replacement by other species.

These features are illustrated in Figure 13.3 which presents modern foraminiferal distributions from two saltmarshes from the east coast of the USA: Pattagansett River marsh in Connecticut (Fig. 13.3a), and Elizabeth River marsh in North Carolina (Fig. 13.3b). These data are then combined with the surface distribution from Hynes Brook marsh, Newfoundland (shown in Fig. 13.2) to produce a composite dataset that spans the Atlantic coast of North America (Fig. 13.3c, d). A general bipartite division exists at all sites, separating a low marsh assemblage dominated by *M. fusca* from an assemblage dominated by high marsh agglutinated taxa such as *B. pseudomacrescens*, *Haplophragmoides* species, *J. macrescens*, and *Trochammina inflata*.

However, while cluster analysis and DCA group the low marsh assemblages from all sites together,

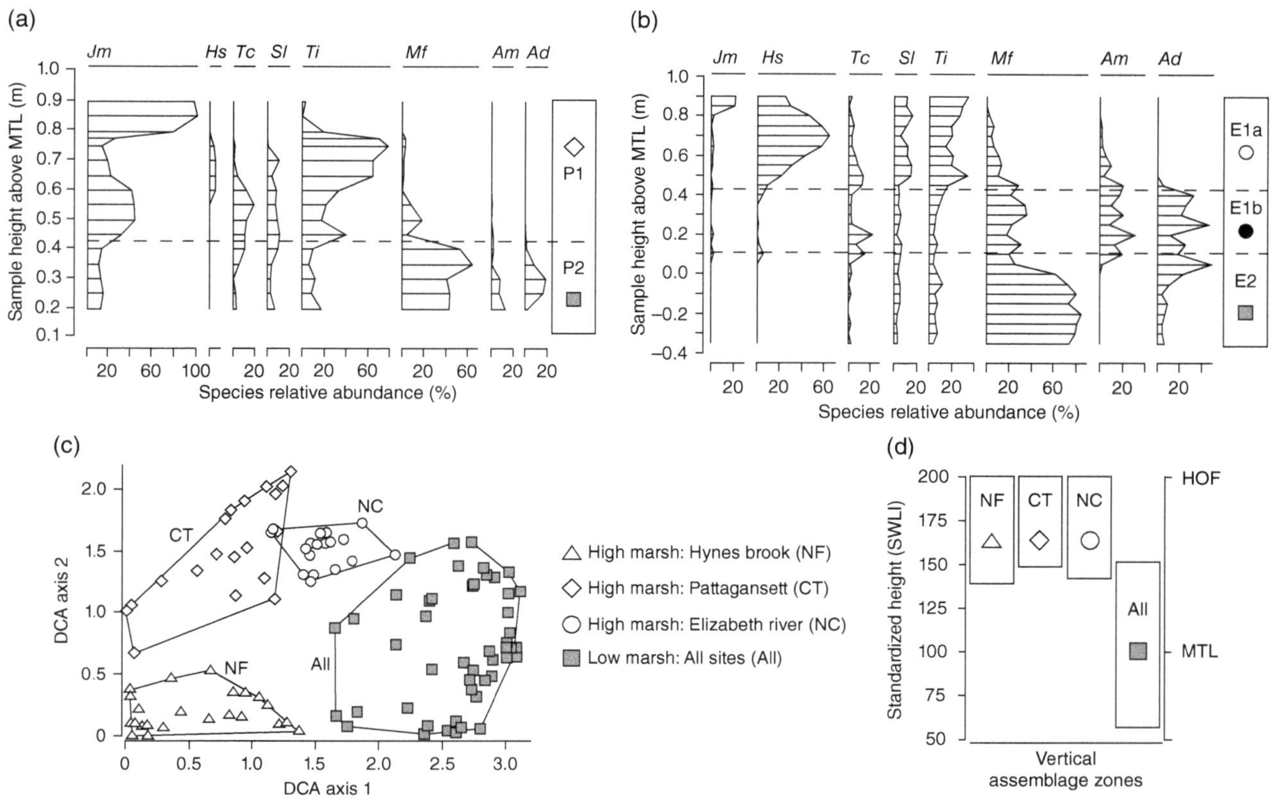

Fig. 13.3. Simplified surface distributions of foraminifera from: (a) Pattagansett River marsh, Connecticut, USA; (b) Elizabeth River marsh, North Carolina, USA. Species abbreviations as in Figure 13.2 with the addition of *Ad*: *Ammobaculities dilitatus*; *Am*: *Arenoparella mexicana*; *Sl*: *Siphotrochammina lobata.* (c) Detrended correspondence analysis of surface data from Pattagansett River, Elizabeth River, and Hynes Brook, showing three distinct, site-specific high marsh zones (CT=P1; NC=E1a+E1b; NF=H1a+H1b) and a single low marsh zone common to all sites (comprising P2+E2+H2). (d) Vertical assemblage zones plotted against height after normalization to account for differences in tidal range. SWLI = standardized water level index, which in this example is defined as (((sample elevation – MTL)/(HOF–MTL))×100)+100, where HOF = highest occurrence of foraminifera and MTL = mean tide level. See Wright et al. (2011) for details.

the high marsh assemblages form distinct clusters with almost no overlap, indicating that assemblage composition varies between sites (Fig. 13.3c). Closer inspection of the data reveals that, while the high marsh in Hynes Brook (NF) is dominated by *B. pseudomacrescens* and *J. macrescens* (Fig. 13.2, Zone H1a), *B. pseudomacrescens* is replaced by *T. inflata* in Pattagansett River (Fig. 13.3a, Zone P1). In Elizabeth River (NC) neither *B. pseudomacrescens* nor *J. macrescens* are dominant, and instead the high marsh is characterized by a mixed assemblage of *Haplophragmoides* species, *T. inflata*, and *Siphotrochammina lobata* (Fig. 13.3b, Zone E1a). Significantly, this kind of finer-scale spatial variation is not only evident when comparing marshes from widely separated regions (e.g., Newfoundland and North Carolina), but is also found in saltmarshes from the same geographical region (e.g., Connecticut; Edwards et al., 2004a), and can even exist within some large, individual marsh systems (e.g., Great Marshes, Massachusetts; de Rijk and Troelstra, 1997).

One obvious implication of this spatial variability is that tidal elevation is not the sole "environmental variable" determining foraminiferal distribution. This is unsurprising since species distribution is the product of numerous, complex interactions among organisms and their environment. While identifying controlling factors and understanding their relative importance is the primary objective of ecologists, paleoecologists seek to simplify and reduce complexity in an attempt to explain observed patterns in terms of their relationship to factors of interest (sometimes termed "target variables"). While it may be ecologically

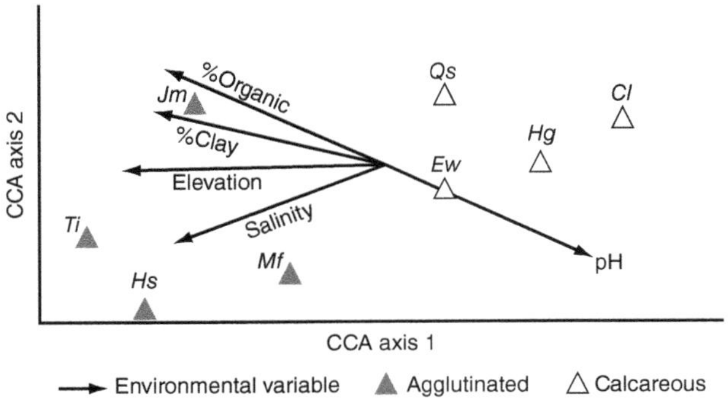

Fig. 13.4. A canonical correspondence analysis biplot of species–environment relationships for modern intertidal foraminifera from Cowpen Marsh, UK. Arrows indicate the strength and direction of influence of known environmental variables. *Source*: Adapted from Horton and Edwards, 2006.

implausible that a single environmental variable, such as tidal elevation, controls species distribution, the objective of the paleoecological approach is simply to extract an "elevation signal" from the data (ultimately it is an exercise in correlation). Hence, as long as the "elevation signal" remains detectable above the contaminating "noise" introduced by the distorting influence of other variables, paleotide-level reconstructions are possible.

While the observed spatial variability indicates tidal elevation is not the only controlling variable, the clear correlation between foraminiferal species and elevation reflects the strong environmental gradient that exists across the intertidal zone; this is a result of its location at the interface between terrestrial and marine conditions. In this way, elevation is really a surrogate for underlying controlling variables, and its strong correlation with species distributions is reasonably explained as a consequence of co-variation with these other parameters. For example, the ordination plot in Figure 13.4 shows the relationships among foraminiferal species and a series of environmental variables measured at Cowpen Marsh, UK. The biplot is produced by canonical correspondence analysis; it is essentially the same as the DCA ordinations presented in Figures 13.2 and 13.3, except that in this case it presents species rather than samples. The ordination also has the additional constraint that the axes must correspond to measured environmental variables or linear combinations of them. Each measured variable (e.g., tidal elevation, salinity) appears as an arrow on the biplot, and is essentially a "vector" indicating the strength and direction of its influence on the distribution of species (see ter Braak, 1995; Legendre and Birks, 2012 for more details).

The results show that axis one, which accounts for 84% of the total variance in species–environment relationships, is highly correlated with tidal elevation and percentage clay. Salinity, organic content, and pH are jointly correlated with both axes one and two, which together account for 93% of the species–environment relationships. The biplot therefore demonstrates the strong environmental gradient that extends from the high marsh on the left of the diagram (high elevation, organic content, salinity, and clay fraction; low pH) to the tidal flat on the right. This is echoed by the foraminiferal species which show the agglutinated taxa to the left and the calcareous taxa to the right. Of the six measured environmental variables, elevation accounted for more of the variance in the foraminiferal data than all of the other variables combined (see Horton and Edwards, 2006, p. 30).

While the simplifications associated with the paleoecological approach mean that it is not necessary for elevation to be the sole environmental control on species distributions, this simplification is accompanied by an important caveat: the influence of other variables must remain constant, both in space and over time. This requirement is particularly important in light of the fact that elevation is acting as a surrogate for other variables (see discussion in Juggins, 2013). If this condition is not met, the precise nature of the correlation between species distribution and tidal elevation will be altered with the result that erroneous inferences may be drawn.

In practice, as Murray (2006) notes, different parameters will influence the distribution of a species at different times and in different places, reflecting the fact that variables at the bounds of tolerance exert the strongest limiting effect. Unfortunately, the ecology of intertidal foraminifera is rather poorly understood (as is the case for most microfossil groups used in paleoenvironmental reconstruction). Consequently, much remains to be done to better understand the environmental variables influencing intertidal foraminiferal distributions and the manner in which abiotic factors (e.g., salinity, temperature, substrate) combine with biotic factors (e.g., competition, predation, food) to ultimately fashion the observed assemblages. In the absence of this information, paleoecologists seek to match fossil assemblages with modern counterparts, following the uniformitarian logic that similar assemblages likely formed in analogous environments (see Birks et al., 2010).

13.2.4 Interpreting fossil foraminiferal assemblages

The spatial (and temporal) variability outlined above means that an important first step in any reconstruction is selecting the appropriate surface data from which to derive the species–tidal elevation correlations that will be used to interpret the fossil assemblages of interest. At present, there is no consensus in the literature regarding how best to compile a modern dataset (often referred to as a "training set"). In fact, given that the spatial and temporal scale of study directly affects the validity of certain key assumptions (e.g., constant influence of secondary variables), training sets should be developed with explicit regard to such issues (see Section 13.3). While no single "best practice" procedure exists, one requirement common to all training sets is the need to find surface foraminiferal assemblages that match those recovered in core material (often referred to as "modern analogs").

13.2.4.1 Finding good modern analogs

There are many ways to assess whether modern and fossil data are comparable, including the cluster and ordination methods used to describe modern assemblages (e.g., Figs 13.2 and 13.3). For example, cluster analysis can be performed on a composite dataset containing both modern and fossil material to determine whether groupings contain both surface and core samples. It is often more instructive to perform an ordination (e.g., DCA) on the surface data and then passively project the fossil samples into the ordination space. If fossil and surface data occupy similar parts of the ordination diagram, it is reasonable to suggest the two datasets are comparable. A third popular test is to employ a "modern analog technique" (MAT) that compares assemblage composition and provides a measure of how similar a fossil sample is to those in the training set (e.g., Edwards and Horton, 2000). While MAT quantifies the similarity between fossil and modern assemblages, determining exactly what constitutes a "good analog" is complex (see Birks et al., 2010; Juggins and Birks, 2012; Simpson, 2012) and sea-level researchers have employed different approaches (e.g., Kemp et al., 2009; Woodroffe, 2009; Watcham et al., 2013). In the example below, we consider fossil samples with a distance to closest modern analog value below that of the 10th percentile to have a good match in the training set (Edwards et al., 2004b).

Figure 13.5 illustrates the issue of modern analogs by comparing surface and fossil data from several sites in New England, USA. Figure 13.5a shows a DCA ordination of surface data from four saltmarshes in Connecticut; samples are coded by site, and some key foraminifera species are indicated for reference. The biplot shows that while similar assemblages are found across Connecticut (overlapping samples from different sites), there are still some local variations in the details of their composition (site-specific portions of the ordination space). When fossil assemblages from Pattagansett River marsh are projected into this ordination space, they show excellent agreement with the surface samples from the same site (Fig. 13.5b). Furthermore, MAT indicates that all the fossil samples have good modern analogs in the Connecticut surface dataset. In this instance, the excellent agreement between fossil and surface assemblages means it is reasonable to use the modern species–tidal elevation relationships from the study site as a basis for reconstruction.

In other cases, fossil data do not match local surface assemblages and an alternative to site-specific analysis is required. This is illustrated in Figures 13.5c, and 13.5d for a sediment core recovered from Menunketesuck River marsh in Connecticut (Gehrels and van de Plassche, 1999). Figure 13.5c presents a DCA ordination of surface foraminiferal assemblages from the Great Marshes in Massachusetts (de Rijk, 1995a; de Rijk and Troelstra, 1997). It reveals distinctive assemblages from different areas of the marsh system which, in part, reflect contrasting salinity regimes at the sampling locations (see Section 13.3.1.1 for further details). In Figure 13.5d, fossil assemblages from Menunketesuck are projected onto the Great Marshes ordination, along with the surface distributions from Connecticut presented in Figure 13.5a. In contrast to the Pattagansett core, MAT indicates that the Menunketesuck samples have limited modern analogs in the Connecticut dataset (shown as white squares) and tend to plot outside of the Connecticut ordination space. Instead, the number of good modern analogs is significantly increased when the Great Marshes data are added (gray squares), indicating that more modern analogs for the Menunketesuck material are found in Massachusetts rather than

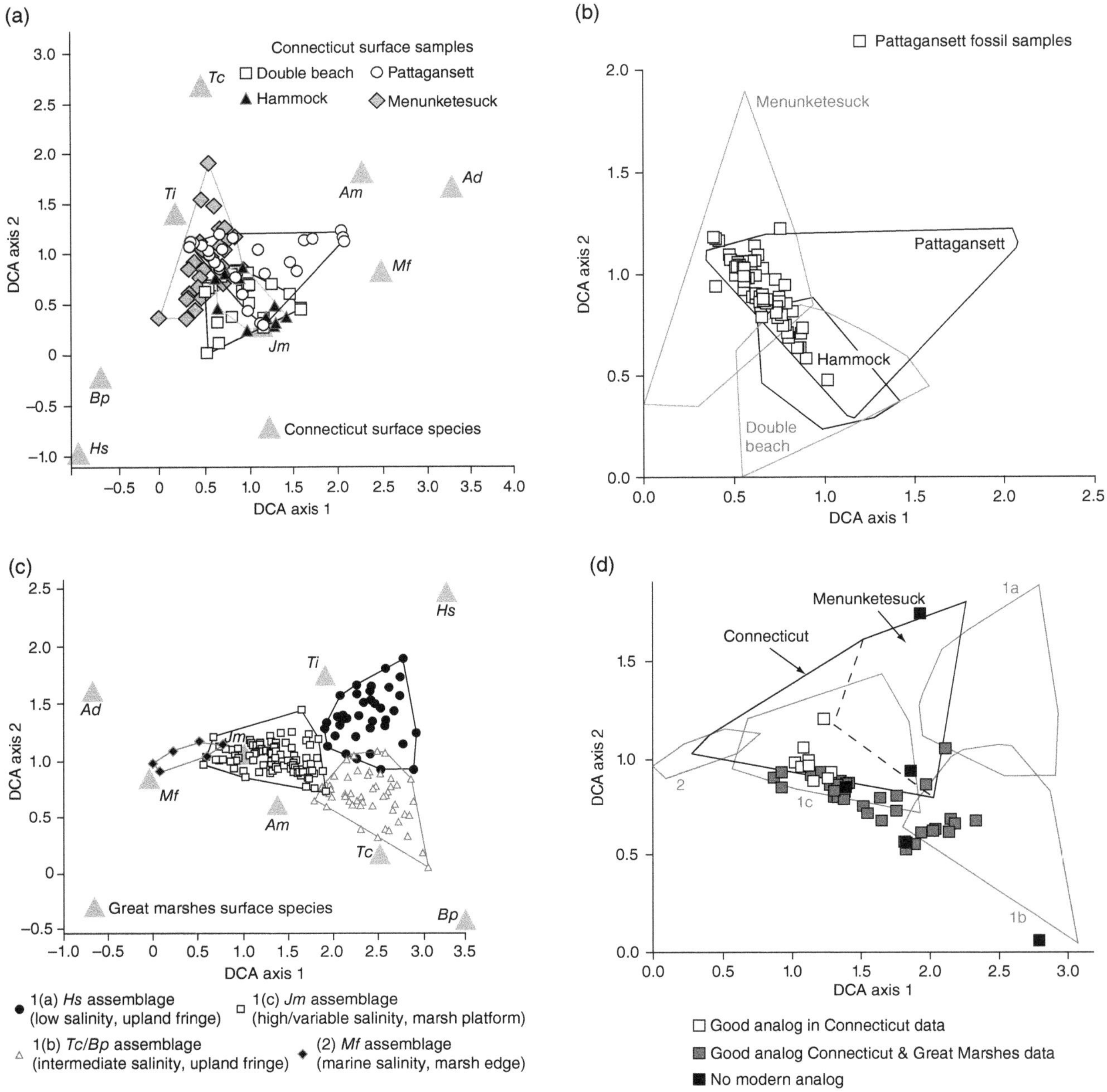

Fig. 13.5. (a) Detrended correspondence analysis (DCA) of surface foraminiferal assemblages from four saltmarshes in Connecticut, showing samples and species. (b) DCA biplot showing fossil samples from Pattagansett, Connecticut (passively) projected into the ordination space of surface distributions from Connecticut. (c) DCA of surface foraminiferal assemblages from the Great Marshes, Massachusetts, showing samples and species. (d) DCA biplot showing fossil samples from Menunketesuck, Connecticut (passively) projected into the ordination space of surface distributions from the Great Marshes, Massachusetts. Outlines of Connecticut surface data shown for reference. *Source*: Data from de Rijk (1995a); de Rijk and Troelstra (1997); Gehrels and van de Plassche (1999); Wright et al. (2011)

Connecticut. Hence, although local, site-specific training sets may be most appropriate for cores that record limited environmental change, more heterogeneous sediments and assemblages may require the combination of data from several sites. Intuitively, the need to introduce additional sites is likely to increase as records become longer, reflecting the greater likelihood that environmental conditions in the more distant past differed from those observed in an area today.

While good modern analog samples are an important prerequisite for reliable paleomarsh-surface elevation reconstruction, it does not necessarily follow that all reconstructions based on good modern analog samples will be reliable. For example, while calcareous taxa are encountered in the intertidal zone, their distributions are not restricted to it. Instead, broadly similar assemblages can be found in subtidal contexts with clearly different vertical relationships to RSL. In such cases, while a fossil sample may have a good analog in a surface intertidal dataset, the resulting reconstruction will be in error if the assemblage was actually of subtidal origin (Woodroffe, 2009). Many saltmarsh studies avoid the use of lower elevation distributions partially as a consequence of this effect, but in other cases low intertidal (and even subtidal) datasets have been employed (e.g., Horton et al., 2007). In some locations, complementary, environmentally diagnostic data (e.g., lithology, alternative biological indicators) can assist in interpreting these assemblages, although this information is not always available (e.g., Woodroffe, 2009).

Although the focus of this chapter is on the use of intertidal foraminifera as precise RSL indicators, it is worth briefly noting that a vertical zonation is also apparent in the depth distributions of subtidal benthic foraminifera (e.g., Phleger, 1965; Sen Gupta, 1977; Rossi and Horton, 2009; Phipps et al., 2010). While the vertical range of these shelf (and slope) biofacies is typically one or two orders of magnitude greater than their intertidal counterparts, they may still be used to constrain past sea level, at least in a qualitative to semi-quantitative manner. For example, subtidal foraminifera from multiple cores have been used in combination with a suite of other data (e.g., marine ostracods, planktonic foraminifera) to infer large (10 m) sea-level changes in the Bonaparte Gulf, Australia following the Last Glacial Maximum (Yokoyama et al., 2000, 2001; De Dekker and Yokoyama, 2009). More quantitative methods have been applied to samples from the northern Adriatic Sea where it is suggested that foraminifera may be able to reconstruct paleo-water depths of between 8 and 100 m with a vertical uncertainty of around 10 m (Rossi and Horton, 2009).

Further work in shallow marine environments clearly has the potential to cast new light on meter-scale RSL changes. However, in the same way that tidal elevation is to some extent a surrogate for other controlling variables, any water depth zonation arises as a consequence of multiple biotic and abiotic factors (e.g., temperature and salinity; substrate; current velocity; availability of oxygen and food). This raises similar modern analog issues to those discussed above for intertidal assemblages. It is also important to emphasize that while the vertical distributions of both intertidal and subtidal assemblages may be plotted relative to sea level, the environmental gradients that produce these distributions are fundamentally different. Furthermore, the contrasting nature of intertidal and subtidal environments means that the resulting fossil foraminiferal assemblages will have experienced markedly different processes during formation. In light of these points, the compilation of intertidal and subtidal assemblages into a composite training set lacks a clear theoretical basis and is best avoided.

13.2.4.2 The genesis of fossil assemblages: some complicating factors

While the absence of a good modern analog may reflect sampling limitations in the surface dataset, an alternative cause could be failure of the fossil dataset to fulfill the requirement that foraminifera are autochthonous and assemblages do not change over time. In an idealized scenario, foraminifera live on the sediment surface (termed "epifaunal") and are immobile so that, upon death, their tests are incorporated into the sediment, buried, and preserved as accumulation progresses. In practice, several phenomena can contribute to deviations away from this idealized situation (see Berkeley et al., 2007 for a useful review).

Firstly, intertidal foraminifera are not exclusively epifaunal (e.g., Goldstein et al., 1995; Ozarko et al., 1997; Saffert and Thomas, 1998). Several studies have reported living foraminifera at depths ranging from a few centimetees up to several decimeters (termed "infaunal"). If infaunal foraminifera occur in sufficient numbers to distort or overprint the original epifaunal component of a sample, a "no modern analog" situation could arise (Goldstein and Harben, 1993). At present, the extent of infaunal influence is poorly quantified, as are the processes responsible for its occurrence. In many cases, infaunal foraminifera are rarely found in large numbers when compared with the accompanying death assemblage, and their influence is often restricted to the top centimeter or so of sediment (e.g., Alve and Murray, 2001; Edwards et al., 2004a; Culver and

Horton, 2005; Horton and Edwards, 2006). While there is some evidence that certain species are more commonly found infaunally (e.g., Duchemin et al., 2005), the paucity of data concerning benthic foraminiferal ecology means it is currently unclear whether they actively seek particular environmental niches found at depth, or whether they are simply more capable of survival when mixed down into the sediment by other processes. In light of this uncertainty, it is advisable to investigate the potential significance of infaunal activity at sites targeted for RSL reconstruction (see Section 13.3).

Sediment mixing (e.g., by bioturbating organisms such as fiddler crabs or reworking by tidal currents) is a second potential cause of "no modern analog" situations and can be grouped with the broader range of processes that modify assemblages by introducing or removing foraminiferal tests (collectively referred to as "taphonomy"). Empty foraminiferal tests behave in a similar fashion to other sediment particles and, while there may be some hydrodynamic differences relating to factors such as test shape, in general the potential for transport will be greatest in areas of highest energy (Murray, 2006). Hence, very sandy sediments often contain a significant allochthonous component, and lower elevation contexts usually have greater potential for reworking due to faster current velocities and the absence of vegetative cover. In contrast, on the vegetated marsh platform, particularly away from creek margins and the lower marsh fringe, energy conditions are low, and sediments are fine (often cohesive) and bound together by organic material. In these sheltered settings, the potential for transport is generally low; this is one of the reasons these environments are favored for RSL reconstruction.

High marsh environments are not completely immune from the impacts of high-energy events such as storms or hurricanes, which may remove material or introduce sediments from lower intertidal or subtidal contexts (e.g., Turner et al., 2006; van de Plassche et al., 2006). However, as outlined in Section 13.2.1, lower elevation contexts are dominated by calcareous taxa while the higher marsh is characterized by agglutinated forms. One of the reasons for this is that calcareous tests rapidly dissolve in the low-pH conditions associated with organic-rich saltmarsh sediments (Scott and Medioli, 1980a). Consequently, in most instances the episodic introduction of exotic calcareous taxa onto the marsh surface will not leave a permanent imprint in the fossil record.

Test destruction by processes such as dissolution is perhaps the most significant taphonomic process operating in saltmarsh (and mangrove) systems. Its influence can vary from altering assemblage composition (e.g., by the selective removal of certain taxa) to completely eliminating all foraminiferal tests from a sample, rendering it useless for RSL reconstruction (Berkeley et al., 2007). Test dissolution is the principal mode of destruction for calcareous taxa, with porcelaneous species perhaps being more susceptible due to the nature of their wall structure. The effects of dissolution range from mild surface etching to complete test removal. In the case of the latter, studies of artificially acid-treated assemblages have demonstrated how anomalous, residual agglutinated assemblages can be produced (e.g., an almost monospecific assemblage of *M. fusca*). While these residual agglutinated assemblages will lack modern analogs, they can retain environmental information and so may still be useful in some applications (Murray and Alve, 1999).

Agglutinated foraminifera are not immune from test destruction which can arise as a consequence of attack to the organic test lining, or to the cement that holds the arrenaceous grains together. Bacterial action has been implicated in this process, and loss of agglutinated tests will therefore be most pronounced in hot, tropical environments. The loss of agglutinated foraminifera is a widely reported issue for mangrove-based studies, and severely restricts the use of mangrove peat in foraminifera-based RSL reconstruction (e.g., Woodroffe, 2009).

Taphonomic processes such as dissolution can produce a mismatch between the composition of the life assemblage (biocoenosis) and the death assemblage (thanatocoenosis) that is ultimately incorporated into the fossil record. For this reason, surface death assemblages are considered by many to be the most appropriate modern analogs for the interpretation of material from sediment cores (e.g., Murray, 2000; Horton and Murray, 2006). There has been considerable debate in the literature as to whether the "total" (life plus death) assemblage is also an appropriate modern analog (e.g., Scott and Medioli, 1980b; Tobin et al., 2005). In many instances the comparative scarcity of living individuals relative to the number of dead tests in a sample renders this more a point of principle than of practice (e.g., Culver and Horton, 2005).

13.2.4.3 Reconstructing paleomarsh-surface elevation

Assigning a tidal elevation value to a sediment sample on the basis of its foraminiferal content (i.e., a reconstruction of paleomarsh-surface elevation) can be achieved in several ways (Fig. 13.6). The ultimate research goal is to reconstruct RSL, and this requires paleomarsh-surface elevation estimates to be combined with other data (e.g., age and stratigraphic information) to produce an accurate picture of change. These additional steps introduce their own sets of considerations (e.g., accurate determination of age, influence of sediment compaction, etc.) which are discussed elsewhere in this handbook (Chapters 2, 11, 23, 30, 32). The following discussion focuses on how paleomarsh-surface elevation can be reconstructed.

One approach is to find a match between fossil and modern data and to assign a paleomarsh-surface elevation on the basis of the range of tidal elevations occupied by the modern analog(s). The vertical assemblage zone approach is perhaps the simplest example of this methodology, either assigning a fossil assemblage to a modern assemblage zone by eye, or using some kind of statistical test as outlined in Section 13.2.4.1 (Fig. 13.6c, f). This approach can be particularly successful when applied to lithological contacts between terrestrial and intertidal sediments (Chapter 2). Where assemblages dominated by marsh agglutinated species are sandwiched between organic-rich samples devoid of foraminifera and minerogenic samples containing calcareous taxa, tidal elevations corresponding to the vegetated marsh can be assigned with confidence. In some circumstances, these paleomarsh-surface estimates can be further refined where characteristic assemblages occupying smaller vertical ranges can be identified (e.g., Fig. 13.1, location 2).

While certain assemblage zones have the potential to fix paleomarsh-surface elevation to within a few centimeters, in many cases assemblage zones are several decimeters thick, especially where tidal ranges are large (e.g., Fig. 13.1, location 9). This vertical resolution is appropriate for examining the types of changes experienced during deglaciation and early in the Holocene (measured in meters to tens of meters), but becomes limiting when the more subtle changes characterizing the late Holocene are considered. The situation is compounded by a methodological shift in higher-resolution RSL studies away from discrete sampling in multiple cores, toward detailed, sometimes contiguous, sampling within an individual core. The assemblage zone approach can only robustly demonstrate tidal elevation change when the shift in foraminiferal composition is sufficient for the sample to be assigned to a different vertical zone. The "quantum" nature of this change is a poor representation of the more subtle, progressive shifts typically observable in the distributions of individual species (Fig. 13.6f). Furthermore, since zone boundaries are limited in occurrence and unevenly spaced, the resulting reconstructions are associated with variable precision and sensitivity which complicates efforts to correlate records.

The desire to resolve finer-scale vertical changes has resulted in a move away from simple "assemblage zone matching" by eye, toward more quantitative, statistically based approaches (Chapter 31). The modern analog technique (MAT) mentioned above is one example of an assemblage-based approach. In simple terms, it assigns a tidal elevation to a fossil sample on the basis of the tidal elevations of the modern samples that it most closely matches (see Birks et al., 2010 for more details). By definition, reconstructions produced by MAT will be sensitive to the "good modern analog" considerations outlined in Section 13.2.4.1.

An alternative approach to "assemblage matching" is to focus on the distributions of the individual species that comprise the assemblage of interest. If the vertical relationship between each species and tidal elevation can be quantified, a consensus estimate of paleomarsh-surface elevation can be made for an assemblage by considering the signatures of its component species, weighted in some way by their relative abundance (Fig. 13.6d, g). This is the essence of several of the transfer-function-based approaches that have become popular with those seeking to develop high-resolution records of recent RSL change (Chapter 31).

Quantifying the details of how individual taxa are distributed across the width of their ecological niches marks a fundamental change in emphasis away from the general features of vertical zonation, toward more subtle, spatially variable, species-specific responses. While this may allow more refined reconstructions of paleomarsh-surface elevation to be made, these come at the cost of increasing the significance of between-site (and within-site) variability which arises from the secondary influences of other environmental variables (Section 13.2.3). Particular care needs to be

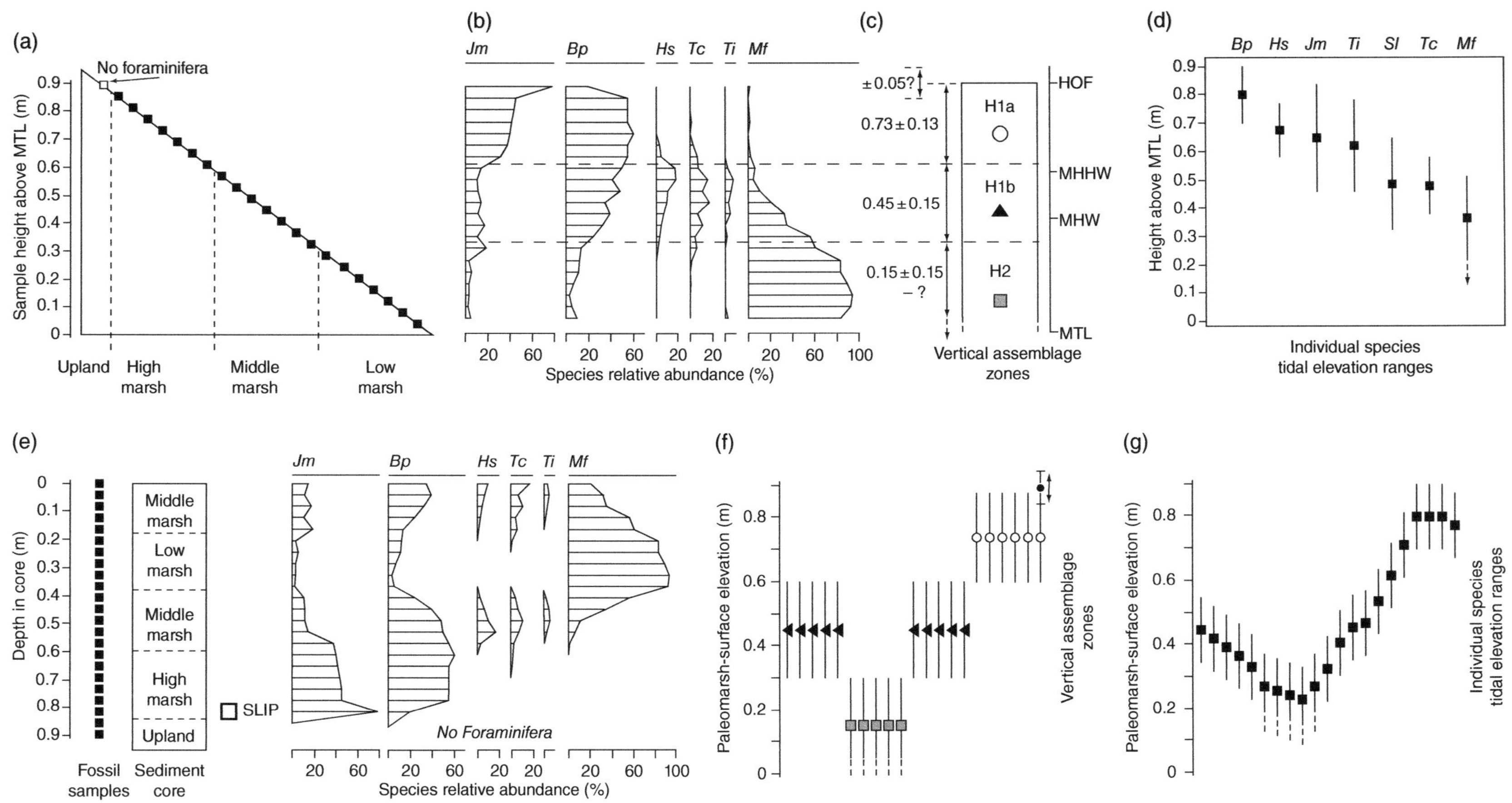

Fig. 13.6. Schematic diagram of paleomarsh-surface elevation reconstruction. (a) Surface samples are collected from the modern marsh to ascertain (b) the vertical distribution of foraminiferal species relative to mean tide level. Tide level information can be determined by (c) an assemblage zone approach or by (d) distilling the height ranges for individual species. (e) Fossil assemblages recovered from core material can then be used to infer paleomarsh-surface elevation either on the basis of (f) matching samples to assemblage zones or (g) by using statistical methods to combine species-specific information. Incomplete sampling of the lower portion of the *M. fusca* distribution means that reconstructions from low marsh contexts are subject to error. In situations where the first occurrence of foraminifera are detected in a core, precise sea-level index points (SLIP) may be established where finer-scale, high marsh zonations are present, for example in (c) and (f). Species abbreviations as in Figure 13.2.

taken to ensure that an appropriate surface dataset is compiled that furnishes good modern analogs while preserving the validity of the simplifying paleoecological assumption that the elevation signal is intact, and remains detectable above the noise introduced by secondary variables (see Wright et al., 2011 for discussion).

13.3 METHODOLOGY

The field and laboratory methods for collecting, preparing, and analyzing saltmarsh foraminifera have changed little since the pioneering work of the 1970s. The methodology described in this section is specifically tailored to studies seeking to use intertidal foraminifera in RSL reconstruction and therefore builds on the rationale and assumptions presented in Section 13.2. The specific nature of these objectives means that, in some instances, sampling and analysis will differ from more general census-type inventory studies, or those seeking to elucidate foraminiferal ecology and wider environmental controls (see Murray, 2006).

13.3.1 Collecting intertidal foraminifera in the field

The primary objectives of surface sampling campaigns conducted in sea-level studies are: (1) to locate modern foraminiferal assemblages that match those encountered in the fossil material used for RSL reconstruction; and (2) to quantify the vertical relationship between these "modern analog" assemblages (or their component taxa) and the tidal frame. In theory, realizing these objectives is a simple matter of collecting surface foraminifera from a range of intertidal environments and measuring the corresponding tidal elevation at each sampling station. In practice, the manner in which a sampling campaign is devised and executed can influence the resulting reconstructions and interpretations.

13.3.1.1 The importance of sampling design

In many instances, sediment cores are recovered from active intertidal environments and the surface sampling campaign will logically commence by surveying the modern intertidal distributions at the study site. Since tidal elevation is the target variable, samples should theoretically be collected at equal vertical intervals across the full range of intertidal contexts. The simplest way to achieve this is to sample along transects established across the intertidal zone, extending from the tidal flats to above the upper limit of marine influence. Transects need not be linear and should avoid areas of modern disturbance or other settings that would not be targeted in RSL reconstruction (e.g., pond holes, creek beds, etc).

Where possible, it is preferable for samples to be collected along multiple transects, since this provides some replication of data (i.e., repeat sampling of the same elevation but at a different location) while reducing the potential influence of spatial autocorrelation. This can be particularly instructive across relatively flat saltmarsh platforms where different sub-environments may be encountered at comparable elevations. As the strong environmental gradients responsible for the vertical zonation of foraminifera also exert an influence on plant distributions, surface sampling can be informed by the transitions between different floral zones. Hence, while statistical methods generally perform best when samples are equally spaced in the vertical, practical and logistical considerations may result in some relaxation of this idealized requirement.

It is important for sampling to extend beyond the upper limit of marine influence in order to capture the distinctive disappearance of foraminifera above the highest water mark. A vertical spacing of around 5 cm is generally optimal for delimiting this transition, although practical constraints may necessitate broader sampling intervals in lower elevation settings where assemblage zones tend to be thicker (especially in macrotidal areas). While coring usually targets sediments that accumulated in the highest marsh zone, it is still important to collect modern samples from lower elevation contexts as this can assist in more reliably delimiting the tidal elevation preference/tolerances of individual species (Wright et al., 2011).

The significance of surface sampling strategy is illustrated by the Great Marshes foraminiferal distributions summarized in Figure 13.5c. Here, more than 180 surface samples were collected from multiple locations within the spatially extensive Barnstable system, with the aim of identifying modern analogs for different stages of marsh development (see de Rijk, 1995a, b; de Rijk and Troelstra, 1997 for details). The DCA analysis shows distinct high marsh assemblages adjacent to the upland fringe, with *Haplophragmoides* associated with low-salinity conditions (black circles), and *T. comprimata/B. pseudomacrescens* more commonly encountered in intermediate-salinity areas (white triangles). On the basis of this spatial variability, the original studies concluded that elevation was not the dominant control on the foraminiferal distribution of the Great Marshes, instead highlighting the significance of factors such as salinity (de Rijk, 1995a, b).

Although this work has been frequently cited as illustrating the absence of a vertical zonation in some systems, it is important to note that the sampling design employed by de Rijk was focused on capturing spatial (horizontal) variability rather than collecting a representative dataset of vertical sub-environments. The expression of this is shown in Figure 13.7 which compares the vertical distribution of samples and associated species' relative abundances from the Great Marshes (Fig. 13.7b), with similar data from Connecticut (Fig. 13.7a). The data have been standardized to account for differences in tidal range such that a value of 200 equates to the upper limit of marine influence/foraminiferal occurrence, while a value of 100 equals mean tide level.

While both distributions exhibit long "tails" of sampling in the lower elevation contexts, the Connecticut dataset possesses a reasonably uniform vertical distribution of samples across the range of saltmarsh environments. In contrast, the initial Great Marshes dataset (de Rijk, 1995b) which emphasized the control of salinity on the observed distribution patterns, has samples that are tightly concentrated on a comparatively small vertical interval in the high marsh (black points/bars in Fig. 13.7b). The full Great Marshes dataset (de Rijk and Troelstra, 1997) reduces this sampling bias to some extent, but is still strongly focused on a 60 cm zone within the high marsh (gray points/bars in Fig. 13.7b). The net result of this is that the Great Marshes dataset highlights

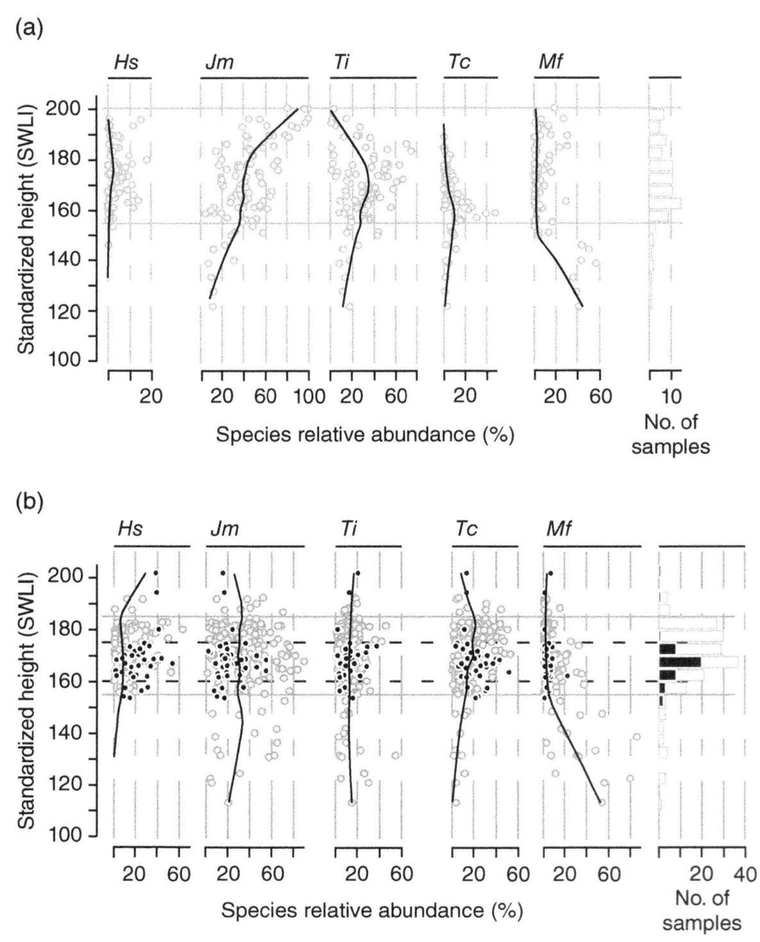

Fig. 13.7. Relative abundances of principal foraminiferal species in surface samples from: (a) Connecticut; and (b) Great Marshes, Barnstable, Massachusetts. Samples plotted against height after standardization to account for differences in tidal range. SWLI: standardized water level index (as defined in Wright et al., 2011). Histograms indicate the number of samples in the dataset distributed in vertical bins of 5 SWLI units.

spatial variability within a narrow height range, but lacks the vertical sampling to reliably capture the upper and lower limits of species distribution. Evenly distributed vertical sampling is especially critical for statistical approaches that use averaging methods to identify the preferred tidal elevation of particular species (e.g., the popular WA-PLS transfer function approach), since any bias in vertical sampling will feed directly to the resulting species optima (Wright et al., 2011; Chapter 31).

13.3.1.2 Methods of sampling modern intertidal foraminifera

Most foraminifera-based RSL studies take surface samples of approximately 10 cm^3 comprising the uppermost 1 cm of sediment. In vegetated areas, it is often helpful to employ a small hand-held corer in concert with a sharp knife to slice through fibrous material. This sampling method is generally effective when recovering the kinds of cohesive sediments that often typify saltmarsh and tidal flat environments. In some instances however (e.g., when sampling clean sands or fluid muds), it is easier to collect a surface sediment scrape using a spoon.

Sediment samples should be placed into appropriately labeled airtight bags or vials for transport to the laboratory. It is important to record site name, sampling station and date/time of collection. It is also useful to photograph the sampling site and record the nature of the substrate and vegetation cover. Many studies collect a larger-volume replicate sediment sample at each station to furnish material for other laboratory analyses (e.g., loss on ignition, particle size, etc.).

In light of the taphonomic processes outlined above, it is important to process and stabilize foraminiferal samples as quickly as possible after collection. Consequently, it is recommended that a preservative is added to each sample in the field and that this is mixed with a protoplasm stain (Rose Bengal) to discriminate between live and dead specimens (Murray and Bowser, 2000). While the distributions of live foraminifera are patchy in time and space, identification of the life assemblage makes the resulting data of wider use to the community interested in foraminiferal ecology. For this reason, it is recommended that surface samples are stained and separate counts of living and dead foraminifera reported, irrespective of whether death or total assemblages are ultimately employed.

Ethanol is an ideal preservative (at least 70% ethanol, buffered by the addition of 1 teaspoon of sodium bicarbonate/litre), and this can be mixed with Rose Bengal (1 teaspoon/litre) before gently shaking or kneading the sample to distribute the solution throughout the sediment. Care should be taken during addition and mixing of the ethanol/Rose Bengal solution to avoid contact with skin or eyes. It should also be noted that ethanol can erase labeling written with permanent markers, so it is advisable to "double bag" all material. In warmer environments (e.g., tropical mangroves) or where a delay between field collection and laboratory sampling is likely, it may also prove beneficial to store material in a cool box until it can be transported to the laboratory.

This sampling protocol assumes that living foraminifera are predominantly epifaunal to shallow infaunal, such that the uppermost centimeter of sediment provides a suitable analog for interpreting subsurface assemblages. Where the potential significance of infaunal foraminifera is to be examined, cores of sediment will need to be recovered and either subsampled and preserved/stained in the field or returned to the laboratory for subsequent processing. The collection of subsurface material by coring is described in more detail in (Chapter 4).

While the time-integrative nature of the foraminiferal death assemblage can be expected to moderate the influence of patchiness to some degree, repeat sampling of surface distributions (e.g., winter, spring, summer, and autumn) can improve transfer function performance (Horton and Edwards, 2003; Horton and Murray, 2006), although this is often not logistically feasible (Culver and Horton, 2005). The limited available data seem to indicate that sampling during the winter months may provide the best alternative if only a single visit is possible.

13.3.2 Processing foraminifera in the laboratory

Compared to many other microfossil groups (e.g., diatoms, pollen) the processing and analysis of intertidal foraminifera is quick, relatively simple, and does not require the use of hazardous chemicals.

On return to the laboratory, sample material should be kept refrigerated below 4 °C to retard biological processes. Samples should be left in the buffered ethanol/Rose Bengal solution for a minimum of 24 hours to ensure complete penetration of the stain into the tests. Where sediment samples are consolidated or extremely fibrous, it can be useful to disaggregate material in a beaker to assist with this process.

Once staining is complete, samples are washed through 500 µm and 63 µm sieves to screen out coarse material, rinse off residual stain and remove the fine sediment fraction. In the case of very fibrous, organic-rich samples, particular care needs to be taken to ensure that foraminiferal tests are flushed from the organic residue which collects on the 500 µm mesh. Intertidal foraminifera are typically smaller than 500 µm, but it is good practice to inspect the coarse residue prior to discarding it in case unusually large specimens remain on the mesh.

Foraminifera are retained on the 63 µm sieve along with fine organic material and sediment grains. It is often helpful to use a fine spray (e.g., from a pressurized garden sprayer) to assist in disaggregating clay-rich samples and rinsing away the finer sediment fraction. The sample material is then washed into a labeled storage bottle, mixed with an equal volume of buffered ethanol, sealed, and stored in a refrigerator.

Sieves should be washed carefully between samples and it is useful to dip them in a solution of methylene blue in order to stain any residual material. In this way, contamination can be easily identified under the microscope.

13.3.3 Indentifying and analyzing foraminifera

Foraminifera are examined under a binocular microscope using reflected light at typical magnifications of × 40–80. The high organic content of saltmarsh sediment means that intertidal assemblages are best analyzed in suspension ("wet picking") rather than from a dried residue as is common practice for fully marine samples. Samples should remain wet at all times, since drying and subsequent re-wetting can destroy fragile agglutinated tests.

It is necessary to count a certain minimum number of foraminiferal tests in order to produce robust data suitable for statistical analysis; the precise figure needed is a function of the relative abundances of the different taxa present, and the required confidence level (Patterson and Fishbein, 1989; Fatela and Taborda, 2002). In species-rich assemblages such as those that typify more fully marine environments, it is common practice to count a minimum of 300 tests per sample. By comparison, intertidal environments are typically species poor and their vertical zonation is expressed as large swings between dominant taxa that typically constitute 50% or more of an assemblage (e.g., Figs 13.2 and 13.3). In these situations it has become common for analysts to count a minimum of around 100 tests, although in some instances counts as low as 40 have been used (e.g., in the highest elevation samples where foraminiferal abundances are extremely low and assemblages can be monospecific).

Table 13.1. Changes in approximate count uncertainties (confidence intervals, %) for a given relative abundance as sample size increases. Results based on calculations in equation 13.1 with Z=1.96 (0.95 confidence level).

Relative abundance (%)	Number of foraminifera counted			
	50	100	200	300
5	± 7	± 5	± 3	± 3
10	± 9	± 6	± 4	± 3
25	± 12	± 8	± 6	± 5
50	± 13	± 10	± 7	± 6
75	± 12	± 8	± 6	± 5
90	± 9	± 6	± 4	± 3
95	± 7	± 5	± 3	± 3
Average	± 10	± 7	± 5	± 4

Ultimately, the choice of minimum count size will be a trade-off between the benefits that greater counts bring (i.e., narrower confidence intervals), and the cost in extra time required to achieve these improvements. As greater emphasis is placed on small changes in relative abundance, the issue of minimum count size becomes increasingly significant. Table 13.1 presents illustrative confidence intervals associated with different count sizes and relative abundances, derived from Equation 13.1 (Mosimann, 1965). While not strictly symmetrical, these can essentially be thought of as error terms associated with the foraminiferal counts (see Maher et al., 2012 for further information).

$$P(c\ \text{limit}) = \frac{\hat{p} + Z^2/2n_p \pm Z\sqrt{\hat{p}(1-\hat{p})/n_p + Z^2/4n_p^2}}{1 + Z^2/n_p} \tag{13.1}$$

where $P(c$ limit) are the confidence limits for the percentage count of a given species; Z is the area of confidence interval under normal curve (in SD units; Z = 1.960 for 0.95 confidence limits); n_p is the total number of foraminifera counted in the sample; and $\hat{p}$ is the percentage relative abundance counted for a given species.

While foraminiferal concentrations vary considerably, it is normal for 10 cm^3 of sediment to contain significantly more than 100 tests; it is therefore advantageous to subdivide the sample to expedite counting. Wet samples can be divided using a laboratory rotary splitter or a "wet-splitter" (see Scott and Hermelin, 1993). Once a sample has been split to the required volume it is transferred onto a spiral-grooved tray using a pipette, where the foraminifera are then identified and counted. It is advisable to spread the sample thinly to facilitate identification, and the entire aliquot should be counted since some size sorting can occur.

Foraminifera are identified on the basis of their test morphology with diagnostic features including the nature of the test wall (agglutinated, calcareous, or porcelaneous), number, shape and arrangement of their chambers, and the number, shape, and position of holes in the test walls (the major holes are termed "apertures"). Agglutinated foraminifera generally appear brown to gray under reflected light and it is sometimes possible to distinguish the sediment grains that comprise their test. In contrast, calcareous tests usually appear white or glassy under the optical microscope with porcelaneous taxa, as their name suggests, having a porcelain-like appearance. Specimens living at the time of collection are distinguished by the red staining of their protoplasm (if processed using Rose Bengal), which is often restricted to the final chambers of the test.

While the number of characteristic intertidal species is comparatively limited, ongoing debate regarding foraminiferal taxonomy and the absence of a single, definitive taxonomic guide complicates the process of identification and analysis. Some of the controversy arises from historical differences between schools of researchers and differing tendencies toward the "lumping" and "splitting" of taxa. Re-evaluation of current taxonomic practices will form an important part of future work seeking to improve our understanding of intertidal foraminiferal distributions and the factors controlling them. In the interim, it is essential that workers are explicit regarding which taxonomic conventions they follow and, where possible, for publications to include plates illustrating which forms are assigned to a particular taxon. This is especially important when data from multiple analysts are combined and analyzed. Although a detailed taxonomic discussion is beyond the scope of this chapter, the Appendix describes some commonly found intertidal foraminifera from the UK and North America, and provides a brief overview of some of the more frequent areas of taxonomic debate.

13.3.4 Presenting foraminiferal data

It is common practice to present intertidal foraminiferal distributions in terms of the relative abundances of the individual species contributing to the total assemblage. These are best plotted against tidal elevation since this is the target variable, although some workers prefer to show changing distributions with horizontal distance when sampling is not equally spaced in the vertical. It is useful to include information on the concentration of tests within a sample, along with the total number of foraminifera counted (the latter is essential when tabulating data). Commonly used programs for analyzing and presenting paleoecological data include C2 (Juggins, 2007), Canoco (ter Braak and Šmilauer, 2002), PAST (Hammer et al., 2001) and TILIA (Grimm, 1992), while new analysis packages are being developed for the R environment (http://www.r-project.org/) and are likely to be widely adopted.

13.4 SUMMARY

Distinct assemblages of benthic foraminifera live at different heights within the intertidal zone. Analysis of modern foraminifera contained within surface sediments permits quantification of species–height relationships, which can be remarkably precise (centimeter to decimeter). Armed with this knowledge, it is possible to infer the tidal elevation at which a sediment sample accumulated from the foraminifera it contains. Fossil assemblages recovered from core material can be used to infer paleomarsh-surface elevation and produce detailed reconstructions of relative sea-level change. Spatial variability in foraminiferal distributions, coupled with the influence of taphonomic processes, can complicate the reliable extraction of tide level information. However, a range of statistical tools are available to assist in the description of foraminiferal data and the assessment of their suitability for RSL reconstruction. The potential precision of foraminifera-based approaches has made them a favored tool for those seeking to produce detailed reconstructions of recent RSL. Future success will depend on further refinement and critical evaluation of reconstruction methods, coupled with an improvement in our understanding of benthic foraminiferal ecology.

APPENDIX: A BEGINNER'S GUIDE TO TAXONOMY OF INTERTIDAL FORAMINIFERA

The use of different taxonomies in foraminifera-based RSL research introduces additional complications and ambiguities when reading the published literature. This brief guide focuses on some of the main intertidal species encountered in the RSL literature, illustrates them with SEM images, and provides brief accompanying descriptions to assist in deciphering how various authors have defined them. The illustrative material is derived from work in the saltmarshes of the UK and USA. This appendix is not intended as a definitive guide, but rather as a useful reference when reading existing work and considering future taxonomic refinements.

Ecophenotypic variation is a notable feature of agglutinated saltmarsh foraminifera. In many instances, a series of gradational forms exist between distinct "end-member taxa" that are often designated as "species". Much of the taxonomic debate in the RSL literature boils down to whether or not members of a gradational series are delimited as distinct species, or are simply classified as "formae" of a single species. In a practical sense, the precise name given to a particular grouping of specimens is less important than the ability to reliably and consistently assign tests to a particular taxon. A balance must therefore be struck between "lumping" forms together to ensure that a robust set of diagnostic criteria can be established, and "splitting" series into distinctive forms which may carry significant paleoecological information.

It is arguably best practice to lean toward a "splitting" approach in the first instance since, while separate forms can ultimately be combined into a single group, the reverse is not possible. When combining data from multiple analysts, an important first step is to standardize the taxonomies employed. The inclusion of illustrative material when publishing data will greatly assist in this process and, in most instances, this can be readily accommodated by the provision of online supplementary images and taxonomic notes.

Principal agglutinated saltmarsh taxa

Arenoparrella mexicana (Kornfeld, 1931)

1931 *Trochammina inflata* (Montagu) var. *mexicana* Kornfeld, p. 86, pl. 13, fig. 5a–c.
1951 *Arenoparella mexicana* (Kornfeld); Andersen, p. 31, fig. 1a–c.

Plate 13.1, Figures 1a–c

Diagnostic features: Tests are agglutinated and pale gray-brown in appearance. Chambers are trochospirally coiled, increase gradually in size, and are separated by radial sutures. Primary aperture is an elongate slit on the face of the final chamber that is approximately parallel to the plane of coiling. Supplementary cribrate aperture may be present near the apex of the final chamber.

Remarks: Arenoparella is distinguished from *Jadammina* by its vertical, slit-like aperture which is usually quite apparent under the optical microscope.

Balticammina pseudomacrescens Brönnimann, Lutz and Whittaker, 1989

1989 *Balticammina pseudomacrescens* Brönnimann, Lutz and Whittaker, p. 169, pl. 1, figs 1–5, pl. 2, figs 1–9, pl. 3, figs 1–4, pls 1–3.

Plate 13.1, Figures 2a–c

Diagnostic features: Tests appear brown in color and are composed of fine detrital grains with abundant organic cement. They are trochospirally coiled but compressed with a wide, shallow umbilicus. Tests possess both a primary interiomarginal aperture and secondary umbilical apertures.

Remarks: In North America, this species has traditionally constituted part of the intragradational series associated with *Trochammina macrescens* by Scott and Medioli (1980a, pl. 3, figs 1–12). de Rijk (1995b) proposed that part of this series (designated *T. macrescens macrescens*) comprised forms with an open umbilicus which should be assigned to *B. pseudomacrescens*, while forms with a closed umbilicus should be assigned to *Jadammina macrescens* (see discussion below). Gehrels and van de Plassche (1999) recommend that a distinction is made between *B. pseudomacrescens* and *J. macrescens* in light of the fact that they exhibit markedly different distributions in both modern and fossil deposits. Similar patterns are evident in the data presented in this chapter (e.g., Figs 13.2, 13.3 and 13.5). These differences in distribution are consistent with the view that the forms are ecophenotypic variants of the same species, and Tobin et al. (2005) suggest that

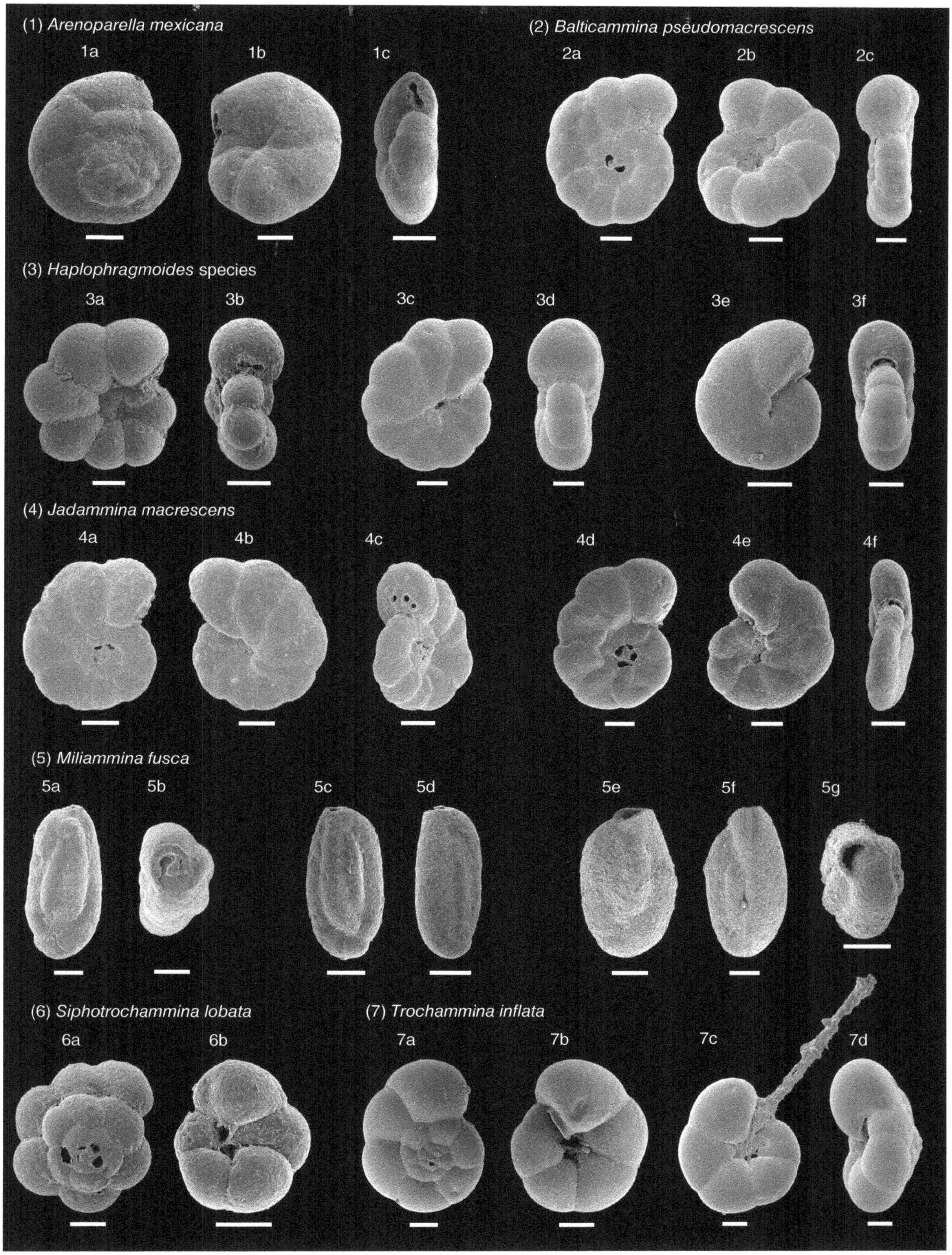

Plate 13.1 Scale bar: 100 μm. 1a–c: *Arenoparella mexicana* (material from Connecticut, Edwards); 2a–c: *Balticammina pseudomacrescens* (material from NE England, Horton); 3a–f: Gradational series of *Haplophragmoides* species (material from (a, b) Connecticut, Edwards; (c, d) NE England, Horton; (e, f) North Carolina, Wright); 4a–f: *Jadammina macrescens* (material from (a–c) NE England, Horton; (d) Connecticut, Edwards; (e, f) Newfoundland, Wright); 5a–g: Gradational series of *Miliammina fusca* (material from (a, b) syntype 2, Natural History Museum London, UK; (c, d) Connecticut, Edwards; (e–g) North Carolina, Wright); 6a–b: *Siphotrochammina lobata* (material from Newfoundland, Wright); 7a–d: *Trochammina inflata* (material from (a, b) Connecticut, Edwards; (c) S England, Edwards; (d) NE England, Horton).

B. pseudomacrescens is best viewed as an end-member of the intergradational series of Scott and Medioli (1980a, pl. 3, figs 1–3) which should be named *T. macrescens pseudomacrescens*.

While the relative merits of whether these taxa should be given the status of distinct species or ecophenotypic "formae" is open to debate, the fact that they may contain significant paleoenvironmental information means that efforts should be made to systematically differentiate them wherever possible. While both *B. pseudomacrescens* and *J. macrescens* have a very low trochospiral test with a primary interiomarginal apertural slit, the former is generally larger and can be distinguished on the basis of its wide, open umbilicus containing supplementary apertures (see Pl. 13.1, Fig. 2b). Similarly, *B. pseudomacrescens* is distinct from *T. inflata*, with the latter having a higher trochospiral test with fewer, but more globular, chambers in the final whorl, and a single interiomarginal aperture.

Haplophragmoides species

1910 Genus *Haplophragmoides* Cushman.

Plate 13.1, Figures 3a–f

Diagnostic features: Tests appear pale brown in color, are composed of fine silt grains held together by organic cement, and have a smooth finish. While superficially resembling the umbilical view of *Jadammina macrescens*, *Haplophragmoides* tests are planispirally coiled (identical on both sides) and possess an interiomarginal aperture.

Remarks: Several different species of *Haplophragmoides* are recognized in the literature on the basis of differences in the form of the periphery (smooth to lobate) and sutures (degree of depression and extent of curvature). The gradational series shown in Plate 13.1 illustrates the spectrum of these forms. Tests with lobate peripheries, inflated chambers, and depressed sutures have been designated as *H. manilaensis* Anderson (1953, p. 22, pl. 4, fig. 8), while those with smooth peripheries and less depressed sutures have been assigned to *H. wilberti* Anderson (1953, p. 21, pl. 4, fig. 7).

Plate 13.1, Figure 3b shows a form with interioareal apertures similar to part of the intergradational series initially identified as *H. manilaensis* by de Rijk and Troelstra (1997, p. 89, pl. 1, figs 1–8), but which they note could be assigned to *Trochamminita salsa* (Cushman and Brönnimann), as redefined by Saunders (1957, p. 6, pl. 1, figs 3–8). These differences can be difficult to distinguish with an optical microscope, especially when the specimens are small and sometimes have organic detritus clogging or obscuring the apertures. Similarly, Horton and Edwards (2006) note that it can be difficult to distinguish *Cribrostomoides jeffreysii* (Williamson, 1858), which has an areal slit just above the base of the final chamber, from true *Haplophragmoides* (see Haynes, 1973, p. 27–31).

Jadammina macrescens (Brady, 1870)

1870 *Trochammina inflata* (Montagu) var. *macrescens* Brady: p. 290, pl. 11, figs 5a–c.

1938 *Jadammina polystoma* Bartenstein and Brand: p. 381, figs 1–3.

1984 *Jadammina macrescens* (Brady); Brönnimann and Whittaker: p. 303–309, figs 1–21 (lectotype erected).

Plate 13.1, Figures 4a–f

Diagnostic features: Tests agglutinated, brown in color, and composed of extremely fine detrital grains with abundant organic cement. Chambers are thin-walled and arranged in a very low trochospiral coil with a shallow umbilicus. The thin and flexible nature of the chambers means they commonly collapse when dried. The primary aperture is a single interiomarginal slit, and supplementary apertures are present as one or more areal pores on the face of the final chamber.

Remarks: Scott and Medioli (1980a) regard the presence of supplementary cribrate openings as ecophenotypic, reflecting high-salinity conditions. While initially recognized as *Jadammina polystoma*, tests bearing these areal pores (e.g., Pl. 13.1, Fig. 4c) were later assigned to part of their gradational series *T. macrescens* as the forma *T. macrescens polystoma* (Scott et al., 1990).

In contrast, de Rijk and Troelstra (1997) assigned all forms with closed umbilical areas to *Jadammina macrescens*, irrespective of the presence of supplementary apertures, since their forms had interiomarginal apertures in peripheral (equatorial) positions (Pl. 13.1, Fig. 4f) as opposed to the more interiomarginal umbilical location characteristic of the genus *Trochammina* Parker and Jones (1859). This convention was later followed by Gehrels and van de Plassche (1999, p. 98, pl. 1, figs 1–5), Edwards et al. (2004a, p. 16, pl. 1, figs 3, 4), Vance et al. (2006, p. 19, pl. 2, fig. 7), Kemp et al. (2009, p. 227, pl. 1, figs 1, 2) and Wright et al. (2011, p. 58, pl. 1, figs 5a–h).

While organic detritus, chamber collapse, and small test size (especially in NW European marshes) can complicate recognition of areal pores, the potential paleoenvironmental information they impart recommends that their presence is recorded wherever possible. In light of the potential confusion that surrounds *Balticammina pseudomacrescens*, *Jadammina macrescens*, and *Trochammina macrescens* in its various forms, it is essential that researchers are explicit with regard to their taxonomic choices when dealing with these species.

Miliammina fusca (Brady, 1870)

1870 *Quinqueloculina fusca* Brady, p. 47, pl. 11, figs 2, 3.
1950 *Miliammina fusca* (Brady); Phleger and Walton, p. 280, pl. 1, fig. 19.

Plate 13.1, Figures 5a–g

Diagnostic features: Tests are agglutinated, pale brown in color and composed of detrital grains in an organic cement. They are elongate, rounded in section and made up of many chambers coiled on a milioline plan. The terminal aperture possesses a tooth.

Remarks: Test morphology is noted to vary with grain size (sand or silt) and some authors differentiate forms with a more elongate test, larger tooth, and finer wall structure as *Miliammina petila* Saunders (Saunders, 1958, p. 87, pl. 1, figs 10, 11). Wright et al. (2011, p. 59, pl. 2, figs 2a–j) present an intraspecfic gradational series which they assign to *M. fusca*, although several of this series lack a clear tooth in their aperture (e.g., Pl. 13.1, Fig. 5g).

Siphotrochammina lobata Saunders, 1957

1957 *Siphotrochammina lobata* Saunders, p. 9, pl. 3, figs 1, 2.

Plate 1, Figures 6a, b

Diagnostic features: Agglutinated test is brown and composed of extremely fine detrital grains with abundant organic cement. Ovate chambers are arranged in a trochospiral coil with deeply depressed sutures, giving a lobate periphery. The aperture comprises a round opening at the end of a siphon that projects into the umbilicus from the final chamber, with the former aperture from the penultimate chamber also extending into the umbilicus.

Remarks: Scott and Medioli (1980a, p. 41, pl. 4, figs 1–3) consider this to be a microspheric form of *Trochammina inflata* (Montagu). Tobin et al. (2005) state that forms of *Siphotrochammina* exist with and without a tube extending into the umbilicus and, where the former occurs alongside *T. inflata*, many intermediate forms can arise.

In contrast, Wright et al. (2011, p. 58, pl. 1, figs 10a–d) separate *S. lobata* from *T. inflata* on the basis that the former possesses a siphonal aperture which extends into the umbilicus, while the latter has a simple interiomarginal slit-like aperture.

Tiphotrocha comprimata (Cushman and Brönnimann, 1948)

1948 *Trochammina comprimata* Cushman and Brönnimann, p. 41, pl. 8, figs 1–3.
1957 *Tiphotrocha comprimata* (Cushman and Brönnimann); emendation Saunders, p. 11, pl. 4, figs 1–4.

Plate 13.2, Figures 8a–g

Diagnostic features: Brown agglutinated test, composed of extremely fine detrital grains with abundant organic cement. Test compressed, trochospirally coiled with depressed, curved sutures. The final chamber has a characteristic lobe that covers the umbilical area to varying degrees, while other chambers have secondary openings into the umbilicus.

Remarks: In emending the genus *Tiphotrocha*, Saunders (1957) illustrates substantial variability in degree of chamber inflation, angle of suture curvature, and development of the T-shaped final chamber. Horton and Edwards (2006, p. 71, pl. 2, figs 7d, e) illustrate "juvenile" forms which lack the complete development of the umbilical lobe (Pl. 13.2, Fig. 8g). When specimens are small or damaged, these juvenile forms can be difficult to distinguish from *J. macrescens* under the optical microscope.

Trochammina inflata (Montagu, 1808)

1808 *Nautilus inflatus* Montagu, p. 81, pl. 18, fig. 3.
1859 *Trochammina inflata* (Montagu); Parker and Jones, p. 347.

Plate 13.1, Figures 7a–d

Diagnostic features: Brown tests, agglutinated, composed of extremely fine detrital grains with an outer organic layer and abundant organic cement. Tests are trochospiral with inflated chambers, the size and extent of globularity varying between individuals. The aperture is an interiomarginal slit with a lip.

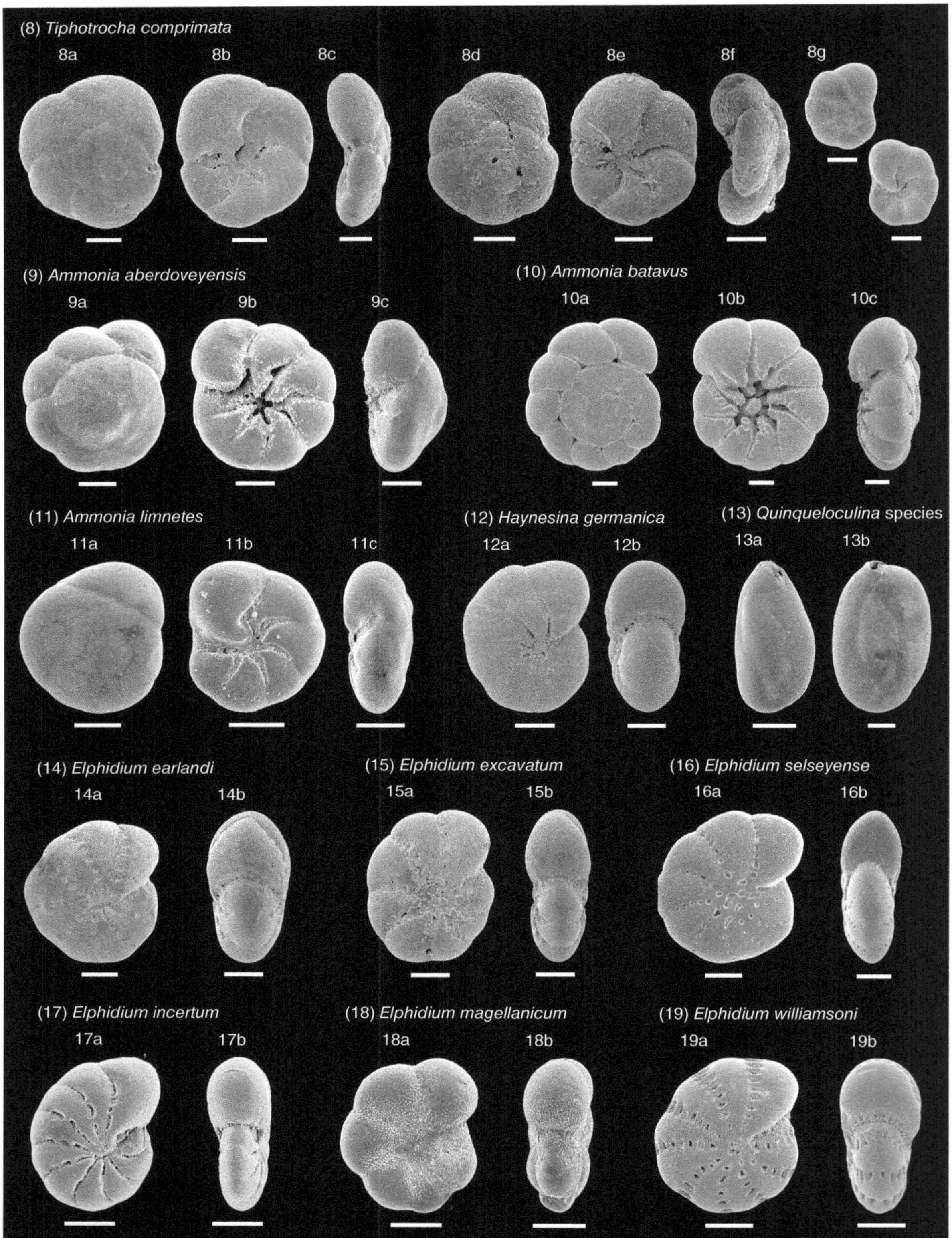

Plate 13.2 Scale bar: 100 μm. 8a–g: *Tiphotrocha comprimata* (material from (a–c, g) S. England, Edwards; (d–f) North Carolina, Wright); 9a–c: *Ammonia aberdoveyensis* (material: paratype 1, Natural History Museum London, UK); 10a–c: *Ammonia batavus* (material from NE England, Horton); 11a–c: *Ammonia limnetes* (material from: (a, c) S. England, Whittaker; (b) paratype, Natural History Museum London, UK); 12a, b: *Haynesina germanica* (material from NE England, Horton); 13a, b: *Quinqueloculina* species (material from S. England, Edwards); 14a, b: *Elphidium earlandi* (material from S. England, Edwards); 15a, b: *Elphidium excavatum* (material from NE England, Horton); 16a, b: *Elphidium selseyense* (material: syntype 2, Natural History Museum London, UK); 17a, b: *Elphidium incertum* (material: paralectotype, Natural History Museum London, UK); 18a, b: *Elphidium magellanicum* (material: paratype2, Natural History Museum London, UK); 19a, b: *Elphidium williamsoni* (material from S. England, Edwards).

Remarks: While *T. inflata* is often distinguished from *S. lobata* due to its lack of siphonal aperture opening into the umbilicus, Horton and Edwards (2006, p. 71, pl. 2, fig. 8d) illustrate a *T. inflata* specimen from the UK with an extended tube emanating from the primary aperture (Pl. 13.2, Fig. 7c). A similar specimen from North Carolina is illustrated by Kemp et al. (2009, p. 227, pl. 1, fig. 5).

Common calcareous intertidal taxa

Ammonia spp

Ammonia aberdoveyensis Haynes, 1973 (Plate 13.2, Figs 10a–c) [1973 *Ammonia aberdoveyensis* Haynes: p. 184, pl. 18, fig. 15, text-figs 38.1–7]
Ammonia batavus (Hofker, 1951) (Plate 13.2, Figs 11a–c)
[1951 *Streblus batavus* Hofker: p. 498, text-figs 340, 341]
Ammonia limnetes (Todd and Brönnimann, 1957) (Plate 13.2, Figs 12a–c)
[1957 *Streblus limnetes* Todd and Brönnimann: p. 38, pl. 10, figs 4a–c]

Plate 13.2, Figures 9a–c, 10a–c, 11a–c

Diagnostic features: Tests are calcitic with radially arranged crystallites and pores. They normally appear glassy or translucent under the binocular microscope but are sometimes brownish in color. Tests are biconvex and subcircular in outline with trochospirally arranged chambers which vary in number depending on species. Interspecific morphological variations include: test size; suture shape and extent of depression; and the development of ornamentation including an umbilical boss.

Remarks: While commonly assigned to *Ammonia beccarii* (Linnaeus, 1758), the molecular systematic work of Hayward et al. (2004) shows that this species is restricted in occurrence to the Mediterranean. Horton and Edwards (2006) differentiate three common species of *Ammonia* in their material from the UK which are illustrated here. *A. limnetes* (Todd and Brönnimann, 1957) is generally associated with lower-salinity conditions (Pl. 13.2, Figs 11a–c), while *A. batavus* (Hofker, 1951) is characteristic of more fully marine conditions (Pl. 13.2, Figs 10a–c). North American material includes a species reported as *Ammonia parkinsoniana* (d'Orbigny, 1839) which, while similar in form to *A. batavus*, has been shown to be distinct at the molecular level (Hayward et al., 2004).

Elphidium williamsoni *Haynes*, 1973

1973 *Elphidium williamsoni* Haynes, p. 207, pl. 24, fig. 7, pl. 25, figs 6, 9, pl. 27, figs 1–3 [see synonymy].

Plate 13.2, Figures 19a, b

Diagnostic features: Calcareous test, rotund with rounded peripheries, comprises 10–12 chambers in the outer whirl. Sutures covered by numerous, long retral processes.

Remarks: Several species of *Elphidium* are associated with intertidal to shallow subtidal environments, and are primarily differentiated on the basis of chamber number/shape, umbilical ornamentation, and the number, length, and shape of retral processes (Pl. 13.2, Figs 14–18). See Horton and Edwards (2006, p. 74–77) for additional remarks.

Haynesina germanica *(Ehrenberg, 1840a,b)*

1840a *Nonionina germanica* Ehrenberg, p. 23 (figured in Ehrenberg, 1840b, pl. 2, figs 1a–g)
1965 *Protelphidium anglicum* Murray, p. 149, 150, pl. 25, figs 1–5, pl. 26, figs 1–6

Plate 13.2, Figures 12a, b

Diagnostic features: Tests radial, calcitic, finely perforate and appear glassy under the optical microscope. They are rotund, planispirally coiled with 5–11 chambers in the last whorl. Sutures are slightly depressed over the periphery but deeply depressed close to the umbilicus. The aperture takes the form of a row of pores at the base of the last chamber.

Remarks: This is a very common lower marsh to mudflat species. It is superficially similar to *Nonion depressulus* (Walker and Jacob, 1798), but differs from it in wall structure, and is generally much less compressed with less-depressed sutures.

Quinqueloculina *spp.* d'Orbigny, 1826

Plate 13.2, Figures 13a, b

Diagnostic features: Porcellaneous tests which appear translucent to opaque under an optical microscope. Chambers are coiled on a quinqueloculine plan, while the aperture usually has a tooth.

Remarks: Several species of *Quinqueloculina* have been reported from intertidal to shallow subtidal environments. While intertidal forms appear to occupy similar height ranges and are often grouped together due to difficulties in reliably distinguishing between them (e.g., Horton and Edwards, 2006), the differentiation of some taxa may be useful in distinguishing between intertidal and subtidal sediments (see Woodroffe, 2009).

REFERENCES

Alve, E., and Murray, J. (2001) Temporal variability in vertical distributions of live (stained) intertidal foraminifera, southern England. The Journal of Foraminiferal Research, 1, 12–24.

Anderson, H.V. (1951) Two new genera of foraminifera from recent deposits in Louisiana. Journal of Paleontology, 25, 31–34.

Anderson, H.V. (1953) Two new species of *Haplophragmoides* from the Louisiana coast. Contributions from the Cushman Laboratory for Foraminiferal Research, 4, 21–22.

Barbosa, C. F., Scott, D. B., Seoane, J. C. S., and Turcq, B. J. (2005) Foraminiferal zonations as base lines for Quaternary sea-level fluctuations in south-southeast Brazilian mangroves and marshes. The Journal of Foraminiferal Research, 35(1), 22–43.

Bartenstein, H., and Brand, E. (1938) Die Foraminiferen Fauna des Jade-Gebiete.1. *Jadammina polystoma* n. g, n. sp. sus dem Jade-Gebiete. Senckenbergiana 20, 381–385.

Berkeley, A., Perry, C. T., Smithers, S. G., Horton, B. P., and Taylor, K. G. (2007) A review of the ecological and taphonomic controls on foraminiferal assemblage development in intertidal environments. Earth Science Reviews, 83(3–4), 205–230.

Berkeley, A., Perry, C. T., Smithers, S. G., Horton, B. P., and Cundy, A. B. (2009) Foraminiferal biofacies across mangrove-mudflat environments at Cocoa Creek, north Queensland, Australia. Marine Geology, 263(1–4), 64–86.

Birks, H., Heiri, O., Seppä, H., and Bjune, A. E. (2010) Strengths and weaknesses of quantitative climate reconstructions based on late-Quaternary biological proxies. The Open Ecology Journal, 3, 68–110.

Brady, H.B. (1870) With analysis and descriptions of foraminifera by H.B. Brady, Part II. In: *The Ostracodes and Foraminifera of Tidal Rivers* (eds Brady, H.B., and Robertson, D.), London, pp. 273–306. Available at http://www.biodiversitylibrary.org/item/93156#page/287/mode/1up (accessed 7 August 2014).

Brönnimann, P., and Whittaker, J.E. (1984) A neolectotype for *Jadammina macrescens* (Brady) and emendation of *Jadammina* Bartenstein and Brand (Protozoa: Foraminiferida). Bulletin of the British Museum, Natural History (Zoology), 46, 303–309.

Brönnimann, P., Lutze, G.F., and Whittaker, J.E. (1989) *Balticammina pseudomacrescens*, a new brackish water Trochamminid from the western Baltic Sea, with remarks on the wall structure. Meyniana, 41, 167–177.

Culver, S., and Horton, B. (2005) Infaunal marsh foraminifera from the outer banks, North Carolina, USA. The Journal of Foraminiferal Research, 2, 148–170.

Cushman, J.A. (1910) *A Monograph of the Foraminifera of the North Pacific Ocean; Part I - Astrorhizidae and Lituolidae*. United States National Museum Bulletin, Washington DC, USA.

Cushman, J.A., and Brönnimann, P. (1948) Some new genera and species of arenaceous foraminifera from brackish water of Trinidad. Contributions from the Cushman Laboratory for Foraminiferal Research, 24(1, 2), 15–21, 37–42.

DeDeckker, P., and Yokoyama, Y. (2009) Micropalaeontological evidence for Late Quaternary sea-level changes in Bonaparte Gulf, Australia. Global and Planetary Change, 66(1–2), 85–92.

de Rijk, S. (1995a) Salinity control on the distribution of saltmarsh foraminifera (Great Marshes, Massachusetts). Journal of Foraminiferal Research 25(2), 156–166.

de Rijk, S. (1995b) Agglutinated foraminifera as indicators of salt-marsh development in relation to late Holocene sea level rise (Great Marshes at Barnstable, Massachusetts). Free University Amsterdam, Netherlands.

de Rijk, S., and Troelstra, S. (1997) Salt-marsh foraminifera from the Great Marshes, Massachusetts: environmental controls. Palaeogeography, Palaeoclimatology, Palaeoecology, 130, 81–112.

d'Orbigny, A.D. (1826) *Tableau Méthodique de la Classe des Céphalopodes*. Crochard, Paris.

d'Orbigny, A.D. (1839) *Foraminifères. Histoire, Physique, Politique et Naturelle de l'Ile de Cuba*. Arthus Bertrand, Paris.

Duchemin, G., Jorissen, F. J., Redois, F., and Debenay, J.-P. (2005) Foraminiferal microhabitats in a high marsh: Consequences for reconstructing past sea levels. Palaeogeography, Palaeoclimatology, Palaeoecology, 226(1–2), 167–185.

Edwards, R.J., and Horton, B.P. (2000) Reconstructing relative sea-level change using UK salt-marsh foraminifera. Marine Geology, 169, 41–56.

Edwards, R.J., Wright, A.J., and van de Plassche, O. (2004a) Surface distributions of salt-marsh foraminifera from Connecticut, USA: modern analogues for high-resolution sea level studies. Marine Micropaleontology, 51, 1–21.

Edwards, R. J., van de Plassche, O., Gehrels, W., and Wright, A. (2004b) Assessing sea-level data from Connecticut, USA, using a foraminiferal transfer function for tide level. Marine Micropaleontology, 51(3–4), 239–255.

Ehrenberg, C. G. (1840a) *Eine weitere Erläuterung des Organismus mehrerer in Berlin lebend beobachterer Polythalamien der Nordsee*. Bericht über die zur Bekanntmachung geeigneten Verhandlungen der Königlichen Preussischen Akademie der Wissenschaften zu Berlin, v. 1840, p. 18–23.

Ehrenberg, C. G. (1840b) *Über noch Jetzt zahlreich lebende Thierarten der Kreidebildung und den Organismus der Polythalamien*. Physikalische Mathematische Abhandlungen der Königlichen Akademie der Wissenschaften zu Berlin, v. 1839, p. 81–174.

Fatela, F., and Taborda, R. (2002) Confidence limits of species proportions in microfossil assemblages. Marine Micropaleontology, 45, 169–174.

Gehrels, W. R., and Plassche, O. van de. (1999) The use of Jadammina macrescens (Brady) and Balticammina pseudomacrescens Brönnimann, Lutze and Whittaker (Protozoa: Foraminiferida) as sea-level indicators. Palaeogeography, Palaeoclimatology, Palaeoecology, 149, 89–101.

Gehrels, W. R., and Woodworth, P. L. (2012) When did modern rates of sea-level rise start? Global and Planetary Change, 100, 263–277.

Goldstein, S., and Harben, E. (1993) Taphofacies implications of infaunal foraminiferal assemblages in a Georgia salt marsh, Sapelo Island. Micropaleontology, 39, 53–62.

Goldstein, S., Watkins, G., and Kuhn, R. (1995) Microhabitats of salt marsh foraminifera: St. Catherines island, Georgia, USA. Marine Micropaleontology, 26(95), 17–29.

Grant, K. M., Rohling, E. J., Ayalon, A., Ramsey, C. B., Satow, C., and Roberts, A. P. (2012) Rapid coupling between ice volume and polar temperature over the past 150,000 years. Nature, 491(7426), 744–747.

Grimm, E.C. (1992) Tilia and Tilia-graph: pollen spreadsheet and graphics programs. Program and Abstracts, *8th International Palynological Congress*, Aix-en-Provence, France, September 6–12, 1992, p. 56.

Hammer, Ø., Harper, D.A.T., and P. D. Ryan, (2001) PAST: Paleontological Statistics Software Package for Education and Data Analysis. Palaeontologia Electronica, 4(1), 9pp.

Hawkes, A. D., Horton, B. P., Nelson, A. R., and Hill, D. F. (2010) The application of intertidal foraminifera to reconstruct coastal subsidence during the giant Cascadia earthquake of AD 1700 in Oregon, USA. Quaternary International, 221(1–2), 116–140.

Haynes, J. R. (1973) Cardigan Bay Recent Foraminifera (Cruises of the R.V. *Antur*, 1962–1964). Bulletin of the British Museum, Natural History, Zoology Series, Supplement 4, 1–245.

Hayward, B., Grenfell, H., and Scott, D. (1999) Tidal range of marsh foraminifera for determining former sea-level heights in New Zealand. New Zealand Journal of Geology and Geophysics, 42(3), 395–413.

Hayward, B. W., Holzmann, M., Grenfell, H. R., Pawlowski, J., and Triggs, C. M. (2004) Morphological distinction of molecular types in Ammonia – towards a taxonomic revision of the world's most commonly misidentified foraminifera. Marine Micropalaeontology, 50, 237–271.

Hofker, J. (1951) *The Foraminifera of the Siboga Expedition, Part 3.* Siboga-Expeditie, Leiden, v. 4b, p. 1–513.

Horton, B. P. (1999) The distribution of contemporary intertidal foraminifera at Cowpen Marsh, Tees Estuary, UK: implications for studies of Holocene sea-level changes. Palaeogeography, Palaeoclimatology, Palaeoecology, 149, 127–149.

Horton, B. P., and Edwards, R. J. (2003) Seasonal distributions of foraminifera and their implications for sea-level studies, Cowpen Marsh, UK. In *Micropaleontological Proxies for Sea-Level Change and Stratigraphic Discontinuities* (eds Orson, H., and Leckie, R.), SEPM Special Publication No. 75, pp. 21–30.

Horton, B. P., and Edwards, R. J. (2006) *Quantifying Holocene Sea-Level Change Using Intertidal Foraminifera: Lessons from the British Isles.* Cushman Foundation for Foraminiferal Research, Special Publication No. 40.

Horton, B. P., and Murray, J. W. (2006) Patterns in cumulative increase in live and dead species from foraminiferal time series of Cowpen Marsh, Tees Estuary, UK: Implications for sea-level studies. Marine Micropaleontology, 58(4), 287–315.

Horton, B. P., Larcombe, P., Woodroffe, S. A., Whittaker, J. E., Wright, M. R., and Wynn, C. (2003) Contemporary foraminiferal distributions of a mangrove environment, Great Barrier Reef coastline, Australia: implications for sea-level reconstructions. Marine Geology, 198(3–4), 225–243

Horton, B., Whittaker, J., Thomson, K., Harbattle, M., Woodroffe, S., and Wright, M. (2005) The development of a modern foraminiferal data set for sea-level reconstructions, Wakatobi Marine National Park, Southeast Sulawesi, Indonesia. Journal of Foraminiferal Research, 35(1), 1–14.

Horton, B. P., Culver, S. J., Hardbattle, M. I. J., Larcombe, P., Milne, G. M., Morigi, C., Whittaker, J. E., and Woodroffe, S. A. (2007) Reconstructing Holocene sea-level change for the Great Barrier Reef using subtidal foraminifera. Journal of Foraminiferal Research, 37, 327–343.

Jennings, A., Nelson, A., Scott, D., and Aravena, J. (1995) Marsh foraminiferal assemblages in the Valdivia estuary, south-central Chile, relative to vascular plants and sea level. Journal of Coastal Research, 11(1), 107–123

Juggins. S. (2007) C2 Version 1.5 User guide. Software for ecological and palaeoecological data analysis and visualisation. Newcastle University, Newcastle upon Tyne, UK. 73pp.

Juggins, S. (2013) Quantitative reconstructions in palaeolimnology: new paradigm or sick science? Quaternary Science Reviews, 64, 20–32.

Juggins, S., and Birks, H.J.B. (2012) Quantitative environmental reconstructions from biological data. In: *Tracking Environmental Change Using Lake Sediments* (eds Birks, H.J.B., Lotter, A.F., Juggins, S. and Smol, J.P.), Springer, Developments in Paleoenvironmental Research vol. 5, pp. 431–494.

Kemp, A. C., Horton, B. P., and Culver, S. J. (2009) Distribution of modern salt-marsh foraminifera in the Albemarle–Pamlico estuarine system of North Carolina, USA: Implications for sea-level research. Marine Micropaleontology, 72(3–4), 222–238.

Kemp, A. C., Horton, B. P., Donnelly, J. P., Mann, M. E., Vermeer, M., and Rahmstorf, S. (2011) Climate related sea-level variations over the past two millennia. Proceedings of the National Academy of Sciences, 108(27), 11017–11022.

Kornfeld, M.M. (1931) Recent littoral foraminifera from Texas and Louisiana. Stanford University, Department of Geology Contributions 1, 77–101.

Legendre, P., and Birks, H.J.B. (2012) From classical to canonical ordination. In: *Tracking Environmental Change Using Lake Sediments: Data Handling and Numerical Techniques* (eds Birks, H.J.B., Lotter, A.F., Juggins, S., and Smol, J.P.), Springer, Dordrecht, 201–248.

Linnaeus, C. (1758) Tomus I. *Syst. nat., ed. 10.* Holmiae, Laurentii Salvii(1–4), 1–824.

Maher, L.J., Heiri, O., and Lotter, A.F. (2012) Assessment of uncertainties associated with palaeolimnological laboratory methods and microfossil analysis. In: *Tracking Environmental Change Using Lake Sediments* (eds Birks, H.J.B., Lotter, A.F., Juggins, S. and Smol, J.P.), Springer, Developments in Paleoenvironmental Research vol. 5, pp. 143–166.

Montagu, G. (1808) *Testacea Britannica*, Supplement. S. Woolmer, Exeter, England.

Mosimann, J.E. (1965) Statistical methods for the pollen analyst: multinomial and negative multinomial techniques. In: *Handbook of Paleontological Techniques*

(eds Kummel, B., and Raup, D.), Freeman, San Francisco, pp. 636–673.
Murray, J. W. (1965) Two new species of British Recent Foraminiferida. Contributions from the Cushman Foundation for Foraminiferal Research, 16, 148–150.
Murray, J. W. (2000) The enigma of the continued use of total assemblages in ecological studies of benthic foraminifera. The Journal of Foraminiferal Research, 3, 244–245.
Murray, J. W. (2006) *Ecology and Applications of Benthic Foraminifera*. Cambridge University Press, Cambridge.
Murray, J. W., and Alve, E. (1999) Taphonomic experiments on marginal marine foraminiferal assemblages: how much ecological information is preserved? Palaeogeography, Palaeoclimatology, Palaeoecology, 149(1–4), 183–197.
Murray, J.W., and Bowser, S.S. (2000) Mortality, protoplasm decay rate, and reliability of staining techniques to recognize living foraminifera: A review. Journal of Foraminiferal Research, 30, 177–191.
Ozarko, D. L., Patterson, R. T., and Williams, H. F. L. (1997) Marsh Foraminifera from Nanaimo, British Columbia (Canada); implications of infaunal habitat and taphonomic biasing. The Journal of Foraminiferal Research, 27(1), 51–68.
Parker, W.K., and Jones, T.R. (1859) On the nomenclature of the Foraminifera; Part II - On the species enumerated by Walker and Montagu. Annual Magazine of Natural History.
Patterson, R.T., and Fishbein, E. (1989) Re-examination of the statistical methods used to determine the number of point counts needed for micropaleontological quantitative research. Journal of Paleontology, 63, 245–248.
Phipps, M., Kaminski, M., and Aksu, A. (2010) Calcareous benthic foraminiferal biofacies along a depth transect on the southwestern Marmara shelf (Turkey). Micropaleontology, 56(2004), 377–392
Phleger, F. (1965) Depth patterns of benthonic Foraminifera in the eastern Pacific. Progress in Oceanography, 3, 273–287.
Phleger, F.B., and Walton, W.R. (1950) Ecology of marsh and bay foraminifera, Barnstable, Massachusetes. American Journal of Science, 248, 274.
Rohling, E. J., Grant, K., Bolshaw, M., Roberts, A. P., Siddall, M., Hemleben, C., and Kucera, M. (2009) Antarctic temperature and global sea level closely coupled over the past five glacial cycles. Nature Geoscience, 2(7), 500–504.
Rossi, V., and Horton, B. P. (2009) The application of a subtidal foraminifera-based transfer function to reconstruct Holocene paleobathymetry of the Po Delta, northern Adriatic Sea. The Journal of Foraminiferal Research, 39(3), 180–190.
Saffert, H., and Thomas, E. (1998) Living foraminifera and total populations in salt marsh peat cores: Kelsey Marsh (Clinton, CT) and the Great Marshes (Barnstable, MA). Marine Micropaleontology, 33, 175–202.
Saunders, J.B. (1957) *Trochamminidae and certain Lituolidae (Foraminifera) from the recent brackish-water sediments of Trinidad, British West Indies*. Smithsonian Miscellaneous Collection.
Saunders, J.B. (1958) Recent foraminifera of mangrove swamps and river estuaries and their fossil counterparts in Trinidad. Micropaleontology, 4, 79–92.
Scott, D. B., and Medioli, F. (1978) Vertical zonations of marsh foraminifera as accurate indicators of former sea-levels. Nature, 272, 528–531.
Scott, D. B., and Medioli, F. S. (1980a) Quantitative studies of marsh foraminiferal distributions in Nova Scotia: implications for sea-level studies. Cushman Foundation for Foraminiferal Research, Special Publication No. 17.
Scott, D. B., and Medioli, F. S. (1980b) Living vs. total foraminifera populations: Their relative usefulness in paleoecology. Journal of Paleontology, 54, 814–831.
Scott, D. B., and Medioli, F. S. (1986) Foraminifera as sea-level indicators. In: *Sea-Level Research: A Manual for the Collection and Evaluation of Data* (ed. van de Plassche, O.), Geo Books, Norwich, pp. 435–456.
Scott, D. B., and Hermelin, J. (1993) A device for precision splitting of micropaleontological samples in liquid suspension. Journal of Paleontology, 67(1), 151–154.
Scott, D. B., Schafer, C. T., and Medioli, F. S. (1980) Eastern Canadian estuarine foraminifera; a framework for comparison. The Journal of Foraminiferal Research, 10(3), 205–234.
Scott, D. B., Williamson, M., and Duffett, T. (1981) Marsh foraminifera of Prince Edward Island: their recent distribution and application for former sea level studies. Maritime Sediments and Atlantic Geology, 17, 98–129.
Scott, D.B., Schnack, E.J., Ferrero, L., Espinosa, M., and Barbosa, C.F. (1990) Recent marsh foraminifera from the east coast of South America: Comparison to the northern hemisphere. In: *Palaeoecology, Biostratigraphy, Palaeoceanography and Taxonomy of Agglutinated Foraminifera* (eds Hemleben, C., Kaminski, M.A., Kuhnt, W., and Scott, D.B.), Kluwer, Dordrecht, 717–737.
Scott, D., Collins, E., and Duggan, J. (1996) Pacific Rim marsh foraminiferal distributions: implications for sea-level studies. Journal of Coastal Research, 12(4), 850–861.
Sen Gupta, B.K. (1977) Depth distribution of modern benthic foraminifera on continental shelves of the world. Indian Journal of Earth Science, 4, 60–83.
Simpson, G.L. (2012) Analogue methods in palaeolimnology. In: *Tracking Environmental Change Using Lake Sediments* (eds Birks, H.J.B., Lotter, A.F., Juggins, S. and Smol, J.P.), Springer, Developments in Paleoenvironmental Research vol. 5, pp. 495–522.
ter Braak, C.J.F. (1995) Ordination. In: *Data Analysis in Community and Landscape Ecology* (eds Jongman, R.H.G., ter Braak, C.J.F., and van Tongeren, O.F.R.), Cambridge University Press, Cambridge, pp. 91–173.
ter Braak, C.J.F., and Šmilauer, P. (2002) CANOCO Reference Manual and CanoDraw for Windows User's Guide: Software for Canonical Community Ordination (version 4.5) Ithaca, NY, USA (www.canoco.com). Microcomputer Power.
Tobin, R., Scott, D.B., Collins, E.S., and Medioli, F.S. (2005) Infaunal benthic foraminifera in some North American marshes and their influence on fossil assemblages. Journal of Foraminiferal Research, 35, 130–147.
Todd, R., and Brönnimann, P. (1957) Recent foraminifera and thecamoebina from the eastern Gulf of Paria. Cushman Foundation for Foraminiferal Research, Special Publication, 3, 1–43.
Turner, R. E., Baustian, J. J., Swenson, E. M., and Spicer, J. S. (2006) Wetland sedimentation from hurricanes Katrina and Rita. Science, 314(5798), 449–452.

van de Plassche, O., Erkens, G., Van Vliet, F., Brandsma, J., Van der Borg, K., and De Jong, A. F. M. (2006) Salt-marsh erosion associated with hurricane landfall in southern New England in the fifteenth and seventeenth centuries. Geology, 34(10), 829.

Vance, D.J., Culver, S.J., Corbett, D.R., and Buzas, M.A. (2006) Foraminifera in the Albemarle Estuarine System, North Carolina: Distribution and recent environmental change. Journal of Foraminiferal Research 36, 15–33.

Waelbroeck, C., Labeyrie, L., Michel, E. et al. (2002) Sea-level and deep water temperature changes derived from benthic foraminifera isotopic records. Quaternary Science Reviews, 21(1–3), 295–305.

Walker, G., and Jacob, E. (1798) An arrangement and description of minute and rare shells. In: *Essays on the Microscope*, 2nd Edition (with considerable additions and improvements by F. Kanmacher) (ed. Adams, E.), Dillon and Keeting, London, 712 p.

Watcham, E. P., Shennan, I., and Barlow, N. L. M. (2013) Scale considerations in using diatoms as indicators of sea-level change: lessons from Alaska. Journal of Quaternary Science, 28(2), 165–179.

Williamson, W.C. (1858) *On the Recent Foraminifera of Great Britain*. Ray Society, London, England.

Woodroffe, S. A. (2009) Recognising subtidal foraminiferal assemblages: implications for quantitative sea-level reconstructions using a foraminifera-based transfer function. Journal of Quaternary Science, 24(3), 215–223.

Woodroffe, S. A., Horton, B. P., Larcombe, P., and Whittaker, J. E. (2005) Intertidal mangrove foraminifera from the central Great Barrier Reef Shelf, Australia: implications for sea-level reconstruction. The Journal of Foraminiferal Research, 35(3), 259–270.

Wright, A. J., Edwards, R. J., and van de Plassche, O. (2011) Reassessing transfer-function performance in sea-level reconstruction based on benthic salt-marsh foraminifera from the Atlantic coast of NE North America. Marine Micropaleontology, 81(1–2), 43–62.

Yokoyama, Y., Lambeck, K., De Deckker P, Johnston, P., and Fifield, L. (2000) Timing of the Last Glacial Maximum from observed sea-level minima. Nature, 406(6797), 713–6.

Yokoyama, Y., De Deckker, P., Lambeck, K., Johnston, P., and Fifield, L. (2001) Sea-level at the Last Glacial Maximum: evidence from northwestern Australia to constrain ice volumes for oxygen isotope stage 2. Palaeogeography, Palaeoclimatology, Palaeoecology, 165(3–4), 281–297.

Chapter 14
Pollen and spores of terrestrial plants

CHRISTOPHER E. BERNHARDT AND DEBRA A. WILLARD
US Geological Survey, Eastern Geology and Paleoclimate Science Center, Reston, VA, USA

14.1 INTRODUCTION

Pollen and spores are valuable tools in reconstructing past sea level and climate because of their ubiquity, abundance, and durability as well as their reciprocity with source vegetation to environmental change (Cronin, 1999; Traverse, 2007; Willard and Bernhardt, 2011). Pollen is found in many sedimentary environments, from freshwater to saltwater, terrestrial to marine. It can be abundant in a minimal amount of sample material, for example half a gram, as concentrations can be as high as four million grains per gram (Traverse, 2007). The abundance of pollen in a sample lends it to robust statistical analysis for the quantitative reconstruction of environments. The outer cell wall is resistant to decay in sediments and allows palynomorphs (pollen and spores) to record changes in plant communities and sea level over millions of years. These characteristics make pollen and spores a powerful tool to use in sea-level research.

This chapter describes the biology of pollen and spores and how they are transported and preserved in sediments. We present a methodology for isolating pollen from sediments and a general language and framework to identify pollen as well as light micrographs of a selection of common pollen grains. We then discuss their utility in sea-level research.

14.2 POLLEN AND SPORES

Unlike other biological proxies used to reconstruct sea level (such as marine foraminifera, ostracodes, diatoms), pollen is the by-product of plants that may or may not be living in the setting analyzed. Regardless, pollen assemblages can provide complex physical information about the depositional setting, including sea level. It is therefore important to define exactly what pollen is.

Pollen is a functional rather than morphological definition, where it is defined as the microspore wall of a seed plant including the microgametophyte that develops within the wall (Traverse, 2007). Pollen develops from the inner mother cells of the anther (the description of fertilization below is derived from the classic text by Esau, 1977). The mother cells undergo meiosis producing a tetrad of haploid cells. A pollen grain represents a plant generation, or a haploid male seed plant. Mature pollen grains typically range in size from ~5 to ~100 μm. Individual cells then separate and sporopollenin production in the exine is increased. In some families however, such as the Ericaeae, tetrads remain together (Erdtman, 1952; Traverse, 2007). Once the pollen is mature, it is transferred via some vector (insect, wind, water, mammal, and others) to the carpel (Proctor et al., 1996). Pollen germinates on the stigma, where a pollen tube is formed. The pollen tube exits the pollen grain through a pore, colpus, or a weakening in the exine. A vegetative nucleus passes into the tube followed by two gametes, which are formed by division of the generative nucleus during germination. The pollen tube grows between the tissue cells in the stigma and the style entering into the ovary. The pollen tube enters the ovule through the micropyle. The two male gametophytes enter the embryo-sac, where one fuses with the egg and the other with the diploid fusion nucleus. The fertilized egg develops to form the embryo and the male gametophyte fused with the diploid nucleus forms the endosperm.

Spores from ferns and fern allies (collectively called pteridophytes) are common in sediments and are used in tandem with pollen when interpreting past vegetation changes; they are therefore included in this chapter. There are two main life cycles in pteridophytes: homosporous and heterosporous. In homosporous pteridophytes, the diploid sporophytic plant produces tetrads of one type of haploid spore (an isospore). From the spore, a gametophyte, with both male and female sex organs, will germinate. Fertilization occurs when spermatozoids are

Handbook of Sea-Level Research, First Edition. Edited by Ian Shennan, Antony J. Long and Benjamin P. Horton.

released by the male organs and swim to a female part of a gametophyte. A diploid pteridophtye forms from the fertilized egg. In heterosporous pteridophytes however, two types of haploid spores are produced (a microspore and a megaspore). Each germinates into a separate gametophyte where the microgametophyte produces the spermatozoid. This spermatozoid swims to a megagametophyte to fertilize and produce an egg that will give rise to the diploid stage of the pteridophyte. An important commonality for both types of life cycles is that water is needed for fertilization.

14.2.1 Pollen/spore wall

The cell wall of pollen and spores is termed the exine, which is composed of an organic compound called sporopollenin that is highly resistant to chemical degradation. This characteristic allows pollen and spores to be preserved in sediments for hundreds of millions of years and for them to be powerful proxies of past sea level and vegetation communities. This resilience means that pollen is subject to post-depositional reworking, however (see Section 14.2.3). The exact organic structure of sporopollenin is unknown because procedures to break it down for analysis result in simple sugars and compounds that do not represent the original molecule. However, it is thought to be similar to the structure of rubber (Traverse, 2007). Even though sporopollenin is an inert organic compound, it is sensitive to extreme oxidation. Sporopollenin's susceptibility to oxidation can result in pollen grains autoxidizing in older microscope slides and oxidizing in coarse grain sediments (Traverse, 2007). In extreme cases, this sensitivity is species specific (e.g., *Juncus*) and results in it almost never being found in pollen assemblages (Peglar et al., 2001).

14.2.2 Pollination strategy and influence on pollen distribution

Understanding that pollen is rarely an *in situ* proxy is critical for ecological and climatic interpretations based on the pollen record from sediments. The sedimentary record is biased toward plants with specific pollination strategies. The two most common pollination syndromes are anemophily (wind) and entomophily (insect) (Proctor et al., 1996). Anemophilous plants produce large amounts of pollen; a single floret can produce over 50,000 grains, with the average pollen rain from a single plant as high as 2100 grains/cm^2/year (Proctor et al., 1996). These pollen grains can sometimes be carried hundreds of kilometers (Proctor et al., 1996). For example, pine pollen can be transported as far as 750 km (Erdtman, 1952). However, the majority of anemophilous pollen is deposited closer to the source plant (Proctor et al., 1996). The concentration of individual taxa in the pollen rain is influenced by the abundance of those taxa in the landscape and the response to precipitation or other environmental factors (Proctor et al., 1996).

Entomphilous pollen is not readily dispersed by wind because it is not aerodynamic and flower architecture only allows pollen to be removed from the anthers by insect manipulation (Proctor et al., 1996). Once pollen is removed by an insect, the dispersal of pollen becomes heavily influenced by density and size of plant population (Bernhardt et al., 2008).

The sedimentary pollen record is biased towards anemophily because it is produced in large quantities and dispersed over broad areas, whereas pollen from insect and animal pollinated plants is less abundant and rarely deposited in sediments. Understanding the fact that pollen and spores are dispersed over a range of distances from the source plant is critical for accurate interpretation of pollen assemblage data, as well as requiring extensive surface sample calibration between standing vegetation and sedimentary pollen assemblages.

14.2.3 Pollen transport to sediments

Pollen is deposited through several mechanisms, including atmospheric pollen rain and aquatic transport. Calibration studies of pollen rain are useful to reconstruct the depositional environment (e.g., Ayyad et al., 1992). Depending on pollen size and shape, it will settle out of the air; the differing velocities affect the distance it will settle out of the air from the source plant (Jackson and Lyford, 1999). Airborne pollen can be sorted by "interception" where taller vegetation traps airborne pollen more than shorter vegetation (Roe and van de Plassche, 2005).

Pollen can be sorted by water currents and redistributed by marine-estuarine processes (Brush and Brush, 1994; Holmes, 1994). Reworking also occurs in coastal settings due to the erosion of basal layers due to tidal currents (Traverse, 1994; Freund et al., 2004). Reworked grains are sometimes visually differentiated due to corroded or damaged exines or because the taxa observed do not belong in the geologic time period of the sample analyzed. For example, extinct Miocene taxa observed in Holocene-aged sediment would indicate reworking. Certain

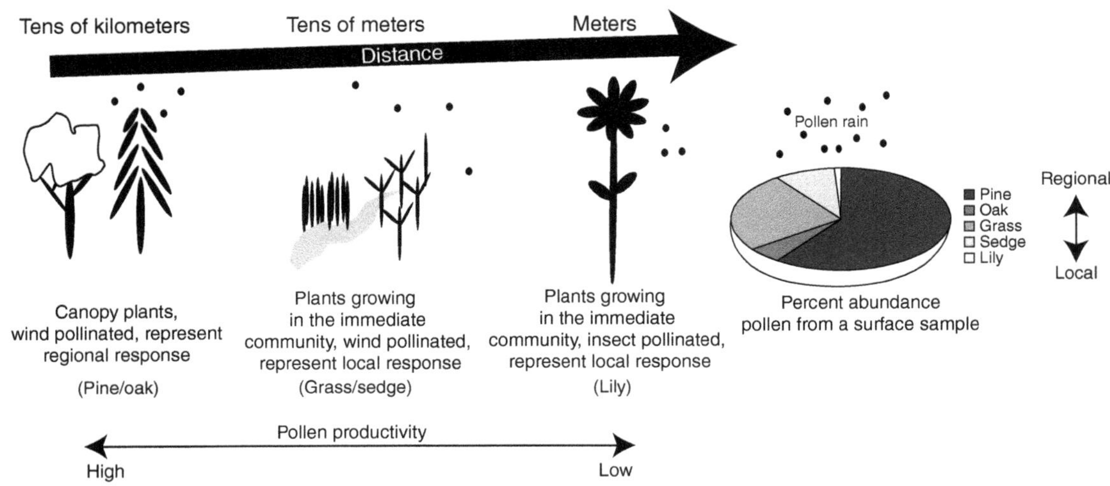

Fig. 14.1. Dispersal factors influencing a hypothetical surface sample. The pie chart comprises five pollen taxa (pine, oak, grass, sedge, and lily). Pollen from wind-pollinated plants with high pollen productivity can dominate assemblages but are transported in from tens of kilometers and represent a regional signature. Insect-pollinated plants rarely reach high abundances in an assemblage, and are usually indicative of very localized conditions.

depositional environments such as deltas transport pollen great distances, and the assemblage found there could represent an entire drainage basin (Chmura and Liu, 1990).

To illustrate the complexity described above, we present a hypothetical pollen assemblage from a surface sample from the Everglades and explain possible pollen sources (Fig. 14.1). The assemblage could have extremely localized pollen, such as that from a nearby insect pollinated plant (e.g., lily pollen). This pollen can be rare because so little of it makes it into the pollen rain. Pollen in the assemblage can represent the local environment on a scale of tens of meters, like grass and sedge pollen (Fig. 14.1). In patterned wetlands, such as the ridge and slough landscape in the Everglades, pollen assemblages from two wetland community types just meters apart can be statistically differentiated based on the local pollen contribution (Bernhardt and Willard, 2009). Regional pollen rain can represent how the greater community is responding to changes. Extralocal pollen comes from distances of tens of kilometers and provides a context of regional changes; however, it may not provide information about changes at the specific site being examined. Cores from marshes in the central Everglades contain high amounts of pine pollen, but pines do not grow in the marshes (Bernhardt and Willard, 2009).

14.3 METHODOLOGY

14.3.1 Isolating pollen from sediments

Pollen is isolated from sediments by digesting clastic and non-palynomorph components and concentrating organic microfossils on microscope slides. Although differences in sediment type (carbonate, silicate, organic) require minor modifications of processing techniques, we outline common methodologies used to isolate palynomorphs from sediment (Traverse, 2007). Splits of wet sediment collected for pollen should be dried at 50 °C. In general, highly organic, fine-grained sediments (i.e., peats, organic muds) contain the highest concentrations of pollen, and oxidized coarse-grained sediments (i.e., sands, limestones) contain the lowest. The amount of sample needed for analysis will therefore vary. Sample requirements range from 0.5 g for peat, to 8–15 g of organic, clay-rich sediment, to well over 60 g of sandy sediments.

For each sample, a tablet of *Lycopodium* spores (or other type of marker such as *Eucalyptus*) should be added to calculate palynomorph concentration (grains g^{-1}). Concentration is determined by the following calculation (Maher, 1981):

$$\text{Pollen concentration} = \frac{\dfrac{\text{Total pollen}}{\text{No. } \textit{Lycopodium} \text{ spores counted}} \times \text{No. } \textit{Lycopodium} \text{ spores per tablet}}{\text{Sample weight}}.$$

Samples are digested in beakers with hydrochloric acid (HCl) and hydrofluoric acid (HF) to remove carbonates and silicates, respectively. In general, samples are left to sit in HCl overnight before neutralizing with deionized water. Dilute HCl (2%) is preferred if the samples have a high percentage of carbonate material. This process is slower with the addition of more washes with dilute HCl to dissolve the carbonate fraction, but the reaction is more controlled using this method. The neutralized samples are covered by HF overnight before a second neutralization with deionized water washes. The remaining residue is transferred to a centrifuge tube, and glacial acetic acid is added. This step ensures sample dehydration before acetolysis. The samples are centrifuged and decanted. Cellular contents are then removed during acetolysis, in which the samples are treated with 1 part sulfuric acid: 9 parts acetic anhydride in a boiling water bath for 10 minutes. Glacial acetic acid is added to stop the reaction, and the sample is centrifuged, decanted, and neutralized with deionized water. Neutralized samples then are treated with 10% KOH for 15 minutes in a water bath at 70 °C. After neutralization with deionized water, residues are sieved with 149 μm and 10 μm nylon mesh to remove the coarse and clay fractions, respectively. When necessary, samples are swirled in a watch glass to remove mineral matter. After staining palynomorphs with Bismarck Brown or another stain, residues are mounted on microscope slides in glycerin jelly. Usually two slides provide a sufficient amount of material for analysis. Other mounting media, such as silicon oil or Canada balsam, may be used. These have different requirements for final processing (outlined in Doher, 1980). Mounting media have different affects on palynomorph size, slide permanence, and mobility of grains on the microscope slide, and the goals of individual studies largely govern the choice of mounting medium. Any additional residue remaining after mounting can be stored and archived in a "residue vial". Pipette remaining residue into a glass vial with a screw cap with a mixture of a 3–4 mL of 25% ethanol (prevents fungal growth) and 2 drops of glycerin jelly. We provide a detailed processing sheet that can be modified and used for tracking laboratory procedures performed on a sample (Fig. 14.2). Detailed notes such as these are essential for keeping track of which step the analysis is at throughout processing, for replicating processing techniques when filling intervals from a previously analyzed core, or resampling a site. They also provide a baseline to return to if procedures need to be adjusted in order to improve the quality of the final slides.

Using a light microscope with at least 400× magnification, at least 300 pollen grains and spores are counted from each sample to determine percent abundance and concentration of palynomorphs.

14.3.2 Tools for pollen and spore identification

Morphological distinction of pollen types was pioneered by Potonié in the early 20th century (Traverse, 2007). Pollen morphology is important because taxa are distinct (Fig. 14.3) and can be taxonomically traced to the "mother" plant taxon. Pollen is primarily identified by the ornamentation of the exine, number and type of apertures, and size of the grain (Fig. 14.3). However, even with this information, some taxa are only discernible to a familial level, which can hinder environmental interpretations. For example, Freund et al. (2004) discuss how the inability to distinguish between the different Cyperaceae generally makes it is difficult to determine marsh elevation based only on pollen.

Many volumes describe the nomenclature for pollen and spore ornamentation and morphology (Erdtman, 1943, 1952; Tschudy and Scott, 1969; Kremp, 1965; Kapp, 2000; Traverse, 2007; Hesse et al., 2009). Herein, we do not attempt to address the finer complexities of morphological descriptions (i.e., defining a furrow v. a colpus); rather we provide the basic terminology and illustrations needed to understand descriptions of pollen and spores. Because spores of lower vascular plants such as ferns also are encountered during analysis of pollen assemblages, we provide basic spore-specific terminology (Fig. 14.3; Tables 14.1 and 14.2).

Detailed reference collections and pollen atlases are vital for confident identification of pollen from sediments. Pollen reference collections should be made from vouchered herbarium sheets to ensure correct identification of source plants. The details of building a reference collection are laid out in Traverse (2007). Many pollen atlases have been compiled, and Hooghiemstra and van Geel (1998) provide a comprehensive list

Sediment Sample Processing Sheet

Field/Core number:_______________ **Location**:_______________________
Depth:______________________
Analysis start date:_________________
Special instructions:____

1. Set up

A. Lithology:______________________________ — Circle One
B. Sample weight (grams):____________________ — dry / moist / wet
C. Physical disaggregation method: — morter/ pestal / hand
D. Lycopodium tablet added: — yes / no

Batch #:__________________________

*place sample in a labeled 250, 400, 04 1000 ml polyethylene beaker

2. HCl

A. Add HCl: Dilute (2%) or Concentrate (10%)

*enough to cover sample
*allow to sit overnight (~12 hours)

B. DI water wash to neutral pH:__________ times

*allow sample to settle between washes

C. Once pH is neutral, pour/siphon off water

3. HF

A. Add HF (49%)

*enough to cover sample
*allow to sit overnight (~12 hours)

B. DI water wash to neutra pH:__________ times

*allow sample to settle between washes

C. Once pH is neutral, pour/siphon off water

*transfer sample to labeled 50ml centrifuge tube

4. Glacial Acetic Acid Wash

2 times:_____________

*centrifuge at 1500 RPM for 10 minutes between washes
*decant liquid between washes

5. Acetolysis

A. Add acetolysis solution

*9 parts Acetic Anhydride : 1 part Sulfuric Acid
*make fresh immediately prior to using
*add enough to cover sample plus 5ml

B. Heat 10 minutes at 100° C in hot water bath
C. Remove from hot water bath

*immediately go to step 6

6. Glacial Acetic Acid wash

A. Add glacial to sample in acetolysis solution
B. Centrifuge at 1500 RPM for 10 minutes
C. Decant liquid

7. DI water wash to neutral:__________ times

*centrifuge at 1500 RPM for 10 minutes between washes
*decant liquid between washes

8. 10%KOH

A. Heat 15 minutes at 100° C in hot water bath

*add enough to cover sample plus 5ml
*immediately go to step 9

9. DI water wash to neutral:__________ times

*centrifuge at 1500 RPM for 10 minutes between washes
*decant liquid between washes

10. Sieve

Screen sizes:_____________ microns
Save coarse fraction: — yes / no

*transfer sieved sample to labelled 25ml centrifuge tube
*centrifuge at 1500 RPM for 10 minutes

11. Stain — yes / no

Type:____

*three to five drops, let sit for approximately 2 minutes
*fill centrifuge tube with deionized water or 25% ethanol

12. Swirl — yes / no

*swirl and separate heavy and light material
*pipette light material and return to centrifuge tube
*centrifuge at 1500 RPM for 10 minutes
*decant liquid

13. Number of slides produced:_______________________________________
Mounting medium:____

14. Residue 3 dram storage vial — yes / no

Processing Comments:____
____
____

Fig. 14.2. Example processing sheet

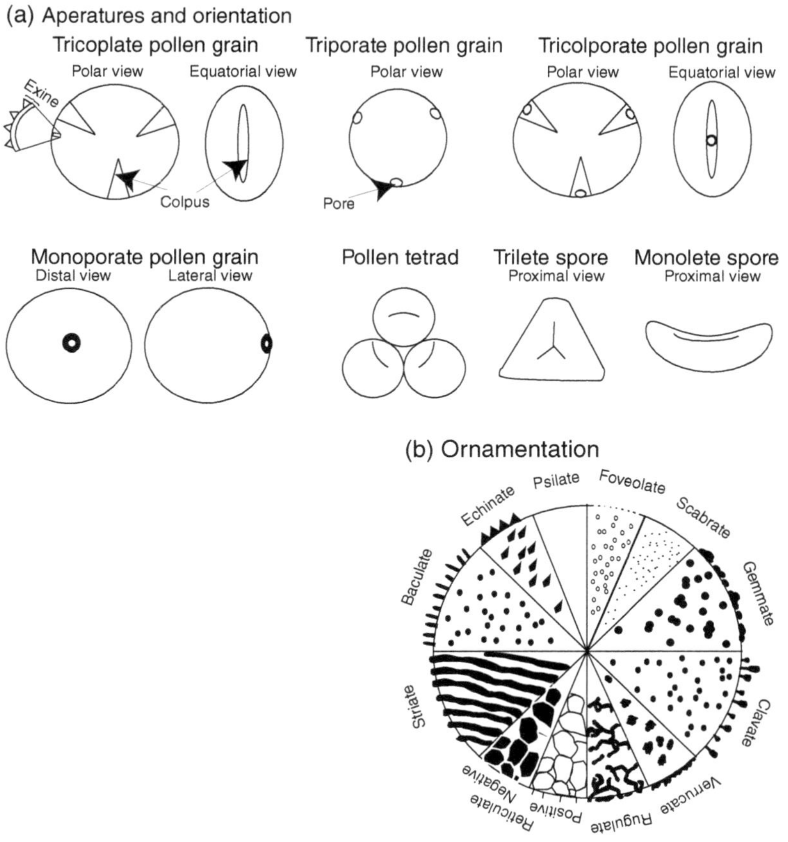

Fig. 14.3. Characteristics used in identification of pollen grains and fern spores. (a) Several of the major aperture types are illustrated in two orientations (polar/equatorial for pollen): tricolporate, triporate, tricolpate, monoporate, as well as a pollen tetrad. The two major spore aperture types are illustrated in the proximal orientation: trilete and monlete. (b) Eleven major ornamentation types are illustrated: psilate, foveolate, scabrate gemmate, clavate, verrucate, rugulate, reticulate (positive and negative), striate, baculate, and echinate. Illustrations are modified from Traverse (2007) and Kremp (1965). *Source*: Photographs by Christopher Bernhardt.

Table 14.1. Common aperture types

Type	General definition	Common example
Bisaccate	Two air sacs	*Pinus*
Inaperaturate	No distinct apertures	*Tsuga*
Ulcerate	Weakly defined aperture, a thinning in the exine	*Carex*
Monoporate	One pore	*Zea*
Dicolpate	Two colpi	*Pontedaria*
Tricolpate	Three colpi	*Acer*
Tricolporate	Three colpi and three pores	*Rhizophora*
Stephanocolporate	More than three colpi and pores equatorially arranged	*Batis*
Diporate	Two pores	*Itea*
Triporate	Three pores	*Betula*
Stephanoporate	More than three pores equatorially arranged	*Pterocarya*
Periporate	Numerous pores scattered across the surface	*Amaranthus*
Fenestrate	Large window-like patterning	*Polygonum*
Sulcate	Elongate distal aperture	*Nymphaea*
Tetrad	A group of four or more grains	*Typha*
Monolete	Spores with one lasura	*Polypodium*
Trilete	Spores with three lasura	*Acrostichum*

Table 14.2. Ornamentation (modified from Faegri and Iversen, 1964)

Term	General definition	Common example
Psilate	Even surface or pitted with pits smaller than 1 μm in diameter	*Zea*
Foveolate	Surface pitted, diameter of pits greater than 1 μm	*Amaranthus*
Scabrate	Radial projections	*Carex*
Gemmate	Sculpturing element rounded, and lower part constricted and diameter greater than its height	*Taxodium*
Clavate	Constricted lower part of the sculpturing element and the element is rounded and higher than largest diameter	*Ilex*
Verrucate	The lower part of the element is not constricted and it is wider than it is tall	*Liriodendron*
Baculate	The lower part of the element is not constricted and it is taller than it is wide	*Nymphaea*
Echinate	A pointed sculpturing element	*Ambrosia*
Rugulate	Irregulary distributed elements that are elongated	*Ulmus*
Striate	Regularly distributed sculpturing elements	*Acer rubrum*
Reticulate	Sculpturing elements for a reticular pattern	*Polygonum*

of pollen atlases from around the world. Two atlases of particular relevance to sea-level research were published after Hooghemiestra and van Geel (1998). Mao et al. (2012) provide detailed color micrographs of mangrove pollen and other taxa relevant to southeast Asian coasts. Willard et al. (2004) covers mangrove pollen as well as wetland plants from Everglades which include taxa common along coastlines of the southeastern US and Caribbean Islands. Online pollen and spore atlases and databases are also useful references as they usually contain more images than a print atlas and are continuously updated with new taxa. A drawback of relying on atlases found online is that the website can be removed from the web. Two online resources for pollen and spore images are PalDat Palynological Database (Buchner and Weber, from 2000) and the Australasian Pollen and Spore Atlas (APSA Members 2007). In this chapter, we present images of select palynomorphs that are typically encountered in wetland sediments, exemplify specific morphological features for identification, and/or are examples of taxa used in age control (Figs 14.4 and 14.5).

14.3.3 Pollen morphology

Generally, pollen identification is based on: aperture type and number, exine ornamentation, and size and orientation (Fig. 14.3; Tables 14.1 and 14.2). These apertures (pores and/or colpi) are a weakening in the exine where the pollen tube emerges. The number of each and their arrangement can be diagnostic of specific plant families, genera, or species. For example, pollen grains with only one pore are generally found within the Poaceae (Figs 14.3 and 14.4). Specific differences in the morphology of an aperture can be used to distinguish similar pollen grains. For example, *Betula*'s characteristically distinct vestibulate pore distinguishes it from other triporate grains (Fig. 14.4). The number of apertures between grains within a genus can occasionally allow identification to the species level. For example, pore frequency can be used to differentiate between four-pore and five-pore types of *Alnus* (Martin, 1958). Table 14.1 provides a list of the terminology used for describing aperture type, number, and orientation.

Another characteristic used to identify pollen is sculpturing of the exine (Fig. 14.3). In several

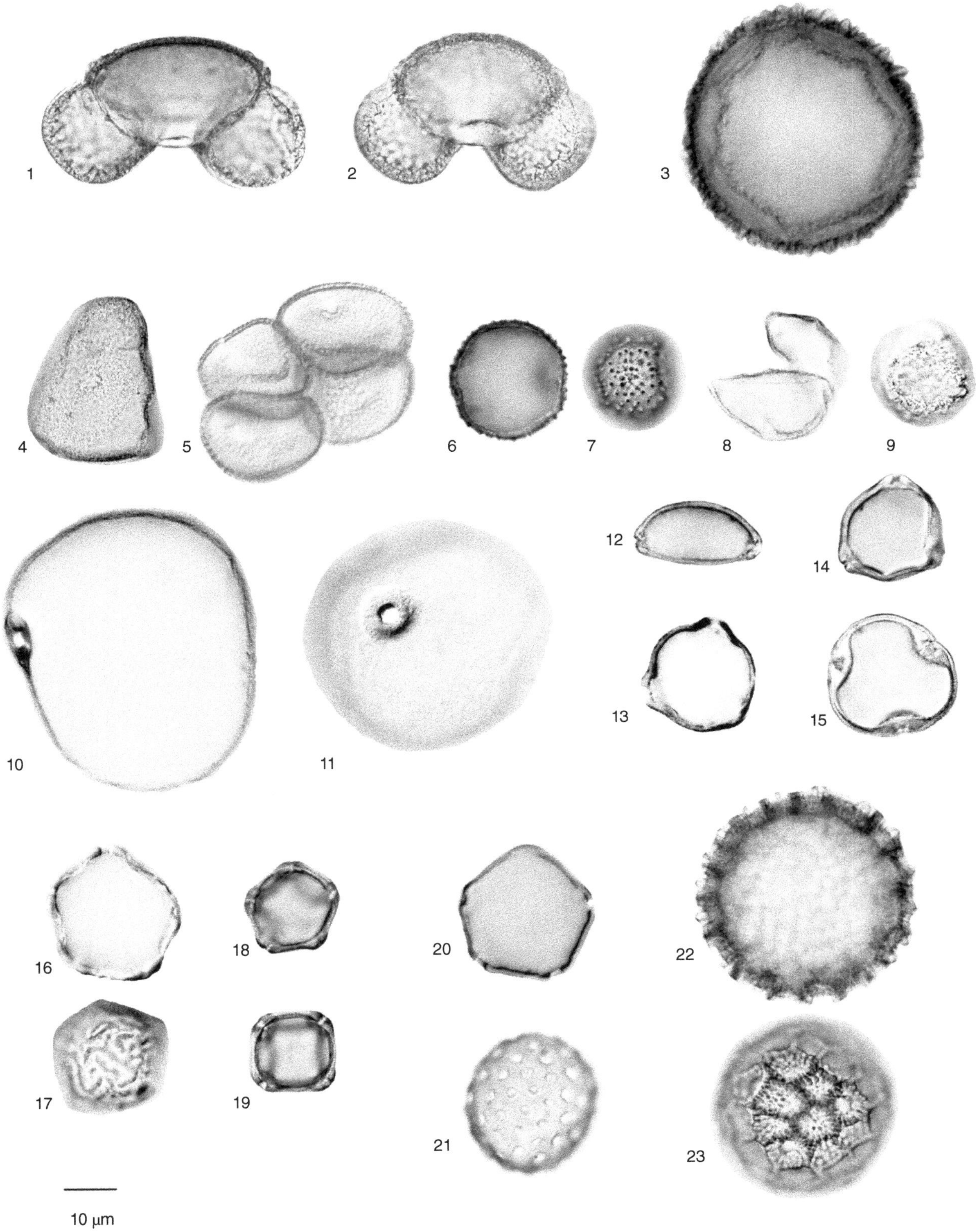

Fig. 14.4. Some of the major pollen grains found in coastal wetland sediments or used in age control. The selection also includes common aperture and sculpture types. Images are photographed at 1000× using a light microscope. The slide medium is glycerin jelly. The scale bar is 10 μm. 1, 2. *Pinus* sp.; 3. *Tsuga candensis*; 4. *Carex* sp.; 5. *Typha latifolia*; 6, 7. *Sagittaria* sp; 8, 9. *Taxodium distichum*; 10, 11. *Zea mays*; 12. *Itea* sp.; 13. *Casuarina equisetifolia*; 14. *Betula sp.*; 15. *Tilia sp.*; 16, 17. *Ulmus sp.*; 18, 19. *Alnus sp.*; 20. *Pterocarya sp.*; 21. *Amaranthus* sp.; 22, 23. *Polygonum sp. Source*: Photographs by Christopher Bernhardt.

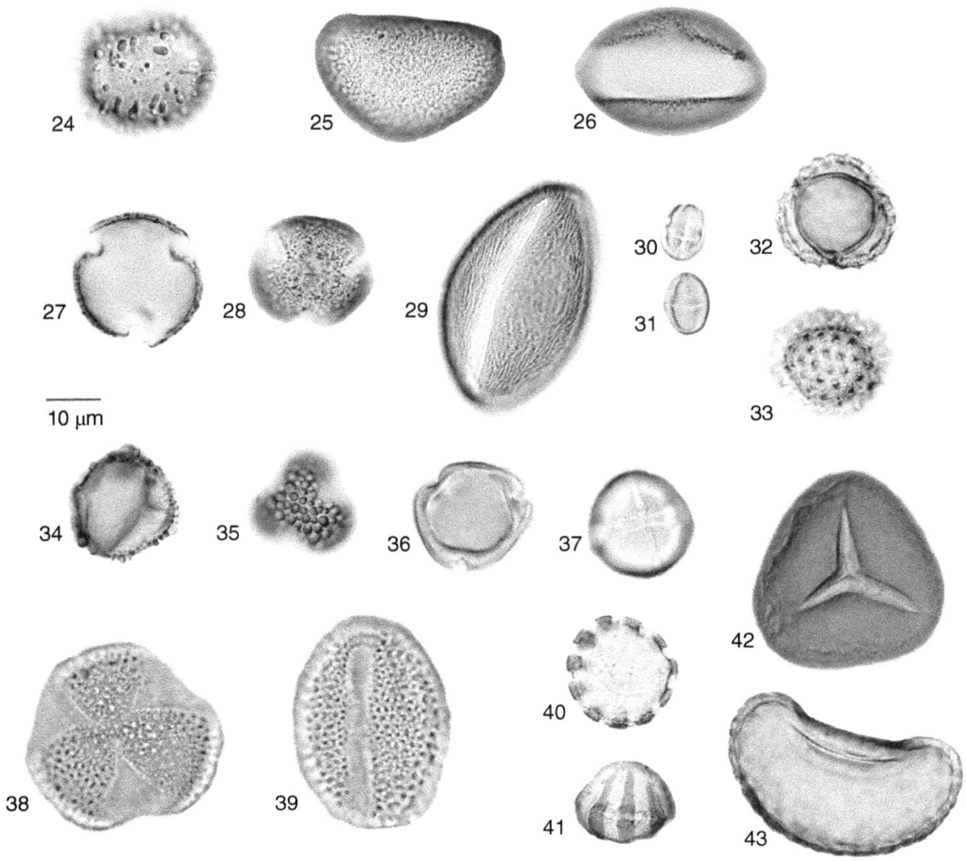

Fig. 14.5. Pollen images continued. 24. *Nymphaea* sp; 25, 26. *Liriodendron tulipifera*; 27, 28. *Quercus*; 29. *Acer rubrum*; 30, 31. *Castanea*; 32, 33. *Ambrosia*; 34, 35. *Ilex* sp.; 36, 37. *Rhizophora* sp.; 38, 39. *Avicennia* sp.; 40, 41. *Utricularia* sp.; 42. *Acrostichum*; 43. *Polypodium* sp. *Source*: Photographs by Christopher Bernhardt.

cases, the features on the exine alone can be diagnostic of specific pollen types. For example, echinate pollen grains generally fall within the Asteraceae (Fig. 14.5). A list of sculpturing types with basic definitions is listed in Table 14.2.

Species level identification can be made for some taxa based on size measurements. For example, in eastern North America, three *Nyssa* species (which are identical morphologically) can be distinguished based on mean polar and equatorial diameters. Size measurements on individual features, such as body or bladder size for pine pollen, is also a potential means of identifying grains to species level (Martin, 1958). Factors such as slide medium type (glycerin jelly v. silicon oil) and *in situ* preservation conditions can however influence pollen size.

Finally, pollen and spores have distinct axes of orientation: polar and equatorial for pollen; and distal and proximal for spores. The orientation on the slide will affect how morphological features are viewed (Figs 14.3–14.5).

14.3.4 Quantitative analysis of pollen assemblages

Pollen data are usually displayed graphically in a pollen diagram with percent abundance of taxa along the *x*-axis and either depth or age on the *y*-axis. Software typically used for analyzing and graphing pollen data include: psimpoll and pscomb (Bennett, 1994); TILIA and TILIA*GRAPH (Grimm, 1990); and C2 (Juggins, 2007). Pollen zones are identified using a combination of visual inspection and multivariate analyses such as cluster analysis (Grimm, 1987) and broken stick model to determine the number of zones to assign (Bennett, 1996). To facilitate a more holistic interpretation of vegetation dynamics through time, palynologists often group taxa into "functional groups" that integrate information on the life history, pollination biology, and taphonomy. This technique combines all pollen taxa counted into discreet ecological groups and is meant to ease interpretation of the traditional pollen diagram, where each taxon is plotted separately, and also to include taxa of <2% abundance that are excluded from the original pollen diagram (Bernhardt and Willard, 2009).

Further numerical statistics can be applied to interpret downcore changes in pollen assemblages (see Birks, 1995). The number of individual pollen grains can vary from sample to sample within a core and between cores, and simple comparisons of diversity can become biased by more taxa being found in samples with more individuals counted or overall higher total pollen sum (Weng et al., 2006). To account for these phenomena, statistical techniques such as rarefaction can be used to calculate sample diversity (Birks and Line, 1992). Rarefaction allows for comparison of diversity by creating a standardized pollen sum and calculating an expected number of taxa encountered in that sample. However, taphonomy and pollination strategy (wind v. insect) can affect species evenness and bias the detection of rare taxa, consequently affecting the rarefaction results (Peros and Gajewski 2008).

Vegetation change can be compared statistically within and between sites using detrended correspondence analysis (DCA), or detrended canonical correspondence analysis (DCCA). Results from DCA of pollen assemblages can be used to discern changes in vegetation communities, for example an intertidal flat that gradually converts to a raised bog through time (Shennan et al., 1995). DCCA can be used to interpret vegetation turnover and variability through time (Birks, 2007). Detrended canonical correspondence analysis reports the amount of vegetation turnover in standard deviation units (SD). The turnover (SD) calculated by DCCA is equivalent to beta-diversity and therefore illustrates the usefulness of the SD metric as an ecologically interpretable unit. Values of 0 indicate no vegetation turnover while values of 4 indicate complete community turnover. Further statistical comparison of fossil assemblages to modern assemblages is made using the modern analog technique (Overpeck et al., 1985) or transfer functions (Engelhart et al., 2007).

14.4 APPLICATION OF POLLEN TO SEA-LEVEL RESEARCH

14.4.1 Dating: pollen biostratigraphy

Robust age models are increasingly more essential as the sample resolution and accuracy of sea-level reconstruction improve. The use of pollen from sediment cores for sea-level research can help confirm and support age models as well as provide age information where outside the range of other geochronologic techniques (Roe and van de Plassche, 2005; Gehrels, 2007). Godwin (1940) used changes in relative abundances between several key plant taxa to establish age control in the absence of radiometric dates, in order to reconstruct Holocene sea level in the English Fenlands. While used frequently for relatively recent sediments (the last ~300 years), this type of biostratigraphy has a long history and has been applied to much older sediments (e.g., sea-level reconstructions for the Eemian interglacial in the Netherlands and North Sea; Zagwijn, 1983). Pollen "chronohorizons", a specific age date associated with the change in abundance of specific pollen taxa, are determined in several ways. Table 14.3 provides a select list of examples of major chronohorizons.

The combination of the written historical record of land use, radiometric dating, and the ecologic response of specific plant taxa is used to establish age control for specific regions. In eastern North America, the increase in *Ambrosia* pollen is tied to land clearance. Based on the local historical record, this increase in *Ambrosia* pollen can be tied to specific dates (Brush, 1989; Cooper et al., 2004). In sea level research, Kemp et al. (2009) used this technique to fill the age gap between ^{210}Pb and ^{14}C dating to construct their age model. The planting of specific tree species such as pine, which increases their contribution to the general pollen rain, has been successfully used as a chronohorizon in sea-level research (Long et al., 1999).

Chronohorizons also are associated with the introduction of new species to a region. Plants can be introduced to a region by natural migration or by humans. For example, *Casuarina* is a non-native plant to North America and was introduced to South Florida in the early 1900s. The presence of its pollen indicates that sediments that contain *Casuarina* are not older than ~1930 AD when the pollen became common in sediments (Willard et al., 2001). The introduction of cereal crops is a potential chronohorizon, but the timing varies

Table 14.3. Examples of pollen chronohorizons

Pollen	Change	Location	Time period	Reference
Castanea	Disappearance	Mid-Atlantic US	~1930 AD	Brush (1989)
Castanea	Decrease	Mid-Atlantic US	~1910 AD	Brush (1989)
Casuarina	Appearance	Florida/Egypt	20th century	Wingard et al. (2003) Mehringer et al. (1979)
Ambrosia	Increase	Eastern US	Post-colonial (exact time varies between locations)	Brush (1989)
Tsuga	Decrease	Eastern US	5.4 ka	Bennett and Fuller (2002)
Pteracarya	Disappearance	US	Plio-Pleistocene boundary	Various
Ulmus	Decrease	Europe	5 ka	Peglar and Birks (1993)
Ceralia	Appearance	Global	Varies with location	Various

depending on location because some cultures established agriculture at different periods. Extinction or expatriation events also provide age control points. *Pterocarya*, which is still extant in China, became extinct in North America at the end of the Pliocene and serves as a good age marker for the Plio-Pleistocene border. More recently, the disappearance of *Castanea* in the eastern United States is a chronohorizon for ~1930 AD (Brush, 1989).

Abrupt decreases in percent abundances due to plant disease characterize other chronohorizons. The decrease in *Tsuga* pollen in the eastern US and the decrease in *Ulmus* pollen in Europe mark mid-Holocene events (Peglar and Birks, 1993; Bennett and Fuller, 2002). The chestnut blight of the early 1900s AD is evident in sediments from North America (Brush, 1989).

Finally, the ratio between two taxa is used to define approximate chronohorizons. The change in the ratio of *Pinus* to *Quercus* pollen (*Quercus* becoming dominant) in central to northeast North America places the sediments age to at least 10.5 ka (Watts, 1979).

14.4.2 Application of pollen evidence to sea-level reconstruction

The relationship of pollen with its source vegetation, and that vegetation's response to changing environmental characteristics, has been used to reconstruct sea level for many years (Godwin, 1940). Godwin's (1940) pioneering work was the first to use pollen for sea-level reconstruction. Today, a common means to use pollen to reconstruct sea level is to create a transfer function (Engelhart et al., 2007; Gehrels, 2007). However, using pollen alone can lead to large errors in vertical elevation (Gehrels, 2007). The utility of pollen is perhaps in its ability to allow researchers to differentiate different vegetation zones, such as upland versus low marsh (Roe and van de Plassche, 2005) and to differentiate different peat types, for example salt versus freshwater peat (Davies, 1980; Freund et al., 2004). Used in conjunction with other proxy data, pollen can be a powerful tool for reconstructing sea level (Freund et al., 2004).

One example of an environment where pollen is successful in reconstructing past sea level is mangrove swamps. Mangrove is a community definition for plants growing in brackish to saline environments, which spans several plant families and divisions. Mangroves have very specific habitat requirements influenced by salinity and, as a result, exhibit strong zonation. Further, in most cases the pollen (and spores) from mangroves is not transported far from the source plant, making mangrove pollen a useful indicator of sea level. Changes in mangrove pollen zonation and the alternation of mangrove and freshwater taxa indicate changes in salinity regime and are related to long-term changes in sea level (Ellison, 2005; Cohen et al., 2008). However, there are many factors that influence the percent abundance of mangroves in surface samples that are unrelated to sea level (Urrego et al., 2010). Well-calibrated surface sample datasets can account for this variability and produce sea-level transfer functions with errors as precise as ±0.22 m (Engelhart et al., 2007).

The mangrove marshes that fringe the southern Everglades and the buried peats of Florida Bay act as a practical setting to demonstrate the past 4.0 ka of sea-level variability and the pollen record. We collected a piston core from a marsh on Cape

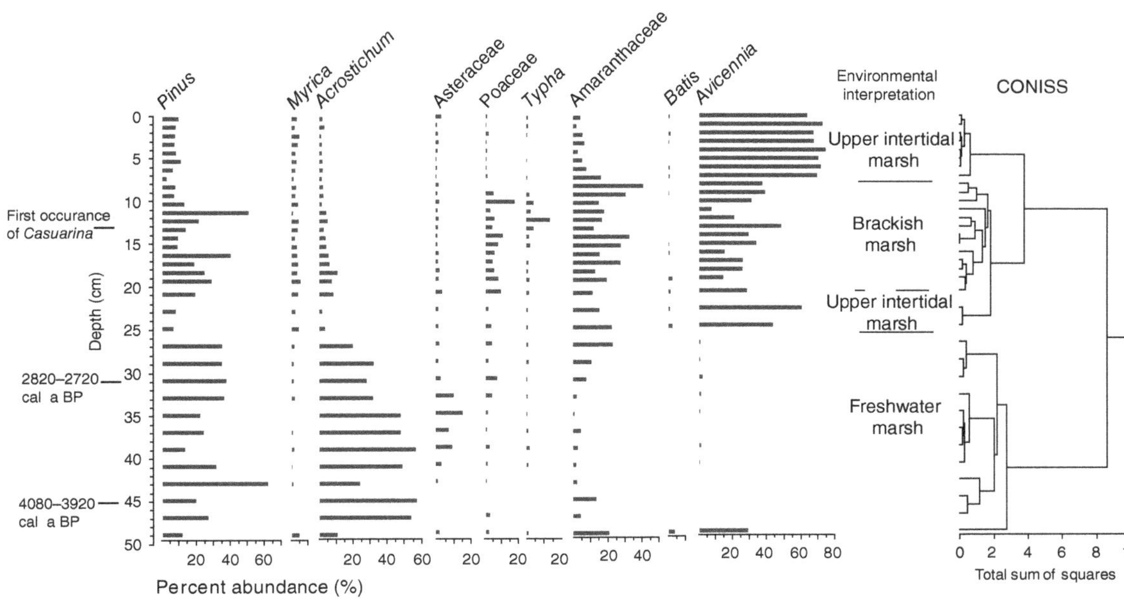

Fig. 14.6. Pollen diagram of select taxa from a sediment core collected on Cape Sable, Fl. Environmental interpretations for each zone are determined from calibration with modern surface samples (Willard et al., 2001).

Sable, Florida. Pollen was isolated, identified, and analyzed as discussed above. The basal section of the core, representing 4–2.8 ka, comprised fern spores and *Pinus* pollen; this was interpreted as freshwater marsh infrequently flooded by high tides (Fig. 14.6; Willard et al., 2001). This is consistent with geologic records of regional sea level. Early Holocene sea level was ~6.2 m below mean sea level (Wanless et al., 1994), leaving Florida Bay exposed. During this time freshwater peats were accumulating on the Florida Platform (Scholl et al., 1969). As rates of sea-level rise slowed by 3.5 ka, mangrove forests began to establish along the coastlines (Wanless, 1989; Wanless et al., 1994). In the Cape Sable record, at ~3 ka there is an abrupt shift from a less-frequently flooded, less-salt-tolerant marsh dominated by *Acrostichum* to an upper intertidal marsh dominated by *Avicennia* pollen. *Avicennia* quickly declines and a brackish marsh characterized by Amaranthaceae, most likely *Salicornia*, Batis, Poaceae, and *Avicennia* pollen, is established. Roe and van de Plassche (2005) found that, based on pollen size (specifically, diameter), *Phragmites*, *Spartina patens*, and *Spartina alfterniflora* can be differentiated from one another. They could then use the differentiation of these taxa to make interpretations about marsh elevation. We did not find significant differences between sizes and did not attempt to differentiate them here.

We use the first occurrence of *Casuarina* as a chronohorizon for ~1910–1930 AD. *Avicennia* pollen abruptly increases during this time period, indicating increased salinity (Fig. 14.6). Regionally, the rate of sea level increased during this time period which could have influenced the transition to more saline marsh conditions (Wanless and Vlaswinkel, 2005). However, during this same time period, canals were cut across Cape Sable that allowed more saline water to enter from Florida Bay (Wanless and Vlaswinkel, 2005).

This example demonstrates the utility of pollen in sea-level research: (1) changes in the relevant abundance of pollen and spores from specific taxa provide information on sea level; (2) if a detailed pollen assemblage-elevation calibration dataset for the area is available, precise elevation estimates can be generated; and (3) the first occurrence of *Casuarina* pollen added age control where radiometric data (e.g., ^{210}Pb) were unavailable.

14.5 CONCLUSION

This chapter provides a concise framework for using pollen and spores from terrestrial plants in sea-level research. Concepts presented early on in the chapter, such as a plant's pollination strategy and pollen transport and deposition, are critical to interpreting assemblages observed in sediments.

This chapter provides a standard methodology to process most sediments encountered during sea-level research. We provide illustrations and images of key morphological features to aid in the identification of major palynomorphs relevant to sea-level research. Citations to print and online pollen atlases are provided as well as references to classic texts on palynology.

Pollen and spores allow for robust statistical examination changes in the regional and local vegetation community, and several common statistical methods have been discussed. Regional changes in vegetation, such as the appearance/disappearance of a plant species, can act as a chronohorizon and support age models from cores used in sea-level research. The chapter ends with a case study from South Florida to illustrate a practical example of how pollen can be used to study sea level.

14.6 DISCLAIMER

Any use of trade, firm, or product names is for descriptive purposes only and does not imply endorsement by the US Government.

REFERENCES

APSA Members (2007) The Australasian pollen and spore atlas V1.0. Australian National University, Canberra. Available at http://apsa.anu.edu.au/ (accessed 22 July 2014).

Ayyad, S.M., Moore, P.D., and Zahran, M.A. (1992) Modern pollen rain studies of the Nile Delta, Egypt. New Phytologist, 121, 663–675.

Bennett, K.D. (1994) "psimpoll" version 2.23: a C program for analyzing pollen data and plotting diagrams. INQUA Commission for the study of the Holocene: Working Group on Data Handling Methods, Newsletter, 11, 4–6.

Bennett, K.D. (1996) Determination of the number of zones in a biostratigraphical sequence. New Phytologist, 132, 155–170.

Bennett, K.D., and Fuller, J.L. (2002) Determining the age of the mid-Holocene *Tsuga canadensis* (hemlock) decline, eastern North America. The Holocene, 12, 421–429.

Bernhardt, C.E., and Willard, D.A. (2009) Response of the Everglades ridge and slough landscape to climate variability and 20th century water management. Ecological Applications, 19, 1723–1738.

Bernhardt, C.E, Mitchell, R. J., and Michaels, H. J. (2008) Effect of *Lupinus perennis* population size and local density on pollinator behavior. International Journal of Plant Sciences, 169, 944–953.

Birks, H.J.B. (1995) Quantitative paleoenvironmental reconstructions. In *Statistical Modeling of Quaternary Science Data* (eds Maddy, D., and Brew, J.S.). Quaternary Research Association, England, pp. 161–254.

Birks, H.J.B. (2007) Estimating the amount of compositional change in late-Quaternary pollen-stratigraphical data. Vegetation History and Archaeobotany, 16, 197–202.

Birks, H.J.B., and Line, J.M. (1992) The use of rarefaction analysis for estimating palynological richness from Quaternary pollen-analytical data. The Holocene, 2, 1–10.

Brush, G.S. (1989) Rates and patterns of estuarine sediment accumulation. Limnology and Oceanography, 34, 1235–1246.

Brush, G.S., and Brush, L.M. (1994) Transport and deposition of pollen in an estuary: signature of the landscape. In: *Sedimentation of Organic Particles* (ed. Traverse, A.), Cambridge University Press, Cambridge, pp. 33–46.

Buchner, R., and Weber, M. (from 2000) PalDat – a palynological database: Descriptions, illustrations, identification, and information retrieval. Available at http://paldat.org (accessed 22 July 2014).

Chmura, G.L., and Liu, K.-B. (1990) Pollen in the Lower Mississippi River. Review of Palaeobotany and Palynology, 64, 253–261.

Cohen, M.C.L., Lara, R.J., Smith, C.B., Angélica, Dias, B.S., and Pequeno, T. (2008) Wetland dynamics of Marajó Island, northern Brazil, during the last 1000 years. Catena, 76, 70–77.

Cooper, S.R., McGlothlin, S.K., Madritch, M., and Jones, D.L. (2004) Paleoecological evidence of human impacts on the Neuse and Pamlico estuaries of North Carolina, USA. Estuaries, 27, 617–633.

Cronin, T.M. (1999) *Principles of Paleoclimateology*. Columbia University Press, New York.

Davies, T.D. (1980) Peat formation in Florida Bay and its significance in interpreting the recent vegetational and geological history of the bay area. PhD thesis, Department of Biology, the Pennsylvania State University, University Park.

Doher, L.I. (1980) Palynology preparation procedures currently used in the paleontology and stratigraphy laoratories, US Geological Survey, Geological Survey Circular 830.

Ellison, J. (2005) Holocene palynology and sea-level change in two estuaries in Southern Irian Jaya. Palaeogeography, Palaeoclimatology, Palaeoecology, 220, 291–309.

Engelhart, S.E., Horton, B.P., Roberts, D.H., Bryant, C.L., and Corbett, D.R. (2007) Mangrove pollen of Indonesia and its suitability as a sea-level indicator. Marine Geology, 242, 65–81.

Erdtman, G. (1943) *An Introduction to Pollen Analysis*. Chronica Botanica Company, Waltham Mass., 239 pp.

Erdtman, G. (1952) *Pollen Morphology and Plant Taxonomy. Angiosperms*. Almqvist and Wiksell, Stockholm, 539 pp.

Esau, K. (1977) *Anatomy of Seed Plants*, 2nd edition. John Wiley & Sons, New York.

Faegri, K, and Iversen, J. (1964) *Textbook of Pollen Analysis*. Hafner Publishing Company, New York.

Freund, H., Gerdes, G., Streif, H., Dellwig, O., and Watermann, F. (2004) The indicative meaning of diatoms, pollen and botanical macro fossils for the reconstruction of paleoenvironments and sea-level fluctuations along the coast of Lower Saxony; Germany. Quaternary International, 112, 71–87.

Gehrels, W.R. (2007) Sea level studies: Microfossil reconstructions. In: *Encyclopedia of Quaternary Sciences* (ed. Elias, S.), Elsevier, pp. 3015–3024.

Godwin, H. (1940) Studies in the post-glacial history of British vegetation, III: fenland pollen diagrams, IV: post-glacial changes of relative land and sea level in the English Fenland. Philosophical Transactions of the Royal Society of London B, 230, 239–303.

Grimm, E.C. (1987) CONISS: a Fortran 77 program for stratigraphically constrained cluster analysis by the method of incremental sum of squares. Computers & Geosciences, 13, 13–35.

Grimm, E.C. (1990) TILIA and TILIA*GRAPH. PC spreadsheet and graphics software for pollen data. INQUA Commission for the study of the Holocene: Working Group on Data Handling Methods, Newsletter, 4, 5–7.

Hesse, M., Halbritter, M., Zetter, R., Weber, M., Buchner, R., Frosch-Radivo, A., and Ulrich, S. (2009) *Pollen Terminology: An Illustrated Handbook*. Springer, Vienna.

Holmes, P.L. (1994) The sorting of spores and pollen by water: experimental and field evidence. In: *Sedimentation of Organic Particles* (ed. Traverse, A.), Cambridge University Press, Cambridge, pp. 9–32.

Hooghiemstra, H., and van Geel, B. (1998) World list of Quaternary pollen and spore atlases. Review of Palaeobotany and Palynology, 104, 157–182.

Jackson, S.T., and Lyford, M.E. (1999) Pollen dispersal models in Quaternary plant ecology: assumptions, parameters, and prescriptions. The Botanical Review, 65, 39–75.

Juggins, S. (2007) *C2 Version 1.5 User guide*. Software for ecological and palaeoecological data analysis and visualization. Newcastle University, Newcastle upon Tyne, UK, 73 pp.

Kapp, R.O. (2000) *Pollen and Spores*, 2nd edition. The American Association of Stratigraphic Palynologists Foundation Publication.

Kemp, A. C., Horton, B.P., Culver, S.J., Corbett, D.R., van de Plassche, O., Gehrels, W.R., Douglas, B.C., and Parnell, A.C. (2009) Timing and magnitude of recent accelerated sea-level rise (North Carolina, United States). Geology, 37, 1035–1038.

Kremp, G.O.W. (1965) *Morphologic Encyclopedia of Palynology: An Illustrated Collection of Definitions and Illustrations of Spores and Pollen*. University of Arizona Press, Tuscon, Program in Geochronology no. 100.

Long, A. J., Scaife, R.G., and Edwards, R.J. (1999) Pine pollen in intertidal sediments from Poole Harbour, UK; implications for late-Holocene sediment accretion rates and sea-level rise. Quaternary International, 55, 3–16.

Maher, L.J. Jr. (1981) Statistics for microfossil concentration measurements employing samples spiked with marker grains. Review of Palaeobotany and Palynology, 32, 153–191.

Mao, L., Batten, D.J., Fujiki, T., Li, Z., Dai, L., and Weng, C. (2012) Key to mangrove pollen and spores of southern China: an aid to palynological interpretation of Quaternary deposits in the South China Sea. Review of Palaeobotany and Palynology, 176–177, 41–67.

Martin, P.S. (1958) Taiga-tundra and the full-glacial period in Chester County, Pennsylvania. American Journal of Science, 256, 470–502.

Mehringer, P.J., Jr., Petersen, K.L., and Hassan, F.A. (1979) A pollen record from Birket Qarun and the recent history of the Fayrum, Egypt. Quaternary Research, 11, 238–256.

Overpeck, J.T., Webb III, T., and Prentice, I.C. (1985) Quantitative interpretation of fossil pollen spectra: dissimilarity coefficients and the method of modern analogs. Quaternary Research, 338, 87–108.

Peglar, S.M., and Birks, H.J.B. (1993) The mid-Holocene *Ulmus* fall at Diss Mere, South-East England, disease and human impact. Vegetation History and Archaeobotany, 2, 61–68.

Peglar, S.M., Birks, H.H., Birks, H.J.B., Appleby, P.G., Faithi, A.A., Flower, R.J., Kraïem, Patrick, S.T., and Ramdani, M. (2001) Terrestrial pollen record of recent land-use changes around nine North African lakes in the CASSARINA Project. Aquatic Ecology, 35, 431–448.

Peros, C.P., and Gajewski, K. (2008) Testing the reliability of pollen-based diversity estimates. Journal of Paleoliminology, 40, 357–368.

Proctor M., Yeo, P., and Lack, A. (1996) *The Natural History of Pollination*. Timber Press, USA.

Roe, H.M., and van de Plassche, O. (2005) Modern pollen distribution in a Connecticut saltmarsh: Implications for studies of sea-level change. Quaternary Science Reviews, 24, 2030–2049.

Scholl, D.W., Craighead, F.C. Sr., and Stuiver, M. (1969) Florida submergence curve revised: its relation to coastal sedimentation rates. Science, 163, 562–564.

Shennan, I., Innes, J.B., Long, A.J., and Zong, Y. (1995) Holocene relative sea-level changes and coastal vegetation history at Kentra Moss, Argyll, northwest Scotland. Marine Geology, 124, 43–59.

Traverse, A. (ed.) (1994) *Sedimentation of Organic Particles*. Cambridge University Press, Cambridge.

Traverse, A. (2007) *Paleopalynology*, 2nd edition. Springer, Dordrecht.

Tschudy, R.H., and Scott, R.A. (1969) *Aspects of Palynology*. Wiley, New York.

Urrego, L.E., González, C., Urán, G., and Polanía, J. (2010) Modern pollen rain in mangroves from San Andres Island, Colombian Caribbean. Review of Palaeobotany and Palynology, 162, 168–182.

Wanless, H.R. (1989) The inundation of our coastlines: past, present and future with a focus on south Florida. Sea Front, 35, 264–271.

Wanless, H.R., and Vlaswinkel, B.M. (2005) Coastal landscape and channel evolution affecting critical habitats at Cape Sable, Everglades National Park, Florida. Final Report to Everglades National Park, National Park Service, U.S. Department of Interior.

Wanless, H.R., Parkinson, R.W., and Tedesco, L.R. (1994) Sea level control on stability of Everglades wetlands. In: *Everglades: The Ecosystem and its Restoration* (eds Davis, S.M., and Ogden, J.C.), St Lucie Press, Delray Beach, pp. 198–223.

Watts, W.A. (1979) Late Quaternary vegetation of Central Appalachia and the New Jersey Coastal Plain. Ecological Monographs, 49, 427–469.

Weng , C., Hooghiemstra, H., and Duivenvoorden, J.F. (2006) Challenges in estimating past plant diversity from fossil pollen data: statistical assessment, problems, and possible solutions. Diversity and Distributions, 12, 310–318.

Willard, D.A., and Bernhardt, C.E. (2011) Impacts of past climate and sea level change on Everglades wetlands: placing a century of anthropogenic change into a late-Holocene context. Climatic Change, 107, 59–80.

Willard, D.A., Weimer, L.M., and Riegel, W.L. (2001) Pollen assemblages as paleoenvironmental proxies in the Florida Everglades. Review of Palaeobotany and Palynology, 113, 213–235.

Willard, D.A., Bernhardt, C.E., Weimer, L., Cooper, S.R, Gamez, D., and Jensen, J. (2004) Atlas of pollen and spores of the Florida Everglades. Palynology, 28, 175–227.

Wingard, G. L., Cronin, T. C., Dwyer, G. S., Ishman, S. E., Willard, D. A., Holmes, C. W., Bernhardt, C. E., Williams, C. P., Marot, M. E., Murray, J. B., Stamm, R. G., Murray, J. H., and Budet, C. (2003) Ecosystem history of southern and central Biscayne Bay; summary report on sediment core analyses. Open-File Report: US Geological Survey, Report: OF 03-0375. Available at http://sofia.usgs.gov/publications/ofr/03-375/ (accessed 22 July 2014).

Zagwijn, W.H. (1983) Sea-level changes in the Netherlands during the Eemian. Geologie en Mijnbouw, 62, 437–450.

Chapter 15
Diatoms

YONGQIANG ZONG[1] AND YUKI SAWAI[2]
[1]*Department of Earth Sciences, the University of Hong Kong, Hong Kong SAR, China*
[2]*Institute of Earthquake and Volcano Geology (IEVG), National Institute of Advanced Industrial Science and Technology (AIST), Tsukuba, Ibaraki, Japan*

15.1 INTRODUCTION

Diatoms commonly live in aquatic, naturally illuminated environments from hot-springs to wet soils, intertidal mudflats to the open ocean. Their ecological preferences to certain environmental conditions such as water salinity, temperature, and nutrient level make diatoms a useful tool for environmental reconstructions. In coastal areas, many habitats including saltmarsh, tidal flat, and estuarine waters are suitable for diatoms to grow. Their siliceous frustules, or valves, are not susceptible to oxidation and are resistant to depositional processes; diatoms are therefore well preserved in fine-grained sediments. Together with their different shapes and sizes, the intricate patterns of apertures perforated on diatom valves (Round et al., 2007) are the most important characteristics for their identification to species or subspecies level. As different species of diatoms tend to occupy different niches of the environment, changes in diatom assemblages have been used to infer changes in the coastal environment from freshwater to marine water conditions or vice versa (van de Plassche, 1986). For sea-level studies, diatoms have been an important proxy for the evaluation and estimation of the indicative meaning of sea-level index points (Horton and Sawai, 2010). Recently, statistical methods have also been applied to quantitative reconstructions of past sea level.

15.2 CHARACTERISTICS OF COASTAL DIATOMS

Diatoms are microscopic, unicellular algal organisms mostly ranging from 5 μm to 200 μm in size. According to the morphological characteristics of diatom's siliceous valves, a useful classification system (among several others) was developed by Round et al. (1990). Based on this system, Hartley et al. (1996) compiled drawings and published an atlas of British diatoms, and this system was updated by Round et al. (2007). Under this system, diatoms are divided into three main classes and a number of subclasses. Class I (Coscinodiscophyceae) is a group of disc-shaped diatoms, mostly circular and some irregular. Frustules are often connected to each other forming a chain. The arrangement of areola on the valve surface and edge are key features for identification of this type of diatoms (Fig. 15.1). Class II (Fragilariophyceae) are diatoms of an elongate box shape (Fig. 15.2). For Fragilariophyceae, structure and markings of the valves including lines of apertures and width of axial area provide the clues for identification. Class III (Bacillariophyceae) includes pennate diatoms. For identification of Bacillariophyceae, the size and shape of diatom valves are important (Fig. 15.2). Other important features can be found from the arrangement of lines of apertures and the central part (or edge for some genera, such as *Nitzschia* and *Hantzschia*) and two ends of the axial area of diatom valves. A useful guide to the identification of diatoms is an article by Rumeau and Coste (1988) which provides an outline of terminology for description of the structure and features of diatom valves, as well as a step-by-step chart to help identify diatoms to genus level. Reference literature for the identification of coastal diatoms to species level include van der Werff and Huls (1958–1974), Patrick and Reimer (1966, 1975), and Hartley et al. (1996).

Diatoms are useful for environmental reconstructions because of their preferences to a number of environmental factors. For sea-level studies, diatoms' preferences to salinity, substrate, and

Handbook of Sea-Level Research, First Edition. Edited by Ian Shennan, Antony J. Long, and Benjamin P. Horton.

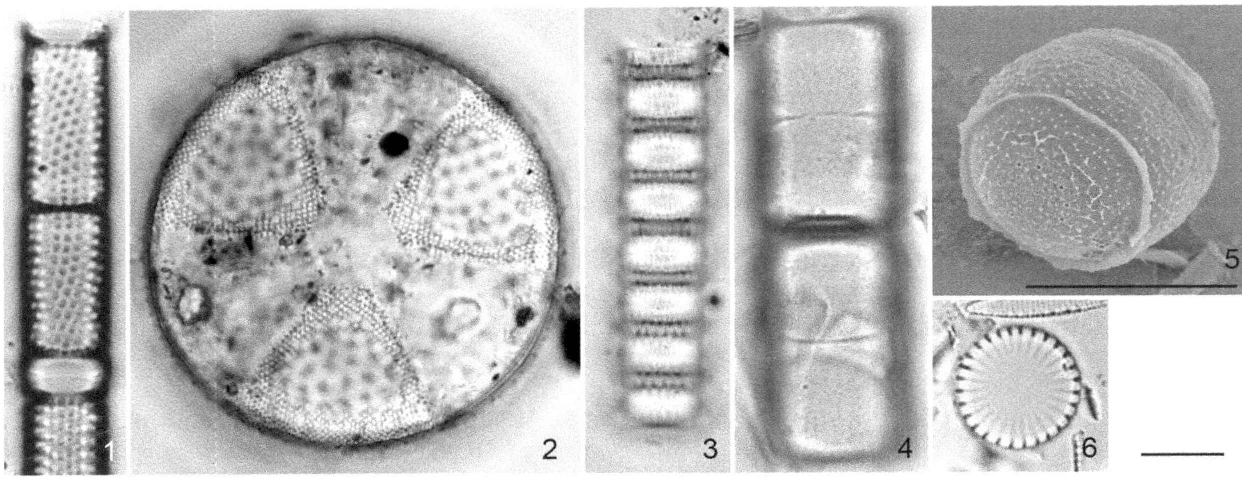

Fig. 15.1. Images of some Coscinodiscophyceae-type diatoms found in coastal environments. *Source*: Reproduced with permission of Sawai and Nagumo, 2003a, and Niigata; Sawai et al. 2001). Reproduced with permission of T. Nagumo, A. Momohara, T. Fujiki, and H. Nasu. 1. *Aulacoseira granulata* (Ehrenberg) Simonsen (Niigata, Japan); 2. *Actinoptychus senarius* (Ehrenberg) Ehrenberg (Alaska, USA); 3. *Paralia sulcata* (Ehrenberg) Cleve (Alaska, USA); 4. *Melosira lineata* (Dillwyn) C. Agardh (Hokkaido, Japan); 5. *Melosira nummuloides* C. Agardh (Chile); 6. *Cyclotella meneghiniana* Kützing (Niigata, Japan). Scale bars: 10 μm.

duration of tidal inundation are particularly valuable. According to these associations, specific environments can be inferred from diatom data. A greater level of understanding of these preferences has been gained over the past few decades, and many diatom species are categorized into salinity groups and life forms (e.g., Denys, 1991–1992; Vos and de Wolf, 1993) as follows.

- *Salinity classifications*: A. Polyhalobous, marine water greater than 30‰ of salinity; B. Mesohalobous, brackish water ranging from 2‰ to 30‰ of salinity; C. Oligohalobous halophile, freshwater but tolerable to salinity up to 2‰; D. Oligohalobous indifferent, freshwater; and E. Halophobous, pure freshwater.
- *Life forms*: 1. Planktonic, free floating in water column; 2. Episammic, attached on sand grains; 3. Epiphytic, attached on plant structure; 4. Epipelic, living on wet sediment surface; and 5. Aerophilous, living on wet sediment surface and able to survive when exposed to air for a period of time.

There are also associations between the shape of diatoms and their life forms. The disc-shape and chain forms of Coscinodiscophyceae Class are usually related to its planktonic life form. The box shape of Fragilariophyceae type is dominantly epiphytic or episammic. Finally, the pennate type of Bacillariophyceae is mostly epipelic and aerophilous.

Note that there are other classification schemes which have been used in the past, including those listed in Denys (1991–1992). These classifications originated from some local studies in northwest Europe, with more recent work carried out in places across the globe showing some species have wider tolerances. There is a therefore need for modern datasets from many regions worldwide. Despite the fact that some regional datasets have lower precision, any new ecological information related to coastal diatoms are useful in addition to the above categorization. For reference, common coastal diatom species and their ecological preference are listed in Appendix 1. Methods of using these classifications in sea-level studies are explained in Section 15.4.

15.3 FIELD AND LABORATORY METHODS

Sediment samples or cores should be sealed in the field and kept at low temperature, preferably in a fridge at c. 4 °C. Procedures of sample preparation and making diatom slides are listed in Appendix 2. Materials and equipment needed include: glass test tubes or plastic centrifuge tubes in a rack; water bath; hydrogen peroxide (H_2O_2) at 20% concentration; hydrochloric acid (HCl) at 10% concentration; distilled water; hotplate; pipettes (3 mL); glass slides; glass coverslips; and mountant media such as Naphrax (with a refractive index of 1.65), Pleurax, or Zrax (with a refractive index greater than 1.70). Several steps are needed for preparation of diatom

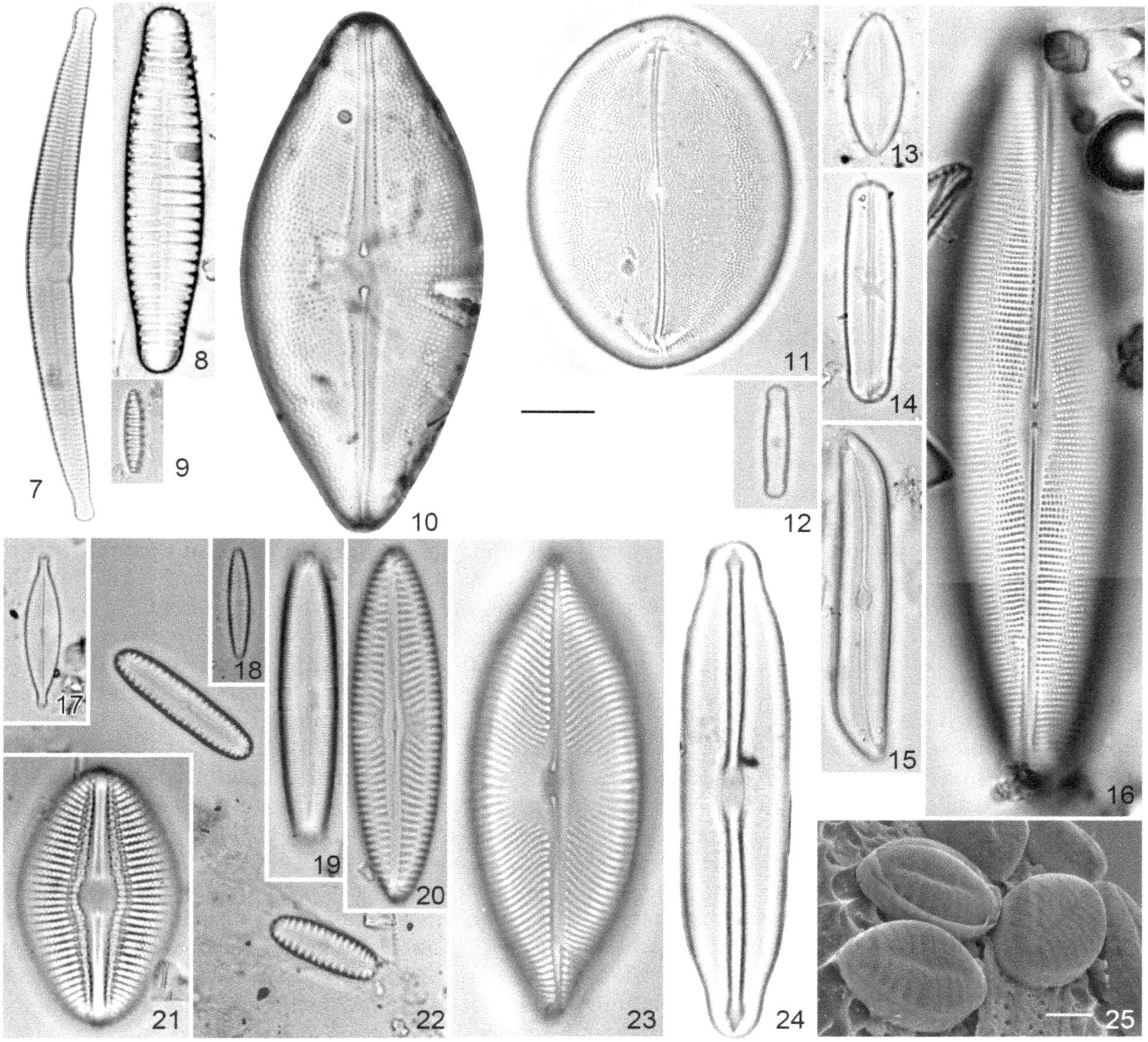

Fig. 15.2. Images of some Fragilariophyceae-type and Bacillariophyceae-type diatoms found in coastal environments. *Source*: Reproduced with permission of Sawai and Nagumo, 2003a, and Sawai and Nagumo, 2003b). Reproduced with permission of T. Nagumo. 7. *Hannaea arcus* var. *arcus* (Ehrenberg) R.M. Patrick in Patrick and Reimer (Oregon, USA); 8. *Diatoma vulgare* Bory (Oregon, USA); 9. *Denticula subtilis* (Alaska, USA); 10. *Cosmioneis delawarensis* (Grunow ex Cleve) D.G. Mann (Oregon, USA); 11. *Cocconeis pseudomarginata* var. *intermedia* (Oregon, USA); 12. *Diadesmis contenta* var. *biceps* (Grunow) P.B. Hamilton (Alaska, Oregon); 13. *Fallacia pygmaea* (Küting) Stickle and Mann (Alaska, USA); 14. *Sellaphora pupula* (*Navicula pupula*) (Kützing) Mereschkovsky (Alaska, USA); 15. *Gyrosigma eximium* (Thwaites) C.S. Boyer (Alaska, USA); 16. *Scolioneis tumida* (Brébisson ex Kützing) D.G. Mann (Hokkaido, Japan); 17. *Navicula gregaria* Donkin (Alaska, USA); 18. *Navicula salinicola* Hustedt (Alaska, USA); 19. *Caloneis bacillum* (Grunow) Cleve (Alaska, USA); 20. *Navicula digitoradiata* (Gregory) Ralfs (Hokkaido, Japan); 21. *Diploneis smithii* (Brébisson) Cleve (Hokkaido, Japan); 22. *Pinnularia lagerstedtii* (Cleve) Cleve-Euler (Alaska, USA); 23. *Pinnunavis elegans* (*Navicula elegans*) (W. Smith) Okuno (Hokkaido, Japan); 24. *Frustulia vulgaris* (Hokkaido, Japan); 25. *Planothidium delicatulum* (Kützing) Round & Bukhtiy (Oregon, USA). Scale bar: 10 μm. Scale bar in 25: 2 μm.

samples as listed in Appendix 2. Under a light microscope with 600–1000 magnifications, most features of a diatom valve can be easily viewed. At these magnifications, many apertures or perforations for pennate types may only be seen as thin lines, but these are sufficient for identification of the majority of taxa.

To aid qualitative and quantitative interpretation of diatom results, it is advisable to collect modern surface sediment samples across a coastal environmental gradient, such as along a cross-transect of intertidal mudflat through saltmarsh onto freshwater wetland environment (e.g., Zong and Horton, 1998) or from marine to freshwater environments of an estuary (e.g., Zong et al., 2010). For each sampling site, environmental variables including ground elevation, sediment particle size, organic content, and water salinity should also be collected.

15.4 APPLICATIONS TO SEA-LEVEL STUDY

15.4.1 Marine transgression and regression

Sea-level change is an important driving force for changes in coastal environments. A period of sea-level rise can cause coastal habitats to retreat landwards and create an accommodation space for tidal and sub-tidal sedimentation, leaving a transgressive sedimentary archive in coastal areas. Likewise, a period of sea-level fall or stagnation can lead to terrestrial habitats colonizing former marine environments, resulting in a regressive sedimentary sequence deposited in coastal areas. As diatoms are found living in these habitats and well preserved in coastal sediment sequences, they are frequently used as a proxy for environmental reconstructions and inference of sea-level change.

In near-field sites, many small rock depressions or basins have experienced submergence and emergence due to glacio-isostatic processes (Fig. 15.3a). Marine inundation took place while they were submerged, while marine regression occurred as relative sea level fell. Called isolation basins, such depressions are found at various altitudes within the coastal landscapes of near-field locations. Diatoms obtained from the transgressive and regressive sediment sequences of these basins have been used to infer sea-level change. A diatom record from one of the basins, Loch nan Eala in northwest Scotland (Shennan et al., 1994), demonstrates such changes. The 5.5 m of sediment from the centre of this basin contains rich assemblages of diatoms which can be divided into four phases of sea-level change (Fig. 15.3b). Phase I is inferred as the basin being inundated by marine water, because the diatom assemblages in this sediment section are dominated by marine and brackish water taxa. As the majority of diatom taxa in this phase is either epiphytic or epipelic forms, water depth in the basin was not deep during this phase of marine inundation. In other words, relative sea level was not much higher than the rock sill. Phase II saw marine and brackish water taxa disappear completely, implying no marine inundation to the basin even during high spring tides. Initially, the diatom community was dominated by epiphytic species, particularly *Cocconeis placentula* and *Fragilaria* sp. These are freshwater taxa tolerable to low levels of salinity and pioneer species to colonize former marine/brackish water environments (Zong, 1997). Afterwards, epipelic forms of freshwater diatoms increased. The diatom assemblages in Phase III are much more variable. Firstly, marine planktonic diatoms (particularly *Paralia sulcata*) appeared strongly in the assemblages (Zong, 1997), suggesting a sudden sea-level rise in the basin. After the initial increase in marine influence, the abundance of marine diatoms decreased first and then increased again in the later part of this phase, indicating a fluctuating sea level. A sharp increase in freshwater diatoms in Phase IV, including the pioneer species *Fragilaria pinnata*, marks the end of marine inundation to this basin (Shennan et al., 1994). Together with the height of the rock sill (6.27 m above sea level) and the local tidal levels (highest astronomical tide is 2.9 m), the changes in diatom assemblages in Loch nan Eala record a sea-level history. Decreasing marine diatoms from Phase I to Phase II (~10,500 ^{14}C a BP) suggest a sea-level fall of at least 2.9 m (Fig. 15.3b). Sea level again rose as indicated by the increase in marine diatoms from Phase II to Phase III (~8310 ^{14}C a BP). The highly mixed assemblages in Phase III imply a fluctuating sea level which was not much higher than the rock sill. Finally, freshwater diatoms dominate as sea level fell below the rock sill by ~4010 ^{14}C a BP.

At mid-latitudes, sea-level change has been recorded in many saltmarsh sequences. In Cumbria, UK a diatom record from the estuary of River Mite (Lloyd et al., 2013) demonstrates both transgressive and regressive processes during the middle Holocene. In this record (Fig. 15.4), all the dominant taxa (over 10% of total diatoms) belong to Class III, that is, epipelic and aerophilous types, while planktonic species only occur up to 3.5%. This information immediately suggests an environment around the upper part of the intertidal zone which is exposed twice-daily during low tides. The change from dominantly freshwater assemblages at 65 cm to high numbers of brackish water diatoms at 50 cm indicates a transgressive process. The high numbers of

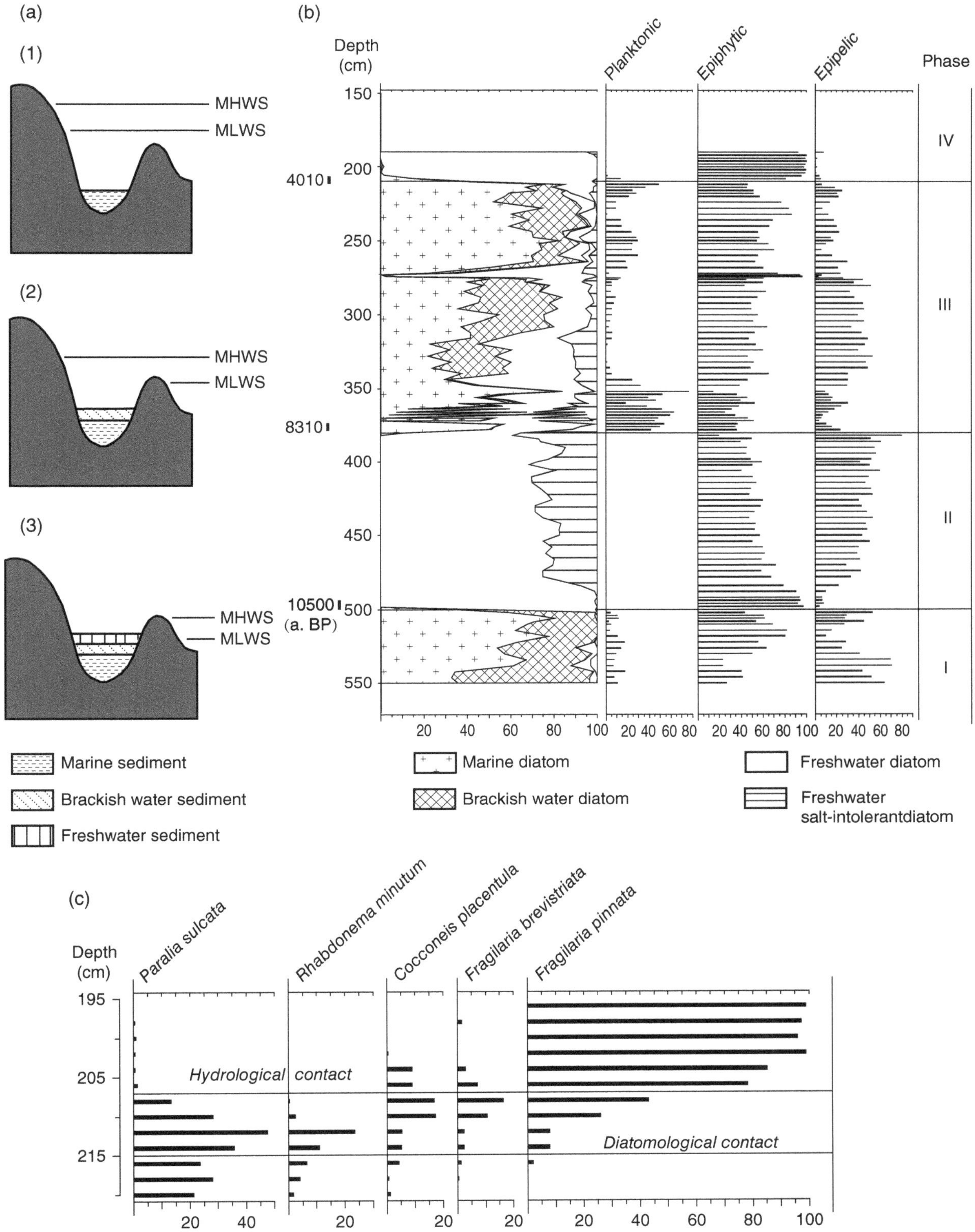

Fig. 15.3. (a) Changes in sea level and the resulting sedimentary sequences in an isolation basin: (1) the basin is inundated at all stages of the tide including mean high water of spring tides (MHWS) and mean low water of spring tides (MLWS) and marine sediments accumulate; (2) the basin is inundated only during high tides and brackish water sediments are deposited; (3) the basin is isolated from the sea and filled with freshwater. (b) Diatom assemblages from Loch nan Eala (after Shennan et al., 1994), northwest Scotland are grouped according to their life forms and preferences to salinity. (c) A close look into the changes in diatom assemblages (after Zong, 1997) showing a diatomological contact at 2.15 m and a hydrological contact at 2.07 m. This case demonstrates a marine transgression in these basins takes place when sea level rises above the height of the rock sill. When sea level falls below the rock sill, a regressive sedimentary sequence is formed.

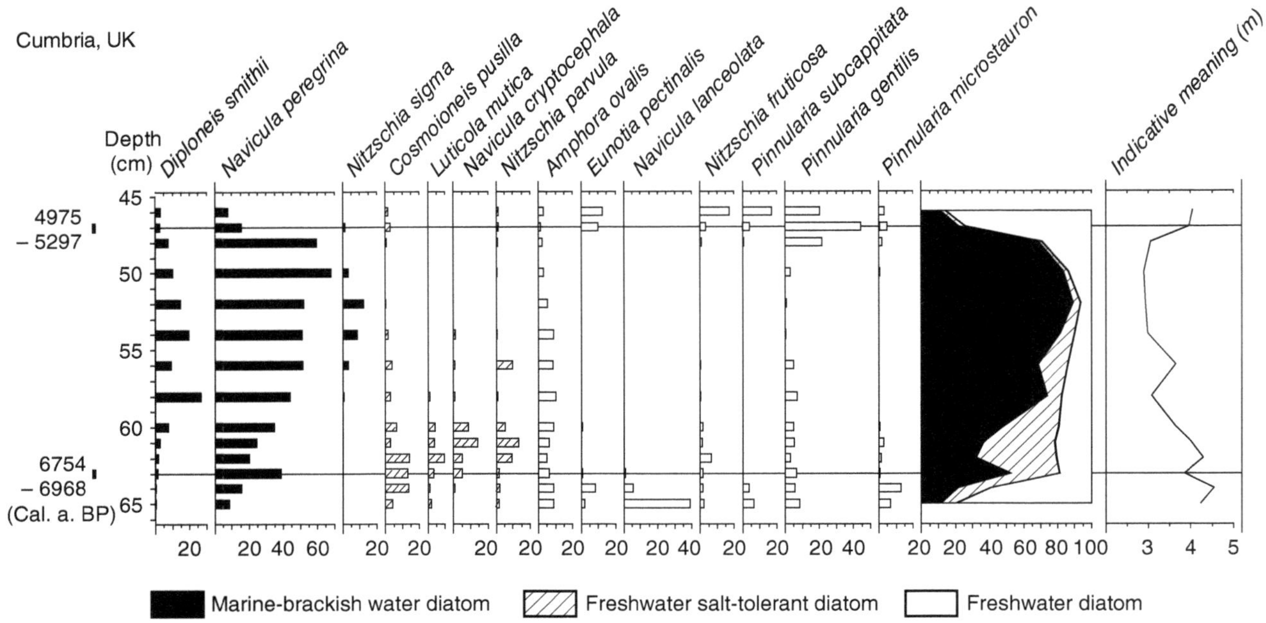

Fig. 15.4. Key diatom species from a saltmarsh sequence at the Cumbrian coast of the UK. *Source*: Adapted from Lloyd et al., 2013 show marine transgression and regression as a result of sea-level rise and fall during the middle Holocene. The curve on the far right shows paleo-ground-level changes reconstructed using a diatom-based transfer function, from which the indicative meanings for the two sea-level index points can be obtained.

Navicula lanceolata, three *Pinnularia* species, and *Eunotia pectinalis* around 65 cm represent freshwater marsh conditions. The assemblages also include low numbers of brackish water diatoms, indicating infrequent inundation by tides. A gradual increase in tidal inundation is indicated by the *Navicula–Pinnularia–Eunotia* assemblages being replaced by *Navicula peregrina*, *Diplonei smithii*, *Nitzschia sigma*, and a group of oligohalobous halophilous taxa (*Cosmioneis pusilla*, *Navicula cryptocephala*, *Luticola mutica*, and *Nitzschia parvula*) which is commonly found in saltmarsh environments. The diatom data therefore record a habitat change from freshwater marsh to saltmarsh and then tidal mudflat, which corresponds well with the lithostratigraphy. Finally, a regressive process takes place from 50 cm to 45 cm, indicating a change from tidal mudflat back to freshwater marsh. This data therefore confirm that relative sea level reached its highest altitude at this location soon after 6754–6982 cal a BP and started falling before 4975–5297 cal a BP.

In low-latitude or far-field regions, an old marine sedimentary unit is commonly found in coastal areas (e.g., Zong et al., 2009). This unit was formed during the previous interglacial sea-level highstand (Marine Isotope Stage 5) and incised by rivers during the last glacial period. The surface of this marine sequence lies at around −15 m to −20 m, and many incised valleys cut down to −40 m, leaving a T-shape accommodation space in the mouth of many rivers (Fig. 15.5a). Diatoms from the postglacial sedimentary sequences provide detailed information of the sea-level history. A record from one such incised valley in the northern South China Sea (Fig. 15.5b) reveals three stages of coastal change in relation to the far-field sea-level history. The record is from a core with a surface elevation of 3.5 m drilled in the middle of the Pearl River delta plain (Zong et al., 2012). In stage I, the diatom data indicate a dominantly freshwater environment during the early part of the early Holocene when sea level rose rapidly from −40 m to −16 m. The assemblages comprise high percentages of freshwater benthic diatoms, indicating strong freshwater outflow. The rapid sea-level rise supplied a large amount of sediment to the limited accommodation space and resulted in a very high sedimentation rate. The sedimentation kept pace with the rise in sea level as the high percentages of freshwater benthic diatoms imply. During stage II, sea level continued to rise over the shoulders of the incised valley. The sea then flooded the much wider accommodation space as the sedimentation was

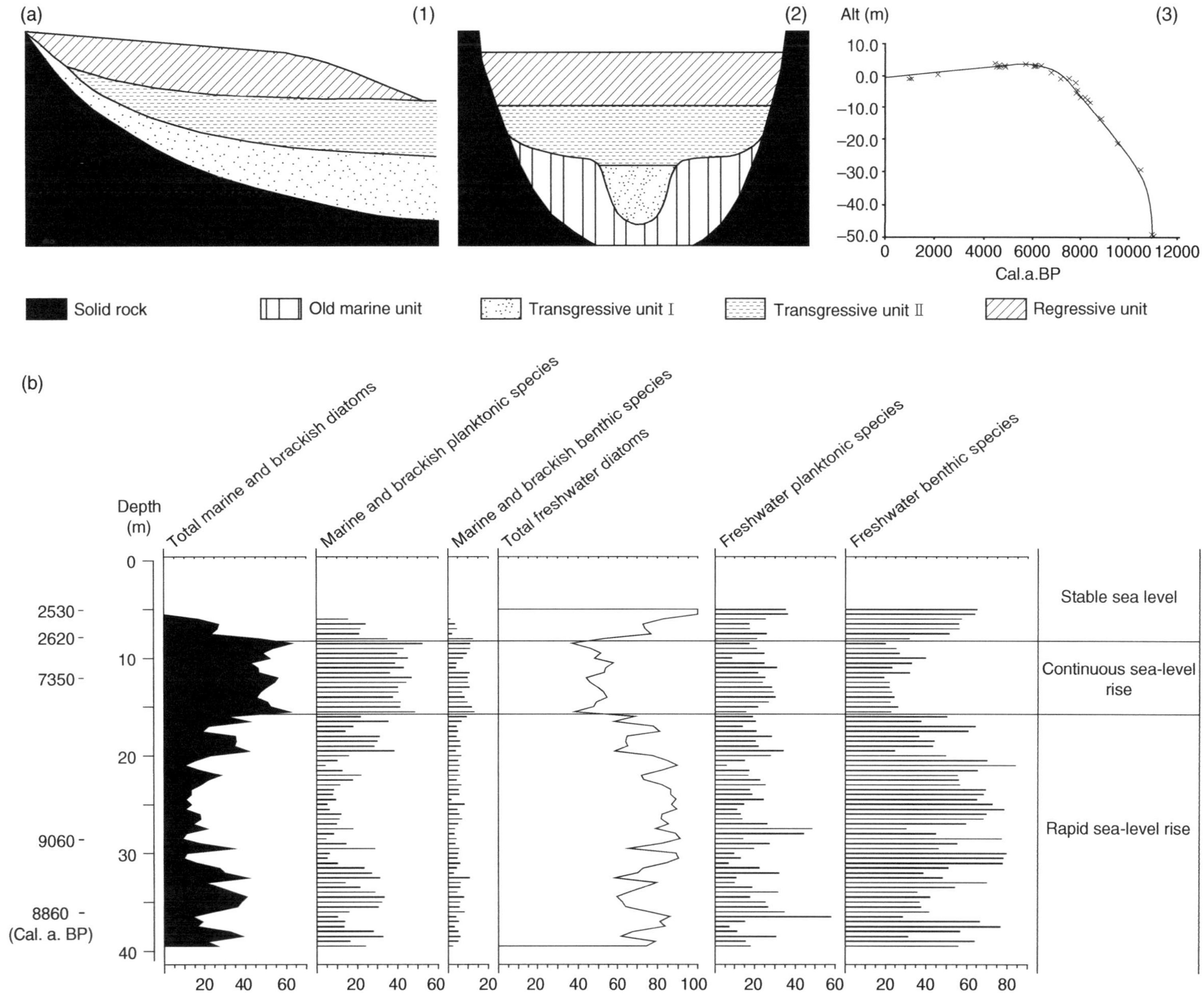

Fig. 15.5. (a) Schematic diagrams showing the impacts of sea-level change on coastal sedimentation in far-field sites: (1) a longitudinal transect of an estuary describing the three sedimentary units that deposited during the Holocene; (2) A cross transect of an estuary displaying the typical paleo-landscape in far-field coasts and the three units of sediment; and (3) the sea-level history of far-field sites (after Zong, 2007). (b) Diatom assemblages from a sediment core from the Pearl River deltaic basin, southern China demonstrate the effects of sea-level change during the Holocene on coastal sedimentation (after Zong et al., 2012).

no longer in pace with sea-level rise, resulting in river mouth retreat. The diatom assemblages between 8 m and 16 m of the core are characterized by high abundances of marine and brackish planktonic species, suggesting stronger marine influence and deeper water conditions. Sea level stabilized during stage III (the past 7000 years), which allowed the deltaic shoreline to advance. The diatom data show an increase in freshwater species, particularly benthic taxa, implying the deltaic shoreline advanced past the core site around 2500 years ago.

15.4.2 Indicative meaning of sea-level index points

One important task in sea-level study is to determine the indicative meaning of sea-level index points. The indicative meaning is the vertical distance between a datable sedimentary horizon or formation and its local, contemporary, mean sea level (Chapter 2). Without the aid of microfossils, the indicative meaning can only be estimated by observing the relationship between tidal levels and coastal sediment; this involves large errors (commonly over ±1.0 m) in coasts with mesotides. With the help of microfossils, the indicative meaning of a sea-level index point can be established qualitatively or quantitatively. Three examples are presented in the following to explain how to use diatoms to estimate indicative meanings.

The depositional regime within isolation basins is governed by the altitudinal relationship between the rock sill and sea level. In Loch nan Eala (Fig. 15.3b), the modern mean spring tidal range is 4.3 m. This means the isolation process may have taken a certain length of time to complete as the sill passed through the tidal range during each period of relative sea-level fall. Generally speaking, an isolation process includes two important moments, each of them identifiable by diatom data as an isolation contact in sedimentary sequences. The first important event occurred when there was a change from fully marine conditions as the basin was connected to the sea at all stages of the tide, to brackish water conditions as the basin was only partially connected to the sea. This change is marked by the occurrence of freshwater diatoms in the sedimentary sequence at 215 cm of core depth, and is known as the diatomological contact (Fig. 15.3c). This contact implies that the local mean low water of spring tides (modern day −1.90 m from mean sea level) may have fallen below the altitude of the rock sill. The second important moment is when the basin became free from tidal inundation, which is marked by the diatom data at 207 cm of core depth (Fig. 15.3c). This contact is called the hydrological contact, which suggests the local mean high water of spring tides (+2.40 m) fell below the altitude of the rock sill. In this case, the high-resolution diatom record from Loch nan Eala has the ability to produce sea-level index points at both 207 and 215 cm core depth, and their precisions are ±0.25 m (Shennan et al., 1994).

As was the case in the Cumbria study, an understanding of the modern relationship between diatoms and ground altitudes can greatly enhance the precision of the estimates of indicative meaning of sea-level index points. In Kentra Bay, modern diatoms were obtained from locations across the mudflat and saltmarsh (Zong and Horton, 1998; Fig. 15.6). The results show two important changes in the diatom assemblages. First, a change from marine-brackish diatoms to freshwater diatoms occurs close to the local mean high water of spring tides, which coincides with a change from pioneer marsh to low marsh. Second, across the local highest astronomical tide, the salt-tolerant freshwater diatoms (such as *Cosmioneis pusilla*) disappear, while salt-intolerant species (such as *Eunotia valida*) emerge. Using these details, the indicative meaning for the dated horizon at 47 cm of the Cumbria core (Fig. 15.4) can be estimated as close to the local mean high water of spring tides (4.02 m; e.g., Lloyd et al., 2013). Further improvement in precision of indicative meaning estimates requires statistical analyses of diatom and environmental data.

Quantitative estimation of the indicative meaning requires development of a diatom-based tidal-level transfer function (Chapter 31). This method was tested by Zong and Horton (1999) in the UK, and applied to studies in North America (e.g., Horton et al., 2006) and Asia (e.g., Sawai et al., 2004a, b; Horton et al., 2007). In the UK study, modern diatom data from as many paleo-environmental settings as possible was merged to produce a comprehensive diatom database. Statistical examinations of the modern diatom data and the associated environmental data confirmed that the relationship between diatoms and tidal inundation expressed as ground altitudes is very strong, and a diatom-based tidal level transfer function could be developed. Furthermore, the predictive ability of the transfer function was highly satisfactory. Using this transfer function, the precision of

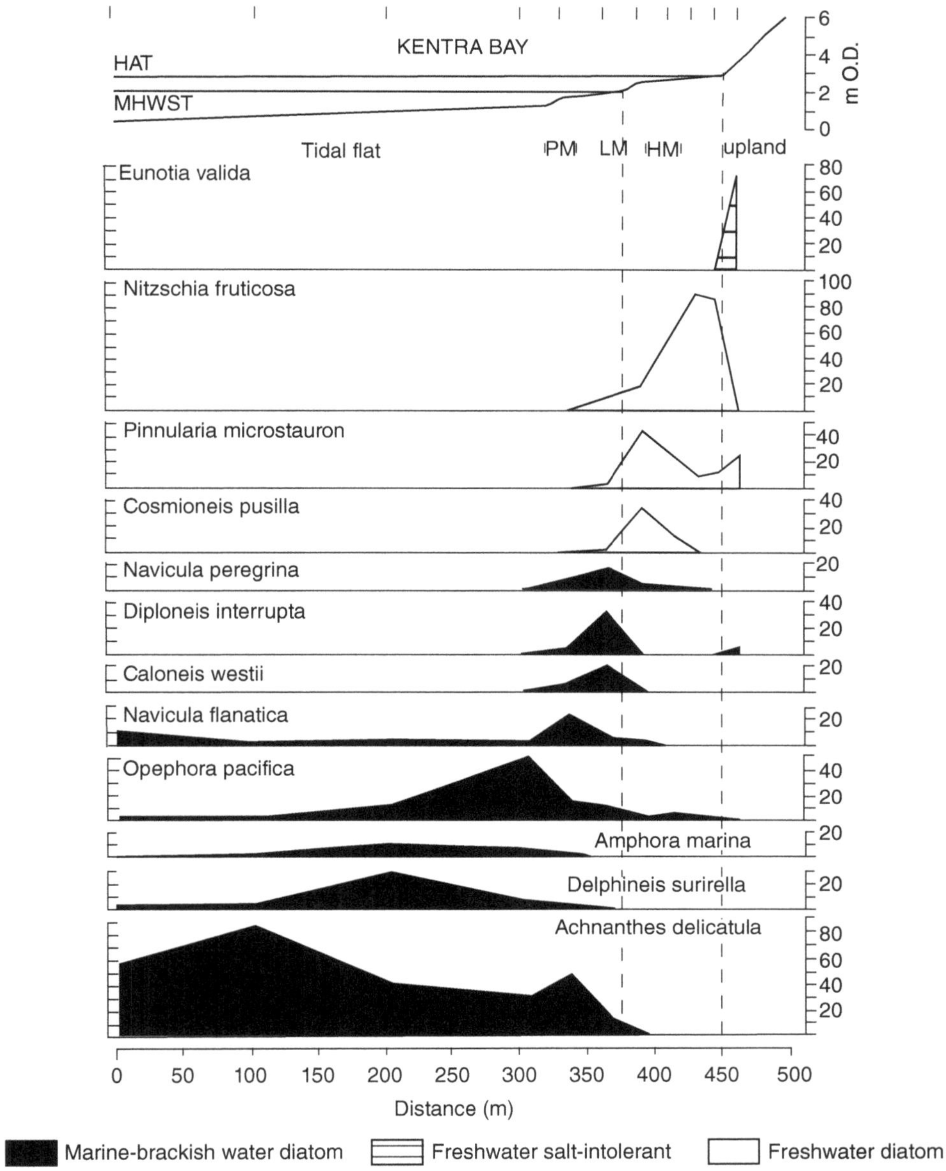

Fig. 15.6. The distribution of modern diatoms across the mudflat and saltmarsh at Kentra Bay. *Source*: Adapted from Zong and Horton, 1998, 1999, northwest Scotland. PM: pioneer marsh (*Salicornia maritima, Armeria maritima*); LM: low marsh (*Puccinellia maritima, Plantago maritima, Armeria maritima*); HM: high marsh (*Puccinellia maritima*, Gramineae); and the relationship to local mean high water of spring tides (MHWST) and highest astronomical tide (HAT).

vertical estimates of the indicative meaning is reduced to 10% of local tidal range according to the root mean square of error for prediction (RMSEP; Chapter 31). When this transfer function is applied to the Cumbria core (Fig. 15.4), the paleo-sea level of the dated horizon at 47 cm is predicted as 3.95 m above modern mean tide level. The prediction error (RMSEP) is ±0.31 m; this error is very small (only 8.2% of the tidal range) because the local tidal range is over 7.56 m between mean high water of spring tides and mean low water of spring tides.

Similar to saltmarsh environment, diatoms from mangrove flats also show a good relationship with ground level. In a mangrove flat of Malaysia, marine planktonic diatoms dominate the mudflat environment (Zong and Kamaludin, 2004). Brackish water benthic species increase from the local mean high water of neap tides upwards. Freshwater benthic taxa, both salt-tolerant and indifferent, emerge from the local mean high water of spring tides upwards as both marine and brackish water diatoms decline. Despite the effects of a tidal creek which resulted in an increase in marine and brackish water diatoms at two sites (ground level around 1.3–1.5 m), freshwater diatoms become dominant around the local extreme high water. The strong relationship between diatom assemblages and ground levels in mangrove coasts was further tested statistically by Horton et al. (2007). In Indonesia, they examined two mangrove flats and confirmed that the relationship is sufficiently strong; diatom-based transfer functions can therefore be developed to aid in estimating indicative meanings of sea-level index points obtained from such environments.

The above examples illustrate that an essentially qualitative approach to interpreting coastal evolution and sea-level change is based on the researcher's knowledge of diatom ecology, particularly the classifications in salinity and life form. For quantitative reconstructions of environmental parameters, such as elevation, this ecological knowledge must be combined with additional numerical analyses (Chapters 2 and 31).

15.4.3 Abrupt sea-level change

Short-term, or abrupt, changes in sea level are commonly caused by earthquakes, tsunami, and storms. Diatoms have provided a useful proxy for sudden sea-level change. There are several ways by which diatoms can help in reconstructing the amount of land-level change relative to sea level and the extent of tsunami or flood water.

Land-level change caused by earthquakes can be reconstructed using the close relationship of diatoms to tidal levels (e.g., Shennan et al., 1996; Atwater and Hemphill-Haley, 1997; Sherrod et al., 2000; Sherrod, 2001; Cisternas et al., 2005; Nelson et al., 2007). Especially near subduction zones, recent, historical, and prehistoric land-level changes are used to reconstruct earthquake cycles to understand the mechanics of plate-boundary ruptures. At subduction zones the coastal areas above the trench on the plate-boundary subside slowly during/before large earthquakes as strain accumulates, and uplift abruptly during earthquakes. In contrast, the coastal areas further inland from the rupture behave in the opposite way.

An example from Alaska shows that ground subsidence during the 1964 earthquake is reflected by a sharp change in diatom assemblages from a marsh community to a tidal flat community. Using a diatom-based tidal-level transfer function, the amount of coseismic subsidence was estimated as 1.8 m at Girdwood (Zong et al., 2003), close to the observed amount of subsidence of 1.7 m. With confidence in the diatom methods, Shennan and Hamilton (2006) extended the reconstruction back to five older events. Their reconstructions of coseismic land movements achieved a very high precision, about ±0.3 m or 6% of the local tidal range. There is also detailed diatom-based land-level reconstruction from estuarine deposits in Hokkaido near the southeast of the Kuril trench. Diatom analyses suggest tidal flats in a small basin changed into freshwater upland in decades after a 17th century tsunami. The tidal flat gradually rose about 1.5 m as a decadal post-seismic deformation process caused by fault slip at a location deeper than the seismogenic zone (Sawai et al., 2004b). This coastal uplift in Hokkaido was therefore comparable in magnitude to that of the recent giant (*M* 9 class) earthquake in 2011 and that in Alaska (1964) and Chile (1960).

The use of diatoms to reconstruct tsunami and storm floods is common because of the robust diatom valves that can resist rough depositional processes. Both tsunami inundation and storm surges are caused by a sudden rise in sea level and marine incursion to low-lying coastal landscapes. Diatoms are found in tsunami/flood deposits and provide details of the run-up, extent, and origin of the deposits. There has recently been an increase in the number of case studies on diatom assemblages within modern tsunami deposits. An understanding of the common nature of the diatom assemblages is still lacking, however. For example, the composition of diatom assemblages in tsunami deposits should have many marine diatoms as for the case of the 2004 tsunami deposits (Sawai et al., 2009), because they are transported by currents associated with rapid marine incursions. However, many freshwater specimens are often found in historical and prehistoric tsunami deposits as a tsunami crosses coastal and inland areas eroding, transporting, and redepositing freshwater diatoms (Dawson et al., 1996; Dawson and Smith, 2000; Smith et al., 2004; Bondevik et al., 2005; Sawai et al., 2009). As a result, diatom valves in tsunami deposits are often highly fractured, reflecting the nature of rapid turbulent marine incursion of tsunami events (Dawson and Smith, 2000; Sawai, 2002; Dawson,

2007). Although some recent tsunami deposits show pristine preservation (Hemphill-Haley, 1996; Sawai et al., 2009, 2012). Additionally, diatoms have been widely used to delineate the landward extent of tsunami run-up, beyond the extent of the heavier sandy sediment. In a study of Washington State, USA (Hemphill-Haley, 1995), marine diatoms were found at sites at least 1 km inland of the landward limit of the tsunami sand.

15.5 COMMON PROBLEMS

Despite the fact that diatoms are a useful proxy for sea-level reconstructions, there are some intrinsic problems. When using diatoms for sea-level study, several matters must be kept in mind. Firstly, diatoms need to be identified to species level. Some species share very similar morphological features. Greater care must therefore be taken in the identification of diatoms. Secondly, there exists the problem of taxonomy created by different authors, leading to a situation of the same species of diatoms having different names. Harmonization of diatom data from multiple data sources may sometime be needed when combining several datasets. Thirdly, the ecological classification of diatoms involves uncertainties, because diatom distributions are controlled by multiple environmental factors. For example, diatoms appear closely associated with ground altitude across intertidal zones, but different substrates are also associated with certain taxa (e.g., Zong and Horton, 1998). In this case, quantitative analysis of diatom assemblages and their associated environmental parameters (e.g., Zong and Horton, 1999) can help measure the strength between diatom data and environmental variables.

The fourth problem is the difficulty of identifying diatoms that live in the very spot of the sampling site (called autochthonous); these must be differentiated from those which are transported from outside the sampling location (referred to as allochthonous) in the same modern sediment sample or fossil sedimentary horizon. It is well known that the amount of allochthonous valves in coastal environments is generally higher than in calm environments, because daily tidal currents can easily transport diatom valves from one place to the other. During such transportation, some weaker valves may break and leave stronger valves to be deposited and preserved in sediment, resulting in preferential preservation of certain taxa over others. Several methods have been suggested to distinguish allochthonous diatoms from fossil assemblages (e.g., Sawai, 2001). One of the most classical approaches is to consider a life form of each diatom species. For instance, planktonic diatoms can be regarded as allochthonous components in fossil diatom assemblages of tidal flat coastal deposits because planktonic forms can be transported by tidal currents from deep-water areas onto tidal flats, while benthic (e.g., epiphytic, epipelic, and epipsammic types) are considered as autochthonous (Vos and de Wolf, 1993). Fragmentation of diatom valves can also be used as an indicator of long-distance transportation (Vos and de Wolf, 1993). However, fragmentation is sometimes complicated by other processes including natural chemical dissolution, compaction, and sample treatment.

Another partial solution to the autochthonous/allochthonous problem is to exclude potential allochthonous species from the data interpretation. A few potential allochthonous components that may require exclusion have been suggested based on observation of modern diatom assemblages. For example, *Paralia sulcata* is commonly found living in the lower column of coastal waters (Hemphill-Haley, 1995). Its valves are very robust, and they often form in chains. The robust long-chained specimens can be easily floated and transported by tidal currents and deposited in up-estuarine locations (Hemphill-Haley, 1995) and into isolation basins (Zong, 1997). Another example is *Cocconeis scutellum*, an epiphytic species living on macrophytes to which the diatom frustules are strongly attached onto the surface of macro-algae through their raphid valves. After their death however, non-attached rapheless valves become separated from the macrophytes whereas raphid valves remain attached by the mucilage. The separated rapheless valves can then be transported over long distances selectively by tidal currents. Consequently, rapheless valves can be found in sediment across the entire tidal environment, although its habitat is limited to the macrophytes zone (Sawai, 2001). From this observation, Sawai (2001) suggested that rapheless valves of *Cocconeis scutellum* should be excluded in reconstructing coastal environments.

Finally, the fifth problem, which is the non-modern-analog situation, is the most serious. This situation occurs when environmental conditions have changed significantly over time in a locality. Many coastal sites have experienced changes in hydrology, sedimentary regime, and morphology between the period of marine transgression in the early Holocene and the period of regression during the middle-late Holocene. Consequently, the diatom assemblages from a modern environment do not match those in a fossil one in the same locality (Wilson and Lamb, 2012). If this happens, it is not appropriate to apply transfer function models for reconstructions using local modern diatom assemblages (Watcham et al., 2013). Even if the non-modern-analog situations do not exist in an analysis, one should also take into account issues such as spatial autocorrelation which may artificially overestimate the performance of a transfer function model (e.g., Telford and Birks, 2005, 2009).

APPENDICES

Appendix 1: Ecological preferences of common coastal diatoms

Ecological preferences of common coastal diatoms (modified from Denys, 1991–1992; Vos and de Wolf, 1993) arranged under the three classes (Hartley, 1996; see Section 15.2 for definitions). Valuable online resources with diatom images and their ecological information include: European diatom database (http://craticula.ncl.ac.uk/Eddi/jsp/); PMN diatom image gallery (http://products.coastalscience.noaa.gov/pmn/image_gallery_diatom.aspx); and UCL Microfossil Image and Circulation for Learning and Education (http://www.ucl.ac.uk/GeolSci/micropal/diatom.html).

Taxonomy	Reference	Life form	Salinity
Class I: Coscinodiscophyceae			
Actinocyclus normanii	Hustedt	1	B
Actinocyclus octonarius	Ehrenberg	1	B
Actinoptychus senarius	Ehrenberg	1	A
Actinoptychus splendens	Ralfs	1	A
Aulacoseira granulata	Simonsen	1	D
Chaetoceros radians	Schutt	1	A
Coscinodiscus blandus	Cleve	1	B
Coscinodiscus divisus	Grunow	1	B
Coscinodiscus lineatus	Ehrenberg	1	A
Coscinodiscus nodulifer	Schmidt	1	A
Coscinodiscus obscurus	Schmidt	1	A
Cyclotella comta	Kutzing	1	D
Cyclotella kutzingiana	Thwaites	1	D
Cyclotella meneghiniana	Kutzing	1	C
Cyclotella radiosa	Lemmerman	1	D
Cyclotella striata	Grunow	1	B
Cyclotella stylorum	Brightwell	1	B
Cymatosira belgica	Grunow	1	A
Odontella biddulphiana	Boyer	1	A
Odontella rhombus	Kutzing	1	A
Paralia sulcata	Cleve	1	A
Plagiogramma sp.		3	A
Podosira stelligera	Mann	1	A
Skeletonema costatum	Cleve	1	A
Stephanodiscus sp.		1	D
Thalassiosira eccentrica	Cleve	1	A
Triceratium sp.		1	A
Class II: Fragilariophyceae			
Delphineis surirella	Andrews	2	A
Diatoma tenue	Agardh	3	C
Diatoma vulgare	Bory	3	B
Fragilaria construents	Brunow	3	D
Fragilaria pinnata	Ehrenberg	3	D
Grammatophora oceanica	Ehrenberg	3	A
Opephora martyi	Heribaud	3	D
Opephora pacifica	Petit	3	A
Rhabdonema minutum	Kutzing	3	A

Taxonomy	Reference	Life form	Salinity
Rhaphoneis amphiceros	Ehrenberg	3	A
Rhaphoneis belgica	Grunow	3	A
Synedra acus	Kutzing	3	D
Synedra fasciculata	Kutzing	3	B
Synedra pulchella	Ralfs	3	B
Synedra ulna	Ehrenberg	3	D
Tabellaria fenestrata	Kutzing	3	E
Tabellaria flocculosa	Kutzing	3	E
Thalasionema nitzschiodes	Grunow	1	A
Trybiblioptychus cocconeiformis	Hendey	3	B
Class III: Bacillariophyceae			
Achnanthes brevipes	Agardh	3	B
Achnanthes delicatula	Grunow	2	B
Achnanthes lanceolata	Grunow	3	D
Achnanthes minutissima	Kutzing	3	D
Amphora coffeaeformis	Kutzing	4	B
Amphora ovalis	Kutzing	4	D
Amphipleura pellicuda	Kutzing	4	E
Anomoeoneis sphaerophora	Pfitzer	4	C
Caloneis amphisbaena	Cleve	4	D
Caloneis bacillum	Cleve	5	D
Caloneis westii	Hendey	4	B
Cocconeis pediculus	Ehrenberg	3	C
Cocconeis placentula	Ehrenberg	3	B
Cocconeis scutellum	Ehrenberg	3	B
Cosmioneis pusilla	W. Smith	5	D
Cymbella sp		3 or 5	D
Denticula subtilis	Grunow	4	D
Diadesmis contenta	Mann	5	E
Diploneis bombus	Ehrenberg	4	A
Diploneis didyma	Cleve	4	B
Diploneis elliptica	Cleve	4	C
Diploneis interrupta	Cleve	4	B
Diploneis ovalis	Cleve	4	D
Diploneis smithii	Cleve	4	A
Encyonema perpusillum	Mann	3	D
Epithemia sorex	Kutzing	3	D
Epithemia turgida	Kutzing	3	D
Eunotia sp		3 or 5	D or E
Fallacia subhamulata	Mann	4	D
Fallacia pymaea	Mann	4	B
Fallacia forcipata	Mann	4	B
Frustulia linkei	Hustedt	4	B
Frustulia rhomboides	De Toni	5	E
Gomphonema acuminatum	Ehrenberg	5	D
Gomphonema parvulum	Kutzing	3	D
Gyrosigma acuminatum	Rabenhorst	4	D
Gyrosigma balticum	Rabenhorst	4	B
Hantzschia amphioxys	Grunow	5	D
Hantzschia virgata	Grunow	4	B
Luticula mutica	Mann	5	C
Lyrella lyra	Mann	4	A
Masogloia smithii	W. Smith	5	D
Navicula cari var. cincta	Ralfs	4	C
Navicula cryptocephala	Kutzing	4	C
Navicula halophila	Grunow	4	B
Navicula lanceolata	Ehrenberg	4	D
Navicula minima	Grunow	4	C
Navicula peregrina	Kutzing	4	B

(*continued*)

(*Continued*)

Taxonomy	Reference	Life form	Salinity
Navicula pupula	Kutzing	4	D
Navicula radiosa	Kutzing	4	D
Navicula viridula	Ehrenberg	4	D
Nitzschia fonticola	Grunow	4	D
Nitzschia frustulum	Grunow	4	C
Nitzschia fruticosa	Hustedt	5	D
Nitzschia granulata	Grunow	4	B
Nitzschia hungarica	Grunow	4	B
Nitzschia navicularis	Grunow	4	B
Nitzschia parvula	Lewis	4	C
Nitzschia sigma	W. Smith	4	B
Nitzschia sinuata	Grunow	4	E
Nitzschia tryblionella	Hantzsch	4	C
Pinnularia borealis	Ehrenberg	5	D
Pinnularia gentlis/nobilis	Ehrenberg	5	D
Pinnularia microstauron	Cleve	5	D
Pinnularia subcapitata	Gregory	5	D
Pleurosigma sp		4	A
Rhopalodia gibberula	O. Muller	3	C
Stauroneis anceps	Ehrenberg	5	D
Stauroneis phoenicenteron	Ehrenberg	4	D
Surirella ovalis	Brebisson	4	B
Trachyneis aspera	Cleve	4	A

Appendix 2: Diatom sample preparation procedure

Sample preparation	
1	Place 0.5 g wet sediment in a tube
2	Add 20 mL of 20% H_2O_2 to each tube, covered with aluminum foil, and put the whole rack into the hot water bath with water temperature up to 80 °C. This step is to remove organic matter from the sediment samples. Higher water temperature can help accelerate the oxidation. If the organic content of a sample is very high, add more H_2O_2 until reactions totally subside. Do not allow the samples to dry during the process.
3	Remove tubes from hot water bath. Add 1–2 drops of 10% HCL to each tube to eliminate remaining H_2O_2 and any carbonates.
4	Top up each tube with distilled water, centrifuge tubes at 4000 rpm for 4 minutes and decant the supernatant liquid. Add a small amount of distilled water to each tube (up to a quarter of the tube). The sediment samples are now ready for mounting on glass slides.
Making diatom slides	
5	Place 22 × 22 mm glass cover-slips onto a warm (c. 40 °C) hotplate.
6	Disturb each sample fully, and allow sediment to settle for about 5 seconds (letting coarse material to settle out first). This leaves finer material (including diatoms) in suspension, from which an aliquot can be taken up into the pipette from just below the liquid surface. Place the liquid sample onto glass cover-slips.
7	Allow the cover-slips to dry away from dust, which may take a few hours.
8	Heat the hotplate in a fume cupboard to temperature 100–130 °C, place a glass slide on the hotplate, and add a small drop of Naphrax (or Pleurax and Zrax) mountant onto the slide. Carefully turn a cover-slip over and mount it onto the slide, allowing the mountant to boil and release its solvent. This boiling may take a few seconds. Once all solvent is out, lift the slide from the hotplate and cool it for labeling.

REFERENCES

Atwater, B.F., and Hemphill-Haley, E. (1997) Recurrence intervals for great earthquakes of the past 3500 years at northeastern Willapa Bay, Washington. US Geological Survey Professional Paper 1576, US Government Printing Office, Washington.

Bondevik, S., Mangerud, J., Dawson, S., Dawson, A.G., and Lohne, O. (2005) A record of three tsunami events in the Shetland Islands during the last 8,000 cal years. Quaternary Science Reviews, 24, 1757–1775.

Cisternas, M., Atwater, B.F., Torrejon, F., Sawai, Y., Machuca, G., Lagos, M., Eipert, A., Youlton, C., Salgado,

I., Kamataki, T., Shishikura, M., Rajendran, C.P., Malik, J.K., Rizal, Y., and Husni, M. (2005) Predecessors of the giant 1960 Chile earthquake. Nature, 437, 404–407.

Dawson, A.G., Shi, S., Dawson, S., Takahashi, T., and Shuto, N. (1996) Coastal sedimentation associated with the June 2nd and 3rd, 1994 tsunami in Rajegwesi, Java. Quaternary Science Reviews, 15, 901–912.

Dawson, S. (2007) Diatom biostratigraphy of tsunami deposits: examples from the 1998 Papua New Guinea tsunami. Sedimentary Geology, 200, 328–335.

Dawson, S., and Smith, D.E. (2000) The sedimentology of mid-Holocene tsunami facies in northern Scotland. Marine Geology, 170, 69–79.

Denys, L. (1991–1992) A check-list of the diatoms in the Holocene deposits of the western Belgium coastal plain with a survey of their apparent ecological requirements. Professional Paper 1991/2 No. 246, De Lescluzestraat, 68, B-2600 Berchem, Belgium.

Hartley, B., Barber, H.G., and Carter, J.R. (1996) *An Atlas of British Diatoms*. Biopress Limited, UK.

Hemphill-Haley, E. (1995) Diatom evidence for earthquake-induced subsidence and tsunami 300 yr ago in southern coastal of Washington. Geological Society of America Bulletin, 107, 367–378.

Hemphill-Haley, E. (1996) Diatoms as an aid in identifying late-Holocene tsunami deposits. The Holocene, 6, 439–448.

Horton, B.P., and Sawai, Y. (2010) Diatoms as indicators of former sea levels, earthquakes, tsunamis, and hurricanes. In: *The Diatoms: Applications for the Environmental and Earth Sciences*, 2nd edition (eds Smol, J.P., and Stoermer, E.F.), Cambridge University Press, Cambridge, pp. 357–372.

Horton, B.P., Corbett, R., Culver, S.J., Edwards, R.J., and Hillier, C. (2006) Modern saltmarsh diatom distributions of the Outer Banks, North Carolina, and the development of a transfer function for high resolution reconstructions of sea level. Estuarine, Coastal and Shelf Science, 69, 381–394.

Horton, B.P., Zong, Y., Hillier, C., and Englehart, S. (2007) Diatoms from Indonesian mangroves and their suitability as sea-level indicators for tropical environments. Marine Micropaleontology, 63, 155–168.

Lloyd, J.M., Zong, Y., Fish, P., and Innes, J.B. (2013) Holocene and Lateglacial relative sea-level change in north-west England: implications for glacial isostatic adjustment models. Journal of Quaternary Science, 28, 59–70.

Nelson, A.R., Sawai, Y., Jennings, A., Bradley, L., Sherrod, B., Sabean, J., and Horton, B.P. (2007) Great-earthquake paleogeodesy and tsunamis of the past 2000 years at Alsea Bay, central Oregon coast, USA. Quaternary Science Reviews, 27, 747–768.

Patrick, R., and Reimer, C.W. (1966) *The Diatoms of the United States, Volume 1*. Monographs of the Academy of Natural Sciences, Philadelphia.

Patrick, R., and Reimer, C.W. (1975) *The Diatoms of the United States, Volume 2*. Monographs of the Academy of natural Sciences, Philadelphia.

Round, F.E., Crawford, R.M., and Mann, D.G. (1990) *The Diatoms: Biology and Morphology of the Genera*. Cambridge University Press, Cambridge, UK.

Round, F.E., Crawford, R.M., and Mann, D.G. (2007) *The Diatoms: Biology and Morphology of the Genera*. Cambridge University Press, Cambridge, UK.

Rumeau, A., and Coste, M. (1988) Introduction into the systematic of freshwater diatoms: for a useful generic diatomic index. Knowledge and Management of Aquatic Ecosystems (former Bulletin Français de la Pêche et de la Pisciculture), 309, 1–69.

Sawai, Y. (2001) Distribution of living and dead diatoms in tidal wetlands of northern Japan: relations to taphonomy. Palaeogeography, Palaeoclimatology, Palaeoecology, 173, 125–141.

Sawai, Y., Momohara, A., Fujiki, T., Nasu, H. (2001) Emergence events inferred from fossil diatom assemblages in the deposits of Lake Sagata, Niigata Prefecture, central Japan. Diatom, 17, 91–100. (in Japanese with English abstract)

Sawai, Y. (2002) Evidence for 17th century tsunamis generated on the Kuril-Kamchatka subduction zone, Lake Tokotan, Hokkaido, Japan. Journal of Asian Earth Science, 20, 903–911.

Sawai, Y. and Nagumo, T. (2003a) Diatom (Bacillariophyceae) flora of salt marshes along the Pacific coast of eastern Hokkaido, northern Japan. Bulletin of the Nippon Dental University General Education, 32, 93–108.

Sawai, Y., and Nagumo, T. (2003b) Diatoms from Alsea Bay, Oregon, USA. Diatom, 19, 33–46.

Sawai, Y., Horton, B.P., and Nagumo, T. (2004a) Diatom-based elevation transfer function along the Pacific coast of eastern Hokkadio, Northern Japan: an aid in paleoseismic study along the coasts near Kurile subduction zone. Quaternary Science Reviews, 23, 2467–2483.

Sawai, Y., Satake, K., Kamataki, T., Nasu, H., Shishikura, M., Atwater, B.F., Horton, B.P., Kelsey, H.M., Nagumo, T., and Yamaguchi, M. (2004b) Transient uplift after a 17th-century earthquake along the Kuril subduction zone. Science, 306, 1918–1920.

Sawai, Y., Jankaew, K., Martin, M.E., Prendergast, A., Choowong, M., and Charoentitirat, T. (2009) Diatom assemblages in tsunami deposits associated with the 2004 Indian Ocean tsunami at Phra Thong Island, Thailand. Marine Micropalaeontology, 73, 70–79.

Sawai, Y., Shishikura, M., Namegaya, Y., Fujii, Y., Miyashita, Y., Kagohara, K., Fujiwara, O., and Tanigawa, K. (2012) Diatom assemblages in tsunami deposits on a paddy field and paved roads from Ibaraki and Chiba prefectures, Japan, generated with the 2011 Tohoku tsunami. Diatom, 28, 1–8.

Shennan, I., and Hamilton, S. (2006) Coseismic and pre-seismic subsidence associated with great earthquakes in Alaska. Quaternary Science Reviews, 25, 1–8.

Shennan, I., Innes, J.B., Long, A.L., and Zong, Y. (1994) Late Devensian and Holocene relative sea-level changes at Loch nan Eala, near Arisaig, northwest Scotland. Journal of Quaternary Science, 9, 261–283.

Shennan, I., Long, A.J., Rutherford, M.M., Green, F.M., Innes, J.B., Lloyd, J.M., Zong, Y., and Walker, K.J. (1996). Tidal marsh stratigraphy, sea-level change and large earthquakes, I: A 5000 year record in Washington, USA. Quaternary Science Reviews, 15, 1023–1059.

Sherrod, B.L. (2001) Evidence for earthquake-induced subsidence about 1100 yr ago in coastal marshes of southern Puget Sound, Washington. Geological Society of America Bulletin, 113, 1299–1311.

Sherrod, B.L., Bucknam, R.C., and Leopold, E.B. (2000) Holocene relative sea-level changes along the Seattle Fault at Restoration Point, Washington. Quaternary Research, 54, 384–393.

Smith, D.E., Shi, S., Cullingford, R.A., Dawson, A.G., Dawson, S., Firth, C.R., Foster, I.D.L., Fretwell, P., Haggart, B.A., Holloway, L.K., and Long, D. (2004) The Holocene Storegga slide tsunami in the United Kingtom. Quaternary Science Reviews, 23–24, 2291–2321.

Telford, R.J., and Birks, H.J.B. (2005) The secret assumption of transfer functions: problems with spatial autocorrelation in evaluating model performance. Quaternary Science Reviews, 24, 2173–2179.

Telford, R.J., and Birks, H.J.B. (2009) Evaluation of transfer functions in spatially structured environments. Quaternary Science Reviews, 28, 1309–1316.

van de Plassche, O. (ed.) (1986) *Sea-level Research: A Manual for the Collection and Evaluation of Data.* Geobooks, Norwich, UK.

van der Werff, H., and Huls, H. (1958–1974) *Diatomeeenflora van Nederland.* 8 parts, published privately, De Hoef, The Netherlands.

Vos, P.C., and de Wolf, H. (1993) Diatoms as tool for reconstruction of sedimentary environments in coastal wetlands: methodological aspects. Hydrobiologia, 269–270, 285–296.

Watcham, E.P, Shennan, I., and Barlow, N.L.M. (2013) Scale considerations in using diatoms as indicators of sea-level change: lessons from Alaska. Journal of Quaternary Science, 28, 165–179.

Wilson, G.P., and Lamb, A.L. (2012) An assessment of the utility of regional diatom-based tidal-level transfer functions. Journal of Quaternary Science, 27, 360–370.

Zong, Y. (1997) Implications of *Paralia sulcata* abundance in Scottish isolation basins. Diatom Research, 12, 125–150.

Zong, Y. (2013) Late Quaternary relative sea-level changes in the tropics. In: *The Encyclopedia of Quaternary Science*, Volume 4 (ed. Elias S.A.), pp. 495–502. Elsevier, Amsterdam.

Zong, Y., and Horton, B.P. (1998) Diatom zones across intertidal flats and coastal saltmarshes in Britain. Diatom Research, 13, 375–394.

Zong, Y., and Horton, B.P. (1999) Diatom-based tidal-level transfer functions as an aid in reconstructing Quaternary history of sea-level movements in the UK. Journal of Quaternary Science, 14, 153–167.

Zong, Y., and Kamaludin, bin H. (2004) Diatom assemblages from two mangrove tidal flats in peninsular Malaysia. Diatom Research, 19, 329–344.

Zong, Y., Shennan, I., Combellick, R.A., Hamilton, S.L., and Rutherford, M.M. (2003) Microfossil evidence for land movements associated with the AD 1964 Alaska earthquake. The Holocene, 13, 7–20.

Zong, Y., Yim, W.W-S., Yu, F., and Huang, G. (2009) Late Quaternary environmental changes in the Pearl River mouth region, China. Quaternary International, 206, 35–45.

Zong, Y., Kemp, A.C., Yu, F., Lloyd, J.M., Huang, G., and Yim, W.W-S. (2010) Diatoms from the Pearl River estuary, China and their suitability as water salinity indicators for coastal environments. Marine Micropaleontology, 75, 38–49.

Zong, Y., Huang, K., Yu, F., Zheng, Z., Switzer, A.D., Huang, G., Wang, N., and Tang, M. (2012) The role of sea-level rise, monsoonal discharge and the palaeo-landscape in the early Holocene evolution of the Pearl River delta, southern China. Quaternary Science Reviews, 54, 77–88.

Chapter 16
Ostracods and sea level

THOMAS M. CRONIN
US Geological Survey, Reston, VA, USA

16.1 INTRODUCTION

Ostracods are small (~0.5–2 mm length) meiofaunal Crustacea whose low-magnesium calcite ($CaCO_3$) shells are preserved as microfossils in lacustrine, estuarine, and marine sediments. A total of approximately 65,000 ostracod species had been described as of 2005 (Ikeya et al., 2005) and, although estimates vary, there may be 20,000 species living today (Rodriguez-Lazaro and Ruiz-Muñoz, 2012). Ostracods inhabit virtually all types of non-marine and marine aquatic environments, including intertidal and subtidal zones in coastal marshes, mudflats, estuaries, bays, and coral reef complexes. A few species even live in moist supratidal terrestrial habitats (Horne et al., 2004). Many species have physiological limits on their survival and/or reproduction to specific temperature and salinity ranges, and most species are adapted to certain substrate types (i.e., mud, sand, submerged aquatic vegetation). Consequently, marginal marine ostracods have been used for reconstructing changes in relative sea level (Cronin, 1987; Penney, 1987; De Deckker and Yokoyama, 2009) and they have been used in numerous studies of coastal zones, mostly in northern Europe, the Mediterranean, parts of North America, and the Indo-Pacific region (Boomer and Eisenhauer, 2002; Frenzel and Boomer, 2005).

This chapter discusses biological and ecological attributes that make ostracods unique sea-level indicators, distinct from other biological proxies, and provides examples where they have been applied to sea-level reconstruction. This focus is on the practical application of species' ecology to paleo-sea-level reconstruction using assemblages (also called associations, biofacies) from coastal sediment records. Although lack of space prevents a discussion of ostracod shell chemistry, it is noteworthy that pioneering studies of magnesium/calcium and strontium/calcium ratios in marginal marine ostracods were first applied to document changes in regional sea level (De Deckker et al., 1988; see also Dettman and Dwyer, 2012; Holmes and De Deckker, 2012).

16.2 TAXONOMY

Correct taxonomic identification of marginal marine species is the foundation of any study of sea level using species' ecology and faunal assemblages. The most useful introduction to the taxonomy of major ostracod groups is that of Horne et al. (2002); Ikeya et al. (2005) also provide a useful review of taxonomic schemes for ostracods. Regarding higher taxonomic groups, ostracod biologists often use "soft-tissue" features (i.e., appendages, copulatory apparatus) to distinguish groups, but these are not typically preserved in sediments. In paleo-sea-level studies, ostracod specialists therefore identify genera and species from morphological features of their $CaCO_3$ shells, notably species-specific patterns of sieve and radial pores, muscle scars, the calcified inner lamella and vestibule, hinge morphology along the dorsal margin, and carapace surface ornamentation, among others (Fig. 16.1). This requires knowledge of the taxonomy and inter- and intra-population morphological variability, which in some species can vary with environmental conditions such as salinity, hydrochemistry, and temperature.

16.3 REPRODUCTION, GROWTH AND SHELL MORPHOLOGY

Ostracods reproduce through a variety of mechanisms (cloning, parthenogenesis, brood care of eggs and/or young) but most species used in

Handbook of Sea-Level Research, First Edition. Edited by Ian Shennan, Antony J. Long, and Benjamin P. Horton.

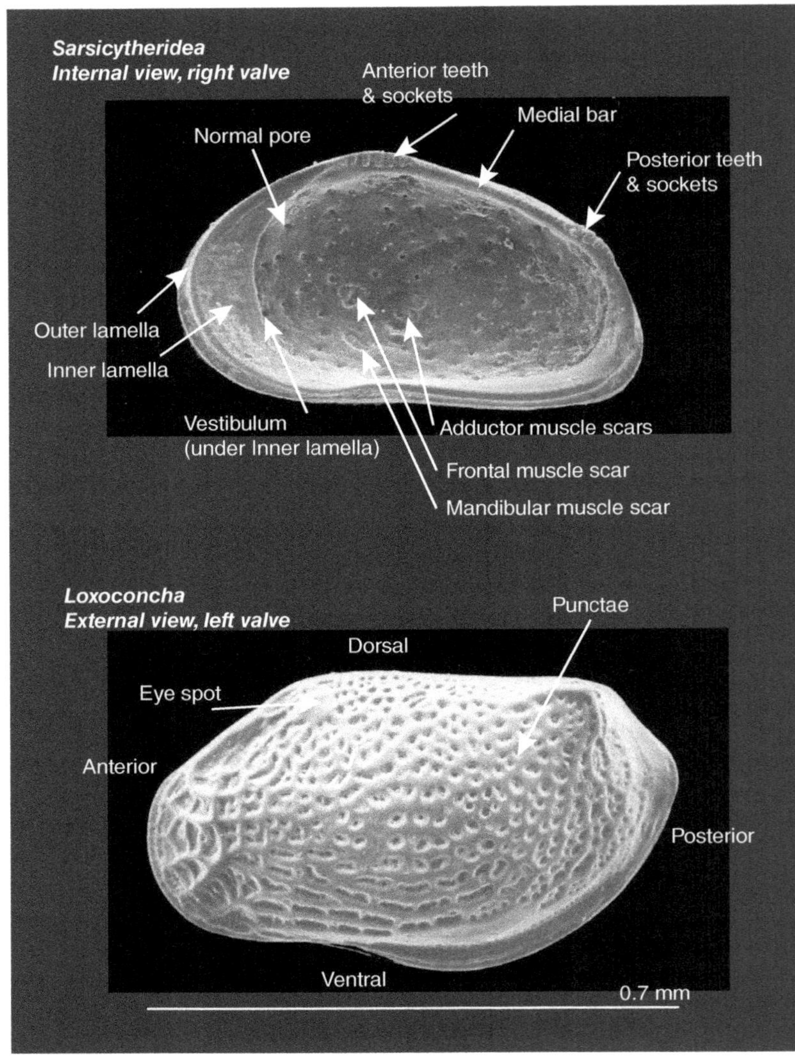

Fig. 16.1. Scanning electron photos of benthic ostracods *Sarsicytheridea* (top, internal view, right valve) from Late Pleistocene Champlain Sea, Quebec, Canada, and *Loxoconcha* (bottom, external view, left valve) from the continental shelf off North Carolina, US. Morphological features used in genus and species identification are shown. For discussion of basic ostracod morphology and taxonomy see Horne et al. (2002) and Rodriguez-Lazaro and Ruiz-Muñoz (2012).

sea-level studies are benthic in habitat, reproduce sexually, and either retain or deposit their eggs. With the exception of those taxa whose eggs are dispersed by birds (*Cyprideis*), or passively by floating marine algae, the lack of a mobile larval stage results in a high level of endemism for ostracods inhabiting coastal zones.

Intra-specific shell morphological variability comes from ostracod ontogeny, sexual dimorphism, and valve asymmetry (Rodriguez-Lazaro and Ruiz-Muñoz, 2012). Ostracods grow in discreet increments by molting (ecdysis) (Kesling, 1951), usually 7–9 molt stages, such that fossil assemblages consist of both adult and juvenile specimens. Shells of adults and 3–4 pre-adult molt stages are often represented in sediment from the >125 micron size fraction used in many microfaunal analyses. Furthermore, sexual dimorphism in carapace size and shape and valve overlap (either the right or left valve is larger and/or a different shape than the other) must be considered in taxonomic and ecological study of modern or fossil ostracods. Although shell morphology (e.g., sieve pore morphology, noding, carapace ornamentation, etc.) and chemistry (stable isotopes, minor element ratios) are used for paleosalinity reconstruction, faunal assemblage analysis (see the following section) is the most common approach for sea-level reconstruction. A useful online reference for further information on ostracode morphology is University College London's website (http://www.ucl.ac.uk/GeolSci/micropal/ostracod.html), and Cohen et al. (2007) provide a list of morphological terms for living ostracods.

16.4 PRESERVATION IN SEDIMENTS

Oceanographic and sedimentological processes affect the preservation of microfossil assemblages in sediments and potentially introduce biases, such as selective transport of juveniles or smaller species, mixing (time-averaging) of populations from several years or longer, and shell dissolution. Burrowing and bioturbation can also mix formerly distinct assemblages up to several centimeters in the sediment. The taphonomy of fossil ostracod assemblages – that is, the degree to which fossil assemblages represent life assemblages – involves the assessment of these biases (Lord et al., 2012). In practice, stratigraphic intervals covering a freshwater to marine transition caused by sea-level rise often contain a sequence characterized by non-marine assemblages, "mixed" assemblages (including both non-marine and marine species), and brackish to marine assemblages with no non-marine species. Mixing of ecologically distinct assemblages in a freshwater to marine transition zone is influenced by the rate of sea-level rise, sources and rates of clastic sediment input, geomorphology of coastal habitats, tidal range, and other factors. In addition to understanding species' ecology, it is necessary to understand local coastal processes to interpret patterns and rates of sea-level change from sediment records.

Dissolution of ostracod calcitic shells can alter the original biocoenoses (life assemblages) either *in situ* (i.e., within the sediment), or after taking sediment cores but before laboratory processing of sediment. This problem applies especially to tidal marsh and estuarine environments. While *in situ* dissolution in sediments is difficult to assess, rapid refrigeration of sediment cores (usually at 2–4 °C) can inhibit dissolution; laboratory processing (washing/sieving) of marsh sediments as soon as possible after taking sediment cores is preferred and should eliminate or minimize dissolution artifacts.

Biological characteristics discussed above distinguish ostracods from other groups, such as diatoms and foraminifera, as valuable tools in paleo-sea-level studies (see De Deckker, 2002). For example, the preservation of a large number of carapaces (two valves still articulated) rather than disarticulated valves depends on morphological factors (dorsal hinge structure, ligament, adductor muscles), as well as environmental conditions (waves, currents, sedimentary processes). As a general rule, preservation of carapaces is more common in low-energy environments when little or no transport has occurred. Population structure – the relative proportions of adults and juveniles of a species – can be an important element in sea-level research, especially in the reconstruction of rapid sea-level rise over centuries or less. In the idealized case where an individual with 8 molt stages reached adulthood, the individual would contribute 14 juvenile valves (two valves, disarticulated during ecdysis, for each juvenile molt) and either 2 adult valves or 1 adult carapace to a sediment sample (assuming all size fractions were studied). In the case of high juvenile mortality, there will be fewer adults than juveniles and perhaps articulated juvenile carapaces that died before reaching adulthood. In any case, knowing the population structure of key species tells the researcher whether or not mixing of assemblages from different habitats occurred.

16.5 BIOGEOGRAPHY AND ECOLOGY

At large spatial scales such as shallow marine biogeographic provinces, ocean temperature required for species' reproduction and/or survival is a major factor controlling the distribution of genera and species on continental shelves (Hazel, 1970; Cronin and Dowsett, 1990; Ikeya and Cronin, 1993; Wood et al., 1993). Conversely, at regional and local spatial scales along coasts, ecological and habitat-related factors controlling the abundance and distribution of species are salinity (including daily, seasonal and interannual salinity variability), substrate, and other resources such as food. Most coastal regions exhibit a high degree of habitat heterogeneity and strong, often fluctuating, environmental gradients. Coastal ostracods from Europe, the Arctic, North America, and Asia shown in Figures 16.2 and 16.3 exemplify the wide range of species and morphological diversity. In addition, even when two or more species co-occur in the same salinity zone within a bay, lagoon, or other aquatic system, they can have distinct, seasonally varying life cycles (King and Kornicker, 1970; Horne, 1983), or habitat preferences, such as sand versus phytal dwelling (Kamiya, 1988).

The vertical resolution for paleo-sea-level estimates derived from microfossil assemblages is often based on species' preferred depth ranges or habitats in relationship to high or mean tide levels. Establishing an "error bar" for past sea levels is a critical aspect of paleoecological analyses of all fossil groups used in sea-level studies (benthic foraminifera, diatoms, corals, and ostracods; Barlow et al., 2013). Typically, preferred depth ranges of

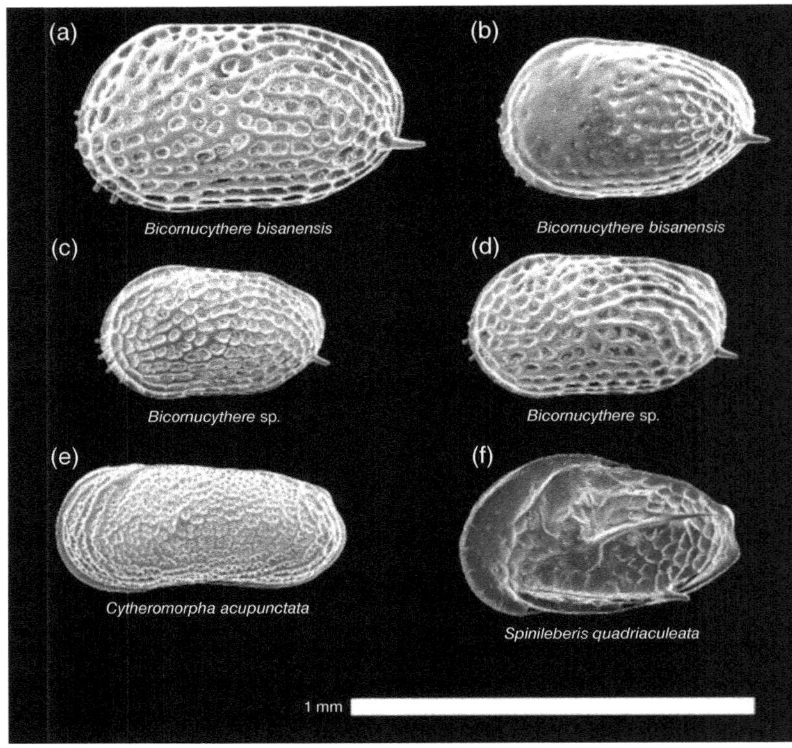

Fig. 16.2. Scanning electron micrographs of ostracode fossils from core HIR94-3 (Japan). Scale bar: 1 mm. All specimens are lateral views of left valves. (a) *Bicornucythere bisanensis*, adult, female, sample no. H36. (b) *Bicurnucythere bisanensis*, A-1 instar, sample no. H36. (c) *Bicurnucythere* sp., A-1 instar, sample no. H52. (d) *Bicornucythere* sp., adult, female, sample no. H52. (e) Cytheromorpha acupunctata, adult, male, sample no. H49. (f) *Spinileberis quadriaculeata* adult, male, sample no. H52. *Source*: Photograph by M. Yasuhara. Reproduced with permission.

coastal ostracod species are less than 10 m water depth (e.g., Yasuhara and Seto, 2006), but many tidal marsh species are dominant in water less than a few meters deep, in some cases due to substrate preferences. For example the depth limits of phytal species, such as those in the genus *Loxoconcha*, are limited by light penetration (usually a few meters in estuaries) which controls seasonal growth of the host vegetation (the sea grass *Zostera*; Kamiya, 1988; Vann et al., 2004). Macrobenthic algae also host ostracod species living within a few meters of sea level, especially in subtropical and tropical regions (e.g., Triantaphyllou et al., 2005).

In sum, understanding the complex ecology of coastal ostracods requires field studies of the distribution and abundance of living populations across salinity gradients and in relation to substrate, submerged aquatic vegetation, and tidal patterns (Boomer, 1998).

16.6 QUANTITATIVE FAUNAL ANALYSES

Due to high endemism and habitat heterogeneity discussed above, species diversity of non-marine, brackish, and marine species can be quite high for any coastal region. A few examples illustrate this point: 105 species in lagoon and carbonate platform environments off Belize (Teeter, 1975); 129 species in the Baltic Sea (Frenzel et al., 2010); 35 species along the mangrove coast of southwest Florida (Keyser, 1975a, 1975b); 35–39 species in Texas bays and lagoons (King and Kornicker, 1970; Garbett and Maddocks, 1979); 81 species in Osaka Bay, Japan (Yasuhara et al., 2004); and about 30 species in Chesapeake Bay (Tressler and Smith, 1948; Cronin and Vann, 2003).

High diversity and complex niche-partitioning among species has led ostracod workers to apply multivariate faunal analyses such as transfer functions, modern analog technique (MAT), cluster analyses, detrended correspondence analysis (DCA) (Viehberg and Mesquita-Joanes, 2012) to characterize modern and fossil ostracode assemblages. These methods involve analysis of species' relative frequencies (RF, percent abundance of each species out of the total ostracod assemblage) in relation to environmental parameters. Modern assemblages are then applied to analysis of fossil assemblage composition from sediments. A total of 300 individuals per sample is a standard used in quantitative microfaunal analysis, but identifying statistically significant temporal assemblage

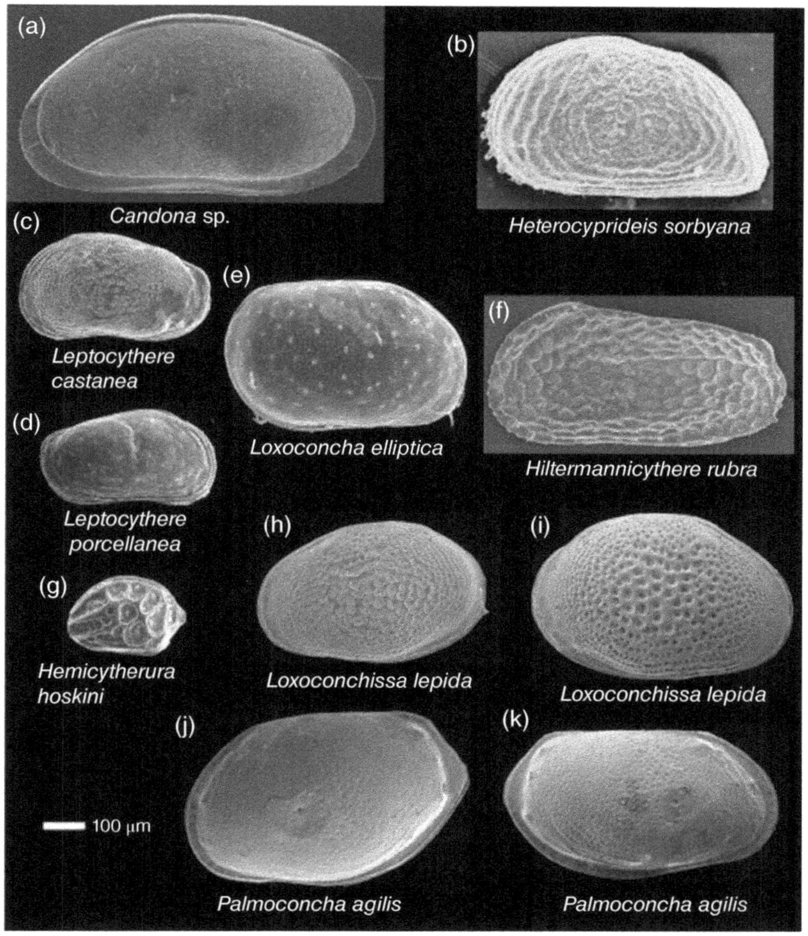

Fig. 16.3. Scanning electron micrographs of ostracode fossils. All external view except (a). Scale bar: 100 μm. (a) *Candona* sp internal view, right valve. *Source*: Rayburn et al., 2011. Reproduced with permission of Elsevier; (b) *H. sorbyana*, external view, right valve; (c) *Leptocythere castanea* external view, left valve; (d) *Leptocythere porcellanea* external view, right valve; (e) *Loxoconcha elliptica* external view, right valve; (f) *Hiltermannicythere rubra* external view, left valve, male; (g) *Hemicytherura hoskini* external view, left valve; (h) *Loxoconchissa (Loxocaspia) lepida*, external view, left valve, male; (i) *Loxoconchissa (Loxocaspia) lepida*, external view, right valve, female; (j) *Palmoconcha agilis* external view, left valve, male; (k) *Palmoconcha agilis* external view, right valve, female. (a) from post-glacial Lake Vermont, New York (Rayburn et al., 2011); (b) from Laptev Sea (Stepanova et al., 2007), courtesy A. Stepanova; (c–e, g) from Tina Menor Estuary, Northern Spain, (Martínez García et al., 2013), courtesy J. Rodriguez-Lazaro; (f, h–k) from Black Sea Holocene (Ivanova et al., 2007), courtesy M. Zenina.

changes can be accomplished with smaller sample sizes if there are large temporal changes in RFs of dominant, ecologically sensitive species (Buzas, 1990). In some cases, temporal patterns in a few dominant species can provide convincing evidence for sea-level change.

16.7 CASE STUDIES

Micropaleontological studies have used ostracods preserved in marine and brackish water sediments to infer changes in sea-level changes during the Phanerozoic over timescales ranging from 10^3 to 10^7 years. Cyclostratigraphy of orbital-scale sea-level oscillations, for example, uses marginal marine ostracods as proxies for sea-level positions in Cretaceous (Cenomanian–Turonian) sediments of the Colorado Plateau (Tibert et al., 2003), Messinian (5.4–5.3 Ma) sediments of the Adriatic (Cosentino et al., 2006, 2011), and Quaternary deposits in the eastern US (Cronin et al., 1981).

In the majority of cases, however, ostracods have been applied to late Quaternary (including

Table 16.1. Application of ostracodes to sea-level reconstruction

Region	Ostracode taxa	Sea level event	References
Chesapeake Bay	*Cytheromorpha*	8.2 ka Holocene climate anomaly	Cronin et al. (2007a, b)
British Isles	Various	Various	Athersuch et al. (1989)
Bahamas	*Candona, Limnocythere, Cypridopsis, Perissocytheridea, Cyprideis, Bairdia, Aurila, Loxoconcha, Xestoleberis*	Late Holocene	Teeter (1995)
Lake Champlain, New York	*Candona, Cytheropteron*	Post-glacial Champlain Sea, Younger Dryas	Rayburn et al. (2011); Cronin et al. (2011)
Black Sea	Candonidae, Leptocytheridae and Loxoconchidae	Early Holocene	Boomer et al. (2010)
Osaka Bay	*Loxoconcha, Bicornucythere, Cytheromorpha, Spinileberis* others	Early Holocene	Irizuki et al. (2001); Yasuhara et al. (2004)
Baltic Sea	Candonids, *Metacypris, Ilyocypris, Cyprideis, Cytheromorpha*	*Littorina* Sea, early Holocene	Wastegård et al. (1995); Viehberg et al. (2008)
Mediterranean	*Cyprideis, Ilyocypris, Loxoconcha*	Holocene	Primavera et al. (2011)
Bonaparte Gulf, Australia	*Tanella, Neocytheretta, Leptocythere*	Early deglacial (19.4 ka)	De Deckker and Yokoyama (2009)
East Anglia, UK	Various	Early Holocene	Boomer and Godwin (1993)

Holocene) sea-level reconstruction, including the following examples (Table 16.1):

- Rapid sea-level rise during early deglaciation (post-19.4 ka cal BP) in Bonaparte Gulf, Australia (De Deckker and Yokoyama, 2009).
- Holocene and Pleistocene sea-level changes in the British Isles (Athersuch et al., 1989; Boomer and Godwin, 1993).
- The Younger Dryas-age (~13 ka cal BP) Champlain Sea marine transgression (Rayburn et al., 2011).
- Rapid early Holocene sea-level rise at the end of the last deglaciation (Irizuki et al., 2001; Yasuhara et al., 2004; Yasuhara and Seto, 2006; Cronin et al., 2007a; Viehberg et al., 2008; Yasuhara, 2008).
- Abrupt Holocene flooding of the Black Sea (Boomer et al., 2010) and Tampa Bay (Cronin et al., 2007b).
- Mid- to late Holocene sea-level changes in the Mediterranean (Primavera et al., 2011).
- Small but significant late Holocene sea-level oscillations (Boomer, 1993; Teeter, 1995; Boomer et al., 2009; Mazzini et al., 2011).

One common feature in these studies is that ostracod microfaunal analyses were part of broader, multidisciplinary assessments of sea level that used other proxy methods (foraminifera, diatoms, palynology, geochemistry), lithostratigraphy, geochronology, and/or geophysical analyses of isostatic, tectonic, or other processes. It is therefore fair to say that, given their suitability for paleo-sea-level reconstruction based on their ecology and preservation in coastal sediments, ostracods hold great promise in future studies of past sea-level changes.

ACKNOWLEDGEMENTS

Thanks go to I. Shennan for the invitation to write this chapter and to I. Boomer, G. Dwyer, J. Hazel, J. Holmes, D. Horne, N. Ikeya, A. Lord, J. Rodriguez-Lazaro, R. Whatley, and M. Yasuhara for sharing their expertise on ostracode over the years. Special thanks are due to J. Rodriguez-Lazaro, A. Stepanova, M. Yasuhara and M. Zenina for SEM photographs and L. Gemery, J. Repetski and three excellent anonymous reviewers for critical reviews. This study was funded by US Geological Survey Climate & Land Use Change and Priority Ecosystems Programs.

REFERENCES

Athersuch, J., Horne, D. J., and Whittaker, J. E. (1989) Marine and brackish water ostracods (superfamilies Cypridacea and Cytheracea): keys and notes for the identification of the species. Synopses of the British Fauna (New Series) No. 43, Leiden, New York. Published for the Linnean Society of London and the Estuarine and Brackish-water Sciences Association by E.J. Brill, 343 pp.

Barlow, N.L.M., Shennan, I., Long, A.J., Gehrels, W.R., Saher, M.H, Woodroffe, S.A., and Hillier, C. (2013) Salt marshes as late Holocene tide gauges. Global and Planetary Change, 106, 90–110.

Boomer, I. (1993) Palaeoenvironmental indicators from Late Holocene and contemporary ostracoda of the Aral Sea. Palaeogeography, Palaeoclimatology, Palaeoecology, 103, 141–153.

Boomer, I. (1998) The relationship between meiofauna (Ostracoda, Foraminifera) and tidal levels is modern intertidal environments of North Norfolk: a tool for palaeoenvironmental reconstruction. Bulletin of the Geological Society of Norfolk, 46, 17–29.

Boomer, I., and Godwin, M. (1993) Palaeoenvironmental reconstruction in the Breydon Formation, Holocene of East Anglia. Journal of Micropalaeontology, 12, 35–46.

Boomer, I., and Eisenhauer, G. (2002) Ostracod faunas as palaeoenvironmental indicators in marginal marine environments. In: *The Ostracoda: Applications in Quaternary Research* (eds Holmes, J., and Chivas, A.), American Geophysical Union Monograph 131, 135–149.

Boomer, I., Wunnemann, B., Mackay, A.W., Austin, P., Sorrel, P., Reinhardt, C., Keyser, D., Guichard, F., and Fontugne, M. (2009) Advances in understanding the late Holocene history of the Aral Sea region. Quaternary International, 194, 79–90.

Boomer, I., Guichard, F., and Lericolais, G. (2010) Late Pleistocene to Recent ostracod assemblages from the western Black Sea. Journal of Micropaleontology, 29, 119–133.

Buzas, M. A. (1990) Another look at confidence limits for species proportions. Journal of Paleontology, 64, 842–843.

Cohen, A. C., Peterson, D. E., and Maddocks, R. F. (2007) Ostracoda. In: *The Light & Smith Manual: Intertidal Invertebrates from Central California to Oregon*, 4th edition (ed. Carlton, J. T.), University of California Press, Berkeley and Los Angeles, pp. 417–446.

Cosentino D., Federici I., Cipollari P., and Gliozzi, E. (2006) Environments and tectonic instability in central Italy (Garigliano Basin) during the late Messinian Lago-Mare episode: New data from the onshore Mondragone well (Garigliano Plain, central Italy). Sedimentary Geology, 188/189, 293–317.

Cosentino, D., Bertini, A., Cipollari, P., Florindo, F., Gliozzi, E., Lo Mastro1, S., and Sprovieri, M. (2011) Orbitally forced paleoenvironmental and paleoclimate changes in the late postevaporitic Messinian of the Central Mediterranean Basin. Geological Society of America Bulletin, 124, 499–516.

Cronin, T.M. (1987) Quaternary sea-level studies in the Eastern United States of America: a methodological perspective. In: *Sea-level Changes* (eds Tooley, M. J., and Shennan, I.), Blackwell Ltd, Oxford, pp. 225–248.

Cronin, T.M., and Dowsett, H.J. (1990) A quantitative micropaleontologic method for shallow marine paleoclimatology: Application to Pliocene deposits of the Western North Atlantic Ocean. Marine Micropaleontology, 16, 117–148.

Cronin, T. M., and Vann, C. D. (2003) The sedimentary record of climatic and anthropogenic influence on the Patuxent Estuary and Chesapeake Bay. Estuaries, 26, 196–209.

Cronin, T.M., Szabo, B.J., Ager, T.A., Hazel, J.E., and Owens, J.P. (1981) Quaternary climates and sea levels, US Atlantic Coastal Plain. Science, 211, 233–240.

Cronin, T. M., Vogt, P. R., Willard, D. A., Thunell, R. Halka, J. Berke, M., and Pohlman, J. (2007a) Rapid sea level rise and ice sheet response to 8200-year climate event. Geophysical Research Letters, 34, L20603, 1–6.

Cronin, T. M., Edgar, T. Brooks, G., Hastings, D., Larson, R., Hine, A., Locker, S., Suthard, B., Flower, B., Wehmiller, J., Willard, D., and Smith, S. (2007b) Sea level rise in Tampa Bay. EOS, Transactions of the AGU, 88(10), 117–118.

Cronin, T. M., Rayburn, J. A., Guilbault, J.-P., Thunell, R., and Franzi, D. A. (2011) Stable isotope evidence for glacial lake drainage through the Saint Lawrence Estuary, eastern Canada, ~13.1–12.9 ka. Quaternary International, 260, 55–65.

De Deckker, P. (2002) Ostracod palaeoecology. In: *Applications of the Ostracoda in Quaternary Research* (eds Holmes, J., and Chivas, A.R.), American Geophysical Monograph, 131, pp. 121–134.

De Deckker, P., and Yokoyama, Y. (2009) Micropalaeontological evidence for Late Quaternary sea-level changes in Bonaparte Gulf, Australia. Global and Planetary Change, 66, 85–92.

De Deckker, P., Chivas, A. R., Shelley, J.M.G., and Torgersen, T. (1988) Ostracod shell chemistry: A new paleoenvironmental indicator applied to a regressive/transgressive record in the Gulf of Carpentaria, Australia. Palaeogeography, Palaeoclimatology, Palaeoecology, 66, 231–241.

Dettman, D. L., and Dwyer, G. S. (2012) Chapter 9: The calibration of environmental controls on elemental ratios in ostracod shell calcite: a critical assessment. In: *Ostracoda as Proxies for Quaternary Climate Change* (eds Horne, D. J., Holmes, J. A., Rodriguez-Lazaro, J., and Viehberg, F. A.), Elsevier, Developments in Quaternary Sciences, 17, pp. 145–163.

Frenzel, P., and Boomer, I. (2005) The use of ostracods from marginal marine, brackish waters as bioindicators of modern and Quaternary environmental change. Palaeogeography, Palaeoclimatology, Palaeoecology, 225, 68–92.

Frenzel, P., Keyser, D., and Viehberg, F. A. (2010) An illustrated key and (palaeo)ecological primer for Postglacial to Recent Ostracoda (Crustacea) of the Baltic Sea. Boreas, 39, 567–575.

Garbett, E. C., and Maddocks, R. F. (1979) Zoogeography of holocene Cytheracean ostracodes in the bays of Texas. Journal of Paleontology, 53, 841–919.

Hazel, J. E. (1970) Atlantic continental shelf and slope of the United States – Ostracode zoogeography in the southern Nova Scotian and northern Virginian faunal provinces. US Geological Survey Professional Paper 529-E, pp. E1–E21.

Holmes, J. A., and De Deckker, P. (2012) Chapter 8: The chemical composition of ostracod shells: applications in Quaternary palaeoclimatology. In: *Ostracoda as Proxies for Quaternary Climate Change* (eds Horne, D. J., Holmes, J. A., Rodriguez-Lazaro, J., and Viehberg, F. A.), Elsevier, Developments in Quaternary Sciences 17, 131–143.

Horne, D. J. (1983) Life-cycles of podocopid Ostracoda, a review (with particular reference to marine and brackish water species). In: *Applications of Ostracoda* (ed. Maddocks, R. F.), Department of Geosciences, University of Houston, pp. 581–590.

Horne, D. J., Cohen, A., and Martens, K. (2002) Taxonomy, morphology, and biology of Quaternary and living Ostracoda. In: *The Ostracoda: Applications in Quaternary Research* (eds Holmes, J., and Chivas, A.), American Geophysical Union Monograph, 131, 5–36.

Horne, D.J., Smith, R.J., Whittaker, J.E., and Murrary, J. (2004) The first British record and a new species of the superfamily Terrestricytheroidea (Crustacea, Ostracoda): morphology, ontogeny, lifestyle and phylogeny. Zoological Journal of the Linnean Society, 142, 253–288.

Ikeya, N., and Cronin, T.M. (1993) Quantitative analysis of Ostracoda and water masses around Japan: application to Neogene paleoceanography. Micropaleontology, 39, 263–281.

Ikeya, N., Tsukagoshi, A., and Horne, D. J. (2005) Preface: The phylogeny, fossil record and ecological diversity of ostracod crustaceans. Hydrobiologia, 538, vii–xiii.

Irizuki, T., Masuda, F., Miyahara, B., Hirotsu, A., Ueda, S., and Yoshikawa, S. (2001) Vertical changes of Holocene ostracodes in bore hole cores from off Kobe, related to the opening on straits and relative sea-level change in western Japan. The Quaternary Research, 40, 105–120.

Ivanova, E., Murdmaa, I. O., Chepalyga, A. L., Cronin, T. M., Pasechnik, I. V., Levchenko, O. V., Howe, S. S., Manushkina, A. V., and Platonova, E. A. (2007) Holocene sea-level oscillations and environmental changes on the Eastern Black Sea shelf. Palaeogeography, Palaeoclimatology, Palaeoecology, 246, 228–259.

Kamiya, T. (1988) Contrasting population ecology of two species of Loxoconcha (Ostracoda, Crustacea) in recent Zostera (eelgrass) beds: adaptive differences between phytal and bottom-dwelling species. Micropaleontology 34, 316–331.

Kesling R.V. (1951) The morphology of ostracod molt stages. Illinois Biological Monographs, 21, 1–324.

Keyser, D. (1975a) Ostracode of the mangroves of south Florida, their ecology and biology: Bulletin American Paleontology, 65(282), 489–499.

Keyser, D. (1975b) Ostracoden aus den Mangrovegebieten von sudwest-Florida. Abhandlung und Verhandlungen des Naturwissenschaften, 18/19, 255–290.

King, C. E., and Kornicker, L. S. (1970) Ostracoda in Texas bays and lagoons: an ecologic study. Smithsonian Contributions to Zoology, 24, 1–92.

Lord, A. R., Boomer, I. Brouwers, E., and Whittaker, J. E. (2012) Chapter 3: Ostracod taxa as paleoclimatic indicators in the Quaternary. In: *Ostracoda as Proxies for Quaternary Climate Change* (eds Horne, D. J., Holmes, J. A., Rodriguez-Lazaro, J., and Viehberg, F. A.), Elsevier, Developments in Quaternary Sciences 17, p. 37–46.

Martínez García, B., Pascual, A., Rodríguez-Lázaro, J., Martín-Rubio, M., and Rofes, J. (2013) The Ostracoda (Crustacea) of the Tina Menor estuary (Cantabria, southern Bay of Biscay): Distribution and ecology. Journal of Sea Research, 83, 111–122.

Mazzini, I., Faranda, C., Giardini, M., Giraudi, C., and Sadori, L. (2011) Late Holocene palaeoenvironmental evolution of the Roman harbour of Portus, Italy. Journal of Paleolimnology, 46, 243–256.

Penney, D.N. (1987) Application of Ostracoda to sea-level studies. Boreas, 16, 237–247.

Primavera, M., Simone, O. Fiorentino, G., and Caldara, M. (2011) The palaeoenvironmental study of the Alimini Piccolo lake enables construction of Holocene sea-level changes in southeast Italy. The Holocene, 21, 553–563.

Rayburn, J. A., Cronin, T. M., Franzi, D. A., Knuepfer, P. L. K., and Willard, D. A. (2011) Timing and duration of glacial lake discharges and the Younger Dryas climate reversal. Quaternary Research, 75, 541–551.

Rodriguez-Lazaro, J., and Ruiz-Muñoz, F. (2012) Chapter 1: A general introduction to ostracods: morphology, distribution, fossil record and applications. In: *Ostracoda as Proxies for Quaternary Climate Change* (eds Horne, D. J., Holmes, J. A., Rodriguez-Lazaro, J., and Viehberg, F. A.), Elsevier, Developments in Quaternary Sciences 17, p. 1–14.

Stepanova, A., Taldenkova, E., Simstich, J., and Bauch, H. A. (2007) Comparison study of the modern ostracod associations in the Kara and Laptev seas: ecological aspects. Marine Micropaleontology, 63, 111–142.

Teeter, J. W. (1975) Distribution of Holocene marine Ostracoda from Belize. In: *Belize Shelf-Carbonate Sediments, Clastic Sediments, and Ecology* (eds Wantland, K.F., and Pusey, C., III), American Association of Petroleum Geologists, Studies in Geology No. 2, pp. 400–498.

Teeter, J. W. (1995) Holocene saline lake history, San Salvador Island, Bahamas. Geological Society of America, Special Paper 300, 117–124.

Tibert, N. E., Leckie, R. M., Eaton, J. G., Kirkland, J. I., Colin, J.-P., Leithold, E., and McCormic, L. E. (2003) Recognition of relative sea level change in upper Cretaceous coal- bearing strata: A paleoecological approach using agglutinated foraminifera and ostracodes to detect key stratigraphic surfaces, In: *Microfossils As Proxies For Sea Level Change and Stratigraphic Discontinuities* (eds Olson, H., and Leckie, R. M.), SEPM (Society for Sedimentary Geology), Special Publication 75, 263–299.

Tressler, W. L., and Smith, E. M. (1948) An ecological study of seasonal distribution of Ostracoda, Solomons Island, Maryland, region. Chesapeake Biological Laboratory Publication No. 71. Solomons Island, Maryland.

Triantaphyllou, M. V., Tsourou, T., Koukousioura, O., and Dermitzakis, M. D. (2005) Foraminiferal and ostracod ecological patterns in coastal environments of SE Andros Island (Middle Aegean Sea, Greece). Revue de Micropaléontologie, 48, 279–302.

Vann, C. D., Cronin, T. M., and Dwyer, G. S. (2004) Population ecology and shell chemistry of a phytal ostracode species (Loxoconcha matagordensis) in the Chesapeake Bay watershed. Marine Micropaleontology, 53, 261–277.

Viehberg, F. A., and Mesquita-Joanes, F. (2012) Quantitative transfer function approaches in palaeoclimatic reconstruction using quaternary ostracods. In: *Ostracoda as Proxies for Quaternary Climate Change* (eds Horne, D.J., Holmes, J.A., Rodgriguez-Lazaro, J., and Viehberg, F.A.), Elsevier, Developments in Quaternary Science, 17, 47–64.

Viehberg, F. A., Frenzel, P., and Hoffmann, G. (2008) Succession of late Pleistocene and Holocene ostracode assemblages in a transgressive environment: A study at a coastal locality of the southern Baltic Sea (Germany). Palaeogeography, Palaeoclimatology, Palaeoecology, 264, 318–329.

Wastegård, S., Andrén, T., Sohlenius, G., and Sandgren, P. (1995) Different phases of the Yoldia Sea in the north-western Baltic Proper. Quaternary International, 27, 121–129.
Wood, A. M., Whatley, R. C., Cronin, T. M., and Holtz, T. (1993) Pliocene paleotemperature reconstruction for the southern North Sea based on Ostracoda. Quaternary Science Reviews, 12, 747–768.
Yasuhara, M. (2008) Holocene ostracod palaeobiogeography of the Seto Inland Sea, Japan: Impact of opening of the strait. Journal of Micropalaeontology, 27, 111–116.
Yasuhara, M., and Seto, K. (2006) Holocene relative sea-level change in Hiroshima Bay, Japan: a semi-quantitative reconstruction based on ostracodes. Paleontological Research (Japan), 10 (2), 99–116.
Yasuhara, M., Irizuki, T., Yoshikawa, S., Nanayama, F., and Mitamura, M. (2004) Holocene ostracode paleobiogeography in Osaka Bay, southwestern Japan. Marine Micropaleontology, 53, 11–36.

Chapter 17
Mollusca

JESSICA E. PILARCZYK[1,2] AND DONALD C. BARBER[3]

[1]*Institute of Marine and Coastal Sciences, Rutgers, The State University of New Jersey, New Brunswick, NJ, USA*
[2]*Earth Observatory of Singapore, Nanyang Technological University, Singapore*
[3]*Department of Geology, Bryn Mawr College, Bryn Mawr, PA, USA*

17.1 INTRODUCTION

Mollusks dominate marine and freshwater settings and are useful for sea-level reconstruction because they are highly diverse in habitat and often abundant in the sedimentary record (e.g., Beckvar and Kidwell, 1988; Anderson and McBride, 1996; Aguirre et al., 2005; Martinez and Rojas, 2013). Sea-level highstand deposits (e.g., marine terraces) often contain well-preserved shell concentrations that, when combined with chronological data, are useful in delineating former sea-level histories (Aguirre et al., 2005; Muhs et al., 2012). In rare cases, the shells of sessile molluscan bivalves (e.g., *Mytilus edilus*) or gastropods (e.g., Vermetidae) with restricted depth habitats are found in life position, thus providing unusually good sea-level indicators. During episodes of relative sea-level rise, landward migration (transgression) of the shoreface erodes former inshore deposits and leaves a lag of reworked shell material behind. Major challenges in using shell beds for reconstructing sea-level include not only the ability to differentiate between lags and highstands, but the ability to differentiate sea-level deposits from shell accumulations caused by other processes such as storms and tsunamis. These challenges can be addressed by employing a combined paleoecologic (i.e., taxonomic) and taphonomic approach where the surface condition of individual shells is used to assess transport and mode of deposition (e.g., Aguirre and Farinati, 1999; Reinhardt et al., 2006). Shell bed genesis can be classified based on two main genetic processes: (1) biogenic (resulting from the activities of shell producers); and (2) sedimentologic (resulting from sea-level change, extreme coastal events, or tidal channel migration, e.g., Kidwell et al., 1986). In theory, each depositional style creates a unique taphonomic signature preserved within the shell bed (Table 17.1). However, post-depositional reworking (e.g., episodes of exhumation, burial, and bioturbation) can modify the original signature, making identification of a shell bed difficult (e.g., Beckvar and Kidwell, 1988; Davies et al., 1989; Anderson and McBride, 1996).

This chapter begins with a review of the methodological approaches for documenting shell beds in section and in the laboratory. The final sections address the relationship between time-averaging of the fossil record and reconstructing former sea-level position.

17.2 METHODOLOGY

17.2.1 Background

When using shell beds to reconstruct sea-level histories, it is important to properly identify how they formed and the degree to which they have been reworked (e.g., Aguirre and Whatley, 1995). Various methods have been developed to address the paleoenvironmental setting, bioclast source (provenance), degree of time-averaging, and accumulation rate. One of the main challenges in documenting a shell bed is recognizing its mechanism of accumulation (e.g., Kidwell, 1991). In addition to sea-level highstand or transgressive lag deposits, shell accumulations can also result from high-energy events (storms and tsunamis), or highly productive species (biogenic deposits; e.g., Kidwell et al., 1986). The combined paleoecologic and taphonomic approach has enabled researchers to distinguish among shell beds formed by these varying processes (Table 17.1; e.g., Davies et al., 1989; Angulo et al., 2006; Reinhardt et al., 2006; Martinez and Rojas, 2013). Paleoecologic (taxonomic) analysis provides information

Handbook of Sea-Level Research, First Edition. Edited by Ian Shennan, Antony J. Long, and Benjamin P. Horton.

Table 17.1. Mechanisms of shell bed formation and their corresponding sedimentologic and taphonomic characteristics

Shell bed type	Sedimentological characteristics	Taphonomic characteristics
Biogenic concentrations		
	• Irregular basal contact, poorly sorted • massive or graded, bioturbated, mottled • occur in areas with very low sedimentation rates	• Range of shell preservation states, with low preservation dominating; • high encrustation and boring
Sedimentologic concentrations		
Storms (washover)	• Erosional base, gradational upper contact • Generally massive, poorly-sorted, cross-stratified • Medium-scale erosion and transport	• Random bioclasts, nested shells, imbrications • Low articulation, mixed assemblage
Storms (*in situ*)	• Rapid burial of the *in situ* community • Matrix composed of marine sand	• High articulation at base of shell bed, low articulation at top • Moderate fragmentation with both rounded and angular edges • Allochthanous sps. dominate base, mixed assemblage dominates the top • Some articulated shells oriented convex-up (in life position)
Tsunamis	• Erosional base • Matrix composed of marine sand • Massive or graded • Large scale erosion and transport • Rapid burial	• Highly concentrated shells, high articulation, random fabric • High degree of fragmentation, fragments dominantly angular • Presence of allochthanous offshore species • Abundant articulated bivalves oriented out of life position • Bifurcating stress fractures
Transgressive lag (sea-level change)	• Evidence of reworking and transport	• Poorly sorted bioclasts • High encrustation and boring; high corrasion; high edge rounding • Low articulation • Shell assemblage is dominated by autochthanous species
Tidal channel lag	• Erosional base, gradational upper contact • Well sorted, graded	• Poorly preserved bioclasts; heavy abrasion, encrustation, and boring • Articulated shells oriented convex-up (in life position) • Autochthanous assemblage • Tidal currents can concentrate and imbricate shell fabric

concerning the depositional environment (e.g., autochthonous assemblage) and possible transport of mollusks (e.g., mixed assemblage), whereas taphonomic analysis provides further insight into the degree and type of transport and reworking (e.g., Davies et al., 1989; Aguirre et al., 2011).

Shell middens are subaerial cultural deposits comprised primarily of mollusk shells. In most cases, middens are readily distinguished from the types of shell beds considered in this chapter due to the morphology of the shell heaps, the taphonomic signatures resulting from selective harvesting, cooking, and consumption, and the presence of other cultural remains (Stein, 1992). The allochthonous constituents of shell middens do not provide direct sea-level indicators, but the results of archeological studies on coastal shell middens are often integrated into regional sea-level reconstructions (e.g., Fairbridge, 1976; Angulo et al., 2006). In arid environments, deflation of eolian deposits containing numerous small shell middens can produce thin layers of mixed shells (Händel, 2009) that, upon initial examination, may resemble an event shell bed (e.g., storm overwash or tsunami deposit). Other archeological, sedimentologic, and geomorphological evidence will generally reveal the cultural origin of reworked shell middens (Händel, 2009).

17.2.2 Field sampling

For locations where molluscan taxa and taphonomic trends are not well documented, a modern analog is useful in establishing taxonomic distributions, which can then be applied down-core or in stratigraphic section to reconstruct the paleoenvironment (e.g., Aguirre et al., 2011). This is accomplished by determining relevant sub-environments for sampling, and then using a 30 cm × 30 cm metal frame to collect the upper 5 cm of sediment and shells (providing 4500 cm^3 of sample material). Care must be taken to ensure that samples are not taphonomically overprinted through sampling (e.g., fragmentation and disarticulation).

The method for sampling shell beds in the field depends on the stratigraphic context, taxonomic group, age of the deposit, and shell size and concentration (e.g., Reinhardt et al., 2006; Aguirre et al., 2011). When examining shell deposits in the stratigraphic record, researchers can choose between core and trench sampling. Trenches or bank exposures are preferred because larger sample sizes can be obtained, and they allow for the assessment of broadscale changes in contacts, grading, lithology, bioclasts, and bed geometry (e.g., Donato et al., 2008). Broad exposures are particularly useful for assessing shell orientation (e.g., in or out of life position), articulation of bivalves, imbrication, and nesting (stacking of shells inside one another). Further, larger species (e.g., *Anadara* sp., *Tagelus* sp.) will likely not be present in sufficient quantities in cores (e.g., Reinhardt et al., 2012). For bulk shell samples obtained from stratigraphic section, articulated specimens should be removed in the field to avoid disarticulation prior to analysis. Bulk samples can be stored below 5 °C in airtight plastic bags. Where coring is necessary, a minimum core diameter of 10 cm is recommended. Care must be taken to ensure that shell orientation is not affected during core recovery (Lanesky et al., 1979). Cores should be immediately sealed in plastic to prevent the sediment from drying or oxidizing and stored below 5 °C to prevent chemical and microbial alteration of the shell material. When documenting shell beds the following should be noted (where possible) while in the field:

- stratigraphic context (e.g., surrounding units, bed geometry);
- lithology of the sediment matrix (e.g., clay, silt, sand, gravel);
- sedimentary structure (e.g., grading, bedding);
- contacts (e.g., erosional, gradational);
- concentration of shells (% volume of bioclasts; matrix- or bioclast-supported);
- lateral changes in species composition throughout the deposit;
- orientation of shells (e.g., random, imbricated, concave- or convex-up, and whether bivalves are in or out of life position);
- evidence of bioturbation;
- time-averaging (see Section 17.3).

17.2.3 Sample preparation

Sample preparation begins with washing bulk samples over a 1 mm sieve using freshwater to remove sediment (e.g., De Francesco and Hassan, 2008). The main challenge in preparing shell samples for analysis is determining appropriate size classes. Highly fragmented samples, or those with a range of shell sizes, may benefit from multiple size classes (e.g., >2 mm fraction, <2 mm fraction). Fractioned samples can be placed in a drying oven at 25 °C overnight, or until dry. Mollusks are then sorted, identified to species level and counted. Following species identification, taphonomic characters are identified and counted, and the collective dataset is expressed as a percentage of overall mass or volume. Depending on the study, considering a subset of target species within the assemblage may be more appropriate, particularly if the shell bed consists of a range of shell structures (e.g., thick, robust *vs.* thin, fragile). Some studies report varying taphonomic effects within the same deposit when using selected target species *vs.* bulk samples from the same stratigraphic unit (e.g., Aguirre et al., 2011). Aguirre et al. (2011) circumvented the bulk sampling bias in shell beds from Argentina by selecting two target species for taphonomic analysis.

17.2.4 Paleoecologic analysis (taxonomy and identification)

The Phylum Mollusca includes approximately 60,000 living species that span freshwater, brackish, and marine environments. Of all molluscan taxonomic classes, gastropods (e.g., snails) and bivalves (a shell with two hinged parts, e.g., clam) are generally the most common components of shell beds (e.g., Davies et al., 1989; Anderson and McBride, 1996; Lopez et al., 2008). There is no universally accepted system for identifying mollusks, and identification generally relies on anatomical features

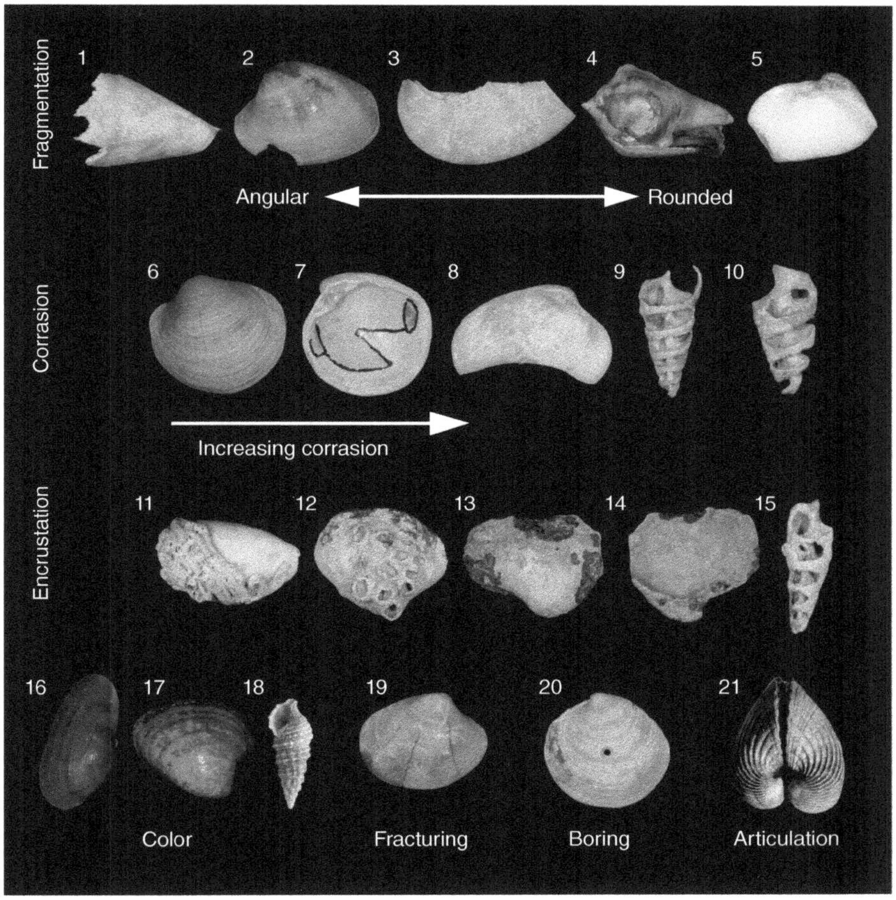

Fig. 17.1. Common taphonomic features of mollusk shells. For color details, please see Plate 30.

such as shell shape and structure, foot, siphon, hinges, pallial line, and muscle scars (e.g., Mikkelsen and Bieler, 2007). Identification at higher taxonomic levels is complicated by morphological variation of these features and the loss of diagnostic soft parts. Although taxonomic effort is generally directed towards the mollusks in a shell deposit, noting the presence of other bioclasts (e.g., echinoids, bryozoans, barnacles, hydrozoans, algae) can be important in reconstructing depositional and paleoenvironment (estuarine *vs.* open marine conditions; e.g., Beckvar and Kidwell, 1988; Reinhardt et al., 2012).

17.2.5 Taphonomic analysis

Taphonomic alteration of a shell is a function of time and environmental conditions after death. Taphonomic analysis, involving the identification of taphonomic characters, enhances paleoenvironmental interpretation (e.g., Davies et al., 1989; Kowalewski, 1996; Kidwell, 1998; Donato et al., 2008; Massari et al., 2009; Aguirre et al., 2011). However, there is no general agreement as to which taphonomic traits to consider (e.g., Aguirre et al., 2011). We therefore recommend that the following taphonomic criteria be used (Fig. 17.1).

- *Articulation*: refers to the decay of the organic ligament that joins the valves together. Disarticulation results from exhumation or exposure by currents and bioturbators. The tendency of bivalves to disarticulate is species dependent and researchers should be selective in choosing which species to enumerate (e.g., avoid species with delicate ligaments that are easily broken).
- *Fragmentation*: refers to the breakage of shell edges. Fragmentation is usually an indication of a high-energy environment (waves, currents, extreme events). Fragments are angular (indicating minimal reworking) or rounded (indicating some degree of reworking) and provide a measure of exposure time before burial.

- *Corrasion*: refers to the degradation of a shell caused by the combined effects of abrasion, dissolution, and bioerosion (Brett and Baird, 1986). Mollusks that have undergone corrasion have chalky and pitted surfaces, loss of ornamentation, and obscured features (e.g., loss of periostracum, indistinct pallial line, and muscle scars). Corrasion is related to subaerial exposure, energy of the environment, and size of the abrasive agent (e.g., Aguirre and Farinati, 1999), with sands and gravels being the most effective abrasives.
- *Encrustation*: refers to the attachment of other organisms onto a mollusk. Encrusters include sponges (*Cliona*), serpulid polychaetes, gastropods (e.g., Vermetidae), coral (*Porites*), red algae (*Lithothamnion*), barnacles, and bryozoans. The percentage of encrusted mollusks provides a measure of the duration of post-mortem exposure of shells on the sea floor, especially encrustation of infaunal shells which are out of reach of most encrusters while the organisms are alive in their burrows. Many encrusters are sensitive to even temporary burial by sediment.
- *Color*: refers to the degradation or bleaching of the original color due to exhumation, exposure, and aging.
- *Fracturing*: refers to the cracking or breaking of a shell. Bifurcating stress fractures are indicative of rapid shell fragmentation caused by turbulent-flow transport, shell-on-shell contact, and impact on hard surfaces such as rock or infrastructure (e.g., Donato et al., 2008).
- *Orientation*: refers to the position of a mollusk in stratigraphic section (random, imbricated, in life position, concave- or convex-up). High percentages of articulated bivalves with random azimuthal orientation are often indicative of exhumation, transport, and rapid burial.

The collective consideration of paleoecologic and taphonomic data can be used to infer not only the origin of a shell bed, but also subsequent reworking or transport events. For example, a shell deposit consisting of both estuarine and marine species, where the estuarine species are less well-preserved than the marine species, indicates that the estuarine species are reworked and relict (e.g., Kidwell, 1998). However, a shell bed containing estuarine and marine species that are equally preserved indicates within-habitat time-averaging as opposed to environmental amalgamation (e.g., Kidwell, 1998). Furthermore, the combined approach enables the differentiation between deposits generated by sea-level change and those created by storms or tsunamis (Table 17.1). For example, an assemblage that consists of mollusks with low articulation but high abrasion and encrustation indicates that the shell material was exposed either continuously or episodically after death. This is often seen in transgressive lag deposits (e.g., Anderson and McBride, 1996). On the contrary, storm deposits contain imbricated or nested shells with moderate fragmentation and minor articulation (Beckvar and Kidwell, 1988; Davies et al., 1989; Williams, 2011); whereas, tsunami shell beds contain high frequencies of shells that are articulated, fragmented, and contain bifurcating stress fractures (Reinhardt et al., 2006; Donato et al., 2008).

17.2.6 Counting and data analysis

When quantifying the taxonomic and taphonomic composition of a shell deposit, a minimum of 300 specimens from each sample should be counted to fully capture the species/traits that are present, but in low abundance (e.g., Aguirre et al., 2011). In the case of bivalves, counts on disarticulated specimens can be divided by two to account for a whole specimen. The ultimate goal in data analysis is to ascribe a depositional origin. Multivariate techniques used to interpret these parameters include: ordination methods (detrended correspondence analysis, non-metric multidimensional scaling), hierarchical clustering, and ternary diagrams (Kowalewski et al., 1995; Scarponi and Kowalewski, 2004; Hammer and Harper, 2006).

17.3 TIME-AVERAGING

A key problem when using molluscan assemblages to reconstruct former sea-level is assessing the degree of time-averaging. Time-averaging is the process by which biological remains (e.g., mollusks) from different time intervals are accumulated together and preserved (Fürsich, 1990; Kidwell, 1998; Muhs et al., 2012). Over time, shell beds can be subjected to repeated cycles of physical reworking of host sediment (e.g., burial/exhumation) and bioturbation. Both of these processes result in time-averaging and a biasing of the geologic record over a broad range of timescales (Kidwell, 1998). Determining the degree of

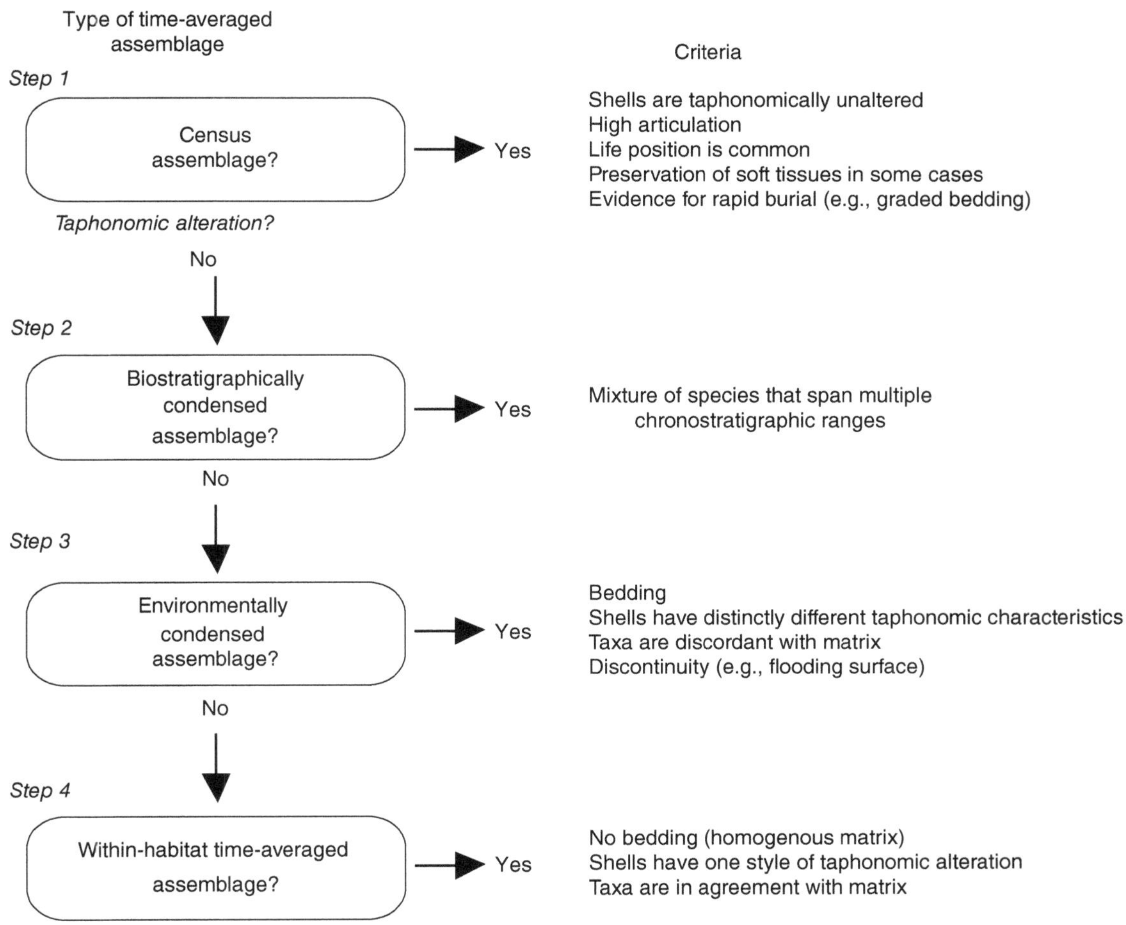

Fig. 17.2. Identifying and interpreting the relative scale of time-averaging in shell beds using stratigraphic, sedimentologic, and paleontologic (taxa and taphonomy) evidence. *Source*: Kidwell, 1998. Reproduced with permission of Elsevier. The extremes, census, and biostratigraphically condensed assemblages are considered first in Steps 1 and 2, followed by the more common assemblage types (environmentally condensed and within-habitat time-averaged) in Steps 3 and 4.

time-averaging is therefore important to interpreting the origin of a shell bed and placing it into proper chronological context.

Assessing the relative scale of time-averaging requires a combination of stratigraphic, sedimentologic, and paleontologic (taxa and taphonomy) evidence and is based on a four-step process of elimination (Kidwell, 1998). This process involves the categorization of shell accumulations into census, within-habitat time-averaged, environmentally condensed and biostratigraphically condensed assemblages (Fig. 17.2; Kidwell and Bosence, 1991). Census assemblages are those that contain minimal time-averaging and record instantaneous or short-term events. Within-habitat time-averaged assemblages are characterized by multiple generations of mollusks that are from a single community, indicating environmental stability. In contrast, environmentally condensed assemblages contain ecologically unrelated species that accumulated over a period of environmental change. Assemblages that are time-averaged over longer timescales (>100 ka), and contain a mixture of species that span multiple chronozones, are referred to as biostratigraphically condensed.

Accumulation history is different from time-averaging because it refers to the time interval over which a shell bed forms. Accumulation histories range from instantaneous (e.g., storm, tsunami) to several thousands of years (e.g., sea-level change). Kidwell (1991) discusses the classification of four types of accumulation histories (from shortest to longest timescales): event; composite;

hiatal; and lag concentrations. Event concentrations (e.g., storm, tsunami) contain uniform preservation and taxonomic composition. Composite concentrations, resulting from multiple events and slow sedimentation, contain variable taxa and preservation states. Hiatal concentrations form during prolonged intervals of sediment starvation, erosion, or reworking and are characterized by variable taxonomic composition and highly variable preservation. Lag concentrations are shell beds that are produced by physical sorting and reworking of pre-existing beds. For example, transgressive lag concentrations contain predominantly autochthonous taxa that are poorly preserved (Fig. 17.3; Table 17.1). Lag concentrations generally have the lowest temporal resolution due to vigorous reworking (e.g., Anderson and McBride, 1996).

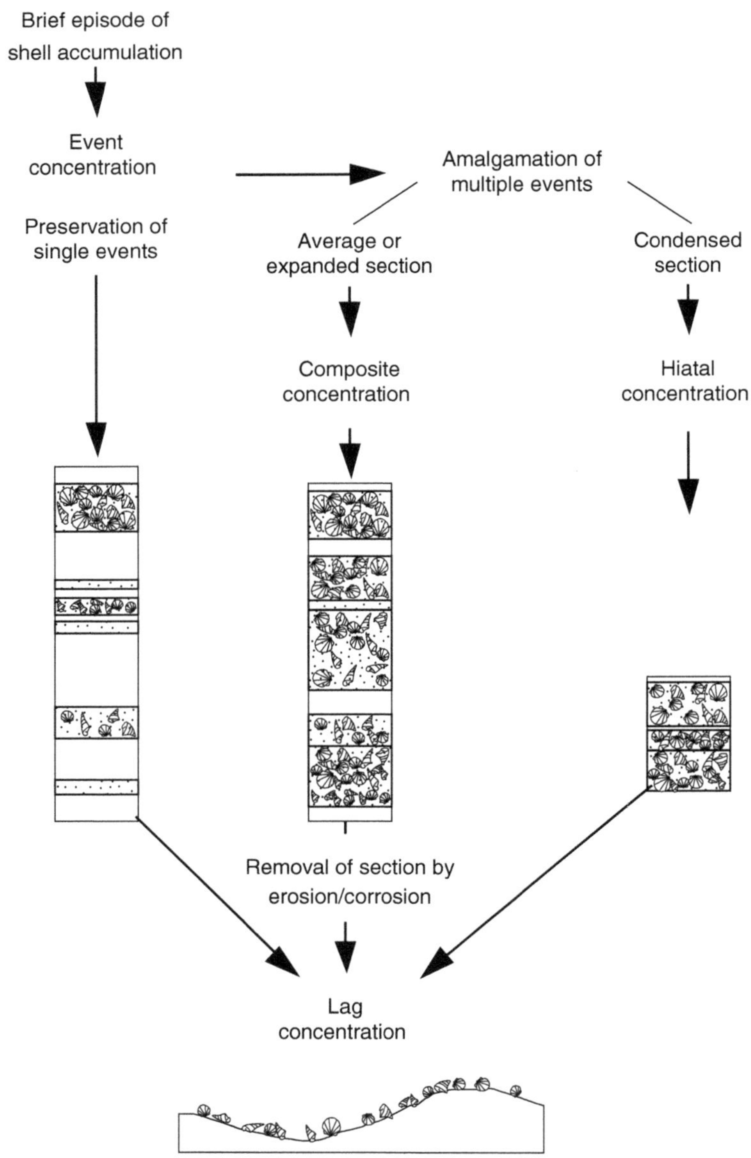

Fig. 17.3. Mechanisms of shell bed formation (also known as accumulation history). Event concentrations, representing ecologically brief episodes, can amalgamate as a result of physical and biological reworking. Amalgamated shell beds that are expanded are referred to as composite concentrations; those that are condensed are hiatal. A lag concentration forms when a portion of a shell bed is truncated by erosion/corrosion. *Source*: Adapted from Kidwell, 1991. Reproduced with permssion of Springer Verlag.

17.4 FOSSIL DATA AND SEA-LEVEL RECONSTRUCTIONS

For molluscan fauna to be used as a sea-level indicator they must possess a systematic and quantifiable relationship to elevation in the tidal frame (van de Plassche, 1986). Many sublittoral mollusks occupy a broad range of water depths (Petersen, 1986), so can only provide marine-limiting sea-level data points, with unconstrained one-way errors. Nevertheless, in some cases, mollusk assemblage data combined with sedimentologic, geomorphologic and chronologic analyses, have helped to provide relatively narrow (2–5 m) paleo-sea-level ranges (e.g., Muhs et al., 2012).

Early studies employing mollusks as sea-level indicators were based on taxonomic analysis with little to no emphasis on taphonomy. Merrill et al. (1965) obtained radiocarbon dates from oyster beds (*Crassostrea virginica*) to construct a Holocene sea-level curve for the US Atlantic coast between Massachusetts and North Carolina. *C. virginica* was used as a qualitative reference to constrain former sea-level position due to its narrow range of habitat, restricted to shallow, brackish lagoons and estuaries. Over a decade later, MacIntyre et al. (1978) re-examined the same oyster beds from North Carolina. High-resolution radiocarbon dates on individual *C. virginica* shells indicated significant time-averaging and shoreward transport within the deposits. Not only did these findings call into question previous sea-level reconstructions (e.g., Curray, 1965; Merrill et al., 1965), they also highlighted the importance in understanding the degree of transport and time-averaging of an assemblage.

The addition of taphonomic analysis to traditional paleoecological analysis on shell beds provided better estimations of accumulation history and the degree of vertical mixing (e.g., Kowalewski, 1996; Kidwell, 1998). This combined approach has been used to assess former sea-levels from Europe (e.g., Gutierrez-Mas, 2011), South America (e.g., Aguirre and Whatley, 1995; Aguirre et al., 2005), and the Gulf of Mexico (e.g., Anderson and McBride, 1996). For example, extensive shell deposits in Argentina record a series of sea-level transgressions and regressions over the late Pliocene to late Quaternary (Aguirre and Whatley, 1995). These shell deposits are predominantly parautochthonous littoral ridges (e.g., beach ridges), but are also represented by tidal flat deposits. Anderson and McBride (1996) showed that transgressive lag deposits, which are subjected to reworking and amalgamation as sea levels rise and fall, can still be recognized on the basis of their combined paleoecologic and taphonomic composition. Shelly transgressive lag deposits from the northeastern Gulf of Mexico are composed of well-preserved shells due to episodically high sedimentation rates during high-energy events.

The use of mollusks in documenting former sea-levels is important deeper in the sedimentary record where post-depositional changes can entirely obscure thin sedimentary records. Episodes of sea-level transgression and regression are archived in shell beds of late Pliocene age (e.g., Aguirre and Whatley, 1995). Older records have the added problem of decreasing preservation; time-averaging, hiatuses, and post-depositional alteration to shells can all lead to incomplete and misinterpreted records (e.g., Kidwell, 1998).

17.5 SUMMARY

Mollusks are useful indicators of sea level because they are abundant and readily recognizable in the stratigraphic record, and their habitat spans freshwater to marine environments. Sea-level highstand and lag deposits form during cycles of sea-level transgression and regression and, when combined with chronological data, can be used to estimate former sea level. The resulting shell accumulations have unique taxonomic and taphonomic signatures that can be distinguished from one another. Through time, these shell beds can become overprinted by physical (e.g., storms, tides, waves) and biological (e.g., bioturbation) processes, thereby confounding the interpretation of past sea levels. Careful evaluation and determination of the type of shell bed under consideration, the characteristics of the bed, and its individual shell components provides a framework for using molluscan shell deposits for sea-level reconstruction. A combined taxonomic and taphonomic approach has been successfully used on shell accumulations to assess the origin (i.e., accumulation history), degree of time-averaging, and overall applicability of the deposit for sea-level studies. This technique has been used to reconstruct sea-level histories from several locations such as Europe, South America, and the Gulf of Mexico.

ACKNOWLEDGEMENTS

This work was supported by the National Science Foundation (Award Nos. EAR-1144537 and EAR-1322658).

REFERENCES

Aguirre, M.L., and Whatley, R.C. (1995) Late Quaternary marginal marine deposits and palaeoenvironments from northeastern Buenos Aires Province, Argentina: a review. Quaternary Science Reviews, 14, 223–254.

Aguirre, M.L., and Farinati, E.A. (1999) Taphonomic processes affecting late Quaternary molluscs along the coastal area of Buenos Aires Province (Argentina, Southwestern Atlantic). Palaeogeography, Palaeoclimatology, Palaeoecology, 149, 283–304.

Aguirre, A., Negro Sirch, Y., and Richiano, S. (2005) Late Quaternary molluscan assemblages from the coastal area of Bahía Bustamante (Patagonia, Argentina): paleoecology and paleoenvironments. Journal of South American Earth Sciences, 20, 13–32.

Aguirre, A., Richiano, S., Farinati, E., and Fucks, E. (2011) Taphonomic comparison between two bivalves (*Mactra* and *Brachidontes*) from Late Quaternary deposits in northern Argentina: which intrinsic and extrinsic factors prevail under different palaeoenvironmental conditions? Quaternary International, 233, 113–129.

Anderson, L.C., and McBride, R.A. (1996) Taphonomic and palaeoenvironmental evidence of Holocene shell-bed genesis and history on the northeastern Gulf of Mexico shelf. Palaios, 11, 532–549.

Angulo, R.J., Lessa, G.C., and de Souza, M.C. (2006) A critical review of mid- to late-Holocene sea-level fluctuations on the eastern Brazilian coastline. Quaternary Science Reviews, 25, 486–506.

Beckvar, N., and Kidwell, S.M. (1988) Hiatal shell concentrations, sequence analysis, and sealevel history of a Pleistocene coastal alluvial fan, Punta Chueca, Sonora. Lethaia, 21, 257–270.

Brett, C.E., and Baird, G.C. (1986) Comparative taphonomy: a key to paleoenvironmental interpretation based on fossil preservation. Palaios, 1, 207–227.

Curray, J.R. (1965) Late Quaternary history, continental shelves of the United States. In *The Quaternary of the United States* (eds Wright, H.E., and Frey, D.C.), Princeton University Press, Princeton, NJ, 723–735.

Davies, D.J., Powell, E.N., and Stanton, R.J. (1989) Taphonomic signature as a function of environmental process: shells and shell beds in a hurricane-influenced inlet on the Texas coast. Palaeogeography, Palaeoclimatology, Palaeoecology, 72, 317–356.

De Francesco, C.G., and Hassan, G.S. (2008) Dominance of reworked fossil shells in modern estuarine environments: implications for paleoenvironmental reconstructions based on biological remains. Palaios, 23(1), 14–23.

Donato, S.V., Reinhardt, E.G., Boyce, J.I., Rothaus, R., and Vosmer, T. (2008) Identifying tsunami deposits using bivalve shell taphonomy. Geology, 36(3), 199–202.

Fairbridge, R.W. (1976) Shellfish-eating Preceramic Indians in Coastal Brazil. Science, 191, 353–359.

Fürsich, F.T. (1990) Fossil concentrations and life and death assemblages. In *Palaeobiology* (eds Briggs, D.E.G., and Crowther, P.R.), Blackwell Scientific Publications, London, p. 235–239.

Gutierrez-Mas, J.M. (2011) Glycymeris shell accumulations as indicators of recent sea-level changes and high-energy events in Cadiz Bay (SW Spain). Estuarine, Coastal and Shelf Science, 92, 546–554.

Hammer, O., and Harper, D. (2006) *Paleontological Data Analysis*. Blackwell Publishing, Oxford, 351 pp.

Händel, M. (2009) Al-Hamriya (Sharjah, UAE): approaching an extensive shell midden site. BioArchaeologica, 5, 103–112.

Kidwell, S.M. (1991) The stratigraphy of shell concentrations. In: *Taphonomy: Releasing the Data Locked in the Fossil Record* (eds Allison, P.A., and Briggs, D.E.G.), Plenum Press, New York, pp. 212–290.

Kidwell, S.M. (1998) Time-averaging in the marine fossil record: overview of strategies and uncertainties. Geobios, 30, 977–995.

Kidwell, S.M., and Bosence, D.W.J. (1991) Taphonomy and time-averaging of marine shelly faunas. In: *Taphonomy: Releasing the Data Locked in the Fossil Record* (eds Allison, P.A., and Briggs, D.E.G.), Plenum Press, New York, pp. 115–209.

Kidwell, S.M., Fürsich, F.T., and Aigner, T. (1986) Conceptual framework for the analysis and classification of fossil concentrations. Palaios, 1, 228–238.

Kowalewski, M. (1996) Time-averaging, overcompleteness, and the geologic record. Journal of Geology, 104, 317–326.

Kowalewski, M., Flessa, K.W., and Hallman, D.P. (1995) Ternary taphograms: triangular diagrams applied to taphonomic analysis. Palaios, 10, 478–483.

Lanesky, D.E., Logan, B.W., Brown, R.F., and Hine, A.C. (1979) A new approach to portable vibracoring underwater and on land. Journal of Sedimentary Petrology, 49, 654–657.

Lopez, R.A., Penchaszadeh, P.E., and Marcomini, S.C. (2008) Storm-related strandings of mollusks on the northeast coast of Buenos Aires, Argentina. Journal of Coastal Research, 24(4), 925–935.

MacIntyre, I.G., Pilkey, O.H., and Stuckenrath, R. (1978) Relict oysters on the United States Atlantic continental shelf: a reconsideration of their usefulness in understanding late Quaternary sea-level history. Geological Society of America Bulletin, 89, 277–282.

Martinez, S., and Rojas, A. (2013) Relative sea level during the Holocene in Uruguay. Palaeogeography, Palaeoclimatology, Palaeoecology, 374, 123–131.

Massari, F., D'Alessandro, A., and Davaud, E. (2009) A coquinoid tsunamite from the Pliocene of Salento (SE Italy). Sedimentary Geology, 221(1–4), 7–18.

Merrill, A.S., Emery, K.O., and Rubin, M. (1965) Ancient oyster shells on the Atlantic continental shelf. Science, 147, 398–400.

Mikkelsen, P.M., and Bieler, R. (2007) *Seashells of Southern Florida: Living Marine Mollusks of the Florida Keys and Adjacent Regions: Bivalves*. Princeton University Press, Princeton, NJ.

Muhs, D.R., Simmons, K.R., Schumann, R.R., Groves, L.T., Mitrovica, J.X., and Laurel, D. (2012) Sea-level history during the Last Interglacial complex on San Nicolas Island, California: implications for glacial isostatic adjustment processes, paleozoogeography and tectonics. Quaternary Science Reviews, 37, 1–25.

Petersen, K.J. (1986) Marine molluscs as indicators of former sea-level stands. In *Sea-level Research: A Manual for the Collection and Evaluation of Data* (ed. van de Plassche, O.), GeoBooks, Norwich, pp. 129–155.

Reinhardt, E.G., Goodman, B.N., Boyce, J.I., Lopez, G., van Hengstum, P., Rink, W.J., Mart, Y. and Raban, A. (2006) The tsunami of December 13, 115 A.D., and the destruction of Herod the Great's harbour at Caesarea Maritima, Israel. Geology, 34, 1061–1064.

Reinhardt E.G., Pilarczyk, J.E., and Brown, A. (2012) Probable tsunami origin for a shell and sand sheet from marine ponds on Anegada, British Virgin Islands. Natural Hazards, 63, 101–117.

Scarponi, D., and Kowalewski, M. (2004) Stratigraphic paleoecology: bathymetric signatures and sequence overprint of mollusk associations from upper Quaternary sequences of the Po Plain, Italy. Geology, 32(11), 989–992.

Stein, J.K. (1992) *Deciphering a Shell Midden*. Academic Press, New York.

van de Plassche, O. (ed.) (1986) *Sea-level Research: A Manual for the Collection and Evaluation of Data*, GeoBooks, Norwich.

Williams, H.L. (2011) Shell bed tempestites in the Chenier Plain of Louisiana: late Holocene example and a modern analogue. Journal of Quaternary Science, 26(2), 199–206.

Chapter 18
Fixed biological indicators

ALESSIO ROVERE[1,2], FABRIZIO ANTONIOLI[3], AND CARLO NIKE BIANCHI[4]

[1] *Sea Level and Coastal Changes Group, MARUM, University of Bremen & ZMT, Leibniz Center for Tropical Marine Ecology, Bremen, Germany*

[2] *Lamont Doherty Earth Observatory, Columbia University, Palisades, NY, USA*

[3] *ENEA-UTMEA-TER, Roma, Italy*

[4] *DISTAV, Università degli Studi di Genova, Genoa, Italy*

18.1 INTRODUCTION

Fixed biological indicators (FBIs) are organisms that live fixed to hard substrates and their living range, or the upper part of it, is located near or at sea level (Fig. 18.1; Laborel and Laborel-Deguen, 2005). When relative sea level (RSL) changes, FBIs experience a modification of their living environmental conditions; they therefore either initiate mechanisms of biological adaptation to keep up with changing RSL or, if the pace of RSL rise (or fall) is too high, they die leaving fossil remains. Bioconstructions, bioerosive morphological traces, and upper limits of elevated remains of non-building fixed plants or invertebrates are commonly used as FBIs (Laborel and Laborel-Deguen, 1996). The main principle to be taken into account when using FBIs as sea-level index points is that, on rocky shores, fauna and vegetation develop in quasi-horizontal belts parallel to the sea surface, defining the marine zonation (Pérès, 1982).

While the subdivision into supralittoral, midlittoral, and infralittoral is based on changes in biotic assemblages, the terms supratidal, intertidal, and subtidal explicitly refer to physical limits, i.e., the tidal properties of an area. The two zonation systems overlap and are sometimes used interchangeably, but do not match closely. In macrotidal seas, the intertidal may extend upward to include the lower part of the supralittoral and downward to include the upper part of the infralittoral. In microtidal seas, the height of the waves and the irregular variations of the sea due to winds and atmospheric pressure are more important than tides in setting the boundaries of the biological zones. For reference, in Figure 18.1 we report the relationship between marine zonation and the most commonly used tidal datums in sea level studies. Nevertheless, in this chapter reference is made to biological zones rather than to tidal levels. At the practical level, the relationship between marine zonation and tidal datums should be derived specifically for a study area comparing the modern marine zonation with tide gauge data.

Different authors have described marine zonation, giving different names to various zones (e.g., Stephenson and Stephenson, 1949; Lewis, 1961; Pérès and Picard, 1964; see Laborel, 1986 for a comparison between the different zonation schemes and for a rationale of their differences). Pérès (1982) defines three main zones:

(1) *Supralittoral* zone, located above the level of the highest tides of calm waters and therefore normally emerged, even if more or less regularly reached by seawater spray. The dominating process is weathering, and the first cyanobacteria appear.
(2) *Midlittoral* zone, submersed at regular intervals by waves or tides. It is subdivided into an upper subzone, only wet by the waves, and a lower one, more directly influenced by the tide as well as waves and barometric seiches. Here, limpets and chitons may be common. Rock-building organisms (coralline rhodophytes, vermetids) also occur.
(3) *Infralittoral* zone, nearly constantly submerged, which extends from just below the lower limit of the midlittoral to tens of meters depth, depending on water transparency. The infralittoral zone is characterized by brown algae, coralline rhodophytes, vermetid gastropods, oysters, annelids, cirripeds, and eroding agents such as clionaid sponges, sea-urchins, and rock-boring mollusks, as well as corals.

Handbook of Sea-Level Research, First Edition. Edited by Ian Shennan, Antony J. Long, and Benjamin P. Horton.

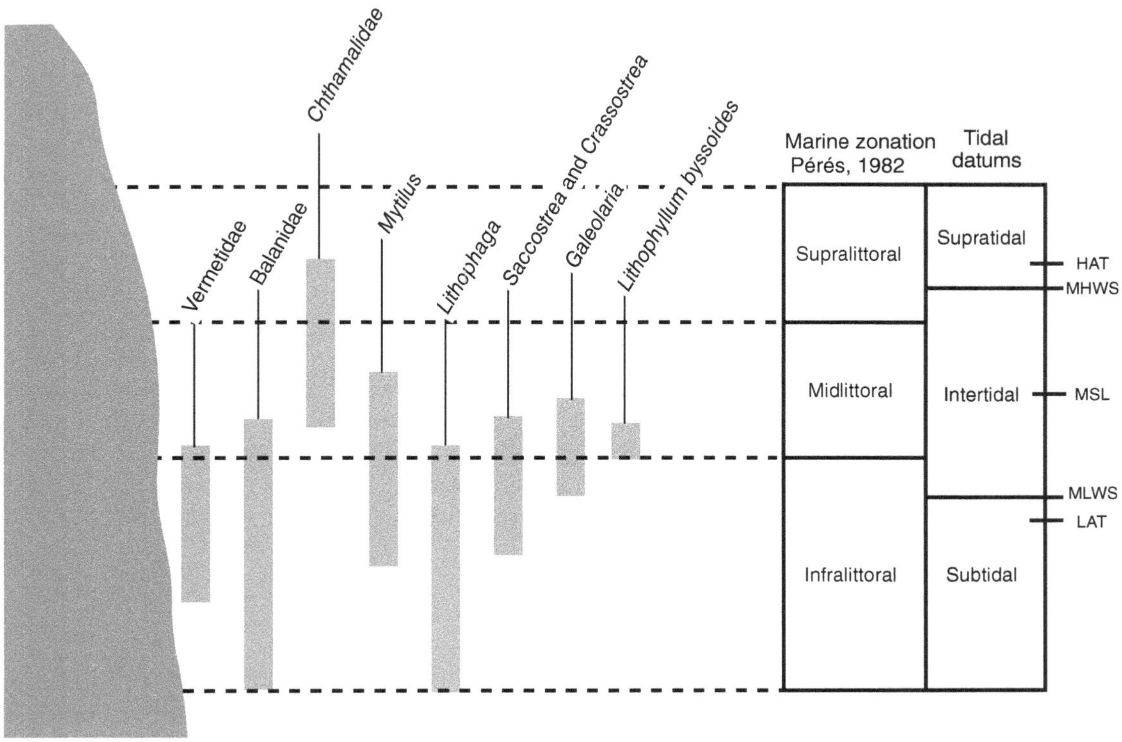

Fig. 18.1. Position of the main FBIs described in the text with respect to the marine zonation (Pérès, 1982) and tidal datums commonly used in sea-level studies. HAT and LAT: highest and lowest astronomical tides; MHWS and MLWS: mean high and mean low water springs; MSL: mean sea level. The relationship between tidal datums and the marine zonation is empirical and varies according to different tidal regimes (see text for details).

Within this zone, many organisms have a clear distribution limit. A few species are restricted to the upper margin of that zone, but many display a clear-cut limit at that level. Some organisms build various reef-like structures (also called bioherms or biostromes), or develop bioconstructed levels that protect the hard substrate from further erosion (Laborel and Laborel-Deguen, 1996).

The limit between the midlittoral and infralittoral zones is defined as biological mean sea level or biological zero, which is characterized by a sharp change in biotic composition (Morri et al., 2004). Species living near the biological mean sea level are particularly important as FBIs, and their significance is increased when used in association with other sedimentological, geological, or archeological markers (e.g., tidal notches or archeological remains; Marriner and Morhange, 2007). Another peculiar characteristic of FBIs is that they provide dateable deposits to constrain the timing of RSL histories assessed with different types of markers, that is, tidal notches or shore platforms.

Depending on the area and environmental conditions, including exposure, tidal range, climate, and physico-chemical properties of water (Morri et al., 2000), several organisms can be used as FBIs. In general, to be used in relative sea-level reconstructions, a FBI must be as close as possible to the following specifications (Laborel and Laborel-Deguen, 1996):

(1) It must provide reliable information upon the general direction of the relative displacement of sea level.
(2) It must allow an accurate and reproducible measurement of the displacement.
(3) It must provide biological material datable by radiochronology.
(4) It must provide valuable data about the direction of sea-level change, from very slow (tectonic, eustatic, isostatic) to rapid (coseismic).
(5) Results obtained should be coherent with those obtained by converging multidisciplinary approaches (morphology, archeology, geology).

Moreover, fossil FBIs can provide a reliable method to assess RSL histories if a suitable indicator is available, an accurate study of local conditions of preservation and sedimentation is performed, and a correlation with other types of sea-level index points is possible.

FBIs have been used mainly to assess Holocene (Stiros et al., 2009; Scheffers et al., 2012) or late Pleistocene (i.e., last interglacial, Ferranti et al., 2006) RSL histories, but in particular conditions older FBIs can be preserved. As for Holocene FBIs, their importance is particularly high for the study of past coseismic events (Palyvos et al., 2008; Maouche et al., 2011).

A broad range of FBIs can be employed as proxies for past RSLs; the following sections describe those used most frequently in the literature, which are also listed in Table 18.1.

18.2 VERMETIDAE

Vermetidae is a family of fixed Gastropoda, in the clade Littorinomorpha. Vermetids are often found as thin incrustations, but may grow in association with coralline rhodophytes to form prominent reef-like structures (Fig. 18.2a, b). As an example, in the Mediterranean the vermetid gastropod *Dendropoma petraeum*, in association with the coralline algae *Neogoniolithon* sp. or *Porolithon* sp., can build important reefs or platforms. Most species of vermetids have a large living range within the infralittoral zone, and should therefore be used as FBIs following species-related considerations (e.g., *Serpulorbis annulatus*, and indo-pacific species limited to depths less than 3 m) and in association with other RSL markers. On the contrary, many species of the genera *Dendropoma* (subgenus *Novastoa*) and *Petaloconchus* (subgenus *Macrophragma*) live close to sea level, in the lower part of the midlittoral zone. Laborel (1986) described the importance of surf and tides in the development of vermetid reefs, highlighting that difference in elevation of more than 1 m can be observed for the same reef measured in sheltered and exposed areas. Observing vermetid reefs in sheltered areas in the Mediterranean, Caribbean, and Brazil, Laborel (1986) estimated that the upper limit of vermetid reefs develop in the lower quarter of the tidal range, i.e., between low water neap tides and low water spring tides.

The morphology of vermetid reefs can be highly variable, and can be divided into four typologies (Laborel, 1986; Antonioli et al., 1999): (1) bench or platform type, in which the biogenic structure overlies the bedrock, usually a shore platform, and is cut by wave action; (2) microatoll type (Safriel, 1975), in which erosion and bioconstruction interplay to form an annular structure with a built-up rim and a "lagoon" inside it; (3) "cornice" type, which is a simple rim perpendicular to the slope of a cliff; and (4) mushroom-like pillars.

The clearest examples of the use of vermetid reefs to reconstruct RSL histories can be found in the Mediterranean Sea (Pirazzoli et al., 1994; Sanlaville et al., 1997; Antonioli et al., 1999) and in the Atlantic Ocean (Van Andel and Laborel, 1964; Jones et al., 1991; Angulo et al., 1999). Pleistocene vermetid reefs are rare due to their fragile nature (Ferranti et al., 2006).

18.3 CIRRIPEDIA

Barnacles belong to the subclass Cirripedia of the phylum Arthropoda, subphylum Crustacea. They are sessile organisms that live mostly in shallow waters; many species can be found at sea level.

18.3.1 Balanidae

Balanidae is a family of barnacles (Figs 18.2c, f, g and Fig. 18.3d). Balanids develop in single-layered populations or in small rims. Among the 19 genera of this family, the species of the genera *Balanus*, *Tetraclita* and their relatives are more often used as sea-level indicators.

The vertical living range of balanids extends from sea level to the infralittoral zone, but their upper limit is significant as it begins near MSL and is limited to the lower midlittoral. Their accuracy in indicating sea level is therefore within centimeters, provided that a linear upper limit can be identified along a cliff or a vertical upper surface (Laborel and Laborel-Deguen, 1994). As they can live in brackish water, balanids have also been used extensively in association with archeological remains, such as ancient harbors (Morhange et al., 2001). Due to the good preservation of their calcareous shells, often more preserved in crevices or cracks of the bedrock, balanids have been used as FBIs for at least 50 years (e.g., Donner, 1963; Laborel and Laborel-Deguen, 1994; Ferranti et al., 2007) and were used extensively to map the extent of coseismic land movements resulting from the 1964 M_w 9.2 earthquake in Alaska (Plafker, 1969).

Table 18.1. Fixed biological indicators described in this chapter, main studies using or describing their use in sea level studies and key points to be considered for each FBI.

FBI	Refs	Key points
Vermetidae	1, 2, 3, 4, 5, 6, 21	• Many species of *Dendropoma* and *Petaloconchus* have their upper limit in the midlittoral zone • Different kinds of reefs • Difference in elevation can be observed for the same reef measured in sheltered and exposed areas.
Balanidae	7, 8, 21, 22, 24	• Upper limit begins at MSL • Upper limit must be identified along a cliff or a vertical upper surface • Used successfully in association with archeological remains • Variations of isotopic ratio in brackish waters may cause problems for radiocarbon dating
Chthamalidae	9, 10, 11, 21, 22	• Vertical living range limited to the splash zone, that is, between the upper midlittoral and lower supralittoral zones • Upper limit varies following local conditions of surf exposure, tidal amplitude and topography • Fossil population must be compared with the modern population • Better accuracies using chthamalids as FBIs have been obtained using fossils in semi-enclosed environments • Shells of small size, difficult dating. Possible deposition of calcareous material inside the shells, affecting age determinations
Mytilus	12	• *Mytilus edulis* complex living in midlittoral–upper infralittoral (0–5 m) • Preservation in this status is often compromised by the fragility of the byssum • Limitations for use in Arctic seas
Lithophaga	13, 14, 15, 21, 22	• Living range is from the midlittoral down to deeper than 30 m • Upper limit coincides closely with sea level, should be observed along a continuous rock surface • Difficult preservation of shells for radiocarbon dating, possible settlement on secondary generations into already-carved boreholes
Saccostrea and *Crassostrea*	16, 21, 22, 24	• Ostreidae typically live in shallow water and species of the genera *Saccostrea* and *Crassostrea* may colonize the midlittoral zone, but in general oyster vertical accuracy in indicating SL is variable to species and regions • Due to their resistance to variable environmental conditions, they can be used in brackish areas and harbors • Require detailed mineralogical analyses before dating to assess shell composition
Serpulidae (*Galeolaria*)	17, 18, 23, 24	• Their reefs are mostly found in enclosed embayments and/or brackish estuaries and lagoons • Some genera, such as *Galeolaria*, colonize the midlittoral • Growth of serpulids on speleothems has been used to constrain and date marine ingression in several limestone caves
Coralline algae	19, 20, 21, 22, 25	• Rims of coralline algae living near the present sea level have been largely used as indicators of ancient sea levels • Necessity to clean samples from younger incrustations, i.e., borings infilled by younger material, before radiocarbon dating

1. Laborel (1986), 2. Antonioli et al. (1999), 3. Van Andel and Laborel (1964), 4. Sivan et al. (2010), 5. Vescogni et al. (2008), 6. Jones and Hunter (1995), 7. Morhange et al. (2001), 8. Donner (1963), 9. Scicchitano et al. (2011), 10. Pirazzoli et al. (1985), 11. Ferranti et al. (2007), 12. Petersen (1986), 13. Antonioli and Oliverio (1996), 14. Shaw et al. (2010), 15. Vacchi et al. (2012), 16. Pirazzoli et al. (1991), 17. Baker et al. (2003), 18. Dutton et al. (2009b), 19. Adley (1986), 20. Laborel (2005), 21. Laborel and Laborel-Deguen (1994), 22. Laborel and Laborel-Deguen (1996), 23. Baker and Haworth (1997), 24. Baker et al. (2001), 25. Faivre et al. (2013).

18.3.2 Chthamalidae

Chthamalidae (Fig. 18.2d, e) are found along almost all coasts of the northern hemisphere, as well as many regions in the southern hemisphere. Their vertical living range is limited to the splash zone, that is, between the upper midlittoral and lower supralittoral zones, and their upper limit varies following local conditions of surf exposure, tidal amplitude, and topography. Their vertical living range is therefore irregular, and may vary from a few centimeters to several meters (Laborel and Laborel-Deguen, 1996). In addition, different species may live at different shore heights in the same area, with little overlap: in the Mediterranean, for instance,

Fig. 18.2. (a) *Dendropoma petraeum* platform reef, developing upon a shore platform at San Vito Lo Capo, Sicily, south Italy. (Antonioli et al., 1999. Photographs by Alessio Rovere.) (b) Mushroom-like pillars of *Dendropoma petraeum* at San Vito Lo Capo, Sicily. (Antonioli et al., 1999. Photographs by Fabrizio Antonioli.) (c) Living balanids (arrow) on intertidal boulders, Western Australia. (Photographs by Fabrizio Antonioli.) (d) Fossil *Chthamalus stellatus* on an uplifted shore platform at Scilla, Calabria, south Italy, Late Holocene (Ferranti et al., 2007. Photographs by Fabrizio Antonioli.) (e) Living *Chthamalus stellatus* band in the splash zone, Palmi, south Italy. (Photographs by Fabrizio Antonioli.) (f) Fossil balanids of Pliocene–Early Pleistocene age, Hondeklipbaai, South Africa. (Photographs by Alessio Rovere.) (g) Living balanids (1) and, in the lower part of the boulder, oysters (2) at low tide, Magnetic Island, West Australia. (h) Living *Mytilus galloprovincialis* (on coralline algae) attached in the lower part of a tidal notch, Gaeta, central Italy. For color details, please see Plate 31.

Euraphia depressa extends into the supralittoral zone, while *Chthamalus stellatus* and *C. montagui* are restricted to the midlittoral (Bianchi et al., 2004).

Due to the possible effects of different exposures to wave action on the vertical distribution of chthamalids, it is imperative that their fossil population is compared with the modern population (Scicchitano et al., 2011) at the same sampling site. Better accuracies using chthamalids as FBIs have been obtained using fossil populations in semi-enclosed environments (e.g., small coastal caves). This allows the possible source of error highlighted by Laborel and Laborel-Deguen (1996) to be avoided, who argued that former periods of intense storms may have led to the development of "energy-elevated" fossil populations which

Fig. 18.3. (a) Last interglacial *Lihtophaga lithophaga* boreholes in north Sicily, Italy. (Photographs by Fabrizio Antonioli.) (b) Close-up of last interglacial *Lithophaga lithophaga* in the Bergeggi marine cave, Liguria, north Italy. (Photographs by Alessio Rovere.) (c) Balanids growing upon oysters in a Pliocene–Early Pleistocene outcrop at Hondeklip Bay, South Africa. (Photographs by Alessio Rovere.) (d) Fossil *Perforatus perforatus* on a Archaic-Greek-age harbor boulder (at Lechaion, Greece). (Photographs by Fabrizio Antonioli.) (e) Fossil Holocene *Galeolaria* reef (dashed line: upper band), West Australia. (Photographs by Fabrizio Antonioli.) (f) Section of (1) speleothem showing serpulid overgrowth on (2) top of continental deposition (Argentarola cave, Italy; Antonioli et al., 2004). (Photographs by Fabrizio Antonioli.) (g) Rim (or "trottoir") of *Lithophyllum byssoides* (arrow) at low tide, Gaeta central Italy. (Photographs by Fabrizio Antonioli.) (h) Measurement of the lower part of a shell deposit using GPS RTK system, Central Florida, USA. (Photograph by P. J. Hearty.) (i) Measurement of the relative position of fossil Holocene *Chthamalus stellatus* with respect to that of the living population (arrow), Scilla, Calabria, Italy. (Photographs by Fabrizio Antonioli.) For color details, please see Plate 32.

might be erroneously interpreted as elevated sea levels. Chthamalids can gather major significance as FBIs if associated with the analysis of other SL indicators (e.g., geomorphological elements, such as shore platforms or tidal notches). Where dating with radiocarbon techniques is employed, caution should be exercised as deposition of calcareous material inside the shells is frequent (Laborel and Laborel-Deguen, 1994).

18.4 MYTILIDAE

Mytilidae is a family of marine bivalve mollusks (mussels) in the order Mytiloida. They live in the lower midlittoral to infralittoral zones. Some species live attached to hard substrates by means of elastic and strong thread-like structures called byssal threads, secreted by byssal glands. In some areas, clustered *Mytilus* attached to each other or to the hard substratum can form banks. Other Mytilidae are endolithic, boring galleries in calcareous rocks through glandular secretion and mechanical action. In the following sections, we describe two representative genera: *Mytilus*, mussels that grow on rocky shores, and *Lithophaga*, a typical endolithic bivalve.

18.4.1 Mytilus

Of particular interest as FBIs are species of *Mytilus* pertaining to the *Mytilus edulis* complex, three closely related species (Varvio et al, 1988). If fossils are found in a living position on a rocky shore (i.e., attached to the substratum), the upper level of *Mytilus* banks can be generally considered as a good SL indicator as it is located near MTL (Fig. 18.2h). Their preservation in this status is often compromised by the fragility of the byssum, which causes their post-mortem detachment from the rocky surface. For this reason, the use of *Mytilus* shells as SL indicators is always subject to considerations of their former living position, and this FBI is often used in association with other SL markers (e.g., geomorphological or archeological markers). It is worth noting that the use of *Mytilus edulis* as FBI in Arctic zones is limited by the fact that this mollusk lives below the intertidal zone because of ice abrasion of the sea floor in winter (Hansen et al., 2011). Hjort et al. (2007) also highlighted that in the Svalbard Islands this bivalve lived 10–20 m below the contemporaneous sea level during the Holocene Climatic Optimum.

18.4.2 Lithophaga

Lithophaga (date mussels or boring mussels) are more or less ubiquitous limestone-boring lamellibranches (Laborel and Laborel-Deguen, 1994). *Lithophaga* living range is from the upper limit of the infralittoral zone down to deeper than 30 m although the highest concentrations are found within the top few meters, and their upper limit may coincide closely with sea level (Laborel and Laborel-Deguen, 1994).

These mussels are homogeneous and abundant not only on cliffs but also inside crevices and cave environments. In general, *Lithophaga* is an accurate FBI if surfaces of burrowed rock with a distinct linear upper limit are preserved (Fig. 18.3a, b). When using *Lithophaga* borehole upper bands (and associated shells) as sea-level index points, Laborel and Laborel-Deguen (1994) recommended: (1) preference should be given to *Lithophaga* populations with a clear-cut horizontal upper limit (or linked with a fossil erosion notch); (2) holes selected should be dug into the bedrock far from sources of fine sediment such as a beach or a river mouth; (3) preference should be given to shell remains not included in a sedimentary matrix; (4) holes near karst freshwater sources must be avoided; and (5) elevated bedrock surface must not be covered by a layer of dead endobiontic invertebrates (foraminifers, serpulids, and barnacles), which indicates that *Lithophaga* did not develop on a cliff surface but in a cryptic environment such as a crevice which was later opened by erosion.

In general, boring mussel shells inside boreholes are rapidly destroyed, but under certain conditions the shell can be sampled and dated (see Shaw et al., 2010 for discussion on their radiocarbon dating). The upper limit of *Lithophaga lithophaga* boreholes has been widely used to date both rapid and slow RSL changes (Lambeck et al., 2004; Ferranti et al., 2006). In some cases, *L. lithophaga* was used as a sea-level indicator on submerged speleothems, as they colonized the very earliest stages of submersion (Antonioli and Oliverio, 1996).

Other than *Lithophaga*, other mollusks produce dense bands of bores in the infralittoral zone. As an example, *Gastrochaena dubia* (family Gastrochaenidae) can occur in high densities and is widespread throughout the Mediterranean and the eastern Atlantic Ocean, where it bores eight-shaped holes into soft rock. Another important borer is the mollusk *Pholas dactylus* (family Pholadidae), which create conical-shaped burrows that have a narrow entrance and a larger rounded chamber.

18.5 OSTREIDAE

Ostreidae (oysters) is a family of bivalves pertaining to the phylum Mollusca, and live fixed on hard substrates. In general, Ostreidae live in shallow water. Members of the genera *Saccostrea* and *Crassostrea* may have their upper limit in the lower midlittoral zone, and have been used as accurate sea-level indicators in different areas (Kelletat, 1988; Beaman et al., 1994; Baker et al., 2001, 2003). On the contrary, the genus *Ostrea* lives continually immersed (Petersen, 1986).

Oysters can act as important builders, and their living range extends between the surface and depths of 20–30 m. Due to their resistance to variable environmental conditions, oysters can be used in brackish areas and harbors. As the resistance of oyster shells to erosion is high, biological remains can often be used for dating. Laborel and Laborel-Deguen (1996) highlighted that oyster vertical accuracy in indicating SL is variable according to species and regions; their use as FBIs therefore requires a preliminary biological survey of the studied area (Fig. 18.3c).

18.6 SERPULIDAE

The serpulids constitute a large (nearly 300 species) family of sedentary Polychaetes (Bianchi et al., 1995). Having a calcareous habitation tube, they may be important encrusters in many marine communities. By filling crevices and cementing rubble in coral or coralline-algae reefs, serpulids may play the role of binders or secondary builders (Fagerstrom, 1987). However, several species are known to be able to act as primary builders, forming reef-like structures, that is, tubes that grow vertical to the substrate in clumps that are cemented to each other (Ten Hove and Weerdenburg, 1978). The major serpulid reefs are found in quiet, enclosed embayments and/or brackish estuaries and lagoons (Bianchi and Morri, 1996, 2001), although some genera such as *Galeolaria* can build reefs on exposed shores.

The family Serpulidae is composed of very different genera, and in general the bathymetric range of these animals is very large, varying with individual species. The morphology of the bioconstructions can also vary according to physical processes (e.g., wave action and bioerosion) to which they are subjected. As a result, serpulids can be found either as reef complexes or as single animals.

Fossil remains of serpulids, better preserved inside cracks or small cavities, can be used as dateable material for other geomorphological or archeological markers (Vacchi et al., 2012). Some species, such as those of the genus *Galeolaria*, form distinct reef-like structures in the midlittoral zone (Fig. 18.3e), although isolated tubes can be found at any level (also above the highest tide, in the splash zone). The species *Galeolaria caespitosa* has been extensively used as an accurate RSL marker along the Australian shorelines (Baker et al., 2003). This serpulid indicates SL position with a high accuracy, especially if associated with the analysis of the lower limits of cirriped populations (Baker et al., 2001; Sloss et al., 2007).

Growth of serpulids on speleothems (Fig. 18.3f) has been used in the last decade to constrain and date marine ingression in several limestone caves, particularly in the Mediterranean (Antonioli et al., 2004; Dutton et al., 2009a, b) and in tropical coastlines (Dill et al., 1998).

18.7 CORALLINE ALGAE

Coralline algae belong to the Order Corallinales of the Phylum Rodhophyta, and are characterized by a hard tallus formed by calcareous deposits contained within the cell walls. Coralline algae colonize rocky substrata, and are widespread along the world's rocky shores. Although different species of coralline algae live in different depths of water, ranging from the midlittoral to the depth of maximum penetration of light, some species can be used as very reliable SL indicators (Laborel, 2005). Rims of coralline algae living near present sea level (Fig. 18.3g) have been largely used as indicators of ancient sea levels, if found in positions different from their living range (Laborel, 1987). In the Mediterranean, fossil rims of *Lithophyllum byssoides* (formerly *L. lichenoides*) are usually considered a good proxy of past sea levels (Laborel and Laborel-Deguen, 1996; Faivre et al., 2013) because their living range is constrained around sea level. *Lithophyllum byssoides* have the narrowest vertical biological range of 30–50 cm and the rim, which occurs in the lower midlittoral zone, is the highest biogenic build-up in the Mediterranean. Rims over 1 m wide may occur in places exposed to strong wave action.

Also in the tropics, it is generally possible to determine the water depth of fossil remains of coralline encrustations or build-ups with a good degree of accuracy (Cabioch et al., 1999). In the Atlantic and Pacific, species of the genus *Porolithon* form large reefs on midlittoral and wave-exposed algal ridges (Littler and Doty, 1975). Other algal ridges, such as those formed by *Lythophyllum congestum*, develop within a few meters below the MSL (Adley, 1986).

18.8 CORALS

Corals are one of the most extensively used FBIs, although their use is greatly limited to tropical and sub-tropical areas. In particular, corals of the genus *Acropora* (e.g., *Acropora palmata*, a major reef framework builder in the wider Caribbean) have been extensively used to reconstruct Holocene SL history (e.g., Bard et al., 1990). Some species of corals such as *Porites* maintain a living position up to the mean low water mark, and morphologies such as microatolls are considered precise sea-level indicators (Woodroffe and McLean, 1990).

The use of corals as sea-level indicators is described in Chapters 7 and 8 on coral reefs and microatolls, respectively. Here we highlight that some temperate corals, such as the Mediterranean *Cladocora caespitosa*, may be important as they provide dateable material to assess the age of more accurate SL indicators such as marine terraces or tidal notches (Hearty, 2002). *C. caespitosa* has also been used as paleotemperature proxy (Peirano et al., 2004; Silenzi et al., 2005).

18.9 BIOEROSION MARKERS

Bioerosion markers represent a particular type of biological indicator, as they are the result of the action of mechanical erosion and biochemical dissolution by different types of organisms on rocky shores. Marks can be produced either by sessile or mobile organisms. As dateable remains are rare on bioerosion markers, they lack one of the main characteristics to be fully considered as FBIs (point (3) in the Introduction: "It must provide biological material datable by radiochronology"). Nevertheless, bioerosion markers have often been used to assess the elevation of paleo-sea levels dated with other markers, or to evaluate the stratigraphy of bioerosion (Bromley, 2004).

The best-known examples of sessile bioerosion markers are boring mollusks, such as *Lithophaga* or *Pholas*. Within mobile organisms, grazers (mainly sea urchins) have a great role as bioeroders along coral reef coastlines (Peyrot-Clausade et al., 2000; Spencer and Viles, 2002). The echinoid *Paracentrotus lividus* is distributed throughout the Mediterranean Sea and in the Northeast Atlantic, from Scotland and Ireland to southern Morocco and the Canary Islands, including the Azores Islands (Boudouresque and Verlaque, 2001). It bores in the bedrock depressions that vary from cup-shaped to deep pockets having a narrow entrance opening. Rasping traces left on the bedrock by modern echinoids have been used to assess the origin of relict traces attributed to the ichnofossil *Gnathichnus pentax* of Pliocene age (Martinell, 1982).

Other organisms leaving evident bioerosion marks along rocky coasts are Clionaidae, a family of boring sponges. Some clionaid species with zooxantellae can chemically dissolve the carbonates deposited by coralline algae (Calcinai et al., 2000; Cerrano et al., 2001; Azzini et al., 2006). Other important bioeroders include endolithic Cyanobacteria (Wisshak and Tapanila, 2008) in the supralittoral zone, and limpets (*Patella* spp.) and chitons in the midlittoral zone (Laborel and Laborel-Deguen, 1996).

Bioerosion and bioerosion traces are a complex and large subject. For more in-depth information we direct the reader to the following authors: Viles (1984), Bromley et al. (2008), Wisshak and Tapanila (2008), Kazmer and Taborosi (2012), and Naylor et al. (2012).

18.10 FIELD METHODS

Fixed biological indicators have been used extensively over the last 50 years to measure vertical RSL movements along the world's coasts. In order to obtain accurate measurements, the elevation of the fossil population should always be referenced to its modern analog rather than to the water level. In order to do so it is necessary to carry out ecological analyses of the modern intertidal zone, establishing where organisms used as FBIs, or their modern analog, are located within it Laborel and Laborel-Deguen (2005) highlighted that the best results are obtained when the uppermost limits of both fossil and living populations are well delineated. The difference between the height on the shore of the fossil population and that of the modern population allows the vertical shift between present and former sea level to be established. One of the advantages of this method makes any direct reference to the actual water level unnecessary, whether observed or calculated from tide tables (Laborel and Laborel-Deguen, 2005).

As some kinds of FBIs can have important vertical variations due to local hydrodynamics, the solution of selecting a unique modern benchmark for several fossil sites might lead to inaccurate reconstructions. If operatively possible, the recommendation is to use the modern analog for each fossil site surveyed in an area. This procedure makes one fundamental assumption, that is, that the tidal range has not altered substantially and the degree of wave exposure has not changed; this assumption may prove false if climate has changed significantly. The best accuracy can therefore be obtained in areas where the upper limit is preserved and where the transition boundaries between species in the intertidal sequence are preserved (Baker and Haworth, 1997).

It is useful to replicate the elevation measurements of the same fossil unit at the same site. In general, a site is considered as an area of coastline of about 10–20 m length, where the fossil FBI crops out at several points. Three different elevation measurements of the same fossil unit at different points at the same site are a minimum requirement to exclude large variations in the elevation of the population on a small scale, which could be due to erosion of part of the unit or variations of wave energy within the site.

If more than one FBI species is present at a single site it is good practice to sample the largest possible species diversity, and in quantities sufficient for more than one radiometric dating. We highlight here that some types of FBI can present dating problems. As an example, small shell size and the possibility of post-mortem calcareous material often prevent cyrripeds and balanids from yielding accurate radiocarbon ages. The presence of the latter in brackish water with varied isotopic ratios may also cause additional uncertainties in radiocarbon ages. For a synopsis of the suitability of different kinds of FBIs as dateable materials, the reader is referred to Chapters 23–27 and Laborel and Laborel-Deguen (1994) and Shaw et al. (2010).

Regarding the techniques used to measure the elevation of fossil FBIs, it is worth highlighting that in most cases the environmental conditions on the coast do not allow a precise measurement of both the fossil and modern population with accuracies greater than 10–50 cm. The most-used technique to measure FBIs consists of obtaining the difference in elevation between fossil and living populations with a measuring tape or surveyor's level and staff. If the difference in elevation is less than 3 m (Fig. 18.3i), the accuracy of this technique is usually constrained within a few centimeters. In the last ten years differential GPS systems have been used to measure the elevation of sea-level index points, in particular in saltmarshes (see Chapters 3, 4 and 10 for details). With the exclusion of high-energy environments, where the instrument can be damaged by seawater, RTK or Omnistar GPS instruments allow centimeter or sub-centimeter accuracy of the measurement of FBIs (Fig. 18.3 h).

ACKNOWLEDGMENTS

AR thanks PLIOMAX NSF OCE-1202632 (www.pliomax.org) for support and for allowing the use of Figures 18.2f and 18.3c, h, i. MEDFLOOD INQUA project (www.medflood.org) allowed AR and FA to meet and share earlier versions of the MS with other Mediterranean scientists. AR's research is funded by the Institutional Strategy of the University of Bremen, German Excellence Initiative, and by ZMT Center for Tropical Marine Ecology, Bremen. This chapter is dedicated to the memory of Professor J. Laborel. He inspired three generations of sea-level scientists with his work on fixed biological indicators around the world, including the authors of this chapter.

REFERENCES

Adley, W.H. (1986) Coralline algae as indicators of sea level. In: *Sea-Level Research: a Manual for the Collection and Evaluation of Data* (ed. van de Plassche, O.), Geo Books, Norwich, pp. 229–280.

Angulo, R.J., Giannini, P.C.F., Suguio, K., and Pessenda, L.C.R. (1999) Relative sea-level changes in the last 5500 years in southern Brazil (Laguna–Imbituba region, Santa Catarina State) based on vermetid 14C ages. Marine Geology, 159, 323–339.

Antonioli, F., and Oliverio, M. (1996) Holocene sea-level rise recorded by a radiocarbon-dated mussel in a submerged speleothem beneath the Mediterranean Sea. Quaternary Research, 45, 4–4.

Antonioli, F., Chemello, R., Improta, S., and Riggio, S. (1999) *Dendropoma* lower intertidal reef formations and their palaeoclimatological significance, NW Sicily. Marine Geology, 161, 155–170.

Antonioli, F., Bard, E., Potter, E.K., Silenzi, S., and Improta, S. (2004) 215-ka history of sea-level oscillations from marine and continental layers in Argentarola cave speleothems (Italy). Global and Planetary Change, 43, 57–78.

Azzini, F., Bavestrello, G., and Cerrano, C. (2006) Boring sponges from the Ha Long Bay, Tonkin Gulf, Vietnam. Zoological Studies, 45(2), 201–212.

Baker, R.G.V., and Haworth, R.J. (1997) Further evidence from relic shellcrust sequences for a late Holocene higher sea level for eastern Australia. Marine Geology, 141, 1–9.

Baker, R.G.V., Haworth, R.J., and Flood, P.G. (2001) Intertidal fixed indicators of former Holocene sea levels in Australia: a summary of sites and a review of methods and models. Quaternary International, 83, 257–273.

Baker, R.G.V., Davis, A.M., Aitchison, J.C., Flood, P.G., Morton, B.S., and Haworth, R.J. (2003) Comment on

"Mid-Holocene higher sea level indicators from the south China coast" by W.W.-S. Yim and G. Huang [Mar. Geol. 182 (2002) 225–230]: a regional perspective. Marine Geology, 196, 91–98.

Bard, E., Hamelin, B., and Fairbanks, R.G. (1990) U–Th ages obtained by mass spectrometry in corals from Barbados: sea level during the past 130, 000 years. Nature, 346, 456–458.

Beaman, R., Larcombe, P., and Carter, R.M. (1994) New evidence for the Holocene sea-level high from the inner shelf, central Great Barrier Reef, Australia. Journal of Sedimentary Research, 64, 882–885.

Bianchi, C.N., and Morri, C. (1996) *Ficopomatus* 'reefs' in the Po River Delta (Northern Adriatic): their constructional dynamics, biology, and influences on the brackish-water biota. Marine Ecology, 17 (1–3), 51–66.

Bianchi, C.N., and Morri, C. (2001) The battle is not to the strong: Serpulid reefs in the Lagoon of Orbetello (Tuscany, Italy). Estuarine, Coastal and Shelf Science, 53, 215–220.

Bianchi, C.N., Aliani, S., and Morri, C. (1995) Present-day serpulid reefs, with reference to an on- going research project on *Ficopomatus enigmaticus*. Publications du Service Geologique du Luxembourg, 29, 61–65.

Bianchi, C.N., Boero, F., Fraschetti, S., and Morri, C. (2004) The wildlife of the Mediterranean. In: *Wildlife in Italy* (eds Minelli, A., Chemini, C., Argano, R., and Ruffo, S.), Touring Editore, Milan, and Italian Ministry for the Environment and Territory, Rome, pp. 248–335.

Boudouresque, C.F., and Verlaque, M. (2001) Ecology of *Paracentrotus lividus*. In: *Edible Sea Urchins: Biology and Ecology* (ed. Lawrence, J.M.) Elsevier, Amsterdam, Developments in Aquaculture and Fisheries Science, pp. 177–216.

Bromley, R.G. (2004) *A Stratigraphy of Marine Bioerosion*. Geological Society, London, Special Publications 228, 455–479.

Bromley, R.G., Hankel, N.M., and Asgaard, U. (2008) Shallow marine bioerosion: the results of an experimental study. Bulletin of the Geological Society of Denmark, 38(1–2), 85–99.

Cabioch, G., Montaggioni, L.F., Faure, G., and Ribaud-Laurenti, A. (1999) Reef coralgal assemblages as recorders of paleobathymetry and sea level changes in the Indo-Pacific province. Quaternary Science Reviews, 18, 1681–1695.

Calcinai, B., Cerrano, C., and Sarà, M. (2000) Boring sponges (Porifera, Demospongiae) from the Indian Ocean. Italian Journal of Zoology, 67(2), 203–219.

Cerrano, C., Bavestrello, G., Bianchi, C.N., Calcinai, B., Cattaneo-Vietti, R., Morri, C., and Sarà M. (2001) The role of sponge bioerosion in the Mediterranean coralligenous accretion. In: *Structure and Processes in the Mediterranean Ecosystems* (eds Faranda, F.M., Guglielmo L., and Spezie, G.), Springer Verlag, Milano, pp. 235–240.

Dill, R.F., Land, L.S., Mack, L.E., and Schwarcz, H.P. (1998) A submerged stalactite from Belize: petrography, geochemistry, and geochronology of massive marine cementation. Carbonates and Evaporites, 13, 189–197.

Donner, J.J. (1963) The Late- and Post-glacial raised beaches in Scotland. Annales Academiae Scientiarum Fennicae, 68, 1–13.

Dutton, A., Bard, E., Antonioli, F., Esat, T.M., Lambeck, K., and McCulloch, M.T. (2009a) Phasing and amplitude of sea-level and climate change during the penultimate interglacial. Nature Geoscience, 2, 355–359.

Dutton, A., Scicchitano, G., Monaco, C., Desmarchelier, J.M., Antonioli, F., Lambeck, K., Esat, T.M., Fifield, L.K., McCulloch, M.T., and Mortimer, G. (2009b) Uplift rates defined by U-series and 14C ages of serpulid-encrusted speleothems from submerged caves near Siracusa, Sicily (Italy). Quaternary Geochronology, 4, 2–10.

Fagerstrom, J.A. (1987) *The Evolution of Reef Communities*. John Wiley and Sons, New York.

Faivre S., Bakran Petricioli, T., Horvatincic N., and Sironic, A. (2013) Distinct phases of relative sea level changes in the central Adriatic during the last 1500 years – influence of climatic variations? Palaeogeography, Palaeoclimatology, Palaeoecology, 369, 163–174.

Ferranti, L., Antonioli, F., Mauz, B., Amorosi, A., Dai Pra, G., Mastronuzzi, G., Monaco, C., Orrù, P., Pappalardo, M., Radtke, U., Renda, P., Romano, P., Sansò, P., and Verrubbi, V. (2006) Markers of the last interglacial sea-level high stand along the coast of Italy: Tectonic implications. Quaternary International, 145–146, 30–54.

Ferranti, L., Monaco, C., Antonioli, F., Maschio, L., Kershaw, S., and Verrubbi, V. (2007) The contribution of regional uplift and coseismic slip to the vertical crustal motion in the Messina Straits, southern Italy: Evidence from raised Late Holocene shorelines. Journal of Geophysical Research, 112, B06401.

Hansen, J., Hanken, N.-M., Nielsen, J.K., Nielsen, J.K., and Thomsen, E. (2011) Late Pleistocene and Holocene distribution of Mytilus edulis in the Barents Sea region and its palaeoclimatic implications. Journal of Biogeography, 38, 1197–1212.

Hearty, P.J. (2002) New data on the pleistocene of Mallorca. Quaternary Science Reviews, 6, 245–257.

Hjort, C., Mangerud, J., Adrielsson, L., Bondevik, S., Landvik, J.Y., and Salvigsen, O. (2007) Radiocarbon dated common mussels Mytilus edulis from eastern Svalbard and the Holocene marine climatic optimum. Polar Research, 14, 239–243.

Jones, B., and Hunter, I.G. (1995) Vermetid buildups from Grand Cayman, British West Indies. Journal of Coastal Research, 11(4), 973–983.

Jones, D.S., MacFadden, B.J., Webb, S.D., Mueller, P.A., Hodell, D.A., and Cronin, T.M. (1991) Integrated geochronology of a classic Pliocene fossil site in Florida: linking marine and terrestrial biochronologies. The Journal of Geology, 99(5), 637–648.

Kazmer, M., and Taborosi, D. (2012) Bioerosion on the small scale–examples from the tropical and subtropical littoral. Hantkeniana, 7, 37–94.

Kelletat, D. (1988) Zonality of modern coastal processes and sea-level indicators. Palaeogeography, Palaeoclimatology, Palaeoecology, 68, 219–230.

Laborel, J. (1986) Vermetid gastropods as sea-level indicators. In: *Sea-Level Research: a Manual for the Collection and Evaluation of Data* (ed. van de Plassche, O.), Geo Books, Norwich, pp. 281–310.

Laborel, J. (1987) Marine biogenic constructions in the Mediterranean-a review. Scientific Reports of the Port-Cros National Park, 13, 97–126.

Laborel, J. (2005) Algal rims. In: *Encyclopedia of Coastal Science* (ed. Schwartz, M.), Wiley, New York, pp. 24–25.

Laborel, J., and Laborel-Deguen, F. (1994) Biological indicators of relative sea-level variations and of co-seismic displacements in the Mediterranean region. Journal of Coastal Research, 10(2), 395–415.

Laborel, J., and Laborel-Deguen, F. (1996) Biological indicators of Holocene sea-level and climatic variations on rocky coasts of tropical and subtropical regions. Quaternary International, 31, 53–60.

Laborel, J., and Laborel-Deguen, F. (2005) Sea-level indicators, biologic. In: *Encyclopedia of Coastal Science* (ed. Schwartz, M.), Wiley, New York, pp. 833–834.

Lambeck, K., Antonioli, F., Purcell, A., and Silenzi, S. (2004) Sea-level change along the Italian coast for the past 10,000 yr. Quaternary Science Reviews, 23, 1567–1598.

Lewis, J.R. (1961) The littoral zone on rocky shores: a biological or physical entity? Oikos, 12, 280–301.

Littler, M.M., and Doty, M.S. (1975) Ecological components structuring the seaward edges of tropical Pacific reefs: the distribution, communities and productivity of Porolithon. The Journal of Ecology, 72(1), 117–129.

Maouche, S., Meghraoui, M., Morhange, C., Belabbes, S., Bouhadad, Y., and Haddoum, H. (2011) Active coastal thrusting and folding, and uplift rate of the Sahel Anticline and Zemmouri earthquake area (Tell Atlas, Algeria). Tectonophysics, 509(1), 69–80.

Marriner, N., and Morhange, C. (2007) Geoscience of ancient Mediterranean harbours. Earth Science Reviews, 80, 137–194.

Martinell, J. (1982) Echinoid bioerosion from the Pliocene of NE Spain. Geobios, 15, 249–253.

Morhange, C., Laborel, J., and Hesnard, A. (2001) Changes of relative sea level during the past 5000 years in the ancient harbor of Marseilles, Southern France. Palaeogeography, Palaeoclimatology, Palaeoecology, 166, 319–329.

Morri, C., Cattaneo-Vietti, R., Sartoni, G., and Bianchi, C.N. (2000) Shallow epibenthic communities of Ilha do Sal (Cape Verde Archipelago, Eastern Atlantic). Arquipélago, Life and Marine Sciences, Suppl. 2 (Part A), 157–165.

Morri, C., Bellan-Santini, D., Giaccone, G., and Bianchi, C. N. (2004) Principles of bionomy: definition of assemblages and use of taxonomic descriptors (macrobenthos). Biologia Marina Mediterranea, 11 (suppl. 1), 573–600.

Naylor, L.A., Coombes, M.A., and Viles, H.A. (2012) Reconceptualising the role of organisms in the erosion of rock coasts: A new model. Geomorphology, 157–158, 17–30.

Palyvos, N., Lemeille, F., Sorel, D., Pantosti, D., and Pavlopoulos, K. (2008) Geomorphic and biological indicators of paleoseismicity and Holocene uplift rate at a coastal normal fault footwall (western Corinth Gulf, Greece). Geomorphology, 96(1), 16–38.

Peirano, A., Morri, C., Bianchi, C.N., Aguirre, J., Antonioli, F., Calzetta, G., Carobene, L., Mastronuzzi, G., and Orrù, P. (2004) The Mediterranean coral *Cladocora caespitosa*: a proxy for past climate fluctuations? Global and Planetary Change, 40, 195–200.

Pérès, J.M. (1982) Zonations and organismic assemblages. In: *Marine Ecology* (ed. Kinne, O.), John Wiley & Sons, Chichester, pp. 9–576.

Pérès, J.M., and Picard, J. (1964) *Nouveau Manuel de Bionomie Benthique de la Mer Méditerranée.* Aix-en-Provence.

Petersen, K.S. (1986) Marine mollusks as indicators of former sea level stands. In: *Sea-Level Research: a Manual for the Collection and Evaluation of Data* (ed. van de Plassche, O.), Geo Books, Norwich, pp. 129–193.

Peyrot-Clausade, M., Chabanet, P., Conand, C., Fontaine, M.F., Letourneur, Y., and Harmelin-Vivien, M. (2000) Sea urchin and fish bioerosion on La Reunion and Moorea reefs. Bulletin of Marine Science, 66(2), 477–485.

Pirazzoli, P.A., Delibrias, G., Kawana, T., and Yamaguchi, T. (1985) The use of barnacles to measure and date relative sea-level changes in the Ryukyu Islands, Japan. Palaeogeography, Palaeoclimatology, Palaeoecology, 49, 161–174.

Pirazzoli, P.A., Laborel, J., Saliège, J.F., Erol, O., Kayan, I., and Person, A. (1991) Holocene raised shorelines on the Hatay coasts (Turkey): Palaeoecological and tectonic implications. Marine Geology, 96, 295–311.

Pirazzoli, P.A., Stiros, S.C., Laborel, J., Laborel-Deguen, F., Arnold, M., Papageorgiou S., and Morhange, C. (1994) Late-Holocene shoreline changes related to palaeoseismic events in the Ionian Islands, Greece. The Holocene, 4(4), 397–405.

Plafker, G. (1969) Tectonics of the March 27, 1964, Alaska earthquake. US Geological Survey Professional Paper, 543-I, p. 74.

Safriel, U.N. (1975) The role of vermetid gastropods in the formation of Mediterranean and Atlantic reefs. Oecologia, 20(1), 85–101.

Sanlaville, P., Dalongeville, R., Bernier, P., and Evin, J. (1997) The Syrian coast: a model of Holocene coastal evolution. Journal of Coastal Research, 13(2), 385–396.

Scheffers, A., Brill, D., Kelletat, D., Bruckner, H., Scheffers, S., and Fox, K. (2012) Holocene sea levels along the Andaman Sea coast of Thailand. The Holocene, 22(10), 1169–1180.

Scicchitano, G., Spampinato, C.R., Ferranti, L., Antonioli, F., Monaco, C., Capano, M., and Lubritto, C. (2011) Uplifted Holocene shorelines at Capo Milazzo (NE Sicily, Italy): evidence of co-seismic and steady-state deformation. Quaternary International, 232, 201–213.

Shaw, B., Jackson, J.A., Higham, T.F.G., England, P.C., and Thomas, A.L. (2010) Radiometric dates of uplifted marine fauna in Greece: Implications for the interpretation of recent earthquake and tectonic histories using lithophagid dates. Earth and Planetary Science Letters, 297, 395–404.

Silenzi, S., Bard, E., Montagna, P., and Antonioli, F. (2005) Isotopic and elemental records in a non-tropical coral (*Cladocora caespitosa*): Discovery of a new high-resolution climate archive for the Mediterranean Sea. Global and Planetary Change, 49, 94–120.

Sivan, D., Schattner, U., Morhange, C., and Boaretto, E. (2010) What can a sessile mollusk tell about neotectonics? Earth and Planetary Science Letters, 296, 451–458.

Sloss, C.R., Murray-Wallace, C.V., and Jones, B.G. (2007) Holocene sea-level change on the southeast coast of Australia: a review. The Holocene, 17, 999–1014.

Spencer, T., and Viles, H. (2002) Bioconstruction, bioerosion and disturbance on tropical coasts: coral reefs and rocky *limestone* shores. Geomorphology, 48, 23–50.

Stephenson, T.A., and Stephenson, A. (1949) The universal features of zonation between tide-marks on rocky coasts. The Journal of Ecology, 37(2), 289–305.
Stiros, S.C., Pirazzoli, P.A., and Fontugne, M. (2009) New evidence of Holocene coastal uplift in the Strophades Islets (W Hellenic Arc, Greece). Marine Geology, 267, 207–211.
Ten Hove, H.A., and Weerdenburg, J. (1978) A generic revision of the brackish-water serpulid Ficopomatus Southern, 1921 (Polychaeta: Serpulidae), including Mercierella Fauvel, 1923, Sphaeropomatus Treadwell, 1934, Mercierellopsis Rioja, 1945 and Neopomatus Pillai, 1960. The Biological Bulletin, 154, 96–120.
Vacchi, M., Rovere, A., Zouros, N., Desruelles, S., Caron, V., and Firpo, M. (2012) Spatial distribution of sea-level markers on Lesvos Island (NE Aegean Sea): Evidence of differential relative sea-level changes and the neotectonic implications. Geomorphology, 159–160, 50–62.
Van Andel, T.H., and Laborel, J. (1964) Recent high relative sea level stand near Recife, Brazil. Science, 145, 580–581.
Varvio, S.L., Koehn, R.K., and Vainola R. (1988) Evolutionary genetics of the Mytilus edulis complex in the North Atlantic region. Marine Biology, 98, 51–60.
Vescogni, A., Bosellini, F.R., Reuter, M., and Brachert, T.C. (2008) Vermetid reefs and their use as palaeobathymetric markers: New insights from the Late Miocene of the Mediterranean (Southern Italy, Crete). Palaeogeography, Palaeoclimatology, Palaeoecology, 267, 89–101.
Viles, H.A. (1984) Biokarst: review and prospect. Progress in Physical Geography, 8, 523–542.
Wisshak, M., and Tapanila, L. (2008) *Current Developments in Bioerosion*. Springer-Verlag, Berlin.
Woodroffe, C., and McLean, R. (1990) Microatolls and recent sea level change on coral atolls. Nature, 344, 531–534.

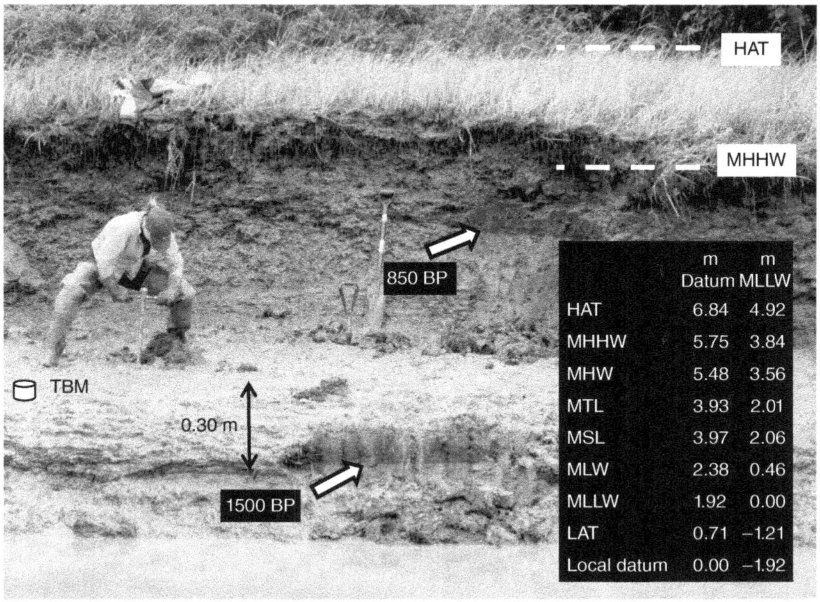

Plate 1 Marsh sequence, 10 km east of Cordova. Two buried peat layers, dated at ~850 and 1500 cal a BP separated by intertidal silt. A temporary bench mark (TBM) was surveyed to 4.10 m above local zero datum, and the lower peat is 0.30 m below the TBM (values for illustration only). No modern equivalent peat forming at the site, but studies elsewhere indicate the indicative meaning for the formation of peat adjacent to tidal flat is a reference water level of midway between MHHW and HAT and an indicative range of ±0.25 m. *Source*: Photograph by Ian Shennan.

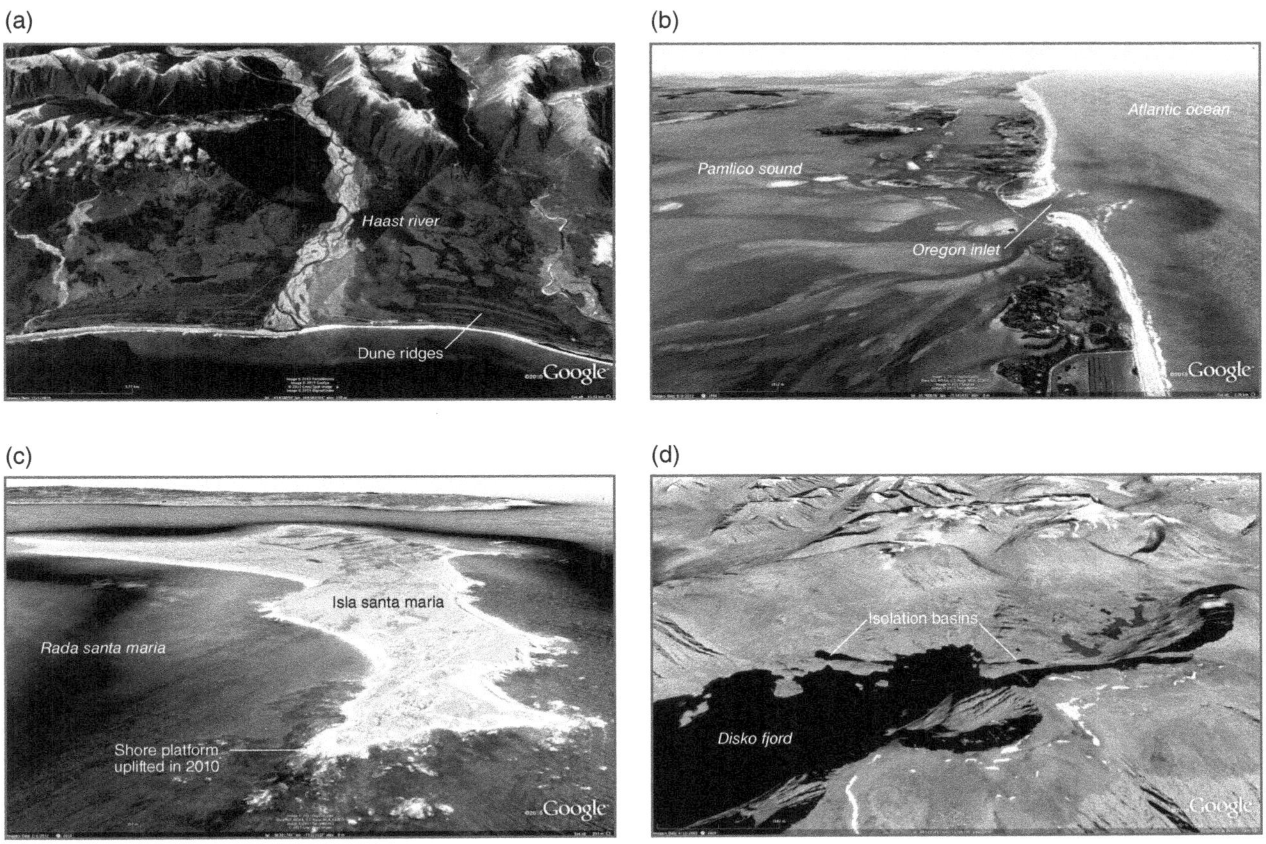

Plate 2 Examples of coastal landscapes viewed with Google Earth. (a) Prograding coastal dune ridges in South Westland, New Zealand mark progressive seaward shifts of the shoreline in response to increased sediment loads in the Haast River following earthquakes on the Alpine Fault (Wells and Goff, 2007). (b) Late Holocene sea-level rise controls the configuration of Oregon Inlet and the barrier islands of the Outer Banks, North Carolina, USA. (c) Sudden tectonic uplift during the 2010 Maule earthquake in central Chile exposed shore platforms surrounding Isla Santa Maria near Concepción. (d) Sediments of isolation-basin lakes around the shores of Disko Fjord in western Greenland record relative sea-level changes related to glacial isostatic adjustment caused by changes in mass of the Greenland ice sheet (Long et al., 2011). *Source*: Google Earth images (© 2010 Google) include data from (a) TerraMetrics, GeoEye, Cnes/Spot Image, and DigitalGlobe; (b) DigitalGlobe, GeoEye, and TerraMetrics; (c) DigitalGlobe, NOAA, US Navy, NGA, GEBCO, TerraMetrics, and Cnes/Spot Image; and (d) DigitalGlobe, GeoEye, US Geological Survey, and IBCAO.

Handbook of Sea-Level Research, First Edition. Edited by Ian Shennan, Antony J. Long, and Benjamin P. Horton.

Plate 3 (a) Uplifted marine terraces of the Yakataga coast, near Icy Bay, Alaska revealed by a digital elevation model (DEM) derived from LiDAR topographic data. Post-processing the LiDAR data removed first returns from vegetation to produce a "bare-earth" DEM with 1 m horizontal and 0.2 m vertical resolution. Thin black lines depict 4 m elevation contours. Plafker et al. (1982) identified four marine terraces T1–T4 (from oldest to youngest) that record tectonic uplift related to collision and subduction of the Yakatat microplate in southern Alaska. The LiDAR image reveals details of the marine terrace geomorphology, including eroded seacliffs along abandoned terrace backedges, and alluvial fans deposited on terrace surfaces that have been subsequently dissected by stream and river channels as a result of progressive uplift. *Source*: Derived from LiDAR topographic data. Reproduced with permission of Ron Bruhn. (b) Photograph of marine terraces flanking the channel and alluvial fan of Little River, near Icy Bay, Alaska. *Source*: Photograph by I. Shennan.

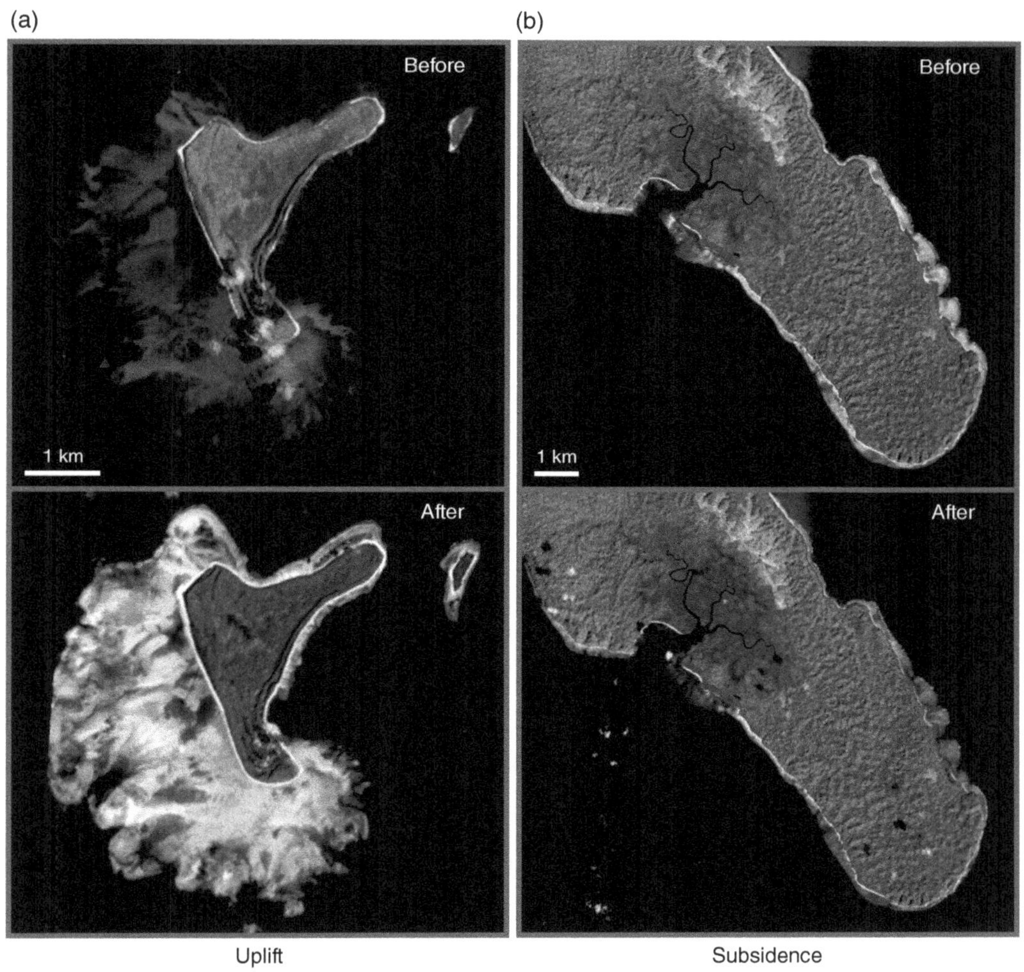

Plate 4 Advanced Spaceborne Thermal Emission and Reflection Radiometer (ASTER) images of North Reef Island (a) before (top) and after (bottom) the great 2004 Sumatra–Andaman Islands earthquake. Submerged coral reefs fringe the island before the earthquake. Emergent reefs, uplifted during the earthquake, surround the island in post-earthquake images. (b) ASTER images of the southeast tip of Havelock Island show shallow coral reefs and beaches exposed before (top) the earthquake and the same shoreline features submerged by deeper water after (bottom) the earthquake as a result of tectonic subsidence. *Source*: Reproduced with permission of Aron Meltzner.

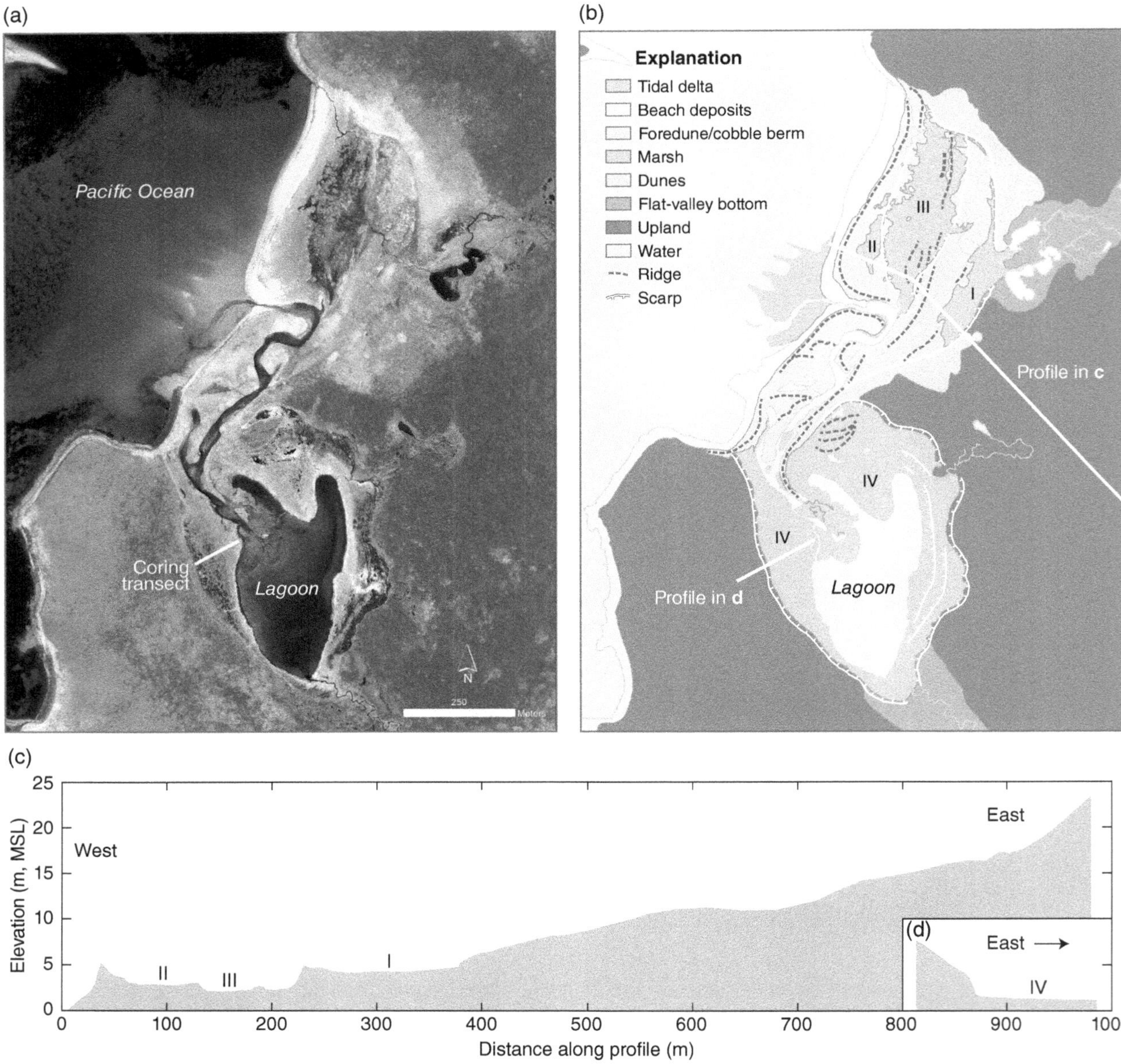

Plate 5 Geomorphological map of an intertidal lagoon on Simeonof Island near Sand Point, Alaska. (a) 2009 satellite image of study area. Cores along the transect near the lagoon indicated that the basin is filled with 1–1.5 m of freshwater peat. (b) Interpreted coastal geomorphology. *Source*: GeoEye image of Simeonof Island from May 2009. © 2009 DigitalGlobe, NextView. (c) Topographic profile corresponding to transect line in image and map above. The map and profile delineate scarps that may reflect recessional shorelines related to progressive draining of coastal lakes rather than to a series of emergent marine strandlines. Slowly rising sea level over the past 3500 years or stable sea level with severe coastal erosion allowed the sea to invade the lakes (Witter et al., 2012).

Plate 6 Comparison of shoreline data from 1927 to 2009 at Neskowin, Oregon, USA. (a) 1927 NOS T-sheet (#4337) depicting the high water line (HWL) as a bold shoreline backing a stippled beach. *Source*: LiDAR data reproduced with permission of Jon Allan. (b) 1953 black and white aerial photography acquired by the US Geological Survey showing the HWL coincident with the boundary between wet and dry beach sand. *Source*: Photographs by Jon Allan. Reproduced with permission. (c) Shaded relief map derived from LiDAR data acquired by the State of Oregon in 2009. The blue line traces the mean shoreline position for the period 2007–2013 measured by GPS survey. Dashed line marks one standard deviation about the mean (unpublished data by Jon Allan). LiDAR data reproduced with permission of Jon Allan. (d) 2009 digital ortho-imagery acquired by the US Department of Agriculture's National Agriculture Imagery Program (NAIP). (e) Late Holocene tree stumps exhumed in the surf zone near Proposal Rock. *Source*: Photograph by Jon Allan. Reproduced with permission.

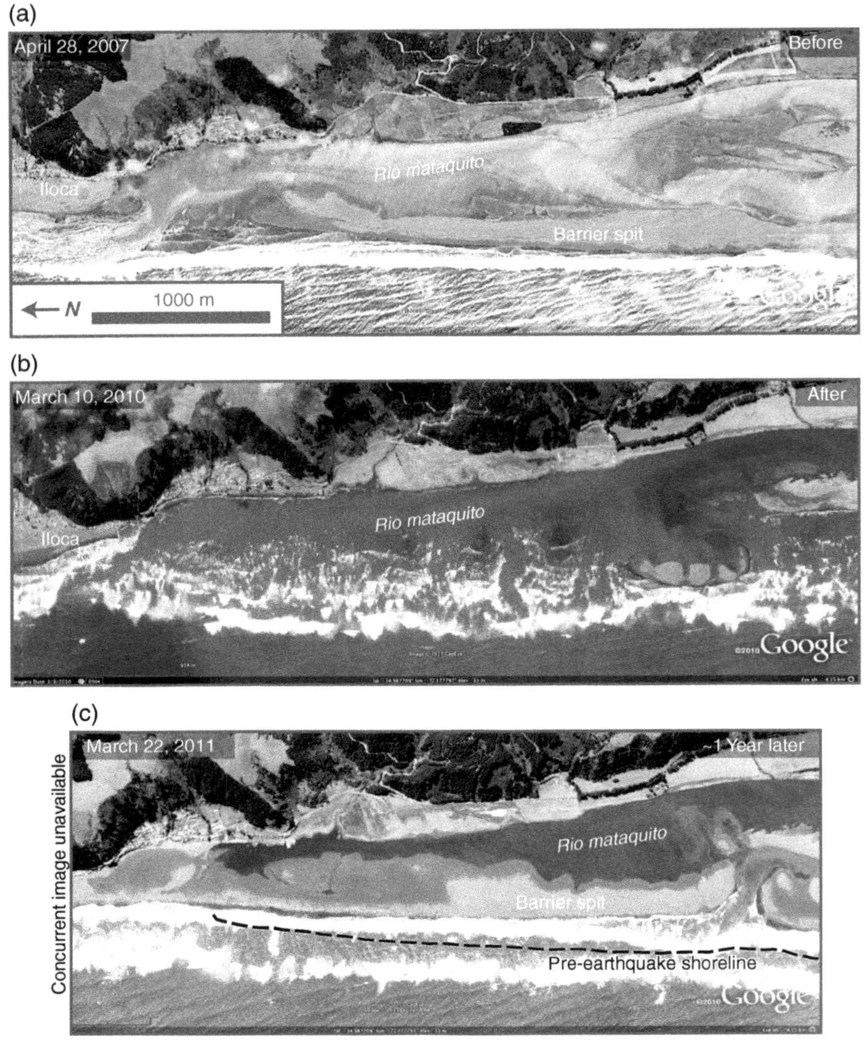

Plate 7 Changes in the configuration of the barrier spit at the mouth of Rio Mataquito following the 27 February 2010, *M* 8.8 Maule, Chile earthquake shown by Google Earth images (© 2010 Google). (a) The long, unbroken barrier spit in April 2007 diverted the mouth of Rio Mataquito to the north near the village of Iloca. Google Earth images from September 2009 show a similar spit configuration with little shoreline change prior to the 2010 earthquake. (b) Post-earthquake images from March 2010 show the disappearance of the spit due to wave erosion. (c) In March 2011, the reconstructed barrier placed the river mouth about 4 km south of Iloca with the shoreline 150 m landward of its 2009 position (dashed shoreline).

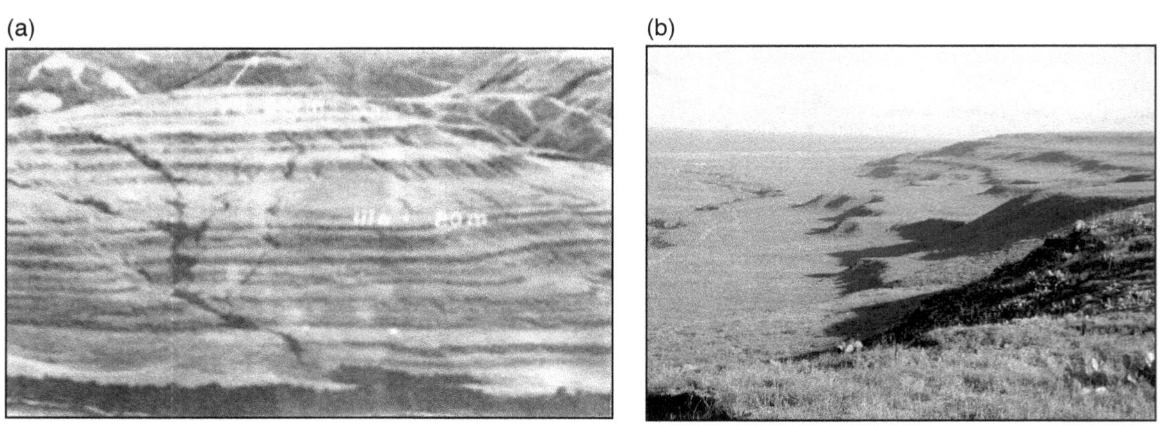

Plate 8 (a) Coralline terraced landscape of Huon Peninsula, New Guinea (6°05′ N, 147°34′ E). *Source*: Reproduced from Chappell (1974, fig. 2). The white lettering on the image is the original annotation of Chappell (1974) and refers to his terrace identifying number and the elevation of the terrace in meters. *Source*: Reproduced with permission of the Geological Society of America. (b) Erosional marine terraced landscape exemplified by marine terraces on the west coast of San Clemente Island, California (32°53′ N, 118°31′ W) looking northward (Muhs et al., 2002). The most extensive terrace, in the foreground and middle ground on the left of the photo, is the c. 120 ka marine terrace. The youngest and lowest terrace, limited in extent and minimally exposed along the embayed coastline in the middle ground, is the 80–100 ka marine terrace. The higher and older terraces, in the center and right on the photograph, are mid-Pleistocene marine terraces (200 ka and older). *Source*: Photograph by Daniel Muhs. Reproduced with permission.

Plate 9 The seacliff–shore platform junction at Drake's Beach, Point Reyes Peninsula, California at low tide (38°01.5' N, 122°58' W). The platform is prominently displayed, as are the tide drainage channels cutting the platform. Platform and seacliff eroded in siltstone and sandstone of the Miocene–lower Pliocene Purisima Formation. The wide, prominent modern platform, the lack of any older uplifted platforms in the landscape, and the low uplift rates along this section of the coast adjacent to the San Andreas fault all allow for the possibility that the platform may have been originally formed in the late Pleistocene and then inherited and reoccupied in the late Holocene. *Source*: Photograph by S. M. Cashman. Reproduced with permission.

Plate 10 Microatolls that were killed, wholly or in part, due to coseismic uplift during the December 2004 Aceh Sumatra earthquake. The three sites, located in the vicinity of 2°50' N, 95°49' E, are described in detail in Meltzner (2010). (a) Microatoll with cup-shaped geometry that implies that rapid submergence was occurring before the abrupt coseismic uplift event that emerged, and killed, the microatoll. Langi site, northern Simeulue Island, Sumatra. *Source*: Photograph by D. H. Natawidjaja. Reproduced with permission of Aron Meltzner. (b) Uneroded concentric ridges and swales characterize this microatoll and attest to annual growth outward once the coral reached the highest level of survival (HLS) dictated by low tide. Outward growth terminated when the coral was killed by coseismic uplift. This specimen is from the Lewak site, northern Simeulue, but similar specimens can be found in the circum-Pacific in tropical regions. *Source*: Photograph by D. H. Natawidjaja. Reproduced with permission of Aron Meltzner. (c) Most of this microatoll was abruptly killed by abrupt relative sea-level fall, although the lowest 5 cm of the colony is still alive in this photo. The fall in relative sea level is measured as the elevation difference between the microatoll's dead flat top (the former HLS) and the highest living corallites today (the new HLS, a few centimeters above water level in this photo). Photo was taken by J. Galetzka at the Lewak site, northern Simeulue, on 5 February 2005. *Source*: Photograph by J. Galetzka. Reproduced with permission of Aron Meltzner.

(a) (b) (c)

Plate 11 Modern examples of the shore platform inner edge, which is also called the platform–seacliff junction or the shoreline angle. (a) Platform inner edge eroded into highly erosive, interbedded (3–6 cm thick) middle Miocene sandstone and argillite, northern California, USA. People are astride the inner edge. Note that the well-rounded beach cobbles, which partially cover the platform, are a distinctly different, much more resistant massive sandstone (40°06' N, 124°06.5 W). (b) Close-up view of modern platform inner edge, La Paz Peninsula, Mexico. Platform and seacliff consist of andesitic lahar deposits (24°19' N, 110°12' W). (c) Modern shore platform and platform inner edge at Arisaig, northern coast of Nova Scotia (45°46' N, 62°09' W). The platform is cut in landward-dipping Silurian sandstone and siltstone (Porter et al., 2010). *Source*: Photograph (a) by Harvey Kelsey and photographs (b) and (c) by A. S. Trenhaile. Reproduced with permission.

Plate 12 (a) Modern tidal notch in tidal zone, La Paz Peninsula, Mexico; notch is eroded into andesitic lahar deposits (24°19' N, 110°12' W). (b) Uplifted paleo-tidal notch, Barbados Island, western Caribbean; notch is eroded into coralline limestone (13°10' N, 59°33' W). *Source*: Photograph (a) by A. S. Trenhaile and photograph (b) by Anthony Long. Reproduced with permission.

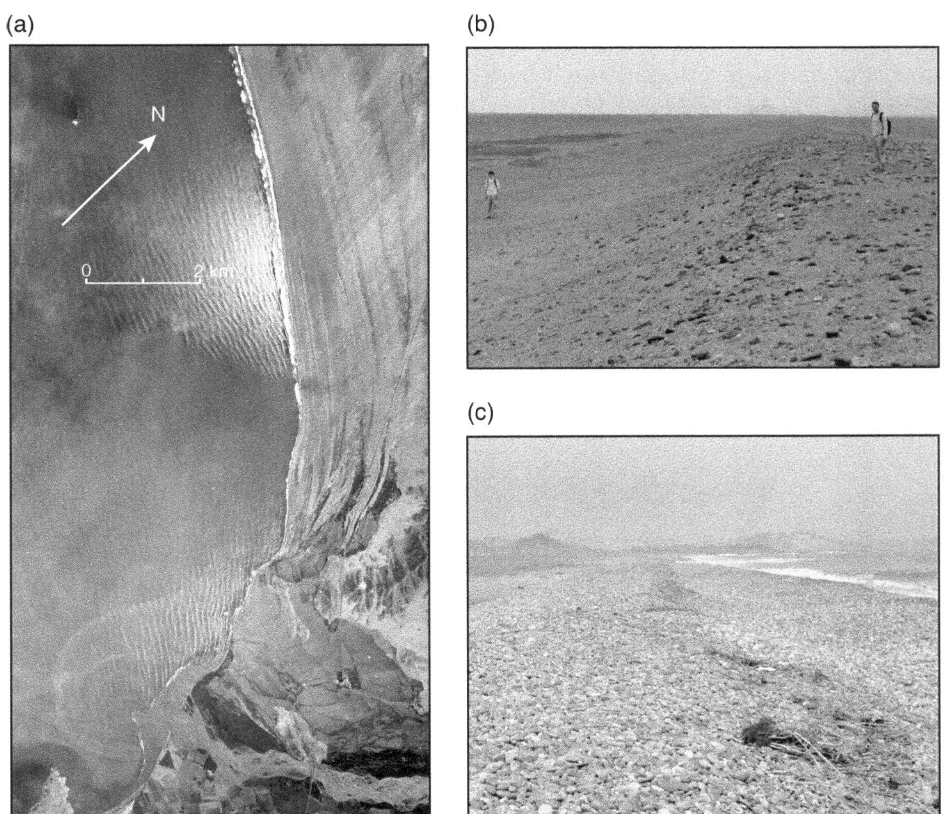

Plate 13 (a) Mosaic of three vertical aerial photographs depicting beach ridges on the progradational beach-ridge plain of the Santa beach ridge complex, latitude 8° S, Peru. Santa River empties to Pacific Ocean to the southeast of the beach-ridge plain. Note the plume of sediment at the river mouth that is being transported north. Source of photograph: Servicio Aerofotográfico Nacional (Peru), Project 147-66, 20 March 1967. *Source*: Photographs mosaicked and contributed by D. Sandweiss. Reproduced with permission. (b) Abandoned inactive beach ridge, Santa beach ridge complex, latitude 8° S, Peru. View to north-northwest. (c) Modern beach ridge, seaward edge of Santa River progradational plain, latitude 8° S, Peru. View to south-southeast. *Source*: Photographs (b) and (c) by David Reid. Reproduced with permission.

Plate 14 (a) Uplifted beach ridges at Cape Turakirae, South Island, New Zealand (41°26' S, 174°55' E). Uplift of all ridges is inferred to be associated with coseismic displacement on the nearby Wairarapa fault, and the most recent, historic beach ridge uplift occurred during coseismic displacement on this fault in 1855. Individual uplifts range in magnitude from 3.5 to 7.1 m (McSaveney et al., 2006; Little et al., 2009). The two stair-stepped treads of marine terraces in the middle ground are oxygen isotope stage 5 (80–125 ka) and stage 7 (~200 ka) uplifted shore platforms (McSaveney et al., 2006). *Source*: Photograph by Lloyd Homer. Reproduced with permission of GNS Science, New Zealand. (b) Late Holocene uplifted beach ridges along the central-east coast of Isla Mocha, which is an island at latitude 38° S off the coast of Chile (see Nelson and Manley, 1992). Isla Mocha sits above the Chilean subduction zone. *Source*: Photograph by Alan R. Nelson. Reproduced with permission.

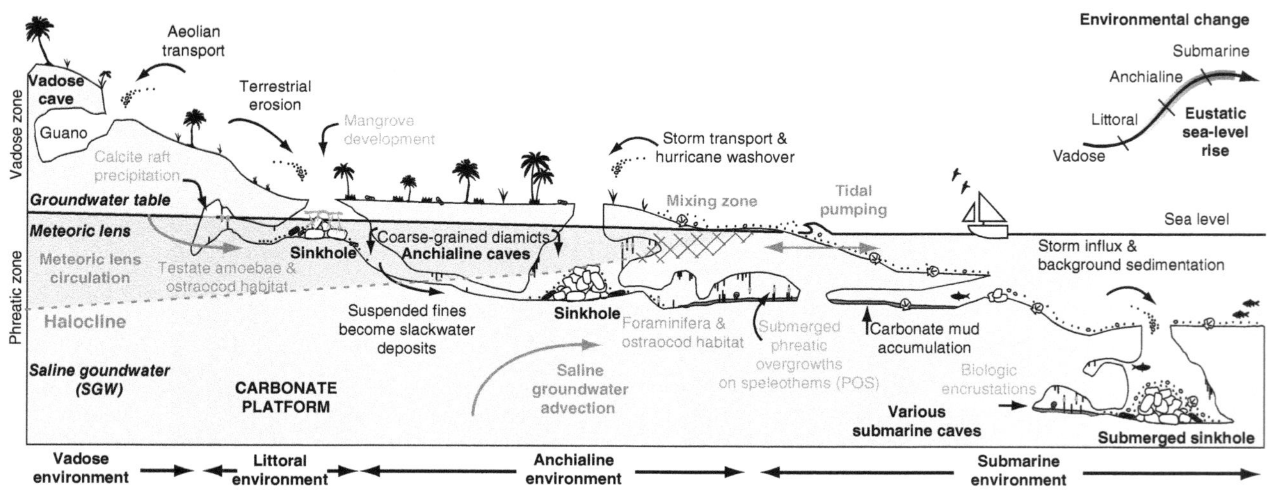

Plate 15 Coastal karst basins (CKBs) provide accommodation space for unique environments, sedimentation patterns, and ecosystems that are distinct from all other coastal environments. All four environmental categories can be observed with increasing distance from the coastline, and sea-level rise causes predictable environmental succession in CKBs (top right corner). The distinctive environmental change that occurs during sea-level oscillations is preserved in the geologic record through speleothems and sediments in CKBs. Green: microfossil remains; red: sea-level indicators; blue: groundwater hydrography and flow; black: common sedimentary processes.

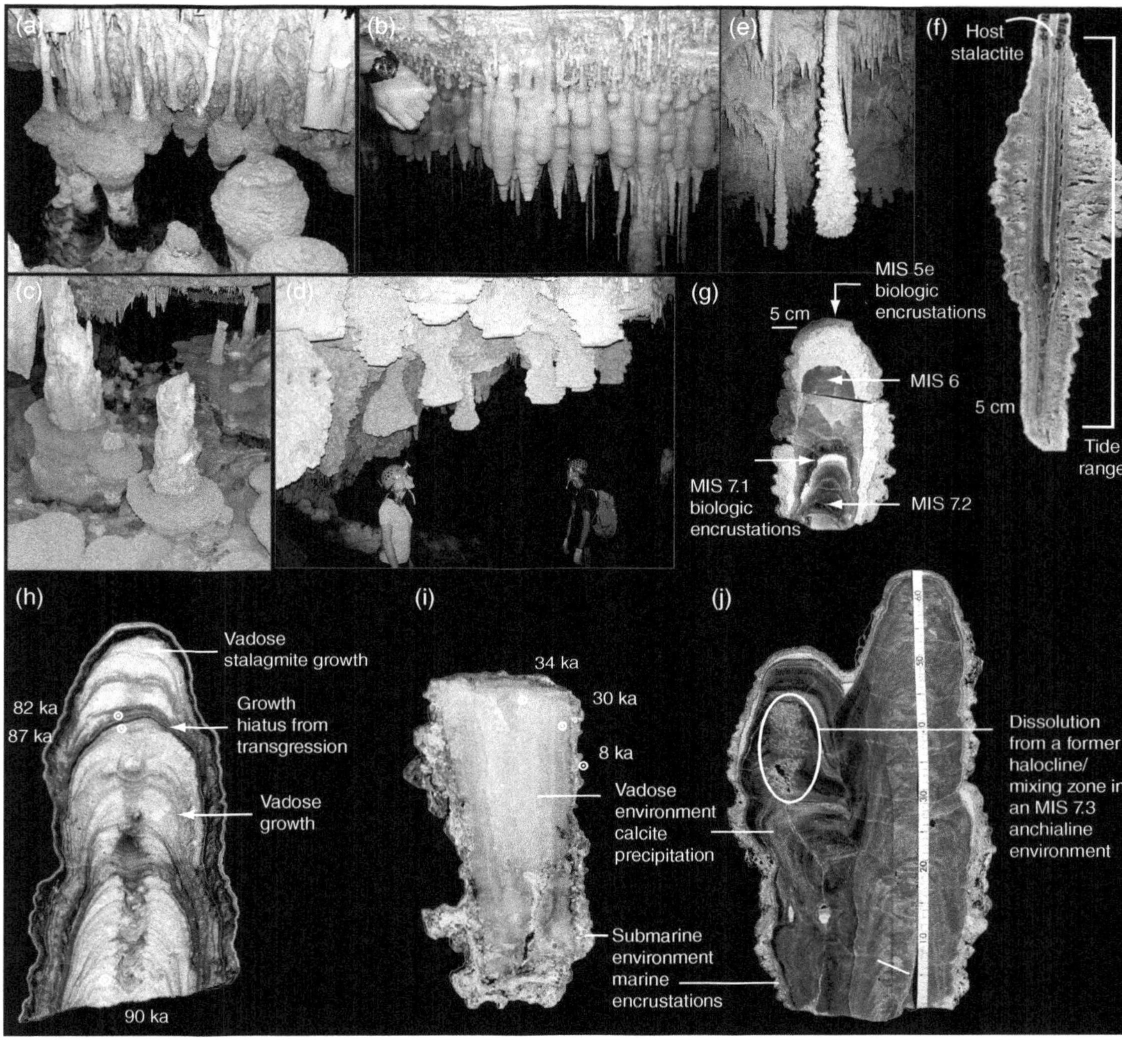

Plate 16 Speleothem sea-level indicators. (a–c) Recent phreatic overgrowths on speleothems (POS) from Mallorcan littoral cave environments. (d, e) Pleistocene POS in Mallorca indicative of prior sea-level highstands. (f) POS developed on a stalactite where thickest point represents mean tide level. (g) Stalagmite ASI from Argenterola Cave (Italy), indicating two cycles of drowning by sea-level rise when the cave became a submarine environment suitable for encrusting marine invertebrates (Bard et al., 2002); *Source*: Photograph by Fabrizio Antonioli. Reproduced with permission. (h) Stalagmite K14 from U Vode Pit (Croatia) depicting a classic growth hiatus from a drowning event. *Source*: Surić et al., 2009. Reproduced with permission of Elsevier). (i) Stalactite P-23 from Tilhovac Bay Cave (Croatia) depicting biologic encrustations as cave flooded by Holocene sea-level rise. *Source*: Surić et al., 2005. Reproduced with permission of Elsevier. (j) Stalagmite ASN from Argentarola Cave (Italy) depicting typical vadose environment calcite precipitation, calcite dissolution from a halocline in an anchialine environment during MIS 7.3, and biologic encrustations developed in a subsequent submarine environment (Dutton et al., 2009). *Source*: Photograph by and reproduced with permission of Fabrizio Antonioli. MIS: Marine Isotope Stage.

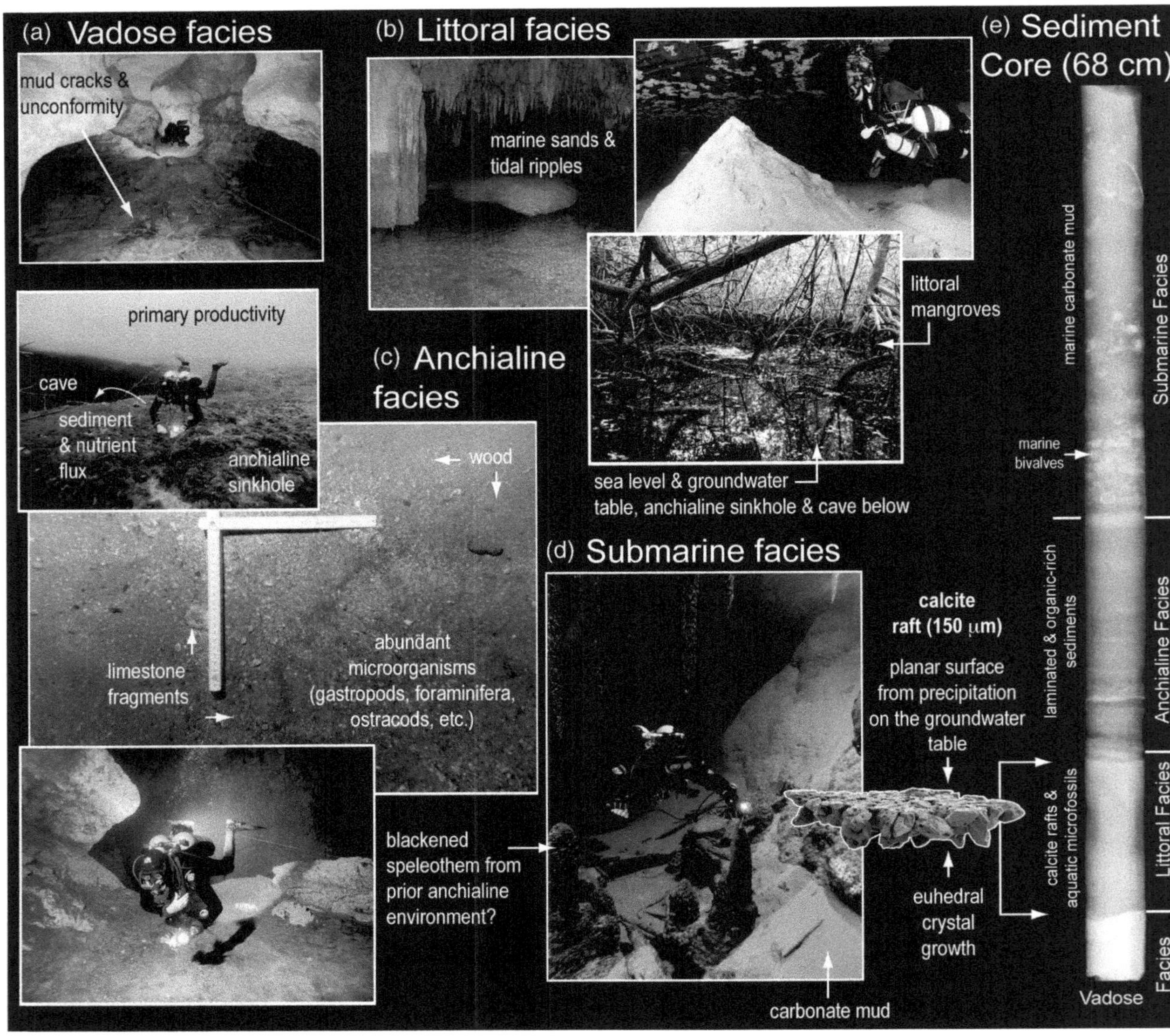

Plate 17 Sediments and facies in CKBs. (a) Mud cracks on sediments that were originally deposited in an underwater cave during a previous vadose environment, and a subsequent unconformity from non-deposition (Mallorca, photograph by Joan Fornos, used with permission). (b) Sediment deposited in littoral caves that are open to the ocean can be primarily marine in character and can even have tidal indicators such as ripples (Bermuda, photograph by Peter van Hengstum), but calcite rafts can precipitate at the water table in isolated littoral cave environments that are not physically connected to the ocean (Mallorca, photograph by Joan Fornos, used with permission). In sinkholes, mangroves commonly grow at sea level (Whiskey Sinkhole, Bermuda). (c) Anchialine facies often comprise both allochthonous sediments eroding into sinkholes and caves, as well as autochthonous sediments from *in situ* cave processes (Mexico and Bermuda, photographs by Daniel Riordan and Peter van Hengstum, used with permission). (d) Carbonate mud and marine invertebrates are common in submarine environments that are well-circulated with the ocean. (e) A sediment core documenting the succession of Green Bay Cave (GBC5) through all four CKB environments in response to Holocene sea-level rise (van Hengstum et al., 2011). *Source*: Photographs by Peter van Hengstum.

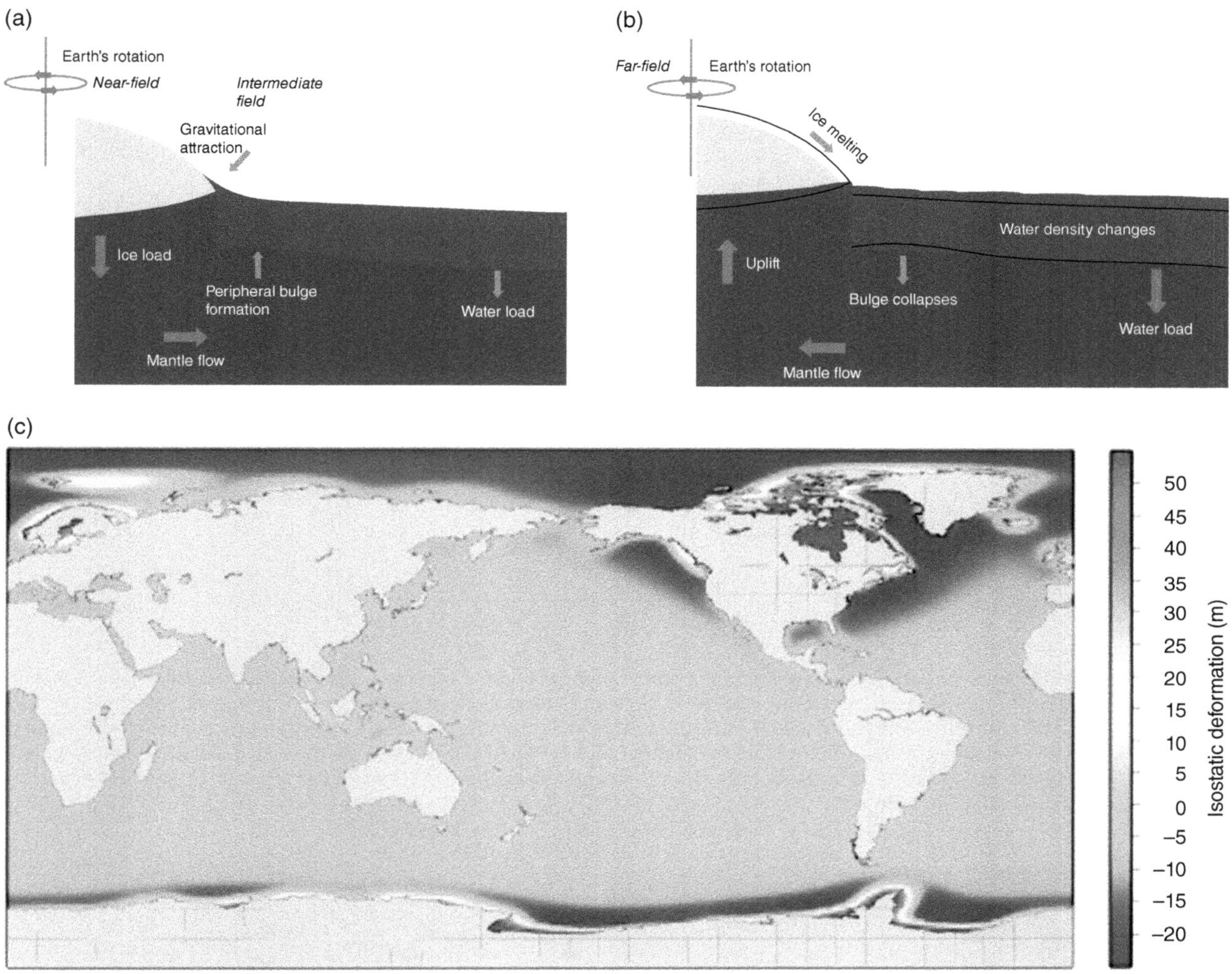

Plate 18 Solid Earth deformation due to growth and decay of large ice sheets. (a) Near the large ice sheet during the ice growth phase, loading of the lithosphere deforms the mantle causing flow in mantle materials away from the center of the ice sheet. (b) Melting of the large ice mass produces extra water in the oceans causing a rise in sea-level. At far-field locations, where most of coral reefs are situated, the additional water loading in ocean basins tends to cause uplift of land masses. (Yokoyama & Esat, 2011. Reproduced with permission of The Oceanography Society). (c) Calculated sea-level change during the last ice age with the eustatic component removed (Yokoyama et al., 2012. Reproduced with permission of John Wiley & Sons). Low-latitude far-field sites are relatively unaffected by deformations induced by former ice sheets. Coral reefs, which respond to all of these changes, therefore provide a unique archive of paleo-sea levels.

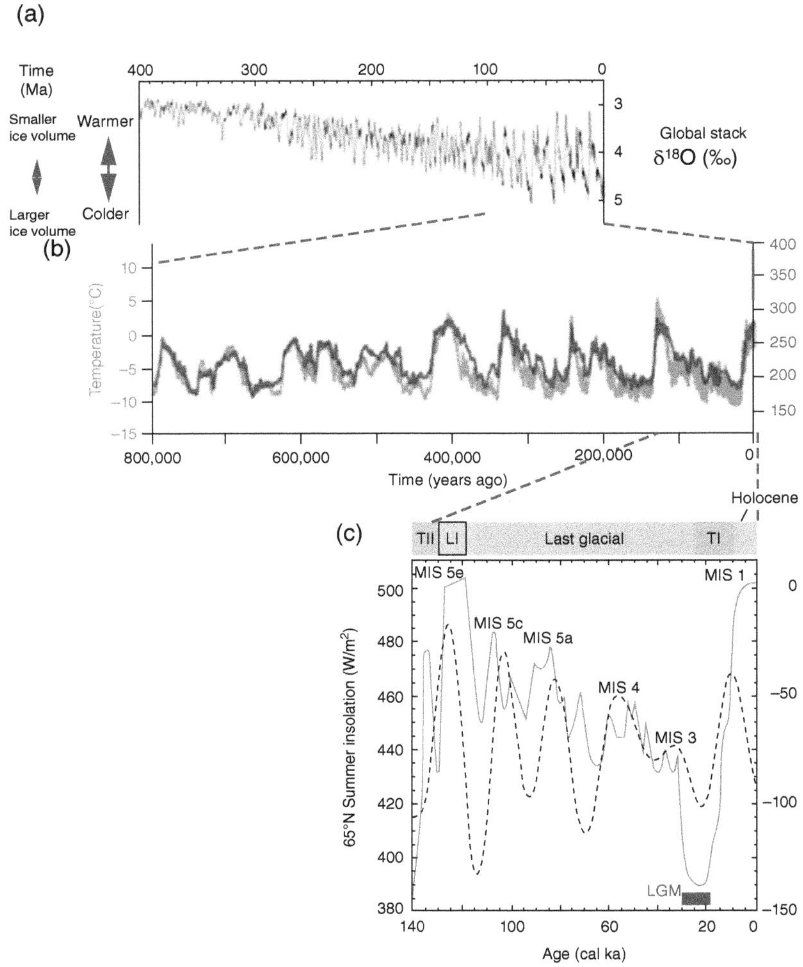

Plate 19 (a) A 4 Ma record of paleo-sea-level reconstructions using stacked deep-sea sediment archives (data from Lisiecki and Raymo, 2005); (b) comparisons between Antarctic surface temperatures and atmospheric CO_2 records from Vostok (data from EPICA Community Members, 2004) and (c) Northern Hemisphere high-latitude insolation curve with mainly coral-based sea-level reconstructions (data from Yokoyama and Esat, 2011). Several distinct climate time windows are identified such as TII (termination II, namely penultimate deglaciation, 129–150 ka); TI (termination I, the last deglaciation, 10.4–24 ka); and LGM (19–26 ka).

Plate 20 Images of corals from Huon Penninsula, Papua New Guinea. (a) Present-day nearshore seafloor around 3 m water depth shows shallow-water-living corals. (b) A wall of Holocene uplifted reef includes robust branching Acropora corals as well as platy and encrusting Acropora. (c) A photo taken offshore from Bobongara section showing a full section ranging from the Holocene reef to the last interglacial reef. (d) Kanzarua section looking up from 30 ka reef to MIS 4 and MIS 5 terraces. *Source*: photographs by Yusuke Yokoyama.

Plate 21 Present-day example of (left) shallow-living coralgal algae (*Hidrolithon* (*Porolithon*) *onkodes*) and (right) *Acropora* at Heron Island, Great Barrier Reef. *Source*: Photographs by Yusuke Yokoyama.

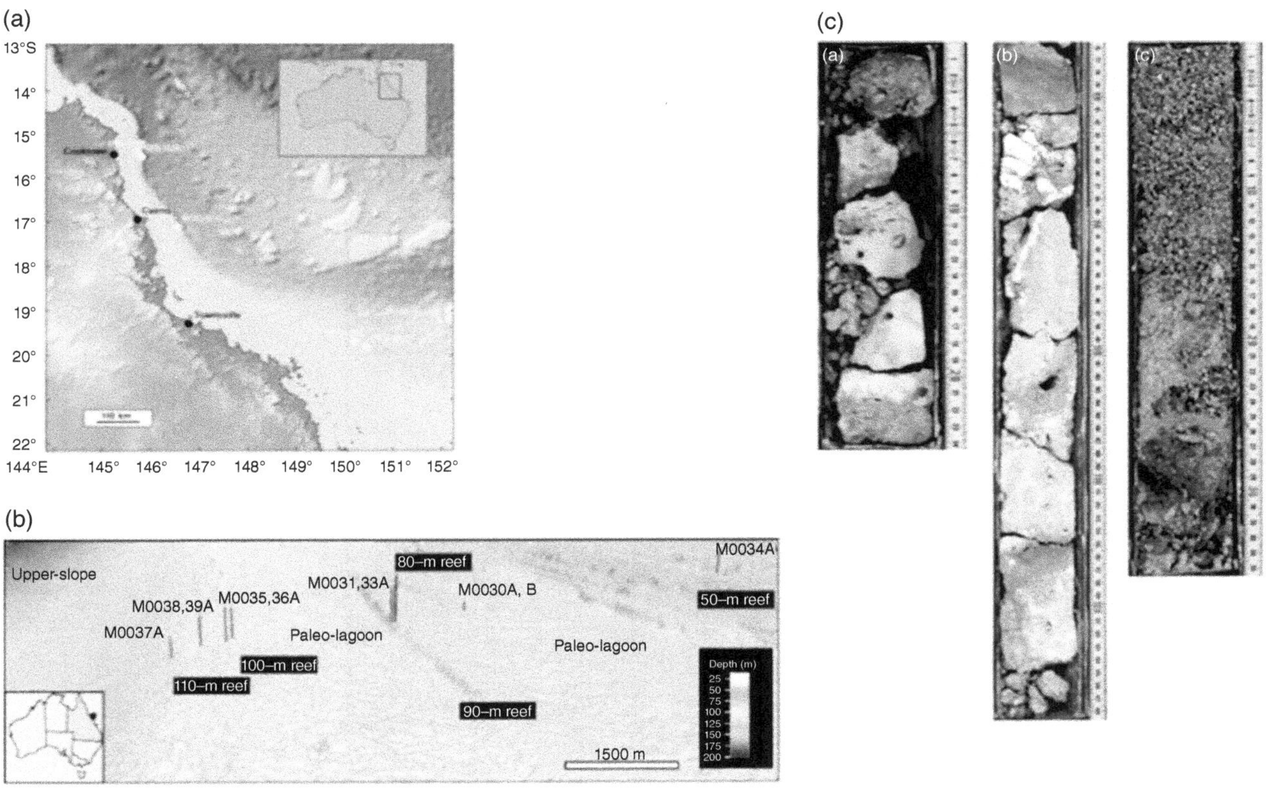

Plate 22 A study of Quaternary sea levels with corals drilled from the Great Barrier Reef, Eastern Australia. *Source*: Yokoyama et al., 2011. (a) Location of drilling sites of IODP Expedition 325; (b) a high-resolution multi-beam image of Hydrographer's Passage transect HYD-0C; and (c) representative lithology in cores. Section A is a coralgal boundstones typically found at the top of the reef structure. Coralgal-microbialite boundstones and unlithfied sediments can be seen in sections B and C respectively.

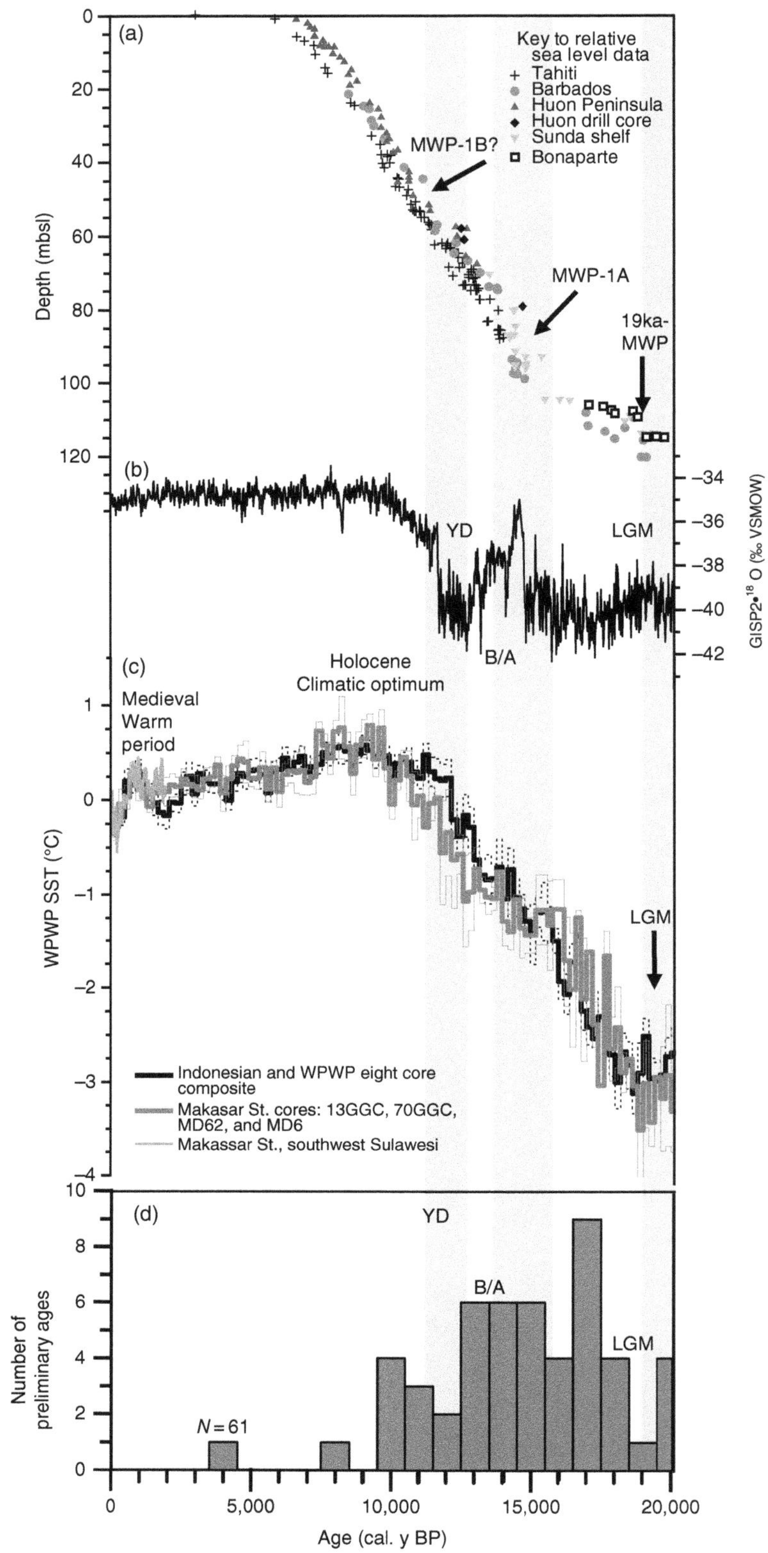

Plate 23 (a) Comparisons between previously reported sea levels during the past 20 ka; (b) Greenland ice core $\delta^{18}O$; (c) sea-surface temperature history of the Western Pacific Warm Pool; and (d) preliminary dating results of GBR IODP Expedition 325 samples. Reef materials obtained from the GBR nicely capture the key intervals of the past climate events including several meltwater pulses, Younger Dryas cold event and the LGM. *Source*: Yokoyama et al., 2011.

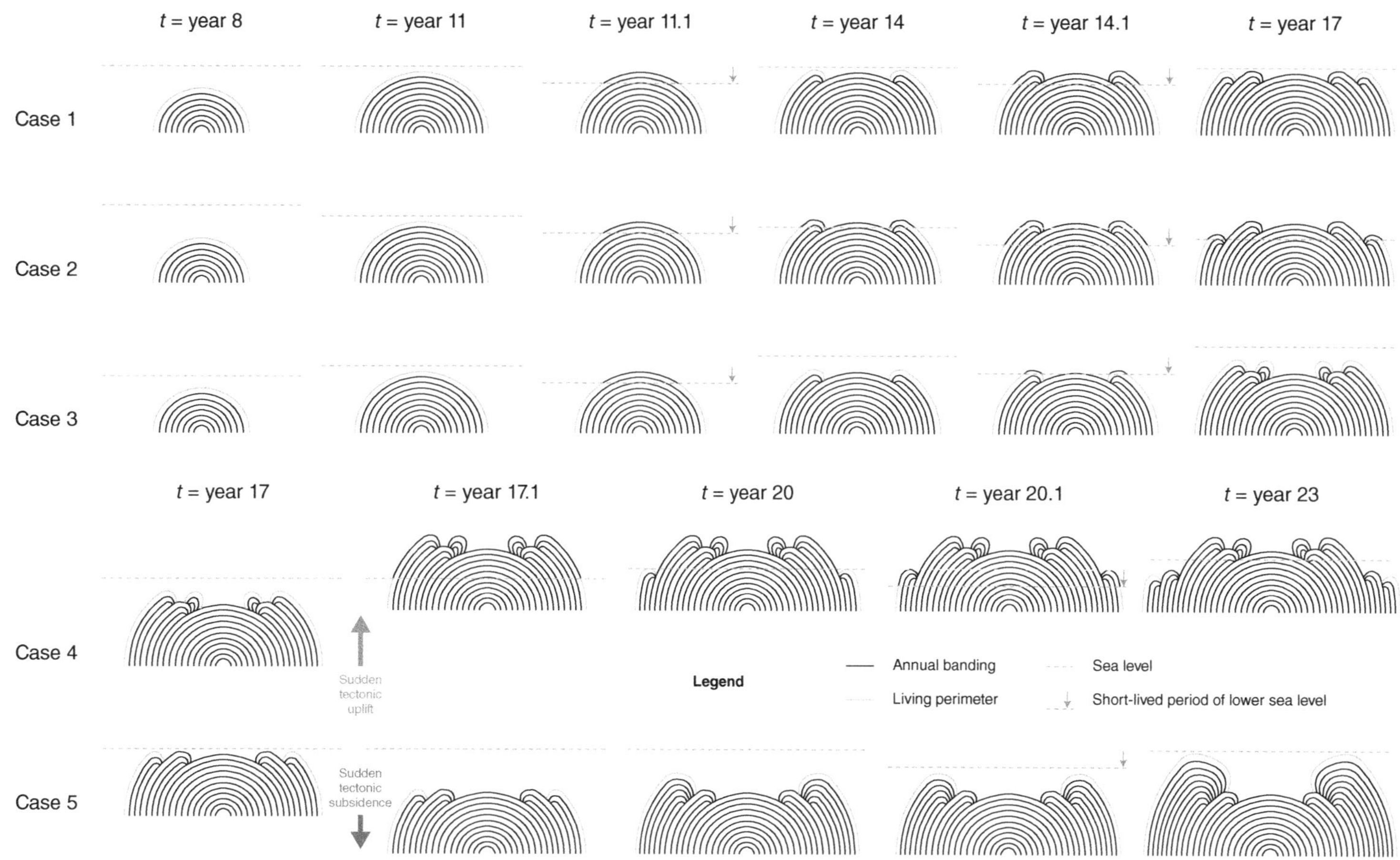

Plate 24 Upper: microatoll development under different relative sea-level history scenarios: under stable sea-level conditions (Case 1); under gradually falling sea level (Case 2); and under gradually rising sea level (Case 3). In all three cases, we superimpose on the long-term trend a realistic interannual variability: at year 11.1 and 14.1, we simulate temporary local sea-level lowerings such as those that tend to occur during La Niña events in the central Pacific, during El Niño events in the western Pacific, or during positive IOD events in the eastern Indian Ocean. Note that concentric annuli form in open-water (unmoated) environments simply from year-to-year fluctuations in the low water level and highest level of survival. Also note the "out-of-sequence" interior annuli that form where relative sea level is rising rapidly enough; this overgrowth protects the underlying inner surface of the microatoll from erosion. Lower: microatoll development affected by sudden (tectonic) changes in land level. Case 4 illustrates the microatoll from Case 3, followed by coseismic uplift at year 17.1; Case 5 illustrates the microatoll from Case 1, followed by coseismic subsidence at year 17.1. In each case, the long-term trend is superimposed on a typical ENSO or IOD cycle, with an additional short-lived period of lower local sea level at year 20.1. In Case 4, the uplift must have been sudden at year 17.1, but if we found the microatoll in Case 5, we could not distinguish between sudden coseismic subsidence at year 17.1 and rapid interseismic subsidence (at an average rate exceeding the coral's growth rate) beginning at year 17.1.

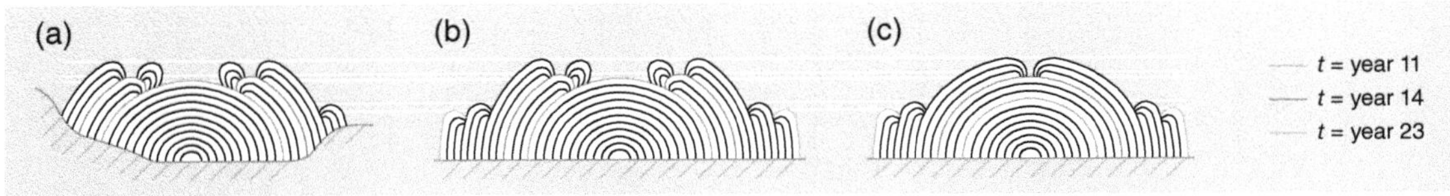

Plate 25 Potential variability in contemporaneous microatolls. Microatolls (a–c) all experienced the relative sea-level history depicted in Cases 3 and 4 of Figure 8.1, having formed under gradually rising sea level with a modest uplift event at year 17.1. Microatoll (b) is the specimen from Case 4 in Figure 8.1. Microatoll (a) started growing at the same time in a depression but at the same basal elevation as microatoll (b), but irregular topography resulted in less accommodation space. The substrate to the left of the coral is sufficiently high that the microatoll's left side did not survive the diedown at year 17.1. The substrate to the right of the coral is slightly lower, which allowed the right side to survive the year-17.1 diedown, but not the lower diedown at year 20.1. Microatoll (c) grew from the same basal elevation as microatoll (b) but started growing only in year 4; its highest level of growth was therefore lower at year 11.1 and it did not die down then. After year 14.1, microatoll (c) recorded a similar history as microatoll (b). White horizontal lines denote elevations of annuli that could be surveyed in the field.

Plate 26 Recognizing the highest living corallites (highest level of survival or highest level of growth) on a microatoll is normally easy to do, if you know what to look for. (a) Close-up and (b) wider view of highest level of survival ~3 cm above the water line on a coral at the PPY-A site of Meltzner et al. (2012), northeastern Simeulue, Sumatra, Indonesia. This coral experienced a diedown of ~40 cm due to uplift in the December 2004 and March 2005 earthquakes, but the coral survived around its base and continued to grow outward and upward from below its highest level of survival. Three years later, in an earthquake in February 2008, the microatoll experienced another diedown of ~2 cm. Above the 20 cm mark on the ruler, the coral is dead and covered by 4 years' accumulation of algae. At the 20 cm mark ~5 cm above the water line, a lip circumscribes the microatoll, marking the coral's highest level of growth from early 2005 to early 2008. The uppermost ~2 cm of this lip, from ~20 cm to ~18 cm, is thinly coated by algae; these are the polyps that died in 2008. Below ~18 cm, the coral is alive and polyps are not coated by algae. An incipient outer lip is forming ~2 cm below the previous lip, although this is not obvious in the photographs (taken in February 2009). When considering the scale, note the optical illusion that only the upper ~5 cm, and the reflection of those ~5 cm, are visible due to the bright reflection of the metallic ruler. (c) Stark contrast between dull, algae-covered, dead polyps above the highest level of survival, and bright, living polyps below on recently uplifted microatolls along the coast of northern Tetepare Island, Western Province, Solomon Islands (June 2012). (d) Field of living microatolls in a moated pool, Mapur Island, Riau Islands, Indonesia (April 2012). Again, note the contrast between the bright living corallites on the rim and perimeter of the colony, and the dull, fine sediment-covered inner upper surface. *Source*: Photographs by Aron Meltzner.

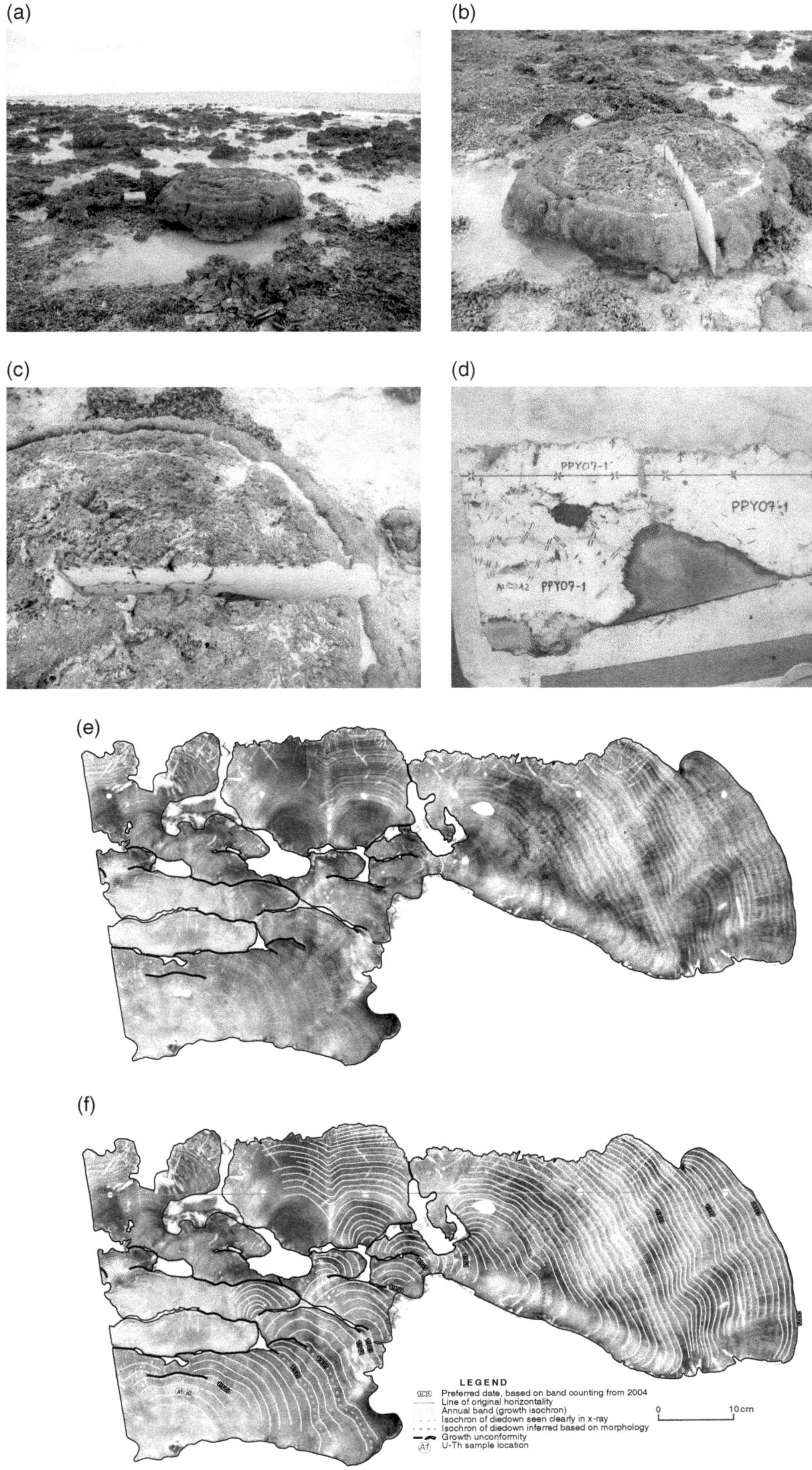

Plate 27 PPY-1 microatoll and slab. (a–c) Views of the microatoll in the field. Removal of the slab allowed us to view below the inner surface of the microatoll, but prior to the slab's removal, only the overgrowth (the flat inner surface of the microatoll) was visible. (d) Photograph of the slab after removal and initial processing. The slab was subsequently X-rayed, and the X-rays were (e) mosaicked and (f) annotated. Although some erosion of the low interior annuli occurred and is evident on the X-rays, the erosion was minor. Once the low interior annuli were overgrown, they were protected from additional erosion. *Source*: Photographs by Aron Meltzner.

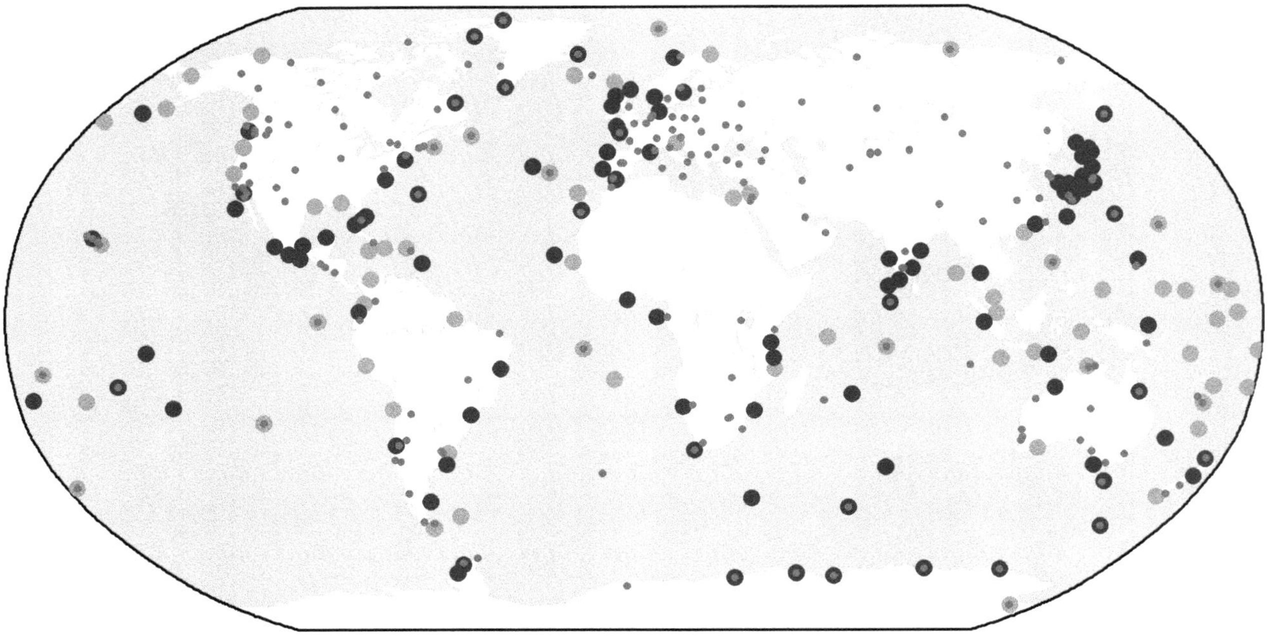

Plate 28 Map of core IGS sites (232 dots), TIGA sites with GPS less than 10 km from tide gauge (224 light circles) and TIGA sites with GPS less than 1 km from the tide gauge (113 dark circles).

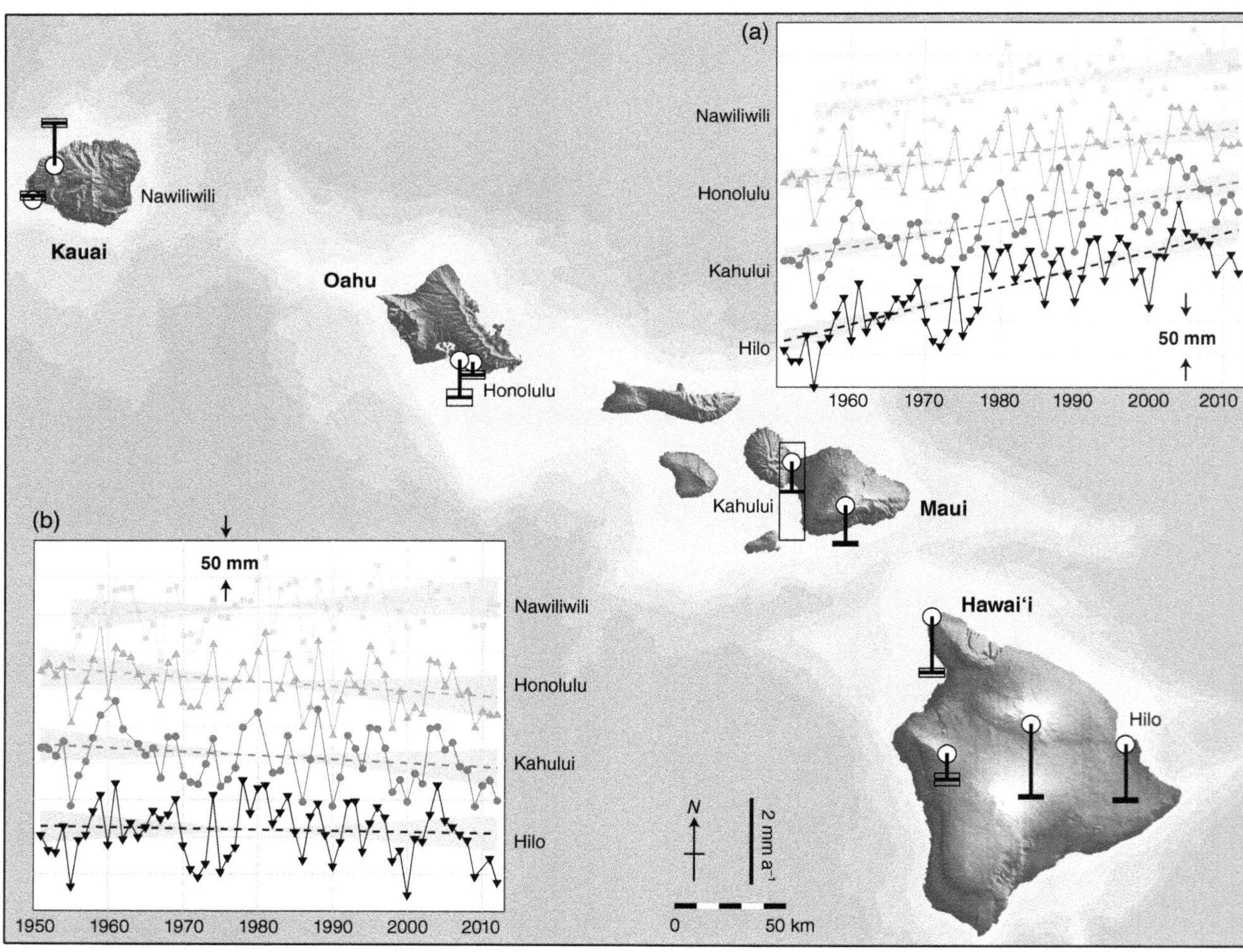

Plate 29 Long-term vertical rates (vertical lines) and error bars (thin lines) for stable continuous GPS sites (circles) in Hawai'i. The increasing mass of the active volcanoes on Hawai'i Island are causing it and Maui to sink with respect to Oahu and Kauai, which are thought to be stable. (a) Annual sea-level changes for tide gauges in Hawai'i, with best-fit slopes (dashed lines) and 95% confidence limits (gray zones). Differences in the sea-level rise rates mirror the GPS vertical velocities. (b) Annual sea-level changes after the mean sea-level rise rate for Nawiliwili, Kauai, and Honolulu, Oahu, is removed, and corrections for the GPS vertical velocities applied. The residual slopes are not significantly different from zero.

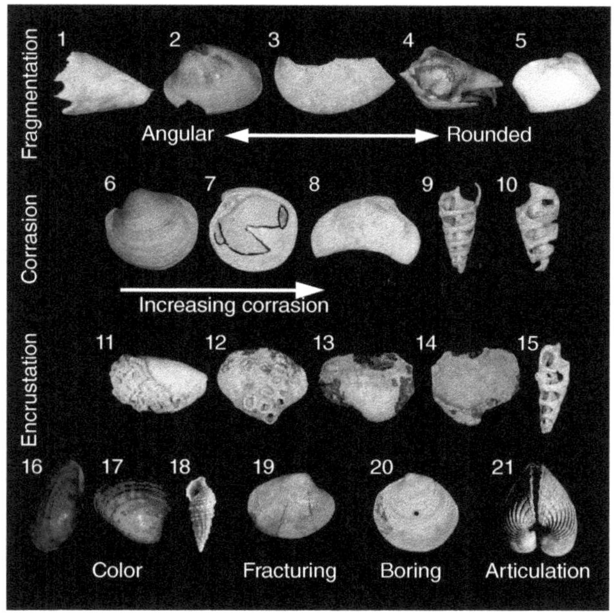

Plate 30 Common taphonomic features of mollusk shells.

Plate 31 (a) *Dendropoma petraeum* platform reef, developing upon a shore platform at San Vito Lo Capo, Sicily, south Italy. (Antonioli et al., 1999. Photographs by Alessio Rovere.) (b) Mushroom-like pillars of *Dendropoma petraeum* at San Vito Lo Capo, Sicily. (Antonioli et al., 1999. Photographs by Fabrizio Antonioli.) (c) Living balanids (arrow) on intertidal boulders, Western Australia. (Photographs by Fabrizio Antonioli.) (d) Fossil *Chthamalus stellatus* on an uplifted shore platform at Scilla, Calabria, south Italy, Late Holocene (Ferranti et al., 2007. Photographs by Fabrizio Antonioli.) (e) Living *Chthamalus stellatus* band in the splash zone, Palmi, south Italy. (Photographs by Fabrizio Antonioli.) (f) Fossil balanids of Pliocene–Early Pleistocene age, Hondeklipbaai, South Africa. (Photographs by Alessio Rovere.) (g) Living balanids (1) and, in the lower part of the boulder, oysters (2) at low tide, Magnetic Island, West Australia. (h) Living *Mytilus galloprovincialis* (on coralline algae) attached in the lower part of a tidal notch, Gaeta, central Italy.

Plate 32 (a) Last interglacial *Lihtophaga lithophaga* boreholes in north Sicily, Italy. (Photographs by Fabrizio Antonioli.) (b) Close-up of last interglacial *Lithophaga lithophaga* in the Bergeggi marine cave, Liguria, north Italy. (Photographs by Alessio Rovere.) (c) Balanids growing upon oysters in a Pliocene–Early Pleistocene outcrop at Hondeklip Bay, South Africa. (Photographs by Alessio Rovere.) (d) Fossil *Perforatus perforatus* on a Archaic-Greek-age harbor boulder (at Lechaion, Greece). (Photographs by Fabrizio Antonioli.) (e) Fossil Holocene *Galeolaria* reef (dashed line: upper band), West Australia. (Photographs by Fabrizio Antonioli.) (f) Section of (1) speleothem showing serpulid overgrowth on (2) top of continental deposition (Argentarola cave, Italy; Antonioli et al., 2004). (Photographs by Fabrizio Antonioli.) (g) Rim (or "trottoir") of *Lithophyllum byssoides* (arrow) at low tide, Gaeta central Italy. (Photographs by Fabrizio Antonioli.) (h) Measurement of the lower part of a shell deposit using GPS RTK system, Central Florida, USA. (Photograph by P. J. Hearty.) (i) Measurement of the relative position of fossil Holocene *Chthamalus stellatus* with respect to that of the living population (arrow), Scilla, Calabria, Italy. (Photographs by Fabrizio Antonioli.)

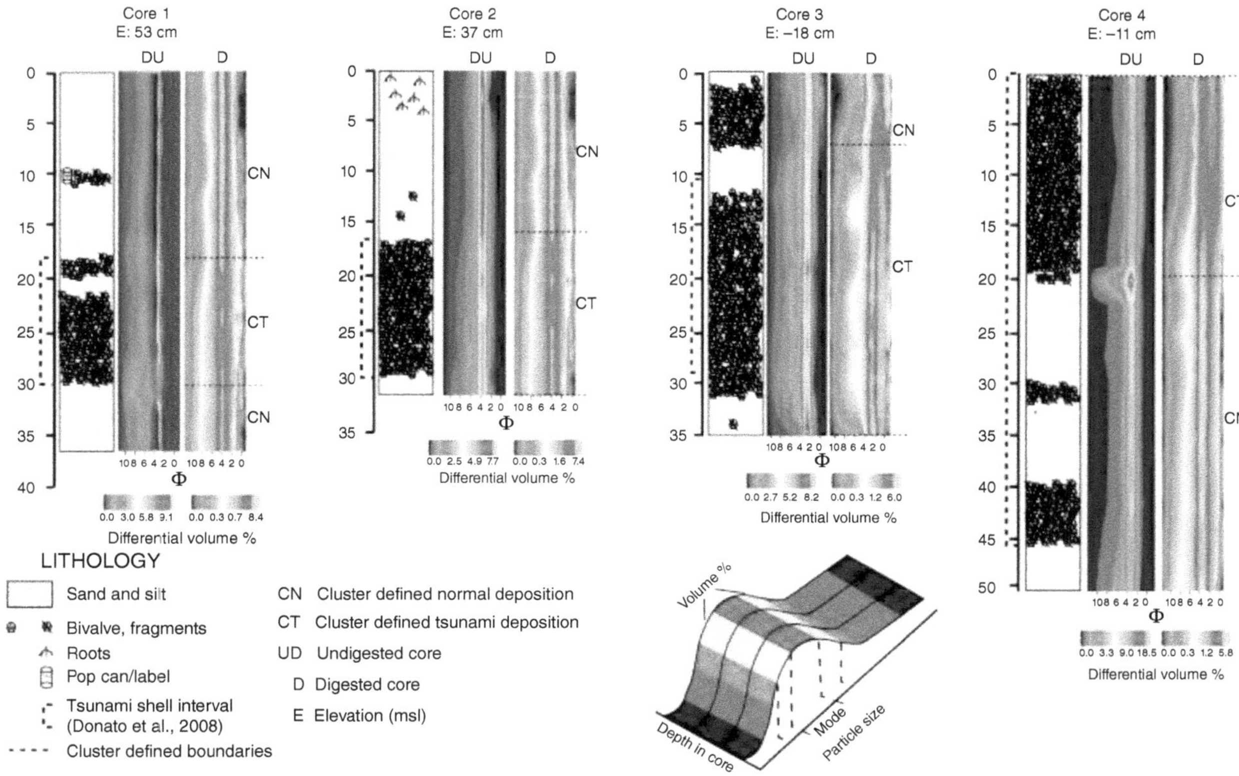

Plate 33 Lithologic logs for four sediment cores collected in Sur Lagoon, Oman. *Source*: Donato et al., 2009. Reproduced with permission from Elsevier. In this case high-resolution particle-size distribution plots were plotted from samples that were analyzed using a Coulter LS 230 instrument. Data are shown as φ surface plots for both treated (digested, D) and untreated (undigested, UD) sediment samples from each core. Differential volumes by percentage color code: blue: absent to minimal; red–pink: common–most common. This technique was found to present the unimodal sediments much more effectively than the use of traditional grain size plots.

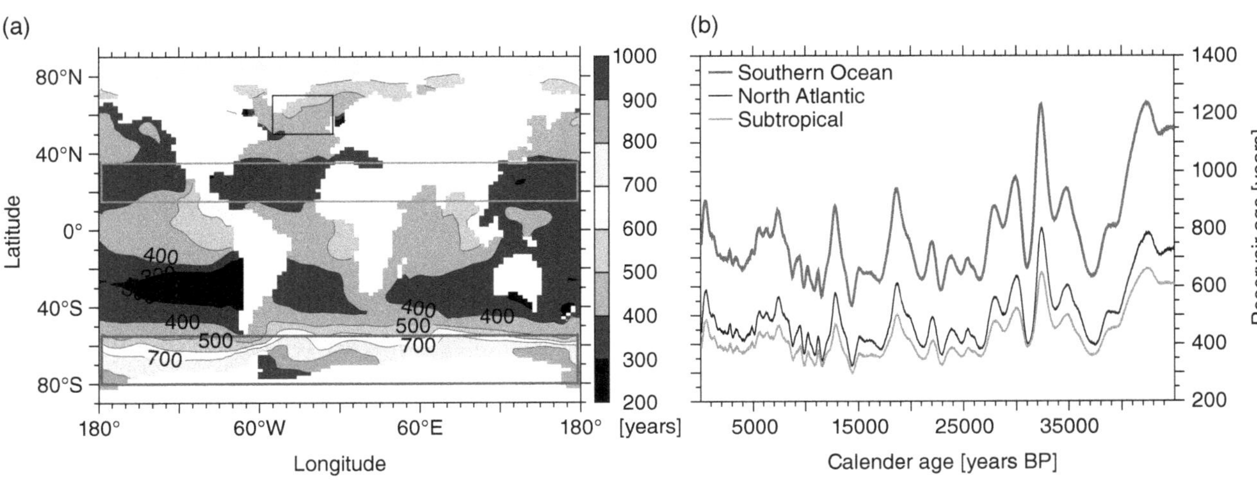

Plate 34 Modeled variability in reservoir ages over the past 45 cal ka (Franke et al., 2008). (a) Spatial pattern of global R variations in surface ocean waters (upper 36 m; J. Franke, pers. comm., 2013). (b) Regional mean temporal R variations for the areas indicated by equally colored rectangles in the map. Note that this is a simulation with present-day boundary conditions in terms of ocean circulation; conditions of reduced Atlantic meridional overturning circulation such as those during the Last Glacial Maximum would lead to higher reservoir ages.

Plate 35 Photos of field measurements, tools, and samples. (a) Hand-held battery-powered drill to recover small drill core samples; (b) larger petrol-powered drill with wider and longer core barrel; (c) surface of coral before drilling; (d) same coral as in (c) but shown with sample and number; (e) surveying position of coral heads into local sea level; and (f) Faviid coral head encrusted with coralline algae that can also be sampled to help assess paleodepth interpretation. *Source*: Photographs by Andrea Dutton.

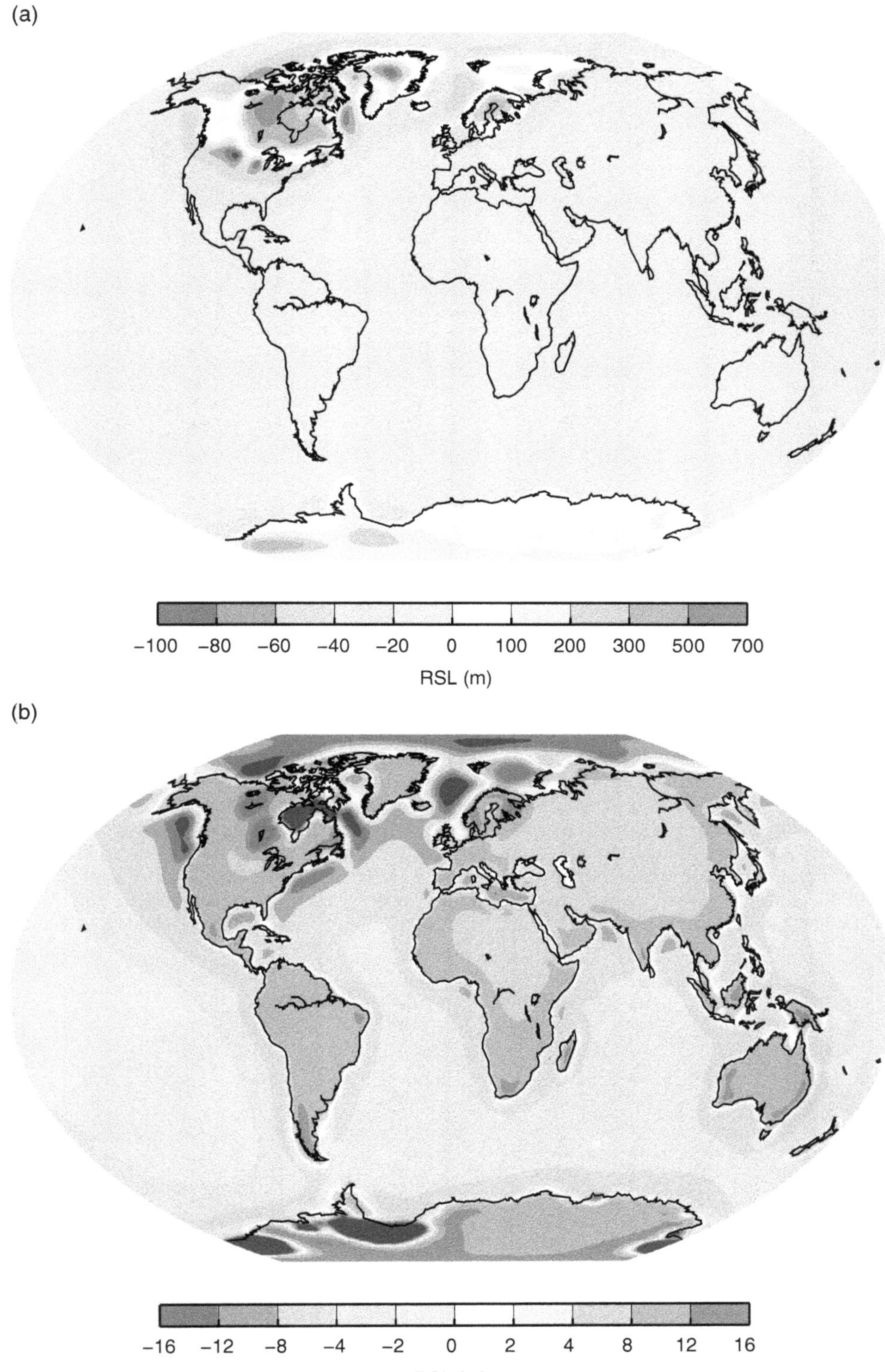

Plate 36 (a) RSL at 10 ka before present due to ice loading changes only. (b) RSL at 10 ka BP due to ocean loading changes only. Note that the contribution of these effects to a globally uniform RSL change through the syphoning process (see Section 28.2) is not included. These model results are based on the ICE5G model of global ice evolution and the VM2 Earth viscosity model (with a 90 km thick lithosphere) (Peltier, 2004).

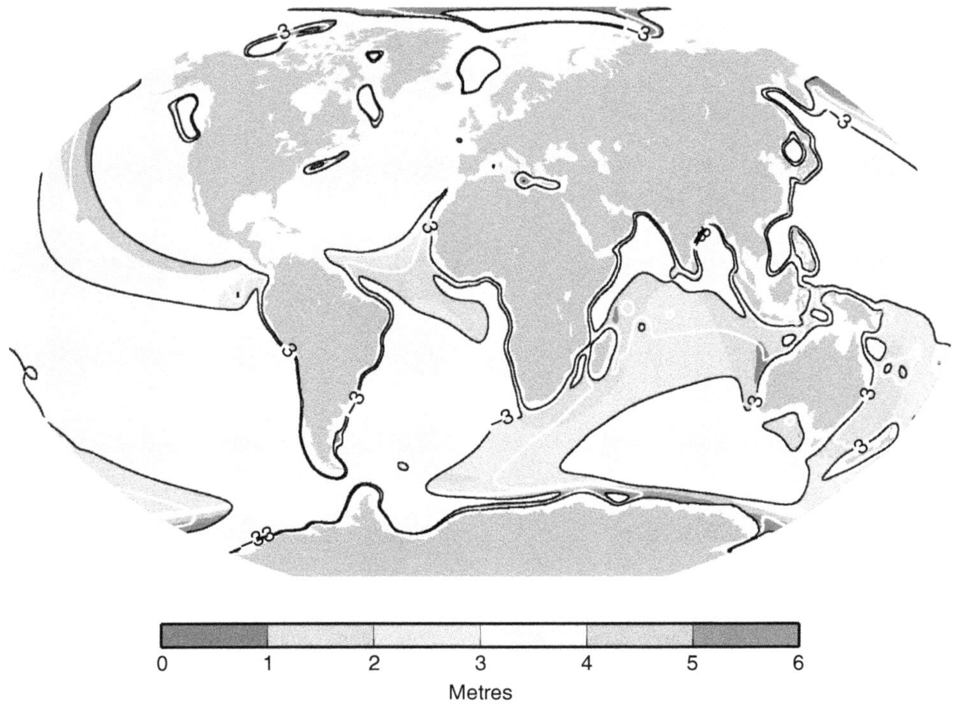

Plate 37 Model results showing locations that would be well suited to estimating ESL from RSL field reconstructions at the last glacial maximum. Ideal locations are where: (1) the modeled RSL values are close to the model ESL value; and (2) the model output is relatively insensitive to uncertainty in the input Earth model. The first criterion is satisfied in areas between the black contours which indicate where RSL is within 3 m of ESL (they are equal along the white line). The second criterion relates to the color shading which indicates the spread (standard deviation) in model output due to variations in Earth model viscosity structure. Blue areas are therefore ideal locations for inferring ESL from RSL data, as the model correction will be relatively small and insensitive to uncertainty in Earth model viscosity structure. These model results are based on the ICE5G model of global ice evolution and a large suite of Earth viscosity models. See Milne and Mitrovica (2008) for further details. *Source*: From Milne and Mitrovica (2008). Reproduced with permission of Elsevier.

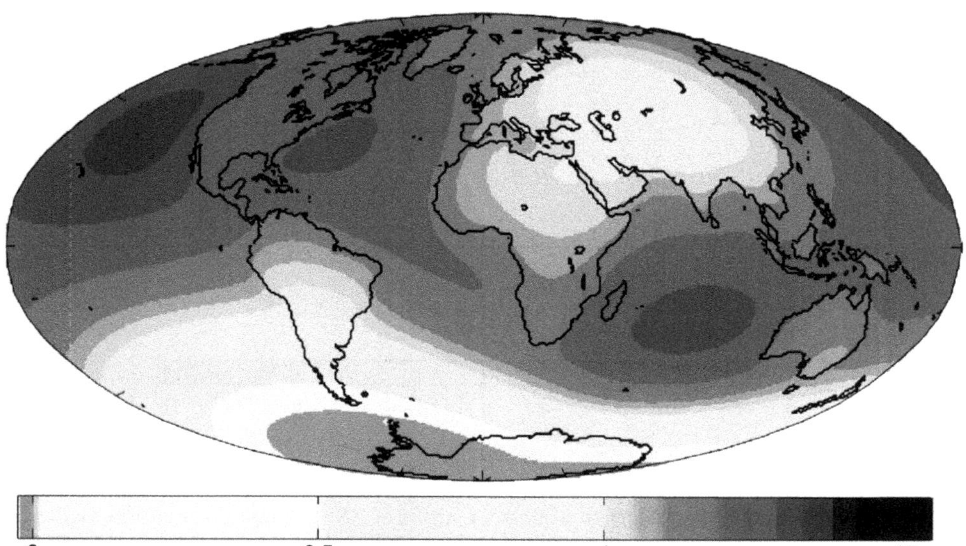

Plate 38 A GIA model prediction of sea-level change associated with a collapse of the West Antarctic ice sheet. The sea-level change is normalized to the eustatic value. Note that a sea-level fall is predicted in the near-field of ablating ice. These computations are based on an elastic Earth model and so assume that the collapse is relatively rapid (of the order 100 years). *Source*: Gomez et al. (2010). Reproduced with permission of John Wiley & Sons.

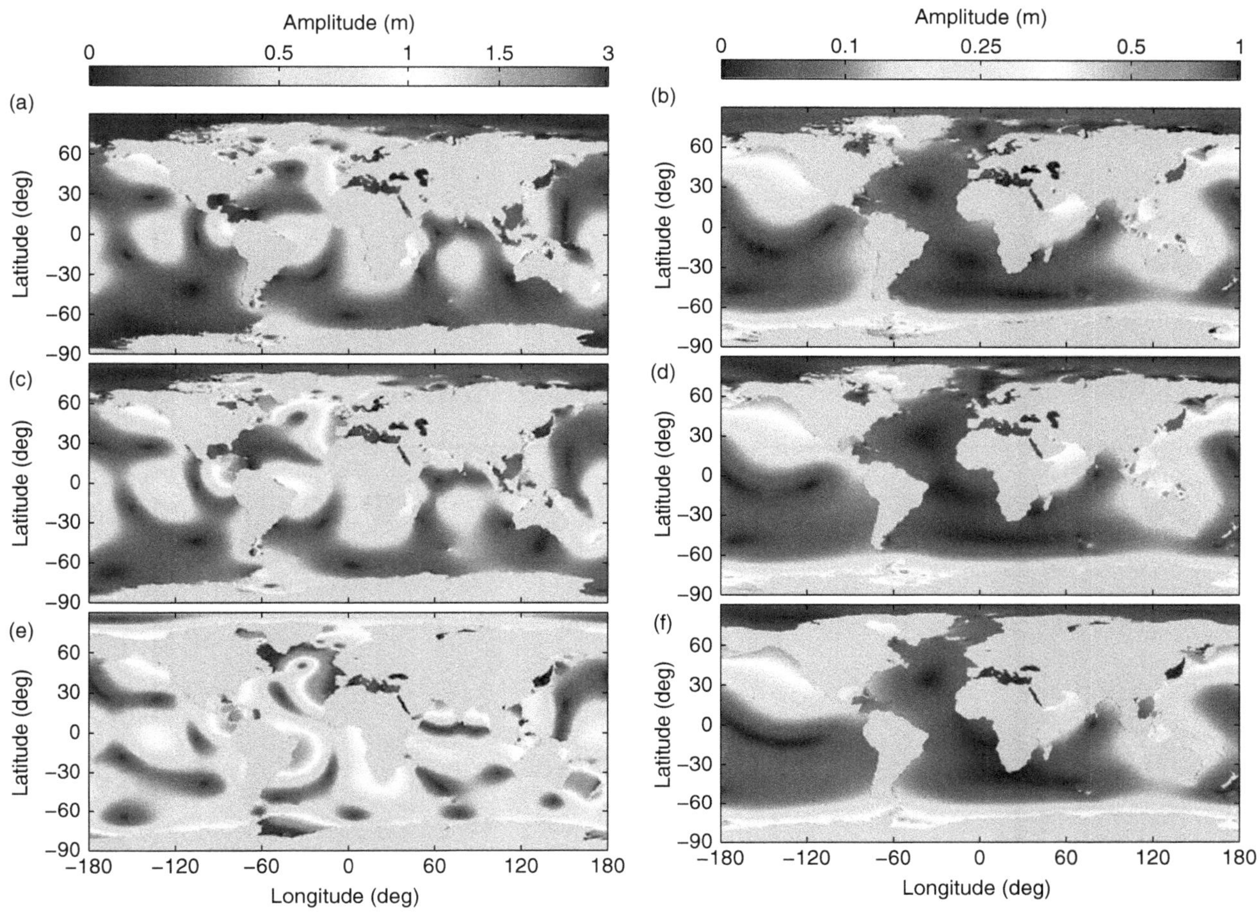

Plate 39 Tidal amplitudes for the M_2 (left) and K_1 (right) constituents. (a, b) As determined by the TPXO 6.2 dataset, described by Egbert and Erofeeva (2002). (c, d) From a global numerical model (Section 29.2). (e, f) A prediction for tides at the LGM from the same global numerical model (see Section 29.4). *Source*: Reproduced with permission from Griffiths and Peltier (2009).

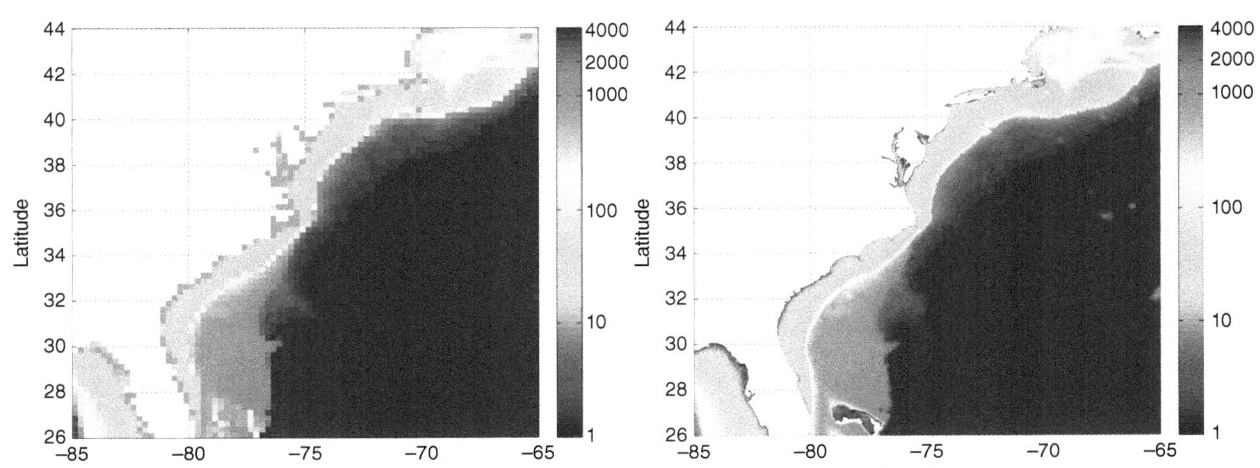

Plate 40 Left: low-resolution bathmetry from the TPXO7.2 global inverse tidal solution (Egbert et al. 1994. *Source*: Reproduced with permission of John Wiley & Sons). Right: high-resolution bathymetry from the EC2001_V2D tidal database (Mukai et al., 2002).

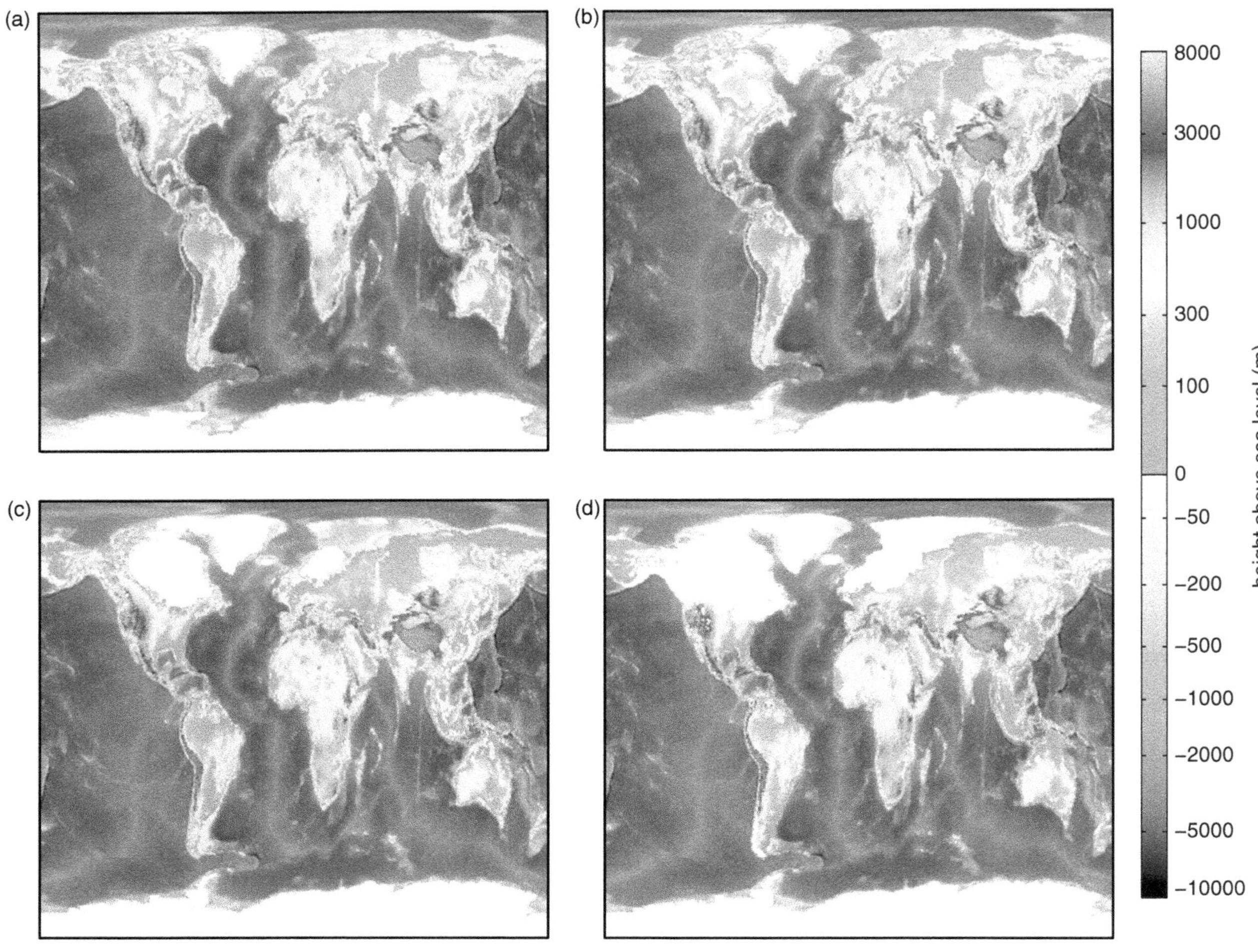

Plate 41 Surface height (m) and ice location according to the ICE-5G v.1.3 dataset, an updated version of that described by Peltier (2004) (a) present day, (b) 8 ka, (c) 12 ka, and (d) 26 ka (LGM).

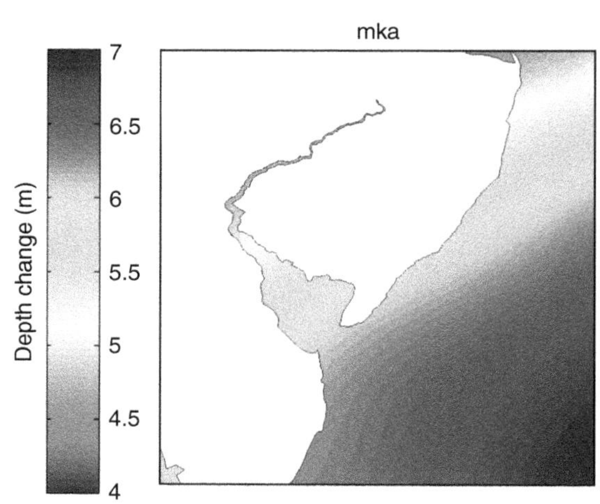

Plate 42 GIA adjustment from 3 ka to present day for the Delaware Bay, USA.

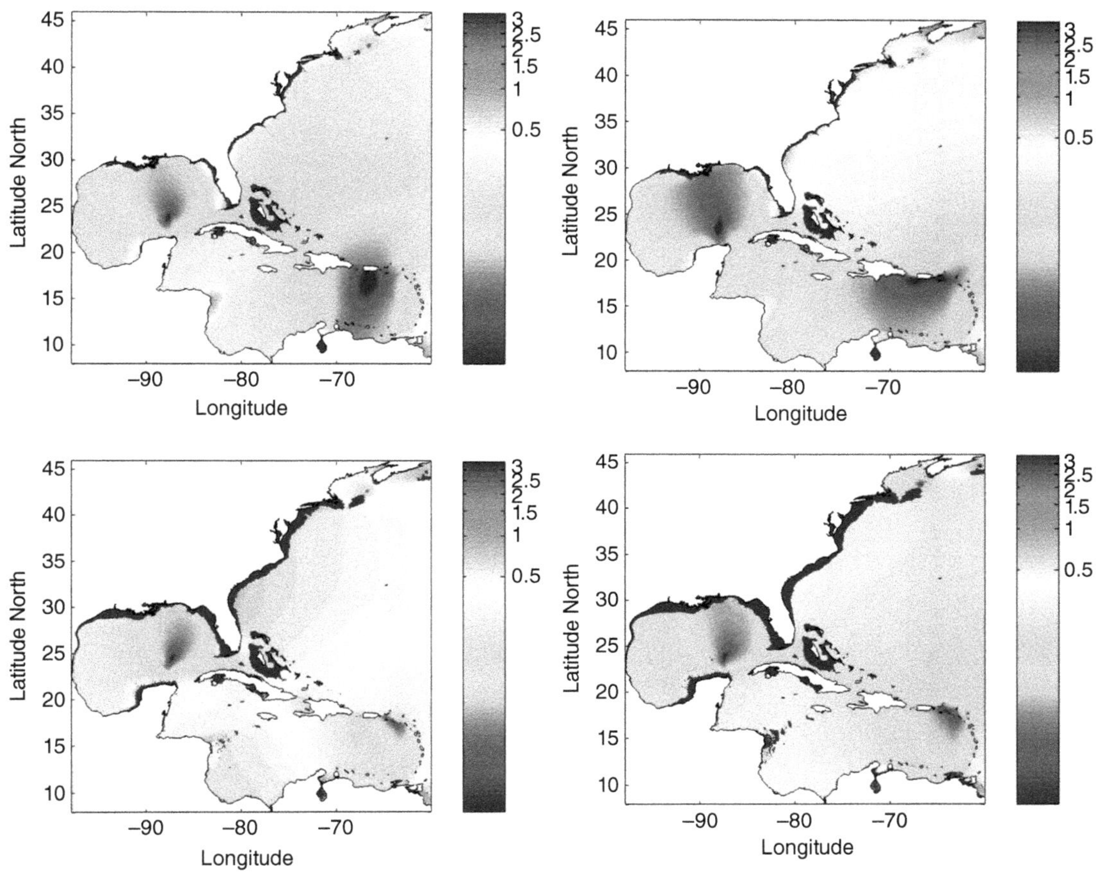

Plate 43 Variation of M_2 tidal amplitude during the early Holocene. Panels (a–d) correspond to 7, 8, 9, and 10 ka, respectively. *Source*: Hill et al. (2011). Reproduced with permission.

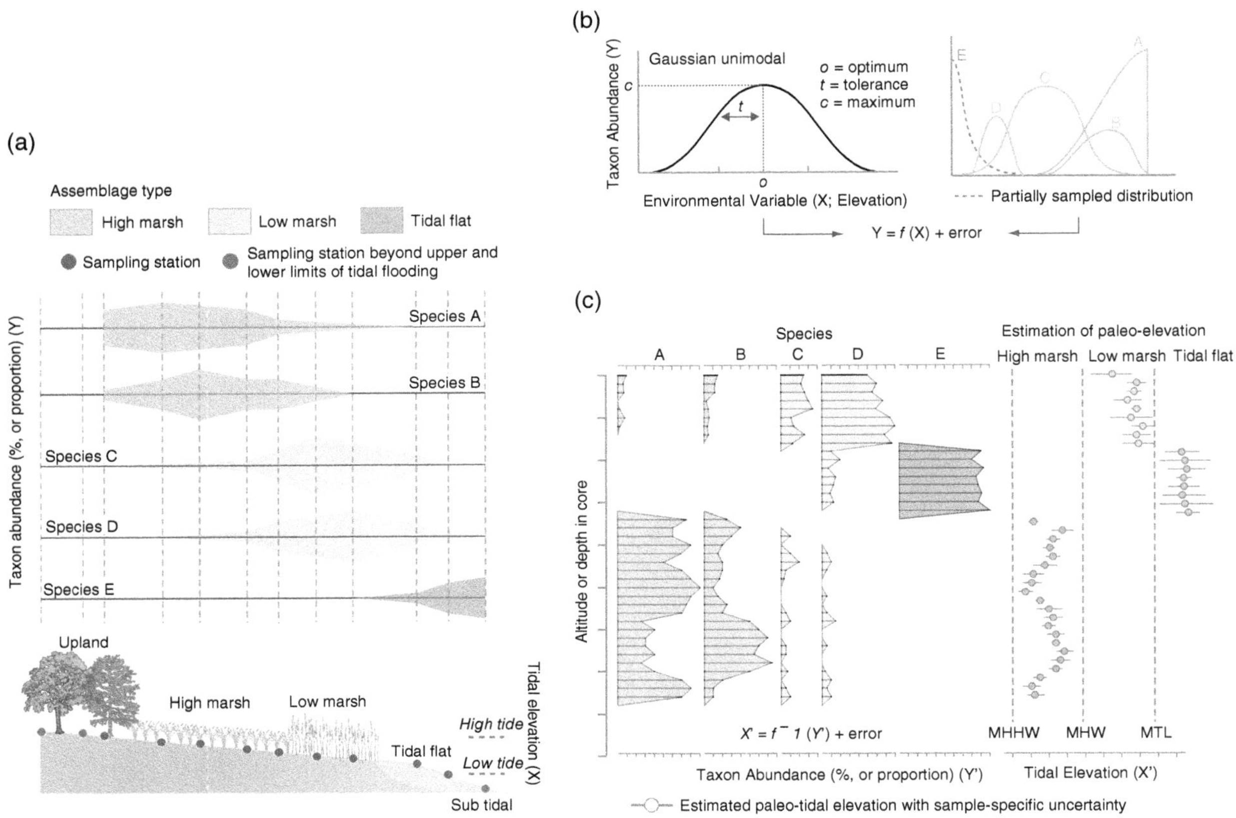

Plate 44 Use of transfer functions to reconstruct relative sea level using intertidal microfossils. (a) A training set comprises paired field observations of species abundance (Y) and tidal elevation (X). Samples are collected across the environmental gradient of interest (here it is tidal elevation in a saltmarsh and tidal flat). Sampling extends into supra- and subtidal settings to ensure that the full environmental gradient is sampled. (b) An appropriate numerical technique quantitatively relates Y and X. This is the transfer function. (c) Application of the transfer function to microfossil assemblages (Y') preserved in buried sediment to estimate the paleotidal elevation where the assemblage formed. A sample-specific uncertainty (1σ) is generated for each fossil sample by bootstrapping, shown by horizontal lines through circles. *Source*: Horton et al., 2013. *Reproduced with permission of Elsevier.* Symbols courtesy of the Integration and Application Network, University of Maryland Center for Environmental Science.

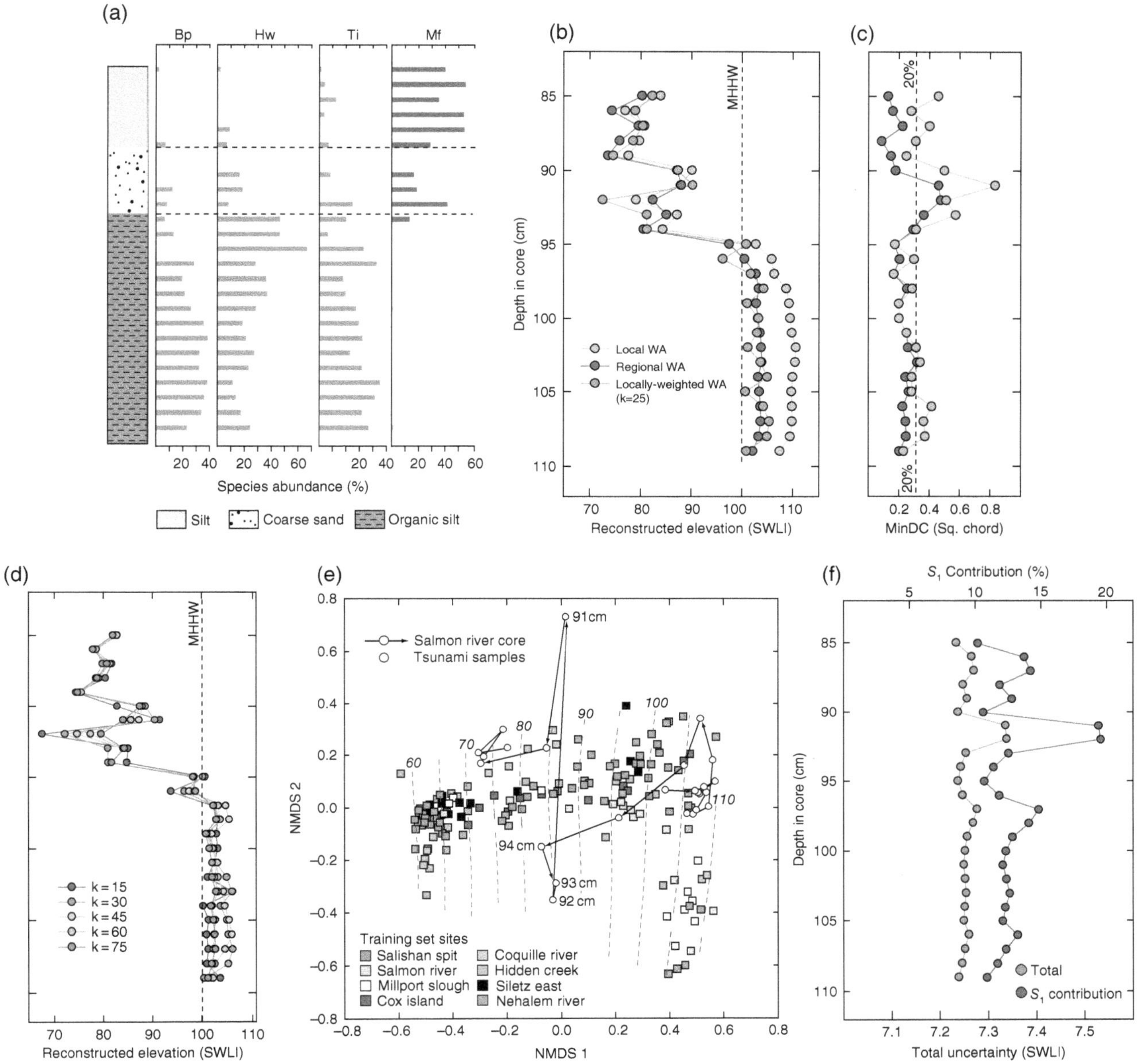

Plate 45 Reconstruction of relative sea-level change caused by coseismic subsidence in Oregon, USA. (a) Lithology of the Salmon River core and assemblages of foraminifera preserved in core samples. Species typically from high-saltmarsh environments are shown in blue and those from low-marsh environments in red. Bp: *Balticammina pseudomacrescens*; Hw: *Haplophragmoides wilberti*; Ti: *Trochammina inflate*; Mf: *Miliammina fusca*. (b) Reconstructed elevation in SWLI units for three transfer functions (WA: weighted averaging, with inverse deshrinking and bootstrap cross-validation) developed using the Oregon training set (152 samples, eight sites) from (Engelhart et al., 2013). The local model only includes samples from the Salmon River site and the regional model includes all samples. (c) Use of analog matching to identify samples lacking a modern analog. Pair-wise comparison of all modern samples identified the 20th percentile threshold (vertical dashed line). For each core sample, the dissimilarity to the closest single modern analog (MinDC) was calculated using the Chord-squared distance metric. Samples exceeding the 20% threshold were assumed to lack a modern analog. (d) Elevation reconstructed using a locally weighted weighted-averaging (LW-WA) transfer function based on increasing values of *k*. (e) Non-metric multidimensional scaling (NMDS) of the modern training set and Salmon River core. Sites with modern samples are represented by colored squares. The samples from the Salmon River core are represented by open circles with a trajectory shown by solid lines with arrows beginning from the bottom of the sequence. Core samples lying outside of the training set cluster lack a modern analog. The four samples from a tsunami-deposited sand layer at 91–94 cm are highlighted. Dashed lines mark the approximate position of standardized water level index (SWLI) elevations. Elevation parallels NMDS axis 1, indicating a strong environmental control on modern species distributions. (f) Decomposition of reconstruction errors for the regional transfer function applied to the Salmon River core.

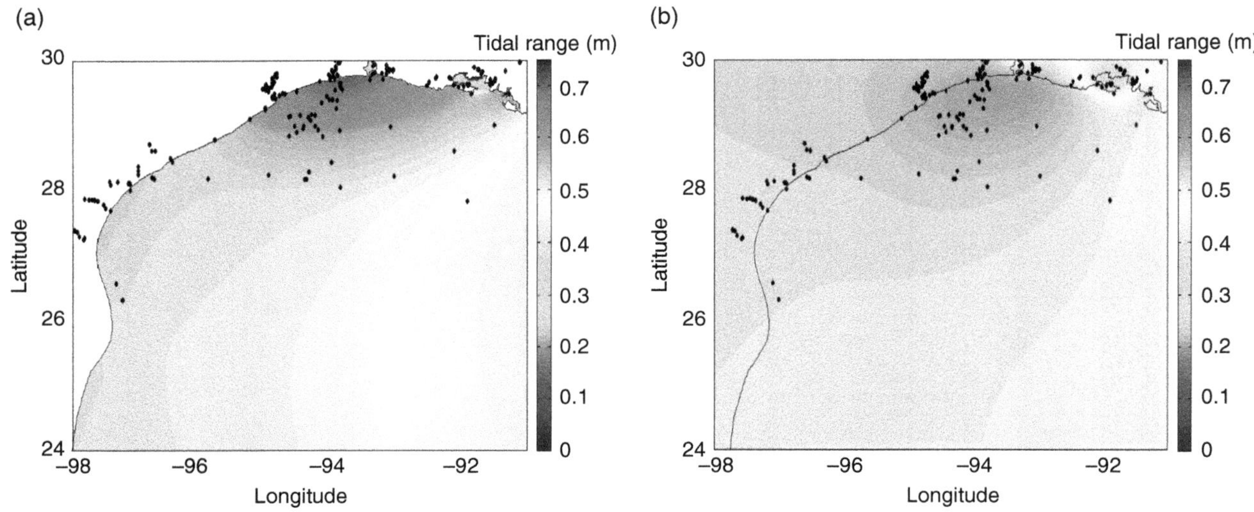

Plate 46 (a) Modeled present-day tidal range (MHHW – MLLW) in the NW Gulf of Mexico. Black dots show the location of SLIPs and limiting data. (b) Tidal range for the entire block with extrapolated values for the onshore portions, enabling the calculation of tidal parameters for sampling sites on land. The offshore part of (b) does not match (a) exactly due to slightly different smoothing parameters. This was necessary because in the data for (a) there were some non-realistic values in regions where nodes are occasionally dry. For those nodes, traditional harmonic analysis is not feasible (see also Chapter 29). To prevent those nodes from influencing the extrapolation, some smoothing was performed on (a) before extrapolation; the panels are therefore not identical for the offshore area.

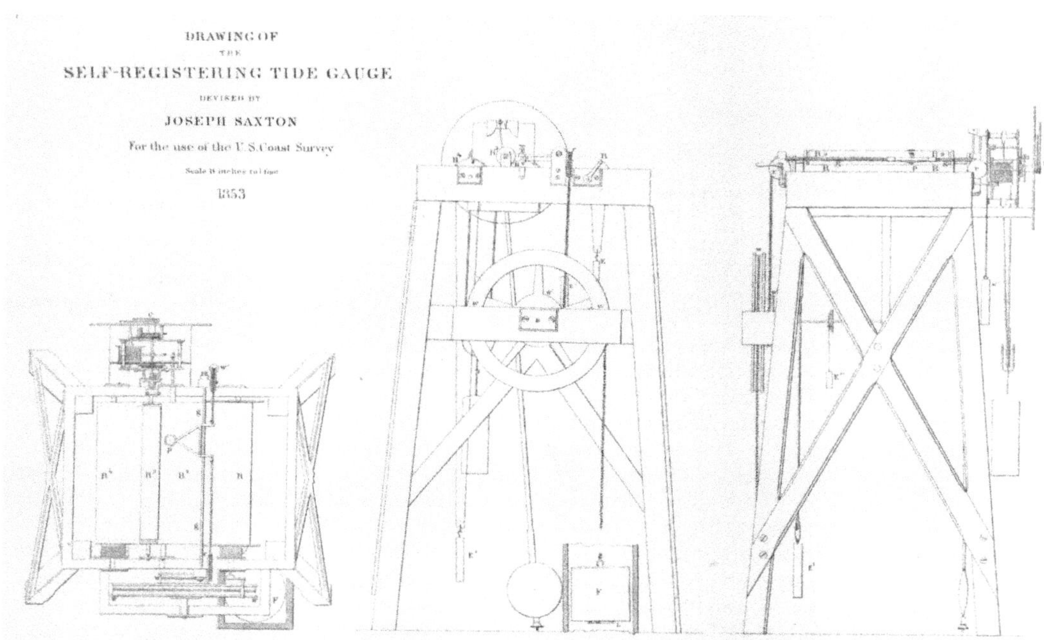

Plate 47 Diagram of the self-registering tide gauge devised by Joseph Saxton for the US Coast Survey in 1851. *Source*: From the National Oceanic and Atmospheric Administration photo library, www.photolib.noaa.gov. The Saxton gauges produced a continuous trace on very long (~20 m) paper scrolls which have been subjected to reanalysis by Talke and Jay (2013).

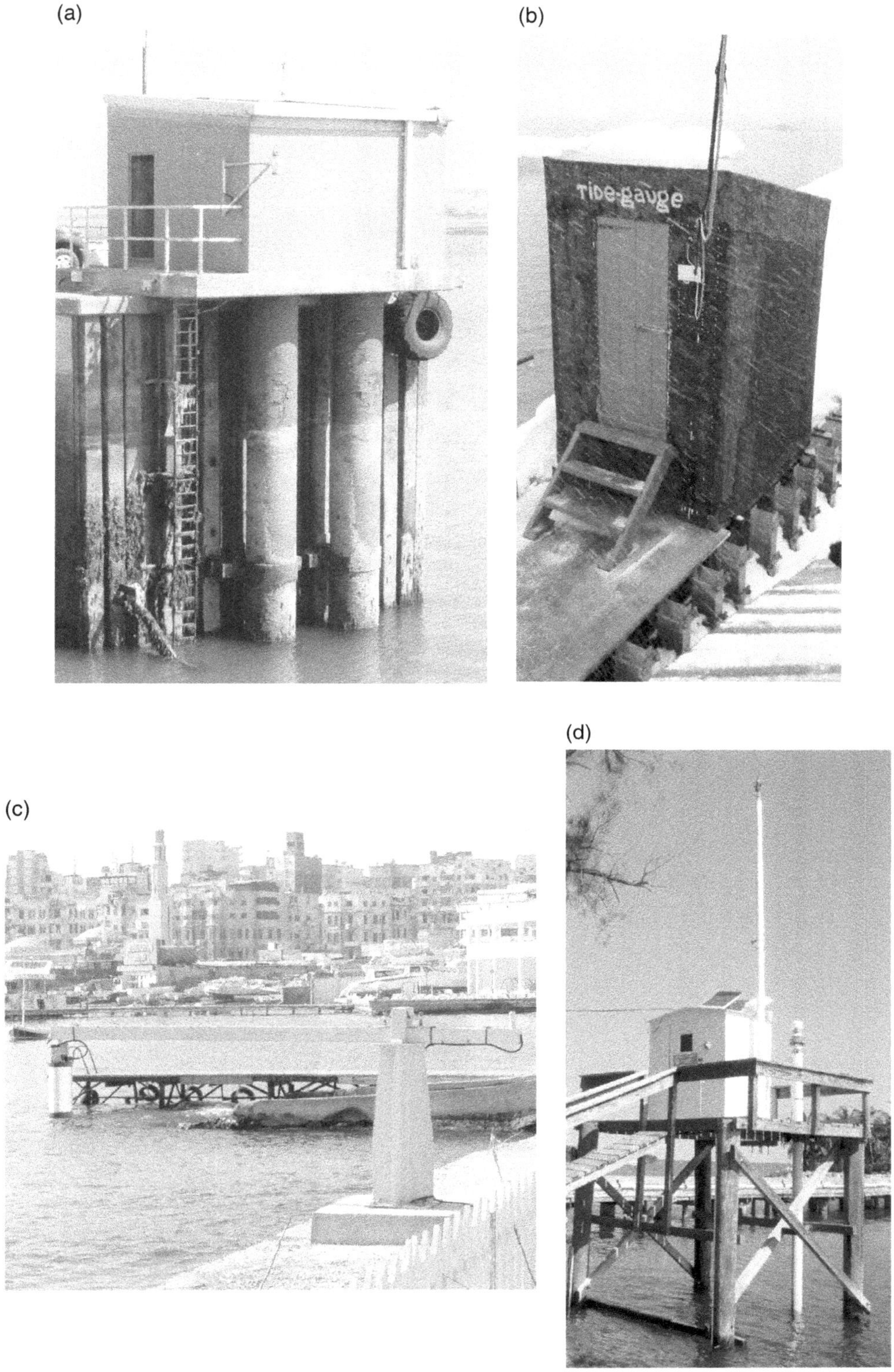

Plate 48 Different types of tide gauge. (a) A classical float and stilling-well gauge at Holyhead, UK. This site has two large wells. (b) The float gauge at Vernadsky, the site of the longest sea-level record in Antarctica. *Source*: Photographs for (a) and (b) by National Oceanography Centre. (c) A radar tide gauge at Alexandria, Egypt. *Source*: Photograph by Thorkild Aarup, Intergovernmental Oceanographic Commission. (d) An acoustic gauge at Vaca Key, Florida. Acoustic gauges now form the majority of the US sea-level network. Photograph also indicates meteorological sensor, solar panels, and communication systems. *Source*: Photograph by NOAA.

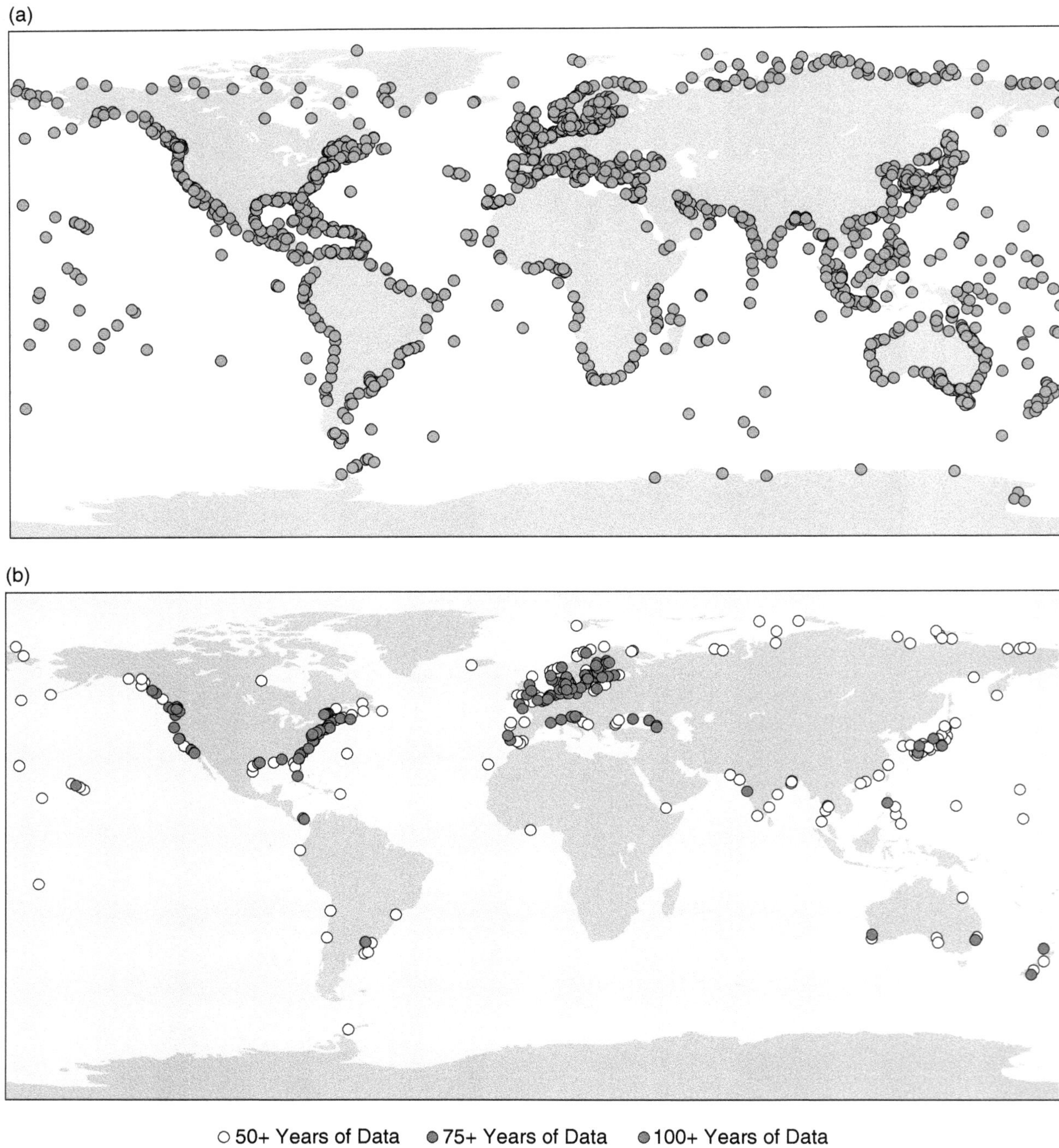

Plate 49 Distributions of (a) all PSMSL stations including both research-quality (RLR) and non-datum controlled stations (known as "metric only" stations) and (b) long, research-quality records in the PSMSL RLR subset. *Source*: From Holgate et al., 2013.

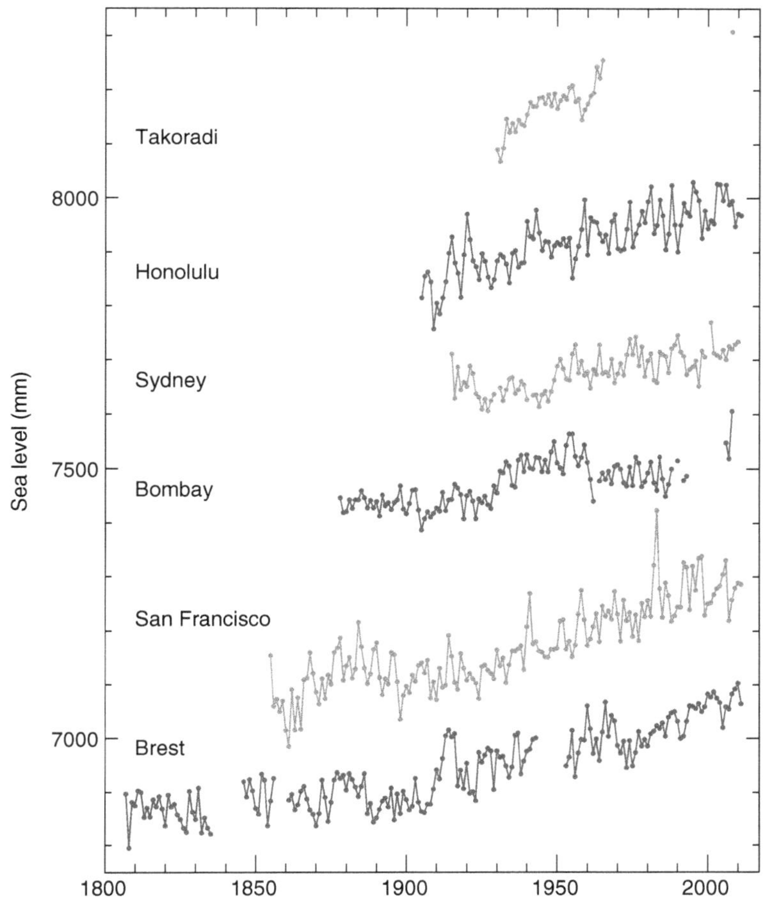

Plate 50 Typical long-term records in the PSMSL dataset from different regions of the world. The Takoradi (Ghana) record, one of the longest in Africa, has recently been restarted after many years. Each record has an arbitrary offset for presentation purposes. *Source*: From Pugh and Woodworth (2014). Reproduced with permission of Cambridge University Press.

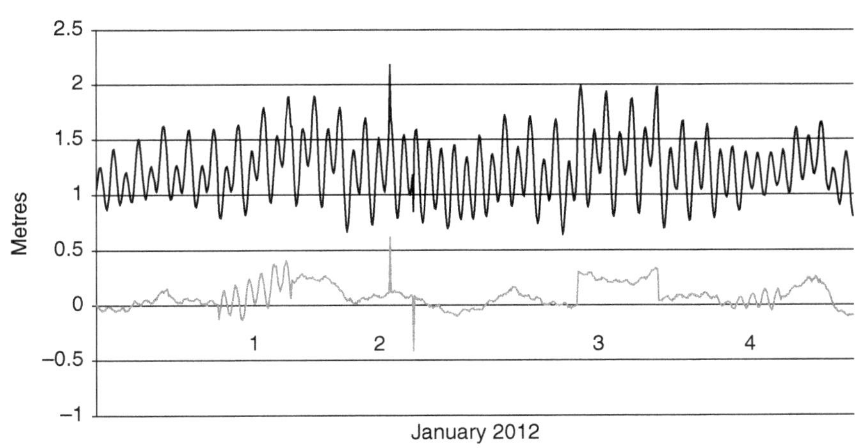

Plate 51 Example of a time series of non-tidal residuals (red) showing characteristic errors due to: (1) a missed 1 hr value; (2) 0.5 m positive and negative spikes; (3) a datum shift of 0.3 m; and (4) a slow reduction in gauge sensitivity due to blockage. The original record, shown by the black curve, is from Spring Bay, Tasmania and includes low-frequency natural variations due to weather effects. The errors are artificially introduced. *Source*: From Pugh and Woodworth (2014). Reproduced with permission of Cambridge University Press.

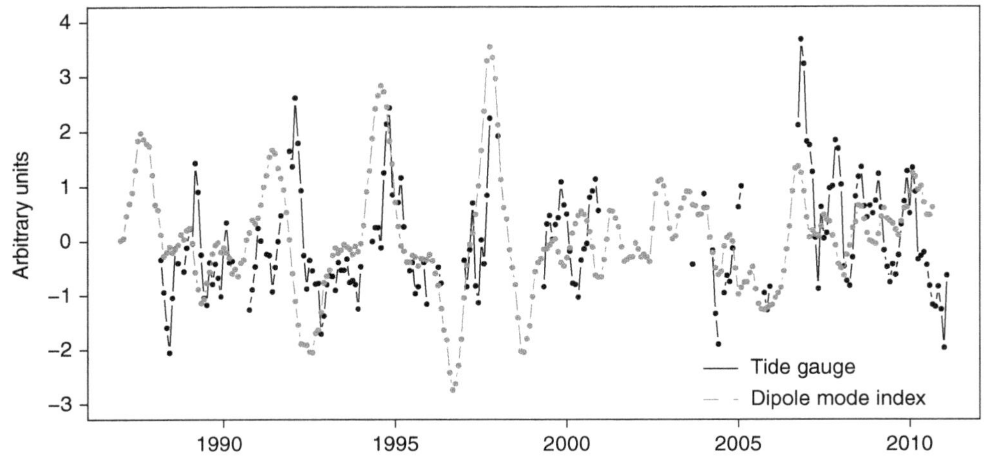

Plate 52 An example of interannual variability in MSL, showing a comparison of monthly MSL since 1988 from Diego Garcia, in the Chagos Archipelago in the western tropical Indian Ocean, to the Indian Ocean Dipole (IOD) index. The IOD is a measure of the oscillation between sea-surface temperatures in the western and eastern Indian Ocean, with positive IOD corresponding to higher temperatures in the west. The sea-level time series has been scaled to have the same standard deviation as the IOD index. *Source*: From Dunne et al. (2012). Reproduced with permission of Elsevier.

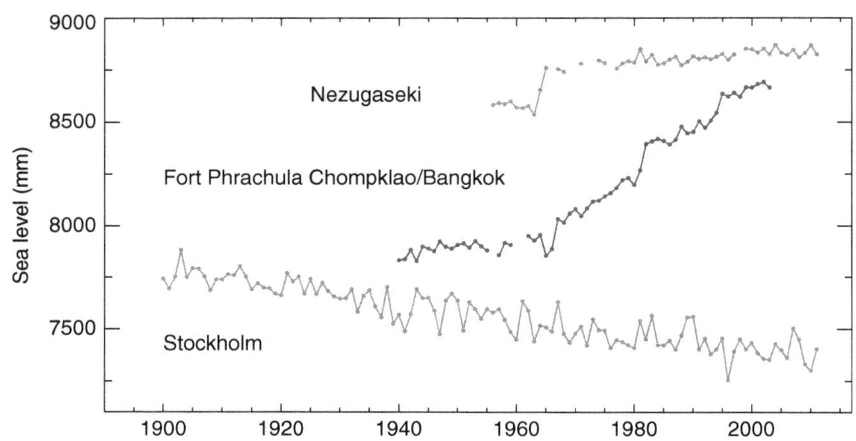

Plate 53 Time series from: Stockholm, Sweden indicating the contribution of GIA to a MSL record; Nezugaseki, Japan indicating an earthquake; and Bangkok, Thailand showing local subsidence under a major Asian city. Data from the PSMSL. *Source*: From Pugh and Woodworth (2014). Reproduced with permission of Cambridge University Press.

Chapter 19
Testate amoebae

DAN J. CHARMAN
Department of Geography, College of Life and Environmental Sciences, University of Exeter, Exeter, UK

19.1 INTRODUCTION

Testate amoebae are small (typically 20–200 μm), single-celled eukaryotic organisms (protists) that form a shell (the test) around the cell. They have also been referred to as rhizopods, arcellaceans, and thecamoebians in various parts of the literature at different times. The term "rhizopods" was largely used by terrestrial biologists and paleoecologists in the 1950s to 1990s, "arcellaceans" is predominantly used in parts of the paleolimnological literature, and "thecamoebians" has mostly been used within the pre-Quaternary, paleolimnological and paleo-sea level community. The term "testate amoebae" is more widely used than any of these other terms in contemporary biological, ecological, and paleoecological literature, and I will therefore use this term throughout this chapter.

Testate amoebae are found throughout a wide range of soils and sediments, especially lake sediments, wetland soils, and peat, but they also occur in much drier soils and even in temporary microhabitats such as moss patches on buildings and trees. They occur in every continent and most terrestrial habitats from warm tropical regions to the ice-free areas of the Arctic and Antarctic (Smith, 1992). They also occur in coastal areas, especially saltmarshes, but also in less obvious locations such as on mudflats (First and Hollibaugh, 2008) and within the interstitial spaces of beach sands (e.g., Golemansky and Todorov, 2004). In other words, there are very few places in soils and sediments where testate amoebae cannot be found. Consequently, they have been applied to a range of ecological and paleoenvironmental problems ranging from the now widely used reconstruction of peatland water tables to more unusual circumstances such as forensic science (Mitchell et al., 2008). Although the occurrence of testate amoebae in coastal sediments has been noted regularly for over 30 years (Medioli et al., 1990), their application in sea-level research has been relatively slow to emerge and is still very much under development.

The first part of this chapter briefly summarizes the history and development of testate amoebae research in sea-level studies. This is followed by a review of the most commonly used methodological approaches to field and laboratory analyses. The final two sections deal with the relationship between sea level and testate amoebae populations and the application of testate amoebae analysis to sea-level reconstruction. The chapter concludes by summarizing the current state-of-the-art and identifying key future areas to develop the methodology and application of the technique.

19.1.1 Development of testate amoebae and sea-level studies

The vast majority of reports of testate amoebae in coastal sediments and modern samples are from studies that were primarily aimed at foraminifera analyses. There are numerous publications that mention finds of testate amoebae in surface sediments from saltmarshes and estuaries during the course of foraminifera analyses. Scott et al. (1977) is one of the earliest of these, and other work by the Dalhousie group in the 1980s showed that testate amoebae are widespread geographically and occur consistently in the highest parts of saltmarshes (see Medioli and Scott, 1983, 1988; Medioli et al., 1990, for reviews of this earlier work). In fact, in almost every marsh surface study that included samples from the uppermost intertidal area, testate amoebae have been recovered. There is not enough space here to mention all of these studies individually; it is however clear that testate amoebae are ubiquitous, including for example tropical regions (Riveiros et al., 2007) and coastal mangroves (Duleba and Debenay, 2003; Barbosa et al., 2005), temperate Southern Hemisphere marshes (Hayward

Handbook of Sea-Level Research, First Edition. Edited by Ian Shennan, Antony J. Long, and Benjamin P. Horton.

et al., 1996; Scott et al., 1990), and the Pacific Rim (Scott et al., 1995).

Because these studies were typically aimed at foraminifera analyses, they focused on the typical size fraction used to separate foraminifera from sediments (often 63–500 μm, but occasionally as low as 37 μm) and used wet-splitting and examination under relatively low magnification for counting. Furthermore, a relatively conservative approach to taxonomy was usually adopted (see the following section). Both of these approaches meant that, while testate amoebae were usually recovered in surface studies, they rarely demonstrated a zonation within the part of the marsh where they were found and were mostly used simply as general indicators of the uppermost marsh zones and relatively freshwater conditions. An alternative approach, based on similar methods to those used in terrestrial peat deposits (Warner, 1988; Hendon and Charman, 1997), involved examining a smaller size fraction (15–300 μm) and counting samples on microscope slides under higher-power magnification; greater diversity and a distinct vertical zonation of testate amoebae assemblages was demonstrated (Charman et al., 1998). Subsequent studies have demonstrated the prevalence of these patterns on saltmarsh surfaces in the United Kingdom and eastern North America (Gehrels et al., 2001, 2006; Charman et al., 2002, 2010; Fig. 19.1), but work using this approach remains relatively limited at present.

While there have been numerous studies reporting on assemblages from surface samples, there are far fewer records of testate amoebae in fossil sediments (see the following section). One aspect of past sea-level reconstruction where large size fraction testate amoebae have been used to good effect is in detecting and quantifying seismic uplift and subsidence related to historical earthquake activity in regions including Alaska and Chile (Shennan et al., 1999; Reinhardt et al., 2010). They have also been recovered from older sediments associated with seismic activity in the Mediterranean (Nixon et al., 2009). Small size fraction testate amoebae have also been recovered from a range of sediments in the UK and eastern North America (Charman et al., 2010), but not without difficulty (see the following section; Roe et al., 2002). Finally, there are a number of studies in which testate amoebae in lake sediments have been used to determine freshwater conditions following sea-level fall in isolation basins (Lloyd, 2000) and sea-level fluctuations in lake sediments in deltaic settings (Farooqui and Naidu, 2010). Indirect evidence of sea-level rise has also been detected using testate amoebae in cave sediments (van Hengstum et al., 2009). The following sections concentrate on the use of testate amoebae in saltmarsh sediments, as that is where the main developments in techniques have been recently. However, many of the principles of these approaches could be adapted for use in the other sediments and settings mentioned above.

19.2 METHODOLOGY

19.2.1 Background

From the brief overview of the development of testate amoebae studies related to sea-level reconstruction, it is clear that there have essentially been two approaches to methods for analysis. The first is based on methods used for foraminifera and the second is based on methods designed specifically to detect the full size range of testate amoebae. Both approaches have advantages and disadvantages and I will refer to each of them here but, because the former approach is essentially the same as foraminifera analysis, I will not discuss the details of preparation procedures (but see Chapter 13). There are four main aspects to testate amoebae methodology: (1) field sampling; (2) subsampling and laboratory preparation; (3) counting and identification; and (4) data analysis. This section focuses primarily on parts (2) and (3) of this process; parts (1) and (4) are covered in Section 19.3 and Chapter 31, respectively.

19.2.2 Field sampling

In terms of field sampling, because testate amoebae are restricted to the uppermost parts of saltmarshes, it is of course necessary to sample this part of the marsh whether the intention is to use surface samples to establish indicative meaning of assemblages, or to use core samples to reconstruct past changes. Furthermore, the vertical zonation of testate amoebae is usually quite tightly constrained with a strong but short gradient. This means that samples for surface sediments should be closely spaced with small vertical distances (5 cm or less) between them. Surface samples are usually from the top 1 cm of sediment as there is no evidence to suggest that testate amoebae are infaunal to any great extent.

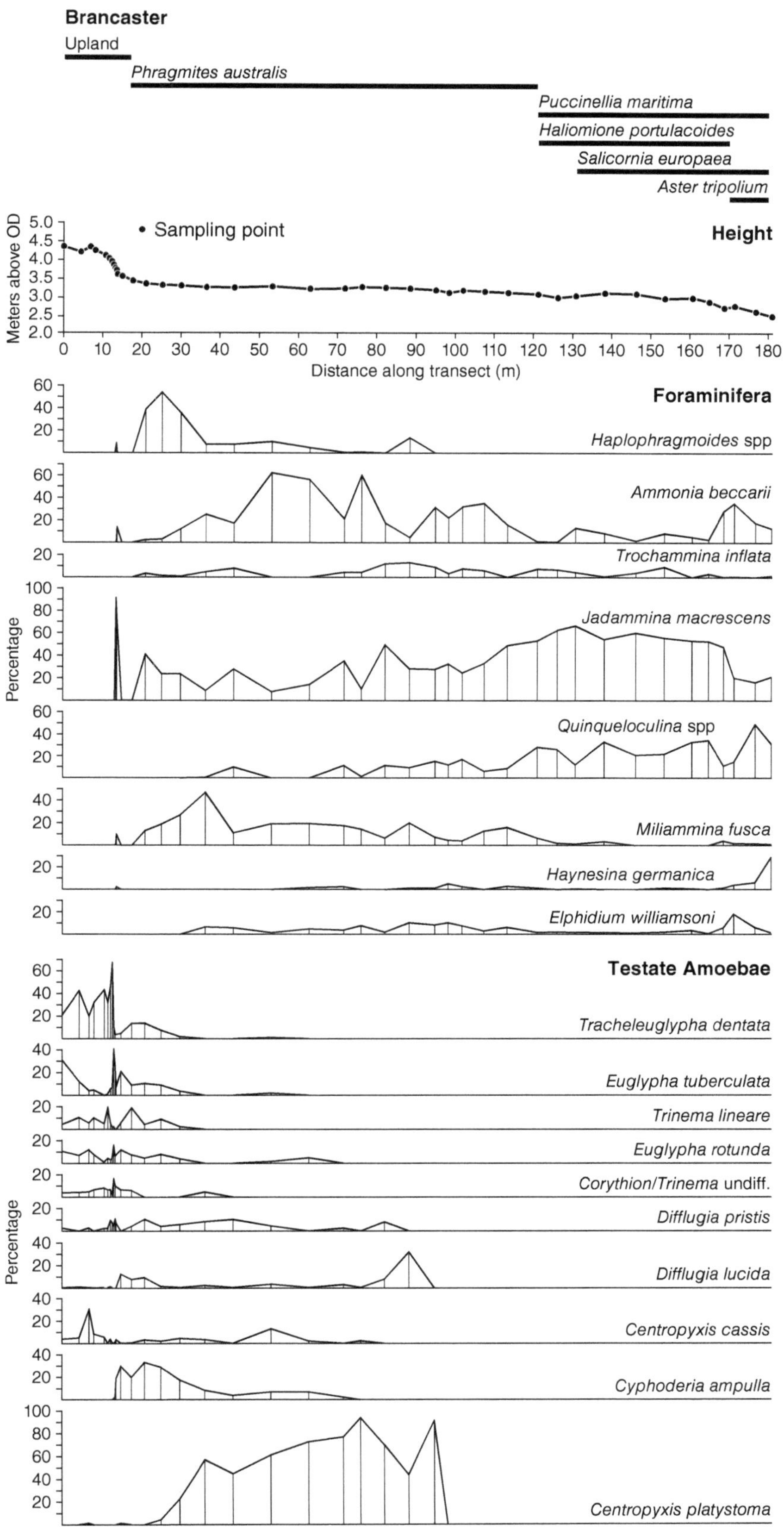

Fig. 19.1. Testate amoebae and foraminifera assemblages for a transect across a saltmarsh at Brancaster, Norfolk, England. *Source*: Gehrels et al., 2002. Reproduced with permission of John Wiley & Sons., showing the overlap in occurrence between these two groups of proxy sea-level indicators and the zonation in relation to elevation.

19.2.3 Sample preparation

There are two main challenges in extracting and counting testate amoebae in saltmarsh sediments. First, the concentration of testate amoebae in the sediment may be very low. Saline conditions appear to be sub-optimal for testate amoebae and, although tolerance varies between taxa (hence the zonation in relation to elevation), in most cases testate concentrations decrease rapidly from highest astronomical tide (HAT), mostly disappear below mean high water springs (MHWS), and are completely absent below mean high water neaps (MHWN). Second, the size range and particle composition of a large proportion of the sediment is very similar to that of the testate amoebae. Testate amoebae construct their tests from a wide range of materials including organic secretion, silica plates, and both organic and minerogenic particles from the sediment in which they live. Hence, a mixture of fine organic and silt particles such as that found in saltmarsh sediment is probably the most difficult matrix from which to isolate testate amoebae. In fully organic sediments such as terrestrial peats, a large proportion of the sediment matrix can be removed by sieving out the more fibrous and larger plant fragments. In saltmarsh sediment, sieving removes rather less material so that the tests are relatively thinly distributed in the sediment residue remaining on the slide. Both low concentration and retention of large amounts of sediment on the slide can make for slow and difficult counting. Several different approaches have been taken for sample preparation to mitigate these problems, and techniques are still very much under development. Here I will summarize the two main approaches and provide a recommended protocol for sample preparation for dedicated testate amoebae analysis.

19.2.3.1 Combined testate amoebae and foraminifera analysis

The advantage of preparing samples in the same way for foraminifera and testate amoebae is that it is quicker both in preparation and counting time (see also Chapter 13). Time savings are made because only one preparation is made and counting is carried out for both groups of organisms simultaneously. The main disadvantage is that it will only account for a small proportion of the total testate amoebae community because the majority of the testate amoebae population is smaller than the standard 63 µm mesh size used for foraminifera (although a number of more recent studies now use 45 µm or even 37 µm sieves, which helps reduce this problem). Charman *et al.* (1998) found that numbers of testate amoebae in the 15–63 µm fraction were sometimes more than two orders of magnitude higher than the >63 µm fraction and numbers of taxa were much higher in the smaller size fraction. Typically there were 1–2 taxa present in the large size fraction, but up to 17 taxa in the small size fraction. This could be improved by using a smaller mesh size in the foram sample preparation.

Practices have varied a little in foraminifera preparations but many foraminifera studies have used 45 µm (e.g., Nixon et al., 2009). The smallest sieve size used is 37 µm (Roe and Patterson, 2006), although rather few of the samples in this study were from saline conditions. No other direct comparisons of concentration of tests have been made in any of these studies, but Roe and Patterson (2006) found 11 additional taxa in the 15–37 µm size fraction when compared with the >37 µm fraction. While smaller sieve sizes helps retention of a larger number of tests, there is a practical limit to the size of particles that can be easily observed under low-power microscopy; with more small particles, the efficiency of counting of the larger foraminifera is reduced. Furthermore, because many testate amoebae taxa have sizes in the range 20–60 µm, this approach is unlikely to retain the full testate amoebae assemblage.

If the aim is simply to identify relatively freshwater conditions, then counting the large size fraction of testate amoebae is adequate in most circumstances. However, if the intention is to develop more precise estimates of past sea level using assemblage zonations or transfer functions, then full counts of the complete assemblage will be needed. Using smaller sieve sizes and high-power microscopy is the only way to achieve this.

19.2.3.2 Preparation of testate amoebae samples

The main challenge for sample preparation of testate amoebae samples is to concentrate the tests as much as possible by disaggregating sediment particles and tests and removing as much of the sediment as possible. This is achieved by strong chemical treatments in pollen and diatom analysis, but this is not appropriate for testate amoebae as they are much less resistant to chemical treatment than pollen or diatoms (Hendon and

Charman, 1997). Consequently, standard preparation procedures for terrestrial peatland samples generally recommend gentle disaggregation in water, followed by sieving to isolate the size range of sediment that contains the testate amoebae (Tolonen, 1986; Warner, 1988; Charman et al., 2000). Although earlier work on terrestrial peats used quite large size fractions, it has now become standard to analyze the 15–300 μm fraction (Charman et al., 2000); this procedure was also applied in the first attempts at saltmarsh testate amoebae analysis (Charman et al., 1998, 2002).

Given the difficulty of sufficiently concentrating tests for counting in some samples, mild chemical treatment has been tested as part of the preparation process. Hendon and Charman (1997) found that a range of chemical treatments routinely used in pollen analysis destroyed or damaged many of the tests in terrestrial peat samples. However, experiments on saltmarsh samples with a weaker alkali (5% KOH compared to 10%) for a short period of time appear to improve the preparation procedure by helping disperse and disaggregate sediment particles (Charman et al., 2010). This means more of the <15 μm sediment particles pass through the smaller sieve, so there is less of this material on the slide. In disaggregating larger clumps of sediment more effectively, it may also "release" more of the testate amoebae from the sediment, increasing the concentration of tests on the slide. Charman et al. (2010) found that including the alkali treatment produced higher concentrations, allowing increased counts with generally less analysis time, with no significant change in taxon assemblages recorded. New work suggests a similar result for samples from Norwegian saltmarshes (Barnett et al., 2013). One recent study on lacustrine sediments has even used a 10% KOH treatment and recovered a significant number and diversity of tests in the 10–150 μm size fraction (Farooqui and Naidu, 2010), although this is not recommended as a routine treatment.

Based on the studies carried out to date the following protocol is recommended for most saltmarsh sediments, but it could also be used for lacustrine or other fine-grained sediments where test concentrations are low and sediment clumping is a problem. The main difference between this and standard freshwater preparations (Tolonen, 1986; Charman et al., 2000) is the addition of the weak alkali step. Because this has still not been extensively tested in a range of sediments it is recommended that, before using it for a full set of samples, some tests are performed on its effect by comparing preparations with and without KOH.

19.2.3.3 A protocol for sample preparation

All equipment should be thoroughly clean and the treatment should be carried out in fume hood; normal health and safety rules should be adhered to.

(1) Weigh or measure volumetrically by displacement, samples of 0.5 g (or 0.5 mL) wet sediment. Sample sizes are not critical and anything in the range 0.2–1.0 g (0.2–1.0 mL) is usually suitable.
(2) Add to 100 mL distilled water. For surface samples where separate counts of living and dead cells are required, add a few drops of Rose Bengal stain and leave to soak for 24 hours.
(3) Add one tablet of *Lycopodium clavatum* L. as an exotic marker (Stockmarr, 1971; currently available from Lund University, http://www.geol.lu.se/kvg/eng/lycopodium.pdf). Other types and methods of exotic spike can be used instead. This step is essential to calculate test concentrations. It can be omitted if only percent data are to be derived, but this is not recommended as concentrations are a useful additional indicator of the suitability of the (paleo)habitat for testate amoebae.
(4) Heat at 80–100 °C for 5–10 min, stirring gently occasionally to aid disaggregation, and then leave to soak overnight or for approximately 12 hours. Heating promotes disaggregation, but also kills any mobile micro-organisms that can be a problem in viewing slides.
(5) Sieve samples through 15–300 μm sieves (see note at end of list) and retain the 15–300 μm size fraction.
(6) Add 25 mL distilled water to the sample and then add 4 mL solution of 5% KOH. Warm at approximately 80 °C for 1–2 min and sieve again at 15 μm to remove more fine detritus.
(7) Store samples in small vials with distilled water in cold conditions (4 °C) until ready for counting. Samples should be analyzed as quickly as possible; although they may often last for months or even several years if stored well, they can deteriorate more quickly.
(8) For counting, samples can be mounted in glycerol or water on a microscope slide with a 22 × 40 mm coverslip sealed with nail varnish.

The best optical quality is normally in water, but this often dries out quite quickly.

Note: Regarding step (5) above, there may be some advantage to undertaking the KOH treatment before any sieving has been done (Barnett et al., 2013). This may aid disaggregation of any sediment clumps larger than 300 μm before they are lost in the first sieving step. This is recommended if clumping of sediment appears to be a particular problem. The upper sieve size is to allow the largest tests to pass through the sieve; 300 μm is conservative for many samples. Smaller upper sieve sizes could be used and most of the data that we have from saltmarsh samples suggests that a smaller sieve size would retain the same number of tests while filtering out a larger proportion of sediment. In most samples a sieve size of 200–250 μm could be used without loss of tests; 212 μm has been used in a recent study (Barnett et al., 2013). If a smaller sieve size is used however, it is recommended that tests with a larger sieve size are carried out to check whether larger specimens are present.

19.2.4 Counting and data analysis

Most testate amoebae analysts have typically aimed to count at least 150 tests per sample (Charman et al., 2000), as this is thought to capture the majority of taxa present even in more diverse assemblages (Warner, 1988; Woodland et al., 1998). However, it is often not practical to count 150 tests in saltmarsh samples because of the low concentration of tests and the large amount of sediment on the slides. In many cases, counts of 50–100 have been used to characterize assemblage zonation and develop transfer functions (Gehrels et al., 2001, 2006; Charman et al., 2002; Ooms et al., 2011). This is not ideal but any microfossil analysis is a tradeoff between count levels and numbers of analyses given time constraints of analytical time. A general recommendation would be to aim for counts of >100 and not to use counts of <50 for any statistics. Counts from foram-based preparations are much more variable because any total count normally includes forams as well as testate amoebae. In deriving transfer functions, percentages of total testate amoebae counts are normally used. The count total should exclude any non-testate amoebae microfossils, even if they are potentially useful sea-level indicators.

It is important to be able to calculate test concentrations because these can be a useful indicator of the potential of the sample to yield useful data. Furthermore, because there are still only a small number of datasets available, it would be useful to see how concentrations vary between different settings. Because counting can take a long time for some samples, an estimate of test concentration during counting can indicate whether this is because of an inherently low concentration of tests or whether it is because of a large amount of sediment particles on the slide. If concentration is very low, it may not be worth attempting to continue the count unless the data from those samples are especially vital to interpretation. The calculation is a simple one and similar to that for any samples where exotic spikes are added:

$$\text{Test concentration} = \frac{T_c \times S_t}{S_c \times M} \left(g^{-1}\right)$$

where T_c is the number of tests counted, S_t is the number of spores (or other exotic marker) added to the sample, S_c is the number of spores counted, and M is the mass of the sample. If volumetric samples are used instead of samples by mass, M is replaced by V (volume in mL) and concentrations are expressed as tests mL^{-1}.

19.2.5 Taxonomy and identification

There is no standard taxonomic approach to the identification of testate amoebae that is accepted and used by all analysts. The basic problem is that the biological species concept does not apply to testate amoebae, which reproduce almost entirely by cell division. The details of this issue are discussed further by Medioli and Scott (1983), but the central problem is that all taxon definitions are based on morphology, which may or may not reflect real species differences. Furthermore, many taxa show a significant degree of morphological plasticity, so that differentiation of some taxa can be difficult. DNA work is gradually clarifying genetic relationships among some taxa and showing that morphology is not always a good guide to genetic difference (e.g., Gomaa et al., 2012), but for practical purposes paleoenvironmental studies will be limited to morphological criteria for some time. The most important qualities of a taxonomy for paleoenvironmental analysis are: (1) that the identifications are repeatable by different observers, that is, the criteria are clearly and preferably quantitatively defined; and (2) that the taxa used have sufficiently different ecological niches to be of use as paleoenvironmental indicators. Because there are two different traditions in the analysis of testate amoebae in paleoecology, two main approaches to identification have arisen and both are in use.

Samples that are prepared using sieving and wet splitting approaches used for foraminifera are normally wet-counted under a low-power dissecting microscope in larger size fractions (mostly >63 µm, see above). Medioli and Scott (1983) established a taxonomy for samples prepared this way and many studies have since followed this route; this method is recommended for combined foraminifera and testate amoebae analysis of large size fractions, where it is possible to see these larger tests in three dimensions. The system essentially uses a conservative species nomenclature but allows the designation of informal infraspecific strains; these are regarded as being from the same species but show sufficient and consistent morphological variation that makes them useful as environmental indicators (Kumar and Dalby, 1998).

Samples prepared for analysis of the 15–300 µm size fraction under high-power microscopy are normally identified using the method of Charman et al. (2000) as a basis, but also referring to the wider taxonomic literature to build up a taxonomy incrementally for use on saltmarsh samples (e.g., see descriptions and illustrations in Charman et al., 1998, 2002; Gehrels et al., 2006). Because the use of testate amoebae on saltmarsh sediments is still developing and there are still relatively few studies that have systematically analyzed either surface or fossil samples, it is recommended that studies include photomicrographs for reference. This is especially important where identifications are uncertain or specimens that may or may not be testate amoebae have been encountered. There are several unknown microfossil "types" that have been repeatedly observed in saltmarsh samples, and these may be useful additional paleoenvironmental indicators even if they prove not to be testate amoebae (e.g., see fig. 9 of Charman et al., 2002). Some of the more commonly encountered taxa are shown in Figure 19.2.

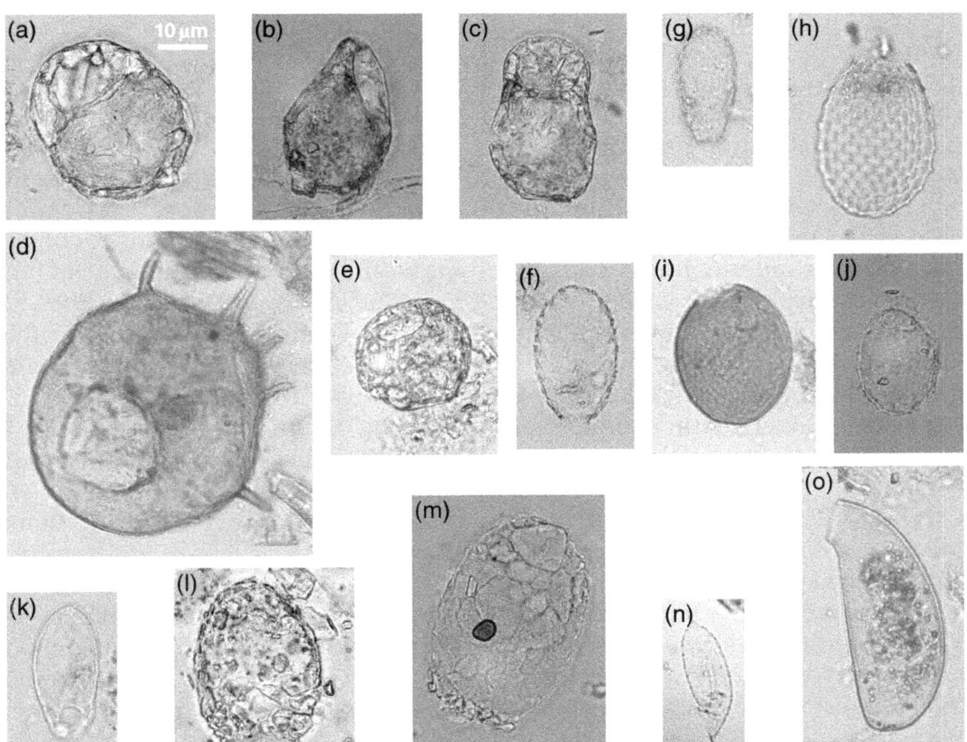

Fig. 19.2. Testate amoebae taxa commonly encountered in saltmarsh samples. All specimens are shown at the same relative size according to the scale bar on (a), unless otherwise specified. (a) *Centropyxis cassis* type front view; (b) *Centropyxis cassis* type side view; (c) *Centropyxis platystoma* type; (d) *Centropyxis aculeatea* type; (e) *Cyclopyxis arcelloides* type; (f) *Tracheleuglypha dentata;* (g) *Euglypha rotunda* type; (h) *Euglypha tuberculata* type, (i) *Assulina muscorum;* (j) *Corythion-Trinema* type; (k) *Trinema lineare;* (l) *Difflugia pristis* type; (m) *D. lucida* type; (n) *Campascus* type; and (o) *Cyphoderia ampulla.*

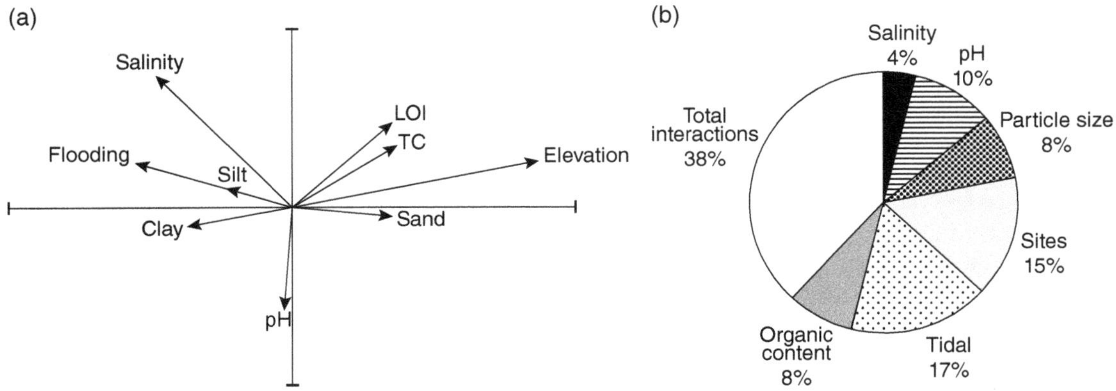

Fig. 19.3. (a) Results from constrained ordination analysis showing the strength and direction of relationships between testate amoebae assemblages and a range of environmental variables from 52 samples from three saltmarshes in the UK *Source*: Charman et al., 2002. Reproduced with permission of John Wiley & Sons. The alignment of the arrows indicates the relationship with the two axes, and its length represents the strength of this relationship. (b) Pie chart expressing the variation of the testate amoebae assemblage explained by components representing the unique contributions of the main variables and the interactions between variables. The total species variance explained by all environmental variables for this dataset is 49%. The tidal factors (flooding and elevation) and organic content (%LOI and total carbon) were grouped for this analysis. "Sites" expresses the variation due to differences between different saltmarsh locations. See Charman et al. (2002) for details and further analysis.

19.3 MODERN DISTRIBUTIONS AND ECOLOGY

The use of biological proxies (testate amoebae, diatoms, foraminifera) for sea-level reconstruction assumes that the main environmental control on species distribution is elevation in relation to mean sea level. Elevation is a convenient proxy for the actual controls on species distribution, partly because it is the target variable for reconstruction but also because repeatable and accurate measurements can be made very easily in the field. Elevation determines the frequency and period of tidal inundation, which in turn determines the degree of waterlogging and salinity conditions in the surface sediment. Other parameters such as vegetation, organic matter content, and particle size may also influence the microbial communities living in the sediment, and these often also correlate with elevation. In practice then, it is difficult to be sure exactly what the testate amoebae are responding to. One approach is to ignore this as a problem; since we wish to reconstruct elevation, we can use that relationship to construct a transfer function for paleo-elevation directly. However, it is important to establish what the underlying relationships with the proximal variables are because an assumption of applying any reconstruction is that the relationships with other variable have not significantly influenced assemblage composition in the past.

The only extensive survey of saltmarsh testate amoebae and multiple environmental variables in the intertidal zone shows that the strongest relationships are between testate amoebae and elevation, with elevation being strongly autocorrelated with flooding duration and, to a lesser extent, salinity (Charman *et al.*, 2002; Fig. 19.3a). Other variables have significant but secondary relationships with testate amoebae assemblages, including pH, organic matter content, and particle size. Partialling out the influence of environmental variables on assemblage variance (Borcard et al., 1992) confirms this generally subsidiary influence of these other variables, but shows that they can still have significant influence on taxon composition (Fig. 19.3b). These variables also tend to show more differences in their relationship with testate amoebae assemblages between sites than flooding duration and salinity. Although there is a general decrease in particle size and organic matter content lower in the tidal gradient, there is therefore some variation in this pattern between sites (Charman et al., 2002). There are also some site-specific factors that were not accounted for by the measured variables, suggesting there are other influences on taxon composition unrelated to elevation (Fig. 19.3b). In general, it is possible to explain more of the variation in taxon assemblages within an individual site (up to 79% in this case).

More studies of this kind are required to test whether these are universal trends and relationships, but some data do at least confirm the strong relationship with elevation (Charman et al., 1998; Gehrels et al., 2006; Barnett et al., 2013) and these can be used to summarize the relationship between elevation and testate amoebae taxa (Fig. 19.4). From the existing data, it is clear that there are very similar patterns in testate amoebae zonation on different saltmarshes. There are relatively few taxa that are tolerant of the marsh areas that are inundated every high tide (i.e., below MHWN), and the most diverse faunas occur around HAT and MHWS (Fig. 19.4). The zonation extends into the supratidal zone (i.e., above HAT), suggesting that there is perhaps an influence of occasional storm events that bring flooding with seawater to areas outside of the normal tidal range.

There are several key taxa that appear to be relatively common and abundant in saltmarshes and these taxa have the strongest potential for sea-level reconstruction. One of the most abundant groups in all saltmarsh studies is the *Centropyxis* and *Cyclopyxis* taxa. *Cyclopyxis arcelloides* type is common in supratidal samples and *Centropyxis cassis* type is probably one of the most abundant taxa at a range of elevations, but especially in the upper intertidal zone. *Centropyxis platystoma* type consistently occurs at lower elevations and can be quite abundant in the lowest samples that contain testate amoebae. A diverse range of taxa occur in the uppermost samples in the supratidal zone. The assemblages here are more variable than at lower elevations and presumably partly depend on the nature of the upland habitats immediately adjacent to the marsh. *Euglypha* taxa are commonly encountered along with *Corythion* and *Trinema* taxa, but their relative position on the tidal gradient is not firmly established because of the relatively small sample numbers available. Of these small taxa, *Euglypha rotunda* type appears most commonly in the lowest position with larger *Euglypha* taxa higher up. One taxon that seems to have particular affinity for saltmarshes is *Tracheleuglypha dentata* and has been recovered from almost every marsh sampled so far, occurring in the marsh just below or around HAT. *Difflugia* taxa are also very common on most marshes and, although this group is best known from lacustrine environments, they reach significant abundances in the zones below HAT and MHWS. The sequence in which they occur is the same in both North American and British marshes. In order of elevation, the taxa found are *D. oblonga* type, *D. lucida* type, and *D. pristis* type, with the latter taxon one of the most abundant in the lowest samples. Two other taxa are worthy of mention as they are regularly encountered on the main open marsh area. *Cyphoderia* taxa and *Campascus* type are found in a number of sites with the former occupying lower elevations than the latter. In conclusion, despite limited data availability, a number of key taxa are emerging as being the most useful indicators for past sea level. Perhaps more importantly, many of these appear to consistently occur at similar relative positions in the tidal gradient at different sites that are widely separated geographically.

19.4 FOSSIL DATA AND RECONSTRUCTIONS

If the ultimate aim is to reconstruct past sea level from biological proxies, an obvious pre-requisite is to be able to retrieve fossil assemblages from sediments that have been affected by changing sea levels. There is a very large and still growing literature on the distribution of sea-level proxy organisms in modern environments (see Chapters 12–18), and the drive to develop and test transfer functions has added impetus to this work. For testate amoebae, almost all the available data are from modern samples. Clearly, while these data are an essential underpinning for interpretation of all biological fossil data, the existence of relationships in contemporary environments does not mean that the analysis methods will be applicable to fossil samples and sediments. Can the potential be turned into actual application?

The most important limitation to applying any proxy-based reconstruction is that the fossil organisms have to be present in the sediment. For testate amoebae, unlike for many freshwater sediments, a key limitation is that testate amoebae are only found in a restricted part of the intertidal and supratidal zones (see Section 19.3 above). Testate amoebae will only be present in sediments if they were derived from these settings. A second potential limitation is preservation in sediments. Many testate amoebae form relatively fragile tests and are perhaps more susceptible to both physical and chemical damage *in situ*. The intertidal zone is a more dynamic environment than most freshwater peatland and lake conditions, and there is therefore a risk that tests will be broken up on the surface of the marsh. Furthermore, the chemical

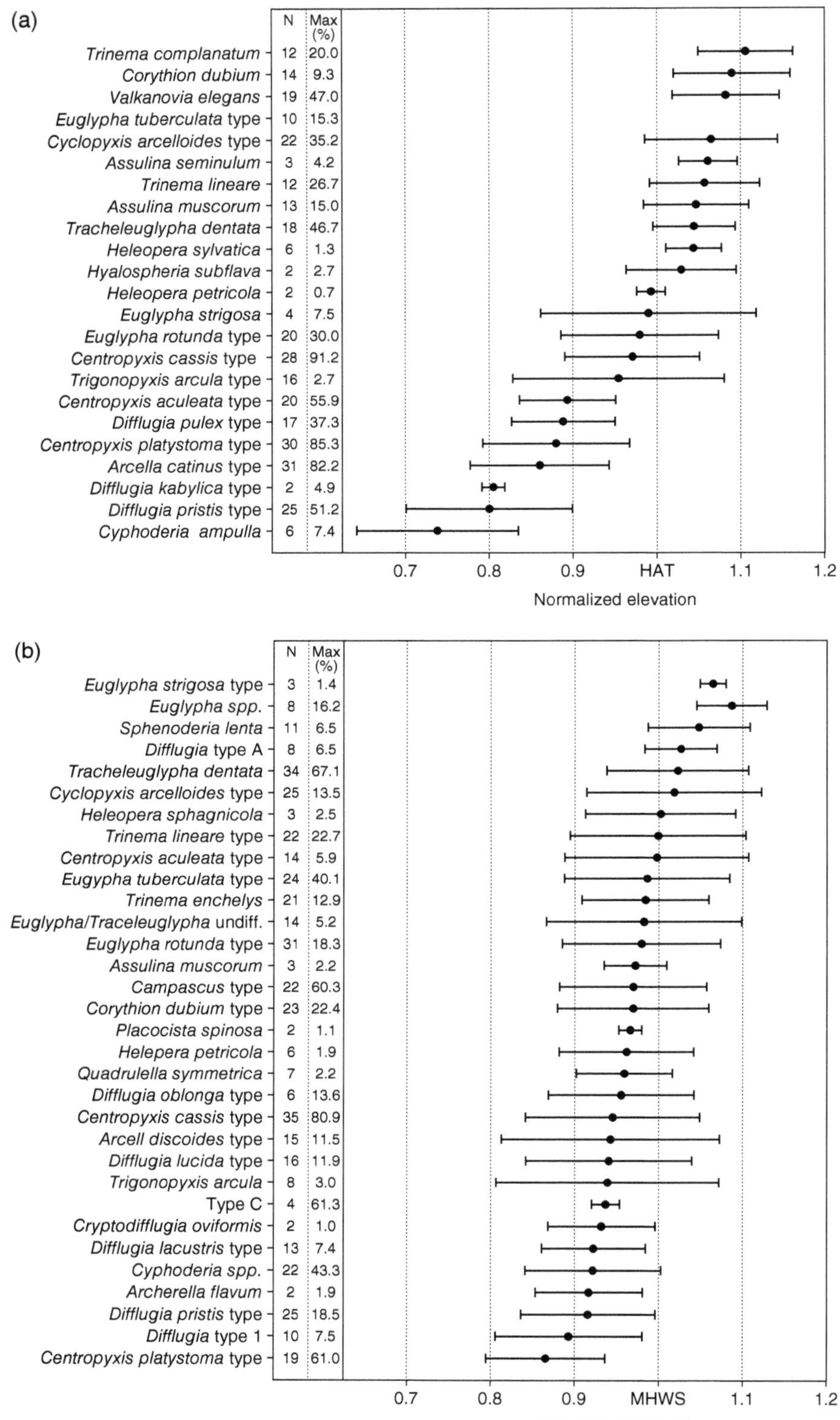

Fig. 19.4. Optima and tolerances for taxa from (a) three marshes in the northeast USA (Gehrels et al., 2006) and (b) three marshes in the United Kingdom (the data used in Charman et al., 2002). Note that the normalized elevations used in these datasets are different. For (a) the normalization is between MTL and HAT, and in (b) the normalization is between MTL and MHWS.

conditions within marsh sediments are probably less conducive to preservation of tests than the organic-rich, often acid conditions found in peat and lake sediments. The first challenge is therefore to locate and retrieve fossil specimens in sufficient numbers to reliably infer past sea level. A second challenge is to be able to apply the assemblage zonations and transfer functions derived from studies of modern assemblage distributions to the fossil data to generate more detailed information on past sea-level change.

As mentioned above, there are now a number of studies that have retrieved fossil testate amoebae from coastal sediments using foraminifera preparation methods for larger size fraction taxa. These have been used to indicate transgressive and regressive sea-level change by their indication of more freshwater or upper tidal range conditions (e.g., Shennan et al., 1999; Lloyd, 2000; Reinhardt et al., 2010). These studies indicate that testate amoebae occurrence is sporadic mainly because there have only been suitable conditions for testate amoebae for part of the period of sediment deposition. Potential problems of preservation are more of a concern. Roe *et al.* (2002) sampled a wide range of fossil coastal sediments and found that they were certainly present in a range of settings including recent saltmarsh sediments and a range of sediments transitional to freshwater environments such as isolation basins and raised bog peat. Recovery from mid-Holocene saltmarsh deposits was more limited, but specimens were present despite the fact that the sediments may have been sub-optimal for testate amoebae because of being lower in the tidal frame. Sediments from supratidal freshwater marsh phases had mostly very low numbers or no testate amoebae present, despite being from an environment favorable for testate amoebae. However, the analyses were performed on samples that had been in storage for longer periods of time and it was suggested that there had been decay and loss of tests from dehydration between sampling and analysis (Roe et al., 2002). More experiments and data are needed to assess whether test preservation is a problem in both sediments and storage. Nevertheless, it is clear that sediments do exist where a range of taxa can be found, including groups such as the idiosomic taxa that are often thought to be more susceptible to decay (Roe et al., 2002; Charman et al., 2010).

While it is clear that it is possible to identify phases of freshwater or low salinity in coastal sediments from the presence of testate amoebae, there is little value in establishing detailed zonation in relation to surface elevation in contemporary samples unless similar detail can be observed in fossil assemblages. Roe et al. (2002) showed that small size fraction testate amoebae show distinct down-core zonation in an isolation basin sediment and also extend well beyond the depth range of the larger size fraction previously reported (Lloyd, 2000). Furthermore, studies on saltmarsh sediments from the last few hundred years also provide assemblages and temporal variations that allow more detailed sea-level records to be established (Roe et al., 2002; Charman et al., 2010).

Figure 19.5 shows the results of testate amoebae analysis from a core from Chezzetcook, Nova Scotia, from which foraminifera had already been analyzed and used for a sea-level reconstruction (Gehrels et al., 2005). This allowed the reconstruction to be tested against another proxy but, because the sediment covers the last 100 years, it was also compared with the local tide gauge record from Halifax (Charman et al., 2010). Both this core and another sequence from the nearby coast of Maine show down-core changes in testate amoebae and assemblages that are suitable for the application of transfer functions from modern samples. The reconstructions suggest that the trends and magnitude of sea-level change estimated from testate amoebae are very similar to those of the tide gauges. In the case of Chezzetcook, the testate amoebae suggest a long term rise of $3.70 \pm 0.70\,mm\ a^{-1}$ compared to the tide gauge measurement of $3.3 \pm 0.1\,mm\ a^{-1}$. At Wells, Maine, the testate amoebae reconstruction showed a rise of 1.62 ± 0.17 and $2.20 \pm 0.25\,mm\ a^{-1}$, depending on the transfer function used, compared with $1.86 \pm 0.12\,mm\ a^{-1}$ for the tide gauge data. The less-conclusive results at Maine are partly a result of low sample numbers because of a low sediment accumulation rate, but both cores suggest that the testate amoebae-based estimates of sea level are accurate. They also concur with the foraminifera-based estimates of sea-level change.

19.5 CONCLUSIONS: POTENTIAL AND FUTURE WORK

In a review of testate amoebae applications in 2001, I identified two areas of future work before testate amoebae could be used in a meaningful way in sea-level reconstructions. These were: (1) "the replicability of zonation in response to tidal parameters at different geographical locations and

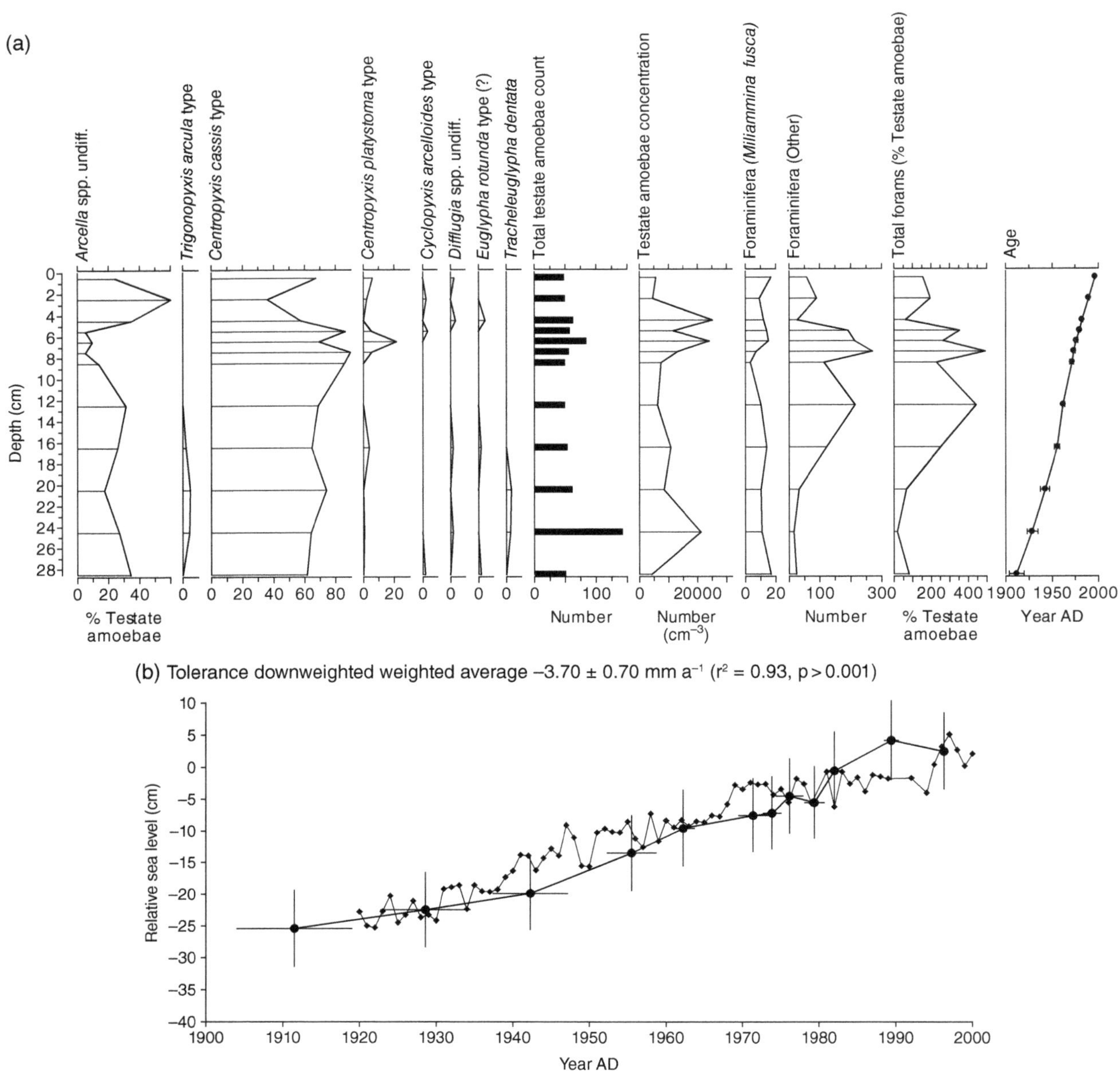

Fig. 19.5. Results of testate amoebae analysis from a saltmarsh core from Chezzetcook Marsh, Nova Scotia, Canada. (a) Testate amoebae taxa with total counts of testates and foraminifera and age based on ^{210}Pb analysis. (b) Reconstructed sea level and transfer function errors (2σ), based on testate amoebae analysis as compared to the tide gauge record at Halifax, NS. *Source*: Charman et al. 2010. Reproduced with permission of Elsevier.

within varied plant communities"; and (2) "the occurrence and detectability of fossil testate amoebae in coastal sediments." Initial work has suggested that this may be much harder than for terrestrial peats and lake sediments (Charman, 2001, p. 1762). There has been some progress in both of these areas although further work is still needed. In terms of establishing zonation, it now seems clear that testate amoebae zonation on saltmarshes is likely to be replicable at a large regional scale, at least covering coasts on both sides of the North Atlantic (Charman et al., 2010). Recent work in Norway also confirms this idea (Barnett et al., 2013). Although the assemblages are less diverse, studies based on foram sampling also suggest there are some taxa that always occur in the same zones for marshes throughout the world (Scott et al., 2001). Although no assumptions can be made concerning testate amoebae zonation outside of the North Atlantic region, it seems very likely that similar patterns will occur given the general cosmopolitanism of testate amoebae. Further work is needed to test this idea.

There has been less progress on the second issue. Certainly nothing has dispelled the idea that analysis in saltmarshes is more difficult than for other sediments, but the work on developing preparation techniques suggests that the situation has improved (Charman et al., 2010; Barnett et al., 2013). There are almost certainly further improvements that could be made by refining the protocol described here further.

Identifying and sampling sediments optimal for testate amoebae occurrence is one further consideration here. Where cores are sampling from lower than the uppermost parts of the marsh, they inevitably have lower concentrations of testate amoebae. For successful core records, it may be much more appropriate to sample the highest areas of the marsh but also to accept that, in many cases, sediments containing testate amoebae will only have accumulated during part of the sedimentation history. Coring non-saltmarsh locations in the upper estuary areas may also be a good way to obtain samples with more reliable testate amoebae content (Ooms et al., 2011). In conclusion, the prospects are good for further development of the application of testate amoebae to sea-level reconstruction. I strongly encourage those who have the opportunity to take up the challenge and work on testate amoebae, alongside other sea-level indicators, whenever the opportunity arises!

ACKNOWLEDGEMENTS

I would like to thank all the contributors to past work on testate amoebae in coastal deposits and also those who have provided samples and data in connection with the results summarized here. Much of the developmental work reported here was made possible by NERC grant GR9/03426 and a University of Plymouth DevR grant, both to Roland Gehrels and me. Rob Barnett provided some of the photographs in Figure 19.2.

REFERENCES

Barbosa, C. F., Scott, D. B., Seoane, J. C. S., and Turcq, B. J. (2005) Foraminiferal zonations as base lines for Quaternary sea-level fluctuations in south-southeast Brazilian mangroves and marshes. Journal of Foraminiferal Research, 35, 22–43.

Barnett, R., Charman, D.J., Gehrels, W.R., Marshall, W., and Saher, M. (2013) Testate amoebae as sea-level indicators in northwestern Norway: preparatory and analytical developments. Acta Protozoologica, 52, 115–128.

Borcard, D., Legendre, P., and Drapeau, P. (1992) Partialling out the spatial component of ecological variation. Ecology, 73, 1045–1055.

Charman, D. J. (2001) Biostratigraphic and palaeoenvironmental applications of testate amoebae. Quaternary Science Reviews, 20(16–17), 1753–1764.

Charman, D. J., Roe, H. M., and Gehrels, W. R. (1998) The use of testate amoebae in studies of sea-level change: a case study from the Taf Estuary, south Wales, UK. Holocene, 8, 209–218.

Charman, D. J., Hendon, D., and Woodland, W. A. (2000) *The Identification of Peatland Testate Amoebae*. Quaternary Research Association Technical Guide, London.

Charman, D. J., Roe, H. M., and Gehrels, W. R. (2002) Modern distribution of saltmarsh testate amoebae: regional variability of zonation and response to environmental variables. Journal of Quaternary Science, 17, 387–409.

Charman, D. J., Gehrels, W. R., Manning, C., and Sharma, C. (2010) Reconstruction of recent sea-level change using testate amoebae. Quaternary Research, 73, 208–219.

Duleba, N., and Debenay, J. P. (2003) Hydrodynamic circulation in the estuaries of Estacao Ecologica Jurtia-Itatins, Brazil, inferred from foraminifera and thecamoebian assemblages. Journal of Foraminiferal Research, 33, 62–93.

Farooqui, A., and Naidu, T. Y. (2010) Thecamoebians and palynological assemblage in Gautami-Godavari River mouth, India: Environment and sea level since 3000 years. Journal of the Geological Society of India, 75, 841–850.

First, M. R., and Hollibaugh, J. T. (2008) Protistan bacterivory and benthic microbial biomass in an intertidal creek mudflat. Marine Ecology Progress Series, 361, 59–68.

Gehrels, W. R., Roe, H. M., and Charman, D. J. (2001) Foraminifera, testate amoebae and diatoms as sea-level indicators in UK saltmarshes: a quantitative multiproxy approach. Journal of Quaternary Science, 16, 201–220.

Gehrels, W. R., Roe, H. M., and Charman, D. J. (2002) Foraminifera, testate amoebae and diatoms as sea-level indicators in UK saltmarshes: a quantitative multiproxy approach. Journal of Quaternary Science, 17, 285–285.

Gehrels, W. R., Kirby, J. R., Prokoph, A., Newnham, R. M., Achterberg, E. P., Evans, H., Black, S. and Scott, D. B. (2005) Onset of recent rapid sea-level rise in the western Atlantic Ocean. Quaternary Science Reviews, 24, 2083–2100.

Gehrels, W. R., Hendon, D., and Charman, D. J. (2006) Distribution of testate amoebae in salt marshes along the North American East Coast. Journal of Foraminiferal Research, 36, 201–214.

Golemansky, V., and Todorov, M. (2004) Shell morphology, biometry and distribution of some marine interstitial testate amoebae (Sarcodina: rhizopoda). Acta Protozoologica, 43, 147–162.

Gomaa, F., Todorov, M., Heger, T. J., Mitchell, E. A. D., and Lara, E. (2012) SSU rRNA phylogeny of Arcellinida (Amoebozoa) reveals that the largest Arcellinid genus, *Difflugia* Leclerc 1815, is not monophyletic. Protist, 163, 389–399.

Hayward, B. W., Grenfell, H., Cairns, G., and Smith, A. (1996) Environmental controls on benthic foraminiferal and thecamoebian associations in a New Zealand tidal inlet. Journal of Foraminiferal Research, 26, 150–171.

Hendon, D., and Charman, D. J. (1997) The preparation of testate amoebae (Protozoa: Rhizopoda) samples from peat. Holocene, 7, 199–205.

Kumar, A., and Dalby, A. P. (1998) Identification key for Holocene lacustrine arcellacean (thecamoebian) taxa. Palaeontologia Electronica, 1, 33.

Lloyd, J. (2000) Combined foraminiferal and thecamoebian environmental reconstruction from an isolation basin in NW Scotland: Implications for sea-level studies. Journal of Foraminiferal Research, 30, 294–305.

Medioli, F. S., and Scott, D. B. (1983) Holocene Arcellacea (Thecamoebians) from eastern Canada. Cushman Foundation Special Publication, 21, 1–63.

Medioli, F. S., and Scott, D. B. (1988) Lacustrine thecamoebians (mainly arcellaceans) as potential tools for palaeolimnological interpretations. Palaeogeography, Palaeoclimatology, Palaeoecology, 62, 361–386.

Medioli, F. S., Scott, D. B., Collins, E. S., and McCarthy, F. M. G. (1990) Fossil thecamoebians: present status and prospects for the future. In: *Paleocology, Biostratigraphy, Paleoceanography and Taxonomy of Agglutinated Foraminifera* (ed. Hemleben, C., Kaminski, M. A., Kuhnt, W., and Scott, D. B.), Kluwer, Amsterdam, 813–839.

Mitchell, E. A. D., Charman, D. J., and Warner, B. G. (2008) Testate amoebae analysis in ecological and paleoecological studies of wetlands: past, present and future. Biodiversity and Conservation, 17, 2115–2137.

Nixon, F. C., Reinhardt, E. G., and Rothaus, R. (2009) Foraminifera and tidal notches: Dating neotectonic events at Korphos, Greece. Marine Geology, 257, 41–53.

Ooms, M., Beyens, L., and Temmerman, S. (2011) Testate amoebae as estuarine water-level indicators: modern distribution and the development of a transfer function from a freshwater tidal marsh (Scheldt estuary, Belgium). Journal of Quaternary Science, 26, 819–828.

Reinhardt, E. G., Nairn, R. B., and Lopez, G. (2010) Recovery estimates for the Rio Cruces after the May 1960 Chilean earthquake. Marine Geology, 269, 18–33.

Riveiros, N. V., Babalola, A. O., Boudreau, R. E. A., Patterson, R. T., Roe, H. M., and Doherty, C. (2007) Modern distribution of salt marsh foraminifera and thecamoebians in the Seymour-Belize Inlet Complex, British Columbia, Canada. Marine Geology, 242, 39–63.

Roe, H. M., and Patterson, R. T. (2006) Distribution of thecamoebians (testate amoebae) in small lakes and ponds, Barbados, West Indies. Journal of Foraminiferal Research, 36, 116–134.

Roe, H. M., Charman, D. J., and Gehrels, W. R. (2002) Fossil testate amoebae in coastal deposits in the UK: implications for studies of sea-level change. Journal of Quaternary Science, 17, 411–429.

Scott, D. B., Medioli, F. S., and Schafer, C. T. (1977) Temporal changes in foraminiferal distributions in Miramichi River estury, New Brunswick. Canadian Journal Of Earth Sciences, 14, 1566–1587.

Scott, D. B., Schnack, E. J., Ferrero, L., Espinosa, M., and Barbosa, C. F. (1990) Recent marsh foraminifera from the east coast of South America: Comparision to the Northern Hemisphere. In: *Paleocology, Biostratigraphy, Paleoceanography and Taxonomy of Agglutinated Foraminifera* (ed. Hemleben, C., Kaminski, M. A., Kuhnt, W., and Scott, D. B.), Kluwer, Amsterdam, 717–737.

Scott, D. B., Hasegawa, S., Saito, T., Ito, K., and Collins, E. (1995) Marsh foraminifera and vegetation distributions in Nemuro Bay wetland areas, eastern Hokkaido. Transactions and Proceedings of the Palaeontological Society of Japan, 180, 282–295.

Scott, D. B., Medioli, F. S., and Schafer, C. T. (2001) *Monitoring of Coastal Environments using Foraminifera and Thecamoebian Indicators*. Cambridge University Press, Cambrige.

Shennan, I., Scott, D. B., Rutherford, M., and Zong, Y. Q. (1999) Microfossil analysis of sediments representing the 1964 earthquake, exposed at Girdwood Flats, Alaska, USA. Quaternary International, 60, 55–73.

Smith, H. G. (1992) Distribution and ecology of the testate rhizopod fauna of the continental Antarctic Zone. Polar Biology, 12, 629–634.

Stockmarr, J. (1971) Tablets with spores used in absolute pollen analysis. Pollen et Spores, 13, 615–621.

Tolonen, K. (1986) Rhizopod analysis. In *Handbook of Holocene Palaeoecology and Palaeohydrology* (ed. Berglund, B. E.), John Wiley & Sons, 645–666.

van Hengstum, P. J., Reinhardt, E. G., Beddows, P. A., Schwarcz, H. P., and Gabriel, J. J. (2009) Foraminifera and testate amoebae (thecamoebians) in an anchialine cave: Surface distributions from Aktun Ha (Carwash) cave system, Mexico. Limnology and Oceanography, 54, 391–396.

Warner, B. G. (1988) Methods in Quaternary Ecology #5. Testate amoebae (Protozoa). Geoscience Canada, 15, 251–260.

Woodland, W. A., Charman, D. J., and Sims, P. C. (1998) Quantitative estimates of water tables and soil moisture in Holocene peatlands from testate amoebae. Holocene, 8, 261–273.

Chapter 20

Stable carbon isotope and C/N geochemistry of coastal wetland sediments as a sea-level indicator

NICOLE S. KHAN[1,2], CHRISTOPHER H. VANE[3], AND BENJAMIN P. HORTON[2,4]

[1] *Sea Level Research, Department of Earth and Environmental Science, University of Pennsylvania, Philadelphia, PA, USA*

[2] *Sea Level Research, Department of Marine and Coastal Sciences, Rutgers University, USA*

[3] *British Geological Survey, Environmental Science Centre, Keyworth, Nottingham, UK*

[4] *Earth Observatory of Singapore and Division of Earth Sciences, Nanyang Technological University, Singapore*

20.1 INTRODUCTION

The stable carbon isotope geochemistry of organic remains preserved in tidal wetland sediments can provide valuable information about relative sea-level change (e.g., Törnqvist et al., 2004a, b; Wilson et al., 2005a, b; Kemp et al., 2012, 2013; Engelhart et al., 2013a, b). Stable carbon isotopes ($\delta^{13}C$) and the ratio of organic carbon to total nitrogen (C/N) together are able to differentiate sources of organic matter that accumulate in coastal depositional sequences, in particular between C_3 and C_4 vegetation and freshwater and marine organic matter. They hold value as sea-level indicators because modern intertidal depositional environments (e.g., tidal flat, low marsh, high marsh, and freshwater environments) with distinct bulk sediment and plant $\delta^{13}C$ and C/N values occur at specific elevation ranges relative to the tidal frame (Chmura and Aharon, 1995; Lamb et al., 2006; Wilson et al., 2005a, b; Kemp et al., 2012; Fig. 20.1).

The range of $\delta^{13}C$ and C/N values of end-members (flora or fauna incorporated in sediments) has been established from studies in temperate and tropical settings (Fig. 20.1a). Further, the general pathway of $\delta^{13}C$ and C/N values from plant tissue to bulk sediments has been identified (Figs 20.1b and 20.2). Bulk sediment $\delta^{13}C$ and C/N values are the result of autochthonous inputs from wetland vegetation and allochthonous sources, such as algae and particulate organic matter (Fry et al., 1977; Chmura and Aharon, 1995; Middelburg et al., 1997). However, application to paleoenvironmental studies may be complicated by accumulation of allochthonous marine or riverine organic matter sources or early degradation of *in situ* organic matter (Fig. 20.2). The early degradation of organic matter causes only a minor change in $\delta^{13}C$ values, although C/N values may change greatly (t_0 to t_1; Kemp et al., 2011). Diagenesis of sediments over Holocene timescales may cause additional deviation in sediment $\delta^{13}C$ and C/N from their original values at time of deposition (t_1 to t_2; Wilson et al., 2005b). Because surface sediment $\delta^{13}C$ and C/N values represent a time averaging of the various sources of organic matter to sediments, surface sediments, rather than contemporaneous vegetation end-members, are the more suitable analog for fossil samples. The bulk sediment $\delta^{13}C$ and C/N values of intertidal depositional environments are recognizable in sedimentary archives.

There are numerous applications of $\delta^{13}C$ and C/N in coastal environments (see Lamb et al., 2006), although most of these studies investigate the provenance of organic matter in coastal archives (e.g., Thorton and McMacnus, 1994; Pilarczyk et al., 2012), carbon dynamics within coastal ecosystems (e.g., Showers and Angle, 1986; Andrews et al., 2000), or broad-scale paleoenvironmental change (e.g., Byrne et al., 2001; Zong et al., 2006). The development of $\delta^{13}C$ and C/N as a sea-level indicator is in its infancy. Among the small number of studies that examine the use of $\delta^{13}C$ and/or C/N in sea-level reconstruction, only a few published datasets exist that specifically examine the relationship of $\delta^{13}C$ and C/N

Handbook of Sea-Level Research, First Edition. Edited by Ian Shennan, Antony J. Long, and Benjamin P. Horton.

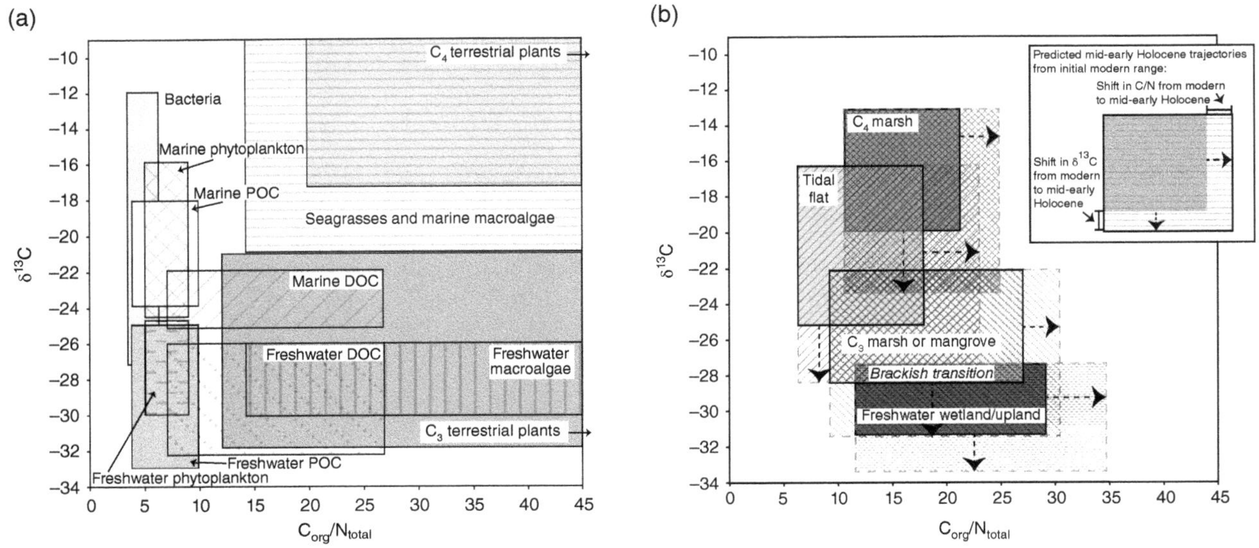

Fig. 20.1. (a) The range of $\delta^{13}C$ and C/N values reported in the literature for characteristic organic matter sources (i.e., living flora and fauna) that accumulate in sedimentary sequences of intertidal wetlands. From Lamb et al. (2006), and updated with values of marine/freshwater macroalgae and seagrasses from Hemminga and Mateo (1996), Ishihi et al. (2001), and Prado and Heck (2011). (b) The range of $\delta^{13}C$ and C/N values for tidal flat, C_4 marsh, C_3 marsh or mangrove, brackish transitional and freshwater/upland environments. Ranges are based on values reported by Chmura et al. (1987), Ember et al. (1987), Middelburg et al. (1997), Andrews et al. (2000), Malamud-Roam and Ingram (2001, 2004), Bouillon et al. (2003), Wilson et al. (2005a, b), Gebrehivet et al. (2008), Kristensen et al. (2008), Kemp et al. (2010, 2012), Engelhart et al. (2013a). Changes in the range of $\delta^{13}C$ and C/N values due to post-depositional alteration on mid–early Holocene timescales is indicated by black arrows and based on observations from Andrews et al. (2000) and Wilson et al. (2005a). The ranges defined in (a) and (b) are more qualitative in nature and not rigorously defined by statistics.

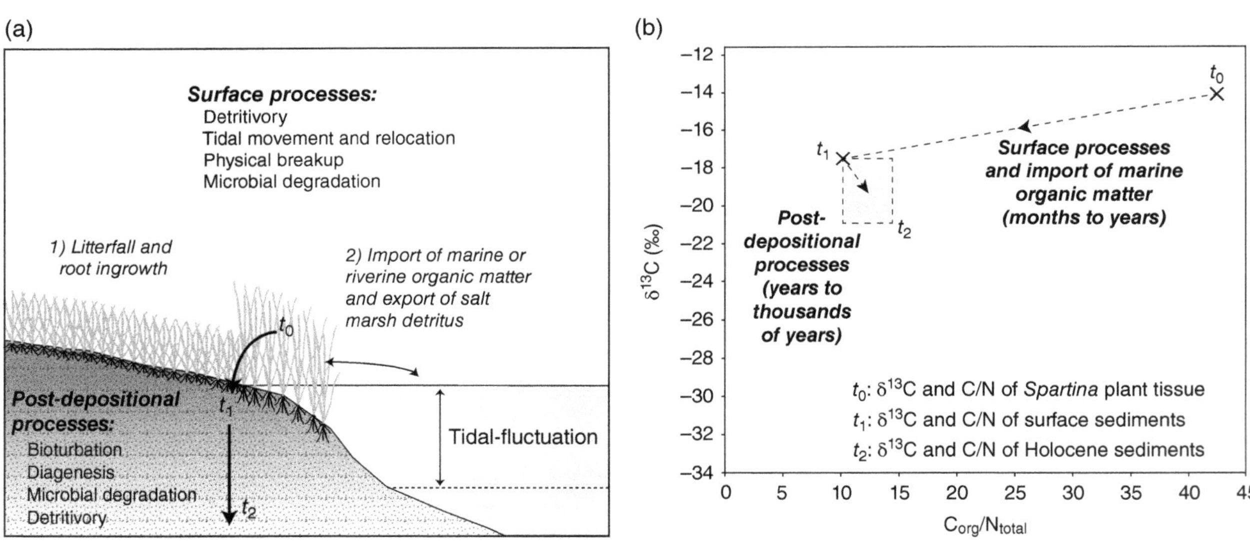

Fig. 20.2. Factors influencing $\delta^{13}C$ and C/N values in a *Spartina*-dominated marsh. (a) Sources of organic matter and processes affecting its accumulation and preservation in saltmarsh sedimentary sequences (after Wooller et al., 2003). The major end-member sources are the (1) autochthonous vascular plant material (including roots) from *Spartina* litter and (2) allochthonous organic matter derived in particular from particulate organic matter from marine sources through tidal delivery or riverine sources in estuarine or deltaic settings. (b) The change in $\delta^{13}C$ and C/N of *Spartina* litter and sediments from fresh plant material (t_0) to burial in Holocene sedimentary sequences (t_2) related to the factors listed in (a). The change in values was estimated from studies by Ember et al. (1987), Benner et al. (1987, 1991), White and Howes (1994) and Klap et al. (1999).

to elevation with respect to the tidal frame (Wilson et al., 2005b; Lamb et al., 2007; Kemp et al., 2010, 2011, 2013; Engelhart et al., 2013a), which is a key determinant in its utility as a sea-level indicator. In a study of the intertidal zone of the southern Atlantic coast of New Jersey, Kemp et al. (2012) showed that $\delta^{13}C$ values, along with the absence/presence of agglutinated foraminifera, distinguished between elevation-dependent environmental zones, while the C/N values remained uniform (Fig. 20.3a, b). Thus, stable carbon isotope geochemistry has been revealed to be most powerful when used in combination with simple metrics of microfossil assemblages (e.g., absence/presence of saltmarsh foraminifera) in a multi-proxy approach.

Here we focus on studies employing $\delta^{13}C$ and C/N to quantitatively reconstruct relative sea level; we direct the reader to Lamb et al. (2006) for a comprehensive review of the use of $\delta^{13}C$ and C/N in coastal paleoenvironmental applications. In this chapter, we describe the factors controlling $\delta^{13}C$ and C/N of wetland plants and sediments and outline basic methods for its use in sea-level reconstruction. We highlight key studies employing $\delta^{13}C$ and C/N, draw attention to strengths and weaknesses of its use, and offer suggestions for future research directions to further develop this technique.

20.2 $\delta^{13}C$ AND C/N OF SOURCES OF ORGANIC MATTER TO TIDAL WETLANDS

There are significant variations in $\delta^{13}C$ and C/N in the sources of organic matter incorporated into tidal wetland sediments (Fig. 20.1). The relatively large variations in the $\delta^{13}C$ of C_3 v. C_4 organic matter (~14‰), terrestrial C_3 v. marine organic matter (5–8‰), seagrass v. planktonic marine organic matter (10‰), and downstream changes in river-estuarine gradients (3–7‰) should predominate over smaller variations in isotopic composition over time caused by seasonal effects or other environmental changes such as local temperature or humidity (Tyson, 1995). On the basis of $\delta^{13}C$ alone, there is a strong overlap of C_4 marsh vegetation and marine-derived organic matter values which may confound paleoenvironmental interpretations. The C/N ratio can help with the distinction between these two sources (Lamb et al., 2006).

20.2.1 Carbon isotopic fractionation

There are three naturally occurring carbon isotopes: ^{12}C, ^{13}C, and ^{14}C. ^{12}C and ^{13}C comprise the stable carbon isotopes, while ^{14}C is radioactive (with a half-life of 5730 years) and is used to determine the age of organic remains preserved in stratigraphic sequences. The ratio of ^{13}C to ^{12}C is expressed in $\delta^{13}C$ notation:

$$\delta^{13}C = \left(\frac{^{13}C/^{12}C_{\text{sample}}}{^{13}C/^{12}C_{\text{standard}}} - 1 \right) \times 10^3,$$

given in parts per thousand (‰) with reference to the standard Pee Dee Belemnite (PDB). ^{12}C has a much higher natural occurrence than ^{13}C (98.9% v. 1.1%, respectively). Slight variations in the ratio of the two isotopes are therefore recorded in organic materials due to isotopic fractionation during physical, chemical, and biological processes (Dawson et al., 2002).

20.2.1.1 Terrestrial carbon

Most of the natural variation in the isotopic composition of organic matter occurs as a consequence of photosynthesis (Wickman, 1952). Terrestrial plants are divided into three major groups on the basis of their photosynthetic pathway: C_3, C_4, and Crassulacean acid metabolism (CAM). The different ways in which these plants fix carbon result in distinct isotopic variations intrinsic to each plant type. The variation in $\delta^{13}C$ is due to the biochemical properties of the primary CO_2-fixing enzymes and limitations of CO_2 diffusion into the leaf (O'Leary, 1988). Plants using the C_3-pathway reduce CO_2 to the 3-C compound phosphoglycerate by use of ribulose bisphosphate carboxylase (Rubisco), which discriminates against the heavier ^{13}C isotope (Calvin and Benson, 1949). Plants with this photosynthetic pathway have $\delta^{13}C$ values ranging from −32‰ to −21‰ (Deines, 1980).

Plants following the C_4-pathway reduce CO_2 to the 4-C compounds of aspartic or malic acid by the enzyme phosphoenol pyruvate carboxylase (PEP) (Slack and Hatch, 1967). The discrimination of PEP against ^{13}C is relatively less than that of Rubisco and results in higher $\delta^{13}C$ values than C_3 terrestrial plants. C_4 plants range in $\delta^{13}C$ from −17‰ to −9‰ (Chmura and Aharon, 1995) and have distinct, mutually exclusive values of $\delta^{13}C$ with respect to C_3 plants (Smith and Epstein, 1971).

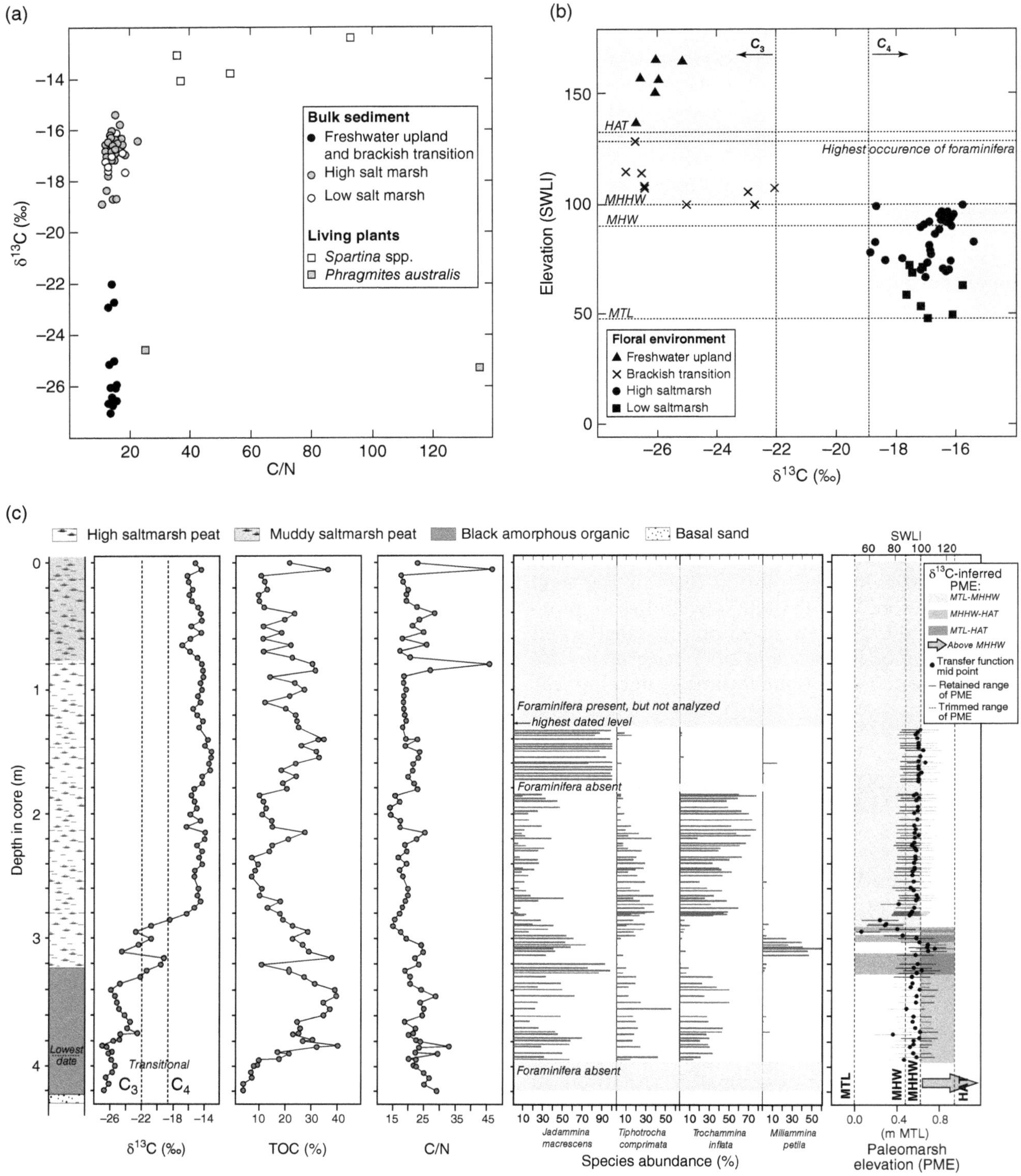

Fig. 20.3. The application of $\delta^{13}C$ and foraminifera to reconstruct Late Holocene vegetation and relative sea-level history in Leeds Point, New Jersey. (a) $\delta^{13}C$ and C/N values of all bulk sediment and living plant samples in the study, classified by depositional environment (open circles: low saltmarsh; gray circles: high saltmarsh; black circles: freshwater upland;and brackish transition; open squares: plant species, *Spartina* spp.; gray squares: *Phragmites australis*). $\delta^{13}C$ values clearly delineate between plants and sediments derived from C_3 v. C_4 plant types. C/N of bulk sediments converges to values between 10 and 20, irrespective of depositional environment. (b) Range in $\delta^{13}C$ and elevation, given in standardized water

CAM plants can minimize water loss by nighttime CO_2 fixation via PEP (Ting et al., 1985). While most CAM plants have a resulting $\delta^{13}C$ closer to C_4 plants, under some environmental conditions, facultative CAM species are able to utilize the C_3 pathway during daytime using Rubisco (O'Leary, 1981). The $\delta^{13}C$ of facultative CAM plants will have a value dependent on the proportion of carbon fixed by Rubisco v. PEP (Osmond et al., 1976). The $\delta^{13}C$ for these plants covers the range of −28‰ to −10‰ (Deines, 1980).

Of all terrestrial plant species, most are C_3. These plant types are dominant in temperate forest and freshwater aquatic/wetland environments, although C_4 and CAM plants have a significant presence in warm, arid environments (Osmond et al., 1982). High salinity imposes a similar stress on plants as water stress in arid environments; C_4 and CAM plants are therefore well adapted to saline intertidal environments (Malamud-Roam and Ingram, 2004).

In addition to photosynthetic pathway, variation in plant $\delta^{13}C$ can occur due to environmental factors. Water availability and internal leaf CO_2 controls the "openness" of the stomatal aperture, expressed as stomatal conductance ($cm\,s^{-1}$) (Shlesinger, 1997). Decreased stomatal conductance due to increased salinity stress (e.g., Guy et al., 1980) or higher atmospheric CO_2 (Park and Epstein, 1960) causes more CO_2 inside the leaf to react with CO_2-fixing enzymes. Less fractionation of isotopes therefore occurs and plant $\delta^{13}C$ increases (Farquhar et al., 1982; Guy and Reid, 1986).

Changes in the source of CO_2 to the atmosphere due to climatic and anthropogenic factors (e.g., Keeling, 1979; Leuenberger et al., 1992) have not remained constant over time and may also affect the $\delta^{13}C$ of terrestrial plants. One important anthropogenic influence is the addition of CO_2 derived from fossil fuels (coal: −25‰; natural gas: −40‰, petroleum: −30‰) that have a combined average $\delta^{13}C$ of −26‰ since the onset of the Industrial Revolution (Tans, 1981; Boutton, 1991). Pre-1850 $\delta^{13}C$ composition of the atmosphere was calculated to be −6.4‰ (Francey et al., 1999), which has decreased over time to the present-day $\delta^{13}C$ of around −8.0‰ (Cuntz, 2011). McCarroll and Loader (2006) suggest the isotopic ratio of atmospheric CO_2 has declined by 1.7‰ since 1850 in two distinct phases: a slow decline from 1850 to 1961 (at a rate of 0.0044‰ a^{-1}) and a more rapid fall from 1962 to 1980 (at a rate of 0.0281‰ a^{-1}), with a comparable rate to the present. Several paleolacustrine records have demonstrated a decrease in $\delta^{13}C$ values from 1850 to present (Verberg, 2007; Castañeda et al., 2011; Jiang et al., 2011), although this effect in coastal sedimentary archives is undocumented. The effect of small variations in atmospheric CO_2 on resulting plant $\delta^{13}C$ should be considered when comparing modern to pre-industrial core material.

20.2.1.2 *Aquatic carbon*

Aquatic photosynthetic organisms (phytoplankton, algae, seagrasses) use the C_3 pathway, but have $\delta^{13}C$ values that differ from terrestrial C_3 plants. This dissimilarity can be explained by slower diffusion of CO_2 in water and ambient CO_2 availability (Fogel et al., 1992). Phytoplankton (diatoms, dinoflagellates, green algae, euglenoids) will preferentially take up dissolved CO_2 until the source is exhausted, and the low availability of CO_2 (−8‰) in the oceans will cause uptake of HCO_3^- (~0‰). As a result, phytoplankton have values ranging from −30‰ to −18‰ (Degens et al., 1968; Deines, 1980; Anderson and Arthur, 1983). Variation occurs in marine (−24‰ to −18‰) and freshwater phytoplankton $\delta^{13}C$ (−30‰ to −25‰) due to higher dissolved CO_2 concentrations compared to HCO_3^- in river water. Similar to phytoplankton, the $\delta^{13}C$ of algae (freshwater: −30‰ to −26‰; marine: −23‰ to −16‰) and sea grasses (−21‰ to −6‰; Hemminga and Mateo, 1996) reflects its use of CO_2 relative to HCO_3^-. Zooplankton retain the signature of their diet, although slight enrichment of ^{13}C occurs along food chains (Fry, 2008).

Fig. 20.3. (*Continued*) level index (SWLI) units, of bulk sediment samples classified by environment from which the sample was collected (triangles: freshwater upland; X: brackish transition; circles: high saltmarsh; squares: low saltmarsh). The elevation of tidal datums and the highest occurrence of foraminifera are indicated by horizontal dashed lines. The ~3‰ demarcation between C_3 and C_4-derived sediment types is indicated by vertical dashed lines, and gray shading is applied to indicate the distinct range in $\delta^{13}C$ and elevation that C_3 and C_4-derived sediment types occupy. (c) Lithology, $\delta^{13}C$, TOC, C/N, and species abundance of key foraminifera taxa (represented by horizontal bars) and reconstructed paleomarsh elevation (PME) in SWLI units and elevation relative to local MTL from a core collected at Leeds Point. Paleomarsh elevation was reconstructed using a transfer function applied to foraminifera preserved in the core (filled circles and error bars) and $\delta^{13}C$-based indicative meanings (gray shading). Dashed lines illustrate the error that was trimmed from the final PME reconstruction on the basis of $\delta^{13}C$ values. Modified from Kemp et al. (2012, 2013).

Particulate organic carbon (POC) is a term that describes suspended organic matter with a diameter >0.45 μm and includes phytoplankton, zooplankton, and plant detritus. Marine POC values lie between −21‰ and −16‰ and freshwater POC values, which strongly resemble terrestrial carbon sources, range between −33‰ and −25‰ (Lamb et al., 2006). A gradient in $\delta^{13}C$ values has been observed with increasing salinity along numerous estuaries in northeast Europe (Wilson et al., 2005b; Middelburg and Herman, 2007). Dissolved organic carbon (DOC) describes dissolved organic matter with a diameter of <0.45 μm. In marine settings, DOC consists primarily of soluble products of plankton decomposition. In fluvial settings, DOC has an isotopic signature comparable to phytoplankton and terrestrial humic substances, all of which should reflect the $\delta^{13}C$ of their precursor (i.e., phytoplankton or C_3/C_4 plant).

20.2.1.3 *Microbial carbon*

Organic matter recycling by microbes (mainly bacteria and fungi) is prevalent in intertidal environments due to the high productivity and availability of organic matter (Blum et al., 2004). Depending on their origin, $\delta^{13}C$ of bacteria ranges from −27‰ to −12‰, which tends to lower bulk sediment $\delta^{13}C$ (Coffin et al., 1989). However, estimates of the proportion of bacterial biomass comprising the total organic carbon of sediment are small (<4% in a North Carolina saltmarsh; Rublee, 1982) and should have minimal effect on the resulting sediment $\delta^{13}C$ (Ember et al., 1987). A similar logic may be applied to the incorporation of fungal mycelia into sediments in that the $\delta^{13}C$ of fungi are about 3‰ higher than the vegetation being decayed (Vane et al., 2003), but represent a small proportion of the overall biomass.

20.2.2 C/N values

C/N ratios were the most common chemical method used to characterize the source and degree of biological and diagenetic alteration of organic matter before the development of stable C isotope methods (Tyson, 1995). C/N values are useful in distinguishing aquatic from terrestrial sources because aquatic organic matter has a significantly higher bulk N content than that of terrestrial organic matter (Tyson, 1995).

20.2.2.1 *Terrestrial organic matter*

Terrestrial plants contain a large fraction of refractory structural biopolymers devoid of N, such as hemicellulose, cellulose, and lignin (Vane et al., 2005, 2006, 2013a). These components make up 57–77% of herbaceous and woody plant tissues (Benner et al., 1987). On the other hand, the amount of protein (C/N ~ 3; Goodell, 1972), a principal N-containing compound, is low in terrestrial organic matter (~1–8%; Romankevich, 1990). C/N ratios in woody materials range from ~150 to 500 (Vane, 2003; Vane et al., 2003, 2006), while the C/N ratio of leaves and herbaceous plants is lower at 20–100 (Tyson, 1995; Vane et al., 2013a, b).

Soil organic matter is a complex mixture of plant, animal, or microbial remains and can be categorized into primary compounds inherited from plant and animal residues entering the soil and secondary compounds synthesized from the breakdown of primary structures. Humic substances are an important variety of secondary compounds, which are resistant to further decay and can make up to 7–25% of soil organic matter in terrestrial soils (Given, 1984). Humic substances typically have C/N values that range from 10 to 34 (Tyson, 1995). The resulting C/N values of soil organic matter range from 13 to 33 (Aitkenhead and McDowell, 2000) and can be as high as 70 in peats (Vane et al., 2013b).

20.2.2.2 *Aquatic organic matter*

Phytoplankton predominantly consist of labile protein, carbohydrates, and lipids (Parsons and Strickland, 1961) and contain no cellulose or lignin (Meyers, 1994). Due to their greater protein content (~34%; Romankevich, 1990), phytoplankton have C/N values in the range 6–8, which is 2–20 times lower than terrestrial organic matter (Tyson, 1995). Seagrasses and macroalgae (seaweed) (C/N ~15 to 50) often have elevated C/N ratios relative to phytoplankton because they contain higher levels of structural carbohydrates (lignin and cellulose; 8–10%) and slightly lower amounts of protein (10–15%; Prado and Heck, 2011).

Particulate organic matter (POM) and dissolved organic matter (DOC) have somewhat variable values. However, C/N of marine surface waters typically range from 3 to 9, while in fluvial environments and within estuaries and deltas, POC and DOC C/N values are usually >14 (Tyson, 1995) due to the admix of aquatic biomass and terrestrial runoff to rivers, which primarily consists of leaf debris and soil organic matter (Kendall et al., 2001).

20.2.2.3 Microbial organic matter

Bacteria have low C/N values of 4–6 (Tyson, 1995) due to their high protein content (15%; Prado and Heck, 2011). Fungal C/N values (10–35) are somewhat higher than those reported for bacteria (Cundell et al., 1979; Valiela, 1984; Rayner and Boddy, 1988), likely reflecting the presence of N containing components chitin, chitosan, and proteins in combination with N-deficient polysaccharides (Vane et al., 2003, 2005).

20.2.3 Accumulation and preservation of organic matter

20.2.3.1 Organic matter accumulation in relation to elevation

Coastal environments represent the convergence of terrestrial and marine systems and, accordingly, the $\delta^{13}C$ and C/N values of these environments will depend on the relative proportions of carbon derived from marsh detritus, phytoplankton, seagrasses, algae, and POC transported by rivers and tides (Chmura and Aharon, 1995; Middelburg et al., 1997; Fig. 20.2). A strong environmental gradient exists along the transition from subtidal, intertidal and supratidal zones. The accumulation of autochthonous and allochthonous sources vary along this gradient in relation to duration of tidal flooding (Cahoon and Reed, 1995), and hence elevation with respect to the tidal frame. Mean high water depth has a first-order control on the sediment deposition rate (French, 1993; Allen, 1995) and the amount of allochthonous organic matter deposited on the wetland surface. Just as delivery of tidal-derived POC increases with decreasing elevation, riverine POC accumulation will vary with proximity to a river (Luternauer et al., 1995), with locations closest to river mouths (e.g., estuarine and deltaic wetlands) receiving greater amounts of riverine POC.

20.2.3.2 Post-depositional alteration and decomposition

The decomposition of organic matter is dependent on a number of factors, including the chemical composition and amount of organic matter, nutrient and oxygen availability, temperature, soil moisture, the depth within soil, and the local fauna and microorganisms (Hemminga and Buth, 1991; Sanderman and Amundson, 2003). There are three predominant phases in the degradation of organic matter in tidal wetland sediments (Valiela et al., 1985). First, soluble compounds are leached and a rapid loss of vascular plant detritus occurs within days or months of deposition. Second, microbial degradation of organic matter occurs, principally by bacteria and to a lesser degree by fungi (Benner et al., 1984). During this phase, which may last up to a year, rates of decomposition are variable. Enhanced nutrient quality of detritus can increase the rate of decomposition, but anaerobic conditions promote slower decay rates than aerobic conditions. In the third phase, which may last an additional year, only relatively refractory material remains and decomposition rates are very slow (Fig. 20.2).

The effect of decomposition processes on $\delta^{13}C$ and C/N varies among depositional environments. In the upper intertidal/supratidal zone, there is a net decrease in C/N values due to C loss from microbial respiration, leaching of soluble compounds and decay products, and particulate transport and a potential increase in N due to fixation of external nitrogen sources (Valiela et al., 1985). In this zone, bulk $\delta^{13}C$ values tend to decrease. Plant litter has isotopically distinct biochemical fractions (cellulose, hemicellulose, and lignin) that decompose at different rates (Vane et al., 2003). Compared to bulk plant material, cellulose and hemicellulose are 1–2‰ higher in $\delta^{13}C$ and lignin is 2–6‰ lower in $\delta^{13}C$ (Benner et al., 1987). Because lignin decomposes more slowly than the cellulose fractions, the $\delta^{13}C$ of litter may decrease during decomposition to values closer to that of plant lignin (Fig. 20.2). In the lower intertidal to subtidal zone, organic matter is mostly derived from POC of marine or riverine provenance. Terrestrial material should already have been extensively degraded upstream and is relatively resistant to further degradation. Phytoplankton is nitrogen-rich and therefore an attractive food source for consumers. Considerable degradation in the water column occurs and only a small proportion of phytoplankton is incorporated into the sediment. When planktonic sources are incorporated, the change in C/N values is different from that of vascular plants. N, instead of C, is lost in greater amounts and the C/N ratio increases in sediments (Valiela et al., 1995). The $\delta^{13}C$ of plankton may increase from end-member to sediment values due to preferential degradation of carbohydrates and proteins (enriched in ^{13}C) relative to lipids in the water column (Lamb et al., 2006).

The effects of decomposition on C/N ratios confound its use in paleoenvironmental interpretations. Due to the contrasting change in direction of values depending on the dominant organic matter type being degraded (e.g., phytoplankton v. vascular plants), surface sediment values have the tendency to converge on uniform values between 10 and 20. The changes in $\delta^{13}C$ are not great enough to bias the distinction between C_3 and C_4 vegetation, but may be problematic when making smaller distinctions between terrestrial and marine sources (Wilson et al., 2005a).

20.3 APPROACHES TO SAMPLE COLLECTION, CORE STORAGE, PREPARATION, AND MASS SPECTROMETRY

20.3.1 Sample collection

Sea level can be reconstructed using $\delta^{13}C$ and C/N values when the indicative meaning of a sample (its vertical relationship to contemporary tide levels) is defined. The floral character of tidal wetlands is controlled by elevation relative to the tidal frame (Allen and Pye, 1992), but it is further influenced by location, local hydrographic conditions, salinity regime, and substrate, which can cause local to regional variability in the $\delta^{13}C$ and C/N of plants and bulk sediment. Unless a dataset that describes the relationship of plant or bulk sediment $\delta^{13}C$ and C/N values relative to tide levels exists for a given location (e.g., Chmura et al., 1987; Törnqvist et al., 2004a), sampling of the contemporary environment is necessary in addition to collection of core material at the site for paleoenvironmental and sea-level reconstruction.

To determine the indicative meaning of $\delta^{13}C$ and C/N values, surface sediment and vegetation samples from each depositional environment along an elevational gradient should be collected and analyzed. Surface sediments (top 0–1 cm in depth) are often collected along shore-normal transects (Fig. 20.4; Kemp et al., 2010), and the elevation of each sampling point should be leveled to a common datum. Dominant vegetation at each sampling point should be noted, as well as the visibility of imported macro-detritus (e.g., algae or seagrass) on the sediment surface. Some studies have incorporated mixing models (e.g., Chmura et al., 1987) or other types of mass balance approach (e.g., Middelburg et al., 1997) to better understand surface sediment $\delta^{13}C$ and C/N. In these cases, recording the percent cover of plant taxa more rigorously using a quadrat (e.g., Malamud-Roam and Ingram, 2001) or taking measurements of sediment accumulating on the surface using sediment traps (e.g., Wheelock, 1997), is useful.

To understand sources and processes contributing to bulk sediment $\delta^{13}C$ and C/N values, it is recommended to collect potential end-members that contribute to surface sediments: above- and below-ground components of all dominant wetland vegetation; suspended sediment samples; and algae or seagrass if present (Vane et al., 2013a). Consideration should be given to the plant components or organs collected for analysis (Fig. 20.4). Because variations occur in $\delta^{13}C$ and C/N among plant components, above- and below-ground components will often be analyzed separately. The time of year when the vegetation is collected, as well as whether newly produced or year-old dead foliage is collected, is also an important consideration. Variations in $\delta^{13}C$ of 1–4‰ and changes of up to ~40 in C/N values may occur due to these factors (Cloern et al., 2002).

Plant and sediment material should be sampled in sufficient quantity to obtain reliable instrumental measurements. To avoid the loss of organic matter and prevent post-collection alteration of $\delta^{13}C$ and C/N values, all samples (plant and sediment) should be kept in a cool box in the field and subsequently refrigerated until samples can be processed (Fig. 20.4).

20.3.2 Sample storage

Plant samples should be dried immediately if possible or frozen until drying can occur (Ehleringer, 1991). Samples may be freeze-dried or dried in an oven at a moderate temperature (40–50 °C). Once dried, the plant sample can be stored without alteration to isotopic composition, although very long storage and slow drying of the living material should be avoided (Ehleringer, 1991).

Sediment samples and cores should be stored at low temperatures (~4 °C) in darkness to limit microbial activity and negate photo-oxidation of the organic components. Freezing or freeze-drying of sediments can also halt organic matter alteration, although this is somewhat impractical for large volumes of sediment; further, if sediments are also to be analyzed for microfossils, these

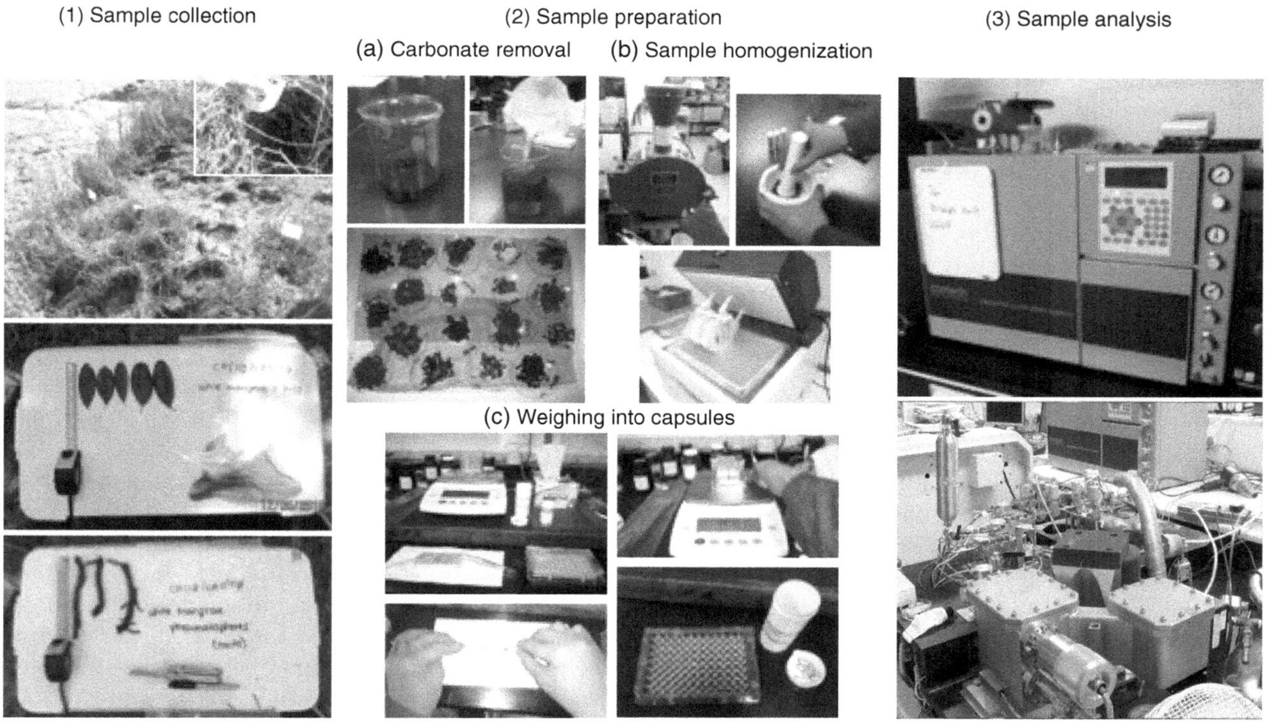

Fig. 20.4. Collection, processing, and instrumental analysis of sediment and vegetation samples for $\delta^{13}C$ and C/N analysis. (1) Top: Collection of surface sediment samples along a transect extending through the transition from tidal flat to low marsh environment from Dartford Creek, southeast England. Center, bottom: Vegetation collected from *Laguncularia racemosa* in southeast Puerto Rico. Plant components (leaves and pneumatophores) were sampled and stored separately in plastic bags on ice in a cooler in the field. (2) Example of methods for processing samples for elemental and isotopic analysis, which include carbonate removal, sample homogenization, and weighing samples in aluminum or silver capsules and crimping them before instrumental analysis. (a) Top: The "rinse" method for removing carbonates (Brodie et al., 2011a, b) where sediment samples are treated with 5% HCl overnight and are rinsed 3 times with at least 1500 mL of deionized water. Bottom: Plant samples following acid pretreatment, laid out to be put in the oven to dry before grinding occurs. (b) Various methods of homogenizing dried sediment and plant samples, including a Wiley mill (top left), pestle and mortar (top right), and freezer mill (bottom). (c) An "in-house" standard being weighed out into a tin capsule using a microbalance. Top left: Equipment and tools necessary for weighing samples into capsules. Top right, bottom left: Forceps used to transfer capsule to microbalance and fold and crimp tin containing weighed sample to remove all air. Bottom right: Plastic tray used to organize and store samples before elemental/isotopic analysis. (3) Instruments used to perform $\delta^{13}C$ and C/N analysis. Top: Costech Elemental Analyzer used for C/N measurement. Bottom: Optima dual-inlet mass spectrometer used for $\delta^{13}C$ analysis.

processes may be destructive. Comparison of $\delta^{13}C$ profiles from Holocene sediments analyzed c. 9 years apart showed that storage at ambient temperatures shifted the $\delta^{13}C$ by up to +0.9‰ (Lamb et al., 2007). In addition, cores should be sealed in an impervious material, such as polyurethane or polycarbonate (plastic) lining, to avoid dehydration and air contamination.

20.3.3 Sample preparation

Sample preparation introduces the most error into elemental and isotopic measurements, often non-systematic and an order of magnitude greater than instrumentation (Boutton, 1991; Brodie et al., 2011a, b). At all steps of preparation, care must be taken when handling samples in the laboratory. Nitrile or Latex gloves should been worn when handling samples to avoid contamination by skin oils (Boutton, 1991). All instruments (e.g., forceps or spatulas) or glassware used to handle samples should be thoroughly cleaned and dried before contact with samples. Best practices include rinsing instruments with weak acid, followed by washing with laboratory cleaning solution, and a final rinse with deionized water before firing in a combustion oven (appropriate for Pyrex or glassware) or rinsing with methanol (suitable for stainless steel instruments). At the very minimum, instruments/glassware should be washed in a laboratory cleaning solution, rinsed with deionized water, and dried before contact with sample.

When subsampling plants and sediment for sample pretreatment, it is necessary to allocate enough material to obtain accurate instrument measurements; conservatively, this is ~500 µg C (Brodie et al., 2011b). In general, it is assumed that carbon comprises ~40% of plant matter, so at least 1.25 mg of plant sample is needed for pretreatment. With sediments, the amount of sample required is dependent on the amount of organic matter in the sample. The color of a sediment sample can give a rough indication of the organic content of a sample, with dark brown samples containing relatively high amounts of organic matter and light gray samples containing little (Steinhardt and Franzmeier, 1979).

Inorganic carbon has an isotopic composition that varies greatly from that of organic carbon sources, with values in general close to 0 ± 4‰ (Boutton, 1991). Its removal from samples is necessary before isotopic analysis of organic matter. Recent investigations by Brodie et al. (2011a, b) indicate that the method used to remove carbonate material from samples may bias resulting $\delta^{13}C$ and C/N values; this method should therefore be kept consistent on all samples within a study, and caution should be taken when comparing values from multiple datasets using varying preparation methods.

A final consideration in minimizing error during sample pretreatment is ensuring sample homogeneity. Because variation in $\delta^{13}C$ and C/N can occur within bulk plant and sediment material, it is important for samples to be ground to a fine powder, ideally able to pass through at least a 40-mesh screen (Ehleringer, 1991). Finely ground material also burns more uniformly during combustion, which helps to reduce "carry-over" effects that lead to inaccuracy in $\delta^{13}C$ and C/N values between measurements. Before grinding sediment samples, large plant fragments should be removed. In general, sediment samples with a high mineral content are easily ground to a fine-consistency powder using a pestle and mortar, although plant samples and sediment samples containing a greater degree of plant fragments should be processed in a laboratory mill (Wiley, ball, freezer, etc.) which is thoroughly cleaned between samples (Fig. 20.4).

20.3.4 Instrumentation

Most commercially available isotope ratio mass spectrometers (IRMS) are fully automated and permit quick $\delta^{13}C$ analysis. Most systems possess six basic components: a sample introduction system; an electron ionization source; a magnetic sector analyzer; a pumping system that can maintain vacuum pressure during the gas analysis; a Faraday-collector detector array; and a computer-controlled data acquisition system. An elemental analyzer, used to measure the amount of C and N in a sample, is commonly coupled with a mass spectrometer (EA-IRMS) to introduce samples into the system.

First, dried homogenized samples are weighed using a microbalance, placed in tin or silver capsules, and folded and crimped to remove all air. Sample capsules are lowered into a combustion furnace, often through an auto-sampler carousel. The sample is combusted into CO_2 under flow of oxygen at elevated temperatures. A helium gas stream carries the combusted sample into a reduction chamber where nitrous oxides are converted into N_2 and excess O_2 is removed. The sample is next carried through a chemical trap to remove water that was produced during combustion, and then into the gas chromatograph where CO_2 is separated from N_2 and both are measured by a thermal conductivity detector, which sequentially generates a signal of each element that is proportional to the amount in each sample. The sample is then carried from the elemental analyzer through an interface to the IRMS system for $^{13}C/^{12}C$ analysis alongside a reference gas standard (or working gas).

Following transfer of CO_2 to the mass spectrometer, molecules are ionized by interaction with an electron beam. The ions are then accelerated through a magnetic field at high voltage. The strength of the magnetic field and voltage applied will control the trajectory of ions to enter the Faraday cups. For measurement of carbon isotope ratio, the cups will be spaced to measure ions with mass to charge ratio of 44, 45, and 46, which corresponds to CO_2 molecules with various combinations of ^{12}C, ^{13}C, ^{16}O, ^{17}O, and ^{18}O. Sharpe (2007) provides a good review of computations involved in deriving $\delta^{13}C$ values from instrumental readings.

Isotope ratios are measured relative to international standards (primary materials; Vienna PDB or VPDB for carbon isotopes); it is therefore necessary to use well-characterized standards (both working gas and/or solid material) whose isotope ratios have been determined against the primary calibration materials. Primary and secondary reference materials (compounds calibrated to primary materials within small uncertainties) are of limited commercial availability, and "in-house" standards are often produced at many isotope facilities. Analysis of blanks and solid material standards placed throughout a run of samples of unknown $\delta^{13}C$ and C/N values can be used to provide indication of instrumental error. Blanks (empty capsules folded as if containing a sample) can show signs of carry-over effects from incomplete combustion of samples or can be used to correct for the contribution of atmospheric gases introduced by an autosampler. In-house standards are used as a quality control measure and can help to estimate the measurement error of unknown samples.

20.4 APPLICATION OF $\delta^{13}C$ AND C/N IN RELATIVE SEA-LEVEL RECONSTRUCTION

The following examples illustrate the use of stable carbon isotope geochemistry in five contrasting environments.

20.4.1 Comparison of $\delta^{13}C$ and C/N from modern and Holocene intertidal sediments to produce sea-level index points, Mersey Estuary, UK

Wilson et al. (2005a, b) examined relations between plant and bulk sediment $\delta^{13}C$ and C/N values from samples collected along a transect extending through intertidal environments in the Mersey Estuary, UK. They found the C_3 marsh vegetation and bulk sediment C/N had little to no relationship to elevation with respect to the tidal frame. However, bulk sediment $\delta^{13}C$ values did show a relationship with elevation, which was due to increased delivery of estuarine POC with decreasing marsh elevation (Wilson et al., 2005a, b). Wilson et al. (2005a) compared the modern bulk sediment $\delta^{13}C$ and C/N to values from Holocene subtidal (channel), intertidal (reed swamp/saltmarsh, saltmarsh, tidal flat) and supratidal (oak-hazel woodland, alder carr, poor fen) depositional environments, which were independently characterized based on diatom, pollen, and grain-size analyses. They found Holocene saltmarsh $\delta^{13}C$ was 3.7‰ lower than modern deposits, and early–middle Holocene saltmarsh sediment C/N (13.3) was higher than modern saltmarsh sediment C/N (10.6). The results show a pattern of decreasing $\delta^{13}C$ and increasing C/N in saltmarsh/tidal flat sediment with age that is consistent with the concentration of refractory organic compounds due to below-ground decomposition of organic matter. Decomposition such as this could complicate proper identification of paleoenvironments in the absence of other indicators (Wilson et al., 2005a). In these instances, application of molecular-level techniques such as analytical pyrolysis-GC/MS or chemolysis-GC/MS could be used to fully understand the extent of decay in selected samples and thus refine the $\delta^{13}C$ and C/N-based interpretations (Vane et al., 2006; Lamb et al., 2007).

20.4.2 The application of plant and sediment $\delta^{13}C$ and C/N to reconstruct Late Holocene relative sea levels from US Atlantic coast saltmarshes

Kemp et al. (2010, 2012) defined the indicative meaning of $\delta^{13}C$ and C/N values they observed at sites along the North American Atlantic coast. In the Outer Banks of North Carolina, low marsh environments are dominated by C_4 *Spartina alterniflora*, while the high marsh is dominated by C_3 *Juncus roemerianus*. Kemp et al. (2010) found that $\delta^{13}C$ values were able to distinguish *Spartina*-dominated low marsh from high marsh *Juncus* zones, but the presence of C_4 vegetation (*Distichlis spicata*, *Spartina cynusoroides*, *Spartina patens)* in the high marsh complicated the use of $\delta^{13}C$, and consequently limited the precision of the technique. Furthermore, *Juncus*-dominated high marsh environments were indistinguishable from freshwater environments on the basis of $\delta^{13}C$ alone, which is problematic for paleoenvironmental reconstructions. As Wilson et al. (2005a, b) observed, sediment C/N values converged to a uniform value across environments and were of little use in discriminating between floral zones.

In contrast to the results from North Carolina, Kemp et al. (2012) were able to define indicative meanings for sediment $\delta^{13}C$ values in combination with the absence/presence of agglutinated foraminifera along the southern Atlantic coast of New Jersey (Fig. 20.3). While $\delta^{13}C$ values were able to distinguish between elevation-dependent environmental zones, C/N values were uniform among these environments and were of little use in distinguishing between environmental zones. Unlike North Carolina, the high marsh zone of New Jersey is dominated by C_4 *Spartina patens* and short-form *Spartina alterniflora*, with minimal occurrence of C_3 species. Indicative meaning was defined for four paleomarsh environments with a vertical resolution of approximately ±0.2–0.5 m or ~30–90% of tidal range. Kemp et al. (2013) further refined the $\delta^{13}C$/foraminifera-based indicative meaning using a multi-proxy approach that combined a foraminiferal transfer function and the bulk sediment $\delta^{13}C$ values to reconstruct paleo-marsh elevation with an average vertical uncertainty of ±0.12 m, which reduced the $\delta^{13}C$ (average ±0.3 m) and foraminiferal transfer function-based (average ±0.18 m) uncertainties by ~60% and ~30%, respectively (Fig. 20.3).

20.4.3 Holocene sea-level reconstruction using macrofossil and sediment $\delta^{13}C$ of basal peats in the Mississippi River Delta, USA

Törnqvist et al. (2004a) used the $\delta^{13}C$ of radiocarbon-dated samples from a series of basal peats to establish sea-level index points in the Lutcher–Gramercy area on the eastern margin of the Mississippi River Delta. Average surface sediment $\delta^{13}C$ values of saline (mean −16.2‰, range −18.0‰ to −14.3‰), brackish (mean −16.9‰, range −20.1‰ to −15.1‰), intermediate (mean −22.1‰, range −25.1‰ to −18.2‰), and freshwater (mean −27.8‰, range −29.4‰ to −25.4‰) environments on the Mississippi River Delta, established by Chmura et al. (1987), were used to compare plant macrofossils (mostly charcoal, determined to be primarily C_4 *Spartina* spp.) found in basal peat from the cores. Törnqvist et al. (2004a) used these $\delta^{13}C$ values to provide evidence that the basal peats accumulated under saline to intermediate conditions, which would suggest that the peat formed between mean sea level and mean spring high water. Based on the microtidal regime (tidal range <0.6 m), sea-level index points were assigned an indicative range of 0.3 m with additional error terms added for leveling and depth measurements within cores for the final precision of ±0.29–0.38 m (range in vertical error of index points).

20.4.4 The application of sediment $\delta^{13}C$ and C/N in the construction of sea-level index points from isolation basins in northwest Scotland, UK

Mackie et al. (2005, 2007) compared $\delta^{13}C$ and C/N-based paleo-salinity estimates to previous foraminiferal and diatom analyses performed on cores collected from isolation basins in northwest Scotland. Over Holocene timescales, the $\delta^{13}C$ and C/N data was in general agreement with the microfossil-based paleo-salinities that previous studies had used to construct sea-level index points (Lloyd, 2000; Shennan et al., 2000). Mackie et al. (2005, 2007) indicated that the $\delta^{13}C$ and C/N-based approach can be used to identify relative sea-level change in northwest Scotland from isolation basins, but in a manner inconsistent with a simple model of paleo-salinity change where $\delta^{13}C$ values would merely increase and C/N values decrease with increasing salinity along the isolation boundary. Instead, changes in water temperature and organic nutrient supply accompanying the final isolation event in turn affect aquatic productivity and stratification within the basins to cause deviations in $\delta^{13}C$ and C/N values along the contact. Together with biological (diatom, foraminifera) proxies, $\delta^{13}C$ and C/N values were able to provide a clear depiction of basin dynamics. In contrast, in late glacial and older-aged sediments, environmental variables independent of the isolation process such as atmospheric CO_2 concentration, poor vegetation development, and temperature had a greater influence on $\delta^{13}C$ and C/N values.

20.4.5 The use of $\delta^{13}C$ and C/N to infer rapid seismic vertical motion along subduction zones

Along tectonically active coastlines, archives of tidal wetland sediments record accounts of rapid vertical displacements of the wetland surface during earthquake events. To overcome issues with microfossil preservation in tropical settings, Dura et al. (2011) used $\delta^{13}C$ and C/N to identify two coseismic subsidence events from western Sumatra along the Mentawai segment of the Sunda subduction zone. Along the Cascadia subduction zone of the North American Pacific coast, Hawkes et al. (2011) used $\delta^{13}C$ values to confirm that the absence of foraminifera in the organic-rich peat unit found below the tsunami sand from the Great Cascadia earthquake of 1700 AD in South Slough, Oregon resulted from formation within a freshwater environment above the tidal limit rather than from poor preservation of foraminiferal tests. Also in Cascadia, Engelhart et al. (2013a, b) compared $\delta^{13}C$ and C/N and foraminiferal reconstructions from a core containing the stratigraphic contact marking the 1700 AD earthquake. The study indicates that $\delta^{13}C$ values may be able to accurately quantify amounts of relative sea-level change associated with coseismic subsidence, although it is somewhat less precise than the foraminiferal-based methods.

20.5 SUMMARY

The stable carbon isotope geochemistry of sedimentary organic matter can be used to reconstruct relative sea level and paleoenvironmental change and has been shown to be most powerful when used in combination with simple metrics of microfossil assemblages (e.g., absence/presence of saltmarsh foraminifera) in a multi-proxy approach (Kemp et al., 2012, 2013). The success of the technique is likely to vary regionally (e.g., Kemp et al. 2010, 2012), where the character of floral zones or delivery of allochtonous organic matter sources is conducive to the $\delta^{13}C$ and C/N-based approach.

The strengths of the $\delta^{13}C$ and C/N-based approach are that analysis is quick and relatively cost-effective. When submitting samples to accelerator mass spectrometer (AMS) labs for ^{14}C analysis, clients are often given the option to complete $\delta^{13}C$ analysis at little to no additional cost, making it easy to obtain these data. Importantly, organic matter is continuously preserved in coastal sedimentary sequences. $\delta^{13}C$ and C/N values can provide paleoenvironmental or relative sea-level information when microfossils are absent or poorly preserved.

The weakness of the approach lies in the fact that $\delta^{13}C$ and C/N values are not solely a reflection of *in situ* vegetation; changes in bulk sediment $\delta^{13}C$ and C/N can be caused by changing sedimentation rate or source, rather than directly to factors related to sea level (e.g., inundation frequency and thus wetland elevation). In addition, organic matter degradation may alter modern $\delta^{13}C$ and C/N values of plants and surface sediments and complicate interpretations of $\delta^{13}C$ and C/N values from Holocene sediment cores. In particular, C/N values are sensitive to decay processes and, in modern intertidal settings, tend to converge to values between 10 and 20, making a distinction between environmental zones difficult on that basis alone. To date, the accuracy of $\delta^{13}C$-based paleo-marsh elevation reconstructions has been shown to be comparable to that of foraminiferal-based reconstructions on late Holocene timescales, although slightly less precise (Kemp et al., 2010, 2012; Engelhart et al., 2013a). The accuracy of $\delta^{13}C$ values may be compromised when applying them to early Holocene samples due to post-depositional changes of up to ~4‰ from modern values (Wilson et al., 2005b). Trade-offs between the accuracy and precision of reconstructions and time of analysis are an important consideration to this approach.

To continue to develop this technique, a better understanding of the pathways of carbon from plants/living organisms to buried sediments and the effect of these transformations on $\delta^{13}C$ and C/N values is needed. Advancement in other bulk organic geochemical analyses such as Rock-Eval pyrolysis may facilitate the distinction between labile and residual organic matter (Carrie et al., 2012; Leng et al., 2013). In addition, molecular-level analytical techniques may provide insight into organic matter transformations or enable use of organic compounds minimally affected by decomposition processes (e.g., lignin, tannins, suberins; Vane et al., 2003, 2006, 2013a; Beramendi-Orasco et al. 2006).

REFERENCES

Aitkenhead, J.A., and McDowell, W.H. (2000) Soil C: N ratio as a predictor of annual riverine DOC flux at local and global scales. Global Biogeochemical Cycles, 14, 127–138.

Allen, J.R.L. (1995) Salt-marsh growth and fluctuating sea level: Implications of a simulation model for Flandrian coastal stratigraphy and peat-based sea-level curves. Sedimentary Geology, 100, 21–45.

Allen, J.R.L., and Pye, K. (1992) *Saltmarshes. Morphodynamics, Conservation and Engineering Significance*. Cambridge, Cambridge University Press.

Anderson, T.F., and Arthur, M.A. (1983) Stable isotopes of oxygen and carbon and their application to sedimentologic and paleoenvironmental problems. In: *Stable Isotopes in Sedimentary Geology* (eds Arthur, M.A., and Anderson, T.F.), Society of Paleontologists and Mineralogists, Tulsa, Oklahoma, 151 p.

Andrews, J.E., Samways, G., Dennis, P.F., and Maher, B.A. (2000) Origin, abundance and storage of organic carbon and sulphur in the Holocene Humber estuary: emphasising human impact on storage changes. In: *Holocene Land–Ocean Interaction and Environmental Change around the North Sea* (eds Shennan, I., and Andrews, J.E.), Geological Society of London, Special Publication, 166, 145–170.

Benner, R., Newell, S.Y., Maccubbin, A.E., and Hodson, R.E. (1984) Relative contributions of bacteria and fungi to rates of degradation of Lignocellulosic detritus in salt-marsh sediments. Applied and Environmental Microbiology, 48, 36–40.

Benner, R., Fogel, M.L., Sprague, E.K., and Hodson, R.E. (1987) Depletion of ^{13}C in lignin and its implications for stable carbon isotope studies. Nature, 329, 708–710.

Benner, R., Fogel, M.L., Sprague, E.K., and Hodson, R.E. (1991) Diagenesis of belowground biomass of Spartina alterniflora in salt-marsh sediments. Limnological Oceanography, 36, 1358–1374.

Beramendi-Orosco, L.E., Vane, C.H., Cooper, M., Sun, C.G., Large, D.J., and Snape, C.E. (2006) Evaluation of errors associated with δ^{13}C analysis of lignin-derived TMAH thermochemolysis products by gas chromatography-combustion-isotope ratio mass spectrometry. Journal of Analytical and Applied Pyrolysis, 76, 88–95.

Blum, L.K., Roberts, M.S., Garland, J.L., and Mills, A.L. (2004) Distribution of microbial communities associated with the dominant high marsh plants and sediments of the United States east coast. Microbial Ecology, 48, 375–388.

Bouillon, S., Dahdouh-Guebas, F., Rao, A.V.V.S., Koedamn, N., and Dehairs, F. (2003) Sources of organic carbon in mangrove sediments: variability and possible ecological implications. Hydrobiologia, 495, 33–39.

Boutton, T.W. (1991) Stable carbon isotope ratios of natural materials: I. Sample preparation and mass spectrometric analysis. In: *Carbon Isotope Techniques* (eds Coleman, D.C., and Fry, B.), San Diego, Academic Press, Inc., 155–170.

Brodie, C.R., Casford, J.S.L., Lloyd, J.M., Leng, M.J., Heaton, T.H.E., Kendrick, C.P., and Zong, Y.Q. (2011a) Evidence for bias in C/N, δ^{13}C and δ^{15}N values of bulk organic matter, and on environmental interpretation, from a lake sedimentary sequence by pre-analysis acid treatment methods. Quaternary Science Reviews, 30, 3076–3087.

Brodie, C.R., Leng, M.J., Casford, J.S.L., Kendrick, C.P., Lloyd, J.M., Zong, Y.Q., and Bird, M.I. (2011b) Evidence for bias in C and N concentrations and δ^{13}C composition of terrestrial and aquatic organic materials due to pre-analysis acid preparation methods. Chemical Geology, 282, 67–83.

Byrne, R., Ingram, B. L., Starratt, S., Malamud-Roam, F., Collins, J. N., and Conrad, M. E. (2001) Carbon-isotope, diatom, and pollen evidence for late Holocene salinity change in a brackish marsh in the San Francisco Estuary. Quaternary Research, 55, 66–76.

Cahoon, D.R., and Reed, D.J. (1995) Relationships among marsh surface-topography, hydroperiod, and soil accretion in a deteriorating Louisiana salt-marsh. Journal of Coastal Research, 11, 357–369.

Calvin, M., and Benson, A.A. (1949) The path of carbon in photosynthesis: The identity and sequence of the intermediates in sucrose synthesis. Science, 109, 140–142.

Carrie, J., Sanei, H., and Stern, G. (2012) Standardisation of Rock-Eval pyrolysis for the analysis of recent sediments and soils. Organic Geochemistry, 46, 38–53.

Castañeda, I.S., Werne, J.P., Johnson, T.C., and Powers, L.A. (2011) Organic geochemical records from Lake Malawi (East Africa) of the last 700 years, part II: Biomarker evidence for recent changes in primary productivity. Palaeogeography, Palaeoclimatology, Palaeoecology, 303, 140–154.

Chmura, G.L., and Aharon, P. (1995) Stable carbon isotope signatures of sedimentary carbon in coastal wetlands as indicators of salinity regime. Journal of Coastal Research, 11, 124–135.

Chmura, G.L., Aharon, P., Socki, R.A., and Abernethy, R. (1987) An inventory of ^{13}C abundances in coastal wetlands of Louisiana, UA - Vegetation and sediments. Oecologia, 74, 264–271.

Cloern, J.E., Canuel, E.A., and Harris, D. (2002) Stable carbon and nitrogen isotope composition of aquatic and terrestrial plants of the San Francisco Bay estuarine system. Limnology and Oceanography, 47, 713–729.

Coffin, R.B., Fry, B., Peterson, B.J., and Wright, R.T. (1989) Carbon isotopic compositions of estuarine bacteria. Limnology and Oceanography, 34, 1305–1310.

Cundell, A.M., Brown, M.S., Stanford, R., and Mitchell, R. (1979) Microbial degradation of Rhizophora mangle leaves immersed in the sea. Estuarine Coastal Marine Science, 9, 281–286.

Cuntz, M. (2011) Carbon cycle: A dent in carbon's gold standard. Nature, 477, 547–548.

Dawson, T.E., Mambelli, S., Plamboeck, A.H., Templer, P.H., and Tu, K.P. (2002) Stable isotopes in plant ecology. Annual Review of Ecology and Systematics, 33, 507–559.

Degens, E.T., Guillard, R.R., Sackett, W.M., and Hellebus, J. (1968) Metabolic fractionation of carbon isotopes in marine plankton. I. Temperature and respiration experiments. Deep-Sea Research, 15, 1–9.

Deines, P. (1980) The isotopic composition of reduced organic carbon. In: *Handbook of Environmental Isotope Geochemistry* (eds Fritz, P., and Fontes, J.C.), Elsevier, Amsterdam, The Terrestrial Environment, Volume 1, 329–406.

Dura, T., Rubin, C.M., Kelsey, H.M., Horton, B.P., Hawkes, A., Vane, C.H., Daryono, M., Pre, C.G., Ladinsky, T., and Bradley, S. (2011) Stratigraphic record of Holocene coseismic subsidence, Padang, West Sumatra. Journal of Geophysical Research: Solid Earth, 116, B11.

Ehleringer, J.R. (1991) ^{13}C/^{12}C fractionation and its utility in terrestrial plant studies. In: *Carbon Isotope Techniques* (eds Coleman, D.C., and Fry, B.), San Deigo, Academic Press, Inc., 187–200.

Ember, L.M., Williams, D.F., and Morris, J.T. (1987) Processes that influence carbon isotope variations in salt-marsh sediments. Marine Ecology Progress Series, 36, 33–42.

Engelhart, S.E., Horton, B.P., Vane, C.H., Nelson, A.R., Witter, R.C., Brody, S.R., and Hawkes, A.D. (2013a) Modern foraminifera, δ^{13}C, and bulk geochemistry of central Oregon tidal marshes and their application in paleoseismology. Palaeogeography, Palaeoclimatology, Palaeoecology, 377, 13–27.

Engelhart, S.E., Horton, B.P., Nelson, A.R., Hawkes, A.D., Witter, R.C., Wang, K., Wang, P.-L. and Vane, C.H. (2013b) Testing the use of microfossils to reconstruct great earthquakes at Cascadia: Geology, doi: 10.1130/G34544.1.

Farquhar, G.D., O'Leary, M.H., and Berry, J.A. (1982) On the relationship between carbon isotope discrimination and

the intercellular carbon dioxide concentration in leaves. Australian Journal of Plant Physiology, 9, 121–137.
Fogel, M.L., Cifuentes, L.A., Velinsky, D.J., and Sharp, J.H. (1992) Relationship of carbon availability in estuarine phytoplankton to isotopic composition. Marine Ecology Progress Series, 82, 291–300.
Francey, R.J., Allison, C.E., Etheridge, D.M., Trudinger, C.M., Enting, I.G., Leuenberger, M., Langenfelds, R.L., Michel, E., and Steele, L.P. (1999) A 1000-year high-precision record of $\delta^{13}C$ in atmospheric CO_2. Tellus B, 51, 170–193.
French, J.R. (1993) Numerical-simulation of vertical marsh growth and adjustment to accelerated sea-level rise, North Norfolk, UK. Earth Surface Processes and Landforms, 18, 63–81.
Fry, B. (2008) *Stable Isotope Ecology*. Springer, New York.
Fry, B., Scalan, R. S., and Parker, L. (1977) Stable carbon isotope evidence for two sources of organic matter in coastal sediments: seagrasses and plankton. Geochimica et Cosmochimica Acta, 41, 1875–1877.
Gebrehiwet, T., Koretsky, C.M., and Krishnamurthy, R.V. (2008) Influence of Spartina and Juncus on saltmarsh sediments. III. Organic geochemistry. Chemical Geology, 255, 114–119.
Given, P.H. (1984) An essay on the organic geochemistry of coal. Coal Science, 3, 63–252.
Goodell, H.G. (1972) Carbon/nitrogen ratio. In: *The Encyclopedia of Geochemistry and Environmental Sciences* (ed. Fairbridge, R.W.), Van Nostrand Reinhold, New York, 136–142.
Guy, R.D., and Reid, D.M. (1986) Photosynthesis and the influence of CO_2 enrichment on $\delta^{13}C$ values in a C3 halophyte. Plant Cell and Environment, 9, 65–72.
Guy, R.D., Reid, D.M., and Krouse, H.R. (1980) Shifts in carbon isotope ratios of two C3 halophytes under natural and artificial conditions. Oecologia, 44, 241–247.
Hawkes, A.D., Horton, B.P., Nelson, A.R., Vane, C.H., and Sawai, Y. (2011) Coastal subsidence in Oregon, USA, during the giant Cascadia earthquake of AD 1700. Quaternary Science Reviews, 30, 364–376.
Hemminga, M.A., and Buth, G.J.C. (1991) Decomposition in salt-marsh ecosystems of the SW Netherlands - the effects of biotic and abiotic factors. Vegetatio, 92, 73–83.
Hemminga, M.A., and Mateo, M.A. (1996) Stable carbon isotopes in seagrasses: Variability in ratios and use in ecological studies. Marine Ecology Progress Series, 140, 285–298.
Ishihi, Y., Yamada, Y., Ajisaka, T., and Yokoyama, H. (2001) Distribution of stable carbon isotope ratio in Sargassum plants. Fisheries Science, 67, 367–369.
Jiang, S., Liu, X., Sun, J., Yuan, L., Sun, L., and Wang, Y. (2011) A multi-proxy sediment record of late Holocene and recent climate change from a lake near Ny-Ålesund, Svalbard. Boreas, 40, 468–480.
Keeling, C.D. (1979) The Suess effect: ^{13}carbon-^{14}carbon interrelations. Environment International, 2, 229–300.
Kemp, A.C., Vane, C.H., Horton, B.P., and Culver, S.J. (2010) Stable carbon isotopes as potential sea-level indicators in salt marshes, North Carolina, USA. Holocene, 20, 623–636.
Kemp, A.C., Vane, C.H., Horton, B.P., Engelhart, S.E., and Nikitina, D. (2012) Application of stable carbon isotopes for reconstructing salt-marsh floral zones and relative sea level, New Jersey, USA. Journal of Quaternary Science, 27, 404–414.
Kemp, A.C., Horton, B.P., Vane, C.H., Bernhardt, C.E., Corbett, D.R., Engelhart, S.E., Anisfeld, S.C., Parnell, A.C., and Cahill, N. (2013) Sea-level change during the last 2500 years in New Jersey, USA. Quaternary Science Reviews, 81, 90–104.
Kendall, C., Silva, S.R., and Kelly, V.J. (2001) Carbon and nitrogen isotopic compositions of particulate organic matter in four large river systems across the United States. Hydrological Processes, 15, 1301–1346.
Klap, V.A., Louchouarn, P., Boon, J.J., Hemminga, M.A., and van Soelen, J. (1999) Decomposition dynamics of six salt marsh halophytes as determined by cupric oxide oxidation and direct temperature-resolved mass spectrometry. Limnology and Oceanography, 44(6), 1458–1476.
Kristensen, E., Bouillon, S., Dittmar, T., and Marchand, C. (2008) Organic carbon dynamics in mangrove ecosystems: a review. Aquatic Botany, 89, 201–219.
Lamb, A.L., Wilson, G.P., and Leng, M.J. (2006) A review of coastal palaeoclimate and relative sea-level reconstructions using $\delta^{13}C$ and C/N ratios in organic material. Earth-Science Reviews, 75, 29–57.
Lamb, A.L., Vane, C.H., Wilson, G.P., Rees, J.G., and Moss-Hayes, V.L. (2007) Assessing $\delta^{13}C$ and C/N ratios from organic material in archived cores as Holocene sea level and palaeoenvironmental indicators in the Humber Estuary, UK. Marine Geology, 244, 109–128.
Leuenberger, M., Siegenthaler, U., and Langway, C. (1992) Carbon isotope composition of atmospheric CO_2 during the Last Ice Age from an Antarctic ice core. Nature, 357, 488–490.
Leng, M.J., Wagner, B., Boehm, A., Panagiotopoulos, K., Vane, C.H., Snelling, A., Haidon, C., Woodley, E., Vogel, H., Zanchetta, G.R., and Baneschi, I. (2013) Late Quaternary climate and carbon cycling change in the Mediterranean from geochemical and isotope data. Quaternary Science Research, 66, 123–136.
Lloyd, J. (2000) Combined foraminiferal and thecamoebian environmental reconstruction from an isolation basin in NW Scotland: Implications for sea-level studies. Journal of Foraminiferal Research, 30, 294–305.
Luternauer, J.L., Atkins, R.J., Moody, A.I., Williams, H.F.L., and Gibson, J.W. (1995) Salt marshes. In: *Geomorphology and Sedimentology of Estuaries* (ed. Perillo, G.M.E.), Elsevier, Amsterdam, Developments in Sedimentology, 307–332.
Mackie, E.A.V., Leng, M.J., Lloyd, J.M., and Arrowsmith, C. (2005) Bulk organic $\delta^{13}C$ and C/N ratios as palaeosalinity indicators within a Scottish isolation basin. Journal of Quaternary Science, 20, 303–312.
Mackie, E.A.V., Lloyd, J.M., Leng, M., Bentley, M.J., and Arrowsmith, C. (2007) Assessment of $\delta^{13}C$ and C/N ratios in bulk organic matter as palaeosalinity indicators in Holocene and Lateglacial isolation basin sediments, northwest Scotland. Journal of Quaternary Science, 22, 579–591.
Malamud-Roam, F., and Ingram, B.L. (2001) Carbon isotopic compositions of plants and sediments of tide marshes in the San Francisco Estuary. Journal of Coastal Research, 17, 17–29.
Malamud-Roam, F., and Ingram, B.L. (2004) Late Holocene $\delta^{13}C$ and pollen records of paleosalinity from tidal

marshes in the San Francisco Bay estuary, California. Quaternary Research, 62, 134–145.

McCarroll, D., and Loader, N.J. (2006) Isotopes in tree rings. In *Isotopes in Palaeoenvironmental Research* (ed. Leng, M.J.), Springer, Netherlands.

Meyers, P.A. (1994) Preservation of elemental and isotopic source identification of sedimentary organic-matter. Chemical Geology, 114, 289–302.

Middelburg, J.J., and Herman, P.M.J. (2007) Organic matter processing in tidal estuaries. Marine Chemistry, 106, 127–147.

Middelburg, J.J., Nieuwenhuize, J., Lubberts, R.K., and van de Plassche, O. (1997) Organic carbon isotope systematics of coastal marshes. Estuarine Coastal and Shelf Science, 45, 681–687.

O'Leary, M.H. (1981) Carbon isotope fractionation in plants. Phytochemistry, 20, 553–567.

O'Leary, M.H. (1988) Carbon isotopes in photosynthesis. Bioscience, 38, 328–336.

Osmond, C.B., Bender, M.M., and Burris, R.H. (1976) Pathways of CO2 fixation in the CAM plant Kalanchoe daigremontiana: III. Correlation with $\delta^{13}C$ value during growth and water stress. Australian Journal of Plant Physiology, 3, 787–799.

Osmond, C.B., Winter, K., and Ziegler, H. (1982) Functional significance of different pathways of CO2 fixation in photosynthesis. In: *Encyclopedia of Plant Physiology. Physiological Plant Ecology II. Water Relations and Carbon Assimilation* (eds Lange, O.L., Nobel, P.S., Osmond, C.B., and Ziegler, H.),, Berlin, Springer-Verlag, 479–547.

Park, R., and Epstein, S. (1960) Carbon isotope fractionation during photosynthesis. Geochimica et Cosmochima Acta, 21, 110–126.

Parsons, T.R., and Strickland, J.D.H. (1961) On the production of particulate organic carbon by heterotrophic processes in sea water. Deep-Sea Research, 8, 211–222.

Pilarczyk, J.E., Horton, B.P., Witter, R.C., Vane, C.H., Chagué-Goff, C., and Goff, J. (2012) Sedimentary and foraminiferal evidence of the 2011 Tōhoku-oki tsunami on the Sendai coastal plain, Japan. Sedimentary Geology, 282, 78–89.

Prado, P., and Heck, K.L. (2011) Seagrass selection by omnivorous and herbivorous consumers: determining factors. Marine Ecology Progress Series, 429, 45–55.

Rayner, A.D.M., and Boddy, L. (1988) *Fungal Decomposition of Wood; Its Biology and Ecology*. Chichester, Wiley.

Romankevich, E.A. (1990) Biogeochemical problems of living matter of the present-day biosphere. In: *Facets of Modern Biogeochemistry* (eds Ittekot, V., Kempe, S., Michaelis, W., and Spitzy, A.), Springer-Verlag, Berlin, Festschrift for E.T. Degens, pp. 39–51.

Rublee, P.A. (1982) Seasonal distribution of bacteria in salt-marsh sediments in North Carolina. Estuarine Coastal and Shelf Science, 15, 67–74.

Sanderman, J., and Amundson, R. (2003) Biogeochemistry of decomposition and detrital processing. In: *Treatise on Geochemistry* (eds Holland, H.D., and Turekian, K.K.), Pergamon, Oxford, 249–316.

Schlesinger, W.H. (1997) *Biogeochemistry: An Analysis of Global Change*, 2nd edition. Academic Press, San Diego.

Sharpe, Z. (2007) *Principles of Stable Isotope Geochemistry*. Pearson Prentice Hall, Upper Saddle River, NJ, 360 p.

Shennan, I., Lambeck, K., Horton, B., Innes, J., Lloyd, J.M., McArthur, J., Purcell, T., and Rutherford, M. (2000) Late Devensian and Holocene records of relative sea-level changes in northwest Scotland and their implications for glacio-hydro-isostatic modelling. Quaternary Science Reviews, 19, 1103–1135.

Showers, W. J., and Angle, D. G. (1986) Stable isotopic characterization of organic carbon accumulation on the Amazon continental shelf. Continental Shelf Research, 6, 227–244.

Slack, C.R., and Hatch, M.D. (1967) Comparative studies on the activities of carboxylases and other enzymes in relation to the new pathway of photosynthetic CO2 fixation in tropical grasses. Biochemical Journal, 103, 660–665.

Smith, B.N., and Epstein, S. (1971) Two categories of $^{13}C/^{12}C$ ratios for higher plants. Plant Physiology, 47, 380–384.

Steinhardt, G.C., and Franzmeier, D.P. (1979) Comparison of organic-matter content with soil color for silt loam soils of Indiana. Communications in Soil Science and Plant Analysis, 10, 1271–1277.

Tans, P. (1981) $^{13}C/^{12}C$ of industrial CO_2. In: *Carbon Cycle Modelling* (ed. Bolin, B.), John Wiley & Sons, New York, pp. 127–129.

Thornton, S. F., and McManus, J. (1994) Application of organic carbon and nitrogen stable isotope and C/N ratios as source indicators of organic matter provenance in estuarine systems: evidence from the Tay Estuary, Scotland. Estuarine, Coastal and Shelf Science, 38, 219–233.

Ting, I.P., Bates, L., Sternberg, L.O., and Deniro, M.J. (1985) Physiological and isotopic aspects of photosynthesis in Peperomia. Plant Physiology, 78, 246–249.

Törnqvist, T.E., Gonzalez, J.L., Newsom, L.A., van der Borg, K., de Jong, A.F.M., and Kurnik, C.W. (2004a) Deciphering Holocene sea-level history on the US Gulf Coast: A high-resolution record from the Mississippi Delta. Geological Society of America Bulletin, 116, 1026–1039.

Törnqvist, T.E., Bick, S.J., González, J.L., Van der Borg, K. and De Jong, A.F.M. (2004b) Tracking the sea-level signature of the 8.2 ka cooling event: New constraints from the Mississippi Delta. Geophysical Research Letters, 31, L23309, doi:10.1029/2004GL021429.

Tyson, R.V. (1995) *Sedimentary Organic Matter: Organic Facies and Palynofacies*. Chapman and Hall, London.

Valiela, I. (1995) *Marine Ecological Processes*, 2nd edition. Springer-Verlag, New York.

Valiela, I., Wilson, J., Buchsbaum, R., Rietsma, C., Bryant, D., Foreman, K., and Teal, J. (1984) Importance of chemical-composition of salt-marsh litter on decay-rates and feeding by detritivores. Bulletin of Marine Science, 35, 261–269.

Valiela, I., Teal, J.M., Allen, S.D., Vanetten, R., Goehringer, D., and Volkmann, S. (1985) Decomposition in salt-marsh ecosystems: the phases and major factors affecting disappearance of above-ground organic-matter. Journal of Experimental Marine Biology and Ecology, 89, 29–54.

Vane, C.H. (2003) Monitoring decay of black gum (*Nyssa sylvatica*) wood during growth of the shiitake mushroom (*Lentinula edodes*) using DRIFT spectroscopy. Applied Spectroscopy, 57, 514–517.

Vane, C.H., Drage, T.C., and Snape, C.E. (2003) Biodegradation of oak (*Quercus alba*) wood during growth of the shiitake mushroom (*Lentinula edodes*): A molecular approach. Journal of Agricultural and Food Chemistry, 51, 947–956.

Vane, C.H., Drage, T.C., Snape, C.E., Stephenson, M.H., and Foster C. (2005) Decay of cultivated apricot wood (*Prunus armeniaca*) by the ascomycete *Hypocrea sulphurea*, using solid state ^{13}C NMR and off-line TMAH thermochemolyis with GC-MS. International Biodeterioaration and Biodegradation, 55, 175–185.

Vane, C.H., Drage, T.C., and Snape, C.E. (2006) Bark decay by the white-rot fungus *Lentinula edodes*: Polysaccharide loss, lignin resistance and the unmasking of suberin. International Biodeterioration & Biodegradation, 57, 14–23.

Vane, C.H., Kim, A.W., Moss-Hayes V, Snape C.E, Castro-Diaz, M., Khan, N.S., Engelhart S.E. and Horton, B.P. (2013a) Mangrove tissue decay by arboreal termites (*Nasutitermes acajutlae*) and their role in the mangrove C cycle (Puerto Rico). Chemical characterisation and organic matter provenance using bulk δ^{13}C, C/N, alkaline CuO oxidation-GC/MS and solid-state ^{13}C NMR. Geochemsistry, Geophysics, Geosystems, doi: 10.1002/ggge.20194.

Vane, C.H., Rawlins, B.G., Kim, A.W., Moss-Hayes, V.M., Kendrick, C., and Leng, M.J. (2013b) Sedimentary transport and fate of polycyclic aromatic hydrocarbons (PAH) from managed burning of moorland vegetation on a blanket peat, South Yorkshire, UK. Science of the Total Environment, 449, 81–94.

Verburg, P. (2007) The need to correct for the Suess effect in the application of δ^{13}c in sediment of autotrophic Lake Tanganyika, as a productivity proxy in the Anthropocene. Journal of Paleolimnology, 37, 591–602.

Wheelock, K. (2003) Pulsed river flooding effects on sediment deposition in Breton Sound Estuary, Louisiana. Louisiana State University, Baton Rouge.

White, D.S., and Howes, B.L. (1994) Long-term δ^{15}N-nitrogen retention in the vegetated sediments of a New England salt marsh. Limnological Oceanography, 39, 1878–1892.

Wickman, F.E. (1952) Variations in the relative abundance of the carbon isotopes in plants. Geochimica et Cosmochimica Acta, 2, 243–254.

Wilson, G.P., Lamb, A.L., Leng, M.J., Gonzalez, S., and Huddart, D. (2005a) δ^{13}C and C/N as potential coastal palaeoenvironmental indicators in the Mersey Estuary, UK. Quaternary Science Reviews, 24, 2015–2029.

Wilson, G.P., Lamb, A.L., Leng, M.J., Gonzalez, S., and Huddart, D. (2005b) Variability of organic δ^{13}C and C/N in the Mersey Estuary, UK and its implications for sea-level reconstruction studies. Estuarine Coastal and Shelf Science, 64, 685–698.

Wooller, M., Smallwood, B., Scharler, U., Jacobson, M., and Fogel, M. (2003) A taphonomic study of δ^{13}C and δ^{15}N values in Rhizophora mangle leaves for a multi-proxy approach to mangrove palaeoecology. Organic Geochemistry, 34, 1259–1275.

Zong, Y., Lloyd, J. M., Leng, M. J., Yim, W. S., and Huang, G. (2006) Reconstruction of Holocene monsoon history from the Pearl River Estuary, southern China, using diatoms and carbon isotope ratios. The Holocene, 16, 251–263.

Chapter 21
Loss on ignition and organic content

ANDREW J. PLATER[1], JASON R. KIRBY[2], JOHN F. BOYLE[1], TIMOTHY SHAW[1], AND HAYLEY MILLS[3]

[1]*Department of Geography and Planning, School of Environmental Sciences, University of Liverpool, Liverpool, UK*
[2]*School of Natural Sciences and Psychology, Department of Geography, Liverpool John Moores University, Liverpool, UK*
[3]*British Oceanographic Data Centre, National Oceanography Centre, Liverpool, UK*

21.1 INTRODUCTION

In this chapter we examine the importance of loss-on-ignition (LOI) and organic matter (OM) data in sea-level research, methodological issues present in the determination of OM content in the field and the laboratory, and recent developments in the consideration of OM provenance and OM autocompaction, adding to our toolkit for well-resolved sea-level reconstruction and coastal evolution.

The use of LOI and OM data in sea-level research did not receive specific attention in van de Plassche (1986), although the broader significance of these parameters was recognized across several chapters. In relation to organic matter, the efficacy of biological remains as indicators of sea level was reviewed in chapters on marine molluscs, corals, foraminifera, ooids, diatoms, and ostracods. Biological material was also reviewed indirectly with respect to obtaining the essential chronological evidence for sea-level reconstructions in the form of radiocarbon dating and dendrochronology. Linking explicitly to the more routine use of LOI and OM in sea-level studies as both direct and indirect proxies of botanical remains, two chapters in Orson van de Plassche's manual examined submerged forest beds and macrofossil remains. In the first, Heyworth (1986) investigated the elevational relationship between coastal forests and different tree species assemblages with respect to limiting tidal levels. The evidence considered here included macrofossils of wood, roots, and bark and (indirectly) fruits, seeds, and of course pollen, as well as evidence for associated coastal wetlands such as alder carr, fen peat, sphagnum bog, saltmarsh, and brackish water silts and clays. Although attention was paid to the environmental conditions and coastal settings that limit/enable the establishment and/or development of vegetation in coastal wetland settings, the treatment of LOI and OM as proxies for identifying or quantifying these transitions was scant. Behre (1986) also reviewed the use of plant macrofossils with peat beds again being a focus or, more explicitly, the range of plant communities to be found in proximity to local MHW. There was a clear emphasis on the establishment of plant cover as a sea-level indicator, linked primarily to salinity and hydroperiod (French, 1993) in addition to the advantages of macro-remains over pollen evidence. The organic/inorganic stratigraphic transition was examined in some detail, as well as the efficacy of intercalated and basal peat stratigraphy, but the emphasis was largely on the resolution of coastal (and riverine) wetland plant communities as opposed to a gross indicator of organic content. Indeed, LOI or OM content would fall very short of providing the kind of detailed information discussed in Behre (1986).

In this respect, it would appear that LOI and OM as either proxies or crude indicators of trajectories of coastal vegetation succession are of limited value in sea-level research. Furthermore, there are geographical limits to the extent to which data from one example can be directly translated to another location where there are significant differences in tidal range, climate and intertidal ecology. In many examples, however, trends in LOI and OM content provide valuable evidence of coastal change, much of it as the first or last indications of "terrestrialization" or evidence of

Handbook of Sea-Level Research, First Edition. Edited by Ian Shennan, Antony J. Long, and Benjamin P. Horton.

subtle or temporary changes that do not necessarily lead to any large-scale environmental change that is expressed in the litho- or biostratigraphy. Here we present a very different perspective: one that illustrates the critical and current importance of accurate LOI and OM content analyses.

21.2 ORGANIC MATTER AND LOI

21.2.1 LOI and related physical analyses

Berglund (1986) includes the determination of organic content (including organic carbon) among the physical analyses listed as a minimum demand for paleoecological reference sites for the study of past environmental changes in lakes and mires. Similarly, Digerfeldt (1986) includes the determination of ignition residue of coarse matter as a metric for assessing changes in lake level. Such physical analyses (including LOI, moisture content, dry bulk density, etc.) yield data on accumulation rates, ecosystem production, decomposition, shrinkage, compaction/autocompaction, water-level changes, and carbon sequestration/cycling.

LOI is described by Aaby (1986) as the content of inorganic material of dried sediment that remains after being ignited at 550 °C, calculated as a percentage of the dry weight. Additional determinations of physical properties of mire deposits that are closely linked to the organic content include moisture content, density or dry bulk density, and degree of humification (Aaby, 1986). Bengtsson and Enell (1986) recommend that these physical properties, along with carbonate content, should be determined from a sequence of air drying of a known sample volume for c. 12 hours at 105 ± 2 °C (moisture content and dry bulk density), ignition for 2 hours at 550 °C (LOI), with potentially an additional 30 minutes ignition at 550 °C after adding 1–2 drops of 20% w/v NH_4NO_3 solution if the first ignition appears incomplete (black with carbon), and then ignition for 4 hours at 925 °C (carbonate content).

LOI is usually applied as a measure of the organic content of the material, but ignition also releases chemically bound water. Under certain circumstances, LOI can also be used to estimate the organic carbon content whereby the organic C is 12/30 of the organic content (assuming a composition of $(CH_2O)_n$; Mackereth, 1966; Håkansson and Jansson, 1983). However, Digerfeldt (1972) recognized that different conversion factors are required for organic C from LOI depending upon the composition of the material. Furthermore, it is crucial that the ignition temperature is regulated to avoid complications due to the volatilization of salts, structural water, and ammonia (Bengtsson and Enell, 1986).

In terms of the determination of LOI, it would also be rather inappropriate to overlook the semi-quantitative field-based determination of organic content, as exemplified by the sediment classification scheme of Troels-Smith (1955) whereby an approximate assessment of organic content, and indeed the composition of this organic matter, can be made based on objective classification of the physical characteristics and component parts of a deposit. This is still a key tool in understanding the macro- to meso-scale pattern of past changes in coastal sedimentary environments, and indeed the micro-scale evidence from follow-up laboratory analysis (Aaby and Berglund, 1986).

Studies of coastal change often overlook the importance of subtle changes in organic facies, yet Bos et al. (2012) stress that such changes can be important in studies of delta evolution for understanding paleoenvironmental conditions and differential susceptibility to compaction (e.g., van Asselen, 2011). They recently proposed a new field method for distinguishing different organic facies found in temperate delta plains based on sedimentary characteristics, that is, the distinction of gyttya from peat based on the presence of lamination or clastic lenses. Peat can then be subdivided according to the botanical remains present. A summary of this field method is given in Table 21.1. With respect to the scope of this chapter, it is interesting to note that Bos et al. (2012) further characterize their organic facies based on typical ranges and average values for LOI: wood peat average LOI 54%, range 30–83%; and sedge and reed-sedge peat average LOI 85%, range 76–92%. Indeed, they suggest that a LOI value of 75% may be used to support field-based differentiation between reed peat (<75%) and reed-sedge peat (>75%).

21.2.2 LOI, OM, structural water and sediment composition: Considerations for ignition temperature

In this section, we consider whether loss-on-ignition (LOI) provides a suitable measure of organic matter (OM) and, if so, then how is it best undertaken. To achieve this it is necessary to examine the principles that underlie LOI

Table 21.1. Classification key for the classification of organic deposits preserved in temperate delta plains.

Applicable for freshly cut material		
	The colour of freshly cut material is grey, brown-grey, yellow-grey	CLASTIC FACIES
	The colour of freshly cut material is brown, green-brown, grey-brown, yellow-brown, white-brown, red-brown, or black	ORGANIC FACIES
For identification of organic facies, the material should be pulled loose		
1.0	The material lacks sedimentary structures and recognizable plant remains, is often coloured black	**Amorphous organics**
2.0	The material is (finely) laminated, contains abundant plant remains and may strongly react to 5% HCl, and may contain shells, shell fragments, ostracods and other small aquatic animal remains and clastic laminae/lenses (sand/silt/clay), breaks open following the laminations.	**Gyttja**
	i. The deposits consist of predominantly coarse plant remains (for example wood pieces, twigs and leaves).	Coarse detrital gyttja
	ii. The deposits consist of a fine-textured mass, generally green-brown, and may contain abundant very small plant remains (<1 mm), may have a gummy-like appearance.	Fine detrital gyttja
	iii. The deposits predominantly consist of white-brown material, which may appear as brittle concretions or as a very fine textured mass, abundant shell fragments, opercula or Chara stems and strongly reacts to HCl.	Calcareous gyttja
	iv. The deposits predominantly consist of yellow-brown material, usually represented as stains, which may appear as brittle concretions or as a very fine (Ø < 1 mm) textures mass, and reacts to HCl, although only at higher temperatures (>30°C).	Siderite gyttja
3.0	The material is composed of plant remains and is strongly rooted.	**Peat**
	i. The material consists of wood fragments, which may be embedded in a fine-textures matrix.	Wood peat
	ii. The dominant components are 5 to 10 mm wide and elongated, curly, yellowish plant remains (reed culms, rhizomes and roots), which are generally accompanied by fine roots (Ø < 1 mm).	Reed peat
	iii. The dominant components are flattened, 0.5 to 2.0 mm wide, elongated, light-brown, curly plant remains (sedge roots), often accompanied by 1.0 to 2.0 mm round orange-brown (or black when oxidized) seeds of *Menyanthes.*	Sedge peat
	iv. The dominant components are either: (i) fine moss leaves and stems, small moss layers can be peeled; (ii) highly fibrous, rope-like (*Eriophorum);* (iii) 1 cm long, lancet-shaped soft plant remains (*Sphagnum);* or (iv) red curly *Ericales* roots.	Oligotrophic peat

Source: After Bos et al., 2012.

measurement and identify issues that affect its accuracy. It should be noted that the question framed here is quite different from that most commonly asked: whether LOI provides a satisfactory measure of organic carbon in soil and sediment.

The standard laboratory procedure for LOI determination is:

(1) Heat the sample at 105 °C in an oven to drive off interstitial moisture. Remove from oven and allow to cool in a desiccator.
(2) Weigh an empty porcelain or platinum crucible to determine original crucible mass.
(3) Place an aliquot of the dry sample in the crucible and reweigh. The mass of the crucible + sample minus the mass of the crucible gives the mass of the dry sample ($mass_{105°C}$).
(4) Place the crucible in a muffle furnace. Ignite the sample for 3 hours at 475 °C. Allow to cool.
(5) Reweigh the crucible with the ignited sample. This mass minus the original mass of the crucible gives the mass of the sample after ignition ($mass_{475°C}$).
(6) LOI is determined from the following:

$$\text{LOI } \%_{w/w} = \frac{\text{mass}_{105°C} - \text{mass}_{475°C}}{\text{mass}_{105°C}} \times 100.$$

This assumes that the ignition at 475 °C for 3 hours causes complete conversion of organic matter into gases (see Section 21.2.1 and the following), and results in no mass change in the mineral matter. LOI is then considered equivalent to the percentage OM by weight ($\%_{w/w}$). It has however long been known that soils and sediments do not usually behave like this. While it is true that only a negligible part of the OM is non-volatile, many minerals lose mass on heating due mainly to dewatering and dehydroxylation (Ball, 1964). The two most widely

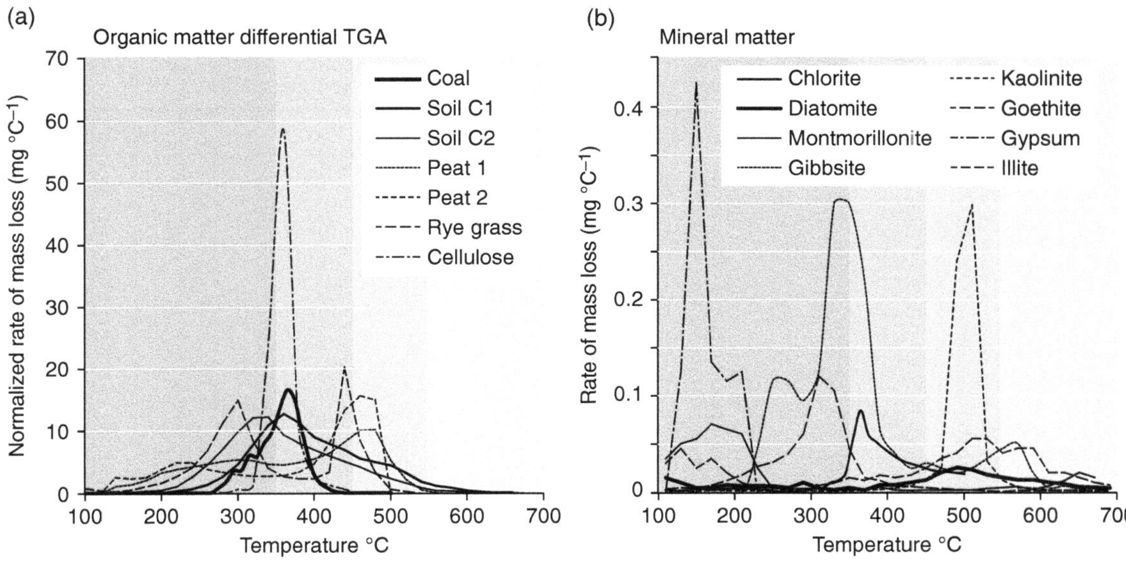

Fig. 21.1. DTGA curves for (a) organic matter and (b) mineral matter. The shaded fields shown delimit three widely used temperatures for LOI determination.

studied materials are the carbonate and clay mineral families. The former was a particular issue in the early days of LOI analysis (see Ball, 1964) because of the common practice of igniting at 850 °C. Reducing the ignition temperature solves this problem readily because very little organic matter remains in material ignited to 600 °C, while none of the carbonate will have pyrolized. On the other hand, clays de-water substantially and in a more complex manner such that non-OM mass loss must be taken into account. To allow for such effects we must know how much non-OM mass loss can be expected, and at what temperatures this will be manifested. Thermogravimetric analysis (TGA), in which the weight of a small sample is measured continuously while being heated in an oven, provides this information.

Figure 21.1 shows the differential thermogravimetric analysis (DTGA) curves for a variety of natural materials, clearly showing the temperature intervals over which mass loss takes place. The organic matter samples (Fig. 21.1a), including rye grass, cellulose, peat and coal, show a drying stage up to c. 150 °C and a broad mass-loss spectrum over the temperature interval 300–600 °C, with peaks throughout the temperature range. It is clear that 375 °C fails to burn much of the OM, while at 550 °C all but a small part is fully ignited.

The mineral matter samples (Fig. 21.1b) also show mass loss across the temperature range 105–700 °C, but with substantial differences between mineral types. Most clays undergo most dehydroxylation at high temperatures, between 500 and 650 °C, and some even higher. However, montmorillonite also has a substantial low-temperature dewatering stage (Mielenz et al., 1953). Importantly, the oxyhydroxides and hydroxides of Al and Fe lose all of their mass loss over exactly the same temperature range as organic matter. Other commonly abundant components of sediment, such as biogenic silica, also dewater within the interval of maximum OM mass loss (Fig. 21.1). It is clear that using an ignition temperature of 550 °C maximizes the water loss from minerals. At 375 °C this problem is minimized.

The total mass loss due to dehydroxylation of common minerals is variable but, in some cases, high (Table 21.2). While the framework silicates typically have LOI values below 1%, mica LOI values fall within the range 1–5% and clays 5–15%. The Fe and Al hydroxides (*sensu lato*) are higher still, and crucially dewater with a similar temperature profile to OM. Other minerals such as gypsum and carbonate show still greater LOI, but typically lose their mass outside the natural organic matter temperature profile.

There are essentially two approaches taken to address this problem: minimization and correction. In both cases it is essential to take a targeted approach based on good knowledge of the thermal properties of the material being studied (Boyle, 2001, 2004).

Table 21.2. Structural water in common minerals, defined as water release on heating above 110 °C. Data from Deer et al. (1966) except where otherwise shown. The values represent the LOI value (wt%) that would be measured by heating from 105 °C to 1000 °C. For most minerals this is close to what would be measured if heating to only 550 °C. The wide range of values within each mineral type shows the danger of applying generalized correction factors.

	N	Log mean H_2O^+ wt%	Minimum H_2O^+ wt%	Maximum H_2O^+ wt%
Gibbsite[a]		34.6		
Gypsum[a]		20.9		
Zeolite	9	14.2	9.47	22.04
Serpentine	3	13.2	12.18	13.68
Chlorite	8	11.4	10.31	13.19
Clay	7	10.6	5.98	17.42
Goethite[a]		10.1		
Opal[b]			6	10
Tourmaline	5	3.2	2.22	4.16
Mica	9	2.7	0.38	7.21
Staurolite	2	1.9	1.92	1.98
Epidote	5	1.4	0.57	2.13
Amphibole	18	1.3	0.19	2.6
Sphene	3	0.7	0.56	0.93
Pyroxene	12	0.3	0.1	1.27
Garnet	4	0.2	0.13	0.48
Feldspar	19	0.2	0.01	0.84
Olivine	3	0.1	0.05	0.33
Quartz[c]		0.0		

[a]Calculated release based on stoichiometry
[b]Range of value for opal-A. No mean quoted.
[c]Assumed representative value.

21.2.2.1 Minimization approach

A large part of the OM mass loss takes place in the temperature interval 200–350 °C, while structural water in many clays is mainly released at higher temperatures. Ball (1964) stated that clay dewatering takes place between 450 and 600 °C, concluding that limiting the ignition temperature to 375 °C entirely avoids interference from clay. Such an interpretation is often repeated; however, the situation is not so simple for three reasons. First, many clay minerals undergo substantial dewatering at temperatures below 375° (Fig. 21.1; Table 21.3). Second, variable but often large amounts of OM remain at 375 °C; ignition to at least 600 °C is required to achieve full loss (Table 21.4). Third, at 375 °C the burn is sluggish, and the extent of OM loss varies with time, precise temperature, and position in the oven (Heiri et al., 2001). Consequently, ignition to low temperatures is prone to reproducibility problems, and must be undertaken with great care. Minimization by restriction of temperature is therefore a compromise solution that must be undertaken with full knowledge of the thermal properties of the material investigated (Boyle, 2004), and with a thorough understanding of potential errors introduced. On the other hand, if specific problem minerals are present – as might be expected for intertidal sediment successions where OM is often present in association with tidal silts and clays – then the effect of these may be largely avoidable by altering the temperature window. For example, if gypsum were present then increasing the "drying" temperature to 200 °C would greatly reduce any interference with OM measurement.

21.2.2.2 Correction approach

While many minerals dewater, it is the clay minerals that are both commonly abundant and rich in structural water. Measurement of total clay content in soils and sediments is a common practice, and the information provides a basis for producing clay-corrected LOI values. Such corrections should not be regarded as universal; the water content of specific clay minerals is highly variable, as is the proportion of the clay size fraction of soil that comprises clay minerals. "Local" correction factors have been favored for some time however, and new formulations still appear. De Vos et al. (2005) report a clay-corrected method for

Table 21.3. Percentage of total LOI (1000 °C) released at three different ignition temperatures.

	375 °C	450 °C	550 °C	Source
Muscovite	11.1	13.3	22.2	Mielenz et al. (1953)
Biotite	72.7	72.7	72.7	Mielenz et al. (1953)
Kaolinite	9.0	12.4	33.8	Mielenz et al. (1953)
Gibbsite	74.4	81.8	91.5	Mielenz et al. (1953)
Montmorillonite	63.4	65.6	67.9	Mielenz et al. (1953)
Montmorillonite	68.6	69.9	73.2	Mielenz et al. (1953)
Montmorillonite	72.3	75.7	80.3	Mielenz et al. (1953)
Illite	35.4	46.3	74.4	Mielenz et al. (1953)
Muscovite	25.0	39.6	60.4	Alexiades and Jackson (1967)
Kaolinite	2.9	5.0	92.8	Alexiades and Jackson (1967)
Chlorite	8.0	16.0	37.6	Alexiades and Jackson (1967)
Diatomite	30.1	39.7	60.3	Chaisene and Rangsriwatananon (2004)
Gypsum	100.0	100.0	100.0	Anon (1985)
Opal-A	30.8	53.8	76.9	Hatipoğlu et al. (2010)
Opal-A	2.2	43.5	76.1	Hatipoğlu et al. (2010)
Opal-A	26.2	45.2	69.0	Hatipoğlu et al. (2010)
Opal-A	27.5	56.9	82.4	Hatipoğlu et al. (2010)

Table 21.4. Percentage of total LOI (1000 °C) released at three different ignition temperatures, with organic matter included.

	375 °C	450 °C	550 °C	Source
Ledo coal	16.0	38.3	97.5	Saikia et al. (2009)
Tikak coal	10.0	30.0	81.3	Saikia et al. (2009)
Cellulose	100	100	100	Fernández et al. (2012)
Soil OC 1	48	77	96	Fernández et al. (2012)
Soil OC 2	66	88	98.5	Fernández et al. (2012)
Soil OC 3	55.3	79.9	96.9	Pallisser et al. (2013)
Rye grass 1	65.0	84.6	100	Sharma et al. (2012)
Rye grass 2	59.8	73.3	100	Sharma et al. (2012)
Rye grass 3	67.9	93	100	Sharma et al. (2012)
Surface peat	45	70.0	100	Kļaviņš et al. (2010)
Sub-surface peat	35.7	52.5	100	Kļaviņš et al. (2010)

predicting organic carbon from LOI for some Belgian soils. While this method is developed for carbon measurement, it lends itself well to estimation of OM because it aims primarily to avoid inclusion of mineral mass loss in the LOI value.

To achieve reliable LOI measurement of OM it is therefore necessary to address the following questions:

(1) What is the mass–temperature profile of the inorganic fraction of the sediment?
(2) Can peaks in the inorganic mass-loss profile be avoided by careful choice of the temperature window? Specifically, (a) What is a suitable drying temperature? 105 °C is conventional, but this may lie on a slope on the mass–temperature curve for the material, and may lie just below a dewatering step for an interfering mineral. (b) What is the best ignition temperature? The three most widely used values are 375 °C, 450 °C, and 550 °C. If (i) there are negligible interfering inorganic phases, or (ii) these are not correlated with the OM, then the higher temperature is likely the best choice, minimizing measurement time (1 hour being sufficient) and improving precision. Alternatively, are the benefits of using a lower temperature sufficient to justify the disbenefits of reduced OM recovery and reduced precision?
(3) Is "clay" correction either necessary or appropriate? The depends on (a) the level of interference; (b) the amount of OM present (the benefits of clay correction being greatest for very low LOI); and (c) whether the "clay" content of the sediment has been quantified.

(4) Is an independent OM measurement for a subset of samples needed to verify the LOI data? If accurate and precise OM measurements are required, this is an essential step.

21.2.3 Thermogravimetric analysis: methodological considerations

In essence, TGA comprises a balance stem positioned in an oven or reaction chamber. The temperature of the chamber is controlled to increase at specified rates, or to hold at particular temperatures. The balance is tared with a crucible in place, a sample is added, and the weight is recorded at specified intervals for the duration of the measurement run. The main technical challenge is to prevent heat from interfering with the operation of the balance. The main variations in technical specification relate to: (1) the size and weight capacity of the crucible (typically some tens of μL and thus tens of μg, but some devices are much larger); (2) temperature measurement, which may be theoretical only (based on heat input to the oven), may measure chamber temperature directly or may measure both this and the temperature of the crucible (differential thermal analysis is a related technique that depends on this approach to quantify heat production or consumption by the sample); (3) control of gas composition in the oven (typically air or N_2); and (4) capture and measurement of gases released by the sample (if CO_2 is measured, then direct assessment of organic carbon can be made).

Here we describe some common practical considerations in the use of DTGA to obtain LOI measurements, and the thermogravimetric properties of the sample.

If the sample is heated in air, oxygen reacts with organic matter. This allows complete conversion of most organic matter to CO_2 over a temperature interval of c. 200–600 °C, and therefore weight estimation of this organic matter by LOI. However, heating in air allows the possibility of combustion rather than ignition, where heating initiates a self-sustaining oxidation reaction or burn. This has two undesirable consequences. First, smoke emission caused by the burn may cause measurable loss of non-organic particulate matter, thereby overestimating the LOI. Second, the weight loss curve is no longer a record of the pyrolytic decomposition of the sample, and thus contains far less information about the nature of the organic matter. Furthermore, the onset of such burning is erratic, leading to highly variable weight loss records.

The issue of sample burning can be avoided by heating the sample in an inert gas, usually N_2. This leads to more stable and repeatable thermogravimetric data, but does not convert all of the organic matter to CO_2. Rather, a proportion of the organic matter remains in the crucible in a more thermally stable form, chiefly elemental carbon. Weight loss curves collected like this therefore underestimate the total sediment organic matter content. To allow for this process, air may be admitted into the reaction chamber at the end of the measurement run, pyrolizing any elemental carbon. This combined approach allows both measurement of total LOI, and repeatable assessment of the thermogravimetric properties of the sediment organic matter.

Example procedure:

(1) Place empty 50 μL ceramic crucibles in the sample changer, and start a tare cycle.
(2) Add about 20 μg of dry sample to each crucible, and start a run cycle.
(3) The run cycle comprises the following steps. (a) Under N_2, heat the chamber to 150 °C at 20 °C min^{-1}. (b) When 150 °C is reached, hold the temperature for 5 min (to provide a reliable "dry" weight measurement). (c) Continue to heat at 20 °C min^{-1} until a temperature of 900 °C is reached. (d) Change gas feed to the chamber from N_2 to air, and hold at 900 °C for 5 min (to pyrolize the elemental carbon). (e) Cool (active cooling) before the instrument automatically moves to the next sample.
(4) LOI between 150 °C and 550 °C is given by:

$$\mathrm{LOI}_{550-150^\circ}\ (\mathrm{wt\%}) = 100 \times \frac{\left(W_{150^\circ\mathrm{C}} - W_{550^\circ\mathrm{C}}\right) + \left(W_{900^\circ\mathrm{C,N_2}} - W_{900^\circ\mathrm{C,air}}\right)}{W_{150^\circ\mathrm{C}}}$$

where W is the sample weight.
(5) Analyses of the thermal decomposition profile is undertaken to characterize the sediment organic matter.

21.3 ORGANIC MATTER AND COASTAL SEDIMENTATION: INDICATORS OF SEA-LEVEL

It is useful to first consider how accessibility to the sedimentary record of past change in coastal settings acts to reduce the importance of LOI and OM determinations in lithostratigraphic correlation. In depositional environments where the sediments are accessed remotely from a boat or raft, Dearing (1986) suggests that there is considerable potential for using organic matter determination by "rapid, cheap and easy to perform" LOI as a robust correlation tool in waterlain and especially lake sediment sequences (e.g., Tolonen et al., 1975; Davis, 1976). Here, the number of cores collected tends to be limited and the morphology of the lake bed may be poorly constrained; core correlation is therefore much more challenging. It is doubtful whether the potential for LOI for core correlation has ever been truly revealed in studies of past sea-level change, mainly because data can be readily obtained on the stratigraphic succession and, where the evidence is unclear, an intermediate core can readily be drilled. Furthermore, the relative and absolute altitudes of each core can be determined with ease. In such circumstances, the field determination of OM is normally applied to select a stratotype core, and LOI is reserved for follow-up laboratory analysis. There are few situations in which LOI would be determined on a full suite of cores where the field data provides more than satisfactory stratigraphic evidence of environmental change, although there are obviously circumstances where the stratigraphic evidence is not sufficiently sensitive, for example in isolation basins.

Acknowledging the importance of field-based assessments of OM, the age–altitude reconstruction of Holocene sea-level change from sediment records is based upon the collection of sea-level index points (SLIPs) or sea-level indicators (SLIs) derived from transitional stratigraphic boundaries between freshwater and marine deposits, that is, at the interface between land and sea. The lithological and biological characteristics of such transitional sedimentary environments are closely linked with position within the tidal frame, broadly manifest in the organic content of the sediment. For example, in saltmarsh and associated perimarine ecosystems, organic-rich, peat sediment forms at the highest elevations, at and above the upper limits of tidal influence, and progressively more inorganic sediment forms lower down the intertidal slope where tidal inundation is more frequent and prolonged, which ultimately prevents vegetation survival. These sub-environments can easily and rapidly be identified by their sedimentary character, particularly organic content. The position within the tidal frame where organic sedimentation, and specifically a *Phragmites* or monocotyledonous peat, gives way to intertidal minerogenic deposits (termed transgressive overlap) in a humid, temperate macrotidal setting, has been shown to approximate a reference water level of MHWST–20 cm with a range of c. 20 cm (Shennan, 1986). Similarly, where the marine influence is decreasing, minerogenic sedimentation (i.e., a saltmarsh deposit) is replaced by organic deposits (termed regressive overlap). If the organic deposits are again *Phragmites* or monocotyledonous peat, the reference water level is M^1–20 cm (where M^1 is the average of HAT and MHWST); if it is fen wood peat above a saltmarsh deposit, this occurs around M^1. Clearly, consideration needs to be given to ground-truthing these stratigraphic relationships with tidal levels in meso- and microtidal settings.

In the mangrove ecosystems of lower latitudes, the principles of stratigraphic succession are largely similar but may be complicated by: (1) macroscale basement topography (e.g., continental shelf v. oceanic island position) and mesoscale environmental context (e.g., river-dominated, tide-dominated, and "carbonate" settings) (Woodroffe, 1990); (2) site morphology and paleogeography; and (3) spatial patterns of sedimentation across a range of variably organic, minerogenic, and calcareous facies (e.g., Woodroffe, 1981; Parkinson, 1989; Larcombe and Carter, 1998). Stratigraphic indicators of former sea level are generally obtained from mangrove peats, for which Woodroffe (1981) quotes LOI values of 50–80%, and terrestrial muds with wood remains deposited *in situ* beneath mangrove forests.

Assuming no significant paleotidal changes, and with careful consideration of the present-day relationship between observed changes in sedimentation, sea level, and tidal range, the altitude at which any of these stratigraphic transitions occurs within the succession can be taken as the altitude of the appropriate reference water level at time of deposition. In this way, a time-altitude plot of sea level may be obtained, corrected appropriately according to the indicative meaning of each SLI (e.g., Shennan, 1986).

In a very basic application, LOI can be used to identify lithostratigraphic changes which can be used to identify and verify SLIs. This is particularly useful in situations where the cut-off between organic and minerogenic sedimentation is clear and/or rapid (e.g., isolation basin contacts, stratigraphic changes resulting from vertical coseismic subsidence in association with earthquakes), and

Table 21.5. Partial redundancy analysis (pRDA) results for all Mersey saltmarsh datasets with only significant environmental variables included (Mills et al., 2013).

Variable	Oglet Bay		Decoy Marsh		Combined	
	Total inertia (%)	Constrained inertia (%)	Total inertia (%)	Constrained inertia (%)	Total inertia (%)	Constrained inertia (%)
Total explained	52	–	73	–	50	–
Distance	6	11	–	–	–	–
Elevation	4	8	–	–	2	4
Organic matter	2	5	25	42	3	7
pH	13	25	–	–	–	–
Phragmites spp.	–	–	–	–	3	6
Salinity	–	–	–	–	3	6
Sand %	–	–	–	–	2	3
Transect	14	27	–	–	4	8
Clay %	–	–	24	40	2	3
Intercorrelation	13	26	11	18	32	64

therefore where high-resolution sampling and determination of LOI can provide greater vertical accuracy in the determination of SLI altitude. In studies of RSL in NW Scotland, SLIs obtained from isolation basin contacts show LOI values rising from 4–8% in transitional marine organic clay sediments to 33–52% in overlying freshwater deposits in <5 cm (Shennan et al., 1996). Similarly, Shennan et al. (1995) use LOI to define and characterize different depositional environments within the modern intertidal zone ranging from sandflat close to MHWNT (<1%), low saltmarsh below MHWST (19–28%), high saltmarsh between MHWST and HAT (>28–72%), and supratidal peat >72%. These LOI values are used along with biostratigraphic information to determine appropriate indicative meanings for the SLIs collected from core material.

The pattern of coastal change associated with land movement prior to and after an earthquake is also clearly manifest in LOI data. For example, Zong et al. (2003) identify a short period of preseismic land subsidence where LOI values decline from c. 40% to 10–20% as coastal marshes become submerged prior to the sudden coseismic subsidence associated with the 1964 earthquake event in Alaska.

The last decade has seen a marked expansion in the statistical characterization of the contemporary coastal wetland continuum for the quantitative reconstruction of former sea level. This transfer function approach (see Chapter 13 and 31) has developed the well-established indicative meaning of stratigraphic transitions with respect to tidal level (cf. Shennan, 1986) by statistically constraining the relationship between saltmarsh fauna and flora and elevation within the intertidal frame. The research community has therefore become accustomed to emerging evidence of recent sea-level change that bridges the sedimentological (10^1–10^4 years) and instrumental (10^{-3}–10^2 years) timescales. Transfer function reconstructions have focused on microfossils such as diatoms, foraminifera, and testate amoebae because they usually preserve well in Holocene sediments. However, as saltmarsh foraminiferal assemblages may be controlled by a number of variables (e.g., salinity, temperature, substrate conditions, sediment composition, etc.) that may have no direct relationship to elevation (e.g., Murray, 1971; Patterson, 1990; de Rijk, 1995; de Rijk and Troelstra, 1999), the significance of elevation as a primary control must be tested in order to determine whether the modern dataset is appropriate to use for sea-level reconstruction. Importantly, because the organic content of saltmarsh surface sediments generally increases with decreasing hydroperiod, there is often a strong correlation between LOI and elevation. Consequently, the statistical significance of elevation in determining foraminiferal zonation across a mid- to high-latitude saltmarsh may be reduced in favor of organic content or the intercorrelation of environmental variables. For example, a partial redundancy analysis of environmental and foraminiferal data from saltmarshes in the macrotidal Mersey estuary (Mills et al., 2013) reveals a significant OM control on zonation on Decoy Marsh (42% of

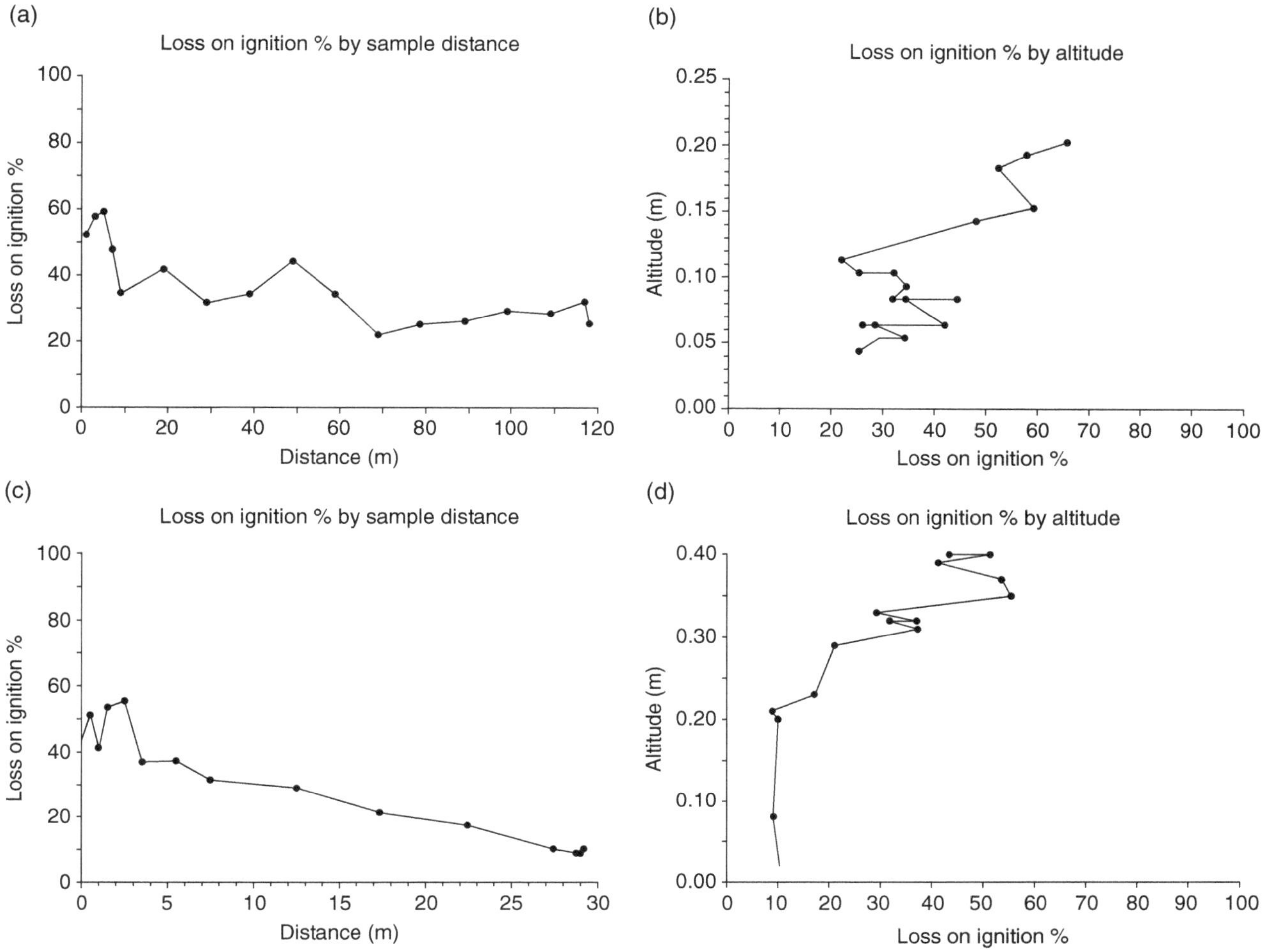

Fig. 21.2. Surface LOI% for (a, b) Jadrtovac saltmarsh in Morinje Bay and (c, d) Blace saltmarsh on the Neretva delta, Croatia. (a, c) Distance from the limit of highest high water and (b, d) altitude (note that some sample elevations are duplicated). The upper and lower saltmarshes, as confirmed by vegetation assemblage and foraminiferal zonation (Shaw et al., 2012), exhibit characteristic LOI% values of c. 60% and 25–40%, respectively. Downcore trend in LOI% in (e) core JD1 from Jadrtovac saltmarsh in Morinje Bay, and (f) BL from Blace saltmarsh on the Neretva delta, Croatia. In comparison with the contemporary LOI% data (a–d), the step changes at 19–20 cm and then at 8–11 cm for JD1 are interpreted, respectively, as environmental transitions from mudflat to lower saltmarsh at c. 1880 AD and then lower to upper saltmarsh at c. 1980–1990. The temporal trend preserved in core BL is much more of a gradual transition from a mudflat to an upper saltmarsh through the upper 5–6 cm of the core.

explained variance) and a substantial influence from environmental variable intercorrelation across Oglet Bay saltmarsh and for the combined dataset (Table 21.5). The determination of OM content by LOI is therefore of considerable importance for establishing the accuracy and reliability of ecological transfer functions of tidal level. Significantly, the fact that LOI as a quantitative proxy for OM content may be compromised in the presence of clay minerals (as one would expect for intertidal sediments) highlights the importance of the methodological considerations illustrated in Section 21.2.

While diatoms and foraminifera, as well as testate amoebae, show a strong relationship with tidal level (e.g., with respect to MHWS), it is also apparent that the organic content of the saltmarsh substrate has an important bearing on species distribution and zonation. Although organic matter may be regarded as a confounding factor when it comes to establishing an ecological transfer function with respect to elevation, it may in itself offer an additional means for reconstructing former sea levels. For example, Shaw et al. (2012) have examined LOI% with elevation and distance across microtidal saltmarshes on the Adriatic coast of Croatia.

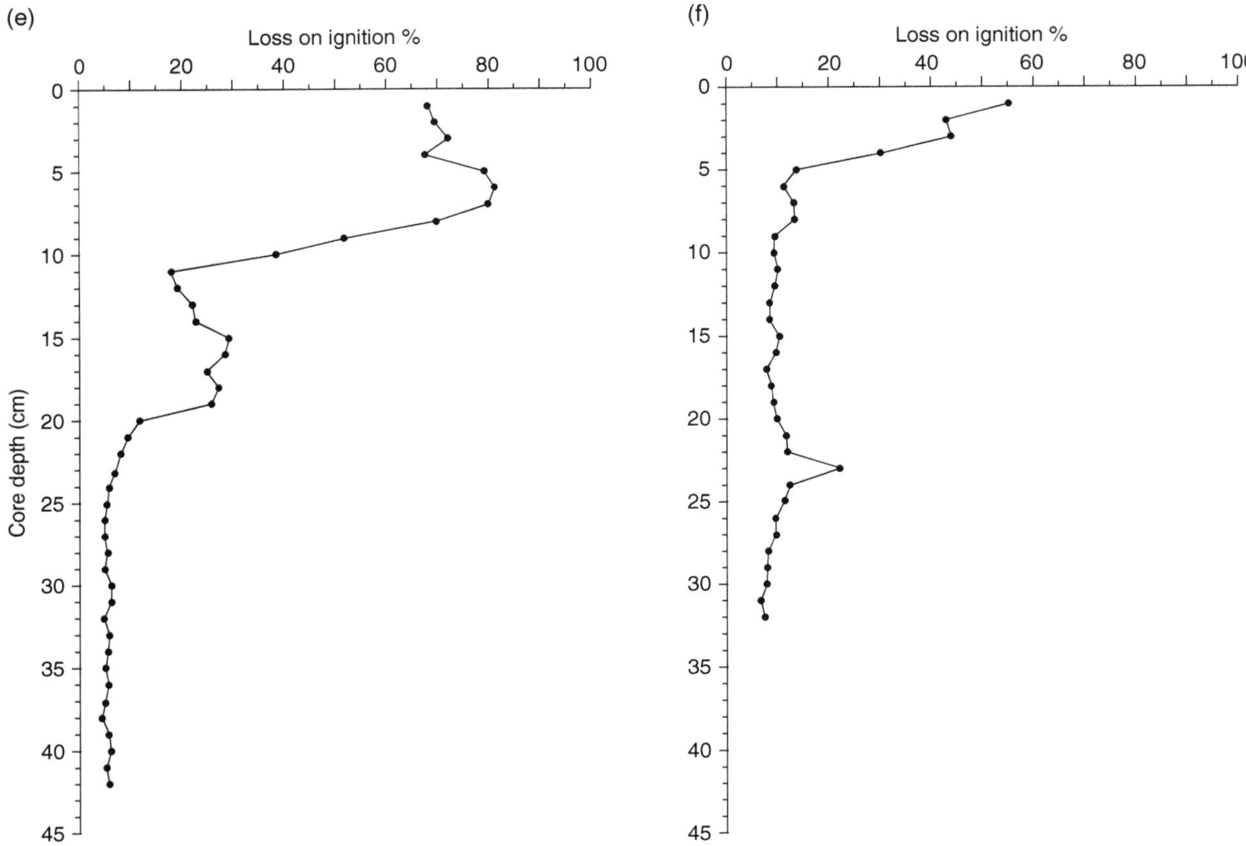

Fig. 21.2. *(Continued)*

Inspection of surface LOI (Fig. 21.2a) reveals a decreasing trend towards the seaward edge in Morinje Bay (Jadrtovac) and the Neretva delta (Blace), showing a clear distinction between upper and lower saltmarsh environments, which exhibit characteristic LOI values of c. 60% and 25–40%, respectively. Moving up through the core taken from the same saltmarshes (Fig. 21.2b), LOI values are very low up to c. 20 cm (<10%) where they then increase to about 25% between 20 and 11 cm, and then to 70–80% between 8 and 0 cm. Here it is argued that these step changes potentially represent first the transition from mudflat to lower saltmarsh at 20–19 cm, and then from lower to upper saltmarsh at 11–8 cm. However, care is needed in interpreting "fossil" LOI data alone due the diagenetic effects of OM decomposition and leaching (see Chapter 20). In this case, the interpretation of step changes is corroborated by foramineral data, suggesting that the LOI provides at least supporting evidence for indicators of tidal level during the last 100–200 years (Shaw et al., 2012).

21.4 AUTOCOMPACTION AND THE USE OF LOI DATA FOR "DECOMPACTION"

An underpinning assumption in the reconstruction of sea-level trends is that the sedimentary or paleoecological evidence of former sea level (i.e., sea-level index points or sea-level indicators) is preserved at the same altitude as it was deposited. This may well be true for "basal peats" that directly overlie an incompressible substrate (e.g., Jelgersma, 1961; van de Plassche, 1982; Gehrels et al., 1996; Shennan and Horton, 2002; Donnelly et al., 2004; Törnqvist et al., 2004), but not thick sequences of unconsolidated Holocene sediments, especially those characterized by significant thicknesses of organic deposits which are more prone to compression than minerogenic sediments (Head, 1988; Allen, 2000). Woodroffe (1990) recognizes that this phenomenon applies equally to mangrove sediments, although less so where the bulk density is high and root penetration low. Indeed, Aaby

(1986) cites a considerable body of evidence highlighting the critical role of compaction in thick organic deposits, and that compressibility varies between different peat types.

Several physical and chemical processes act to reduce sediment volume post-depositionally, and hence lead to compaction or what is often referred to as "autocompaction" (Bloom, 1964; Kaye and Barghoorn, 1964; Haslett et al., 1998; Allen, 1999, 2000; Edwards, 2006; Long et al., 2006; Horton and Shennan, 2009; van Asselen et al., 2009, 2011; Brain et al., 2011, 2012). Autocompaction arises from decreases in pore space, de-watering, structural collapse or biological decay of vegetal remains, and biogeochemical alteration (e.g., oxidation) of organic matter (see van Asselen et al., 2009). Further loss of volume may arise from the nature of coastal change since deposition. For example, both Baeteman et al. (2002) and Long et al. (2006) invoke local erosion by channels and resulting drainage as the main cause of wetland surface lowering. As a consequence of autocompaction, SLIs may be preserved at altitudes that are significantly lower than their depositional position; these are therefore indicative of faster (or accelerated) rates of sea-level rise. Further, Allen (1999) suggests that compaction is more likely to have a greater effect on the elevation of transgressive stratigraphic contacts where minerogenic sediments overlie thick peat beds. While this phenomenon is methodologically challenging when it comes to sea-level studies, autocompaction is also of considerable significance in controlling the fate of coastal wetlands in response to present and future sea-level rise (e.g., Reed, 1995; McKee et al., 2007; Day and Giosan, 2008).

As noted by Brain et al. (2012) the effects of compaction, especially in relation to saltmarsh-based ecological transfer function reconstructions of sea level, are poorly constrained (see Chapter 30). Indeed, they stress that the numerical modeling of compaction for potential decompaction of Holocene stratigraphy (e.g., Pizzuto and Schwendt, 1997; Paul and Barras, 1998; Tovey and Paul, 2002; Massey et al., 2006; Meckel et al., 2007) may not be appropriate for low-energy intertidal successions if based on classical principles of soil mechanics. Furthermore, van Asselen et al. (2011) stress that the geotechnical properties of peat are very difficult to estimate due to its inherently heterogeneous nature, the considerable variations in properties that readily occur over short distances in peat successions, and because many models of syn-depositional compaction do not incorporate time-dependent compaction.

Recognizing that many compression parameters are controlled both directly and indirectly by organic matter content, and thus LOI as an effective proxy, Brain et al. (2012) have developed a numerical model to, in essence, "decompact" typical intertidal stratigraphies based on the analysis of contemporary mid-latitude saltmarsh and intertidal sediments and sequential removal of overlying sedimentary layers.

In a similar vein, van Asselen (2011) applies a combination of LOI and dry bulk density for assessing the accommodation space created by Holocene peat compaction, and van Asselen et al. (2011) also use LOI in the modern calibration set and scenario testing for their numerical model of peat compaction in Holocene deltaic successions. Their results show that subsidence is positively correlated with the thickness and LOI of the peat sequence, as well as the overburden thickness. In addition, van Asselen (2011) notes that variability in compaction within one peat sequence is due to variations in organic content (LOI) and the structure of the vegetal matter.

The results of Brain et al. (2012) reveal that the greatest post-depositional lowering increases linearly to a maximum at about mid-depth in the modeled stratigraphic succession (as expected from the relative balance between underlying compressible material and overlying load; van Asselen et al., 2009), and that this lowering is greater for transgressive successions as the LOI in the uppermost layer decreases. It is therefore concluded that autocompaction may lead to an increase in the apparent rate of sea-level rise of 0.1–0.2 mm a^{-1} in successions with subtle variations in lithostratigraphy, while this has the potential to increase to 0.4 mm a^{-1} in thick, transgressive successions (Fig. 21.3). This is in good agreement with previously published long-term estimates of autocompaction from mid-latitude Holocene coastal stratigraphic successions (e.g., Edwards, 2006: 0.7–1.0 mm a^{-1}; Horton and Shennan, 2009: 0.4 ± 0.3 mm a^{-1}; van Asselen, 2011: 0.1–1.7 mm a^{-1}), but lower than observed centennial to decadal-scale compaction rates (Törnqvist et al., 2008: 5–10 mm a^{-1}).

As further contribution to the potential decompaction of coastal intertidal stratigraphies, Brain

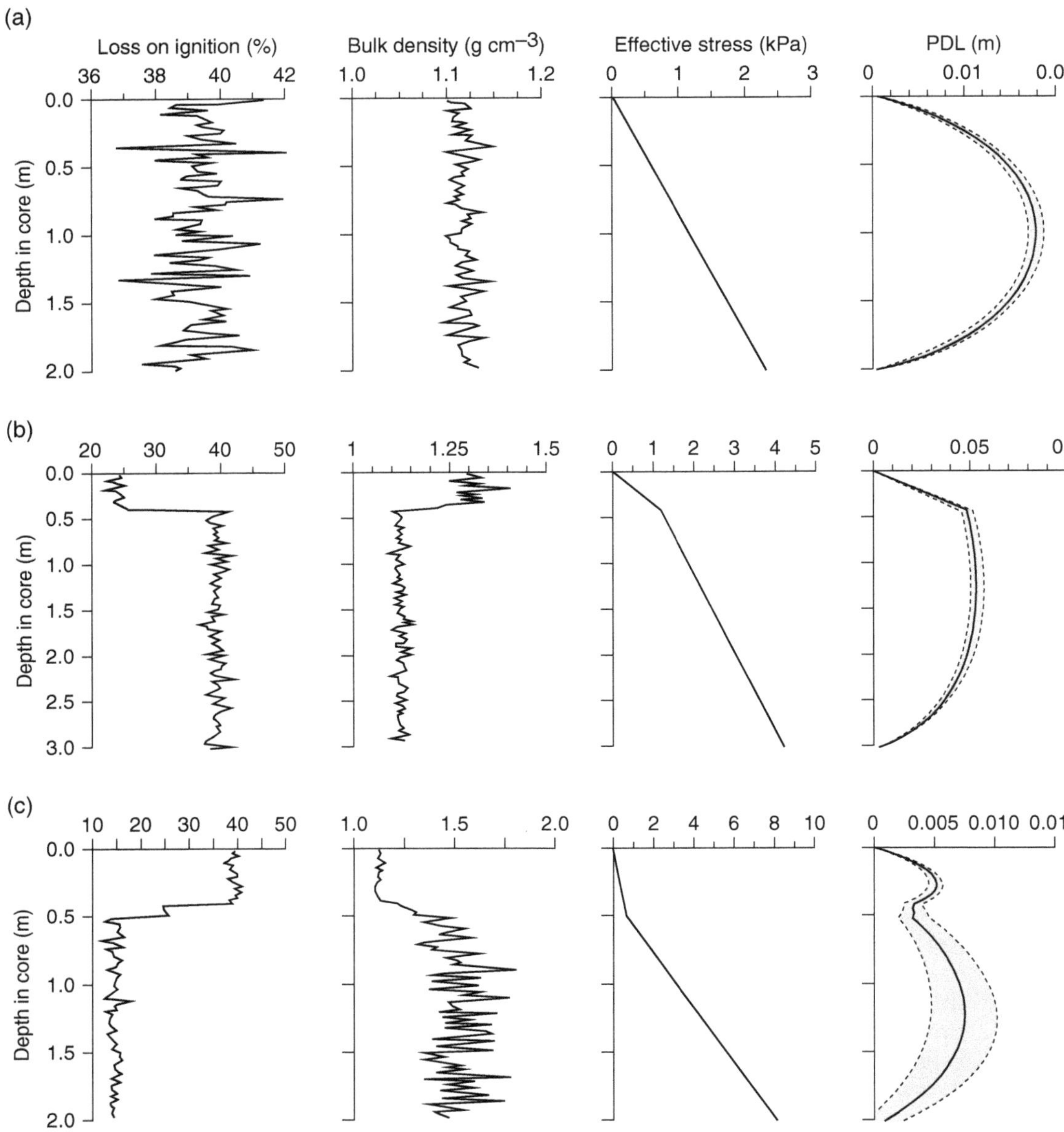

Fig. 21.3. Post-depositional lowering (PDL) profiles based on 100 model runs (mean ± SD) of autocompaction in intertidal sediment successions: (a) uniform 2 m succession; (b) 3 m transgressive succession; and (c) 2 m regressive succession. LOI, dry bulk density and effective stress profiles are used to demonstrate similarity with naturally occurring intertidal successions. *Source*: Adapted from Brain et al. (2012). Reproduced with permission of Elsevier and M. J. Brain.

et al. (2012) recognize that thin sediment successions should be selected for analysis when basal peat successions are not present. Further, it is recommended that LOI and dry bulk density data should be routinely obtained and considered. Indeed, LOI may be preferable to dry bulk density, particularly for short cores that may contain over-consolidated sediments (e.g., Gehrels et al., 2006).

21.5 PROVENANCE OF ORGANIC MATTER: ISOTOPIC AND ELEMENTAL (C&N) DATA

Stable isotopic (C and N) and elemental variations (C/N) in bulk sediment samples from Holocene coastal successions may be applied in the reconstruction of past sea-level trends (Törnqvist et al., 2004; Wilson et al., 2005a, b; Lamb et al., 2006,

2007; Kemp et al., 2010; Chapter 20). Indeed, Wilson and Lamb (2012) recently applied $\delta^{13}C$ and C/N data as an independent validation tool for a diatom-based transfer function reconstruction of Holocene sea level in a mid-latitude macrotidal estuary, although it is important to stress that bulk organic sediment $\delta^{13}C$ and C/N values cannot be applied as quantitative indicators of paleotidal level because sources of organic matter vary both spatially and temporally during the period of deposition, and the isotopic and elemental signatures are subject to post-depositional degradation (see review by Lamb et al., 2006).

The potential for using $\delta^{13}C$ and C/N data in studies of Holocene coastal change in response to changes in relative sea level is further illustrated in depositional settings where there have been significant shifts in the provenance of organic matter, for example in lagoons (e.g., Müller and Mathesius, 1999; Müller and Voss, 1999; Sampei and Matsumoto, 2001), and in isolation basins whereby the drop-off in marine-derived organic matter reflects the critical threshold elevation of the basin's sill and therefore relative sea-level change (e.g., Chivas et al., 2001; Westman and Hedenström, 2002; Mackie et al., 2005). Other applications have included the investigation of seasonal influences on estuary sedimentation (Weiguo et al., 2003; Allen et al., 2007) and longer-term studies of Quaternary coastal evolution (Da Cruz Miranda et al., 2009; Castro et al. 2010).

While this approach shows considerable promise, and indeed proven application, there are other considerations that make the isotopic and elemental detection of Holocene sea-level change problematical (see Chapter 20). First, the technique is restricted to areas where C_3 and C_4 plants have been present during the period of interest (Wilson et al., 2005a). Second, the relative inputs of the different sources of organic matter vary in time and space as a function of river catchment area (Goñi et al., 2003), estuarine hydrodynamics (tidal prism v. freshwater discharge, wind/wave climate; see Fichez et al., 1993; Goñi et al., 2009; Tremblay and Gagne, 2009), saltmarsh/perimarine wetland ecology and productivity (Chmura and Aharon, 1995; Middelburg et al., 1997; Sondi et al., 2008), and distance along and across the estuary (Odum, 1984; Yu et al., 2010). Furthermore, OM undergoes biogeochemical breakdown post-depositionally through dissolution, mineralization, and autolysis, and even during sample storage (Lamb et al., 2006, 2007). Some of these problems may be overcome by isotopic analysis of selected or extracted sediment components as opposed to bulk sediments (e.g., plant macrofossils; Törnqvist et al., 2004), or indeed particular organic compounds (Wang et al., 2003; Tanner et al., 2007).

21.6 CONCLUSIONS

What emerges from this review chapter is the fact that the measurement of LOI has been somewhat taken for granted or indeed virtually ignored in what would be regarded as more established approaches to Holocene sea-level reconstruction. This is perhaps not surprising considering the extent to which field-based investigation of biogenic and minerogenic sedimentation patterns dominates interpretation of the record of coastal change and therefore the sampling strategy for follow-up laboratory analyses. High-resolution sampling for LOI does provide an effective and rapid tool for assessing meso- to micro-scale temporal changes in the marine influence on sedimentation (e.g., studies of isolation basins and patterns of land deformation associated with pre- and coseismic events), but even here LOI is usually superseded by micropaleontological analyses that provide better resolved information on the environment of deposition with respect to sea level (i.e., SLIs). However, the observed LOI of present-day saltmarsh and mangrove deposits still offers potential for interpreting down-core trends in LOI as being linked to changes in sea level.

Even with the emergence of new methodological approaches for sea-level reconstruction, LOI is still measured in a crude, traditional fashion without appropriate consideration of how the results may be interpreted with respect to the presence of mineral matter and structural water contained within. This is of some significance in the reconstruction of recent sea-level trends from mid- and high-latitude saltmarsh sediment records, where LOI is taken as a quantitative proxy for OM when examining the controls on the modern ecological training set (even in the obvious presence of clay minerals). When it comes to correction of ecologically based SLI altitudes for autocompaction, the same applies in using the LOI proxy for OM with respect to evaluating the extent of post-depositional volume loss. As a further consideration, OM decomposition and leaching of organic compounds post-depositionally make it very difficult to differentiate between "real" changes in environment preserved in the sediment record from apparent change due to diagenesis. Whether it be due to LOI being an imperfect proxy for OM or suffering from post-depositional losses, methodological errors may be introduced through which the statistical significance of elevation in controlling ecological zonation is incorrectly evaluated and the extent of autocompaction may be over- or underestimated. The critical imperative to at least address the clay contamination of contemporary samples by using DTGA cannot be stressed enough when the wider scientific community are enthusiastically accepting emerging evidence of historical/modern-day accelerations in sea-level rise (e.g., Kemp et al., 2011; Gehrels and Woodworth, 2013). This perhaps unfairly represents the tremendous advances that are being made in establishing statistically robust records of sea-level change for the last 200 years or so and in ensuring the observed trend is not an artifact of diagenesis.

In recognizing recent developments in the use of stable isotopic and elemental C and N data applications, considerable progress has been made in the last decade in using evidence of varying OM provenance in Holocene coastal sediments for reconstructing sea-level/coastal change. Indeed, the influence of sea level on past coastal change at the macro- and meso-scales can be readily determined from such isotopic and elemental data.

In returning to the opening section, LOI and OM were only considered implicitly in van de Plassche (1986) in association with other proxies of past sea level. Indeed, this is generally true of the wealth of publications that have adopted a lithostratigraphic approach to acquiring the first data on sea-level change and then subsequent laboratory analytical ecological data. This actually does something of a disservice in that Orson van de Plassche was one of the pioneers of the development of carbon isotopic evidence of OM provenance in saltmarshes (e.g., Middelburg et al., 1997) and has played a fundamental role in progressing the ecological transfer function approach to sea-level reconstruction (e.g., Gehrels and van de Plassche, 1999; Edwards et al., 2004a, b; Roe and van de Plassche, 2005). Hence, to be critical of the prominence of LOI and OM in *Sea Level Research: A Manual for the Collection and Evaluation of Data* does not reflect on the wider influence of Orson van de Plassche in developing what we regard as new methods and approaches in sea-level research today.

REFERENCES

Aaby, B. (1986) Palaeoecological studies of mires. In: *Handbook of Holocene Palaeoecology and Palaeohydrology* (ed. Berglund, B.E.), The Blackburn Press, Caldwell, New Jersey, pp.145–164.

Aaby, B., and Berglund, B.E. (1986) Characterization of peat and lake deposits. In: *Handbook of Holocene Palaeoecology and Palaeohydrology* (ed. Berglund, B.E.), The Blackburn Press, Caldwell, New Jersey, pp. 231–246.

Alexiades C.A., and Jackson M.L. (1967) Chlorite determination in clays of soils and mineral deposits. *American Mineralogist*, 52, 1855–1873.

Allen, J.R.L. (1999) Geological impacts on coastal wetland landscapes: some general effects of sediment autocompaction in the Holocene of northwest Europe. Holocene, 9(1), 1–12.

Allen, J.R.L. (2000) Morphodynamics of Holocene salt marshes: A review sketch from the Atlantic and Southern North Sea coasts of Europe. Quaternary Science Reviews, 19(12), 1155–1231.

Allen, J.R.L., Lamb, A.L., and Dark, P. (2007) Seasonality of $\delta^{13}C$ and C/N ratios in modern and mid-Holocene sediments in the Severn Estuary Levels, SW Britain. Holocene, 17(1), 139–144.

Anon (1985) *Thermal Analysis of Gypsum*. Application Brief 22. SII nanoTechnology, pp. 1–3.

Baeteman, C., Scott, D.B., and van Strydonck, M. (2002) Changes in coastal zone processes at a high sea-level stand: a late Holocene example from Belgium. Journal of Quaternary Science, 17, 547–559.

Ball D.F. (1964) Loss-on-ignition as estimate of organic matter and organic carbon in non-calcareous soils. Journal of Soil Science, 15, 84–92.

Behre, K.-E. (1986) Analysis of botanical macro-remains. In: *Sea-level Research: A Manual for the Collection and Evaluation of Data* (ed. van de Plassche, O.), Geo Books, Norwich, pp. 413–433.

Bengtsson, L., and Enell, M. (1986) Chemical analysis. In: *Handbook of Holocene Palaeoecology and Palaeohydrology* (ed. Berglund, B.E.), The Blackburn Press, Caldwell, New Jersey, pp. 423–451.

Berglund, B.E. (1986) Palaeoecological reference areas and reference sites. In: *Handbook of Holocene Palaeoecology and Palaeohydrology* (ed. Berglund, B.E.), The Blackburn Press, Caldwell, New Jersey, pp. 111–126.

Bloom, A.L. (1964) Peat accumulation and compaction in a Connecticut salt marsh. Journal of Sedimentary Petrology, 34, 599–603.

Bos, I.J., Busschers, F.S., and Hoek, W.Z. (2012) Organic-facies determination: a key for understanding facies distribution in the basal peat layer of the Holocene Rhine-Meuse delta, The Netherlands. Sedimentology, 59, 679–703.

Boyle J.F. (2001) Inorganic geochemical methods in palaeolimnology. In: *Tracking Environmental Change Using Lake Sediments. Volume 2: Physical and Geochemical Methods* (eds Last, W.M., and Smol, J.P.), Kluwer Academic Publishers, Dordrecht, Netherlands, pp. 83–141.

Boyle J. (2004) A comparison of two methods for estimating the organic matter content of sediments. Journal of Paleolimnology, 31, 125–127.

Brain, M.J., Long, A.J., Petley, D.N., Horton, B.P., and Allison, R.J. (2011) Compression behaviour of minerogenic low energy intertidal sediments. Sedimentary Geology, 233, 28–41.

Brain, M.J., Long, A.J., Woodroffe, S.A., Petley, D.N., Milledge, D.G., and Parnell, A.C. (2012) Modelling the effects of sediment compaction on salt marsh reconstructions of recent sea-level rise. Earth and Planetary Science Letters, 345–348, 180–193.

Castro, D.F., de Fátima Rossetti, D., and Ruiz Pessenda, L.C. (2010) Facies, $\delta^{13}C$, $\delta^{15}N$ and C/N analyses in a late Quaternary compound estuarine fill, northern Brazil and relation to sea level. Marine Geology, 274, 135–150.

Chaisene, A., and Rangsriwatananon, R. (2004) Effects of thermal and acid treatments on some physico-chemical properties of Lampang diatomite. Suranaree Journal of Science and Technology, 11, 289–299.

Chivas, A.R., Garcia, A., van der Kaars, S., Couapel, M.J.J., Holt, S., Reeves, J.M., Wheeler, D.J., Switzer, A.D., Murray-Wallace, C.V., Banerjee, D., Price, D.M., Wang, S.X., Pearson, G., Edger, N.T., Beaufort, L., De Deckker, P., Lawson, E., and Cecil, C.B. (2001) Sea-level and environmental changes since the last interglacial in the Gulf of Carpentaria, Australia: an overview. Quaternary International, 83–85, 19–46.

Chmura, G.L., and Aharon, P. (1995) Stable carbon isotope signatures of sedimentary carbon in coastal wetlands as indicators of salinity regime. Journal of Coastal Research, 11(1), 124–135.

Da Cruz Miranda, M.C., de Fátima Rossetti, D., and Ruiz Pessenda, C.L. (2009) Quaternary paleoenvironments and relative sea-level changes in Marajó Island (Northern Brazil): Facies, $\delta^{13}C$, $\delta^{15}N$ and C/N. Palaeogeography, Palaeoclimatology, Palaeoecology, 282(1–4), 19–31.

Davis, M.B. (1976) Erosion rates and land use history in southern Michigan. Environmental Conservation, 3, 139–148.

Day, J.W., and Giosan, L. (2008) Geomorphology: survive or subside? Nature Geoscience, 1, 156–157.

de Rijk, S. (1995) Salinity control on the distribution of salt-marsh foraminifera (great-marshes, Massachusetts). Journal of Foraminiferal Research, 25, 156–166.

de Rijk, S., and Troelstra, S. (1999) The application of a foraminiferal actuo-facies model to salt-marsh cores. Palaeogeography Palaeoclimatology Palaeoecology, 149, 59–66.

De Vos, B., Vandecasteele, B., Deckers, J., and Muys, B. (2005) Capability of loss-on-ignition as a predictor of total organic carbon in non-calcareous forest soils. Communications in Soil Science and Plant Analysis, 36, 2899–2921.

Dearing, J.A. (1986) Core correlation and total sediment influx. In: In: *Handbook of Holocene Palaeoecology and Palaeohydrology* (ed. Berglund, B.E.), The Blackburn Press, Caldwell, New Jersey, pp. 247–270.

Deer, W.A., Howie, R.A., and Zussman, J. (1966) *An Introduction to the Rock-forming Minerals*. Longman, Harlow, UK.

Digerfeldt, G. (1972) The post-glacial development of Lake Trummen. Regional vegetation history, water level changes and palaeolimnology. Folia Limnologica Scandinavica, 16, 104.

Digerfeldt, G. (1986) Studies on past lake-level fluctuations. In: *Handbook of Holocene Palaeoecology and Palaeohydrology* (ed. Berglund, B.E.), The Blackburn Press, Caldwell, New Jersey, pp. 127–144.

Donnelly, J.P., Cleary, P., Newby, P., and Ettinger, R. (2004) Coupling instrumental and geological records of sea-level change: evidence from southern New England of an increase in the rate of sea-level rise in the late 19th century. Geophysical Research Letters, 31(5), L05203-1–4.

Edwards, R.J. (2006) Mid- to late-Holocene relative sea-level change in southwest Britain and the influence of sediment compaction. Holocene, 16(4), 575–587.

Edwards, R.J., Wright, A., and van de Plassche, O. (2004a) Surface distributions of salt-marsh foraminifera from Connecticut, USA: Modern analogues for high-resolution sea level studies. Marine Micropalaeontology, 51(1–2), 1–21.

Edwards, R.J., van de Plassche, O., Gehrels, W.R., and Wright, A.J. (2004b) Assessing sea-level data from Connecticut, USA, using a foraminiferal transfer function for tide level. Marine Micropalaeontology, 51(3–4), 239–255.

Fernández, J.M., Peltre, C., Craine, J.M., and Plante, A.F. (2012) Improved characterization of soil organic matter by thermal analysis using CO2/H2O evolved gas analysis. Environmental Science and Technology, 46, 8921–8927.

Fichez, R., Dennis, P., Fontaine, M.F., and Jickells, T.D. (1993) Isotopic and biochemical composition of particulate organic matter in a shallow water estuary (Great Ouse, North Sea, England). Marine Chemistry, 43(1–4), 263–276.

French, J.R. (1993) Numerical simulation of vertical marsh growth and adjustment to accelerated sea-level rise, north Norfolk, UK. Earth Surface Processes and Landforms, 81, 63–81.

Gehrels, W.R., and van de Plassche, O. (1999) The use of *Jadammina macrescens* (Brady) and *Balticammina pseudomacrescens* (Bronnimann, Lutze and Whittaker) (Protozoa : Foraminiferida) as sea-level indicators. Palaeogeography, Palaeoclimatology, Palaeoecology, 149(1–4), 89–101.

Gehrels, W. R., and Woodworth, P.L. (2013) When did modern rates of sea-level rise start? Global and Planetary Change, 100, 263–277.

Gehrels, W.R., Belknap, D.F., and Kelley, J.T. (1996) Integrated high precision analyses of Holocene relative sea-level changes: lessons from the coast of Maine. Geological Society of America Bulletin, 108, 1073–1088.

Gehrels, W.R., Marshall, W.A., Gehrels, M.J., Larsen, G., Kirby, J.R., Eiriksson, J., Heinemeier, J., and Shimmield, T. (2006) Rapid sea-level rise in the North Atlantic Ocean since the first half of the 19th century. Holocene, 16, 948–964.

Goñi, M.A., Teixeira, M.J., and Perkey, D.W. (2003) Sources and distribution of organic matter in a river-dominated estuary (Winyah Bay, SC, USA). Estuarine, Coastal and Shelf Science, 57(5–6), 1023–1048.

Goñi, M.A., Voulgaris, G., and Kim, Y.H. (2009) Composition and fluxes of particulate organic matter in a temperate estuary (Winyah Bay, South Carolina, USA) under contrasting physical forcings. Estuarine, Coastal and Shelf Science, 85(2), 273–291.

Håkansson, L., and Jansson, M. (1983) *Principles of Lake Sedimentology*. Springer Verlag, New York.

Haslett, S.K., Davies, P., Curr, R.H.F., Davies, C.F.C., Kennington, K., King, C.P., and Margetts, A.J. (1998) Evaluating late-Holocene relative sea-level change in the Somerset Levels, southwest Britain. Holocene, 8, 197–207.

Hatipo⊠lu, M., Can, N., and Karali, T. (2010) Effects of heating on fire opal and diaspore from Turkey. Physica B, 405, 1729–1736.

Head, K.H. (1988) *Manual of Soil Laboratory Testing: Permeability, Shear Strength and Compressibility Tests.* Pentech Press, London/Plymouth.

Heiri, O., Lotter, A.F., and Lemcke, G. (2001) Loss on ignition as a method for estimating organic and carbonate content in sediments: reproducibility and comparability of results. Journal of Paleolimnology, 25, 101–110.

Heyworth, A. (1986) Submerged forests as sea-level indicators. In: *Sea-level Research: A Manual for the Collection and Evaluation of Data* (ed. van de Plassche, O.), Geo Books, Norwich, pp. 401–411.

Horton, B.P., and Shennan, I. (2009) Compaction of Holocene strata and the implications for relative sea-level change on the east coast of England. Geology, 37, 1083–1086.

Jelgersma, S. (1961) *Holocene Sea Level Changes in the Netherlands*. Van Aelst, Maastricht.

Kaye, C.A., and Barghoorn, E.S. (1964) Quaternary sea-level change and crustal rise at Boston, Massachussetts, with notes on the autocompaction of peat. Geological Society of America Bulletin, 75, 63–80.

Kemp, A.C., Vane, C.H., Horton, B.P., and Culver, S.J. (2010) Stable carbon isotopes as potential sea-level indicators in salt marshes,North Carolina, USA. The Holocene, 20(4), 623–636.

Kemp, A.C., Horton, B.P., Donnelly, J.P., Mann, M.E., Vermeer, M., and Rahmstorf, S. (2011) Climate related sea-level variations over the past two millennia. Proceedings of the National Academy of Sciences of the United States of America, 108(27), 11017–11022.

Kļaviņš, M., Ansone, L., Tjutrins, J., Silamiķele, O., and Purmalis, O. (2010) Differential thermal analysis of peat and peat humic acids in relation to their origin. In: *Mires and Peat* (ed. Kļaviņš, M.), University of Latvia Press, Riga, pp. 207–214.

Lamb, A.L., Wilson, G.P., and Leng, M.J. (2006) A review of coastal palaeoclimate and relative sea-level reconstructions using δ^{13}C and C/N ratios in organic material. Earth Science Reviews, 75, 29–57.

Lamb, A.L., Vane, C.H., Wilson G.P., Rees, J.G., and Moss-Hayes, V.L. (2007) Assessing δ^{13}C and C/N ratios from organic material in archived cores as Holocene sea level and palaeoenvironmental indicators in the Humber Estuary, UK. Marine Geology, 244, 109–128.

Larcombe, P., and Carter, R.M. (1998) Sequence architecture during the Holocene transgression: an example from the Great Barrier Reef shelf, Australia. Sedimentary Geology, 117, 97–121.

Long, A.J., Waller, M.P., and Stupples, P. (2006) Driving mechanisms of coastal change: peat compaction and the destruction of late Holocene coastal wetlands. Marine Geology, 225, 63–84.

Mackereth, F.J.J. (1966) Some chemical observations of post-glacial lake sediments. Philosophical Transactions of the Royal Society of London, 250, 167–213.

Mackie, E.A.V., Leng, M.J., Lloyd, J.M., and Arrowsmith, C. (2005) Bulk organic δ^{13}C and C/N ratios as palaeosalinity indicators within a Scottish isolation basin. Journal of Quaternary Science, 20, 301–408.

Massey, A.C., Paul, M.A., Gehrels, W.R., and Charman, D.J. (2006) Autocompaction in Holocene coastal back-barrier sediments from south Devon, southwest England, UK. Marine Geology, 226, 225–241.

McKee, K.L., Cahoon, D.L., and Feller, I.C. (2007) Caribbean mangroves adjust to rising sea level through biotic controls on changes in soil elevation. Global Ecology and Biogeography, 16, 545–556.

Meckel, T.A., ten Brink, U.S., and Williams, S.J. (2007) Sediment compaction rates and subsidence in deltaic plains: numerical constraints and stratigraphic influences. Basin Research, 19, 19–31.

Middelburg, J.J., Nieuwenhuize, J., Lubberts, R.K., and van de Plassche, O. (1997) Organic carbon isotope systematics of coastal marshes. Estuarine, Coastal and Shelf Science, 45(5), 681–687.

Mielenz, R.C., Schieltz, N.C., and King, M.E. (1953) Thermogravimetric analysis of clay and clay-like minerals. Clays and Clay Minerals, 2, 285–314.

Mills, H., Kirby, J., Holgate, S., and Plater, A. (2013) The distribution of contemporary saltmarsh foraminifera in a macrotidal estuary: an assessment of their viability for sea-level studies. Journal of Ecosystem and Ecography, 3, 131, doi: 10.4172/2157-7625.1000131

Müller, A., and Mathesius, U. (1999) The palaeoenvironments of coastal lagoons in the southern Baltic Sea: I. The application of sedimentary C_{org}/N ratios as source indicators of organic matter. Palaeogeography, Palaeoclimatology, Palaeoecology, 145, 1–16.

Müller, A., and Voss, M. (1999) The palaeoenvironments of coastal lagoons in the southern Baltic Sea: II. δ^{13}C and δ15N ratios of organic matter: sources and sediments. Palaeogeography, Palaeoclimatology, Palaeoecology, 145, 17–32.

Murray, J.W. (1971) Living foraminiferainiferids of tidal marshes; a review. The Journal of Foraminiferal Research, 1, 153–161.

Odum, W.E. (1984) Dual-gradient concept of detritus transport and processing in estuaries. Bulletin of Marine Science, 35(3), 510–521.

Pallisser, R., Minasny, B., and McBratney, A.B. (2013) Soil carbon determination by thermogravimetrics. *PeerJ*, 15, https://peerj.com/articles/6/.

Parkinson, R.W. (1989) Decelerating Holocene sea-level rise and its influence on southwest Florida coastal evolution: a transgressive/regressive stratigraphy. Journal of Sedimentology, 59(6), 960–972.

Patterson, R.T. (1990) Intertidal benthic foraminifera biofacies on the Fraser River Delta, British Columbia. Micropaleontology, 36, 229–244.

Paul, M.A., and Barras, B.F. (1998) A geotechnical correction for post-depositional sediment compression: examples from the Forth Valley, Scotland. Journal of Quaternary Science, 13(2), 171–176.

Pizzuto, J.E., and Schwendt, A.E. (1997) Mathematical modeling of autocompaction of a Holocene

transgressive valley-fill deposit, Wolfe Glade, Delaware. Geology, 25(1), 57–60.
Reed, D. (1995) The response of coastal marshes to sea-level rise: survival or submergence? Earth Surface Processes and Landforms, 20, 39–48.
Roe, H.M., and van de Plassche, O. (2005) Modern pollen distribution in a Connecticut saltmarsh: Implications for studies of sea-level change. Quaternary Science Reviews, 24(18–19), 2030–2049.
Saikia, B.K., Boruah, R.K., Gogoi, P.K., and Baruah, B.P. (2009) A thermal investigation on coals from Assam (India). Fuel Processing Technology, 90, 196–203.
Sampei, Y., and Matsumoto, E. (2001) C/N ratios in a sediment core from Nakaumi Lagoon, southwest Japan: usefulness as an organic indicator. Geochemical Journal, 35, 189–205.
Sharma, H.S.S., Mellon, R.M., Givens, D.I., Park, R.S., Archer, J., Lyons, G., and Fletcher, H. (2012) Evaluation of perennial ryegrass (*Lolium perenne L.*) for digestibility using thermogravimetry. Animal Feed Science and Technology, 177, 30–39.
Shaw, T., Plater, A., Holgate, S., and Kirby, J. (2012) Reconstruction of historical sea-level trends for the Croatian coast of the Adriatic using saltmarsh foraminifera. Poster G31A-0915, AGU Fall Meeting, San Francisco, 3–7 December 2012.
Shennan, I. (1986) Flandrian sea-level changes in Fenland II. Tendencies of sea-level movement, altitudinal changes and local and regional factors. Journal of Quaternary Science, 1, 155–179.
Shennan, I., and Horton, B. (2002) Holocene land- and sea-level changes in Great Britain. Journal of Quaternary Science, 17(5–6), 511–526.
Shennan, I., Innes, J.B., Long, A.J., and Zong, Y. (1995) Holocene relative sea-level changes and coastal vegetation history at Kentra Moss, Argyll, northwest Scotland. Marine Geology, 124, 43–59.
Shennan, I., Green, F., Innes, J., Lloyd, J., Rutherford, M., and Walker, K. (1996) Evaluation of rapid relative sea-level changes in North-West Scotland during the Last Glacial-Interglacial transition: Evidence form Ardtoe and other isolation basins. Journal of Coastal Research, 12, 862–874.
Sondi, I., Lojen, S., Juračić, M., and Prohić, E. (2008) Mechanisms of land-sea interactions - the distribution of metals and sedimentary organic matter in sediments of a river-dominated Mediterranean karstic estuary. Estuarine, Coastal and Shelf Science, 80(1), 12–20.
Tanner, B.R., Uhle, M.E., Kelley, J.T., and Mora, C.I. (2007) C_3/C_4 variations is salt-marsh sediments: an application of compound specific isotopic analysis of lipid biomarkers to late Holocene palaeoenvironmental research. Organic Geochemistry, 38, 474–84.
Tolonen, K., Siiriäinen, A., and Thompson, R. (1975) Prehistoric field erosion sediment in Lake Lojarvi, S. Finalnd and its palaeomagnetic dating. Annales Botanici Fennici, 12, 161–164.
Törnqvist, T.E., deJong, A.F.M., Kurnik, C.W., Gonzalez, J.L., Newsom, L.A., and van der Borg, K. (2004) Deciphering Holocene sea-level history on the U.S. Gulf Coast: A high-resolution record from the Mississippi Delta. Bulletin of the Geological Society of America, 116(7–8), 1026–1039.
Törnqvist, T.E., Wallace, D.J., Storms, J.E.A., Wallinga, J., VanDam, R.L., Blaauw, M., Derksen, M.S., Klerks, C.J.W., Meijneken, C., and Snijders, E.M.A. (2008) Mississippi Delta subsidence primarily caused by compaction of Holocene strata. Nature Geoscience, 1, 173–176.
Tovey, N.K., and Paul, M.A. (2002) Modelling self-weight consolidation in Holocene sediments. Bulletin of Engineering Geology and the Environment, 61(1), 21–33.
Tremblay, L., and Gagne, J.P. (2009) Organic matter distribution and reactivity in the waters of a large estuarine system. Marine Chemistry, 116(1–4), 1–12.
Troels-Smith, J. (1955) Karakterisering af løse jordarter (Characterization of unconsolidated sediments). Danmarks Geologiske Undersøgelse Series IV, 3, 38–73.
van Asselen, S. (2011) The contribution of peat compaction to total basin subsidence: implications for the provision of accommodation space in organic-rich deltas. Basin Research, 23, 239–255.
van Asselen, S., Stouthamer, E., and van Asch, Th.W.J. (2009) Effects of peat compaction on delta evolution: A review on processes, responses, measuring and modelling. Earth Science Reviews, 92, 35–51.
van Asselen, S., Karssenberg, D., and Stouthamer, E. (2011) Contribution of peat compaction to relative sea-level rise within Holocene deltas. Geophysical Research Letters, 38, L24401.
van de Plassche, O. (1982) Sea-level change and water-level movements in the Netherlands during the Holocene. Mededelingen Rijks Geologische Dienst, 36, 1–93.
van de Plassche, O. (ed.) (1986) *Sea-level Research: A Manual for the Collection and Evaluation of Data*. Geo Books, Norwich.
Wang, X.-C., Chen, R.F., and Berry, A. (2003) Sources and preservation of organic matter in Plum Island salt marsh sediments (MA, USA): long-chain *n*-alkanes and stable carbon isotope compositions. Estuarine, Coastal and Shelf Science, 58, 917–928.
Weiguo, L., Zisheng, A., Weijian, Z., Head, M.J., and Delin, C. (2003) Carbon isotope and C/N ratios of suspended matter in rivers: an indicator of seasonal change in C_4/C_3 vegetation. Applied Geochemistry, 18, 1241–1249.
Westman, P., and Hedenström, A. (2002) Environmental changes during isolation processes from the Littorina Sea as reflected by diatoms and geochemical parameters: a case study. The Holocene, 12, 531–540.
Wilson, G.P., and Lamb, A.L. (2012) An assessment of the utility of regional diatom-based tidal-level transfer functions. Journal of Quaternary Science, 27(4), 360–370.
Wilson, G.P., Lamb, A.L., Leng, M.J., Gonzalez, S., and Huddart, D. (2005a) Variability of organic $\delta^{13}C$ and C/N in the Mersey Estuary, UK and its implications for sea-level reconstruction studies. Estuarine, Coastal and Shelf Science, 64, 685–698.
Wilson, G.P., Lamb, A.L., Leng, M.J., Gonzalez, S., and Huddart, D. (2005b) $\delta^{13}C$ and C/N as potential coastal palaeoenvironmental indicators in the Mersey Estuary, UK. Quaternary Science Reviews, 24, 2015–2029.
Woodroffe, C.D. (1981) Mangrove swamp stratigraphy and Holocene transgression, Grand Cayman Island, West Indies. Marine Geology, 41, 271–294.

Woodroffe, C.D. (1990) The impact of sea-level rise on mangrove shorelines. Progress in Physical Geography, 14(4), 483–520.

Yu, F., Zong, Y., Lloyd, J.M., Huang, G., Leng, M.J., Kendrick, C., Lamb, A.L., and Yim, W.W.S. (2010) Bulk organic $\delta^{13}C$ and C/N as indicators for sediment sources in the Pearl River delta and estuary, southern China. Estuarine, Coastal and Shelf Science, 87(4), 618–630.

Zong, Y., Shennan, I., Combellick, R.A., Hamilton, S.L., and Rutherford, M.M. (2003) Microfossil evidence for land movements associated with the AD 1964 Alaska earthquake. The Holocene, 13, 7–20.

Chapter 22
Grain size analysis

ADAM D. SWITZER[1,2] AND JEREMY PILE[1]
[1]*Earth Observatory of Singapore, Nanyang Technological University, Singapore*
[2]*Division of Earth Sciences, Nanyang Technological University, Singapore*

22.1 INTRODUCTION

The measurement of grain size in sediments is a fundamental tool for interpreting sedimentary strata and depositional systems in most environments. Such endeavors are a primary part of many sea-level studies and, although grain size can rarely be used as a direct proxy for sea-level change (with the potential exception of transgressive and regressive system tracts on continental shelves), the measurement of grain size forms an integral part of many proxy studies.

The grain size of sediments reflects relative energy, which is one of the fundamental environmental factors that control the erosion, transport, and deposition of sediment particles. In most landform systems that are of interest to sea-level researchers, the particle size distribution (PSD) of constituent sediments reflects morphological characteristics associated with the physical processes of landform development (Table 22.1). The variety of grain sizes encountered by those interested in sea level, anything from fine clays and silts to large boulders, means that no one technique can be applied to all environments or studies (Fig. 22.1). It is generally desirable to have a representative, repeatable, precise, accurate and, if possible, convenient method by which to conduct sediment grain size measurements when reconstructing paleoenvironments or investigating processes such as sea-level change (e.g., Syvitski et al., 1991a, b; Donato et al., 2009). Of course, the variety of sediments encountered by sea-level researchers dictates that such desires must be traded against logistics (e.g., sample sizes, representativeness of samples collected and results obtained) and the economics of analysis. A myriad of grain size techniques and publications exist that compare different techniques and, in many cases, comment on the limitations (inaccuracies) of one technique with respect to another (see Switzer, 2013 for a complete list).

A combination of methods is usually needed to quantify a grain size population, for example sieving and a sedimentation technique (e.g., pipette analysis) or sieving and laser particle size analysis (PSA). No particular technique can be considered "more accurate" or "more precise" than another, as all techniques have inherent limitations. Generally, researchers presenting in sea-level studies use "phi" units or microns along with associated measures of standard deviation, skewness, and kurtosis.

22.2 DEFINING GRAIN SIZE

Providing a definition for "grain size" that everyone and anyone can use is not an easy task. According to the *Dictionary of Geological Terms*, 3rd edition (Bates and Jackson, 1984, p. 370), grain size or particle size is defined as "The general dimensions, such as average diameter or volume, of the particles in a sediment or rock, or of the grains of a particular mineral that make up a sediment or a rock, based on the premise that the particles are spheres or that the measurements made can be expressed as diameters of equivalent spheres".

For those interested in sea level the grains that are measured are usually composed of inorganic mineral grains or rock fragments. In some coastal settings, constituent grains will exhibit relatively molecular homogeneity (e.g., quartz grains on a beach) while others such as coralline beaches will be relatively inhomogeneous in composition. Inhomogeneity is inherently problematic, particularly when conducting grain size analysis on platy sediments, carbonates, evaporites and sediments of highly variable density. One example of particular interest to sea-level researchers

Handbook of Sea-Level Research, First Edition. Edited by Ian Shennan, Antony J. Long, and Benjamin P. Horton.

Table 22.1. Characteristic features of coastal depositional systems according to grain size (after Orton and Reading, 1993). Lithofacies after Miall (1985).

	Gravel	Gravel and sand	Fine sand	Mud/silt
Delta plain				
Size	Small ($<50\,km^2$)	Intermediate ($<1000\,km^2$)	Intermediate ($<50{,}000\,km^2$)	Large ($<500{,}000\,km^2$)+
Gradient	High ($>5\,m\,km^{-1}$)	Intermediate ($0.5–5.0\,m\,km^{-1}$)	Intermediate ($0.1–1.0\,m\,km^{-1}$)	Low ($0–0.1\,m\,km^{-1}$)
Percentage subaerial	High (>90%)	Intermediate (70–90%)	Intermediate (50–70%)	Low (<50%)
Environments	Gravel bar deposits	Braided sandflat, strand plain or beach ridges	Abandoned fluvial channels, sand flats, marsh, aeolian sand dunes, lagoon with storm washover sands	Interdistributary bays, lakes or swamps, tidal flat or chenier plain, fluvial and tidal channels, levees, crevasse splays
Lithofacies	Gravel-massive/ crudely bedded, gravel-matrix supported, gravel-planar crossbeds, sand (medium–very coarse) -trough cross-beds	Gravel-trough crossbeds, sand (medium–very coarse) -trough cross-beds, and planar cross-beds	Sand (medium–very coarse) -trough cross-beds, sand (very fine–very coarse) -horizontal lamination, sand (medium–very coarse) -planar cross-beds, sand (very fine–coarse) ripple marks	Fines (mud/silt) – massive with dessication cracks, fines (mud/silt/sand) –fine lamination, sand (medium–very coarse) -trough cross-beds, sand (very fine–coarse) ripple marks
River mouth				
Mixing behavior	Friction-inertia	Friction	Buoyancy-friction	Buoyancy
Relative density	Homo- and hyperpycnal	Homo- and hyperpycnal	Hypopycnal–hyperpycnal	Hypopycnal flow
Saltwedge intrusion	Never	Occasionally	Sometimes	Commonly
Effluent spreading angle	Wide	Intermediate	Intermediate	Narrow
Distributary pattern	Channel bifurcation, unchannelized	Channel bifurcation	Channel bifurcation	Single channel
Deposits	Radial bars, Gilbert-type foresets, lunate bars	Radial, middle-ground bars	Radial, middle-ground bars	Bar-finger sands
Shoreline				
Shape	Straight	Straight–cuspate	Irregular–lobate	Irregular–elongate
Gradient	Steep ($>50\,m\,km^{-1}$)	Intermediate ($10–50\,m\,km^{-1}$)	Intermediate ($10–50\,m\,km^{-1}$)	Gentle ($0–10\,m\,km^{-1}$)
Morphology	Reflective	Reflective–dissipative	Dissipative–reflective	Dissipative
Width of shoreface	<50 m	50–200 m	50–500 m	>500 m
Important wave processes	Traction currents, wave-winnowing on shoreface	Traction currents, longshore, transverse, and rhythmic bars	Large-scale nearshore circulation patterns involving longshore drift and rip currents	Easy suspension of sediment, slow shoaling of waves, nearshore trapping of sediment
Width of intertidal zone	<500 m	500 m–2 km	500 m–5 km	>5 km
Tidal influence	Low, limited ability of currents to transport sediment intergranular seepage on foreshore prevents formation of tidal channels	Intermediate	Intermediate, tidal inlets are common	High, river channel often as mixed estuary, frequent tidal inlets and ebb-tidal deltas if barrier islands form

(*Continued*)

Table 22.1. (*Continued*)

	Gravel	Gravel and sand	Fine sand	Mud/silt
Depositional features	Non-barred shoreface, except during storms	Barred shoreface, small barrier islands and associated lagoons, ebb-tidal deltas are rare	Barrier islands, barred shoreface, aeolian sanddunes, tidal channels and deltas	Shore-attached mud banks, chenier plain, tidal flats
Lithofacies	Sand (medium–very coarse) -trough cross-beds, 2D wave ripples, gravel lags	Sand (medium–very coarse) -trough cross-beds, sand (medium–very coarse) -planar cross-beds, 2D wave ripples	Sand (medium–very coarse) -trough cross-beds, hummocky cross-stratification, tidal sandwaves	Fines (mud/silt) – massive with dessication cracks, fines (mud/silt/sand) –fine lamination, sand (very fine–coarse) ripple marks
Sediment mobility	Low	Intermediate	Intermediate	High
Long-term coastline stability	Low	Intermediate	Intermediate	High
Subaqueous delta front				
Gradient	High ($50–1000\,m\,km^{-1}$)	Intermediate ($50–500\,m\,km^{-1}$)	Intermediate ($5–200\,m\,km^{-1}$)	Low ($<20\,m\,km^{-1}$)
Organization	Low	Intermediate	Intermediate	High
Frequency of re-sedimentation events	High	Intermediate	Intermediate	Low
Magnitude of flows	Small	Intermediate	Intermediate	Large
Depositional features	Blocky talus cones, distal avalanche blocks	Braided sand chutes	Branching or meandering sand chutes	Persistent mudflow gully or channel with levees, mud diapirs, growth faults
Supply mechanism	Viscous debris flows	High density turbidity currents	Turbidity currents, deposition from suspension	Mudflows, slumps, deposition from suspension

is the analysis of carbonate grains; the vastly different densities and shapes found in biogenic sediments mean that measurement devices and techniques will respond differently to the same particle (e.g., Folk and Robles, 1964; Kench and McLean, 1997).

This chapter is not an exhaustive review of the problems and issues of grain size analysis. Problems that arise with particular techniques or types of analysis will be considered throughout the chapter. Although carbonate coasts are important for sea level, this work will primarily deal with an assumption that the basic analysis needs of sea-level researchers pertain to near-spherical grains of homogenous composition (e.g., quartz sands).

22.2.1 Grain size analysis, grain sizing and grain size

The broad-ranging text edited by Syvitski (1991) shows very clearly that although there are many methods for grain size, none have been adopted as a comprehensive standard in Earth sciences. It could be argued that the use of International Standards Organization (ISO) or American Society for Testing and Materials (ASTM) protocols are appropriate standards (e.g., ISO 1999 a, b, 2008, 2009; ASTM, 2010). It is however apparent that these standards are primarily used by engineers and food technologists, and are the result of a desire for consistent and comparable reporting procedures when dealing with industrial or food-grade powders and particles (e.g., powder coatings).

Grains commonly present a different cross-section depending on their orientation and this is one of many measurement problems faced when determining grain size. A second is that the cross-section of an irregularly shaped particle has different "diameters" depending on where the chord is drawn (Webb, 2008). This limitation is overcome to some extent if sphericity is assumed and, as such, an equivalent spherical value is produced as all techniques aim to provide a single number

km	m	mm		phi	Udden (1914)	Wentworth (1922)	Blair & McPherson (1999)	Blott & Pye (2012)
1075				−30			Very coarse megalith	
538				−29			Coarse megalith	
269				−28			Medium megalith	
134				−27			Fine megalith	
67.2				−26			Very fine megalith	
33.6				−25			Very coarse monolith	
16.8				−24			Coarse monolith	
8.4				−23			Medium monolith	
4.2				−22			Fine monolith	
2.1				−21			Very fine monolith	
1.0	1048.6			−20			Very coarse slab	
0.5	524.3			−19			Coarse slab	
0.26	262.1			−18			Medium slab	
	131.1			−17			Fine slab	
	65.5			−16			Very coarse block	
	32.8			−15			Coarse block	
	16.4			−14			Medium block	
	8.2			−13			Fine block	
	4.1	4096		−12			Very coarse boulder	Megaclasts
	2.0	2048		−11			Coarse boulder	Very large boulder
	1.0	1024		−10		Bowlder gravel	Medium boulder	Large boulder
	0.5	512		−9			Fine boulder	Medium boulder
	0.25	256		−8	Large boulders	Cobble gravel	Coarse cobble	Small boulder
		128		−7	Medium boulders		Fine cobble	Very small boulder
		64		−6	Small boulders	Pebble gravel	Very coarse pebble	Very coarse gravel
		32		−5	Very small boulders		Coarse pebble	Coarse gravel
		16		−4	Very coarse gravel		Medium pebble	Medium gravel
		8		−3	Coarse gravel		Fine pebble	Fine gravel
		4		−2	Gravel	Granule gravel	Granule	Very fine granule
		2		−1	Fine gravel	Very coarse sand	Very coarse sand	Very coarse sand
		1	µm	0	Coarse sand	Coarse sand	Coarse sand	Coarse sand
		1/2	500	1	Medium sand	Medium sand	Medium sand	Medium sand
		1/4	250	2	Fine sand	Fine sand	Fine sand	Fine sand
		1/8	125	3	Very fine sand	Very fine sand	Very fine sand	Very fine sand
		1/16	62.50	4	Coarse silt or dust	Silt	Very coarse silt	Very coarse silt
		1/32	31.25	5	Medium silt or dust		Coarse silt	Coarse silt
		1/64	15.63	6	Fine silt or dust		Medium silt	Medium silt
		1/128	7.81	7	Very fine silt or dust		Fine silt	Fine silt
		1/256	3.91	8	Coarse clay	Clay	Clay	Very fine silt
		1/512	1.95 nm	9	Medium clay			Very coarse clay
		1/1024	977	10	Fine clay			Coarse clay
		1/2048	488	11				Medium clay
		1/4096	244	12				Fine clay
		1/8192	122	13			↓ ?	Very fine clay

Fig. 22.1. Comparison of particle size scales and terminology commonly used in geology and sedimentology. *Source*: Udden (1914) and Wentworth (1922), with Blair and McPherson (1999) and new particle sizes/terminology proposed by Blott and Pye (2012).

that is indicative of grain size. Most sizing techniques therefore assume that the material being measured is spherical and report the particle size as the diameter of the "equivalent sphere". In most cases, the equivalent sphere approximation for an irregularly shaped particle is dependent on the physical property measured by the chosen technique (e.g., laser diffraction techniques provide a diameter of the sphere that yields an equivalent light scattering pattern to the particle being measured).

22.2.2 Choosing the correct grain size technique/instrument

The grain size instrument and technique must be suitable for the material to be measured (in this case, the sediment being analyzed) and must also provide data to meet the specific needs of the sea-level researcher. For a given particle, all different particle sizing techniques are likely to produce a different "size" and, in most cases, all of them are likely to be technically "correct". The best instrument for the application (the best size definition) may be the one that most closely relates the "particle size" to the application of the user. In sea-level research, determining grain size is rarely the ultimate objective; determining the process of deposition or landform development and inferring its link to sea level is usually the reason for the measurement. That stated, it is still clear that grain size analysis remains one of the key components in landscape reconstructions in the coastal zone and, as such, is of significant relevance to sea-level studies (e.g., Van Hengstum et al., 2011).

22.3 GRAIN SIZE ANALYSIS SAMPLE PREPARATION

In most sea-level studies many samples will be of interest at any one time and a first step is to systematically check all material for identification numbers and labels (e.g., are they legible?). Laying out samples sequentially and ensuring that external sample labels are readable is an essential practice that also allows the examination of the sampling sequence so that any gaps can be identified.

Sample preparation for grain size analysis depends on the technique to be used for the size determination exercise. For example, sandy sediment obtained from oceanic or brackish environments can contain small quantities of salt within their pore spaces that, during drying, can become deposited on grains, cementing adjacent particles together, and thus generating misleading analyses (see Loring and Rantala, 1992).

Some techniques (e.g., sieve/pipette) require a user to separate the sand and mud fractions by wet sieving them, so they can be examined separately using different techniques. In such cases it is important that fine separation be carried out before drying of the sediment, because silts and clays can produce crusts or durable pellets on heating (McManus, 1988). Although these crusts or pellets may break down to individual particles during analysis, there is no guarantee that individual flakes of clay would become separated by purely physical means. Combining these techniques may require dividing the sample into two fractions, to allow later recombination of the different analytical results.

22.3.1 Sediment treatment 1: Removing carbonates

Carbonate material in the form of micro- and macro-organisms can bias the grain size distribution if they occur in significantly high concentrations. They are commonly removed for grain size analysis because they commonly form *in situ*; they are therefore not considered to be hydraulically representative of the depositional environment from a textural standpoint (e.g., Reineck and Singh, 1975). Their presence therefore alters the grain size and other textural data (e.g., shape) and complicates interpretation. The removal of biogenic carbonate (calcite or aragonite) may be selectively dissolved from the bulk sample using cold, dilute (10%) hydrochloric acid (HCl). In the case of low clay content, ~10% HCl solution can be added to the samples before they are left for a day. Extra HCl is added to check for completeness of the reaction. The samples can then be carefully rinsed with deionized water and decanted, avoiding the loss of sediment as much as possible (centrifuging may also help here). For samples with significant clay content, use Morgan's solution (a solution of deionized water and acetic acid, which is buffered to pH5 with sodium acetate) in place of HCl to avoid damaging clay minerals. Other considerations for the use of HCl include alterations to sediments that may affect subsequent geochemical analysis (e.g., elemental determination, Fernandes and Krull, 2008; and carbon isotope studies, Komada, et al., 2008).

22.3.2 Sediment treatment 2: Removing organics

Organic matter (OM) is an important cementing agent in many sedimentary environments and sediments; those which contain high amounts of OM generally have good aggregation (i.e., the particles are stuck together by the OM). Removal of OM is therefore an important pretreatment in textural or particle size analysis as well as in the evaluation of mineralogy, particle shape, or micropaleontology.

Dilute (<10%) hydrogen peroxide (H_2O_2) is the usual reagent of choice. It was first used in the 1920s (Robinson, 1927) to destroy soil organic matter and remains the most widely used chemical reagent for removing OM in soil and sediments. However, there is some literature indicating that it may not be a good reagent for that purpose due to some unwanted effects on the mineral properties of clay particles. In one study, Mikutta et al. (2005) compared the performance of the three most accepted reagents for OM removal: hydrogen peroxide, sodium hypochlorite (NaOCl), and disodium peroxodisulfate ($Na_2S_2O_8$). The authors found that the use of hydrogen peroxide to remove OM should be avoided for the determination of mineral particle properties (including grain size), since the treatment may promote organic-assisted dissolution of poorly crystalline minerals at low pH, disintegration of expandable clay minerals, and transformation of vermiculite into mica-like products due to ammonium (NH_4) fixation.

They concluded that sodium hypochlorite and disodium peroxodisulfate are less harmful to soil minerals than hydrogen peroxide; prolonged heating to >40 °C during any pretreatment may transform poorly crystalline minerals into more crystalline minerals. Sodium hypochlorite can be used at 25 °C, therefore preventing heat-induced mineral changes.

22.3.3 Salts and clays in saline environments

Fine-grained (silt and clay) samples collected from saline environments (e.g., estuaries) can be problematic as the relative concentration and composition of salts in the interstitial waters partially affects the behavior of the finest size fractions. The problem is that whether clays remain as discrete particles or cluster together (forming flocs or aggregates) is partially dependent on the hydrochemistry. This affects grain size, as flocs or aggregates will have diameters much larger than the individual component particles. In such cases the grain size analysis may not be a true indicator of the relationship between the particles and their depositional environment, as samples from saline environments that have been rinsed will have a different concentration of salt which may induce separation or aggregation of the particles that, in the natural system, would have been transported in a different form. Although this issue is widely recognized, most conducting grain size for sea-level research will remove salts before subjecting samples to water-based treatments (e.g., pipette, hydrometer, or laser PSA) or drying for camera-based techniques (e.g., Camsizer). The issue of removing salts is succinctly discussed in the landmark paper on treating marine sediments by Loring and Rantala (1992). Some fine-grained sediments are transported and deposited as flocs and, in such cases, the issue of transport medium is an important consideration. If the user feels that the most likely transport was as individual particles then the electro-chemical bonding forces that hold the flocs together can be partially overcome by adding a dispersing agent/deflocculant (e.g., sodium hexametaphosphate) to assist the separation of individual particles. A caveat exists however, in that the mode and method of transport is assumed, and this should be acknowledged by the user.

22.4 ANALYTICAL TECHNIQUES

Table 22.2 summarizes the common techniques applied for the analysis of materials of different grain sizes in sea-level studies. For most techniques a detailed step-by-step process is not described. In contrast, we have focused on four common techniques applied. For others, references are supplied to key studies where step-by-step explanations can be found.

Gravel, cobble, and boulder size material is rarely examined in sea-level studies. This is a good thing as the task of obtaining accurate grain size data from coarse sediments >64 mm can be time-consuming and challenging (see comments in Adams, 1979 and Switzer, 2013) and is usually conducted through the direct measurement of individual pebbles, by pebble counts and, more recently, by camera-based analysis (Warrick et al., 2007, 2009) or differential global positioning system (DGPS) surveys (Engel and May, 2012).

Table 22.2. Grain size techniques used in coastal science

Technique	Description	Reference
Sieving	This can be either wet or dry. A weighed sample is placed at the top of a sieve stack, with sieve mesh sizes decreasing down the stack. The sample is then agitated for 15–20 minutes. Weights of fractions of sample left in each sieve are then used to calculate particle distribution. The technique is examined in more detail in Section 22.4.1.	McManus (1988); Dalsgaard (1991); Switzer (2013)
Sedimentation	Used for analysis of sands, muds and clays, these related techniques are based on Stokes' Law of settling (Section 22.4.2). • *Settling column.* The sediment sample is placed in the sedimentation tube or settling column (c. 2 m tall), and the weight of sample settled at the base of the column measured at regular time intervals. This technique is now becoming less common, due to the increase in use of laser- and camera-based instrumentation.	Emery (1938); McCave and Syvitski (1991); Switzer (2013); Zeigler and Gill (1959); Zeigler et al. (1960)
	• *Pipette.* Used for analysis of fine-grained sediments. Samples are taken by pipette at a given depth below the water surface at timed intervals. According to Stokes' Law, repeated sampling will yield finer sediment fractions as time progresses. Tables of recommended times and sampling depths are available.	Krumbein and Pettijohn (1961); Switzer (2013)
	• *Hydrometer.* Used for analysis of fine-grained sediments. Based on Stokes' Law for falling spheres in a viscous liquid. Grain diameter can be calculated from knowledge of distance and falling time. Specific gravity of the liquid containing the sediment (in suspension) can be used to calculate percentage of particles of a certain equivalent diameter.	McCave and Syvitski (1991); Switzer (2013)
X-ray attenuation (SediGraph)	Based on X-ray absorption and sedimentation theory, that is, Stokes' and Beer-Lambert laws. Used for determining the relative mass distribution of a sample by particle size. X-rays are attenuated by particles settling in a liquid medium. The measurement of attenuation determines the concentration of particles at decreasing sedimentation depths.	McCave and Syvitski (1991); Switzer (2013)
Scanning electron microscopy (SEM),	A time-intensive technique, SEM can be used to count the number of particles in different size fractions. Automated grid-based counting techniques and image analysis can be used to provide a repeatable characterization of particle size and shape.	Cheetham et al. (2008); Switzer (2013)
Laser diffraction analysis (Mastersizer)	The technique involves suspending sediment in a liquid and firing a laser through a detection cell containing the liquid sediment suspension. Laser light passing through the cell is diffracted by the particles and particle size is characterized by the diffraction angle of the laser beam. See Section 22.4.3.	Switzer (2013)
Camera-based measurements (Camsizer)	Dry particulates are dropped through the image path of two CCD cameras, which record images at 25 frames per second. These images are analyzed and the equivalent spherical grain diameters of the sediment are reported. See Section 22.4.4.	

Coastal landforms of interest to sea-level researchers are commonly composed of sand, silt, and clay material, which often require the routine application of grain size analysis. Several methods of obtaining particle size distribution data are presently available for use in sandy/muddy sediments. These include sieve/hydrometer (combined), X-ray attenuation, scanning electron microscopy (SEM), sedimentation, laser diffraction techniques, and camera-based measurements (Table 22.2).

22.4.1 Sieving

Sieving can be divided into wet and dry applications and is perhaps the oldest and most traditional of the analytical techniques for grain size analysis. With an appropriate stack of sieves, particles between 0.002 and 250.0 mm in size may be separated into regular size class intervals for analysis. Although it is possible to sieve silt particles, sieving is primarily used for size determination of sand-sized material or greater (McManus, 1988).

Fig. 22.2. Analytical sieves. *Source*: Photograph by J. Pile, 2013.

Sieve screens for dry sieving are often made of strong wire mesh of stainless steel or brass with finer wire meshes used for smaller particles (Fig. 22.2). In contrast, wet sieving equipment is usually stainless steel or plastic with Nylon sieves.

In both wet and dry sieving the coarsest sieve required is placed at the top of the stack in which the square screen openings become progressively smaller down the stack. Sieving generally gives the intermediate measure of a particle because of the way particles orient themselves to pass through the mesh. The particle sieve size can either be defined as the smallest sieve size through which a particle can pass (D_{pass}) or as the largest sieve size through which the particle did not pass, that is, the retaining sieve size (D_{ret}) The number of sieves reflects the number of size fractions and hence the relative level of analytical detail. Once the sample is in place, the sieve stack is usually agitated by a mechanical shaker for a predetermined time interval, usually 15–20 min. The material retained in each sieve is commonly emptied onto a sheet of paper by tapping gently in a direction diagonal to the mesh and sweeping with appropriate sieve brushes to release particles which are stuck in the mesh (McManus, 1988). Each fraction of sediment obtained is weighed (usually to ~0.01 g). The sieve mesh sizes, raw weights, weight percentages, and cumulative percentages finer or coarser than the specific sieve can then be calculated.

Dry sieving is subject to a few potential errors such as: particle aggregation where aggregates form clusters considerably larger than the original single component grains (this can be checked by binocular microscope); incomplete mesh cleaning; and overloading of sieves (this may cause mesh distortion, which also introduces error). For sand samples, ~100 g of material is usually adequate for sieving but larger weights are required for coarse gravelly deposits (McManus, 1988). There is also considerable disagreement on the standardization of sieving time although a reproducible analysis is commonly obtained after 20 min (Dalsgaard et al., 1991).

Since many types of coastal sediment contain mixtures of coarse and fine particles, the dry sieving technique may not be appropriate as finer particles require other methods of analysis (e.g., pipette analysis) for which the separation of coarse and fine fractions was customarily made at the silt/sand boundary. Wet sieving commonly starts with drying the whole sediment sample to a constant mass at ~110 °C before resaturating in water containing a dispersant such as sodium hexametaphosphate. The sample is then periodically stirred in water for ~1 hour before being washed through an appropriate stack of sieves until the water, draining from the base of the stack, runs clear. The total content of different fractions can be determined as the difference between the initial mass and the retained material mass.

22.4.2 Sedimentation

Before the advent of laser particle size analysis and high-resolution camera techniques, the grain size analysis of silts and clays was primarily based upon indirect computations of diameters based on observation of the grains in fluids or the response of the fluids to grain-induced displacement based on Stokes' law. These methods, which include pipette analysis, are collectively known as sedimentation methods and are based on the speed at which particles settle through fluids. Such information yields settling velocities from which equivalent grain diameters are then computed. Those who primarily use, or are interested in, sedimentation techniques can refer to Emery (1938), and Krumbein and Pettijohn (1961), or the reviews of Syvistki et al. (1991a) and Switzer (2013).

22.4.3 Laser diffraction analysis

Laser diffraction analysis (LDA) has become increasingly popular as a method for particle size distribution analysis of sediments and soils. A variety of instruments from different manufacturers (e.g., Beckman-Coulter, Malvern, Retsch, Horiba) are available, and they vary in terms of the size range of analysis and the diversity of options for data processing. Standard operating procedures for laser diffraction analysis have also been published (e.g., ISO, 1999a, b; ASTM, 2010), although they are fairly generalized and aimed

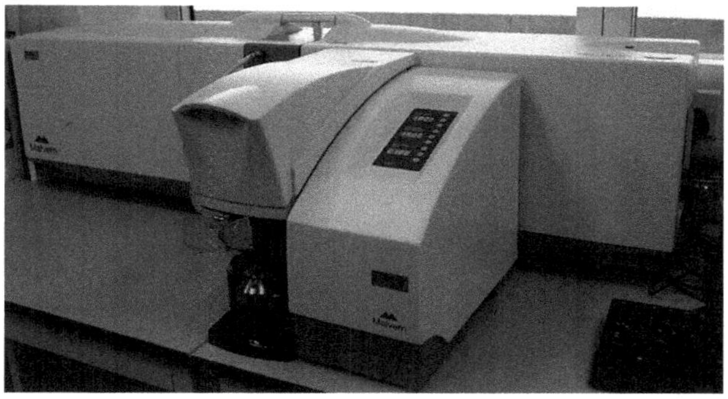

Fig. 22.3. The Malvern Mastersizer 2000 at EOS/DES environmental sedimentology labs. *Source*: Photograph by J. Pile, 2013.

mainly at industrial powders rather than the sediments analyzed in sea-level studies. In LDA, a laser is directed through a cell and the light is diffracted by grains that pass through the beam. As the diffraction angle is directly proportional to the grain size angle, the intensity of laser light scattered by a suspended sediment sample is measured and converted to a volume distribution based primarily on the Mie optical theory; some machines (e.g., Malvern Mastersizer 2000; see Fig. 22.3) can however be set to use Fraunhofer (de Boer et al., 1987). Previous work has shown that, if appropriate sample preparation and handling procedures are employed, laser diffraction provides a precise method for the analysis and comparison of most sediments, soils, and similar material (Pye and Blott, 2004; Blott and Pye, 2006). However, less attention has been given to the question of "accuracy", although it has long been recognized that there will be significant differences between results obtained using laser diffraction compared with other methods (Agrawal et al., 1991; Agrawal and Pottsmith, 2000). Many authors have also shown that laser diffraction underestimates the amount of clay compared with pipette and hydrometer analysis (e.g., McCave et al., 1986), and significant differences have also been noted between laser diffraction and sieving results for some sands (e.g., Konert and Vandenberghe, 1997).

Running samples is usually a three-step process. First, background measurements are taken of the suspension medium (usually water). Second, samples are added to the instrument. This is often done using a large beaker or dispersion bath and sediments are agitated by ultrasound until the laser obscuration is in the optimum range (10–20%) for the instrument. If obscuration is too high it results in inter-particle interference, while too-low obscuration leads to poor signal to noise ratio.

Fig. 22.4. The Retsch Camsizer at EOS/DES environmental sedimentology labs. *Source*: Photograph by J. Pile, 2013.

Third, the data output can then be grouped according to standard outputs including standard sieve-size fractions for correlation statistics; percentage volume data can also be used.

22.4.4 Digital image analysis (Camsizer)

The Camsizer is an instrument produced by Retsch of Germany (Figs 22.4 and 22.5). It is an optically based digital image analyzer capable of determining grain size to within ±1% over the range 30 μm to 30 mm (~5φ to −5φ) and also produces particle images. It is operated by means of two charge-coupled device (CCD) cameras, a basic camera (CCD-B) which records large particles, and a zoom camera (CCD-Z) which records the small particles; the systems can be used in tandem or separately. The Camsizer images a falling curtain of sediment at 25 Hz, and then determines the grain size of each particle in the image by determining the cross-sectional area of the particle and then reporting the diameter of a circle of equivalent area. The high-speed cameras perform several million individual measurements, which means that the resulting dataset can be assessed at relatively high resolution.

22.5 GRAIN SIZE SCALES: THE UDDEN-WENTWORTH SCALE

The Udden-Wentworth scale (Fig. 22.1) is a relatively simple geometric progression of sizes based on the phi (φ) scale originally developed by Krumbein (1934). The phi scale uses the

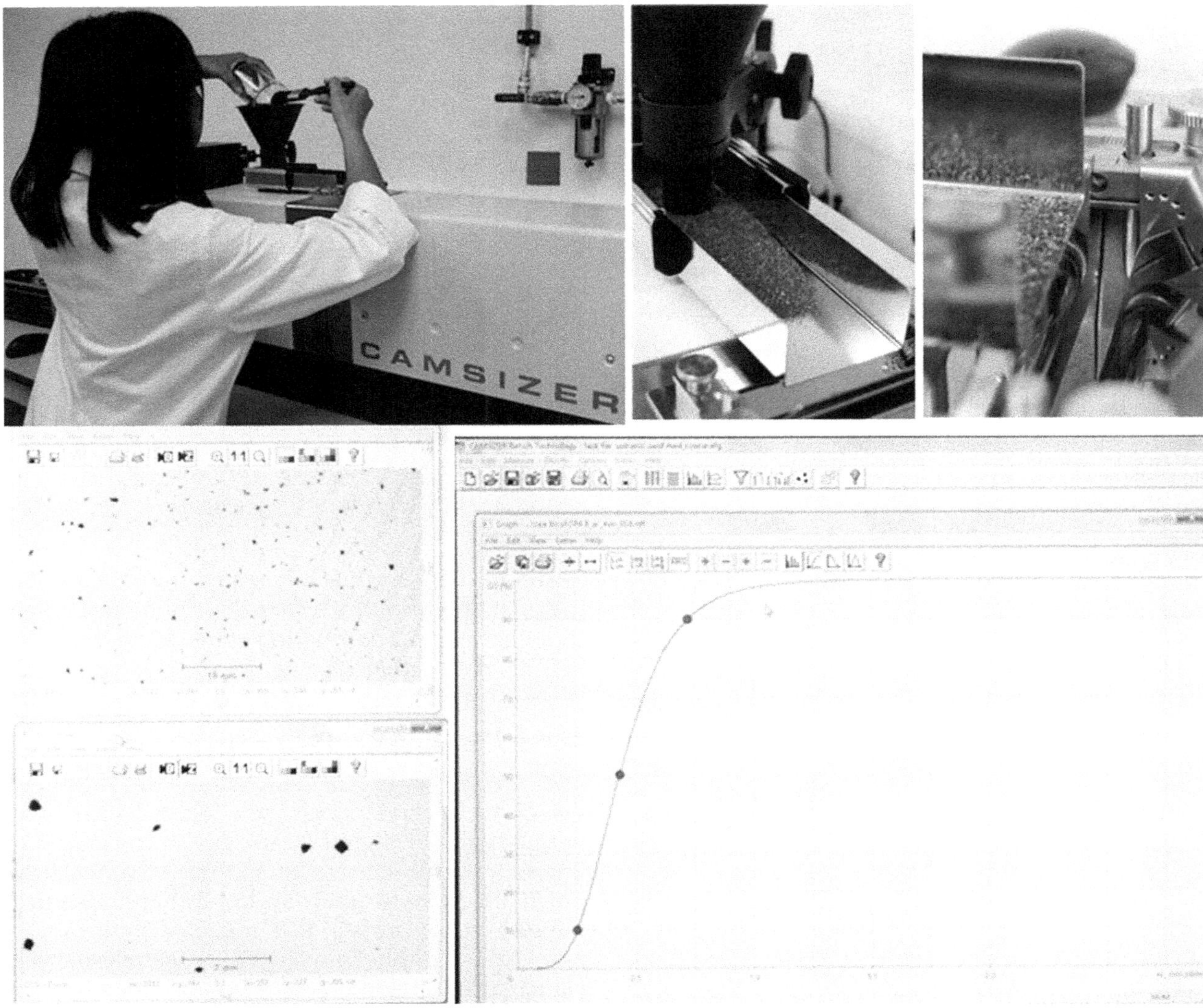

Fig. 22.5. Camsizer in action at EOS/DES environmental sedimentology labs. Top, left to right: dry sedimentary material is fed into the hopper; as the hopper and delivery tray/slide are vibrated the sediment moves slowly toward the analysis chamber; a moving curtain of sediment drops from the delivery mechanism to enter the analysis chamber. Bottom left to right: live images from both CCDs can be displayed on screen, with the PSD generated and updated in real time. *Source*: Photographs and graphic by J. Pile, 2013.

(negative) logarithm to the base 2 of the grain diameter (D, in millimeters) instead of simply the diameter itself, expressed in millimeters as:

$$\varphi = -\log_2 D.$$

Using the phi scale also simplifies the statistical computations (inclusive mean, inclusive standard deviation, skewness, and kurtosis) conducted on grain size data that lead to a sediment textural classification scheme to describe mixtures of mud, sand, and gravel that defined the basic descriptive names for both unconsolidated and lithified sediments (Folk, 1954). A modification of the phi scale in the 1960s generally agreed that by adding a nominal value of $D_0 = 1$ mm, the phi system becomes dimensionless, allowing it to be correctly used for statistical analysis to derive factors described above (McManus, 1963; Krumbein, 1964):

$$\varphi = -\log_2\left(\frac{D}{D_0}\right)$$

where D_0 is the diameter of a 1 mm grain.

The Udden-Wentworth scale was further modified in the late 1990s by Blair and McPherson (1999) who, upon recognizing that boulder sizes in natural environments can range well beyond the 4.1 m maximum of the Udden-Wentworth scale, proposed an expanded size scale to include "grains" up to an astonishing 1075 km in diameter (-30φ) . In their review of the use of particle size scales, Blott and Pye (2012) propose a new classification scheme, again based on the Udden-Wentworth scale, but with five main (first-order) classes which are divided into five subdivisions at 1φ intervals (see Fig. 22.1). Although the original and modified forms of Udden-Wentworth scale remain popular, especially for use in the field, the advent of laser-based technologies (primarily developed from materials used in the engineering and pharmaceutical industries) is reflected in a move to the practice of using microns (μm) (e.g., Pye and Blott, 2004).

22.6 INTERPRETATION OF PARTICLE SIZE DATA

Although the analysis of grain size in sediments is commonly used in the identification of sedimentary environments, landforms, and facies in a variety of depositional settings, there remains a considerable variation in presentation techniques. Pioneering work of the early–middle 20th century (e.g., Udden, 1914; Wentworth, 1922; Krumbein, 1934, 1938; Inmam, 1952; Folk and Ward, 1957) established descriptive statistics that are still commonly used to characterize particle size distributions (i.e., sorting, skewness, and kurtosis; see Table 22.3).

22.6.1 Use of bivariate and ternary plots (scattergraphs, triangular plots)

Numerous approaches have been utilized to interpret grain size data and the associated summary statistics of grain size distributions. Commonly the data are plotted on bivariate scattergrams, from which researchers have identified graphic envelopes within which deposits of particular environments are plotted (e.g., Tanner, 1991a, b; Lario et al., 2002). Several other approaches are reviewed in detail by Syvitski (1991) and Gale and Hoare (1991) and will not be discussed at length here. Despite the usefulness of grain size data, most attempts to determine the environment of deposition based on grain size are usually inadequate as an unequivocal means of paleoenvironmental reconstruction. However, it is clear that particle size data obtained by any means can be used in conjunction with other evidence to determine the environment of deposition of a sedimentary landform and, where that can be tied to sea level, to indicate sea-level changes of the past. For a discussion on bivariate plots the reader is referred to Tanner (1991a), who acknowledged the inherent problems faced when using such plots and examined methods that may assist in the resolution of the environment of deposition.

Grain size data are often also plotted on ternary (or triangular) plots, with either gravel, sand, and mud (GSM) or sand, silt, and clay (SSC) end-members (Blott and Pye, 2012). Discriminating between different depositional settings, using either bivariate or ternary plots, is based on the primary assumption that the statistical parameters reliably reflect differences in sediment transportation and deposition and hence environment (Sutherland and Lee, 1994). Many studies have tried to make environmental sense from bivariate plots of parameters that describe the sample size spectrum; success has been varied for several reasons, such as oversimplified discrimination, for example beach v. river where dunes and other depositional settings are ignored (see Socci and Tanner, 1980; Tanner, 1991a, b). As explained by Griffiths (1967), both mean grain size and sorting are hydraulically controlled in most environments; this results in the best-sorted sediments in almost all environments having a mean grain size in the fine sand size range.

22.6.2 Recent advances in data presentation

Where PSA plays a major role in the analysis, most researchers will find that conventional graphical summary statistics are still the preferred method of displaying grain size data despite the noted limitations of such statistics to adequately characterize the grain size distribution (Blott and Pye, 2001; Fig. 22.6). Recent work in coastal sequences has seen some notable attempts at new analysis and display techniques (e.g., Donato et al., 2009; Fig. 22.7). This is primarily a result of recent advances in instrumentation (e.g., LDA and Camsizer) that now mean it is now possible to analyze large numbers of samples economically and obtain a statistically significant spread of data.

Table 22.3. Grain size parameters and their common indicative meaning

Parameter	Description	Main indicative meaning in PSA
Mode	On a size frequency histogram, the size class in which the greatest percentage is recorded provides the modal class. On the size frequency distribution plot, the highest point on the curve provides the modal value.	The modal size represents the most common grain size in a distribution. The frequency curves with several peaks indicate polymodal distributions, potentially indicating the presence of more than one population of grains.
Median (*Md*)	Half of the grains are coarser and half finer than the median diameter. This is most readily determined from the 50% line of the cumulative distribution curve.	Although useful for unimodal sediments, in polymodal distributions (most natural sediments are polymodal) the median may fall in the tails of two subpopulations of grains.
Mean (*M*)	The measure of average grain size, computed from sizes of particles spread through a range of percentile values. In its simplest form the graphic mean $M = (a + b + c)/3$ where a, b, and c are the 16th, 50th and 84th percentiles in a cumulative frequency curve, which assumes that three values alone are sufficient to give a useful mean. More percentage values may be averaged which approximates the mean of moment statistics.	The mean is the most commonly presented parameter of sediment size and is often presented with a measure of standard deviation to express the basic properties of a sediment (e.g., poorly sorted fine sand)
Sorting (s)	There are two common measures of sorting the graphic standard deviation(s): Inman (1952), which provides a measure of the spread of one standard deviation on either side of the mean; and the inclusive graphic standard deviation (Folk and Ward, 1957) which defines a spread of 1.65 standard deviations either side of the mean.	Sorting relates to the range of grain sizes. Well-sorted samples have a small range (aeolian sediments are well sorted, i.e., grain sizes are very similar; it can therefore be said that winds sort sediments well). Glaciers sort poorly (meaning there is a large spread of grain sizes in glacial deposits).
Skewness (a)	In a normal distribution with a bell-shaped frequency curve the median and mean values coincide. Any tendency for a distribution to lean to one side, that is, to deviate from normality, leads to differences between the median and mean values. These differences are used to characterize the asymmetry or skewness of the curve and are determined from the value of the mean less the median, all divided by the range used in defining the mean.	Skewness has a positive or negative value when more fine or more coarse materials are present. Skewness is a positively or negatively signed dimensionless number; it has neither metric nor phi value and lies within the range -1 to $+1$.
Kurtosis	A measure of the "peakedness" of the distribution. Although frequently calculated, kurtosis rarely receives much attention. It is a measure of the dispersion and the normality of the distribution; it is also a ratio of the spread of the tails and centre of the distribution (and is therefore dimensionless).	Generally kurtosis is a second indicator of sorting where very flat curves equate to poorly sorted sediments or those with bimodal frequency curves are platykurtic; strongly peaked curves are considered leptokurtic and exhibit very good sorting of the central part of the distribution.

22.6.3 Using modern PSD datasets

Grain size analysis statistics were primarily developed for the ease of calculation and characterization of the grain size distribution from graphical cumulative frequency curves, which meant that the derived statistics were simply a convenient way of working with the data.

Modern grain size instrumentation can now provide high-resolution, precisely measured, and easily comparable datasets for statistical analysis using multivariate techniques (e.g., cluster analysis). Additionally, the use of multivariate statistics and 3D surface plots now allows better characterization of sedimentary processes and environments (e.g., Beierle et al., 2002; Van Hengstum et al., 2007; Donato et al., 2009). Combining the grain size data with other relevant datasets (e.g., texture, carbonate content, faunal data) can adequately discriminate landforms and environments and, in suitable environments, infer changes in sea level through time (e.g., Van Hengstum et al., 2011).

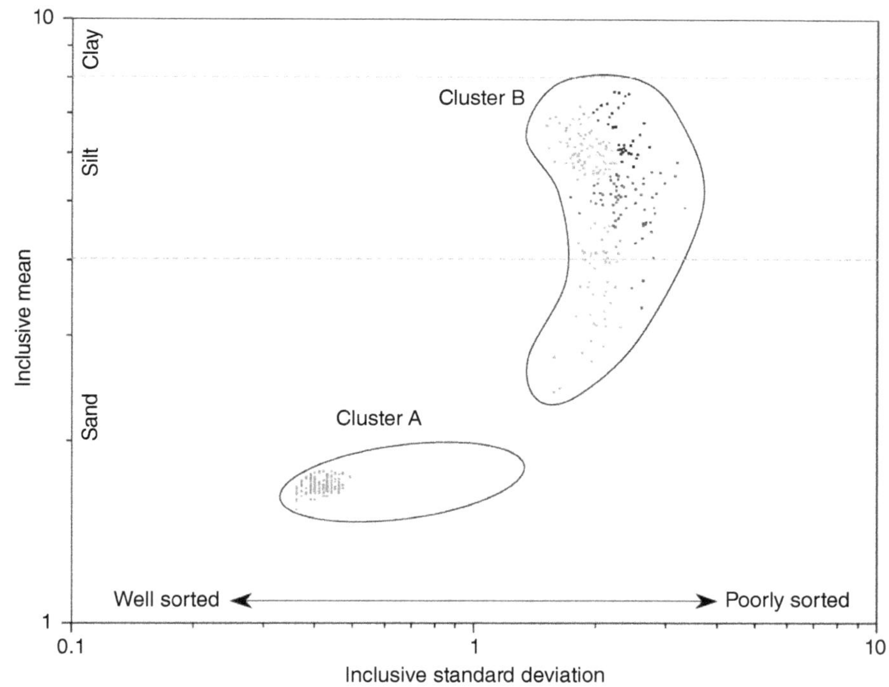

Fig. 22.6. Bivariate plot of inclusive standard deviation against inclusive mean for all samples from vibracores collected at Killalea Lagoon, southeast Australia (Switzer et al., 2006; Switzer and Jones, 2008). Two well-defined clusters are observed. Cluster A is associated with the large sandsheet deposit, is predominantly clean sand and contains two subsets (one of clean sand and a second more poorly sorted sand with a minor silt component). Cluster B is associated with the silty peat deposits that vertically confine the sandsheet. Cluster B exhibits more variability and is predominantly fine sands to silts with small amounts of aeolian sand representative of a back barrier environment dominated by lagoonal sedimentation of peaty muds.

In a recent example, Donato et al. (2009) defined modern environmentally distinct facies distributions determined in Sur Lagoon, Oman using grain size data and multivariate cluster analysis (Q-mode). To do this they used blind clustering of grain size data and compared it to surface topography, remote sensing, and visual field surveys. This allowed a comparison of the mapped distribution of sedimentary facies to determine the effectiveness of grain size data for determining lagoon sub-environments in the geological record (see previous work by Loizeau and Stanley, 1993). This method is especially useful for distinguishing multiple sedimentary processes, which can appear as additional modes within the grain size dataset (e.g., Pilarczyk et al., 2010; Van Hengstum et al., 2011). Such techniques can now be applied to most depositional environments and landforms at any spatial or temporal resolution and, as Donato et al. (2009) noted, a combination of conventional summary statistics with surface plots of grain size data increases its utility as a paleoenvironmental proxy that, in coastal settings, allows the reconstruction of past sea level (see recent example of Van Hengstum et al., 2011).

22.7 CONCLUSION: COMPARING DIFFERENT TECHNIQUES

So which technique is best? Referring to "correctness" of the results obtained for particle size measurements by two or more different analytical techniques, Webb (2008) stated: "Provided that the instruments used are capable of producing high-quality data, the pertinent questions, then, are, 'was the sample properly prepared and properly presented to the instrument,' and 'were the analytical parameters applied correctly'. If the answer to both is 'yes,' then both analytical results probably are equally correct; they are just expressed in different terms." Natural sediments are usually not perfect spheres of homogenous composition and weight, so the determination of grain size on the same sediment sample by all

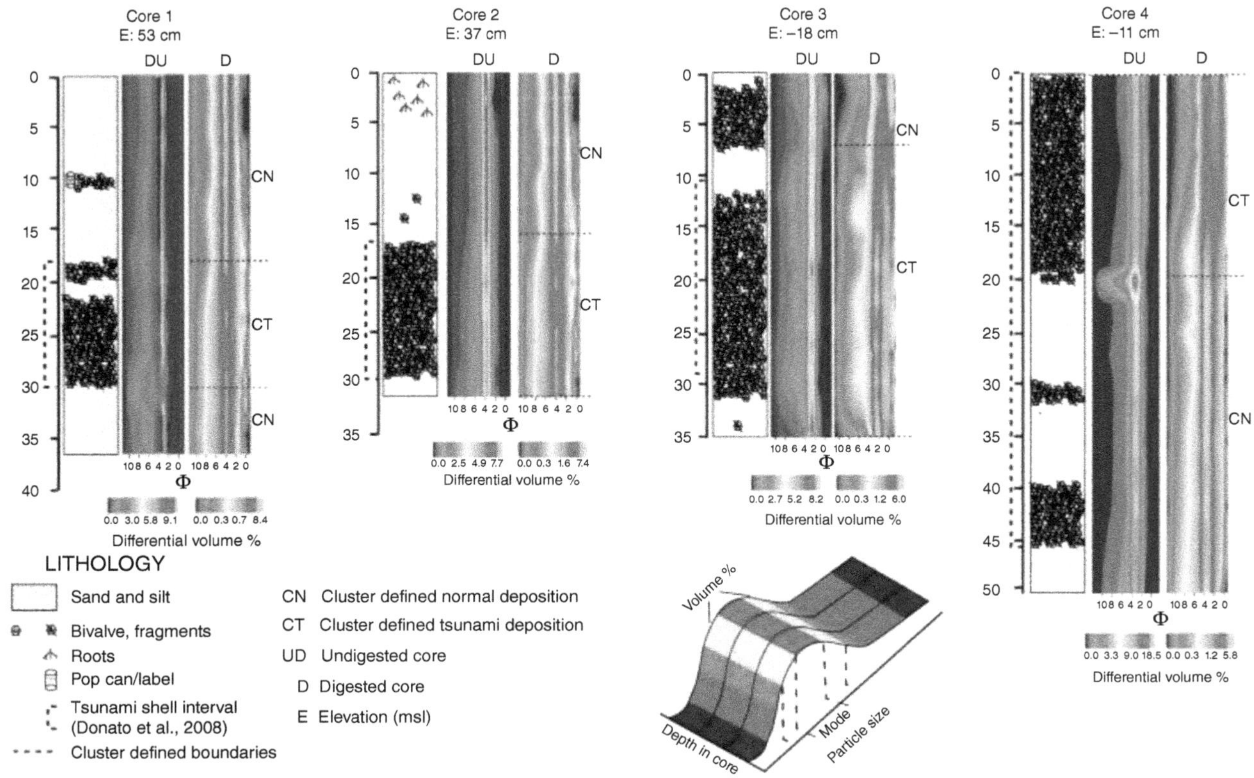

Fig. 22.7. Lithologic logs for four sediment cores collected in Sur Lagoon, Oman. *Source*: Donato et al., 2009. Reproduced with permission from Elsevier. In this case high-resolution particle-size distribution plots were plotted from samples that were analyzed using a Coulter LS 230 instrument. Data are shown as φ surface plots for both treated (digested, D) and untreated (undigested, UD) sediment samples from each core. Differential volumes by percentage color code: blue: absent to minimal; red–pink: common–most common. This technique was found to present the unimodal sediments much more effectively than the use of traditional grain size plots. For color details, please see Plate 33.

techniques will yield different results for mean size, modal size, and quantity distribution by size. So which technique is "correct" or provides the most "accurate" representation of the sediment population? If the sample is prepared properly and the analytical parameters are correctly applied, then all the answers are "correct"; they are simply measuring different things, and researchers must be aware of the limitations of each technique. We therefore have to consider the original question. Why is the analysis being performed? What is the main purpose of the grain size analysis? What type of results is wanted/required/needed for interpretation purposes? Is it necessary to use a specialized sample collection method for a particular analytical method? These questions must be considered preferably before the start of a field campaign, but at the very least before selecting a technique that may in the end prove to be either inappropriate or disproportionate to the task at hand.

REFERENCES

Adams, J. (1979) Gravel size analysis from photographs. ASCE Journal of the Hydraulics Division, 10510, 1247–1255.

Agrawal, Y.C., and Pottsmith H.C. (2000) Instruments for particle size and settling velocity observations in sediment transport. Marine Geology, 168(1–4), 89–114

Agrawal, Y.C., McCave, I.N., and Riley, J.B. (1991) Laser diffraction style analysis. In: *Principles, Methods and Applications of Particle Size Analysis* (ed. Syvitski, J.P.M.), Cambridge University Press, Cambridge, pp. 119–128.

ASTM (2010) Standard E1382–97 (2010) Standard test methods for determining average grain size using semi-automatic and automatic image analysis. ASTM International West Conshohocken, Pennsylvania, USA, doi: 10.1520/E1382-97R10.

Bates, R.L., and Jackson, J.A. (eds) (1984) *Dictionary of Geological Terms*, 3rd Edition. Anchor Books, New York.

Beierle, B.D., Lamoureux, S.F., Cockburn, J.M.H., and Spooner, I. (2002) A new method for visualizing sediment particle size distributions. Journal of Paleolimnology, 27, 279–283.

Blair, T.C. and McPherson, J.G. (1999) Grain-size and textural classification of coarse sedimentary particles. Journal of Sedimentary Research, 691, 6–19.

Blott, S.J., and Pye, K. (2001) Gradistat: a grain size distribution and statistics package for the analysis of unconsolidated sediments. Earth Surface Processes and Landforms, 26, 1237–1248.

Blott, S.J., and Pye, K. (2006) Particle size distribution analysis of sand-sized particles by laser diffraction: an experimental investigation of instrument sensitivity and the effects of particle shape. Sedimentology, 53, 671–685.

Blott, S.J., and Pye, K. (2012) Particle size scales and classification of sediment types based on particle size distributions: Review and recommended procedures. Sedimentology, 59, 2071–2096.

Cheetham, M.D., Keene, A.F. Bush, R.T. Sullivan, L.A. and Erskine, W.D. (2008) A comparison of grain-size analysis methods for sand-dominated fluvial sediments. Sedimentology, 55, 6.

Dalsgaard, K., Jensen, J.L. and Sorensen, M. (1991) Methodology of sieving small samples and calibration of sieve set. In: *Theory, Methods and Applications of Particle Size Analysis* (ed. Syvitski, J.P.M.), Cambridge University Press, New York, pp. 64–75.

de Boer, G.B.J., de Weerd, C., Thoenes D., and Goossens, H.W.J. (1987) Laser diffraction spectrometry: Fraunhofer versus Mie scattering. Particle Characteristics, 4, 14–19.

Donato, S.V., Reinhardt, E.G., Boyce, J.I., Pilarczyk, J.E., and Jupp, B.P. (2009) Particle-size distribution of inferred tsunami deposits in Sur Lagoon, Sultanate of Oman. Marine Geology, 257(1–4), 54–64.

Emery, K.O. (1938) Rapid method of mechanical analysis of sands. Journal of Sedimentary Petrology, 8, 105–111.

Engel, M., and May, S.M. (2012) Bonaire's boulder fields revisited: Evidence for Holocene tsunami impact on the Leeward Antilles. Quaternary Science Reviews, 54, 126–141.

Fernandes, M., and Krull, E. (2008) How does acid treatment to remove carbonates affect the isotopic and elemental composition of soils and sediments? Environmental Chemistry, 5(1), 33–39.

Folk, R.L. (1954) The distinction between grain size and mineral composition in sedimentary rock nomenclature. Journal of Geology, 62(4), 344–359.

Folk, R.L. and Ward, W.E. (1957) Brazos River bar: a study in the significance of grain size parameters. Journal of Sedimentary Petrology, 27, 3–26.

Folk, R.L., and Robles, R. (1964) Carbonate sands of Isla Perez, Alacran Reef complex, Yucatan. The Journal of Geology, 72, 255–292.

Gale, S.J., and Hoare, P.G. (1991) *Quaternary Sediments.* Wiley, Chichester and New York.

Griffiths, J.C. (1967) *Scientific Method in Analysis of Sediments.* McGraw-Hill, New York.

Inmam, D.L. (1952) Measures for describing the size distribution of sediments. Journal of Sedimentary Petrology, 22, 125–145.

International Standard Organization (1999a) ISO 13320-1: Particle size analysis: Laser diffraction method. Part 1. General principles. International Standard Organization, Geneva.

International Standard Organization (1999b) ISO 13320-2: Particle size analysis: Laser diffraction method. Part 2. Validation of inversion procedures. International Standard Organization, Geneva.

International Standard Organization (2008) ISO 9276-6: Representation of results of particle size analysis. Part 6: Descriptive and quantitative representation of particle shape and morphology. International Standard Organization, Geneva.

International Standard Organization (2009) ISO 11277: Soil quality: Determination of particle size distribution in mineral soil material. Method by sieving and sedimentation. International Standard Organization, Geneva.

Kench, P.S., and McLean, R.F. (1997) A comparison of settling and sieve techniques for the analysis of bioclastic sediments. Sedimentary Geology, 109(1–2), 111–119.

Komada, T., Anderson, M.R., and Dorfmeier, C.L. (2008) Carbonate removal from coastal sediments for the determination of organic carbon and its isotopic signatures, δ13C and Δ14C: comparison of fumigation and direct acidification by hydrochloric acid. Limnology Oceanography Methods, 6, 254–262.

Konert, M., and Vandenberghe, J. (1997) Comparison of laser grain size analysis with pipette and sieve analysis: A solution for the underestimation of the clay fraction. Sedimentology, 44(3), 523–535.

Krumbein, W.C. (1934) Size frequency distributions of sediments. Journal of Sedimentary Petrology, 10, 57–69.

Krumbein, W.C. (1938) Size frequency distributions of sediments and the normal phi curve. Journal of Sedimentary Petrology, 8, 84–90.

Krumbein, W.C. (1964) Some remarks on the phi notation. Journal of Sedimentary Petrology, 34, 195–196.

Krumbein, W.C. and Pettijohn, F.L. (1961) *Manual of Sedimentary Petrography.* Appleton-Century-Crofts, New York.

Lario, J., Spencer, C., Plater, A.J., Zazo, C., Goy, J.L., and Dabrio, C.J. (2002) Particle size characterisation of Holocene back-barrier sequences from North Atlantic coasts SW Spain and SE England. Geomorphology, 42, 25–42.

Loizeau, J.L., and Stanley, D.J. (1993) Petrological-statistical approach to interpret recent and subrecent lagoon subfacies, Idku, Nile delta of Egypt. Marine Geology, 1111(2), 55–81.

Loring, D.H., and Rantala, R.T.T. (1992) Manual for the geochemical analyses of marine sediments and suspended particulate matter. Earth Science Reviews, 324, 235–283.

McCave, I.N., and Syvitski, J.P.M. (1991) Principles and methods of geological particle size analysis. In: *Principles, Methods and Applications of Particle Size Analysis* (ed. Syvitski, J.P.M.), Cambridge University Press, Cambridge, pp. 3–21.

McCave, I.N., Bryant, R.J., Cook, H.F., and Coughanowr, C.A. (1986) Evaluation of a laser-diffraction-size analyzer for use with natural sediments. Journal of Sedimentary Research, 56, 561–564.

McManus, D.A. (1963) A criticism of certain usage of the phi notation. Journal of Sedimentary Petrology, 33, 670–674.

McManus, J. (1988) Grain size determination and interpretation. In: *Techniques in Sedimentology* (ed. Tucker, M.), Blackwell, Oxford, UK, pp. 63–85.
Miall, A.D. (1985) Architectural-element analysis: a new model of facies analysis applied to fluvial deposits. Earth Science Reviews, 22, 261–308.
Mikutta, R., Kleber, M., Kaiser, K., and Jahn, R. (2005) Organic matter removal from soil using hydrogen peroxide, sodium hypochlorite, and disodium peroxodisulfate. Soil Science Society of America Journal, 69, 120–135.
Orton, G.J., and Reading, H.G. (1993) Variability of deltaic processes in terms of sediment supply, with particular emphasis on grain size. Sedimentology, 40, 475–512.
Pilarczyk, J.E., Reinhardt, E.G., Boyce, J.I., Schwarcz, H.P., and Donato, S.V. (2011) Assessing surficial foraminiferal distributions as an overwash indicator in Sur Lagoon, Sultanate of Oman. Marine Micropaleontology, 80(3–4), 62–73.
Pye, K., and Blott, S.J. (2004) Particle size analysis of sediments, soils and related particulate materials for forensic purposes using laser granulometry. Forensic Science International, 144, 19–27.
Reineck, H.E., and Singh, L.B. (1975) *Depositional Sedimentary Environments*. Springer-Verlag, Berlin, Germany
Robinson, W.O. (1927) The determination of organic matter in soil by means of hydrogen peroxide. Journal Agricultural Research, 34, 339–356.
Socci, A., and Tanner, W.F. (1980) Little known but important papers on grain-size analysis. Sedimentology, 27, 231–232.
Sutherland, R.A., and Lee, C.-T. (1994) Discrimination between coastal subenvironments using textural characteristics. Sedimentology, 416, 1133–1145.
Switzer, A.D. (2013) Measuring and analyzing particle size in a geomorphic context. In: *Treatise on Geomorphology* (ed. Shroder, J.F.), Academic Press, San Diego, Vol. 14, pp. 224–242.
Switzer, A.D., and Jones, B.G. (2008) Large-scale washover sedimentation in a freshwater lagoon from the southeast Australian coast: sea-level change, tsunami or exceptionally large storm? The Holocene, 18(5), 787–803.
Switzer, A.D., Bristow, C.S., and Jones, B.G. (2006) Investigation of large-scale washover of a small barrier system on the southeast Australian coast using ground penetrating radar. Sedimentary Geology, 183(1), 145–156.
Syvitski, J.P.M. (ed.) (1991) *Principles, Methods and Application of Particle Size Analysis*. Cambridge University Press, Cambridge.
Syvitski, J.P.M., Asprey, K.W., and Clattenburg, D.A. (1991a) Principles, design and calibration of settling tubes. In: *Principles, Methods and Application of Particle Size Analysis* (ed. Syvitski, J.P.M.), Cambridge University Press, Cambridge, pp. 45–63.
Syvitski, J.P.M., LeBlanc, K.W.G., and Asprey, K.W. (1991b) Inter-laboratory, inter-instrument calibration experiment. In: *Principles, Methods and Application of Particle Size Analysis* (ed. Syvitski, J.P.M.), Cambridge University Press, Cambridge, pp. 174–193.
Tanner, W.F. (1991a) Suite statistics: the hydrodynamic evolution of the sediment pool. In: *Principles, Methods and Application of Particle Size Analysis* (ed. Syvitski, J.P.M.), Cambridge University Press, Cambridge, pp. 225–236.
Tanner, W.F. (1991b) Application of suite statistics to stratigraphy and sea-level changes. In: *Principles, Methods and Application of Particle Size Analysis* (ed. Syvitski, J.P.M.), Cambridge University Press, Cambridge, pp. 283–292.
Udden, J.A. (1914) Mechanical composition of clastic sediments. Geological Society of America Bulletin, 25, 655–744.
Van Hengstum, P.J., Reinhardt, E.G., Boyce, J.I., and Clark, C. (2007) Changing sedimentation patterns due to historical land-use change in Frenchman's Bay, Pickering, Canada. Evidence from high-resolution textural analysis. Journal of Paleolimnology, 37(4), 603–618.
Van Hengstum, P.J., Scott, D.B., Gröcke, D.R., and Charette, M.A. (2011) Sea level controls sedimentation and environments in coastal caves and sinkholes. Marine Geology, 286(1–4), 35–50.
Warrick, J.A., George, D.A., Stevens, A.W., Eshleman, J., Gelfenbaum, G., Kaminsky, G.M., Schwartz, A.K., and Bierne, M. (2007) Beach morphology monitoring in the Elwha River littoral cell, 2004–2009. US Geological Survey, Data Series 288, v. 1.1, 59 p. and data folder available at http://pubs.usgs.gov/ds/288/ (accessed 27 July 2014). Revised 2010.
Warrick, J.A., Rubin, D.M., Ruggiero, P., Harney, J.N., Draut, A.E., and Buscombe, D. (2009) Cobble cam: Grain-size measurements of sand to boulder from digital photographs and autocorrelation analyses. Earth Surface Processes and Landforms, 34(13), 1811–1821.
Webb, P.A. (2008) *Interpretation of Particle Size Reported by Different Analytical Techniques*. Micromeritics Instrument Corp., Norcross, GA, USA.
Wentworth, C.K. (1922) A scale of grade and class terms for clastic sediments. Journal of Geology, 30, 377–392.
Zeigler, J.M., and Gill, A. (1959) Tables and graphs for the settling velocity of quartz in water, above the range of Stoke's Law. Report No. 59-36, Woods Hole Oceanographic Institution, MA, USA.
Zeigler, J.M., Whitney, G.G., and Hayes, C.R. (1960) Woods Hole Rapid Sediment Analyzer. Journal of Sedimentary Petrology, 30, 490–495.

Part 3

Dating Methods

Chapter 23
Radiocarbon dating and calibration

TORBJÖRN E. TÖRNQVIST[1], BRAD E. ROSENHEIM[1,2], PING HU[1], AND ALVARO B. FERNANDEZ[1]

[1]*Department of Earth and Environmental Sciences, Tulane University, New Orleans, LA, USA*
[2]*College of Marine Science, University of South Florida, St. Petersburg, FL, USA*

23.1 INTRODUCTION

Reconstructions of late Quaternary sea-level change started in earnest in the 1950s, in large part due to the invention of ^{14}C dating (Libby, 1952). Prior to this, numerical ages were essentially non-existent and sea-level curves were therefore highly qualitative in nature. This chapter offers an overview of the principles of ^{14}C dating, including a brief historical background (more detailed accounts of the history of ^{14}C dating are provided by Taylor, 2000 and Olsson, 2009). An excellent introduction to the principles of ^{14}C dating within the context of sea-level research was provided by Mook and Van de Plassche (1986). However, important developments have taken place since that time, most prominently due to the advent of accelerator mass spectrometry (AMS) that has increasingly replaced the traditional, radiometric approach to ^{14}C dating. Furthermore, considerable progress has been made with ^{14}C calibration, as reflected by the fact that the publication of the precursor of the present volume (Van de Plassche, 1986) coincided with the first Calibration Issue of the journal *Radiocarbon* (Stuiver and Kra, 1986). The sixth Calibration Issue was published in 2013.

23.2 PRINCIPLES OF ^{14}C DATING

The formation of the radioisotope ^{14}C is the result of cosmic ray neutron bombardment of ^{14}N in the upper troposphere and stratosphere, as follows:

$$^{14}N + {}^{1}n \rightarrow {}^{14}C + {}^{1}p^{+}. \tag{23.1}$$

Subsequently, ^{14}C reacts with oxygen to form $^{14}CO_2$ and rapidly mixes throughout the atmosphere. The production of ^{14}C is mediated not only by the influx of cosmic rays, but also by solar activity and the strength of the Earth's magnetic field, all of which have been shown to change through time. As soon as ^{14}C is formed, it is subject to beta decay to ^{14}N. The half-life of ^{14}C, the time needed to reduce the ^{14}C content of a sample by 50%, was originally determined by Engelkemeir et al. (1949). The ultimately agreed-upon half-life of 5730 ^{14}C a (Godwin, 1962) was close to the Engelkemeir et al. (1949) half-life of 5720 ± 47 ^{14}C a. Interestingly, the half-life recommended for use in ^{14}C dating (Libby, 1952) is different (5568 ^{14}C a) and known as the "Libby half-life" (Godwin, 1962). The difference between these half-lives is merely a consequence of the ^{14}C dating method becoming widely applied before geochemists and nuclear physicists had converged on a half-life.

The actual quantity measured in a ^{14}C age determination is the ratio of $^{14}C/^{12}C$. This quantity is related to age through the expression:

$$t = -8033 \times \ln f_M \tag{23.2}$$

where fraction modern f_M is the ratio of $^{14}C/^{12}C$ (^{12}C is stable and also the most abundant carbon isotope, constituting 98.9% of carbon in the environment) in the sample divided by the same ratio in wood from the year 1950 AD, and 8033 is the mean life of ^{14}C (reciprocal of the decay constant λ) based on the Libby half-life. As such, the time before present t is actually defined as the age relative to 1950 AD. Wood from 1950 AD is generally no longer used; ^{14}C laboratories normalize to secondary standards that have been calibrated to 1950 AD wood. The relationship between f_M and ^{14}C age is illustrated in Figure 23.1. Measurements of ^{14}C content that are converted to age using Equation (23.2) are said to have units of ^{14}C years (^{14}C a BP) which are distinct from calendar years because: (1) the Libby half-life is used; and (2) an

Handbook of Sea-Level Research, First Edition. Edited by Ian Shennan, Antony J. Long, and Benjamin P. Horton.

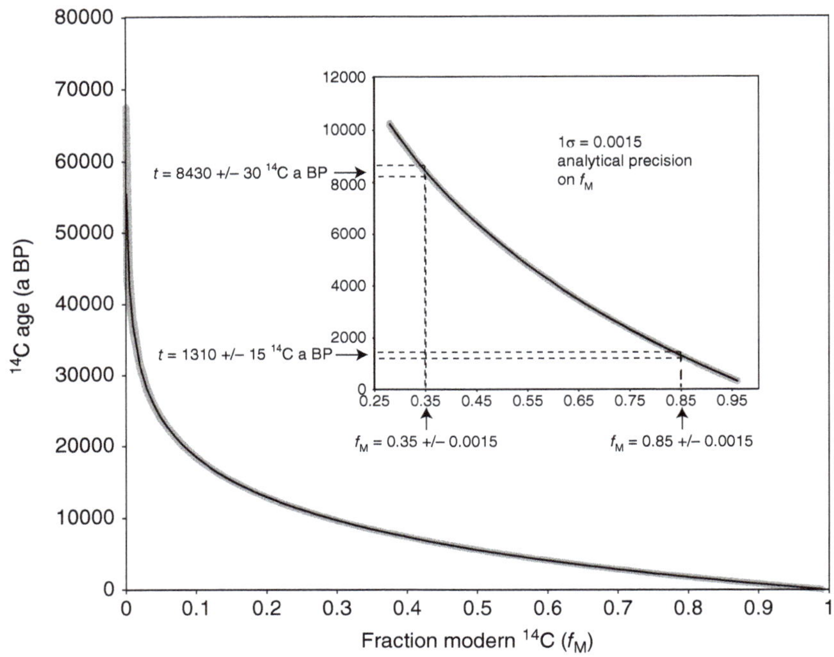

Fig. 23.1. Relationship between f_M and ^{14}C age with an error envelope that reflects the typical analytical precision of f_M of 1.5‰. The inset illustrates in more detail how uncertainties in f_M translate to the analytical ^{14}C age error. Also note that as the curve becomes steeper when f_M approaches zero, this analytical error translates not only to larger but also increasingly asymmetric ^{14}C age errors. The age limit of ^{14}C dating is usually taken to be around 40–45 ^{14}C ka BP.

assumption of constant ^{14}C production in the atmosphere is made. As discussed in more detail in Section 23.7, this assumption is not valid. The true half-life must be used when the ^{14}C content is determined from materials of known age to obtain the ^{14}C deficit ($\Delta^{14}C$):

$$\Delta^{14}C = \left(f_M e^{\lambda(1950-t)} - 1\right) \times 1000 \qquad (23.3)$$

where t is the known age of the sample (in calendar years) and λ is defined relative to the true half-life of ^{14}C (i.e., $\lambda = 1/8267$). The age limit that can be attained with ^{14}C dating is typically in the range of 40–45 ^{14}C ka BP (Fig. 23.1); beyond about 8 half-lives even minute amounts of contamination can produce finite ages from otherwise ^{14}C dead materials. An excellent review of ^{14}C calculation and reporting convention is given by Stuiver and Polach (1977), where the counted activity ratio between a sample and a standard is equivalent to the f_M reported by AMS laboratories.

When carbon is incorporated by an organism that is used as a sea-level indicator, radioactive decay is balanced by the replenishing of ^{14}C due to photosynthesis in plants and the consumption of plant tissue by animals. The ^{14}C clock commences after the death of the organism. Alternatively, when dissolved inorganic carbon (DIC) is utilized to form datable carbonate minerals, decay begins immediately upon crystallization. In aquatic organisms, the CO_2 used in primary production can possess an apparent ^{14}C age compared to terrestrial organisms due to mixing with older DIC. Depending on the mixing rates of water masses and the age of DIC reservoirs within the system this age, known as the reservoir age, can be considerable. As discussed further in Section 23.5, a proper understanding of the reservoir age is of fundamental importance in sea-level studies.

Isotopic fractionation can also affect ^{14}C ages. Processes that transfer carbon between different carbon pools (e.g., photosynthesis, crystallization) can discriminate for either heavier or lighter isotopes. Generally, the two stable isotopes of carbon, ^{12}C and ^{13}C, are measured and normalized to an international standard:

$$\delta^{13}C = \left[\frac{\left(\frac{^{13}C}{^{12}C}\right)_{sam} - \left(\frac{^{13}C}{^{12}C}\right)_{std}}{\left(\frac{^{13}C}{^{12}C}\right)_{sam}}\right] \times 1000. \qquad (23.4)$$

Here, isotope ratios of samples (sam) and standards (std) (known relative to international standards) are measured as ratios of the rare isotope to the abundant isotope, and the difference in these ratios relative to the standard is multiplied by 1000 to yield a value in per mil. In a similar fashion, ^{14}C is fractionated in nature proportionally to ^{13}C. The stable isotopic composition is relatively easy to measure by conventional mass spectrometry and requires much less sample material than a typical ^{14}C measurement. In order to know how the fraction modern relates to the ^{14}C age independent of fractionation, f_M must be corrected for fractionation by multiplying with the approximated term (Stuiver and Polach, 1977):

$$\left[1-\frac{2\left(25+\delta^{13}C\right)}{1000}\right]. \tag{23.5}$$

To clarify the importance of this correction in ^{14}C dating, we consider a shell ($CaCO_3$) in a body of water with zero reservoir age and compare it with coeval terrestrial plant material. If these two materials die and are deposited at exactly the same time, they should have the same ^{14}C age. However, they will have different $\delta^{13}C$ values due to the differences in fractionation into the organic carbon pool v. the inorganic pool. If the plant material had a typical terrestrial plant $\delta^{13}C$ value of –27‰ and the shell had a value of 0‰, and f_M was simply measured without determining $\delta^{13}C$ values, the apparent age difference would be 420 ^{14}C a (note that this number was obtained from a more precise calculation than the approximate Equation (23.5), as detailed further in Section 23.4). The plant material would always seem older because it is isotopically lighter and discriminates against heavy isotopes, including ^{14}C.

As with nearly all numerical dating methods, ages are expressed by a central value and an associated error (Fig. 23.1) which, in the case of ^{14}C, is derived from a variety of analytical errors (e.g., contamination during sample preparation, analytical uncertainty in the AMS) and expressed as the standard deviation (1σ level). Laboratories generally report f_M with a 1σ uncertainty based on their blank determinations and analytical uncertainty. If pre-preparation steps were undertaken prior to submission of material to a ^{14}C laboratory, the submitter should make every effort to determine what the blank contribution to the sample from those steps was. These combined uncertainties are also known as the precision of the ^{14}C age, which should not be confused with dating accuracy. Dating accuracy is largely dependent on how well the sample reflects the phenomenon that is the target of dating. As discussed further in Section 23.3, a prime example is a ^{14}C age of bulk sedimentary organic material. Such an age may be analytically precise to as little as 15 ^{14}C a; however, it likely reflects the mean age of a mixture of organic carbon from many different sources. At best, this age will be far less precise than the analytical precision suggests; at worst, if the sample included material irrelevant to the phenomenon targeted for dating, it will be significantly offset from the true age.

Many laboratories round ^{14}C ages to the nearest 5 or 10 ^{14}C years, although others round values to the nearest year (this is particularly common for younger ages). It has also been argued that ^{14}C ages are best represented by age ranges, rather than being reported primarily by a central value (i.e., the mean or median). As shown in more detail in Chapter 34, the 2σ or 95% confidence interval (we do not recommend using the 1σ confidence interval) is commonly used to express the age uncertainty of sea-level index points. However, it should be noted that the central value remains relevant, in particular for the calculation of relative sea-level trend curves that rely on single values for each index point. Chapter 34 offers further details on the graphical representation of age errors.

When multiple ^{14}C measurements are available for the same phenomenon, it may be justifiable to calculate a weighted mean ^{14}C age with an associated error, according to the following two equations:

$$\bar{t} = \frac{\sum_{i=1}^{n} \frac{t_i}{\sigma_{a_i}^2}}{\sum_{i=1}^{n} \frac{1}{\sigma_{a_i}^2}} \tag{23.6}$$

and

$$\sigma_a = \frac{1}{\sqrt{\sum_{i=1}^{n} \frac{1}{\sigma_{a_i}^2}}} \tag{23.7}$$

where t is the conventional ^{14}C age (a BP) and σ_a is the standard deviation of the ^{14}C age. Caution should be exercised however, since such calculations result in reduced errors by definition. This procedure is therefore only justifiable if multiple

^{14}C ages exhibit only minor scatter that can be ascribed entirely to random errors.

The radiocarbon community deserves credit for being exceptionally well organized with wide consensus on protocols (Stuiver and Polach, 1977) and the convention, adopted at the very outset, to report unique laboratory numbers associated with each sample. Furthermore, the radiocarbon community has a long history of ^{14}C laboratory intercomparison studies (International Study Group, 1982) that continue today (Scott et al., 2010). The high participation rates in these studies have been a major asset to ensure quality control. Intercomparison studies have highlighted that unexplained variability exists between different laboratories. Clearly, such efforts are critical for fields such as sea-level research where the compilation of large databases (Chapter 34) is greatly facilitated by this type of documentation.

23.3 RADIOMETRIC V. AMS ^{14}C DATING

The global ^{14}C stock amounts to no more than about 1 ton, resulting in a natural $^{14}C/^{12}C$ ratio of the order of magnitude of 10^{-12}. The radiometric techniques for determining ^{14}C concentrations rely on the measurement of beta decay. This places significant constraints on sample size: about 1 g of carbon is typically needed to perform a radiometric age measurement with sufficient precision in a reasonable amount of time. After it was recognized that measuring ^{14}C concentrations directly was feasible in principle by means of high-energy accelerators (Muller, 1977), AMS enjoyed rapid development during the 1980s. With respect to early expectations, AMS ^{14}C dating has been unable to push back the age limit (mainly due to the challenges associated with sample contamination) but the sample size reduction has exceeded initial expectations. Furthermore, measuring time has been significantly reduced – from days to minutes – depending on the sample size and the desired analytical precision of the measurement. Arguably, AMS has been the single-most important advance in ^{14}C dating since its inception.

AMS ^{14}C dating has opened a vast array of new opportunities in sea-level research. ^{14}C dated materials vary widely and include everything from well-defined, short-lived specimens (e.g., seeds of plants, articulated single shells) to samples that are mixtures of a wide range of elements with potentially widely variable ages (e.g., bulk peat, shell hash, beach rock). Due to sample size demands, radiometric ^{14}C dating nearly always concerned the latter of these categories. The progression from radiometric methods to AMS has increased the potential for dating small and hence better-defined samples; however, it should be noted that it is not uncommon for AMS ^{14}C dating to be applied to bulk samples, while in some cases radiometric dating has been possible for single specimens (e.g., wood). In the case of organic-rich sediments (notably peat), a number of studies have demonstrated that plant macrofossils extracted from the sediment provide more accurate ages than bulk peat samples (Törnqvist et al., 1992; Björck et al., 1998; Nilsson et al., 2001) and dating of short-lived, well-identified materials is therefore now the preferred approach. A recent statistical analysis (Hu, 2010) of ^{14}C ages from bulk peat and macrofossils (Törnqvist et al., 1992) revealed that an additional error of ±100 ^{14}C a should be applied to bulk peat samples to account for uncertainties associated with various types of sample contamination. It is reasonable to assume that similar uncertainties should be considered in bulk carbonate samples (also see Chapter 34).

In view of the above (but see also Section 23.5 on reservoir effects), well-identifiable terrestrial macrofossils – commonly plant remains – are perhaps the most ideal material for ^{14}C dating in sea-level studies. Proper identification of such materials is non-trivial and often requires the involvement of paleobotanists, paleoentomologists, or related experts. Materials for dating are commonly extracted from bulk sediments by means of careful wet sieving (a 250–500 μm screen is preferable). Sample storage in acidified, de-ionized water at a temperature of <4 °C is generally adequate, provided that samples are submitted for dating as soon as possible after collection in the field. Some studies (Wohlfarth et al., 1998) have identified problems due to fungal growth on macrofossils in wet storage, and therefore recommend drying as the optimal solution for longer-term sample storage.

23.4 ISOTOPIC FRACTIONATION

While the correction of ^{14}C ages for isotopic fractionation (based on $\delta^{13}C$ measurements) as discussed above has been a routine procedure for a considerable time (Stuiver and Polach, 1977), this was not the case in the earlier days of ^{14}C dating. Many sea-level data from the 1970s or before carry potentially large uncertainties since this effect, if unaccounted for, could lead to age offsets of the order of hundreds of ^{14}C years. Equation (23.5) is the approximate term commonly used to correct ^{14}C ages for isotopic fractionation. However, a more precise ^{14}C age correction (Δt_i), shown graphically in Figure 23.2, is as follows:

$$\Delta t_i = 15262.7 \times \ln\left(1 + \frac{\delta^{13}C}{1000}\right) + 386.418. \qquad (23.8)$$

Table 23.1 shows the age correction and associated error ($\Delta t_i \pm \sigma_i$) based on Equation (23.8) for a range of carbonaceous and calcareous materials commonly used in sea-level studies that can be applied to samples where isotopic fractionation correction has not been performed. It should be noted that this table relies primarily on data from the US Gulf Coast that may not always be appropriate for other regions. Clearly, such corrections should be based on regional data wherever possible. It should also be pointed out that $\delta^{13}C$ data such as those listed in Table 23.1 can be used to determine paleoenvironments (e.g., by capitalizing on the large isotopic contrast between C_3 and C_4 plants and the associated paleo-salinities) and therefore constrain the indicative range of sea-level index points. These issues are discussed at length in Chapter 20.

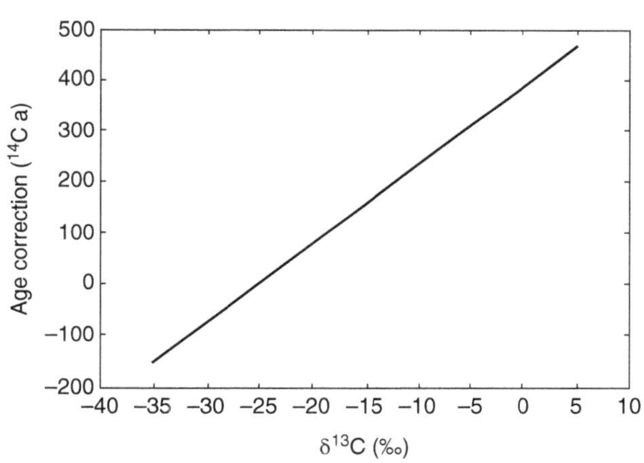

Fig. 23.2. Relationship between $\delta^{13}C$ and ^{14}C age correction due to isotopic fractionation (Δt_i) based on Equation (23.8).

A different approach must be followed when it is unknown whether isotopic fractionation correction has been performed. Since this correction was not adopted by every ^{14}C laboratory simultaneously, it is sometimes difficult to ascertain whether this correction was in fact carried out. In such cases, we define the isotopic fractionation error σ_I as:

$$\sigma_I = |\Delta t_i| + |\sigma_i|. \qquad (23.9)$$

All the uncertainties associated with isotopic fractionation are fully incorporated in state-of-the-art sea-level databases, as illustrated in Chapter 34.

23.5 RESERVOIR EFFECTS

A fundamental assumption of ^{14}C dating is that any organism used for dating was in equilibrium with the atmosphere – in terms of its $^{14}C/^{12}C$ ratio – at the time of death (the same principle would apply to non-biogenic carbonates). However, it was quickly demonstrated (Deevey et al., 1954) that this assumption is often not valid. In particular, water bodies such as lakes, estuaries, and the open ocean exchange carbon with the atmosphere at rates slow enough that they exhibit "apparent" ^{14}C ages. As a result, organisms that assimilate or crystallize carbon (i.e., DIC) from ambient waters acquire a $^{14}C/^{12}C$ ratio that may be significantly

Table 23.1. Range of $\delta^{13}C$ values and ^{14}C age correction for materials relevant to sea-level studies, largely based on data from the US Gulf Coast.

Material	$\delta^{13}C$ (‰)	$\Delta t_i \pm \sigma_i$ (^{14}C a)
Wood (Walker, 2005)	−29 to −23	−20±50
Organic-rich sediment from freshwater environments (Chmura et al., 1987)	−28.5 to −25.5	−30±20
Organic-rich sediment from intermediate environments (Chmura et al., 1987)	−25 to −18	+60±60
Organic-rich sediment from brackish environments (Chmura et al., 1987)	−20 to −15	+120±40
Organic-rich sediment from saline environments (Chmura et al., 1987)	−18 to −14	+140±30
Undifferentiated peat (González, 2008)	−28 to −14	+60±110
Estuarine carbonates (Milliken et al., 2008b, 2008c)	−7 to +5	+370±90
Marine carbonates (Walker, 2005)	−3 to +3	+390±50

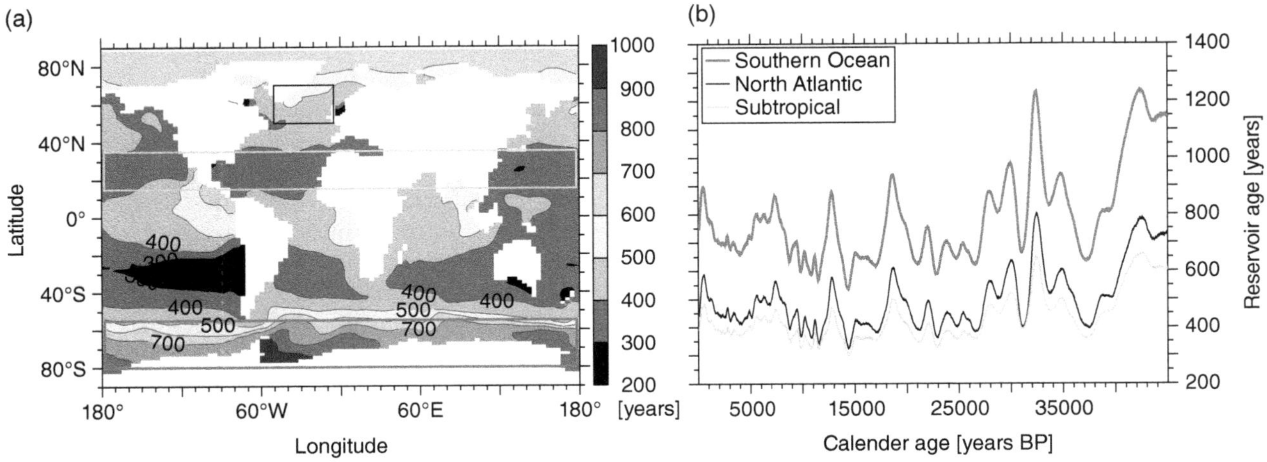

Fig. 23.3. Modeled variability in reservoir ages over the past 45 cal ka (Franke et al., 2008). (a) Spatial pattern of global *R* variations in surface ocean waters (upper 36 m; J. Franke, pers. comm., 2013). (b) Regional mean temporal *R* variations for the areas indicated by equally colored rectangles in the map. Note that this is a simulation with present-day boundary conditions in terms of ocean circulation; conditions of reduced Atlantic meridional overturning circulation such as those during the Last Glacial Maximum would lead to higher reservoir ages. For color details, please see Plate 34.

lower than that of the atmosphere. This phenomenon is known as the reservoir effect (*R*, also referred to as *R*(*t*)), which indicates the deviation (in ^{14}C a) of a sample subject to a reservoir effect from a coeval sample that is in equilibrium with the atmosphere.

In surface waters of the open ocean *R* is often assumed to be around 400 ^{14}C a; recent calibration protocols (Hughen et al., 2004b; Reimer et al., 2009) use a reservoir correction of 405 ± 22 ^{14}C a. However, it is well documented that this is merely a global average; it may be a reasonable approximation at low latitudes, but considerably larger reservoir effects occur in areas of upwelling and/or closer to the poles. For example, the Southern Ocean around Antarctica is widely known to possess much higher reservoir ages. To account for this spatial variability, the deviation of a local reservoir effect from the global average, commonly referred to as ΔR (Stuiver et al., 1986), needs to be known. In addition to spatial variability, temporal variability cannot be ruled out (Waelbroeck et al., 2001). Recent modeling (Franke et al., 2008) has investigated the role of variations in ^{14}C production rate and changes in ocean circulation in the spatiotemporal variability of *R*. This work demonstrates (Fig. 23.3) that over the past 45 cal ka no single location within the global ocean, including the tropics, has experienced a range of reservoir ages <200 ^{14}C a. Clearly, this observation demands the addition of a significant error for any ^{14}C age derived from marine carbonates.

Estuarine reservoir effects are even more problematic, because these environments are influenced by the level of mixing between marine waters and freshwater that may be significantly depleted in ^{14}C (depending, for example, on the presence of ^{14}C dead hinterland carbonate rocks). Alternatively, as recently demonstrated for the Baltic Sea (Lougheed et al., 2013), reservoir effects in low-salinity coastal waters may be well below the global average. This presents a particular challenge for sea-level studies that often rely on ^{14}C dated carbonates (e.g., mollusks, gastropods, foraminifera) from estuarine environments. To account for the large spatial variability of *R* in coastal waters, a marine reservoir correction database is available as an online resource (http://calib.qub.ac.uk/marine/; Reimer and Reimer, 2001). However, an awareness of its limitations is important; many regions are not well covered by data (for example, the Gulf of Mexico currently features only five data points). Furthermore, it is often underappreciated how variable *R* can be on a local to regional scale. For example, an assessment of the variability of the modern *R* along the US Atlantic Coast yields values for ΔR that range from +394 to −280 ^{14}C a ($n = 58$; mean $\Delta R = 55 \pm 97$ ^{14}C a). Particularly troubling is the observation that very large variations can occur over very short distances (e.g., in Long Island Sound). Finally, this database does not address temporal variability in reservoir ages.

We further illustrate the problem of estuarine reservoir effects by means of recently obtained

Table 23.2. Estuarine reservoir effect illustrated by ^{14}C data from the Mississippi Delta.

Laboratory number	Sample name[a]	Material dated	^{14}C age (a BP)	Reference age[b]
UCIAMS-101369	Bayou Sale VIII-1a	1 *Macoma mitchelli*	7630 ± 20	7400 ± 6
UCIAMS-101370	Bayou Sale VIII-1b	1 *Macoma mitchelli*	7595 ± 15	7400 ± 6
UCIAMS-101371	Bayou Sale VIII-1c	1 *Macoma mitchelli*	7640 ± 20	7400 ± 6
UCIAMS-101372	Bayou Sale VIII-1d	*Elphidium* spp.	7570 ± 15	7400 ± 6

[a]All samples collected from 14.34–14.44 m below the land surface; UTM coordinates (NAD 83, zone 15R) 643950/3278980; elevation +0.1 m (NAVD 88);
[b]Weighted mean ^{14}C age from nine terrestrial macrofossil samples that capture the final Lake Agassiz drainage event (Li et al., 2012).

early Holocene RSL data from the US Gulf Coast. A paired wood-barnacle sample from a sediment core in Corpus Christi Bay, Texas (Simms et al., 2008) yielded ^{14}C ages of 7670 ± 45 and 8430 ± 45, respectively. The age difference (760 ^{14}C a) represents the local reservoir effect that clearly exceeds the widely used global average of ~400 ^{14}C a (note that this would be a minimum age difference in the event that the wood to which the barnacle was attached was dead). In contrast, recently obtained data from mollusks and calcareous foraminifera from a lagoonal deposit in the Mississippi Delta were used for comparison with a directly associated, distinct event: the final drainage of proglacial Lake Agassiz ^{14}C dated at 7400 ± 6 (weighted mean age of nine terrestrial subsamples; Li et al., 2012). The coeval carbonates (Table 23.2) provide a weighted mean ^{14}C age of 7601 ± 8, indicating a local reservoir effect of ~200 ^{14}C a. In other words, ΔR varies from about +360 to −200 ^{14}C a along the US Gulf Coast. These findings are consistent with an inferred southwest to northeast gradient along the Texas coast (Milliken et al., 2008a), which has been linked to diminishing bicarbonate levels in estuarine waters. Since this is currently a small dataset that exhibits considerable spatial variability, and temporal variability cannot be ruled out, the most rational approach to interpreting ^{14}C ages from estuarine carbonates in this region would be to adopt a reservoir correction of 500 ± 300 ^{14}C a (ΔR = 100 ± 300 ^{14}C a) that captures the full range of currently observed local reservoir effects. Clearly, this is a large error source that on its own can potentially explain disparate sea-level index points (also see Chapter 34).

23.6 ERROR CALCULATION

The wide range of error sources that can potentially affect ^{14}C ages must be fully addressed prior to any subsequent analysis. The commonly used approach to adding errors such as those discussed in Sections 23.3–23.5 is the following:

$$\sigma_t = \sqrt{\sigma_a^2 + \sigma_b^2 + \sigma_i^2 + \sigma_r^2} \quad (23.10)$$

where σ_t is total age error; σ_a is analytical (measurement) error; σ_b is bulk error; σ_i is isotopic fractionation error (to be substituted by σ_I whenever appropriate, as discussed in Section 23.4); and σ_r is reservoir correction error.

As an illustration of some of the age corrections and errors discussed in Sections 23.3–23.5, consider the following example that is discussed in more detail in Chapter 34. A brackish peat sample from the Mississippi Delta was dated at 1725 ± 105 ^{14}C a BP (Coleman and Smith, 1964) at a time when isotopic fractionation correction was not yet commonly used. Given the known $\delta^{13}C$ range for brackish peat from the US Gulf Coast of −15 to −20‰ (Table 23.1), an isotopic fractionation correction of +120 ± 40 ^{14}C a is used. In addition, a bulk error of ± 100 ^{14}C a must be applied. Using Equation (23.10), the corrected age therefore amounts to 1845 ± 150 ^{14}C a BP. This example shows that fully accounting for all age errors can not only reduce the temporal resolution of sea-level data, but also shift the central value of ^{14}C ages relative to how they were originally published.

23.7 RADIOCARBON CALIBRATION: ^{14}C YEARS V. CALENDAR YEARS

While the initial, implicit assumption was that ^{14}C ages would provide true ages in calendar years, it was soon demonstrated (De Vries, 1958) that this is in fact not the case. However, it took a long time for the necessary data to accumulate to enable the conversion of ^{14}C ages into calendar ages over geologically meaningful timescales; arguably, this has been one of the primary efforts of the international

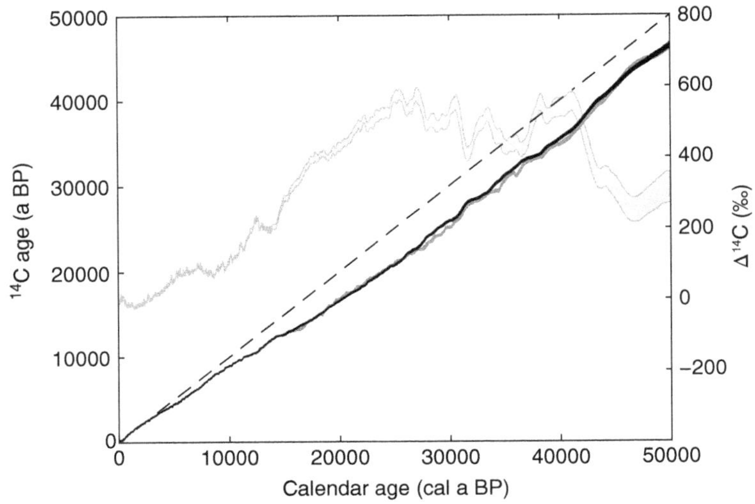

Fig. 23.4. IntCal09 (dark gray) and IntCal13 (black) calibration curves and associated $\Delta^{14}C$ values (light gray, derived from the IntCal13 data) with 1σ envelopes (modified after Reimer et al., 2009, 2013). Dashed line indicates where ^{14}C ages equal calendar ages. The IntCal09 and IntCal13 calibration curves are essentially similar during the Holocene, but deviate by up to ~1000 ^{14}C a during the last glacial, particularly for ages >27 cal ka BP.

radiocarbon community. Early-generation calibration curves (Suess, 1970) that were sufficiently long to capture millennial-scale features demonstrated that, going back in time, ^{14}C ages increasingly deviate (i.e., become progressively younger) compared to dendrochronologically obtained calendar ages. These early studies recognized variations in the Earth's magnetic field as a major control of long-term ^{14}C variations, but also identified smaller (multi-decadal to century scale) phenomena (Suess, 1970), commonly referred to as "wiggles", that proved more elusive.

Despite the early recognition of the non-linear nature of the ^{14}C timescale, the road toward widely accepted calibration curves has been a long one. This painstaking effort initially depended to a large extent on the development of tree-ring chronologies, mainly from bristlecone pine in the western US and oaks in central Europe. High-density and high-precision radiometric ^{14}C dating allowed for calibration curves to be developed, but it was only when precision was pushed close to 1‰ that the existence of medium-term ^{14}C variations (wiggles) could be firmly demonstrated (De Jong et al., 1979). Another major advance was the development of high-precision techniques for U–Th dating (Chapter 26) based on mass spectrometry (akin to the transition from radiometric to AMS ^{14}C dating) which was a major impetus to push calibration curves further back in time (Bard et al., 1990). Finally, AMS ^{14}C dating of terrestrial macrofossils from annually laminated lake sediments (Lotter, 1991) led to the discovery of larger, multi-centennial flat portions of the calibration curve (known as "plateaus"), including one near 10 ^{14}C ka BP, that is, the Pleistocene–Holocene transition.

Subsequent work (Hughen et al., 2004a) has confirmed that millennial-scale, long-term ^{14}C variations are largely due to changes in the intensity of the Earth's magnetic field (and associated variability in the amount of "shielding" of the upper atmosphere from cosmic rays), as well as changes in ocean circulation (notably changes in the ventilation rate of the deep ocean). Medium-term ^{14}C variations exhibit a strong correlation with records of other cosmogenic isotopes (^{10}Be) and are therefore likely associated with fluctuations in solar activity (Bard et al., 1997), where increases in the solar wind reduce ^{14}C production rates. Short-term ^{14}C variations largely reflect the 11-year sunspot cycle (Lingenfelter, 1963). As discussed in Section 23.2 (Equation (23.3)), the deviation of ^{14}C ages from calendar ages is usually expressed as $\Delta^{14}C$; an illustration of changes in this variable throughout the past 50 cal ka is given in Figure 23.4.

From the perspective of sea-level studies, the long-term ^{14}C variations (Fig. 23.4) are of great importance because efforts to determine past rates of relative sea-level change require the use of calendar ages rather than ^{14}C ages. Clearly, merely relying on conventional ^{14}C ages would typically yield rates that are too high. With the increased sophistication of sea-level studies and current attempts to extract decadal- to century-scale relative sea-level fluctuations from the geologic record, fully resolving medium-term ^{14}C variations has also become critical. Only the short-term (sunspot) cycles largely remain beyond the resolution of ^{14}C-based sea-level studies.

Given the historical developments outlined above, it is easy to see that arriving at an internationally accepted calibration curve has taken some time. This occurred with the publication of the first Calibration Issue of the journal *Radiocarbon* (Stuiver and Kra, 1986). This initial calibration curve was entirely based on ^{14}C-dated tree rings but it captured only the past ~7 cal ka. The most recent Calibration Issues (the fifth and sixth iterations) have extended the calibration curve in order to capture the full time window of ^{14}C dating back to 50 cal ka BP (Reimer et al., 2009, 2013; Fig. 23.4). Continuous dendrochronological records have now been extended to ~14 cal ka BP; beyond this time range calibration data consist of a combination of ^{14}C-dated calcareous foraminifera from annually laminated marine sediments, corals subjected to combined U–Th and ^{14}C dating, as well as plant macrofossil ^{14}C ages from annually laminated lake sediments.

The current, internationally adopted, calibration curves (Reimer et al., 2013) are known as IntCal13 and Marine13, intended for terrestrial and marine ^{14}C age calibration, respectively. In other words, both calibration curves are relevant to sea-level studies, depending on the specific nature of the material used for ^{14}C dating. It is important to note that these calibration curves remain subject to change (Fig. 23.4). For example, the most recent major contribution is a detailed new record with 651 ^{14}C ages for the time interval 11.2 to 52.8 cal ka BP from the annually laminated lake sediments of Lake Suigetsu, Japan (Bronk Ramsey et al., 2012) which plays a major role in the IntCal13 effort (Reimer et al., 2013). This new record is important because it is based entirely on terrestrial macrofossils and therefore circumvents problems associated with uncertain marine reservoir effects that affect most previously available calibration data from the last glacial. However, the younger portion of the IntCal13 curve (<14 cal ka BP, i.e., including all of the Holocene) that is entirely based on tree-ring chronologies is less likely to see major modifications in the future.

23.8 CALIBRATION PROCEDURES: PITFALLS AND OPPORTUNITIES

In association with the development of standardized calibration curves, calibration software has proliferated and, particularly in the 1980s, this led to a plethora of programs as reviewed by Aitchison et al. (1989). However, differences between these programs were generally rather small. Over time, many of the early calibration programs have been discontinued and currently the "calibration market" is dominated by CALIB (which traces back to Minze Stuiver's group at the University of Washington and is currently led by Paula Reimer at Queen's University in Belfast; Stuiver and Reimer, 1993) and OxCal (spearheaded by Christopher Bronk Ramsey at the University of Oxford; Bronk Ramsey, 1995). Both programs are freely available and can be accessed at http://calib.qub.ac.uk/calib/ and http://c14.arch.ox.ac.uk/, respectively. While the mathematical principles that underpin the basic calibration procedures of the two programs are largely similar, the two packages offer some differences in terms of their interface, versatility, and specific options. For example, OxCal provides numerous options for the calibration of stratigraphic datasets that contain independent constraints on the relative age of ^{14}C data (Bronk Ramsey, 2009); tools such as these have been used as additional constraints on ^{14}C-based time series in sea-level studies (e.g., González and Törnqvist, 2009; Li et al., 2012). More sophisticated techniques for determining age models are discussed in Chapter 32.

The ^{14}C calibration curve tends to widen calendar age ranges after the conversion of ^{14}C ages into calendar ages. This is due to several factors. First, the calendar timescale is longer than the ^{14}C timescale (Fig. 23.4); for example, the beginning of the Holocene at 10 ^{14}C ka BP is 11.7 cal ka BP in calendar years (Rasmussen et al., 2006). Second, the presence of medium-term ^{14}C variations (Fig. 23.5) often leads to the conversion of Gaussian ^{14}C age probability distributions into extended and irregular probability distributions for calendar ages. For example, a recently obtained highly precise age for a sea-level jump associated with the final drainage of proglacial Lake Agassiz (Table 23.2) was 7400 ± 12 ^{14}C a BP (2σ), but the presence of a flat portion of the calibration curve that nearly coincides with this particular age yields a bimodal distribution and a 2σ calendar age range of 8.18–8.31 cal ka BP (Li et al., 2012; Fig. 23.5).

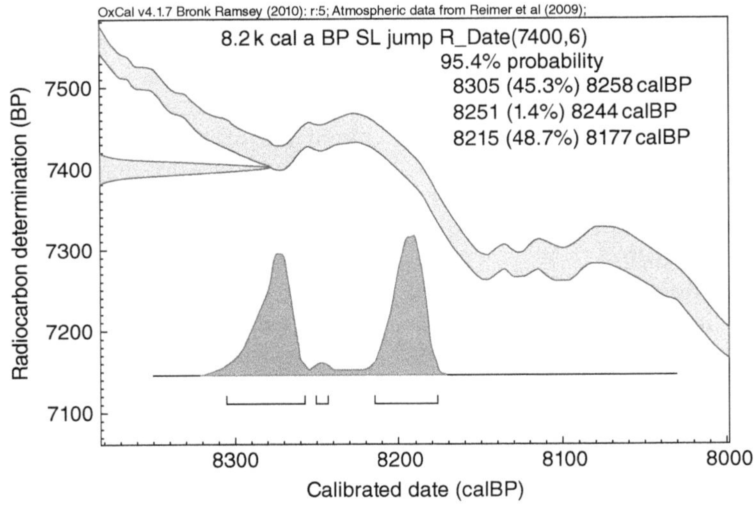

Fig. 23.5. Calibration of a ^{14}C age (Table 23.2) representing a decimeter-scale sea-level jump in the Mississippi Delta associated with the final drainage of proglacial Lake Agassiz (Li et al., 2012) with the OxCal program using the IntCal09 calibration curve (the IntCal13 curve yields a nearly identical result). Note the very tight Gaussian probability distribution of the ^{14}C age that is converted into a much wider, bimodal distribution due to the flat portion in the calibration curve. The calibration curve from ~8.0–8.4 cal ka BP exhibits the characteristic amplitude and period of medium-term ^{14}C variations.

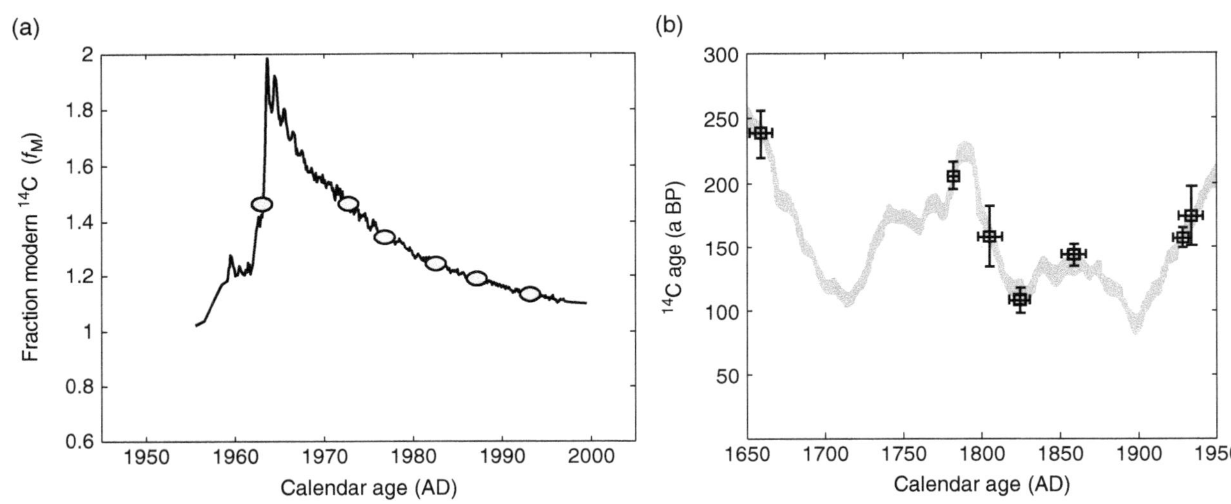

Fig. 23.6. Wiggle matching of ^{14}C data from a coastal marsh over (a) decadal to (b) centennial timescales (modified after Marshall et al., 2007). Note that ^{14}C measurements in (a) are expressed in f_M.

In principle, it is possible to take advantage of the irregularities of the calibration curve using a technique known as "wiggle matching". This involves dense series of highly precise ^{14}C ages from a continuous stratigraphic section that are matched with the calibration curve to obtain better-resolved chronologies. A number of quantitative techniques have been developed for this purpose (e.g., Blaauw et al., 2003). So far, wiggle matching has been carried out in sea-level studies targeting the past few centuries (Marshall et al., 2007; Fig. 23.6) that exhibit particularly large ^{14}C variations due to a combination of factors, including the industrial release of ^{14}C dead CO_2 (known as the "Suess effect") and a distinct peak from 1963 associated with nuclear bomb testing. Studies along these lines are often performed in conjunction with other age markers (Chapters 24 and 25). The CALIB software discussed above includes a module known as CALIBomb (http://calib.qub.ac.uk/CALIBomb/) that has the specific purpose of developing ^{14}C chronologies within the "post-bomb" time window.

23.9 FINAL REMARKS

Since its inception just over 60 years ago, ^{14}C dating has taken remarkable strides. In particular, dating accuracy and precision have increased substantially, in part due to the transition from radiometric dating to AMS and the compilation of sophisticated calibration curves. While some of this progress has led to a better understanding of the limitations of the method (e.g., the increased recognition that reservoir effects are a major challenge indeed, not in the least for sea-level studies), numerous exciting new opportunities have presented themselves. For example, wiggle matching has so far only been used in a few cases in sea-level studies focusing on the past few centuries. There is great potential to apply techniques such as these to older sea-level records in order to investigate subtle, decimeter-scale relative sea-level fluctuations and their possible links with climate change.

REFERENCES

Aitchison, T.C., Leese, M., Michczynska, D.J., Mook, W.G., Otlet, R.L., Ottaway, B.S., Pazdur, M.F., Van der Plicht, J., Reimer, P.J., Robinson, S.W., Scott, E.M., Stuiver, M., and Weninger, B. (1989) A comparison of methods used for the calibration of radiocarbon dates. Radiocarbon, 31, 846–862.

Bard, E., Hamelin, B., Fairbanks, R.G., and Zindler, A. (1990) Calibration of the ^{14}C timescale over the past 30,000 years using mass spectrometric U–Th ages from Barbados corals. Nature, 345, 405–410.

Bard, E., Raisbeck, G.M., Yiou, F., and Jouzel, J. (1997) Solar modulation of cosmogenic nuclide production over the last millennium: comparison between ^{14}C and ^{10}Be records. Earth and Planetary Science Letters, 150, 453–462.

Björck, S., Bennike, O., Possnert, G., Wohlfarth, B., and Digerfeldt, G. (1998) A high-resolution ^{14}C dated sediment sequence from southwest Sweden: age comparisons between different components of the sediment. Journal of Quaternary Science, 13, 85–89.

Blaauw, M., Heuvelink, G.B.M., Mauquoy, D., Van der Plicht, J., and Van Geel, B. (2003) A numerical approach to ^{14}C wiggle-match dating of organic deposits: best fits and confidence intervals. Quaternary Science Reviews, 22, 1485–1500.

Bronk Ramsey, C. (1995) Radiocarbon calibration and analysis of stratigraphy: The OxCal program. Radiocarbon, 37, 425–430.

Bronk Ramsey, C. (2009) Bayesian analysis of radiocarbon dates. Radiocarbon, 51, 337–360.

Bronk Ramsey, C., Staff, R.A., Bryant, C.L., Brock, F., Kitagawa, H., Van der Plicht, J., Schlolaut, G., Marshall, M.H., Brauer, A., Lamb, H.F., Payne, R.L., Tarasov, P.E., Haraguchi, T., Gotanda, K., Yonenobu, H., Yokoyama, Y., Tada, R., and Nakagawa, T. (2012) A complete terrestrial radiocarbon record for 11.2 to 52.8 kyr BP. Science, 338, 370–374.

Chmura, G.L., Aharon, P., Socki, R.A., and Abernethy, R. (1987) An inventory of ^{13}C abundances in coastal wetlands of Louisiana, USA: vegetation and sediments. Oecologia, 74, 264–271.

Coleman, J.M., and Smith, W.G. (1964) Late Recent rise of sea level. Geological Society of America Bulletin, 75, 833–840.

Deevey, E.S., Jr., Gross, M.S., Hutchinson, G.E., and Kraybill, H.L. (1954) The natural C^{14} contents of materials from hard-water lakes. Proceedings of the National Academy of Sciences of the United States of America, 40, 285–288.

De Jong, A.F.M., Mook, W.G., and Becker, B. (1979) Confirmation of the Suess wiggles: 3200-3700 BC. Nature, 280, 48–49.

De Vries, H. (1958) Variation in concentration of radiocarbon with time and location on earth. Proceedings of Koninklijke Nederlandse Akademie van Wetenschappen, Series B, 61, 94–102.

Engelkemeir, A.G., Hamill, W.H., Inghram, M.G., and Libby, W.F. (1949) The half-life of radiocarbon (C^{14}). Physical Review, 75, 1825–1833.

Franke, J., Paul, A., and Schulz, M. (2008) Modeling variations of marine reservoir ages during the last 45 000 years. Climate of the Past, 4, 125–136.

Godwin, H. (1962) Half-life of radiocarbon. Nature, 195, 984.

González, J.L. (2008) Exploring the late Holocene sedimentary record of the Mississippi Delta for climate/sea level connections. PhD thesis, Tulane University.

González, J.L., and Törnqvist, T.E. (2009) A new Late Holocene sea-level record from the Mississippi Delta: evidence for a climate/sea level connection? Quaternary Science Reviews, 28, 1737–1749.

Hu, P. (2010) Developing a quality-controlled postglacial sea-level database for coastal Louisiana to assess conflicting hypotheses of Gulf Coast sea-level change. MSc thesis, Tulane University.

Hughen, K.A., Baillie, M.G.L., Bard, E., Beck, J.W., Bertrand, C.J.H., Blackwell, P.G., Buck, C.E., Burr, G.S., Cutler, K.B., Damon, P.E., Edwards, R.L., Fairbanks, R.G., Friedrich, M., Guilderson, T.P., Kromer, B., McCormac, G., Manning, S., Bronk Ramsey, C., Reimer, P.J., Reimer, R.W., Remmele, S., Southon, J.R., Stuiver, M., Talamo, S., Taylor, F.W., Van der Plicht, J., and Weyhenmeyer, C.E. (2004a) Marine04 marine radiocarbon age calibration, 0-26 cal kyr BP. Radiocarbon, 46, 1059–1086.

Hughen, K., Lehman, S., Southon, J., Overpeck, J., Marchal, O., Herring, C., and Turnbull, J. (2004b) ^{14}C activity and global carbon cycle changes over the past 50,000 years. Science, 303, 202–207.

International Study Group (1982) An inter-laboratory comparison of radiocarbon measurements in tree rings. Nature, 298, 619–623.

Li, Y.-X., Törnqvist, T.E., Nevitt, J.M., and Kohl, B. (2012) Synchronizing a sea-level jump, final Lake Agassiz drainage, and abrupt cooling 8200 years ago. Earth and Planetary Science Letters, 315–316, 41–50.

Libby, W.F. (1952) *Radiocarbon Dating.* University of Chicago Press, Chicago.

Lingenfelter, R.E. (1963) Production of carbon 14 by cosmic-ray neutrons. Reviews of Geophysics, 1, 35–55.

Lotter, A.F. (1991) Absolute dating of the late-glacial period in Switzerland using annually laminated sediments. Quaternary Research, 35, 321–330.

Lougheed, B.C., Filipsson, H.L., and Snowball, I. (2013) Large spatial variations in coastal ^{14}C reservoir age – a case study from the Baltic Sea. Climate of the Past, 9, 1015–1028.

Marshall, W.A., Gehrels, W.R., Garnett, M.H., Freeman, S.P.H.T., Maden, C., and Xu, S. (2007) The use of 'bomb spike' calibration and high-precision AMS ^{14}C analyses to date salt-marsh sediments deposited during the past three centuries. Quaternary Research, 68, 325–337.

Milliken, K.T., Anderson, J.B., and Rodriguez, A.B. (2008a) A new composite Holocene sea-level curve for the northern Gulf of Mexico. Geological Society of America Special Paper, 443, 1–11.

Milliken, K.T., Anderson, J.B., and Rodriguez, A.B. (2008b) Record of dramatic Holocene environmental changes linked to eustasy and climate change in Calcasieu Lake, Louisiana, USA. Geological Society of America Special Paper, 443, 43–63.

Milliken, K.T., Anderson, J.B., and Rodriguez, A.B. (2008c) Tracking the Holocene evolution of Sabine Lake through the interplay of eustasy, antecedent topography, and sediment supply variations, Texas and Louisiana, USA. Geological Society of America Special Paper, 443, 65–88.

Mook, W.G., and Van de Plassche, O. (1986) Radiocarbon dating. In: *Sea-Level Research: a Manual for the Collection and Evaluation of Data* (ed. van de Plassche, O.), Geo Books, Norwich, pp. 525–560.

Muller, R.A. (1977) Radioisotope dating with a cyclotron. Science, 196, 489–494.

Nilsson, M., Klarqvist, M., Bohlin, E., and Possnert, G. (2001) Variation in ^{14}C age of macrofossils and different fractions of minute peat samples dated by AMS. The Holocene, 11, 579–586.

Olsson, I.U. (2009) Radiocarbon dating history: Early days, questions, and problems met. Radiocarbon, 51, 1–43.

Rasmussen, S.O., Andersen, K.K., Svensson, A.M., Steffensen, J.P., Vinther, B.M., Clausen, H.B., Siggaard-Andersen, M.-L., Johnsen, S.J., Larsen, L.B., Dahl-Jensen, D., Bigler, M., Röthlisberger, R., Fischer, H., Goto-Azuma, K., Hansson, M.E., and Ruth, U. (2006) A new Greenland ice core chronology for the last glacial termination. Journal of Geophysical Research, 111, D06102.

Reimer, P.J., and Reimer, R.W. (2001) A marine reservoir correction database and on-line interface. Radiocarbon, 43, 461–463.

Reimer, P.J., Baillie, M.G.L., Bard, E., Bayliss, A., Beck, J.W., Blackwell, P.G., Bronk Ramsey, C., Buck, C.E., Burr, G.S., Edwards, R.L., Friedrich, M., Grootes, P.M., Guilderson, T.P., Hajdas, I., Heaton, T.J., Hogg, A.G., Hughen, K.A., Kaiser, K.F., Kromer, B., McCormac, F.G., Manning, S.W., Reimer, R.W., Richards, D.A., Southon, J.R., Talamo, S., Turney, C.S.M., Van der Plicht, J., and Weyhenmeyer, C.E. (2009) IntCal09 and Marine09 radiocarbon age calibration curves, 0–50,000 years cal BP. Radiocarbon, 51, 1111–1150.

Reimer, P.J., Bard, E., Bayliss, A., Beck, J.W., Blackwell, P.G., Bronk Ramsey, C., Buck, C.E., Cheng, H., Edwards, R.L., Friedrich, M., Grootes, P.M., Guilderson, T.P., Haflidason, H., Hajdas, I., Hatté, C., Heaton, T.J., Hoffmann, D.L., Hogg, A.G., Hughen, K.A., Kaiser, K.F., Kromer, B., Manning, S.W., Niu, M., Reimer, R.W., Richards, D.A., Scott, E.M., Southon, J.R., Staff, R.A., Turney, C.S.M., and Van der Plicht, J. (2013) IntCal13 and Marine13 radiocarbon age calibration curves 0–50,000 years cal BP. Radiocarbon, 55, 1869–1887.

Scott, E.M., Cook, G.T., and Naysmith, P. (2010) The Fifth International Radiocarbon Intercomparison (VIRI): An assessment of laboratory performance in Stage 3. Radiocarbon, 52, 859–865.

Simms, A.R., Anderson, J.B., Rodriguez, A.B., and Taviani, M. (2008) Mechanisms controlling environmental change within an estuary: Corpus Christi Bay, Texas, USA. Geological Society of America Special Paper, 443, 121–146.

Stuiver, M., and Polach, H.A. (1977) Reporting of ^{14}C data. Radiocarbon, 19, 355–363.

Stuiver, M., and Kra, R. (eds) (1986) *Calibration Issue*, Radiocarbon, 28, 805–1030.

Stuiver, M., and Reimer, P.J. (1993) Extended ^{14}C data base and revised CALIB 3.0 ^{14}C age calibration program. Radiocarbon, 35, 215–230.

Stuiver, M., Pearson, G.W., and Braziunas, T. (1986) Radiocarbon age calibration of marine samples back to 9000 cal yr BP. Radiocarbon, 28, 980–1021.

Suess, H.E. (1970) Bristlecone-pine calibration of the radiocarbon time-scale 5200 BC to present. In: *Radiocarbon Variations and Absolute Chronology* (ed. Olsson, I.U.), Almqvist & Wiksell, Stockholm, pp. 303–311.

Taylor, R.E. (2000) Fifty years of radiocarbon dating. American Scientist, 88, 60–67.

Törnqvist, T.E., De Jong, A.F.M., Oosterbaan, W.A., and Van der Borg, K. (1992) Accurate dating of organic deposits by AMS ^{14}C measurement of macrofossils. Radiocarbon, 34, 566–577.

Van de Plassche, O. (ed.) (1986) *Sea-Level Research: a Manual for the Collection and Evaluation of Data.* Geo Books, Norwich.

Waelbroeck, C., Duplessy, J.-C., Michel, E., Labeyrie, L., Paillard, D., and Duprat, J. (2001) The timing of the last deglaciation in North Atlantic climate records. Nature, 412, 724–727.

Walker, M. (2005) *Quaternary Dating Methods.* John Wiley, Chichester.

Wohlfarth, B., Skog, G., Possnert, G., and Holmquist, B. (1998) Pitfalls in the AMS radiocarbon-dating of terrestrial macrofossils. Journal of Quaternary Science, 13, 137–145.

Chapter 24

^{210}Lead and ^{137}Cesium: establishing a chronology for the last century

D. REIDE CORBETT[1,2] AND J.P. WALSH[1,2]

[1] *East Carolina University, Greenville, NC, USA*

[2] *UNC Coastal Studies Institute, Wanchese, NC, USA*

24.1 INTRODUCTION

Marine, freshwater, and intertidal sediments can offer a continuous record of environmental change. However, without some scale of time, a core is merely a column of sediment reflecting possible variations, but with no means to relate to processes. ^{210}Pb and ^{137}Cs are two radionuclides that have been commonly used to quantify sediment accumulation rates and establish a deposition chronology for the past 100–150 and 50 years, respectively. Developing a modern geochronology in any fresh or marine system requires a fundamental understanding of the sedimentation in the area, and specifically with some constraints on the source and supply of material that controls sediment accumulation with time. With this knowledge and suitable conditions, the application of these two radionuclides has proven to be useful in many different areas and environments around the globe, including freshwater, marine, and intertidal systems. Once a chronology for a system has been determined, a wide range of sedimentological and environmental questions can be addressed (e.g., changes in water quality, deposition history of heavy metals, or trapping of contaminants). In this chapter, we provide the scientific background to the application and utility of these two tracers in developing accurate geochronologies in fresh, marine, and intertidal environments, including the most common models applied and potential methodological pitfalls.

24.2 ANTHROPOGENIC FALLOUT OF ^{137}CS: ESTABLISHING A POINT IN TIME

Cesium-137 is an anthropogenic radionuclide, a byproduct of fission reactions. The first hydrogen bomb was tested on 1 November 1952 on Enewetak Atoll in the Marshall Islands. Atmospheric ^{137}Cs ($t_{1/2}$ = 30.1 years) derived from nuclear weapons testing was first detectible in sediments in approximately 1953, with the maximum fallout occurring in 1963 in the northern hemisphere (Fig. 24.1; UNSCEAR, 2000). A similar temporal pattern, although with significantly reduced activities, is present in the Southern Hemisphere, making this technique much less useful south of the equator. The historic annual flux of ^{137}Cs provides the possibility of two dated horizons in particular, one reflecting the initial onset of deposition in the early 1950s and the 1963 peak in ^{137}Cs fallout; researchers have used both. Some Northern Hemisphere areas had notable additional input in 1986 following the Chernobyl incident (Aarkrog, 1988), thus generating a secondary peak. The annual variation allows ^{137}Cs to be a viable chronostratigraphic marker in a wide variety of depositional environments including reservoirs, lakes, wetlands, coastal areas, and floodplains (Pennington et al., 1973; Chmura and Kosters, 1994; Neubauer et al., 2002).

24.2.1 Experimental methods

Measurement of ^{137}Cs is easily achieved through gamma spectroscopy. Like ^{210}Pb (see Section 24.3), a high-resolution germanium gamma detector is used; this makes gamma spectroscopy especially attractive as ^{137}Cs, ^{210}Pb, and ^{226}Ra can be simultaneously measured (Table 24.1). The ^{137}Cs photopeak lies at 661.7 keV and has an 85.1% emission yield. This area of the spectrum has a fairly linear energy efficiency and a relatively low background. If an older, low-resolution detector is used for analysis, the user should be aware of possible interference by the low yielding ^{214}Bi photopeak at 665.5 keV.

Handbook of Sea-Level Research, First Edition. Edited by Ian Shennan, Antony J. Long, and Benjamin P. Horton.

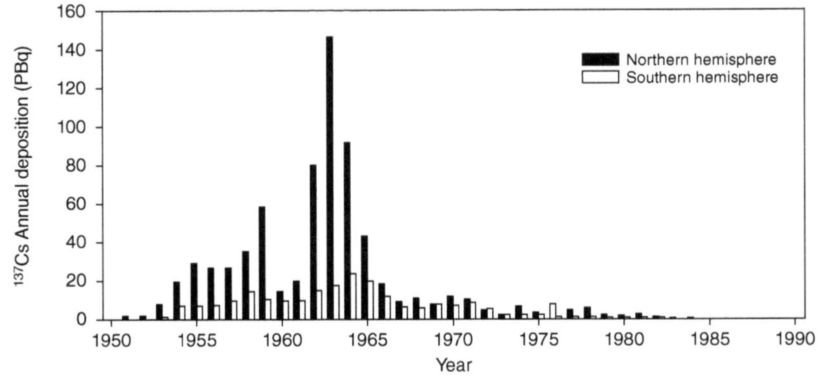

Fig. 24.1. Annual deposition of ^{137}Cs produced in atmospheric nuclear testing (UNSCEAR, 2000).

Table 24.1. The advantages and disadvantages should be considered when deciding which analytical method to use for ^{210}Pb analysis.

α Spectrometry	
Pro	• Low background, high sensitivity • Small amounts of sample can be used (<1.0 g dry sediment) • Detectors are less costly, easier to maintain • Shorter counting time, increased sample throughput
Con	• Requires time-consuming and involved chemical separation • Measures total ^{210}Pb via daughter product (^{210}Po), assume secular equilibrium • Does not evaluate $^{210}Pb_{supported}$, must measure independently or make assumption • Requires a measure of chemical yield for each measurement, adding additional error • Destructive procedure
γ Spectrometry	
Pro	• Measure multiple isotopes simultaneously (i.e., ^{210}Pb, ^{137}Cs) • Non-destructive, simple sample preparation • Measure supported ^{210}Pb simultaneously
Con	• Higher background, lower sensitivity • Self-absorption correction necessary • Longer counting time • Difficult efficiency calibration required

24.2.2 Establishing a chronology

Sediment accumulation rates are calculated by relating the initial appearance or maximum peak ^{137}Cs activity in the strata to the years 1953 and 1963, respectively. However, four assumptions must be made when applying ^{137}Cs in the calculation of sediment accumulation rates in aquatic systems: (1) ^{137}Cs entering the water body is quickly adsorbed to suspended material in the water column; (2) where ^{137}Cs is deposited in the catchment area of a water body, the radionuclide is firmly adsorbed on soil particles and any redistribution results chiefly from erosion and deposition; (3) once deposited in the sediment column, the radionuclide is immobile and not affected by reworking or diffusional movement up or down the sediment column (Hutchinson, 1995); and (4) the depth of the physical and/or biologically mixed surface should be factored into the calculation. ^{137}Cs is typically associated with clay minerals and associated coatings of organic matter and sesquioxides (Goodbred and Kuehl, 1998; Alverez-Iglesias et al., 2007). Due to this affinity, ^{137}Cs can become mobile in anoxic sediments, leading to upward or downward migration and invalidating assumption (3) (Crusius and Anderson, 1995; Christiansen et al., 2002; Bondavalli, 2003). It is often more accurate and advantageous to use the maximum peak in ^{137}Cs (e.g., 1963) rather than the initial appearance of the isotope, due to detection limit constraints and the possible influence of mixing and nuclide migration. In these cases, it has been shown that the use of ^{239,240}Pu is much more effective as this nuclide is less prone to mobilization (Crusius and Anderson, 1995), but the disadvantage is that the analysis of Pu is considerably more involved and costly.

Measuring the downcore activity of ^{137}Cs provides a date stamp for the core and is fairly straightforward in establishing a single point within a cores chronology. However, when applying this radionuclide to calculating sedimentation rates, it is important to understand that only an average rate over the entire period since the date used (e.g., 1953 or 1963) can be estimated and assumes no change in rate with time. The average sedimentation rate is simply calculated by dividing the depth (cm or g cm^{-2} for vertical accretion or mass accumulation rates, respectively) of maximum activity (or first appearance) by the years before collection that the event occurred.

24.3 ^{210}PB TRACER OF SEDIMENT ACCUMULATION

Lead-210 is a naturally occurring radioisotope that is part of the ^{238}U decay series (Fig. 24.2) and is often used for developing a modern (last c. 100 years) sediment chronology (Krishnaswamy et al., 1971; Oldfield and Appleby, 1984). ^{210}Pb is continually produced in the atmosphere and in the Earth's surface. Applying ^{210}Pb as a geochronometer is based on the disequilibrium that occurs between ^{210}Pb and its longer-lived parent isotope, ^{226}Ra ($t_{1/2}$ = 1602 years), by diffusion of a short-lived intermediate isotope ^{222}Rn ($t_{1/2}$ = 3.82 days). ^{226}Ra is found in most rocks and soils, thus providing a continuous source of ^{222}Rn and therefore ^{210}Pb. Once ^{226}Ra in soils, rocks, and sediments decays and produces ^{222}Rn, the inert gas (^{222}Rn) diffuses into the atmosphere, eventually decaying through a series of short-lived daughters (e.g., milliseconds to c. 30 min) to ^{210}Pb (Fig. 24.3). This rapid production of ^{210}Pb in the atmosphere is quickly removed by wet and dry deposition, ^{210}Pb is then removed from solution onto particulate matter by direct adhesion on the interface and by adsorption on inorganic particles or on organic matter within a watershed or water bodies (Appleby and Oldfield, 1992). Once sequestered on sediment, ^{210}Pb is essentially geochemically immobile and can therefore serve as a useful tracer of sediment processes, providing a temporal context through its decay. Given a half-life of 22.3 years, ^{210}Pb is used for dating sediments of up to approximately 100 years of age (i.e., after five halves 3.1% remains, roughly equivalent to the measurement error).

The total activity of ^{210}Pb measured in the sediments at any time is the sum of two components: (1) supported ^{210}Pb; and (2) unsupported ^{210}Pb. Supported ^{210}Pb refers to the amount of ^{210}Pb associated with the sediment, that is, generated by ^{226}Ra in or on the particles. It is often assumed that ^{226}Ra and supported ^{210}Pb in the sediments are in secular equilibrium due to their relatively long and short half-lives, respectively. The unsupported or excess ^{210}Pb is the activity above that produced by the sediment particles, which were scavenged onto the particle before burial. This excess is quantified by subtracting the supported ^{210}Pb activity (i.e., from the ^{226}Ra of the sediment particles) from the total ^{210}Pb measured for a given sample. Ultimately, this excess ^{210}Pb represents the amount of ^{210}Pb added to the sediment via atmospheric fallout or through scavenging from the water (e.g., within the water column; Fig. 24.3).

$$^{210}Pb_{Total} = {}^{210}Pb_{Excess} + {}^{210}Pb_{Supported} \qquad (24.1)$$

where $^{210}Pb_{Supported}$ is equal to ^{226}Ra, assuming secular equilibrium.

$$^{238}U \xrightarrow{4.51\times10^9\ y} {}^{226}Ra \xrightarrow{1620\ y} {}^{222}Rn \xrightarrow{3.82\ d} {}^{210}Pb \xrightarrow{22.26\ y} {}^{210}Po \xrightarrow{138.4\ d} {}^{206}Pb$$

Fig. 24.2. An abbreviated ^{238}U decay series, including their associated half-lives, with the principal radionuclides in the production of ^{210}Pb and the final stable Pb daughter in the chain. Dashed lines indicate several intermediate steps not included for simplicity, while a solid line is direct decay, from parent to daughter. Note the long half-life of ^{226}Ra versus ^{222}Rn and ^{210}Pb.

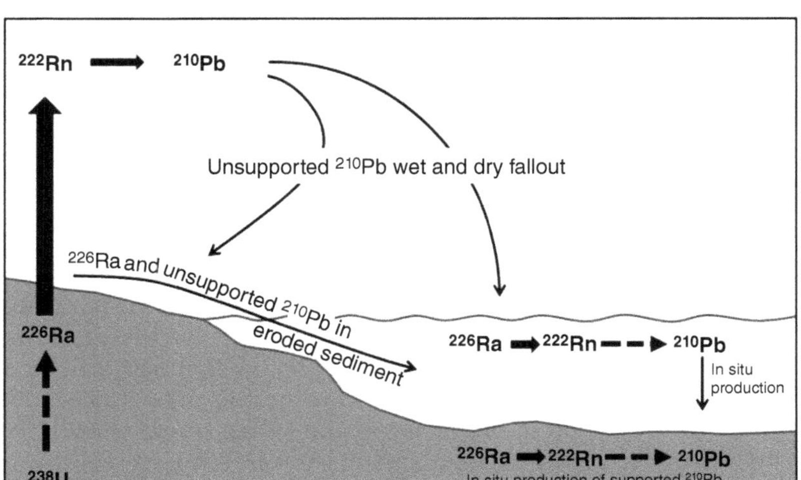

Fig. 24.3. Pathways for ^{210}Pb into fresh, marine, and intertidal systems. *Source*: Adapted from Oldfield and Appleby, 1984.

The potential application of ^{210}Pb as a chronometer was first described by Goldberg (1963) in a seminal paper to determine snow accumulation in Greenland. Krishnaswamy et al. (1971) applied this same approach to a lake core, introducing a new tool in understanding sediment dynamics and quantifying a geochronology. Over the following 40 years, most of the same analytical methods and models described in these original papers and others (e.g., Koide et al., 1971; Pennington et al., 1976; Appleby and Oldfield, 1978; Robbins, 1978; Nittrouer et al., 1979) are still being used today without significant modifications. Application of this tracer has expanded well beyond Greenland snow and lake cores to include wetlands (floodplain and saltmarsh), glacial environments, and marine systems (coastal to deep sea).

24.3.1 Experimental methods

Analysis of ^{210}Pb is typically conducted in one of two ways: direct counting via gamma spectroscopy, or indirectly via alpha spectroscopy using its daughter ^{210}Po. There are advantages and disadvantages of each method (Table 24.1), and often the analysis employed is dependent on the instrument availability of the laboratory. For the alpha procedure, a small amount of homogenized sediment (1.0–1.5 g) is spiked with a yield determinant (e.g., ^{209}Po) and digested in a strong acid solution (e.g., 8 N HNO_3) in an open vessel on a hot plate (Nittrouer et al., 1979) or a vented vessel in a microwave. The $^{210,\,209}Po$ solution is then electrodeposited onto silver or nickel planchets in a dilute acid solution (modified from Flynn, 1968). Samples are then counted in an alpha spectrometer (e.g., Ortec Octet) to determine the activities of ^{209}Po and ^{210}Po activities; counting times can range from 24 to 72 hours, depending on sample activity. Counting error follows a Poisson distribution, and thus one standard deviation is equal to the square root of the counts. Total ^{210}Pb activities are calculated assuming secular equilibrium with the quantified ^{210}Po.

Analyzing ^{210}Pb via gamma spectroscopy involves little sample preparation relative to the chemical separation required for alpha detection. Samples are typically dried, homogenized by grinding, and packed into standardized vessels. Sample size falls within the ranges c. 2–25 g dry weight, depending on the counting geometry. In order to calculate excess ^{210}Pb, samples must be held for more than three weeks to accurately assess the ^{226}Ra activity. The total ^{210}Pb activity is measured directly at the 46.5 keV photopeak, while ^{226}Ra is determined indirectly by counting the gamma emissions of its granddaughters, ^{214}Pb (295 and 351 keV) and ^{214}Bi (609 keV). Activities at the low energy of ^{210}Pb should be corrected for self-adsorption. Correction methods using an experimental direct transmission correction (Cutshall et al., 1983; Cable et al., 2001), as well as Monte Carlo simulations have been successfully applied (Sima and Dovlete, 1997).

Finally, before downcore activities can be used to assess sediment accumulation rates and define a geochronology, the excess ^{210}Pb must be determined. This is straightforward when gamma spectroscopy is employed. Using this method, the total ^{210}Pb and supported ^{210}Pb (e.g., ^{226}Ra) are quantified simultaneously; simply subtract the latter from the former (Equation 24.1). However, if alpha spectroscopy is used, ^{226}Ra is not measured so an assumption must be applied. In this case, the supported activity is typically estimated to be equal to the constant background level found at depth in the total ^{210}Pb profile. This approach assumes that the supported value of ^{210}Pb does not change throughout the core. Care is required when making this assumption as changes in sediment composition or grain-size distribution can influence the total and supported activity. For instance, several studies have shown a direct relationship between grain size, surface area, and/or organic content with excess ^{210}Pb activity (Chanton et al., 1983; Livens and Baxter, 1988; He and Walling, 1996; Kirchner and Ehlers, 1998; Alvarez-Iglesias et al., 2007). However, comparing the two techniques (i.e., gamma and alpha spectrometry) show good agreement in the absolute values of excess ^{210}Pb and the derived sediment accumulation rates (Tanner et al., 2000; Zaborska et al., 2007), although the total activity is often different.

24.3.2 Models for establishing accumulation rates and sediment age with ^{210}Pb

An ideal downcore ^{210}Pb activity profile for age dating will have an exponential decrease with depth if there is no deep sediment mixing and sediment accumulation is steady over the life of the measurable excess ^{210}Pb (c. 100 years). However, there are many environmental processes that can lead to a less-than-ideal downcore profile, including variations in sediment accumulation, physical and biological mixing, and changes in sediment source or composition. Because of these potential environmental effects, a number of different models can be applied when trying to interpret excess ^{210}Pb data from a core. The most suitable model depends on environmental conditions (e.g., sedimentation processes) and requires some knowledge of the area studied. Although many models have been developed and applied since the first application of ^{210}Pb in sedimentation studies, there are three models that are most widely used and described here: (1) constant flux–constant sedimentation (CFCS) model, also referred to as the "Simple" model; (2) constant rate of supply (CRS); and (3) constant initial concentration (CIC). Although these three models each have specific assumptions, they are all predicated on the following suppositions: ^{210}Pb is quickly removed from the atmosphere and water column and sequestered in the sediments; ^{210}Pb is immobile once deposited; and there are no interruptions (i.e., hiatal periods) in sedimentation through time. The selection and use of a specific model should be based on the nature and constancy of the activity and sediment accumulation (Table 24.2).

This section of the chapter will explore selection and use of the different models for stratigraphic studies. For illustrative and comparison purposes, data from a single core collected in the Albemarle Sound, North Carolina, USA is used for each model calculation (Table 24.3; Fig. 24.4). This core was collected in this wind-tide-dominated estuarine system in 2001 near the confluence of two river mouths, the Roanoke and Chowan rivers. The accumulation rates calculated were used to evaluate the decadal-scale sedimentation, and these results coupled with other geochemical measurements (e.g., stable isotopes) highlight environmental changes that had taken place over the last century (Corbett et al., 2007). Note that the the standard SI units of Becquerels per kilogram are used for the activities (Table 24.3; Fig. 24.4). Also, a surface mixed layer is not present; as a result, this is not taken into account in any of the models described in the following sections.

Table 24.2. Summary of assumptions for the three most commonly used models to interpret downcore ^{210}Pb activities when determining a sediment accumulation rate or developing an age–depth relationship.

Model	Specific Activity	Accumulation Rate	Flux of $^{210}Pb_{excess}$
CFCS	Constant	Constant	Constant
CRS	Variable	Variable	Constant
CIC	Constant	Variable	Variable

24.3.2.1 *Constant flux–constant sedimentation (CF/CS or Simple model)*

Assuming a constant accumulation rate, constant flux of unsupported ^{210}Pb, and a resulting constant initial activity upon deposition, the CFCS model can be applied to downcore profiles. In this simplest of the models, the depth of burial z is related to the elapsed time since burial t through the rate of sedimentation S. By linking depth with time, the exponential decay law of ^{210}Pb, and any other radionuclide, can be applied to the downcore activity such that:

$$A_z = A_0 e^{-\lambda t} \quad (24.2)$$

Table 24.3. Data summary of downcore radionuclide activities and grain size (% <63 μm) for a core collected at the head of Albemarle Sound.

Depth (cm)	Porosity	% <63 μm	Mass accumulation (g cm^{-2})	Excess ^{210}Pb (Bq kg^{-1})	^{137}Cs (Bq kg^{-1})
1	0.86	95.6	0.67	146 ± 25	26 ± 1
3	0.84	97.4	1.44	92 ± 22	26 ± 1
5.5	0.83	94.9	2.66	95 ± 34	30 ± 2
8.5	0.81	93.1	4.03	80 ± 32	41 ± 2
11.5	0.80	91.8	5.40	80 ± 21	46 ± 1
14.5	0.82	94.1	6.77	64 ± 19	63 ± 2
17.5	0.81	94.8	8.14	65 ± 19	87 ± 2
20.5	0.81	95.1	9.50	42 ± 17	91 ± 2
23.5	0.81	96.4	10.87	32 ± 15	13 ± 1
26.5	0.80	97.1	12.24	19 ± 22	0 ± 1
29.5	0.81	97	13.61	15 ± 13	0 ± 2
32.5	0.79	96.5	14.98	12 ± 13	0 ± 2
38.5	0.81	97	17.71	9 ± 14	0 ± 2

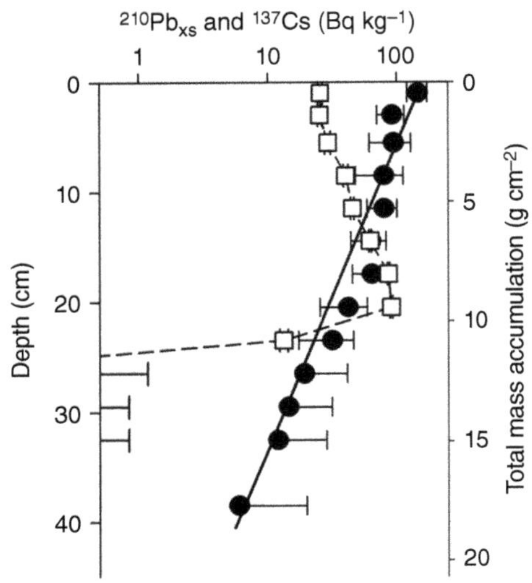

Fig. 24.4. Downcore profile of ^{210}Pb (●) and ^{137}Cs (□) for a core collected in Albemarle Sound. *Source*: Corbett et al., 2007.

$$t = z / S \tag{24.3}$$

and therefore

$$A_z = A_0 e^{-\lambda(z/S)} \tag{24.4}$$

where A_z is the $^{210}Pb_{excess}$ activity at depth z, A_0 is the initial $^{210}Pb_{excess}$ activity upon deposition, and λ is the ^{210}Pb decay constant (0.03114 a^{-1}). The sedimentation rate can then be calculated using the slope of a least-square fit derived from the linear regression of the natural log of $^{210}Pb_{excess}$ (A_z) versus sample depth (z):

$$S = \lambda / b \tag{24.5}$$

where S is the vertical accretion rate (cm a^{-1}) and b is the slope of the best-fit line. The age of the sediments at depth z is simply calculated by dividing the depth by the sedimentation rate (Equation 24.3). In fact, to account for variations in sediment porosity caused by the compaction of sediment after burial or during core collection, one can simply use the cumulative dry-mass of sediment (m) in the above calculations rather than depth (z). This approach provides a mass accumulation rate ($g_{dry\ sediment}$ cm^{-2} a^{-1}), which is also useful when estimating fluxes of sediment-bound constituents (e.g., carbon, trace metals) to the seabed.

Applying the CFCS model to the Albemarle Sound core data estimates a ^{210}Pb accretion rate of 0.43 ± 0.04 cm a^{-1} (mass accumulation = 0.19 ± 0.02 g cm^{-2} a^{-1}). The ^{137}Cs peak corresponding to 1963, 38 years prior to sampling, occurs at a depth of c. 19 cm, giving a calculated accretion rate of 0.50 ± 0.08 cm a^{-1}. The Simple model and ^{137}Cs data correspond fairly well, providing confidence in the model and assumptions used (Fig. 24.5).

24.3.2.2 Constant rate of supply (CRS model)

In many aquatic systems, the sedimentation rate is not constant with time, contradicting a fundamental assumption of the Simple model above. First developed by Goldberg (1963), then later applied by Appleby and Oldfield (1978) and Robbins (1978), the CRS model accounts for changes in sedimentation rate, but still assumes a constant flux of ^{210}Pb (F) to the sediments (Table 24.2). The initial specific activity is variable and dependent on the mass accumulation rate. The CRS model is thought to be the most widely used and accepted model (particularly in lake systems, but widely used in coastal systems) and is governed by the assumption of constant direct atmospheric fallout (Appleby et al., 1990).

Given the assumptions above, the excess ^{210}Pb activity integrated with depth z or the cumulative dry-mass of sediment m is equal to the ^{210}Pb flux F over the same time interval. If F is constant, the residual ^{210}Pb in the sediments of age t or greater is therefore:

$$I_z = \frac{F}{\lambda} e^{-\lambda t} \tag{24.6}$$

at $t = 0$ (present time), the ^{210}Pb inventory of the whole record (I_0) is equal to

$$I_0 = F / \lambda \tag{24.7}$$

and therefore:

$$I_z = I_0 e^{-\lambda t} \tag{24.8}$$

where I_z (dpm cm^{-2}) is the cumulative residual excess ^{210}Pb activity beneath the sediments of depth z and I_0, also noted above, is the cumulative excess ^{210}Pb activity in the whole sediment column. The inventory I values are calculated by numerical integration of the activity versus depth in the downcore profile. The age of the sediments at depth z is then:

$$t = \frac{1}{\lambda} \ln\left(\frac{I_0}{I_z}\right) \tag{24.9}$$

and the mass accumulation rate M is:

$$M = \lambda \frac{I_z}{A_z}. \tag{24.10}$$

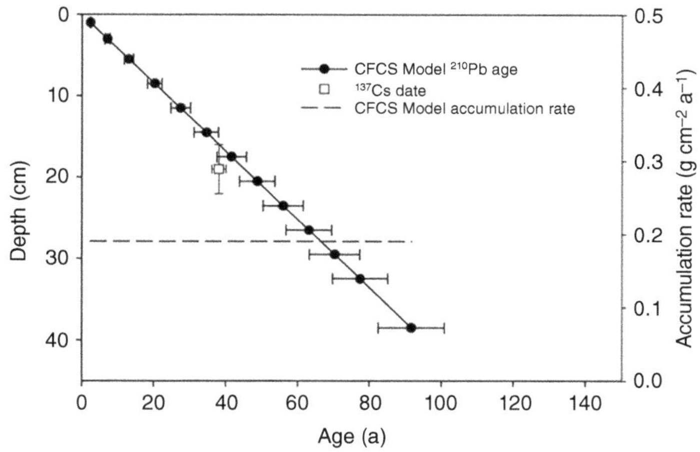

Fig. 24.5. Sediment age dates and accumulation rates in the Albemarle Sound core calculated using the CFCS model. Also plotted is the 1963 ^{137}Cs peak for comparison.

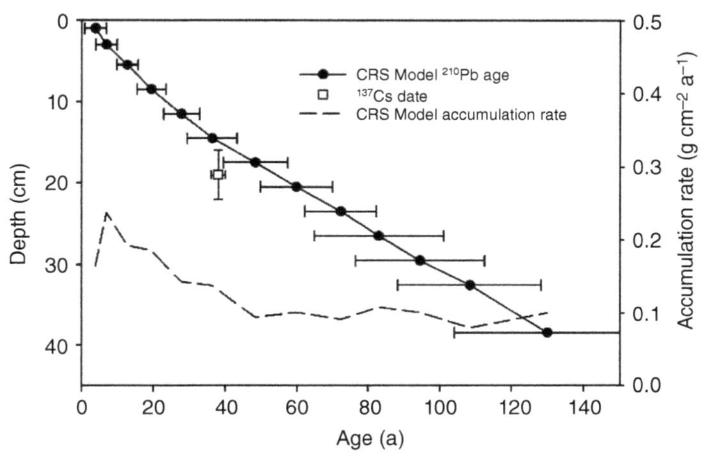

Fig. 24.6. Sediment age dates and accumulation rates in the Albemarle Sound core calculated using the CRS model. Also plotted is the 1963 ^{137}Cs peak for comparison.

Applying the CRS model to the Albemarle Sound core (Table 24.2) suggests significant change in the sediment mass accumulation rate over time, that is, c. 0.1 g cm^{-2} a^{-1} increasing to 0.2 g cm^{-2} a^{-1} in the last 40 years (Fig. 24.6). However, the ^{137}Cs peak is slightly deeper than predicted by the ^{210}Pb model. This may be associated with incorrect assumptions associated with this model or could be a product of downward mobilization of ^{137}Cs due to the high organic content and anoxic sediments. However, the calculated chronology is within the error of the calculated ages for both tracers.

24.3.2.3 Constant initial concentration (CIC model)

The CIC model, also referred to as the constant specific activity model, assumes that the initial ^{210}Pb activity will be constant regardless of sediment accumulation rates (Table 24.2). An increased flux of sediments will therefore remove proportionately more ^{210}Pb from the water column, leading to the same ^{210}Pb activity on the sediments deposited. Under these assumptions, the age t of the sediments at depth z can be calculated using Equation 24.2 if the activity at depth A_z and the initial activity A_0 are known.

As for the CFCS model, the accumulation rate measured for the CIC model can be calculated using the slope of the ^{210}Pb activity against depth. The difference between the CIC and the CFCS models is that the non-linear depth profile, indicative of a non-steady state accumulation rate, can be divided into multiple linear segments and the slope from each segment can be used to calculate the sediment accumulation rate for that section of the core. Unlike the CFCS, the CIC model does not require a linear downcore relationship with the natural log of $^{210}Pb_{excess}$ and allows for the calculation of changes in sedimentation through time (Appleby and Oldfield, 1992). Applying this model to the Albemarle Sound data suggests considerable variation in the mass accumulation rate

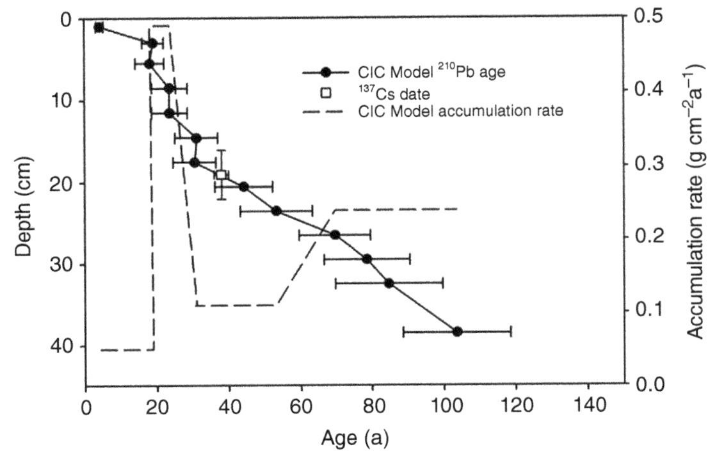

Fig. 24.7. Sediment age dates and accumulation rates in the Albemarle Sound core calculated using the CIC model. Also plotted is the 1963 ^{137}Cs peak for comparison.

($0.05–0.49\,g\,cm^{-2}\,a^{-1}$; Fig. 24.7), particularly during 20–40 years prior to sampling (2001). Although the depth of the ^{137}Cs peak agrees quite well with this model, it is important to note the step-like profile of the age with depth near the top of the core. This is a result of equal activities at certain sampling depths, possibly a product of rapid accumulation.

24.3.2.4 Selecting a model

The application of ^{210}Pb and ^{137}Cs in furthering our understanding of sediment dynamics in aquatic systems (e.g., freshwater, marine, and intertidal) has increased dramatically since it was first introduced by Goldberg (1963). Carroll and Lerche (2003) noted that Robbins and Herche (1993) estimated the number of citations related to the use of ^{210}Pb as a geochronometer exceeded 2000. This has no doubt dramatically grown since that time. A simple Science Direct search (www.sciencedirect.com) for "210Pb and sediment" gave more than 17,000 results. A recent bibliographic summary highlighted more than 4200 articles using ^{137}Cs in studies of erosion and sediment deposition (Ritchie and Ritchie, 2008). In most of these studies, one of the three models described above had been implemented, likely due to their simplicity. However, users should be careful of the model employed to interpret the downcore profiles, as sediment character (e.g., grain sizes, sediment composition), dynamics (e.g., deposition, focusing, stability) and other potential processes (e.g., organic matter decay) may influence the ^{210}Pb functioning in the system. The model used to calculate age and sedimentation rate should therefore be based on an understanding of the processes driving accumulation and an evaluation of the ^{210}Pb data.

In most cases, the CFCS model is initially applied to the downcore ^{210}Pb activity, particularly those cores with a near exponential decrease with depth. However, slight deviations from this trend can lead to significant errors in the age approximation due to small variations in sediment accumulation, particularly for the older sediments. Cores with non-exponential ^{210}Pb activity decline with depth are more problematic and require careful considerations. In practice, the model employed is subjective and may hinge on the best agreement with an independent chronometer (i.e., ^{137}Cs, stable Pb, pollen, etc.). More often than not, the CRS model is applied as it often yields more reasonable results than that of the CIC in both fresh and marine environments (Oldfield et al., 1978; Blais et al., 1995; Appleby, 2008, Leorri et al., 2010). However, there are a couple of simple guidelines that should be followed: (1) the CIC model should only be used if the downcore ^{210}Pb activity decreases monotonically; (2) the CRS model should not be used in cases where the core excess ^{210}Pb inventory is excessively high or low relative to what is expected for atmospheric flux to the area (often established in a nearby area of known constant accumulation; e.g., peat bog, closed lake, etc.); and (3) the CRS model is likely not valid in locations with significant hydrologic changes or the presence of hiatal surfaces in the sediment record (Appleby, 2001). Holding true to these few "rules" and the underlying assumptions of the individual models has provided solid, reliable rates of sediment accumulation and age estimates in scores of environmental studies.

Although the three models described above are almost exclusively applied in most geochronological studies, numerical methods have recently been developed to help interpret downcore tracer data in systems where sediment accumulation, ^{210}Pb flux, and the specific activity vary with time (refer to Table 24.2), or biological and/or physical mixing cannot be ignored. For example, the sediment isotope tomography (SIT) model was recently introduced to evaluate non-linear ^{210}Pb profiles (Liu et al., 1995; Carroll et al., 1995). Although this numerical approach has not been widely used, synthetic data has demonstrated the SIT model to be a reliable alternative to the CRS and CIC models (Carroll and Lerche, 2003). The SIT model uses inverse numerical analysis techniques in combination with a predictive activity module to reproduce the excess ^{210}Pb downcore profile. Non-exponential changes in the downcore activity, caused by either change in the sedimentation rate and/or ^{210}Pb flux, are modeled by a Fourier cosine series (Carroll et al., 1999). Through a non-linear iterative process, the values of the Fourier coefficients are determined for the measured downcore ^{210}Pb data (Carroll and Lerche, 2003). This model can be further constrained by introducing one or more independent time markers (e.g., ^{137}Cs), essentially eliminating many combinations of potential model parameters and increasing the accuracy of the calculated chronology.

Finally, several researchers have developed numerical models of radionuclide distributions in sediments. These numerical models typically treat sedimentation as an advective process and often incorporate bioturbation/physical mixing as a diffusional process. These models provide another approach to quantitatively interpret downcore ^{210}Pb profiles in steady- and non-steady-state conditions (Peng et al., 1979; Santschi et al., 1980; Boudreau, 1986; Robbins, 1986; Crusius et al., 2004; Crusius and Kenna, 2007). Like the SIT model, these numerical methods are not constrained by the same assumptions as the CFCS, CRS, and CIC models (Table 24.2) described here. They are used much less frequently however, likely due to their numerical complexity and less-simple constraints. As for all models, analytical and numerical alike, users must take care to avoid going beyond the bounds of the model parameters or inherent assumptions.

24.4 POTENTIAL PITFALLS

There are many factors that may influence the accuracy of the geochronology beyond the laboratory analyses and model choices described above. Some consideration must be given to core location, material disturbance and/or compaction during core collection, water content of the sediments, radon loss from sediments, downcore variations in grain size, and post-depositional changes (e.g., erosion, bioturbation, physical mixing). Many studies have shown that these tracers are most accurate when applied in quiescent water bodies with sediments that are not mixed, with little to no change in sedimentation over time (Appleby, 1997; Noller, 2000). However, there are many examples (including those described here) of studies that have developed reliable geochronologies from more complicated depositional environments.

A good chronology often starts with careful core collection and sample processing. A core should be collected in a region that is consistently depositing sediments and, ideally, with modest physical reworking and a reasonably well-understood benthic community. Core collection should be conducted in such a way as to avoid significant alteration of the sediment column (e.g., compaction) or loss of the surface layer (Baxter et al., 1981; Crusius and Anderson, 1991; Crusius and Kenna, 2007). Once collected, the core should be carefully sectioned to avoid cross-contamination (e.g., downcore smearing; Chant and Cornett, 1991) and at a resolution that is sufficient to resolve decay in the depositional environment. Ducat and Kuehl (1995) suggest that a 2 cm sampling interval throughout a core is sufficient to capture the nature of ^{210}Pb behavior in some regions (e.g., shelf areas with high sediment input). Crusius and Kenna (2007) provide a more thorough discussion of the importance to sampling resolution with examples related to the ^{137}Cs peak. They clearly show a decrease in the measured activity of the ^{137}Cs peak with coarser sampling resolution. This decrease and broadening of the peak can lead to a misidentification of the peak and ultimately the age–depth model. Although detector time, funds, and personnel time often limit a scientist's ability to increase sample throughput in any laboratory, sampling resolution should be considered prior to core collection.

Finally, it is important to understand and accept that all cores cannot be used for geochronology and learn to recognize when a core cannot be dated accurately. There are many reasons that may lead to an incorrect or inconclusive age–depth model. Areas that have slow accumulation rates and/or intense physical and biological mixing are difficult to date as the decay of excess ^{210}Pb may be unsolvable or overprinted by other processes. Depositional environments with complex sedimentation that results in significant variations in grain size and irregular deposition (e.g., haital surfaces or erosion) may also be difficult or impossible for calculating accumulation rates. However, grain size variations may be overcome through normalization of the ^{210}Pb data to clay fraction or Al concentration, often used as a geochemical proxy for grain size (Chanton et al., 1983; Andersen et al., 2000; Alvarez-Iglesias et al., 2007). Although an independent chronometer can provide additional information into the depositional history and should be used to verify ^{210}Pb-modeled sedimentation, a willingness to acknowledge that not all cores or areas have other chronological tools available (e.g., ^{137}Cs) is required; in these situations, users must be especially critical. The greatest mistake a scientist can make is extrapolating an age–depth model for a core that should be considered undateable.

24.5 CONFIDENCE IN THE GEOCHRONOLOGY

It is vital that confidence can be demonstrated in the dates assigned to the depths in a core. As Noller (2000) points out, confidence in the chronological results based on ^{210}Pb is dependent on the analytical methods, the model employed, statistical treatment of the data, and ultimately the reporting of the results. Care must be taken from core collection to data publication, avoiding the possible problems outlined above, and procedural and other error should be carefully estimated and reported where possible. Increased confidence in any ^{210}Pb chronology often falls upon the use of an independent time-stratigraphic marker (e.g., ^{137}Cs) to confirm model results. Even some of the original modeling efforts demonstrated how an exponential downcore decrease in ^{210}Pb activity can be generated through post-depositional sediment mixing (Robbins, 1978). Validation using at least one independent tracer should therefore be a goal for age–depth models. Additionally, if possible the ^{210}Pb data and model interpretation along with the calculated geochronology should be published or cited. Based on a 2001 editorial by Smith (2001), the *Journal of Environmental Radioactivity* implemented the requirement of acceptance that any manuscripts that use ^{210}Pb geochronology must include figures of the data or a hard copy appendix containing the data and description of the methods. All authors should follow suit regardless of the journal. Others will have greater confidence in the science published if the data are accessible and verifiable.

REFERENCES

Aarkrog, A. (1988) Radiological impact of Chernobyl debris compared with that from nuclear weapons fallout Journal of Environmental Radioactivity, 6(2), 151–162.

Alvarez-Iglesias, P., Quintana, B., Rubio, B., and Perez Arlucea, M. (2007) Sedimentation rates and trace metal history in intertidal sediments from San Simon Bay (Ria de Vigo, NW Spain) derived from ^{210}Pb and ^{137}Cs chronology. Journal of Environmental Radioactivity, 98 229–250.

Appleby, P.G. (1997) Sediment records of fallout radionuclides and their application to studies of sediment-water interactions. Water, Air and Soil Pollution, 99 573–586.

Appleby, P.G. (2001) Chronostratigraphic techniques in recent sediments. In *Tracking Environmental Change using Lake Sediments, Basin Analysis, Coring, and Chronological Techniques, Development in Paleoenvironmental Research* (ed. Last, W.L., and Smol J.P.), Kluwer Academic Publishers, Dordrecht.

Appleby, P.G. (2008) Three decades of dating recent sediments by fallout radionuclides: a review. The Holocene 18(1), 83–93.

Appleby, P.G., and Oldfield, F. (1978) The calculation of lead-210 dates assuming a constant rate of supply of unsupported ^{210}Pb to the sediment. Catena, 5, 1–8.

Appleby, P.G., and Oldfield, F. (1992) Application of lead-210 to sedimentation studies. In *Uranium-Series Disequilibrium: Application to Earth, Marine, and environmental Sciences* (ed. Ivanovich, M., and Harman R.S.), Clarendon Pressm, Oxford.

Appleby, P.G., Richardson, N., Nolan, P.J., and Oldfield, F. (1990) Radiometric dating of the United Kingdom SWAP sites. Philosophical Transactions of the Royal Society of London, 327(B), 233–238.

Andersen, T.J., Mikkelsen, O.A., Møller, A.L., and Pejrup, M. (2000) Deposition and mixing depths on some European intertidal mudflats based on ^{210}Pb and ^{137}Cs activities. Continental Shelf Research, 20, 1569–1591.

Baxter, M.S., Farmer, J.G., McKinley, I.G., Swan, D.S., and Jack, W. (1981) Evidence of the unsuitability of gravity coring for collecting sediment in pollution and sedimentation rate studies. Environmental Science and Technology, 15, 843–846.

Blais, J.M., Kalff, J., Cornett, R.J., and Evans, R.D. (1995) Evaluation of ^{210}Pb dating in lake sediments using stable Pb, Ambrosia pollen, and ^{137}Cs. Journal of Paleolimnology, 13, 169–178.

Bondavalli, C. (2003) Effect of eutrophication on radionuclide dynamics in the Sacca di Goro lagoon (Po River Delta, Italy): a combined field, experimental, and modeling study. Environmental Pollution, 120, 433–446.

Boudreau, B.P. (1986) Mathematics of tracer mixing in sediments: II. Nonlocal mixing and biological conveyor-belt phenomena. American Journal of Science, 268, 199–238.

Cable, J., Burnett, W., Moreland, S., and Westmoreland, J. (2001) Empirical assessment of gamma ray self-absorption in environmental analyses. Radioactivity and Radiochemistry, 12, 30–39.

Carroll, J., and Lerche, I. (2003) *Sedimentary Processes: Quantification Using Radionuclides*. Elsevier Science, Amsterdam.

Carroll, J., Lerche, I., Abraham, J.D., and Cisar, D.J. (1995) Model determined sediment ages from Pb-210 profiles in unmixed sediments. Nuclear Geophysics, 9, 553–565.

Carroll, J., Lerche, I., Abraham, J.D., and Cisar, D.J. (1999) Sediment ages and flux variations from depth profiles of Pb-210: lake and marine examples. Applied Radiation and Isotopes, 50, 793–804.

Chant, L.A., and Cornett, R.J. (1991) Smearing of gravity core profiles in soft sediments. Limnology and Oceanography, 36(7), 1492–1498.

Chanton, J.P., Martens, C.S., and Kipphut, G.W. (1983) Lead-210 sediment geochronology in a changing coastal environment. Geochimica et Cosmochimica Acta, 47, 1791–1804.

Chmura, G.L., and Kosters, E.C. (1994) Storm deposition and 137Cs accumulation in fine-grained marsh sediments of the Mississippi Delta Plain. Estuarine, Coastal and Shelf Science, 39, 33–34.

Christiansen, C., Kunzendorf, H., Emeis, K.C., Endler, R., Struck, U., Neumann, T., and Sivkov, V. (2002) Temporal and spatial sedimentation rate variabilities in the eastern Gotland Basin, the Baltic Sea. Boreas, 31, 65–74.

Corbett, D.R., Vance, D., Letrick, E., Mallinson, D., and Culver, S. (2007) Decadal-scale sediment dynamics and environmental change in the Albemarle estuarine system, North Carolina. Estuarine, Coastal and Shelf Science, 71(3–4), 717–729.

Crusius, J., and Anderson, R.F. (1991) Core compression and surficial sediment loss of lake sediment of high porosity caused by gravity coring. Limnology and Oceanography, 36, 1021–1031.

Crusius, J., and Anderson, R.F. (1995) Evaluating the mobility of ^{137}Cs, $^{239+240}$Pu and 210Pb from their distributions in laminated lake sediments. Journal of Paleolimnology, 13, 119–141.

Crusius, J., and Kenna, T.C. (2007) Ensuring confidence in radionuclide-based sediment chronologies and bioturbation rates. Estuarine, Coastal and Shelf Science, 71(3–4), 537–544.

Crusius, J., Bothner, M., and Sommerfield, C.K. (2004) Bioturbation depths, rates and processes in Massachusetts Bay sediments inferred from modeling of ^{210}Pb and $^{239+240}$Pu profiles. Estuarine, Coastal and Shelf Science, 61(4), 643–655.

Cutshall, N.H., Larsen, I.L., and Olsen, C.R. (1983) Direct analysis of ^{210}Pb in sediment samples: self-adsorption corrections. Nuclear Instruments and Methods, 206, 309–312.

Dukat, D.A., and Kuehl, S.A. (1995) Non-steady-state ^{210}Pb flux and the use of ^{228}Ra/^{226}Ra geochronometer on the Amazon continental shelf. Marine Geology, 125, 329–350.

Flynn, W.W. (1968) The determination of low levels of polonium-210 in environmental materials. Analytica Chimica Acta, 43, 221–227.

Goldberg, E. D. (1963) Geochronology with ^{210}Pb. In *Radioactive Dating*. International Atomic Energy Agency, Vienna, pp. 121–131.

Goodbred, S.L., and Kuehl, S.A. (1998) Floodplain processes in the Bengal Basin and the storage of Ganges Brahmaputra river sediment: an accretion study using 137Cs and 210Pb geochronology. Sedimentary Geology, 121, 239–258.

He, Q., and Walling, D.E. (1996) Interpreting particle size effects in the adsorption of ^{137}Cs and unsupported ^{210}Pb by mineral soils and sediments. Journal of Environmental Radioactivity, 30, 117–137.

Hutchinson, S.M. (1995) Use of magnetic and radiometric measurements to investigate erosion and sedimentation in a British upland catchment. Earth Surface Processes and Landforms, 20, 293–314.

Kirchner, G., and Ehlers, H. (1998) Sediment geochronology in changing coastal environments: potentials and limitations of the ^{137}Cs and ^{210}Pb methods. Journal of Coastal Research, 14, 483–492.

Koide, M., Soutar, A., and Goldberg, E.D. (1971) Marine geochronology with Pb-210. Earth and Planetary Science Letters, 14, 442–446.

Krishnaswamy, S., Lal, D., Martin, J.M., and Meybeck, M. (1971) Geochronology of lake sediments. Earth and Planetary Science Letters, 11, 401–414.

Leorri, E., Cearreta, A., Corbett, D.R., Blake, W., Fatela, F., Gehrels, R., and Irabien, M.J. (2010) Identification of suitable areas for high-resolution sea-level studies in SW Europe using commonly applied ^{210}Pb models. Geogaceta, 48, 35–38.

Liu, J., Carroll, J.L., and Lerche, I. (1991) A technique for disentangling temporal source and sediment variations from radioactive isotope measurements with depth. Nuclear Geophysics, 5, 31–45.

Livens, F.R., and Baxter, M.S. (1988) Particle size and radionuclide levels in some west Cumbrian soils. Science of the Total Environment, 70, 1–17.

Neubauer, S.C., Anderson, I.C., Constantine, J.A., and Kuehl, S.A. (2002) Sediment deposition and accretion in a Mid-Atlantic tidal freshwater marsh. Estuarine, Coastal and Shelf Science, 54, 713–727.

Nittrouer, C.A., Sternberg, R.W., Carpenter, R., and Bennett, J.T. (1979) The use of ^{210}Pb geochronology as a sedimentological tool: an application to the Washington continental shelf. Marine Geology, 31, 297–316.

Noller, J.S. (2000) Lead-210 geochronology. In: *Quaternary Geochronology: Methods and Applications* (eds Noller, J., Sowers, J., and Lettis, W.), American Geophysical Union, Washington, DC.

Oldfield, F., and Appleby, P.G. (1984) Empirical testing of ^{210}Pb-dating models for lake sediments. In: *Lake Sediments and Environmental History* (eds Hayworth, E.Y., and Lund, J.W.G.), Leicester University Press, Leicester, pp. 93–124.

Oldfield, F., Appleby, P.G., and Battarbee, R.W. (1978) Alternative dating results from the New Guinea Highlands and Lough Erne. Nature, 271, 339–342.

Peng, T.-H., Broecker, W.S., and Berger, W.H. (1979) Rates of benthic mixing in deep-sea sediment as determined by radioactive tracers. Quaternary Research, 11, 141–149.

Pennington, W., Cambray, R.S., and Fisher, E.M. (1973) Observations on lake sediments using fallout ^{137}Cs as a tracer. Nature, 242, 324–326.

Pennington, W., Tutin, T.C., Cambray, R.S., Eakins, J.D., and Harkness, D.D. (1976) Radionuclide dating of recent sediments of Blelham Tarn. Freshwater Biology, 6, 317–331.

Ritchie, J., and Ritchie, C. (2008) *Bibliography of Publications of ^{137}Cesium Studies Related to Erosion and Sediment Deposition.* USDA-ARS; HRSL-2008-02.

Robbins, J.A. (1978) Geochemical and geophysical applications of radioactive lead isotopes. In *Biogeochemistry of Lead* (ed. Nriago, J.P.), North Holland, Amsterdam, pp. 285–393.

Robbins, J.A. (1986) A model for particle-selective transport of tracers in sediments with conveyor belt deposit feeders. Journal of Geophysical Research, 91, 8542e8558.

Robbins, J.A., and Herche, L.R. (1993) Models and uncertainty in ^{210}Pb dating of sediments. Verhandlung des Internationalen Verein Limnologie, 25, 217–222.

Santschi, P.H., Li, Y.-H., Bell, J.J., Trier, R.M., and Kawtaluk, K. (1980) Pu in coastal marine environments. Earth and Planetary Science Letters, 51, 248–265.

Sima O., and Dovlete, C. (1997) matrix effects in the activity measurement of environmental samples implementation of specific corrections in a gamma-ray spectrometry analysis program. Applied Radiation and Isotopes, 48, 59–69.

Smith, J.N. (2001) Why should we believe 210Pb sediment geochronologies? Journal of Environmental Radioactivity, 55, 121–123.

Tanner, P.A., Pan, S.M., Mao, S.Y., and Yu, K.N. (2000) γ-ray spectrometric and α-counting method comparison for the determination of Pb-210 in estuarine sediments. Applied Spectroscopy, 54(10), 1443–1446.

UNSCEAR (2000) Sources and effects of ionizing radiation. Sources, Annex C: Exposures from man-made sources of radiation, vol. I. United Nations Committee on the Effects of Atomic Radiation. Available at http://www.unscear.org/unscear/en/publications/2000_1.html (accessed 28 July 2014).

Zaborska, A., Carroll, J., Papucci, C., and Pempkowiak, J. (2007) Intercomparison of alpha and gamma spectrometry techniques used in ^{210}Pb geochronology. Journal of Environmental Radioactivity, 93(1), 38–50.

Chapter 25

Chronohorizons: indirect and unique event dating methods for sea-level reconstructions

WIL MARSHALL

School of Geography, Earth and Environmental Sciences, University of Plymouth, Plymouth, UK

25.1 INTRODUCTION

Coastal sediments may be used to provide information about past events and changes in the intertidal zone. For example, saltmarsh sediments contain subtle signals of marsh surface evolution and past sea levels. If we can link this information to a robust chronology, we can use these natural archives to establish the timing and the rate of sea-level change in the past. Under ideal conditions the traditional "direct dating" methods examined in Chapters 23 and 24 can be used to produce the deposition date of coastal sediments. This notwithstanding, it is now evident that both ^{14}C and ^{210}Pb dating suffer from a number of limitations which can restrict the applicability of these methods for high-resolution late Holocene sea-level studies. For specific information on the issues which can be encountered when using ^{14}C and ^{210}Pb dating for sea-level studies, the reader is directed to Chapters 23 and 24. Some sediment sequences are very minerogenic and often do not contain sufficient organic carbon material suitable for high-density ^{14}C dating. Likewise, the ^{210}Pb dating method can be problematic, particularly in marine sediments (Smith, 2001; Thomson et al., 2002; Abril, 2004), and when it can be used successfully it only covers the last 100–150 years. Fortunately, in many situations it is possible to obtain deposition dates for sediments using indirect methods to establish "chronohorizons".

We can provide deposition dates for coastal sediments using signals of unique events with established timings. These may be events which have occurred in the historical period and were recorded at the time, or more ancient events for which an age has been established using a robust direct dating method, for example ^{14}C in a "master sequence". These signals can be the product of an outstanding natural event such as a volcanic eruption, or the legacy of unique anthropogenic event. For example, the signal may be a subtle change in the local pollen rain reflecting a change in land use, or the sudden fallout of debris produced by a catastrophic fire or explosion.

In all cases to use the unique event dating approach we need: (1) an event to have occurred which produced a detectable signal of its occurrence; (2) knowledge of the date at which that signal was produced; and (3) the ability to identify the signal in the sediments of interest. It follows that:

- the spatial extent of the signal (1) must encompass the location of our study site;
- the precision of the initial age-control (2) of the event will be pivotal to the uncertainty we must attach to the any deposition dates we eventually obtain; and
- the reliable identification of the signal (3) in the sediments will be influenced by the accumulation rate, lithology, and geochemistry of the sequence.

To demonstrate the application of this approach we review some examples of unique event dating methods. These methods are the use of metal pollution histories, pollen and charcoal records, the fallout of industrial black carbon or spheroidal carbonaceous particles (SCPs), and tephrochronology. This list of methods must not be regarded as exclusive or comprehensive, however. Furthermore, we can only present an introduction to the examples in this chapter. For more technical details, instrumentation, and sample preparation the reader is directed to the specialist literature on the individual methods cited in each section. The examples presented are from studies involving saltmarsh sediments, but the approach can be used in other types of coastal sediments. However, it must be noted that different sites will

Handbook of Sea-Level Research, First Edition. Edited by Ian Shennan, Antony J. Long, and Benjamin P. Horton.

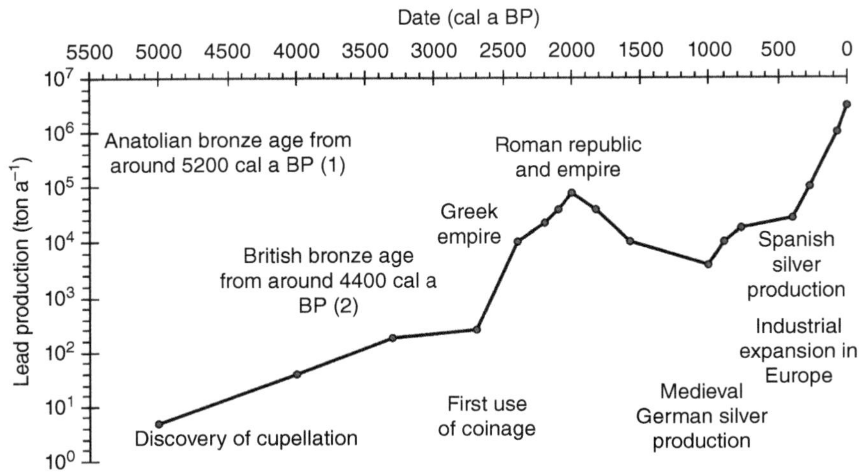

Fig. 25.1. World Pb production. Adapted from Reuer and Weiss (2002) with additions from (1) Şahoğlu (2005) and (2) Northover (1999).

offer "unique events", and what will work in one region will be useless elsewhere. There is no guaranteed method which will work every time. In addition, the geographical location of the site under examination and the time-period of interest will dictate which unique event dating methods we can successfully use to establish a deposition chronology. In some situations we may be able to use more than one dating method, and a local opportunity may present itself which will not work in other locations.

25.2 METAL POLLUTION HISTORIES

The anthropogenic release of metals into the environment can produce widespread signals which are often preserved in suitable sediments. Saltmarsh sediments are natural archives of heavy-metal deposition, and mining and other industrial activity in local catchments contribute to the heavy-metal load in intertidal sediments. If we know the timing of any outstanding pollution events which have contributed to the metal loading of our sediment sequence, we can use this information to date sediment deposition. However, we need to be aware that the retention of metal in coastal sediments is related to grain size. There is a tendency for finer sediments to be better at retaining metal particles (Loring, 1991; Lee and Cundy, 2001). This means that if a saltmarsh is clay-poor it will not retain metals as well as a marsh which has received a significant supply of clay-rich sediments (Cundy et al., 2005).

In addition to utilizing point-source waterborne metal pollution, we can use regionally synchronous records of anthropogenic atmospheric fallout. This fallout is the product of activities which may have been a single short-term event or something which continued over decades or even centuries (Renberg et al., 2001). For example, we can find signals of the emergence of Old World Bronze Age metal working, the rise and fall of the Roman Empire, and the expansion of industrial activities during the 19th century (Bollhöfer and Rosman 2001). Later during the 20th century we can detect the rise, and subsequent decline, in the use of lead as a vehicle fuel additive. This metal is noteworthy from a dating point of view. Lead has been emitted as an atmospheric pollutant since the dawn of metallurgy (Reuer and Weiss, 2002). It has a long history of being worked as an independent metal and, because lead is found in trace concentrations in polymetallic ore bodies, it is emitted during the refining and processing of other metals such as tin, gold, copper, and silver (Fig. 25.1). Lead is also emitted when coal is burnt and by many metal working industries, for example during steel production. Once emitted into the atmosphere, particulate lead can travel long distances. Hong et al. (1994) found a lead enrichment signal in the Greenland Summit GRIP ice core that corresponds to the timing of the maximum industrial activity in Europe during the Roman Empire.

Lead can also be defined using its stable isotopes and in some cases it can be traced back to the orebody from which it was mined (Begemann et al., 1999; Klein et al., 2004). If the source of the lead

Table 25.1. Isotopic values of the major world Pb ore bodies exploited in the late 20th century.

Country	Tonnage produced in 1972	Main mining district	Host rock age	$^{206}Pb/^{204}Pb$	$^{206}Pb/^{207}Pb$	$^{208}Pb/^{206}Pb$
USA	619,000	Missouri and Mississippi	Pennsylvanian	21.78	1.385	1.872
		Idaho	Precambrian	16.45	1.052	2.21
Australia	451,000	Mount Isa	Precambrian	16.08	1.037	2.236
		Broken Hill				
Canada	419,000	British Columbia	Precambrian	16.68	1.064	2.194
		New Brunswick	Ordovician	18.39	1.160	2.096
Peru	208,000	Cerro de Pasco	Miocene	18.86	1.200	2.076

Source: Adapted from Chow et al. (1975).

can be determined, we can better start to make judgments as to when it was produced. Furthermore, if we can identify the source of the lead ore it is possible to distinguish between anthropogenic lead and that which is the product of natural weathering (Shotyk et al., 1998). Like many elements, natural lead contains a number of different isotopes and these ratios can be used to "fingerprint" a sample and establish the ore body that was used to produce the metal. The original primordial isotope is ^{204}Pb, but the radiometric decay, via intermediate daughters of ^{232}Th, ^{235}U, and ^{238}U, has created the three radiogenic stable lead isotopes: ^{206}Pb, ^{207}Pb, and ^{208}Pb. Once radiometric stability is achieved the isotopic ratio is fixed in the ore and, under normal circumstances, no fractionation occurs when the metal is extracted and processed (Rohl, 1995). This means that once entombed in a sedimentary sequence the isotopic ratio of fallout lead will not change, and the metal will faithfully preserve a record of its geological origin.

As the industrial use of lead has increased over the last 2000 years, the location, and isotopic composition, of the exploited ore deposits in the world have changed from being mainly European to those located in the Americas and the Australian deposits in the Broken Hill mining district (Table 25.1). The isotopic signature of the signal in the sedimentary record has shifted accordingly and, if we know when these changes occurred in the source of the polluting lead, we can use these as "unique events". In the last 200 years the isotopic signature of atmospheric lead became progressively more radiogenic until federal controls on lead emissions were implemented in both the USA and Europe during the 1970s and 1980s. These controls produced an abrupt shift to less radiogenic values (Reuer and Weiss, 2002), and this signal can be detected in many sedimentary records.

The start, increase, and the subsequent phasing out of anti-knock fuel additives containing lead has been successfully monitored using stable lead isotopes. Blais (1996) showed that the USA was the main source of atmospheric lead over eastern Canada, and Véron et al. (1994) mapped the contamination of the North Atlantic using isotopic fingerprinting. In Europe the petrol additives used were derived mainly from Australian ore which has a lower $^{206}Pb/^{207}Pb$ than the Mississippi Valley ores commonly used for fuel additives in the USA (Weiss et al., 1999; Reuer and Weiss, 2002). We also know that the first leaded fuel was sold in the USA in February 1923 (Nriagu, 1990), and introduced in the UK during January 1928 but not marketed commercially by British Petroleum (BP) until 15 April 1931 (Bamberg, 1994). If we can detect a signal of lead from vehicle emissions in a North American saltmarsh sequence, we therefore know that the sediments were deposited after AD 1923.

By the early 1970s petrol additives comprised 65–85% of all anthropogenic atmospheric Pb emissions (Reuer and Weiss, 2002). The maximum world consumption occurred in the late 1960s and early 1970s. During this time, over 10^5 metric tons of Pb metal were being processed per year to supply the worldwide demand for leaded petrol, of which 80% was burned solely in the USA (Nriagu, 1983). In most cases the peak deposition of petrol lead occurs in European records some 10–20 years after the absolute maximum of total lead deposition from the atmosphere. This timing can be detected in many European lead isotope records as a shift towards higher $^{206}Pb/^{207}Pb$ and lower $^{208}Pb/^{206}Pb$. Since 1970, the USA and western Europe have progressively reduced the use of leaded fuel and emissions have consequently fallen significantly (Fig. 25.2); in some regions

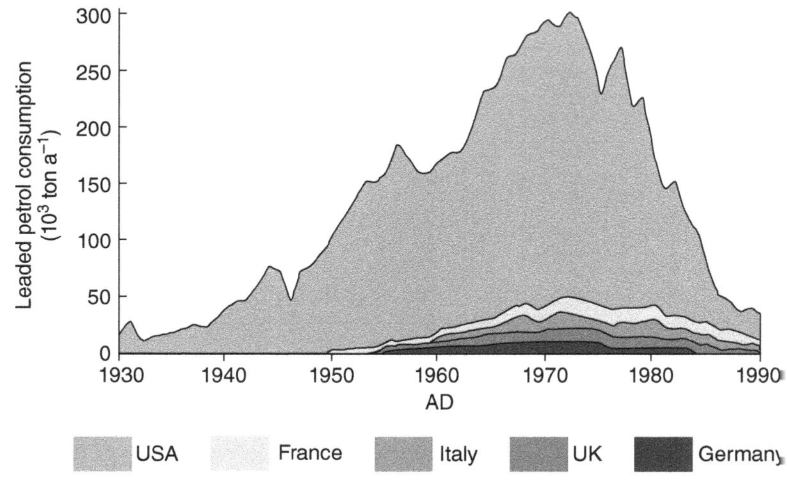

Fig. 25.2. Leaded petrol consumption in the USA and the four countries with the highest use in western Europe, 1930–1990. *Source*: Adapted from Reuer and Weiss (2002).

however the regular use of leaded fuel has persisted into the 21st century (Reuer and Weiss, 2002). For example, India phased out leaded fuel after 1999 (Singh and Singh, 2006), China was still using leaded fuel in 2001 (Li et al., 2012) and leaded fuel was only taken off the South African market in January 2006 (Monna et al., 2004). Similar unique "local" histories of the persistence of leaded fuel can be found in parts of South America and eastern Europe, and these records can be used to provide chronological markers for sites in those regions.

To use metal pollution signals to estimate deposition dates, we must use some form of geochemical analysis on our saltmarsh sediments. There are a number of different analytical methods available to determine metal concentrations in saltmarsh sediments, for example atomic absorption spectroscopy (AAS), X-ray fluorescence (XRF), and inductively coupled plasma mass spectrometry (ICP-MS). They all have advantages and disadvantages, and the sensitivity and precision of the analysis will influence the uncertainty of any chronomarkers derived from the geochemical data. The sample density (vertical resolution) and the analysis quality will influence the accuracy and precision of any dates obtained by this method. We must also remember that a temporary change in the sediment lithology or grain size may change the apparent concentration of metals in our samples, and we should view trends and excursions in the raw concentration curves with caution.

Once the raw metal concentration data is obtained, one of the simplest methods of quantifying the excess lead or other metal in sediments is the traditional normalizing approach of relating the concentration of the element of interest to a physical property such as grain size. Alternatively we can use the concentration in the sample of a conservative element as a proxy for changes in the local minerogenic contribution (Loring, 1991). In its simplest form this is applied as a ratio (e.g., pollution metal/standard) but if the local natural background, that is, the pre-anthropogenic ratio, can be established using samples from the lower sections of a sequence, then more sophisticated methods can be used, for example: $(\text{metal/standard})_{\text{sample}}$ divided by $(\text{metal/standard})_{\text{background}}$ (Shotyk et al., 1998). Aluminum has commonly been used as a conservative element to normalize metal concentrations, for example Weiss et al. (1999) and Schettler and Romer (2006), but other elements have been shown to be more applicable in some circumstances (Loring, 1991). Some examples include lithium (Aloupi and Angelidis, 2001), scandium (Monna et al., 2004) and titanium (Marshall et al., 2009).

If lead isotope data are available for our sediments we can use the "fingerprint approach" to identify different phases of pollution and, in some circumstances, attribute individual events to lead derived from a specific ore-body. This information can be used to confirm our interpretation of the pollution signals and refine the chronology derived from the lead concentration data. One way to use lead isotope data is to plot the samples as an isotopic mixing line as shown in Figure 25.3. This shows lead isotope data from Viðarhólmi saltmarsh in western Iceland (Marshall et al.,

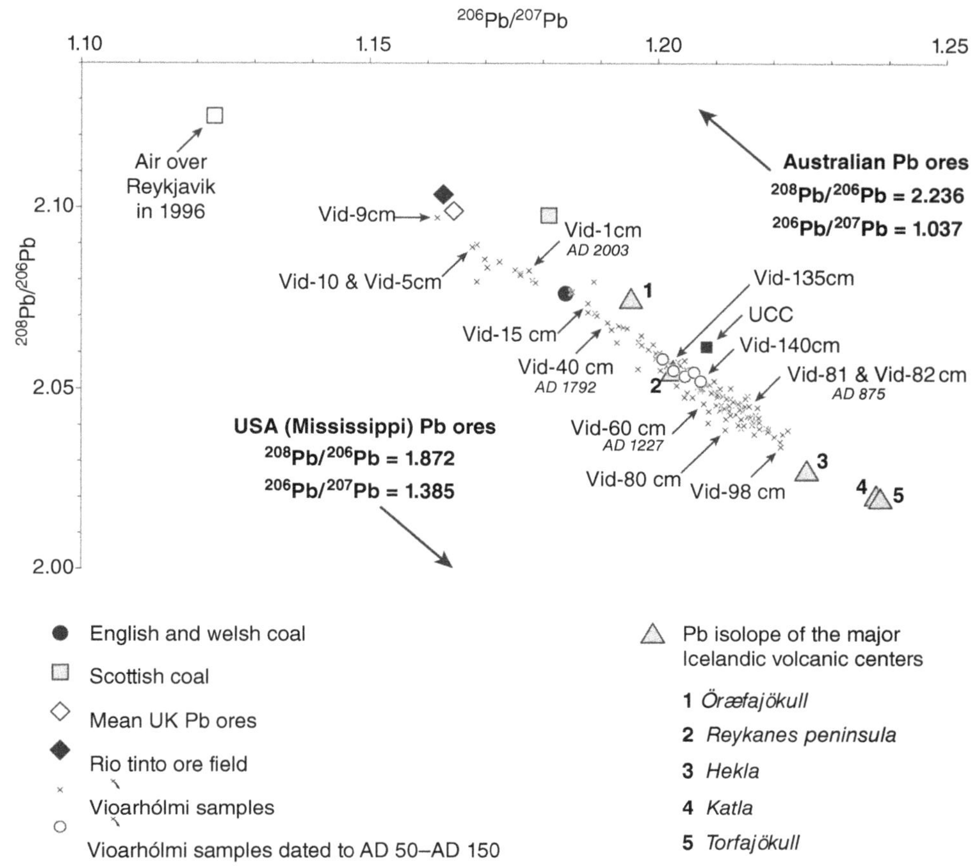

Fig. 25.3. Plot of the Viðarhólmi (Iceland) saltmarsh, $^{208}Pb/^{206}Pb$ versus $^{206}Pb/^{207}Pb$ and other relevant lead isotope values. UCC: mean upper continental crust lead isotope ratios. *Source*: From Marshall et al. 2009, fig. 6.

2009). When these data are plotted as $^{206}Pb/^{207}Pb$ values versus $^{208}Pb/^{206}Pb$ values, they produce a well-defined linear array ($r^2 = 0.958$) with the most radiogenic samples plotting in the lower right-hand sector. The plot shows how the isotopic signature of the sediments has significantly changed through time as different phases of lead pollution were active in Europe. The extreme radiogenic end-member is from 98 cm which was deposited c. AD 500 but the oldest samples, from 135 cm to 140 cm, plot significantly toward the less radiogenic section of the mixing line. This indicates some significant contribution in these lower levels from a lead source depleted in radiogenic isotopes.

If we examine the distribution of the younger samples along the mixing line in Figure 25.3 we can see how during the post-Roman period lead emissions in northern Europe stopped influencing the fallout on the saltmarsh until the start of the social expansion and technical development associated with the Medieval Period. Likewise, the isotopic signature of the samples above 40 cm (AD 1792) shifts rapidly toward the quadrant enriched in heavy isotopes. This signals the start of the industrial use of imported Australian metal in Europe. Later during the 20th century we can see how, following the removal of lead from European vehicle fuel, the isotopic signature of the youngest sediments moved away from the most depleted end-member (Vid-9 cm) in the left-hand top corner toward the pre-industrial section of the mixing line.

The concept of a metal "pollution-based chronology" is not new and has been successfully applied to intertidal sediments in a number of studies, for example those by Varekamp et al. (1992) in Connecticut, USA and Berry and Plater (1998) in northeast England. An example of the use of local historic metal pollution to estimate the rate of recent saltmarsh accretion is that by Price et al. (2005) in southwest England. In

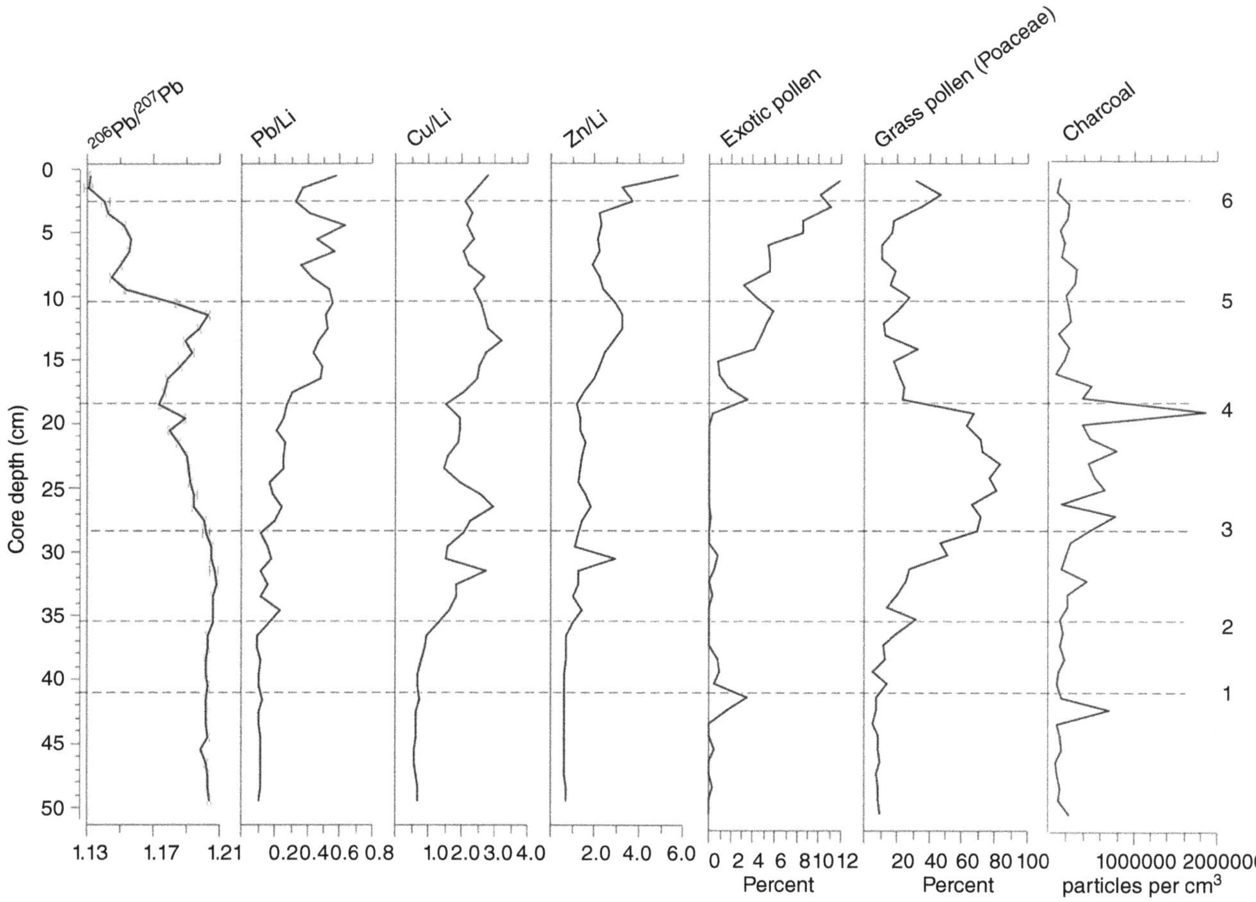

Fig. 25.4. Lead isotope ratios, Pb/Li, Cu/Li, and Zn/Li values, and pollen and charcoal curves from Little Swanport on the eastern Tasmanian coast. Also shown are six chronomarkers derived from these data. *Source*: Adapted from Gehrels et al. 2012. The chronomarkers identified by Gehrels et al. (2012) from these data are: (1) AD 1830: first appearance of exotic pollen and charcoal peak marking the arrival of significant European settlers. (2) AD 1880: signals of catchment disturbance, lead enrichment, and start of widespread land clearance. (3) AD 1895: start of $^{206}Pb/^{207}Pb$ decline as regional anthropogenic lead emissions increase. (4) AD 1937: start of large-scale zinc mining and increase in the use of coal for smelting. (5) AD 1948: start of post-WW2 $^{206}Pb/^{207}Pb$ decline. (6) AD 1988: unleaded petrol introduction.

this study AAS was used to identify sediment horizons enriched in copper, zinc, and lead. These signals were then matched to the documented history of local mining in the 19th century to obtain deposition dates for the metal enriched horizons. This study involved the analysis of predominantly waterborne metal pollution signals that were confined, in the main, to one catchment and produced by local, relatively short-term, anthropogenic industrial activity. However, Price et al. (2005) considered the total changes in the concentration data obtained by the analysis of bulk sediment. They did not consider the "relative enrichment" of the individual metals in the sediments, and this limitation makes detecting subtle short-term changes or attributing province to the pollution difficult. A more complex approach, using lead isotopes and normalization, was used by Gehrels et al. (2012) to date sea-level changes in the south Pacific. They used lithium to normalize lead, copper, and zinc to provide chronomarkers for a Tasmanian saltmarsh sequence. In this study, changes in the downcore concentrations of heavy metals in the sediments, measured using ICP-MS, were matched to local and regional records of mining and vehicle lead emissions (Fig. 25.4). Stable lead ratios ($^{206}Pb/^{207}Pb$) were used to confirm the subtle changes in the lead source through time. These independent chronomarkers were integrated with ^{14}C and ^{210}Pb dates to build a composite chronology of saltmarsh accumulation and sea-level changes on the eastern Tasmanian coast.

25.3 POLLEN AND CHARCOAL AS CHRONOLOGICAL MARKERS

Pollen analysis is one of the longest-established methods of investigating past changes in vegetation cover and landscape evolution. It is often used as a proxy for climate change and its basic principles are covered by a wealth of literature. Chapter 14 introduces and discusses the application of pollen analysis to sea-level studies. However, in addition to being an important paleoenvironmental proxy, under some circumstances pollen can be used as a means to estimate the deposition age of sediment horizons. If a well-dated pollen record exists for an area, we can use this to date unique events and outstanding changes in vegetation cover. For example, the widespread and synchronous signal around 5 cal ka BP of the European decline of *Ulmus* (elm) can be used to mark this time in all pollen diagrams which span this event (Parker et al., 2002; Lamb and Thompson, 2005). Alternatively, there may be anthropogenic events documented which created disturbances in the local pollen rain, for example changes in agricultural practices involving the introduction of a new species of plant.

The rise in *Ambrosia* (ragweed) pollen, marking deforestation following the European settlement of eastern North America in the 18th century, has been successfully used in western North Atlantic sea-level studies (e.g., Gehrels et al., 2005). Gehrels et al. (2012) used the appearance of exotic pollen in their sediments to signal the establishment of European settlements in Tasmania (Fig. 25.4). Long et al. (1999) used *Pinus* (pine) and *Spartina anglica* (cord grass) pollen to help constrain the history of sea-level change in Poole Harbour, UK, during the 17th, 18th, and 19th centuries. Conifers were relatively uncommon in the 17th century landscape around Poole Harbour. The subsequent establishment of conifer plantations in Hampshire and Dorset during the last 300 years resulted in a rapid increase in the representation of *Pinus* pollen in the Poole Harbour pollen sum. Long et al. (1999) were able to match this increase, and later fluctuations, in their pollen diagram to the documented records of afforestation. The other species successfully used by Long et al. (1999), *Spartina anglica*, is a hybrid pioneer saltmarsh grass. This plant emerged around the middle of the 19th century in Southampton Water. The first confirmed formal identification of this species was in AD 1892 near Lymington in Hampshire (Thompson et al., 1991). This invasive plant spread south along the English Channel coast, and was recorded in Poole Harbour between AD 1890 and AD 1900. Subsequently, *Spartina anglica* has invaded numerous saltmarshes in the British Isles and other parts of the world (e.g., Daehler and Strong, 1996; Cottet et al., 2007). In addition, *Spartina anglica* has been propagated and introduced deliberately in some locations, for example China in 1963 (Li et al., 2009), to encourage sediment accumulation and saltmarsh development. Although this is not easy pollen to differentiate from other large *Poaceae* (grass) grains (e.g., cereals), its arrival and subsequent increase in the landscape is therefore closely documented in many regions such as Tasmania (Hedge and Kriwoken, 2000) and San Francisco Bay (Daehler and Strong, 1996), and it has great potential as an independent chronomarker for the 19th and 20th centuries.

Changes in land cover are sometimes associated with fires, and in some regions charcoal deposition can used to signal known episodes of landscape change and historic burning events. Charcoal is produced from the incomplete combustion of wood and other organic material, and can be produced by wildfires or anthropogenic burning (Power et al., 2010). It is grouped here because it is often found as a pollen-sized grain, it is dispersed by similar mechanisms, and it can be quantified during pollen analysis. Although lacking the chronological precision of other fallout histories, sudden increases in charcoal in sediments can be used to support other methods. For example, Gehrels et al. (2012) used charcoal as an additional chronomarker in their Tasmanian sea-level reconstruction (Fig. 25.4). Although this chronomarker was of relatively low precision, it could be used as an additional independent check to the chronology. Charcoal has a number of physical features which contribute to its potential as a chronomarker. For example, it is not affected by oxidation or microbial activity and so it can have excellent preservation in sediments. It is also resistant to most strong reagents, and the smaller particles (10–50 μm) can be counted from pollen analysis preparations using normal high-power microscopy. These small grains can be dispersed by the wind, and their analysis yields a widespread signal which can be correlated with established regional records. Larger grains which are normally dispersed by water and are typically >100 μm in length will only reflect relatively local events, for example fires on or near the shore of an estuary or its fluvial system.

25.4 SPHEROIDAL CARBONACEOUS PARTICLES

In some regions spheroidal carbonaceous particles (SCPs) can be found in saltmarsh sediments. As for charcoal, these are also combustion modified carbon; in contrast to charcoal however, they have no natural sources and therefore provide an unambiguous indicator of industrial activity (Rose, 2001). SCPs are formed from the incomplete combustion of fossil fuels at high temperature. Along with inorganic ash spheres (IAS), produced by the fusion of minerals present within the fuel, SCPs form fly-ash and can be transported long distances (Rose et al., 1999; Yang et al., 2001). Once entombed they are not susceptible to post-depositional alteration, degradation, or movement within a sediment column as a result of changing pore-water conditions, for example pH or redox (Rose et al., 2003). SCPs can be processed and counted as part of pollen analysis or extracted using a number of normal laboratory techniques (Rose, 2001). As yet, they have not been widely applied to sea-level studies but offer potential chronomarkers in the period from the middle 19th to the late 20th century. Marshall et al. (2007) included SCPs in their composite chronology for the saltmarsh site originally studied by Long et al. (1999) in Poole Harbour. In the Marshall et al. (2007) study, SCPs were counted alongside pollen grains and quantified using *Lycopodium clavatum* as an exotic marker.

Once released into the atmosphere, SCPs are rapidly dispersed and can be deposited many hundreds of kilometers from their original source. They have been found in sediments in most areas of the world, both remote and industrialized, for example Scandinavia (Larsen, 2003), eastern Europe (Solovieva et al., 2002), North Africa and the Middle East (Rose et al., 2003), China (Wu et al., 2005), North America (Kralovec et al., 2002), and South America (Chirinos et al., 2006). The record of SCP deposition is longest, and most extensive, in northern Europe and the UK, reflecting the long history of heavy industry and the high-temperature burning of fossil fuels in this region (Yang et al., 2001; Rose and Rippey, 2002; Rose and Appleby, 2005).

There is an underlying regional SCP chronology for the UK and northwest Europe, but with local variations in amplitude and short-term timings (Rose and Appleby, 2005; Muri et al., 2006). A typical European SCP record will start around 1830–1850. SCPs will subsequently increase rapidly as the burning of coal by industries involving blast furnaces, and later electrical power generation, expanded and responded to the social demands of the 19th and 20th centuries. In the post-1945 era, most European sediment records show a rapid increase in SCP deposition as post-war rebuilding and industrial expansion drove the industrial use of fossil fuels upward. In response to growing environmental concerns in the 1960s and 1970s, the enactment in a number of countries of various national legislative measures compelled a widespread reduction in smoke-stack emissions in Europe (e.g., the UK Clean Air Act of 1968). This resulted in an apparent instantaneous fall to 1945, or lower levels, in many UK SCP records in the late 1970s and early 1980s (Rose, 2001).

25.5 TEPHROCHRONOLOGY

Volcanic eruptions are outstanding natural events and their ash falls, or tephra layers, can be regarded as synchronous over a widespread region. If this material is found in a sedimentary sequence, and the date of eruption known, it can be used as a chronomarker (Edwards et al., 1994; Newton et al., 2007). Tephra can be defined as the air-fall component of the magma ejected during a volcanic eruption. The larger material will rapidly fall back to Earth and will form a deposit proximal to the eruption site, but the smaller shards of volcanic glass, crystallized magma, and ash may be transported great distances by the wind (Rhoades, 2002). The thickness of the resulting deposit and the grain size of the tephra decrease with the distance from the source (Bonadonna et al., 1998; Koyaguchi and Ohno, 2001). Far-travelled tephra deposit grains are typically less than 100 μm in size, and an individual layer may be less than 3 mm thick. Different volcanic systems will produce different tephra and different eruptions, or even different phases of the same eruption from the same volcano, can differ in physical characteristics and geochemistry. If these variables can be quantified it is possible to delimit individual tephra layers, map their spatial disruption, and identify their probable source (Newton et al., 2007).

Iceland and other countries surrounding the North Atlantic have been at the forefront of the development and use of tephra as a dating tool (Newton et al., 2007). However, tephra stratigraphies are constantly being developed, refined, and published for other regions, for example New Zealand (Shane, 2000; Lowe et al., 2008) and southern South America (Wastegård et al., 2013). Once-isolated tephra layers can be described using a number of parameters such as particle size, color, and composition. For example, the silica-rich rhyolitic tephras from the Katla volcano on Iceland have a distinct gray-green appearance (Larsen et al., 2001) when compared to the blacker basaltic tephras from this system (Wastegård et al., 2003). A more quantified approach is the analysis of the tephra geochemistry, and it has proved possible to fingerprint a particular eruption if its geochemical signature is sufficiently unique. The most widely used technique for determining tephra geochemistry is electron microprobe analysis (Hall and Pilcher, 2002), but XRF (Schmid et al., 2000) and ICP-MS (Eastwood et al., 1998) have both been used with some success.

There have been a number of important advances during the last ten years in the detection of tephra layers in sediments, the use of tephrochronology to date sequences, and the analytical methods used to define individual tephra shards (see Lowe, 2011; Pearce et al., 2011; Hayward, 2012). Unfortunately the distribution of saltmarshes containing suitable tephra-rich sediments is limited to those coasts within the fallout zones of the world's volcanic systems; even if an eruption takes place and a fall of ash occurs, there may not be preservation of a viable layer in the sediment. If the eruption is small, or occurs subglacially, it may not produce any measurable tephra fall (Larsen et al., 2001). A deposit may be spatially discontinuous even at sites proximal to an eruption source, and the timing of the eruption with regard to local weather conditions, for example wind or snow cover (Begemann et al., 1999), is an important factor in tephra preservation. A further potential complication at an intertidal site will be the tidal state and sea conditions during the tephra fall.

Tephrochronology can provide a robust independent check for ^{14}C-based chronologies, and be used to reliably correlate between dispersed sites. A small number of studies have successfully used tephrochronology to help date saltmarsh sediment deposition. For example, Hamilton and Shennan (2005) were able to use an Augustine tephra to confirm the ^{14}C dates obtained for the deposition of sediments at Cook Inlet, Alaska. In Iceland, Gehrels et al. (2006) used the Landnám tephra or Settlement Layer (AD 875 ± 6; Wastegård et al., 2003) as an independent check to support ^{14}C dates. Likewise, in New Zealand Hayward et al. (2006) used three rhyolitic tephras (Taupo, Waimihia, Whakatane) to provide robust age control in their study of Holocene coastal environments at Ahuriri Inlet, southern Hawke's Bay. However, this study demonstrated that extra care is needed when using tephra at sites proximal to volcanic systems. In addition to discrete *in situ* primary layers of tephra shards, Hayward et al. (2006) found lenses of re-worked tephra shards present in some sections of the Ahuriri Inlet stratigraphy. A similar situation was found by Gehrels et al. (2006) in Iceland. Under these circumstances care must be taken to ensure that the provenance of the tephra shards in the layer being used as a chronohorizon is reliable, that is, the shards have not been reworked from older deposits.

25.6 DISCUSSION AND EVALUATION

We have shown the potential of unique deposition events to produce dated chronohorizons in saltmarsh sediment sequences. The examples presented in this chapter demonstrate how this approach can be used in isolation to estimate the deposition timing for a single layer, or integrated with other dates as part of a composite chronology for a sequence. Furthermore, we have shown how indirect dating methods can be used to establish carbon-independent chronomarkers when ^{14}C dating is difficult or inconclusive. For example, the use of tephrochronology by Hamilton and Shennan (2005), Hayward et al. (2006), and Gehrels et al. (2006) are good examples of how an indirect dating method can be used to confirm a chronology based on ^{14}C dates. In addition, when it is possible to obtain robust ^{14}C ages, indirect dating methods can be used to resolve problematic sections of the ^{14}C calibration curve, for example we can use an independent chronohorizon to establish a "limiting age".

If we consider the typical calibration for 234 ± 20 ^{14}C a BP (e.g., using CALIB 5.0 in Marshall et al., 2007), age ranges of AD 1643–1669, AD 1780–1798 and AD 1945–1951 are produced. These possible solutions span over 300 years, which is less than ideal. We can use statistical methods to indicate which age range is the most probable solution, but there is always the possibility that this can be wrong. However, if we can establish that the sediments in the horizon related to the ^{14}C age were deposited sometime before AD 1740 by using an independent outstanding event (e.g., the appearance of specific tree pollen or a dated metal pollution event), we can exclude two of the three solutions and significantly reduce the chronological uncertainty. This works even if the chronological uncertainty on the timing of the unique event is large, say ±40 years. As long as the unique event signal in the sediment sequence is stratigraphically higher, and thus post-dates the horizon the ^{14}C date is associated with, we can confidently disregard all ^{14}C calibrations younger than the timing of the unique event.

An important time period for sea-level research is the transition from the Little Ice Age to our present-day climate. Unfortunately, dating sediment deposition during this period can be difficult with ^{14}C because of the widespread use of fossil fuels and other factors (Chapter 23). When a ^{14}C age can be produced, in many cases the calibrated chronological uncertainty is too large for the result to be meaningful. Furthermore, the critical transition period in the 19th century is beyond the reliable use of ^{210}Pb dating (Chapter 24). However, indirect dating methods can be used to produce deposition ages in the 18th, 19th, and 20th centuries. The expansion of existing industry and the development of new processes during this period, for example high-temperature coal-fired blast furnaces, introduced a vast array of environmental pollutants into the environment which we can use as chronmarkers. For example, atmospheric lead fallout, and SCPs can both be used to track the rise of industrial coal burning during the last 200 years. Likewise, during the 18th, 19th, and 20th centuries many areas of the world experienced the arrival and expansion of new agricultural practices and land clearance; these events can be tracked using pollen. There are therefore a number of indirect dating methods which can be used to "bridge the gap" between ^{210}Pb dating and the youngest reliable use of ^{14}C dating in the historical period.

We can conclude that indirect dating methods are very useful tools for sea-level studies, and have the potential to contribute to high-resolution composite chronologies. However, the fundamental requirements for these methods to work will not be present at all sites. For example, has the sediment for which we are trying to obtain a deposition age been reworked? We need to check that the sediments being sampled are *in situ*. In common with ^{210}Pb dating, if a sequence has disturbed horizons or missing sections it may be impossible to date using some methods. In addition, there are a number of problems which may prevent specific indirect dating methods from being successful. The site may be too remote for metal pollution to be present at measurable levels, or it may not receive regular fallout from an active volcanic system. Low numbers of preserved pollen, or SCPs, in a sediment sequence will prevent or severely constrain an analysis. The mechanical degrading of both pollen grains and SCPs is possible if the sediments are very minerogenic, especially if the sedimentation rate is low. In addition, tephra shards can be damaged and fragmented if subjected to re-working by marsh surface water flows. This notwithstanding, in less than ideal circumstances it is often possible for useful chronological information to be obtained using indirect dating methods. In the case of pollen and SCPs, results can be obtained using normal laboratory equipment. Furthermore, these particles do not become entrained in the pore-water so post-deposition migration is not an issue once the material has been entombed. In addition, all the indirect dating methods discussed here are independent of the need for organic carbon, and the stability of a signal in the sediments is, in the main, not influenced by the geochemistry of the sediments. However, without a means of converting the data into a chronological age, the geochemical, pollen, or SCPs analysis is of limited value. The availability of a suitable well-dated "master record" from which to obtain an age is a fundamental component of all unique event-dating methods.

REFERENCES

Abril, J.M. (2004) Constraints on the use of ^{137}Cs as a time-marker to support CRS and SIT chronologies. Environmental Pollution, 129, 31–37.

Aloupi, M., and Angelidis, M.O. (2001) Geochemistry of natural and anphropogenic metals in coastal sediments of the island of Lesvos, Aegean Sea. Environmental Pollution, 113, 211–219.

Bamberg, J.H. (1994) *The History of The British Petroleum Company. Volume 2. The Anglo-Iranian Years, 1928–1954.* Cambridge University Press, Cambridge.

Begemann, F., Kallas, K., Schmitt-Strecker, S., and Pernicka, E. (1999) Tracing ancient tin via isotope analyses. In: *Proceedings of the International Conference 'The Beginnings of Metallurgy'* (eds Hauptmann, A., Pernicka, E., Rehren, T. and Yalcin, U.), Deutsches Bergbau-Museum, Bochum, 277–290.

Berry, A., and Plater, A.J. (1998) Rates of tidal sedimentation from records of industrial pollution and environmental magnetism: The Tees Estuary, North-East England. Water, Air, and Soil Pollution, 106, 463–479.

Blais, J.M. (1996) Using isotopic tracers in lake sediments to assess atmospheric transport of lead in eastern Canada. Water, Air, and Soil Pollution, 92, 329–342.

Bollhöfer, A., and Rosman, K.J.R. (2001) Isotopic source signatures for atmospheric lead: The Northern Hemisphere. Geochimica et Cosmochimica Acta, 65, 1727–1740.

Bonadonna, C., Ernst, G.G.J., and Sparks R.S.J. (1998) Thickness variations and volume estimates of tephra fall deposits: the importance of particle Reynolds number. Journal of Volcanology and Geothermal Research, 81, 173–184.

Chirinos, L., Rose, N. L., Urrutia, R., Munõz, P., Torrejón, F., Torres, L., Cruces, F., Araneda, A., and Zaror, C. (2006) Environmental evidence of fossil fuel pollution in Laguna Chica de San Pedro lake sediments (Central Chile). Environmental Pollution, 141, 247–256.

Chow, T.J., Snyder, C.B., and Earl, J.L. (1975) Isotopic sources of lead as pollutant source indicators. In: Isotopes as pollutant source and behaviour indicators. United Nations FAO and International Atomic Energy Association Symposium, April, 1975, Vienna, Austria, Vol. 1, IAEA-SM-191/4.

Cottet, M., de Montaudouin, X., Blanchet, H., and Lebleu, P. (2007) Spartina anglica eradication experiment and in situ monitoring assess structuring strength of habitat complexity on marine macrofauna at high tidal level Estuarine. Estuarine, Coastal and Shelf Science, 71, 629–640.

Cundy, A.B., Hopkinson, L. Lafite, R., Spencer, K., Taylor, J.A., Ouddane, B., Heppell, C.M., Carey, P.J., Charman, R., Shell, D., and Ullyott, S. (2005) Heavy metal distribution and accumulation in two Spartina sp.-dominated macrotidal salt marshes from the Seine estuary (France) and the Medway estuary (UK). Applied Geochemistry, 20, 1195–1208.

Daehler, C.C., and Strong D.R. (1996) Status, prediction and prevention of introduced cordgrass Spartina spp. invasions in Pacific estuaries. Biological Conservation, 78, 51–58.

Eastwood, W.J., Pearce, N.J.G., Westgate, J.A., and Perkins, W.T. (1998) Recognition of Santorini (Minoan) Tephra in lake sediments from Golhisar Golu, southwest Turkey by laser ablation ICP-MS. Journal of Archaeological Science, 25, 677–687.

Edwards, K.J., Buckland, P.C., Blackford, J.J., Dugmore, A.J., and Sadler, J.P. (1994) The impact of Tephra: proximal and distal studies of Icelandic eruptions. In: *Environmental Change in Iceland* (eds Stotter, J., and Wilhelm, F.), Munchener Geographische Abhandlungen, Reihe B, Band B12, 76–99.

Gehrels, W.R., Kirby, J.R., Prokoph, A., Newnham, R.M., Achterberg, E.P., Evans, H., Black, S., and Scott, D. B. (2005) Onset of recent rapid sea-level rise in the western Atlantic Ocean. Quaternary Science Review, 24, 2083–2100.

Gehrels, W.R., Marshall, W.A., Gehrels, M.J., Larsen, G., Kirby, J.R., Eiriksson, J., Heinemeier, J., and Shimmield, T. (2006) Rapid sea-level rise in the North Atlantic Ocean since the first half of the, 19th century. Holocene, 16, 948–964.

Gehrels, W.R., Callard, S.L., Moss, P.T., Marshall, W.A. Blaauw, M., Hunter, J., Milton, J.A., and Garnett, M.H. (2012) Nineteenth and twentieth century sea-level changes in Tasmania and New Zealand, Earth Planet. Science Letters, 315, 94–102.

Hall, V.A., and Pilcher J.R. (2002) Late Quaternary Icelandic tephras in Ireland and Great Britain: detection, characterization and usefulness. Holocene, 12, 223–230.

Hamilton, S., and Shennan I. (2005) Late Holocene relative sea-level changes and the earthquake deformation cycle around upper Cook Inlet, Alaska. Quaternary Science Review, 24, 1479–1498.

Hayward B.W., Grenfell, H.R., Sabaa, A.T., Carter, R., Cochran, U., Lipps, J.H., Shane, P.R., and Morley, M.S. (2006) Micropaleontological evidence of large earthquakes in the past 7200 years in southern Hawke's Bay, New Zealand. Quaternary Science Review, 25, 1186–1207.

Hayward, C.L. (2012) High spatial resolution electron probe microanalysis of tephras and melt inclusions without beam-induced chemical modifications. Holocene, 22, 119–125.

Hedge, P., and Kriwoken, L.K. (2000) Evidence for effects of Spartina anglica invasion on benthic macrofauna in Little Swanport estuary, Tasmania. Austral Ecology, 25, 150–159.

Hong, S., Candelone, J.-P., Patterson, C.C., and Boutron, C.F. (1994) Greenland ice evidence of hemispheric lead pollution two millennnia ago by Greek and Roman civilizations. Science, 265, 1841–1843.

Klein, S., Lahaye, Y., Brey, G.P., and Kaenel, H.-M. (2004) The early Roman Imperial AES coinage II: Tracing the copper sources by analysis of lead and copper isotopes: copper coins of Augustus and Tiberius. Archaeometry, 46, 469–480.

Koyaguchi, T., and Ohno, M. (2001) Reconstruction of eruption column dynamics on the basis of grain size of tephra fall deposits 2. Application to the Pinatubo, 1991 eruption. Journal of Geophysical Research, 106, 6513–6533.

Kralovec, A.C., Christensen, E.R., and Van Camp, R.P. (2002) Fossil fuel and wood combustion as recorded by carbon particles in Lake Erie sediments 1850–1998. Environmental Science and Technology, 36, 1405–1413.

Lamb, H., and Thompson, A. (2005) Unusual mid-Holocene abundance of Ulmus in western Ireland – human impact in the absence of a pathogen? Holocene, 15, 447–452.

Larsen, G., Newton, A.J., Dugmore, A.J., and Vilmundardottir, E. (2001) Geochemistry, dispersal, volumes and chronology of Holocene silicic tephra layers from the Katla volcanic system, Iceland. Journal Quaternary Science, 16, 119–132.

Larsen, J. (2003) Size distributions and concentrations of spheroidal carbonaceous fly-ash particles (SCPs) in lake sediments as an aid to detecting locally deposited atmospheric pollution. Water, Air, and Soil Pollution, 149, 163–175.

Lee, S.V., and Cundy, A.B. (2001) Heavy metal contamination and mixing processes in sediments from the Humber Estuary, Eastern England. Estuarine, Coastal and Shelf Science, 53, 619–636.

Li, H., Zhia, Y., Ana, S., Zhaoa, L., Zhoua, C., Denga, Z. and Gua, S. (2009) Density-dependent effects on the dieback of exotic species Spartina anglica in coastal China. Ecological Engineering, 35, 544–552

Li, Q., Cheng, H., Zhou, T., Lin, C., and Guo, S. (2012) The estimated atmospheric lead emissions in China, 1990–2009. Atmospheric Environment, 60, 1–8.

Long, A.J., Scaife, R.G., and Edwards, R.J. (1999) Pine Pollen in intertidal sediments from Poole Harbour, UK; implications for late-Holocene sediment accretion rates and sea-level rise. Quaternary International, 55, 3–16.

Loring, D.H. (1991) Normalization of heavy-metal data from estuarine and coastal sediments. Journal of Marine Science, 48, 115–115.

Lowe, D.J. (2011) Tephrochronology and its application: A review. Quaternary Geochronology, 6, 107–153.

Lowe, D.J., Shane, P.A.R., Alloway, B.V., and Newnham, R.M. (2008) Fingerprints and age models for widespread New Zealand tephra marker beds erupted since 30,000 years ago: a framework for NZ-INTIMATE. Quaternary Science Review, 27, 95–126.

Marshall, W.A., Gehrels, W.R., Garnett, M.H., Freeman, S.P.H.T., Maden, C., and Xu, S. (2007) The use of 'bomb spike' calibration and high-precision AMS ^{14}C analyses to date salt-marsh sediments deposited during the past three centuries. Quaternary Research, 68, 325–337.

Marshall, W.A., Clough, R., and Gehrels, W.G. (2009) The isotopic record of atmospheric lead fall-out on an Icelandic salt marsh since AD 50. Science of the Total Environment, 407(8), 2734–2748.

Monna, F., Galop, D., Carozza, L., Tual, M., Beyrie, A., Marembert, F., Chateau, C., Dominik, J., and Grousset, F.E. (2004) Environmental impact of early Basque mining and smelting recorded in a high ash minerogenic peat deposit. Science of the Total Environment, 327, 197–214.

Muri, G., Wakeham, S.G., and Rose, N.L. (2006) Records of atmospheric delivery of pyrolysis-derived pollutants in recent mountain lake sediments of the Julian Alps (NW Slovenia). Environmental Pollution, 139, 461–468.

Newton, A.J., Dugmore, A.J., and Gittings, B.M. (2007) Tephrabase: tephrochronology and the development of a centralized European database. Journal of Quaternary Science, 22, 737–743.

Northover, J.P. (1999) The earliest metalworking in southern Britain. In: *Proceedings of the International Conference 'The Beginnings of Metallurgy'* (eds Hauptmann, A., Pernicka, E., Rehren, T., and Yalcin, U.), Deutsches Bergbau-Museum, Bochum, 211–225.

Nriagu, J.O. (1990) The rise and fall of leaded gasoline. Science of the Total Environment, 92, 13–28.

Parker, A.G., Goudie, A.S., Anderson, D.E., Robinson, M.A., and Bonsall, C. (2002) A review of the mid-Holocene elm decline in the British Isles. Progress in Physical Geography, 26, 1–45.

Pearce, N.J.G., Perkins, W.T., Westgate, J.A., and Wade, S.C. (2011) Trace element microanalysis by laser ablation ICP-MS: the quest for comprehensive chemical characterisation of single sub-10um volcanic glass shards. Quaternary International, 246, 57–81.

Power, M.J., Marlon, J.R., Bartlein, P.J., and Harrison, S.P. (2010) Fire history and the Global Charcoal Database: A new tool for hypothesis testing and data exploration Palaeogeography, Palaeoclimatology, Palaeoecology, 291, 52–59.

Price, G., Winkle, K., and Gehrels, W.R. (2005) A geochemical record of the mining history of the Erme Estuary, south Devon, UK Gregory D. Marine Pollution Bulletin, 50, 1706–1712.

Renberg, I., Bindler, R., and Brännvall, M.-L. (2001) Using the historical atmospheric lead-deposition record as a chronological marker in sediment deposits in Europe. Holocene, 11, 511–516.

Reuer, M.K., and Weiss, D.J. (2002) Anthropogenic lead dynamics in the terrestrial and marine environment. Philosophical Transactions of Royal Society of London, A360, 2889–2904.

Rhoades, D.A., Dowrick, D.J., and Wilson, C.J.N. (2002) Volcanic hazard in New Zealand: Scaling and attenuation relations for tephra fall deposits from Taupo Volcano. Natural Hazards, 26, 147–174.

Rohl, B.M. (1995) Application of lead isotope analysis to Bronze Age metalwork in the British Isles. Unpublished PhD thesis, Oxford University.

Rose, N.L. (2001) Fly ash particles. In: *Tracking Environmental Change Using Lake Sediments, Vol. 2. Physical and Chemical Techniques* (eds Last, W.M., and Smol, J.P.), Kluwer Academic Publishers, Dordrecht, the Netherlands.

Rose, N.L., and Rippey, B. (2002) The historical record of PAH, PCB, trace metal and fly-ash particle deposition at a remote lake in north-west Scotland. Environmental Pollution, 117, 121–132.

Rose, N.L., and Appleby, P.G. (2005) Regional applications of lake sediment dating by spheroidal carbonaceous particle analysis I: United Kingdom. Journal of Paleolimnology, 34, 349–361.

Rose, N.L., Juggins, S., and Watt, J. (1999) The characterisation of carbonaceous fly-ash particles from major European fossil-fuel types and applications to environmental samples. Atmospheric Environment, 33, 2699–2713.

Rose, N.L., Flower, R.J., and Appleby, P.G. (2003) Spheroidal carbonaceous particles (SCPs) as indicators of atmospherically deposited pollutants in North African wetlands of conservation importance. Atmospheric Environment, 37, 1655–1663.

Şahoğlu, V. (2005) The Anatolian trade network and the Izmir region during the Early Bronze Age. Oxford Journal of Archaeology, 24, 339–361.

Schettler, G., and Romer, R.L. (2006) Atmospheric Pb-pollution by pre-medieval mining detected in the sediments of the brackish karst lake An Loch Mór, western Ireland. Applied Geochemistry, 21, 58–82.

Schmid, P., Peltz, C., Hammer, V.M.F., Halwax, E., Ntaflos, T., Nagl, P., and Bichler, M. (2000) Separation and analysis of Theran volcanic glass by INAA, XRF and EPMA. Microchimica Acta, 133, 143–149.

Shane, P. (2000) Tephrochronology: a New Zealand case study. Earth-Science Reviews, 49, 223–259.

Shotyk, W., Weiss, D., Appleby, P.G., Cheburkin, A., Frei, R., Gloor, M., Kramers, J., Reese, S., and Van Der Knaap, W.O. (1998) History of atmospheric lead deposition since 12,370 ^{14}C yr BP from a peat bog, Jura Mountains, Switzerland. Science, 281, 1635–1640.

Singh, A.K., and Singh, M. (2006) Lead decline in the Indian environment resulting from the petrol-lead phase-out programme. Science of The Total Environment, 368(2–3), 686–694.

Solovieva, N., Jones, V.J., Appleby, P.G., and Kondratenok, B.M. (2002) Extent, environmental impact and long-term trends in atmospheric contamination in the USA basin of East-European Russian Arctic. Water, Air, and Soil Pollution, 139, 237–260.

Smith, J.N. (2001) Why should we believe ^{210}Pb sediment geochronologies? Journal of Environmental Radioactivity, 55, 121–123.

Thompson, J.D., McNeilly, T., and Gray, A.J. (1991) Population variation in Spartina anglica C.E. Hubbard. New Phytologist, 117, 115–128.

Thomson, J., Dyer, F.M., and Croudace, J.W. (2002) Records of radionuclide deposition in two salt marshes in the United Kingdom with contrasting redox and accumulation conditions. Geochimica et Cosmochimica Acta, 66, 1011–1023.

Varekamp, J.C., Thomas, E., and van de Plassche, O. (1992) Relative sea-level rise and climate change over the last 1500 years. Terranova, 4, 293–304.

Véron, A.J., Church, T.M., Patterson, C.C., and Flegal, A.R. (1994) Use of stable lead isotopes to characterize the sources of anthropogenic lead in North Atlantic surface waters. Geochimica et Cosmochimica Acta, 58, 3199–3206.

Wastegård, S., Hall, V.A., Hannon, G.E., van den Bogaard, C., Pilcher, J.R., Sigurgerisson, M.Á., and Hermanns-Auðardóttir, M. (2003) Rhyolitic tephra horizons in northwestern Europe and Iceland from the AD 700s–800s: a potential alternative for dating first human impact. Holocene, 13, 277–283.

Wastegård S., Veresa, D., Kliemd, P., Hahnd, A., Ohlendorfd, C., and Zolitschkad, B. (2013) The PASADO Science Team 1 Towards a late Quaternary tephrochronological framework for the southernmost part of South America – the Laguna Potrok Aike tephra record. Quaternary Science Review, 71, 81–90.

Weiss, D., Shotyk, W., Appleby, P.G., Kramers, J.D., and Cheburkin, A.K. (1999) Atmospheric Pb deposition since the Industrial Revolution recorded by five Swiss peat profiles: Enrichment factors, fluxes, isotopic composition, and sources. Environmental Science and Technology, 33, 1340–1352.

Wu, Y.H., Wang, S.M., Xia, W.L., and Liu, J. (2005) Dating recent lake sediments using spheroidal carbonaceous particle (SCP). Chinese Science Bulletin, 50, 1016–1020.

Yang, H., Rose, N.L., and Battarbee, R.W. (2001) Dating of recent catchment peats using spheroidal carbonaceous particle (SCP) concentration profiles with particular reference to Lochnagar, Scotland. Holocene, 11, 593–597.

Chapter 26
Uranium-thorium dating

ANDREA DUTTON
Department of Geological Sciences, University of Florida, Gainesville, FL, USA

26.1 INTRODUCTION

Over the past several decades U–Th dating, also known as U-series disequilibrium dating, has been applied to geologic archives of past sea-level position such as fossil coral reefs, submerged speleothems, aragonite carbonate bank sediments, and fossil mollusks in beach deposits. These archives differ in terms of their ability to record and subsequently preserve the primary chemistry during deposition that is ultimately measured to calculate a radiometric age. As for any radiometric dating method, one of the conditions that must be met is that the archive has remained a closed system, meaning that it has not exchanged isotopes with the surrounding environment after deposition. Of the aforementioned archives, molluscan carbonate is the most questionable as a valid technique to extract an age of the timing of growth, in part because most of the uranium in these shells is taken up post-depositionally so it is clear that they behave as an open system. U–Th dating of submerged speleothems and aragonite carbonate bank sediments provide more robust benchmarks for past sea-level position, but have been less commonly applied to studies of past sea level than coral reefs. Accordingly, this chapter emphasizes the processes and issues pertaining to the U-series dating of fossil corals, which is the most widespread application of U-series dating in studies of past sea-level position. U-series dating of speleothems is also addressed, but is given secondary emphasis to allow for expanded discussion of the issues involved in obtaining and interpreting U-series dates of fossil coral reefs.

Initially, U-series dating on fossil coral reefs was largely focused on ascertaining whether the timing of sea-level highstands correlated with Milankovitch orbital cycles (e.g., Broecker et al., 1968). To achieve this goal, a geochemist could go to a fossil reef on a coastline or on an uplifted terrace to collect a few coral samples to take back to the lab for analysis without having paid much attention to the precise elevation, stratigraphy, and facies context. In the past several decades, however, instrumental advances have led to increases in the typical precision of U-series ages (easily an order of magnitude in some cases) and have led to the ability to date older samples than was previously possible (Andersen et al., 2008). This has been coupled with increased ease and accessibility of precise elevation measurements that have changed the landscape of the types of questions that researchers can address regarding sea-level history using the U-series dating technique. At the same time, more comprehensive studies that have sampled multiple corals from the same terrace or multiple subsamples of the same coral have revealed that subtle diagenesis can lead to significant open-system behavior of the U-series decay chain (e.g., Gallup et al., 1994; Thompson et al., 2003). The complication of open-system behavior that is particularly common in older corals has significantly reduced the clarity of interpreting coral U-series ages. All of these factors come into play when planning a field expedition to collect coral samples for U-series dating in terms of developing a sampling strategy (how much, what, and where to collect). In particular, one of the major developments in terms of conducting fieldwork to collect samples for U-series analysis, aside from the smaller sample size that is now required for analysis, is the increased attention that must be given to complementary data that will significantly enhance the final analysis and value of the geochemical data.

This chapter is designed to provide an overview of how U-series dating in sea-level applications is carried out, including the basic theory behind the method, field techniques, assessment of sample

Handbook of Sea-Level Research, First Edition. Edited by Ian Shennan, Antony J. Long, and Benjamin P. Horton.

integrity, data reporting, and relevant considerations for data interpretation.

26.2 BASIC PREMISE

26.2.1 Corals

Fossil corals that lived near the sea surface can be used to track the position of sea level through time given the availability of accurate and precise measurements on both their age and elevation (Chapter 7). Because corals live below the sea surface, they provide a minimum estimate of the past position of sea level where the major source of elevation uncertainty arises from the assumed paleodepth habitat of the coral. Mass spectrometric U-series dating can achieve extremely accurate and precise measurements; providing that the assumptions of the method have been satisfied (Section 26.3), coral U–Th age and elevation data have the potential to deliver a highly resolved record of the past position and timing of sea-level change.

26.2.2 Speleothems

Stalagmites or flowstones in coastal caves are usually the preferred cave formations for U-series dating in sea-level reconstruction studies (Chapter 6). Because these features form in caves that are sub-aerially exposed, dating the time of their growth provides an upper bound to the position of sea level, which must have been lower than that of the cave. Records of intermittent cave submergence and emergence from seawater in coastal caves are more common in presently submerged caves, making these records rare yet extremely valuable archives of past sea-level oscillations. In comparison to corals, speleothems are less susceptible to diagenesis and therefore can provide a more reliable record that can be more easily extended farther back in time.

26.3 U-SERIES GEOCHEMISTRY

The basis of the theory underpinning U-series dating of carbonates is summarized here to introduce key concepts that have important bearing on project design and data interpretation. Detailed treatment of U-series dating of carbonates can be found elsewhere (Edwards et al., 2003a, b; Richards and Dorale, 2003).

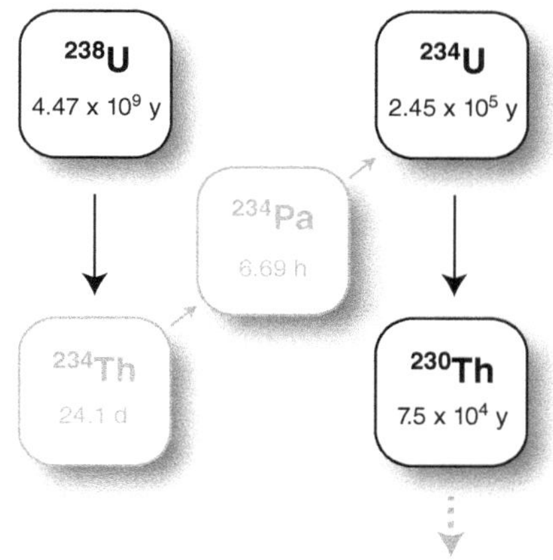

Fig. 26.1. Isotopes and their associated half-lives for the beginning of the ^{238}U decay chain, highlighting the three isotopes (^{238}U, ^{234}U, and ^{230}Th) that form the basis of the U-series dating technique. The half-lives of ^{234}Th and ^{234}Pa are so short that they can be ignored in carbonate dating applications.

A portion of the ^{238}U decay chain is represented in Figure 26.1 to illustrate the isotopes of interest for this radiometric dating technique. U-series disequilibrium arises from any process that fractionates members of the decay chain to create a non-steady-state condition. Geochemists use the term "secular equilibrium" to describe the steady-state situation that is eventually reached when the system remains closed for a time period that is sufficiently long compared to the half-lives of the intermediate daughter isotopes. It is important to recognize that, at secular equilibrium, the activity of all the nuclides in the decay chain is the same, meaning that the activity ratios of the nuclides in the decay chain are all equal to 1, where activity A is defined:

$$A = \lambda N \tag{26.1}$$

where λ is the decay constant, and N is the number of nuclides.

There are two important processes that set up disequilibrium between ^{238}U, ^{234}U, and ^{230}Th in natural waters (Fig. 26.2). The first of these is a fractionation (separation) between U and Th that results from strong differences in solubility: Th is very insoluble and tends to adsorb to the surfaces of particles (e.g., clay minerals) whereas U is very soluble, and hence transportable, in oxidized solutions such as seawater or cave drip water.

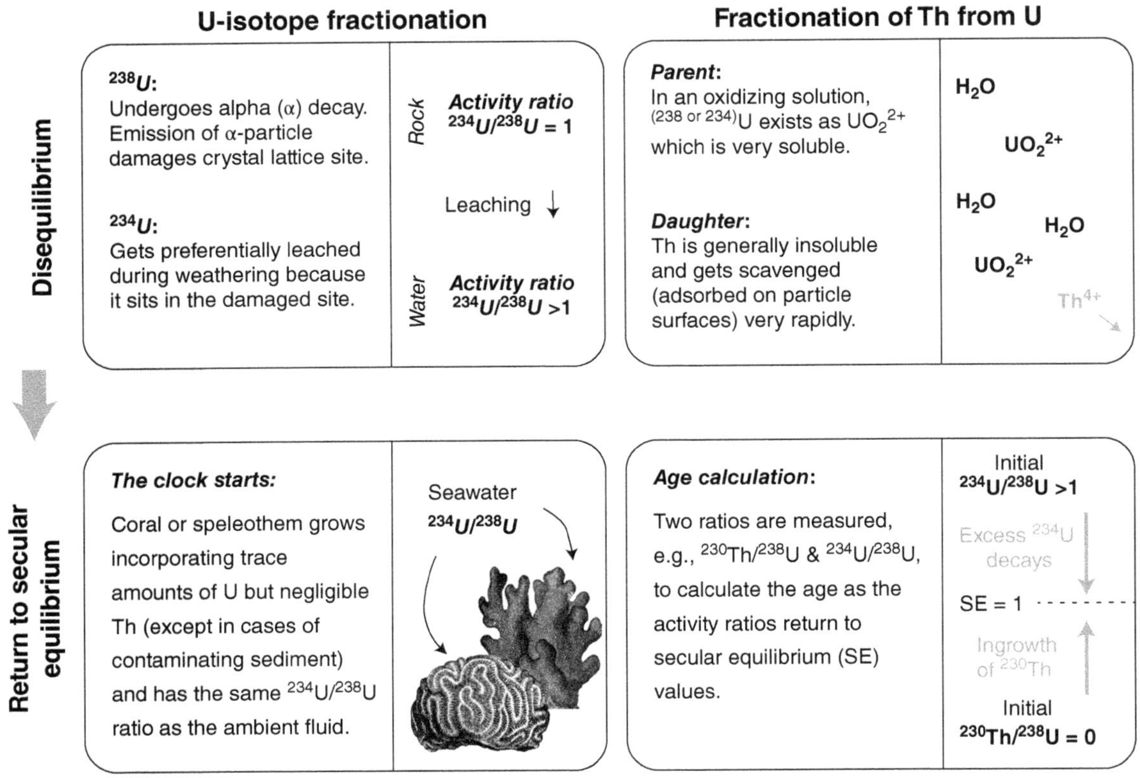

Fig. 26.2. General premise of the U-series dating technique, based on fractionation of isotopes and return to secular equilibrium.

Additional fractionation between the uranium isotopes (^{234}U and ^{238}U) in natural waters arises because ^{234}U is preferentially leached relative to ^{238}U from crystal lattice sites that are damaged during the alpha-decay process.

Uranium in natural waters is co-precipitated during coralline aragonite formation and speleothem deposition. Aragonitic corals contain ~3 ppm U whereas calcitic speleothems typically have much lower concentrations, making them more challenging to date. Once the carbonate lattice is formed, the U–Th radiometric clock starts to tick: any ingrowth of ^{230}Th is the result of the decay of ^{234}U, and the ^{230}Th content increases systematically with time. In U-series disequilibrium dating, it is the return to secular equilibrium that allows for the calculation of time based on the ingrowth relationships between ^{238}U, ^{234}U, and ^{230}Th (Fig. 26.3).

To calculate an age two ratios must be measured, for example $^{230}Th/^{238}U$ and $^{234}U/^{238}U$, to characterize the present composition of the carbonate (Fig. 26.3). The uranium isotope activity ratio is often reported in delta notation relative to the secular equilibrium value of 1, as follows:

$$\delta^{234}U = \left[\left(\frac{^{234}U}{^{238}U}\right)_{\text{sample}} - 1\right] \times 1000 \qquad (26.2)$$

where the term in parentheses denotes the activity ratio.

Presently, the most common method used to measure these isotope ratios is mass spectrometry, which allows for extremely accurate and precise age determination. Ultimately, the accuracy of these ages is dependent upon: (1) how well the decay constants for these three nuclides are known; (2) the condition that no ^{230}Th was initially present in the sample at the time of growth; and (3) the condition that the U-series decay chain has remained closed to isotopic exchange with the surrounding environment since formation. Screening for initial, or detrital, ^{230}Th is accomplished through simultaneous measurement of the more abundant and longer-lived isotope, ^{232}Th. If ^{232}Th is detected, there are various approaches

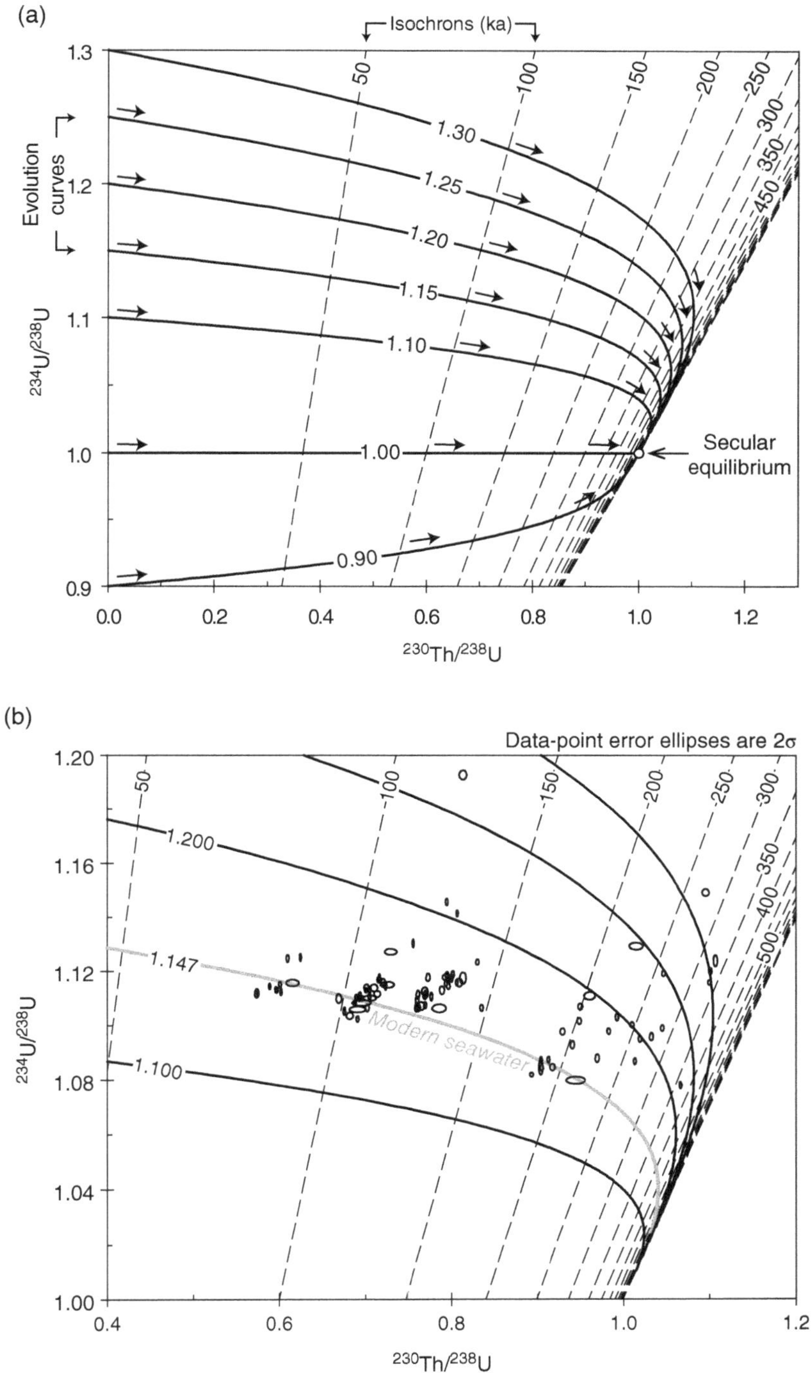

Fig. 26.3. U-series evolution diagrams. (a) Arrows represent evolution of isotope activity ratios through time towards secular equilibrium. Evolution curves are labeled with the initial uranium isotope activity ratio. Isochrons are shown with dashed lines and labeled with age in ka. (b) Measured coral compositions from multiple fossil reef terraces in Barbados reported in Thompson et al. (2003). Modern seawater has a $^{234}U/^{238}U$ activity ratio of 1.147 ($\delta^{234}U$ = 147‰), so corals that grew under such conditions should all plot on the 1.147 evolution curve (gray, bold curve). The array of data extending to the upper right and away from this curve represents samples interpreted as affected by open-system diagenesis. Both diagrams were generated using Isoplot.

to correcting the measured isotope ratios to account for the influence of the initial Th (Section 26.4.2). Detecting and correcting for open-system behavior is more difficult and contentious, and is addressed in more depth in the following sections.

26.4 ASSESSING SAMPLE INTEGRITY

A critical aspect of carrying out a U-series study is to perform a robust assessment of sample integrity, or degree of geochemical alteration. The better this step of the analysis is, the more meaningful the conclusions that can be drawn regarding the interpretation and reliability of the final dataset.

In general, dense calcitic speleothems are much less susceptible to post-depositional alteration (diagenesis) than coralline aragonite; the latter is both porous and metastable, meaning that it is prone to alteration through conversion to calcite. Assessment of speleothem alteration can generally be accomplished through visual inspection of a hand sample and thin-section to assess evidence for dissolution, pitting, or recrystallization. However, diagenesis in coralline aragonite can be much more subtle, requiring additional steps to evaluate sample integrity.

26.4.1 Mineralogy

X-ray diffraction (XRD) analysis is an effective means to assess the degree of conversion to calcite that has been employed in many studies. XRD analysis should be a routine step in screening coral samples before U-series dating is performed. Wider adoption of reporting XRD data associated with coral samples that undergo U-series measurement will improve our understanding of the link between mineralogical composition and anomalous U-series compositions. This is complicated by the fact that the XRD sample sizes are larger than those required for most mass spectrometric analyses, so the XRD approach is not able to resolve sample heterogeneity at the same scale as the U-series data.

Trace element concentrations such as Sr and Mg can have significantly different concentrations in aragonite versus calcite, hence providing another avenue towards evaluating the geochemical preservation of pristine aragonite. This approach has been used in studies that examine diagenesis in detail, but has not been widely applied as a tool to screen for well-preserved samples.

26.4.2 Detrital thorium

Ideally, the ^{230}Th content of a coral or speleothems is entirely derived from radioactive decay within the carbonate lattice. In some cases, specimens may contain additional ^{230}Th derived from the incorporation of mineral or organic detrital material during carbonate deposition. In this situation, the initial ^{230}Th contribution can be subtracted out by measuring the concentration of ^{232}Th and using an assumed value for the initial ^{230}Th/^{232}Th activity ratio (such as the average value of the upper continental crust) or by determining this ratio directly using the isochron technique (Richards and Dorale, 2003). The isochron approach entails the measurement of multiple coeval samples to establish an array of data from which to extrapolate the initial ^{230}Th/^{232}Th activity ratio. Measurement of a modern sample growing in the same environment is yet another means to establish the initial ^{230}Th/^{232}Th activity ratio, assuming that the contaminant source has remained constant through time.

26.4.3 Open-system behavior

Disequilibrium in the initial ^{234}U/^{238}U ratio was aptly termed by Edwards et al. (2003b) as a "happy complication", in part because it allows for back-calculation of the initial uranium isotope ratio at the time of growth once an age has been calculated for a sample. This is an important constraint that can be used to test for open-system behavior of the U-series isotopes in corals that precipitate in seawater. This diagenetic screening tool is based on the observation that seawater δ^{234}U is homogeneous in the open ocean and is further based on the premise that the oceans have had a reasonably constant δ^{234}U value over the past several hundred thousand years (within 15‰ of the modern seawater value of 147‰; Henderson, 2002). Because there is no fractionation of uranium isotopes during carbonate precipitation, coralline aragonite should have the same δ^{234}U value as the ambient seawater. Hence, the initial δ^{234}U value is routinely calculated and used to test for open-system diagenesis that often manifests with unusually elevated δ^{234}U values (relative to modern seawater) and anomalously old ages (Fig. 26.3b). This test is not used for speleothems because groundwater has a more variable range of compositions than seawater.

Another approach that has been used to characterize and detect open-system behavior is to simultaneously determine the ^{231}Pa age (Edwards et al., 2003b). These data together with the ^{230}Th–^{234}U–^{238}U data perhaps constitute the most rigorous test for age accuracy and detection of open-system behavior. This approach has not been widely adopted however, in part due to the technical challenges associated with ^{231}Pa techniques.

26.5 FIELD SAMPLING TECHNIQUES

To build a strategy of the best plan for collecting samples in the field, the best place to start is to work backwards. First, envision (1) what the final goal of the project is (e.g., to determine the peak position of sea level, the timing of peak sea level, both timing and position, assess the duration of an interglacial, etc.) and (2) what kind of information needs to be communicated to deliver that message (e.g., consider the figures that will convey this information and what will be needed to construct them). We then consider what implications this has for the number and type of samples that need to be collected. It is also relevant to consider what the standards are for other similar studies, so that the data collected can be effectively integrated and compared to other datasets well into the future. Collecting data that can be effectively integrated into and compared with the pre-existing framework developed from other studies is desirable and ensures more value and longevity of the dataset. More is said on this topic in Sections 26.6 and 26.7.2.

26.5.1 Logistics of collecting samples

In the case of caves or coral reefs, permission is usually required to conduct the fieldwork, whether on public or private land. The stipulations of a permit may constrain the localities that can be visited or how much material may be removed, particularly because there are often issues related to preservation of sites that have some recreational, historical, social, or economic value. In certain areas, noise or damage associated with drilling can be unacceptable to local authorities, which has ramifications on how samples are collected. Particularly in caves, it may be helpful to conduct a reconnaissance trip if possible to drill a small plug from the base (and potentially the top) of a stalagmite to assess if the time period of growth is appropriate for the study before removing a specimen from a cave. This of course requires multiple trips and significantly more time, but is an extremely effective strategy. If required, drill holes from sampling corals or speleothems can be filled in with concrete to satisfy permitting agencies that may be worried about preservation of the site.

Although obtaining a sample from a fossil coral head can easily be obtained with a geological field hammer, drilling samples can be preferable because it is less damaging to the outcrop and it penetrates past the more weathered, and potentially altered, surface into the interior of the coral. Using a hammer is also more likely to preferentially sample the outer portion and to require removal of more material to penetrate into the better-preserved part of the coral head. A small, battery-powered, hand-held drill fitted with a drill press swivel and a diamond-tipped drill core barrel (Fig. 26.4a) can be used. Alternatively, a petrol-driven drill (similar to those used by paleomagnetists), which can accommodate a wider-diameter core assembly and can be fitted with extensions to penetrate up to a meter or more, can be used if desired (Fig. 26.4b).

26.5.2 Sampling strategy

In terms of choosing which corals to sample and how many samples to collect, there are several considerations. The following is a list of recommended practices.

(1) *Understand the modern reef.* First and foremost, to do a good job of understanding and sampling the fossil reef it is imperative to get out there and explore the modern reef. This can be done through the literature, but time spent snorkeling and observing key features of the modern reef is also a wise – and generally pleasant – investment of time. Being informed about the taxonomic diversity, 3D morphology, reef facies, depth of the modern reef crest, depth habitat of various taxa, etc. will be incredibly useful to help target sampling of the fossil reef and will also help with the data interpretation stage. To this end, detailed, statistically based studies of depth and facies distributions of coral taxa and their growth morphologies are needed across geographically disparate regions. This information would enable significantly better

Fig. 26.4. Photos of field measurements, tools, and samples. (a) Hand-held battery-powered drill to recover small drill core samples; (b) larger petrol-powered drill with wider and longer core barrel; (c) surface of coral before drilling; (d) same coral as in (c) but shown with sample and number; (e) surveying position of coral heads into local sea level; and (f) Faviid coral head encrusted with coralline algae that can also be sampled to help assess paleodepth interpretation. *Source*: Photographs by Andrea Dutton. For color details, please see Plate 35.

reconstruction of sea-level position through time using corals.

(2) *Paleodepth.* Generally the preference is to target coral taxa that are known to have a shallow paleodepth, and ideally a narrow one, to most closely approximate the position of sea level. Paleodepth can also be inferred from the reef facies (e.g., reef crest versus reef slope) so it is important to understand the reef architecture to target sampling locations and to estimate paleodepth. Additional indicators of paleodepth such as the taxonomic composition of the associated fauna (coralline algae, etc.) should be noted, documented, and collected where possible to provide a more robust interpretation.

(3) *Preservation potential.* Selecting coral taxa with thicker wall structures may ensure better geochemical preservation. Clearly, this sampling preference may not be possible depending on the taxonomy and biodiversity of the study site.

(4) *Sample size.* Collect plenty of material to allow for multiple analyses, as well as any complementary analyses such as XRD and elemental analyses. Sample sizes required for U-series analysis vary by laboratory, so having this knowledge to hand is helpful. In general, larger sample sizes are desirable for younger material because of its lower ^{230}Th content. Because coral heads can be large, it is usually not a problem to collect ample material. It is useful, at least in some cases, to collect multiple drill samples from the same coral to evaluate geochemical heterogeneity. Multiple subsamples can also be taken from a single core, but obviously reflects a different spatial scale than collecting multiple drill cores. Both are useful. For speleothems, there is a tradeoff between sample size where a larger sample translates to a higher-precision date but smaller samples can provide increased temporal resolution.

(5) *Modern comparison.* It is common practice to analyze a modern coral from the same setting to test assumptions about the initial uranium isotope composition and initial Th. Be aware that importing and exporting modern coral samples usually requires special permission.

(6) *Spatio-temporal relationships.* To test hypotheses about the 3D growth of the reef, it is useful to collect samples along transects, both vertical and horizontal, where possible. In particular, collecting coral samples that are directly superimposed, that is, where there is an unambiguous temporal relation implied by the stratigraphy, is very useful as an external test on the accuracy of the ages.

26.6 COMPLEMENTARY FIELD DATA AND OBSERVATIONS

Collecting supporting information regarding each coral sample that will be dated by U-series will enable a far richer context from which to draw conclusions and evaluate the robustness of the data. Although earlier studies of coral reefs emphasized the timing of reef growth for comparison with orbital frequencies, the community is now asking questions that require a more detailed understanding to define the exact magnitude, timing, and rates of sea-level change in the past. It has therefore become increasingly important to observe and describe the 3D setting of the samples within the reef architecture, the precise elevation, and additional supporting observations of paleodepth indicators and sedimentary evidence for sea-level oscillations (e.g., exposure or erosional surfaces).

It should be emphasized that to reconstruct past sea level fundamentally requires two pieces of information: the age and elevation of the sample. While U-series dating studies go to great length to develop a precise and accurate chronology, surprisingly little emphasis has been devoted to precise and accurate surveying to define the other half of the data required to discuss past sea-level position and rates of change. Given the ease with which elevation data can now be collected (Chapter 10), there is no reason to ignore the elevation component or to rely upon estimations of elevation from contour maps, which has been surprisingly common practice. As a case in point, even though sea level during the last interglacial period (125,000 years ago) has been extensively studied, only 68% of the 847 U-series coral data from the last interglacial period that have been compiled have any elevation data associated with them (database updated from Dutton and Lambeck, 2012). Furthermore, most of the existing elevation data that do exist in this database were collected in a fashion that was not documented and with no defined elevation benchmark; in many cases, both the precision and accuracy of these elevation values is quite unclear.

The following elements are recommended to enable construction of a more valuable dataset that will have more longevity and use within the discipline.

(1) *Surveying.* This component of the fieldwork is essential. Methods for ensuring robust measurement of elevation are covered in Chapters 10 and 11. It is desirable to define the elevation of every sample collected, not just one point on the outcrop. Because the corals are all minimum estimates for the past position of sea level, the measurement should reflect the uppermost surface of the coral head. If the coral has been planed off through erosion, an estimate of its former height should be made. Traditional optical surveying equipment can be used (Fig. 26.4e) or differential global positioning system (DGPS) technology such as the real-time kinematic (RTK) systems, but in either case care must be taken to reference the elevation measurement to a local tidal datum (Chapter 11). In the case of corals, it is most

useful to use a benchmark of low mean water springs, which is generally taken as the highest point to which a coral can grow. If the data is collected using a different benchmark such as mean sea level, then the offset between mean sea level and mean low water springs should be reported. If the modern reef does not grow up to mean low water springs, it is helpful to have an estimate of the depth of the modern reef crest to establish the appropriate modern analog for paleodepth. The uncertainty associated with the elevation measurements should also be reported.

(2) *Stratigraphic context and facies analysis.* Documenting and reporting the 3D stratigraphic context of each sample is useful for making interpretations regarding the robustness of the data and also for drawing interpretations with respect to sea-level position through time. This point is taken up in more depth in the following section on data interpretation. For the purposes of carrying out the fieldwork, any steps that can be taken towards documenting the position of the sample within the reef will prove helpful later on. This includes making sketches of the outcrop, drawing cross-sections and stratigraphic columns, describing the sedimentary facies and taxonomic composition of the surrounding area, documenting any evidence for flooding surfaces or exposure surfaces that might indicate sea-level oscillations, and evidence for multiple occupations of sea level at the same locality. On a more practical level, documenting everything with digital photography, such as close-ups of sample location before and after the sampling has been conducted (Fig. 26.4), are helpful to have after the laboratory analysis is complete and the data can be considered in the context of the field observations. Although a lot of attention is given to the outcrop being sampled, walking the outcrop (and documenting it) on a larger scale to understand the context within a bigger picture is also useful.

26.7 BEYOND THE FIELDWORK: FROM SAMPLE TO DATA

Laboratories will have their own recommended procedures for preparing samples for chemistry. Usually coral samples are subsampled using a hand-held dental drill into small chunks. Several studies have documented the benefit of drilling out the finer skeletal material that is more susceptible to geochemical alteration. Areas with obvious visual evidence of staining, pitting, recrystallization, or chalky texture should generally be avoided. Solid chunks of carbonate material are also desirable over powdered samples that have high electrostatic attraction for dust contamination. These solid chunks can then be cleaned (sonicated) before dissolution.

Extracting solid pieces from a spelothem sample is usually more destructive to the sample than drilling out powder from the surface, but is a recommended approach where possible to avoid contamination issues. As with the coral samples, areas exhibiting evidence of post-depositional alteration or significant sediment contamination should be avoided. Care must be taken to rigorously assess the exact position along the growth axis of the sample that will be critical in the construction of an age model of the speleothems. Stalagmites are usually sectioned and polished along the growth axis. Samples for dating are taken on or near the growth axis where detrital Th contamination is less of an issue than near the edges of the stalagmite. If extracting solid samples rather than powdered point samples, attention must also be given to the orientation of curvilinear layers so that sampling is conducted in a parallel fashion.

26.7.1 Chemical preparation and measurement

In general, the chemical preparation of samples involves dissolution, removal of organic contaminants, addition of an isotopic spike, and isolation of the U and Th isotopes using column chemistry in a clean lab. Ion-exchange resins such as TRU-Spec or UTEVA are commonly used to isolate the desired isotopes. There are many, sometimes subtle, variations on this process that are used by different laboratories. Analysis is then carried out on the U and Th fractions, either separately or simultaneously, on a mass spectrometer to quantify the ratios of ^{232}Th, ^{230}Th, ^{234}U, and ^{238}U (and usually ^{235}U) as well as the tracer isotopes in the spike. This combination of isotopes presents an analytical challenge of precisely measuring isotopes with large differences in abundances combined with significant background corrections that may be required to subtract the tail of the large ^{238}U peak. However, advances in both instrumental technology and analytical methodology have given rise to even higher-precision measurements that have extended the time range of the dating technique to >500 ka (Andersen et al., 2008).

One issue that has recently come to light regarding the accuracy of U-series mass spectrometric measurements is the recognition that the $^{238}U/^{235}U$ ratio of natural materials can vary at the epsilon level (Stirling et al., 2007; Weyer et al., 2008). This is important because it has been common practice to assume a fixed value of 137.88 for $^{238}U/^{235}U$ and use it to correct for the significant isotopic fractionation (known as mass bias) that occurs during sample introduction to the mass spectrometer. Because most laboratories have typical analytical precisions that exceed the epsilon level, this is not a significant concern unless a high-precision technique such as the Faraday–Faraday measurement protocol is employed (Andersen et al., 2004). However, the recognition that modern seawater and associated Quaternary marine carbonates precipitated from seawater have a slight ^{235}U-enrichment would introduce a small systematic bias to measurements made under the assumption of a $^{238}U/^{235}U$ ratio of 137.88 (Hiess et al., 2012). Cave precipitates appear to display more variable $^{238}U/^{235}U$ compositions than corals (Stirling et al., 2007), which would preclude the application of any procedure that adopts a fixed composition of $^{238}U/^{235}U$ to account for mass bias in high-precision measurements of speleothem samples.

Additional issues regarding the accuracy of U-series measurements include the revision of the decay constants of ^{234}U and ^{230}Th by Cheng et al. (2000) as well as small variations between different aliquots of the Harwell uraninite (HU-1) secular equilibrium standard that has been used to calibrate measurements in many laboratories around the world. It is difficult to fully account for all the small, systematic errors in age accuracy of published data that are introduced by the combination of all these factors. In light of these complications, and also given that the advance in analytical capabilities is beginning to generate U-series ages with analytical precisions that outstrip their accuracy, community-wide efforts are presently underway to address global standardization in interlaboratory calibration procedures.

The chemical preparation and analytical measurement of U-series samples is involved, so usually laboratories only train those people who will be running large numbers of samples to do the technique themselves. If it is not possible to carry out the laboratory component of the project, then it is important to try to observe or otherwise understand the process as completely as possible to appreciate the nuances of the procedure that may later have a bearing on data interpretation.

26.7.2 Data reporting

Although there are unwritten accepted practices of data reporting in the field, codified recommendations are notably lacking. This causes problems, particularly in making data comparisons between studies and in developing regional and global syntheses of data. In particular, reporting the age without listing the measured data makes it impossible to evaluate open-system behavior, which is increasingly recognized as important to document and understand before making interpretations. Additional complications in data comparisons arise from the use of different decay constants for ^{230}Th and ^{234}U. Because of this change, it is necessary to recalculate the ages and initial $\delta^{234}U$ values for older data that use the formerly adopted decay constants of ^{230}Th and ^{234}U before making comparisons to measurements that adopt the revised constants.

In particular, it is important to report all measurements and associated errors and to detail not only the decay constants used in the calculations but the method of spike calibration (to a gravimetric or secular equilibrium standard), since this has bearing on subsequent calculations that may be needed to normalize data to a common set of decay constants. Aside from the U-series measurements, reporting the specifics of the sample elevation, elevation benchmark, and stratigraphic context should be included with publication of the U–Th age data. There is a clear need to establish guidelines for the required and suggested parameters that should be reported for U-series measurements. Wide adoption of such guidelines and the online archival of data to an open-access database such as EarthChem/Geochron (http://www.earthchem.org) will ensure the utility and longevity of U-series data and, in so doing, ultimately enhance our analysis of past sea-level position through time.

26.8 DATA INTERPRETATION OF CORAL U–TH AGES

In some studies, there is a tendency to acquire U-series data, plot the age versus elevation of fossil corals, and then draw a squiggly line of sorts

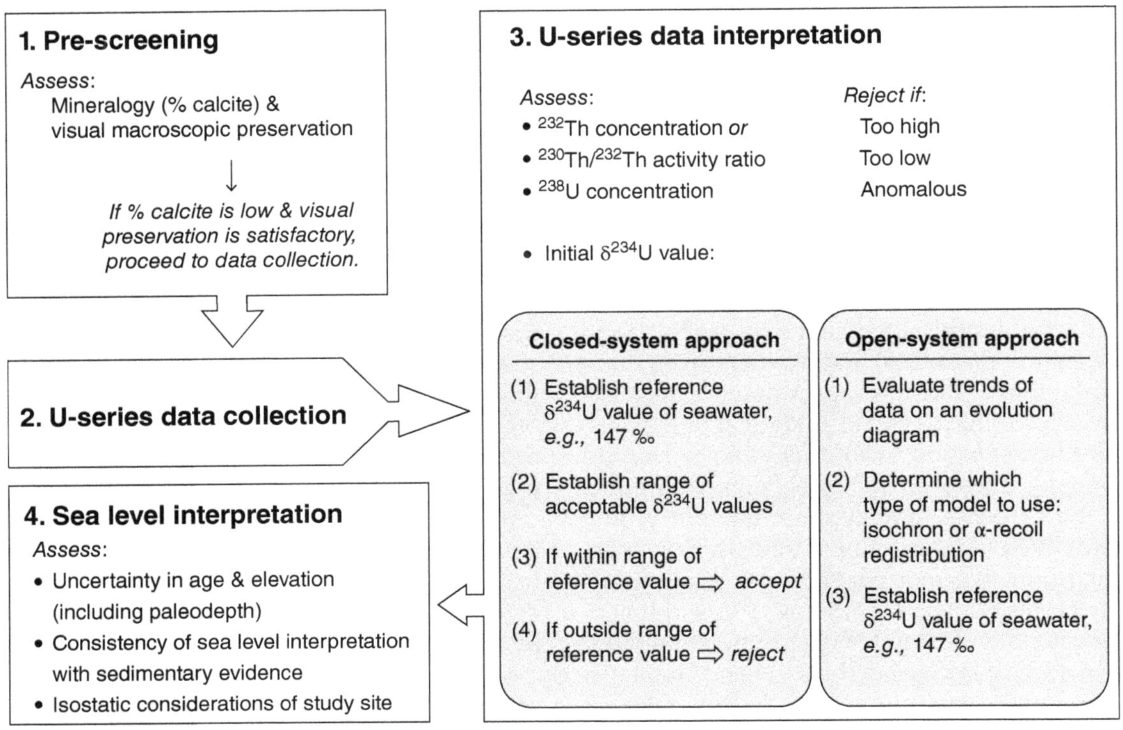

Fig. 26.5. Flow chart of procedure for U-series data collection and interpretation.

through the data to arrive at a sea-level interpretation. The discipline has now advanced to a point where this treatment of the data, in the absence of the consideration of complementary data that supports or refutes such an interpretation, is not helpful in advancing our state of knowledge. The following represents a partial listing of factors that should be considered when making interpretations on sea-level behavior based on the final age–elevation dataset. In addition, a flowchart to guide interpretation of coral U-series data is also provided (Fig. 26.5).

26.8.1 Sedimentary and stratigraphic evidence

Evaluation of sedimentary and stratigraphic evidence of former sea-level position and oscillations is increasingly – and appropriately – playing an important role in the interpretation of U-series data from fossil corals. In fact, it would be good practice for all studies to discuss the sedimentary features hand-in-hand with the U-series data.

Speed and Cheng (2004) and Blanchon et al. (2009) provide excellent examples of documenting the larger-scale stratigraphy as well as the localized outcrop-scale features that are important to the final data interpretation. To interpret U-series age–elevation data in the absence of such considerations can result in misleading interpretations. For example, if a handful of data points for corals are collected from a single outcrop and then several rapid sea-level oscillations are inferred based on the curve drawn through those data points, this conclusion should be backed up with consistent sedimentary evidence in the outcrop. If on the other hand the corals seem to have grown together in a patch reef that suggests no evidence for multiple re-occupations of sea level, then this interpretation should be reconsidered.

As another example, assessing the stratigraphic consistency of the data by plotting the U-series age–elevation data to reject data based on the presence of stratigraphic inversions is not acceptable practice for fossil reefs. Coral reefs do not grow in the same way as gravity-controlled deposition of sediments in layers. Reefs can accrete vertically as well as laterally, with nooks and crannies that can be reoccupied at a later stage. For this reason – unlike a sediment core that should generally have a predictable relation between increasing elevation and younger ages – a coral sample that has a lower elevation in an outcrop is not necessarily older, particularly in the case of a reef that may be prograding seaward during a relative stillstand in sea level. A corollary of this statement is that corals at the same elevation may have grown at different times; they need not all be the same age just because they are at the same elevation. This type of behavior is exemplified by the Holocene development of portions of the Great Barrier Reef, where seaward progradation has been well-documented at many sites as manifest by ages at the surface that progressively get younger from the landward to the seaward margin of the reef (Hopley et al., 2007). In contrast, other reefs that are growing rapidly in the vertical sense to keep up or catch up with sea-level rise (Chapter 7) may indeed display approximately the same age along elevation contours if there has been no re-occupation of sea level that is superimposed.

Finally, important information about sea-level position, including submergence and exposure history, can be interpreted from sedimentary, stratigraphic, and petrologic analysis even in the absence of U-series data. For example, Montaggioni and Hoang (1988) present an analysis of marine limestones from the Seychelles, paired with some U–Th α-dates that have precisions of ±10 ka. In this case the chronology is not terribly precise, but the model of reef accretion and sea-level behavior that they ultimately interpret rests mainly on their careful analysis of carbonate facies, fabrics, associated fauna, and carbonate petrography. With the advance of instrumental capability in enhancing the precision and accuracy of U-series ages, it is unfortunate that some of these extremely valuable complementary tools have been given less attention. Ultimately, studies that can integrate all of these facets will prove the most valuable.

26.8.2 Open-system diagenesis

As U-series studies of fossil corals began to proliferate, it became clear that, particularly in many older samples, there was a correlation with anomalously old ages and elevated initial δ^{234}U values, which is the result of what appears to be excess ^{230}Th and ^{234}U introduced into the system (Bard et al., 1991; Hamelin et al., 1991; Henderson et al., 1993). This pattern was systematically described by Gallup (1994) and several models have subsequently been proposed to explain the mechanisms behind open-system diagenesis of aragonitic sediments and corals (Henderson and Slowey, 2000; Thompson et al., 2003; Villemant and Feuillet, 2003). This suite of models is based on the concept of α-recoil mobilization and redistribution of uranium daughter products that appear to approximate the array of open-system U-series compositions observed in multiple settings (e.g., Figs 26.3b or 26.6a). This means that following α-decay of ^{238}U, particle reactive nuclides of Th are either ejected from the crystal lattice and then locally rapidly re-adsorbed, or adsorbed following brief advective transport. Coupled addition of these α-recoil nuclides – both ^{230}Th and ^{234}Th (that rapidly decays to ^{234}U with a half-life of 24.1 days) – could then explain the simultaneous increase in both ^{234}U and ^{230}Th that often characterizes corals that have been altered by open-system diagenesis.

Unfortunately, none of these models completely explain all aspects of open-system diagenesis that have been observed in U and Th isotopes in corals, yet some are very effective in fitting some of the empirically observed trends. Notably, it is clear that open-system diagenesis can follow a number of different pathways, and not just those pathways predicted by the existing models (Scholz et al., 2007). This observation supports an alternative strategy to modeling the initial composition of open-system samples that involves linear-regression, also referred to as the isochron method (Potter et al., 2004; Scholz et al., 2007). In this case, rather than using a mechanistic model to extrapolate the initial composition of an altered sample, the empirical trend is used to define the extrapolation (Fig. 26.6b).

The prevalence of open-system behavior gives rise to several important considerations and options that can be employed in the analysis and interpretation of U-series coral data, as described in the following sections.

26.8.2.1 Closed-system screening versus open-system model

At present, there are two dominant approaches to interpret absolute ages from coral U-series data. One is to apply an open-system model to calculate an age; the second is to define an acceptable range of initial δ^{234}U values and then screen the data based on this criterion (Fig. 26.5). The merits of each of these approaches are briefly discussed, and the issue of why these two methods can result in the interpretation of different absolute ages is addressed in the following section on assumptions about seawater δ^{234}U values.

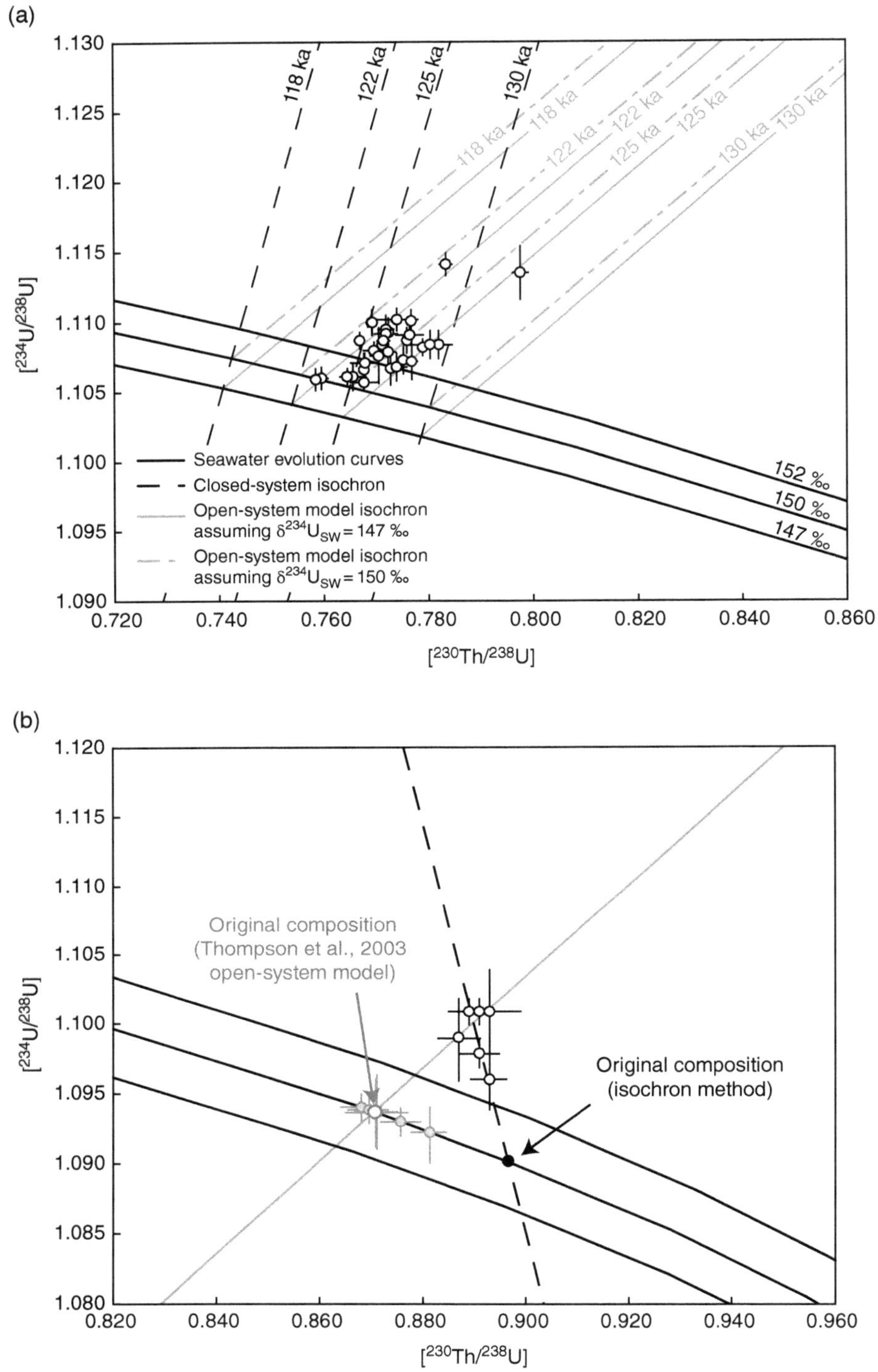

Fig. 26.6. Evolution diagrams showing activity ratios of $^{234}U/^{238}U$ versus $^{230}Th/^{238}U$. (a) Data from Bahamian fossil corals from the last interglacial period in 'Reef I' (Thompson et al., 2011). Using the Thompson et al. (2003) open-system model, these data points were extrapolated backwards to the 147‰ seawater evolution curve and averaged to yield an age of 123.1±1.2 ka. If seawater $\delta^{234}U$ during the last interglacial period were slightly higher, say 150‰, then the modeled open-system age would be correspondingly older (see isochrons above). Closed-system ages for these data largely fall between 125 and 130 ka, consistent with the ages reported in Chen et al. (1991). (b) Data for subsamples of coral BB02-4-1 from the Barbados MIS 6.5 terrace (Scholz et al., 2007). These subsamples display a trend defined by the black dashed line (ELWS regressions). Extrapolations of the data using the isochron method, as in Scholz et al. (2007), and an open-system model assuming a $\delta^{234}U$ of 150‰, yield significantly different compositions and ages.

Closed-system screening. A common approach to interpreting U-series data is to first reject any samples that did not meet acceptable preservation criteria (e.g., samples with significant amounts of calcite, high ^{232}Th concentrations, anomalous ^{238}U concentrations) and to then screen the remaining data according to the calculated initial $\delta^{234}U$ value. This is done by accepting all samples that fall within a predefined range, such as ±4‰ (Gallup et al., 1994), of modern seawater (147‰). Different studies employ different tolerances to this screening process, in some cases accepting data up to 10‰ within the modern seawater value. Gallup et al. (1994) estimate that an altered sample with an initial $\delta^{234}U$ value that is 4‰ higher than the marine value will have an age that is ~1 ka too old. The bias in accuracy of the closed-system age interpretation will therefore depend upon the frequency distribution of initial $\delta^{234}U$ values within the sample pool. If most of the samples have initial values that closely approximate that of the initial seawater composition then the age bias will be minimal; if most of the samples have initial $\delta^{234}U$ values that are significantly different than the seawater in which they grew however, then the age bias will be greater.

It is important to note that the $\delta^{234}U$ value for modern seawater is approximately 3‰ lower using the revised decay constants for ^{234}U and ^{230}Th of Cheng et al. (2000) compared to the value when calculated using the "old" decay constants. This becomes an important distinction when setting parameters for screening closed-system data based on initial $\delta^{234}U$ values of the corals, and also highlights the need to recalculate the measured isotope ratios (where a gravimetric spike was used) as well as the age and initial $\delta^{234}U$ (regardless of the spike calibration method) to normalize all of the data to the same set of decay constants before making comparisons between datasets.

Open-system model. With progressively older age, there is an increasing amount of coral U-series data that exhibit anomalously old ages coupled with higher initial $\delta^{234}U$ values that give rise to an approximately linear trend (e.g., Fig. 26.3b). If only the data having initial $\delta^{234}U$ values close to that of modern seawater are accepted in the final analysis, this means having to throw out large amounts of data. This is where open-system modeling holds great promise, providing a potential means to extrapolate the measured, open-system U–Th isotope compositions back to their initial unaltered values. This translates to being able to retain the majority of the data in the final analysis, creating a potentially statistically more robust result. A pre-requisite to applying this technique is that at least two assumptions are met: (1) that the model accurately describes or at least approximates the process that produced the array of geochemically modified data; and (2) that the initial value of seawater $\delta^{234}U$ is known. If the open-system model is applied to a system that has undergone a different type of diagenesis, then it may extrapolate the data to produce a different interpretation (Fig. 26.6b). In the second case, because the initial seawater $\delta^{234}U$ value is effectively the point at which the model "stops" and calculates an acceptable age, the accuracy of this value will directly translate into the accuracy of the interpreted absolute age (Fig. 26.6a).

At this stage, the validity of applying an open-system model (e.g., Thompson et al., 2003) to interpret U-series data from samples exhibiting anomalous closed-system ages is still debated within the community (Stirling and Andersen, 2009). Given the failure of macroscopic screening techniques to identify a large number of samples that have behaved as an open system in combination with the costly and time-consuming procedure in acquiring a U-series age, it is tempting to apply such a model to data that do not fall within the expected range of the seawater evolution curve. Open-system modeling remains controversial however, so there are no widely adopted criteria for assessing when and where the application of this model may be appropriate. Given that some corals produce different trends from those predicted by the open-system models (e.g., Fig. 26.6b), applications of such a model to data where each sample is represented by a single data point are dubious unless it can be independently established that there is an expectation that the sample has been modified in a way that is consistent with the open-system model prediction. In other words, multiple subsamples of each coral are extremely valuable to establish the diagenetic trends before models are applied to the data.

It has been shown that different diagenetic patterns exist between different localities or even between different corals within a locality. These observations indicate that multiple diagenetic pathways are possible (Scholz et al., 2007; Andersen et al., 2010). As researchers are beginning to perform multiple analyses of single coral heads more routinely, new observations are coming to light with respect to patterns in open-system behavior. For example, in corals from the Bahamas, Thompson et al. (2011) observed a trend that is consistent with the Thompson et al. (2003) open-system model when comparing compositions between different coral heads, but the same pattern does not hold for subsamples within a single coral. Scholz et al. (2007) describe arrays of U-series data for single coral specimens that show different trends between different individual specimens. Such observations suggest that open-system models based on the α-recoil redistribution process may under-represent the diagenetic processes occurring (Stirling and Andersen, 2009). For this reason, applying an open-system model may introduce large errors to the age accuracy. Until the open-system diagenetic process is better understood, it is difficult to assess the true reliability of the modeled open-system ages. In the meantime, it remains valuable to analyze data using both the closed-system and open-system approaches.

26.8.2.2 *Seawater $\delta^{234}U$ assumptions*

Regardless of whether conventional closed-system ages or modeled open-system ages are employed, the assumption that past seawater $\delta^{234}U$ is the same as that of the modern ocean will affect the accuracy of the interpreted age. For example, if the initial seawater $\delta^{234}U$ value was different than present, then this adjusted value should be employed to screen the closed-system ages or to calculate the modeled open-system age, instead of employing the value of modern seawater. In both cases, the age bias should be similar where a tight screening has been conducted on the initial $\delta^{234}U$ values of the closed-system data, regardless of which technique is used.

It is not clear exactly how much seawater $\delta^{234}U$ has varied in the past several hundred thousand years. Given the relatively long residence time of U (~400 ka, Ku et al., 1977) and the homogeneity of open-ocean seawater samples, it was initially assumed that seawater $\delta^{234}U$ was invariant on these timescales. However, modeling of potential changes in weathering on glacial–interglacial timescales suggests the possibility of permil–percent level change in seawater $\delta^{234}U$ (Chen et al., 1986; Hamelin et al., 1991; Richter and Turekian, 1993; Henderson, 2002; Robinson et al., 2004; Esat and Yokoyama, 2006). In addition, there is an increasing amount of data that indicate a lowering of seawater $\delta^{234}U$ during the last glacial maximum by c. 5–10‰ (Cutler et al., 2003, 2004; Hughen et al., 2004; Esat and Yokoyama, 2006). Presently, there is no effective technique to independently reconstruct seawater $\delta^{234}U$, so the uncertainty surrounding past secular variation continues to contribute to the overall uncertainty in coral U-series age accuracy.

26.8.2.3 *High-intensity subsampling*

As noted earlier, open-system diagenesis of U and Th isotopes can be subtle in that even if several methods are used to assess diagenesis, a seemingly well-preserved sample may still be compromised. For example, corals that present with 100% aragonite mineralogy and initial $\delta^{234}U$ values close to that of modern seawater may still have experienced open-system diagenesis. Blanchon et al. (2009) report on several coral samples from the Yucatan, some of which have "acceptable" initial $\delta^{234}U$ values according to closed-system screening; however, three samples taken from the same coral head display an age range of more than 10 ka, so clearly some alteration has taken place. This type of open-system behavior can be difficult to detect, and is a testament to the value of performing multiple analyses for each coral as an additional check on the robustness of the age data. Although tedious, costly, and time-consuming, it is clear that performing multiple analyses on every coral sample may become a requisite procedure, particularly for older samples where open-system behavior is more of an issue. Multiple measurements of the same coral may also help to clarify the type of open-system diagenesis, and whether it is consistent with some of the models that can be used to interpolate ages.

26.8.3 Isostatic effects

When making interpretations about the timing of past sea-level changes it is tempting to amalgamate data from many regions to assess the agreement between sites. It is however important to remember that because of local non-eustatic contributions to relative sea-level records, these data need to be viewed through the lens of glacial isostatic adjustment (GIA) processes (Chapter 28). This is most effectively done when directly combined with isostatic modeling to aid interpretation, but there are certainly valuable conclusions that can be reached even in the absence of direct modeling. For example, there may be intervals of time where the rate of eustatic sea-level change is so great as to overwhelm the isostatic or non-eustatic components of relative sea-level change, such that the timing of sea-level change will be synchronous between sites. Alternatively, a general knowledge of the nature of the isostatic response in a given region, even in the absence of direct calculations, can be helpful to interpret the significance of the U–Th age data. In general however, it is not advisable to superimpose U–Th age and elevation data from different geographic sites in a single diagram to make interpretations regarding the timing and magnitude of past sea-level change, due to differential isostatic corrections that must be applied to relative sea-level data derived from different sites. For example, during interglacial periods, the absolute timing as well as the duration of the sea-level highstand will vary between sites (Dutton and Lambeck, 2012; Lambeck et al., 2012). These considerations apply equally to coral U–Th ages as well as those derived from speleothems.

26.9 DATA INTERPRETATION OF SPELEOTHEM U–TH AGES

Interpretations of sea-level behavior based on speleothem U-series measurements are relatively straightforward in comparison to the case of corals, partly because of the lack of need to deal with issues related to open-system behavior (usually). The manifestation of cave submergence in speleothems and considerations for interpreting U–Th ages on either side of an unconformity or other marker of sea-level submergence are discussed in more detail by van Hengstrum et al. (Chapter 6). In general, it is important to recognize that U–Th dates and elevations derived from speleothems that grew in a vadose environment provide an upper limit to the past position of sea level. Additionally, U–Th dates derived from material that bound the period of time that the cave was submerged provide a maximum estimate of the duration of seawater inundation.

26.10 SUMMARY

U-series data offer an opportunity to ascribe rigorous chronologies to climate and sea-level change in the past. Importantly, the value of such data lies not only in acquiring high-quality measurements of U-series isotopic ratios but in the strength of the complementary measurements and analysis of sample preservation, sedimentary context, paleodepth reconstruction, glacio-hydro isostatic history, and, of course, elevation. Finally, it is clear that even once these parameters are well defined, there is still room for interpretation of the significance of and even (in cases where open-system diagenesis is an issue) the assignment of an absolute age to the sample. Sea-level reconstructions based on intercomparisons of U–Th data from different studies and regions of the globe will benefit from careful amalgamation and normalization of data, and consideration of issues related to open-system behavior, seawater $\delta^{234}U$ history, and glacial isostatic processes. The construction of open-access databases of U–Th sea-level data will play an important role in facilitating such synoptic efforts in the future. In light of this, future development of and adherence to data reporting guidelines are a critical vehicle to establishing such databases, and will undoubtedly improve the quality and transparency of the data in addition to providing a richer context from which to establish the past history of sea level.

ACKNOWLEDGEMENTS

I would like to acknowledge my colleagues for discussions that contributed to many of the ideas presented here, in particular participants of the PALSEA and PALSEA2 working group, funded by PAGES and INQUA.

REFERENCES

Andersen, M.B., Stirling, C.H., Potter, E.-K., and Halliday, A.N. (2004) Toward epsilon levels of measurement precision on $^{234}U/^{238}U$ by using MC-ICPMS. International Journal of Mass Spectrometry, 237, 107–118.

Andersen, M.B., Stirling, C.H., Potter, E.-K., Halliday, A.N., Blake, S.G., Mcculloch, M.T., Ayling, B.F., and O'Leary, M.J. (2008) High-precision U-series measurements of more than 500,000 year old fossil corals. Earth and Planetary Science Letters, 265, 229–245.

Andersen, M.B., Stirling, C.H., Potter, E.-K., Halliday, A.N., Blake, S.G., McCulloch, M.T., Ayling, B.F., and O'Leary, M.J. (2010) The timing of sea-level high-stands during Marine Isotope Stages 7.5 and 9: Constraints from the uranium-series dating of fossil corals from Henderson Island. Geochimica et Cosmochimica Acta, 74, 3598–3620.

Bard, E., Fairbanks, R.G., Hamelin, B., Zindler, A., and Hoang, C.T. (1991) Uranium-234 anomalies in corals older than 150,000 years. Geochimica et Cosmochimica Acta, 55, 2385–2390.

Blanchon, P., Eisenhauer, A., Fietzke, J., and Liebetrau, V. (2009) Rapid sea-level rise and reef back-stepping at the close of the last interglacial highstand. Nature, 458, 881–885.

Broecker, W., Thurber, D.L., Goddard, J., Ku, T.-L., Matthews, R.K., and Mesolella, K.J. (1968) Milankovitch hypothesis supported by precise dating of coral reefs and deep-sea sediments. Science, 159, 297–300.

Chen, J.H., Edwards, R.L., and Wasserburg, G.J. (1986) ^{238}U, ^{234}U and ^{232}Th in seawater. Earth and Planetary Science Letters, 80, 241–251.

Chen, J.H., Curran, H.A., White, B., and Wasserburg, G.J. (1991) Precise chronology of the last interglacial period: ^{234}U–^{230}Th data from fossil coral reefs in the Bahamas. Geological Society of America Bulletin, 103, 82–97.

Cheng, H., Edwards, R.L., Hoff, J., Gallup, C.D., Richards, D.A., and Asmerom, Y. (2000) The half-lives of uranium-234 and thorium-230. Chemical Geology, 169, 17–33.

Cutler, K.B., Edwards, R.L., Taylor, F.W., Cheng, H., Adkins, J., Gallup, C.D., Cutler, P.M., Burr, G.S., and Bloom, A.L. (2003) Rapid sea-level fall and deep-ocean temperature change since the last interglacial period. Earth and Planetary Science Letters, 206, 253–271.

Cutler, K.B., Gray, S.C., Burr, G.S., Edwards, R.L., Taylor, F.W., Cabioch, G., Beck, J.W., Cheng, H., and Moore, J. (2004) Radiocarbon calibration and comparison to 50 kyr BP with paired 14C and 230Th dating of corals from Vanuatu and Papua New Guinea. Radiocarbon, 46, 1127–1160.

Dutton, A., and Lambeck, K. (2012) Ice volume and sea level during the Last Interglacial. Science, 337, 216–219.

Edwards, R., Cutler, K.B., Cheng, H., and Gallup, C.D. (2003a) *Geochemical Evidence for Quaternary Sea-Level Changes*. Elsevier, Treatise on Geochemistry, 6, pp. 343–364.

Edwards, R., Gallup, C.D., and Cheng, H. (2003b) Uranium-series dating of marine and lacustrine carbonates. In: *Uranium-Series Geochemistry* (eds Bourdon, B., Henderson, G.M., Lundstrom, C.C., and Turner, S.P.), Mineralogical Society of America, Washington, DC, pp. 363–406.

Esat, T., and Yokoyama, Y. (2006) Variability in the uranium isotopic composition of the oceans over glacial-interglacial timescales. Geochimica et Cosmochimica Acta, 70, 4140–4150.

Gallup, C.D., Edwards, R.L., and Johnson, R.G. (1994) The timing of high sea levels over the past 200,000 years. Science, 263, 796–800.

Hamelin, B., Bard, E., Zindler, A., and Fairbanks, R.G. (1991) $^{234}U/^{238}U$ mass spectrometry of corals: How accurate is the U-Th age of the last interglacial period? Earth and Planetary Science Letters, 106, 169–180.

Henderson, G.M. (2002) Seawater (234U/238U) during the last 800 thousand years. Earth and Planetary Science Letters, 199, 97–110.

Henderson, G.M., and Slowey, N.C. (2000) Evidence from U-Th dating against Northern Hemisphere forcing of the penultimate deglaciation. Nature, 404, 61–66.

Henderson, G.M., Cohen, A.S., and O'nions, R.K. (1993) $^{234}U/^{238}U$ ratios and ^{230}Th ages for Hateruma Atoll corals: implications for coral diagenesis and seawater $^{234}U/^{238}U$ ratios. Earth and Planetary Science Letters, 115, 65–73.

Hiess, J., Condon, D., Mclean, N.M., and Noble, S. (2012) 238U/235U systematics in terrestrial uranium-bearing minerals. Science, 335, 1610–1614.

Hopley, D., Smithers, S.G., and Parnell, K.E. 2007. *The Geomorphology of the Great Barrier Reef: Development, Diversity and Change*. Cambridge University Press, Cambridge.

Hughen, K.A., Baillie, M.G.L., Bard, E., Beck, J.W., Bertrand, C.J.H., Blackwell, P.G., Buck, C.E., Burr, G.S., Cutler, K.B., Damon, P.E., Edwards, R.L., Fairbanks, R.G., Friedrich, M., Guilderson, T.P., Kromer, B., Mccormac, G., Manning, S., Ramsey, C.B., Reimer, P.J., Reimer, R.W., Remmele, S., Southon, J.R., Stuiver, M., Talamo, S., Taylor, F.W., Van Der Plicht, J., and Weyhenmeyer, C.E. (2004) Marine04 marine radiocarbon age calibration, 0–26 cal kyr BP. Radiocarbon, 46, 1059–1086.

Ku, T.-L., Knauss, K.G., and Mathieu, G.G. (1977) Uranium in open ocean: concentration and isotopic composition. Deep-Sea Research, 24, 1005–1017.

Lambeck, K., Purcell, A., and Dutton, A. (2012) The anatomy of interglacial sea levels: The relationship between sea levels and ice volumes during the Last Interglacial. Earth and Planetary Science Letters, 315–316, 4–11.

Montaggioni, L.F., and Hoang, C.T. (1988) The last interglacial high sea level in the granitic Seychelles, Indian Ocean. Palaeogeography, Palaeoclimatology, Palaeoecology, 64, 79–91.

Potter, E.-K., Esat, T.M., Schellmann, G., Radtke, U., Lambeck, K., and McCulloch, M.T. (2004) Suborbital-period sea-level oscillations during marine isotope substages 5a and 5c. Earth and Planetary Science Letters, 225, 191–204.

Richards, D.A., and Dorale, J.A. (2003) Uranium-series chronology and environmental applications of speleothems. In: *Uranium-Series Geochemistry* (eds Bourdon, B., Henderson, G.M., Lundstrom, C.C., and Turner, S.P.), The Mineralogical Society of America, Washington, DC.

Richter, F.M., and Turekian, K.K. (1993) Simple models for the geochemical response of the ocean to climatic and tectonic forcing. Earth and Planetary Science Letters, 119, 121–131.

Robinson, L.F., Henderson, G.M., Hall, L., and Matthews, I. (2004) Climatic control of riverine and seawater uranium-isotope ratios. Science, 305, 851–854.

Scholz, D., Mangini, A., and Meischner, D. (2007) U-redistribution in fossil reef corals from Barbados, West Indies, and sea-level reconstruction for MIS 6.5. Developments in Quaternary Sciences, 7, 119–139.

Speed, R.C., and Cheng, H. (2004) Evolution of marine terraces and sea level in the last interglacial, Cave Hill, Barbados. Geological Society of America Bulletin, 116, 219–232.

Stirling, C.H., and Andersen, M.B. (2009) Uranium-series dating of fossil coral reefs: Extending the sea-level record beyond the last glacial cycle. Earth and Planetary Science Letters, 284, 269–283.

Stirling, C.H., Andersen, M.B., Potter, E.-K., and Halliday, A.N. (2007) Low-temperature isotopic fractionation of uranium. Earth and Planetary Science Letters, 264, 208–225.

Thompson, W.G., Spiegelman, M.W., Goldstein, S.L., and Speed, R.C. (2003) An open-system model for U-series age determinations of fossil corals. Earth and Planetary Science Letters, 210, 365–381.

Thompson, W.G., Curran, H.A., Wilson, M.A., and White, B. (2011) Sea-level and ice-sheet instability during the Last Interglacial: new Bahamasw evidence. Nature Geoscience, 4, 684–687.

Villemant, B., and Feuillet, N. (2003) Dating open systems by the 238U-234U-230Th method: application to Quaternary reef terraces. Earth and Planetary Science Letters, 210, 105–118.

Weyer, S., Anbar, A.N., Gerdes, A., Gordon, G.W., Algeo, T.M., and Boyle, E.A. (2008) Natural fractionation of 238U/235U. Geochimica et Cosmochimica Acta, 72, 345–359.

Chapter 27
The application of luminescence dating in sea-level studies

MARK D. BATEMAN
Geography Department, University of Sheffield, Sheffield, UK

27.1 INTRODUCTION

Luminescence has been widely applied to coastal and marine sediments for as long as the technique has been used (e.g., Wintle and Huntley, 1980). It is a radiometric technique which is based on the accumulation of signal in sediment while that sediment is buried. There are significant advantages of luminescence when compared to other chronometric techniques, as the material it can be applied to is almost ubiquitous and durable in sedimentary environments. Whereas radiocarbon is reliant on the preservation of biogenic material (e.g., shells, carbonate material, plant remains) and uranium series dating is largely limited to carbonate sediments (e.g., corals and mollusk shells), luminescence can be applied to quartz and feldspar minerals which are widely available both in offshore marine and onshore coastal sediments. Luminescence dating also has a wide age range, allowing decadal through to multi-millennial depositional records within coastal sediments to be obtained. Luminescence ages in excess of 200 ka have been published (e.g., Banerjee et al., 2001; Bateman et al., 2011), and older ages are possible with the development of new technique protocols (e.g., Jacobs et al., 2011). Finally, by giving a burial age, luminescence dating also allows direct age determination of sedimentation within marine and coastal settings.

The basic principles of the luminescence technique including sampling, preparation, and measurement are outlined in this chapter. This is followed by sections illustrating how this technique has been applied to a variety of coastal and offshore sediment from intertidal flats and beach ridges through to offshore cores. Each section is used to highlight particular issues regarding the application of the technique and how these may be avoided or overcome.

27.2 BACKGROUND

27.2.1 Basic principles

Thermoluminescence (TL), using heat to stimulate luminescence, was originally developed to date fired artifacts and natural sediments. Optically stimulated luminescence (OSL), using specific wavelengths of light to stimulate luminescence, was subsequently developed to date natural sediments (see Aitken, 1998). Both TL and OSL rely on the premise that buried sediments containing quartz or feldspar are subjected to low-level ionizing radiation found in most environments. This radiation displaces electrons into traps caused by imperfections within the crystal lattices of the feldspar and quartz minerals. During burial, these displaced electrons accumulate through time. Trapped electrons can be stimulated with heat or light into being released from the traps back to a more stable state. In doing so, photons are released causing luminescence which can be measured. In other words, for two samples subjected to the same low-level ionizing radiation, the sample emitting large amounts of luminescence is mostly likely to be older than the other sample. The key for this to be a viable dating technique applicable to naturally deposited sediments is that every time feldspar or quartz grains are exposed to sunlight (or heat) the accumulated luminescence signal is removed (bleached) away. If sunlight bleaching is sufficient to zero the luminescence signal at deposition, the amount of luminescence is related to burial and not an antecedent geological event. Full zeroing of the luminescence signal can occur in less than 30 seconds of full sunlight exposure, which needs to occur at some point between sediment erosion, transportation, and deposition. Once further sunlight is blocked by a few millimeters of overlying sediment, accumulation of the luminescence signal, induced by background radiation, can resume.

Handbook of Sea-Level Research, First Edition. Edited by Ian Shennan, Antony J. Long, and Benjamin P. Horton.

If the rate at which background radiation is dosing a sample can be measured and luminescence used to establish the total accumulated dose since burial, the time elapsed since burial can be obtained. Age by luminescence is defined:

$$\text{Age (ka)} = \frac{\text{Total accumulated dose (Gy)}}{\text{Dose rate (Gy / ka)}}$$

27.2.2 Sampling and sample preparation

While luminescence sampling can be performed at night to minimize the potential for light contamination, with care and efficiency this is not a necessity. Sediment samples for luminescence are generally collected either as carved-out intact blocks (~20 cm^3) where sediment is indurated in light-tight tubes (~5 cm × 15 cm) from freshly exposed sediment or with a barrel-type auger (Fig. 27.1). The technique can be applied to sediments which are dominated by fine–medium sand (90–250 µm) or by the fine silt fraction (~4–11 µm). Stratigraphic breaks and close proximity to soil horizons and the ground surface are generally avoided unless critical to the research, as they potentially introduce heterogeneity to the background radiation field and/or changes in this field through time. Once sampled, care has to be taken to ensure samples are not exposed to sunlight or heat prior to their arrival at a luminescence dating facility. To this end, care is required to ensure that blocks do not break up, allowing sun-exposed sediment from block surfaces to mix with unexposed sediment from the centre, and that samples in tubes are tightly packed to prevent sediment movement. It is also a requirement for all luminescence samples that the depth from surface is measured on site as the amount of sediment overburden affects the contribution of cosmogenic radiation to the background radiation field. Where pre-existing vertical sections are sampled which are thought to have existed for the majority of burial history, for example an ancient sea-cliff formed relatively soon after barrier–dune construction, this should be noted as it also affects the contribution of cosmogenic radiation dose rate.

Sample preparation needs to be undertaken within dedicated luminescence dating facilities under controlled lighting conditions. For sand-size grains, sample preparation is designed to

Fig. 27.1. Examples of three common forms of OSL sampling. (a) A series of opaque 5 cm diameter plastic tubes driven into a freshly extracted vertical exposure of sediment to collect luminescence samples. Note the opaque end-caps in place. An end-cap should be fitted immediately on the opposite end as the tube is extracted from the section. (b) For indurated sediment, chiseling out a large intact block is common for OSL sampling. Note the size is controlled by the necessity to collect c. 200 cm^3 of appropriately sized sand material following removal of up to 2 cm of light-contaminated material from all block surfaces. Care must be taken to avoid any cracks within the block, as this renders the sample useless. In this case, a sub-tidal beach rock has been sampled as a block by chiseling off a large overhang. (c) Collection of luminescence samples using an auger. In this case, coastal dunes samples are being extracted using a dormer sand auger fitted with a luminescence sampling barrel. *Source*: Photographs by Mark Bateman.

remove carbonate and organic material, reduce the grain size to within 90–250 μm, and then separate out and clean either feldspars or quartz grains (e.g., Bateman and Catt, 1996). For silt-size grains, carbonate and organic material are also removed during sample preparation but different minerals are not isolated due to their small grain size making density or chemical separation difficult.

27.2.3 Measurement

Luminescence ages for sediment require two key measurements. First, the measurement of the total accumulated luminescence since the sediment was last exposed to sunlight (i.e., when it was buried; Fig. 27.2a) is required. Measurements are usually made with prepared minerals mounted as a monolayer on 9.6 mm diameter metal discs referred to as aliquots, and many aliquots are measured to produce a single age. Depending on the grain size of the sample and the amount of surface of the disc coated with sample, each luminescence measurement is an average of tens to thousands of grains. This produces a good signal to background ratio (Fig. 27.2a).

Where grains of different ages are mixed within a single sample however, for example due to partial sample bleaching prior to burial or post-depositional disturbance, this averaging effect can lead to age determination inaccuracies (Bateman et al., 2003; Duller, 2008; Arnold and Roberts, 2009). More recent developments now allow the measurement of individual grains of sand to overcome this (e.g., Roberts et al., 1999; Duller and Murray, 2000; Olley et al., 2004; Duller, 2008). As all quartz or feldspars have different sensitivities to dose, all measurements require the establishment of laboratory-based dose-luminescence calibration curves (known as growth curves) in order to equate the naturally accumulated luminescence signal to absorbed radiation dose (Fig. 27.2b). This is known as the equivalent dose (De or ED), reported in Grays (Gy) and now most frequently measured by the single aliquot regeneration (SAR) protocol (Murray and Wintle, 2003). For feldspars, originally heat (TL) and now infrared light wavelengths are used to stimulate luminescence (referred to as infrared stimulated luminescence or IRSL). For quartz OSL, blue-green light is used for stimulation. Typically a minimum of 24 replicate measurements at the single aliquot level of measurement and a minimum of 50 grain replicates which met the quality assurance criteria would be made.

The second critical measurement needs to establish the amount of radiation dose per year for that particular sediment/site. This dose is a result of alpha, beta, and gamma irradiation, is referred to as the dose rate, and is often reported in Gy ka^{-1} or μGy a^{-1}. For environmental samples the background radiation primarily stems from normally low levels of naturally occurring uranium (U), thorium (Th), and potassium (K), with some dose contribution (depending on site location and sample depth) from cosmic sources and also from rubidium (Rb). A range of different approaches are commonly adopted to measure the background dose rate.

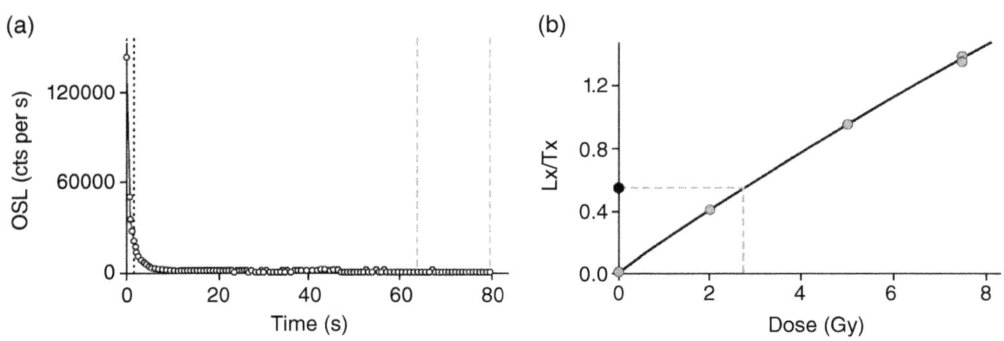

Fig. 27.2. (a) A typical OSL shine down curve in counts (cts) per second (s) showing how, when quartz is exposed to light (in this case, laboratory-based blue-green diodes), luminescence is emitted. As exposure continues, this OSL signal reduces exponentially until all stored luminescence has been emitted. Note the dashed vertical lines bracketing the data used as OSL signal and background for the construction of the growth curve. (b) A typical OSL growth curve where the naturally acquired OSL signal is measured (filled circle) and then the OSL response to known laboratory doses are measured (gray circles). Initially OSL signal grows linearly with radiation dose but, at higher doses, it becomes exponential until all OSL traps are filled; at this point the growth curve is flat and saturated. Equivalent doses are obtained by interpolating the naturally acquired OSL signal onto the growth curve and seeing what this equates to in terms of radiation dose. Note the vertical axis Lx/Tx scaling refers to the OSL dose which has been corrected for sensitivity changes as per the SAR protocol of Murray and Wintle (2003).

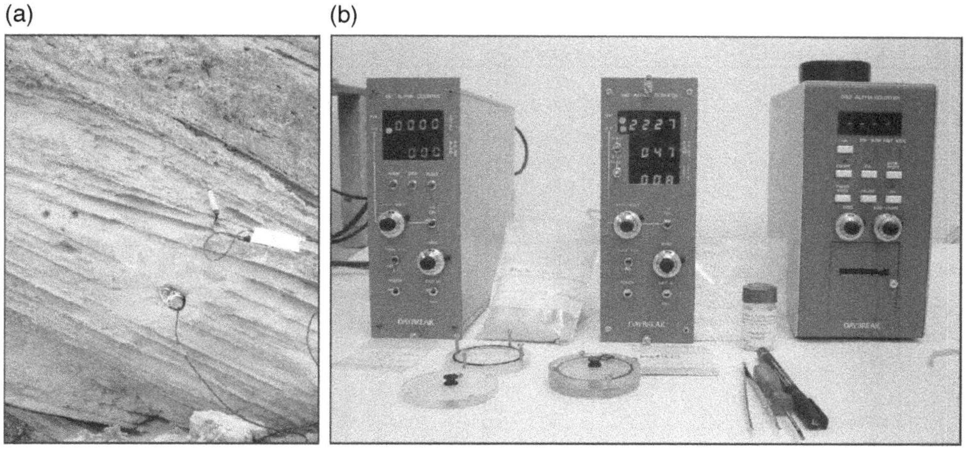

Fig. 27.3. Different measurement techniques used to establish the dose rate. (a) Collection of background dose rate using a portable gamma spectrometer inserted into the exact same position as samples from luminescence dating. Photograph taken at an exposure within a barrier dune in the southern Cape, South Africa. (b) Thick source alpha counters used for measurement of samples once they are at the luminescence dating facility. *Source*: Photographs by Mark Bateman.

As gamma radiation is high energy, its contribution to the dose rate can come from quite a distance from the sediment sampled. For example, in a sediment sequence with typical crustal concentrations of U, Th, and K, where the unit of interest is 20 cm thick, and was sampled in the middle, only 71% of the gamma dose would come from this unit. A further 26% would come from the sediments 20 cm above and below this unit, and the final 4% from sediments 40 cm above and below it. Clearly if sedimentary units above the unit of interest had higher concentrations of U, Th, and K, then a higher proportion of the gamma dose rate would come from them. Especially in sites with heterogeneous sediments, the gamma dose component of the dose rate is most accurately measured at the point of sampling either with a long-term dosimeter but more commonly using a portable gamma-spectrometer (Fig. 27.3a). The beta component of the dose rate can be based on the sediment sampled for luminescence and is commonly determined by thick source alpha counting (for U and Th; Fig. 27.3b), beta counting, or based on concentrations determined by geochemical methods such as inductively coupled plasma mass spectroscopy. The alpha dose rate is only applicable for fine-grained samples, as sand samples are commonly etched with hydrofluoric acid during sample preparation in order to remove any part of the grain subjected to alpha irradiation.

As potassium feldspars contain potassium within their crystal matrix they also self-irradiate, so the concentration of K within the feldspar is required to determine the dose rate. This can be measured on feldspars separated out from a sediment sample by geochemical techniques. The cosmic dose rate contribution is nearly always calculated based on an algorithm by Prescott and Hutton (1994) which corrects for sediment overburden, site altitude, longitude, and latitude.

27.2.4 Age calculation

Luminescence ages are always presented in years from the year of measurement and not using the term BP (which pertains to radiocarbon dating only). Quoted age uncertainties are normally at the one standard deviation level unless stated otherwise. Perhaps the main limiting factor in the application of luminescence dating to marine and coastal sediments and understanding sea-level histories is that, in comparison with other chronometric techniques, these errors are much higher, typically in the range 5–10% of the derived age. However, no recalibration of ages is required and the errors incorporate not only instrumental error but the propagation in quadrature of all errors associated with measured quantities. These include uncertainties from laboratory beta source calibration, counting statistics of the measurements, dose rate conversion, and attenuation factors, as well as uncertainty associated with moisture content (Armitage et al., 2006). As each age is based on multiple replicate aliquots/grains, much of the age uncertainty comes from how much inter-grain–inter-aliquot variability in the De value exists within a sample. As discussed later in Sections

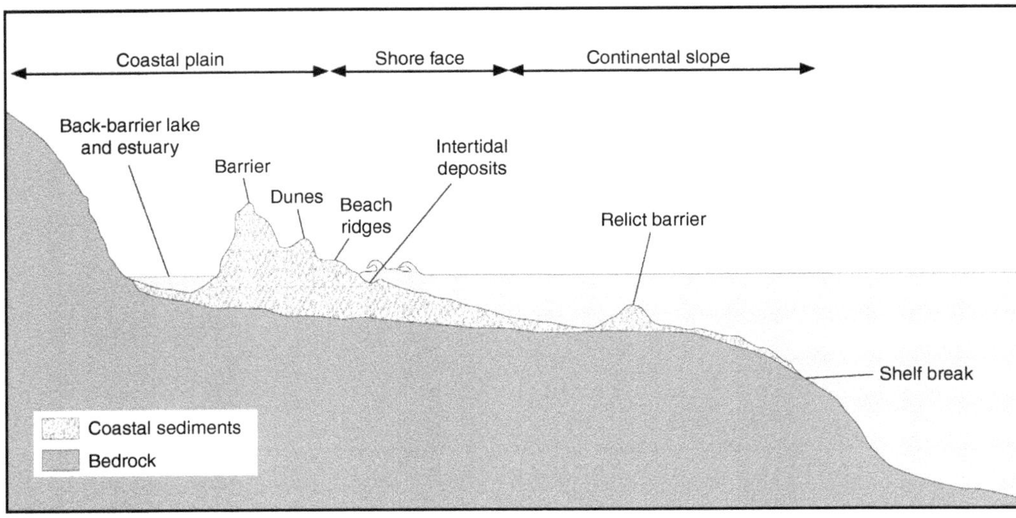

Fig. 27.4. Schematic cross-section of the typical depositional environments which might be present on the terrestrial–marine margin.

27.3.3 and 27.3.4, samples which are not fully reset prior to burial will have grains containing a variety of De values, and thus an average De from multiple replicates will have a high uncertainty.

27.3 APPLICATIONS OF LUMINESCENCE DATING IN COASTAL AND MARINE CONTEXTS

Luminescence dating has been widely applied to a large range of coastal and marine environments in order to better understand past sea-level histories. This section seeks to illustrate how it has been applied in different contexts, and any particular considerations required in these contexts. In the following sections, the terrestrial marine transition has been subdivided into four main depositional environments: coastal dunes; beaches and beach ridges; intertidal/marsh/back-barrier; and offshore sites (Fig. 27.4).

27.3.1 Coastal dunes

Sand dunes, including those found at the coast, are viewed as the optimum depositional environments for the application of luminescence dating. This is primarily due to them being almost entirely mineral thus providing ample sediment to sample, and because aeolian transport (particularly saltation) provides a very high probability of sunlight exposure prior to burial. Dunes also have the advantage of accumulating rapidly, thereby minimizing potential re-exposure to sunlight, and providing thick sections with similar dose histories. Where dunes become lithified to form aeolianite, preservation is good and the chances of accidental sunlight exposure or mixing with different-aged sediment are low. As a consequence, luminescence dating has been widely applied to coastal dune sequences throughout the world (e.g., Murray and Clemmensen, 2001; Murray-Wallace et al., 2002; Ballarini et al., 2003; Sivan and Porat, 2004; Clemmensen et al., 2007; Giannini et al., 2007; Porat and Botha, 2008; Szkornik et al., 2008; Thomas et al., 2008; Fornos et al., 2009; Roberts et al., 2009; Rink and Lopez, 2010; Bateman et al, 2011; Reiman et al., 2011).

One of the earlier major studies of coastal dune systems was that of the Coorong coastal plain in south-eastern Australia as reviewed by Murray-Wallace (2002). Here a series of TL ages, subsequently supplemented with OSL ages (Banerjee et al., 2003), were collected from a series of shore-parallel barrier ridge dunes located between the present-day coastline and approximately 60 km inland. From a luminescence perspective, Australian quartz works very well being not only bright (i.e., high luminescence counts per unit dose received) but able to store a very large dose, allowing chronologies to extend back further before saturation is encountered. The work on the dunes on the Coorong plain dated the barriers to >700 ka. The work was able to show not only increasing antiquity with distance from the present-day coast, but to relate barrier formation to interglacial highstands. It therefore constitutes perhaps the most complete record of Pleistocene highstand barriers spanning the past 11 interglacials. Given the tectonic stability of the region, its far field situation, and the positions and relative elevations, the work also showed that over this time period sea-level maxima during highstands did not deviate by more than 6 m from that of the present.

Similar work has been conducted on the Wilderness barrier dune system on the Southern Cape of South Africa (see Bateman et al., 2011). In this instance, although the dose rates are on average only moderately higher than those reported from the Coorong (0.7 Gy ka^{-1} as opposed to 0.5 Gy ka^{-1}), the maximum luminescence age reported before the samples went into saturation was only 250 ka. This illustrates how quartz from different geological sources in different areas can have a profound effect on the maximum limit of the quartz single aliquot regenerative (SAR) technique. Despite covering only the last three interglacials, the work from the Wilderness barrier dunes was able to show a strong relationship between sea-level highstands and barrier dune formation. The research also demonstrated the controlling influence of paleo-coastline configurations, in affecting whether barrier dunes accreted vertically or formed wider or more numerous ridges.

Working further west along the South African coastline, Jacobs et al. (2011) reported a longer-term record from coastal dunes with luminescence ages back to marine isotope stage (MIS) 11 (~370 ka). This study used a more recently developed luminescence measurement method called thermally transferred optically stimulated luminescence (TT-OSL) to avoid the saturation problems encountered with old dune samples, and provided internally consistent ages in line with independent stratigraphical evidence. Due to low signal, sensitivity changes, and large residuals (eee the review of Duller and Wintle, 2012 for further details), its application has largely been restricted to aeolian sediments of considerable antiquity.

At the other end of the age range, luminescence dating has been successfully applied to dunes of historical and modern ages. Bateman and Murton (2006) reported an OSL age of just 9 ± 3 years from a modern coastal dune found in northwest territory, Canada. The work of Ballarini et al. (2003) reported a detailed study of coastal dunes found on an island just offshore of the northern Netherlands. Using the quartz SAR OSL method they were not only able to date dunes, but also to compare the resultant ages to independent

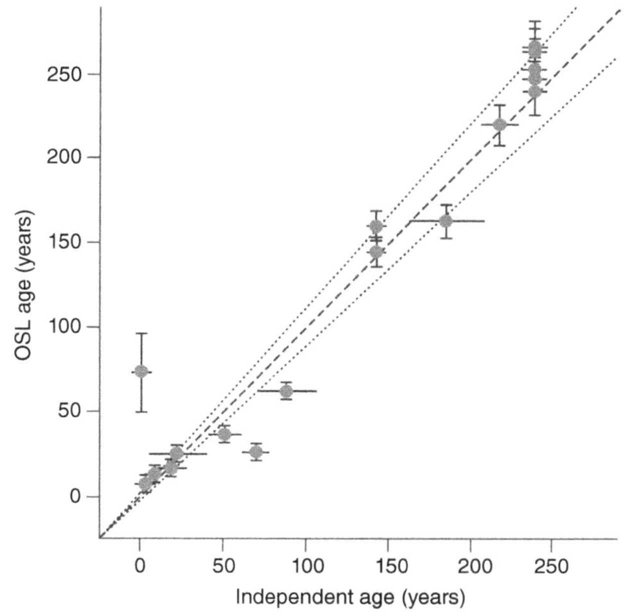

Fig. 27.5. Results of the application of quartz SAR OSL to very young and modern dunes compared to their known historical age, showing the applicability of OSL dating even in quite young contexts. *Source*: Adapted from Ballarini et al., 2003. Note that dotted lines indicate 10% from unity, and age errors are shown at one standard deviation.

historical records (primarily maps). This showed a high degree of concordance of the generated OSL ages with their known ages in the age range of 10 years back to 260 years (Fig. 27.5). It also showed the potential to achieve decadal accuracy when luminescence-dating coastal dunes.

One of the few studies to directly use luminescence dating in the construction of a relative sea-level curve is that of van Heteren et al. (2000). This study used coarse-grained feldspar IRSL dating of a dune–beach facies boundary from a barrier in Massachusetts, US. By correcting for anomalous fading of the feldspar IRSL signal and allowing for elevational differences between the sediments dated with luminescence and the material dated with radiocarbon, good agreement between the two methods was achieved. Elevation data of the dune–beach facies boundary was used with the IRSL chronology to construct index points relative to the datum used and to the modern dune–beach boundary found at 2.3–2.7 m above datum. As a result, a luminescence base sea-level curve spanning the last 6000 years was produced.

Some minor considerations might be required when applying luminescence to coastal dune deposits. Firstly, because the luminescence signal can be reset very rapidly with exposure to sunlight, any dune reworking (as found in van Heteren et al., 2000) or dune movement might mean that the resultant luminescence age does not reflect dune emplacement. Second, as with any sampling of sediment archives, the mode of accumulation must be considered, for example gradual accretion or episodic rapid accretion. Avoidance of sampling on active crests or slip faces may avoid sediments which have been recently reset. Careful consideration of internal sedimentary bedding, where available, may help focus sampling to best reconstruct accumulation modes. Alternatively, paired sampling within a dune unit and/or multiple samples from the surface down the dune plinth may allow the identification of erroneous ages.

Third, as the cosmogenic component of the dose rate is strongly controlled by the depth of sediment overburden, where coastal dunes have accreted vertically over a long period of time some consideration should be given to average burial depths of samples rather than depths found at sampling. For example, a sedimentary unit of age 120 ka which was buried under 2 m of sediment at 50 ka and a further 1 m at 10 ka has only been buried to a depth of 1 m between 50 ka and 10 ka and to a depth of 3 m from 10 ka to the present. The average overburden which should be used to calculate the cosmic dose contribution should therefore be 1.4 m and not 3 m. In this example, for a dose rate based on typical crustal values and for a mid-latitude site at low elevation, this would reduce the dose rate by 2%, making a difference in the age of 2.76 ka. Such an approach was taken by Carr et al. (2010) when OSL was applied to a sequence of periodically aggrading coastal aeolianites which spanned from 113 ka to 3 ka. In order to overcome this issue, Burdette et al. (2012) successfully developed a model to derive the average burial depth for the entire history of deposition.

Other complications arise when coastal dune sand has become lithified as aeolianite. In such a scenario, some consideration should be given to the affect that cementation may have had on the dose rate. Carbonate material will have a different intrinsic dose rate to the mineral host and, by filling pore spaces, attenuate the dose rate received by grains. Work undertaken by Nathan and Mauz (2008) on coastal dune sediments showed that carbonate cement could change the dose rate by up to 30% with a corresponding change in age of 15%. However, if it is assumed or shown that cementation occurred rapidly after burial (as has been the assumption for the Wilderness dunes dated by Bateman et al., 2011 or in Sardinia by Andreucci et al., 2010), then the dose rate as measured when sampling should reflect the dose rate received by the sand during burial. Alternatively, if it is assumed (or if sedimentary evidence suggests) that cementation happened only recently, then the burial dose rate may be estimated on the sediment once cementing material has been removed (e.g., Andreucci et al., 2012). Where cement is thought to have gradually accumulated then the model proposed by Nathan and Mauz (2008) can be used to account for these dose-rate changes. Changes in sediment moisture content with time can also affect dose rates due to water attenuating the beta dose component (see Section 27.3.3). However, as dunes are generally free draining and in the vadose zone, unless there is clear sedimentological or other evidence to suggest past changes in moisture, values as measured at the present-day are generally used.

A final consideration for coastal dune samples for OSL dating is that of post-depositional disturbance, either by human activity or by plants and animals. Post-depositional disturbance is problematic from a luminescence perspective as it can potentially mix together older-aged sediment with younger material (Bateman et al., 2003, 2007). Alternatively, burrowing by animals for example can cause sediment exhumation (allowing sunlight resetting) and re-burial (Bateman et al., 2003, 2007). Sampling sedimentary facies which have pristine primary bedding avoids accidentally sampling post-depositionally disturbed sediment, but bedding is not always present or preserved.

In depositional contexts where post-depositional disturbance might be an issue, two strategies may be employed. First, OSL sampling may be relocated away from the present-day surface, paleo-surfaces, or the disturbed unit to directly adjacent sediment. This approach was successfully adopted by Roberts et al. (2008) and Jacobs and Roberts (2009) when trying to date fossilized footprints found within coastal aeolian sediments. Alternatively, measurement of OSL ages using the single grain method potentially allows for recently zeroed or much older sand grains to be identified and excluded from the subsequent age calculations. Using single grain OSL dating and a comparison to standard aliquot OSL measurements and a radiocarbon date, Bateman et al. (2008) were able to illustrate that coastal dune sediment just below a Late Stone Age midden deposit located in South Africa had been disturbed. Single aliquot ages over-estimated the true burial age due to the incorporation of older sediment into the sample (assumed to be human trampling while the midden was active). Component analysis of the single grain OSL measurements was able to identify this old component and exclude it from age calculations.

27.3.2 Beaches and beach ridges

Beach ridges are narrow, elongate, sand or gravel shore-parallel mounds generally formed by wave action near the high tide level, sometimes with subsequent additional aeolian sedimentation. They provide a geomorphic divide between the modern shore zone and the coastal hinterland (Rink and Forest, 2005; Otvos, 2012). In places where the coastline has prograded, they have formed into extensive sequences so that their geometry and depositional age have been used to indicate past relative sea levels, shoreline position, and shape. Irrespective of whether beach ridges are defined as being purely built by wave action or a combination of wave and aeolian action, in both beaches and beach ridges the chances of sediment being exposed to sunlight and the luminescence signal reset prior to sediment burial is high. Additionally, while the sediment size might often be coarser than that of coastal dunes, there is usually plenty of sediment in the correct size range for the application of luminescence dating. Beaches (relict, raised, or otherwise) and beach ridges are therefore seen as good coastal geomorphic features to which to apply OSL dating; luminescence dating has been applied to beaches and beach ridges from various parts of the world (e.g., Bristow and Pucillo, 2006; Bjørnsen et al., 2007; Brooke et al, 2008; Lopez and Rink, 2008; Benedetti et al., 2009; Nott et al., 2009; Carr et al., 2010; Burdette et al., 2012; Dörschner et al., 2012).

Nielsen et al. (2006) successfully established OSL-based chronologies for one of the world's largest beach ridge plains, the Jerup beach ridge plain in Denmark. Use of the quartz SAR method to 20 samples from the beach ridge plain enabled the establishment of a chronology from 2.7 ka to the present day, consistent with independently derived radiocarbon dates and recent aerial photography. The OSL chronology also showed an increase in age of beach ridges with distance from the present coast and good consistency in ages obtained from separate localities on the same ridge. From this study it was established that a beach ridge formed roughly every 15 years and that average lateral migration was of the order 2 m a^{-1}.

Rink and Forest (2005) applied course-grained quartz SAR OSL dating to beach ridges found on the Cape Peninsular, Florida, USA. Their study showed that shells incorporated within these ridges were recycled, resulting in radiocarbon dates that were 2000–3000 years too old when compared to the OSL burial ages. The derived OSL-based chronology for all the ridges showed that storm surge events forming beach ridges occurred every ~80 years, with coastline progradation occurring at the rate of ~135 m per century.

Elsewhere, Forsyth et al. (2010, 2012), working in Queensland, Australia, used quartz SAR OSL dating to establish beach ridge records for the last 5 ka. These were used not only to work out coastal progradation rates but also, based on ridge heights, to model the magnitude and frequency of landfall of tropical cyclones.

Tamura et al. (2012) undertook an intensive OSL dating program of beach ridges on the Mekon Delta, Vietnam. Resulting from this work was an OSL-based chronology spanning the last 5 ka up to the present day, which fitted with available radiocarbon ages and historical observations. Tamura et al. (2012) used this to establish the initiation of the lower plain delta at 3.5 ka, and to relate its propagation and asymmetry to enhanced long-shore sediment drift caused by the winter monsoon.

In their study of North Carolina, USA, Mallinson et al. (2011) included OSL dating of back beach and overwash sediments situated above either coastal swamp peats or estuarine sediments. Using coarse-grained quartz OSL, these samples were dated to fall within marine isotope stages (MIS) 5a and they were able to show relative sea level was at or higher than present. After making a glacio-isostatic adjustment, these sea levels were still 10 m higher than those recorded on the Houn peninsular coral reef terraces with the difference attributed to uncorrected hydro-isostatic loading. A parallel study based on beach ridges was able to show that sea level had regressed during MIS 1. OSL dates showed that the coastline was >5 km further west (inland) of its present position 2000–3000 years ago.

Benedetti et al. (2009) undertook a coarse-grained quartz SAR OSL study of raised beaches from the Estremadura coastline of Portugal.

Results from this showed good internal consistency of the OSL ages and, for the older samples, agreement with radiocarbon ages (younger radiocarbon ages are thought to have suffered from problems associated with podzolization). Raised beaches were dated to MIS 6 to MIS 3, although the latter were most widespread. By assuming that sea level at the time of formation of the beaches was the same as the global eustatic average, elevations from the MIS 3 OSL-dated raised beaches were used to calculate rapid localized tectonic uplift rates of 0.4–4.3 mm a^{-1}.

Many of the considerations outlined for coastal dunes apply to beaches and beach ridges. Both can be subsequently reworked by aeolian processes (e.g., Clemmensen et al., 2007). Establishing that sediment sampled is within primary beach/beach ridge sediment is important if the OSL ages are to provide information about sea level and coastal processes. Where beach deposits are thin or beach ridges form only low geomorphic forms, sampling for OSL below soil and sediment disturbed by plants and animals may be more challenging. However, as both tend not to vertically accrete through time or be active for long periods, consideration pertaining to changing overburden through time or phases of deposition are less of a concern.

27.3.3 Subtidal, intertidal, marsh and back-barrier lagoons

As intertidal, marsh, and back-barrier sediments tend to be more variable in terms of size, both coarse- and fine-grained approaches have been adopted when applying luminescence dating to these types of deposits. Critical to the successful application is whether exposure to sunlight of sediment took place prior to burial. Compared to aeolian and beach deposits, this is less certain due to diminished light penetration through the water column when fine grains are suspended or sediment emplaced *en masse*, for example overwash deposits. For tidal sediment, Mauz et al. (2010) suggested that fine-grained samples should theoretically be better bleached; when this was tested however, both coarse- and fine-grained bleaching was found to be highly variable. Measurement at the single grain or small aliquot level with appropriate statistical analysis to try to exclude incompletely bleached gains mitigates this problem and has had some successes. For example, Mauz et al. (2010) had good agreement with independent age controls for tidal samples dated with OSL.

In active subtidal and intertidal zones, marshes, and lagoons, clearly the other challenge is to collect a luminescence sample without exposure to light and without mixing it with sediment of other ages. This is done by using opaque core liners and not overly aggressive core catchers, although paired coring, enabling core logging to take place and OSL samples to be extracted from the most suitable sediment and units, is helpful.

Early studies in applying luminescence dating to intertidal deposits met with mixed success due to incomplete bleaching, as demonstrated with TL by Plater and Poolton (1992) and with IRSL and OSL by Richardson (2001). However, more recent studies have shown success in the application of luminescence in these environments. For example, Boomer and Horton (2006) dated intertidal deposits cored with opaque liners from off the coastline of Norfolk, UK. Using the fine-grained approach and IRSL they were able to produce ages between ~7.9 ka and 2.7 ka which fitted with the associated radiocarbon chronology, suggesting sufficient bleaching had taken place for these samples. This chronology and transfer functions based on foraminifera enabled these authors to produce new sea-level index points and a new relative sea-level curve for this region. Similarly, research performed by Carr et al. (2010) in South Africa using quartz SAR OSL dating, dated preserved tidal inlet, storm berm, and shoreface sedimentary units. In conjunction with OSL-dated dune sediments overlying the raised beaches, this provided the chronology for sea-level limiting and index points within MIS 5e and showed that sea level during MIS5e was 6–8 m above present-day mean sea level.

In another study, Davids et al. (2010) targeted overwash sand layers associated with hurricane strikes preserved in saltmarsh deposits in order to establish a luminescence-based chronology. Such sandy layers were shown to have been bleached prior to burial as quartz OSL and feldspar IRSL ages were found to be in stratigraphic order and consistent within errors with radiocarbon ages and with each other.

Fruergaard et al. (2011) successfully demonstrated how the application of OSL to cores from a tidal lagoon behind the barrier island of Fanø, Denmark could be used to reconstruct the paloenvironmental evolution of the coastal lagoon and its relationship to sea-level changes. In this study, cores were collected from tidal flats and sandy facies targeted for OSL dating using quartz and

SAR. Previously, when working in the same region, Madsen et al. (2005, 2007) indicated that modern sub-aqueous coarse-grained sediments such as these could be well bleached on deposition. Resultant ages, which ranged from 40 years to 5.5 ka, had low uncertainties, suggesting good age accuracy. Results were able to show slow vertical aggradation of the sand flats (1.5–2 mm a^{-1}) within the lagoon and the preservation of a millennial-scale sediment archive due to the rising sea level. Centurial timescale sediments were only preserved where unaffected by the lateral migration of channels.

The application of OSL dating to intertidal deposits requires a range of considerations. One of these is sediment bioturbation. In intertidal zones, marshes, and lagoons, sediments are clearly prone to bioturbation as well as water disturbance (i.e., resuspension) with vertical sediment mixing being of particular concern from a luminescence perspective. This was demonstrated by Madsen et al. (2011) with a very-high-resolution record collected from sandy tidal flats in the Danish part of the Wadden Sea and dated with the coarse-grained quartz SAR approach. Results clearly showed a jump in ages associated with the bottom of the mixing layer (at around 20 cm depth). Above this, sediments were young as they were presumed to have been recently exhumed by lugworms. An age offset of 620 years was reported, the time taken for sufficient sediment to accumulate above a sample to move it out of the mixing zone. Even with more solid samples, such as submerged beachrock and aeolianite, burrows and larger voids can allow light penetration within blocks carved for luminescence dating (Fig. 27.6).

Fig. 27.6. Block sample of aeolianite collected from the intertidal zone, southern Cape, South Africa by Dr Adam Dunajko, which was revealed to contain numerous borings created by marine organisms and therefore problematic for luminescence purposes. *Source*: Photograph by Mark Bateman.

This group of sediments can also suffer from changes in dose rate through time through the dissolution of carbonate cement (see Section 27.3.1), changes in concentration of radioactive elements (notably K and U), and also from changes in water content. For most of these depositional contexts, the sediments will spend varying amounts of time both in the vadose and phreatic/saturated zones. Use of modern analogs to determine water content are therefore not always appropriate and present-day water contents of sediments may not reflect average moisture contents. The problem is compounded as, with further sedimentation, autocompaction occurs and water content (porosity) will likely decrease. Richardson (2001) exemplified this problem, demonstrating that very high and variable water contents for fine-grained subaqueous sediments had a large effect on their age determination. The effect of moisture on OSL age calculations can be illustrated with a hypothetical coarse-grained sample with a dose rate based on typical crustal values and a true average moisture content of 5% by weight. If the moisture content was erroneously doubled to 10% by weight, the calculated sample age would overestimate true age by 6%. Use of coarser gains mitigated this to some degree, as saturated sand can only have a moisture content range of the order 25% by weight (compared to values of up to 86% reported by Richardson, 2001 for fine-grained sediments).

Mauz and Bugenstock (2007) applied fine-grained quartz OSL dating to a mesotidal back-barrier tidal flat system to test a methodology for establishing sea level in these depositional contexts. Using multiple core samples and detailed facies analysis, the relationship of sedimentation to sea level was established. Issues of moisture content for the luminescence ages were corrected for by using a model proposed by Roberts and Plater (2007) which varied the moisture content according to whether the sediments were above or beneath mean tidal level and for average sea-level rise. The consistency of the OSL ages with expected time ranges suggests this correction worked well. As a result, each sample was converted to an index point and a relative sea-level curve, spanning the last 2800 years, was constructed showing variations in levels between −200 cm and +60 cm.

Compaction may also take place during sampling, again affecting moisture and dose rates if no corrections were applied. The work of Madsen et al. (2005) attempted to modify measured dose rates in order to account for the effects of sampling compaction by looking at the relation of sample bulk density with depth. By using a linear accumulation model, they also corrected for the effects of sampling very shallow estuarine samples (less than 60 cm depth) which would have received a non-infinite matrix gamma-dose for a considerable percentage of the burial history. As a result, Madsen et al. (2007) were able to demonstrate good agreement between OSL ages and those derived from ^{210}Pb for tidally derived sediments which ranged in age from a few years up to one thousand years.

An alternative and novel attempt to circumvent some of the issues with natural and sampling-induced changes in dose rates within marsh sediment was undertaken by Davids et al. (2010). They dated both feldspars and quartz and attributed the difference between the paleodoses to the internal K dose within feldspars, which is unlikely to have changed through time. Where the paleodose was over 1 Gy, ages measured by this subtraction method were consistent with independent radiocarbon ages, although the errors associated with the former were much larger than the more conventional luminescence ages.

27.3.4 Offshore sites

Some of the initial work looking at establishing luminescence as a dating technique was carried out on marine sediments from Atlantic and Pacific deep-sea cores (e.g., Wintle and Huntley, 1980). Establishing whether sediment in this depositional context could have been exposed to sunlight at some point before final burial on the ocean floor was key. Working on Arctic ocean cores using TL and IRSL measurements of fine-grained material, Berger (2006) found that bleaching was highly variable. As a result, he reported many results overestimating those determined by other independent methods. This has not been the case everywhere, however. Bitinas et al. (2001) reported IR-OSL results from marine samples from the Baltic which showed good agreement with intercalated radiocarbon ages. This was attributed to sand-dominated sedimentation taking place under low-energy conditions where full bleaching of the luminescence signal could take place.

Using the multi-grain SAR approach on coarse-grained samples from the central Arctic Ocean, Jakobsson et al. (2003) were successful in obtaining MIS 5 ages (~110 ka) from some of the samples in line with expectations. However, some samples clearly overestimated their expected stratigraphic ages. Likewise, using single grains of quartz extracted from a core collected offshore of Western Australia in the Indian Ocean, Olley et al. (2004) produced OSL ages in the range of 1.8–51 ka which are in good agreement with radiocarbon ages. Even with this study however some samples were found to have been incompletely bleached within this core, despite the fact that the Australian quartz is very well behaved (see Section 27.3.1) and thought to have originated from aeolian sources prior to deposition offshore.

Working in the English Channel using OSL on fine-sand-sized quartz grains, Mellet et al. (2012) were able to show, through analysis of replicates, that while coastal sediments (nearshore beach and washover fan sediments) were incompletely bleached (with large scatter on the replicates), samples taken from shallow marine settings were well bleached (with high reproducibility). This was against expectations as the high swash–backwash processes found in the English Channel were expected to have fully bleached the coastal sediments. Bleaching away of the luminescence signal prior to burial of sediments in marine settings would appear not only to be variable in time and space but also unpredictable based on stratigraphic or sedimentological interpretations of depositional environments. Having multiple downcore samples, as well as performing replicate luminescence measurements on individual samples, may help to isolate poorly bleached samples.

While offshore luminescence samples can be assumed to be 100% water saturated, thereby avoiding some of the issues outlined above, moisture contents through time may still change due to compaction (see Section 27.3.3) and, depending on accumulation rates, bioturbation. In addition, luminescence samples from offshore cores may also have suffered from time-dependent changes in the dose rate. This is due to the well-known disequilibria in the U^{238} decay series in carbonaceous marine sediments with increases in the amount of Th^{230} and decay of U^{234} over time. Using high-resolution gamma spectrometry, Jakobsson et al. (2003) reported little excess of Th^{230} in samples from the Arctic. In contrast, Olley et al. (2004)

did find disequilibria, but used an iterative model to successful overcome time-dependent changes in Th^{230} and U^{234}. This approach decreased the estimated sample dose rates by 1–5% when compared to those measured at the present day.

In summary, luminescence ages are possible for offshore sediments, but additional care is required both in terms of extra measurements to establish whether uranium disequilibria is an issue as well as extra corrections for this and moisture and overburden changes. Note that, even with additional care, samples may still overestimate true age due to incomplete bleaching prior to burial.

27.4 CONCLUSIONS

A considerable number of studies have now shown the utility of the application of luminescence dating to coastal and marine sediments in order to better understand the timing of sea-level changes. While coastal dunes, beaches, and beach ridges are the most optimal environments for the technique, successful application to intertidal and offshore records has also been achieved. Methodological advances are still pushing the limits of luminescence dating back to older dates. Critical to all applications of luminescence which desire accurate chronological information is the understanding how different marine and coastal settings may have affected the luminescence signal and dose rate, and how these may be avoided, mitigated against, or overcome.

REFERENCES

Aitken, M.J. (1998) *An Introduction to Optical Dating: The Dating of Quaternary Sediments by the use of Photo-Stimulated Luminescence.* Oxford Science Publication, Oxford.

Andreucci, S., Clemmensen, L.B., Murray, A.S., and Pascucci, V. (2010) Middle to late Pleistocene coastal deposits of Alghero, northwest Sardinia (Italy): Chronology and evolution. Quaternary International, 222, 3–16.

Andreucci, S., Bateman, M.D., Zucca, C., Kapur, S., Akşit, I., Dunajko, A., and Pascucci, V. (2012) Evidence of Saharan dust in Upper Pleistocene reworked palaeosols of Northwest Sardinia, Italy: Palaeoenvironmental implications. Sedimentology, 59, 917–938.

Armitage, S.J., Botha, G.A., Duller, G.A.T., Wintle, A.G., Rebêlo, L.P., and Momadeet, F.J. (2006) The formation and evolution of the barrier islands of Inhaca and Bazaruto, Mozambique. Geomorphology, 82, 295-308.

Arnold L.J., and Roberts, R.G. (2009) Stochastic modelling of multi-grain equivalent dose (De) distributions: Implications for OSL dating of sediment mixtures. Quaternary Geochronology 4, 204–230.

Ballarini, B., Wallinga, J., Murray, A.S., van Heteren, S., Oost, A.P., Bos, A.J.J., and van Eijk C.W.E. (2003) Optical dating of young coastal dunes on a decadal time scale. Quaternary Science Reviews, 22, 1011–1017.

Banerjee, D., Hildebrand A.N., Murray-Wallace, C.V., Bourman, R.P., Brooke, B.P., and Blaira, M. (2003) New quartz SAR-OSL ages from the stranded beach dune sequence in south-east South Australia. Quaternary Science Reviews, 22, 1019–1025.

Bateman, M.D., and Catt, J.A. (1996) An absolute chronology for the raised beach and associated deposits at Sewerby E. Yorkshire, England. Journal of Quaternary Science, 11, 389–395.

Bateman, M.D., and Murton, J.B. (2006) The chronostratigraphy of late Pleistocene glacial and periglacial aeolian activity in the Tuktoyaktuk coastlands, NWT, Canada. Quaternary Science Reviews, 25, 2552–2568.

Bateman, M.D., Frederick, C.D., Jaiswal, M.K., and Singhvi, A.K. (2003) Investigations into the potential effects of pedoturbation on luminescence dating. Quaternary Science Reviews, 22, 1169–1176.

Bateman, M.D., Boulter, C.H., Carr, A.S., Frederick, C.D., Peter, D., and Wilder, M. (2007) Detecting Post-depositional sediment disturbance in sandy deposits using optical luminescence. Quaternary Geochronology, 2, 57–64.

Bateman, M.D., Carr, A.S., Murray-Wallace, C.V., Roberts, D.L., and Holmes, P.J. (2008) A dating intercomparison study on Late Stone Age coastal midden deposits, South Africa. Geoarchaeology, 23, 715–741.

Bateman, M.D., Carr, A.S., Dunajko, A., Holmes, P.J., Roberts, D.L., McLaren, S.J., Bryant, R.G., Marker, M.E., and Murray-Wallace, C.V. (2011) The evolution of coastal barrier systems: A case study of the Middle-Late Pleistocene Wilderness barriers, South Africa. Quaternary Science Reviews, 30, 63–81.

Benedetti, M.M., Haws, J.A., Funk, C.L., Daniels, J.M., Hesp, P.A., Bicho, N.F., Minckley, T.A., Ellwood, B.B., and Forman, S.L. (2009) Late Pleistocene raised beaches of coastal Estremadura, central Portugal. Quaternary Science Reviews, 28, 3428–3447.

Berger, G.W. (2006) Trans-arctic-ocean tests of fine-silt luminescence sediment dating provide a basis for an additional geochronometer for this region. Quaternary Science Reviews, 25, 2529–2551.

Bitinas, A., Damušyte, A., Hütt, G., Jaek, I., and Kabailiene, M. (2001) Application of the OSL dating for stratigraphic correlation of Late Weichselian and Holocene sediments in the Lithuanian Maritime Region. Quaternary Science Reviews, 20, 767–772.

Bjørnsen, M., Clemmensen, L.B., Murray, A., and Pedersen, K. (2008) New evidence of the Littorina transgressions in the Kattegat: Optically Stimulated Luminescence dating of a beach ridge system on Anholt, Denmark. Boreas, 37, 157–168.

Boomer, I., and Horton, B.P. (2006) Holocene relative sea-level movements along the North Norfolk Coast, UK. Palaeogeography, Palaeoclimatology, Palaeoecology, 230, 32– 51.

Bristow, C.S., and Pucillo, K. (2006) Quantifying rates of coastal progradation from sediment volume using GPR and OSL: the Holocene fill of Guichen Bay, south-east South Australia. Sedimentology, 53, 769–788.

Brooke, B., Ryan, D., Pietsch, T., Olley, J., Douglas, G., Packett, R., Radke, L., and Flood, P. (2008) Influence of climate fluctuations and changes in catchment land use on Late Holocene and modern beach-ridge sedimentation on a tropical macrotidal coast: Keppel Bay, Queensland, Australia. Marine Geology, 251, 195–208.

Burdette, K.E., Rink, W.J., Lopez, G.I., Mallinson, D.J., Parham, P.R., and Reinhardt, E.G. (2012) Geological investigation and optical dating of Quaternary siliciclastic sediments near Apalachicola, North-west Florida, USA. Sedimentology, 59, 1836–1849.

Carr, A.S., Bateman, M.D., Roberts, D.L., Murray-Wallace, C.V., Jacobs, Z., and Holmes, P.J. (2010) The last interglacial sea-level high-stand on the southern Cape coastline of South Africa: optically stimulated luminescence and amino acid racemization chronologies. Quaternary Research, 73, 351–363.

Clemmensen, L.B., Bjørnsen, M., Murray, A., and Pedersen, K. (2007) Formation of aeolian dunes on Anholt, Denmark since AD 1560: A record of deforestation and increased storminess. Sedimentary Geology, 199, 171–187.

Davids, F., Duller, G.A.T., and Roberts, H.M. (2010) Testing the use of feldspars for optical dating of hurricane overwash deposits. Quaternary Geochronology, 5, 125–130.

Dörschner, N., Reimann, T., Wenske, D., Lüthgens, C., Tsukamoto, S., Frechen, M., and Böse, M. (2012) Reconstruction of the Holocene coastal development at Fulong Beach in north-eastern Taiwan using optically stimulated luminescence (OSL) dating. Quaternary International, 263, 3–13.

Duller, G.A.T. (2008) Single-grain optical dating of Quaternary sediments: why aliquot size matters in luminescence dating. Boreas, 37, 589–612.

Duller, G.A.T., and Murray, A.S. (, 2000) Luminescence dating of sediments using individual mineral grains. Geologos, 5, 88–106.

Duller, G.A.T., and Wintle, A.G. (2012) A review of the thermally transferred optically stimulated luminescence signal from quartz for dating sediments. Quaternary Geochronology, 7, 6–20.

Fornos, J.J., Clemmensen, L.B., Gomez-Pujol, L., and Murray, A.S. (2009) Late Pleistocene carbonate aeolianites on Mallorca, Western Mediterranean: a luminescence chronology. Quaternary Science Reviews, 28, 2697–2709.

Forsyth, A.J., Nott, J., and Bateman, M.D. (2010) Beach ridge plain evidence of a variable late-Holocene tropical cyclone climate, North Queensland, Australia. Palaeogeography, Palaeoclimatology, Palaeoecology, 297, 707–716.

Forsyth, A.J., Nott, J., Bateman, M.D., and Beaman, R.J. (2012) Juxtaposed beach ridges and foredunes within a ridge plain, Wonga Beach, northeast Australia. Marine Geology, 307–310, 111–116.

Fruergaard, M., Andersen, T.J., Nielsen, L.H., Madsen, A.T., Johannessen, P.N., Murray, A.S., Kirkegaard, L., and Pejrup, M. (2011) Punctuated sediment record resulting from channel migration in a shallow sand-dominated micro-tidal lagoon, Northern Wadden Sea, Denmark. Marine Geology, 280, 91–104.

Giannini, P.C.F., Sawakulchi, A.O., Martinho, C.T., and Tatumi, S.H. (2007) Eolian depositional episodes controlled by Late Quaternary relative sea level changes on the Imbituba-Laguna coast (southern Brazil) Marine Geology, 237, 143–168.

Jacobs, Z., and Roberts, D.L. (2009) Last Interglacial Age for aeolian and marine deposits and the Nahoon fossil human footprints, Southeast Coast of South Africa. Quaternary Geochronology, 4, 160–169.

Jacobs, Z., Richard, G., Roberts, R.G., Lachlan, T.J., Karkanas, P., Marean, C.W., and Roberts, D.L. (2011) Development of the SAR TT-OSL procedure for dating Middle Pleistocene dune and shallow marine deposits along the southern Cape coast of South Africa. Quaternary Geochronology, 6, 491–513.

Jakobsson, M., Backman, J., Murray, A.S., and Løvlie, R. (2003) Optically stimulated luminescence dating support central Arctic Ocean cm-scale sedimentation rates. Geochemistry, Geophysics, Geosystems, 4, Article no. 1016.

Lopez, G.I., and Rink, W.J. (2008) New quartz optical stimulated luminescence ages for beach ridges on the St. Vincent Island Holocene strandplain, Florida, United States. Journal of Coastal Research, 24, 49–62.

Madsen, A.T., Murray, A.S., Andersen, T.J., Pejrup, M., and Breuning-Madsen, H. (2005) Optically stimulated luminescence dating of young estuarine sediments: a comparison with ^{210}Pb and ^{137}Cs dating. Marine Geology, 214, 251–268.

Madsen, A.T., Murray, A.S., Andersen, T.J., and Pejrup, M. (2007) Temporal changes of accretion rates on an estuarine salt marsh during the late Holocene: Reflection of local sea level changes? The Wadden Sea, Denmark. Marine Geology, 242, 221–23.

Madsen, A.T., Murray, A.S., Jain, M., Andersen, T.J., and Pejrup, M. (2011) A new method for measuring bioturbation rates in sandy tidal flat sediments based on luminescence dating. Estuarine, Coastal and Shelf Science, 92, 464–471.

Mallinson, D.J., Smith, C., Mahan, S., Culver, S., and McDowell, K. (2011) Barrier island response to late Holocene climate events, North Carolina, USA. Quaternary Research, 76, 46–57.

Mauz, B., and Bungenstock, F. (2007) How to reconstruct trends of late Holocene relative sea level: A new approach using tidal flat clastic sediments and optical dating. Marine Geology, 237, 225–237.

Mauz, B., Baeteman, C., Bungenstock, F., and Plater, A.J. (2010) Optical dating of tidal sediments: Potentials and limits inferred from the North Sea coast. Quaternary Geochronology, 5, 667–678.

Mellett, C.L., Mauz, B., Plater, A.J., Hodgson, D.M., and Lang, A. (2012) Optical dating of drowned landscapes: A case study from the English Channel. Quaternary Geochronology, 10, 201–208.

Murray, A.S., and Clemmensen, L.B. (2001) Luminescence dating of Holocene aeolian sand movement, Thy, Denmark. Quaternary Science Reviews, 20, 751–754.

Murray A.S., and Wintle A.G. (2003) The single aliquot regenerative dose protocol: potential for improvements in reliability. Radiation Measurements, 37, 377–381.

Murray-Wallace, C.V. (2002) Pleistocene coastal stratigraphy, sea-level highstands and neotectonism of the southern Australian passive continental margin: a review. Journal of Quaternary Science, 17, 469–489.
Murray-Wallace, C.V., Banerjee, D., Bourman, R.P., Olley, J.M., and Brooke, B.P. (2002) Optically stimulated luminescence dating of Holocene relict foredunes, Guichen Bay, South Australia. Quaternary Science Reviews, 21, 1077–1086.
Nathan, R.P., and Mauz, B. (2008) On the dose-rate estimate of carbonate-rich sediments for trapped charge dating. Radiation Measurements, 43, 14–25.
Nielsen, A., Murray, A.S., Pejrupa, M., and Elberling, B. (2006) Optically stimulated luminescence dating of a Holocene beach ridge plain in Northern Jutland, Denmark. Quaternary Geochronology, 1, 305–312.
Nott, J., Smithers, S., Walsh, K., and Rhodes, E. (2009) Sand beach ridges record 6000 year history of extreme tropical cyclone activity in northeastern Australia. Quaternary Science Reviews, 28, 1511–1520.
Olley, J.M., De Deckker, P., Roberts, R.G., Fifield, L.K., Yoshida, H., and Hancock, G. (2004) Optical dating of deep-sea sediments using single grains of quartz: A comparison with radiocarbon. Sedimentary Geology, 169, 175–189.
Otvos, E.G. (2012) Coastal barriers: Nomenclature, processes, and classification issues. Geomorphology, 139, 39–52.
Plater, A.J., and Poolton, N.R. (1992) Interpretation of Holocene sea level tendency and intertidal sedimentation in the Tees estuary using sediment luminescence techniques: a viability study. Sedimentology, 39, 1–15.
Porat, N., and Botha, G. (2008) The luminescence chronology of dune development on the Maputaland coastal plain, southeast Africa. Quaternary Science Reviews, 27, 1024–1046.
Prescott, J.R., and Hutton, J.T. (1994) Cosmic ray contributions to dose rates for luminescence and ESR dating: large depths and long-term time variations. Radiation Measurements, 2/3, 497–500.
Reimann, T., Tsukamoto, S., Harff, J., Osadczuk, K., and Frechen, M. (2011) Reconstruction of Holocene coastal foredune progradation using luminescence dating: An example from the Swina barrier (southern Baltic Sea, NW Poland). Geomorphology, 132, 1–16.
Richardson, C.A. (2001) Residual luminescence signals in modern coastal sediments. Quaternary Science Reviews, 20, 887–892.
Rink, W.J., and Forrest, B. (2005) Dating evidence for the accretion history of beach Ridges on Cape Canaveral and Merritt Island, Florida, USA. Journal of Coastal Research, 21, 1000–1008.
Rink, W.J., and Lopez, G.I. (2010) OSL-based lateral progradation and aeolian sediment accumulation rates for the Apalachicola Barrier Island Complex, North Gulf of Mexico, Florida. Geomorphology, 123, 330–342.
Roberts, R.G., Galbraith, R.F., Olley, J.M., Yoshida, H., and Laslett, G.M. (1999) Optical dating of single and multiple grains of from Jinmium rock shelter, northern Australia: Part II. Results and implications. Archaeometry, 41, 365–395.
Roberts, D.L, Bateman, M.D., Murray-Wallace, C.V., Carr, A.S., and Holmes, P.J. (2008) Last Interglacial fossil elephant trackways dated by OSL/AAR in coastal aeolianites, Still Bay, South Africa. Palaeogeography, Palaeoclimatology, Palaeoecology, 257, 261–279.
Roberts, D.L., Bateman, M.D., Murray-Wallace, C.V., Carr, A.S., and Holmes, P.J. (2009) West coast dune plumes: Climate driven contrasts in dunefield morphogenesis along the western and southern South African coasts. Palaeogeography, Palaeoclimatology, Palaeoecology, 271, 24–38.
Roberts, H.M. and Plater, A.J. (2007) Reconstruction of Holocene foreland progradation using optically stimulated luminescence (OSL) dating: an example from Dungeness, UK. The Holocene, 17, 495–505.
Sivan, D., and Porat, N. (2004) Evidence from luminescence for Late Pleistocene formation of calcareous aeolianite (kurkar) and paleosol (hamra) in the Carmel Coast, Israel. Palaeogeography, Palaeoclimatology, Palaeoecology, 211, 95–106.
Szkornik, K., Gehrels, W.R., and Murray, A.S. (2008) Aeolian sand movement and relative sea-level rise in Ho Bugt, western Denmark, during the 'Little Ice Age'. Holocene, 18, 951–965.
Tamura, T., Saito, Y., Lap Nguyen, V., Oanh Ta, T.K., Bateman, M.D., Matsumoto, D., and Yamashita, S. (2012) Origin and evolution of interdistributary delta plains; insights from the Mekong River delta. Geology, 40, 303–306.
Thomas, P.J., Murray, A.S., Granja, H.M., and Jain, M. (2008) Optical dating of late quaternary coastal deposits in northwestern Portugal. Journal of Coastal Research, 24, 134–144.
van Heteren, S., van de Plassche, O., and Lubberts, R.K., (2000) Optical dating of dune sand for the study of sea-level change. Geology, 28, 411–414.
Wintle, A.G., and Huntley, D.J. (1980) Thermoluminescence dating of ocean sediments. Canadian Journal of Earth Sciences, 17, 348–360.

Part 4

Modeling

Chapter 28
Glacial isostatic adjustment

GLENN A. MILNE
Department of Earth Sciences, University of Ottawa, Ottawa, Ontario, Canada

28.1 INTRODUCTION

A variety of processes contribute to sea-level change (see Chapter 2). On 1–100 ka timescales during the Quaternary, large-scale changes in the spatial distribution of land ice through the growth and melting of continental-scale ice sheets were the primary control on relative sea-level (RSL) change. Reconstructed sea-level changes from this period reflect fluctuations in the volume of land ice as well as the response of the solid Earth and gravity field to the climate-driven surface ice-water mass redistribution. The latter is known as glacial isostatic adjustment (GIA) (e.g., Milne and Shennan, 2013) and is the focus of this chapter.

The influence of GIA on past sea levels is illustrated in Figure 28.1 which shows RSL reconstructions from three localities, each distinct in terms of its distance from a major glaciation centre of the last glacial maximum (LGM). The influence of GIA is most evident at locations that were once ice covered (so-called "near-field" regions), where the ice-induced deformation is largest. This is illustrated by the RSL reconstruction from Angerman River in Sweden, which shows a monotonic fall in sea level associated mainly with seafloor uplift following the deglaciation of the Fennoscandian ice sheet. This type of RSL curve is typical for locations located near the centre of previously glaciated regions such as the Gulf of Bothnia in Fennoscandia and Hudson Bay in Canada. Peripheral to these regions, the influence of GIA is still relatively large but the rates of change can be close to those associated with glacial meltwater influx, which can lead to more complex RSL curves (e.g., that for Arisaig, Scotland in Fig. 28.1). The more complex form of this curve is largely a consequence of local seafloor uplift associated with the deglaciation of the British–Irish ice sheet dominating during the early and late parts of the record causing a sea-level fall and meltwater-driven sea-level rise dominating between ~11 and 6 cal. ka BP (Lambeck, 1993). At greater distances from past and present glaciation centers, the influence of GIA becomes less evident, particularly that associated with ice loading. For example, the sea-level curve from Barbados shows a monotonic sea-level rise following the end of the LGM, which is believed to be dominated by the influx of glacial meltwater to the oceans. Even though the meltwater signal dominates in areas located thousands of kilometers from major glaciation centres (so-called "far-field" regions), GIA can contribute significantly to the observed signal (as shown in the following section), making the interpretation of these data for changes in land ice volume non-trivial.

Given the influence of GIA on sea-level changes during the Quaternary, an understanding of this process is required in order to interpret the RSL changes reconstructed from field observations. As in most scientific disciplines, this understanding is advanced and tested by comparing the output from models that represent the physics of the underlying processes to observations that are sensitive to these processes. In the case of GIA, reconstructions of past sea-level changes are the dataset that has been the most commonly used for this purpose. In Section 28.2, some underlying theory and aspects of GIA models are briefly presented. Section 28.3 focuses on data-model comparison and the consideration of how GIA sea-level models have been applied to interpret RSL reconstructions, test model accuracy, and infer model parameters.

28.2 THEORY AND MODEL COMPONENTS

To construct a model of RSL change caused by volume fluctuations in land ice, it is first necessary to define RSL and consider the relevant underlying processes that would cause changes in this quantity. At any position in the oceans, RSL is the

Handbook of Sea-Level Research, First Edition. Edited by Ian Shennan, Antony J. Long, and Benjamin P. Horton.

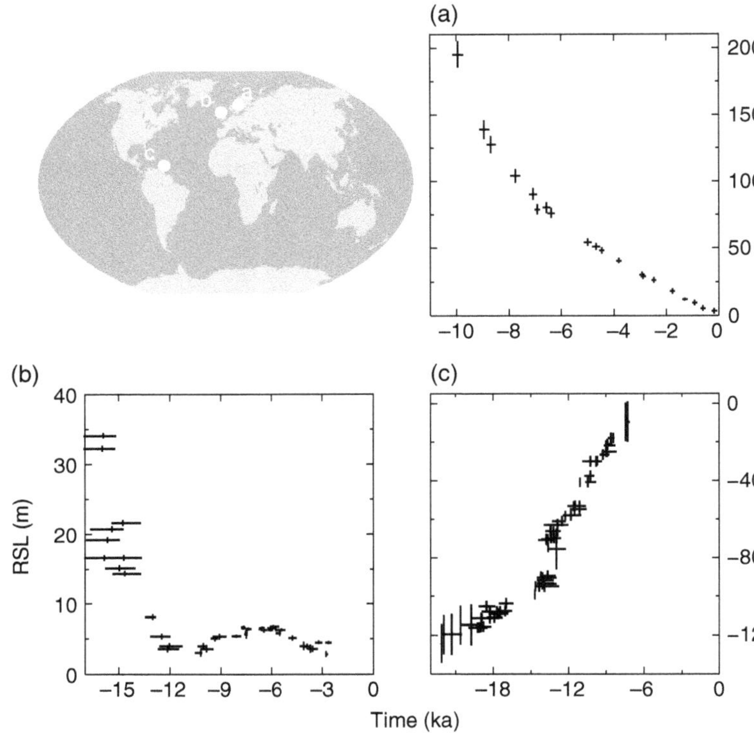

Fig. 28.1. RSL reconstructions from three locations: (a) Angerman River, Sweden (Lambeck et al., 1998); (b) Arisaig, Scotland (Shennan et al., 2005); and (c) Barbados (Fairbanks, 1989; Bard et al., 1990). These locations illustrate how the sea-level response is a strong function of distance from ancient glaciation centers.

height of the ocean surface relative to the ocean floor. RSL is therefore zero at coastal areas where these two surfaces intersect. Of course, this intersection is central to reconstructing changes in RSL from the geological record. By convention, these reconstructions define a change in RSL from a past time relative to the present at a specific location. In order to simulate these reconstructions, a GIA sea-level model must compute changes in both ocean surface and ocean floor height through time due to the mass loading associated with changes in land ice extent and sea-level change. A brief discussion of the relevant GIA processes and how they are modeled is presented in the following.

Changes in the vertical position of the ocean floor are directly due to isostatic deformation of the solid Earth. Changes in the height of the ocean surface due to volume changes in land ice are produced by two processes: (1) changes in the Earth's gravity field due to the redistribution of surface ice/water mass and solid Earth deformation associated with this surface mass redistribution; and (2) the influx of glacial meltwater to the oceans. The former is a result from the fact that the equilibrium ocean surface must lie on an equipotential of the Earth's gravity field. The latter leads to changes in the amount of water in the oceans and, when averaged over the ocean basins, defines what is known as glacio-eustatic (henceforth eustatic) or ice-equivalent sea-level change:

$$\Delta SL_{\mathrm{EUS}}(t) = -\frac{\rho_{\mathrm{I}} V_{\mathrm{I}}(t)}{\rho_{\mathrm{W}} A_{\mathrm{O}}} \qquad (28.1)$$

where $V_{\mathrm{I}}(t)$ is the volume change in grounded continental ice at time t relative to some reference time, A_{O} is ocean area and ρ_{I} and ρ_{W} are the density of ice and water, respectively. A more accurate version of eustatic sea-level change includes the variation in ocean area over time (e.g., Lambeck et al., 2000):

$$\Delta SL_{\mathrm{EUS}}(t) = \int_0^t \frac{\dot{V}_{\mathrm{I}}(t')}{A_{\mathrm{O}}(t')} \mathrm{d}t'. \qquad (28.2)$$

where the dot above V_{I} indicates the time derivative of this quantity. Over relatively short timescales (several decades), the meltwater influx also influences RSL through perturbing ocean density structure and thus the dynamical (flow) response (e.g., Kopp et al., 2010). This short-term component of the RSL response is not simulated in GIA models.

Farrell and Clark (1976) were the first to present a sea-level theory that accounted for both Earth deformation and changes in the gravity field caused by mass loss/gain by land ice. They presented this theory in the form of an integral equation (an equation in which the solution depends on itself), which has become known as the "sea-level equation":

$$\Delta SL(\theta,\psi,t)=\left(\frac{\rho_{\mathrm{I}}}{g}\Phi_{\mathrm{I}}*I\right)+\left(\frac{\rho_{\mathrm{W}}}{g}\Phi_{\mathrm{O}}*\Delta SL\right)+C_{SL}(t) \quad (28.3)$$

where g is the gravitational acceleration, θ is the co-latitude, and ψ is the longitude of the sea-level change. I and ΔSL are functions that define changes in grounded ice thickness and ocean depth, respectively, and Φ is a so-called sea-level Green's function that, when convolved in space and time (represented by *) with a surface load, defines the contributions of that load history to ocean depth change. The subscripts I and O indicate that the spatial convolutions are performed over ice and ocean area, respectively. The final term on the right-hand side of Equation 28.3, $C_{SL}(t)$, represents a spatially uniform height shift of the ocean surface and is added to ensure conservation of ice-ocean water mass. Integrating Equation 28.3 over ocean area (assumed constant in Farrell and Clark, 1976) and then solving for $C_{SL}(t)$ yields:

$$C_{SL}(t)=-\frac{\rho_{\mathrm{I}}V_{\mathrm{I}}(t)}{\rho_{\mathrm{W}}A_{\mathrm{O}}}-\frac{1}{A_{\mathrm{O}}}\left\langle\left(\frac{\rho_{\mathrm{I}}}{g}\Phi_{\mathrm{I}}*I\right)+\left(\frac{\rho_{\mathrm{W}}}{g}\Phi_{\mathrm{O}}*\Delta SL\right)\right\rangle. \quad (28.4)$$

The first term in Equation 28.4 is the eustatic change as defined in Equation 28.1; the second term is the contribution of GIA-driven changes in sea surface and ocean floor height integrated over the ocean area. This latter contribution is termed "syphoning" (Mitrovica and Milne, 2002).

In order to solve the sea-level equation, two model inputs must be defined: (1) space–time evolution of grounded land ice; and (2) the density and rheology of the solid Earth (which are required to compute the sea-level Green's function Φ in Equation 28.3). Regarding the latter, information on the density and elastic properties of the Earth are adopted from seismic models (e.g., Dziewonski and Anderson, 1981). The non-elastic component of the solid Earth deformation, governed by the viscous structure of the Earth model, is often treated as a free parameter set to be inferred when fitting GIA-related data such as RSL (see the following section). In the majority of previous GIA analyses, the Earth model response to the ice-ocean surface loading was computed by considering relatively simplistic, spherically symmetric, Maxwell Earth models (Peltier, 1974). These models are simplistic in the sense that they do not incorporate lateral variations in Earth properties or departures from linear (stress proportional to strain rate) deformation of mantle rock, both of which are known to exist in the actual Earth. Some implications related to these Earth model simplifications are discussed in Section 28.3.

The ice model component of GIA models is developed by considering a number of different constraints, including inferences of lateral and vertical ice extent from the interpretation of erosional and depositional features in the landscape, as well as the application of analytical and numerical models of ice-sheet shape and evolution. The past reconstruction of ice sheets is described more fully in Section 28.3. Figure 28.2 illustrates the key elements of a GIA sea-level model.

A few different methods have been applied to solve the sea-level equation. Early analyses employed a finite element approach by representing the ice and ocean loads as a sum of disks (e.g., Tushingham and Peltier, 1991, figs 1 and 2) in order to compute the spatial convolutions in the sea-level equation. However, an alternative approach known as the pseudo-spectral method, which involves representing spatial fields as a weighted sum of spherical harmonic functions, is now the most commonly applied for the case of spherically symmetric Earth models (see Mitrovica and Peltier, 1991 for details). In recent years, Earth models have been applied that incorporate lateral variations in model parameters. These are based on finite element or finite volume methods and so the convolutions in the sea-level equation are computed in the space domain (e.g., Wu and van der Wal, 2003).

The original sea-level equation proposed by Farrell and Clark (1976) has been extended to include lateral shoreline migration due to the advance and retreat of ice margins grounded below sea level as well as marine regression and transgression in ice-free areas. A full discussion of this extension and comparison of different approaches applied to implement it are found in Mitrovica and Milne (2003). A second extension includes the effect of Earth rotation on GIA-induced RSL change. GIA produces changes in Earth rotation and these changes feedback on RSL and other GIA observables such as present-day land motion (e.g., Milne and Mitrovica, 1998). This feedback is dominated by a rotational change known as True Polar Wander (e.g., Sabadini et al.,

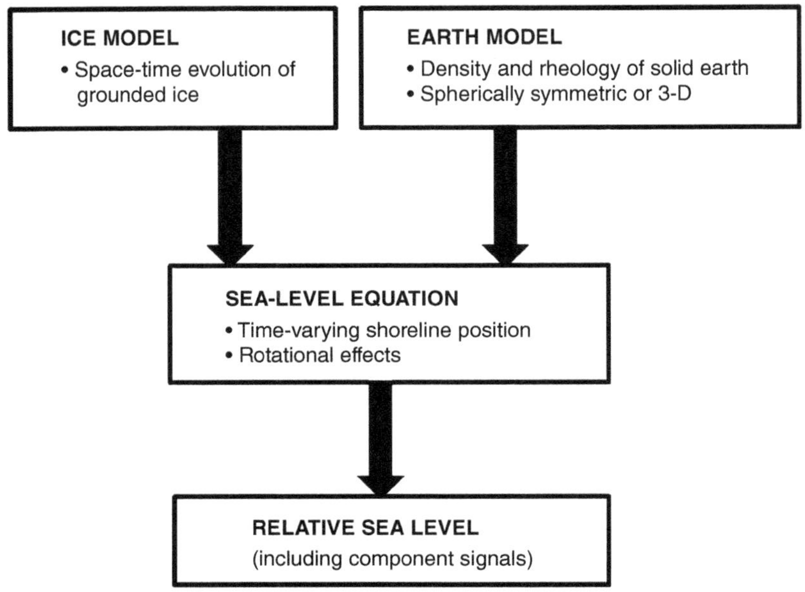

Fig. 28.2. Diagram illustrating basic components of a GIA sea-level model. There are two key inputs: a model of grounded ice extent in space and time and a model of solid Earth properties and structure. Once these two inputs are known, the sea-level equation can be solved, with or without a time-varying shoreline or rotational feedback, to compute changes in relative sea level as well as contributions to this signal (e.g., sea-surface height change or influence of ice-loading only).

1982) which results in a very long wavelength pattern on RSL change (described by a second-degree, first-order spherical harmonic function; Milne and Shennan, 2013, fig. 2). This rotation-induced signal is relatively small in amplitude (few meters or less) during deglaciation, but can be relatively large compared to other component GIA signals when considering the RSL response to ice mass changes over short timescales (decades to centuries). These short-timescale changes are known as sea-level fingerprints (e.g., Mitrovica et al., 2001).

Figure 28.3 depicts the model output of RSL at the three localities, for which reconstructed curves are shown in Figure 28.1. The modeled curves simulate the observed curves relatively well. They also provide the opportunity to better understand the processes leading to the observed changes. For example, the solid gray lines in each frame represent the contribution of vertical motion of the ocean floor to RSL. As would be expected, this contribution is largest at the site located near the center of the Fennoscandian ice sheet (Angerman River). The contribution of ocean floor height change at the other sites is considerably less. At Arisaig, it dominates early in the record but changes in sea-surface height also play an important role at this site, particularly during periods of rapid ice melting when RSL change becomes positive (e.g., at ~14 and ~11 ka). It is important to note that the sea-surface height changes (gray dashed lines) are not simply the eustatic curve; they also include the contributions from syphoning and changes in gravity due to surface ice-water mass redistribution and solid Earth deformation. For example, syphoning leads to a RSL fall of a few tens of meters during deglaciation, as does ice melting in areas near and adjacent to the major ice sheets due to the influence of mass loss on the regional gravity field. These processes are largely responsible for the relatively small sea-surface height changes at all three sites considered (including Barbados, which is not strictly a far-field site due to its relative proximity to the North American ice sheets). Indeed, it is quite surprising to note that the modeled RSL change at Barbados includes a larger contribution from ocean-floor subsidence rather than sea-surface rise since the LGM.

28.3 MODEL RESULTS AND APPLICATIONS

Clark et al. (1978) were the first to compare predictions of sea-level change based on the sea-level equation to a global distribution of observations.

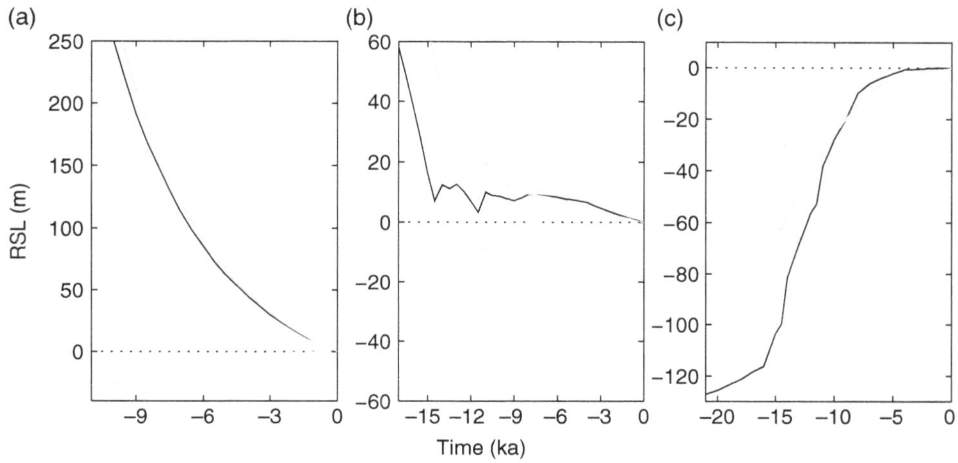

Fig. 28.3. Model output of RSL at the three localities considered in Fig. 1: (a) Angerman River; (b) Arisaig; and (c) and Barbados. The black lines show RSL at each site and the gray solid and gray dashed lines show the contributions to the RSL signal from relative land height change and sea-surface height change, respectively. Note that negative relative land height change is shown to directly indicate the contribution of vertical motion of the Earth's solid surface to RSL (e.g., land uplift contributes to a RSL fall). These model results are based on the ICE5G model of global ice evolution and the VM2 Earth viscosity model (with a 90 km thick lithosphere) (Peltier, 2004).

Their primary aim in doing so was to determine how well the model could account for the spatial variation in the reconstructed RSL curves. In this regard, the comparison was successful as the predictions captured many aspects of the complex variation recorded in the observations. This was an important result as it provided strong support that the GIA process has dominated secular sea-level changes during the Quaternary. Based on their model results, Clark et al. (1978) mapped out six regions, each characterized by a distinct sea-level history (Fig. 28.4); this map has served as a useful resource for field scientists for many decades.

The data-model comparisons in Clark et al. (1978) and other early papers (e.g., Peltier et al., 1978) mark the beginning of modern GIA analyses that solve the sea-level equation on global (spherical) Earth models. Even though some of the data-model residuals are large in these early studies, the results demonstrated the potential of GIA models to simulate the observed RSL changes by varying model parameters, specifically, the Earth model viscosity structure and ice-loading history. GIA sea-level modeling is one of the primary methods used to infer information on large-scale Earth viscosity structure and the spatio-temporal evolution of the late Quaternary ice sheets which play a key role in our understanding of solid Earth evolution (i.e., plate tectonics) and past climate change, respectively. Data-model RSL comparisons to infer model parameter values are the focus of the remainder of this chapter. Earth viscosity structure, regional ice-sheet histories, global ice volume, geographic sources of rapid melt events, and projections of future sea-level change are discussed in the following sections.

28.3.1 Earth viscosity structure

Even prior to the modern era of GIA modeling which began in the 1970s, RSL observations had been used to infer the viscosity of the Earth's mantle (e.g., Haskell, 1935). These analyses estimated RSL change by assuming that it resulted from vertical ocean floor motion only based on the loading of a half-space Earth model; model accuracy was therefore considerably poorer compared to that for models which solve the sea-level equation on spherically symmetric Earth models. It is therefore quite remarkable that the inferences obtained in some of these early studies are similar to those obtained relatively recently (Mitrovica, 1996).

Inferences of Earth model viscosity structure have primarily been based on RSL reconstructions from previously glaciated regions, where vertical motion of the Earth's solid surface dominates the total RSL response (see Fig. 28.3). This structure is commonly defined as an outer layer of very high viscosity (to simulate the Earth's most mechanically competent region: the lithosphere) and two or more deeper layers to define viscosity variation

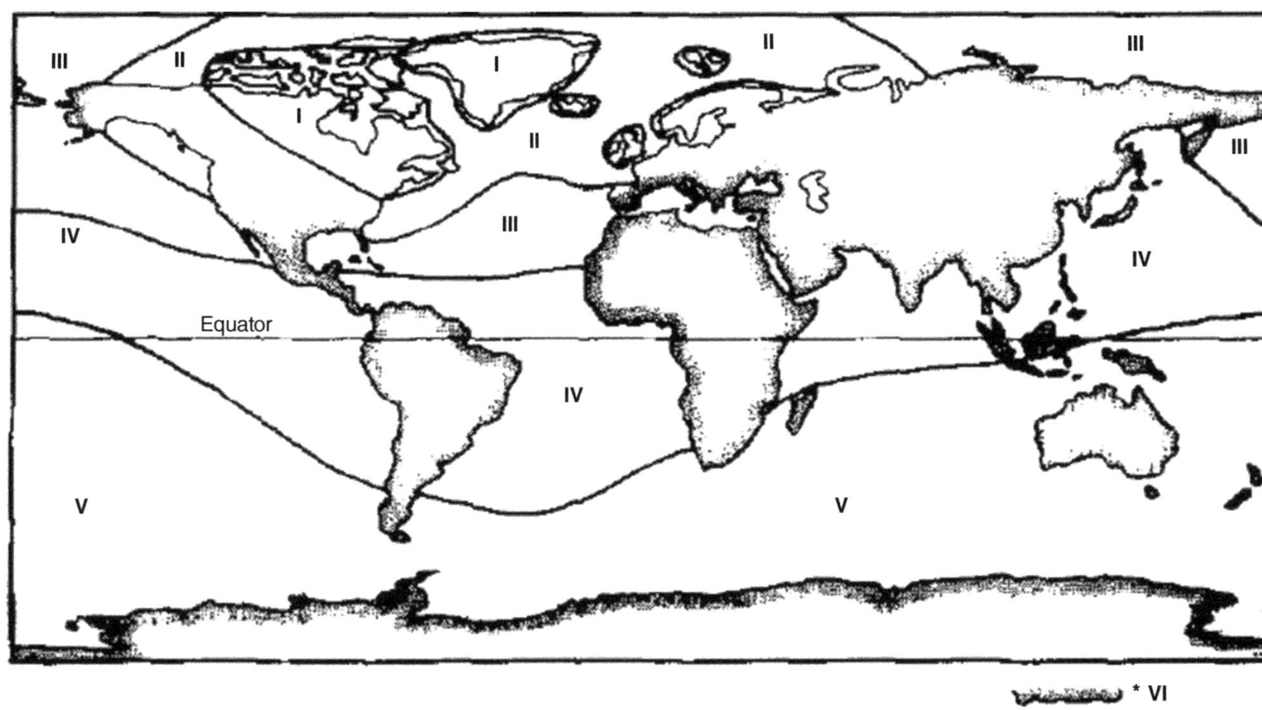

Fig. 28.4. Map illustrating the six zones in which the modeled RSL variations display similar forms. See Clark et al. (1978) for a discussion of the characteristic RSL curves for zones I-VI. *Source*: Clark et al. (1978). Reproduced with permission of Elsevier.

between the model lithosphere and the core–mantle boundary. One issue that was identified early on is that of non-uniqueness, that is, the possibility of fitting the data with more than one set of ice and Earth model parameters. This is a very common issue in the majority of geophysical modeling problems. In order to produce a robust inference of Earth viscosity structure, either the ice history must be known to a good degree of accuracy or the data must be relatively insensitive to uncertainty in this model input. One example of the latter is the so-called "inverse decay time" data parameterization (e.g., Walcott, 1980). This involves representing the RSL curves that display the characteristic exponential-like sea-level fall (e.g., Angerman River in Fig. 28.1) by a simple analytical equation with three parameters: A, τ, and C:

$$\mathrm{RSL}(t) = A\left[\exp\left(-\frac{t}{\tau}\right) - 1\right] + C. \qquad (28.5)$$

It has been shown that the parameter τ shows little sensitivity to changes in the ice-loading model; modeling this parameter, as opposed to the actual RSL curve, therefore produces a relatively robust inference of Earth viscosity structure (e.g., Mitrovica, 1996).

RSL data from mid- and low-latitude locations have also been used to infer Earth viscosity structure. At most of these locations, the isostatic deformation is dominated by the ocean load. In a manner similar to the non-uniqueness problem in modeling data from near-field regions, this application of far-field data requires that the eustatic function is well known (as this has a large impact on far-field RSL changes) or that the data being considered are insensitive to this component of the RSL change. This insensitivity can be achieved by considering spatial differences in RSL which are dominated by the ocean-floor (Earth deformation) contribution to the signal (e.g., Nakada and Lambeck, 1989).

When estimating the viscosity structure of the Earth through modeling RSL data, it is important to consider how well the model structure is actually constrained by the data. This relates to the non-uniqueness issue introduced above and is an important consideration in all geophysical inversion problems that aim to delineate spatial variations in subsurface Earth properties. It has been show that RSL data do not resolve viscosity structure particularly well, and that 2–3 layers beneath the lithosphere to depths of ~1200 km represent the limit of what these data can delineate (Mitrovica, 1996). This resolving power is governed primarily by the extent of the ice sheet and the viscosity structure within the mantle; for example, a larger ice sheet induces deformation at greater depths and so sea-level data from Hudson Bay in Canada have more sensitivity to deeper mantle structure than those from Fennoscandia (Mitrovica, 1996). In order to improve constraints on deep Earth structure, additional datasets such as plate velocities and large-scale gravity anomalies have been inverted simultaneously with RSL data (e.g., Mitrovica and Forte, 2004). While there remains considerable uncertainty in defining this function, most results indicate that the viscosity averaged over depth and latitude/longitude in the upper mantle (base of the lithosphere to 670 km depth) lies in the range ~$3–8 \times 10^{20}$ Pas and that the average in the lower mantle (670 km to the core–mantle boundary) is greater than this by up to two orders of magnitude.

Lateral variations in Earth viscosity structure are known to exist, based on differences in the inference of this structure from different regions and the results of inversions for 3D seismic velocity structure (e.g., Romanowicz, 2003). Regarding the former, the most extreme examples are studies from regions near active plate boundaries such as Iceland (e.g., Sigmundsson, 1991) and the northwest Pacific (e.g., James et al., 2009). The upper mantle viscosities inferred in these regions is generally 1–2 orders of magnitude lower than that inferred in cratonic regions (e.g., the Canadian Shield). This reflects the influence of geodynamic features such as mantle plumes (Iceland) and subducting oceanic lithosphere (northwest Pacific) on local and regional viscosity structure (see above references for a more detailed discussion). Lateral variations in seismic wave speeds are at least partly governed by variations in temperature (a primary control on the viscosity of rock). Spherical Earth models that can accommodate 3D viscosity structure in the mantle have been used to simulate RSL since the early 2000s. These studies have focused on quantifying the influence of lateral viscosity structure on simulating RSL changes (e.g., Wu and van de Wal, 2003) as well as the corresponding bias or error made when estimating radial viscosity structure using a 1D (spherically symmetric) Earth model (Paulson et al., 2007). While these more complex Earth models enable a higher degree of predictive accuracy, their application is limited due to the current uncertainty in defining viscosity variations in 3D. A common approach involves converting models of 3D seismic velocity structure to models of viscosity structure by applying a series of scaling relationships to go from perturbations in velocity to density, temperature, and finally viscosity (e.g., Latychev et al., 2005).

These scaling relationships generally ignore potential effects due to compositional variations. When this limitation and the fact that the 3D velocity structure is not well constrained in certain parts of the mantle are considered, it is clear that accurate and precise 3D models of mantle viscosity structure are some time away from being realized.

As noted above, the majority of previous RSL modeling studies assumed a linear (Maxwell) rheology for simulating mantle flow in which the deviatoric stress and strain rate are proportional. However, laboratory deformation experiments indicate that mantle materials often exhibit non-linear flow where the strain rate is proportional to the stress to some power (Karato, 2008). Departures from non-linear flow have been explored via the application of finite element models that can simultaneously incorporate linear and non-linear flow in a so-called composite rheology (van der Wal et al., 2010). Such studies indicate that adopting a composite rheology enables a statistically improved fit to RSL data when considering a small subset of ice models. While this increase in model complexity is supported by a large number of deformation experiments on Earth materials, it remains to be seen if GIA-related data are able to unequivocally demonstrate the dominant occurrence of non-linear flow.

An additional departure from linear Maxwell flow is transient creep, in which there is a period of transition between short-timescale elastic deformation and steady-state creep during which the viscosity of the material increases. It has been suggested that, due to the relatively short timescales and small strains associated with GIA, the deformational response is within this transient regime (Karato, 2008). A small number of studies have considered the influence of this process by modeling the GIA response of a more general rheological model (a standard linear solid or Burgers body) (e.g., Peltier et al., 1980). These studies demonstrate that this effect can be significant, but the GIA data are not able to discriminate between these more general models and those based on a simpler Maxwell rheology due to the uncertainty in rheological parameters.

The take-home message is that the spherically symmetric Maxwell viscoelastic models most commonly employed in GIA modeling of RSL data are a simplification of reality. Inferences of parameters such as mantle viscosity based on these models are therefore most likely biased.

28.3.2 Regional ice-sheet histories

As noted above, RSL data also provide valuable information on the past distribution and volume of grounded ice. Figure 28.5 shows the contribution of (a) ice loading and (b) ocean loading to RSL at 10 ka before present. A comparison of these indicates that ice loading is the dominant signal in previously glaciated regions; RSL reconstructions from these regions have therefore been commonly used to infer or reconstruct past ice-sheet histories.

Early ice-sheet reconstructions were largely based on morphological evidence of lateral ice extent (e.g., ICE1, Peltier and Andrews, 1976) with comparisons to RSL data only used as a check on the accuracy of the model (with no subsequent model updating). Tushingham and Peltier (1991) published the first global ice reconstruction (known as ICE-3G) that was tuned to fit both geomorphological observations of ice extent as well as RSL observations using an assumed Earth viscosity profile. This model was made available to the wider community and was adopted in a large number of GIA modeling studies throughout the 1990s and early 2000s until the release of the ICE-5G model (Peltier, 2004).

In contrast to the global approach illustrated by the ICE-xG sequence of models, other groups have focused on the development of higher-resolution models at the regional scale. For example, Kurt Lambeck and colleagues produced model reconstructions of the British–Irish and Fennoscandian ice sheets during the 1990s (e.g., Lambeck, 1993; Lambeck et al., 1998). The general approach taken is to first develop a starting ice model based on geomorphological constraints of margin extent and chronology and determine the thickness distribution by applying analytical expressions for the ice height profile assuming that the ice sheet was frozen to its bed (Cuffey and Paterson, 2010). This assumption usually results in an ice sheet that is too thick. RSL predictions based on the starting ice model are then generated using a range of different Earth viscosity models to examine data-model residuals. Based on these, a regional ice thickness scaling parameter is then used to find a final combination of ice model and Earth

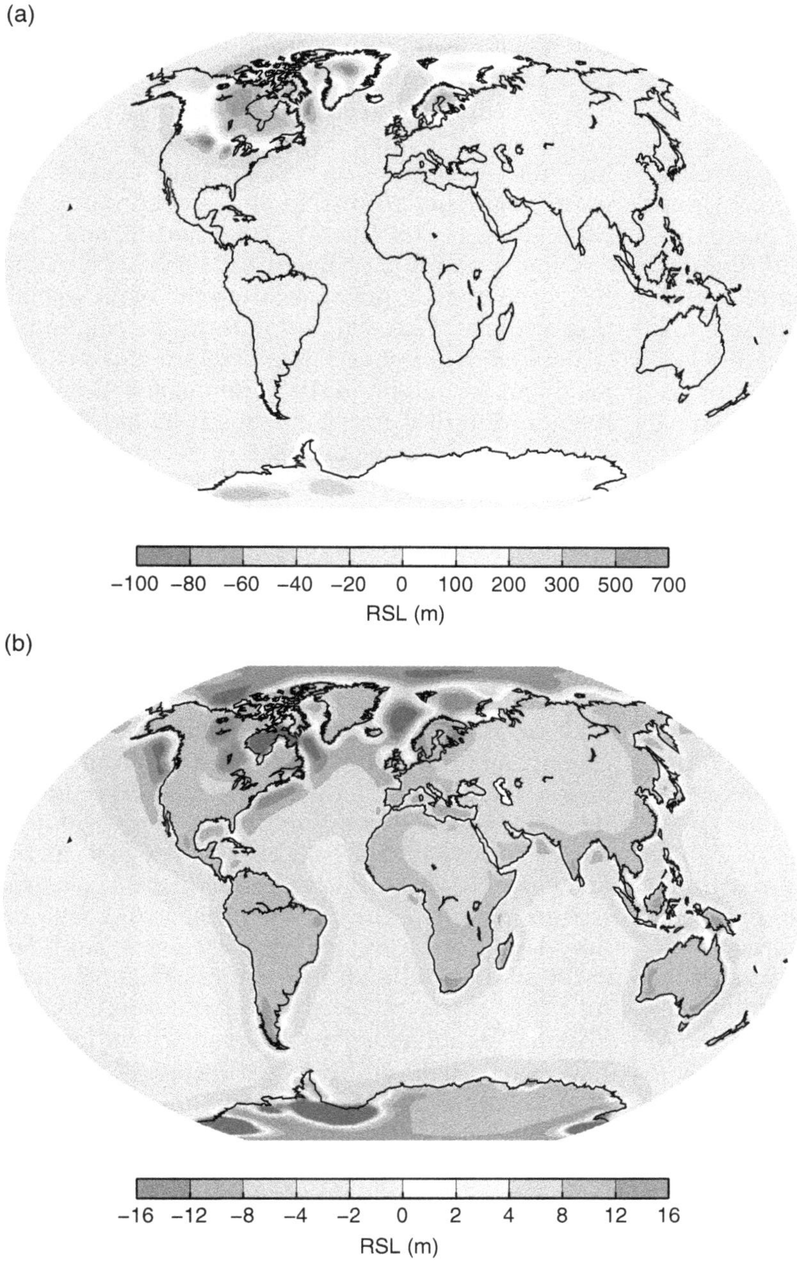

Fig. 28.5. (a) RSL at 10 ka before present due to ice loading changes only. (b) RSL at 10 ka BP due to ocean loading changes only. Note that the contribution of these effects to a globally uniform RSL change through the syphoning process (see Section 28.2) is not included. These model results are based on the ICE5G model of global ice evolution and the VM2 Earth viscosity model (with a 90 km thick lithosphere) (Peltier, 2004). For color details, please see Plate 36.

model parameters that produce an optional fit to the RSL data. Through careful sensitivity testing and examination of data-model residuals, trade-offs between Earth and ice model parameters can be identified and, in some cases, minimized.

There are common elements in these two approaches to producing ice-sheet reconstructions; for example, the regional approach implicitly requires the application of a global model in which to embed the regional ice model in order to compute regional RSL curves. While each approach has its strengths and weaknesses, a clear advantage of the regional approach is the ability to tune Earth as well as ice model parameters in seeking an optimal fit. While it can be argued that employing different Earth model parameters in different regions is inconsistent with the application of a spherically symmetric (1D) Earth model, and including the regional viscosity structure as an additional set of variable parameters leads to increased non-uniqueness, this approach has become the most common in the past decade or so.

A relatively recent advance in modeling RSL data to constrain regional ice-sheet reconstructions is the application of glaciological models that simulate the accumulation and ablation of ice in response to climate change as well as the temperature, rheology, and flow of ice (e.g., Huybrechts and de Wolde, 1999). The first study to consider this approach focused on the deglacial history of the Greenland ice sheet (Tarasov and Peltier, 2002). While this approach is more complex and time-consuming in that there are more parameters to consider with regard to tuning the ice model, thus increasing the non-uniqueness of the problem, the main benefits are that the resulting solution is compatible with the physics of ice sheets (as represented in the applied model) and an opportunity is provided to test glaciological ice models and better understand how ice sheets respond to climate change. This approach has become more widely adopted in recent years (e.g., Simpson et al., 2009; Whitehouse et al., 2012).

With the application of more realistic ice models to GIA modeling problems, the issue of non-uniqueness becomes more acute. With the increase in model parameters, it becomes increasingly difficult to explore parameter sensitivities manually and determine the range of model parameters that satisfy the data (to within uncertainty). For this reason, automated or semi-automated procedures are the logical way forward. A recent study focusing on reconstructing the North American ice sheets defines the state-of-the-art in this approach (Tarasov et al., 2012). The procedure applied in this case is based on the production of a large number (several thousand) of model simulations that sample a range of specified ice model parameters. These results are then used within a Bayesian Neural Network scheme to identify a new suite of parameters that would result in improved data–model fits. This second parameter subset is then used to generate a new (smaller) ensemble of model output and the process is repeated until convergence on an optimal parameter set is obtained.

28.3.3 Global ice volume

Figure 28.5a indicates that the influence of ice loading is much reduced in locations removed from major glaciation centers. For this reason, far-field RSL data have been commonly used to provide constraints on changes in global ice volume or eustatic sea level, which serves as an important target that regional ice-sheet reconstructions must sum to at a given time. Interpreting far-field data for changes in global ice volume is non-trivial however, due to the significant influence of, primarily, ocean loading (Fig. 28.5b) and the syphoning effect described in Section 28.2. One of the most surprising results in the classic study by Clark et al. (1978) (see Fig. 28.4 and related discussion) was the significant influence of GIA on RSL even at great distances from the main glaciation centers. This led the authors to conclude, "There are no 'stable' regions where eustatic sea level can be measured, because deglaciation and the addition of water to the ocean basins deform the Earth and change the observer's point of reference." Other processes, such as tectonics or sediment compaction, can also be important at some locations. GIA and non-GIA signals must be estimated and removed from the observations before inferring ice volume changes. Since the GIA RSL signal cannot be estimated from empirical information, a model is commonly used to estimate and remove it from observations.

Constraining grounded ice volume at the last glacial maximum (LGM) has been a common application of far-field RSL reconstructions since it captures the large ice mass loss during Termination 1 and provides a measure of this quantity that can be compared to those derived from marine oxygen isotope records (Waelbroeck et al., 2002). Due to the depth of LGM sea levels in the far-field (>100 m), only a handful of records exist that capture this sea-level lowstand. Two examples are RSL reconstructions from Barbados (Peltier and Fairbanks, 2006) and Bonaparte Gulf (Yokoyama et al., 2000). The former is based on fossil corals and the latter on microfossil assemblages of fauna. When these observations are corrected for GIA (and tectonics in the case of Barbados), eustasy at LGM is estimated to be −130 to 135 m (Bonaparte Gulf; Yokoyama et al., 2000) and c. −120 m (Barbados; Peltier and Fairbanks, 2006). Note that this latter estimate is based on assuming a fixed ocean area equal to that at present since the LGM (i.e., application of Equation 28.1 using present-day ocean area for A_0); this assumption leads to an underestimate in the calculated eustatic rise and so reduces the discrepancy between these two ice volume estimates by about 5%. Furthermore, these differences can likely be reconciled when uncertainties in the reconstructed RSL and model parameters (e.g., viscosity structure) are accounted for (e.g., Milne et al., 2002).

From considering geomorphological land forms related to glacial erosion and deposition as well as plausible ice surface profiles, the CLIMAP Project (CLIMAP, 1981) put forward two models of LGM ice extent to represent minimum (127 m eustatic sea level) and maximum (163 m ESL) scenarios. The recent estimates of LGM eustatic sea level discussed above clearly support the minimum CLIMAP estimate. Recent reconstructions of the Antarctic ice sheets at the LGM give eustatic sea level estimates of ~7–11 m (Whitehouse et al., 2012; Ivins et al., 2013; note that these values take into account that a large proportion of ice was grounded below mean sea level and so did not contribute to eustatic sea level). These estimates are a factor of 2–3 smaller than in most previous studies, largely reflecting the improvement in data constraints from this continent over the past 5–10 years. Some relatively recent estimates of eustatic sea-level contributions at LGM from the major ice-covered regions lead to the following ranges: 68–74 m for North America (Peltier, 2004; Tarasov et al., 2012); 13–22 m for Eurasia (Siegert et al., 2001; Lambeck et al., 2010); and 3–5 m for Greenland (Fleming and Lambeck, 2004; Simpson et al., 2009). When summing these values, a range of 91–112 m eustatic sea level is obtained. The contribution from smaller bodies of ice will increase these numbers by around 6 m at most (Clark and Mix, 2002) and so, at the time of writing, the LGM eustatic sea-level budget is not closed.

The mid-to-late Holocene is also a common time period to estimate eustatic sea level as it provides an important context for rates of ice volume loss associated with global warming in the past ~150 years. Mid-Holocene sea-level highstands found in many far-field locations mark a significant deceleration in land ice melting associated with the end of the main phase of Termination 1. As the rates of eustatic sea-level rise decreased, isostatic effects – specifically ocean loading and syphoning – became dominant, resulting in a change from RSL rise to fall in many far-field locations (e.g., Mitrovica and Milne, 2002). Estimating and removing the GIA signal suggests a net eustatic sea-level change of a few meters between ~7 and ~2 ka BP (Fleming et al., 1998) which is believed to be dominated by changes in the Antarctic ice sheets (Nakada and Lambeck, 1988). During the past few millennia there is no evidence for any change in eustatic sea level exceeding a few tenths of a millimeter per year sustained over millennial timescales (Lambeck et al., 2004), indicating that current rates of global mean sea level (GMSL) rise (2–3 mm a^{-1}) estimated from tide gauge records are consistent with a response of the climate system to recent warming. However, it is important to note that GMSL cannot be directly compared to eustatic sea level, particularly for recent times, due to the large contribution (~50%) of ocean warming and expansion to GMSL rise. In contrast, the contribution of this process to millennial-scale GMSL following the LGM is estimated to be at the few percent level (McKay et al., 2011).

Based on their model results, Clark et al. (1978) concluded that there are no places where eustatic sea level can be measured directly, even when only considering the influence of GIA. Their results also indicate that the departure from eustatic sea level due to GIA is time and space dependent, with the largest differences occurring in near-field regions. Model output can therefore be used to indicate where the GIA correction would be relatively small for a given time interval and therefore where the conversion from RSL to eustatic sea level would be more accurate. This model application was pursued by Milne and Mitrovica (2008), who presented model output that indicates optimal locations for the purpose of reconstructing RSL to estimate eustatic sea level, based on reducing the contamination due to GIA. Given that model output will depend on the specific parameter set used, Milne and Mitrovica (2008) considered two published deglaciation models and several hundred Earth viscosity models to map out some of this uncertainty. The results of their analysis are summarized in Figure 28.6 for the LGM.

The estimation of eustatic sea level from GIA-corrected RSL reconstructions has been predominantly based on 1D spherically symmetric Earth models. As outlined above, Earth models that can incorporate 3D Earth structure have been developed and used to simulate sea-level change. One recent application of these more complex models considered the impact of lateral structure on modeling Barbados sea levels during and following the LGM (Austermann et al., 2013). A novel aspect of this analysis was the explicit treatment of subducted lithosphere and its influence on the regional GIA response which, as noted above, has been demonstrated in studies from convergent tectonic margins (e.g., James et al., 2009). Comparing predictions of a 1D Earth model to those of a 3D Earth model show that modeled RSL is shallower at the LGM by greater than 7 m due to

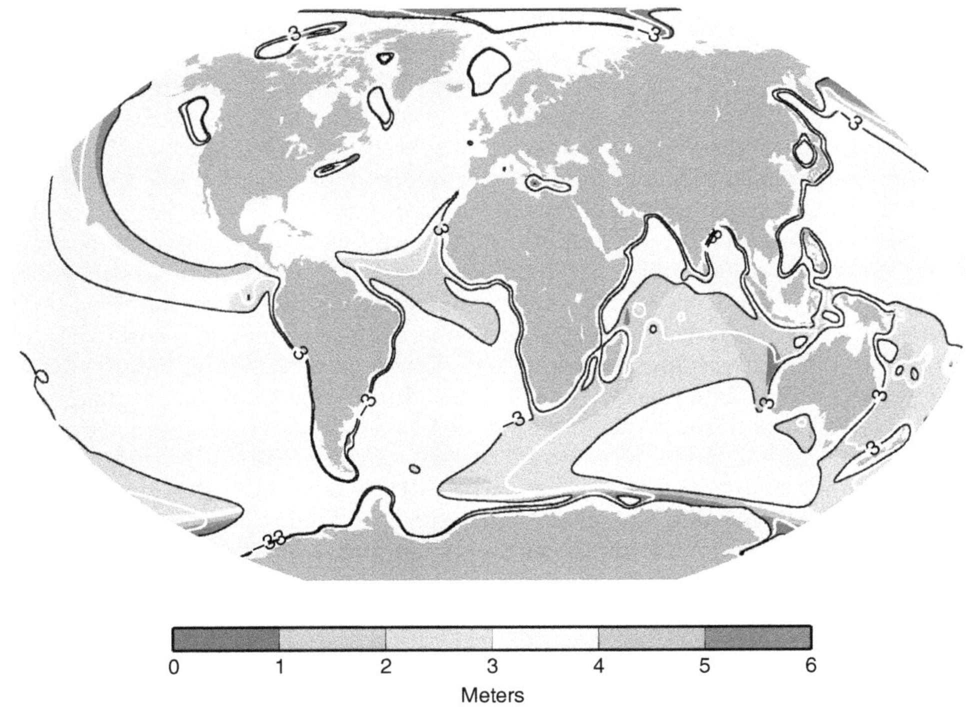

Fig. 28.6. Model results showing locations that would be well suited to estimating ESL from RSL field reconstructions at the last glacial maximum. Ideal locations are where: (1) the modeled RSL values are close to the model ESL value; and (2) the model output is relatively insensitive to uncertainty in the input Earth model. The first criterion is satisfied in areas between the black contours which indicate where RSL is within 3 m of ESL (they are equal along the white line). The second criterion relates to the color shading which indicates the spread (standard deviation) in model output due to variations in Earth model viscosity structure. Blue areas are therefore ideal locations for inferring ESL from RSL data, as the model correction will be relatively small and insensitive to uncertainty in Earth model viscosity structure. These model results are based on the ICE5G model of global ice evolution and a large suite of Earth viscosity models. See Milne and Mitrovica (2008) for further details. *Source*: Milne and Mitrovica (2008). Reproduced with permission of Elsevier. For color details, please see Plate 37.

the influence of lateral viscosity variations (and primarily those associated with the high-viscosity subducted lithosphere). Application of their more complex 3D Earth model therefore resulted in eustatic sea level estimates from Barbados corals that are ~7 m more negative compared to those based on the more common 1D Earth models (i.e., more land ice is required to push LGM sea levels lower). This result could account for the above-noted discrepancy between eustatic sea level estimates from Barbados and Bonaparte Gulf, as the latter is not located above subducted lithosphere.

Given current concern regarding the stability of the Greenland and Antarctic ice sheets and their possible contributions to future sea-level rise, a focus of recent research has been on estimating eustatic sea level during past times when the climate was warmer than at present. The last interglacial (LIG, ~130–115 ka BP) is the most recent period when polar temperatures and sea levels in many areas were significantly greater than at present. The procedure of estimating and removing the GIA contribution from RSL reconstructed during the last interglacial is more challenging than doing so for the LGM or late Holocene, given that the model correction in this case is sensitive to the global ice history from the penultimate glacial maximum through to the present day (Lambeck et al., 2012). Nevertheless, some recent studies have applied GIA models and estimated LIG eustatic sea levels to be 7.2 ± 1.3 m (67% confidence; Kopp et al., 2009) and 5.5–9 m (Dutton and Lambeck, 2012). These results indicate a significant response from the Greenland and Antarctic ice sheets to warmer last interglacial temperatures. Note that these estimates are for GMSL rather than eustatic sea level given the possible contribution from thermal expansion of the oceans. However, this is estimated to be relatively small during the LIG (McKay et al., 2011). Reconstructions of RSL during the last interglacial have also been used to estimate rates of RSL change, with the most recent results indicating that millennial average rates of GMSL rise were likely (67% probability) less than or equal to 7 m ka^{-1} (Kopp et al., 2013).

The mid-Pliocene has also been identified as a part analog for future climate due to the elevated temperatures and CO_2 levels during this period. RSL reconstructions vary considerably (~10–40 m) and many have been interpreted directly as changes in GMSL. It was recently demonstrated that GIA can contribute significantly (up to several tens of meters) to Pliocene RSLs (Raymo et al., 2011), so must be included when interpreting these data for ice volume relative to that at present. When interpreting RSLs over long timescales (>1 Ma), the influence of other solid Earth processes such as sediment redistribution and the corresponding isostatic response and vertical land motion due to mantle flow (Moucha et al., 2008; Rowley et al., 2013) must also be considered.

28.3.4 Sources of rapid melt events

There is strong field evidence to support the occurrence of exceptionally rapid sea-level changes (rates of several tens of millimeters per year) during Termination 1. The largest event, termed Meltwater Pulse 1A (Fairbanks, 1989), occurred between 14.65 and 14.31 ka BP with rates of eustatic sea-level rise likely exceeding 4 m per century (Deschamps et al., 2012). These events provide useful information on how quickly ice sheets can deliver meltwater to the global oceans. They are also central to understanding millennial-scale climate change due to their potential influence on ocean circulation. Both of these applications require that the geographical distribution of meltwater sources is well defined; this is often difficult to do however, as illustrated in the case of MWP-1A (Carlson and Clark, 2012). In addition to considering near-field evidence of ice extent and RSL to constrain the contribution of a given ice sheet to MWP-1A (e.g., Tarasov et al., 2012; Whitehouse et al., 2012), the technique known as "sea-level fingerprinting" which considers far-field data has also provided useful information.

Sea-level fingerprinting involves modeling the spatial variation in RSL across a relatively short time interval such that the sea-level response is dominated by gravitational effects and elastic deformation of the solid Earth. An early study considered spatial variations (fingerprints) in RSL associated with changes in the Antarctic ice sheets (Clark and Lingle, 1977). More recently, it has been demonstrated that spatial variations in 20th century sea-level change are compatible with significant mass loss of the Greenland ice sheet (Mitrovica et al., 2001), although this analysis did not consider the possible contribution from ocean dynamic/steric changes. The fingerprinting technique is well suited to studying meltwater pulses since these are large and rapid events with a relatively small ocean dynamic/steric overprint.

Clark et al. (2002) were the first to apply this technique to consider the source distribution of MWP-1A. They interpreted the RSL records at Barbados and Sunda Shelf as indicating a MWP-1A amplitude of ~25 m at each site. They then proceeded to demonstrate that this constraint was not compatible with a sole or dominant Laurentide source, but was compatible with a sole or dominant Antarctic source. Subsequent studies have also suggested that the far-field RSL data are more compatible with a dominant Antarctic as opposed to Laurentide source (Bassett et al., 2005; Deschamps et al., 2012). However, there remains considerable disagreement regarding the interpretation of field data from Antarctica and the plausibility of a large retreat across the MWP-1A time window (Carlson and Clark, 2012).

28.3.5 Projecting future sea-level change

An important application of GIA sea-level models is to project future sea-level changes. GIA associated with past changes in land ice and ocean loading will continue into the future due to the ongoing viscous response of the solid Earth. This signal will make a significant contribution to sea-level change, particularly in near-field areas, and will add to the signal associated with contemporary climate change: ocean steric changes and land ice melting. The latter will include both a GIA component (e.g., Gomez et al., 2010; see Fig. 28.7) and an ocean dynamic component associated with the influence of meltwater influx on ocean density structure and thus circulation (e.g., Stammer et al., 2011). Clearly, all of these component signals have inherent spatial variability; projections of future sea-level change that consider them will therefore exhibit this variability (e.g., Slangen et al., 2012).

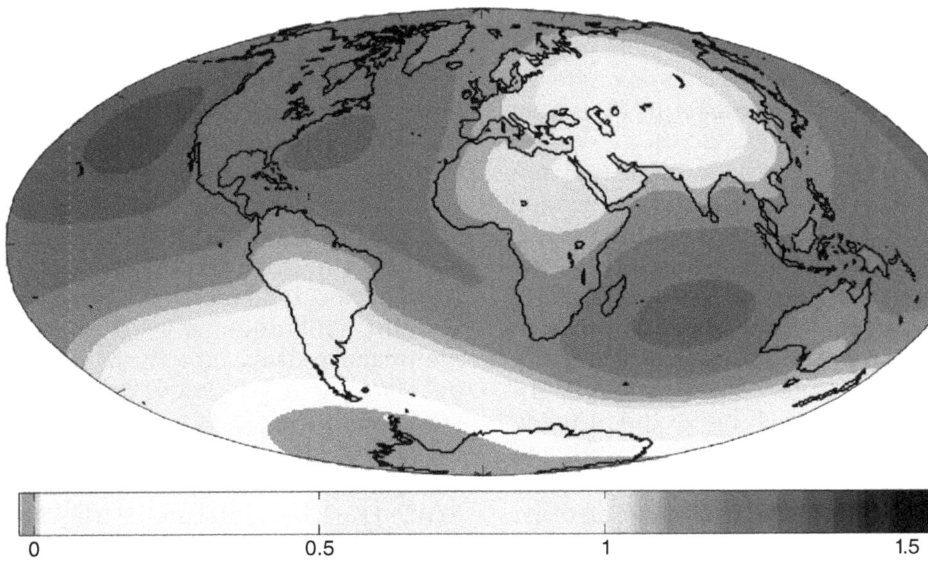

Fig. 28.7. A GIA model prediction of sea-level change associated with a collapse of the West Antarctic ice sheet. The sea-level change is normalized to the eustatic value. Note that a sea-level fall is predicted in the near-field of ablating ice. These computations are based on an elastic Earth model and so assume that the collapse is relatively rapid (of the order 100 years). *Source*: Gomez et al. (2010). Reproduced with permission of John Wiley & Sons. For color details, please see Plate 38.

28.4 SUMMARY

GIA is the dominant contributor to RSL changes during the late Quaternary as it explains the space–time variation in RSL curves reconstructed from the geological record. Accurate models of GIA-driven RSL were first produced in the 1970s through the development and application of the sea-level equation (Farrell and Clark, 1976) and global, visco-elastic Earth models (Peltier, 1974). Since the late 1970s, GIA sea-level model output has been compared to RSL observations with two primary aims: (1) to improve understanding and knowledge of solid Earth viscosity structure and thus the long-term evolution of the Earth's interior and surface geology; and (2) late Quaternary ice-sheet evolution in response to climate change. The former is arguably more challenging due to the complexity of solid Earth rheology and the difficulty in sampling material from the Earth's deep interior. While this application of RSL data has led to improved constraints on Earth viscosity structure, it has become clear that RSL data are only able to delineate this structure within a few (~2–3) layers extending to the shallow lower mantle (~1200 km depth; Mitrovica, 1996). While the recent addition of 3D (as opposed to spherically symmetric 1D) Earth models is an important advance, the application of these models is somewhat limited due to uncertainty in defining Earth structure in three dimensions.

The second aim of constraining past ice-sheet evolution has become more common in recent years due to the current and projected climate warming and concern regarding the potential contributions of the Greenland and Antarctic ice sheets to future sea-level change. The use of ice models that simulate the relevant glaciological processes is an important step forward in this regard, as they provide the opportunity to relate the inferred ice extent changes to a climate forcing (Tarasov and Peltier, 2002). While these models also bring additional glaciological constraints to the problem, the increase in (ice) model parameters leads to a higher degree of solution non-uniqueness. The use of these models is therefore much more effective through the application of automated and semi-automated inversion schemes (Tarasov et al., 2012).

With regard to future GIA model development, a key aspect will be the continued production of new RSL reconstructions for areas and times where/when current data control is sparse (e.g., Antarctica), and the development of regional databases using consistent methodologies and quality assessment criteria (Chapter 34). Making these data easily accessible to the modeling community is also an important criterion. Of course, the flow of information should not be in one direction only; as discussed above, modeling studies provide useful information on potential field sites to target in order to better constrain key model parameters (e.g., Milne and Mitrovica, 2008). It is therefore equally important for model output to be made available to the field community to stimulate more interaction and, ultimately, enhance the rate of progress in this fascinating and challenging research area.

REFERENCES

Austermann, J., Mitrovica, J.X., Latychev, K., and Milne, G.A. (2013) Barbados based estimate of ice volume at Last Glacial Maximum effected by subducted plate. Nature Geoscience, 6, 553–557.

Bard, E., Hamelin, B., Fairbanks, R.G., and Zindler, A. (1990) Calibration of the 14C timescale over the past 30,000 years using mass spectrometric U–Th ages from Barbados corals. Nature, 345, 405–410.

Bassett, S.E., Milne, G.A., Mitrovica, J.X., and Clark, P.U. (2005) Ice sheet and solid earth influences on far-field sea-level histories. Science, 309, 925–928.

Carlson, A.E., and Clark, P.U. (2012) Ice-sheet sources of sea-level rise and freshwater discharge during the last deglaciation. Reviews of Geophysics, 50, doi: 10.1029/2011RG000371.

Clark, J.A. and Lingle, C.S. (1977) Future sea-level changes due to West Antarctic ice sheet fluctuations. Nature, 269(5625), 206–209.

Clark, J.A., Farrell, W.E., and Peltier, W.R. (1978) Global changes in postglacial sea level: a numerical calculation. Quaternary Research, 9, 265–287.

Clark, P.U., and Mix, A.C. (2002) Ice sheets and sea level of the Last Glacial Maximum. Quaternary Science Review, 21, 1–7.

Clark, P.U., Mitrovica, J.X., Milne, G.A., and Tamisiea, M. (2002) Sea-level fingerprinting as a direct test for the source of global meltwater pulse 1A. Science, 295, 2438–2441.

CLIMAP Project Members (1981) Seasonal Reconstruction of the Earth's Surface at the Last Glacial Maximum. Geological Society of Amerirca, Map and Chart Series, Vol. 36, 18.

Cuffey, K.M., and Paterson, W.S.B. (2010) *The Physics of Glaciers*, 4th edition. Academic Press.

Deschamps, P., Durand, N., Bard, E., Hamelin, B., Camoin, G., Thomas, A.L., Henderson, G.M., Okuno, J., and Yokoyama, Y. (2012) Ice-sheet collapse and sea-level rise at the Bolling warming 14,600 years ago. Nature, 483(7391), 559–564.

Dutton, A., and Lambeck, K. (2012) Ice volume and sea level during the last interglacial. Science, 337, 216–219.

Dziewonski, A.M., and Anderson, D.L. (1981) Preliminary reference earth model (PREM). Physics of the Earth and Planetary Interiors, 25, 297–356.

Fairbanks, R.G. (1989) A 17,000-year glacio-eustatic sea level record: influence of glacial melting rates on the Younger Dryas event and deep-ocean circulation. Nature, 342, 637–642.

Farrell, W.E., and Clark, J.A. (1976) On postglacial sea-level. Geophysical Journal of the Royal Astronomical Society, 46, 647–667.

Fleming, K., and Lambeck, K. (2004) Constraints on the Greenland ice sheet since the Last Glacial Maximum from sea-level observations and glacial-rebound models. Quaternary Science Reviews, 23(9–10), 1053–1077.

Fleming, K., Johnston, P., Zwartz, D., Yokoyama, Y., Lambeck, K., and Chappell, J. (1998) Refining the eustatic sea-level curve since the Last Glacial Maximum using far and intermediate-field sites. Earth and Planetary Science Letters, 163, 327–342.

Gomez, N., Mitrovica, J. X., Tamisiea, M. E., and Clark, P. U. (2010) A new projection of sea level change in response to collapse of marine sectors of the Antarctic Ice Sheet. Geophysical Journal International, 180, 623–634.

Haskell, N.A. (1935) The motion of a fluid under a surface load, 1, Physics, 6, 265–269, 1935.

Huybrechts, P., and de Wolde, J. (1999) The dynamic response of the Greenland and Antarctic ice sheets to multiple-century climatic warming. Journal of Climate, 12, 2169–2188.

Ivins, E.R., James, T.S., Wahr, J., Schrama, E.J.O, Landerer, F.W., and Simon, K.M. (2013) Antarctic contribution to sea level rise observed by GRACE with improved GIA correction. Journal of Geophysical Research (Solid Earth), 118, 3126–3141.

James, T.S., Gowan, E.J., Wada, I., and Wamg, K. (2009) Viscosity of the asthenosphere from glacial isostatic adjustment and subduction dynamics at the northern Cascadia subduction zone, British Columbia, Canada. Journal of Geophysical Research (Solid Earth), 114, B04405.

Karato, S. (2008) *Deformation of Earth Materials: Introduction to the Rheology of the Solid Earth*. Cambridge University Press, Cambridge.

Kopp, R.E., Simons, F.J., Mitrovica, J.X., Maloof, A.C., and Oppenheimer, M. (2009) Probabilistic assessment of sea level during the last interglacial stage. Nature, 462, 863–867.

Kopp, R.E., Mitrovica, J.X., Griffies, S.M., Yin, J., Hay, C.C., and Stouffer, R.J. (2010) The impact of Greenland melt on local sea levels: a partially coupled analysis of dynamic and static equilibrium effects in idealized water-hosing experiments. Climate Change, 103, 619–625.

Kopp, R.E., Simons, F.J., Mitrovica, J.X., Maloof, A.C., and Oppenheimer, M. (2013) A probabilistic assessment of sea level variations within the last interglacial stage. Geophysical Journal International, 193, 711–716.

Lambeck, K. (1993) Glacial rebound of the British Isles. II: a high resolution, high-precision model. Geophysical Journal International, 115, 960–990.

Lambeck, K., Smither, C., and Johnston, P. (1998) Sea-level change, glacial rebound and mantle viscosity for northern Europe. Geophysical Journal International, 134, 102–144.

Lambeck, K., Yokoyama, Y., Johnston, P., and Purcell, A. (2000) Global ice volumes at the Last Glacial Maximum. Earth and Planetary Science Letters, 181, 513–527.

Lambeck, K., Anzidei, M., Antonioli, F., Benini, A., and Esposito, A. (2004) Sea level in Roman time in the Central Mediterranean and implications for recent change. Earth and Planetary Science Letters, 224, 563–575.

Lambeck, K., Purcell, A., Zhao, J. and Svensson, N.-O. (2010) The Scandinavian Ice Sheet: from MIS 4 to the end of the Last Glacial Maximum. Boreas, 39, 410–435.

Lambeck, K., Purcell, A., and Dutton, A. (2012) The anatomy of interglacial sea levels: The relationship between sea levels and ice volumes during the Last Interglacial. Earth and Planetary Science Letters, 315–316, 4–11.

Latychev, K., Mitrovica, J.X., Tromp, J., Tamisiea, M.E., Komatitsch, D. and Christara, C.C. (2005) Glacial isostatic adjustment on 3-d earth models: a finite-volume formulation. Geophysical Journal International, 161, 421–444.

McKay, N.P., Overpeck, J.T., and Otto-Bliesner, B.L. (2011) The role of ocean thermal expansion in Last Interglacial sea level rise. Geophysical Research Letters, 38, L14605.

Milne, G.A., and Mitrovica, J.X. (1998) Postglacial sea-level change on a rotating Earth. Geophysical Journal International, 133, 1–19.

Milne, G., and Mitrovica, J. (2008) Searching for eustasy in deglacial sea level histories. Quaternary Science Reviews, 27, 2292–2302.

Milne, G.A., and Shennan, I. (2013) Isostasy: glaciation-induced sea-level change. In Encyclopedia of Quaternary Science, volume 3, Elsevier, Oxford, p. 452–459.

Milne, G.A., Mitrovica, J.X., and Schrag, D.P. (2002) Estimating past continental ice volume from sea-level data. Quaternary Science Review, 21, 361–376.

Mitrovica, J.X. (1996) Haskell [1935] revisited. Journal of Geophysical Research, 101, 555–569.

Mitrovica, J.X., and Peltier, W.R. (1991) On post-glacial geoid subsidence over the equatorial oceans. Journal of Geophysical Research, 96, 20053–20071.

Mitrovica, J.X., and Milne, G.A. (2002) On the origin of postglacial ocean syphoning. Quaternary Science Review, 21, 2179–2190.

Mitrovica, J.X., and Milne, G.A. (2003) On post-glacial sea level - I. General theory. Geophysical Journal International, 154, 253–267.

Mitrovica, J.X., and Forte, A.M. (2004) A new inference of mantle viscosity based upon joint inversion of convection and glacial isostatic adjustment data. Earth and Planetary Science Letters, 225, 177–189.

Mitrovica, J.X., Tamiseia, M.E., Davis, J.L., and Milne, G.A. (2001) Recent mass balance of polar ice sheets inferred from patterns of global sea-level change. Nature, 409, 1026–1029.

Moucha, R., Forte, A. M., Mitrovica, J. X., Rowley, D. B., and Quéré, S. (2008) Dynamic topography and long-term sea-level variations: There is no such thing as a stable continental platform. Earth and Planetary Science Letters, 271, 101–108.

Nakada, M., and Lambeck, K. (1988) The melting history of the late Pleistocene Antarctic ice sheet. Nature, 33, 36–40.

Nakada, M., and Lambeck, K. (1989) Late Pleistocene and Holocene sea-level change in the Australian region and mantle rheology. Geophysical Journal International, 96, 497–517.

Paulson, A., Zhong, S., and Wahr, J. (2007) Limitations on the inversion for mantle viscosity from postglacial rebound. Geophysical Journal International, 168, 1125–1209.

Peltier,W.R. (1974) The impulse response of a Maxwell Earth. Reviews of Geophysics, 12, 649–669.

Peltier, W.R. (2004) Global glacial isostasy and the surface of the ice-age earth: The ICE-5G (VM2) model and GRACE. Annual Review of Earth and Planetary Sciences, 32, 111.

Peltier, W.R., and Andrews, J.T. (1976) Glacial isostatic adjustment I: the forward problem. Geophysical Journal of the Royal Astronomical Society, 46, 605–646.

Peltier, W.R., and Fairbanks, R.G. (2006) Global glacial ice volume and Last Glacial Maximum duration from an extended Barbados sea level record. Quaternary Science Review, 25, 3322–3337.

Peltier, W.R., Farrell, W.E., and Clark, J.A. (1978) Glacial isostasy and relative sea level: A global finite element model. Tectonophysics, 50, 81–110.

Peltier, W.R., Yuen, D.A., and Wu, P. (1980) Postglacial rebound and transient rheology. Geophysics Research Letters, 7(10), 733–736.

Raymo, M. E., Mitrovica, J. X., O'Leary, M. J., DeConto, R. M., and Hearty, P. J. (2011) Departures from eustasy in Pliocene sea level records. Nature Geoscience, 4, 328–332.

Romanowicz, B. (2003) Global mantle tomography: progress status in the last 10 years. Reviews of Geophysics and Space Physics, 31(1), 303–328.

Rowley, D.B., Forte, A.M., Moucha, R., Mitrovica, J.X., Simmons, N.A., and Grand, S.P. (2013) Dynamic topography change of the eastern United States since 3 million years ago. Science, 340, 1560–1563.

Sabadini, R., Yuen, D.A., and Boschi, E. (1982) Polar wander and the forced responses of a rotating, multi-layered, viscoelastic planet. Journal of Geophysical Research, 87, 2885–2903.

Shennan, I., Hamilton, S., Hillier, C., and Woodroffe, S. (2005) A 16,000-year record of near-field relative sea-level changes, northwest Scotland, United Kingdom. Quaternary International, 133–134, 95–106.

Siegert, M. J., Dowdeswell, J. A., Hald, M., and Svendsen, J. I. (2001) Modelling the Eurasian Ice Sheet through a full (Weichselian) glacial cycle. Global and Planetary Change, 31, 367–385.

Sigmundsson, F. (1991) Post-glacial rebound and asthenosphere viscosity in Iceland. Geophysical Research Letters, 18(6), 1131–1134.

Simpson, M. J. R., Milne, G. A., Huybrechts, P., and Long, A. J. (2009) Calibrating a glaciological model of the Greenland ice sheet from the Last Glacial Maximum to present-day using field observations of relative sea level and ice extent. Quaternary Science Review, 28, 1631–1657.

Slangen, A.B.A., Katsman, C.A., van de Wal, R.S.W., Vermeersen, L.L.A., and Riva, R.E.M. (2012) Towards regional projections of twenty-first century sea-level change based on IPCC SRES scenarios. Climate Dynamics, 38, 1191–1209.

Stammer, D., Agarwal, N., Herrmann, P., Kohl, A., and Mechoso, C.R. (2011) Response of a coupled ocean-atmosphere model to Greenland ice melting. Surveys in Geophysics, 32, 621–642.

Tarasov, L., and Peltier, W.R. (2002) Greenland glacial history and local geodynamic consequences. Geophysical Journal International, 150, 198–229.

Tarasov, L., Dyke, A.S., Neal, R.M., and Peltier, W.R. (2012) A data-calibrated distribution of deglacial chronologies for the North American ice complex from glaciological modeling. Earth and Planetary Science Letters, 315–316, 30–40.

Tushingham, A.M., and Peltier, W.R. (1991) Ice-3G: a new global model of late Pleistocene deglaciation based upon geophysical predictions of post-glacial relative sea-level change. Journal of Geophysical Research, 96, 4497–4523.

van der Wal, W., Wu, P., Wang, H., Sideris, M.G. (2010) Sea levels and uplift rate from composite rheology in glacial isostatic adjustment modeling. Journal of Geodynamics, 50(1), 38–48.

Waelbroeck, C., Labeyrie, L., Michel, E., Duplessy, J.C., McManus, J.F., Lambeck, K., Balbon, E., and Labracherie, M. (2002) Sea-level and deep water temperature changes derived from benthic foraminifera isotopic records. Quaternary Science Reviews, 21(1–3), 295–305.

Walcott, R.I. (1980) Rheological models and observational data of glacio-isostatic rebound. In: *Earth Rheology, Isostasy and Eustasy* (ed. Morner, N.-A.), John Wiley & Sons, New York, pp. 3–10.

Whitehouse, P.L., Bentley, M.J., and Le Brocq, A.M. (2012) A deglacial model for Antarctica: geological constraints and glaciological modelling as a basis for a new model of Antarctic glacial isostatic adjustment. Quaternary Science Reviews, 32, 1–24.

Wu, P., and van der Wal, W. (2003) Postglacial sea-levels on a spherical, self-gravitating viscoelastic earth: effects of lateral viscosity variations in the upper mantle on the inference of viscosity contrasts in the lower mantle. Earth and Planetary Science Letters, 211, 57–68.

Yokoyama, Y., Lambeck, K., De Deckker, P., Johnston, P., and Fifield, L.K. (2000) Timing of the Last Glacial Maximum from observed sea-level minima. Nature, 406, 713–716.

Chapter 29
Tidal modeling

STEPHEN D. GRIFFITHS[1] AND DAVID F. HILL[2]
[1]*Department of Applied Mathematics, University of Leeds, Leeds, UK*
[2]*School of Civil and Construction Engineering, Oregon State University, Corvallis, OR, USA*

29.1 INTRODUCTION

Tides are periodic oscillations in sea-surface height and currents, which owe their origin to the rotation of the Earth and the gravitational attraction of the Sun and Moon. The amplitude of the surface oscillations is about 50 cm over much of the open ocean and about 1 m along many coastlines, but local resonances can lead to tides of over 5 m in special coastal locations (e.g., Garrett 1972; Arbic et al. 2007). These large coastal oscillations lead to large fluxes of mass, momentum, and energy, which have relevance to coastal inundation, sediment transport, and other processes. Additionally, the variations in these oscillations over time mean that tides play an important role in determining sea-level history from coastal observations.

The nature of the tidal signal is well understood at many coastal locations, sometimes based upon observations lasting hundreds of years (Bowen, 1972; Cartwright, 1972). Historically, corresponding information in the open ocean was sparse, but satellite altimeter data (notably from the TOPEX/Poseidon mission, launched in 1992) have led to accurate mapping of tides over much of the globe (Egbert et al., 1994). At any location a set of periodic *tidal constituents* of distinct frequencies that are determined by the orbital geometry of the Earth, Moon, and Sun are found. Some constituents are directly astronomically forced, while others are harmonics (or overtides) that are generated through non-linear interactions in the ocean. Details of the largest directly forced constituents, all close to diurnal or semi-diurnal, are given in Table 29.1. By far the largest constituent, when measured by globally integrated energy of the oceanic response, is the principal lunar tide which is denoted by the abbreviation M_2.

If the tidal signal is assumed to consist of N constituents each with frequency ω_j, then the ocean depth perturbation h at a given location may be written

$$h(t) = \mathrm{Re}\left(\sum_{j=1}^{N} \hat{h}_j \, e^{-i\omega_j t}\right) \tag{29.1}$$

where $\hat{h}_j$ is a complex tidal amplitude, which determines the amplitude (through $|\hat{h}_j|$) and phase (through the complex argument of $\hat{h}_j$) of each constituent. Maps of $|\hat{h}|$ are shown in Figure 29.1a, b for the most energetic semi-diurnal and diurnal tidal constituents, according to satellite altimetry. Note how the (semi-diurnal) M_2 tide is largest in the North Atlantic, while the (diurnal) K_1 tide is largest in most of the Pacific.

From a modeling perspective, what is required are predictions of $\hat{h}$ for the largest tidal constituents; exactly how many constituents are required will depend upon the accuracy desired and the location in question. Extensive use of observational data allows various kinds of assimilative models to be derived, such as the TPXO 6.2 model used to generate Figure 29.1a, b. However, here we are interested in dynamical numerical modeling approaches that use little (or no) tidal data, and instead rely upon the (known) astronomical forcing, datasets of ocean bathymetry and stratification, and properties of the elastic Earth. Although such dynamical models are not as accurate as assimilative models, their lack of reliance upon data means they can be used to examine how tides may have changed in the past in response to sea-level and bathymetric variations (perhaps due to glacial cycles), or how they may change in the future. They are therefore of special interest in studies of sea-level change.

Handbook of Sea-Level Research, First Edition. Edited by Ian Shennan, Antony J. Long, and Benjamin P. Horton.

Table 29.1. Main directly forced tidal constituents, ordered by energy of the oceanic response. The first column gives the commonly used Darwin symbol, or tidal abbreviation. The tidal period and forcing amplitude A (see Equation (29.4)) are taken from Cartwright (1977), with the latter being modified by the Love numbers taken from tables 11 and 12 of Dehant et al. (1999). The total energy (kinetic plus potential) is taken from table 1 of Egbert and Ray (2003), based on their TPXO.5 assimilative solution

Tidal constituent	Period (h)	Energy (PJ)	A (cm)
M_2	12.42	312.26	16.98
K_1	23.93	49.92	10.51
S_2	12.00	49.87	7.90
O_1	25.82	24.87	7.07
N_2	12.66	14.11	3.25
P_1	24.07	4.78	3.34
K_2	11.97	4.06	2.15
Q_1	26.87	1.16	1.35

Two main challenges can be distinguished within dynamical tidal modeling:

(1) *Global modeling*: with no tidal data, can the entire global tidal response be modeled?
(2) *Regional and local modeling*: given information about the tidal amplitudes in the deep ocean, can the coastal tidal response be modeled?

The resolution (computational cost) limitations associated with global modeling mean that sometimes these challenges are combined in a nested approach, with a purely dynamical global model forcing a higher-resolution local model. We examine each of these modeling challenges in turn, and how they have been applied to study paleotides.

As described extensively by Cartwright (1999), the modern physical and mathematical description of tides dates back to Laplace (1776). His famous tidal equations remain the starting point for all dynamical numerical modeling, although they have been supplemented in the last century with important additional terms relating to turbulent bottom friction, internal tides, and deformation of the elastic Earth. Most of the modern-day challenges relate to efficient implementation of numerical schemes, due to the many physical processes and wide range of spatial scales that must be accounted for.

29.2 GLOBAL MODELING

29.2.1 Depth-averaged models

Most global tidal models are based upon the single-layer shallow-water equations, which model the depth-averaged dynamics of motions that are long relative to the ocean depth:

$$\frac{\partial \mathbf{u}}{\partial t} + (\mathbf{u}\cdot\nabla)\mathbf{u} + \mathbf{f}\times\mathbf{u} = -g\nabla\left(h - h_{\text{sal}} - h_{\text{eq}}\right) - \frac{\mathbf{D}}{\rho(H+h)} \tag{29.2}$$

$$\frac{\partial h}{\partial t} + \nabla\cdot\left((H+h)\mathbf{u}\right) = 0, \tag{29.3}$$

where $\mathbf{u}$ is the depth-averaged horizontal flow, h is the perturbed water depth, H is the undisturbed water depth, $\mathbf{f}$ is the Earth's rotation vector oriented towards the local vertical, g is the acceleration due to gravity, ρ is the average density of seawater, t is time, and ∇ is the horizontal gradient operator. The remaining terms, discussed further in the following, are the equilibrium tide h_{eq}, the self-attraction and loading potential h_{sal}, and parameterized drag $\mathbf{D}$. Laplace's famous tidal equations are Equations (29.2) and (29.3) in spherical geometry, with the non-linear terms and drag neglected (Laplace, 1776).

The equilibrium tide h_{eq} accounts for the direct astronomical forcing of the ocean, and also for the deformation of the elastic Earth under the same forcing. These latter effects appear in the momentum equation (rather than the height equation) when h is taken to be the perturbed water depth, rather than the perturbed sea-surface height. Although some tidal models calculate h_{eq} directly given the positions of the Sun and Moon (e.g., Müller et al., 2012), most tidal models use a harmonic decomposition of h_{eq}. By accounting for N directly forced constituents, h_{eq} then takes the form:

$$h_{\text{eq}} = \text{Re}\left(\sum_{j=1}^{N}\hat{h}_{\text{eq},j}\text{e}^{-i\omega_j t}\right),$$

$$\text{where } \hat{h}_{\text{eq},j} = \begin{cases} A_j\cos^2\theta\text{e}^{-2i\phi} & \text{for semi-diurnal tides,} \\ A_j\sin 2\theta\text{e}^{-i\phi} & \text{for diurnal tides,} \end{cases} \tag{29.4}$$

where θ is latitude, ϕ is longitude, and the amplitude A_j for each of the larger tidal constituents is given in Table 29.1. Further details of this theory may be found in Cartwright (1977), Hendershott (1981), or Arbic et al. (2004a).

The self-attraction and loading potential h_{sal} accounts for three additional effects. First, the perturbed water column induces a gravitational field – the oceanic self-attraction – leading to a weak force towards regions of sea-surface elevation. The Earth also deforms slightly under the perturbed water column, leading to a modification of the perturbed ocean depth h, and an

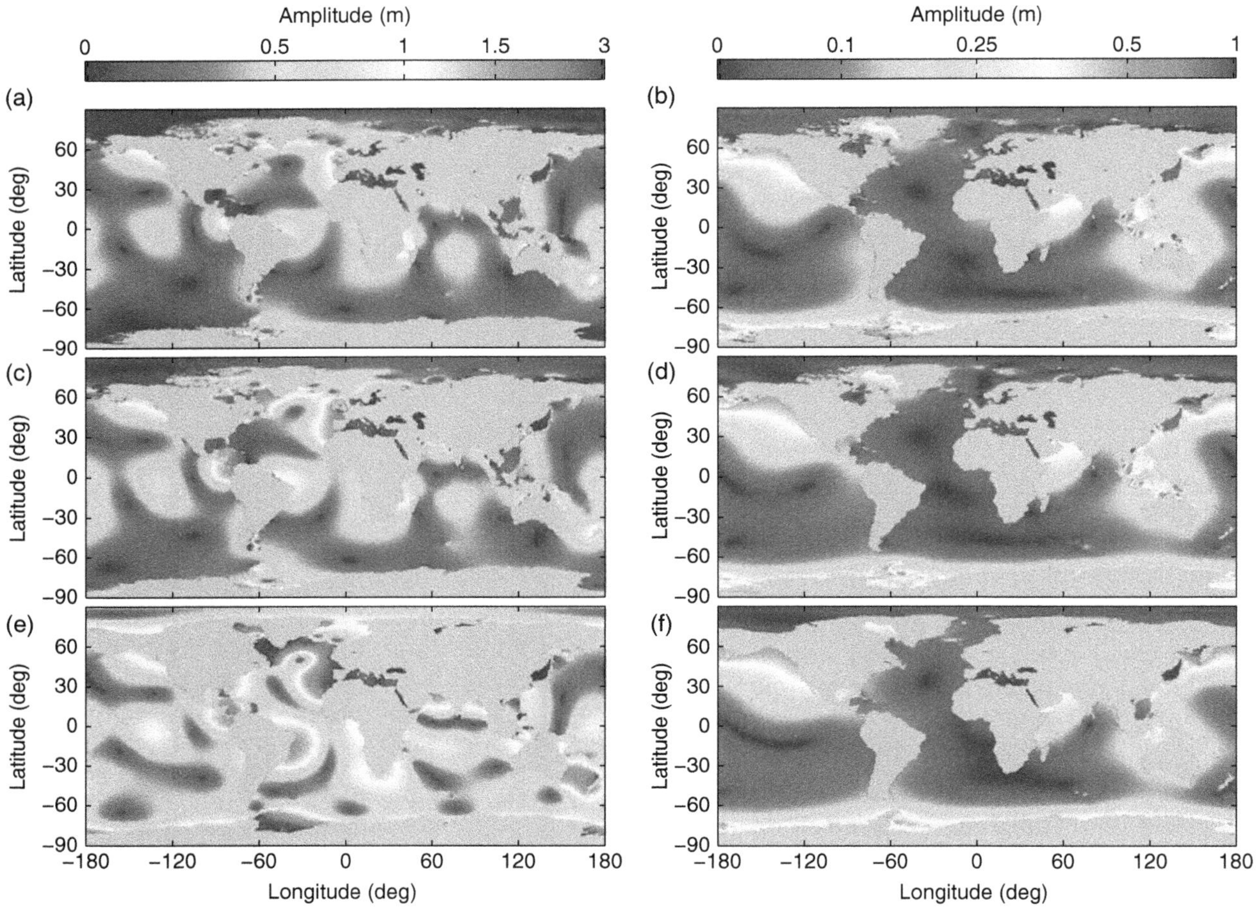

Fig. 29.1. Tidal amplitudes for the M_2 (left) and K_1 (right) constituents. (a, b) As determined by the TPXO 6.2 dataset, described by Egbert and Erofeeva (2002). (c, d) From a global numerical model (Section 29.2). (e, f) A prediction for tides at the LGM from the same global numerical model (see Section 29.4). *Source*: Reproduced with permission from Griffiths and Peltier (2009). For color details, please see Plate 39.

additional gravitational potential associated with these displacements. These three effects are accounted for through the single term h_{sal}, which may be expressed in terms of h using a spherical harmonic transform. Full details of the implementation are given by Ray (1998). Although h_{sal} is generally small compared with h_{eq}, ∇h_{sal} can be comparable to ∇h_{eq}, since h_{sal} has a smaller characteristic lengthscale (set by the tides) than that of h_{eq}.

Inclusion of some drag, denoted by the stress $\mathbf{D}$ in Equation (29.2), is vital to achieve realistic tides. These stresses are thought to arise from two distinct mechanisms. The first is due to the presence of a turbulent bottom boundary layer at the seafloor, which will be strongest in shallow marginal seas where the horizontal currents are strongest. Following the pioneering study of Taylor (1920), the implied stresses are often parameterized using a classical drag law $\mathbf{D}_{BL} = \rho c_d \mathbf{u} \mid \mathbf{u} \mid$, where $c_d \approx 0.0025$. This is of course a simple parameterization; see Marchuk and Kagan (1989) for a more detailed discussion. The second contribution to the drag is associated with the generation of internal waves of tidal frequency as the tide flows over topography. These internal tides may propagate away from the generation site, exerting a local pressure drag at the ocean floor and extracting energy from the horizontal flow; background theory and observations are reviewed by Garrett and Kunze (2007). The implied drag $\mathbf{D}_{IT}$ must be parameterized in depth-averaged models, which are unable to resolve any internal wave dynamics. Unfortunately, $\mathbf{D}_{IT}$ is a non-local function of the surrounding flow and topography, and different scaling laws are expected for small and large topography. Various parameterization schemes of differing complexity have been used in tidal models; see Green and Nycander (2013). In the present-day ocean, it is thought that bottom boundary layer drag and internal tide drag extract energy from the depth-averaged tides with the same order of magnitude when globally averaged (Egbert and Ray, 2003). For example, for the M_2 constituent, observations place about 1.6 TW of dissipation in coastal zones (presumably due to bottom drag on continental shelves, and some internal tide generation at continental slopes) and about 0.8 TW in the deep ocean (presumably due to internal tide generation).

29.2.2 Nature of the response

Many features of the tidal response can be understood using simplified versions of Equations (29.2) and (29.3). There are two important ideas. First, in the absence of h_{sal}, h_{eq}, $\mathbf{D}$, and the non-linear terms, free-wave solutions of the equations can be sought, either in spherical or local Cartesian geometry. A range of waves may be identified at tidal frequencies such as inertia-gravity waves, coastally trapped Kelvin waves, topographic Rossby waves, and planetary waves, as discussed by Hendershott (1981). Inertia-gravity waves and Kelvin waves are particularly relevant to tides. In the simplest models of these waves, the frequency ω and horizontal wavenumber k are linked through $\omega / k \approx \sqrt{gH}$. At tidal frequencies with $\omega \approx 10^{-4}\,\text{s}^{-1}$, waves of lengthscales $k^{-1} \approx 500\,\text{km}$ are expected on the continental shelves (with $H \approx 200\,\text{m}$) and of about 2000 km in the deep ocean (with $H \approx 4000\,\text{m}$). That is, the waves excited at tidal frequencies are large scale.

The second important idea relates to the existence of "normal modes" of ocean basins, which oscillate with discrete natural frequencies. Of interest for tidal modeling are those modes with periods close to diurnal or semi-diurnal, which are expected to be strongly excited by tidal forcing. On a regional scale, Garrett (1972) has shown how the large semi-diurnal tides of the Bay of Fundy can be explained as the near-resonant excitation of a gravity wave mode with a natural period of about 13 hours. On a global scale, a set of normal modes can be calculated for the world ocean. It is thought that the North Atlantic has a normal mode with a period close to semi-diurnal, while the Pacific has a normal mode with a period close to diurnal, potentially explaining the different amplification patterns visible in Figure 29.1. Full details may be found in Platzman (1984); updated calculations have been made by Müller (2007).

29.2.3 Numerical modeling

Numerical solutions of Equations (29.2) and (29.3) at high resolution are required to produce realistic tidal solutions. In addition to expressions for h_{eq} and elastic constants for the Earth in order to evaluate h_{sal}, datasets for stratification and bathymetry H are required. The former, which are required for parameterization of internal tide drag, can be estimated on a 1 degree global grid using the World Ocean Atlas dataset (Antonov et al., 2006; Locarnini et al., 2006), for example. There are a number of global bathymetric datasets freely available online, such as the ETOPO1 1-minute dataset and the GEBCO_08 30 arc-second dataset, both of which rely upon the work of Smith and Sandwell (1997) and others.

Previous studies suggest that global models need to be run at spatial resolutions of at least 1/4 degree to obtain realistic solutions without introducing artificially large parameterized drag (see Egbert et al., 1994, section 3.1). Various spatial discretizations are possible. Regular latitude–longitude rectangular grids are often used, coupled with low-order finite differences. Such models face the problem of a polar singularity, which is avoided by either removing a portion of the Arctic Ocean (e.g., Arbic et al., 2004a; Egbert et al., 2004), rotating the axis of the numerical coordinate system to run through Greenland and Antarctica (e.g., Griffiths and Peltier, 2009), or using a different coordinate system near the poles (e.g., Simmons et al., 2004). Finite-element discretizations have also been used for tidal modeling (e.g., Lyard et al., 2006; Stuhne and Peltier, 2009), which more naturally offer the possibility of unstructured meshes, allowing for higher resolution around the coastlines.

Most depth-averaged tidal models are based on time-stepping of Equations (29.2) and (29.3). Starting from rest, the equations are evolved until equilibrium is reached (typically after about 20 days), which is followed by a further period of evolution to allow for spectral analysis to distinguish between the tidal constituents. If many tidal constituents are included in the solution, this latter period may need to be about 200 days. This is computationally expensive, because the barotropic wave speed c in the deep ocean is so large ($c \approx 200\,\text{m}\,\text{s}^{-1}$). For example, on a regular latitude–longitude grid of 1/4 degree resolution, which implies a minimum east–west grid spacing $\Delta x \approx 4\,\text{km}$ due to convergence of meridians towards the poles, a time-step $\Delta t < 20$ s is required to satisfy the CFL condition $c\Delta t \leq \Delta x$. Simulations with just one semi-diurnal and one diurnal constituent perhaps require just a 30 day simulation, and thus a fraction of the computational time. Indeed, given the globally integrated energies in Table 29.1, many global solutions are calculated for M_2 and K_1 alone.

In a time-stepping model, calculation of the self-attraction and loading potential at every time-step is regarded as too computationally expensive, due to the cost of the spherical harmonic transforms that are normally used. Various approximate treatments are used. The simplest of these is the so-called scalar approximation, which involves setting $h_{sal} = \beta h$, where β is a constant. In a globally averaged sense, observations suggest $\beta \approx 0.08$ for semi-diurnal tides and $\beta \approx 0.12$ for diurnal tides, although there are large differences between the optimal values for shallow and deep water. Although widely used because of its simplicity, the scalar approximation cannot be justified in modern tidal models, as explained by Ray (1998). Fortunately, iterative approaches are available (e.g., Arbic et al., 2004a; Griffiths and Peltier, 2009), in which an approximation to h_{sal} is determined every few tidal periods (using a full spherical harmonic transform) and then updated periodically.

An alternative to time-stepping models is to work in the frequency domain. Because tides are approximately linear over much of the ocean, the dominant frequencies of the response are known *a priori* as equal to those of the forcing. Assuming that the solution takes the form of Equation (29.1), the time-dependence of the problem is removed and, in the absence of non-linear terms, an elliptic partial differential equation is obtained to solve for each tidal constituent on the surface of the sphere. Given a low-order discretization of the spatial operators, Equations (29.2) and (29.3) may be written as a large sparse matrix system that may be inverted directly. This approach is much faster than time-stepping methods, but is memory intensive. It has been implemented (globally) by Lyard et al. (2006) and Hill et al. (2011), and the same ideas are implicit in the assimilative numerical schemes of Egbert and Erofeeva (2002). Note that non-linear bottom friction must be dealt with approximately or iteratively, as must self-attraction and loading, which corresponds to a dense matrix operation.

With appropriate tuning, both classes of models lead to globally averaged errors of 5–10 cm for the M_2 tide and about 2–3 cm for the K_1 tide (e.g., Arbic et al., 2004a, 2010; Egbert et al., 2004; Griffiths and Peltier, 2009). These are root-square-sum errors that take account of errors in both amplitude and phase, and largely reflect accuracy in the deep ocean. Errors on the continental shelves will be larger than those in the deep ocean, so global models (in isolation) are of limited use for studying tidally induced changes in sea-level records. A typical set of model results is shown in Figure 29.1c, d, which was produced using the model of Griffiths and Peltier (2009).

29.2.4 Three-dimensional ocean models

Depth-averaged models are unable to resolve the stratified dynamics of the ocean interior. We have already noted that such dynamics are central to the generation of internal tides, which is an important process in setting global tidal amplitudes through the feedback of internal tide drag on the depth-averaged flow. In a depth-averaged model, this internal tide drag must be parameterized. However, as computing power increases, it is becoming possible to model both the surface and internal tides simultaneously using three-dimensional ocean models. Although such tidal modeling is routinely performed on a regional scale, at the present time significant challenges remain for global modeling.

The main demands of three-dimensional modeling are associated with spatial resolution. In addition to the extra demands of vertical resolution and discretization, the horizontal scale of internal tides poses a serious problem. Because internal waves are driven by much weaker restoring forces than surface waves, they have much slower wave speeds (c. 1 m s^{-1} in the deep ocean, and less in shallow seas). Thus, at a given frequency, they have a much smaller wavelength than surface waves. At tidal frequencies, the first few internal modes have wavelengths of 10–100 km in the deep ocean and 1–10 km in shallow seas. A grid spacing of perhaps 1 km will therefore be required to resolve the dynamics of the first few internal wave modes over much of the ocean. However, the slow internal wave speeds can work in favor of the modeler, since much larger time-steps may be taken for the internal dynamics, if they can be separated from those of the depth-averaged dynamics.

The first such three-dimensional tidal model (Arbic et al., 2004a) took the extreme case of a

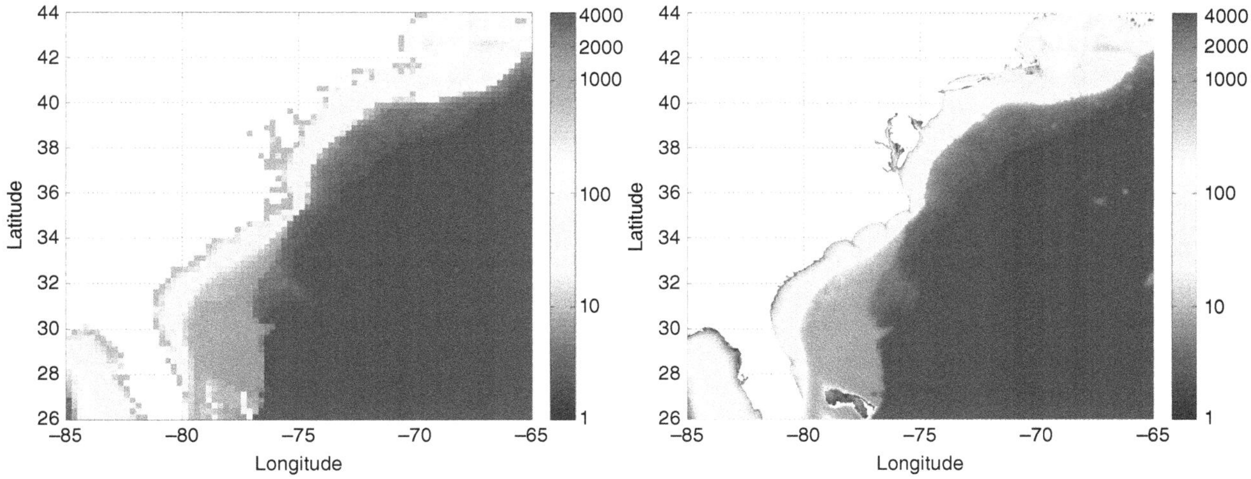

Fig. 29.2. Left: low-resolution bathmetry from the TPXO7.2 global inverse tidal solution (Egbert et al. 1994. *Source*: Reproduced with permission of John Wiley & Sons). Right: high-resolution bathymetry from the EC2001_V2D tidal database (Mukai et al., 2002). For color details, please see Plate 40.

two-layer ocean solved at 1/4 degree resolution, with implied limitations for the integrity of the internal tides. There have been steady improvements in horizontal and vertical resolution. Now state-of-the-art ocean general circulation models are being used to model tides: recent results have been obtained with 32 layers at 1/12.5 degree resolution (Arbic et al., 2010), and 40 layers at 1/10 degree resolution (Müller et al. 2012). However, these models do not produce surface tides that are more accurate than those obtained from depth-averaged models (in which internal tide drag may be tuned). Nevertheless, three-dimensional models are the future of global tidal modeling, although even higher resolution is required to resolve all the internal tide drag. A further benefit of using ocean general circulation models is that additional physics may be incorporated easily, such as the interaction of tides with eddies, the effects of sea ice, or seasonal changes in stratification.

29.3 REGIONAL AND LOCAL MODELING

Regional and local modeling trade horizontal coverage for increased spatial resolution. Here, *regional* suggests a domain on the order of 1000 km in extent and *local* suggests a domain on the order of 100 km. For brevity we will use the term regional to encompass both of these scales. The need for higher resolution than that afforded by typical 1/4 degree global models is demonstrated in Figure 29.2. Shown are low- and high-resolution bathymetries for the western Atlantic Ocean. The colorbar has been saturated at 4000 m and a logarithmic scale used in order to better highlight the differences in shallow areas. It is clear that a coarse grid filters out many features of interest to sea-level studies including bays, rivers, and barrier islands, among others. Simple extrapolation of results from global model output to these unresolved regions will be generally unsatisfactory due to non-linear interactions, wetting and drying of intertidal regions and other processes unresolved by a global approach. Lyard et al. (2006) conclude their review of recent tidal modeling advances by noting that improvements in coastal and shelf regions are needed and that regional models are the first step in this process.

29.3.1 Governing equations and strategies

In principle, the formulations discussed in Section 29.2 can also be applied to regional scales. The primary difference between global and regional applications is the presence of one or more open ocean boundaries along which various boundary conditions must be prescribed. In the case of depth-averaged regional models, only water surface elevations or depth-averaged currents along the open boundary are required. These data, in the form of constituent amplitudes and phases, abound. Example databases include those based upon altimetric data (e.g., GOT99.2; Ray, 1999), purely dynamical models (e.g., FES95.2; Le Provost et al., 1997), and dynamical models that also assimilate data (e.g., TPXO, Egbert et al., 1994; FES2004, Lyard et al., 2006). A common and straightforward modeling strategy (e.g., Hinton, 1996; Shennan et al., 2003; Uehara et al., 2006; Hill et al., 2011) is that of "nesting", where global tidal analysis is used to determine tidal constituents on a coarse grid; these results are interpolated and used to force the open boundary of a regional grid. This is most often done in a "decoupled" fashion in which the global model and regional model are not dynamically linked. The open boundary of the regional model should, when possible, be placed in deep water. Most global tidal databases contain a limited set (8–10) of constituents that often omit the overtides generated in shallow water.

In practice however, regional models are often implemented with a smaller set of included physical processes. For example, it is common to conduct local tidal modeling studies without directly including internal tide drag, self-attraction, and loading effects. The implicit assumption is that the effects of those processes have been captured in the global tidal model results, which are in turn being used to force the smaller domain. As a second example, the relatively small horizontal scale, particularly for a local study, often permits the use of projected coordinates and an f-plane approximation.

The relatively shallow water of regional tidal model domains often requires modification of Equations (29.2) and (29.3). As one example, quadratic friction laws with a spatially constant friction coefficient can be replaced by hybrid friction laws that specify a deep-water coefficient and then transition to larger values in shallow water. As a second example, in domains that have large intertidal areas, the wetting and drying (Bunya et al., 2009) of these regions will produce a significant amount of dissipation of energy and should be included.

The extremely shallow water of local studies brings with it other considerations. Bathymetry data are typically given relative to a low water datum such as mean low water (MLW) or mean lower low water (MLLW). Topographic data are typically given relative to a geodetic datum such as NAVD88. Combined bathymetric/topographic grids, which are needed for wetting/drying and for inundation studies, require adjustment of one of the datasets to the datum of the other. Additionally, the bathymetric offset from MLLW to mean sea level (MSL) needs to be applied. In deep-water (global) applications, these offsets are a negligible fraction of the water column depth and often ignored. In shallow water applications (e.g., Bunya et al., 2010), they can be significant. Traditionally, these adjustments were made using coarse interpolations from a limited number of coastal gauging stations. Modern products, such as VDatum (NOAA), use high-resolution regional models to calculate these offsets instead.

For sea-level studies, there is generally considerable interest in tidal datums (mean higher high water; MHHW, etc.) in nearshore regions, including estuaries, rivers, and barrier islands. Tidal model output in the form of constituent amplitudes and phases is easily converted to information on datums in one of two ways. First, time series water elevation output may be post-processed to obtain the datums. For a large domain, it is far more computationally efficient to use algorithms (Mofjeld et al., 2004) that instead convert constituent amplitudes and phases directly to datums.

29.3.2 Numerical modeling

Again, much of the discussion from Section 29.2 remains relevant on smaller spatial scales; here we seek only to point out some notable differences in approach and notable consequences of the smaller scale. First, most global ocean models have traditionally used regular latitude–longitude rectangular grids. In addition to the difficulties this choice presents in the vicinity of the poles, it yields poor representation of the complex boundary formed by the shoreline (Fig. 29.2). Some global models (e.g., Lyard et al., 2006; Stuhne and Peltier, 2009) and many regional models (Luettich and Westerink, 1991; Fringer et al., 2006; Chen et al., 2007) instead use an unstructured finite-element grid. These grids are variable in resolution, allowing for refinement as needed to meet various grid quality and stability criteria. Grid generation has seen recent advances (Hagen et al., 2001) that seek optimal node placement based upon the minimization of truncation errors.

Second, the computational demands of a local model can be surprising. A regional model with the same number of nodes as a global model will require more computational time due to the smaller grid spacing and the stability constraints on a temporally explicit numerical scheme. As an example, a local grid with a minimum spacing of

30 m and a water depth of 10 m will require a time-step of less than 3 s; in contrast a time-step of about 20 s is required for a global model with a grid spacing of about 4 km (as detailed in Section 29.2).

Appropriately tuned regional models can be very successful in predicting tidal constituents at the coastline. As an example, Spargo et al. (2004) developed a regional model database for the north-eastern Pacific Ocean and compared their results against data from 91 stations along the west coast of North America. The overall average error (averaged over 8 constituents and 91 stations) in constituent amplitude was approximately 7%. At stations where more than one data source was available, it was shown that the average difference between the data records was on the order of 5%. Much of the difference between model output and data may therefore be ascribed to variability in the data itself.

29.4 PALEOTIDAL MODELING

Tidal amplitudes are sensitive to the frequency of the astronomical forcing (i.e., the orbital configuration), ocean depth, density stratification, and coastal configuration. All of these have changed considerably in the past, and the implied changes in ocean tides are of interest for several reasons as follows:

(1) Estimates of sea-level change during the Holocene are often derived from measurements of MHHW, which is of course influenced by tidal amplitudes. Changes in tidal amplitudes may therefore need to be removed from such records before sea-level changes can be properly deduced.
(2) Changes in nearshore tidal currents may lead to changes in mixing, with ecological implications, and changes in erosion rates or sediment transport pathways, with implications for bed evolution.
(3) In the deep ocean, internal tides are thought to lead to vertical mixing that plays a role in setting the large-scale temperature and circulation of the oceans (e.g., Wunsch, 2000). The idea that changes in the magnitude of internal tide generation could have led to changes in ocean circulation and global climate has been explored in idealized models by Wunsch (2005), and in coupled atmosphere–ocean models by Green et al. (2009).
(4) Tidal dissipation controls the rate at which the Moon recedes from the Earth and the corresponding slowing of the Earth's rotation (e.g., Lambeck, 1977). Observational evidence and model results related to paleotidal dissipation are reviewed by Kagan (1997).

Of course, the details of such issues can only be resolved with the kind of dynamical tidal modeling that we have examined in Sections 29.2 and 29.3, with appropriately modified input datasets and boundary conditions. Changes in the strength of the astronomical forcing from present-day are of the order of 0.01% per century (Cartwright, 1977), and can therefore be neglected for studies over the past 50 ka; see Munk and Bills (2007) for speculation of some longer-period effects. The remaining factors to consider are changes in stratification and bathymetry.

29.4.1 Global modeling considerations

Ocean stratification is an important input for global models, but is poorly constrained for past climates. Estimates of salinity and temperature at the seafloor are available from sediments (e.g., Adkins et al., 2002), which may permit some crude estimates of stratification. Stratification can also be estimated from coupled atmosphere–ocean climate simulations, although the structure of the ocean interior in such models is often sensitive to mixing parameterizations. This uncertainty is a potentially serious issue, since it is likely that internal tide drag (rather than bottom boundary layer drag) was the dominant dissipation mechanism under glacial conditions, when the shallow marginal seas of the present-day were absent. For further discussion, see Griffiths and Peltier (2009), or the sensitivity analysis of Egbert et al. (2004).

There have been significant changes to bathymetry over the last 100 ka, associated with quasi-periodic glacial climates during which large continental ice sheets existed in the northern hemisphere and sea level was reduced by 100 m or more, on average. There is a well-established theory of glacial isostatic adjustment (GIA, see Chapter 28) which leads to topographic reconstructions that take into account the water taken up by the ice sheets, the deformation of the Earth beneath the ice sheets, and the slow rebound once the ice sheets have been removed. Important inputs are the spatial and temporal location of the ice sheets, their volume, and the assumed viscoelastic structure of the Earth. The most widely used reconstructions are those of Peltier; his ICE-4G (Peltier, 1996) and ICE-5G (Peltier, 2004) reconstructions have both been widely used in paleotidal models (e.g., Thomas and Sündermann, 1999; Egbert et al., 2004; Green et al., 2009). Paleobathymetry according to the ICE-5G v. 1.3 reconstruction is shown in Figure 29.3. Note how there are rapid changes in the geometry of the North Atlantic with the blockage of the Hudson Strait and the English Channel between around 8 ka and 12 ka. At 26 ka, most of the shallow marginal seas of the present-day ocean have disappeared.

In addition to changes in bathymetry, changes in the amount of sea-ice and floating ice shelves can also be considered for some tidal models. However, the extent of these changes is poorly constrained.

29.4.2 Regional modeling considerations

Regional studies of paleotides confront several important issues. First, bathymetric depths must be properly adjusted from present-day values. Many previous studies (Scott and Greenberg, 1983; Gehrels et al., 1995; Hinton, 1995, 1996; Uehara et al., 2002; Leorri et al., 2011) have adopted spatially uniform water surface declines (a so-called "bathtub" model). In many cases, the decline was tied to a particular time in the past through use of local sea-level curves based upon field data. Other studies (Shennan et al., 2003; Uehara et al., 2006; Hill et al., 2011; Hall et al., 2013) used GIA models to apply spatially variable adjustments from present-day bathymetric depths. In some cases, these adjustments can have strong spatial gradients that are important on local scales. For example, shown in Figure 29.4 is the GIA

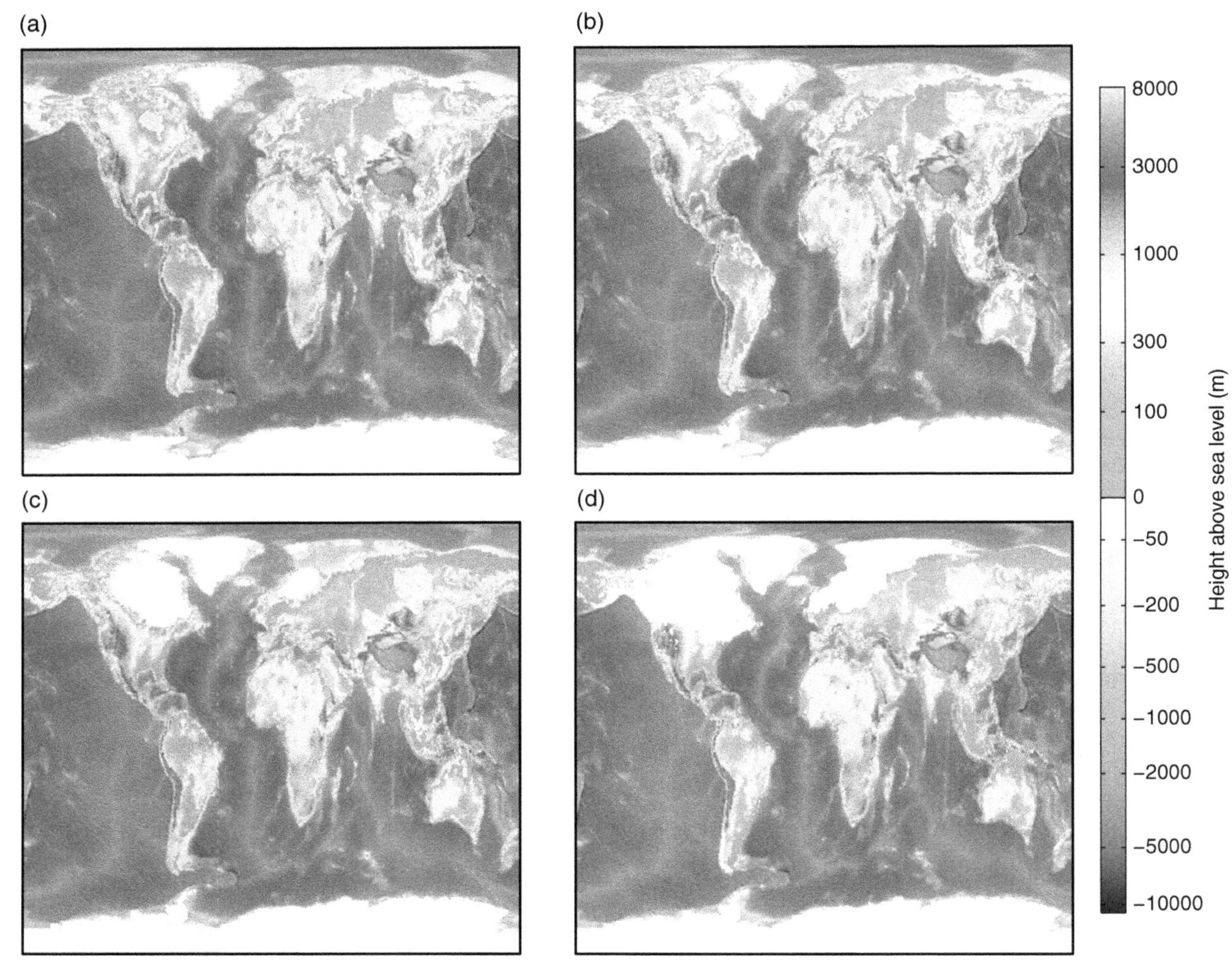

Fig. 29.3. Surface height (m) and ice location according to the ICE-5G v.1.3 dataset, an updated version of that described by Peltier (2004). (a) present day, (b) 8 ka, (c) 12 ka, and (d) 26 ka (LGM). For color details, please see Plate 41.

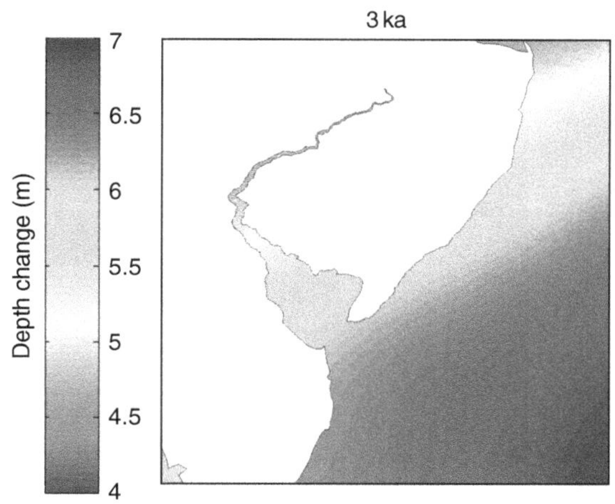

Fig. 29.4. GIA adjustment from 3 ka to present day for the Delaware Bay, USA. For color details, please see Plate 42.

adjustment (using ICE-6G/VM5B) from 3 ka to the present day in the Delaware Bay, USA. Between the lower and upper portions of the bay (separated by only 100 km) there is a difference of 1.5 m in the size of the GIA signal.

Second, regional models of paleotides must decide how to treat the temporal variability of the open boundary forcing. Many studies (Scott and Greenberg, 1983; Gehrels et al., 1995; Hinton, 1995, 1996; Shennan et al., 2003; Leorri et al., 2011) assumed, without justification, that open ocean tides would not change during the Holocene. Other studies (Uehara et al., 2002; Hall et al., 2013) justified their decisions to hold open boundary forcing fixed in time through the use of global tidal models that showed minimal changes over the period of their studies. Uehara et al. (2006) conducted parallel studies, including one in which open boundary forcing was kept constant and a second in which it was varied according to simulations from a global tidal model.

29.4.3 Behavior during the Holocene

There have been several numerical modeling studies of tides over the Holocene (11.5 ka to the present day). From about 7 ka to the present day, there appear to have been only small changes at the basin scale (Uehara et al., 2006; Hill et al., 2011). However, large changes at a regional scale have been demonstrated, usually associated with changes in the resonance characteristics of a particular estuary or nearshore area that accompanied rising sea level. For example, the nested simulations of Hill et al. (2011) performed at 1 ka intervals show small changes at the basin scale from 7 ka to the present day, but temporally resolve a large amplification of tides in the Gulf of Maine and Bay of Fundy at around 4 ka. This latter finding is consistent with earlier studies (e.g., Gehrels et al., 1995).

Model results suggest there were more significant changes in both global and regional tides in the early Holocene, defined as 11.5–7 ka, in line with the significant topographic changes shown in Figure 29.3b, c. Thomas and Sündermann (1999), Egbert et al. (2004), Uehara et al. (2006), Arbic et al. (2008) and Green et al. (2009) all found that the amplitude and global dissipation of the M_2 tide is considerably higher at 10 ka than at present. Hill et al. (2011) found a substantial basin-scale amplification of semi-diurnal tidal consitituents (and, correspondingly, tidal range) in the western North Atlantic that was constrained to occur at around 9 ka (Fig. 29.5). Numerous theories have been advanced to explain these phenomena, all of which are tied to the significant rise in sea level that occured during the early Holocene. For example, this rise and the retreat of ice sheets leading to the opening of seas such as Hudson Bay (Arbic et al. 2008) greatly increased the areal extent of shallow marginal seas, increasing dissipation and reducing tidal amplitudes.

29.4.4 Behavior at the Last Glacial Maximum

Topographic changes at the Last Glacial Maximum (LGM), which is thought to have occurred at about 26 ka, were larger again, as shown in Figure 29.3d. There have been several modeling studies of tides for this topographic configuration, which was presumably typical of quasi-periodic glacial climates further back in time. All of these studies (e.g., Egbert et al., 2004; Uehara et al., 2006) have found globally energized semi-diurnal tides at the LGM, with a strong amplification in the North Atlantic. The dominant dissipation mechanism switches from drag in shallow seas to internal tide generation in the deep ocean. However, changes in diurnal tides appear to have been less pronounced on a global scale.

Of special interest are tides in the polar regions, due to their possible interaction with adjoining ice shelves and ice streams. Arbic et al. (2004b) were the first to speculate that the amplified tides of the Labrador Sea under glacial climates may have interacted with the Laurentide ice sheet to trigger discharges of ice into the North Atlantic (so called Heinrich events). More detailed studies of polar

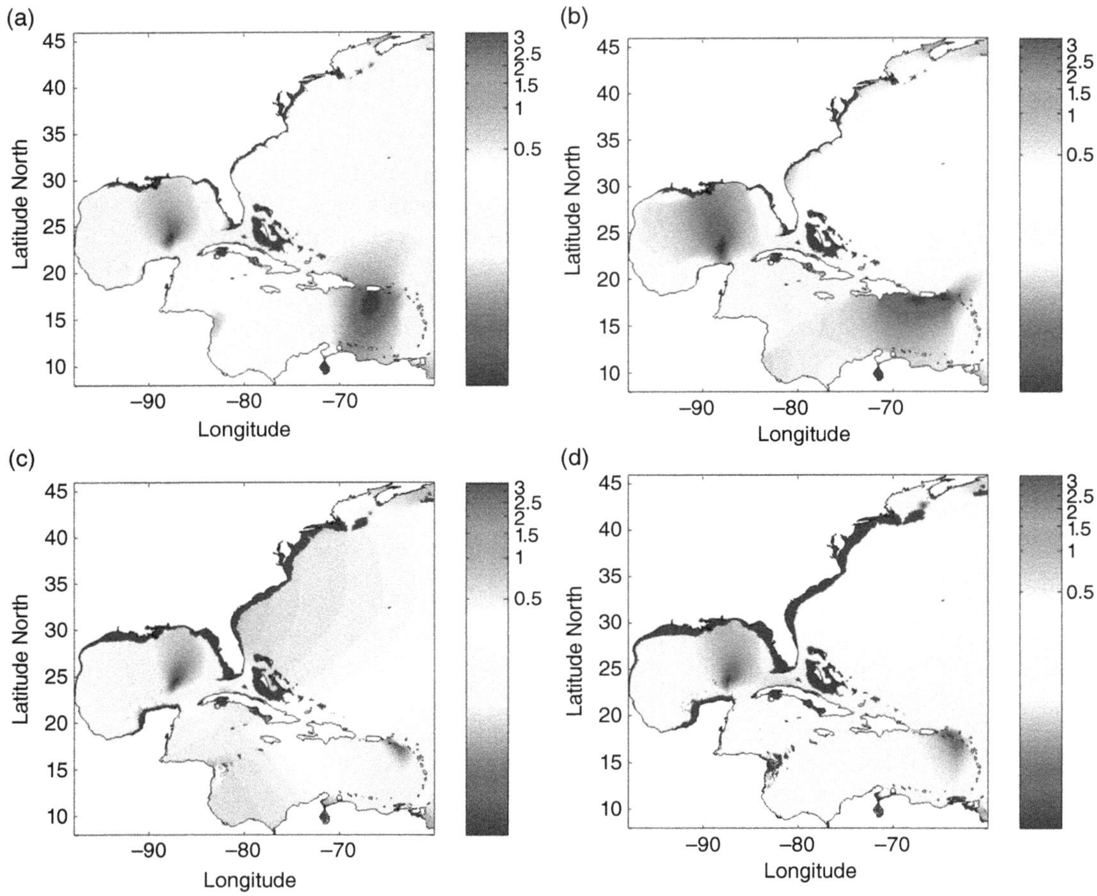

Fig. 29.5. Variation of M_2 tidal amplitude during the early Holocene. Panels (a–d) correspond to 7, 8, 9, and 10 ka, respectively. *Source*: Hill et al. (2011). Reproduced with permission. For color details, please see Plate 43.

tides have also shown a semi-diurnal resonance in the Arctic Ocean, which is almost a closed basin at the LGM (Griffiths and Peltier, 2008), and amplifications of Antarctic tides (Griffiths and Peltier, 2009). However, these polar amplifications are sensitive to the assumed location of the grounding lines of coastal ice sheets, which leads to the possibility of interactions between tides and climate forcing.

29.4.5 Application to sea-level index points

Relative sea level (RSL) integrates eustatic, isostatic, tectonic, and local processes (Horton et al., 2013). Relevant local processes include sediment compaction and change in tidal range. There is a clear physical path by which RSL changes can lead to tidal range changes (Hill et al., 2011) since RSL changes alter water depth and shelf width, both of which help to control tidal wave amplification and dissipation. There is also a feedback, where tidal range changes affect RSL change estimates. Sea-level "index points" are defined as reconstructions of sea level at a particular location, with quantified uncertainties in age and in vertical position. Establishment of a sea-level index point requires knowledge of the indicative range (Shennan, 1986), which is the elevational range occupied by a particular sea-level indicator, and the vertical relationship between that indicator and some reference water level. As an example (van de Plassche, 1986), marsh plants, saltmarsh foraminifera, diatoms, and barnacles are all found consistently near the high tide level and can therefore be used to relocate former high tide. Knowledge of tidal range can then be used to relocate other tidal datums, such as mean tide level (MTL). If tidal range has changed over time, however, an adjustment between tidal datums based on modern values will not be applicable at earlier times. As a result, the established RSL

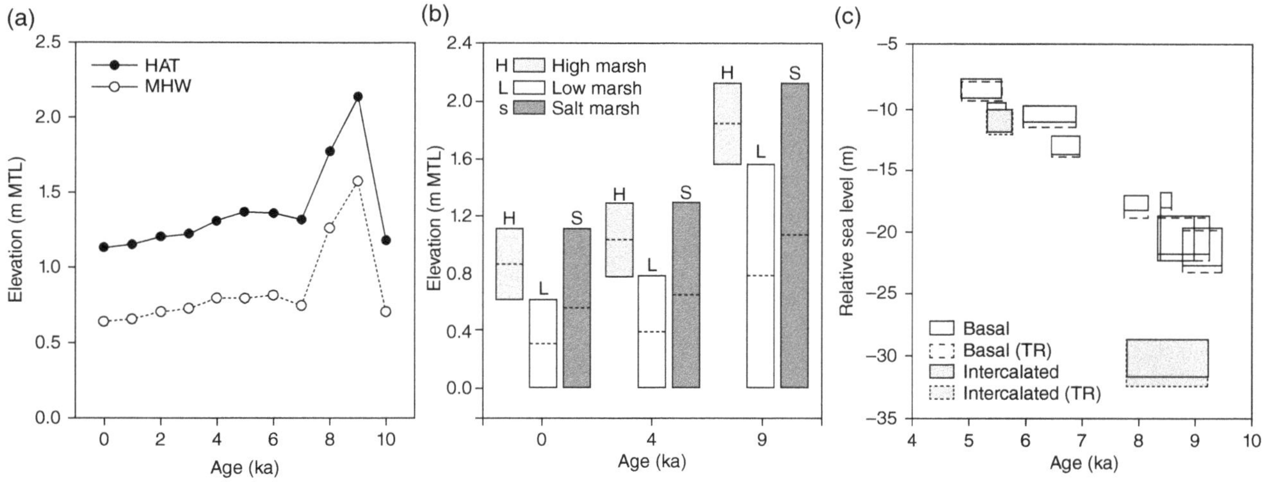

Fig. 29.6. (a) Tidal data used to generate the indicative meanings (HAT, highest astronomical tidea; MHW, mean high water; MTL, mean tide level) during the Holocene for Brigantine, New Jersey, USA. (b) Reference water level and indicative range of saltmarsh, high marsh, and low marsh at 0, 4, and 9 ka for Brigantine, New Jersey, USA. (c) Sea-level index points with and without tidal range change correction in the middle and early Holocene. *Source*: Reproduced with permission from Horton et al. (2013).

curve will differ from the "true" RSL curve (Gehrels et al. 1995).

An illustrative example is provided in Figure 29.6 (Horton et al., 2013). The left panel shows paleotidal model results at Brigantine, New Jersey, USA from 10 ka to the present day. This time series result shows the same significant amplification demonstrated across the domain in Figure 29.5. The center panel shows the reference water levels (dotted lines) and indicative range (box heights) of high, low, and saltmarsh indicators at 0, 4, and 9 ka for the same location. Finally, the right-hand panel shows RSL curves without (solid) and with (dotted) tidal range corrections.

29.5 CONCLUDING REMARKS

The physics of the tidal system is thought to be well understood, but numerical modeling of the full tidal system remains a challenge. At a global level, purely dynamical models (with no data assimilation) are unable to adequately resolve the dissipation of the barotropic tide; all or part of the dissipation due to turbulent bottom drag in shallow seas and internal tide generation in the deep ocean has to be parameterized. Three-dimensional stratified global tidal models can now resolve a good portion of the internal tide generation, but higher horizontal resolution is still required.

Coastal tides from purely dynamical global models are generally not of sufficient accuracy for use in studies of sea level. This deficiency can be reduced somewhat by using a model with enhanced resolution on the continental shelves, but typically some kind of regional modeling (using open ocean boundary conditions from observed tides or a constrained global model) is required for acceptable nearshore results. Since any changes in coastal tides need to be accounted for when interpreting sedimentary records used to construct relative sea-level history, there is a particular need for improved accuracy in purely dynamical global paleotidal models, which can be used to force regional paleotidal models.

Due to the importance of tidal resonances, even relatively small changes in sea-level or coastal configuration can lead to non-trivial changes in local tides, as has been shown by various analytical and numerical tidal models. There is therefore an obvious challenge to model how tides might change in the future in response to rising sea levels and changes in stratification associated with climate change.

REFERENCES

Adkins, J., McIntyre, K., and Schrag, D. (2002) The salinity, temperature, and $\delta^{18}O$ of the glacial deep ocean. Science, 298, 1769–1773.

Antonov, J., Locarnini, R., Boyer, T., Mishonov, A., and Garcia, H. (2006) World Ocean Atlas 2005, Volume 2: Salinity. NOAA Atlas NESDIS 62 ed., US Government Printing Office, Washington, DC.

Arbic, B., Garner, S., Hallberg, R., and Simmons, H. (2004a) The accuracy of surface elevations in forward global barotropic and baroclinic tide models. Deep-Sea Research, 51, 3069–3101.

Arbic, B., MacAyeal, D., Mitrovica, J., and Milne, G. (2004b) Palaeoclimate: Ocean tides and Heinrich events. Nature, 432, 460.

Arbic, B., St-Laurent, P., Sutherland, G., and Garrett, C. (2007) On the resonance and influence of the tides in Ungava Bay and Hudson Strait. Geophysical Research Letters, 34, 2007.

Arbic, B., Mitrovica, J., MacAyeal, D., and Milne, G. (2008) On the factors behind large Labrador sea tides during the last glacial cycle and the potential implications for Heinrich events. Paleooceanography, 23, 1–14.

Arbic, B., Wallcraft, A., and Metzger, J. (2010) Concurrent simulation of the eddying general circulation and tides in a global ocean model. Ocean Modelling, 32, 175–187.

Bowen, A. J. (1972) The tidal regime of the River Thames; long-term trends and their possible causes. Philosophical Transactions of the Royal Society of London. Series A, Mathematical and Physical Sciences, 272, 187–199.

Bunya, S., Kubatko, E., Westerink, J., and Dawson, C. (2009) A wetting and drying treatment for the Runge-Kutta discontinuous Galerkin solution to the shallow water equations, Computational Methods in Applied Mathematics and Engineering, 198, 1548–1562.

Bunya, S., Dietrich, J.C., Westerink, J.J. et al. (2010) A high-resolution coupled riverine flow, tide, wind, wind wave and storm surge model for southern Louisiana and Mississippi. Part 1: Model development and validation. Monthly Weather Review, 138, 345–377.

Cartwright, D. (1972) Secular changes in the oceanic tide at Brest. Geophysical Journal Royal Astronomical Society, 30, 433–449.

Cartwright, D. (1977) Oceanic tides. Reports in Progress in Physics, 40, 665–708.

Cartwright, D. (1999) *Tides: A Scientific History*. Cambridge University Press, Cambridge.

Chen, C., Huang, H., Beardsley, R., Liu, H., Xu, Q., and Cowles, G. (2007) A finite volume numerical approach for coastal ocean circulation studies: comparisons with finite difference models. Journal of Geophysical Research, 112, C03018.

Dehant, V., Defraigne, P., and Wahr, J. (1999) Tides for a convective Earth. Journal of Geophysical Research, 104, 1035–1058.

Egbert, G., and Erofeeva, S. (2002) Efficient inverse modeling of barotropic ocean tides. Journal of Atmospheric and Ocean Technology, 19, 183–204.

Egbert, G., and Ray, R. (2003) Semi-diurnal and diurnal tidal dissipation from TOPEX/Poseidon altimetry. Geophysical Research Letters, 30(17), doi: 10.1029/2003GL017676.

Egbert, G., Bennett, A., and Foreman, M. (1994) TOPEX/Poseidon tides estimated using a global inverse model. Journal of Geophysical Research, 99, 24,821–24,852.

Egbert, G., Ray, R., and Bills, B. (2004) Numerical modeling of the global semidiurnal tide in the present day and in the Last Glacial Maximum. Journal of Geophysical Research, 109, doi: 10.1029/2003JC001973.

Fringer, O., Gerritsen, M., and Street, R. (2006) An unstructured-grid and finite-volume nonhydrostatic, parallel coastal ocean simulator. Ocean Modelling, 14, 139–178.

Garrett, C. (1972) Tidal resonance in the Bay of Fundy and Gulf of Maine. Nature, 238, 441–443.

Garrett, C., and Kunze, E. (2007) Internal tide generation in the deep ocean. Annual Review of Fluid Mechanics, 39, 57–87, 2007.

Gehrels, W., Belknap, D., Pearce, B., and Gong, B. (1995) Modeling the contribution of M2 tidal amplification to the Holocene rise of mean high water in the Gulf of Maine and the Bay of Fundy. Marine Geology, 124, 71–85.

Green, J., Green, C., Bigg, G., Rippeth, T., Scourse, J., and Uehara, K. (2009) Tidal mixing and the strength of the Meridional Overturning Circulation from the Last Glacial Maximum. Geophysical Research Letters, 36, doi: 10.1029/2009GL039309.

Green, M., and Nycander, J. (2013) A comparison of tidal conversion parameterizations for tidal models. Journal of Physical Oceanography, 43, 104–119.

Griffiths, S., and Peltier, W. (2008) Mega-tides in the Arctic Ocean under glacial conditions. Geophysical Research Letters, 35, doi: 10.1029/2008GL033263.

Griffiths, S., and Peltier, W. (2009) Modeling of polar ocean tides at the Last Glacial Maximum: amplification, sensitivity, and climatological implications. Journal of Climate, 22, 2905–2924.

Hagen, S., Westerink, J., Kolar, R., and Horstman, O. (2001) Two-dimensional unstructured mesh generation for tidal models. International Journal for Numerical Methods in Fluids, 35, 669–686.

Hall, G., Hill, D., Horton, B., Engelhart, S., and Peltier, W. (2013) A high resolution study of tides in the Delaware Bay: past conditions and future scenarios. Geophysical Research Letters, 40(2), 338–342.

Hendershott, M. (1981) Long waves and ocean tides. In *Evolution of Physical Oceanography* (eds Warren, B., and Wunsch, C.), MIT Press, pp. 292–341.

Hill, D., Griffths, S., Peltier, W., Horton, B., and Tornqvist, T. (2011) High-resolution numerical modeling of tides in the western Atlantic, Gulf of Mexico, and Caribbean Sea during the Holocene. Journal of Geophysical Research, 116, C10014.

Hinton, A. (1995) Holocene tides of The Wash, UK: the influence of water-depth and coastline-shape changes on the record of sea-level change. Marine Geology, 124, 87–111.

Hinton, A. (1996) Tides in the northeast Atlantic: considerations for modeling water depth changes. Quaternary Science Reviews, 15, 873–894.

Horton, B., Engelhart, S., Hill, D., Kemp, A., Nikitina, D., Miller, K., and Peltier, W. (2013) Influence of tidal-range change and sediment compaction on Holocene relative sea-level change in New Jersey, USA. Journal of Quaternary Science, 28, 403–411.

Kagan, B. (1997) Earth–Moon tidal evolution: model results and observational evidence. Progress in Oceanography, 40, 109–124.

Lambeck, K. (1977) Tidal dissipation in the oceans: astronomical, geophysical and oceanographic consequences. Philosophical Transactions of Royal Society of London, 287, 545–594.

Laplace, P.-S. (1776) *Recherches sur Plusieurs Points du Systeme du Monde.* Memoires de l'Academie Royale Sciences de Paris, Paris, pp. 187–264.

Le Provost, C. L., Genco, M., Lyard, F., Vincent, P., and Canceil, P. (1997) Spectroscopy of the ocean tides from a finite element hydrodynamic model. Journal of Geophysical Research, 99, 24,777–24,797.

Leorri, E., Mulligan, R., Mallinson, D., and Cearreta, A. (2011) Sea-level rise and local tidal range changes in coastal embayments: an added complexity in developing reliable sea-level index points. Journal of Integrated Coastal Zone Management, 241, 307–314.

Locarnini, R., Mishonov, A., Antonov, J., Boyer, T., and Garcia, H. (2006) *World Ocean Atlas 2005,* Volume 1: Temperature. NOAA Atlas NESIS 61 ed., US Government Printing Office, Washington, DC.

Luettich, R., and Westerink,J. (1991) A solution for the vertical variation of stress, rather than velocity, in a three-dimensional circulation model. International Journal for Numerical Methods in Fluids, 12, 911–928.

Lyard, F., Lefevre, F., Letellier, T., and Francis, O. (2006) Modelling the global ocean tides: modern insights from FES2004. Ocean Dynamics, 56, 394–415.

Marchuk, G., and Kagan, B. (1989) *Dynamics of Ocean Tides.* Springer.

Mofjeld, H., Venturato, A., Gonzales, F., Titov, V., and Newman, J. (2004) The harmonic constant datum method: options for overcoming datum discontinuities at mixed-diurnal tidal transitions. Journal of Atmospheric and Oceanic Technology, 21, 95–104.

Mukai, A., Westerink, J., Luettich, R., and Mark, D. (2002) Eastcoast 2001, a tidal constituent database for western north Atlantic, Gulf of Mexico and Caribbean Sea. Technical Report ERDC/CHL TR-02-24, US Army Corps of Engineers, 25 pp.

Müller, M. (2007) The free oscillations of the world ocean in the period range 8 to 165 hours including the full loading effect. Geophysical Research Letters, 34, L50606.

Müller, M., Cherniawsky, J., Foreman, M., and von Storch, J.-S. (2012) Global M_2 internal tide and its seasonal variability from high resolution ocean circulation and tide modelling. Geophysical Research Letters, 39, L19607.

Munk, W., and Bills, B. (2007) Tides and the climate: some speculations. Journal of Physical Oceanography, 37, 135–147.

Peltier, W. (1996) Mantle viscosity and ice-age ice sheet topography. Science, 273, 1359–1364.

Peltier, W. (2004) Global glacial isostasy and the surface of the ice-age Earth: The ICE-5G (VM2) model and GRACE. Annual Reviews of Earth and Planetary Science, 32, 111–149.

Platzman, G. (1984) Normal modes of the world ocean. Part IV: Synthesis of diurnal and semidiurnal tides. Journal of Physical Oceanography, 14, 1532–1550.

Ray, R. (1998) Ocean self-attraction and loading in numerical tidal models. Marine Geodesy, 21, 181–192.

Ray, R. (1999) A global ocean tide model from TOPEX/Poseidon altimetry: GOT99, NASA Technical Memorandum 209478, Goddard Space Flight Center, Greenbelt, MD, USA.

Scott, D., and Greenberg, D. (1983) Relative sea-level rise and tidal development in the Fundy tidal system. Canadian Journal of Earth Sciences, 20, 1554–1564.

Shennan, I. (1986) Flandrian sea-level changes in the Fenland, II: Tendencies of sea-level movement, altitudinal changes, and local and regional factors. Journal of Quaternary Science, 1, 155–179.

Shennan, I., Coulthard, T., Flather, R., Horton, B., Macklin, M., Rees, J., and Wright, M. (2003) Integration of shelf evolution and river basin models to simulate Holocene sediment dynamics of the Humber Estuary during periods of sea-level change and variations in catchment sediment supply. Science of the Total Environment, 314–316, 737–754.

Simmons, H., Hallberg, R., and Arbic, B. (2004) Internal wave generation in a global baroclinic tide model. Deep-Sea Research, 51, 3043–3068.

Smith, W., and Sandwell, D. (1997) Global seafloor topography from satellite altimetry and ship depth soundings. Science, 277, 1957–1962.

Spargo, E., Westerink, J., Luettich, R., and Mark, D. (2004) ENPAC 2003: A tidal constituent database for Eastern North Pacific Ocean. Technical Report ERDC/CHL TR-04-12, US Army Corps of Engineers.

Stuhne, G., and Peltier, W. (2009) An unstructured C-grid based method for 3-D global ocean dynamics: Free-surface formulations and tidal test cases. Ocean Modelling, 28, 97–105.

Taylor, G. (1920) Tidal friction in the Irish Sea. Philosophical Transactions of the Royal Society of London, 220, 1–33.

Thomas, M., and Sündermann, J. (1999) Tides and tidal torques of the world ocean since the Last Glacial Maximum. Journal of Geophysical Research, 104, 3159–3183.

Uehara, K., Saito, Y., and Hori, K. (2002) Paleotidal regime in the Changjiang (Yangtze) estuary, the East China Sea, and the Yellow Sea at 6 ka and 10 ka estimated from a numerical model. Marine Geology, 183, 179–192.

Uehara, K., Scourse, J., Horsburgh, K., Lambeck, K., and Purcell, A. (2006) Tidal evolution of the north west European shelf seas from the Last Glacial Maximum to the present. Journal of Geophysical Research, 111, 1–15.

van de Plassche, O. (1986) *Sea-level Research: A Manual for the Collection and Evaluation of Data.* Geobooks, Norwich.

Wunsch, C. (2000) Moon, tides and climate. Nature, 405, 743–744.

Wunsch, C. (2005) Speculations on a schematic theory of the Younger Dryas. Journal of Marine Research, 63, 315–333.

Chapter 30
Compaction

MATTHEW J. BRAIN
Sea Level Research Unit, Department of Geography, Durham University, Science Laboratories, Durham, UK

30.1 INTRODUCTION

Compaction describes a range of syn- and post-depositional diagenetic processes that result in volumetric reductions of sediments and distortion of stratigraphic sequences (see fig. 5 of Allen, 2000). In low-energy intertidal environments, compaction results in settlement and lowering of the depositional surface relative to the intertidal frame, causing a local-scale relative sea-level rise. Compaction also lowers sea-level index points (SLIs) from their depositional altitudes, resulting in errors in the estimation of the magnitude and rate of sea-level change reconstructed from intertidal sediments (Kaye and Barghoorn, 1964). In a rising sea-level sequence, compaction causes the rate of reconstructed sea-level rise to be overestimated if not corrected for (Fig. 30.1a). Where reconstructions of sea level display a simple regressive trend, the rate of sea-level fall is underestimated (Fig. 30.1b).

Studies that quantify compaction-induced rates of lowering demonstrate the potential magnitude of the resultant error in reconstructed sea level (Bloom, 1964). Horton and Shennan (2009) showed that compaction can double the rate of reconstructed sea-level rise. Isolating and correcting for the effects of compaction on sea-level records is therefore an important component of any research that relies on SLIs obtained from intertidal sedimentary successions.

Approaches to limit or correct for the effects of compaction on reconstructed sea level vary in complexity and range from using compaction-free basal peats to more detailed geotechnical modeling efforts. Such approaches are the focus of this chapter.

30.2 TERMINOLOGY

30.2.1 Compaction

Allen (2000, page 1186) defines (auto-)compaction as "...the group of interlinked processes whereby the sediment within a growing stratigraphic column diminishes in volume, on account of burial and self-weight, leading to a rearrangement of the mineral skeleton, and in the case of vegetable matter, a loss of mass as the result of biological and chemical decay...". This definition is favored because it is comprehensive, describing the full range of (bio-) mechanical and (bio-) chemical processes that lead to a volumetric reduction of sediments. Compaction of a sediment layer, an individual stratum or a full stratigraphic column is often expressed as a percentage of its original thickness.

30.2.2 Compression processes and effective stress

Mechanical compaction processes cause compression of sediments and result from the vertical stress (pressure) exerted by overburden materials. Compression is the sum of two key processes: consolidation and creep. Both consolidation and creep are time- and stress-dependent and result in decreased porosity and higher sediment bulk density.

Consolidation describes the time-dependent process of pore water expulsion, causing solid particles to be packed closer together. Consolidation results from an increase in effective stress σ', which is defined as:

$$\sigma' = \sigma - u \tag{30.1}$$

where σ is the total stress resulting from the weight of overburden sediments and u is the pore water pressure. Increases in effective stress result in

Handbook of Sea-Level Research, First Edition. Edited by Ian Shennan, Antony J. Long, and Benjamin P. Horton.

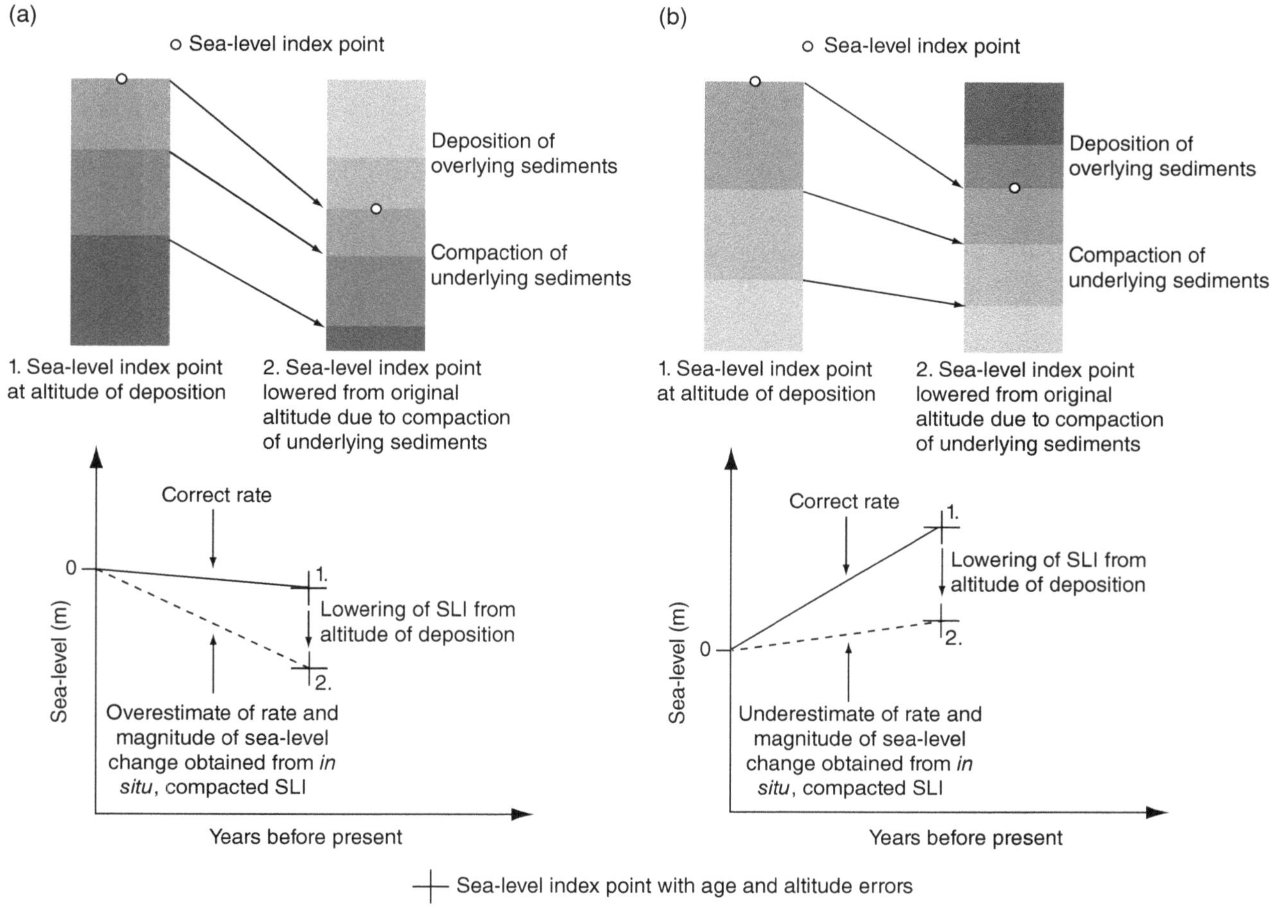

Fig. 30.1. The effects of sediment compaction on sea-level index points within low-energy intertidal stratigraphic successions and the subsequent effects on age-altitude reconstructions of sea level in (a) "transgressive" (deepening-upward) and (b) "regressive" (shallowing-upward) successions. Scales in (a) and (b) are indicative and are not directly comparable. See also Paul and Barras (1998), Horton and Shennan (2009), and van Asselen et al. (2009).

sediment deformation (Head, 1988). The volume change associated with consolidation is termed primary compression.

Creep describes the continued viscous movement and rearrangement of sediment particles. Conventionally, this has been assumed to occur following completion of primary consolidation (i.e., under constant effective stress), though both processes can occur simultaneously in organic sediments. The volume change attributable to creep is termed secondary compression. Creep can result in ongoing, prolonged compaction.

30.2.3 Biochemical processes and biodegradation

Compaction can occur as a result of interacting biological and chemical processes that cause humification of organic matter. Such processes include fragmentation, microbial decay, and chemical transformation, notably oxidation (Lillebø et al., 1999). The effects of humification can be profound and can transform a fibrous organic deposit into an amorphous, homogeneous black peat.

30.2.4 Post-depositional lowering

The net effect of compaction is post-depositional lowering (i.e., settlement), which describes compaction-induced changes in altitude of an SLI since deposition. Post-depositional lowering experienced throughout a sediment column varies with depth and is the sum of compaction in all underlying sediment layers. Post-depositional lowering is the height correction that must be added to the *in situ* altitude of an individual SLI to return it to its depositional altitude (Paul and Barras, 1998; Brain et al., 2012). To do so, a "decompaction" procedure must be undertaken.

30.2.5 Normal and over-consolidation

Sediment that has not experienced an effective stress greater than that generated by the existing overburden is described as normally consolidated. In contrast, sediments that have experienced effective stresses greater than the existing overburden are termed overconsolidated. The increase in density that results from the increase in effective stress renders overconsolidated sediments more resistant to compression than equivalent normally consolidated sediments, until the previous maximum effective stress experienced by the sediment is exceeded. This stress is termed the preconsolidation stress (see Head, 1988).

30.2.6 Basal and intercalated strata

Basal strata directly overlie an incompressible substrate. Samples obtained from such strata that are in direct contact with the incompressible basement (i.e., "base of basal" deposits) cannot be lowered from their depositional altitudes by compaction. However, samples obtained from basal strata that are not in direct contact with the basement surface are likely to have experienced some compaction. In this chapter, the term "basal peat" is used to describe "base of basal" sediments only.

Intercalated strata are both over- and underlain by other compressible strata. SLIs obtained from intercalated strata are most prone to post-depositional lowering. Intercalated SLIs typically plot below same-age basal index points on the vertical (altitude) axis of sea-level plots (Horton and Shennan, 2009).

30.3 MAGNITUDE AND RATE OF POST-DEPOSITIONAL LOWERING

The magnitude of post-depositional lowering experienced by a sediment column can be directly observed using field techniques (Cahoon et al., 1995) or can be estimated by comparing the *in situ* altitudes of intercalated and basal SLIs of the same age (Törnqvist et al., 2008; Horton and Shennan, 2009). Compaction can introduce vertical errors of several meters into reconstructions of sea level obtained from low-energy intertidal sediments. Such errors are considerably greater than other conventionally considered sources of altitude error in reconstructions of sea level (Long et al., 2006).

Division of the observed magnitude of post-depositional lowering by either the observation period or time elapsed since formation of SLIs provides rates of post-depositional lowering. Cahoon et al. (1995) reported lowering rates of a saltmarsh surface in the Mississippi Delta, USA of up to 24.5 mm a^{-1} during a two-year monitoring period. Over millennial timescales, Törnqvist et al. (2008) found post-depositional lowering rates of 5 mm a^{-1} in the Mississippi Delta and suggested decadal or centennial rates of 10 mm a^{-1} or more. Horton and Shennan (2009) reported average post-depositional lowering rates of 0.4 ± 0.3 mm a^{-1} over millennial timescales in eastern England.

The reported variability in the magnitude and rate of post-depositional lowering reflects the non-linearity of many compaction processes, local lithological and stratigraphic controls, and the time period over which average rates of post-depositional lowering are calculated. Compaction rates should not be universally applied or extrapolated through time and space.

30.4 CONTROLS ON COMPACTION AND POST-DEPOSITIONAL LOWERING

30.4.1 Lithology

Considerable differences in compaction potential exist between primarily minerogenic sediments typical of the lower intertidal zone and the organogenic sediments that form in mid- and high-marsh environments. Highly organic peats can experience a 90% reduction in volume, whereas sand strata experience almost negligible compaction (Shennan and Horton, 2002). However, such high values of compaction for peat reflect maximum states that require vertical effective stresses and periods of time that are generally far greater than those experienced in Holocene intertidal stratigraphies (cf. Allen, 2000). Nevertheless, compaction in coastal Holocene stratigraphies can be significant in the context of sea-level reconstruction, even within the relatively low-stress environment (1×10^0 to 1×10^2 kPa) and short geological timeframes (1×10^1 to 1×10^3 years) typical of such stratigraphies.

Large variations in compaction between different lithologies are widely reported in coastal Holocene stratigraphies. High compressibility in wholly or partly organic saltmarsh deposits results largely from the presence of vascular saltmarsh plants that create highly porous sediment structures (Delaune et al., 1994). High-porosity structures are more prone to compaction than sediments forming or deposited lower in the intertidal frame where the relative percentage of organic material decreases. "High-marsh" materials are typically the most organic and therefore the most compressible sediments encountered in low-energy intertidal settings. Their low bulk densities also result in reduced unit weights (i.e., compressive load contributions per unit volume).

Organic deposits typically experience creep to a greater degree than minerogenic sediments because, unlike mineral grains, organic matter is compressible (Head, 1988). Minerogenic sediments typically do not display such significant creep deformation because the sediment particles are not compressible.

Biodegradation of organic matter can result in a change in compressibility through time and can cause a loss of turgidity and compressive strength. This results in collapse in structure and an increase in bulk density without an increase in vertical effective stress (Delaune et al., 1994). Again, such processes are not generally observed in minerogenic sediments.

30.4.2 Subaerial and hydrogeological environment

Low-energy intertidal sediments are subjected to a range of conditions other than those resulting from effective stress increases during/following burial. Desiccation of sediments during exposure to subaerial conditions between tidal flooding events can generate significant suction stresses (negative pore water pressures), particularly in hot, dry conditions (Hawkins, 1984). This can increase effective stresses without increasing total (overburden) stress (Equation 30.1). Similarly, groundwater level falls over diurnal (tidal) or seasonal timescales can also create negative pore water pressures in sediments located above the groundwater table via capillary suction and reduce pore water pressures below the water table. This increases effective stress throughout the sediment column, potentially causing overconsolidation.

Overconsolidation can also result from (bio-) chemical processes. For example, reduction-oxidation ("redox") reactions can cause geochemical zonation in the stratigraphy due to post-depositional redistribution of redox sensitive elements. Diagenetic enrichment of elements may result in increased bulk density and reduced compaction potential as initial depositional structures are cemented and strengthened (Hawkins, 1984).

Groundwater table dynamics also affect the rate of humification in organic soils. Humification occurs most rapidly when dead plant matter is exposed to air or aerated groundwater (Clymo, 1965). Hence, water table fluctuations and aeration within the acrotelm increase the rate of biodegradation of organic matter in this zone.

30.4.3 Time

The reduced compaction experienced by actively accreting, recent, and thin sediment deposits in comparison to older, thicker sequences is not solely a result of differences in the stress environment. Older deposits have undergone time-dependent compaction processes, such as creep and biodegradation, for longer. The rate of operation of individual processes typically reduces non-linearly with time under a given set of stress of environmental conditions.

30.4.4 Stratigraphy

Post-depositional lowering at any specific depth within a sediment column is dependent on the cumulative effect of compaction in all underlying sediments. Hence, stratigraphic configuration plays an important role in controlling the magnitude of post-depositional lowering.

The position of an SLI within a stratigraphic column is important. Uniform stratigraphic successions that show only minor (stochastic) variations in lithology typically display a characteristic, continuously curved post-depositional lowering depth profile characterized by zero post-depositional lowering at the depositional surface and at the base of the core (Fig. 30.2a). While the location of maximum effective stress at the base of the core coincides with the most compacted layer, this is not the depth at which maximum post-depositional lowering occurs (cf. van Asselen et al., 2009). This is because sediments that directly overlie the incompressible basement cannot be lowered in altitude (i.e., basal sediments). Instead, maximum post-depositional lowering occurs at the middle of the sequence where the optimal combination of loading by overburden sediments and lowering as a result of compaction of underlying material occurs (Fig. 30.2a).

In "transgressive" (deepening-upward) stratigraphies, less compressible sediments with greater

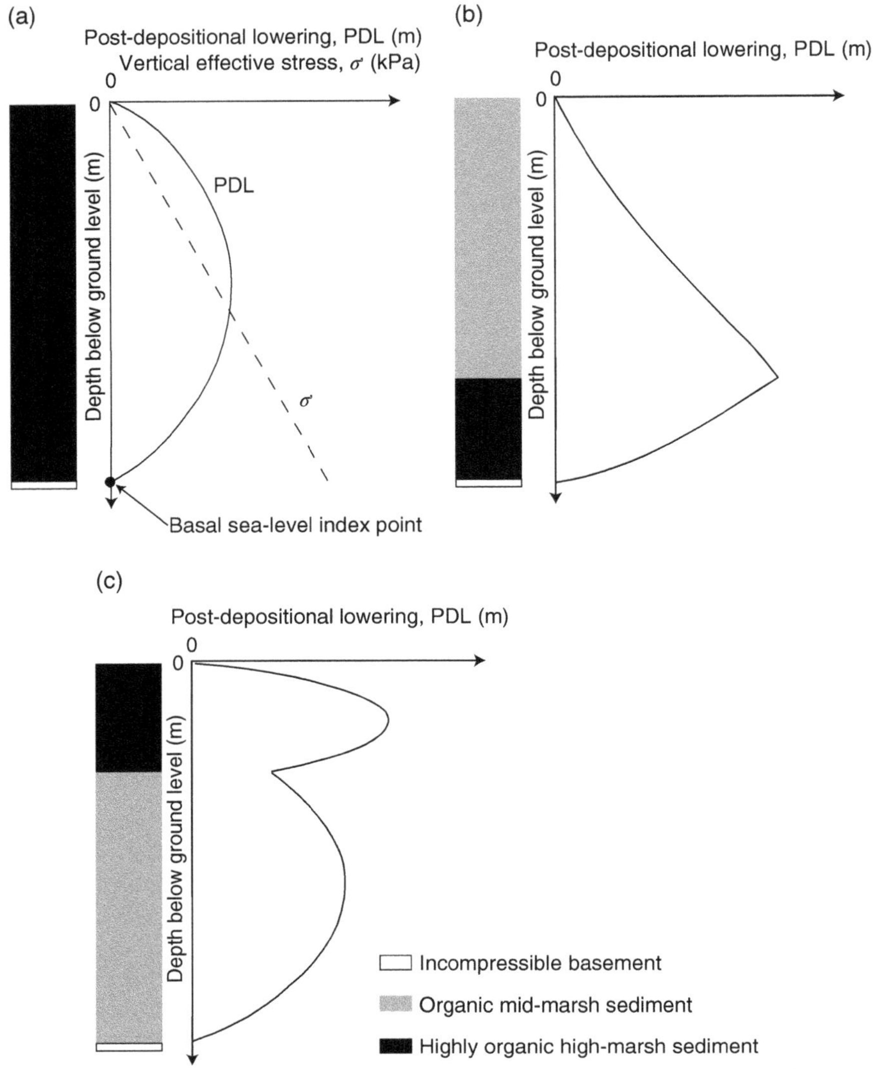

Fig. 30.2. Indicative post-depositional lowering (PDL) profiles for (a) uniform, (b) "transgressive" (deepening-upward) and (c) "regressive" (shallowing-upward) stratigraphic successions. In (a), the effective stress profile is also displayed. *Source*: Adapted from Brain et al. (2012) and used with permission of Elsevier. See also Paul and Barras (1998) and van Asselen et al. (2009).

unit weight exert pressure on lower-density, higher-compressibility materials. This has three main effects (Fig. 30.2b). First, the overall magnitude of post-depositional lowering experienced is greater than in a uniform succession of the same overall thickness. Second, maximum post-depositional lowering may not be in the mid-point of the core. Instead, it typically reflects the point at which the change in stratigraphy occurs and may be towards the depositional surface or base of the core. Third, if the stratigraphic contact is sharp, the resultant post-depositional lowering profile may also demonstrate a sharp inflection.

In contrast, "regressive" (shallowing-upward) stratigraphies are characterized by relatively dense, low-compressibility sediments overlain by high-compressibility sediments with relatively low unit weights. In such successions, values of post-depositional lowering magnitude may be lower than uniform sequences of equivalent thickness. Again, peak post-depositional lowering does not necessarily occur at the mid-point of the stratigraphic column. If stratigraphic contacts are sharp and strata display sufficiently different geotechnical properties, mid-strata peaks in post-depositional lowering may be evident (Fig. 30.2c; Brain et al., 2012).

30.5 BASAL PEATS

Basal peats can provide SLIs that have not been post-depositionally lowered (Fig. 30.2a). As such, they provide the best way to minimize uncertainty in sea-level reconstructions resulting from sediment compaction. However, basal peats are far from ubiquitous through space and time (Shennan and Horton, 2002) and compilations of basal SLIs generally lack the resolution necessary to identify sub-millennial trends in sea level. Hence, basal peats are typically used to constrain long-term (millennial-scale) patterns of sea-level change (cf. Shennan and Horton, 2002). Basal peats generally only record sea-level rises and are less commonly used to detect falls or fluctuations in sea level (e.g., Shennan et al., 2005). A fall in sea level typically produces a SLI of negative tendency that overlies a transgressive, basal SLI. By definition, the regressive SLI is not truly basal and may therefore experience some post-depositional lowering as the underlying layers experience compaction. Basal peats also require sufficient relief (slope) in the underlying, incompressible surface to provide compaction-free sediment over the vertical range of historic sea-level rise of interest. Such underlying topography is not found at all locations.

Where basal peats are not present, or are unsuitable for the reasons outlined above, use of intercalated samples becomes necessary. Intercalated SLIs are prone to post-depositional lowering and so require correction by quantitative modeling methods to return them to their depositional altitudes.

30.6 MODELLING COMPACTION

30.6.1 Decompaction models

An accurate decompaction routine can provide altitudinal corrections for post-depositional lowering in sediment layers at all depths within a stratigraphic column. Decompaction models must consider the full stratigraphic succession above the incompressible basement, even if only a single SLI within the succession requires correction for post-depositional lowering. This is because the post-depositional lowering experienced at any depth is dependent both on the cumulative compaction in all underlying sediments and the weight of overlying material (Section 30.4.4).

Decompaction routines involve either forward- or reverse-modeling approaches, both of which require the sediment column to be artificially split into discrete layers. Within each layer, the lithological and geotechnical characteristics are assumed to be effectively uniform, though probabilistic approaches can be used to address the effects of stochastic variability in lithology. Forward-iterative models simulate sequential deposition of sediment layers on an incompressible substrate until the stress state and age of *in situ* sediments is achieved (Pizzuto and Schwendt, 1997; Tovey and Paul, 2002). In contrast, reverse-modeling begins with the *in situ* (field) condition and involves sequential removal of the uppermost layer(s) (Paul and Barras, 1998). During the addition or removal of a single layer, compaction resulting from both stress- and time-dependent processes is calculated in all underlying layers below that added or removed. By summation of thickness changes in each layer, post-depositional lowering throughout the sediment column can be calculated (Brain et al., 2012).

To date, the majority of models employed in modeling (de)compaction in intertidal sediments describe only the primary compression component. The effects of other compaction processes and how these interact with overconsolidating mechanisms are not well studied in these sediments, though attempts have been made to model their broad effects (cf. Rybczyk et al., 1998).

30.6.2 Model transferability and validity

Soil mechanics and geotechnical engineering provide a useful basis for modeling stress-dependent (mechanical) compaction processes (e.g., Pizzuto and Schwendt, 1997; Paul and Barras, 1998). However, aspects of the theories and models developed in these disciplines are not directly applicable to low-energy intertidal sediments, where lithologies and diagenetic conditions vary considerably from those conventionally considered in soil mechanics (Section 30.4). Hence, classical soil mechanics models may not provide sufficiently accurate descriptions of volume change in saltmarsh and tidal flat deposits. "Black box" application of conventional soil mechanics models to intertidal sediments may produce erroneous predictions of post-depositional lowering and so should be avoided. The following section focuses on some of the most recent developments in modeling compaction processes in low-energy intertidal sediments. Further details of alternative compaction models are summarized by van Asselen et al. (2009).

30.6.3 Model inputs and modern analogs

Successful decompaction requires direct measurements or accurate estimates of compression behavior. However, the volumetric history of a sediment layer cannot be accurately predicted on the basis of the laboratory compaction behavior of *in situ* sediments obtained from depth. In order to overcome this issue, a key assumption is that equivalent contemporary sediments demonstrate the same early-stage compaction behavior as those observed at depth prior to reaching their *in situ* condition. Hence, the physical controls that determine compaction behavior at a particular elevation at a given study site are assumed to be constant through time.

An important prerequisite of this modern analog approach in compaction modeling involves explaining the compaction behavior observed in near-surface sediments in terms of depositional, environmental and early diagenetic processes. This requires comparison of compression properties measured in the laboratory with, for example, physical properties or conditions experienced in the depositional environment at the elevation of formation. Both (semi-) quantitative and qualitative descriptions of the depositional environment are helpful, since they can help to demonstrate that controls on compression behavior are elevation-specific at a particular site and are not variable through time.

30.7 FIELDWORK

30.7.1 Land access and sampling permissions

Prior to undertaking fieldwork, an appropriate risk assessment must be undertaken and permission should be sought from landowners and appropriate regulatory bodies to access sites of interest, particularly when obtaining bulk sediment samples. This is particularly the case on land that is subject to protective environmental legislation. Every effort should be made to minimize the impact that sampling has on surface appearance and conditions.

30.7.2 Identifying sampling locations

Detailed stratigraphic investigation (i.e., high borehole density, high-quality topographic and levelling control, and detailed sediment description and classification) is needed to ensure that modern analogs of all materials encountered within a sediment succession of interest are identified in and sampled from the contemporary intertidal environment. Comparison of buried and contemporary sediments in the field relies on lithological description and, if possible in the field, assessment of plant macrofossil content. A sampling transect throughout the intertidal zone is a useful first step in obtaining modern analogs for sediments encountered in the fossil core. Samples obtained at fixed elevation intervals should be supplemented by those obtained at locations where abrupt changes in lithology and/or vegetation occur because lithology may vary considerably between sampling altitudes. However, the local, modern environment may not provide appropriate analogs for the paleoenvironments recorded throughout the stratigraphic column, particularly in deeper, older stratigraphies. In the absence of suitable modern analogs at a particular study site, reconnaissance and sampling at additional sites within the region is necessary.

30.7.3 Field survey and observations

Sampling locations should be leveled to an appropriate height datum to permit comparison of compression properties with, for example, tidally controlled flooding frequency and duration. At each sampling location and throughout each sediment core selected, a description of both lithology and physical properties should be undertaken. Key characteristics to note include the presence and nature of organic content (fibrosity and decomposition state), sample stiffness and a qualitative description of moisture content, since these provide an initial assessment of both compressibility and stress history (British Standards Institute, 2002). For surface samples, recording the vegetation assemblage and noting the presence, absence, and nature of near-surface biomass are useful observations that can explain the degree of

Fig. 30.3. Collection of undisturbed geotechnical samples. (a) Removal of saltmarsh vegetation to limit disturbance to sediment. (b) Wide-diameter, short-length core sampling tube prior to insertion into the ground. (c) Removal of the undisturbed sediment core following excavation and removal of the surrounding material. In (b) and (c), the sampling tube has a diameter of 150 mm and a height of 200 mm. Photographs by Matthew Brain.

overconsolidation within samples, assisting with use of the modern analog approach (Section 30.6.3).

30.7.4 Sample collection

Geotechnical laboratory analysis of compression behavior requires collection of undisturbed sediment samples. Poor sampling practices can result in adverse compaction, densification, or shearing of sediments. This can result in both over- and underestimations of bulk density and can change sample compressibility.

Application and validation of decompaction models require samples to be obtained from both the depositional surface and from depth. For the former, the ground surface should be appropriately prepared to minimize disturbance to the sediment sample. Excessive movement of saltmarsh plants during sampling can causing tearing of the sample and/or result in an artificial reduction in bulk density as the sediment is inadvertently pulled apart. Carefully and safely removing this vegetation using scissors or pruning shears is an important step in the sampling process (Fig. 30.3a).

Undisturbed sediment samples collected in the field must be of sufficiently large dimensions to allow further subsampling in the laboratory prior to geotechnical testing. Surface samples are ideally collected as wide diameter (approximately 100–150 mm) cores (100–200 mm length). Larger samples may more fully represent the heterogeneous structures of organic sediments.

In soft, non-fibrous materials, thin-walled (<2 mm) core sampling tubes with a beveled lower edge can be pushed into the ground, though considerable care must be taken to insert the tube vertically (Fig. 30.3b). This process must be undertaken slowly to prevent sudden increases in pore water pressure and hence loss of strength. The area around the core tube can subsequently be carefully excavated using hand tools until the base of the sample can be cut from the substrate (Fig. 30.3c).

In densely fibrous organic sediments, it is necessary to cut through plant material without disturbing the sample sediment structure. van Asselen and Roosendaal (2009) describe a near-surface coring device that has sinuous cutting "teeth" that is lowered into the sediment using repeated circular sawing motions (±10 mm). Only minimal downwards pressure should be exerted in order to prevent compaction of susceptible organic sediments.

There is a limit to the depth below ground level to which undisturbed samples can be obtained using such techniques, since detachment of samples from the underlying substrate is not possible at depth (>0.5 m) without significant sample disturbance. Hence, it is again necessary to obtain model inputs by testing the compression properties of samples collected from the near surface and applying these to equivalent lithologies at depth. However, physical and lithological properties, notably loss on ignition (LOI) and bulk density, must still be obtained from depth using coring techniques to allow model calibration and validation. For this purpose, wide diameter (≥40 mm) cores are again preferable to minimize disturbance to sediment.

Samples must be stored appropriately to prevent disturbance from stress relief, moisture content loss, and decomposition. This typically involves keeping samples confined within sampling containers, wrapping them in impervious materials (plastic wrap or "cling film") or sealing them with wax, and storing them in refrigerators at approximately 4 °C prior to testing.

30.8 INITIAL LABORATORY ANALYSES

Prior to geotechnical testing, initial laboratory analysis should be used to characterize the physical properties of the sediments to permit explanation of observed compression behavior and model validation. For surface and core sediments, measurements of bulk density, moisture content, voids ratio (e, the ratio of the volume of voids to the volume of solids, dimensionless), LOI, particle size distribution, and specific gravity (particle density) are required. Samples obtained from sediment cores should be taken at depth intervals appropriate to the test and the required resolution of sea-level reconstruction. The majority of these tests can be undertaken relatively quickly and require relatively small sample masses and volumes to obtain reliable results. These tests can be undertaken at high resolution (<0.02 m depth intervals). In contrast, obtaining measurements of specific gravity requires greater masses of sediment and can be highly time-consuming. It may therefore be more appropriate to obtain specific gravity measurements at coarser resolution (0.10 m depth intervals, for example). A detailed description of the relevant methods is beyond the scope of this chapter but further information is provided in Head (1980) and British Standards Institute (1990).

30.9 CALCULATING EFFECTIVE STRESS PROFILES

Downcore effective stress profiles are important for defining the effective stress range used in geotechnical laboratory tests. Downcore effective stress can be calculated using downcore saturated bulk density (ρ_d, in g cm^{-3}) values, from which the saturated unit weight of the layer can be calculated (γ_{sat}, in kN m^{-3}):

$$\gamma_{sat} = \rho_d g \tag{30.2}$$

where g is the gravitational constant (9.81 m s^{-2}). Beginning with the uppermost sediment layer at the depositional surface, the total stress at the base of this layer can be calculated from:

$$\sigma = \gamma_{sat} t \tag{30.3}$$

where t is the thickness of the layer in meters. Assuming hydrostatic conditions and no near-surface unsaturated or capillary saturated zones (i.e., that the groundwater level is at the ground surface), pore water pressure at the base of the layer can be calculated:

$$u = \gamma_w d \tag{30.4}$$

where γ_w is the unit weight of water (9.81 kN m^{-3}) and d is depth (m) at the base of the layer below the ground surface. Effective stress at the base of this layer can be calculated from Equation 30.1. By summation, effective stress at the base of any layer σ'_i, can then be calculated from:

$$\sigma_i' = \sigma_{i-1}' + \gamma_{\mathrm{sat}(i)} t_i - u_i \quad (30.5)$$

where σ'_{i-1} is the effective stress at the base of the overlying layer, $\gamma_{\mathrm{sat}(i)}$ is the saturated unit weight of the layer, t_i is the thickness of the layer, and u_i is the pore water pressure at the base of the layer (Brain et al., 2012).

30.10 GEOTECHNICAL LABORATORY TESTING

30.10.1 Equipment requirements

Geotechnical laboratory testing can be used to estimate lithology-specific values of parameters describing volume change in response to applied stress and through time for input into compaction models. A range of equipment is available that varies in cost, ease of use, and suitability.

The selected apparatus must be capable of accurately replicating the stress conditions observed in the field, both in terms of the range of stresses experienced, but also the necessity to reproduce one-dimensional (vertical) compression only. Lateral confinement of samples is required, since this replicates the zero lateral strain conditions experienced in natural sedimentary conditions (Powrie, 2004).

Sample size is also a key consideration. Larger samples provide results describing the bulk compression behavior of heterogeneous organic sediments that demonstrate rapid lateral and vertical variability in geotechnical properties. This approach may be suitable in conventional civil engineering tests that consider volumetric response of thick sequences of peat following application of (relatively) high stresses. However, large samples require considerable extension to test durations and may provide data that are not applicable to the thin layers (1×10^{-2} m) that require individual characterization and treatment in Holocene stratigraphies. A more suitable alternative is to test multiple "thin" samples of the same lithology to understand variability in compression behavior (Brain et al., 2011).

The oedometer (Fig. 30.4) fulfils many of these testing requirements, so is frequently used to simulate the one-dimensional compression behavior of low-energy intertidal sediments in the laboratory.

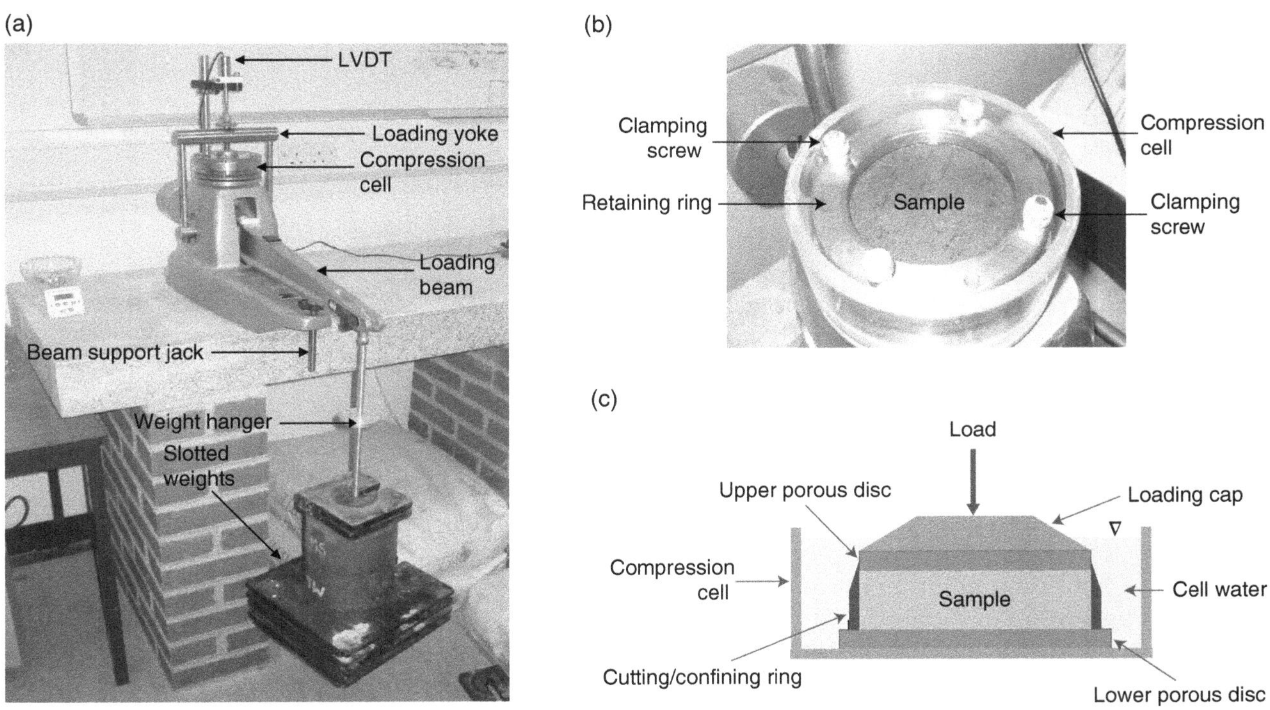

Fig. 30.4. The oedometer compression apparatus. (a) Full laboratory set-up. (b) Location of a sample within the cell body. (c) Simplified cross-section of a typical oedometer compression cell. For a more detailed description of the oedometer compression apparatus, see Head (1988). *Source*: Photographs by Matthew Brain.

30.10.2 The oedometer test

The oedometer test is carried out by applying a sequence of increasing vertical loads to a laterally confined sediment sample that has a height of approximately one-quarter of its diameter (typically 19 mm high, 75 mm diameter; Fig. 30.4). A detailed description of the oedometer test and the procedural stages involved is provided by Head (1988) and is summarized below.

A sample obtained in the field is firstly carefully removed from its sampling container. This can then be cut to a more manageable size using appropriate cutting tools before being carefully trimmed using the oedometer cutting ring, which has a beveled lower edge. Soft cohesive sediments can be quickly and easily prepared with a gentle pushing action using the cutting ring. Sediment that becomes detached from the sample during the trimming processes should gradually be removed. In fibrous organic sediments, the beveled edge of the cutting ring is insufficient to cut through roots and rhizomes without causing disturbance and creating an unfavorable sample geometry. In such sediments, sample preparation can be highly time-consuming and require a range of cutting tools.

Following sample preparation, the sample is placed in the cell body on top of a porous disc, through which the sample can drain when loaded. This is clamped into the cell with a retaining unit ring, and an upper porous disc attached to the loading cap is then placed on top of the sample. The sample is then saturated by adding water to the cell. Any displacement that occurs can be monitored manually through use of a dial gauge, or can be recorded digitally using a linear variable differential transformer (LVDT) in conjunction with a computer and an appropriate software package.

Following saturation, the effective stress acting on the sample is increased using an incremental loading strategy. Masses ("weights") are added to the loading beam which, in turn, applies a (multiplied) load to the sample. The resultant total stress applied to the sample is a function of the sample cross-sectional area to which the load is applied and the loading ratio of the loading beam, which can be varied on some oedometer models.

During application of each loading stage, vertical displacement associated with compression is monitored (Fig. 30.5a). The compression that occurs at the end of each loading stage represents a single point on a plot of a volumetric parameter, typically the voids ratio against vertical effective

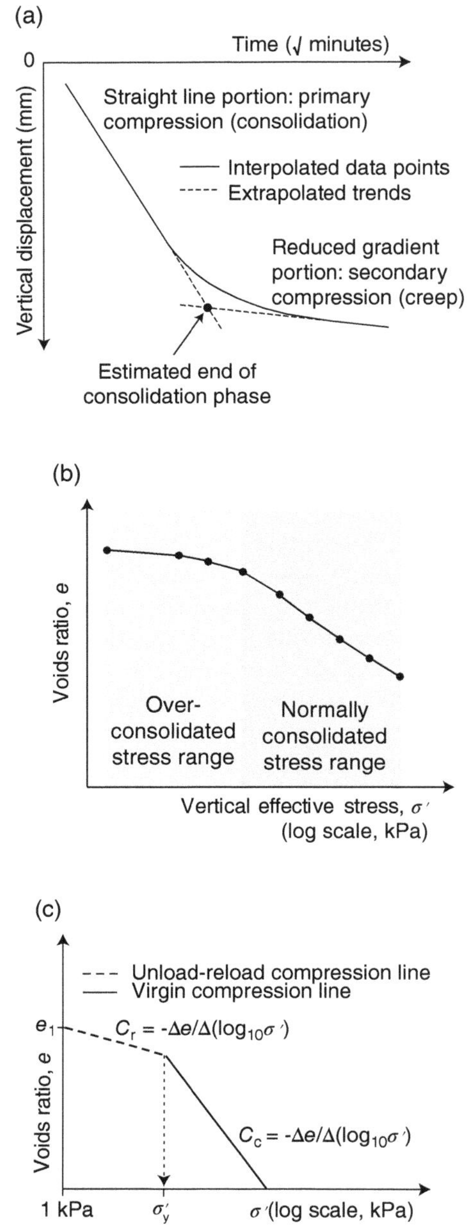

Fig. 30.5. Oedometer test data and model development. (a) Time–displacement plot for a single load increment of a multi-stage oedometer test demonstrating how the end of the primary compression phase can be estimated from the point at which primary and secondary portions of the curve intersect (see Head, 1988). The horizontal axis refers to time in square-root minutes. (b) $e\log_{10}\sigma'$ plot for a multi-stage oedometer test. The reduced gradient over-consolidated portion of the curve is distinct from the normally consolidated portion. Each point on the plot represents a single loading stage (see Head, 1988). (c) Conceptual model of primary compression behavior in $e\log_{10}\sigma'$ space. *Source*: Adapted from Brain et al. (2012) and used with permission of Elsevier.

stress expressed in kPa, on a common logarithmic scale (i.e., an $e\log_{10}\sigma'$ plot, Fig. 30.5b). If only primary compression is of interest, each loading stage must be stopped as consolidation ceases (Fig. 30.5a).

The effective stress range replicated in laboratory compression tests should reflect the effective stress range experienced in the field. The minimum initial stress that can be applied to the sample depends on the lever arm loading ratio and the mass of the loading cap, but ideally should be as close to 0 kPa as possible to simulate (near-) surface field stress conditions. The value of the maximum applied effective stress must equal or exceed that experienced at the base of the stratigraphic column.

30.11 MODELLING PRIMARY COMPRESSION

30.11.1 Model description

Primary compression behavior can be modeled in $e\log_{10}\sigma'$ space using four parameters (Fig. 30.5c):

(1) e_1, a constant (graphical intercept) that describes the value of e at 1 kPa;
(2) C_r, the recompression index, which describes changes in e with effective stress in the reduced compressibility (overconsolidated) phase;
(3) C_c, the compression index, which describes changes in e with effective stress in the normally consolidated phase; and
(4) σ'_y, the yield stress, which describes the effective stress at which compression behavior changes from the reduced compressibility phase described by the recompression line to the greater volumetric decreases associated with the normally consolidated phase. σ'_y describes the effective stress at which structural resistance to compression begins to reduce.

Values of each of the four parameters must be obtained for each lithology and over the full effective stress range experienced within a stratigraphic succession of interest. The applicability of this framework to low-energy intertidal sediments was explored and demonstrated by Brain et al. (2011, 2012) and is an adaptation of compression models used previously in such stratigraphies (Paul and Barras, 1998; Massey et al., 2006). Critically, this four-parameter model considers the low compressibility (overconsolidated) phase, without which decompaction routines may significantly overestimate the magnitude of post-depositional lowering (Brain et al., 2011).

The yield stress as defined here is numerically slightly less than the preconsolidation stress (Section 30.2.5) but is more meaningful when modeling compression because it better describes the behavior of the sediment, rather than an estimation of the previous maximum effective stress experienced (Brain et al., 2011). The yield stress value can be obtained by estimating the effective stress at which the recompression and compression lines intersect. Alternatively, it can be estimated using statistical methods.

Primary compression behavior can be modeled using the following equations:

$$e = e_1 - C_r\left(\log_{10}\sigma' - \log_{10}\sigma'_y\right) \text{ if } \log_{10}\sigma' \le \log_{10}\sigma'_y$$
$$e = e_1 - C_c\left(\log_{10}\sigma' - \log_{10}\sigma'_y\right) \text{ if } \log_{10}\sigma' > \log_{10}\sigma'_y \quad (30.6)$$

Use of the four-parameter compression model assumes that overconsolidation is controlled only by near-surface processes at the time of deposition. Hence, values of σ'_y and e_1 for a particular sediment are assumed to be unchanged from their initial values at the depositional surface. Erosive phases are assumed not to occur, since they may "overwrite" the depositional signature of stress history by increasing the *in situ* yield stress.

The logarithmic functions in Equations 30.6 used to describe voids ratio prevent the use of 0 kPa as the value of effective stress at the depositional surface. An appropriate value must be selected. This should be based on the thickness of the layers used in the modeling procedure which, in turn, is dependent on the downcore resolution of post-depositional lowering corrections required. Massey et al. (2006) used layer thicknesses of 0.1–0.2 m in thick (10–15 m) early–middle Holocene stratigraphies. In contrast, Brain et al. (2012) suggested a layer thickness of 0.02 m for high-resolution late Holocene applications, since this is the typical minimum height at which reliable estimates of compression properties can be obtained in the laboratory. Using this layer thickness, the order of magnitude of the minimum effective stresses observed by Brain et al. (2012) during modeling was 0.01 kPa, though this value may vary by several orders of magnitude with different lithologies and layer thicknesses.

30.11.2 Parameter estimation

To successfully employ the modern analog approach and to overcome the difficulties associated with obtaining undisturbed samples from depth, values of compression model parameters can be estimated using empirical relationships that exist with physical properties or environmental conditions. While some correlations have been explored in civil engineering applications (Paul and Barras, 1998), they were not developed for the unique lithologies of the intertidal environment and they focus largely on C_c (Brain et al., 2011).

Brain et al. (2012) considered relationships between physical and environmental conditions and compression property parameters for a range of UK intertidal sediments. C_r and C_c are intrinsic material properties and are controlled by the lithological characteristics of the sediment, particularly the organic content, as typically estimated by laboratory LOI tests. Brain et al. (2012) observed statistically significant empirical relationships between LOI and both C_r and C_c. These correlations appear to transcend site boundaries, reflecting a potentially universal organic control on compressibility. It therefore seems possible that C_r and C_c can be predicted for sediments at any site on the basis of LOI results using well-constrained regression models, though determining whether such relationships hold at sites with varying environmental and geomorphic characteristics is a key research priority. Similarly, specific gravity (G_s) can also be estimated on the basis of empirical relationships with LOI (Brain et al, 2012; Fig. 30.6).

Compressive yield stress σ'_y is controlled by local ecological, sedimentological, climatic, and hydrographic conditions that cause overconsolidation at the depositional surface (Brain et al., 2011). Brain et al. (2012) demonstrated a statistically significant relationship between σ'_y and standardized elevation within the intertidal zone at two study sites in the UK. This relationship is characterized by an increase in σ'_y with elevation that results from increasing subaerial desiccation associated with a reduction in flooding frequency and duration. Values of σ'_y were observed to peak in the low marsh. At greater elevations, σ'_y declined to a minimum in the high marsh, which Brain et al. (2012) attribute to a change in plant species assemblage and an associated shift from below-ground to above-ground productivity. Increased above-ground biomass produces mats of decaying vegetation, which reduce the potential for desiccation. Given this strong ecological control, it is likely that the relationship between σ'_y and elevation is site specific. For example, limited production or persistence of above-ground biomass may cause a unidirectional increase in σ'_y with elevation and desiccation potential. This further demonstrates the importance of providing a physical explanation for any relationships observed between σ'_y and site-specific environmental conditions.

The reference voids ratio e_1 is a function of both lithology and local environmental conditions. While voids ratio is partly controlled by organic content (Section 30.4.1), the degree of overconsolidation resulting from near-surface processes exerts a site-specific control on e_1.

Since σ'_y and e_1 are influenced to at least some degree by local factors, universally applicable predictive relationships are unlikely to exist. Hence, how σ'_y and e_1 are related to elevation and organic content must first be determined on contemporary samples and explained in terms of environmental processes and lithology prior to prediction of values downcore. To do so, quantification of the indicative meaning throughout sediment cores is necessary, since this will provide an estimate of where the sediment formed within the intertidal frame. In turn, the indicative meaning can be related to σ'_y and e_1 on the basis of patterns observed within the contemporary intertidal zone.

If downcore values of compression parameters are estimated from physical properties or the indicative meaning using regression models, the statistical (prediction) errors must be considered in the modeling procedure to assess and quantify sensitivity to such errors. This can be done using a repeat-iteration, stochastic approach such as Monte Carlo analysis. Predictions of post-depositional lowering will therefore have an associated error range, which can be incorporated into the overall error of an SLI.

30.11.3 Model validation

Initial model validation must be undertaken to assess the accuracy of the primary compression model. Validation can be assessed by statistically comparing observed (downcore) and model-predicted values of bulk density or voids ratio, which are related as follows:

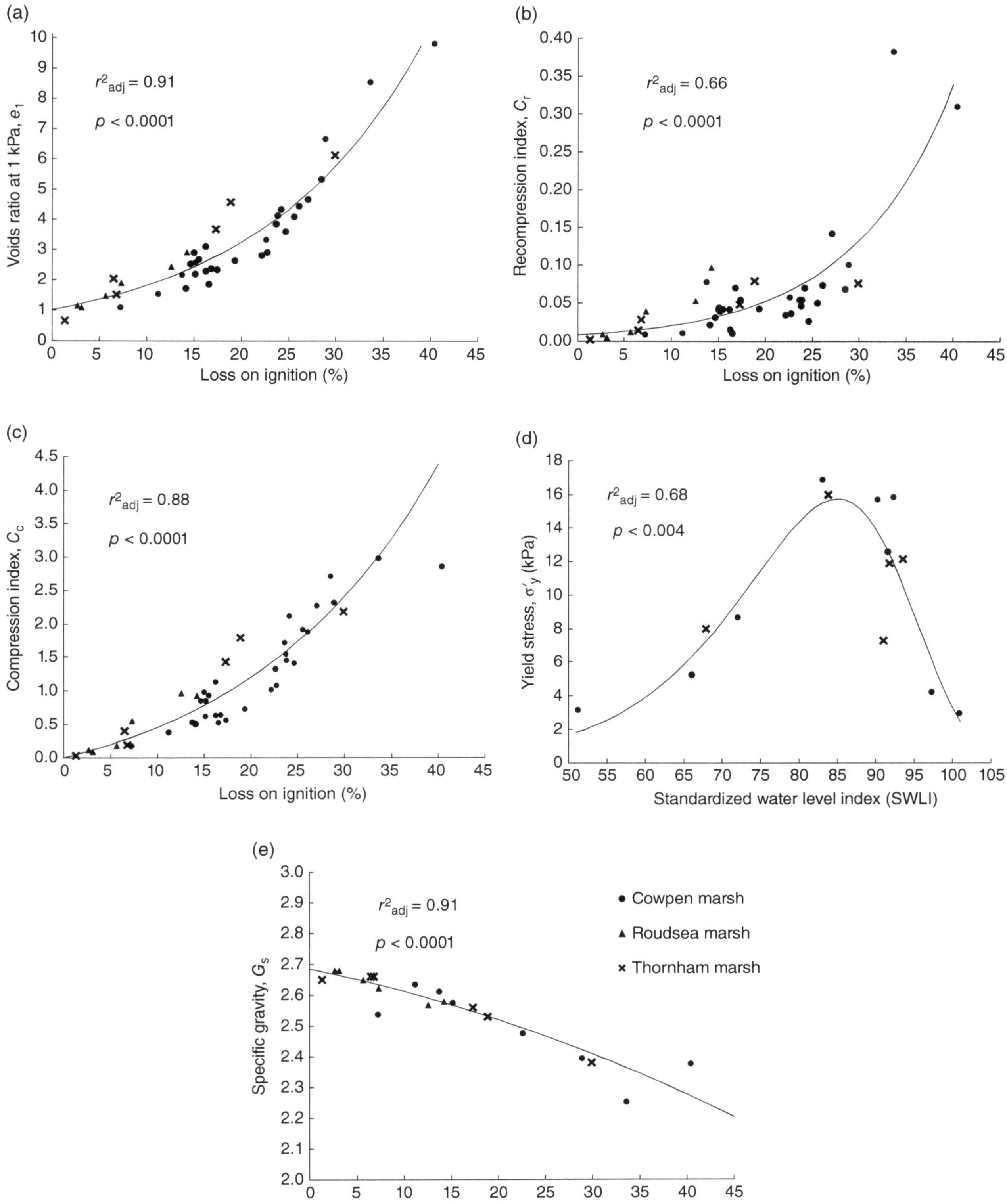

Fig. 30.6. Variations of key compression and physical properties with controlling variables, demonstrating the goodness of fit and statistical significance of predictive regression models. *Source*: Adapted from Brain et al. (2012) and used with permission of Elsevier.

$$\rho_d = \frac{G_s + e}{1 + e} \tag{30.7}$$

Using Equations (30.1)–(30.7) and assuming an appropriate value of effective stress at the depositional surface, model predictions of bulk density and voids ratio throughout the sediment succession of interest can be obtained. If observed and predicted values display broad (within error) parity, the primary compression model can be assumed to be providing an accurate description of volume change within the sediment core. Systematic offsets suggest the influence of other processes in controlling compaction. If the compression model underpredicts observed voids ratios, this may indicate creep deformation or biodegradation. If overprediction occurs, this may suggest incomplete consolidation. Either way, consideration of time-dependent processes may be required.

30.11.4 Decompaction routine

Figure 30.7 demonstrates the division of a stratigraphic column into layers for use in a typical reverse-modeling decompaction routine. Layer L_0 is that for which a post-depositional lowering correction is calculated in a particular model iteration, layer L_{-1} is the immediately overlying layer, and L_N is the lowermost (basal) layer in the column.

To calculate post-depositional lowering at the top of L_0, the value of effective stress at the base of L_{-1} is subtracted from the values of effective stress in layers L_0 to L_N inclusive (i.e., all layers including and underlying L_0). This creates a new effective stress profile, from which new values of voids ratio can be predicted in each layer (Equations 30.6). Individual layer thickness changes in L_0 to L_N inclusive can then be calculated using:

$$\Delta t_i = \frac{t_i\left(e_p - e_s\right)}{1 + e_s} \tag{30.8}$$

where Δt_i is the change in thickness of the ith layer, t_i is the *in situ* ("compacted") thickness of the ith layer (and therefore equal to the user-defined initial layer thickness), e_p is the model-derived ("decompacted") voids ratio of the layer, and e_s is the *in situ* ("compacted") voids ratio. The individual thickness changes in layers L_0 to L_N are then summed to calculate the post-depositional lowering (PDL) experienced at the top of L_0 since deposition of overburden sediments:

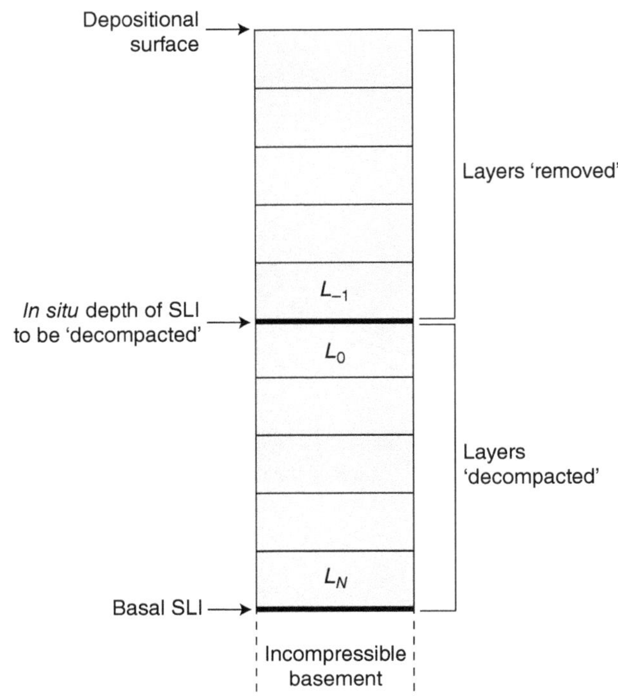

Fig. 30.7. Division of an arbitrary stratigraphic succession into model layers for use in decompaction. L_0 is the layer of interest, at the top of which lies the SLI requiring a post-depositional height correction. L_{-1} is the immediately overlying layer. L_N is the lowermost layer in the succession. See also Tovey and Paul (2002).

$$\mathrm{PDL}_{(L_0)} = \sum_{i=0}^{N} \Delta t_i \tag{30.9}$$

This decompaction routine may then be iterated to calculate the post-depositional lowering for any layer (depth) within the stratigraphic column. When completed for all depths, summary graphs of depth can be produced (Fig. 30.2).

30.11.5 Assessing model performance

Sea-level curves can be corrected for compaction by addition of depth-specific, model-derived estimates of post-depositional lowering to the corresponding *in situ* altitudes of the constituent SLIs. Generalizing the effects of compaction on reconstructed sea level is, however, difficult to do. Such effects are complex and are broadly dependent on the magnitude, rate and direction of sea-change, and stratigraphic factors, as detailed in Section 30.4. In short, compaction effects on sea level are likely to be most profound where the background rate of sea-level change is low and post-depositional lowering is high.

Overall model performance should, where available, be compared to compaction-free records of sea level. For example, basal peats provide an ideal reference against which decompacted intercalated SLIs can be compared. The decompaction procedure can be deemed to be a success if same-age basal and decompacted intercalated SLIs lie, within error, at the same altitude. For studies of recent sea-level change, if tide gauges of appropriate duration are locally available, comparison of saltmarsh reconstructions of sea level in terms of rate, magnitude, and variability can be used to assess the accuracy and validity of the decompaction model.

30.12 TIME-DEPENDENT COMPRESSION PROCESSES

30.12.1 Consolidation

A key assumption of primary compression models is that dissipation of excess (above hydrostatic) pore water pressure effectively occurs synchronously with overburden sedimentation loading over annual to decadal timescales. In other words, consolidation is assumed to keep pace with sedimentation. This means it is not considered necessary to examine the temporal element of primary consolidation. This is a reasonable assumption in thin late Holocene intertidal sediments successions, where sedimentation rates are low, drainage distances are short, and permeability is relatively high (Gibson, 1958).

In deeper, early Holocene sequences, longer drainage lengths and reduced permeability resulting from compaction may mean that full consolidation may not be achieved within an annual loading increment (cf. Tovey and Paul, 2002). Despite well-established numerical models in soil mechanics, modeling the degree of consolidation in such situations has not been fully investigated in low-energy intertidal sediments. This is because of: the organic nature of the sediments, which display a different volumetric behavior through time to that assumed by classical models (Berry and Poskitt, 1972); the mismatch between field loading and drainage conditions and those that are realistic and possible in the laboratory, which prevent appropriate estimates of relevant model input values from being obtained with sufficient accuracy (Head, 1988); and scaling issues that limit the accuracy of applying observations of volume change in small laboratory samples to field situations, where plant macrostructures may expedite dissipation of pore water pressure (Samson and La Rochelle, 1972; Hobbs, 1986). Exploratory modeling and calibration using conventional theories can assist in determining the broad effects of delayed consolidation on post-depositional lowering through time (cf. Pizzuto and Schwendt, 1997). Nevertheless, these issues demonstrate the need for initial validation of primary compression models using *in situ* density data. They also highlight the need for further research into consolidation processes in low-energy intertidal sediments.

30.12.2 Creep

Creep processes in low-energy intertidal sediments have received scant attention. The effects of creep are likely to be reduced in overconsolidated sediments that display denser, more stable structures (Head, 1988). However, the significance of creep in controlling post-depositional lowering is unknown and numerical models that describe how it operates in intertidal stratigraphies are yet to be developed.

Difficulties in modeling creep rates and the associated contribution of secondary compression to post-depositional lowering result from: (1) uncertainties associated with the extrapolation of relatively short-term laboratory test results to the decadal, centennial, and millennial timescales in the field (Lefebvre et al., 1984); (2) the simultaneous operation of consolidation and creep in even partially organic sediments (Bjerrum, 1967); and (3) the effects of strain history, whereby the duration of creep and resultant magnitude of secondary compression in one loading stage will affect creep rate and the magnitude of secondary compression in the subsequent stage (den Haan, 1996). However, considerable advances in modeling creep and secondary compression have been made in cognate disciplines, such that the difficulties associated with modeling creep have been largely addressed (Berry and Poskitt, 1972; den Haan, 1996; van Asselen et al., 2011). However, the applicability of these models to Holocene intertidal stratigraphies has yet to be investigated; mindful of the arguments presented in Section 30.6.2, a description of these models is therefore beyond the scope of this chapter.

Nevertheless, such models are likely to provide a good starting point for application to low-energy intertidal sediments. Obtaining accurate model input parameters for the thin layers involved and the effects of a variable effective stress regime resulting from groundwater dynamics are important future research objectives.

30.13 SUMMARY

This chapter has briefly outlined the effects of compaction (a suite of processes that affect sediment volume) and post-depositional lowering (the effect on a sea-level index point) on reconstructions of sea level obtained from low-energy intertidal sediments. The degree of compaction and post-depositional lowering depends largely on the lithologies present, stratigraphic thickness and configuration and time. The best way to avoid the effects of compaction is through use of compaction-free basal index points that overlie an incompressible basement. However, basal peats are not found everywhere and so use of compaction-prone intercalated deposits becomes necessary.

Geotechnical modeling provides a means to correct sea-level index points for compaction, returning them to their depositional altitudes. The accuracy of such models is dependent on an empirically informed and validated model that has been developed for, or adapted to suit, low-energy intertidal sediments and stress conditions. Model inputs describing sediment compressibility require detailed geotechnical tests on high-quality, undisturbed sediment samples. However, obtaining samples from depth is extremely difficult and so model input values can be predicted on the basis of empirical relationships with physical properties and/or environmental conditions. While recent results suggest some similarity in compression properties between sites, local calibration of compaction models is required to obtain information describing locally controlled compression parameters.

Model accuracy should be assessed by comparing predictions of bulk density with those observed *in situ*. Any systematic offsets may result from the operation of processes not considered by the model, such as consolidation and creep. Appropriate predictive models for these processes have not yet been developed for use in low-energy intertidal sediments. Future work must focus on understanding these processes and developing predictive models to describe their effects.

REFERENCES

Allen, J.R.L. (2000) Morphodynamics of Holocene salt marshes: a review sketch from the Atlantic and Southern North Sea coasts of Europe. Quaternary Science Review, 19, 1155–1231.

Berry, P.L., and Poskitt, T.J. (1972) The consolidation of peat. Géotechnique, 22, 27–52.

Bjerrum, L. (1967) Engineering geology of Norwegian normally-consolidated marine clays as related to settlements of buildings. Géotechnique, 17, 81–118.

Bloom, A.L. (1964) Peat accumulation and compaction in a Connecticut salt marsh. Journal of Sedimentary Petrology, 34, 599–603.

Brain, M.J., Long, A.J., Petley, D.N., Horton, B.P., and Allison, R.J. (2011) Compression behaviour of minerogenic low energy intertidal sediments. Sedimentary Geology, 233, 28–41.

Brain, M.J., Long, A.J, Woodroffe, S.A., Petley, D.N., Milledge, D.G., and Parnell, A.C. (2012) Modelling the effects of sediment compaction on salt marsh reconstructions of recent sea-level rise. Earth and Planetary Science Letters, 345–348, 180–193.

British Standards Institute (1990) BS 1377 Methods of Test for Soils for Civil Engineering Purposes. British Standards Institute, Milton Keynes.

British Standards Institute (2002) BS EN ISO 14688-1: Eurocode 7.Geotechnical investigation and testing: identification and classification of soils. Part 1: Identification and Description. British Standards Institute, London.

Cahoon, D.R., Reed, D.J., and Day Jr, J.W. (1995) Estimating shallow subsidence in microtidal salt marshes of the southeastern US: Kaye and Barghoorn revisited. Marine Geology, 128, 1–9.

Clymo, R.S. (1965) Experiments on breakdown of Sphagnum in two bogs. Journal of Ecology, 53, 747–758.

Delaune, R.D., Nyman, J.A., and Patrick, J.W.H. (1994) Peat collapse, ponding and wetland loss in a rapidly submerging coastal marsh. Journal of Coastal Research, 10(4), 1021–1030.

den Haan, E.J. (1996) A compression model for non-brittle soft clays and peats. Géotechnique, 46, 1–16.

Gibson, R.E. (1958) The progress of consolidation in a clay layer increasing in thickness with time. Géotechnique, 3, 171–182.

Hawkins, A.B. (1984) Depositional characteristics of estuarine alluvium: some engineering implications. Quarterly Journal of Engineering Geology, 17, 219–324.

Head, K.H. (1980) *Manual of Soil Laboratory Testing: Soil Classification and Compaction Tests.* Pentech Press.

Head, K.H. (1988) *Manual of Soil Laboratory Testing: Permeability, Shear Strength and Compressibility Tests.* Pentech Press, London/Plymouth.

Hobbs, N.B. (1986) Mire morphology and the properties and behaviour of some British and foreign peats. Quarterly Journal of Engineering Geology, 19, 7–80.

Horton, B.P., and Shennan, I. (2009) Compaction of Holocene strata and the implications for relative sea-level change on the east coast of England. Geology, 37, 1083–1086.

Kaye, C.A., and Barghoorn, E.S. (1964) Quaternary sea-level change and crustal rise at Boston, Massachusetts, with notes on the autocompaction of peat. Geological Society of America Bulletin, 75, 63–80.

Lefebvre, G., Langlois, P., Lupien, C., and Lavalee, J.-G. (1984) Laboratory testing and in situ behaviour of peat as embankment foundation. Canadian Geotechnical Journal, 21, 322–337.

Lillebø, A.I., Flindt, M.R., Pardal, M.A., and Marques, J.C. (1999) The effect of macrofauna, meiofauna and microfauna on the degradation of Spartina maritima detritus from a salt marsh area. Acta Oecologica, 20, 249–258.

Long, A.J., Waller, M.P., and Stupples, P. (2006) Driving mechanisms of coastal change: peat compaction and the destruction of late Holocene coastal wetlands. Marine Geology, 225, 63–84.

Massey, A.C., Paul, M.A., Gehrels, W.R., and Charman, D.J. (2006) Autocompaction in Holocene coastal back-barrier sediments from south Devon, southwest England, UK. Marine Geology, 226, 225–241.

Paul, M.A., and Barras, B.F. (1998) A geotechnical correction for post-depositional sediment compression: examples from the Forth Valley, Scotland. Journal of Quaternary Science, 13, 171–176.

Pizzuto, J.E., and Schwendt, A.E. (1997) Mathematical modeling of autocompaction of a Holocene transgressive valley-fill deposit, Wolfe Glade, Delaware. Geology, 25, 57–60.

Powrie, W. (2004) *Soil Mechanics: Concepts and Applications*. Spon Press/Taylor and Francis Group, London and New York.

Rybczyk, J.M., Callaway, J.C., and Day, J.W., Jr. (1998) A relative elevation model for a subsiding coastal forested wetland receiving wastewater effluent. Ecological Modelling, 112, 23–44.

Samson, L., and La Rochelle, P. (1972) Design and performance of an expressway constructed over peat by preloading. Canadian Geotechnical Journal, 9, 447–466.

Shennan, I., and Horton, B.P. (2002) Holocene land- and sea-level changes in Great Britain. Journal of Quaternary Science, 17, 511–526.

Shennan, I., Hamilton, S., Hillier, C., and Woodroffe, S. (2005) A 16,000-year record of near-field relative sea-level changes, northwest Scotland, United Kingdom. Quaternary International, 133–134, 95–106.

Törnqvist, T.E., Wallace, D.J., Storms, J.E.A., Wallinga, J., Van Dam, R.L., Blaauw, M., Derksen, M.S., Klerks, C.J.W., Meijneken, C., and Snijders, E.M.A. (2008) Mississippi Delta subsidence primarily caused by compaction of Holocene strata. Nature Geoscience, 1, 173–176.

Tovey, N.K., and Paul, M.A. (2002) Modelling self-weight consolidation in Holocene sediments. Bulletin of Engineering Geology and the Environment, 61, 21–33.

van Asselen, S., and Roosendaal, C. (2009) A new method for determining the bulk density of uncompacted peat from field settings. Journal of Sedimentary Research, 79, 918–922.

van Asselen, S., Stouthamer, E., and van Asch, Th.W.J. (2009) Effects of peat compaction on delta evolution: a review on processes, responses, measuring and modelling. Earth Science Reviews, 92, 35–51.

van Asselen, S., Karssenberg, D., and Stouthamer, E. (2011) Contribution of peat compaction to relative sea-level rise within Holocene deltas. Geophysical Research Letters, 38, L24401.

Chapter 31
Transfer functions

ANDREW C. KEMP[1] AND RICHARD J. TELFORD[2]
[1]*Department of Earth and Ocean Sciences, Tufts University, Medford, MA, USA*
[2]*Department of Biology, University of Bergen and Bjerknes Centre for Climate Research, Bergen, Norway*

31.1 INTRODUCTION

> *Transfer functions are empirically-derived equations for calculating quantitative estimates of past atmospheric and oceanic conditions from paleontological data*
> (Sachs et al., 1977).

Paleoenvironmental research aims to reconstruct past conditions such as sea level or temperature and their changes through time. With the development of advanced numerical techniques and the growing availability of computers and software, a quantitative revolution in paleoenvironmental research has taken place since the early 1970s. The result was augmentation of qualitative descriptions of past environments based on a researcher's judgment with quantitative reconstructions derived from numerical methods. The seminal paper of Imbrie and Kipp (1971) estimated past ocean temperature from planktonic foraminifera, and is widely credited with introducing the principles that underpin quantitative paleoenvironmental reconstructions using micropaleontological assemblages preserved in sediments. At around the same time, other groups developed numerical techniques to reconstruct past environmental conditions from pollen assemblages (e.g., Webb and Bryson, 1972). These and subsequent studies established the varied procedures that are collectively referred to as transfer functions.

The term transfer function does not refer to a single numerical technique, but rather a common approach that draws from a suite of techniques to select those most appropriate for a specific study. It is reasoning by analogy, grounded in the assumption that the environmental associations of species were unchanged through time. Modern observations of an assemblage therefore constitute a suitable analogy for interpreting their counterparts preserved in sedimentary archives. Using a transfer function to make quantitative reconstructions of past environmental conditions is a three-stage process (Juggins and Birks, 2012; Figs 31.1 and 31.2) as follows.

(1) *Development of a modern training set.* Paired observations of micropaleontological assemblages (described by the relative abundances of several species) and environmental variables make up the modern training set (Fig. 31.1a). These empirical field data are collected in modern environments that are expected to be analogous to those preserved in the sedimentary record and later used in paleoenvironmental reconstruction.
(2) *Development of a transfer function.* Without a deterministic model that can entirely explain species distributions as a function of measured environmental variables, it is necessary to create an ecological response function using the modern training set. A variety of numerical techniques are available to perform this step, from which the researcher must select one that is appropriate given the aims of the investigation and nature of available data. The resulting model is the transfer function (Fig. 31.1b), and the principal differences among models are the assumed response of species to their environment (Birks, 1995). Model performance is usually assessed by internal cross-validation.
(3) *Paleoenvironmental reconstruction.* The final stage is to produce a paleoenvironmental reconstruction using the transfer function. Micropaleontological assemblages preserved in the sedimentary record (termed the fossil

Handbook of Sea-Level Research, First Edition. Edited by Ian Shennan, Antony J. Long, and Benjamin P. Horton.

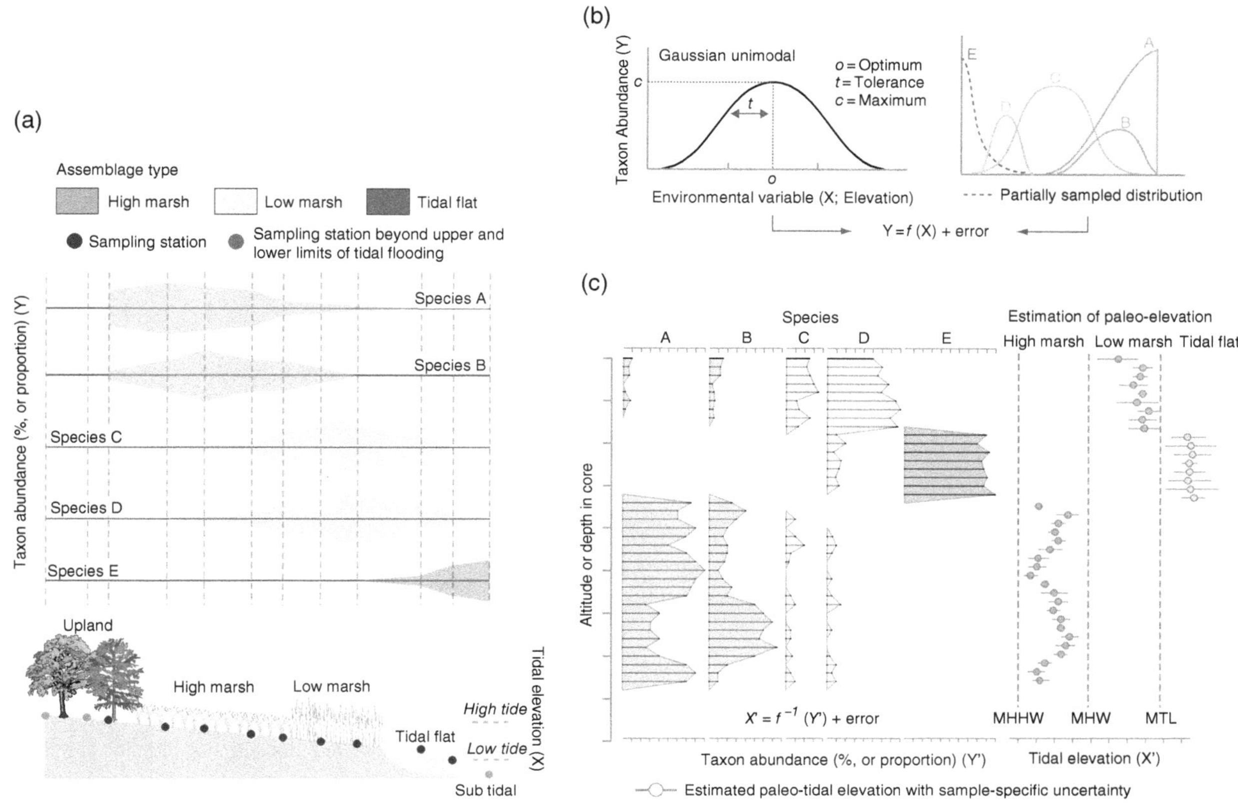

Fig. 31.1. Use of transfer functions to reconstruct relative sea level using intertidal microfossils. (a) A training set comprises paired field observations of species abundance (Y) and tidal elevation (X). Samples are collected across the environmental gradient of interest (here it is tidal elevation in a saltmarsh and tidal flat). Sampling extends into supra- and subtidal settings to ensure that the full environmental gradient is sampled. (b) An appropriate numerical technique quantitatively relates Y and X. This is the transfer function. (c) Application of the transfer function to microfossil assemblages (Y') preserved in buried sediment to estimate the paleotidal elevation where the assemblage formed. A sample-specific uncertainty (1σ) is generated for each fossil sample by bootstrapping, shown by horizontal lines through circles. *Source*: Horton et al., 2013. *Reproduced with permission of Elsevier*. Symbols courtesy of the Integration and Application Network, University of Maryland Center for Environmental Science. For color details, please see Plate 44.

assemblage and described by the relative abundance of species) can be observed and are used as the input to the transfer function. The output of the transfer function is an estimate of past environmental conditions that are the unobservable target of the reconstruction (Fig. 31.1c). A variety of techniques are used to investigate the accuracy and reliability of the reconstruction.

All paleoenvironmental reconstructions made using transfer functions are reliant upon the validity of several implicit assumptions, irrespective of the type of micropaleontological data used, the environmental variable being reconstructed, or the numerical technique selected. Credible reconstructions must satisfy each of these assumptions in modern environments and over the period of time for which reconstructions are made (Juggins, 2013).

(1) Modern taxa are systematically related to their environment.
(2) The environmental variable that is the target of reconstruction is an important ecological determinant of the observed distribution of taxa, or it has a linear relationship to another factor that is ecologically important.
(3) The relationship between taxa and their environment has not changed over the period of time under consideration. Modern and fossil assemblages must be similar in composition, although not all taxa need to be present in both datasets.

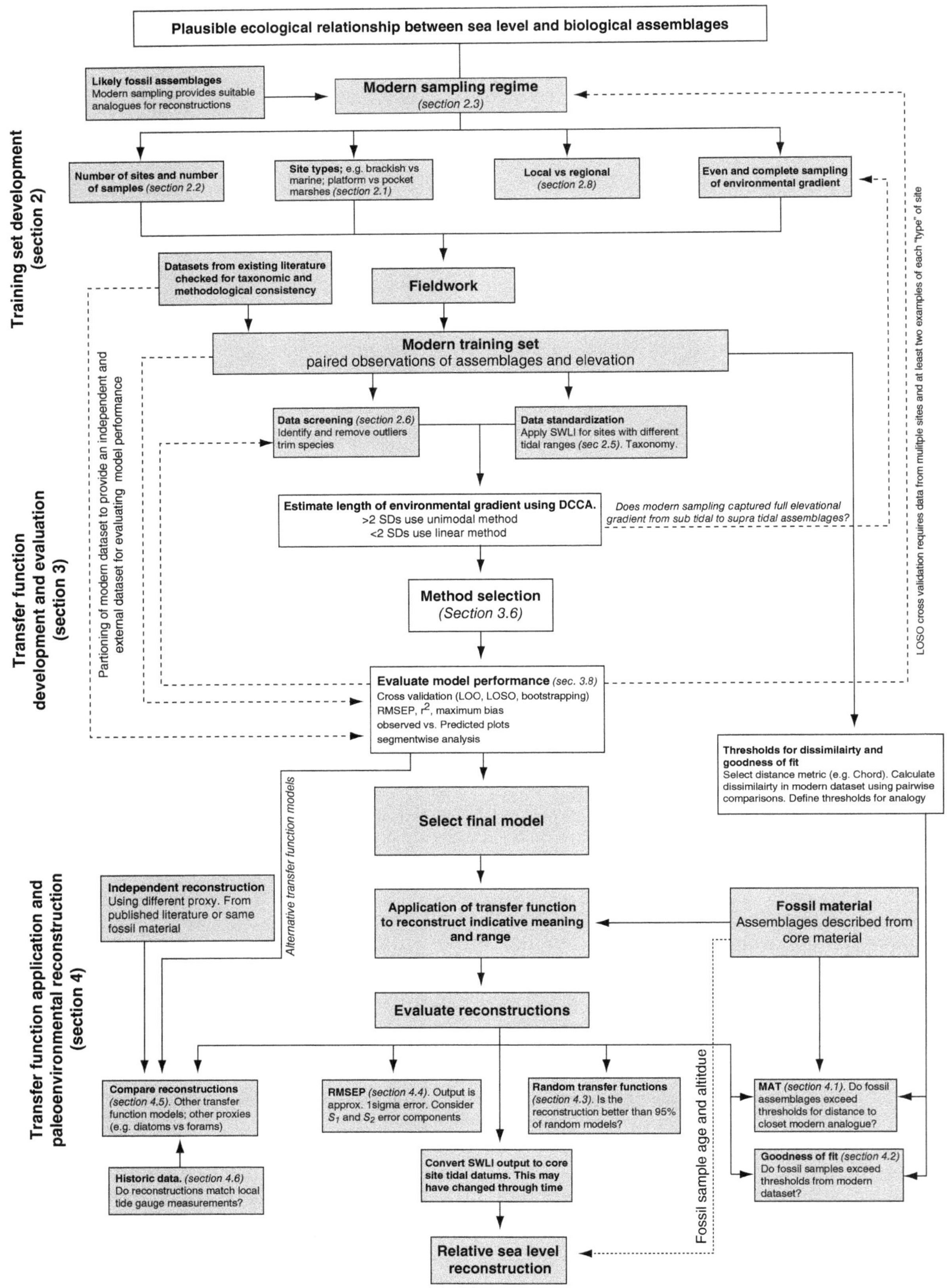

Fig. 31.2. Flow chart showing the process of using a transfer function to reconstruct relative sea level from intertidal microfossils such as foraminifera or diatoms. The division of steps into shaded groups follows the three-stage process of training set development, transfer function development, and evaluation and application of the transfer function to reconstruct indicative meaning and indicative range.

(4) Numerical techniques employed in developing the transfer function produce accurate and unbiased reconstructions.
(5) Environmental variables other than that to be reconstructed have limited effect on taxa distributions, or their relationship to the target variable has not changed over the period of time under consideration.

Transfer functions were first used in sea-level research to reconstruct the amount of coastal subsidence caused by large earthquakes that are manifest in coastal sediments as instantaneous episodes of relative sea-level rise (Guilbault et al., 1995, 1996). They have since become a key part of quantitative sea-level reconstructions that aim to develop Holocene sea-level index points (e.g., Horton and Edwards, 2006; Woodroffe and Long, 2010), produce detailed sea-level histories for the last ~100–3000 years (e.g., Gehrels et al., 2006), and estimate the magnitude of coseismic land-level movements (e.g., Sawai et al., 2004; Hamilton and Shennan, 2005; Hawkes et al., 2010). The motivations for using transfer functions to make quantitative sea-level reconstructions are the same as in other paleoenvironmental research fields such as paleolimnology. Primarily, they provide a quantitative and objective methodology that allows results from different studies to be replicated and directly compared with one another. In sea-level research specifically, transfer functions complement a discrete classification of sea-level indicators (e.g., high-saltmarsh peat) with continuous reconstructions of the indicative meaning (Chapter 2) that are sample specific to improve accuracy and reduce vertical uncertainty (i.e., a smaller indicative range). Despite the quantitative and objective nature of the numerical techniques, there are numerous steps in the process of using a transfer function to reconstruct sea level where the researcher is required to exercise judgment and make decisions. We highlight these steps and provide guidance, although the choices made must reflect the specific aims of each study and the particular datasets employed.

In this chapter we introduce the fundamental concepts behind using transfer functions to reconstruct sea level for field-orientated researchers who may be less familiar with quantitative techniques. There is particular emphasis on microfossils (such as foraminifera and diatoms) from low-energy coastal environments such as tidal marshes, mangroves, and tidal flats because this is currently the most common application of transfer functions in sea-level research, but the concepts are widely applicable (e.g., to pollen, ostracods, chironomids, or testate amoebae). We provide a practical guide to transfer function development, application, and assessment that makes recommendations and identifies common problems specific to sea-level reconstruction rather than providing detailed descriptions of the numerical methods (Sections 31.2–31.4). Examples are provided to illustrate the concepts as they are introduced and we make suggestions regarding software for new users (Section 31.5).

31.2 DEVELOPING MODERN TRAINING SETS

Quantitative sea-level reconstructions are reliant upon the availability of a suitable training set consisting of modern species assemblages (Y) and associated environmental variables (X). This is the empirical field data that the transfer function will be developed from (Fig. 31.1a). The accuracy and precision of any reconstruction necessarily reflects the nature and qualities of the modern training set that it is based upon. The first step in using transfer functions to reconstruct sea level is therefore to develop a suitable modern training set comprising assemblages enumerated from surface sediment samples, where the environmental variable of interest is measured at the same time and/or on the same sample. The most commonly used biological sea-level indicators from low-energy, depositional coastal environments are foraminifera (Chapter 13), diatoms (Chapter 15), and thecameobians (Chapter 19). In sea-level research, the environmental variable is elevation expressed as depth below, or altitude above, tide levels. Often elevation is the only environmental variable measured in developing a modern training set, but this sampling strategy impairs the ability to investigate if the transfer function satisfies assumptions about reconstructing an important environmental variable and the influence of secondary variables (Juggins, 2013). We discuss several pertinent topics that frequently arise when developing a modern training set.

31.2.1 Modern and paleoenvironments

As the transfer function is based on the analogy between modern and fossil assemblages, the training set should comprise samples from environments that are similar to those encountered in the sedimentary record that will be analyzed. In sea-level research, consideration should be given to the prevailing salinity and climate regime, sediment texture, vegetation type, and geomorphic setting when determining the degree of analogy between modern and paleoenvironments. The researcher is therefore required to exercise judgment in interpreting the environment of deposition in which fossil assemblages are preserved. In practice, there are two approaches to ensuring that modern and fossil datasets are from analogous environments. First, a training set is collected for application to a specific fossil sedimentary sequence that was previously described from cores and the likely environment of deposition gauged by sediment character, including the presence or absence of other paleoenvironmental indicators such as plant macrofossils (Chapter 12). In this instance, it may be possible to restrict the collection of modern samples to a particular set of sites or part of the environmental gradient. For example, Gehrels (2000) sampled modern saltmarshes in Maine to compile a training set of foraminifera for reconstructing sea level at one of the sites. Modern samples did not extend to elevations below ~0.5 m above mean tide level that were flooded more than 40% of the time because the core of high-saltmarsh peat chosen for analysis was not deposited in these environments. Gehrels (2000) therefore chose to restrict sampling of the environmental gradient to the interval most likely to be analogous to that preserved in core material. In the second approach, modern transects are collected to provide analogs for environments that are known to be well suited to sea-level reconstruction, even if a specific core has not been collected. Usually this approach emphasizes the collection of modern samples over the complete environmental gradient from elevations below the lowest tides to above the highest tides. For example, Zong and Horton (1999) sampled a wide range of tidal flats and saltmarshes in the United Kingdom to generate a training set relating the distribution of modern diatoms to tidal elevation. They sampled as much of the environmental gradient as possible at a variety of sites to produce a training set with broad geographical and environmental scope that could be applied to an equally varied range of fossil assemblages. In practice modern and core samples are often collected during the same field campaign, and a balance between these approaches is sought by relying on the interpretation of core lithology at the time of collection. Where fieldwork is conducted in difficult-to-reach locations it is prudent to sample a diverse range of modern environments, even if it is not immediately apparent that all samples will be needed to provide modern analogs. To ensure that modern samples come from environments analogous to fossil samples, it may be necessary to collect additional material (from the field or published literature) after an initial round of sampling and analysis.

31.2.2 Training set size

There are few guidelines on the minimum or optimal number of samples required to develop a training set, nor has a consensus emerged from the sea-level literature. For sea-level reconstructions using foraminifera and diatoms, training sets comprising 21 to >200 samples have been used, with an average of approximately 50 (Barlow et al., 2013). Given that many of these transfer functions were applied successfully, 50 may be a reasonable guideline. Watcham et al. (2013) however recommended using a minimum of 100 samples (and preferably 200) for diatoms, based on the range of samples that were needed to provide modern analogs when reconstructing sea level in Alaska during the last ~3700 years from tidal flat and saltmarsh deposits. Using diatoms from North American lakes, Reavie and Juggins (2011) described a methodology for exploring the effect of training-set size and concluded that a minimum of 40–70 samples was necessary. Applying this approach to a training set of intertidal diatoms from Scotland, (Barlow et al., 2013) suggests that 80–100 samples are the minimum required for this example dataset as evidenced by the plateauing of root-mean-square error of prediction values (Fig. 31.3a). Analysis of a training set of intertidal foraminifera from Oregon, USA (Hawkes et al., 2010; Engelhart et al., 2013) indicates that 30–50 samples are necessary to develop a transfer function (Fig. 31.3b), similar to the minimum of 30 offered by Juggins and Birks (2012) as a rule of thumb.

The choice of an appropriate training set size should be informed by several factors including the practical constraints of time. Fewer modern samples are needed to characterize simple, strong

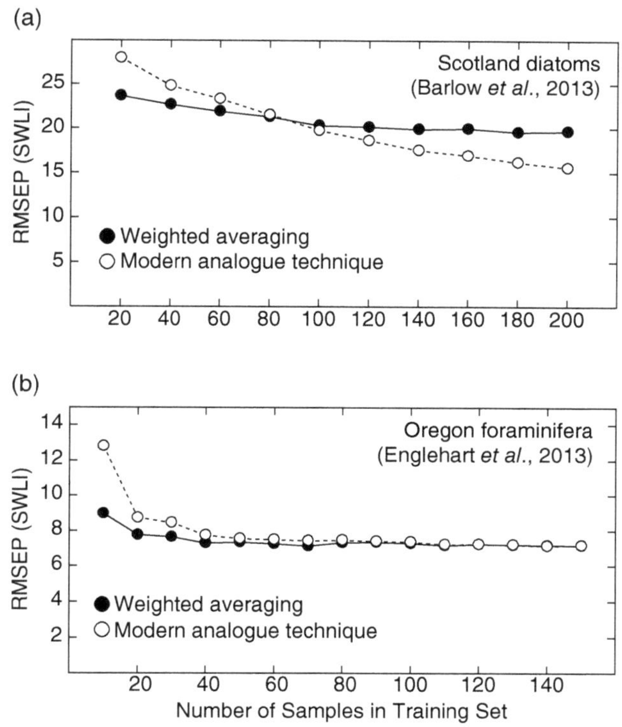

Fig. 31.3. (a, b) Effect of training set size (number of samples) on transfer function performance for weighted-averaging and modern analog technique models applied to example datasets. Root-mean-square error of prediction (RMSEP) was estimated by leave-one-out cross-validation. Plateauing of RMSEP indicates the number of samples needed to construct a training set for the example datasets. SWLI calculations are not the same for both datasets and therefore cannot be compared directly.

environmental gradients than for complex ecological systems, where variables in addition to elevation play an important role in determining species distributions. A transect across a tidal flat, low saltmarsh, high saltmarsh, and into a freshwater upland represents a simple, strong environmental gradient because a high proportion of variability among samples is explained by differences in the duration and frequency of inundation (elevation). For example, foraminifera from high-saltmarsh environments in North America are readily distinguished from low-marsh or tidal-flat populations at the same site using 10–25 samples, because elevation is a simple and strong environmental gradient. In contrast, transects from multiple sites or regions represent a more complicated ecological system because biological variability among sites reflects the influence of additional environmental parameters such as salinity, sediment-type, and prevailing climate. Transects from several sites (different estuaries or water bodies) show that high-saltmarsh assemblages of foraminifera also reflect the influence of other environmental factors that vary among sites, resulting in a number of distinctive assemblages that all occupy high-saltmarsh elevations (e.g., Wright et al., 2011). To capture and characterize this variability requires a greater number of sites and samples.

Sedimentary records that include pronounced environmental changes require a larger and more diverse training set than cores that represent a period of environmental stability. An expanded diversity of modern analogs is required if environmental variables in addition to elevation (e.g., the salinity or climate regime) changed significantly during the period under consideration. These changes are more likely to influence reconstructions spanning relatively long periods of time, but significant changes also occur over short time intervals. Age alone is therefore not always a good indication of how many training set samples are needed. In some cases clear lithostratigraphic changes recognized from exploratory coring show that pronounced environmental change took place, while in others the change may only be recognized after micropaleontological analysis is started. One way to test if sufficient modern analogs are available in the training set is to use analog matching (see Section 31.4.1).

In training sets with large numbers of species, a relatively large number of samples may be required to adequately represent the ecological niches and signal-to-noise properties typical of high-diversity assemblages. Tidal flat and salt-marsh diatoms are commonly more diverse than foraminifera from the same samples, for example. Diatoms also display a greater diversity of assemblages among sites, particularly in higher intertidal environments (e.g., Woodroffe and Long, 2010; Wilson and Lamb, 2012). The site-specific nature of diatom assemblages requires a large number of modern samples to interpret fossil populations, because even relatively small paleoenvironmental changes may have caused significant assemblage changes.

With the increasing numbers of modern samples that have been collected for sea-level reconstruction, data from multiple studies can be merged to conveniently generate larger training sets or provide an adequate training set without having to collect new modern samples. Using existing data requires that taxonomic consistency is achieved by ensuring that the measurement of environmental variables and identification and treatment of species is consistent among datasets. For example, Wright et al. (2011) combined *Trochammina inflata* and *Siphotrochammina lobata* when they compiled a dataset of modern foraminifera because one of the previous studies did not differentiate the two species. When combining datasets it is also important to minimize the influence of secondary environmental gradients such as climate. For example, it is probably not ecologically reasonable to combine datasets from tropical and arctic regions. We encourage presentation of new training sets to include archived or tabulated data to ensure that others can use the data effectively.

31.2.3 Sampling regimes

The numerical techniques used in transfer functions assume that training sets are composed of independent samples (one sample from each site). In sea-level research using intertidal microfossils, this assumption is almost always invalidated by a sampling regime that collects multiple samples along a transect. There is usually one or more transect at each site and there are often multiple sites (e.g., individual salt-marshes or estuaries). Training sets of this type are referred to as "spatially clustered". This sampling regime is necessary because the prevailing environmental gradient is elevation within a single site rather than elevation among multiple sites, making transects an appropriate (and necessary) sampling regime. This approach makes the ecological assumption that microfossil assemblages vary more within a single site because of the strong environmental gradient presented by elevation than they do among nearby sites with similar tidal and physiographic conditions. It is necessary to acknowledge the effect of spatially-clustered training sets when assessing transfer function performance (Section 31.3.8; Payne et al., 2012), although its influence can only be removed by avoiding the type of linear, within-site sampling that is typically used in sea-level research.

Transfer functions require paired biological and environmental data (collected at and representative of the same time and the same place). Sampling of surface sediment provides time-averaged biological data to reduce or remove the influence of seasonality and intra-annual variability. The degree of time averaging reflects the thickness of sample collected and rate of sediment accumulation. Some environmental variables can be measured from the same sediment sample (e.g., grain size or organic content), but sea-level research requires the elevation of each sample to be measured relative to tidal datums. There may be some disparity between the time interval represented by biological assemblages and the period over which tidal datums were established, particularly if tidal data from a different site (e.g., a nearby tide gauge) are used. It is therefore recommended that the period for which water level was derived is investigated. If there is likely to be a mismatch with biological data, it may be necessary to seek alternative water-level data or to collect local water-level measurements using a logger deployed at the field site.

The numerical techniques that underpin transfer functions are sensitive to the distribution of training set samples along the environmental gradient, such as absence of data from the extremes or an overabundance of samples in one part of the gradient at the expense of a less-frequently sampled interval. Uneven sampling of the environmental gradient can cause over-optimistic measures of transfer function performance (Telford and Birks, 2011a). Ideally, samples should be distributed evenly across the full environmental gradient with bracketing samples collected to ensure that a complete ecological response is recorded (e.g., extending intertidal transects to include sub- and supratidal samples; Fig. 31.1a).

Field-based sampling rarely provides systematic, evenly distributed samples because of natural variability and/or sampling difficulties. Researchers (consciously or subconsciously) often skew sampling toward parts of the environmental gradient that are most common in fossil material to maximize the number of modern analogs. In other instances there are practical difficulties that prevent the evenly distributed collection of samples. Datasets of intertidal diatoms from Scotland (Barlow et al., 2013), foraminifera from Oregon, USA (Engelhart et al., 2013), and foraminifera from Connecticut, USA (Wright et al., 2011) demonstrate, to varying degrees, a bias in sampling toward higher elevations (Fig. 31.4). In the case of Connecticut, there is often an absence of low-marsh and tidal-flat environments in Long Island Sound that prevents this part of the elevational gradient being sampled; it is commonly represented by a "step" from the saltmarsh platform to lower elevations. This geomorphological feature is rare in Oregon, which allowed Engelhart et al. (2013) to sample the elevational gradient more evenly, despite a concentration of samples around mean higher high water (MHHW), because of a deliberate effort to capture the highest elevation supporting foraminifera. The dataset of modern diatoms from Scotland has a near-normal distribution with a maximum concentration of samples close to mean high water (MHW; SWLI of 175), which is partly a legacy from combining multiple studies.

Large datasets could be trimmed to ensure even sampling of elevation, but this would involve removing otherwise acceptable data or shortening the length of the gradient sampled and is not recommended. When designing a sampling regime it is suggested that the full length of the environmental gradient is sampled at evenly spaced intervals to minimize bias in estimating transfer function performance (Telford and Birks, 2011a) and to capture complete species responses. The influence of uneven sampling along the environmental gradient is discussed further in Section 31.3.8.

31.2.4 Standardized water level index

Measurements of environmental variables must be comparable among samples and among sites for them to be included in a training set. When the variable is tidal elevation, absolute values relative to a tidal datum such as mean high water do not permit data from sites with different tidal ranges to be combined, because the absolute elevation does not represent the same frequency and duration of flooding as it does at another site with a different tidal range. It is therefore ecologically unreasonable to combine data from multiple sites using absolute elevation. To compensate for differences in tidal range, elevations are normalized and expressed in the training set as a standardized water-level index (SWLI). This allows samples from sites with different tidal ranges to be combined in an ecologically reasonable manner because the SWLI approximates the frequency and duration of tidal inundation. If compiling data from several studies, it is important to recognize that the SWLI calculation may vary; it is therefore recommended that studies provide elevations with respect to tidal datums (and the relationship between tidal datums) so that others can recalculate SWLI as needed.

A number of SWLI calculations have been used since Horton et al. (1999), all of which share the common trait of standardizing elevation between a lower and upper limit. Most commonly the upper limit is mean higher high water (MHHW) or highest astronomical tide (HAT), and the lower limit is mean tide level (MTL) or mean lower low water (MLLW). One example of a SWLI calculation is:

$$\mathrm{SWLI}_n = \frac{100(h_n - h_{\mathrm{MTL}})}{h_{\mathrm{MHHW}} - h_{\mathrm{MTL}}} + 100 \qquad (31.1)$$

where SWLI_n is the standardized water-level index for sample n, h_n is the sample elevation, h_{MTL} is the elevation of mean tide level, and h_{MHHW} is the elevation of mean higher high water (modified from Barlow et al., 2013). All elevations are expressed relative to the same datum and a SWLI of 100 is MTL and MHHW is 200.

According to Wright et al. (2011), using an upper tidal level such as MHHW in SWLI calculations causes poor vertical alignment of samples at higher elevations because of non-linearity between elevation and tidal inundation. To correct for this, they used the highest occurrence of foraminifera (HOF)

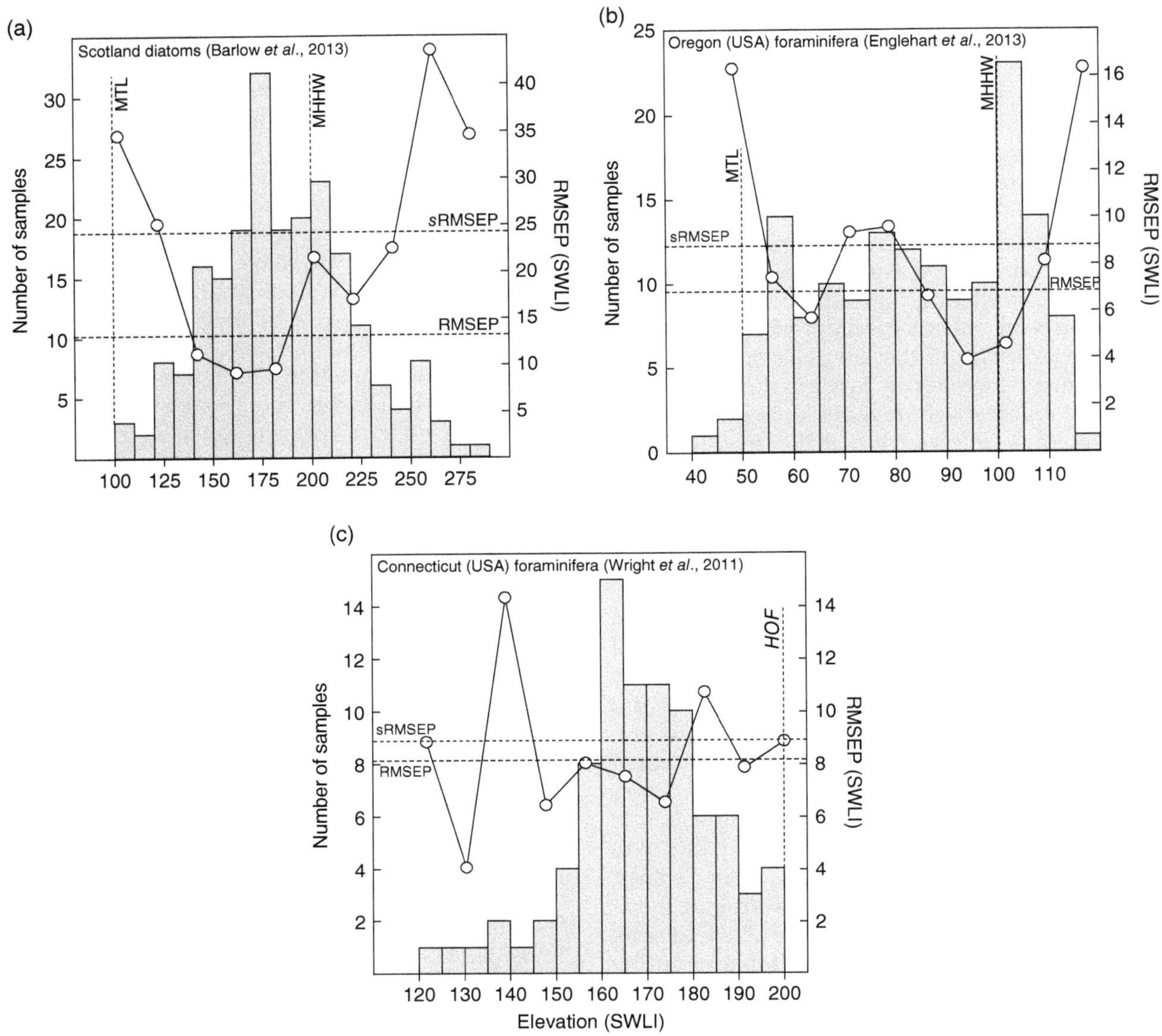

Fig. 31.4. Samples distributed along an elevational gradient from three sea-level studies using: (a) intertidal diatoms from Scotland; (b) intertidal foraminifera from Oregon, USA; and (c) saltmarsh foraminifera from Connecticut, USA. Histograms represent the frequency of samples at elevational intervals measured in standardized water level index (SWLI) units. The SWLI function used in each example is different, but tidal and biological datums are shown by dashed vertical lines. Each example shows, to varying degrees, the uneven sampling of the gradient that is characteristic of almost all training sets. A weighted-averaging transfer function with leave-one- out (LOO) cross-validation was developed for each dataset and the reported root-mean-square error of prediction (RMSEP) is shown by a horizontal dashed line. The transfer functions were also cross-validated (with LOO) by dividing the gradient into ten segments (open circles), where RMSEP is calculated for each segment and for the model as a whole (sRMSEP). Cross-validation under-predicts RMSEP if the environmental gradient is unevenly sampled; this can be accounted for using segment-wise cross-validation. More densely sampled intervals are better constrained and have correspondingly lower sRMSEP. Reconstructions from these intervals are likely to be more precise than those from under-sampled segments.

as the upper limit of marine influence and reported that it improved the alignment of the highest samples and compensated for differences in the quality and availability of tidal data. Using HOF in SWLI calculations relies on capturing the HOF by ensuring that supratidal samples devoid of foraminifera are collected at each site. Failure to capture HOF at a site would prohibit inclusion of those samples in a regional model where HOF is part of the SWLI calculation. It also places considerable emphasis on a single sample and requires some judgment as to where HOF is located if there is a gradual rather than abrupt disappearance of foraminifera. A comparable upper SWLI bound has not been used for diatoms because, unlike foraminifera, diatoms are present in freshwater coastal environments and there is a turnover of species rather than a highest occurrence.

An alternative to expressing the vertical position of intertidal microfossils to tidal datums is to calculate the duration and frequency of flooding for each sample since this, rather than elevation, is the environmental control on species distributions (Section 31.2.7; Horton and Edwards, 2006). Developing a transfer function with flood duration as the environmental variable is therefore an ecologically reasonable alternative to using elevation (Gehrels, 2000) and does not require standardization among sites with differing tidal ranges. However, in most cases it is easier to measure elevation than flood duration by leveling samples to benchmarks. Flood duration fails to take into account assemblages found above the highest astronomical tides as a consequence of the coastal water table, and the non-linearity between elevation and flooding characteristics at high tidal elevations is difficult to calculate without *in situ* water-level measurements for individual samples. Transfer functions based on flood duration require a transformation to elevation after being applied to a fossil assemblage in order to reconstruct sea level. This approach therefore assumes that the relationship between flood duration and elevation has remained unchanged through time. Implicitly, this assumption also underpins transfer functions with elevation as the environmental variable (Section 31.2.7).

31.2.5 Data screening

Field data comprising biological assemblages (*X*) and environmental variables (*Y*) usually include outliers. These atypical observations can be removed from the training set if they meet predetermined criteria for exclusion. This is a step in the transfer function process requiring an informed decision from the researcher. In some cases screening of samples has been a contentious, polarizing issue that has elicited strong responses. It is therefore recommended that researchers carefully weigh the advantages and disadvantages of screening or not screening and the criteria applied. Prior to screening it is prudent to check the outlying sample(s) to ensure that the data are accurate and were entered correctly. There is no consensus on if and how to screen samples, but broadly it can be done to improve transfer function performance or to exclude samples with known ecological limitations. Outliers in the modern training set can be recognized as samples that the transfer function is unable to accurately predict (large residuals) under cross-validation (see Section 31.3.8); these can be removed to improve performance. Juggins and Birks (2012) proposed that samples with residuals exceeding 2 or 2.5 standard deviations under cross-validation be screened to remove modern assemblages that lie outside of the approximately 95% or 99% confidence interval for the relationship between taxa and their environment (see Section 31.3.8). This type of screening should be limited to a single pass, since the effect of multiple iterations is to remove samples that in prior steps were not outliers. It should be recognized that this type of screening artificially improves transfer function performance; it is recommended that all samples are retained, if possible, to capture natural variability in modern populations.

Samples can also be excluded if there are specific justifications to believe that they are ecologically or statistically unreasonable to include. For example, in some locations undergoing sea-level rise and transgression, relict high-saltmarsh peat deposits are buried just below the surface of a modern tidal flat. A sample collected from this environment may inadvertently include older high-marsh foraminifera that it is unreasonable and inaccurate to include in the transfer function, and it should therefore be excluded (e.g., Kemp et al., 2013). Detailed field notes may help to identify such samples during laboratory analysis. Low counts are a statistical reason to screen particular samples if the researcher considers them to be unrepresentative. The decision on how many individuals constitutes a "low" count depends on the type of microfossil being used and the level of species diversity within the sample. Diverse assemblages (e.g., diatoms) require a greater minimum count than less-diverse groups such as foraminifera. Some datasets are trimmed by excluding rare or unusual species (typically those with a maximum abundance of <1–2%). This often results in improved transfer function performance (Juggins and Birks, 2012), reduces the frequency of zero abundances in modern and fossil data, and excludes species where the limited occurrence prevents adequate modeling of species–environment relationships.

31.2.6 Elevation as a surrogate environmental variable

Reconstructions of elevation are the basis for investigating changes in sea level. Elevation is not a true environmental variable because it does not exert a controlling influence on the modern distribution of microfossils. Instead, it is a convenient and useful substitute (a composite or surrogate variable) for the frequency and duration of tidal inundation (and other environmental co-variables) that observations show to be a direct ecological control on the distribution of intertidal microfossils. This surrogacy is acceptable in developing transfer functions where elevation is predictably related to the true controlling environmental variable(s) and where that relationship has remained unchanged for the period of reconstruction (Juggins, 2013). In circumstances where the frequency and duration of inundation changed independently of elevation, an unreliable reconstruction could be generated. If the environmental variable being reconstructed (e.g., elevation) is ecologically important, it follows that the same biological response ought to be observed in different regions and under variable environmental conditions (Juggins, 2013). This can be tested by comparing the environmental optima and tolerance of taxa among different sites and/or regions.

Broadly, "optima" is the elevation at which a species is most abundant and "tolerance" is the range of elevations over which it occurs. If species optima vary significantly in space, it is possible that they also changed over the time for which reconstructions are made and consequently invalidate one of the fundamental assumptions in using transfer functions to make paleoenvironmental reconstructions (Juggins, 2013). This comparison could be made at any spatial scale, but in most instances where transfer functions have been used to reconstruct sea level, the environmental change through time is equivalent to a relatively small spatial change. For example, it would be appropriate to compare intertidal microfossils between the mid US, northeastern US, and Canadian Atlantic coasts if the focus was a Holocene sea-level reconstruction from these regions. It is probably unnecessary to compare distributions from Arctic and tropical regions in different ocean basins, since the modern spatial variability in environmental conditions likely exceeds those encountered during the time period under consideration. The optima and tolerance of common intertidal foraminifera from three studies on the US Pacific northwest, mid-Atlantic, and northeast Atlantic coasts is compared in Figure 31.5. The regions have different climate and oceanographic regimes, but the optima and tolerance of foraminifera are similar for most taxa.

31.2.7 Regional v. local training sets

Training sets composed of spatially-independent samples are inherently regional in scale. However, when sampling regimes use transects it is possible to generate a local training set where all samples are from a single site (even if collected from multiple transects). Alternatively, transects from several sites can be combined to produce a regional training set, although there is ambiguity in the spatial definition of "local" and "regional" among studies. Here we define regional as being multiple sites spread across an area that can reasonably be assumed to share a prevailing set of climate and oceanographic conditions. Since the first application of transfer functions to sea-level reconstructions, the relative merits of local and regional training sets have been debated in the context of accuracy and vertical precision. Transfer functions trained on local datasets produce sea-level reconstructions with smaller uncertainties, because many environmental variables (e.g., salinity regime) at a single site are common to all samples and the influence of factors other than elevation is minimized. Frequently, the local training set is from the vicinity of where core material will be collected in the hope that it is the best single source of analogs for fossil assemblages; this is not necessarily the case however, and a local training set from elsewhere may be employed.

The principle advantage of regional datasets is that they include a greater range of modern environmental conditions, making them applicable to a greater range of paleoenvironments. In addition, they capture some of the natural variability of species distributions among sites. They are particularly useful when paleoenvironmental conditions are distinct from those encountered at

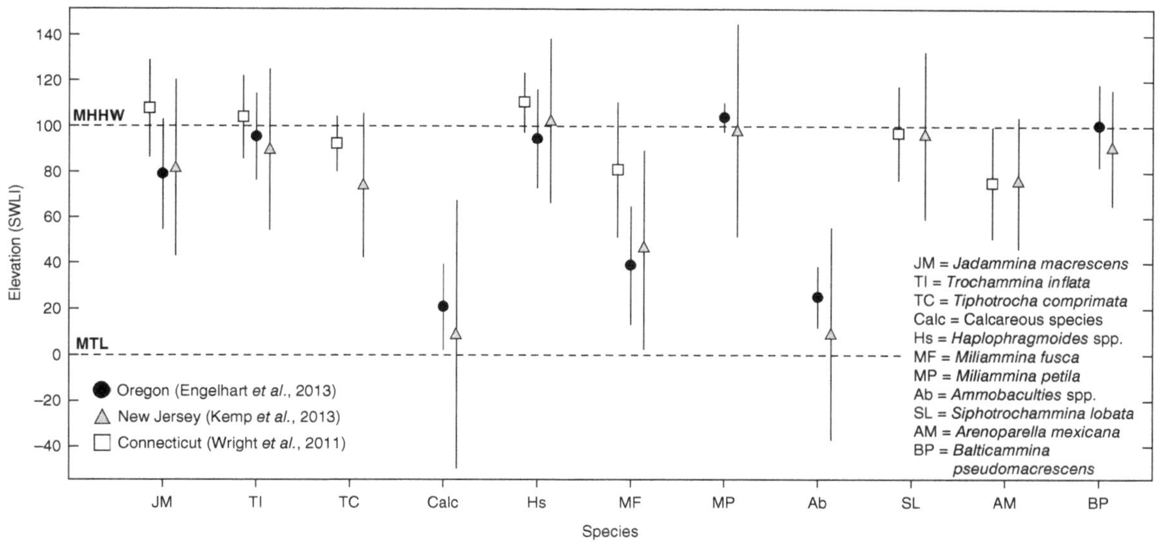

Fig. 31.5. Species optima (circles) and tolerances (vertical lines) for common species of intertidal foraminifera calculated for two regions on the US east coast (Connecticut and New Jersey) and one on the US west coast (Oregon). Optima and tolerances were derived from a weighted-averaging transfer function under leave-one-out cross-validation. The standardized water level index (SWLI) was the same for each dataset, where MTL = 0 and MHHW = 100.

the locality today. Regional compilations can compensate for difficulties associated with local sampling (e.g., absence of certain saltmarsh subenvironments and associated microfossils) to achieve regular and complete sampling of the environmental gradient.

Horton and Edwards (2005) compared the utility of local and regional datasets of intertidal foraminifera in reconstructing sea level and concluded that local training sets resulted in smaller cross-validated errors than regional training sets, but more frequently returned no-analog results that may be associated with less reliable and more uncertain reconstructions. In particular, diatoms may need to utilize large regional training sets because their high species diversity among sites commonly causes fossil assemblages to lack a suitable analog in a local training set (Section 31.4.1, e.g., Nelson et al., 2008; Kemp et al., 2009; Wilson and Lamb, 2012). In Alaska, Watcham et al. (2013) recognized that local training sets of diatoms frequently failed to provide acceptable analogs for reconstructing sea level from fossil assemblages, while regional training sets produced reliable reconstructions when tested against observed changes in sea level. They recommended collecting additional local datasets when the existing regional training set did not include suitable analogs as assessed by analog matching (Section 31.4.1), and favored a local contribution of 20–35% samples in the regional transfer function under the assumption that the local site will be the best single source of analogs for core assemblages. They also concluded that regional transfer functions were important for reconstructing sea-level changes from older sediment (>100 years) because environmental conditions (and therefore diatom assemblages) became increasingly dissimilar to local equivalents over longer periods of time.

An alternative to choosing between a local or regional dataset is to apply a dynamic technique such as locally-weighted transfer functions that generate a unique modern training set for each fossil sample (Juggins and Birks, 2012). This technique seeks to offer the precision of a local dataset in conjunction with the broad range of analogs provided by a regional training set, and is presented in greater detail in Section 31.3.5.

Development and application of local, regional, and locally-weighted transfer functions are illustrated using an example of intertidal foraminifera from Oregon, USA (Fig. 31.6). The modern training set consists of 152 samples from 8 sites with elevations expressed in SWLI units, where 0 is MLLW, ~50 is MTL, and 100 is MHHW (Hawkes et al.,

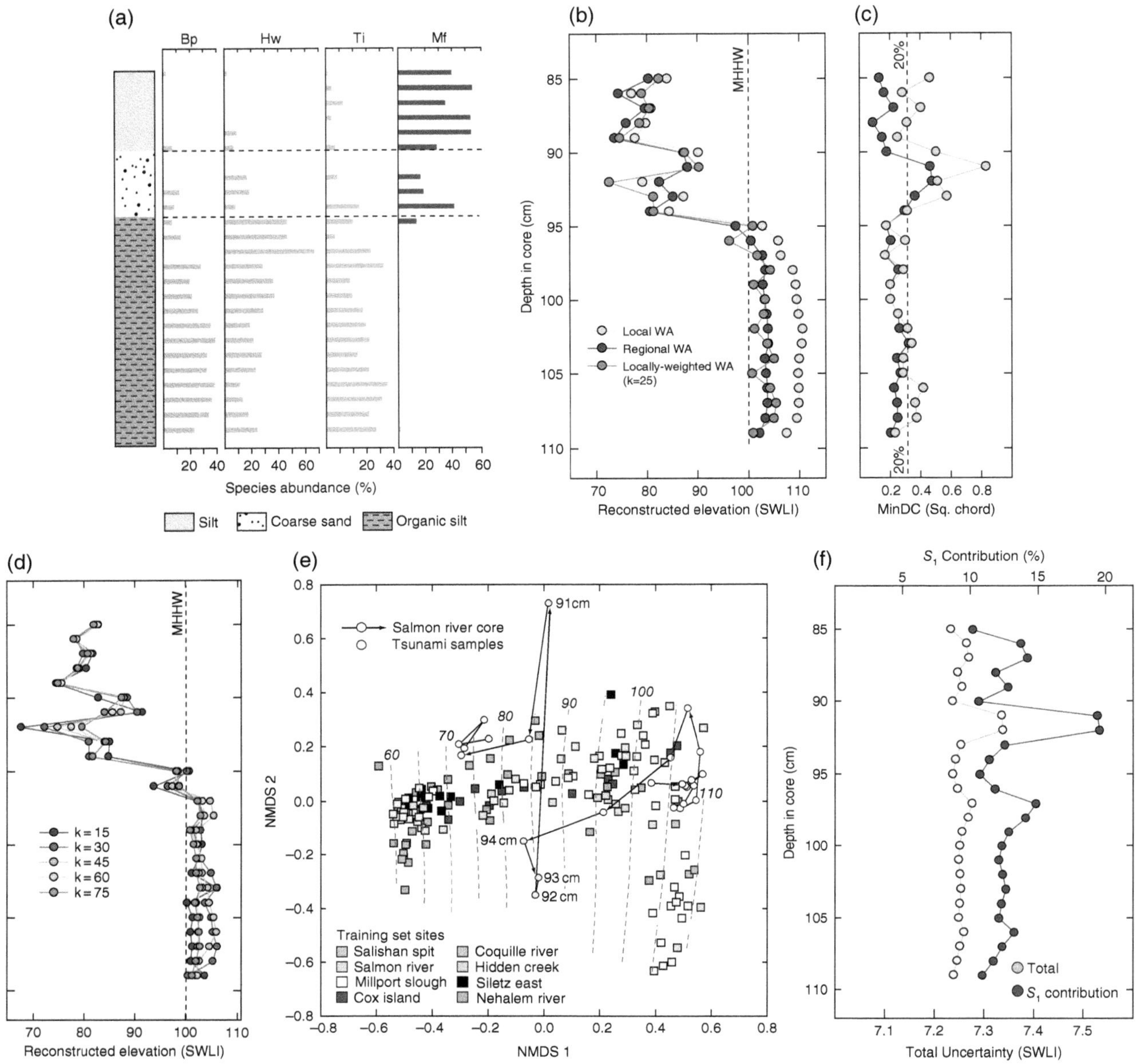

Fig. 31.6. Reconstruction of relative sea-level change caused by coseismic subsidence in Oregon, USA. (a) Lithology of the Salmon River core and assemblages of foraminifera preserved in core samples. Species typically from high-saltmarsh environments are shown in blue and those from low-marsh environments in red. Bp: *Balticammina pseudomacrescens*; Hw: *Haplophragmoides wilberti*; Ti: *Trochammina inflata*; Mf: *Miliammina fusca*. (b) Reconstructed elevation in SWLI units for three transfer functions (WA: weighted averaging, with inverse deshrinking and bootstrap cross-validation) developed using the Oregon training set (152 samples, eight sites) from (Engelhart et al., 2013). The local model only includes samples from the Salmon River site and the regional model includes all samples. (c) Use of analog matching to identify samples lacking a modern analog. Pair-wise comparison of all modern samples identified the 20th percentile threshold (vertical dashed line). For each core sample, the dissimilarity to the closest single modern analog (MinDC) was calculated using the Chord-squared distance metric. Samples exceeding the 20% threshold were assumed to lack a modern analog. (d) Elevation reconstructed using a locally weighted weighted-averaging (LW-WA) transfer function based on increasing values of *k*. (e) Non-metric multidimensional scaling (NMDS) of the modern training set and Salmon River core. Sites with modern samples are represented by colored squares. The samples from the Salmon River core are represented by circles with a trajectory shown by solid lines with arrows beginning from the bottom of the sequence. Core samples lying outside of the training set cluster lack a modern analog. The four samples from a tsunami-deposited sand layer at 91–94 cm are highlighted. Dashed lines mark the approximate position of standardized water level index (SWLI) elevations. Elevation parallels NMDS axis 1, indicating a strong environmental control on modern species distributions. (f) Decomposition of reconstruction errors for the regional transfer function applied to the Salmon River core. For color details, please see Plate 45.

2010; Engelhart et al., 2013). The training set is used to reconstruct relative sea level and specifically changes caused by coseismic subsidence preserved in a sediment core from one of the sites (Salmon River) where modern transects were sampled (Hawkes et al., 2011). A "local" transfer function (weighted averaging with bootstrapping cross-validation and inverse deshrinking) using only the modern samples from Salmon River was applied to the core, as was a regional (all 152 samples) and locally weighted ($k=25$, Chord-squared distance metric) version of the same model (Fig. 31.6b, d). The core includes organic silt overlain by coarse sand and then a silt unit (Fig. 31.6a). This sequence represents a saltmarsh that was subsided by coseismic land-level change and overwashed by a sand-laden tsunami, and the establishment of a tidal flat or low-marsh environment in response to being lowered in the tidal frame (Hawkes et al., 2011). Foraminifera in the lower part of the core comprise species typical of high-saltmarsh environments (e.g., *Haplophragmoides wilberti*), while the silt unit is dominated by the low-marsh and tidal-flat foraminifera *Miliammina fusca*.

The local, regional, and locally-weighted averaging transfer functions all reconstructed a distinct lowering at ~95 cm depth, supporting the lithostratigraphic interpretation (Fig. 31.6b). While the pattern is the same among models, the absolute changes are not. The local model generated a higher estimate of elevation for the samples in the organic silt unit, representing the site-specific distribution of high-marsh species and positioning of samples along the environmental gradient. Cross-validation estimates an uncertainty of 5 SWLI units for the local model, 7 SWLI units for the regional model, and 6 SWLI units for the locally weighted model, indicating that the local model would generate the most precise sea-level reconstruction. The 20% threshold for dissimilarity established by Chord-squared distance for the regional dataset is exceeded for three samples in the regional transfer function (in the tsunami sand) and ten in the local model (Fig. 31.6c), indicating that no analog situations are more likely when using a local model. The locally weighted transfer function was applied with $k=15$, 30, 45, 60, and 75 to show the effect of sample size on the reconstruction (Fig. 31.6d). There is close agreement among the models, suggesting that this particular dataset is robust to the choice of k. The greatest disparity among models is for the tsunami sample at 92 cm with a mixed assemblage of foraminifera. It is unsurprising that the assemblage of foraminifera in the tsunami sand lacks modern analogs under the local and regional models, because no modern equivalent could be sampled and included in the training set.

31.3 TRANSFER FUNCTION DEVELOPMENT

The second stage in using transfer functions to make quantitative sea-level reconstructions is to empirically model the ecological response of biological assemblages (Y) to the environmental variable of interest (X) using the modern observations compiled into a training set (Figs 31.1b and 31.2). This involves selecting and applying an appropriate numerical technique to define the relationship between the chosen sea-level indicator and tidal elevation (expressed as a SWLI where necessary). The particular properties and nature of modern and fossil assemblages dictate what numerical methods are suitable to use. Most microfossil training sets share the following attributes (Birks, 1995; Juggins and Birks, 2012).

- The absence (zero values) of particular species in many samples.
- A large number of species. This is a feature of diatom training sets in particular and, in some instances, there may be more species than samples in the training set.
- Species distributions that are highly correlated or collinear. For example, there are several species of intertidal foraminifera that often occur together in high-saltmarsh environments such as *Jadammina macrescens* and *Trochammina inflata*.
- Many or most species have a non-linear relationship to the environmental variable of interest. Few species of diatoms or foraminifera display relative abundances that increase or decrease linearly with tidal elevation; more often they have a unimodal distribution (Fig. 31.1b). This relationship is in part defined by the length of the environmental gradient that is sampled. A unimodal relationship may be well approximated by a linear response if only a part of the environmental gradient is sampled.
- Species abundances are expressed as proportions or percentages and therefore must total 1 or 100% within a single sample (prior to

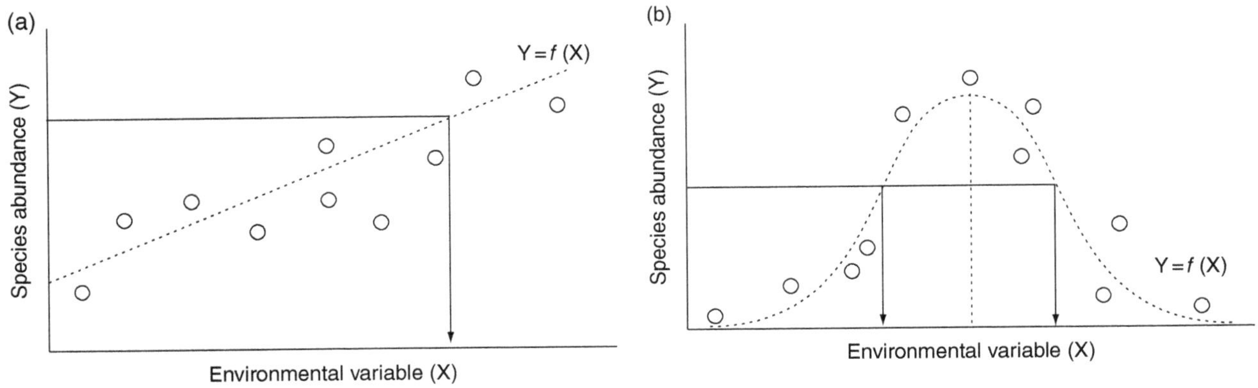

Fig. 31.7. Theoretical response of a species (*Y*) to an environmental variable (*X*) that is a primary control on its distribution in the modern environment. The distribution is derived from empirical field data (open circles) that comprise paired observations of *X* and *Y* and is described by a linear (a) or unimodal response (b).

removing any species from further analysis). This is a constant sum constraint.

- Species abundances are affected by structured noise (e.g., influence of environmental variables other than elevation such as salinity or sediment texture) and unstructured noise (e.g., preservation bias, counting errors).
- Despite the best intentions and careful sampling design, most training sets do not sample the environmental gradient evenly (see Section 31.3.4).

Sea-level researchers must select from the suite of numerical techniques that, to varying degrees, take into account these properties of training set data. We introduce four of the most common techniques and two more recent additions, and discuss their suitability for reconstructing sea level, assumptions, advantages, and disadvantages. The underlying mathematics is not presented; for more detail see Birks (1995) and Juggins and Birks (2012).

31.3.1 Maximum likelihood

The maximum likelihood method is an example of a "classical" approach where the modern assemblage data comprised of many species (*Y*) is regressed on the measured environmental variable (*X*) by linear, non-linear, or multivariate regression, depending on what is most suitable for the dataset at hand (Birks, 1995; Juggins and Birks, 2012). With one environmental variable and one species, classical regression would take the form:

$$Y = f(X) + \text{error}. \qquad (31.2)$$

For a linear relationship there would be one solution for any given species abundance (Fig. 31.7a), but in the case of unimodal relationships between a species and the environmental variable there would be two solutions (Fig. 31.7b). In reality, multiple species make up *Y* and one motivation for employing a transfer function is to utilize the whole assemblage rather than relying on a single indicator species (e.g., Scott and Medioli, 1978). In this instance, response curves are estimated for each species within the constant sum constraint. For classical methods such as maximum likelihood, the modeled species response (*f*) to environmental variables must be inverted to reconstruct an unknown environmental variable using fossil species data. The computational difficulty of this step was formerly a barrier to using classical methods, but this is no longer the case. The inversion takes the form:

$$X' = f^{-1}(Y') + \text{error}, \qquad (31.3)$$

where X' is the paleoenvironmental variable to be reconstructed (elevation), Y' is the fossil microfossil assemblage enumerated from a core that the transfer function is applied to, and f^{-1} is the inverted transfer function. The principal advantages and disadvantages of using maximum likelihoods transfer functions are presented in Table 31.1.

31.3.2 Weighted averaging

Weighted averaging (WA) is an "inverse" method. This group of methods treats the environmental variable (*X*) as the response and species

Table 31.1. Principal advantages and disadvantages of common numerical techniques used to develop transfer functions

	Advantages	Disadvantages
Maximum likelihood (ML)	Ecologically sound as it is based on a realistic species response to environmental conditions; good predictive power; robust to uneven sampling of the environmental gradient	Strongly influenced by outliers and zero values (in modern and fossil datasets), which are more frequent in high-diversity populations.
Weighted averaging (WA)	Statistically robust; relatively simple but results are similar to classical methods; underlying species-environment response is well supported by ecological theory and data; robust performance with test data	Affected by uneven sampling of the elevation gradient, but helped by large datasets; distortion at extremes of elevation gradient (edge effects) causes low samples to be over-predicted and high samples to be under-predicted (structured residuals) (this is particularly true for long environmental gradients >3SD); does not utilize residual correlations unlike WA-PLS
Weighted averaging, partial least squares (WA-PLS)	Reduces the influence of edge effects; exploits residual correlations in the modern training set that remain after fitting a WA model	Risk of over-fitting the transfer function by selecting too many components; affected by uneven sampling of the elevation gradient, but helped by large datasets
Modern analog technique (MAT)	Simple underlying theory; provides a test of ecological plausibility for reconstructions using other methods	May require a large training set; requires subjective decision on number of analogs to use and thresholds for degree of analogy; effected by uneven sampling of the elevation gradient, but helped by large datasets

Source: Modified from Juggins and Birks (2012).

abundance as the predictor (Y), and may therefore be perceived to lack ecological plausibility. However, species are not treated as drivers of environmental conditions; rather, the form of inverse methods is a numerical convenience (Birks, 1995). The transfer function (f) is derived directly from empirical data and does not require the inversion step that is necessary in classical methods. The format of the transfer function developed from the modern dataset is therefore

$$X = f(Y) + \text{error} \tag{31.4}$$

and inserting the fossil assemblage (Y') data allows the paleoenvironmental variable (X') to be reconstructed directly as

$$X' = f(Y') + \text{error.} \tag{31.5}$$

The ecological assumption underlying WA is that a species is more abundant at locations where environmental conditions are favorable; conversely, it is absent or rare where conditions are less favorable. More formally, each species has environmental optima where abundance is at a maximum, and also an environmental tolerance, which is the range over which the species is found (Figs 31.1b, 31.5, and 31.7b). The optimum is calculated as the average of environmental values where a species occurs, weighted by its abundance. The environmental value of a sample where the species is highly abundant is therefore given more importance in calculating the average environmental value than a sample where the species is a minor part of the assemblage. Tolerance is the weighted standard deviation of species abundance (Birks, 1995). The paleoenvironmental variable is reconstructed by taking a weighted average of the species optima in a fossil sample. Calculation of weighted averages therefore twice shrinks the range of the environmental variable compared to the original sampled range. To correct for this, a "deshrinking" step is necessary using either inverse, classical, or monotonic spline regression. The classical approach is more suitable for reconstructions close to the ends of the environmental gradient, while inverse deshrinking is better for reconstructions close to the middle of the gradient (Juggins and Birks, 2012). It is necessary to decide which approach is most suitable for the particular dataset being used to reconstruct sea level. Unlike classical and inverse methods, monotonic deshrinking is a non-linear approach that adds flexibility and is often the method with the best performance statistics (Birks and Simpson, 2013).

It may be desirable to favor paleoenvironmental estimates from species with narrow ecological ranges (small tolerances) in the belief that this will produce a more precise reconstruction. This is achieved by placing less emphasis (down-weighting) on the contribution made by species with broad tolerances, and is termed weighted averaging tolerance downweighting (WA-Tol). In sea-level research, species that occupy a relatively large elevational range are effectively given less importance than those which are found at a restricted range of elevations in modern intertidal environments. Species with a smaller indicative range generate more precise sea-level reconstructions, but accuracy should be carefully assessed, particularly if tolerance downweighting emphasizes the importance of rare or unusual species. However, tolerance downweighting often makes little practical difference because the optima of rare or unusual species are poorly defined by the modern training set. The principal advantages and disadvantages of WA are listed in Table 31.1 (Juggins and Birks, 2012).

31.3.3 Weighted-averaging partial-least-squares (WA-PLS)

WA-PLS, an inverse method, is perhaps the most widely employed transfer function technique in sea-level research. The weighted averaging part of this technique is the same as that described in the previous section. After the application of weighted averaging, some correlations remain between observed and predicted environmental variables that are not taken into account by weighted averaging (Birks, 1995). The partial-least-squares (PLS) addition utilizes these residuals to improve the fitted relationship between taxa and the environmental variable of interest. This is achieved by maximizing the covariance between elevation and linear combinations of species abundances, and therefore takes into account (to some degree) edge effects and the environmental factors other than elevation that influence the distribution of taxa (ter Braak and Juggins, 1993).

WA-PLS generates results for 1–n components, which are linear combinations of the original biological data. Component 1 is equivalent to the WA model with inverse deshrinking. Use of WA-PLS requires a decision to be made on the number of components to include in the model. It is advisable to use the minimum acceptable number, because choosing additional components may overfit the data. In most cases, sea-level reconstructions based on WA-PLS restrict themselves to component 2 since the fitting of additional components improves model performance very little at the expense of increase complexity. Barlow et al. (2013) proposed restricting WA-PLS to three components or fewer, and defined improved performance as >5% reduction in root-mean-square error of prediction (RMSEP) and increase in r^2_{boot} (Section 31.3.6). Researchers could however adopt alternative guidelines such as testing the significance of improved model performance with a randomization *t*-test. The principal advantages and disadvantages of WA-PLS are listed in Table 31.1 (Juggins and Birks, 2012).

31.3.4 Modern analog technique (MAT)

Perhaps the most straightforward approach to reconstruction that can be envisaged is assigning a fossil sample the optima and tolerance of its most similar counterpart in the modern training set. This is termed space-for-time substitution by Juggins and Birks (2012), and is the principal behind MAT. It does not impose a presumed species–environment relationship (Birks, 1995). The numerical method in MAT is calculation of dissimilarity between a fossil sample and each sample in the modern training set. A number of metrics are available to make this calculation, each of which has strengths and weaknesses (Jackson and Williams, 2004; see Section 31.4.1).

The simplest use of MAT would be to select the single closest modern analog (lowest measured dissimilarity) for each fossil sample in core material. A more common approach is to use the average environmental value (weighted or not) from the k closest analogs in the modern dataset, where k is a number chosen by the researcher (commonly between 5 and 10). The choice of k is a tradeoff between bias and variance and should balance the need to employ multiple modern analogs against the size of the training set. For example, using one analog is likely to result a high degree of variability among reconstructions by overemphasizing the importance of a single modern sample. Using a proportion of modern analogs that is unreasonably large may be unreliable or disguise the paleoenvironmental signal by supressing variance and increasing bias.

In datasets with a high degree of spatial autocorrelation, MAT transfer functions commonly perform better (lower RMSEP, higher r^2) than alternatives such as WA or WA-PLS because the fit between species and elevation is localized. It is therefore recommended that MAT is used cautiously with modern datasets displaying a high degree of spatial autocorrelation (Telford and Birks, 2009), which is often a common trait of the training sets developed for reconstructing sea level. In these circumstances, MAT reconstructions should also be compared with reconstructions from other methods (see Section 31.4.5). Furthermore, Birks (1995) states that using MAT transfer functions requires a relatively large training set and, unlike other methods, is unable to extrapolate to values for the environmental variable beyond those captured in the training set.

In addition to being a numerical technique for reconstructing sea level, MAT is also used to assess the ecological plausibility of reconstructions derived from other methods such as WA and WA-PLS, by classifying the degree of analogy between modern and core samples (see Section 31.4.1).

31.3.5 Locally weighted models

Locally weighted transfer functions use the training set dynamically to generate a unique model for individual fossil samples. The motivation of this approach is to strike a balance between the precision of a small, local dataset in reconstructing sea level and the wide range of modern analogs available in a larger, regional training set. For each fossil sample the number of modern samples (k) to be included in the training set is specified; Juggins and Birks (2012) propose 30–50. Those k analogs are then identified by MAT (Section 31.3.4) and used to generate a training set with the additional option of weighting modern samples by distance. A numerical technique such as weighted averaging (Section 31.3.2) is then used to develop each transfer function, resulting in a locally weighted weighted-averaging (LW-WA) transfer function. This approach has been successfully applied to develop transfer functions relating pollen to climatic factors (Lu et al., 2011) and offers the potential to exploit the benefits of local and regional training sets to reconstruct sea level. One advantage of this approach is that large numbers of samples can be assimilated to maximize the variety of analogs available, but potentially without the disadvantage of weakening model performance because samples that are deemed unnecessary to include by MAT play no role in the reconstruction.

31.3.6 Bayesian models

Bayesian modeling offers an alternative and new approach to developing and applying transfer functions for paleoenvironmental reconstruction. It is based on modeling the probabilities associated with the targeted environmental variable, the fossil dataset, and all model parameters (Salonen et al., 2011). The first step is forward modeling to build the relationship between modern taxa and the environmental variable of interest. Backward modeling then generates a reconstruction that is consistent with the fossil assemblage data and has a formalized uncertainty. This approach has not yet been used in sea-level reconstruction.

31.3.7 Model selection

No single type of transfer function can be recommended over all others for reconstructing sea level. Instead, technique selection is a necessary and critical stage in making quantitative paleoenvironmental reconstructions. The first step is to determine if models underpinned by linear or unimodal relationships between biological assemblages and elevation are most appropriate (Birks, 1995). This is achieved by calculating the length of the environmental gradient from the modern training set not as a range in meters, but rather as standard deviation (SD) units of biological turnover. The environmental tolerance of a species is 1 SD. Over a gradient length of 4 SD, the species would therefore appear, increase to a maximum abundance, decline, and become rare. If an elevational range of 4 SD was sampled, there would be few species in common between the lowest and highest samples. Gradient length is estimated by detrended canonical correspondence analysis (DCCA) of the modern training set with elevation as the only environmental variable. Birks (1995) recommended that if the gradient is <2 SD linear methods are used, and if the gradient length is >2 SD unimodal methods are used. If the environmental variable is an important constraint on species distributions it will be strongly correlated to the first unconstrained axis of a detrended correspondence analysis (DCA), and this analysis can be used in lieu of DCCA. This approach does not help in the decision of whether or not to use MAT.

There is less guidance for choosing among the different types of unimodal transfer function. The most common approach is to generate a range of transfer functions using the training set (WA, WA-PLS, ML, LW-WA, MAT, etc.) and to compare their performance. If fossil material is already available, the transfer functions are applied and the results tested for ecological plausibility before selecting a final model. It is important that the researcher is familiar with the ecology and paleoecology of the organisms being used as sea-level indicators so that the validity of model reconstructions can be assessed qualitatively. Some transfer functions will perform well statistically, but generate results that appear contradictory to the species–environment relationship observed when developing a training set. In many cases where intertidal foraminifera and diatoms are used to reconstruct sea level, weighted-averaging methods (WA and WA-PLS) are chosen because they are relatively simple, mathematically robust, ecologically plausible, perform well even under poor-analog conditions, and commonly outperform other methods in comparative studies (Birks, 1995).

An example of transfer functions developed using a suite of numerical methods is presented in Figure 31.8 for the modern training set of Scottish diatoms (Barlow et al., 2013). The dataset comprises 215 samples that were all used in model development. The results are typical of training sets used in sea-level reconstructions where there is no single model that outperforms each of the others by every measure. The lowest RMSEP (16.0 SWLI units) and highest r^2 (0.81) are for the locally weighted weighted-averaging model, but this model has a relatively high maximum bias (43.5 SWLI units) because of systematic misfits below ~150 SWLI and above ~250 SWLI. The lowest maximum bias was returned by the maximum likelihood model (22.4 SWLI units). Comparison of observed and predicted elevations show a tendency for all of the transfer functions to systematically over-predict low elevations and under-predict higher elevations to varying degrees. This trait is also shown by inspecting residuals between observed and predicted elevations for structure.

Qualitatively, the best fit between observed and predicted elevations is from the LW-WA and ML transfer functions. The modern analog model has a clear structure in the residuals indicating a strong and systematic bias which, coupled with its relatively poor measured performance, would lead to this model not being pursued further. Similarly, the WA-PLS model outperforms the WA model which can likely be set aside. There is little to choose between the WA-PLS, LW-WA, and ML models and additional methods of model evaluation may be necessary to select one over another; equally, any (or all) of these models could be used.

31.3.8 Model evaluation

Applying a transfer function to an independent dataset of modern observations is the most robust means of model evaluation. This type of external data is not usually employed because researchers understandably prefer to include all modern samples in the training set given the time and effort that was invested in collecting and analyzing them, although data from existing literature can be used in this capacity. Transfer functions are therefore most commonly evaluated by simulating new data from the modern training set using cross-validation. Cross-validation methods include the following.

- Leave-one-out (LOO) excludes each sample from the dataset in turn and uses the remaining samples to predict the environmental value of the excluded sample.
- Bootstrapping selects a fixed number of samples (n) with replacement to generate a new training set. Some samples will be included more than once because of replacement. The unselected samples form the test set. The procedure is repeated many times (e.g., 1000 or 10,000) to provide a robust estimate of uncertainty.
- Leave-one-site-out (LOSO) excludes all samples from a particular site in turn and uses the remaining samples to predict the environmental value of the excluded samples (see below). The user defines what constitutes a single site and consequently which samples belong to each site. A single site may have more than one modern transect of samples.
- Leave-one-group-out (LGO) excludes samples assigned to a particular group in turn and uses the remaining samples to predict the environmental value for the excluded samples. Groups are specified by the researcher and could, for example, be samples from particular settings such as open coast or back-barrier locations. LOSO is a specific application of LGO cross-validation.

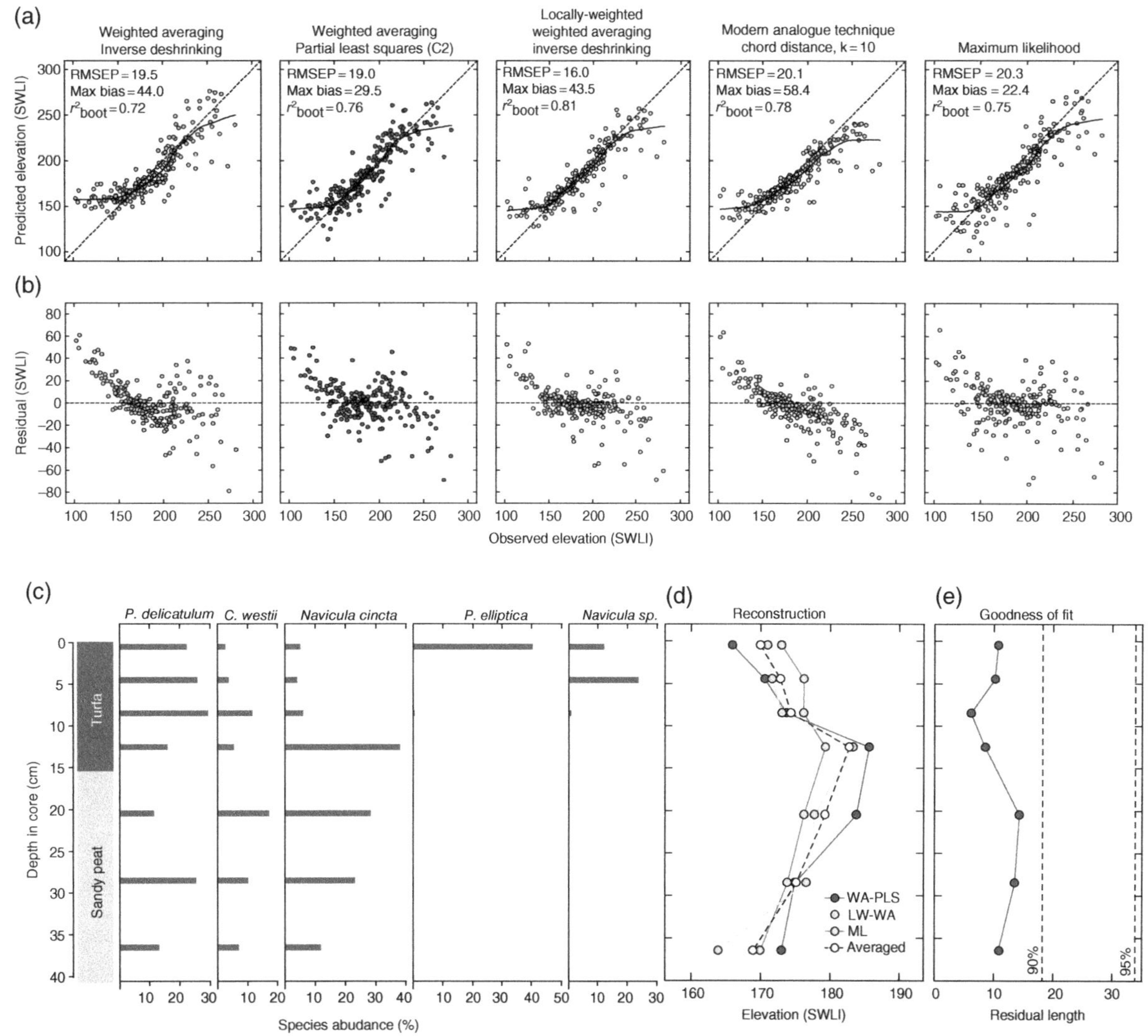

Fig. 31.8. Cross-validated performance and application of transfer functions developed for a training set of Scottish diatoms (Barlow et al., 2013). SWLI = standardized water level index. (a) Plots in the top row of panels are observed and predicted sample elevations with 1000 bootstrap trials for the 215 modern samples. Solid lines represent a local regression (LOESS) smooth; dashed line is 1:1. (b) Lower panels are the residual difference between observed and predicted elevations for each sample; dashed line is zero residual. (c) Fossil diatoms in an example core from Loch Laxford. (d) Reconstructions of elevation from three transfer functions derived by applying each model to the Loch Laxford core. An average of the three models was generated and is presented. (e) Goodness-of-fit calculated for each core sample in the weighted-averaging partial-least-squares model. Thresholds at the 90th and 95th percentile were established from the canonical correspondence analysis of the training set into which core samples were passively projected.

Several terms are used to evaluate and compare transfer function performance. The root-mean-square error of prediction (RMSEP) is the square root of mean squared differences between measured environmental values (often termed observed) and those predicted for the same sample by the transfer function under cross-validation. The units associated with RMSEP are the same as the training set (e.g., SWLI for sea-level research) and it can be represented (Juggins and Birks, 2012) as:

$$\text{RMSEP} = \sqrt{\text{SEP}^2 + \text{MB}^2} \quad (31.6)$$

where SEP is the standard error of the predicted residuals (i.e., observed – predicted) and MB is the mean bias (the mean of those residuals). These random and systematic components are returned by the transfer function and are commonly assessed by plotting observed elevations against residuals from transfer function estimates (Fig. 31.8). If there is structure in the residuals then a systematic bias is inferred. This systematic component is measured by the maximum bias, calculated by dividing the environmental gradient into segments (usually 10) and calculating the mean bias for each, the maximum bias being the largest value. The coefficient of determination (r^2) between observed and predicted environmental values is also calculated during cross-validation. In assessing transfer function performance, a lower RMSEP, absence of visual structure in plots of observed v. predicted environmental values, a lower maximum bias, and higher r^2 are desirable criteria.

Evaluation of transfer functions assumes that samples in the training set are independent of one another. In sea-level research, where the modern samples were collected along transects, this assumption is less robust because of the possibility of spatial autocorrelation among samples (Telford and Birks, 2005, 2009). Spatial autocorrelation is the tendency for samples that are physically close to one another to be more similar in species composition than randomly selected samples. This causes transfer functions to have artificially low RMSEP values and may result in a misguided model choice because some techniques are more sensitive to autocorrelation than others. Qualitatively, this effect can be investigated by comparing MAT and WA or WA-PLS models. A protocol for assessing the influence of spatial autocorrelation was described by Telford and Birks (2009), where the effect of randomly removing samples is compared to the effect of removing geographically ordered samples at distance intervals (e.g., within 5, 10, 50, and 100 km). Sea-level research relies on modern datasets where multiple samples are collected at individual sites separated by many kilometers, referred to as "clustered training sets". The Telford and Birks (2009) method for assessing spatial autocorrelation is not well suited to situations where samples may be just a few meters apart, as they are on intertidal transects. In the case of transect data, one solution is to use leave-one-site-out (LOSO) cross-validation, where all samples from a single site are omitted and data from the remaining sites are used to predict them (Payne et al., 2012).

To illustrate the effect of LOSO cross-validation, the method is applied to three example datasets. The training set of Scottish diatoms (Barlow et al., 2013) includes samples from nine sites. Modern foraminifera in Oregon, USA (Hawkes et al., 2010; Engelhart et al., 2013) were sampled along transects at eight sites, while a similar dataset from New Jersey, USA documented foraminifera at 12 sites representing a diverse range of tidal marshes (Kemp et al., 2012, 2013). A WA-PLS (component 2) transfer function was developed for each training set with LOSO and LOO cross-validation. This model choice was not the preferred model in each case, but was used illustratively and to ensure comparability among datasets. In each case, RMSEP under LOSO cross-validation was greater than using LOO (Fig. 31.9), although only large differences would indicate that model performance was overly optimistic in spatially-clustered datasets of the type frequently used in sea-level reconstruction. The disparity was greater for diatoms than foraminifera (Fig. 31.9), which may reflect the high degree of variability in diatom assemblages among sites reported in several studies (e.g., Woodroffe and Long, 2010; Wilson and Lamb, 2012). In the example from New Jersey one site (Jayne Drive) stands out as having a particularly large difference between RMSEP estimated under LOSO and under LOO cross-validation. This site has a unique (in the dataset) assemblage of foraminifera and illustrates the effect of having a single unusual site in LOSO cross-validation. Since the site was unique, there was not a good analog retained in the dataset during LOSO so it is poorly predicted by the remaining sites. The increased RMSEP under LOSO cross-validation

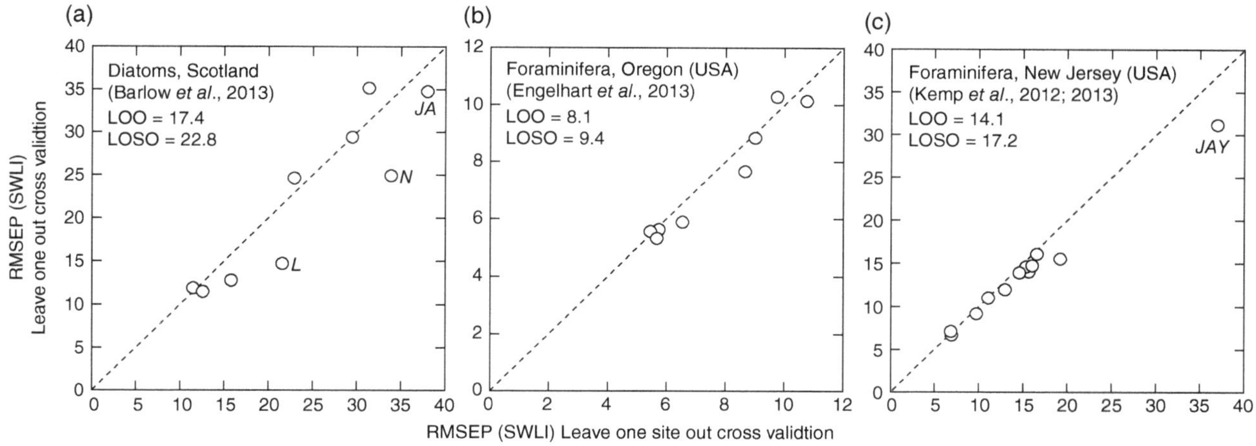

Fig. 31.9. Comparison of root-mean-square error of prediction (RMSEP) for three modern training sets under leave-one-out (LOO) and leave-one-site-out (LOSO) cross-validation. In each case the transfer function is a weighted-averaging partial-least-squares (component 2) model. Data points represent individual sites and reported RMSEP values in text are for the overall model. Elevations are in standardized water level index (SWLI) units and are not directly comparable among the three datasets. (a) Scottish diatoms; JA: Jura; N: Nonach; L: Loch Creran and Loch Linnhe. (b) Oregon foraminifera. (c) New Jersey foraminifera; JAY: Jayne Drive.

does not mean that local transfer functions should not be used to reconstruct sea level, although it does confirm that training sets including multiple sites capture additional variability in species–environment relationships.

Modern sampling for sea-level reconstruction often fails to sample the environmental gradient evenly (e.g., Fig. 31.4). Uneven sampling along an environmental gradient can cause bias in transfer function performance (Telford and Birks, 2011a). The effect of uneven sampling is assessed by segment-wise analysis of the transfer function, where RMSEP is estimated for pieces (segments) of the environmental gradient. The segments are equally sized intervals of the elevational range included in the training set. Apparent performance (RMSEP) is better in the heavily sampled part of the gradient because there are more analogs and these are retained more often in cross-validation. Conversely, the less-frequently sampled parts of the gradient have lower precision as there are fewer analogs. Decreased overall model performance from uneven sampling does not necessarily lead to more uncertain paleoenvironmental reconstructions if fossil samples are analogous to the heavily sampled segments (Telford and Birks, 2011a).

To illustrate segment-wise analysis of transfer function performance, the training sets of Scottish diatoms, Oregon foraminifera, and Connecticut foraminifera were used as an example (Fig. 31.4). The environmental gradient was divided into 10 segments of equal environmental length (range of elevation). Cross-validation was performed for each segment in turn, and these were averaged to generate an estimate of segment-wise RMSEP (sRMSEP). This approach removes the bias of uneven sample distribution, inherent when the full training set is cross-validated at once. In each of the three datasets, RMSEP under segment-wise analysis exceeded RMSEP under cross-validation of the full training set. For individual segments, RMSEP was usually greater for parts of the environmental gradient represented by fewer samples, indicating that cross-validation under estimates RMSEP and that the most precise sea-level reconstructions would come from fossil environments analogous to those sampled most frequently in the training set.

31.4 SEA-LEVEL RECONSTRUCTION (TRANSFER FUNCTION APPLICATION)

Fossil assemblages preserved in sedimentary archives (Y') are used to reconstruct sea level (X'). The transfer function developed from the modern training set is applied to fossil assemblages and generates an output that is reconstructed elevation (indicative meaning) and a sample-specific estimate of uncertainty (indicative range). The units are the same as those used as input for transfer function development; if they are SWLI units they must be back-converted by rearranging the original SWLI equation and using the tidal datums for the core site, which may or may not change through time. In all circumstances, application of a transfer function will generate an output that must be evaluated. Six approaches for evaluating transfer function performance are described.

31.4.1 Modern analog evaluation

Reconstructions are more likely to be reliable if the fossil assemblage that was the basis for the reconstruction has appropriate analogs in the training set (Simpson, 2012). MAT measures dissimilarity between fossil and modern samples using a selected metric (e.g., Bray-Curtis distance). Simple distance metrics (e.g., Euclidean, Manhattan) that weight the abundance of species are dominated by the most common taxa. In contrast, metrics that weigh each species equally are disproportionately influenced by rare or unusual taxa. Signal-to-noise metrics (e.g., Chord-squared distance) offer an alternative, intermediate approach and are the most useful for evaluating modern analogs (Jackson and Williams, 2004). Chord-squared distance performs well with the closed compositional assemblages (percentages or proportions rather than counts) that are typical of microfossil groups used in sea-level reconstructions because it down-weights rare species (noise) to highlight the principal patterns (signal). The χ^2 and information statistic metrics behave similarly to Chord-squared distance and can often be used interchangeably (Simpson, 2012). Bray-Curtis measures absolute distances among samples rather than squared differences. The first step in evaluating modern analogs is to establish thresholds of dissimilarity from samples in the modern dataset by calculating the distance between all possible pairings in the training set (Simpson, 2012). Commonly thresholds or cut-off values are percentiles (e.g., 5%, 10%, or 20%) of dissimilarity from pair-wise comparison of samples in the training set. These values are equated to qualitative descriptions of analogy such as "good" or "close". Often a sliding scale of thresholds is used to identify close, good, acceptable, or poor analogs from increasingly high percentiles. For example, Watcham et al. (2013) determined that for intertidal diatoms from Alaska, a fossil sample exceeding the 20% threshold had a "poor" analogy to modern samples, samples below the 20% cut-off had a "close" analogy, and those below the 5% level had "good" analogs.

The second step is to measure dissimilarity between modern and fossil samples. For each fossil sample, the dissimilarity to its single closest modern analog (minimum dissimilarity or minDC) determines whether or not it falls within or outside of the thresholds set previously. In some studies, fossil samples exceeding a pre-determined threshold are excluded from the resulting sea-level reconstruction, while in others they are retained but interpreted with a degree of caution.

The limitation of this approach is that the selection of percentiles is a subjective decision (e.g., Overpeck et al., 1985; Anderson et al., 1989; Jackson and Williams, 2004). Furthermore, there is no guarantee that a fossil sample exceeding the chosen threshold(s) for dissimilarity is producing an unreliable reconstruction. In a homogenous training set, dissimilarity among modern samples is small and the absolute distance (as measured by a distance metric) for a particular percentile threshold is correspondingly small. A fossil sample that is quite similar in biological composition to the modern training set could therefore exceed the chosen threshold and wrongly be deemed to lack a modern analog. It is therefore recommended that a relatively high percentile (e.g., 20%) is selected as the threshold for analogy when using homogenous modern training sets. In sea-level research, intertidal foraminifera form low-diversity assemblages and where sampling is biased toward high-saltmarsh environments (particularly at one site or a small number of similar sites), modern training sets can be somewhat homogenous. Conversely, in a heterogeneous training set dissimilarity among modern samples is large and the absolute distance for a particular threshold is correspondingly high. As such, a fossil sample with a distinctive biological composition may wrongly be treated as having an acceptable modern analog. It is therefore recommended that a relatively low percentile (e.g., 5%) is selected as the threshold for analogy when using heterogeneous modern training sets. Intertidal diatoms form heterogeneous training sets because species diversity is high within and among sites.

The taxonomic distance between modern and fossil samples can also be visualized and assessed using non-metric multidimensional scaling (NMDS) or a similar ordination method. Fossil samples are included passively on a NMDS ordination plot of modern samples and joined in stratigraphic order by a line representing upcore trajectory and changes in assemblage composition through time. If fossil samples fall within the scatter of modern samples, then there is close taxonomic agreement. The training set and core of intertidal foraminifera from Oregon, USA (Hawkes et al., 2010, 2011; Engelhart et al., 2013) were used to illustrate application of NMDS to reconstructing coseismic relative sea-level change (Fig. 31.6e). NMDS axis one parallels sample elevation, indicating that elevation is an important control on species distributions on the dataset. The core includes a lower organic silt unit, overlain by a tsunami-deposited sand and then a silt unit (Fig. 31.6a). The trajectory of core samples passively inserted into the NMDS ordination show that samples below the sand formed close to MHHW (100 SWLI units). The four samples from a tsunami-deposited sand lie outside of the modern cluster and therefore lack a modern analog, confirming the analysis using MAT. The samples above the sand formed at a lower elevation (approximately 75 SWLI units), confirming the interpretation of coseismic subsidence lowering of the marsh surface in the tidal frame. It is clear from the NMDS ordination that a local transfer function would not provide adequate analogs for interpreting this sedimentary sequence because most core samples lie away from the cluster of modern samples from the same site.

31.4.2 Goodness-of-fit statistics

Evaluating how well transfer functions reconstruct the target environmental variable is an important part of assessing model performance and sea-level reconstructions. Goodness-of-fit is estimated by placing the modern micropaleontological data into a constrained ordination (e.g., canonical correspondence analysis or redundancy analysis) with elevation as the only environmental constraint (Birks, 1995; Simpson and Hall, 2012). The squared residual distance between each sample and its fitted position on the first constrained axis is calculated. As with modern analog evaluation, thresholds are established from the modern dataset. When core samples are passively projected into the ordination, a residual distance is calculated that is compared with the modern thresholds to evaluate goodness-of-fit between fossil samples and elevation. The proposed thresholds for this analysis are the 90th ("poor" fit) and 95th ("very poor" fit) percentiles of the modern training set (Simpson and Hall, 2012). Similarly to MAT, it is possible to calculate and apply any chosen percentile as a threshold, but these values are recommended. To illustrate the application of this method an example is presented in Figure 31.8e. A weighted-averaging partial-least-squares (component 2) transfer function developed from the training set of Scottish diatoms was applied to a short core from Loch Laxford. Residual distances between modern samples and the first axis of a canonical correspondence analysis set the 90th percentile and 95th percentile thresholds. Passive projection positioned all of the seven core samples below the 90th percentile, and they were therefore deemed to have a good fit to elevation.

31.4.3 Tests of reconstruction significance

Telford and Birks (2011b) introduced a method to assess the statistical significance of paleoenvironmental reconstructions, where the amount of variance in fossil data explained by a transfer function is compared to a large number of alternative transfer functions trained on random environmental data. A large number of random transfer functions (e.g., 999) are used to generate a probability distribution ($n = 1000$ with the addition of the actual transfer function). A reconstruction is statistically significant if it explains more variance than 95% of the random models. This test was applied to a core of intertidal foraminifera from Oregon, USA where the transfer function applied was the regional weighted-averaging model used and described previously (Fig. 31.10). Reconstructed elevation explained 61% of the variance observed in the fossil assemblages, which exceeded more than 99% of the random models. In passing this test, the transfer function is deemed significant.

In sea-level research, a core of sediment representing an intertidal environment (such as a high saltmarsh) that kept pace with sea-level rise could have an unchanged microfossil assemblage. This unusual circumstance would cause the reconstruction to unfairly fail the significance test. If downcore assemblages of microfossils do not

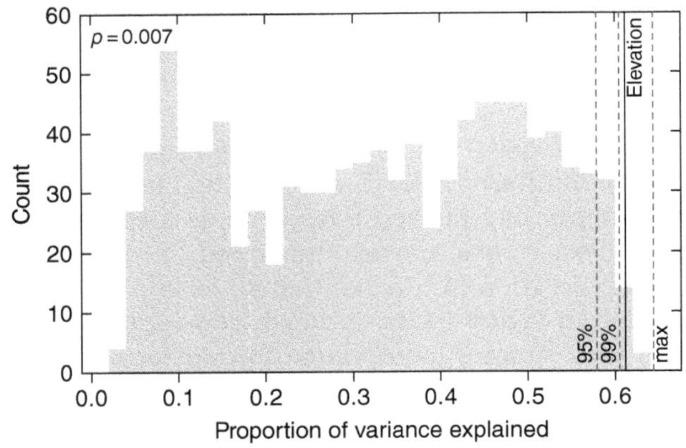

Fig. 31.10. Statistical significance of a weighted-averaging transfer function from Oregon, USA developed from a regional training set of foraminifera and applied to a sediment core from Salmon River. The transfer function explained 61% of variance in the core (solid vertical line marked elevation) compared to a maximum possible of 64% (dashed vertical line marked max). For comparison, 999 weighted-averaging transfer functions developed on normally distributed, random environmental variables were generated (gray histogram). The model trained on elevation explains more variance than >95% of the random models and is therefore deemed significant.

change composition there is minimal variance for any model to explain, including those trained on random environmental variables. As a result, the reconstruction of elevation may not perform any better than reconstructions trained on random and synthetic environmental variables (Kemp et al., 2013). In this instance the reconstruction could fail the significance test, but actually be a robust and useful model.

31.4.4 Sample-specific error

Transfer function output includes an error estimate for each fossil sample that was derived by bootstrapping. To get a ~95% confidence interval for the reconstructions, this value should be multiplied by 1.96. Most studies making a paleoenvironmental reconstruction (including sea level) do not multiply the error term in this way. The reason for this is that it normally generates an uncertainty that is large enough to mask all but the most pronounced paleoenvironmental changes, and no trends can be established since all changes lie within the uncertainty of the reconstruction. If RMSEP is used as the reconstruction uncertainty there is an inherent assumption that the training set adequately captured the relationship between modern assemblages and elevation (i.e., the transfer function is correct) and that no additional error is introduced from a lack of analogy between modern and core samples.

Reconstruction error can be broken down into two parts. S_1 is the uncertainty in estimating model parameters (the species–environment relationship). This is calculated as the standard deviation of reconstructed values for core samples during bootstrapping and is therefore unique to each sample. Larger S_1 values are returned for samples including taxa that are poorly represented in the modern training set. S_2 is the difference between observed and predicted elevations established by cross-validation of the training set, and has the same value for all core samples (Birks et al., 1990; Juggins and Birks, 2012). Commonly, the S_2 factor is dominant and S_1 makes a relatively small (10–20%) contribution. If species in fossil samples are underrepresented in the training set, S_1 may make a larger contribution. Plotting the downcore percentage of total error from the S_1 and S_2 components can help to recognize samples where dissimilarity between modern and core samples contributes an unusually high proportion and which may therefore generate unreliable reconstructions (Juggins and Birks, 2012). In the example from Oregon (Fig. 31.6f), the total reconstruction error for the regional transfer function is approximately 7 SWLI units for each core sample. The S_1 component varies between 10% and 20% of the total, and is greater in the samples of tsunami sand that are known to lack a modern analog as evidenced by MAT and NMDS.

31.4.5 Comparison with other reconstructions

In many instances, measures of cross-validated performance (RMSEP, r^2, maximum bias) and inspection of observed v. predicted plots fail to identify a particular numerical technique as being superior to the alternatives (Fig. 31.8). One means of evaluating the robustness of the reconstruction is to apply other types of transfer function (numerical technique) with different underlying ecological response models to the same training and fossil datasets to produce alternative reconstructions. If the resulting sea-level reconstructions are similar, then there can be a higher degree of confidence than if they are different. Indeed, reconstructions from several methods can be combined (e.g., simple averaging) to generate a composite reconstruction (Juggins and Birks, 2012). Transfer functions using several numerical techniques were developed from the training set of Scottish diatoms (Barlow et al., 2013). Cross-validation indicated that the weighted-averaging and modern analog technique models were inferior. These were discarded from further analysis and the WA-PLS, maximum likelihood, and locally weighted weighted-averaging models were applied to an illustrative core. The core included seven samples from a lower sedimentary unit of sandy peat overlain by turfa peat (Fig. 31.8c). The transfer functions reconstruct elevations between 160 and 190 SWLI units. The downcore trends in reconstructed elevation are similar among all models giving a degree of confidence that the reconstructions are accurate. The average uncertainty for each model is approximately ±18 SWLI units and within this range all of the reconstructions are in agreement. An average of the three models shows that the locally weighted weighted-averaging model is the closest to the inter-model mean for most samples. In this case the choice of any (or all) of the three models would be appropriate.

In addition to making comparisons among reconstructions from several model types, in some cases it may be possible to compare reconstructions made using other proxies. For example, if paired modern and fossil samples of diatoms and foraminifera were available, independent reconstructions could be made using these two sea-level proxies (e.g., Kemp et al., 2009). Similar results would give confidence in the reliability of the sea-level reconstruction and the skill of both microfossil groups and transfer functions in reconstructing sea level.

31.4.6 Validation with historic measurements

Historical records of sea-level variability provide an opportunity to independently check transfer function results. Tide gauges are an instrumental record of past sea-level changes that can be compared directly to nearby reconstructions from transfer functions (e.g., Gehrels et al., 2005). Sea-level reconstructions from coastal sediment naturally filter out the short-lived and annual variability that is obvious in instrumental measurements through the response time of biological assemblages and a sampling regime using slices of core material that span an interval of time (often several years). As such it is unreasonable to expect reconstructions to resolve sea-level changes with the same resolution as tide gauges. The natural filtering is an advantage of proxy reconstructions of sea level because it isolates persistent trends from noise. Comparing trends in tide-gauge records and sea-level reconstructions is therefore an appropriate means to investigate the reliability of the transfer function. However, Juggins and Birks (2012) caution that agreement between measured and reconstructed values validates the reconstruction, but not necessarily other reconstructions that use the same transfer function model.

31.5 SOFTWARE

The most commonly used Windows-based software for developing and applying transfer functions is C2 (Juggins, 2011), which provides an easy-to-use implementation of most types of transfer function and cross-validation in a format that is familiar and intuitive for Windows users. Increasingly, transfer functions are being implemented using the R statistical software. R is a common platform that is customized by downloading and installing additional packages into the R environment to perform particular types of analysis. Several of the techniques described can currently only be applied in R (e.g., segment-wise RMSEP, leave-one-site-out cross-validation). The techniques we described are available in the *vegan*, *analogue*, *rioja*, and *palaeoSig* packages. R is freely available for download from http://www.r-project.org/. The format of R provides greater flexibility and is likely to be the favored format for implementation of new techniques in the future. Picking up R can be time-consuming and challenging, although user guides (e.g., Crawley, 2012), online tutorials, internet help pages, and a growing availability of university training courses make it increasingly accessible and widespread. All R packages are supported by manuals that can be downloaded at no charge. R also provides a single program that implements a range of numerical techniques relevant to sea-level research beyond transfer functions such as age–depth modeling. The R code applied to the example datasets is provided as an appendix.

APPENDIX: R CODE USED IN EXAMPLES

The following code for R (from the *rioja*, *analogue*, *vegan*, and *palaeoSig* packages) was used to generate some of the examples used in this chapter. Other results come from analysis performed in C2.

The object "MODERN" is a training set of modern assemblage data with species abundances expressed as proportions. The object "ELEVATION" is the measured elevation of each sample in the training set. The object "SITE" is a numerical code assigned to each sample in the training set specifying which site it was collected at. The object "CORE" is assemblage data from core samples from which elevation is to be reconstructed.

A weighted-averaging model was generated with LOO, LOSO, and bootstrap cross-validation by:

```
mod<-WA(sqrt(MODERN), ELEVATION$"SWLI")
mod.loo<-crossval(mod, cv.method="loo")
mod.boot<-crossval(mod, cv.method="boot",
  nboot=1000)
mod.loso<-crossval(mod, cv.method="lgo",
  ngroups=SITE$Site)
```

A weighted-averaging partial-least-squares model was generated with LOO, LOSO, and bootstrap cross-validation by:

```
mod<-WAPLS(sqrt(MODERN),
ELEVATION$"SWLI")
mod.loo<-crossval(mod, cv.method="loo")
mod.boot<-crossval(mod, cv.method="boot",
  nboot=1000)
mod.loso<-crossval(mod, cv.method="lgo",
  ngroups=SITE$Site)
```

Investigation of training set size (Figure 31.3) was performed by:

```
spp<-MODERN
env<-ELEVATION$SWLI
subsample<-function(env,n,ndivis=1,...){ #
  stratified subsampling
  extras<-length(env)%%ndivis
  sz<-rep(length(env)%/%ndivis, ndivis)
  ext<-sample(1:ndivis, size=extras)
  sz[ext]<-sz[ext]+1
  cut.at<-cumsum(c(0,sz))
  env.cut<-cbind(1:length(env),cut(rank(env,
    ties.method="random"), cut.at, labels=F))
  maxlen<-max(tapply(env.cut[,1],env.
    cut[,2],length))
  samps<-sapply(sample(1:ndivis),function(n){
    batch<-env.cut[env.cut[,2]==n,]
    b2<-batch[sample(1:nrow(batch)),1]
    b3<-rep(NA,maxlen)
    b3[1:length(b2)]<-b2
    b3
  })
  samps<-as.vector(t(samps))
  samps<-samps[!is.na(samps)][1:n]
  samps
}
cut.to<-c(20,40,60,80,100,120,140,160,180,200)
nrun=50
swap.ss<-sapply(cut.to,function(n){
print("#########################");print(n);
  print("~~~~~~~~~~~~~~~~~~~~")
  res1<-replicate(nrun,{
    ss<-subsample(env=env,n,ndivis=4)
    spp<-spp[ss,]
    spp<-spp[,colSums(spp)>0]
    env<-env[ss]
    wa.rmsep<-performance(crossval(WA(sqrt(s
      pp), env)))$crossval[1,1]
    mat.rmsep<-performance(MAT(spp,
      env))$object[5,1]#5th analogue
    c(wa.rmsep, mat.rmsep)
  })
  apply(res1,1,median)
})
```

Comparison of LOO and LOSO models (Figure 31.9):

```
looloso<-sapply(unique(SITE$Site),function(n)
{N<- SITE$Site==n c(LOO=sqrt(mean((mod.
loo$predict[N,1]-ELEVATION$"SWLI"[N])^2)),
LOSO=sqrt(mean((mod.loso$predict[N,1]-
ELEVATION$"SWLI"[N])^2)))
})
```

Dissimilarity between modern and fossil samples with Bray Curtis dissimilarity (Figure 31.6c):

```
core.diss<- analog(sqrt(MODERN), sqrt(CORE),
  k=5, method="bray")
```

Goodness-of-fit analysis (Figure 31.8e) was performed by:

```
res.length<-residLen(sqrt(MODERN),
  ELEVATION$"SWLI", sqrt(CORE),
  method=c("cca"))
```

Non-metric multidimensional scaling (Figure 31.6e) of modern and core samples:

```
mod<-metaMDS(sqrt(MODERN),
distance="euclidean")
x11()
p<-plot(mod)
ordisurf(mod, ELEVATION$"SWLI", add=TRUE)
sapply(unique(SITE$"site"), function(n){
  x<-scores(mod)[ SITE$"site"==n,]
  polygon(x[chull(x),], border=
    sample(c(1:6,8),1))
  })
identify(p, what=SITE$"site")
all.MODERN<-Merge(MODERN, CORE)
mod<-metaMDS(sqrt(all.MODERN),
  distance="euclidean")
plot(mod, type="n")
points(mod, select=-(1:nrow(MODERN)),
  type="l", col=4)
points(mod, select=1:nrow(MODERN))
ordisurf(mod, c(ELEVATION$"SWLI", rep(NA,
  nrow(CORE))), add=TRUE)
```

The significance test of Telford and Birks (2011b) for a weighted-averaging transfer function:

```
mod.random<-randomTF(sqrt(MODERN),ELEVA
  TION,sqrt(CORE),n=999,fun=WA,col=1)
```


ACKNOWLEDGEMENTS

ACK is grateful for postdoctoral support from the Yale Climate and Energy Institute and the National Oceanic and Atmospheric Administration (Award NA11OAR4310101). Norwegian Research Council FriMedBio project palaeoDrivers (213607) helped support RJT. Natasha Barlow, Simon Engelhart, Andrea Hawkes, and Alex Wright kindly provided the datasets used as examples. We thank two reviewers for their detailed and insightful comments.

REFERENCES

Anderson, P.M., Bartlein, P.J., Brubaker, L.B., Gajewski, K., and Ritchie, J.C. (1989) Modern analogues of Late-Quaternary pollen spectra from the western interior of North America. Journal of Biogeography, 16, 573–596.

Barlow, N.L.M., Shennan, I., Long, A.J., Gehrels, W.R., Saher, M.H., Woodroffe, S.A., and Hillier, C. (2013) Salt marshes as late Holocene tide gauges. Global and Planetary Change, 106, 90–110.

Birks, H.J.B. (1995) Quantitative palaeoenvironmental reconstructions, In: *Statistical Modelling of Quaternary Science Data* (eds Maddy, D., and Brew, J.S.), Quaternary Research Association, Cambridge, pp. 161–254.

Birks, H.J.B., and Simpson, G.L. (2013) Diatoms and pH reconstruction (1990) revisited. Journal of Paleolimnology, 49, 363–371.

Birks, H.J.B., Line, J.M., Juggins, S., Stevenson, A.C., and Braak, C.J.F.T. (1990) Diatoms and pH Reconstruction. Philosophical Transactions of the Royal Society of London. B, Biological Sciences, 327, 263–278.

Crawley, M.J. (2012) *The R Book*, second edition. John Wiley & Sons.

Engelhart, S.E., Horton, B.P., Nelson, A., Hawkes, A.D., Witter, R.C., Wang, K., Wang, P.L., and Vane, C.H. (2013) Testing the use of microfossils to reconstruct great earthquakes at Cascadia. Geology, 41, 1067–1070.

Gehrels, R. (2000) Using foraminiferal transfer functions to produce high-resolution sea-level records from salt-marsh deposits, Maine, USA. The Holocene, 10, 367–376.

Gehrels, W.R., Kirby, J.R., Prokoph, A., Newnham, R.M., Achterberg, E.P., Evans, H., Black, S., and Scott, D.B. (2005) Onset of recent rapid sea-level rise in the western Atlantic Ocean. Quaternary Science Reviews, 24, 2083–2100.

Gehrels, W.R., Marshall, W.A., Gehrels, M.J., Larsen, G., Kirby, J.R., Eiriksson, J., Heinemeier, J., and Shimmield, T. (2006) Rapid sea-level rise in the North Atlantic Ocean since the first half of the nineteenth century. Holocene, 16, 949–965.

Guilbault, J.P., Clague, J.J., and Lapointe, M. (1995) Amount of subsidence during a late Holocene earthquake--evidence from fossil tidal marsh foraminifera at Vancouver Island, west coast of Canada. Palaeogeography, Palaeoclimatology, Palaeoecology, 118, 49–71.

Guilbault, J.P., Clague, J.J., and Lapointe, M. (1996) Foraminiferal evidence for the amount of coseismic subsidence during a late holocene earthquake on Vancouver Island, West Coast of Canada. Quaternary Science Reviews, 15, 913–937.

Hamilton, S., and Shennan, I. (2005) Late Holocene relative sea-level changes and the earthquake deformation cycle around upper Cook Inlet, Alaska. Quaternary Science Reviews, 24, 1479–1498.

Hawkes, A.D., Horton, B.P., Nelson, A.R., and Hill, D.F. (2010) The application of intertidal foraminifera to reconstruct coastal subsidence during the giant Cascadia earthquake of AD 1700 in Oregon, USA. Quaternary International, 221, 116–140.

Hawkes, A.D., Horton, B.P., Nelson, A.R., Vane, C.H., and Sawai, Y. (2011) Coastal subsidence in Oregon, USA, during the giant Cascadia earthquake of AD 1700. Quaternary Science Reviews, 30, 364–376.

Horton, B.P., and Edwards, R.J. (2005) The application of local and regional transfer functions to the reconstruction of Holocene sea levels, north Norfolk, England. Holocene, 15, 216–228.

Horton, B.P., and Edwards, R.J. (2006) Quantifying Holocene sea-level change using intertidal foraminifera: lessons from the British Isles. Cushman Foundation for Foraminiferal Research, Special Publication no. 40, 97.

Horton, B.P., Edwards, R.J., and Lloyd, J.M. (1999) UK

intertidal foraminiferal distributions: implications for sea-level studies. Marine Micropaleontology, 36, 205–223.

Horton, B.P., Engelhart, S.E., Kemp, A.C., and Sawai, Y. (2013) Microfossils in tidal settings as indicators of sea-level change, paleoearthquakes, tsunamis, and tropical cyclones, Treatise on Geomorphology, 14, 292–314.

Imbrie, J., and Kipp, N.G. (1971) A new micropaleontological method for quantiative paleoclimatology: application to a Late Pleistocene Caribbean core. In: *The Late Cenozoic Glacial Ages* (ed. Turekian, K.K.), Yale University Press, New Haven, CT, pp. 77–181.

Jackson, S.T., and Williams, J.W. (2004) Modern analogs in Quaternary paleoecology: Here today, gone yesterday, gone tomorrow? Annual Review of Earth and Planetary Sciences, 32, 495–537.

Juggins, S. (2011) C2, edition 1.7.6. Available at http://www.staff.ncl.ac.uk/staff/stephen.juggins/software/C2Home.htm (accessed 1 August 2014).

Juggins, S. (2013) Quantitative reconstructions in palaeolimnology: new paradigm or sick science? Quaternary Science Reviews, 64, 20–32.

Juggins, S., and Birks, H.J.B. (2012) Quantiative environmental reconstructions from biological data, In: *Data Handling and Numerical Techniques* (eds Birks, H.J.B., Lotter, A.F., Juggins, S., and Smol, J.P.), Springer, pp. 431–494.

Kemp, A.C., Horton, B.P., Corbett, D.R., Culver, S.J., Edwards, R.J., and van de Plassche, O. (2009) The relative utility of foraminifera and diatoms for reconstructing late Holocene sea-level change in North Carolina, USA. Quaternary Research, 71, 9–21.

Kemp, A.C., Horton, B.P., Vann, D.R., Engelhart, S.E., Vane, C.H., Nikitina, D., and Anisfeld, S.C. (2012) Quantitative vertical zonation of salt-marsh foraminifera for reconstructing former sea level; an example from New Jersey, USA. Quaternary Science Reviews, 54, 26–39.

Kemp, A.C., Telford, R.J., Horton, B.P., Anisfeld, S.C., and Sommerfield, C.K. (2013) Reconstructing Holocene sea-level using salt-marsh foraminifera and transfer functions: lessons from New Jersey, USA. Journal of Quaternary Science, 28(6), 617–629.

Lu, H., Wu, N., Liu, K.-b., Zhu, L., Yang, X., Yao, T., Wang, L., Li, Q., Liu, X., Shen, C., Li, X., Tong, G., and Jiang, H. (2011) Modern pollen distributions in Qinghai-Tibetan Plateau and the development of transfer functions for reconstructing Holocene environmental changes. Quaternary Science Reviews, 30, 947–966.

Nelson, A.R., Sawai, Y., Jennings, A.E., Bradley, L.-A., Gerson, L., Sherrod, B.L., Sabean, J., and Horton, B.P. (2008) Great-earthquake paleogeodesy and tsunamis of the past 2000 years at Alsea Bay, central Oregon coast, USA. Quaternary Science Reviews, 27, 747–768.

Overpeck, J.T., Webb, T., and Prentice, I.C. (1985) Quantitative interpretation of fossil pollen spectra: Dissimilarity coefficients and the method of modern analogs. Quaternary Research, 23, 87–108.

Payne, R.J., Telford, R.J., Blackford, J.J., Blundell, A., Booth, R.K., Charman, D.J., Lamentowicz, Å.u., Lamentowicz, M., Mitchell, E.A.D., Potts, G., Swindles, G.T., Warner, B.G., and Woodland, W. (2012) Testing peatland testate amoeba transfer functions: appropriate methods for clustered training-sets. The Holocene, 22, 819–825.

Reavie, E., and Juggins, S. (2011) Exploration of sample size and diatom-based indicator performance in three North American phosphorus training sets. Aquatic Ecology, 45, 529–538.

Sachs, H.M., Webb, T., and Clark, D.R. (1977) Paleoecological transfer functions. Annual Review of Earth and Planetary Sciences, 5, 159–178.

Salonen, J.S., Ilvonen, L., Seppä, H., Holmström, L., Telford, R.J., Gaidamavicius, A., Stancikaite, M., and Subetto, D. (2011) Comparing different calibration methods (WA/WA-PLS regression and Bayesian modelling) and different-sized calibration sets in pollen-based quantitative climate reconstruction. The Holocene, 22, 413–424.

Sawai, Y., Satake, K., Kamataki, T., Nasu, H., Shishikura, M., Atwater, B.F., Horton, B.P., Kelsey, H.M., Nagumo, T., and Yamaguchi, M. (2004) Transient uplift after a 17th-century earthquake along the Kuril Subduction Zone. Science, 306, 1918–1920.

Scott, D.B., and Medioli, F.S. (1978) Vertical zonations of marsh foraminifera as accurate indicators of former sea levels. Nature, 272, 528–531.

Simpson, G.L. (2012) Analogue methods. In: *Data Handling and Numerical Techniques* (eds Birks, H.J.B., Lotter, A.F., Juggins, S., and Smol, J.P.), Springer, pp. 495–522.

Simpson, G.L., and Hall, R.I. (2012) Human impacts: applications of numerical methods to evaluate surface water acidification and eutrophication. In: *Data Handling and Numerical Techniques* (eds Birks, H.J.B., Lotter, A.F., Juggins, S., and Smol, J.P.), Springer, pp. 579–614.

Telford, R.J., and Birks, H.J.B. (2005) The secret assumption of transfer functions: problems with spatial autocorrelation in evaluating model performance. Quaternary Science Reviews, 24, 2173–2179.

Telford, R.J., and Birks, H.J.B. (2009) Evaluation of transfer functions in spatially structured environments. Quaternary Science Reviews, 28, 1309–1316.

Telford, R.J., and Birks, H.J.B. (2011a) Effect of uneven sampling along an environmental gradient on transfer-function performance. Journal of Paleolimnology, 46, 99–106.

Telford, R.J., and Birks, H.J.B. (2011b) A novel method for assessing the statistical significance of quantitative reconstructions inferred from biotic assemblages. Quaternary Science Reviews, 30, 1272–1278.

ter Braak, C.J.F., and Juggins, S. (1993) Weighted averaging partial least squares regression (WA-PLS): an improved method for reconstructing environmental variables from species assemblages. Hydrobiologia, 269–270, 485.

Watcham, E.P., Shennan, I., and Barlow, N.L.M. (2013) Scale considerations in using diatoms as indicators of sea-level change: lessons from Alaska. Journal of Quaternary Science, 28, 165–179.

Webb, T., and Bryson, R.A. (1972) Late and postglacial climatic change in northern Midwest USA: quantitative estimates derived from fossil pollen spectra by multivariate statistical analysis. Quaternary Research, 2, 70–115.

Wilson, G.P., and Lamb, A.L. (2012) An assessment of the utility of regional diatom-based tidal-level transfer functions. Journal of Quaternary Science, 27, 360–370.

Woodroffe, S.A., and Long, A.J. (2010) Reconstructing recent relative sea-level changes in West Greenland:

Local diatom-based transfer functions are superior to regional models. Quaternary International, 221, 91–103.

Wright, A.J., Edwards, R.J., and van de Plassche, O. (2011) Reassessing transfer-function performance in sea-level reconstruction based on benthic salt-marsh foraminifera from the Atlantic coast of NE North America. Marine Micropaleontology, 81, 43–62.

Zong, Y., and Horton, B.P. (1999) Diatom-based tidal-level transfer functions as an aid in reconstructing Quaternary history of sea-level movements in the UK. Journal of Quaternary Science, 14, 153–167.

Chapter 32

Using chronological models in late Holocene sea-level reconstructions from saltmarsh sediments

ANDREW C. PARNELL[1] AND W. ROLAND GEHRELS[2]

[1]*School of Mathematical Sciences (Statistics), Complex and Adaptive Systems Laboratory, University College Dublin, Dublin, Ireland*

[2]*Environment Department, University of York, Heslington, York, UK*

32.1 INTRODUCTION

In the last decade or so, many studies from mid- and high-latitude coastal settings have shown that high-resolution sea-level reconstructions can be derived from chronological and micropaleontological analyses of saltmarsh sediment cores (e.g., Gehrels, 2000; Leorri et al., 2008; Kemp et al., 2009, 2011; Long et al., 2012). The inclusion of uncertainties in the analysis of data is a vital aspect of these sea-level studies. One key uncertainty is that of the age of the sea-level index points (SLIPs), especially when reconstructed from proxy data that may have undergone uncertain dating methods. In this chapter we show how these uncertainties might be estimated appropriately using statistical chronology models, and then used to estimate rates of sea-level change.

We consider a situation where sea-level data have been obtained from a core taken in the upper part of a saltmarsh, close to the highest level of tidal influence. Fossil plant remains from the core have been sent for dating. These samples contain radioactive carbon and are therefore radiocarbon dated. Where feasible, sediments may be dated by a variety of other methods (Gehrels et al., 2006; Marshall et al., 2007; Kemp et al., 2012), including other radiometric methods (e.g., ^{210}Pb), marker horizons (e.g., pollen, stable lead isotopes, ^{137}Cs, or trace metals) or tephra. If all the dates are generated without contamination, then those lower down the core will tend to be older than those at the top of the core. This phenomenon is known mathematically as "monotonicity".

A chronology model uses the depths and dates in the core (however obtained) combined with assumptions about the sedimentation process to produce an age–depth relationship which satisfies the monotonicity constraint. If a statistical model is used then the relationship is probabilistic, such that at each depth in the core a probability distribution can be obtained indicating the probable age of the sediment. Such statistical models can be seen to have three uses. First, they allow for the ages of non-dated sediment to be estimated with uncertainty. Second, the monotonicity constraint allows for the ages of dated sediments to have reduced uncertainty (compared to the age estimate if the monotonicity constraint not been used), as parts of the date probability distribution can be removed. Third, they allow for reduced uncertainty for dated levels in the core where various different methods produce different ages.

If the dates in a core are highly uncertain then the ages estimated from the chronology model will be similarly so. Due to the nature of (especially radiocarbon) dating, the probability distributions of the dates are not normally distributed and may display severe skewness and non-normality. As is previously known (Telford et al., 2004), the mean of such dates (or even the 95% confidence interval) is not a good summary of the age of that sediment. Furthermore, most chronology models show uncertainties that increase considerably away from dated levels, producing an enlarging or "bowing" uncertainty (see Fig. 32.1). If chronology models are required to estimate the ages of SLIPs that are not dated directly, there will therefore be increased temporal uncertainty that must be taken into account. This is an important consideration when working with saltmarsh sediments, as sedimentation rates can fluctuate greatly.

When the output of the chronology model is combined with that of a sea-level reconstruction, each SLIP can be thought of as a bivariate probability distribution which we refer to informally as

Handbook of Sea-Level Research, First Edition. Edited by Ian Shennan, Antony J. Long, and Benjamin P. Horton.

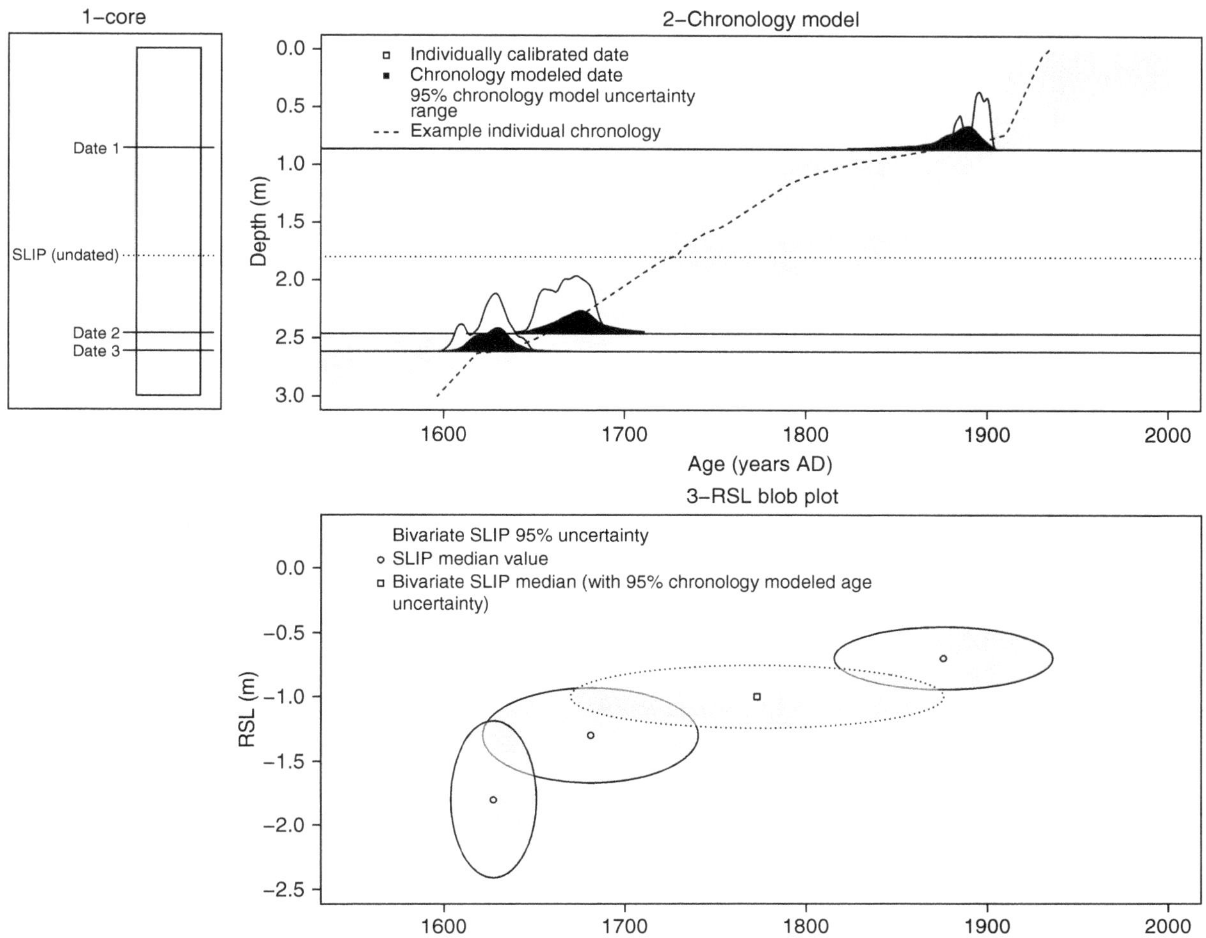

Fig. 32.1. Transformation from a core (panel 1) to a chronology model (the gray shaded area of panel 2) to a relative sea-level plot (panel 3). In panel 1 there are three dates shown at depths of 0.8, 2.4, and 2.6 m as well as a dashed line indicating a SLIP which has not been directly dated (but for which a sea-level reconstruction is possibly based, for example, on its microfossil content). In panel 2 the individual radiocarbon dates are calibrated to produce pdfs and then run through a chronology model which estimates the ages of all depths in a core. The gray shaded region shows the 95% uncertainty from the chronology model. For the three dates in our core, we show the chronology modeled pdfs at those depths in black. We can use the chronology model to estimate the age of the undated SLIPs as shown by the dotted line. In panel 3 we show the relative sea-level values (with 95% uncertainties) and the chronology model uncertainties from panel 2 as ellipses indicating the bivariate uncertainty in each point. For the SLIPs without a date, we use the estimated age (with uncertainty) from the chronology model. The circles (and square) show the median values of each RSL point.

a "blob". Figure 32.1 depicts the process by which ages from a core (panel 1) can be used to produce a chronology model (the gray shaded area of panel 2; note the monotonicity constraint) and thence an RSL plot (panel 3) containing blobs. In panel 3, each individual SLIP is made up of the age probability distribution taken from the chronology model, and the RSL uncertainty taken from the reconstruction. This RSL uncertainty is often assumed to be normally distributed (Parnell, 2005). Note that the presentation of SLIPs as blobs is not a regular occurrence (most commonly plus signs or boxes are used), but one we strongly encourage as it enables a full appreciation of the joint uncertainty.

Often the term "error" is used to describe the numerical value of the uncertainty. However, we caution against this approach since the "error" could represent any of: the standard deviation of the measurement; twice the standard deviation of the measurement (i.e., approximately the half-width of the 95% confidence interval); or four times the standard deviation of the measurement (approximately the width of the 95% confidence interval). Given this confusion, and the potential for widespread mischaracterization of the uncertainty in RSL studies, we use the term standard deviation throughout as a numerical measure of "error" and plot blobs according to the 95% uncertainty interval, which for normally distributed data is the mean plus and minus 1.96 times the standard deviation. For non-normally distributed data we plot the interval created from the 2.5th to the 97.5th percentiles.

Chronological models have been applied in sea-level studies but their use is still quite rare. Gehrels et al. (2005) produced a 1000 year sea-level reconstruction based on a core taken from a saltmarsh on the Atlantic coast of Nova Scotia (Canada). Statistical age–depth modeling was performed using "in-house" software. Ages were calculated from weighted calibrated radiocarbon dates, but each one-sigma calibrated age bracket was assumed to be normally distributed. The same approach was taken by Gehrels et al. (2006) for a late Holocene sea-level reconstruction in Iceland, which is further discussed as one of our case studies in Section 32.4.1. Since software packages have become available, the modeling of probability distributions of calibrated radiocarbon ages have become more realistic (Bronk Ramsey, 2001). Gehrels et al. (2012) used the Bacon software (Blaauw and Christen, 2011) to establish a chronological model for a sea-level reconstruction from Tasmania. We use the same data for our second case study in Section 32.4.2. Probabilistic age modeling was also used in the sea-level reconstructions from North Carolina by Kemp et al. (2009, 2011) based on the Bchron software of Haslett and Parnell (2008). In Holocene paleoecological studies from environments where sedimentation rates are more constant (e.g., peat bogs, lakes, and shallow and deep sea) the use of chronological age modeling is more common (e.g., Parnell et al., 2008).

Once a chronology model has been applied, the goal is often to estimate the long-term rate of sea-level change. This task is made harder because of the multiple uncertainties present in both age and sea level. A standard linear regression (LR) approach will fail here as it will ignore uncertainties and produce overly precise estimates. By contrast, we show that an "errors-in-variables" (EIV) approach can overcome this issue and estimate rates by taking account of all the available uncertainties. These estimated rates will necessarily be more uncertain than the simple linear regression approach, but will be more honest and enable a realistic comparison of rates across different regions where previously they may have been rendered incompatible.

In this chapter we show how chronology models can be created and how they may then be used to calculate suitable uncertainty estimates of the rate of sea-level change. In Section 32.2 we review some of the current range of chronology models and show how they produce estimated ages and uncertainties. In Section 32.3 we discuss how the chronology model can be used to estimate rates of sea-level change, while fully taking account of uncertainties. In Section 32.4 we provide two case studies based on sea-level data from Iceland (Gehrels et al., 2006) and Tasmania (Gehrels et al., 2012), for which we estimate chronology models and calculate rates of sea-level change. There are two appendices which cover the mathematics behind the statistical models we use. The code used to produce the plots and to fit the models makes use of the R package, Bchron. The individual scripts are available at http://mathsci.ucd.ie/~parnell_a/Rpack/Bchron.html.

32.2 CHRONOLOGY MODELS

We start by considering a generic situation whereby we have a set of dates obtained throughout a core. It is likely that, either due to expense or availability of material, not every part of the core will have been dated. Our task is to find the ages of every depth in the core given just these dates. An added complexity is that some of these dates will not be direct estimates of the age of the associated depths in the core, but will be biased to a certain degree. This bias may depend on properties inherent to the dating method, or to the degree with which contamination may have occurred either *in situ* or at the dating lab. The most common form of dating for ancient sediments is that of radiocarbon (Bowman, 1990). Dates created in this fashion are provided in radiocarbon years and require calibration into calendar years before their use in further analysis. The calibration step takes the estimated mean date from the lab together with an estimate of the standard deviation, and involves finding plausible calendar ages which match the radiocarbon age probability distribution (see Chapter 23). The resulting probability distributions of the calendar ages are often skewed and multi-modal. They are therefore not amenable to simple statistical analysis techniques.

Radiocarbon dating is well established as a robust dating method for sediments older than ~300 years (Scott, 2003). A possible complication is that of marine influence, but this only applies to shell dates which generally produce imprecise sea-level estimates and are therefore best avoided. In rare situations, SLIPs have been derived from algal mats whose carbon may partially have a marine source (Gehrels et al., 2012). Dating sediments younger than ~300 years is not straightforward as calibration often results in multiple solutions that can span wide age ranges. High-precision accelerator mass spectrometry (AMS) ^{14}C dating of multiple samples, which can produce radiocarbon ages with 1σ errors of 10 ^{14}C years or better, has the potential to overcome these issues (Marshall et al., 2007). Analyses of radioactive isotopes, such as "bomb" ^{14}C, ^{137}Cs, and ^{210}Pb, provide additional constraints for sediments younger than ~100 years. Stratigraphic marker techniques, such as Pb isotopes and trace metal analyses, can be used to date sediments deposited in recent centuries by comparison with regional and hemispheric pollution records, while pollen markers and charcoal records can reveal vegetation changes that can sometimes be tied to historical events (e.g., Gehrels et al., 2008; Kemp et al., 2009). The use of tephra in sea-level studies is possible in volcanically active regions such as Iceland (Gehrels et al., 2006) and Japan (Sawai et al., 2004).

Assuming enough suitable dates can be found, a chronology model can be used to obtain estimated ages for other layers in the core. The chronology models we describe here contain some common features:

- They create individual "chronologies", each of which is a single set of depths and ages equally probable given the data. A large number of these individual chronologies are often created and then summarized by calculating means or 95% uncertainty intervals by depth. The dashed line in panel 2 of Figure 32.1 shows a single chronology, while the gray area indicates the general 95% uncertainty band when the chronologies are summarized. Some of the chronology models we mention below allow access to the individual chronologies, while others only present the summaries.
- Each individual chronology is consistent with a positive sedimentation rate which is allowed to change (either smoothly or roughly) across the depth range of the core. The size and variability of the sedimentation rate at various points in the core can be estimated by the statistical model or input by the user. The positive sedimentation rate naturally enforces the monotonicity constraint. However, the summaries (e.g., means or medians) might not display monotonic behavior.
- Bayesian methods are used to create the chronologies. The Bayesian method allows for uncertainty in all of the different quantities (the dates themselves, the sedimentation process, the degree to which dates are outliers) to be included naturally as part of the model. External or prior information can also be included if required (though the particular software implementation will allow this to a greater or lesser degree).
- The chronology models are fitted using a technique known as Markov chain Monte Carlo (MCMC), which works by guessing plausible chronologies and accepting or rejecting them depending on their compatibility with the data. Since the initial set of guesses might be very implausible, the model takes time to "converge" to probabilistically consistent estimates. Depending on the software used, the convergence criteria are assessed by the user or automatically by the software. More detail on convergence can be found in Appendix A.
- The nature of the sedimentation rate assumptions produces wider uncertainties where there are large gaps in between dated levels. This can be seen in panel 2 of Figure 32.1 at depth 1.8 m. Conversely, where ages overlap or occupy similar ranges, uncertainty will be reduced as the models are forced to assume that older sediment lies at deeper depths. This can be seen in panel 2 of Figure 32.1 for dates 2 and 3.

A more mathematical description of the construction of the individual chronologies is given in Appendix A.

It is estimated (Scott, 2003) that approximately 5% of all radiocarbon dates are outliers. Such outliers need to be removed or adjusted to enable robust chronology reconstruction. The most common way is to use a "flag and shift" approach first suggested by Christen (1994). When using this approach, the chronology model has the additional task of estimating which radiocarbon dates are outliers (flagging them) and, if so, shifting them into a position that forms a valid set of chronologies. This approach works well when there are relatively few outlying dates, but can perform poorly when many multiple dates are in conflict with the monotonicity assumption. In such circumstances it may be preferable to remove suspect dates by hand and perform multiple runs of the model to determine the sensitivity to their removal. An alternative to the flag and shift approach is to change the probability distribution of the date (which are normally distributed for radiocarbon dates when in radiocarbon years) to one which allows for more extreme values, such as the *t*-distribution (as proposed by Christen and Perez, 2009). Unfortunately this approach similarly struggles where there are multiple outliers, due to none of the dates being actively removed from the chronology model.

Throughout this chapter we use the chronology model Bchron (Haslett and Parnell, 2008), which has been shown to produce suitably robust chronologies in a wide variety of situations (Parnell et al., 2011). However, there are other chronology models which may produce similar results such as Oxcal (Bronk Ramsey, 2008) and Bacon (Blaauw and Christen, 2011). Each of these chronology models has a non-specialist software implementation. Furthermore, they share the common features listed above, but differ in their description of the sedimentation process. A formal comparison of the chronology models can be found in Parnell et al. (2011).

In short, Bchron creates piecewise linear chronologies, where the number of breaks and the variation in sedimentation rate is estimated from the data. Oxcal requires the user to input the value of sedimentation variability (referred to as a k value), though it is somewhat unclear how this value should be estimated (Parnell et al., 2011 propose and test suitable values for this). Bacon has a more complex model where sedimentation rates change smoothly over the depth range of the core. Both Bchron and Bacon are written in the open source statistical programming language R (R Foundation For Statistical Computing Austria, 2011) so allow for all output and computations to be customized by the user. The Oxcal model runs in a web browser and has many more options with regards to dating formats. Both Oxcal and Bchron were found to perform similarly well in a large data-driven comparison exercise (Parnell et al., 2011; although Bacon was not included in this comparison, since it was released after the paper was published). All three allow for outlier handling according to one or more of the methods outlined above and in Appendix A.

32.3 ESTIMATING RATES OF SEA-LEVEL CHANGE WITH A CHRONOLOGY MODEL

We now consider how estimated rates of sea-level change can be computed when a chronology model has been used to calculate ages and their uncertainties. Given that the chronology models described in the previous section are all Bayesian in their formulation, it makes little sense to use traditional frequentist techniques to analyze the data further. Although the models outlined in this section have arisen from the frequentist literature, they are easily adapted into Bayesian models by simply including a prior distribution on the unknown parameters and treating all unknown quantities as probabilitistic.

Given a plot such as that in panel 3 of Figure 32.1, it is tempting to use linear regression to estimate the rate of sea-level change. We may write such a statistical model as $y_i = \alpha + \beta x_i + \epsilon_i$ where y_i is the sea-level measurement for observation i, α is the y-intercept, β is the slope (our main quantity of interest), x_i is the age of observation i, and ϵ_i is a residual term, usually assumed to be normally distributed with constant standard deviation σ. Standard linear regression theory shows that β is normally distributed with mean $\left[\sum(y_i - \bar{y})(x_i - \bar{x})\right] / \sum(x_i - \bar{x})^2$ and standard deviation $\sigma\sqrt{\sum(x_i - \bar{x})^{-2}}$, where $\bar{x}$ and $\bar{y}$ are the means of the x and y measurements. Estimates of the likely rate and its 95% uncertainty interval can be obtained directly from these formulae.

There are several problems with using standard linear regression (as outlined in the previous section) to estimate sea level rates:

- Most importantly, the model assumes that the age variable x is fixed. It does not allow for any uncertainty in its value, or for that uncertainty to change between data points.
- Often the standard deviation of y is calculated as part of the data collection exercise. The above model does not use this information at all.
- Often the data are presented having undergone a correction (e.g., isostatic adjustment) which is similarly uncertain. The data y_i we are presented with are therefore really observations $y_i - \lambda x_i$ where λ is the correction parameter.

All three of these issues will combine to produce estimates of the sea level rate β that are biased and underestimate the true uncertainty. The first issue can be overcome by using an errors-in-variables model which we outline in the following. The second issue requires that we replace our residual standard deviation σ by σ_i which is the observation standard deviation given by the data. The last issue can be overcome provided that we are given the value λ at which the rate was corrected, together with some estimate as to its uncertainty.

To include the uncertainty obtained from the chronologically modeled ages, we can use an errors-in-variables (EIV) model (Fuller, 1987). A first attempt at including the uncertainty would be to write the statistical model as:

$$y_i = \alpha + \beta z_i + \varepsilon_i$$
$$x_i = z_i + \gamma_i$$

where y_i, α, β, and ϵ_i are as before except that RSL y now depends on the *mean* of the age x_i, here denoted as z_i with some residual term γ_i which, at a first step, may be assumed $N(\theta,\gamma_i^2)$. Results are available which give estimates of β for this model (see chapter 1 of Fuller, 1987). If we make the assumption that the chronology model output for each depth is normally distributed, then we can use the posterior sample means and standard deviations of the ages to produce an estimate of the slope β.

A more suitable EIV model can be obtained by relaxing the normality assumption in the above and using Monte Carlo (Robert and Casella, 2005) techniques to sample from the chronology model directly. This avoids the assumption in the above equations that the mean is a suitable value for the central location of a radiocarbon date. In the above equation, this would mean that γ_i is no longer normally distributed and samples of ages x_i are obtained by simulating chronologies (such an example simulated chronology is shown in panel 2 of Fig. 32.1). These are then input into a standard linear regression model. We defer the mathematics of this version to Appendix B. In the case studies discussed in the Section 32.4, the difference between the two EIV models (i.e., Gaussian errors v. non-Gaussian errors) is slight, though we use the latter for mathematical correctness as there is little difference in computational speed between the two versions.

The degree to which the uncertainty in the rates β are affected by uncertainty in the ages is hard to quantify, as it will depend on the number of SLIPs and the size of the age and RSL uncertainties. When these latter uncertainties are small, then a linear regression model may well give equivalent results. A linear regression may also be appropriate when the uncertainties are not small but the number of SLIPs is large.

The statistical models we outline above are implemented in the chronology software Bchron via the function BchronRSL. This function will take SLIP data with standard deviations and output from a chronology model (the default is the Bchron chronology model, though any other can be used provided the data can be put in the same format) to produce estimated regression parameters. The function allows for higher powers than just linear regression so rates of acceleration, for example, can also be estimated. The data can be divided up into age periods so that rates can be compared. A more advanced version which uses Bayesian change point modeling (Carlin et al., 1992) can be found in Kemp et al. (2009). The function BchronRSLplot produces SLI plots such as that found in panel 3 of Figure 32.1. We are not aware of any similar functions for other chronology packages.

32.4 CASE STUDIES

In this section we demonstrate how chronology models can be used with EIV regression to produce estimated sea-level rates for two contrasting sites. We show how the chronology model will yield different estimates of the rates when compared with a method that neglects the uncertainties altogether.

32.4.1 Western Iceland

The sea-level reconstruction for western Iceland (Gehrels et al., 2006) spans the last 2000 years and is based on foraminiferal analyses of an exposed saltmarsh peat section at Viđarhŏlmi, Snæfellsness. The section contains several tephra layers, including the Settlement Layer (~875 AD) and the Medieval Layer (1226/7 AD), which provide very precise age markers in addition to 14 radiocarbon dates. The reconstruction also includes three basal SLIPs from an adjacent saltmarsh, which show that compaction is minimal except at the very bottom of the section. Analyses of ^{137}Cs and Pb levels, $^{206}Pb/^{207}Pb$ ratios, and paleomagnetism provide markers for the last 200 years.

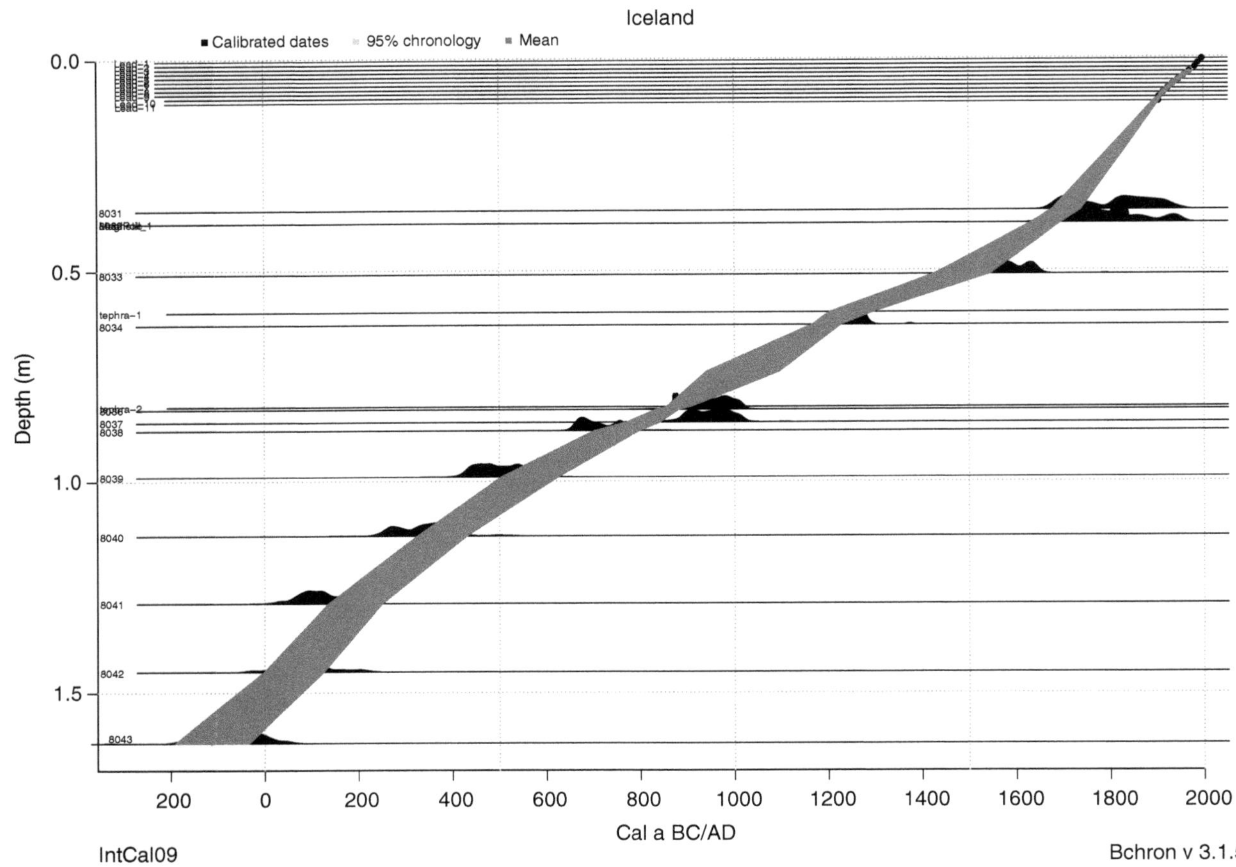

Fig. 32.2. Bchron chronology model output for the Iceland core. *Source*: Gehrels et al., 2006.

A plot of the chronology created for the Iceland core is shown in Figure 32.2. The figure shows the default run for Bchron using one million MCMC iterations. Note that the uncertainty, particularly in the upper part of the core (around 0.3 m), is vastly reduced due to the younger end of the probability distribution functions (pdfs) being rendered incompatible with the modern Pb dates at the top of the core.

From the resulting chronology models, we can create a blob plot such as that shown in Figure 32.3. Here, we supplement the blobs by also including the pdfs to provide a more detailed look at the shape of the uncertainty within each SLIP. Overlaid on the figure are the simple linear regression estimate (created when neglecting all uncertainties), and the EIV linear regression estimate together with the associated 95% uncertainty interval. Note that the EIV regression model uncertainty interval (which takes into account age and RSL uncertainties) is considerably wider than the unrealistically small linear regression estimate. Other chronology models (not shown) will produce similar blob plots and regression parameter estimates.

It is perfectly feasible to extend the EIV model to high powers, for example quadratic regression with an acceleration term included. The details of such models are given in Appendix B. When included, the regression estimates change slightly as there is now competition between the different parameters (the rate and the acceleration) to explain the variation in the data. Table 32.1 gives the numerical values associated with Figure 32.3 and also the quadratic regression model. However, the quadratic regression model does not improve the fit further (which we judge by reduction in the "deviance information criterion"; Spiegelhalter et al., 2002); the standard linear EIV model seems most appropriate. Note that the widths of the uncertainty intervals in Table 32.1 are far larger when the uncertainties are included.

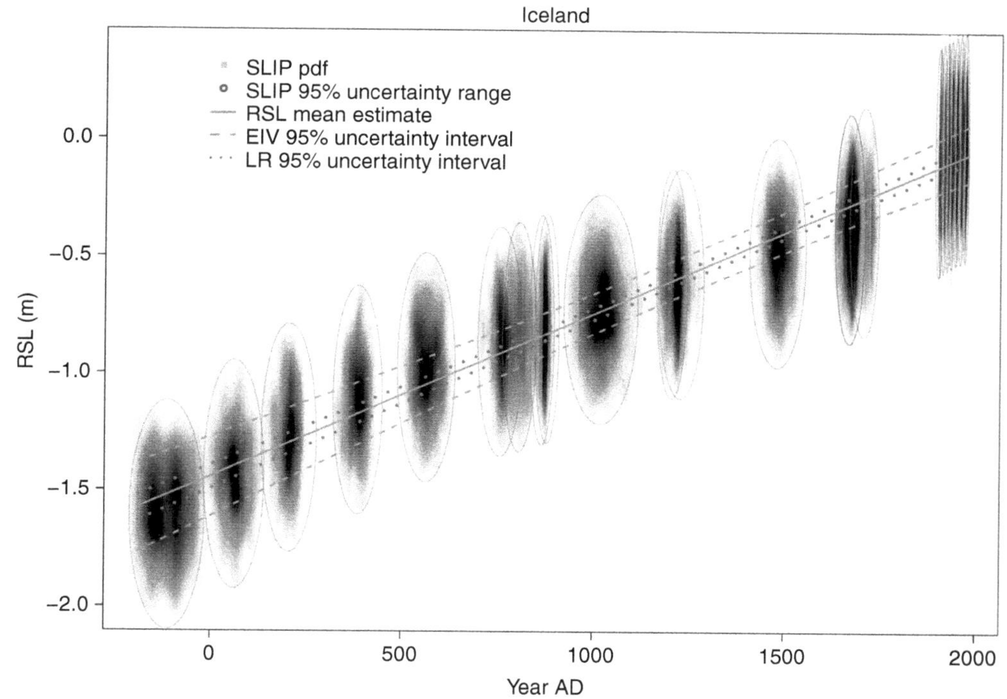

Fig. 32.3. Fitted lines and 95% uncertainty interval for the linear regression and EIV linear regression model fitted to the Iceland data. The SLIPs are shown as grayscale blobs where the darker area indicates higher probability. 95% uncertainty intervals are overlaid as ellipses. The mean fitted line is the solid line, while the dashed and dark lines show the 95% uncertainty intervals for the EIV-LR and LR model respectively. Note that the LR model (which neglects all uncertainties) has an unrealistically small uncertainty interval.

Table 32.1. Rates of sea level change (mm a^{-1}) from Western Iceland under different rate models. The standard linear regression model (top row) is computed by neglecting all uncertainties in the SLIP data. The EIV models (rows 2 and 3) include all uncertainties. Note that the mean estimates are almost identical, but the uncertainty intervals are considerably smaller under the (incorrect) linear regression method compared to the EIV approach. The quadratic model contains both a rate and an acceleration parameter. We report only the rate parameter; the acceleration parameter was approximately zero

Model	Lower 95%	Mean	Upper 95%	95% interval width
Standard LR	0.666	0.702	0.737	0.071
EIV with Bchron	0.578	0.701	0.821	0.243
Quadratic EIV rate	0.558	0.700	0.837	0.279

32.4.2 Tasmania

The Tasmania reconstruction (Gehrels et al., 2012) is a rare Southern Hemisphere example of a high-resolution sea-level reconstruction spanning the last 200 years. The chronology is based on ten ^{210}Pb age estimates, five ^{14}C "bomb" dates, five pollution marker horizons derived from Pb, Zn, and ^{206}Pb/^{207}Pb analyses, a pollen chronohorizon marking the first settlement by Europeans, and seven radiocarbon ages, four of which have 1σ errors of 15 years or less. There are also three middle–late Holocene SLIPs, which are not considered here.

Figure 32.4 shows the chronology for the Tasmania reconstruction, created using the Southern Hemisphere 2004 calibration curve (McCormac et al., 2004) with a Bchron run of 1 million MCMC iterations. The top eight Pb dates point to a period of constant sedimentation over the past 40 years, while the Zn date seems to suggest an increase in sedimentation between 1940 and 1960. The lower dates, as is usual in many cores, suffer from greater uncertainty than those from the more recent past.

RSL change in Tasmania is quite clearly non-linear, so we experiment with a cubic linear regression model (shown in Figure 32.5). Such plots are relatively simple to create using the BchronRSL function. The cubic linear regression plot shows a clear rate rise starting just before 1900 and ending just after 1950. However, there are considerable boundary effects: it is important not to focus too much on the mean lines which

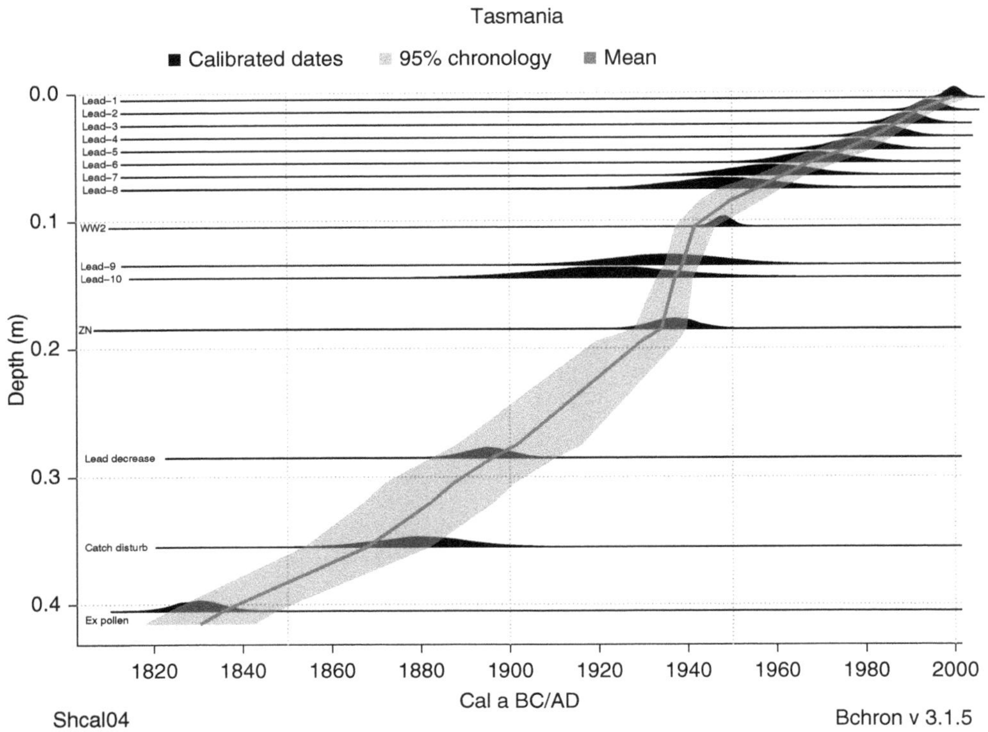

Fig. 32.4. Bchron chronology model output for the Tasmania core. *Source*: Gehrels et al., 2012.

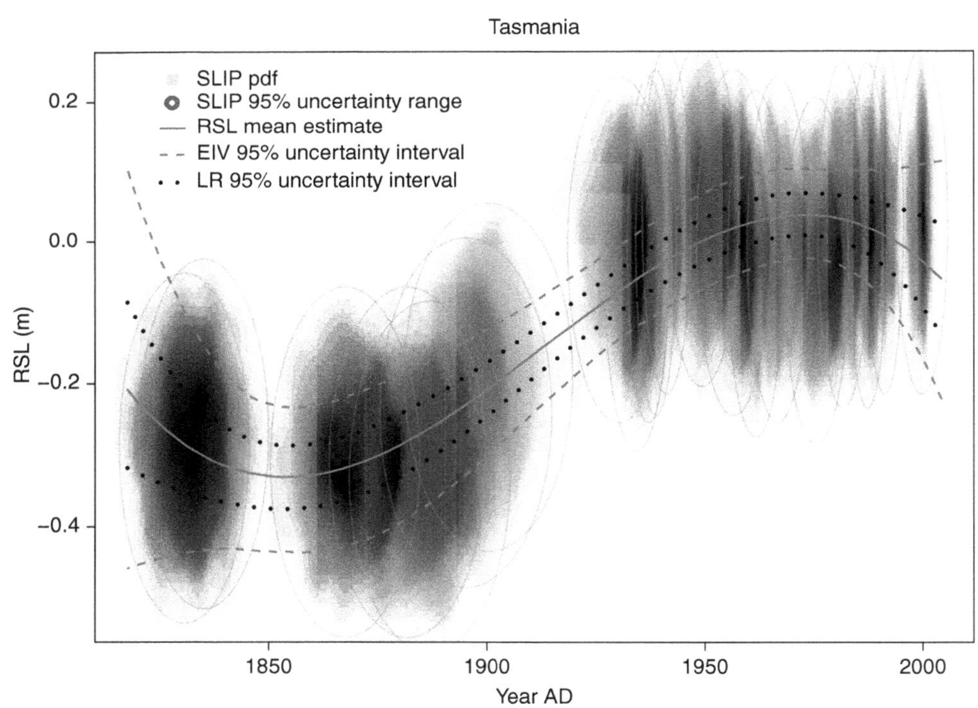

Fig. 32.5. Fitted lines and 95% uncertainty interval for a cubic linear regression and EIV cubic linear regression model fitted to the Tasmania data. The SLIPs are shown as grayscale blobs where the darker area indicates higher probability. 95% uncertainty intervals are superimposed as ellipses. The mean fitted line is the solid line, while the dashed and dotted lines show the 95% uncertainty intervals for the EIV-LR and LR model, respectively. Note that the LR model (which neglects all uncertainties) has an unrealistically smaller uncertainty interval.

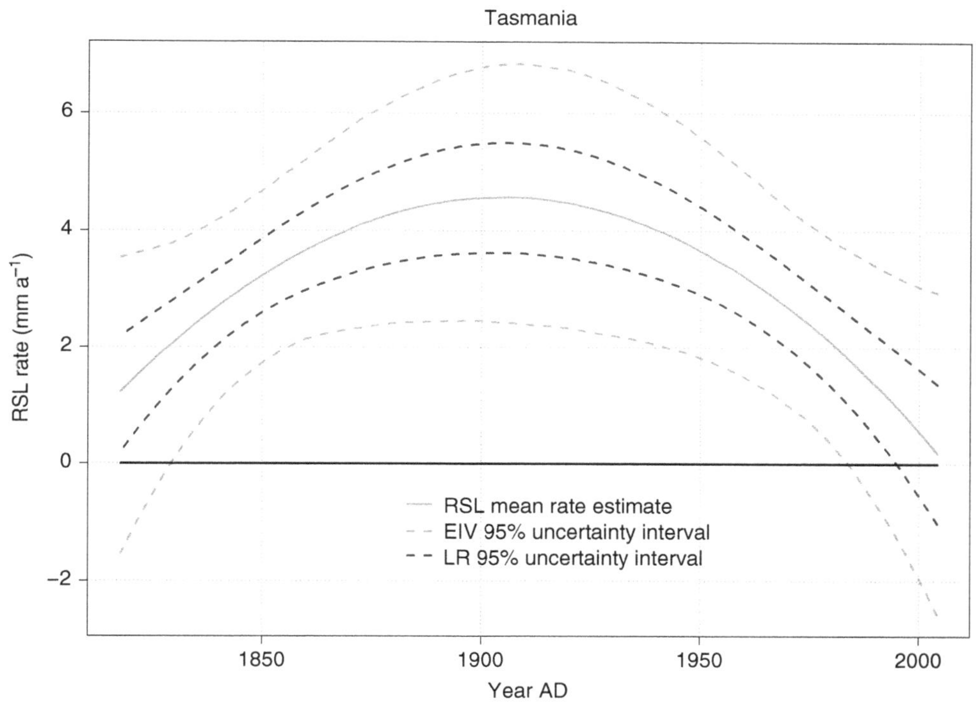

Fig. 32.6. Fitted lines and 95% uncertainty interval for the rates obtained from the cubic linear regression and EIV cubic linear regression models fitted to the Tasmania data. The mean fitted rate is the gray solid line, while the gray and black dashed lines show the 95% uncertainty intervals for the EIV-LR and LR model rates, respectively. Note that a positive rate implies increasing sea level at that particular time point. The LR model (which neglects all uncertainties) again produces unrealistically small uncertainty intervals.

suffer from the usual problems at the edges when using a rigidly parametric model. Figure 32.6 shows the estimates of RSL rate in mm a^{-1} given by the cubic LR and EIV models.

The data resolution at Tasmania is higher (23 SLIPs in ~200 years in Tasmania compared to 24 SLIPs in ~2000 years in Iceland) so the difference between the EIV model and standard linear regression is smaller. However, it is still the case that the linear regression model provides a clear underestimate of the uncertainty in the rate of sea-level change because it ignores the age and RSL uncertainty values. An even richer model would be an EIV change-point model such as that used by Kemp et al. (2009). We do not explore such models further here. However, the Bchron software does allow for the rates to be calculated for differing parts of the core.

32.5 DISCUSSION

In this chapter we have shown how a chronology model can be constructed for a dataset containing SLIPs and then analyzed using EIV regression to give properly quantified estimates of uncertainty in rates of sea-level change. EIV models allow us to take account of uncertainty in both age and RSL. They have the unfortunate but honest side effect of increasing the uncertainty in estimates of RSL rate compared with standard LR.

The chronology model we use throughout is the Bchron monotonic smoothing programme (Haslett and Parnell, 2008) implemented in the statistical software package R. Other chronology models are available which, if used properly, will yield similar results to that of Bchron. The chronology models are constructed by assuming a monotonic relationship between age and depth such that older sediments lie at deeper depths. Bayesian methods are used in their construction so that all uncertainties can be taken into account. The resulting chronologies can be used to predict and possibly constrain the ages of SLIPs.

While linear regression is commonly used to estimate rates of sea-level change, the uncertain estimates of the rate of sea-level change it provides are often large underestimates of the true uncertainty. In our two case studies (Sections

32.4.1 and 32.4.2) we have shown that the uncertainty is underestimated by a factor of approximately 3.4 (Iceland) and 1.4 (Tasmania). Failing to properly account for such uncertainty will lead to inevitable incompatibilities when comparing records between different sites.

When analyzing RSL data using chronological models where the interest is in rates of sea-evel change, we make the following recommendations.

- Use a chronology model (such as Bchron, Bacon or Oxcal) that properly accounts for uncertainty, and allows you to input all of the available information. If you cannot input all the available information, then speak to a chronology modeler who may be able to expand the model to suit your purposes.
- Run the chronology model for a suitable number of iterations (on some packages, e.g., Oxcal, this is done for you) to ensure convergence.
- Use the estimated ages from the chronology model as the uncertainties for the SLIPs that you wish to analyze.
- Plot the SLIPs with appropriate uncertainty using blobs (or even better the raw pdfs) to enable a proper visual inspection of uncertainty.
- Use an appropriate EIV model (e.g., that implemented in the Bchron function BchronRSL) to estimate the rates of change. Ensure you include an appropriately quantified uncertainty interval and not just the mean value. Do not use linear regression and do not ignore important uncertainties.

In the future, the simpler chronology and EIV models presented here will become outdated; the key idea of including all uncertainties will not, however. These future models will most likely better incorporate dependencies between rate estimates and chronology models, possibly by estimating both simultaneously. Information about the RSL rate may therefore feed back into the chronology model and its associated sedimentation rate parameters. Furthermore, simple EIV models such as those presented in this chapter will be replaced with those from the non-parametric smoothing literature (see e.g., Eilers and Marx, 2004, for spline methods). However, they will still need to be incorporated into an EIV framework to take account of age uncertainty. We look forward to these exciting new developments.

APPENDIX

A. Statistical description of chronology models

In a Bayesian statistical model we are concerned with creating a probability distribution of our unknown parameters given our known data. This probability distribution is termed the "posterior distribution". In a chronology modeling scenario, the unknown parameters are the ages of the sediment at all depths, the parameters controlling the sedimentation rate, and any parameters which identify outliers. Our data include the radiocarbon ages and their uncertainties, together with the depths at which they were measured. Bayes' theorem allows us to create a posterior distribution via the relationship:

$$\text{posterior} \propto \text{likelihood} \times \text{prior}$$

which can read as "posterior is proportional to likelihood times prior". The likelihood is the probability of observing the data if the parameters were known, while the prior is the probability distribution of the unknown parameters before the data are seen. Often this relationship is expressed mathematically with parameters written as θ and data as x. Bayes' theorem now reads:

$$p(\theta \mid x) \propto p(x \mid \theta) p(\theta) \tag{34.1}$$

where $p()$ is probability (or, more correctly, probability density), and the vertical bar $\mid$ means "given".

All of the modern Bayesian chronology models use at base a common likelihood where the radiocarbon age is assumed normally distributed with known variance (also given by the lab) and a mean given by the calibration curve (Reimer et al., 2009). We adjust our notation slightly so that x is the radiocarbon age of a sample with standard deviation σ and associated calendar age θ. We write the calibration curve as $r(\theta)$ which returns a radiocarbon age from the calendar age. The likelihood is therefore written:

$$x \mid \theta \sim N\left(r(\theta), \sigma^2\right)$$

The calibration curve itself suffers from uncertainty; it is similarly assumed to be normally distributed and written $r(\theta) \sim N(\mu(\theta), \tau(\theta))$, where $\mu(\theta)$ and $\tau(\theta)$ are the mean and variance of the calibration curve in radiocarbon years associated with calendar age θ. A common confusion that arises in the statistical calibration of radiocarbon ages is that the standard deviation σ is assumed fixed in the likelihood; strictly speaking, all the data should be given a probability distribution given the parameters. This is a modeling assumption and has been relaxed in certain scenarios (see Christen and Perez, 2009 for a more detailed discussion). When dealing with multiple dates, as we always are in a chronology modeling scenario, we have a set x_i, $i=1,\ldots,n$ of radiocarbon ages, each of which has an associated standard deviation σ_i and calendar age θ_i. The likelihood now becomes a product of normal densities:

$$x_i \sim N\left(r(\theta_i),\sigma_i^2\right),$$
$$p\left(x_1,\ldots,x_n \mid \theta_1,\ldots,\theta_n\right)=\prod_{i=1}^{n} p\left(x_i \mid \theta_i\right)$$

The final part of the model is that of the prior distribution. We must now include our depths, written as a continuous positive value d. We now write $x(d_i)$ to represent the radiocarbon date i at depth d_i and $\theta(d)$ as the associated calendar age at depth d. (Note that x is indexed by i through d since there may be multiple dates at the same depth.) The prior distribution represents our beliefs about the nature of sedimentation occurring in the core. A common prior, used by both Oxcal and Bchron, represents the cumulative sedimentation at depth d as a sum of positive random variables:

$$\theta(d)-\theta(d+k)=\sum_{i=1}^{N(k)+1} g_i \qquad (32.2)$$

where k is some positive value, $N(k)$ is the number of sedimentation "events" in the depth range $(d, d+k)$ and g_i is a positive random variable. In words, this equation states that the age difference between depths d and $d+k$ is the sum of a number of sedimentation events each of differing size. Note that this formulation requires that sedimentation is written in terms of units of age rather than units of depth, as the prior distribution needs to be specified on θ. For the Oxcal model (Bronk Ramsey, 2008), $N(k)$ is a Poisson distribution but g_i is fixed. In Bchron, g_i is additionally allowed to vary according to a Gamma distribution. The rate parameter of the Poisson distribution and the extra parameters in the gamma distribution (in the case of Bchron) can be considered extra parameters and estimated as part of the posterior distribution. The prior distribution for the Bacon model is more complicated, but can be written in a similar form to that shown in Equation 32.2.

To complete our model we write out the full posterior distribution:

$$p\left(\theta(d),\phi \mid x(d_1),\ldots,x(d_n)\right)\propto\prod_{i=1}^{n} p\left(x(d_i)\mid\theta(d_i)\right)$$
$$\times\prod_{i=1}^{n-1} p\left(\theta(d_i)-\theta(d_{i+1})\mid\phi\right)\times p(\phi)$$

where ϕ are the parameters governing the sedimentation rate (the Poisson rate and the Gamma parameter in the case of Bchron) which are also required to have a prior distribution. The prior distribution for these parameters is commonly set to be vague.

As stated in Section 32.2, chronology models are usually fitted via Markov chain Monte Carlo (MCMC). This is a trial and error technique that, rather than computing the posterior distribution directly, creates samples from it. We therefore obtain plausible values of the calendar ages at different depths and sedimentation parameters, given the available data. We can then summarize them using standard statistical tools such as means and standard deviations. An outline MCMC algorithm for simulating from the posterior distribution is as follows.

(1) Guess initial values for the calendar ages and sedimentation parameters. Write these as $\phi^{(0)}$ and $\theta^{(0)}(d_1), \ldots, \theta^{(0)}(d_n)$
(2) For iteration $t=1, 2, \ldots,$

   (a) Propose a new set of values of the calendar ages. Write these proposed values as $\theta^*(d_1), \ldots, \theta^*(d_n)$.
   (b) Calculate whether these proposed values match the likelihood and the prior (using the latest set of ϕ values) better than the old values. If they do, then accept the new values and write $\theta^{(t)}(d_i)=\theta^*(d_i)$. If the proposed values do not match the likelihood and prior distribution, then accept them with probability according to the ratio of likelihood times prior for the proposed and old values. If you reject the proposed values, set $\theta^{(t)}(d_i)=\theta^{(t-1)}(d_i)$.
   (c) Propose a new set of sedimentation parameters. Write these proposed values as ϕ^*.
   (d) Calculate whether these proposed values match the likelihood and prior (using the

latest set of $\theta(d_i)$ values) better than the old values. If they do, then accept the new values and write $\phi^{(t)} = \phi^*$. If the proposed values do not match the likelihood and prior distribution, then accept them with probability according to the ratio of likelihood times prior for the proposed and old values. If you reject the proposed values, set $\phi^{(t)} = \phi^{(t-1)}$.

There are many extensions which we do not consider here. The most important of these is the inclusion of outliers which results in a slightly more complex likelihood term $x_i \mid \theta_i \sim N\left(r(\theta_i) + f_i s_i, \sigma_i^2\right)$ with flag parameters f_i and shift parameters s_i. These are given prior distributions, and therefore require extra steps in the MCMC algorithm. Another extension is to include non-radiocarbon dates. We might assume, for example, that there are n radiocarbon dates followed by a further m dates which are assumed normal; we therefore have $x_{n+j} \sim N\left(\theta_{n+j}, \sigma_{n+j}^2\right)$ with their own associated variance. This does not complicate the model further; in fact, it can be seen as a simplification since the likelihood terms are quicker to evaluate (involving no call to the calibration curve).

B. Statistical description of the EIV model

We start with the basic linear regression model:

$$y_i = \alpha + \beta x_i + \varepsilon_i$$

where y_i is the sea-level measurement for observation i, α is the y-intercept, β is the slope (our main quantity of interest), x_i is the age of observation i, and ϵ_i is a residual term, usually assumed to be normally distributed with standard deviation σ_i. Most commonly, the explanatory variables x_i are mean corrected (i.e., their sample mean is subtracted so they have mean zero) to enable more efficient model fitting. The model can readily be extended to incorporate higher-order terms:

$$y_i = \alpha + x_i^T \beta + \varepsilon_i$$

where now $x_i^T = \{x_i, x_i^2, \ldots, x_i^p\}$ is a vector of ages x_i raised to the chosen power p and $\beta = \{\beta_1, \beta_2, \ldots, \beta_p\}$ is a vector of parameters representing the rate, acceleration, etc. It is unusual to fit models with large values of p as the model behavior deteriorates and numerical precision is lost. Where large p is required due to the flexibility of the shape of the data, a non-parametric smoothing model (e.g., a spline) may be preferred.

The linear regression model can be extended to produce an errors-in-variables (EIV) model with the addition of an error term on the explanatory variables:

$$\begin{aligned} y_i &= \alpha + z_i^T \beta + \varepsilon_i \\ x_i &= z_i + \gamma_i \end{aligned}$$

where y_i, α, β, and ϵ_i are all as before except that RSL y now depends on the *mean* of the age x_i, here denoted as z_i with some residual term γ_i which, at a first step, may be assumed $N\left(0, \tau_i^2\right)$. The vector z_i is made up of powers of z_i exactly as in the linear regression version.

The EIV model can be fitted in a Bayesian fashion by noting that the parameters α, β are to be estimated while all other variables in the EIV model are known data (e.g., y, x, σ_i, τ_i). The means z_i of the EIV model are also known, since we know the distribution of the ages from the chronology model. A posterior distribution can therefore be created in the same spirit as that given in Appendix A:

$$p(\alpha, \beta \mid y, x, \sigma, \tau) \propto \prod_{i=1}^{n} p(y_i \mid z_i, \alpha, \beta) \times \prod_{i=1}^{n} p(x_i \mid z_i, \tau_i) \times p(\alpha, \beta)$$

The last term on the right-hand side of the equation above represents prior knowledge of the rates which can be included if required. In our implementation, we use flat prior distributions on these parameters. The model is fitted using MCMC where the steps to update (α, β) are identical to those given in Appendix A.

Our final extension of the EIV model is to allow for non-normally distributed residuals on the age variable. This necessitates a Monte Carlo approach where the likelihood terms above $\left(\prod_{i=1}^{n} p(y_i \mid z_i, \alpha, \beta) \times \prod_{i=1}^{n} p(x_i \mid z_i, \tau_i)\right)$ are replaced by a simulation step that first simulates from the chronology model and then calculates the likelihood of the parameters given this simulation. Simulating from the chronology model is performed repeatedly in order to ensure a full mixing across the possible chronologies for each core. It is this latter method that is implemented in the Bchron R package through the function

BchronRSL. See the website http://mathsci.ucd.ie/~parnell_a/Rpack/Bchron.html for full details of the code used to run this model.

REFERENCES

Blaauw, M. and Christen, J.A. (2011) Flexible paleoclimate age-depth models using an autoregressive gamma process. Bayesian Analysis, 6(3), 457–474.

Bowman, S. (1990) *Radiocarbon Dating*. British Museum Publications Ltd, London.

Bronk Ramsey, C. (2001) Development of the radiocarbon calibration program OxCal. Radiocarbon, 43, 355–363.

Bronk Ramsey, C. (2008) Deposition models for chronological records. Quaternary Science Reviews, 27(1–2), 42–60.

Carlin, B.P., Gelfand, A.E., and Smith, A.F.M. (1992) Hierarchical Bayesian analysis of change point problems. Applied Statistics, 41, 389–405.

Christen, J.A. (1994) Summarizing a set of radiocarbon determinations: a robust approach. Applied Statistics, 43, 489–503.

Christen, J.A. and Perez, E. (2009) A new robust statistical model for radiocarbon data. Radiocarbon, 51(3), 1047–1059.

Eilers, P.H.C., and Marx, B.D. (2004) Splines, knots, and penalties. Wiley Interdisciplinary Reviews Computational Statistics, 2(6), 637–653.

Fuller, W.A. (1987) *Measurement Error Models*. Wiley.

Gehrels, W.R. (2000) Using foraminiferal transfer functions to produce high-resolution sea-level records from salt-marsh deposits, Maine, USA. The Holocene, 10(3), 367–376.

Gehrels, W.R., Kirby, J.R., Prokoph, A., Newnham, R.M., Achterberg, E.P., Evans, H., Black, S., and Scott, D.B. (2005) Onset of recent rapid sea-level rise in the western Atlantic Ocean. Quaternary Science Reviews, 24(18–19), 2083–2100.

Gehrels, W.R., Marshall, W.A., Gehrels, M.J., Larsen, G., Kirby, J.R., Eiriksson, J., Heinemeier, J., and Shimmield, T. (2006) Rapid sea-level rise in the North Atlantic Ocean since the first half of the nineteenth century. The Holocene, 16(7), 949–965.

Gehrels, W.R., Hayward, B.W., Newnham, R.M., and Southall, K.E. (2008) A 20th century acceleration of sea-level rise in New Zealand. Geophysical Research Letters, 35(2), 1–5.

Gehrels, W.R., Callard, S.L., Moss, P.T., Marshall, W.A., Blaauw, M., Hunter, J., Milton, J.A., and Garnett, M.H. (2012) Nineteenth and twentieth century sea-level changes in Tasmania and New Zealand. Earth and Planetary Science Letters, 315–316, 94–102.

Haslett, J., and Parnell, A.C. (2008) A simple monotone process with application to radiocarbon-dated depth chronologies. Journal of the Royal Statistical Society, Series C, 57, 399–418.

Kemp, A.C., Horton, B.P., Culver, S.J., Corbett, D.R., van de Plassche, O., Gehrels, W.R., Douglas, B.C., and Parnell, A.C. (2009) Timing and magnitude of recent accelerated sea-level rise (North Carolina, United States). Geology, 37(11), 1035–1038.

Kemp, A.C., Horton, B.P., Donnelly, J.P., Mann, M.E., Vermeer, M., and Rahmstorf, S. (2011) Climate related sea-level variations over the past two millennia. Proceedings of the National Academy of Sciences of the United States of America 108(27), 11017–11022.

Kemp, A.C., Sommerfield, C.K., Vane, C.H., Horton, B.P., Chenery, S., Anisfeld, S., and Nikitina, D. (2012) Use of lead isotopes for developing chronologies in recent salt-marsh sediments. Quaternary Geochronology, 12, 40–49.

Leorri, E., Horton, B.P., and Cearreta, A. (2008) Development of a foraminifera-based transfer function in the Basque marshes, N. Spain: Implications for sea-level studies in the Bay of Biscay. Marine Geology, 251(1–2), 60–74.

Long, A.J., Woodroffe, S.A., Milne, G.A., Bryant, C.L., Simpson, M.J.R., and Wake, L.M. (2012) Relative sea-level change in Greenland during the last 700 yrs and ice sheet response to the Little Ice Age. Earth and Planetary Science Letters, 315–316, 76–85.

Marshall, W.A., Gehrels, W.R., Garnett, M.H., Freeman, S.P.H.T., Maden, C., and Xu, S. (2007) The use of 'bomb spike' calibration and high-precision AMS 14C analyses to date salt-marsh sediments deposited during the past three centuries. Quaternary Research, 68, 325–337.

McCormac, F.G., Hogg, A.G., Blackwell, P.G., Buck, C.E., Higham, T.F.G., and Reimer, P.J. (2004) SHCal04 Southern Hemisphere calibration, 0-11.0 cal kyr BP. Radiocarbon, 46(3), 1087–1092.

Parnell, A.C. (2005) The statistical analysis of former sea level. PhD thesis, University of Sheffield.

Parnell, A.C., Haslett, J., Allen, J.R.M., Buck, C.E., and Huntley, B. (2008) A flexible approach to assessing synchroneity of past events using Bayesian reconstructions of sedimentation history. Quaternary Science Reviews, 27(19–20), 1872–1885.

Parnell, A.C., Buck, C.E., and Doan, T.K. (2011) A review of statistical chronology models for high-resolution, proxy-based Holocene palaeoenvironmental reconstruction. Quaternary Science Reviews, 30(21–22), 2948–2960.

R Foundation For Statistical Computing Austria, V. (2011) R Development Core Team, R: a language and environment for statistical computing. Available at http://www.r-project.org (accessed 7 August 2014).

Reimer, P.J., Baillie, M.G.L., Bard, E., et al. (2009). INTCAL09 and MARINE09 radiocarbon calibration curves, 0-50,000 years cal BP. Radiocarbon, 51(4), 1111–1150.

Robert, C., and Casella, G. (2005) *Monte Carlo Statistical Methods*. Springer, Springer Texts in Statistics.

Sawai, Y., Satake, K., Kamataki, T., et al. (2004) Transient uplift after a 17th-century earthquake along the Kuril subduction zone. Science, 306(5703), 1918–20.

Scott, E.M. (2003) The fourth international radiocarbon inter-comparison. Radiocarbon, 45, 135–408.

Spiegelhalter, D.J., Best, N.G., Carlin, B.P., and van der Linde, A. (2002) Bayesian measures of model complexity and fit. Journal of the Royal Statistical Society (Series B), 64, 583–639.

Telford, R., Heegaard, E., and Birks, H. (2004) The intercept is a poor estimate of a calibrated radiocarbon age. The Holocene, 14(2), 296–298.

Chapter 33
Paleogeography

GEERT-JAN VIS[1], KIM M. COHEN[1,3,4], WIM E. WESTERHOFF[1], JOHAN H. TEN VEEN[1], MARC P. HIJMA[2,3], AD J.F. VAN DER SPEK[3], AND PETER C. VOS[3]

[1] *TNO, Geological Survey of the Netherlands, Utrecht, The Netherlands*
[2] *Department of Earth and Environmental Sciences, Tulane University, New Orleans, LA, USA*
[3] *Deltares, Applied Geology and Geophysics, Utrecht, The Netherlands*
[4] *Department of Physical Geography, Utrecht University, Utrecht, The Netherlands*

33.1 INTRODUCTION

Paleogeography describes past geological landscapes and relies on a multi-disciplinary approach integrating scattered, fragmentary, and multiform data of various disciplines such as sedimentology, stratigraphy, palynology, and archeology. Dating techniques are used to place reconstructed landscapes in a chronological perspective.

In the context of sea-level research, paleogeographic reconstruction is used in two ways. One way is the use of a well-constrained paleogeographic setting as input for generating high-quality sea-level index points (SLIs). In that context, prospecting, collecting, screening, and categorizing of field evidence is key. Such a study considers the immediate surroundings of a site (10^{-1}–10^{0} km^2). If data availability allows, it also includes the regional-scale estuary, delta, or coast (10^{3}–10^{5} km^2). This approach relies heavily on fieldwork and is the focus of this chapter. Paleogeographic reconstruction on larger scales (continental and global) typically uses sea-level reconstruction as input and provides paleoenvironmental maps as output.

This chapter advocates the use of paleogeography as input for studies collecting data on past sea-level positions. As such it can be considered a method to control and assess the quality of SLIs at regional scales. Applying paleogeographic techniques this way can be considered for three timescales as follows.

(1) *Historical timescale*, covering the last two centuries. To study these timescales, continuous sequences from upper saltmarsh deposits are densely sampled for SLIs using the fossil content. Using present-day and historical maps is sufficient: the paleogeographic setting is known simply because it is so recent. The landscape of the sampled marshes has not usually changed over the last centuries.

(2) *Holocene and/or post-LGM timescale*, covering the last few ten thousand years. At these timescales SLIs are typically collected from deposits overlying paleosurfaces and from within coastal wedges. Samples may come from tens of meters depth, from paleoenvironmental settings unmatched by the current landscape. In this case paleogeography is required to assess the indicative meaning of the SLIs. At this timescale, paleogeographic and sea-level reconstructions benefit from iterations between them. The iterations use data collection and integrated analyses with existing data for best results.

(3) *Pleistocene timescale(s)*, covering up to hundreds of thousands of years. A much more fragmentary record, a more limited amount and resolution of data, and dating-accuracy limitations restrict the application of paleogeographic and sea-level reconstruction techniques at these timescales. Rather than dating single samples and using them as SLIs, complete depositional units are dated. Using one of their sedimentary contacts (top or base), their age is subsequently linked to a sea-level indicator aided by trajectory analysis (Section 33.4.3). At these scales, the iterative process of tuning relative sea level and paleogeographic reconstructions moves forward relatively slowly.

Among settings suited for paleogeographic reconstruction, lowlands in general and coastal and fluvial environments specifically are well

Handbook of Sea-Level Research, First Edition. Edited by Ian Shennan, Antony J. Long, and Benjamin P. Horton.

suited as these areas have sufficient accommodation space to accumulate sediment. This is especially true for the sedimentary record produced during and following times of rising sea level, notably periods of eustatic rise following glacial maxima. Applications of paleogeography to late Pleistocene and Holocene deposits are consequently the most widespread, also because their records are best preserved and omnipresent due to their young age.

In sea-level reconstruction, knowledge of antecedent morphology and paleogeographic setting is key input. Where sea-level rise has caused transgression, the antecedent morphology determines which type of environment and sedimentary facies are established where and when. As well as the aspect of time in paleogeography, aspects of scale are of importance. Paleogeographic differences demonstrate that sea-level rise histories were different within a region. When looking at larger deltaic systems under modest sea-level rise parts of the delta will be prograding, whereas parts receiving less sediment may simultaneously be subsiding and experiencing transgression. This can be seen as apparent sea-level stability in the prograding section and apparent rise in the sediment-starved section. Sea-level reconstruction has to reconcile local differences in apparent signal. Reconstructing past coastal landscapes will overlap with reconstructing past sea levels, and the two are united in requiring a thorough understanding of the paleoenvironment.

An important concept for paleogeographic reconstruction is *Walther's Law of Facies* (1894), which states that facies occurring conformably on top of each other in a vertical succession must be the product of spatially adjacent environments. The facies occurring on top of each other were therefore formed by environments occurring next to each other. It follows that transgressive sequences are good recorders of sea-level rise history. The distribution of facies within them is chiefly governed by the balance between sediment supply and the rate of relative sea-level rise (i.e., the creation of accommodation space). To read and understand sea-level history from sediments, and recover SLIs from them, paleogeographic understanding of the genesis of a landscape in time and space is needed. Using the methods mentioned in Chapters 3–5 a SLI can be derived from numerous features, but what is its indicative meaning (cf. Van de Plassche, 1986) and value? We discuss specific paleogeographic methodologies to help answer this question with respect to the abovementioned Holocene/post-LGM–Pleistocene timescales. We then present key paleogeographic examples from different depositional sedimentary environments and guidelines to interpret SLIs from them.

In the given examples, the main allogenic processes affecting sea-level histories are eustatic sea-level change, vertical land movement, and changes in sediment supply from both the hinterland and the sea. On top of that, various autogenic processes such as subsidence due to compaction of young sediments and coastal currents and tides – causing coastal zones to shift around their depocenters with time – are affecting the sites, modulating the indicative meaning of their contained SLIs.

The examples in this chapter are from selected settings, but they are applicable to various localities on the globe. The importance of basal peat is explained with respect to the determination of SLIs (Section 33.2.3). Correctly using basal peat is vital for the collection of SLIs in any environment. In the methods section we present ways to identify and use storm-surge signatures (Section 33.2.4) to understand the variability of extreme high-water levels. This is a universal problem and approach, which is applicable to any coastal-dune sequence around the globe. The effects of basin/estuary morphology on tidal levels are then discussed using the Westerschelde estuary (Section 33.2.5). The landward-directed variation of tidal range along estuaries and tidal basins is a widespread issue applicable to estuaries such as the Severn, Humber, and Thames (UK), Seine, and Gironde (France), USA east coast, Gambia River (Africa), Yangtze (China), Saemankum (South Korea), and Fitzroy (Australia).

In the section on coastal settings a tidal inlet and basin system in the Netherlands is discussed (Section 33.3.1). The basin was not connected to a river on the landward side and was only connected to the sea via a tidal inlet through the coastal barrier. Such basins develop where antecedent morphology has a low gradient, where no rivers come in from the landward side, and where sufficient sand is available to be turned into barriers by waves to protect the basin. A relatively novel approach is using paleoenvironmental data to time the transition from intertidal to supratidal sedimentation as a proxy for sea level (Section 33.3.2). The approach can be used in any landward onlapping setting of lagoonal systems.

In the absence of dwelling mounds, sedimentology and faunal content may help determine the facies contacts.

Three deltaic settings are covered, of which wide low-gradient deltas (Section 33.4.1) are very common along the world's shelf seas such as the Mississippi River, Po River, Danube River, Ganges–Brahmaputra, and Amazon River deltas. Contrary to low-gradient deltas, confined or bay-head deltas (Section 33.4.2) are found along coasts with an elevated land surface and a narrow continental shelf, enabling deep incision during sea-level lowstands. Confined deltas in incised valleys are a common phenomenon around the globe. Examples can be found along French, Spanish, and Portuguese Atlantic coasts, along the Italian coast, the northern Gulf of Mexico and USA mid-Atlantic margins, Pakarae River (New Zealand), Yangtze River (China), and in Japan along the Sea of Japan. Stacked delta and shelf sequences exhibit multiple SLIs that formed during multiple glacio-eustatic cycles (Section 33.4.3). The large scale of deltas and shelves as opposed to the detailed studies on saltmarsh deposits has the advantage that it provides sea-level trends over long time-scales, with the disadvantage of a limited temporal resolution. This approach relies on seismic data and is therefore generally applied to regions which are petroleum-geologically interesting. Regions to which this method has been applied include the North Sea shelf (Netherlands), the Gulf of Mexico shelf and Mississippi delta (USA), the Po delta and shelf (Italy), the Ebro delta (Spain), the Yangtze delta (China), and the Nile delta (Egypt).

33.2 SPECIFIC METHODOLOGIES

Paleogeographic reconstruction requires a multi-disciplinary approach using facies analysis, architecture, chronology, (paleo)bathymetry, and general sedimentary concepts of deltas, estuaries, and coasts. For facies description, lithological, sedimentological, pedologic, and environmentally indicative biotic criteria should be integrated to facilitate reconstruction of environmental change from logged sections. Paleoenvironmental reconstruction of a single site involves interpretation of sediments in terms of their depositional environment. When this is done for a number of sites in a region in a spatially continuous fashion, connecting all sites and incorporating additional data with a denser spatial coverage, it is a paleogeographic reconstruction.

33.2.1 Data collection

In lowland coastal settings, paleogeographic reconstruction requires the collection and analysis of subsurface data. In late Holocene settings near coastal barriers and along the inland rim of the coastal plain, paleogeographic sea-level reconstruction can be carried out using a series of shallow cores (<5 m, occasionally 10–15 m) recovered either along transects to document lateral sedimentary changes or at specific sites. For these applications, manual coring is most time- and cost-efficient. Edelman augers can be used for the top soil, and gouge, Livingstone piston corer, and Van der Staay suction-corer for the sampling and description of sediments below the groundwater table. Cores should be logged in the field using standard protocols, occasionally taking samples from cores for sedimentological, paleo-biological or chronological analyses.

To understand the spatial distribution of sedimentary environments, their three-dimensional relationships are important. These can be obtained by arranging boreholes in transects perpendicular to the coast or the valley direction, where most changes in sedimentary environment are expected. Depending on the scale of the investigated area, borehole spacing on transects should be around 100 m for a regional study, and denser around sites which are sampled for SLI collection. Once the spatial distribution of sedimentary environments is understood, a sampling campaign revisits the most optimal locations, collects core material, and takes that to a lab for further investigation. Russian-type coring is best for sampling from up to slightly compacted sediments, as it results in better recovery and prevents sediment compaction. More compacted sediments can be sampled using, for example, a Livingstone piston corer or gouge; the latter is however less accurate and reliable.

The use of mechanical boreholes is needed when deeper sediments, for example for early–middle Holocene sea-level rise, are targeted, or for sampling under a thick overburden. To best place such a borehole (e.g., in the thickest sequence), geotechnical information (boreholes, cone-penetration tests) provides valuable insight in sediment thickness and distribution. Detailed paleogeographic reconstruction of deeper buried

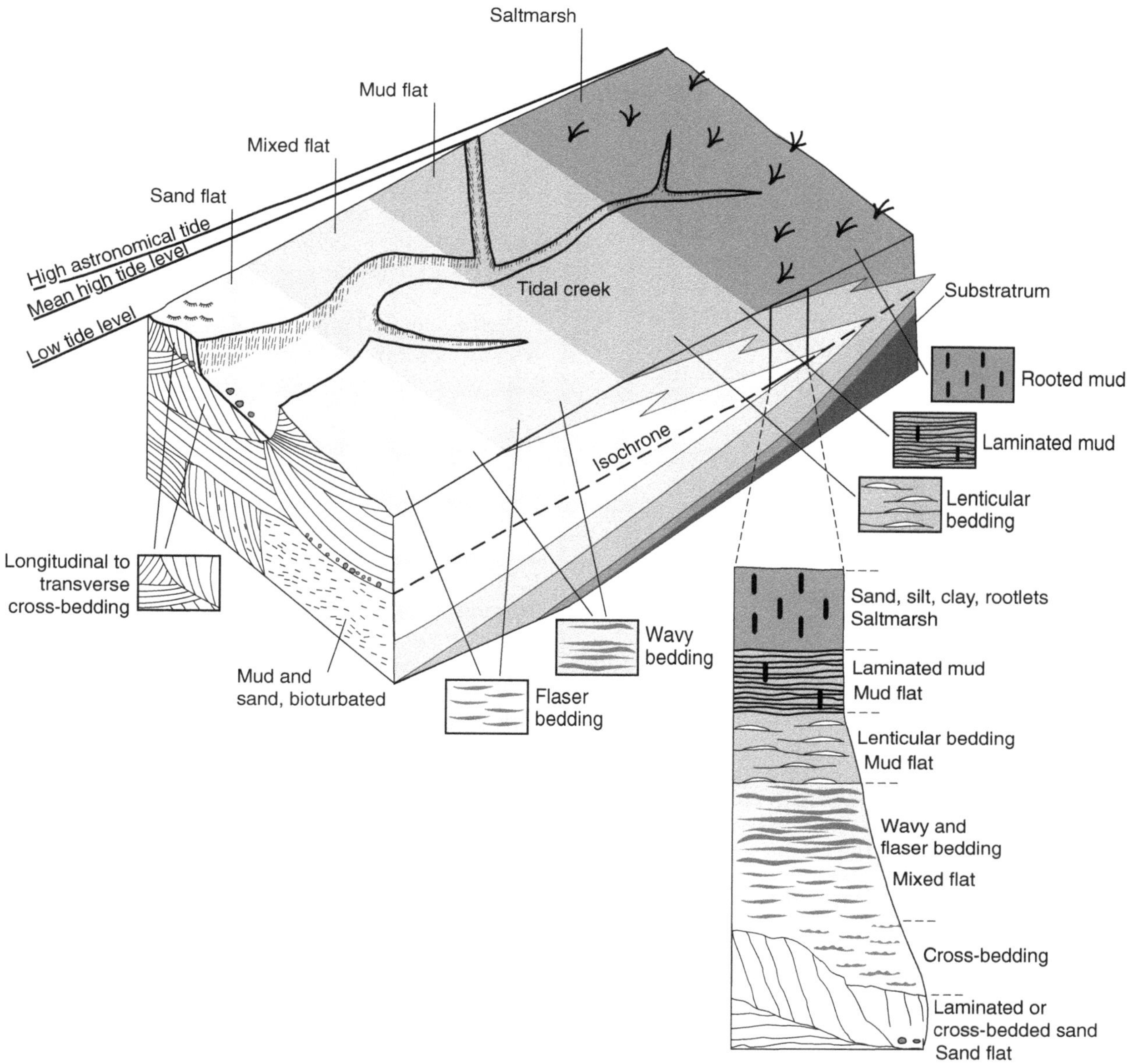

Fig. 33.1. Distribution of facies in a tidal flat area. The column illustrates the sequence of facies resulting from a seaward-prograding back-barrier coastline (local regression). *Source*: Mauz and Bungenstock, 2007. Reproduced with permission of Elsevier.

transgressive environments is possible because of the availability of high-quality but low-density scientific data, integrated with medium-quality high-density engineering data. These data are available because many of such environments are in densely populated urban regions. The use of geo-electric measurements, ground radar, and seismic acquisition may also provide useful information, particularly when deployed as part of a sampling campaign at key sites where SLIs are to be collected.

Outcrops and excavations should be considered for paleogeographic reconstructions as they have the advantage of providing more detail on sedimentary structures than borehole data. It is therefore useful to closely examine sedimentary structures in such locations, which may provide indications for paleo-sea level. Well-known examples come from densely populated areas with much building activity such as Singapore and the Netherlands (Bird et al., 2007; Hijma and Cohen, 2010).

33.2.2 Sedimentology

Sedimentary structures and textures in a coastal setting reflect the sediment transport mechanism, water depth, and flow conditions. They can be used to reconstruct these conditions. To do so requires formal description and analysis of facies sequences in cores and outcrops at individual sites. Paleogeographic reconstruction adds the lateral and temporal setting to sedimentological reconstruction. It allows diachronous features to be identified in three-dimensional mappings of sedimentary facies. The latter is essential to develop space- and time-constrained sedimentation models for local and regional sea-level change studies.

As opposed to high-energy sandy coastal and barrier-island environments, low-energy fine-grained environments in estuaries and back-barrier lagoons are most suited for sea-level reconstruction. These environments can be divided in several zones (based on Schwartz, 2005) as follows.

- *Subtidal zone*: permanently submerged, below (neap) low tide water level; a subaqueous environment.
- *Intertidal zone*: emerged during (neap) low tide, submerged during (spring) high tide.
- *Supratidal zone*: above the highest regular tidal inundation, incidentally flooded only; a terrestrial environment.

The intertidal zone includes tidal mudflats and the tidal marsh fringes. This zone (Fig. 33.1) is most useful in sea-level studies and can be further subdivided as follows.

- *Low tidal flats*: exposed relatively shortly, during mean and extreme low spring tide only, in mixed-sediment systems generally dominated by sandy deposition.
- *Middle tidal flats*: regularly inundated and exposed during every tidal cycle, whether neap or spring. In mixed-sediment systems, these flats have alternating sand and mud deposition. The higher parts may be vegetated (saltmarsh seaward fringe; rooted). Regular exposure of the sediments leaves initial pedogenetic overprints to the intertidal zone (ripened soils, denser textures) relative to subaqueous equivalents.
- *High tidal flats*: briefly inundated during regularly returning higher tides, generally mud-dominated and covered by marsh vegetation or mangrove. In deltaic areas and in upper estuaries of larger rivers, this zone is typically a brackish or freshwater tidal-fluvial environment (reed marsh). Mud deposition dominates with shallower systems due to the inland dissipation of tidal energy and the enhanced trapping of sediment by vegetation and algal biofilms and flocculation of clay particles in brackish waters. Incidental sand layers are present as a result of storms.

Sediments of the intertidal and supratidal zones are the combined product of current and wave action moving sediments, alternating with slack water allowing it to settle (Fig. 33.1). In low-energy mixed-sediment systems, this results in cross-bedding containing mud drapes in higher-energy environments (flaser or flaser bedding) and muddy deposition with lenticular interbeds marking sandy ripples (lenses) (Reineck and Singh, 1980). Often no bedding is visible in the muds topping the sequence, for example due to strong bioturbation and pedogenic alterations. Subsequent deposition of subtidal, intertidal, and supratidal sediments generally results in a fining-upwards sequence (Fig. 33.1). In the subtidal and intertidal zones, typical bivalve mollusk species occur, that can be further zoned (example in Section 33.3.2). In the supratidal zone, the remains of freshwater gastropods may dominate. Saltmarsh vegetation may form thin peat beds. Brackish and freshwater tidal environments (reed marsh, willow swamp, mangrove swamp) may produce thicker clayey-peat beds. Besides environmental sedimentary indication, facies rich in *in situ* shells or terrestrial organics also provide radiocarbon-dating opportunities, helping to chronologically constrain SLIs. The presence of agglutinating foraminifera (Chapter 13) and dinoflagellates in these muds is instrumental for reconstructing the paleoenvironment. Depending on the local tidal range, the thickness of the above-described sedimentary sequence varies, being thickest in macro-tidal (>4 m) regions and thinnest or least-developed in micro-tidal (<2 m) regions along open ocean coasts.

For late Holocene sea-level reconstruction, the alternative to using tidal deposits is utilizing coastal-barrier deposits. The occurrence of the deepest dry aeolian scour and the highest level of marine burrowing and small-scale cross-lamination gives an indication of MHW level. For MLW level, the level of the thickest shell beds, the range of structure-less highly porous sand, the range of low-angle bars and the occurrence of centimeter-thick clay layers are used (Roep and Beets, 1988). These methods come with a relatively high vertical error of decimeters up to a meter. Because this environment usually lacks organic deposits for radiocarbon dating, optical stimulated luminescence dating (OSL) may be used. It is well-suited for this type of environment and has been shown to yield good results (Van Heteren et al., 2000).

The remnants of coastal barriers and tidal channels may be preserved seaward of the present coastline, indicating levels from late glacial and early Holocene transgression stages (which are typically resolved onshore using the basal peat methodology; Section 33.2.3). Barriers continuously migrate landward or remain in place and are finally overstepped, depending on local circumstances. From the Portuguese north coast and the Adriatic Sea in northeast Italy, examples are known demonstrating the seaward occurrence of coastal barriers at the seafloor (e.g., Rodrigues et al., 1991). These barriers may be identified using marine seismic surveying techniques and seafloor sediment sampling. The scour bases and basal fills of major tidal inlet channels which formed during transgression preserve well and are visible on seismic lines (e.g., Rieu et al., 2005). The shallower parts of transgressive systems have a lower potential for preservation; paleogeography and sedimentology are therefore more likely to produce late Holocene SLIs than middle Holocene SLIs.

33.2.3 Basal peat

Transgression causes the progressive drowning of valley floors. In hydrologically suitable cases, this leads to freshwater peat growth on top of the pre-transgression substrate. In areas with a low-gradient substrate, basal peat growth occurs most extensively (Fig. 33.2a–d) as wide areas are flooded and turned into wetlands. In less-wide, low-shouldered transgressed valleys, basal peat growth may occur regionally. In narrow valley systems this occurs only locally, near the valley edge or where small tributaries join as for example in the Lower Tagus Valley (Vis et al., 2008).

In advance of the transgression, near-coastal groundwater rise is often only partially controlled by sea level and care must be taken to differentiate between groundwater-controlled peat and purely sea-level-controlled peat (e.g., Hijma and Cohen, 2010). Traditionally, series of SLIs are based on basal peat samples collected at single sites (Chapters 3, 4, and 23). SLIs collected on spatially separated sites need to be integrated into one dataset to increase the timespan of the sea-level reconstruction, which is also helpful to validate site-attributed indicative meanings. Reliably merging SLIs based on basal peat from various sites requires a good understanding of paleogeography based on thorough three-dimensional reconstruction (e.g., Cohen, 2005). In general, such understanding is crucial to determine which dated basal peat samples represent the most-seaward and lowest-elevation coastal peatlands and hence are SLIs. All basal peat samples reflect groundwater rise, but only some of them are sea-level controlled and can be used to extract SLIs. To select basal peat samples representing sea-level control, the key criteria are as follows.

- They are from the top of a terrestrial organic deposit (i.e., peat formed at the water table) which is covered by transgressive marine deposits (verified using sedimentology and/or faunal content).
- At a given inland position, they are the youngest radiocarbon date from that depth that far inland (within dating uncertainty). This means that no younger sample exists from a deeper level upstream, and no younger sample exists from a deeper level towards the valley sides. In this way, a series of SLIs of decreasing depth with increasing distance upstream is obtained (in a transgressed valley).

Only detailed knowledge of paleogeography allows points to be selected in this way; paleogeography can therefore greatly improve the quality of sea-level reconstructions.

33.2.4 Storm-surge signatures

In addition to the quest for mean sea-level data covering longer timespans (MSL, MHW, MSHW), several sedimentological characteristics can be utilized to identify short-lived extreme sea-level events or storm surges. Studying the frequency and magnitude of storm surges in coastal dunes may considerably improve the understanding of the variability of extreme high-water levels and their relationship with sea-level and climate over the last few thousand years. The frequency of

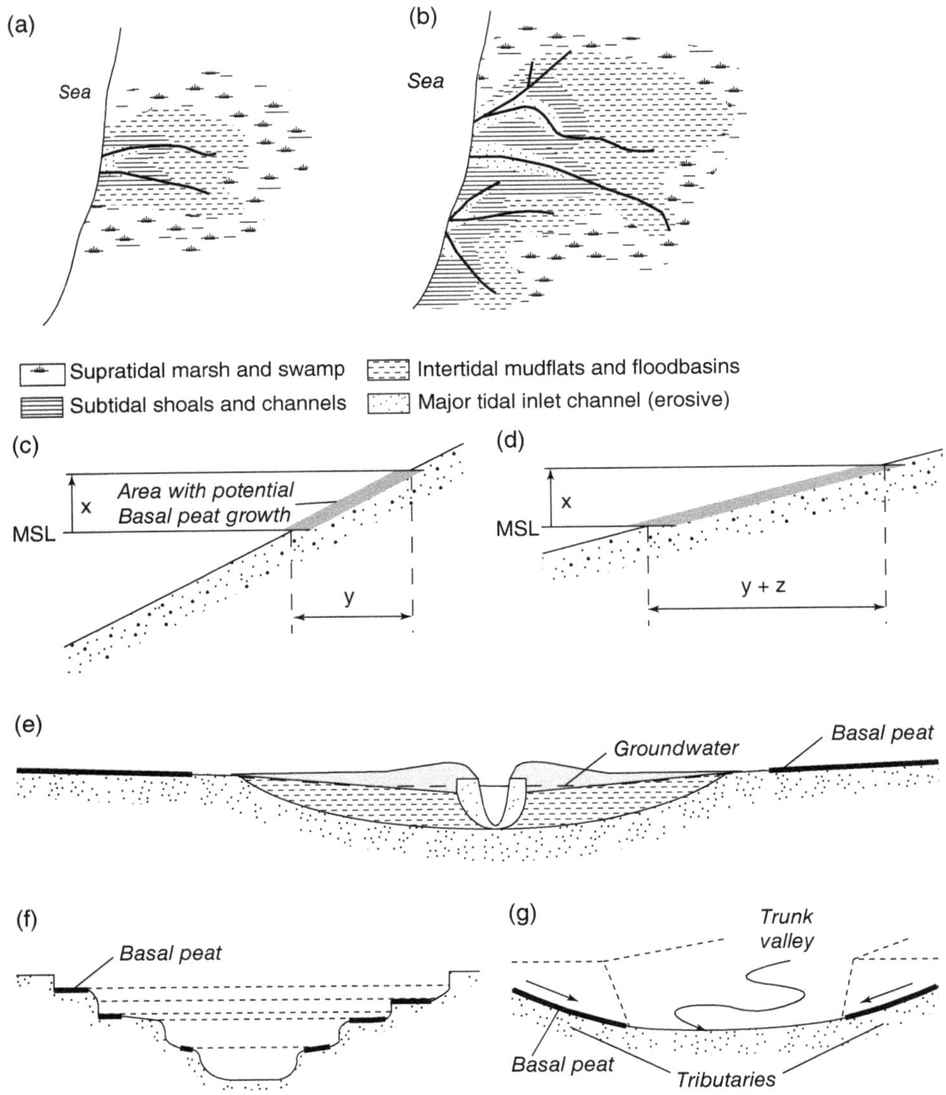

Fig. 33.2. (a, b) Basal peat (marsh) fringes a back-barrier basin. The basal peat is present at the base of the Holocene sequence. Dating the base of the basal peat gives the age of groundwater-driven peat growth announcing transgression; dating the top gives clear drowning which is a stronger sea-level signal, albeit more susceptible to compaction. The distribution of such points in space and time indicates how far inland the transgression reached and at what time. (c, d) Effect of topographic gradient on the areal extent of synchronous basal peat growth. In steeply sloped settings, sea-level rise leads to wetland formation in a relatively narrow zone (or wetlands are even absent), while in low-gradient settings extensive wetlands are formed. (e) Basal peat growth in a low-gradient low-shouldered alluvial valley. (f) Enhanced opportunities for the formation and preservation of groundwater-rise-driven basal peat layers in a valley. (g) Tributary valleys with wet and low-energy environments which favor basal peat growth.

storm surges is usually studied based on sandy overwash deposits in tidal-marsh deposits. Their magnitude can be studied using high dune topographic locations with embedded storm deposits. Such specific information is difficult to reveal from sedimentary sequences in shoreface and back-barrier marsh areas. Moreover, storm deposits in coastal dunes enable the application of OSL dating, by which reliable ages over the last few thousand years can be obtained.

Storm surges are difficult to discriminate from normal events in the sedimentary record of coastal dunes on other criteria than elevation. Their sedimentological preservation potential is limited,

Fig. 33.3. (a) Fresh exposure of foredunes on the beach near Heemskerk (NL) after the storm surge of 9 November 2007. (b) Storm deposit in a sequence of aeolian dune sand. Convolute bedding at the top is highlighted by mollusks and underlain by a slump structure. Euro coin for scale. *Source*: Photographs M.A.J. Bakker, TNO, Geological Survey of the Netherlands. Reproduced with permission.

which tends to make records of storm surges incomplete and discontinuous.

Storm-surge signatures have been studied along the Dutch barrier coast (Fig. 33.3a), in extensive coastal dune complexes that formed in Medieval times (~500–1500 AD). The elevated topography of the dunes facing the beach provides suitable conditions to trap sediments during periods of severe wave action (Roep, 1986; Jelgersma et al., 1995; Cunningham et al., 2011). The sedimentary characteristics of such storm deposits in coastal dunes are as follows (Fig. 33.3b).

(1) Beds with hinged shell couplets or large shells in the convex-side-up position embedded between aeolian sand, indicating their deposition in flowing water and not by wind.
(2) Convolute bedding, resulting from water- and air-escape processes where non-saturated, loosely packed sand was rapidly buried.
(3) Channel formation in the loosely packed dune sand and associated slump structures indicating the sliding of sediments into newly formed channels.
(4) Presence of coarse-grained components such as gravel, pieces of wood, peat, and anthropogenic material such as bricks and coal, derived from drowned and destroyed coastal structures.

Historical data may provide additional means of quantifying storm-surge magnitude and frequency. Such data are typically taken from written records or marks on historical buildings, covering the last few centuries to the last millennium.

33.2.5 Effects of basin/estuary morphology on tidal levels

The tidal range varies along estuaries and tidal basins in a landward direction. The interplay between basin or estuary morphology and the tidal wave results in changes in HW and LW levels and thus in tidal range (Chapter 10). This complicates the reconstruction of coastal MSL on the basis of inland water-level indicators. There is no such thing as *the* HW or LW level; tide levels are average values. In general, the tidal range in an estuary depends on the distortion of the tidal wave by bed friction and landward narrowing of the estuary. If the effect of narrowing dominates the tidal wave is amplified, resulting in higher HW levels. If friction is stronger, the tidal wave is damped and the tidal range decreases. Moreover, abrupt changes in estuary cross-section can induce variation in HW levels, for example relatively narrow passages will induce lowering of upstream HW levels (a so-called "flood-depression"; e.g., Van Veen, 1950; Van Veen et al., 2005).

An increase in average depth of the estuary by scouring of the channels and/or reduction of the intertidal flat area due to siltation of the intertidal flats to supratidal level results in reduced friction, acceleration of the tidal wave celerity, and increasing HW levels. Moreover, the location of maximum HW shifts further upstream. The historic evolution of the Westerschelde estuary illustrates this (Fig. 33.4). In the last 400 years, this estuary changed shape from a basin with large intertidal areas to the present funneled situation. It made the mean tidal range at the inland harbor of Antwerpen increase from 3.2 m in 1650 to 5.2 m in 1967. Over the same interval, the tidal range at the estuary mouth increased by ~0.4 m and MSL rose by ~0.7 m. Simulation of historic tides in the Westerschelde estuary for 1650, 1800 and 1967 (Van der Spek, 1997) shows the inland tidal range increase to be predominantly caused by a rise in HW levels.

On a more local scale, the delay in filling/draining of (sub)basins results in reduced water levels; for HW this is called the "flood-basin effect" (Van Veen, 1950; Van Veen et al., 2005; see also Van de Plassche, 1980 for a discussion of its impact on sea-level reconstructions). The water level in the basin is still rising when at the entrance high tide is already reached. Although water levels at the entrance start to fall, inflow still continues due to inertia of the water motion. Inflow reverses to outflow somewhat later, when entrance water levels drop below those inside the basin. A similar delay occurs around low water slack. Consequently, the HW level in the basin is lower than outside and the LW level is higher. The tidal-range dampening mechanism works for entire tidal basins as well as for smaller units such as tidal-flat areas. It also affects peaty marsh and swamp environments that maintain water tables that connect to HW and the tidal range in the open channels to which they connect, and is therefore of significance for sea-level reconstructions (Van de Plassche, 1980). The effect increases with decreasing cross-sectional area of the feeder channel. Additionally, the composition of the subsurface of the inlet channel is important, since it will influence the channel cross-sectional

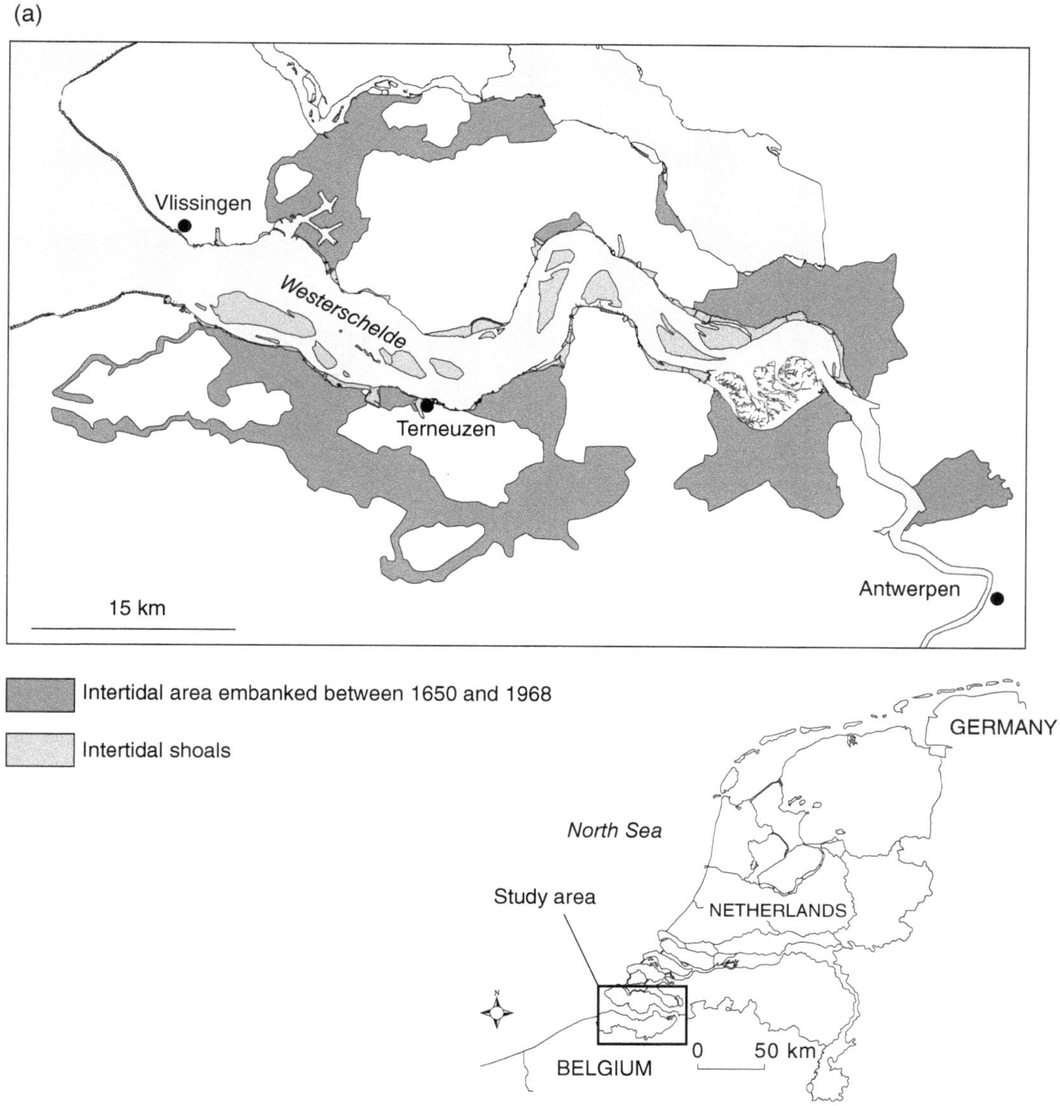

Fig. 33.4. (a) Change of the Westerschelde estuary from a branched tidal basin into a funnel-shaped estuary between 1650 and 1968. *Source*: Adapted from Van der Spek, 1997.

area. In areas with a freely erodible subsurface, the channel cross-section will be in equilibrium with the inflow/outflow discharge and the flood-basin effect will be small. In areas with an erosion-resistant subsurface, the channel will be shallower and/or narrower than its equilibrium dimensions, inducing a significant flood-basin effect.

Widening and/or deepening of the feeder channel will result in a (sudden) increase in tidal range in the basin (Van Veen, 1950; Van Veen et al., 2005). Widening and/or deepening of inner estuary flood basins can cause the tidal range to drop. The reverse is observed in remaining open channels in steadily infilling estuaries and tidal basins. In all these cases, knowledge of the local paleogeography helps to break down observed HW change based on SLIs into components of tidal-range change and mean sea-level change, or to identify those locations in the landscape where HW change is least affected by such processes.

33.2.6 Elevation and compaction

Measurement of sample elevation is crucial for correct correlation of different sites and establishment of interrelationships. For reconstruction of

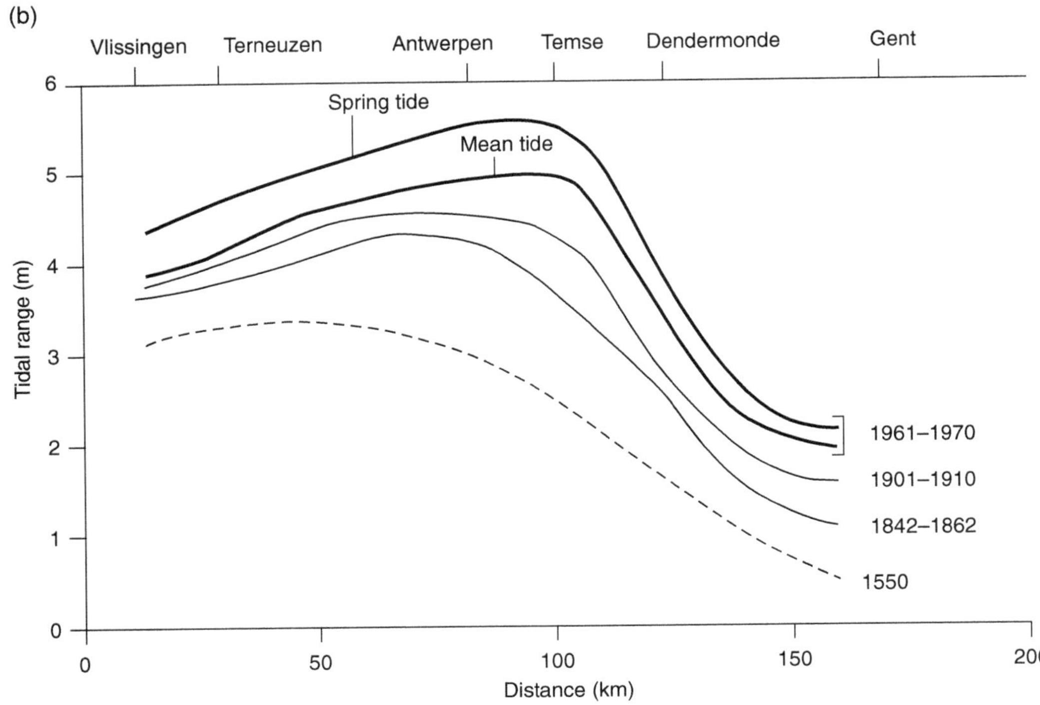

Fig. 33.4. (*Continued*) (b) Changes in mean tidal range over five centuries as a result of changing estuarine morphology. *Source*: Adapted from Coen, 1988.

relative sea level, a detailed and correct measurement of SLI elevation is even more important (Chapters 9 and 10). The samples themselves may be compacted or extended due to the sampling method used. Careful administration of penetration depth of the coring equipment and recovered sample length should resolve this issue, although this may still introduce a relatively large error in the process (see Chapters 30 and 34).

In areas without topographic maps with elevation data and no field elevation marks which are connected to a national reference datum, it is advisable to measure all sampled locations correctly among the sampled sites using the same local reference level such as a building with a strong foundation.

The elevation of a SLI since its time of deposition may have changed as a result of several factors (post-depositional lowering; Chapter 30). Loading of deposits with younger deposits causes porosity reduction and expelling of gas and pore water due to the overburden weight (auto-compaction; Chapter 30). Differential compaction occurs naturally and is a driver of the shifting of sedimentation depocenters over time. This is not to be confused with shifting depocentres due to sea-level change, which may not be identified as such due to a limited study area. Over multiple millennia, the effect is biggest for SLIs from halfway down a compacted sequence. Towards the base of a compacting sequence, compressed rhizomes may give an indication of the amount of compaction from comparing its folded length with its unfolded length. Assuming vertical growth down into the sediment, the uncompressed length represents the original sediment thickness. In other types of deposits, the options to reconstruct the amount of compaction are limited and empirical modeling, which increases vertical error, has to be relied upon.

A way in which paleogeographic understanding of a study area can reduce the compaction problem is by identifying those sites least affected by compaction. This is typically applied in sea-level studies relying on basal peat sampling. Sites with a compaction-free base are regionally found at the base of transgressive (fine-grained) sedimentary sequences, either Pleistocene pedogenetically matured surfaces (paleovalley) or older substrate (bedrock). More locally, the flanks of coastal or inland dunes or coarse sandy alluvial channel belts may provide compaction-free substrate for back-barrier and estuarine sea-level markers formed in the deposits onlapping these features.

In some areas of the world, additional compaction has occurred in the last hundreds of years due to human drainage and dewatering of reclaimed coastal plains (i.e., polders). An additional phenomenon in such areas is direct surface lowering due to oxidation of dewatered organic topsoil (see Van Asselen et al., 2009 for an overview). These phenomena lead to differential compaction and topographic inversion. This negatively affects the quality of the former coastal plain in providing late Holocene sea-level records, but improves opportunities to resolve and understand middle Holocene landscapes and the SLIs obtained from them.

33.2.7 Prehistoric humans

In areas not studied before, the presence of archeological remains may be a useful starting point for the identification of former landscapes (Chapter 9). Human behavior and changes thereof may be indicative of changes in a coastal system, and therefore provide a valuable aid when reconstructing past landscapes. A good example of paleogeographical use of archeological information in sea-level reconstructive context is provided by the Mesolithic shell middens in the Lower Tagus Valley in Portugal. These are presently located along fluvial channels in tributary valleys. Some 8 ka ago however, tidal conditions prevailed in those valleys (Van der Schrieck et al, 2007).

33.2.8 Data organization and visualization

Paleogeographic studies usually involve large datasets which require a well-designed database and visualization or mapping software to enable a clear spatial understanding of the identified facies. Essential for local and regional reconstructions are GIS software (e.g., ESRI ArcGIS, QuantumGIS), paleogeography-dedicated digital map data design, and geological databases (e.g., Berendsen et al., 2007a). Software packages can be used to construct (conceptual) geological models and plot cross-sections (an internet search will show variously priced software packages). Although software is needed to plot lithology or other properties of boreholes on a cross-section, the interpretation of the deposits between them should be performed manually by a geologist based on sedimentological and geological concepts. Further, integrating new data with legacy data and merging of institutional databases is highly recommended for regional paleogeography. The storage of sea-level data in databases is elaborated upon in Chapter 34.

33.3 COASTAL SETTINGS

This section provides some examples of the application of paleogeographic knowledge in coastal sea-level studies.

33.3.1 Tidal inlet and basin system

An example of a large paleotidal basin is located in the northwest of the Netherlands in the Holland coastal plain. Paleogeographic reconstruction has shown this Bergen tidal basin to be a regional tidal basin which was fed by one major tidal inlet and surrounded by peat swamps in the back-barrier part. The basin was never connected to a major fluvial system (Fig. 33.5a) and originated at around 5000–4500 BC behind a transgressive barrier. Rising groundwater led to growth of basal peat (Fig. 33.5b). The transgressive barriers migrated landward until ~2400 BC, when sea-level rise decelerated (Beets et al., 1992; Beets and Van der Spek, 2000). During the following 1.5 ka the barriers rapidly prograded seaward and the tidal basin gradually silted up. A major tidal channel protruded some 40 km from the inlet into the back-barrier area. It formed the main pathway for sediment supply into this regional basin. Sand was mainly deposited in and along the tidal channel and its numerous tributaries, while in between the channels clay and silt were deposited in mudflats. These mudflats were mainly formed under intertidal conditions but in the distal part silting up occurred to above MHW levels. As a result, the most inland part of the tidal basin became suitable for human inhabitance and during the Bronze Age (3000–800 BC) several settlements existed in the area (Roep and Van Regteren Altena, 1988). In the distal part of the tidal channel system, clastic sedimentation was interrupted regularly and peat layers formed (Fig. 33.5b).

The example demonstrates subregional paleogeographic developments in the system, controlled by autogenic processes unrelated to sea-level change. An important aspect with respect to sea-level reconstruction in the basin is the influence of the flood-basin effect

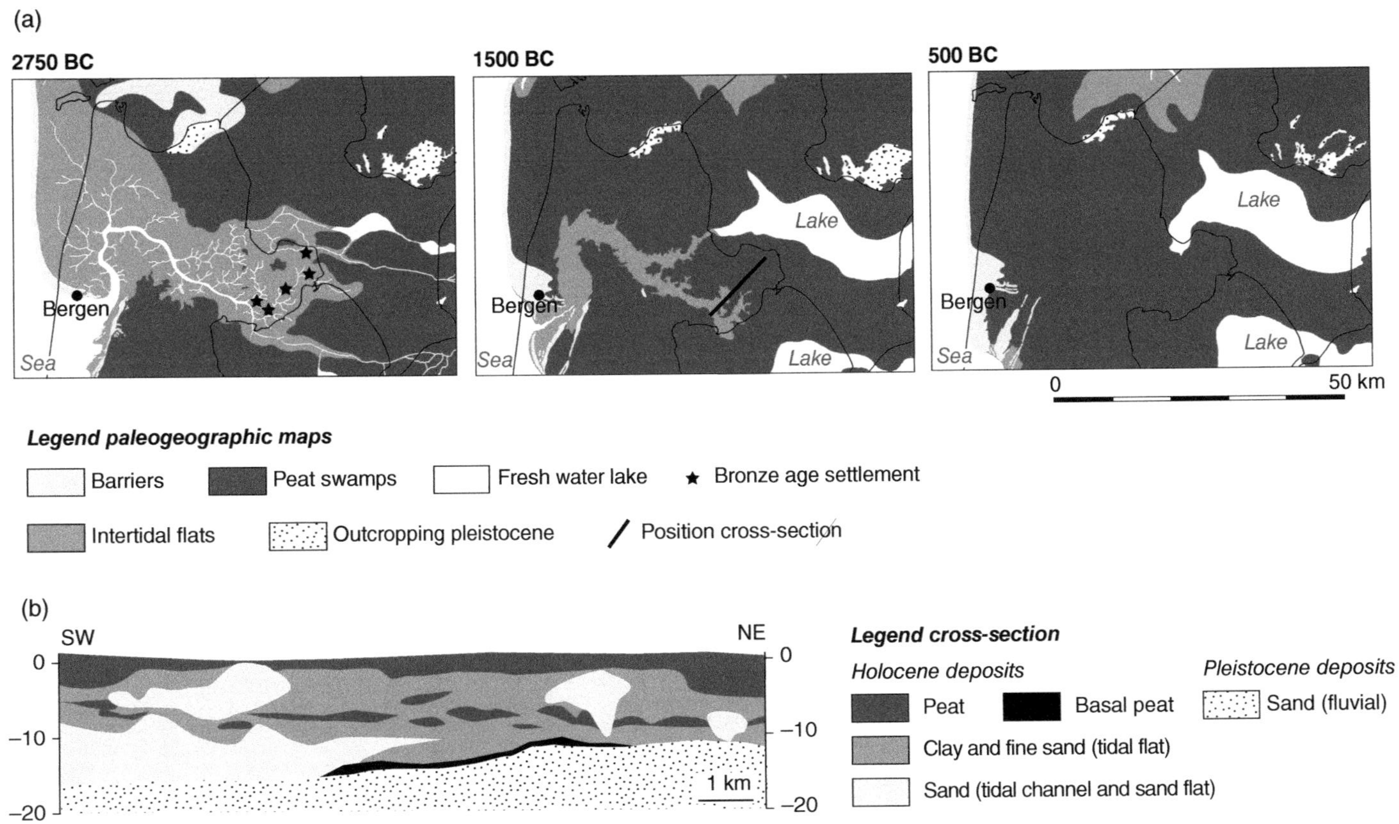

Fig. 33.5. (a) Paleogeography of the tidal basin in NW Netherlands (see Fig. 33.6 for location). (b) Stratigraphic cross-section illustrating tidal deposits which have been covered by peat since the tidal inlet at the town of Bergen was closed.

(Section 33.2.5). Paleotidal levels at the entrance of the tidal basin may have differed considerably from those in the distal part of the basin, thereby affecting sea-level reconstruction. Another factor hampering sea-level reconstruction from the inner part of the basin is related to the muddy sediments formed in the environment: they are highly susceptible to compaction.

The main inlet of the tidal basin closed around 1300 BC (Beets et al., 1996). As a result, the seaward drainage of the freshwater surplus from the back-barrier area was blocked and groundwater in the back-barrier area started rising. From that moment onwards, there has been a full decoupling of the development of the basin from sea-level change. Rising groundwater resulted in the Bronze Age settlements becoming deserted. Eventually, the whole area evolved into extensive peat swamps where thick oligotrophic peat cushions developed, which persisted until ~1000 AD when they were reclaimed for agriculture.

33.3.2 Tidal-facies transitions as SLIs

In this example from the northwest of the Netherlands, a series of paleogeographic maps documents the regional evolution of a tidal-flat landscape (Fig. 33.6a). Paleogeographic reconstruction was aided by archeological excavations which supplied paleoenvironmental information and age control. Most of the key sites consist of several generations of excavated dwelling mounds; the earliest date from ~600 BC. The dwelling mounds were typically built on the relatively higher silted-up parts of the former saltmarshes (Vos and Gerrets, 2005).

This example uses paleoenvironmental data to time the transition from intertidal to supratidal sedimentation as a proxy for sea level. The facies contact between intertidal flat and supratidal saltmarsh (see Section 33.2.2) can be easily distinguished below the dwelling mounds. Towards the top, the intertidal flat sediments consist of clayey, very fine sands. The sediment is almost homogeneous (no layering) due to strong bioturbation by marine organisms. In many cases, the last phases of mudflat deposition were radiocarbon dated from *in vivo* marine shells. The most common species is *Srobicularia plana*, but the bivalve species *Cerastoderma edule* and *Macoma baltica* are also found. The saltmarsh deposits consist of clays with a varying number of thin and crinkly sand or silt layers. The saltmarsh deposits contain occasional beds of marine shells, *ex vivo* shells and shell fragments (such as *Mytilus edule* and *Cerastoderma edule*). The shells originate from subtidal and intertidal environments, and were displaced towards the saltmarsh during storms. Doublets of bivalves were selected for radiocarbon dating. The vertical position of the boundary between the two facies was leveled relative to ordnance datum during the archeological excavation (centimeter-scale accuracy). The age of this boundary was obtained from radiocarbon dates just above and just below the transition: "bracketing dates". The sedimentary boundary can also be given an indicative meaning, relating it to MHW level at the time of deposition. Based on nearby sites, paleo-MHW is estimated to lie about 20 cm above the base of the saltmarsh facies (Roep and Van Regteren Alterna, 1988).

MHW SLIs are calculated for nine archeological sites (Fig. 33.6b). Error assessment was part of the calculation, and the data are presented as 1σ boxes for graphic purposes (2σ boxes are standard). The vertical position follows from the base of the saltmarsh facies, adding 20 cm with 5 cm error on both sides and adding an additional term that corrects for post-depositional compaction. The amount of compaction-induced subsidence of the mudflat deposits due to the loading of saltmarsh deposits and the dwelling mound was explicitly assessed for one of the sites. A range of 2–10 cm was applied as the compaction correction to all nine sites.

It is worth noting that the observed changes in paleo-MHW level should not only be attributed to sea-level change, but also to changes in the geometry of local tidal systems. At present, MHW near the mainland coast is 20 cm higher than in the main tidal inlet and between half and double that difference under past inlet configurations. To fully separate sea-level rise from these morphodynamic effects, the bathymetry of the past tidal system also needs to be paleogeographically assessed. SLIs for the same time period need to be collected from various locations within the same tidal system.

Regional paleogeographic reconstructions suggest that suitable sites for applying this method cover the last 3.5 ka without critical gaps. The method yields information about an otherwise poorly covered period in sea-level reconstructions in the Netherlands. This demonstrates that methods using spatially integrated, paleogeographically

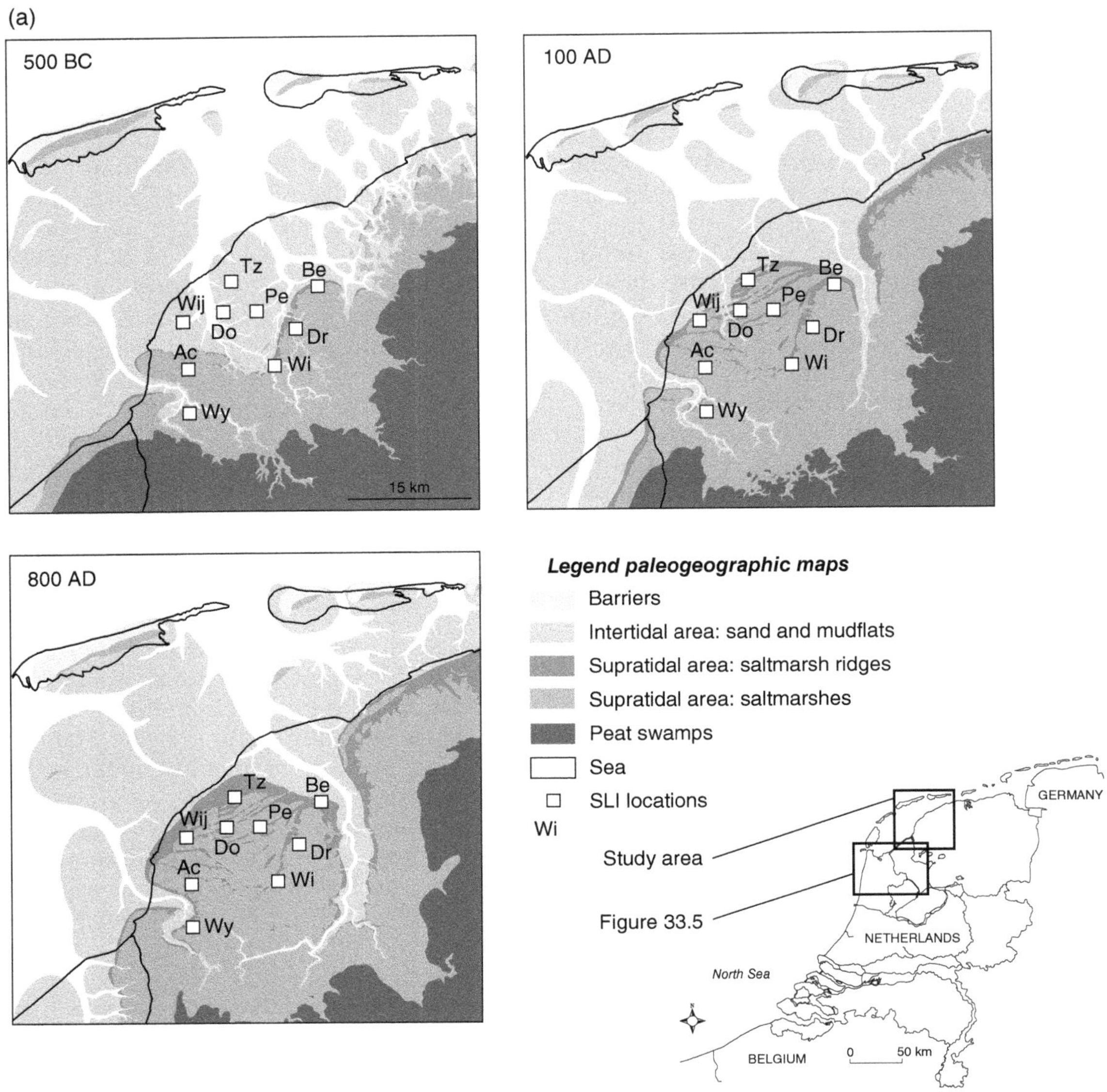

Fig. 33.6. (a) Paleogeographic reconstruction of NW Friesland (500 cal a BC–800 cal a AD). The reconstruction is based on the geological borehole descriptions of the TNO-DINO database (>40,000 boreholes), geomorphologic visualizations of LIDAR data (laser altimetry), and paleoenvironmental data from geoarcheological key sites.

dated sedimentary contacts and facies transitions bridge such gaps in sea-level reconstruction. They also provide information on sedimentation rates and offer constraints on age and paleoenvironment for related disciplines.

33.4 DELTAIC SETTINGS

This section discusses the collection of sea-level data using paleogeographic knowledge from three large-scale deltaic systems.

33.4.1 Wide low-gradient deltas

Many passive-margin shelves are much wider than the global average of ~80 km and hence have relatively low gradients. Consequently, the low-stand valleys are typically much shallower than the classical excavated/filled-in estuaries that have been used to develop sequence-stratigraphic concepts (e.g., Dalrymple and Choi, 2007). During transgression, modestly incised and relatively wide paleovalleys turn into vast wetlands, and the organic accumulation in those environments

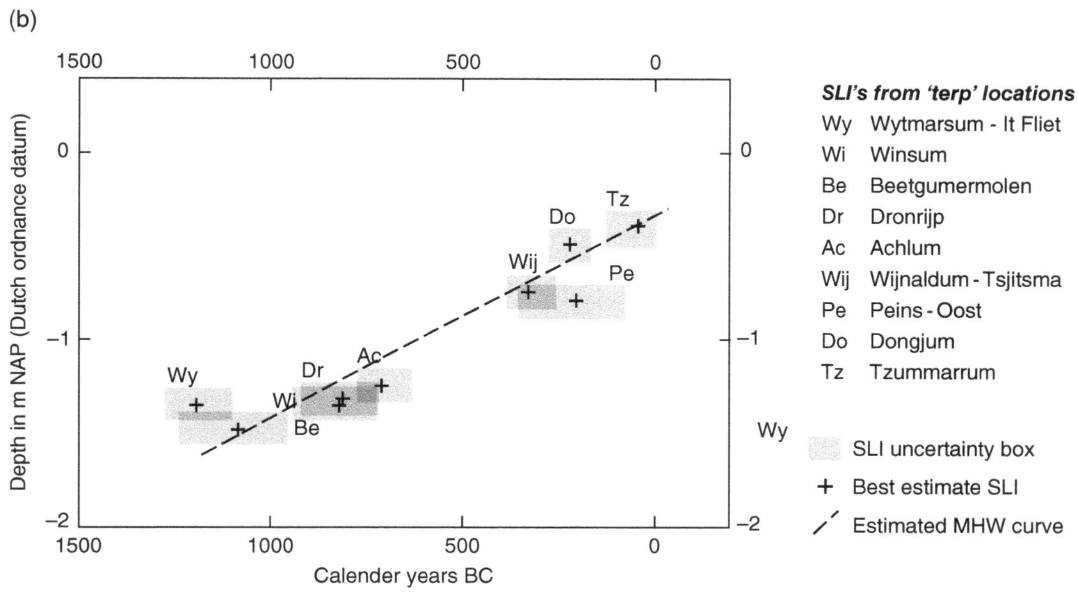

Fig. 33.6. (*Continued*) (b) Reconstructed paleo-MHW SLIs at nine locations in NW Friesland between 1500 cal a BC and 0 AD.

makes wider shelf regions suitable for sea-level research. In addition, the shallow valleys are often filled in well before the end of the transgression; the fluvial system can therefore migrate out of its paleovalley (Fig. 33.2e). A typical example is the Rhine–Meuse back-barrier delta, famous for the pioneering basal-peat-based sea-level research of, for example, Van de Plassche (1982). That and subsequent work (Cohen, 2005; Berendsen et al., 2007b; Hijma and Cohen, 2011) brought important insight in the indicative meaning of tidal-deltaic SLIs, and how to spatially resolve these. This concerns river avulsions, the flood-basin effect (see Section 33.2.5), and the river-gradient effect in open channels and their connected marshes and swamps.

The river-gradient effect tends to raise tidal levels in an upstream direction. This means that an inland indicator will form at a higher level than the same indicator more seaward. In the earliest phase of the transgression, the gradient effect is often substantial and then subsequently drops as a highstand coastal system develops. For example, in the Rhine–Meuse delta, a floodplain gradient of 0.12 m km^{-1} existed between 9000 and 8500 cal a BP and then dropped to a few centimeters after 6500 cal a BP when a barrier-lagoon system developed (Van de Plassche, 1982; Cohen, 2005). These gradients were calculated using vertical series of floodplain peat SLIs of similar age but from different along-stream positions. River avulsions also affect the gradients in open water and in floodplains: (1) in the abandoned estuary, the river gradient diminishes but at the same time the tidal wave will penetrate further inland; and (2) typical groundwater levels maintained in floodplain wetlands will change. Avulsions will also lead to adjustments in the tidal prisms of the tidal/fluvial basins. The flood-basin effect will therefore change, as the various sediment loci and areas experience peat growth.

All the above factors influence local water level and changes induced by these factors can be erroneously interpreted as sea-level changes. For deltas, paleogeographic considerations need to be added to the sea-level reconstruction to disentangle the contributing components to the SLI. For example, a SLI marking MHW may form: at 1 m above MSL close to the coast; just above MSL at a point further inland, away from major rivers (flood-basin effect dominant); and at 1 m above MSL at a third location equally far inland, but close to a major river (river-gradient effect dominant). In most cases only (paleo-) tidal ranges for the open sea are available to calculate indicative ranges (see Chapter 34), so these delta-wide variations should be taken into account. With sufficient paleogeographic information it is possible to: (1) find and predict the region in the delta where, for example, the tidal wave was most dampened and hence MHW very close to

MSL, or to avoid areas where the river-gradient effect would be substantial in a particular time; and (2) estimate the tidal range in the delta and correct the indicative range for each sample. This information can also be used to select the most suitable study sites prior to sampling. If data availability prohibits such analyses, an approach can be taken to increase the indicative range to account for these uncertainties or to add an additional error. This will often result in points of similar age at different elevations, reflecting the variation in water level within the delta (Section 33.2.3). A best-estimate MSL curve is then drawn just below the lowest samples for each time period, as MHW indicators by definition do not form below MSL. With limited data points, this could lead to a slight overestimation of paleo-MSL.

33.4.2 Confined or bayhead deltas

Confined or bayhead deltas are found along narrow continental shelves where deeply incised valleys have formed. In a highstand coastal zone, these valleys turn into estuaries during postglacial sea-level rise. The incisions result from fluvial erosion during sea-level lowstands and reach far inland. When the inland incised valley is transgressed, shallow marine, tidal, and fluvio-deltaic environments are established. The confined nature of the incised valleys prevents their lateral expansion with transgression. Consequently, only limited accommodation space is available for distal flood-basin deposition (e.g., basal peat, lagoonal peat, extensive wetlands), valuable for sea-level reconstruction (Fig. 33.2c, d). When the river supplies sufficient sediment, tidal open water and clastic sedimentation dominate. Although many global coastlines feature confined or bayhead deltas, attention has only recently been paid to them for sea-level reconstruction, possibly due to their less-favorable sedimentological properties. Finding sites for the collection of SLIs requires a multi-environment approach as described in the following.

An undulating valley floor or fluvial terrace remnants imply that, at a given valley cross-section, some levels are flooded later than the thalweg. This provides enhanced opportunities for the formation and preservation of groundwater-rise-driven basal peat layers in the valley (Fig. 33.2f). Dating the base of the peat yields an upper limit for MSL, assuming that basal peat does not grow below MSL (Section 33.2.3). The first occurrence of marine indicators in the peat such as diatoms, agglutinated foraminifera, brackish water plant seeds, and framboidal pyrite crystals marks the supratidal saltmarsh level. Radiocarbon dates of material from this level provide the supratidal range (MHW–HAT) for SLIs.

Where tidal-marsh facies (Section 33.2.2) directly overlie a lowstand fluvial surface, SLIs can be collected by dating the contacts and measuring their elevations relative to ordnance datum (Fig. 33.7b). This can be seen as the incised-valley equivalent of basal peat SLIs in a broad shelf setting. As in basal peats, the preferred material for radiocarbon dating are (flushed in) freshwater terrestrial botanical macrofossils picked from the sediment, such as leaves and seeds. Dating *in vivo* marine shells with estimated reservoir effect may be an alternative. Washed in freshwater, mollusks should be avoided because of their unknown reservoir effect.

Many incised valleys have equally deep incised tributary valleys, which are often relatively narrow. When the tributaries have limited catchments they have a limited sediment supply and therefore act as distal flood basins, perpendicular to the main valley axis. These valleys fill up relatively rapidly with tidal-marsh deposits, and may also contain a basal peat layer (Fig. 33.7b). Samples collected from low-energy wetland environments developed in these tributary valleys can provide SLIs in a similar way as described above (Fig. 33.2g). This type of configuration can be found in many places around the globe, including the massive Lower Amazon system with its tributaries incised in the Amazonian shield. A detailed study of the Lower Tagus Valley in Portugal shows that the tributaries silted up soon after relative sea-level had reached its post-glacial highstand (Vis et al., 2008). Simultaneously, the trunk river valley was still an estuarine environment. As the Tagus River continued supplying sediment, its fluvial wedge prograded downstream, thereby blocking consecutive tributaries. Due to the blockage of the tributaries, swamp conditions were created on top of the tidal marshes which led to growth of freshwater peat (Fig. 33.7c). The age of the base of the peat may be used as an indicator for the transition from tidal to freshwater environments, which is an indicator for the deceleration of sea-level rise. It therefore gives a minimum age of the flattening of the sea-level rise curve (*terminus ante quem*). Pollen analysis of sediments from the transition can help to pinpoint the end of *Chenopodiaceae* species pollen, coming from brackish-environment plants.

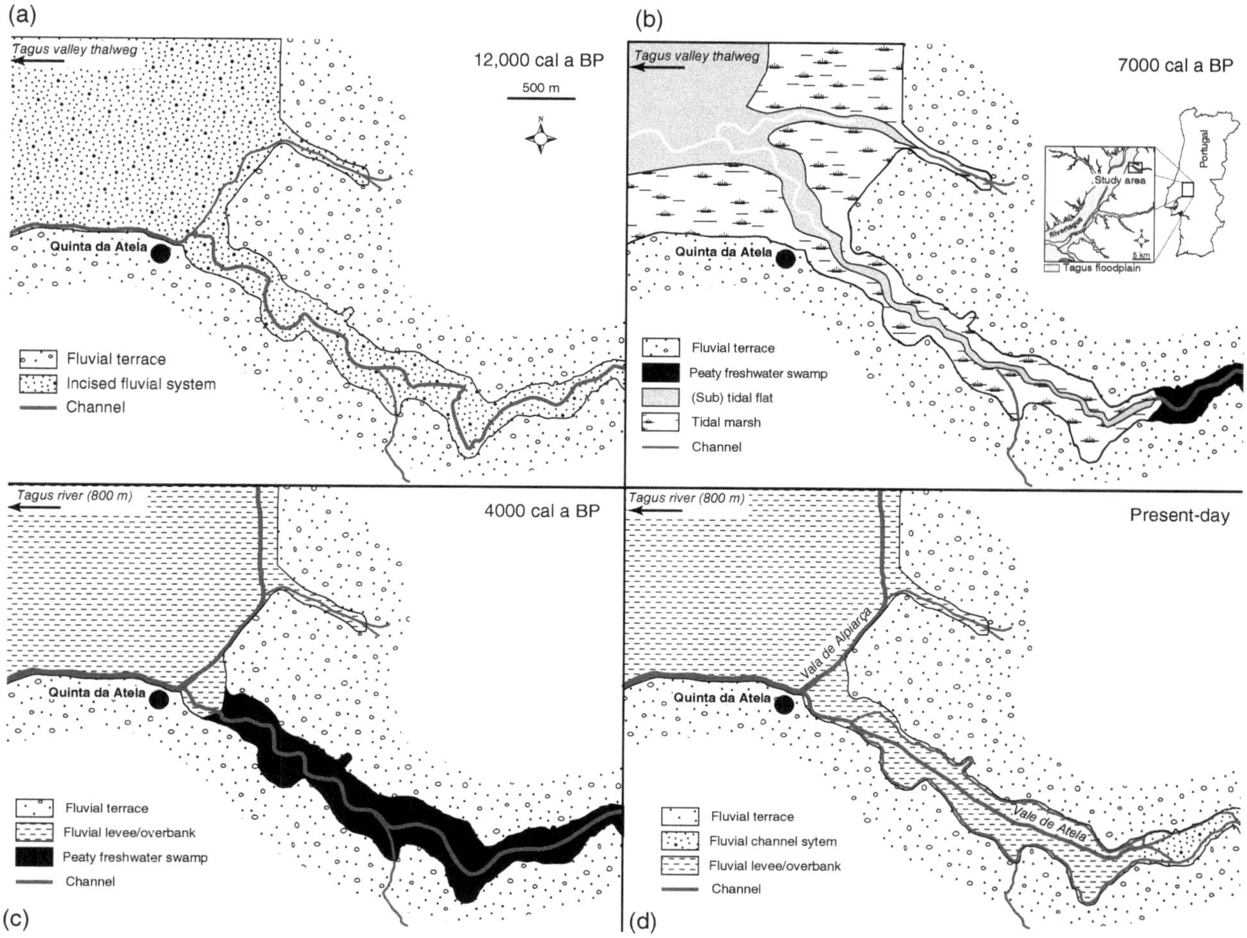

Fig. 33.7. Paleogeography of a tributary valley in the Lower Tagus Valley, Portugal: (a) an incised Tagus River trunk valley and incised tributary valleys; and (b) post-glacial sea-level rise had almost ended; tidal-marsh facies had established and directly overlie the lowstand fluvial surface. The contact between the two facies is ideal for extracting SLIs. (c) As a result of Tagus River downstream progradation its levee blocked the tributary valley, creating swampy conditions ideal for peat formation on top of the tidal deposits. Dating the base of that peat gives a minimum age of the flattening of the sea-level rise curve (*terminus ante quem*). (d) Strongly increased clastic overbank deposition due to human impact in the catchments of both the Tagus River and the tributary Vale de Atela.

Finally, incised valleys may have a narrow bedrock-controlled seaward entrance. In such a case, the width of that entrance may affect the amplification or attenuation of the tide when moving into the tidal basin further inland. This would lead to an altered tidal range (Section 33.2.5).

33.4.3 Deltas and shelves over multiple glacial cycles

Whereas much of the emphasis in previous sections was on the identification of SLIs in different deltaic settings, this section shifts the focus to stacked delta and shelf sequences exhibiting multiple SLIs that formed during multiple glacio-eustatic cycles. In particular, we discuss how three-dimensional seismic data is instrumental in detecting SLIs and outlining paleogeographic information in these stacked delta sequences. Using seismic data to reconstruct sea-level changes is an art referred to as "seismic sequence-stratigraphy". Scattered borehole information, mostly consisting of well-log data and sporadic core material, can be used to constrain the interpretations based on seismic data. Such calibration against physical data is required since seismic data only reveals architectural information and sea-level interpretations are sometimes ambiguous, as demonstrated in the following paragraph. Moreover, subtle sedimentological details of coastal deposits thinner than ~10 m are often not detectable in seismic data because of seismic resolution limitations.

Given sufficient sediment supply and water depth, the basinward progradation of coastal prisms and clastic wedges results in the deposition of clinoforms (Fig. 33.8a).These may form either on a scale of tens of meters as individual shoreface clinoforms, or on a shelf-margin scale (hundreds of meters). Both types of clinoforms are formed by the interplay between sediment supply and sea-level changes (allogenic control) and a range of external influencers. The formation of shoreline clinoforms is importantly controlled by this interplay acting at fourth-, fifth-, and sixth-order scales (100–10 ka), whereas shelf clinoforms are shaped by third-order (0.5–3.0 Ma) changes of relative sea-level and sediment-supply rates. Deltas may form at both scales and either feed the shoreline or the entire shelf.

Trajectory analysis (e.g., Helland-Hansen and Hampson, 2009) determines the migration angle between two successive shoreline or shelf-edge positions and enables a description of the internal deltaic architecture (Fig. 33.8b–e). The method has proven useful in analyzing along-profile seismic sections of deltas where it quantitatively distinguishes between transgressions and normal and forced regressions. The trajectory can be split into vectors where the vertical component represents accommodation (relative sea-level) changes and where the size and direction of the horizontal component is indicative of variations in sediment supply (e.g., Bijkerk et al., 2013). As such, the sea-level trajectory is regarded as a good tool to establish the ratio between accommodation change and sediment supply; it is therefore a valuable aid in sequence-stratigraphic interpretations. The scale and order of the interpretation depends on the clinoform type considered. A weakness in single 2D trajectory analysis is that the effects of lateral avulsion can be easily overlooked or misinterpreted. The strength of multi-2D or 3D trajectory analysis, for instance using 3D seismic data, is that external and internal controls on delta architecture can (theoretically) be more easily disentangled. Autogenic processes such as autocyclicity, autoretreat, lobe-switching, and autostability may locally affect sediment supply. Consequently, the horizontal component of the trajectory may show significant differences depending on the position on the delta. The vertical accommodation component of the trajectory is mainly derived from the two main allogenic controls – tectonics and (glacio-) eustasy – and is expected to be more uniform throughout a delta body studied.

Distribution plots of the vertical and horizontal trajectory components are good visual aids in sensing the role of sea-level change versus autogenic processes, respectively (Fig. 33.8f). Depositional behavior such as progradation, retrogradation, and aggradation may exist next to each other and be laterally transitional.

A trajectory analysis study performed on the Plio-Pleistocene southern North Sea delta (Fig. 33.8g, h) not only shows prolonged delta progradation but also reveals vertical trends in accommodation, which allows a sequence-stratigraphic interpretation. In this case two major sea-level drops are interpreted that are related to Gelasian age glacio-eustasy and which are enhanced in their expression due to concurrently diminishing sediment supply. Without the disentanglement of the vertical and horizontal components of the sea-level trajectory, the relationship between climate effects on sediment delivery and sea-level trends would be easily misread from the stratigraphic record.

33.5 CONCLUDING REMARKS

This chapter has discussed the use of a well-constrained paleogeographic reconstruction as input for generating high-quality sea-level index points. Specific methods have been presented to reconstruct paleoenvironments based on various data sources and to integrate them into a paleogeographic reconstruction. Using examples from coastal and deltaic settings, we have shown the importance of detailed paleogeographic knowledge to correctly extract SLIs from them.

As well as the value of paleogeography for the extraction of SLIs, it can serve the most advanced paleogeographic reconstructions based on modeling at continental and global scales. These models incorporate isostatic land movement, geoidal-eustatic changes, sea-level changes (Chapter 28) and related paleotidal changes (Chapter 29). For the reconstructions the paleogeography of ice-sheet extent is used as input. For coastal regions the models merely use point data on paleo-sea level as input instead of reconstructed paleo-coastlines. Furthermore, present-day bathymetry and surface topography are often used. However, large-scale paleogeographic modeling would benefit strongly from the use of reconstructed paleo-surfaces and paleo-bathymetries

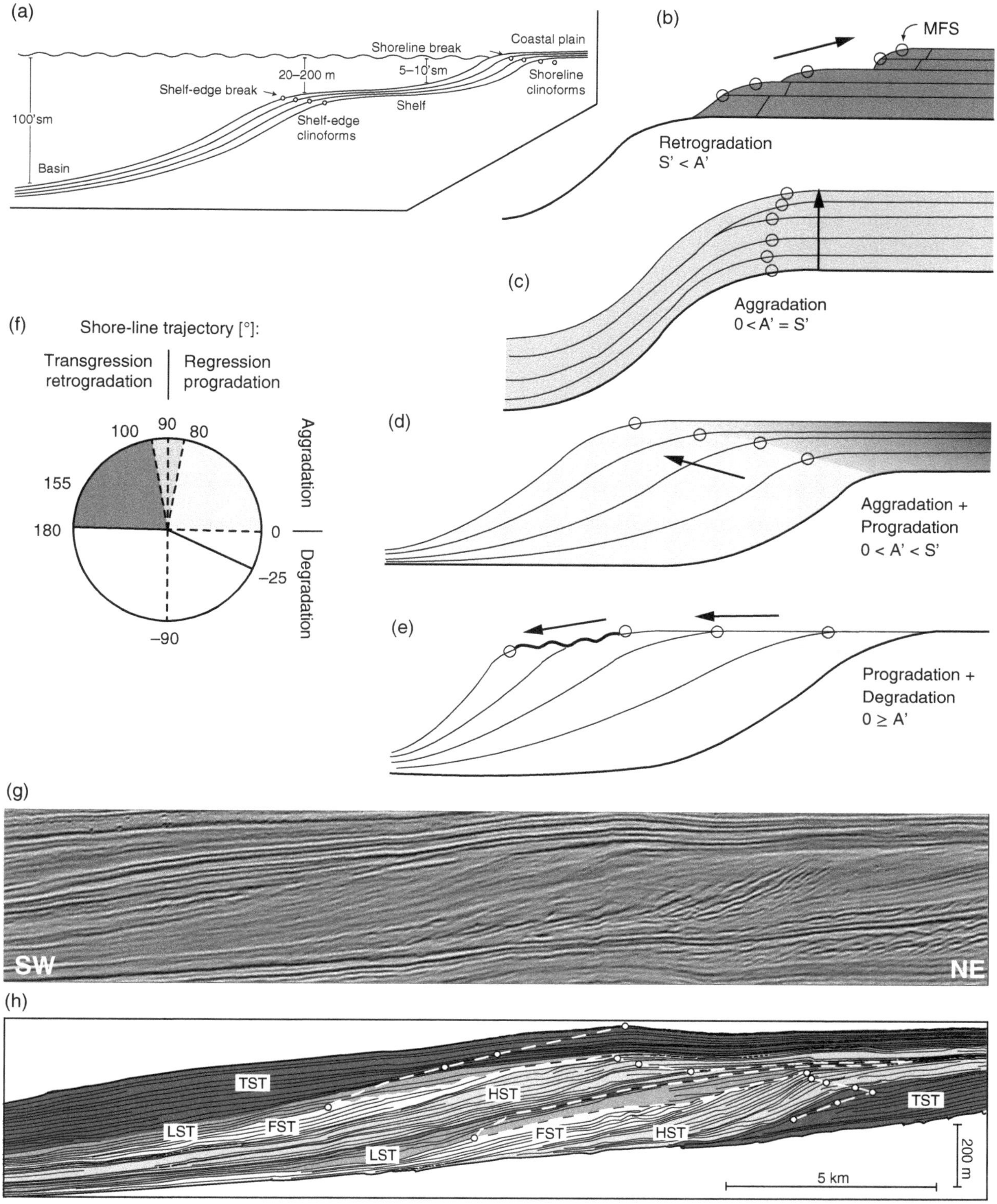

Fig. 33.8. Principle of sea-level trajectory analysis (modified from Helland-Hansen and Hampson, 2009). (a) Compound clinoform set with shoreline- and shelf-edge trajectories that are both indicative of changes in accommodation. (b–e) Cross-sectional representation of the relationship between accommodation change (A′) and sediment supply (S′), the resulting depositional geometry, and shoreline trajectories (for subaqueous settings; MFS: maximum flooding surface). (f) Relationship between trajectory angle and depositional trends; the latter are indicated by gray shading as in (b–e). Positive or negative angles represent aggradational or degradational (erosional) deposition, respectively. Trajectory angles of 90±10° are regarded as purely aggradational, whereas higher or lower angles indicate the land or seaward shift of the clinoform (i.e., retrogradation or progradation). (g) East–west seismic section through the Plio-Pleistocene southern North Sea delta in the northern Dutch offshore. (h) Interpretation of (g) (G. de Bruin, pers. comm., 2013) showing the position of sea-level index points (based on shelf-edge breaks) and the sea-level trajectory (dashed line) that forms the basis for the sequence-stratigraphic interpretation provided (gray shading as in (f)).

(such as those of the base of Holocene coastal wedges), especially at regional scales. At such scales, sedimentation and erosion change the morphology of surfaces over time. This calls for the incorporation of coastal geomorphological modules linked to geophysical models (as implemented for glaciological modules). This link has not yet been made, but it is a likely goal for the near future. In preparation for this, paleogeographic reconstructions at regional to continental scales need to become available in a digital format.

Digital paleogeographic reconstructions are already available for some places in the world, and philosophy and techniques to digitally create and store paleogeographic information exist (e.g., Berendsen et al., 2007a). In general, the Quaternary geological communities along the world's shelves, coasts, and deltas have not yet organized this. However, concerns about expected future global sea-level rise, coastal retreat, and subsidence have led to an increasing amount of raw subsurface data becoming available. The exponential growth of subsurface data gives way to new applications, such as high-resolution 3D voxel models of the subsurface (Stafleu et al., 2011). Building these models is greatly aided by digitally available paleogeographic reconstructions, cancelling out their time-consuming nature. In the end, sea-level studies will benefit from both the voxel models and the digital paleogeographic reconstructions.

REFERENCES

Beets, D.J., and Van der Spek, A.J.F. (2000) The Holocene evolution of the barrier and the back-barrier basins of Belgium and the Netherlands as a function of late Weichselian morphology, relative sea-level rise and sediment supply. Netherlands Journal of Geosciences, 79, 3–16.

Beets, D.J., Van der Valk, L., and Stive, M.J.F. (1992) Holocene evolution of the coast of Holland. Marine Geology, 103, 423–443.

Beets, D.J., Roep, Th.B., and Westerhoff, W.E. (1996) The Holocene Bergen inlet: closing history and related barrier progradation. Mededelingen Rijks Geologische Dienst, 57, 97–131.

Berendsen, H.J.A., Cohen, K.M., and Stouthamer, E. (2007a) The use of GIS in reconstructing the Holocene palaeogeography of the Rhine-Meuse delta, The Netherlands. International Journal of Geographical Information Science, 21, 589–602.

Berendsen, H.J.A., Makaske, B., Van de Plassche, O., Van Ree, M.H.M., Das, S., Van Dongen, M., Ploumen, S., and Schoenmakers, W. (2007b) New groundwater-level rise data from the Rhine-Meuse delta - implications for the reconstruction of Holocene relative mean sea-level rise and differential land-level movements. Netherlands Journal of Geosciences, 86, 333–354.

Bijkerk, J.F., Ten Veen, J., Postma, G., Mikeš, D., Van Strien, W., and De Vries, J. (2013) The role of climate variation in delta architecture: lessons from analogue modeling. Basin Research, 26(3), 351–368.

Bird, M.I., Fifield, L.K., Teh, T.S., Chang, C.H., Shirlaw, N., and Lambeck, K. (2007) An inflection in the rate of early mid-Holocene eustatic sea-level rise: A new sea-level curve from Singapore. Estuarine, Coastal and Shelf Science, 71, 523–536.

Coen, I. (1988) Ontstaan en ontwikkeling van de Westerschelde. Water, 43, 156–162.

Cohen, K.M. (2005) 3D geostatistical interpolation and geological interpretation of palaeo-groundwater rise within a Holocene coastal prism. In: *River Deltas: Concepts, Models, and Examples* (eds Giosan, L., and Bhattacharya, J.P.), SEPM, Special Publication 83, 341–364.

Cunningham, A.C., Bakker, M.A.J., Van Heteren, S., Van der Valk, B., Van der Spek, A.J.F., Schaart, D. R., and Wallinga, J. (2011) Extracting storm-surge data from coastal dunes for improved assessment of flood risks. Geology, 39, 1063–1066.

Dalrymple, R.W., and Choi, K. (2007) Morphologic and facies trends through the fluvial–marine transition in tide-dominated depositional systems: A schematic framework for environmental and sequence-stratigraphic interpretation. Earth-Science Reviews, 81, 135–174.

Helland-Hansen, W. and Hampson, G.J. (2009) Trajectory analysis: concepts and applications. Basin Research, 21, 454–483.

Hijma, M.P., and Cohen, K.M. (2010) Timing and magnitude of the sea-level jump preluding the 8200 yr. event. Geology, 38, 275–278.

Hijma, M.P., and Cohen, K.M. (2011) Holocene transgression of the Rhine river-mouth area, The Netherlands/Southern North Sea: palaeogeography and sequence stratigraphy. Sedimentology, 58, 1453–1485.

Jelgersma, S., Stive, M.J.F., and Van der Valk, L. (1995) Holocene storm surge signatures in the coastal dunes of the western Netherlands. Marine Geology, 125, 95–110.

Mauz, B., and Bungenstock, F. (2007) How to reconstruct trends of late Holocene relative sea level: A new approach using tidal flat clastic sediments and optical dating. Marine Geology, 237, 225–237.

Reineck, H.-E., and Singh, I.B. (1980) *Depositional Sedimentary Environments: With Reference to Terrigenous Clastics.* Springer.

Rieu, R., Van Heteren, S., Van der Spek, A.J.F., and De Boer, P.L. (2005) Development and preservation of a mid-Holocene tidal channel network offshore the western Netherlands. Journal of Sedimentary Research, 75, 413–423.

Rodrigues, A., Magalhães, F., and Dias, J.A. (1991) Evolution of the north Portuguese coast in the last 18,000 years. Quaternary International, 9, 67–74.

Roep, Th.B. (1986) Sea-level markers in coastal barrier sand: examples from the North Sea coast. In: *Sea-Level Research: A Manual for Collection and Evaluation of Data* (ed. Van der Plassche, O.), Geobooks, Norwich, 97–128.

Roep, Th.B., and Beets, D.J. (1988) Sea level rise and paleotidal levels from sedimentary structures in the coastal

barriers in the western Netherlands since 5600 BP. Geologie en Mijnbouw, 67, 53–60.
Roep, Th.B., and Van Regteren Altena, J.F. (1988) Palaeotidal levels in tidal sediments (3800–3635 BP); compaction sea-level rise and human occupation (3275–2620 BP) at Bovenkarspel, NW Netherlands. In: *Tide-Influenced Sedimentary Environments and Facies* (eds de Boer, P.L., van Gelder, A., and Nio, S.D.), Reidel Publishing Company, Dordrecht, 215–231.
Schwartz, M.L. (ed.) (2005) *Encyclopedia of Coastal Science*. Springer, Dordrecht, the Netherlands.
Stafleu, J., Maljers, D., Gunnink, J.L., and Busschers, F.S. (2011) 3D modelling of the shallow subsurface of Zeeland, the Netherlands. Netherlands Journal of Geosciences, 90(4), 293–310.
Van Asselen, S., Stouthamer, E., and Van Asch, Th.W.J. (2009) Effects of peat compaction on delta evolution: A review on processes, responses, measuring and modeling. Earth-Science Reviews, 92, 35–51.
Van de Plassche, O. (1980) Holocene water-level changes in the Rhine-Meuse Delta as a function of changes in relative sea level, local tidal range and river gradient. Geologie en Mijnbouw, 59, 343–351.
Van de Plassche, O. (1982) Sea-level change and water-level movements in The Netherlands during the Holocene. PhD thesis, VU University Amsterdam, Amsterdam, The Netherlands.
Van de Plassche, O. (ed.) (1986) *Sea-Level Research: A Manual for the Collection and Evaluation of Data*. Geobooks, Norwich.
Van der Schriek, T., Passmore, D.G., Stevenson, A.C., and Rolão, J. (2007) The palaeogeography of Mesolithic settlement-subsistence and shell midden formation in the Muge valley, Lower Tagus Basin, Portugal. The Holocene, 17, 369–385.
Van der Spek, A.J.F. (1997) Tidal asymmetry and long-term evolution of Holocene tidal basins in The Netherlands: simulation of palaeo-tides in the Schelde estuary. Marine Geology, 141, 71–90.
Van Heteren, S., Huntley, D.J., Van de Plassche, O., and Lubberts, R.K. (2000) Optical dating of dune sand for the study of sea-level change. Geology, 28, 411–414.
Van Veen, J. (1950) Eb- en vloedschaar systemen in de Nederlandse getijwateren (with English summary). Tijdschrift Koninklijk Nederlands Aardrijkskundig Genootschap, 67, 303–325.
Van Veen, J., Van der Spek, A.J.F., Stive, M.J.F., and Zitman, T. (2005) Ebb and flood channel systems in the Netherlands tidal waters. Journal of Coastal Research, 21, 1107–1120.
Vis, G.-J., Kasse, C., and Vandenberghe, J. (2008) Late Pleistocene and Holocene palaeogeography of the Lower Tagus Valley (Portugal): effects of relative sea level, valley morphology and sediment supply. Quaternary Science Reviews, 27, 1682–1709.
Vos, P.C., and Gerrets, D.A. (2005) Archaeology: a major tool in the reconstruction of the coastal evolution of Westergo (northern Netherlands). Quaternary International, 133–134, 61–75.
Walther, J. (1894) Einleitung in die Geologie als Historische Wissenschaft, Vol. 3. Lithogenesis der Gegenwart, pp. 535–1055.

Chapter 34
A protocol for a geological sea-level database

MARC P. HIJMA[1,2], SIMON E. ENGELHART[3], TORBJÖRN E. TÖRNQVIST[1], BENJAMIN P. HORTON[4,5], PING HU[1], AND DAVID F. HILL[6]

[1]*Department of Earth and Environmental Sciences, Tulane University, New Orleans, LA, USA*
[2]*Deltares, Applied Geology and Geophysics, Utrecht, The Netherlands*
[3]*Department of Geosciences, University of Rhode Island, Kingston, RI, USA*
[4]*Sea Level Research, Department of Marine and Coastal Sciences, Rutgers University, USA*
[5]*Earth Observatory of Singapore and Division of Earth Sciences, Nanyang Technological University, Singapore*
[6]*School of Civil and Construction Engineering, Oregon State University, Corvallis, OR, USA*

34.1 INTRODUCTION

A sea-level index point (SLIP) estimates relative sea level (RSL) at a specified time and place, with an associated uncertainty. In the preceding chapters, numerous examples have been provided detailing how to collect sea-level indicators from different geomorphic settings (Chapters 3–10) and the means to interpret them (Chapters 12–22). Various methods of dating SLIPs (Chapters 23–27) have been discussed, as well as how to use modeling to account for compaction or changes in tidal range (Chapters 29 and 30). In order to compare SLIPs collected by differing techniques, it is necessary to analyze their associated errors in an objective and uniform way. SLIPs that are represented as discrete, errorless data points in age/elevation space may lead to erroneous inferences of RSL fluctuations that often reflect inherent uncertainties in the underlying data. The usefulness of geological sea-level data increases significantly if they are subjected to a rigorous error analysis with well-quantified uncertainties. As a consequence, SLIPs have played a major role in the last decades in estimating future sea-level change by establishing long-term background rates of vertical land motion (e.g., Engelhart et al., 2009) and in refining glacial isostatic adjustment (GIA) models (e.g., Lambeck et al., 1998; Peltier et al., 2002; Milne et al., 2005; Vink et al., 2007).

The database approach is a method for analyzing large numbers of SLIPs with all data stored according to a well-defined error protocol. A sea-level database can elucidate regional variations in past RSL which are of interest to a wide range of topics such as ice-sheet dynamics, archeology, Earth rheology, and future sea-level change. Multiple approaches to database construction have been used worldwide (e.g., Flemming, 1982; Shennan and Horton, 2002; Toscano and Macintyre, 2003; Dutton and Lambeck, 2012; Engelhart and Horton, 2012; Yu et al., 2012), all with unique strategies and emphases. In this chapter we present a comprehensive protocol for analyzing and standardizing sea-level data, including a format for constructing a sea-level database that captures all the relevant variables. In particular, we build on the work initiated by the Durham University group in the 1980s (e.g., Shennan, 1989) that culminated in a comprehensive sea-level database for the UK (Shennan and Horton, 2002). The hallmark of the UK sea-level database is the evaluation in a systematic fashion of a large range of variables to produce SLIPs and limiting data points. Here we expand upon this approach, especially by quantifying dating errors and by incorporating both modern-day and paleotidal modeling. The overarching philosophy is to include as much of the original, "raw" data as possible, and to maximize caution in assigning errors. The latter implies that assigned errors are often larger than in previous analyses of similar data.

The database described here was developed within the framework of studies of RSL change since the Last Glacial Maximum (LGM) along the US Gulf and Atlantic coasts. However, the methods can be applied worldwide as they are suitable for a

Handbook of Sea-Level Research, First Edition. Edited by Ian Shennan, Antony J. Long, and Benjamin P. Horton.

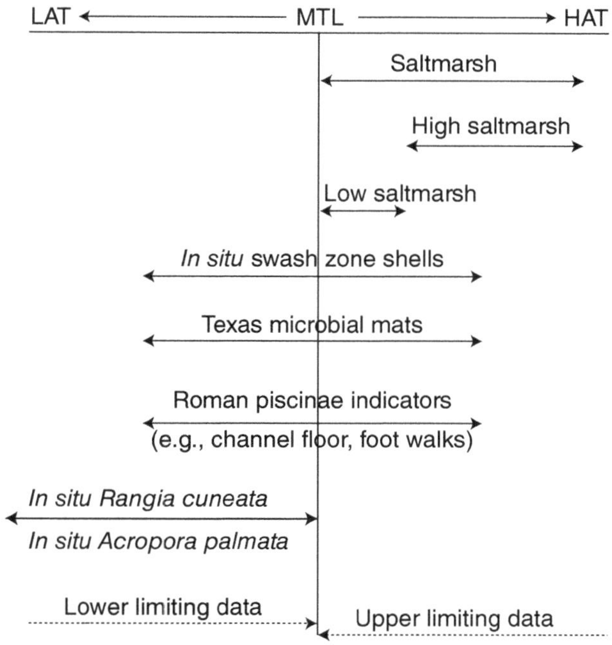

Fig. 34.1. An example of several indicators that can be analyzed using the database protocol. The indicative ranges are provided to illustrate that these samples collectively cover the whole (extreme) tidal range, but as they differ between areas are given in common terms only (HAT: highest astronomical tide; LAT: lowest astronomical tide). For upper and lower limiting data the reference water level is MTL and, as their respective upper and lower limits are unconstrained, they are represented by a dashed line. The indicative ranges of saltmarshes, swash zone shells (e.g., *Donax roemeri*), Texas microbial mats, and Roman piscinae indicators are based on Redfield (1972) and Van de Plassche (1991), Milliken et al. (2008), Livsey and Simms (2013), and Lambeck et al. (2004), respectively. According to LaSalle and Parsons (1985) and Lighty et al. (1982), maximum living depths for *Rangia cuneata* and *Acropora palmata*, respectively, can be set at −5 m MTL but no consensus exists on this issue.

variety of timescales, for both low- and high-resolution dating techniques, and for any type of sea-level indicator, provided that they are analyzed within the framework of the indicative meaning. We realize that some aspects of the protocol, in particular the paleotidal modeling, cannot (yet) be followed in all parts of the world and, for those cases, we offer alternatives. Also, several of the errors that we assign are standardized errors for use when insufficient information is available from the original source. If enough information exists from either publications or archives of researchers, such information always trumps the standardized errors. In order to maximize the usefulness of the protocol for researchers around the globe, the database is currently in an Excel spreadsheet format.

We continue this chapter with a brief review of the indicative meaning of sea-level indicators, along with an explanation of upper and lower limiting data. We describe the database structure, followed by a comprehensive discussion of error calculations. We illustrate our approach with RSL data from the US Gulf Coast by means of Tables 34.1–34.8 that contain 77 variables (columns) that are discussed in the following sections. An abbreviated version of the US Gulf Coast database is available from the NOAA Paleoclimate datasets (http://hurricane.ncdc.noaa.gov/pls/paleox/f?p=519:1:0::::P1_STUDY_ID:16361) with an expanded version in progress. We end the chapter with an example of a re-analyzed sea-level dataset and a call for the creation of a global sea-level database.

34.2 SEA-LEVEL DATA

34.2.1 Sea-level index points and indicative meaning

The most fundamental elevation attribute in RSL reconstruction is the indicative meaning that describes where – with respect to tide levels – the sea-level indicator formed (Shennan, 1982; Van de Plassche, 1986). This relationship should be determined by detailed analysis of the indicator in modern environments. When the indicator is preserved, and with the assumption that this relationship did not change through time, the contemporary tide level can be reconstructed. Many types of indicators can be used with this method, allowing for direct comparison between sea-level data obtained from different environments (Fig. 34.1). The indicative meaning consists of two parameters: the indicative range (IR) and the reference water level (RWL). The IR is the elevational range over which an indicator forms and the RWL is the midpoint of this range, expressed relative to the same datum as the elevation of the sampled indicator (geodetic datum or tide level). The RSL for a SLIP is then calculated from:

$$\mathrm{RSL} = E_s - \mathrm{RWL}_s \tag{34.1}$$

where E_s and RWL_s are the elevation and reference water level of sample s, expressed relative to the same datum, and RSL is relative to local sea level at the time of sampling. E_s is commonly established by measuring the depth of a sample in a core (or outcrop) where the surface elevation was established using surveying methods or estimated

Table 34.1. Structure of the sea-level database illustrated by means of four examples: ID information (NAD: North American Datum from 1927; UTM: Universal Transverse Mercator)

1. Region	2. Location	3. Latitude	4. Longitude	5. Original *x*-coordinate	6. Original *y*-coordinate	7. Original coordinate system	8. Reference(s)
Louisiana USA	St. James Parish	30.0393	−90.7113	3325150	720700	NAD27, 15R (UTM)	Törnqvist et al. (2004)
Louisiana USA	Bird Island Bayou	29.60	−91.87				Coleman and Smith (1964)
Alabama USA	Morgan Peninsula	30.2572	−87.9466				Blum et al. (2003)
Texas USA	Copano Bay	28.120	−97.062				Troiani et al. (2011)

(less precisely) from the environment in which the core was collected (e.g., high saltmarsh). For modern (surface) samples, RWL and E_s are equal and hence modern RSL is zero for all locations. To illustrate these points, we provide an example:

Core X was recovered from a saltmarsh environment where mean high water (MHW) = 1 m above mean tide level (MTL) and the highest astronomical tide (HAT) = +2 m MTL. The surface elevation of +0.6 m MTL was established by leveling to a national geodetic survey benchmark. A high marsh *Spartina patens* rhizome was selected for dating 2.7 m below the surface, hence E_s is −2.1 m MTL. The plant macrofossil suggests that the ^{14}C-dated sample formed in a high-saltmarsh environment. The dated sample is assigned a RWL of the midpoint between MHW and HAT (+1.5 m MTL). Thus, RSL = −2.1 m minus 1.5 m = −3.6 m MTL.

34.2.2 Upper and lower limiting data points

Dated samples from freshwater as well as most estuarine and marine environments cannot be directly related to past tide levels, but they can be used as limiting data points (Shennan and Horton, 2002). Although freshwater samples generally form above the HAT, in some cases they may form within the upper part of the intertidal zone (e.g., Jelgersma, 1961; Van de Plassche, 1982; Shennan et al., 2000) and a conservative RWL of MTL is therefore preferred (Fig. 34.1) in most micro- and mesotidal environments. Under macrotidal conditions, it is unlikely that freshwater samples form close to MTL and another RWL should be used. It is essential that freshwater samples are taken directly above or very close to an unconsolidated substrate and hence do not have significant compaction problems; otherwise, they should preferably not be used as upper limiting data.

Marine and estuarine organisms frequently have very uncertain vertical living ranges. Such organisms can therefore only be used as lower limiting data points. Ideally, these materials should be sampled *in situ*, but a problem is that many marine indicators have been collected offshore by grab samples or with coring techniques that do not allow for this. For open marine organisms this is less of a problem; if they are reworked, the elevational change is likely less than the typically wide living ranges. For organisms living within the intertidal zone or within close proximity to MTL, an *ex situ* occurrence is more problematic as they could have ended up above the contemporaneous MTL. Care is therefore required when using such samples, or samples for which it is not known whether they were obtained *in situ* or *ex situ*. For this reason, and also because the indicative meaning of the lower limiting data is not exactly known, we use a RWL for marine limiting data of MTL rather than lowest astronomical tide (LAT). However, there are several marine indicators where the RWL is sufficiently known to use them as SLIPs. This applies in particular to some coral species (e.g., *Acropora palmata*, *Montastrea annularis*) that have been widely used for sea-level reconstruction (Fig. 34.1 and Chapter 7).

34.3 DATABASE STRUCTURE AND PROTOCOL

34.3.1 "ID" information

The location of all data points (SLIPs, lower and upper limiting data) is described by means of a region, a location name, and latitude and longitude (Table 34.1, columns 1–4). Decimal degrees are preferred as this allows for easy plotting and

Table 34.2. Structure of the sea-level database: sediments (variables that require interpretation have been shaded)

9. Sample name	10. Sampling method	11. Material dated	12. Dated facies	13. Paleo-environmental analysis	14. Sample thickness (m)	15. Estimated sample thickness (m)	16. Corrected sample thickness (m)	17. Overburden facies	18. Underlaying facies (nearest layer)	19. Overburden thickness (m)	20. Depth to consolidated substrate (m)
Lutcher V-1	Hand coring	>10 charcoal fragments	Humic clay		0.03		0.07	Silt and clay	Paleosol on Pleistocene deposit	10.25	0.00
2	Piston core	Peat	Peat		0.18		0.46	Deltaic sediments	Deltaic sed. with peat layers	1.22	
	Vibrocoring	Sand	Swash-zone sediments			0.20	0.20	Beach sediments		2.00	0.00
CB08-02 -10.93	Rotary coring	*Nuculana concentrica* (articulated)	Dark gray mud			0.02	0.02	Open bay, dark gray mud	Dark gray mud	12.93	

Table 34.3. Structure of the sea-level database: elevation (DGPS: differential global positioning system; DEM: digital elevation model; NED: National elevation dataset (Gesch, 2007)

21. Surveying method	22. Land/ subaqueous elevation (m)	23. Depth top sample (m)	24. Depth base sample (m)	25. Depth midpoint sample (m)	26. Reference level	27. Reference level (m NAVD)	28. Elevation top sample (m NAVD)	29. Elevation base sample (m NAVD)	30. Elevation midpoint sample (m NAVD)
Total station; DGPS	2.00	10.25	10.28	10.27	NAVD 88	0.00	−8.25	−8.28	−8.27
MHW mark		1.22	1.40	1.31	MHW	0.29	−0.93	−1.12	−1.02
DEM (NED 1/3)	0.20			2.00	NAVD 88	0.00			−1.80
Read from cross-section	−2.0			10.93	MSL	0.13			−12.80

usage in a wide range of models. The number of decimals is determined by the resolution as provided by the original data source: 4, 3, and 2 decimals indicate that a sampling site can be located within ~10 m, ~100 m, and ~1 km, respectively. Original formats are converted wherever necessary and, if the position is only shown on a map, this map is georeferenced in a GIS program to obtain the coordinates. The original coordinates, the reference(s), and the sample name are also listed (Table 34.1, columns 5–8). Multiple references are sometimes needed if the details of the data are distributed over multiple publications (e.g., a journal article and a PhD thesis). Although a commonly used criterion for a SLIP or limiting data point is that its location should be known to within 1 km (Shennan, 1989), depending on the application (e.g., continental-scale GIA modeling) a lesser accuracy may be acceptable at the discretion of the user. Throughout Tables 34.1–34.4 we have distinguished between fixed variables (i.e., variables that do not require any interpretation, such as laboratory codes) and unfixed variables that involve interpretation (highlighted by shading).

34.3.2 Sediments

The next section of the database contains the dated material, the sample thickness, the sampling method, the dated facies, as well as the facies underlying and overlying the sample. The dated facies (Table 34.2, column 12) is important for three reasons: (1) it is often used to assign an indicative meaning; (2) it can be used to check whether the indicator has been reworked (e.g., a wood fragment within an estuarine facies); and (3) it is used as a criterion whether or not to decompress the thickness of the sample (Section 34.3.3). The dating technique (Table 34.4, column 31) can also play a role in estimating sample thickness (Table 34.2, columns 14–16). When the thickness is unknown, it can sometimes be estimated using the listed thickness of other samples from the same study. Otherwise, a literature review by Hu (2010) can be used to estimate the average thickness for AMS and radiometrically ^{14}C dated samples (0.1 and 0.5 m, respectively). For OSL dated samples we adopt an average value of 0.5 m.

The influence of compaction on sediments below a SLIP (including compaction of the sample itself) has long been recognized as a confounding factor in interpreting sea-level data (Jelgersma, 1961; Kaye and Barghoorn, 1964). As vertical displacement can be as much as several meters (Törnqvist et al., 2008; Horton and Shennan, 2009; Van Asselen et al., 2009), various models have been developed to correct for this (Chapter 30). However, corrections carry very large uncertainties that are hard to quantify. It is therefore not currently feasible to perform such corrections with a high degree of confidence. Instead, key stratigraphic information is documented where available, including the vertical distance of the base of the sample to the top of underlying compaction-free strata, the thickness of the overburden, and the nature of the underlying and overlying facies (Table 34.2, columns 17–20). These stratigraphic data within the database will possibly enable future quantitative corrections once more skillful compaction models become available. However, current understanding justifies estimating the compression of organic and mud-rich samples by increasing the sample thickness by a factor of 2.5 (following Van de Plassche et al., 2005; Van Asselen, 2011). It should be noted that Van de Plassche et al. (2005) also changed the elevation of the sample; here we only adjust sample thickness and use it as an error (see Section 34.4.2). This correction factor could be changed if new data become available or, again, at the discretion of the user. Samples taken at the land surface do not have to be decompressed. The depth to the consolidated substrate can also be used to subdivide the index points into basal and intercalated categories (not done in this chapter). This enables an initial assessment of the influence of compaction. Intercalated samples are derived from easily compressible organic sediment with clastic units above and below in the sedimentary column. Basal samples overlie a relatively incompressible substrate. The influence of sediment compaction for basal samples is therefore minor compared to index points from peat intercalated between thick Holocene clastic sediments (Jelgersma, 1961; Kaye and Barghoorn, 1964). Edwards (2006) and Horton et al. (2013) used linear regression of the residual between intercalated and basal index points to the readily available stratigraphic parameters to "decompact" the intercalated index points.

The availability of paleoenvironmental data (e.g., diatoms, foraminifera, pollen), if any, is also reported (Table 34.2, column 13). Over the past decade, $\delta^{13}C$ values of ^{14}C dated samples have increasingly been used to determine paleoenvironments

(Törnqvist et al., 2004; Mackie et al., 2005; Lamb et al., 2006; Engelhart et al., 2013; see also Chapter 23); $\delta^{13}C$ values are listed in the dating section of the database (Section 34.3.4).

34.3.3 Elevation

One of the most important attributes of a SLIP or a limiting data point is the elevation from which it was sampled, as well as the elevation of the RWL. The sources of error involved in obtaining these are often underappreciated, but it is crucial to account fully for them. In the ideal scenario the elevation of the sampled indicator is measured with high-precision leveling and tied to the nearest geodetic benchmark. This will give the smallest error, although the benchmark elevation error alone can be considerable (e.g., ~0.1 m for most of the US). In some cases however, published sea-level data lack precise elevation information or the data were related to sea level, a tidal datum, or vegetation zones at the time of sampling, leading to even larger errors that can only be assessed after careful analysis of the raw data. Samples taken offshore often have large elevation errors due to corrections for waves and tides and the errors associated with the instruments measuring water depths. The surveying method (Table 34.3, column 21) is therefore an essential piece of information, but frequently not described in journal articles. Depending on the surveying method used, an error is assigned (see Section 34.4.3); if the method remains unknown, a conservative error is introduced. Columns 22–25 in Table 34.3 describe the elevation of the land or subaqueous surface and the depth of the sample below this surface. The used reference level (Table 34.3, column 26) for elevation measurements varies widely (e.g., from local mean sea level or MSL to a geodetic datum). In Table 34.3, column 27 reference levels are converted – wherever necessary – to a geodetic datum (in this case, NAVD 88). One of the advantages of a geodetic/orthometric datum is that it remains constant for long periods of time and is therefore very suitable for archival purposes, whereas a tidal datum changes more frequently. To convert data that were related to tide levels we use the Vertical Datum (VDatum) Transformation tool of NOAA (2011; note that this tool is only valid for the US). With this information the elevation of the sample relative to a geodetic datum is calculated (Table 34.3, columns 28–30). In other places where similar tools are not available and where it is not possible to relate all data to a geodetic datum, the data should at least be related to the same tide level. The Admiralty Tide Tables produced by the United Kingdom Hydrographic Office (http://www.ukho.gov.uk/productsandservices/paperpublications/pages/nauticalpubs.aspx) provide tidal datums for all locations across the globe and are therefore very useful. It should be noted, however, that tidal ranges within estuaries and on saltmarshes are often substantially different from those on open coasts (see also Section 34.3.5).

34.3.4 Dating

The vast majority of published, post-LGM RSL data have been dated by means of ^{14}C (Chapter 23), ^{210}Pb, ^{137}Cs (Chapter 24) or U–Th (Chapter 26). In recent years, optically stimulated luminescence (OSL) dating (Chapter 27) has become more common. The sea-level database reports the 2σ age errors, rounded to the nearest 10 for ^{14}C and OSL dating (different rounding approaches may be more appropriate for other dating techniques). In the literature, OSL and U–Th ages are usually given with their 1σ analytical error so that value should be doubled. For ^{14}C analysis, dating accuracy is at least as important as analytical precision and depends on the nature of the dated material, reservoir effects, and uncertainties associated with isotopic fractionation correction (below and Chapter 23).

The laboratory code (Table 34.4, column 32) provides critical information about where and when the sample was dated. In some cases this is crucial as different laboratories use different protocols. A good example is the correction of ^{14}C ages for isotopic fractionation (based on $\delta^{13}C$ measurements). This had become a standard procedure at most laboratories by the late 1970s (Stuiver and Polach, 1977), but some laboratories have only applied this correction since the mid-1980s. With the laboratory code it can be verified with the laboratories, in principle, whether or not this correction has been applied. Uncorrected sea-level data have potentially large uncertainties as the fractionation effect can – if unaccounted for – lead to age offsets of hundreds of ^{14}C years. The necessary correction depends on the dated material and is detailed in Chapter 23. The $\delta^{13}C$ value of a sample (Table 34.4, column 37) gives important information about the amount of fractionation and, if available, should be listed. However,

Table 34.4. Structure of the sea-level database: dating

31. Dating technique	32. Lab. code	33. No. sub-samples	34. Age (^{14}C a BP)	35. Age error (^{14}C a)	36. Bulk error (^{14}C a)	37. $\delta^{13}C$ (‰)	38. Isotopic fractionation correction (^{14}C a)	39. Isotopic fractionation error (^{14}C a)	40. Corrected age (^{14}C a BP)	41. Corrected age error (^{14}C a)	42. ΔR (^{14}C a)	43. Reservoir age error (^{14}C a)	44. Calendar age 2σ older limit (cal a BP)	45. Calendar age 2σ younger limit (cal a BP)	46. Calendar age weighted mean (cal a BP)
^{14}C	UtC-11212		6490	60		−22.0			6490	60			7510	7270	7390
^{14}C	O		1725	105	100		+120	40	1845	150			2140	1410	1790
OSL	UNL-430												7600	3240	5420
^{14}C	MTC-13319		7241	90		−0.33			7241	90	95	300	8270	6980	7630

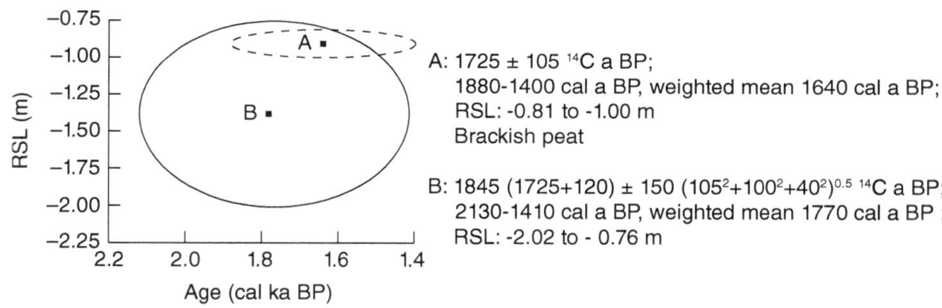

Fig. 34.2. Comparison of error ellipses obtained without (Ellipse A) and with (Ellipse B) a comprehensive error analysis. Ellipse A shows sample 2 (Table 34.1) of Coleman and Smith (1964) based on their reported ^{14}C age and vertical error, and the inferred RWL at MHW. Ellipse B shows the result for the same sample following the database protocol described in this chapter. Because it is a brackish peat, 120±40 ^{14}C a is added to correct for isotopic fractionation (Chapter 23). Furthermore, a bulk error of ±100 ^{14}C a is added. The vertical error increases significantly, because we determine RWL as the midpoint between HAT and MTL, and the uncertainties in the sample elevation (±0.5 m) and the IR (±0.23 m) are relatively large. Error calculations in both the vertical and horizontal direction are based on Equation 34.2. The black squares indicate the points with maximum probability density.

whenever this value is known the correction has almost certainly been carried out. In addition, due to the fact that isotopic fractionation correction was not adopted by all dating laboratories simultaneously, an error should be assigned to samples where it is unknown whether a correction was performed. This error (σ_I) is defined (see also Chapter 23):

$$\sigma_I = |\Delta t_i| + |\sigma_i| \quad (34.2)$$

where Δt_i represents the isotopic fractionation correction for the specific dated material (Table 34.4, column 38) and σ_i represents the associated fractionation correction error (Table 34.4, column 39; Fig. 34.2).

Table 34.4, columns 34 and 35 list the ^{14}C age and the associated analytical error. Sometimes multiple ^{14}C ages are obtained from the same layer (e.g., several dated macrofossils from a thin peat bed) that can be combined to a single ^{14}C age with a smaller error (the calculation of weighted mean ^{14}C ages is discussed in Chapter 23). The number of subsamples is listed in Table 34.4, column 33. Since the development of accelerator mass spectrometry (AMS), the potential for using small ^{14}C samples has increased. Several studies have demonstrated that for organic material (notably peat), AMS ^{14}C dating of plant macrofossils provides better constrained ages than bulk peat samples (Chapter 23) and it is now the preferred method. Based on a recent quantitative analysis (Hu, 2010), a "bulk error" of ±100 ^{14}C a (Table 34.4, column 36) is applied to bulk peat samples (Fig. 34.2); we adopt the same additional error for bulk shell samples. Note that this does not mean that all conventional ages receive this extra bulk error; if a single wood or shell fragment was ^{14}C dated conventionally such an additional error is not necessary.

Finally, ^{14}C dated samples can be subject to a reservoir effect (marine or estuarine; see Chapter 23). Values reported by the original authors on local reservoir effects should be used wherever available (unless new data have become available since publication). It is often underappreciated how variable reservoir effects can be on a regional scale, compared to the global average marine reservoir effect of 405±22 ^{14}C a (Reimer et al., 2009). This should be verified with the online database of Reimer and Reimer (2001) where end-member values are often estuarine rather than marine (and hence particularly relevant to sea-level studies). In estuaries, the reservoir effect is influenced by the level of mixing between marine waters and freshwater that may or may not be significantly depleted in ^{14}C (depending on the nature of source rocks). Also, reservoir effects have in many cases been subject to changes through time (McGregor et al., 2008; Yu et al., 2010). Table 34.4, column 42 lists the ΔR value which is the deviation of the local reservoir effect from the global average (e.g., a reservoir effect of 500 ^{14}C a corresponds to a ΔR of 95 ^{14}C a) and is often an input variable in calibration software. The error associated with the reservoir age (Table 34.4, column 43) can vary widely, depending on the available data.

For calibrating ^{14}C ages, several programs are available (Chapter 23). We prefer OxCal (Bronk

Table 34.5. Structure of the sea-level database: tidal parameters

47. Modern MSL (m NAVD)	48. Modern HAT (m NAVD)	49. Modern MHHW (m NAVD)	50. Modern MHW (m NAVD)	51. Modern MTL (m NAVD)	52. Modern MLW (m NAVD)	53. Modern MLLW (m NAVD)
0.14	0.57	0.32	0.16	0.13	0.12	−0.04
0.13	0.60	0.32	0.16	0.14	0.11	−0.07
0.11	0.56	0.31	0.13	0.09	0.09	−0.08
0.13	0.61	0.33	0.15	0.16	0.11	−0.11

Ramsey, 1995, 2009) as it provides weighted means for the calibrated age. Results from other dating techniques that do not require calibration (e.g., OSL) are listed in Table 34.4, columns 44–46. If a bulk or isotopic fractionation correction is required, a corrected ^{14}C age is calculated prior to calibration. Estuarine and marine carbonate ages are calibrated with the Marine13 curve (Reimer et al., 2013) using the appropriate ΔR value and error, while all other samples are calibrated with the IntCal13 curve (Reimer et al., 2013). Some organisms (e.g., algal mats) can take up both terrestrial and marine carbon and a mix of the Marine13 and IntCal13 curves can be used (e.g., Livsey and Simms, 2013). Many studies in the past did not correct marine carbonates for isotopic fractionation and reservoir effects; because they are of similar magnitude but opposite sign, they approximately cancel each other out. This results in a mean of the calibrated age range that may be accurate, but with a significantly underestimated error.

34.3.5 Tidal parameters

SLIPs are defined with respect to specified tidal datums (e.g., MTL or MSL). The traditional approach (e.g., Shennan, 1982) has been to use a nearby tide gauge record to quantify these datums before application to geological sea-level data. The limitation of this approach is twofold: (1) tide gauges are not always available near the site, meaning that distant or open coast data may have to be relied upon although the sample may have formed further inshore; and (2) tidal range may have changed significantly since the sample was formed, particularly during the middle and early Holocene (Van der Molen and De Swart, 2001; Uehara et al., 2006; Griffiths and Peltier, 2009; Hill et al., 2011). Although for many places around the globe tide gauge data remain the only option, (paleo)tidal modeling (Chapter 29) is an increasingly used and powerful tool. Here, we incorporate the tidal model of Hill et al. (2011). The model resolves the issue of distant tide gauge data as it provides tidal data for any given location off the US Gulf and Atlantic coastline for the past 10 cal ka at 1 cal ka intervals. This model (see Chapter 29 for an indepth description) can accurately reproduce modern observations from tide gauge records.

With this model we calculated tide levels for the modern situation (Table 34.5, columns 47–53) and for the appropriate paleo-time interval for each database entry, relative to a geodetic datum. Since we use modern data to calculate the RWL (see below), we do not list all paleotidal data in the table. Since many sea-level data are collected on land, we had to extrapolate the tidal data onshore (Fig. 34.3). We assume the calculation for 1000 cal a BP to be valid for the period 501–1500 cal a BP and follow this principle throughout, with two exceptions. First, since the modeled change in tidal range between 8000 and 9000 cal a BP is linked to the opening of the Hudson Strait between 8450 and 8300 cal a BP (Hill et al., 2011; Törnqvist and Hijma, 2012), we link all samples dated to 7501–8300 cal a BP to the model output for 8000 cal a BP and samples dated to 8301–9500 cal a BP to the output for 9000 cal a BP. Any future refinement or change in the timing of the opening of the Hudson Strait may require a modification to this exception. Furthermore, all samples dated older than 9501 cal a BP are linked to the 10,000 cal a BP output.

Paleotidal models predict tide levels using bathymetries based on GIA models, which should produce good approximations for open marine tides. However, modeling a large region requires compromise, most notably in under-resolving complex nearshore bathymetry and coastal features. The inland propagation of the tidal wave is therefore not modeled accurately. Also, the models do not account for sedimentation or erosion (Shennan and Horton, 2002), which will have a significant impact on the accuracy of tidal datums in nearshore

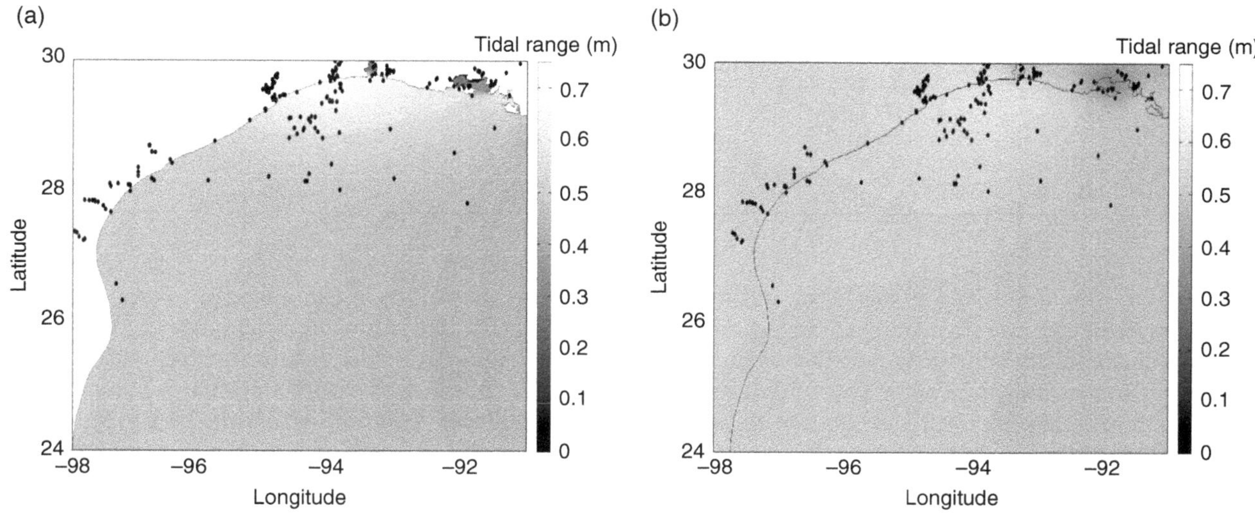

Fig. 34.3. (a) Modeled present-day tidal range (MHHW – MLLW) in the NW Gulf of Mexico. Black dots show the location of SLIPs and limiting data. (b) Tidal range for the entire block with extrapolated values for the onshore portions, enabling the calculation of tidal parameters for sampling sites on land. The offshore part of (b) does not match (a) exactly due to slightly different smoothing parameters. This was necessary because in the data for (a) there were some non-realistic values in regions where nodes are occasionally dry. For those nodes, traditional harmonic analysis is not feasible (see also Chapter 29). To prevent those nodes from influencing the extrapolation, some smoothing was performed on (a) before extrapolation; the panels are therefore not identical for the offshore area. For color details, please see Plate 46.

settings. For databases that cover large areas it is therefore best to use the tidal modeling results for the present-day situation to calculate RWL and IR when calculating paleo-RSL. We then compare the modern and the paleo-IR for each sample and incorporate the difference as an additional error (Section 34.4.1). This means that we restrict the application of paleotidal modeling to an error estimate rather than to adjust the elevation of paleo-RSL. This approach also avoids circularity problems: paleobathymetries used in paleotidal models are often solely based on GIA models that are calibrated and validated with RSL data.

In more local sea-level studies there are often better constraints on the paleogeography, and bathymetries can be estimated from local data rather than GIA models (e.g., Shennan et al., 2003; Hall et al., 2013). In those cases it is possible to fine-tune a tidal model to smaller areas (~100 km across) such that the results are reliable enough to use paleotidal data to correct the elevation of SLIPs and limiting data (Horton et al., 2013). In regions where similar models as that of Hill et al. (2011) exist or are being developed, the same methodology as used here can be applied. Where tidal models do not currently exist, tide gauge data must still be employed. The Admiralty Tide Tables provide this type of information. The tide gauge nearest to the sampling site is useful when it is located, for instance, within the same bay. In other cases the average of the nearest two (one on each side of the sampling site) should be used (e.g., Shennan, 1982), with half the difference as an error.

A substantial number of SLIPs formed inshore (e.g., small bays, marshes) and hence in environments where the tidal datums could have been substantially different (in most cases lower) than at nearshore sites, where large-scale tidal models give the best results and where most tide gauges are situated. This means that in most cases the tidal range error is overestimated (Hijma and Cohen, 2010). In deltas, the inland propagation of the tidal wave is also influenced by the river-gradient effect (Van de Plassche, 1982) that raises the elevation of the tidal datums (see also Chapter 33).

34.3.6 Indicative range, reference water level, and relative sea level

Based on all the paleoenvironmental and stratigraphic information (Table 34.2, columns 13–20) we label each data entry as either a SLIP or an upper or lower limiting data point (Table 34.6, column 54). For each SLIP the IR is calculated (Table 34.6, columns 55–57), based on the paleoenvironment in which the sea-level indicator formed (see also Fig. 34.1). Along the US Atlantic and

Table 34.6. Structure of the sea-level database: reference water level and indicative range

54. Classification	55. RWL	56. Modern RWL (m NAVD)	57. Modern IR (m)	58. Paleo-IR (m)
SLIP	(HAT+MTL)/2	0.35	0.44	0.57
SLIP	(HAT+MTL)/2	0.37	0.46	0.46
Upper limiting	MTL	0.09		
Lower limiting	MTL	0.16		

Gulf coasts, the majority of sea-level indicators formed within saltmarshes or brackish marshes. Within a saltmarsh it is usually possible to observe a clear vertical zonation of plants into high-marsh and low-marsh zones (and in some instances more) that reflect the preferences and tolerances of halophytic species to the frequency and duration of tidal inundation between MTL and HAT. If preserved plant macrofossils or microfossils suggest a high-saltmarsh environment, an IR of MHW to HAT is assigned. For low-saltmarsh environments, the IR applied is MTL to MHW.

As outlined above we use modeled present-day tidal data to calculate RWL (Table 34.6, columns 55–56) and incorporate the paleotidal model results into the error (Section 34.4.1). We do this based on a comparison of the modern and paleo-IR (Table 34.6, columns 57–58). The latter is calculated using the modeled paleotidal data. In studies that apply the highest possible standards to minimize other elevation errors associated with sampling and surveying, the IR typically ends up being the main contributor to the vertical uncertainty. With RWL and Equation 34.1 we calculate paleo-RSL for the SLIPs (Table 34.8, column 74). For limiting data we also use Equation 34.1, but in these cases we label the outcome as limiting-data elevation (Table 34.8, column 75).

34.4 VERTICAL ERRORS

So far, we have described the database structure in general terms, including the steps involved in calculating paleo-RSL or the elevation of limiting data. This section focuses specifically on vertical error calculations. Wherever a publication provides sufficiently specific information on such errors, that information trumps any standardized error estimates as outlined below. Note that not all listed errors apply in every case. For the summation of errors we use the commonly applied expression:

$$E_t = \sqrt{\left(e_1^2 + e_2^2 + \cdots + e_n^2\right)} \qquad (34.3)$$

where $e_1 \ldots e_n$ are component errors and E_t is the total error. This is the same equation as that used for the ^{14}C age error calculation (Chapter 23).

While the 2σ age error is statistically rigorous, the confidence interval associated with the vertical error is less well defined. Nevertheless, we aim for a reasonable approximation of a 2σ error by assessing all vertical errors very conservatively. Total upward and downward errors are computed separately (Table 34.8, columns 76, 77), because the non-vertical drilling error (Table 34.8, column 67) is unidirectional.

To plot SLIPs we prefer to use shapes that approximate ellipses (also referred to as "blobs"; Chapter 32) defined by the 2σ horizontal (age) error range and the 2σ vertical (elevation) error range (Fig. 34.2). Ellipses represent the 2σ error of SLIPs better than the frequently used error boxes. While the SLIPs illustrated here (Figs 34.2 and 34.4b) are represented by ellipses, it should be noted that the point of maximum probability density is commonly not located exactly in their centers (both the horizontal and vertical errors are often asymmetric). In other words, a more accurate geometric representation that accounts for these (often slight) asymmetries would constitute a further improvement.

For limiting data points, several symbols have been used in the literature (e.g., boxes, crosses, T-shaped symbols). Here we use T-shaped symbols with the horizontal bar placed at the extreme end of the vertical 2σ error range. For upper limiting data points the horizontal bar lies at the top, while for lower limiting data points the horizontal bar lies at the base of this range. The width of the horizontal bar is defined by the horizontal 2σ error range. It is important to bear in mind that, in the case of an upper limiting data point, MSL is interpreted to have been below the horizontal line merely at some point along its full width. If information about the length of the vertical error range is to be included, a T-shape should be used for upper limiting data (Fig. 34.4b), while a ⊥-shape should be used for lower limiting data. The length

Table 34.7. Structure of the sea-level database: vertical errors

59. IR error (m)	60. RWL modeling error (m)	61. IR modeling error (m)	62. IR change error (m)	63. VDatum error (m)	64. Sample thickness error (m)	65. Sampling error (m)	66. Core shortening/ stretching error (m)	67. Non-vertical drilling error (m)	68. Tidal error (m)	69. Vegetation zone error (m)	70. DEM/ map/water depth error (m)	71. Leveling error (m)	72. DGPS error (m)	73. Benchmark error (m)
0.22	0.10	0.09	0.07		0.04	0.01	0.05	0.21				0.03	0.04	0.10
0.23	0.10	0.09	0.00	0.05	0.23	0.01	0.15	0.03		0.50				
	0.10				0.10	0.01	0.15	0.04			0.50			
	0.10			0.05	0.01	0.01	0.15	0.22	0.22		0.50			

Table 34.8. Structure of the sea-level database: paleo-RSL calculation

74. Paleo-RSL (m)	75. Limiting-data elevation (m)	76. Total upward error (m)	77. Total downward error (m)
−8.61		0.36	0.29
−1.39		0.63	0.63
	−1.89	0.54	0.54
	−12.96	0.62	0.58

of the vertical bar is then defined by the vertical error range and points (by means of an arrow) in the direction of paleo-RSL; its position is defined by the weighted mean calendar age. Other studies (e.g., Engelhart and Horton, 2012) have used a T-shape for lower limiting data and a ⊥-shape for upper limiting data to avoid confusion with coral-based RSL reconstruction usage (e.g., Peltier and Fairbanks, 2006) where the vertical bar is used to define the living range of coral species and with the RSL curve expected to pass through the bar. The position of the horizontal bar in their analysis is similar as in the present protocol, but the length of the vertical bar is not defined by the magnitude of the vertical error.

34.4.1 Indicative meaning related errors

In the following sections, all errors related to specific parts of the database are discussed using numbered lists.

(1) The IR error is defined as half the IR (Table 34.7, column 59).
(2) The error associated with the modeled RWL is based on a comparison of predictions with the model of Hill et al. (2011) for 40 tide gauges from Texas to Maine using MTL to mean higher high water (MHHW) as the IR (HAT was not always available) and the measured values at those stations. The average difference between the modeled and measured RWL is 0.10 m; we therefore adopt an error of ±0.10 m for the RWL (Fig. 34.5; Table 34.7, column 60). Since the average difference between the modeled and measured IR is 20% of the modern IR, we use this to calculate the IR error (Table 34.7, column 61). In other regions where other models may be used, these errors can be substantially different.
(3) The IR change error accounts for the difference between the modern IR and the paleo-IR. It is defined as half the difference between the modern IR and the paleo-IR (Table 34.7, column 62) and also applies to cases where the paleo-IR is smaller than the modern IR.
(4) For sites that were not linked to a geodetic datum by the original authors, we converted the elevation using VDatum (NOAA, 2011). The average difference between VDatum computations for MSL relative to NAVD 88 at tide gauges and the actual data from those stations is ±0.05 m (Fig. 34.6; Table 34.7, column 63; see also http://vdatum.noaa.gov/docs/est_uncertainties.html). VDatum can also be used to convert elevations that were related to the older National Geodetic Vertical Datum of 1929. In other regions a different approach using Admiralty Tide Tables or local data will change this error.

34.4.2 Sampling related errors

(1) The thickness error (Table 34.7, column 64) is defined as half the thickness of the sample (Shennan, 1986). If the sample was decompressed (Section 34.3.3) we calculate this error using the corrected sample thickness (Table 34.2, column 17).
(2) Sampling errors (Table 34.7, column 65) arise from measuring the sample depth within a core or section. They are often not reported, but can be considered small. We use ±0.01 m, following Shennan (1986).
(3) Core stretching/shortening errors (Table 34.7, column 66) can be quite substantial, especially for rotary coring and vibracoring. According to Morton and White (1997), up to 30% of core shortening can occur within a sampling interval of less than 1 m; on the other hand, alternating unshortened and shortened sampling intervals are commonly observed. We set the error at ±0.15 m for rotary coring and vibracoring (Morton and White, 1997), ±0.05 m for hand coring (Woodroffe, 2006) and ±0.01 m for a Russian sampler (Woodroffe, 2006).
(4) Non-vertical drilling makes the apparent depth of a sample larger than the true depth, with discrepancies increasing with drilling depth. According to Törnqvist et al. (2004), the resulting error may be as large as −0.02 m per meter depth of hand coring. Assuming this error is similar for other coring methods, we assign an error of 0.02 m m^{-1} depth (Table 34.7,

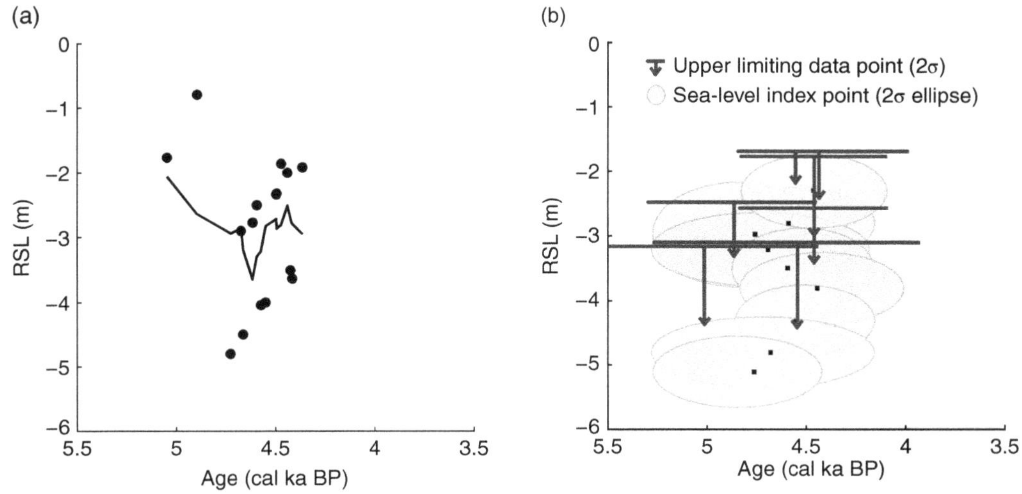

Fig. 34.4. Comparison of the analysis of the same dataset by (a) Balsillie and Donoghue (2004) and (b) the protocol described in the present chapter. Differences are apparent, but the major limitation in the Balsillie and Donoghue analysis is to not include any errors. The period 5500–4000 cal a BP was chosen because it is the only portion where the used data (Spackman et al., 1966; Scholl and Stuiver, 1967; Davies, 1980; Robbin, 1984) are from a well-defined region (southern Florida); using data from the entire US Gulf Coast in one curve would be inappropriate given spatially non-uniform vertical land motions. The horizontal bars of the limiting data points are positioned at the top of the vertical 2σ interval. The black squares in (b) indicate the points with maximum probability density.

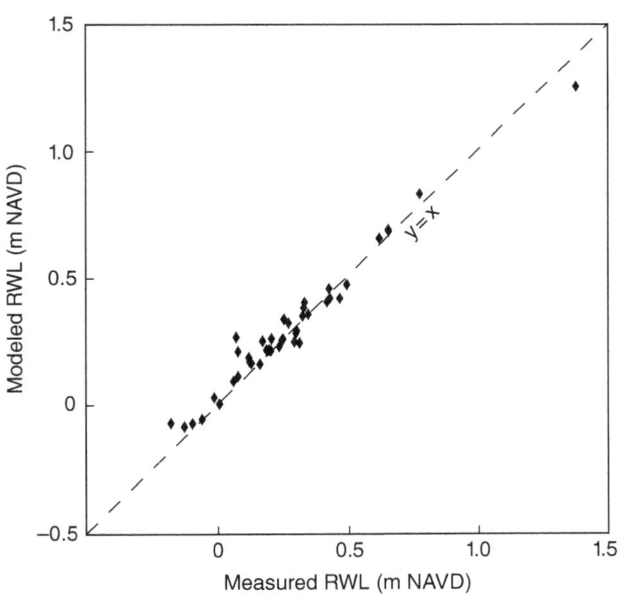

Fig. 34.5. Difference between the measured and modeled values of RWL (MHHW+MTL)/2 relative to NAVD 88 at tide gauges for the modern situation. The average difference is 0.10 m. The tide gauges are located along the US Gulf and Atlantic coasts (from Corpus Christi, Texas to Eastport, Maine).

column 67). This error is unidirectional and makes the upward portion of the total vertical error larger than the downward portion.

34.4.3 Elevation measurement related errors

(1) A tidal error (Table 34.7, column 68) has to be assigned for samples taken offshore with reference to the water surface. As the sampling could have occurred anytime during a tidal cycle, a tidal error of half the tidal range is assigned (Shennan, 1989). This tidal range can be modeled or based on data from nearby tide gauges.

(2) In some studies land-surface elevations were estimated using vegetation zones (e.g., Redfield, 1972; Goodbred et al., 1998; Wright et al., 2005) resulting in elevation errors

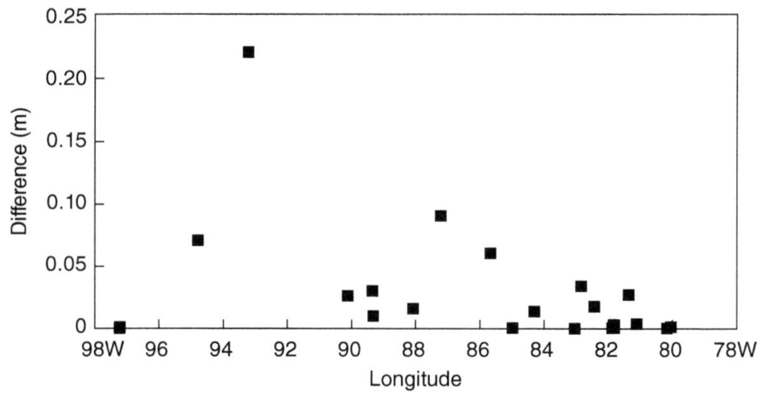

Fig. 34.6. Difference between the measured value of MSL at tide gauges (relative to NAVD 88) and the modeled value by Vdatum. The tide gauges are located along the US Gulf Coast, between Corpus Christi (Texas) and Lake Worth Pier (Florida). The outlier (0.22 m) is the tide gauge at Lake Charles, Louisiana, located 50 km inland of the open coast.

(Table 34.7, column 69) of ±0.5 m. In areas with a high tidal range this error could be larger (Engelhart, 2010).

(3) The elevation of the land surface is frequently not reported but, if the coordinates of the sampling site are known, the elevation can be approximated using high-resolution digital elevation models (DEMs) wherever available. The associated error does not only depend on the vertical accuracy of the DEM (often of the order 0.1–0.2 m), but also on the accuracy of the coordinates. In areas with significant relief, such as strand plains with beach ridges, this matters when the sample was taken from a ridge, but the coordinate points towards a swale. An error of ±0.5 m is assigned (Table 34.7, column 70). If conventional topographic maps are used instead of DEMs, the error is half the contour-line interval (±0.75 m for US topographic maps).
(4) Water depths are typically measured using fathometers, echo sounders, and poles, along with several other techniques. If not specified by the authors, we use an error of ±0.5 m (Table 34.7, column 70).
(5) The error arising during leveling (Table 34.7, column 71) with high-precision (e.g., total station) equipment is, if unknown, set at ±0.03 m (Törnqvist et al., 2004).
(6) The standard error (Table 34.7, column 72) of a (D)GPS base station that is often used during surveying is ±0.1 m, but if carefully used the error can be as low as 0.04 m (Törnqvist et al., 2004). Care must be taken to ensure the accuracy of the D(GPS), especially in regions with variable satellite coverage or restricted reception.
(7) The US National Geodetic Survey (NGS) benchmark elevations are subject to an error (Table 34.7, column 73) of ±0.1 m (Engelhart, 2010) for the most reliable and stable benchmarks. For less reliable benchmarks this error is larger. Note that the error for benchmark elevations can be different in other regions.

Again, we stress here that the standardized errors should only be used if no sufficient information was provided in the original source. If specific information is published or can be obtained by other means (e.g., directly from the author(s)), that information overrules the standardized errors discussed above. This means that all errors should be assessed on a case-by-case basis.

34.5 DISCUSSION

While a long history exists of compiling geological sea-level databases for a variety of applications, relatively few have involved a comprehensive and rigorous assessment of elevation and age errors associated with the underlying data. Here we present the most comprehensive approach to sea-level database construction to date, with a number of error sources (notably associated with dating errors and tidal range uncertainties) that have previously not been considered. Our approach aims to be as inclusive as possible, and we only exclude data that, according to our analysis, have no significance whatsoever with respect to past sea levels. The spirit of our approach is that any data with limited usefulness will be reflected by large vertical and/or horizontal errors. Rather than passing judgment on these data ourselves, such an assessment is then made at the discretion of the user.

By means of illustration, we compare the protocol described above with a sea-level data analysis by Balsillie and Donoghue (2004). The latter authors compiled data (see also Donoghue, 2011) from the US Gulf Coast (separated into offshore and onshore data) and used a seven-point moving average trend line to obtain a RSL curve (Fig. 34.4a) without taking into account any errors or separating between SLIPs and limiting data points. We compared our approach with theirs for the period 5500–4000 cal a BP. Using the protocol described above, it becomes apparent that several of the used indicators are freshwater peats (based on the original descriptions and/or vegetation type) and can therefore only be used as upper limiting data. The vertical 2σ errors of all data are of the order 1 m, precluding the detection of decimeter-scale sea-level fluctuations as suggested by the trend line of Balsillie and Donoghue. Where Balsillie and Donoghue interpret a RSL drop of ~1.75 m between 5000 and 4200 cal a BP, followed by a very rapid RSL rise, our analysis shows that many more interpretations are possible: sea level was somewhere between –6 m and –2 m from 5500 to 3700 cal a BP, most likely with a net RSL rise (Fig. 34.4b). The hallmark of our approach is that assigning conservative errors minimizes the risk of over-interpreting RSL records.

As outlined above, insight into the role of changes in the paleotidal regime is crucial when interpreting sea-level data. Although we use a high-resolution tidal model, we still choose to incorporate any differences between the modern and paleotidal regime as an error instead of adjusting RWL. The main reason for this is that the model covers a very wide region and does not incorporate paleogeographic changes on a local scale which are critical for understanding a local tidal regime. Paleotidal models that are fine-tuned to take into account paleogeographic changes can be used to adjust RWL to get more insight into local RSL changes (e.g., Shennan et al., 2003; Hall et al., 2013).

Finally, when our database protocol is used to analyze published data, newly calculated errors will differ from those in the original publications; in most cases, they will be larger (Fig. 34.2). This could lead to new insights such as revised rates of RSL change, and hence different conclusions. This obviously does not mean that previous studies were executed poorly; it is merely a step forward in the error analysis of sea-level data. Our goal is that the protocol presented here be adopted more widely, and a start is made with the analysis of the wealth of existing geological sea-level data in a uniform way. In order to be widely used, it will become necessary for the database to be stored online. The longer-term goal is for researchers to have the opportunity to directly upload their sea-level data, analyzed using the protocol described here, to this database.

REFERENCES

Balsillie, J.H., and Donoghue, J.F. (2004) High resolution sea-level history for the Gulf of Mexico since the last glacial maximum. Report of investigation. Florida Geological Survey, Tallahassee.

Blum, M.D., Sivers, A.E., Zayac, T., and Goble, R.J. (2003) Middle Holocene sea-level and evolution of the Gulf of Mexico Coast. Gulf Coast Association of Geological Societies, Transactions, 53, 64–77.

Bronk Ramsey, C. (1995) Radiocarbon calibration and analysis of stratigraphy: The OxCal program. Radiocarbon, 37(2), 425–430.

Bronk Ramsey, C. (2009) Bayesian analysis of radiocarbon dates. Radiocarbon, 51(1), 337–360.

Coleman, J.M., and Smith, W.G. (1964) Late recent rise of sea level. Geological Society of America Bulletin, 75(9), 833–840.

Davies, T.D. (1980) Peat formation in Florida Bay and its significance in interpreting the recent vegetational and geological history of the bay area. PhD thesis, Pennsylvania State University.

Donoghue, J.F. (2011) Sea level history of the northern Gulf of Mexico coast and sea level rise scenarios for the near future. Climatic Change, 107(1–2), 17–33.

Dutton, A., and Lambeck, K. (2012) Ice volume and sea level during the last interglacial. Science, 337(6091), 216–219.

Edwards, R.J. (2006) Mid-to late-Holocene relative sea-level change in southwest Britain and the influence of sediment compaction. The Holocene, 16(4), 575–587.

Engelhart, S.E. (2010) Sea-level changes along the U.S. Atlantic coast: implications for glacial isostatic adjustment models and current rates of sea-level change. PhD thesis, University of Pennsylvania.

Engelhart, S.E., and Horton, B.P. (2012) Holocene sea level database for the Atlantic coast of the United States. Quaternary Science Reviews, 54, 12–25.

Engelhart, S.E., Horton, B.P., Douglas, B.C., Peltier, W.R., and Törnqvist, T.E. (2009) Spatial variability of late Holocene and 20th century sea-level rise along the Atlantic coast of the United States. Geology, 37(12), 1115–1118.

Engelhart, S.E., Horton, B.P., Vane, C.H., Nelson, A.R., Witter, R.C., Brody, S.R., and Hawkes, A.D. (2013) Modern foraminifera, δ13C, and bulk geochemistry of central Oregon tidal marshes and their application in

paleoseismology. Palaeogeography, Palaeoclimatology, Palaeoecology, 377, 13–27.
Flemming, N.C. (1982) Multiple regression analysis of earth movements and eustatic sea-level change in the United Kingdom in the past 9000 years. Proceedings of the Geologists' Association, 93(1), 113–125.
Gesch, D. (2007) The National Elevation Dataset. In: *Digital Elevation Model Technologies and Applications: The DEM User Manual*, 2nd edition (ed. Maune, D.), American Society for Photogrammetry and Remote Sensing, Bethesda, Maryland, USA, 99–118.
Goodbred, S.L., Wright, E.E., and Hine, A.C. (1998) Sea-level change and storm-surge deposition in a late Holocene Florida salt marsh. Journal of Sedimentary Research, 68(2), 240–252.
Griffiths, S.D., and Peltier, W.R. (2009) Modeling of polar ocean tides at the last glacial maximum: amplification, sensitivity, and climatological implications. Journal of Climate, 22(11), 2905–2924.
Hall, G.F., Hill, D.F., Horton, B.P., Engelhart, S.E., and Peltier, W.R. (2013) A high-resolution study of tides in the Delaware Bay: Past conditions and future scenarios. Geophysical Research Letters, 40(2), 338–342.
Hijma, M.P., and Cohen, K.M. (2010) Timing and magnitude of the sea-level jump preluding the 8200 yr event. Geology, 38(3), 275–278.
Hill, D.F., Griffiths, S.D., Peltier, W.R., Horton, B.P., and Törnqvist, T.E. (2011) High-resolution numerical modeling of tides in the western Atlantic, Gulf of Mexico, and Caribbean Sea during the Holocene. Journal of Geophysical Research, 116, C10014.
Horton, B.P., and Shennan, I. (2009) Compaction of Holocene strata and the implications for relative sea-level change on the east coast of England. Geology, 37(12), 1083–1086.
Horton, B.P., Engelhart, S.E., Hill, D.F., Kemp, A.C., Nikitina, D., Miller, K.G., and Peltier, W.R. (2013) Influence of tidal-range change and sediment compaction on Holocene relative sea-level change in New Jersey, USA. Journal of Quaternary Science, 28(4), 403–411.
Hu, P. (2010) Developing a quality-controlled postglacial sea-level database for coastal Louisiana to assess conflicting hypotheses of Gulf Coast sea-level change. MSc thesis, Tulane University, New Orleans.
Jelgersma, S. (1961) Holocene sea-level changes in The Netherlands. Mededelingen Geologische Stichting, 7, 1–101.
Kaye, C.A., and Barghoorn, E.S. (1964) Late Quaternary sea-level change and crustal rise at Boston, Massachusetts, with notes on the autocompaction of peat. Geological Society of America Bulletin, 75(2), 63–80.
Lamb, A.L., Wilson, G.P., and Leng, M.J. (2006) A review of coastal palaeoclimate and relative sea-level reconstructions using δ13C and C/N ratios in organic material. Earth-Science Reviews, 75(1–4), 29–57.
Lambeck, K., Smither, S., and Johnston, P. (1998) Sea-level change, glacial rebound and mantle viscosity for Northern Europe. Geophysical Journal International, 134, 102–144.
Lambeck, K., Anzidei, M., Antonioli, F., Benini, A., and Esposito, A. (2004) Sea level in Roman time in the Central Mediterranean and implications for recent change. Earth and Planetary Science Letters, 224(3–4), 563–575.
LaSalle, M.W., and Parsons, J. (1985) Species profiles: life histories and environmental requirements of coastal fishes and invertebrates (Gulf of Mexico). Common Rangia, 82, 1–18.
Lighty, R.G., Macintyre, I.G., and Stuckenrath, R. (1982) Acropora palmata reef framework: A reliable indicator of sea level in the western atlantic for the past 10,000 years. Coral Reefs, 1(2), 125–130.
Livsey, D., and Simms, A.R. (2013) Holocene sea-level change derived from microbial mats. Geology, doi: 10.1130/G34387.1.
Mackie, E.A.V., Leng, M.J., Lloyd, J.M., and Arrowsmith, C. (2005) Bulk organic δ13C and C/N ratios as palaeosalinity indicators within a Scottish isolation basin. Journal of Quaternary Science, 20(4), 303–312.
McGregor, H.V., Gagan, M.K., McCulloch, M.T., Hodge, E., and Mortimer, G. (2008) Mid-Holocene variability in the marine 14C reservoir age for northern coastal Papua New Guinea. Quaternary Geochronology, 3(3), 213–225.
Milliken, K.T., Anderson, J.B., and Rodriguez, A.B. (2008) A new composite Holocene sea-level curve for the northern Gulf of Mexico. In: *Response of Upper Gulf Coast Estuaries to Holocene Climate Change and Sea-Level Rise* (eds Anderson, J.B., and Rodriguez, A.B.), Geological Society of America, Special Paper 443, 1–11.
Milne, G., Long, A., and Bassett, S. (2005) Modelling Holocene relative sea-level observations from the Caribbean and South America. Quaternary Science Reviews, 24, 1183–1202.
Morton, R.A., and White, W.A. (1997) Characteristics of and corrections for core shortening in unconsolidated sediments. Journal of Coastal Research, 13(3), 761–769.
NOAA (2011) Vertical Datum Transformation. Available at http://vdatum.noaa.gov/welcome.html (accessed 4 August 2014).
Peltier, W.R., and Fairbanks, R.G. (2006) Global glacial ice volume and Last Glacial Maximum duration from an extended Barbados sea level record. Quaternary Science Reviews, 25(23–24), 3322–3337.
Peltier, W.R., Shennan, I., Drummond, R., and Horton, B. (2002) On the postglacial isostatic adjustment of the British Isles and the shallow viscoelastic structure of the Earth. Geophysical Journal International, 148, 443–475.
Redfield, A.C. (1972) Development of a New England salt marsh. Ecological Monographs, 42, 201–201.
Reimer, P.J., and Reimer, R.W. (2001) A marine reservoir correction database and on-line interface. Radiocarbon, 43, 461–463.
Reimer, P.J., Baillie, M.G.L., Bard, E. et al. (2009) IntCal09 and Marine09 radiocarbon age calibration curves, 0-50,000 years cal BP. Radiocarbon, 51(4), 1111–1150.
Reimer, P.J., Bard, E., Bayliss, A. et al. (2013) IntCal13 and Marine13 radiocarbon age calibration curves 0-50,000 yr cal BP. Radiocarbon, 55(4), 1869–1887.
Robbin, D.M. (1984) A new Holocene sea level curve for the Upper Florida Keys and Florida reef tract. In: *Environments of South Florida: Present and Past II.* (ed.

Gleason, P.J.), Coral Gables, Miami Geological Society, 437–458.

Scholl, D.W., and Stuiver, M. (1967) Recent submergence of Southern Florida: A comparison with adjacent coasts and other eustatic data. Geological Society of America Bulletin, 78(4), 437–454.

Shennan, I. (1982) Interpretation of Flandrian sea-level data from the Fenland, England. Proceedings of the Geologists' Association, 93(1), 53–63.

Shennan, I. (1986) Flandrian sea-level changes in the Fenland. II: Tendencies of sea-level movement, altitudinal changes, and local and regional factors. Journal of Quaternary Science, 1, 155–179.

Shennan, I. (1989) Holocene crustal movements and sea-level changes in Great Britain. Journal of Quaternary Science, 4(1), 77–89.

Shennan, I., and Horton, B. (2002) Holocene land- and sea-level changes in Great Britain. Journal of Quaternary Science, 17(5–6), 511–526.

Shennan, I., Lambeck, K., Horton, B., Innes, J., Lloyd, J., McArthur, J., and Rutherford, M. (2000) Holocene isostasy and relative sea-level changes on the east coast of England. In: Holocene Land-Ocean Interaction and Environmental Change around the North Sea (eds Shennan, I. and Andrews, J.), Geological Society, London, Special Publications 166(1), 275–298.

Shennan, I., Coulthard, T., Flather, R., Horton, B., Macklin, M., Rees, J., and Wright, M. (2003) Integration of shelf evolution and river basin models to simulate Holocene sediment dynamics of the Humber Estuary during periods of sea-level change and variations in catchment sediment supply. Science of the Total Environment, 314–316, 737–754.

Spackman, W., Dolsen, C.P., and Riegel, W. (1966) Phytogenic organic sediments and sedimentary environments in the Everglades-Mangrove Complex. Part I: Evidence of a transgressing sea and its effects on environments of the shark river area of southwestern Florida. Palaeontographica Abteilung B, 117(4–6), 135–152.

Stuiver, M., and Polach, H. (1977) Discussion; reporting of C-14 data. Radiocarbon, 19(3), 355–363.

Törnqvist, T.E., and Hijma, M.P. (2012) Links between early Holocene ice-sheet decay, sea-level rise and abrupt climate change. Nature Geoscience, 5, 601–606.

Törnqvist, T.E., González, J.L., Newsom, L.A., Van der Borg, K., De Jong, A.F.M., and Kurnik, C.W. (2004) Deciphering Holocene sea-level history on the US Gulf Coast: A high-resolution record from the Mississippi Delta. Geological Society of America Bulletin, 116(7), 1026–1039.

Törnqvist, T.E., Wallace, D.J., Storms, J.E.A., Wallinga, J., van Dam, R.L., Blaauw, M., Derksen, M.S., Klerks, C.J.W., Meijneken, C., and Snijders, E.M.A. (2008) Mississippi Delta subsidence primarily caused by compaction of Holocene strata. Nature Geoscience, 1(3), 173–176.

Toscano, M.A., and Macintyre, I.G. (2003) Corrected western Atlantic sea-level curve for the last 11,000 years based on calibrated Acropora Palmata framework and intertidal mangrove peat. Coral Reefs, 22, 257–270.

Troiani, B.T., Simms, A.R., Dellapenna, T., Piper, E., and Yokoyama, Y. (2011) The importance of sea-level and climate change, including changing wind energy, on the evolution of a coastal estuary: Copano Bay, Texas. Marine Geology, 280(1–4), 1–19.

Uehara, K., Scourse, J.D., Horsburgh, K.J., Lambeck, K., and Purcell, A.P. (2006) Tidal evolution of the northwest European shelf areas from the Last Glacial Maximum to the present. Journal of Geophysical Research, 111, C09025–C09025.

Van Asselen, S. (2011) The contribution of peat compaction to total basin subsidence: implications for the provision of accommodation space in organic-rich deltas. Basin Research, 23, 239–255.

Van Asselen, S., Stouthamer, E., and Van Asch, T.W.J. (2009) Effects of peat compaction on delta evolution: A review on processes, responses, measuring and modeling. Earth-Science Reviews, 92(1–2), 35–51.

Van de Plassche, O. (1982) Sea-level change and water-level movements in the Netherlands during the Holocene. Mededelingen Rijks Geologische Dienst, 36, 1–93.

Van de Plassche, O. (1986) Introduction. In: *Sea-Level Research: A Manual for the Collection and Evaluation of Data.* (ed. Van de Plassche, O.) Geobooks, Norwich, 1–26.

Van de Plassche, O. (1991) Late Holocene sea-level fluctuations on the shore of Connecticut Inferred from transgressive and regressive overlap boundaries in salt-marsh deposits. Journal of Coastal Research, 11, 159–179.

Van de Plassche, O., Bohncke, S.J.P., Makaske, B., and Van der Plicht, J. (2005) Water-level changes in the Flevo area, central Netherlands (5300-1500 BC): implications for relative mean sea-level rise in the Western Netherlands. Quaternary International, 133–134, 77–93.

Van der Molen, J., and De Swart, H.E. (2001) Holocene tidal conditions and tide-induced sand transport in the southern North Sea. Journal of Geophysical Research, C, 106, C5, 9339–9362.

Vink, A., Steffen, H., Reinhardt, L., and Kaufmann, G. (2007) Holocene relative sea-level change, isostatic subsidence and the radial viscosity structure of the mantle of northwest Europe (Belgium, the Netherlands, Germany, southern North Sea) Quaternary Science Reviews, 26(25–28), 3249–3275.

Woodroffe, S.A. (2006) Holocene relative sea-level changes in Cleveland Bay, North Queensland, Australia. PhD thesis, Durham University, Durham, UK.

Wright, E.E., Hine, A.C., Goodbred, S.L., and Locker, S.D. (2005) The effect of sea-level and climate change on the development of a mixed siliciclastic-carbonate, deltaic coastline: Suwannee River, Florida, USA. Journal of Sedimentary Research, 75(4), 621–635.

Yu, K., Hua, Q., Zhao, J.-X., Hodge, E., Fink, D., and Barbetti, M. (2010) Holocene marine 14C reservoir age variability: Evidence from 230Th-dated corals in the South China Sea. Paleoceanography, 25(3), PA3205.

Yu, S.-Y., Törnqvist, T.E., and Hu, P. (2012) Quantifying Holocene lithospheric subsidence rates underneath the Mississippi Delta. Earth and Planetary Science Letters, 331–332, 21–30.

Part 5

Direct Measurements

Chapter 35
Sea-level measurements from tide gauges

PHILIP L. WOODWORTH[1], DAVID T. PUGH[1], AND ANDREW J. PLATER[2]

[1]*National Oceanography Centre, Liverpool, UK*

[2]*Department of Geography and Planning, School of Environmental Sciences, University of Liverpool, Liverpool, UK*

35.1 INTRODUCTION

This chapter describes some aspects of the measurement of sea level using tide gauges. By "sea level" we mean "still water level", the level of the sea surface which occurs due to processes in the ocean such as the tides, storm surges, tsunamis, or mean sea-level changes, but not the high-frequency changes due to wind waves. What is referred to as "sea level" in the UK is usually called "water level" in the USA. By "tide gauge" we mean "sea-level recorder", as the instrument measures changes in level due to all processes and not just the tide. However, the traditional name has been retained.

35.2 THE FIRST TIDE GAUGES

Modern tide gauges can be said to have originated from the proposal of Moray (1665) to use a stilling well to measure sea-level changes. A stilling well is a vertical pipe in which a float moves up and down as the water level inside it changes (Fig. 35.1). The bottom of the pipe has a hole which allows seawater to flow in and out so that the level inside the pipe is, in principle, the same as that outside. However, the hole provides an important function in acting as a low-pass filter and in dampening (stilling) the energetic fluctuations in sea level due to wind waves. In the first stilling wells, one end of a vertical rod would be fixed to the top of the float, with the rod long enough such that its other end was visible above the top of the well, enabling its height to be measured on a vertical graduated scale. In this way, water level values could be obtained from visual measurements.

Stilling-well gauges were used at many ports during the 18th and 19th centuries, with the measured heights and times of high and low waters recorded in the port ledgers. However, the full tidal curve was never (to our knowledge) measured this way, and there seems to have been little scientific interest in the full curve until much later. Captains of the Royal Navy had standing instructions to install temporary stilling wells whenever they could, in order to learn as much as possible about the tide around the world (Herschel, 1849).

A second major development took place in the 1830s with the construction of the first automatic (self-registering) tide gauge. In these devices the rod was replaced by a wire and a system of pulleys and counterweights, with the other end of the wire attached to the pen of a chart recorder so that the full curve of the rise and fall of the tide could be represented on a rotating drum of paper. The resulting charts could be inspected subsequently to provide values of sea level at regular intervals (e.g., hourly), as well as at the high and low "turning points". Methods for calibration of a gauge enabled observed sea levels to be related to the heights of benchmarks on the nearby land (such observed sea level is also known as "relative sea level" in geodesy and oceanography i.e., relative to land level).

The first such tide gauge is often credited to the installation at Sheerness, UK (Palmer, 1831), although the same ideas were being pursued at many places at around the same time (Matthaus, 1972; Woodworth, 2014). The first self-registering tide gauge in the United States was made by Joseph Saxton for the US Coast Survey in 1851 (Fig. 35.2). By the end of the 19th century, similar equipment had been installed in major ports around the world for many practical purposes such as the determination of port and national datums for surveying purposes, and for the acquisition of sufficient data to enable tidal constants to

Handbook of Sea-Level Research, First Edition. Edited by Ian Shennan, Antony J. Long, and Benjamin P. Horton.

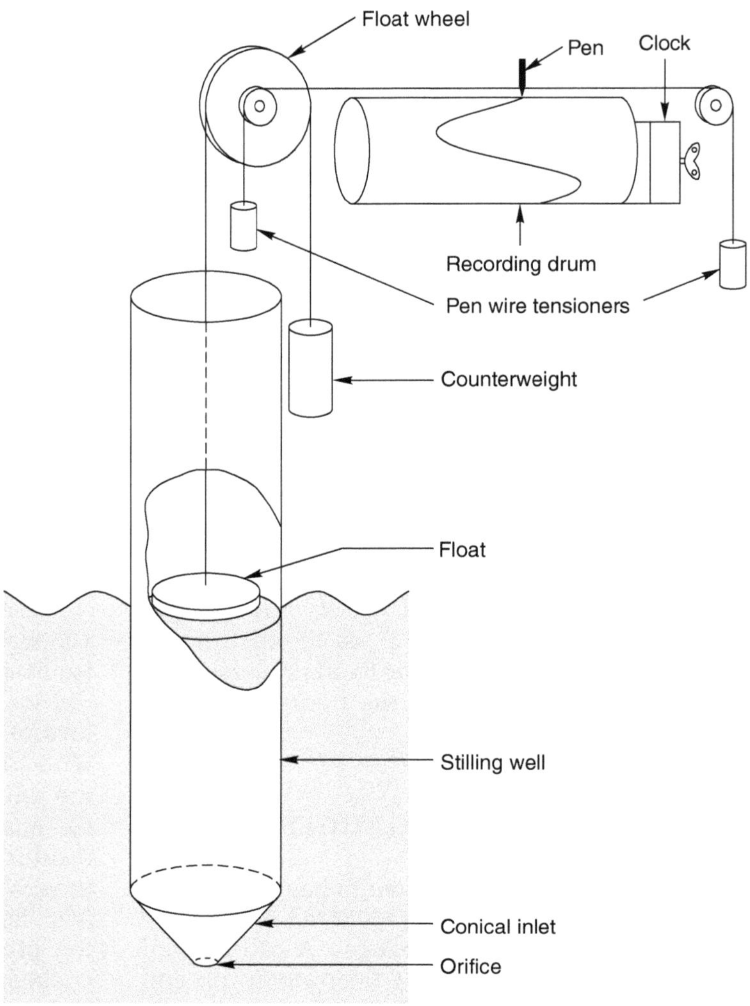

Fig. 35.1. A traditional stilling-well tide gauge with a float and chart recorder. The function of the hole in the conical inlet is to filter the higher-frequency motions due to waves. The well diameter is typically 0.3–0.5 m with a cone orifice diameter typically 10% of the well diameter *Source*: From IOC, 2006.

be determined for tidal prediction. This was at a time when methods of harmonic analysis of the ocean tide were under development, techniques which are still used today (Cartwright, 1999). Gradually, the accumulation of many years of data made possible the construction of long time series of sea-level change for science.

Traditional float and stilling-well gauges are still used in many places, with the clockwork chart recorders and paper charts replaced by digital shaft encoders and electronic data loggers. Many now provide data in near-real time for operational purposes such as harbor navigation, storm surge flood warning, or verification of tsunami alerts (IOC, 2006). It is important to have an appreciation of how they work if the time series of historical sea levels stored by the Permanent Service for Mean Sea Level (PSMSL) and other centers are to be analyzed. Most of the PSMSL dataset from the 19th and 20th centuries was obtained using such technology.

Measurements of sea level were also made throughout the 18th and 19th centuries at a number of locations in northern Europe (e.g., Liverpool, Brest, Stockholm, and Amsterdam) without the benefit of stilling wells. These measurements were made using "tide gauges", graduated markings on the stone walls of docks or "sluices" (gated passages between the sea and inland waters), intended to show water depth over an entrance sill. Alternatively, wooden measuring rods called "tide poles" or "tide staffs" were used. In some cases (e.g., Liverpool), only the heights and times of high water were recorded (Woodworth et al., 2011a). Dock personnel always had to be on duty at high tide when ships passed in and out of the dock; measurements of low waters would have

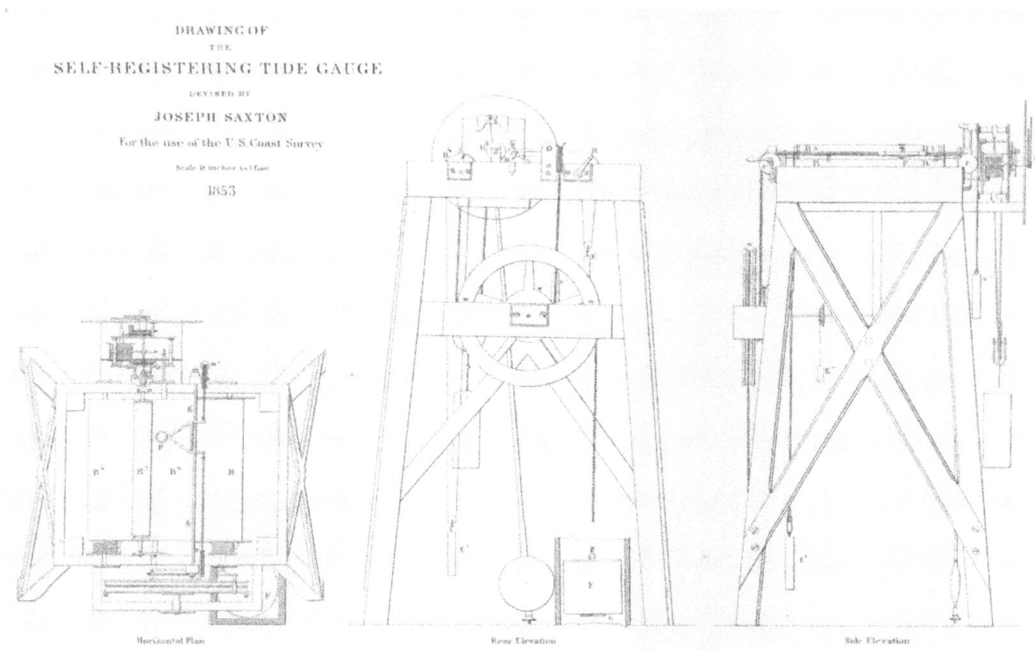

Fig. 35.2. Diagram of the self-registering tide gauge devised by Joseph Saxton for the US Coast Survey in 1851. *Source*: From the National Oceanic and Atmospheric Administration photo library, www.photolib.noaa.gov. The Saxton gauges produced a continuous trace on very long (~20 m) paper scrolls which have been subjected to reanalysis by Talke and Jay (2013). For color details, please see Plate 47.

required a separate tide gauge and were of little operational interest.

35.3 TYPES OF TIDE GAUGE

Although stilling-well gauges can still be found around the world, in most cases they have been replaced by different technologies. These include subsurface pressure, acoustic, and radar gauges (Fig. 35.3).

There are several types of pressure gauge. "Bubbler gauges" have no active electronics under the water. Air is passed at a metered rate along a small-bore tube to a "pressure point" cylinder fixed underwater below the low tide level. A small "bleed hole" is drilled half-way down the pressure point through which air bubbles, and the hole is the effective datum of the pressure measurements. Air pressure is measured at the other (landward) end of the tube by a pressure transducer (Pugh, 1972). Bubbler tide gauges have been used successfully for several decades, having replaced float and stilling-well gauges at many locations (e.g., at the 44 sites in the UK network). A second type of pressure gauge has the pressure transducer itself installed in the sea, connected by a signal cable to a data logger on land. A sophisticated variant comprises three pressure transducers in a "B gauge", which provides effective datum control to the resulting sea-level time series (Woodworth et al., 1996).

Another type of tide gauge is based on acoustic range measurements. The time taken by a pulse of sound to travel from the source to a reflecting surface and back again is a measure of the distance from the source to the reflector. The sound source, an acoustic transducer, is mounted inside a vertical sounding tube, the bottom end of which is in the sea. The travel time of the pulse t_p is given by:

$$t_p = \frac{2l_z}{C_a}$$

where l_z is the distance to be measured between transducer and water surface, and C_a is the speed of sound in air. For dry air (at 10 °C and one atmosphere pressure), $C_a = 337.5\,\mathrm{m\,s^{-1}}$ so that a change in l_z of 0.01 m corresponds to a change in t_p of 59 μs. Corrections must be made for the variations of C_a with air temperature, pressure, and humidity. For example, if the temperature falls to 0 °C, C_a decreases to $331.5\,\mathrm{m\,s^{-1}}$ and it is necessary to

Fig. 35.3. Different types of tide gauge. (a) A classical float and stilling-well gauge at Holyhead, UK. This site has two large wells. (b) The float gauge at Vernadsky, the site of the longest sea-level record in Antarctica. *Source*: Photographs for (a) and (b) by National Oceanography Centre. (c) A radar tide gauge at Alexandria, Egypt. *Source*: Photograph by Thorkild Aarup, Intergovernmental Oceanographic Commission. (d) An acoustic gauge at Vaca Key, Florida. Acoustic gauges now form the majority of the US sea-level network. Photograph also indicates meteorological sensor, solar panels, and communication systems. *Source*: Photograph by NOAA. For color details, please see Plate 48.

reduce the calibration factor of the gauge by 1.8% to compensate. One way of making these adjustments is to record the travel time of a sound pulse over a known standard distance. The US National Oceanic and Atmospheric Administration (NOAA) and authorities in Australia and other countries installed extensive networks of such gauges in the early 1990s and this technology became something of a standard in the global network until recently. A second type of acoustic gauge has the pulse transmitted in the open air, without a sounding tube. However, these instruments have proved to be less accurate.

Radar gauges are now a familiar sight around the world as they are relatively cheap and easy to install and maintain; modern radar gauges draw much less power than earlier devices and can be readily operated from solar panels. Some radars measure changes in sea level by monitoring the time-of-flight of a radar pulse from the transmitter/receiver unit to the surface and back to the unit, while others use a frequency-modulated continuous wave (FMCW) system, in which transmitted radar waves are mixed with signals reflected from the surface to determine the phase shift between the two waves and thereby the range. As radar travels at $299.792 \times 10^6\,m\,s^{-1}$ in air (at 10 °C and a pressure of one atmosphere), timing to an accuracy of 0.6×10^{-10} seconds is necessary to resolve a level difference of 0.01 m.

Radar gauges can be positioned several meters above the surface of the sea, and be operated in the open air without a stilling well (Fig. 35.3c). The radar beam has a typical beam-width of 5°. When installed on harbor walls, they must be mounted at the end of a cantilevered arm extending far enough over the water so that no reflections are obtained from the wall. In principle, they can also be installed inside a large stilling well, although in this case radar systems that transmit microwave pulses down a waveguide tube are preferable to the open-air type.

Radar gauges offer several advantages over float, pressure, and acoustic gauges. The main advantage is the ease of installation and maintenance, requiring neither extensive fixings to a harbor wall or pier (as for a stilling well), nor the involvement of divers (for underwater pressure gauges). Unlike acoustic systems, radar range measurements are not affected by either temperature or vertical temperature gradients. In addition, as far as we know, most radar gauges have little instrumental drift and so are suitable for both temporary and long-term installations. However, there is an initially unknown offset in their range measurements, which must be determined either in the laboratory prior to deployment (e.g., using a target at a known distance from the gauge), or in the field using simultaneous tide pole measurements.

Consequently, radar gauges have become a cost-effective choice for new or refurbished sea-level stations, but with some caveats. Experience of them so far has been limited. In particular, there are concerns about ocean wave bias in the radar sea-level measurements, with different bias characteristics in different systems. Such effects are yet to be fully understood. Other disadvantages include their potential exposure to damage during major storms or tsunamis, including the possibility that the water level in such events may even exceed the height of the radar sensor, or that floating debris or boats may pass under the beam, resulting in false measurements.

Most modern gauges can provide sea level at any required sampling, such as the 1 min values required for tsunami monitoring (IOC, 2006). Pressure and some radar gauges can even sample fast enough to measure wave spectra (several Hertz). Such capabilities are certainly adequate for all the applications described in this chapter. For more technical details of tide gauges, see the Intergovernmental Oceanographic Commission (IOC) manuals on sea-level measurement and interpretation (IOC, 2002, 2006).

35.4 SEA-LEVEL DATA AND DATABANKS

There are in general three streams of data from tide gauges as follows.

(1) *Real or near-real time data.* These are data which are required for operational purposes such as port operations, storm-surge flood warning, or tsunami identification. In these cases, there is no requirement for, or possibility of, rigorous quality control of the sea-level data. Such data are usually monitored around the clock by experienced personnel who will be able to judge whether any data anomalies are real or due to instrument malfunction.
(2) *Fast data.* These are data which are required on timescales of days and so which may have been subjected to some quality control. Fast data are required for applications such as satellite altimeter data validation.

(3) *Delayed mode.* These are data which have been inspected thoroughly and have been given flags to show a subsequent user whether they are good, suspect, or bad. The main application of delayed mode data is scientific research.

Displays of real-time data are available locally (e.g., in a port operations building) and are often provided by national authorities on websites. The only international real-time center is that operated at the Flanders Marine Institute on behalf of the Intergovernmental Oceanographic Commission (IOC) (http://www.ioc-sealevelmonitoring.org/). Fast data for many sites are provided by the University of Hawaii Sea Level Center (UHSLC, http://uhslc.soest.hawaii.edu/). Delayed mode data for sites in the Global Sea Level Observing System (GLOSS) may be obtained from the research-quality dataset of the UHSLC and the British Oceanographic Data Centre (BODC, www.bodc.ac.uk), while data from many other sites can be obtained from national authorities (see list in www.psmsl.org/links/sea_level_contacts/).

All three streams of data are known as "higher-frequency" data (Section 35.6.1), that is, higher than the monthly and annual mean sea level (MSL) values (Section 35.6.2). As regards delayed mode information, the most common data type has historically comprised spot values of sea level every hour, obtained from inspection of the ink trace on a tide gauge chart. An analyst would smooth through any high-frequency variability in the trace due to seiching before estimating the hourly values. Nowadays, the data loggers of tide gauges provide their data electronically. These data can be either spot values, integrated (averaged) values over specified periods (e.g., 6 min), or integrated over a specified period within a longer sampling period (e.g., averaged over 3 min every 6 min).

Monthly and annual MSL can be computed in different ways. The PSMSL recommends that all higher-frequency data be filtered to provide hourly values, hourly values then filtered to make so-called "daily mean values", followed by an arithmetic averaging of the daily means to derive monthly and annual values of MSL. The choice of the first filter, to provide hourly values from the tide gauge measurements, depends on the particular way that the measurements were obtained. The most common filters used to convert hourly values to daily means are the Doodson-X_0 and University of Hawaii filters (IOC, 2002). Whichever filter is employed, or even if the hourly values are simply averaged arithmetically into monthly and annual means, monthly MSL values are usually the same to within several millimeters.

MSL information derived from delayed mode data is archived by the PSMSL (Holgate et al., 2013). In addition to data, the PSMSL web site (www.psmsl.org) contains a large amount of training and technical information on sea-level measurements. The web page for each tide gauge provides links to the data held at other centers (e.g., UHSLC) and to Global Positioning System (GPS, or generically Global Navigation Satellite System, GNSS) data, if available, archived at the Système d'Observation du Niveau des Eaux Littorales (SONEL) center at the University of La Rochelle (www.sonel.org). In principle, the GPS time series from SONEL can be used to derive rates of vertical crustal movement at the gauge sites (Santamaría-Gómez et al., 2012). The IOC Manuals (IOC, 2002, 2006) and Chapter 10 provide details on the monumentation and operation of GPS at tide gauge sites.

Mean tide level (MTL) is a similar parameter to MSL. It is obtained by averaging values of high and low water into monthly and annual mean high and low waters (MHW and MLW) and then by averaging those two quantities into MTL. Tabulations of MTL are often found in archives when port operators recorded only the highs and lows of the tide rather than the full tidal curve. There is a systematic offset between MTL and MSL which depends upon the shallow-water tide (Pugh, 2004). Consequently, time series of MTL must be stored in databanks separately from those of MSL in order to avoid confusion between them, and an allowance for the offset must be made when the two quantities are combined in research.

The PSMSL is the main databank for long-term sea-level information from tide gauges. It was founded in 1933 and operates under the auspices of the International Council for Science. Its datasets are used intensively in research including the regular research assessments of the Intergovernmental Panel on Climate Change (IPCC). The PSMSL currently holds over 60,000 station-years of monthly and annual means of sea level, of which the earliest dates from 1807. Station records in the Revised Local Reference (RLR) subset are

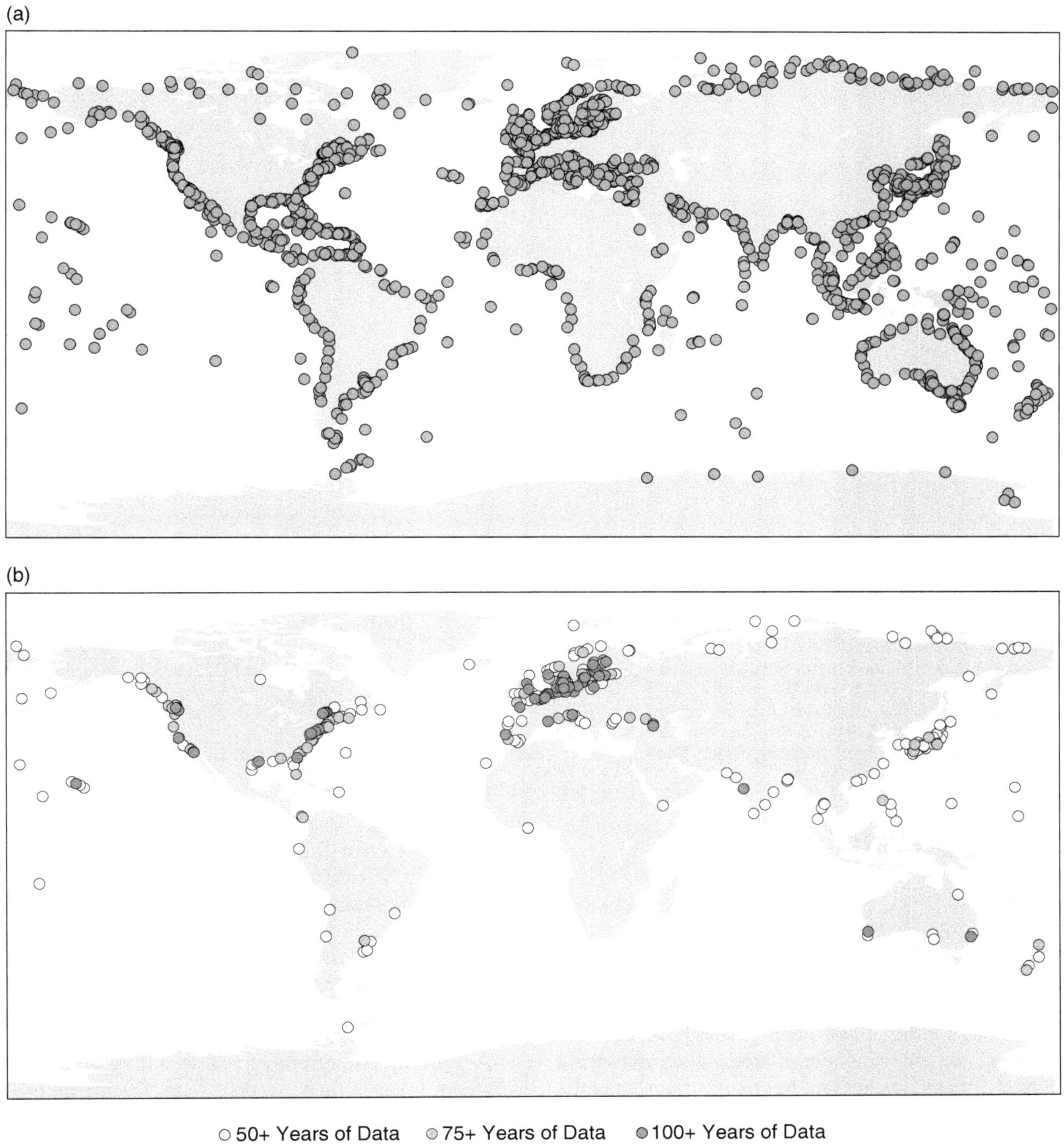

Fig. 35.4. Distributions of (a) all PSMSL stations including both research-quality (RLR) and non-datum controlled stations (known as "metric only" stations) and (b) long, research-quality records in the PSMSL RLR subset. *Source*: From Holgate et al., 2013. For color details, please see Plate 49.

measured to a consistent datum (benchmark height) and can therefore be analyzed to determine sea-level trends and variability. Figure 35.4a shows a map of all stations represented in the PSMSL dataset, while Figure 35.4b shows stations with long records in the RLR subset. Figure 35.5 shows long records from different regions, indicating long-term sea-level rise.

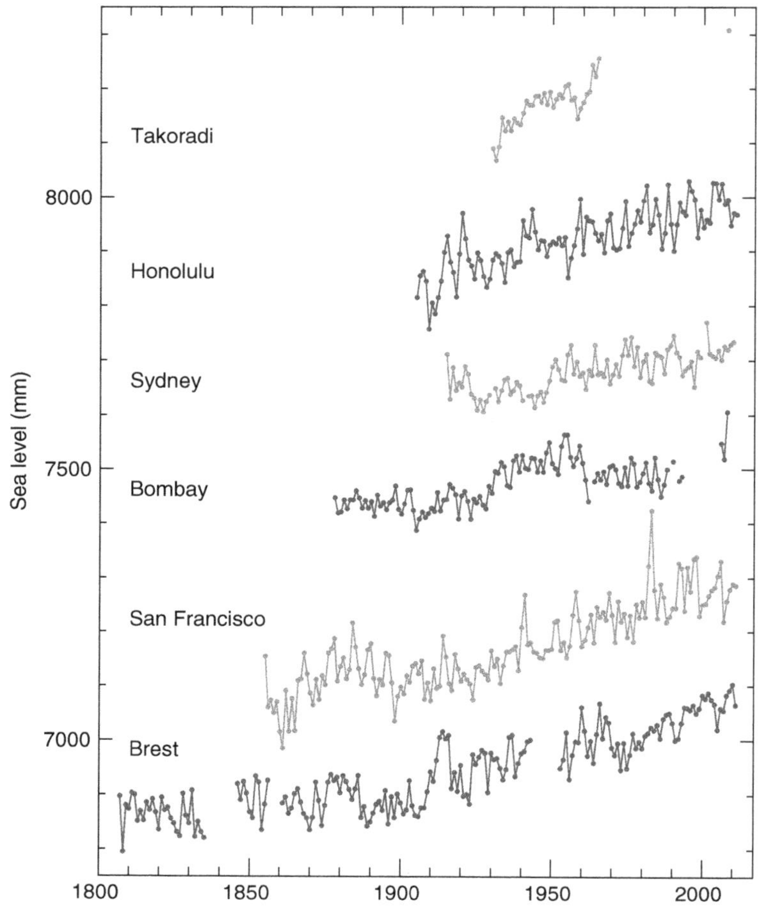

Fig. 35.5. Typical long-term records in the PSMSL dataset from different regions of the world. The Takoradi (Ghana) record, one of the longest in Africa, has recently been restarted after many years. Each record has an arbitrary offset for presentation purposes. *Source*: From Pugh and Woodworth (2014). Reproduced with permission of Cambridge University Press. For color details, please see Plate 50.

35.5 TIDE GAUGE DATA QUALITY CONTROL

Before higher-frequency data recorded by a tide gauge are ready for any scientific analysis, they must go through a quality control procedure called data reduction. Data obtained from international centers will often have been through this procedure. However, the associated documentation should always be checked to ensure that this is the case. Data obtained in the field (e.g., from a temporary tide gauge) will require proper checking.

A first important consideration is whether the times in the dataset are correct. The timing in modern tide gauges is obtained from GPS clocks, and loggers are usually set to record in either local time or Universal Time (Greenwich Mean Time). Poor documentation can result in misunderstandings on the time zone used. However, errors of an integer number of hours can be easily identified by comparison of tidal constants obtained from the dataset with those obtained previously. The possibility of timing errors must always be considered when analyzing older data, especially due to slippage in the clockwork mechanisms of traditional float gauges. When a reported sea level is based on the average of many individual readings (for example, to average out waves), the reported time must be the middle of the averaging period.

A second consideration is to know the units that sea level is measured in (e.g., feet or meters, or mbar in the case of pressure sensors). The time series of sea level must then be checked for errors, a process which requires some experience. For example, an isolated sea level 10 m higher than anything previously observed almost certainly represents an error. However, smaller "spikes" should be considered with care and only rejected if there is independent evidence of malfunction. Otherwise, there is a danger of disregarding real but unexpected events (e.g., small tsunamis). It is always useful to plot the observations to see

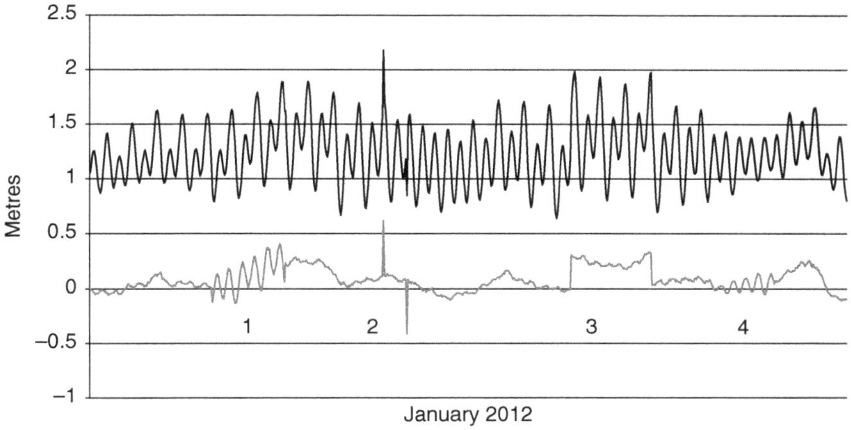

Fig. 35.6. Example of a time series of non-tidal residuals (red) showing characteristic errors due to: (1) a missed 1 hr value; (2) 0.5 m positive and negative spikes; (3) a datum shift of 0.3 m; and (4) a slow reduction in gauge sensitivity due to blockage. The original record, shown by the black curve, is from Spring Bay, Tasmania and includes low-frequency natural variations due to weather effects. The errors are artificially introduced. *Source*: From Pugh and Woodworth (2014). Reproduced with permission of Cambridge University Press. For color details, please see Plate 51.

which appear as isolated spikes, or lie outside an expected range. Values may be tested to see whether they differ by more than a specified amount from a curve interpolated between neighboring values. For example, a 7-point Lagrangian fit requires each sea-level value (x_t) at time t (hours) to satisfy the condition:

$$\begin{aligned}|-0.0049x_{t-7}+0.0410x_{t-5}-0.1709x_{t-3}+0.6836x_{t-1}\\ +0.5127x_{t+1}-0.0684x_{t+3}+0.0068x_{t+5}-x_t|<t\end{aligned}$$

where τ is a specified tolerance. Where tides are dominant, a simple method is to check against values at similar states of the tide on previous and subsequent days:

$$\left|\frac{1}{6}\left(-x_{t-50}+4x_{t-25}+4x_{t+25}-x_{t+50}\right)-x_t\right|<t.$$

When records are dominated by the astronomical tides, a more sensitive test can be made by subtracting tidal predictions from observed values, and by plotting the residual values. Some common errors in tide gauge operation have well-defined characteristics in these plots. Figure 35.6 shows a plot of residuals arising from: (1) a missed 1 hr value; (2) 0.5 m positive and negative spikes; (3) a datum shift of 0.3 m; and (4) a slow reduction in gauge sensitivity due to blockage.

"Buddy checking" involves the comparison of sea-level records from nearby stations, using either the original (e.g., hourly) tide gauge data or daily, monthly, or annual mean sea-level values. Inspection of the differences between the time series, or of the residuals, provides a powerful means for spotting data errors.

Some records may have gaps extending from a few hours to several days, usually due to instrument failure. There might also be small gaps between successive records that need to be filled to obtain a complete series. Gaps of a few hours may be interpolated by eye. Fitting a function through the observed values on either side, if the gap is not too large, can fill larger gaps. The 25-hour function given above can be used for gaps up to a day. Gaps should only be interpolated if there is a special need (e.g., in the application of a numerical filter with a broad window when treating the original observations as two separate series would lose too much data). A more elaborate interpolation gap-filling method is to add tidal predictions to interpolations of the residuals either side of the gap. Whichever method is used, it is important that gaps are flagged clearly in the resulting dataset.

More detailed information on the quality control of tide gauge data may be found in the IOC manuals (IOC, 2002, 2006, 2014).

35.6 SEA-LEVEL SCIENCE

35.6.1 Higher-frequency data

"Higher-frequency" data refers to sea levels (real-time, fast, or delayed-mode) sampled every hour or more frequently (i.e., not MSL). These are the data needed for the study of processes such as ocean tides, storm surges, extreme sea levels, etc. (Pugh, 2004; WMO, 2011). Most countries now provide data with 15, 10, or 6 minute sampling, and 1 minute values are now becoming standard where tsunamis, meteotsunamis, and seiches are of interest. Higher-frequency data are needed if tidal constants are to be determined for the purposes of tidal prediction and determination of tidal datums such as mean high water springs (MHWS) (Chapter 11).

Unfortunately, while national agencies have been willing to exchange MSLs in the past, they have been less disposed to provide higher-frequency sea levels, for different reasons. This situation is slowly improving, but one consequence is that datasets of extreme sea levels, and other quantities related to higher-frequency data, are less copious than those of MSL.

"Daily sea-level values" fall in between the sampling of sea level provided by higher-frequency data and by MSL. Daily means can provide a compact dataset (one value per day) which are more suitable than monthly means for the study of processes such as El Niño–Southern Oscillation (ENSO) events and, at some island locations, of the ocean mesoscale (Firing and Merrifield, 2004).

35.6.2 MSL records

MSL records, such as those available from the PSMSL, allow many aspects of sea-level variability to be studied, depending on the length of the particular record. Seasonal variability refers to changes through the year, with annual variability referring to the 12 month component. Interannual variability refers to changes from year to year or on timescales of several years, while decadal variability refers to changes from one decade to the next. Secular sea-level change refers to the linear rate of change of sea level between specified dates. Some of these topics are mentioned briefly below; for further details, see Pugh and Woodworth (2014).

35.6.2.1 The seasonal cycle of MSL

A seasonal cycle is an obvious feature in all MSL records, with an amplitude at some locations comparable to that of the tide. The reasons for seasonal variations differ between low and high latitudes, between shallow waters and the deep ocean and, in many cases, can depend on very local factors. Nevertheless, there are several well-understood reasons for such variations. These include air pressure changes and winds both in shallow-water areas and in the deep ocean, local surface heating density effects, and advected density changes due to the seasonality of the ocean circulation. Glaciological and hydrological contributions to the cycle will also be present, including seasonal melting of ice at high latitudes and run-off for gauges located near to large rivers. Wave setup in small harbors may also contribute to the observed seasonal cycle, as may a variety of instrumental effects, depending on tide gauge technology. Maps of the amplitude and phases of the annual and semi-annual components derived from PSMSL data are available (e.g., Tsimplis and Woodworth, 1994). While they are primarily forced by oceanographic and meteorological processes, they contain small long-period astronomical tidal contributions (S_a and S_{sa}). These have respective amplitudes of 1.1 and 6.7 mm at the equator (double at the poles).

35.6.2.2 The pole tide

The angular displacements of the Earth's axis of rotation around its mean geographic pole contain a component with a period of 14 months (435 days), with an amplitude that can reach 0.3 arc-seconds (10 m). This is called the "Chandler Wobble". It induces a 14 month pole tide in the ocean with a maximum amplitude of about 5 mm (assuming an Equilibrium response) at 45° N/S. This signal is observable in MSL records that do not contain significant additional variability around the same frequency (e.g., Boston). Larger signals with the same period, several times larger than Equilibrium expectations, have been observed in the North and Baltic seas. These are interpreted as being due to wind forcing, although the reasons for such a forcing in regional meteorology are not well understood (Tsimplis et al., 1994).

35.6.2.3 The nodal tide

The nodal tide is an astronomical long-period tide in MSL with a period of 18.61 years. There are good reasons for believing that the nodal tide has an amplitude and spatial variation similar to that expected from an Equilibrium response to the tidal potential (Woodworth, 2012). However, attempts to extract nodal tide signals from individual MSL records have had limited success due to their small amplitude compared to those of other decadal-timescale ocean processes.

35.6.2.4 Air-pressure-related sea-level variability

Over timescales of several days or longer, sea level will rise (fall) in response to a drop (increase) in local air pressure by approximately $\Delta P_A/\rho g$ where P_A is air pressure, ρ is water density and g is the acceleration due to gravity. This expression corresponds to 1 cm mbar^{-1} to within about 0.5%. This is called the local inverse barometer (LIB) and, in this isostatic adjustment of the sea surface, the consequence is no change in subsurface pressure (pressure that would be measured by a sensor at a level in the ocean just below low tide) or bottom pressure (pressure measured by a sensor on the seabed). Oceanographers usually prefer the inverse barometer (IB) model rather than the LIB. This model has an additional term that accounts for the average air pressure over the global ocean, that is, ΔP_A is replaced by $\Delta P_A - \Delta P_A^{av}$. Global ocean average air pressure ΔP_A^{av} has a root-mean-square variation of approximately 0.6 mbar and contains a large annual component that peaks in July–August.

However, air pressure changes are not the only meteorological forcing responsible for sea-level variability. Winds blowing over the shallow waters of continental shelves result in wind setup (or "surge") which is, of course, most important on timescales of hours and days when coastal flooding can result, but which can also be an important signal in MSL on timescales of weeks or months. Meteorological effects in tide gauge records can be accounted for in two ways. The first method adopts an empirical approach and parameterizes a record in a form such as:

$$\mathrm{MSL}(t) = at + b\left(\Delta P_A - \Delta P_A^{av}\right) + c\tau_x + d\tau_y + \varepsilon(t)$$

where t is time; τ_x and τ_y are local wind stress in the long- and cross-shore directions; a, b, c, and d are parameters determined by linear regression; and $\varepsilon(t)$ represents the residuals to the regression fit (e.g., Thompson, 1986). The *at* term accounts for secular trends due to long-term changes in the ocean and to vertical land movement of the land upon which the tide gauge is located, while the terms involving b, c, and d parameterize the meteorological effects on MSL. Often a choice is made to work with the mean seasonal cycle removed from each quantity, so that the regression parameters are more representative of a spectrum of variability rather than simply of the sometimes-dominant seasonal cycle. Other examples of such regression fits involve the use of both local and distant wind and wind stress, wind stress curl (a measure of the angular momentum of the wind field which relates to the "spinning up" of the ocean circulation), air temperature (which is frequently correlated with wind direction), sea surface temperature (associated with heat content), and rainfall (related to river flow).

A second method of accounting for meteorology in MSL records is provided by regional or global barotropic (two-dimensional or depth-averaged) ocean numerical models, similar to those used for operational storm surge prediction. The models require long time series of high-spatial-resolution air pressure and wind information, and detailed maps of ocean bathymetry, in order to provide time series of MSL throughout the model grid, including at the grid points near to tide gauge locations. In this way, time series of model MSL can be used to simulate the contribution of air pressures and winds to the MSL record from the tide gauge.

While the first method remains an important tool for understanding relationships between parameters, the second method tends to be more powerful as it takes into account both local and distant meteorological forcings in a rigorous way, as well as providing information on the spatial scales of variability. Whichever method is used, once the contributions from air pressure, winds, and other factors are accounted for, research can progress to studying the residuals of the regression fit which may contain interesting information on other ocean processes.

35.6.2.5 Large-scale patterns of interannual variability

The Earth's climate system contains a number of major modes of variability, which are evident in global datasets of climate parameters (air and sea temperatures, air pressure, sea level, etc.) (Trenberth et al., 2007). The most energetic mode occurs primarily in the Pacific when the trade winds relax. Equatorial Kelvin waves transfer water from the western tropical "warm pool" to the eastern side of the basin. These are called ENSO events and they occur approximately every 3–7 years. At the eastern boundary, the equatorial waves are converted into coastally trapped poleward-propagating Kelvin waves which can be observed clearly along the entire American Pacific coastline. The ENSO event is then followed by La

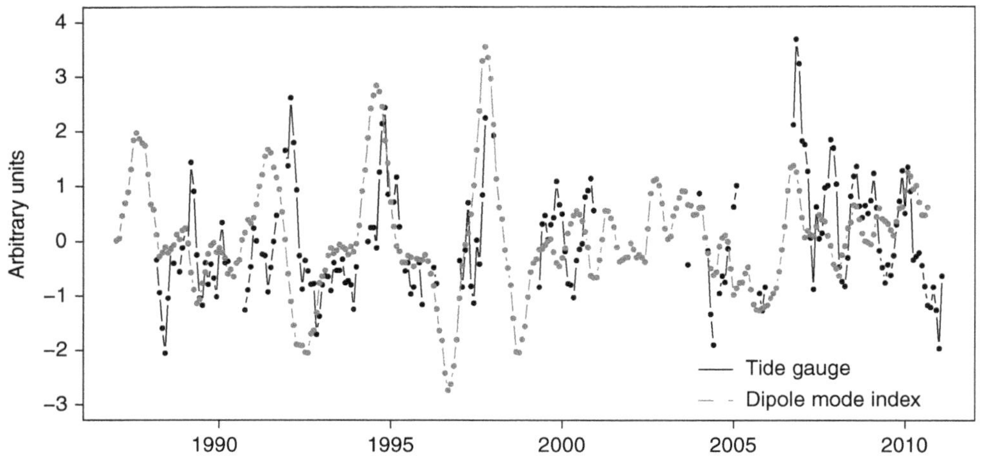

Fig. 35.7. An example of interannual variability in MSL, showing a comparison of monthly MSL since 1988 from Diego Garcia, in the Chagos Archipelago in the western tropical Indian Ocean, to the Indian Ocean Dipole (IOD) index. The IOD is a measure of the oscillation between sea-surface temperatures in the western and eastern Indian Ocean, with positive IOD corresponding to higher temperatures in the west. The sea-level time series has been scaled to have the same standard deviation as the IOD index. *Source*: From Dunne et al. (2012). Reproduced with permission of Elsevier. For color details, please see Plate 52.

Niña conditions, during which the Pacific recovers its normal state until the next event. Other major climatic modes include the Pacific Decadal Oscillation (PDO), Indian Ocean Dipole (IOD, Fig. 35.7), North Atlantic and Arctic Oscillations (NAO and AO), Southern Annular Mode (SAM), and Atlantic Multidecadal Oscillation (AMO). Sea-level variability is associated with each mode.

Any MSL record will contain variability on timescales of more than one year due to variability on similar timescales in the forcing factors (temperatures, air pressures, etc.), whether that variability is associated with climate modes or not. There will also be some low-frequency variability due to dynamical modes of variation in the ocean itself. Whatever the source of this variability it means that, in a time series of MSL annual means, any one value cannot be considered as independent of earlier or later years. This feature is called "serial correlation" and means that, when calculating a rate of sea-level change from a time series, then a standard error of the rate which is calculated by an ordinary least-squares procedure that treats each annual mean as independent will be underestimated.

35.6.2.6 Secular trends in MSL

In addition to the variability on various timescales discussed above, most MSL records will exhibit a low-frequency long-term trend – often called a "secular trend" – which may be linear or quadratic in character. The PSMSL website has a "trend viewer" which can provide a first assessment of trends at particular sites.

These secular trends can come from two sources. The first is vertical crustal movement of the land on which the tide gauge is located; recall that a gauge measures relative sea level where "relative" means with respect to the height of the nearby land. A second source is long-term change in ocean volume, resulting from changes in heat content or from changes in the volume of freshwater in the ocean due to melting of ice in glaciers and ice sheets. Further reasons for secular trends at particular sites will arise from changes in the ocean circulation. (There are many related aspects which we do not have space for here. See Pugh and Woodworth, 2014 for more details.)

Long-term MSL change due to vertical crustal movement. Figure 35.8 gives three examples of MSL records being affected by geological processes. Unlike the sea-level rise experienced during the 20th century at most locations around the world, the MSL record at Stockholm in Sweden shows a sea-level fall of approximately 4 mm a^{-1}, which is a consequence of the land on which the tide gauge is situated experiencing a high rate of crustal uplift due to glacial isostatic adjustment (GIA). The record from Nezugaseki shows an example of a near-instantaneous change of MSL of about 20 cm due to the 1964 Niigata earthquake off the west coast of Japan. Both of these examples are due to natural processes in the solid Earth. The third example is of a change in land level (and

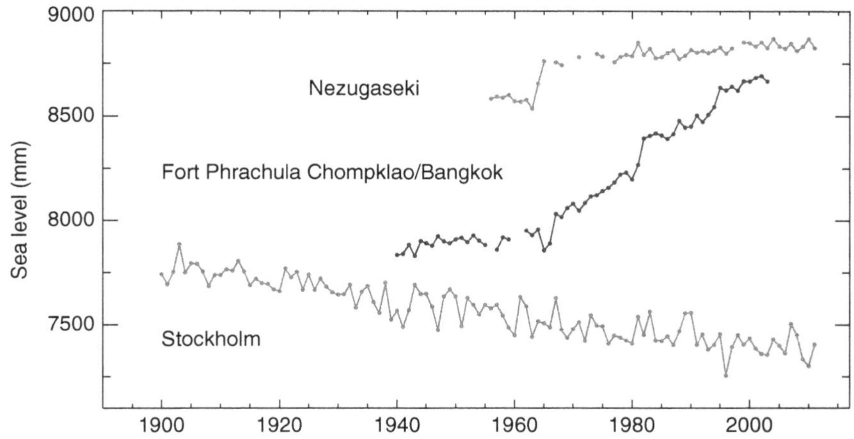

Fig. 35.8. Time series from: Stockholm, Sweden indicating the contribution of GIA to a MSL record; Nezugaseki, Japan indicating an earthquake; and Bangkok, Thailand showing local subsidence under a major Asian city. Data from the PSMSL. *Source*: From Pugh and Woodworth (2014). Reproduced with permission of Cambridge University Press. For color details, please see Plate 53.

therefore relative sea level) due to an anthropogenic process, in this case groundwater pumping under Bangkok, Thailand.

Most tide gauge analysts would like to adjust their records in order to remove the contributions of vertical land movements, resulting in a time series reflecting only ocean change. However, this is a more difficult issue than might be appreciated. When copious geological data exist from near to the tide gauge (and with a requirement for there to have been no major earthquakes), then it is possible to use the former to determine a long-term baseline of relative sea-level change which can be subtracted from the tide gauge record, thereby estimating any sea-level changes that may be due to recent climate change (e.g., Shennan and Woodworth, 1992). An alternative approach is to make use of modern geodetic techniques, primarily GPS, at the gauge site in order to monitor vertical land movement, irrespective of the geological reasons for it. Much progress has been made in using GPS information to adjust tide gauge records in this way (Santamaría-Gómez et al., 2012; Chapter 10), although there are two important caveats in such work. The first is that it requires an assumption that the vertical land movement trend measured by GPS, typically over a decade or so, is representative of that during the entire tide gauge record, which may span more than a century. The second is that GPS does not fully "correct" the tide gauge record, because vertical land movement is always accompanied by associated changes in the gravitational field (Shennan et al., 2012).

Without geological or GPS data, the only recourse for analysts of the global MSL dataset wishing to account for land movements is to employ predictions of relative sea level obtained from geodynamic models of GIA (Peltier, 2004). Similarly, analysts of GPS time series can compare trends in their records to GIA model predictions of vertical land movements; GIA predictions of relative sea level are not simply the negative of vertical land movement because of the associated changes in the gravitational field (Shennan et al., 2012). GIA is the only geological process for which suitable global models exist. However, it is clear that their use will account for only one component of overall vertical land movement, albeit a large one in the case of a location such as Stockholm.

Secular MSL changes on multi-decadal and century timescales. Sea-level change is a topic of considerable scientific and public interest. Three of the main complementary methods currently used to study this topic are based on saltmarsh, tide gauge, and space (altimetry and space gravity) data, providing knowledge of how sea level has changed on typical timescales of several hundreds, tens–hundreds, and tens of years, respectively. The fact that saltmarsh records can span several centuries means that they are particularly interesting in providing an idea of the rate of sea-level change before the instrumental era when tide gauge measurements became available. (See other chapters in this volume for discussion of other sea-level measurement techniques.)

The three techniques provide a validation of each other when their time series overlap.

Saltmarsh and tide gauge datasets, both examples of relative sea-level measurements, have been compared for several sites with favorable results (Gehrels and Woodworth, 2013). Satellite altimetry measurements are geocentric, with sea level measured with respect to the center of the Earth or to an Earth-centered standard ellipsoid. However, their time series (now spanning more than two decades since the early 1990s) compare favorably to tide gauges when both are properly adjusted (e.g., Woodworth et al., 2005). "Sea level" for the altimeter community usually means the measured sea surface height corrected for the inverse barometer effect, yielding a parameter akin to subsurface pressure. On the other hand, for the tide gauge community "sea level" has always meant the physical "still water level". These different conventions have caused confusion at times, and it is always important to verify whether any reported "sea level" has been adjusted for air pressure or not.

Many authors have attempted to combine individual PSMSL records such as those of Figure 35.5 into global-average curves (sometimes called "reconstructions") in spite of the spatial and temporal limitations of the dataset evident in Figure 35.4. The reconstructed global-average changes since the end of the 19th century have been largely monotonic with an average rate of rise of approximately 1.8 mm a^{-1}. For further discussion of trends and accelerations in MSL during the 19th and 20th centuries using saltmarshes, tide gauges, and altimetry, the reader is referred to Woodworth et al. (2011a, b), Pugh and Woodworth (2014) and to the IPCC Research Assessments (IPCC, 2013).

35.7 TEMPORARY TIDE GAUGES

35.7.1 General considerations

Most sea-level science takes place with the use of data from permanent tide gauge installations which nowadays are usually equipped with multiple sensors, real-time telemetry, and continuous GPS receivers to provide a monitor of vertical land movements. The international GLOSS program has specified standards for the operation of permanent gauges, together with related requirements such as the establishment of a local benchmark network (IOC, 2006).

However, there is frequently a need for temporary tide gauges in short-term deployments of a few days to several months. Determining which type of gauge to use depends on the intended application. For example, for hydrographic surveying of a river estuary, a temporary installation with equipment (e.g., radar gauge) that was not very different to that of a permanent site would probably be required. A similar choice might apply for determining a MSL over an extended period as precisely as possible, although a satisfactory alternative might be provided by a relatively cheap differential pressure sensor (one which measures subsurface pressure minus air pressure, which can be converted to sea level by dividing by density times acceleration due to gravity), together with a visual tide pole to provide a datum connection to local benchmarks.

If high-frequency sea-level phenomena such as seiches (Lowry et al., 2008) or even tides (e.g., Pugh et al., 2011) are to be studied, then a compact absolute pressure sensor (one which records subsurface pressure without removal of air pressure) should be adequate. In the case of seiches, the sea level (or pressure) signals will usually occur over timescales much shorter than air pressure variations. In the case of tides, the air pressure contributions will provide only a background noise continuum to the well-defined line spectrum of tides, so will not significantly affect the determination of harmonic constants. (An exception concerns the determination of the S_2 constituent, "radiational S_2" being present in air pressure.)

Different types of pressure sensors (either differential or absolute) are often employed in temporary deployments (e.g., Pugh, 1978; Woodworth et al., 2010). However, it is important to keep in mind that all pressure transducers have an instrumental drift, especially the older and cheaper models. Any linear drift with time may either be unimportant to the application (e.g., in the case of study of seiches), or needs to be measured and removed from the resulting dataset. A large drift can be determined with the use of visual measurements by a nearby tide pole, or by more technically sophisticated methods (e.g., see comparison of tide pole, radar, and pressure tide gauge and GPS-buoy measurements by Martín Míguez et al., 2012).

35.7.2 Temporary gauges in geological fieldwork

Temporary tide gauges are often required in geological fieldwork, such as that concerned with the reconstruction of sea level from fossil indicators in saltmarsh sediments (Edwards and Horton,

2006; Gehrels and Woodworth, 2013). The objective is always to establish the relationship between the altitudinal distribution of the indicator in question (e.g., foraminifera in a saltmarsh core) and a reference tidal level such as MSL or MHWS. In this case, the notation MSL (or MHWS, etc.) refers to values of these quantities averaged over an extended period, so defining reference tidal levels (or "hydrographic datums"), rather than to monthly and annual mean values (Chapter 11).

In an ideal case, the geological indicator is obtained from a site in close proximity to a permanent tide gauge. The altitudes of the indicator samples can then be referred directly, by means of conventional leveling or differential GPS (dGPS; see following paragraphs), to those of the tide gauge benchmarks and thereby to the reference tidal levels at the site. The reference tidal levels will have been determined from many years of sea-level measurements, and will often be published in national tide tables (usually referenced to Chart Datum; Chapter 11). However, such an ideal case is unusual. It is more likely that the tide gauge is many kilometers away. The question then arises of spatial variation in the tide and variation in the relationship between the various reference tidal levels, especially where there are known to be large spatial gradients in tidal characteristics along that coast (e.g., differences in tidal range or tidal asymmetry; see Pugh, 2004), or where there are large rivers.

In such circumstances, it is best to install a temporary tide gauge (e.g., tide pole or bottom pressure sensor). This ensures that, after the tidal data are analyzed, an accurate relationship can be established between the altitudes of the geological indicators and those of the local reference tidal levels. Depending upon the analysis, the altitudes of the reference tidal levels might be expressed either as heights relative to local MSL, as heights in a national leveling system, or as heights with respect to a geodetic reference surface obtained with dGPS (Chapter 11). Whichever method is chosen, the essential point is that the altitudes of the indicator samples and those of the sea levels provided by the temporary gauge (and thereby the local reference tidal levels) can all be referenced to the same datum when analyzed in combination.

If national leveling system heights are used, then there has to be a connection by means of conventional leveling to national survey benchmarks whose heights are known in the system. However, there are practical constraints on using such benchmarks in fieldwork, including their proximity, accuracy, and maintenance (as well as availability). Inspection of available benchmark catalogs can reveal that the nearest might be several kilometers from the study site and that many are no longer reliable or, indeed, extant. Although a survey team may be very skilled, there may be large closure errors in leveling when using distant national survey benchmarks (leveling from benchmark to site and then back again), even when using the most modern leveling equipment.

The now-routine use of dGPS offers a solution in that the altitude data are readily obtained relative to a geodetic reference surface (Chapter 11). dGPS can be used to determine the altitude of an indicator sample, or to establish the altitude of a temporary local benchmark from which conventional leveling can be made to the samples. dGPS data conversion to the most accurate local altitudes requires a degree of post-processing using data from reference stations obtained at the same time as the field measurements. The need for this post-processing can make the use of dGPS seem like something of a "black box", especially when outsourced to specialist geodesists. Similarly, the collection of dGPS data in the field offers little means of independent check other than "seeming about right", in contrast to the more transparent use of conventional leveling. Nevertheless, there is no doubt that the use of dGPS has revolutionized altitude determination in this area of research as it has elsewhere.

The decision whether to use a temporary tide gauge has to be based on an assessment of the likely improvement in accuracy of the geological field data, given that there will always be other sources of uncertainty. For example, there might be unusually high or low air pressure during the fieldwork, resulting in anomalously low or high sea level due to the inverse barometer (see Section 35.6.2.4). If the gain in accuracy in relating indicator altitudes to reference tidal levels from using a temporary gauge was thought to be small, then greater consideration might be given to meteorological monitoring. Nevertheless, in practice it is always difficult to estimate potential sources of error beforehand, and the use of a temporary tide gauge is often a sensible and practical addition to any fieldwork.

The advantages of a tide pole for determining local tidal characteristics are that the rise and fall can be monitored by a readily understood low-level technology, and the heights and times of individual high and low slack waters (turning points) identified. The tide pole has to be located so that its base (zero sea level on the pole, effectively its "zero datum") is below low water level and so that it is not fully submerged at high tide. In the case of the latter, a staircase of tide poles may be deployed, each one leveled and referenced in altitude to the others. Siting a tide pole underwater brings it own challenges, as does ensuring it remains upright during the fieldwork. Similarly, reading a tide pole requires some degree of skill, which is more difficult for open water sites where waves are present. However, tide poles provide a good "belt-and-braces" approach, can be inexpensive and easily constructed (e.g., a broomstick graduated with a felt tip pen), can be readily deployed when on-site collecting samples, and work well in microtidal inlets and lagoons.

Bottom pressure sensors provide measurements of water level over more extended periods than tide pole measurements possibly can, giving direct evidence of any local tidal characteristics including possible significant tidal asymmetry within inlets or estuaries, thereby enabling the determination of periods of inundation and exposure on the different parts of a saltmarsh or beach. A pressure sensor is required to be installed so that it is submerged at all likely occurrences of low water, and therefore has the same installation challenges as a tide pole. Submerged permanent structures (quayside, jetty, pier) offer good sites for anchoring a pressure gauge, especially as these can be used time and again. However, a heavy object may act as a sufficient anchor and as an effective reference point for leveling. Depending on the particular instrumental specifications, it may be best to install the pressure sensor with its long axis horizontally above the bed to prevent the instrument becoming clogged with particulate matter. Other factors to be considered include obtaining atmospheric pressure data for the site over the period of observation and, if there are likely to be significant changes in water density due to tidal flushing, salinity data should be obtained. Instrument drift should be small for short-duration deployments and the bottom pressure tide gauges can be pre-programmed for a period of observation, and left *in situ* without an observer having to be present, recording data at high frequency (and can therefore be used to examine the local tidal characteristics in detail). Again, the "black box" aspect of the deployment may mean that it cannot be readily checked that the pressure transducer is recording without removing it and downloading the data. It would therefore be reassuring to deploy both a pressure sensor and a tide pole when working in remote areas, where revisiting the site at a later date would be problematic or costly. In both cases, the leveling of the zero datum for the tide pole and the pressure sensor must be undertaken with respect to the same altitude datum as for the geological indicator samples.

35.8 CONCLUDING REMARKS

Tide gauges might be thought of as old technology, but there have been many important technical developments in recent years. Tide gauges complement saltmarshes for the measurement of long-term sea-level change. They also complement altimetry for measuring changes in sea level and ocean circulation over decadal and shorter timescales. Altimetry is an excellent technique for measuring sea-level changes offshore, while gauges measure changes at the coast where people live. Variations in the geostrophic surface currents of the ocean circulation are provided by both techniques in terms of sea-level gradients; altimetry is again the more important in the deep ocean, while pairs of gauges provide continuous measurements across straits. Tide gauges are also used in altimeter data validation.

However, the status of tide gauge recording is not completely satisfactory. The GLOSS global network remains to be completed, especially in Africa and South America, while some regional tsunami networks are far from complete. GPS equipment is still to be installed at many tide gauge sites (a GLOSS requirement). Access is required to restricted time series of historical tide gauge data in some countries. Greater efforts are needed in "data archeology", whereby historical sea-level records in archives around the world are converted into computer form for scientific research (e.g., Talke and Jay, 2013). The objective in all this work is to have the best possible datasets for the study of global and regional MSL change and, of greater interest to coastal dwellers, of changes in extreme sea levels.

ACKNOWLEDGEMENTS

We are grateful to Ian Shennan and Antony Long for the invitation to write this chapter. We thank Simon Holgate for discussions on temporary gauges. More details of the topics discussed here can be found in IOC publications and Pugh and Woodworth (2014). This work was partly funded by the UK Natural Environment Research Council.

REFERENCES

Cartwright, D.E. (1999) *Tides: A Scientific History.* Cambridge University Press, Cambridge.

Dunne, R.P., Barbosa, S.M., and Woodworth, P.L. (2012) Contemporary sea level in the Chagos Archipelago, central Indian Ocean. Global Planetary Change, 82–83, 25–37.

Edwards, R.J., and Horton, B.P. (2006) Developing detailed records of relative sea-level change using a foraminiferal transfer fuction: an example from North Norfolk, UK. Philosophical Transactions of Royal Society of London, A, 364, 973–991.

Firing, Y.L., and Merrifield, M.A. (2004) Extreme sea level events at Hawaii: the influence of mesoscale eddies. Geophysical Research Letters, 31, L24306.

Gehrels, W.R., and Woodworth, P.L. (2013) When did modern rates of sea-level rise start? Global Planetary Change, 100, 263–277.

Herschel, J.F.W. (ed.) (1849) *A Manual of Scientific Enquiry; Prepared for the Use of Her Majestys Navy; and Travellers in General.* John Murray, London. 488pp.

Holgate, S.J., Matthews, A., Woodworth, P.L., Rickards, L.J., Tamisiea, M.E., Bradshaw, E., Foden, P.R., Gordon, K.M., Jevrejeva, S., and Pugh, J. (2013) New data systems and products at the Permanent Service for Mean Sea Level. Journal of Coastal Research, 29, 493–504.

IOC (2002) Manual on sea-level measurement and interpretation. Volume III: Reappraisals and recommendations as of the year, 2000 (ed. P.L. Woodworth). Intergovernmental Oceanographic Commission, Manuals and Guides, No. 14, 47 pp. UNESCO/IOC, Paris. Available at http://www.psmsl.org/train_and_info/training/manuals/ (accessed 5 August, 2014).

IOC (2006) Manual on sea-level measurement and interpretation. Volume IV: An update to, 2006 (eds Aarup, T., Merrifield, M., Perez, B., Vassie, I., and Woodworth, P.). Intergovernmental Oceanographic Commission, Manuals and Guides, No. 14, 80 pp. UNESCO/IOC, Paris. Available at http://www.psmsl.org/train_and_info/training/manuals/ (accessed 5 August, 2014).

IOC (2014) Manual on quality control of sea level observations (eds Garcia, M.J., Martín Míguez, B., Perez Gonez, B. et al.), in press. UNESCO/IOC, Paris.

IPCC (2013) Climate Change 2013: The Physical Science Basis. Contribution of Working Group I to the Fifth Assessment Report of the Intergovernmental Panel on Climate Change (eds Stocker, T.F., Qin, D., Plattner, G.-K., Tignor, M., Allen, S.K., Boschung, J., Nauels, A., Xia, Y., Bex, V., and Midgley, P.M.), Cambridge University Press, Cambridge, United Kingdom and New York, NY, USA, 1535 pp, doi: 10.1017/CBO9781107415324.

Lowry, R., Pugh, D.T., and Wijeratne, E.M.S. (2008) Observations of seiching and tides around the islands of Mauritius and Rodrigues, Western Indian Ocean. Journal of Marine Science, 7, 15–28.

Martín Míguez, B., Testut, L., and Wöppelmann, G. (2012) Performance of modern tide gauges: towards mm-level accuracy. Advances in Spanish Physical Oceanography, Scientia Marina, 76(S1), 221–228.

Matthaus, W. (1972) On the history of recording tide gauges. Proceedings of the Royal Society of Edinburgh, B, 73, 26–34.

Moray, R. (1665) Considerations and enquiries concerning tides; likewise for a further search into Dr. Wallis's newly publish't hypothesis. Philosophical Transactions of Royal Society of London, 1, 298–301, doi:10.1098/rstl.1665.0113.

Palmer, H.R. (1831) Description of a graphical register of tides and winds. Philosophical Transactions of Royal Society of London, 121, 209–213, doi:10.1098/rstl.1831.0013.

Peltier, W.R. (2004) Global glacial isostasy and the surface of the ice-age Earth: the ICE-5G (VM2) model and GRACE. Annual Review of Earth and Planetary Science, 32, 111–149, doi:10.1146/annurev.earth.32.082503.144359.

Pugh, D.T. (1972) The physics of pneumatic tide gauges. International Hydrographic Review, 49, 71–97.

Pugh, D.T. (1978) *Techniques for the measurement of sea level around atolls.* In: Coral Reefs: Research Methods (eds. D.R.Stoddart and R.E.Johannes), Monograph on Oceanographic Methodology, No.5. UNESCO, Paris.

Pugh, D.T. (2004) *Changing Sea Levels: Effects of Tides, Weather, and Climate.* Cambridge University Press, Cambridge, 280 pp.

Pugh, D.T., and Woodworth, P.L. (2014) *Sea-level Science: Understanding Tides, Surges, Tsunamis and Mean Sea-Level Changes.* Cambridge University Press, Cambridge.

Pugh, D.T., Woodworth, P.L., and Bos, M.S. (2011) Lunar tides in Loch Ness, Scotland. Journal of Geophysical Research, 116, C11040, doi:10.1029/2011JC007411.

Santamaría-Gómez, A., Gravelle, M., Collilieux, X., Guichard, M., Martín Míguez, B., Tiphaneau, P., and Wöppelmann, G. (2012) Mitigating the effects of vertical land motion in tide gauge records using a state-of-the-art GPS velocity field. Global Planetary Change, 98–99, 6–17, doi:10.1016/j.gloplacha.2012.07.007.

Shennan, I., and Woodworth, P.L. (1992) A comparison of late Holocene and twentieth century sea-level trends from the UK and North Sea region. Geophysical Journal International, 109, 96–105, doi:10.1111/j.1365-246X.1992.tb00081.x.

Shennan, I., Milne, G., and Bradley, S. (2012) Late Holocene vertical land motion and relative sea-level changes: lessons from the British Isles. Journal of Quaternary Science, 27, 64–70, doi:10.1002/jqs.1532.

Talke, S.A., and Jay, D.A. (2013) Nineteenth century North American and Pacific tidal data: lost or just forgotten? Journal of Coastal Research, 29(6a), 118–127, doi: 10.2112/JCOASTRES-D-12-00181.1.

Thompson, K.R. (1986) North Atlantic sea-level and circulation. Geophysical Journal of the Royal Astronomical Society, 87, 15–32, doi: 10.1111/j.1365-246X.1986.tb04543.x.

Trenberth, K.E., Jones, P.D., Ambenje, P., Bojariu, R., Easterling, D., Klein Tank, A., Parker, D., Rahimzadeh, F., Renwick, J.A., Rusticucci, M., Soden, B., and Zhai, P. (2007) Observations: surface and atmospheric climate change. In: *Climate Change, 2007: The Physical Science Basis.* Contribution of Working Group I to the Fourth Assessment Report of the Intergovernmental Panel on Climate Change (eds Solomon, S., Qin, D., Manning, M., Chen, Z., Marquis, M., Averyt, K.B., Tignor, M., and Miller, H.L.), Cambridge University Press, Cambridge.

Tsimplis, M.N., and Woodworth, P.L. (1994) The global distribution of the seasonal sea level cycle calculated from coastal tide gauge data. Journal of Geophysical Research, 99, C8, doi: 10.1029/94JC01115.

Tsimplis, M.N., Flather, R.A., and Vassie, J.M. (1994) The North Sea pole tide described through a tide-surge numerical model. Geophysics Research Letters, 21, 6, doi: 10.1029/94GL00181.

WMO (2011) Guide to storm surge forecasting. WMO Report No. 1076. World Meteorological Organization, Geneva. Available from http://library.wmo.int/.

Woodworth, P.L. (2012) A note on the nodal tide in sea level records. Journal of Coastal Research, 28, 316–323, doi: 10.2112/JCOASTRES-D-11A-00023.1.

Woodworth, P.L. (2014) Tidal measurements. In: *History of Cartography* (ed. Monmonier, M.), Volume 6. University of Chicago, Chicago.

Woodworth, P.L., Vassie, J.M., Spencer, R., and Smith, D.E. (1996) Precise datum control for pressure tide gauges. Marine Geodesy, 19(1), 1–20.

Woodworth, P.L., Pugh, D.T., Meredith, M.P., and Blackman, D.L. (2005) Sea level changes at Port Stanley, Falkland Islands. Journal of Geophysical Research, 110, C06013, doi: 10.1029/2004JC002648.

Woodworth, P.L., Pugh, D.T., and Bingley, R.M. (2010) Long term and recent changes in sea level in the Falkland Islands. Journal of Geophysical Research, 115, C09025, doi: 10.1029/2010JC006113.

Woodworth, P.L., Menéndez, M., and Gehrels, W.R. (2011a) Evidence for century-timescale acceleration in mean sea levels and for recent changes in extreme sea levels. Surveys in Geophysics, 32, 603–618, doi: 10.1007/s10712-011-9112-8.

Woodworth, P.L., Gehrels, W.R, and Nerem, R.S. (2011b) Nineteenth and twentieth century changes in sea level. Oceanography, 24(2), 80–93, doi: 10.5670/oceanog.2011.29.

Index

accuracy, 9, 34–35, 44, 46, 50–51, 57, 60, 62, 66, 70, 75, 97, 106, 111, 114, 118, 127, 138–141, 143–144, 147, 151–152, 157–162, 164–170, 172, 178–179, 201, 216, 219, 227, 269–272, 274–277, 288, 291, 304, 307, 313, 318, 320–321, 331, 336, 351–352, 359–362, 364, 369, 376, 387–388, 391, 393, 395, 397, 399–400, 407, 409, 413, 415, 421–422, 425, 427–428, 431, 434, 438, 442–443, 449–450, 457–458, 464, 466–468, 471, 473, 479, 481, 486, 495, 498, 514, 516, 527, 540, 544, 550, 561, 571, 573
acidification, 88, 345, 352, 498
Acropora, 16–17, 102, 107–111, 113–116, 118, 122, 127, 144, 276, 537–538, 552–553
Aegean, 82, 148, 256, 280, 383
aeolian / aeolianite, 76–77, 84, 332–333, 342–343, 408–417, 518, 521–522
Africa, 231, 272–273, 308, 376, 380, 384, 407, 409, 411–413, 415–417, 515, 564, 572
Alaska, 4, 9–10, 12, 22, 24–25, 29, 31, 34, 36–40, 42, 45–46, 52, 80, 178, 180, 217, 234–235, 242, 247–248, 270, 279, 282, 294, 330, 381, 383, 474, 481, 492, 498–499
Aleutian, 24, 37–39, 44, 46
allochthonous, 93–95, 98, 100, 200, 243, 259, 295–296, 301, 307
altimetry, 129, 158, 169–170, 438, 450–451, 528, 569–570, 572
AMS, 111, 124, 183, 190, 307, 349–352, 356, 359–360, 384, 440, 503, 513, 540, 543; see also radiocarbon / radiocarbon dating
Andaman Islands, 35, 37, 45–46, 81, 145, 279
Antarctica, 106, 116, 120–122, 124, 216, 281, 294, 309, 354, 431–437, 441, 448, 560
aquatic, 94, 97, 99, 184, 188–189, 219, 231, 233, 247, 249, 251–252, 299–300, 306, 308–309, 314, 350, 362, 366, 368, 498
aquifer, 84, 87, 100–102, 149, 161
aragonite, 87–88, 98, 112, 335, 386, 388, 390, 397, 400
Arctic, 79–82, 179, 251, 271, 274, 281, 385, 414–416, 441, 448, 450, 476, 480, 568
Argentina, 260, 265–266
Asia, 31, 111, 224, 240, 247, 251, 569
astronomical, 11, 13, 81, 104, 115, 122, 124, 165, 172, 174, 179, 236, 240–241, 269, 284, 435–436, 438–439, 445, 449–450, 477, 479, 537–538, 566, 573
Atlantic / Atlantic Ocean, 30, 33, 63, 88, 96–97, 102, 107, 114, 122–123, 146, 179, 194, 216–217, 228, 255, 265–266, 270, 274–276, 279–280, 292–293, 297, 305, 326–327, 345, 354, 360, 371, 375, 379, 381, 383, 385, 414, 438, 441, 443, 446–447, 450–451, 468, 480–481, 497, 499, 502, 513, 516, 536, 544–545, 549, 551–553, 568, 573
atoll, 70, 100, 121, 125, 128–129, 131, 134, 136, 138–141, 145, 175–176, 190, 280, 361, 402
Australia, 25, 52, 81, 101–102, 104, 112–113, 115, 117, 120–121, 123–125, 140, 144, 154, 179, 193, 199, 214–215, 217, 224, 230, 254–255, 272–273, 275, 277–279, 309–310, 327–328, 343, 346, 375, 377, 408, 411, 414–417, 436, 515, 553, 561
autocompaction, 312–313, 322–326, 328, 469, 552
autocorrelation, 203, 244, 248, 288, 346, 486, 490

Bahamas, 85–88, 92, 94, 96, 98–103, 112, 121, 254, 256, 398, 400, 402–403
Baltic, 189, 214, 252, 254–257, 328, 354, 360, 371, 417, 566
Barbados, 16–17, 24, 70, 75, 81, 96, 101, 113–116, 119–120, 122–123, 278, 294, 389, 398, 402, 421–422, 424–425, 430–433, 435–436, 552
barrier / barrier coast, 13, 30, 33–34, 39–40, 43–46, 69–70, 79–80, 108, 111, 117, 120, 122, 124–125, 140–142, 144, 179–180, 189, 215, 217, 278, 328, 332–333, 343, 345–346, 397, 402, 405, 407–409, 412–413, 415–417, 443–444, 484, 488, 515–520, 522, 525–529, 534–535
Bayesian, 359, 430, 487, 498, 503–505, 509–510, 512–513, 551
beach, 9, 31–32, 34–35, 37, 39–42, 44–46, 66–67, 69, 73, 76–82, 87, 198, 231, 265, 274, 278, 281, 331–332, 341, 346, 352, 386, 404–405, 408–409, 411–412, 414–417, 521–522, 539, 550, 572
bedding, 58, 60, 63, 260, 263, 410, 517–518, 521–522
Belgium, 186, 189, 247, 294, 317, 326, 523, 528, 534, 553
Belize, 96, 100, 185, 190, 252, 256, 278, 294
benchmark, 13, 136, 147, 152, 157–159, 161, 165–167, 174–179, 276, 386, 393–395, 479, 538, 541, 547, 550, 557, 563, 570–571
Bengal, 205, 207, 285, 371
Bermuda, 85–86, 88, 91, 94–96, 98, 100–102, 171, 173

Handbook of Sea-Level Research, First Edition. Edited by Ian Shennan, Antony J. Long, and Benjamin P. Horton.

bias, 17, 41, 66, 149, 159–160, 167, 204–205, 216, 219, 227, 251, 260, 262, 302, 304, 308, 335, 395, 399–400, 427–428, 472, 477, 484, 486, 488–492, 494, 502, 505, 561
bioerosion, 72, 74–75, 131, 134, 152–153, 262, 268, 275–276, 278–280
bioturbation, 98, 200, 251, 258–262, 265, 296, 369, 371, 413–414, 517–518
bivalves, 97–98, 258–262, 266, 274, 527
Bonaparte Gulf, 25, 115, 119, 121, 199, 214, 254–255, 430, 432
Brazil, 76, 80, 96, 99, 214, 230, 266, 270, 277, 280, 293, 327, 416
Britain / British / British Isles, 8, 24–25, 82, 101–102, 155, 188–190, 214–217, 231, 233, 248, 254, 256, 266, 278, 289, 294–295, 312, 326–329, 371, 374–375, 379, 383–384, 421, 428, 435, 458, 460, 468–469, 498, 513, 551–553, 562, 573

calcite, 84–92, 95–97, 99, 101–102, 112, 213, 249, 251, 255, 335, 388, 390, 396, 399
calibration, 8, 16, 45, 58, 60, 111, 121, 163, 219, 229, 305, 323, 345–346, 349, 351, 353–360, 362, 381–382, 384, 395, 399, 402, 406–407, 435, 460, 467–468, 498, 502–503, 507, 510–513, 543–544, 551–552, 557, 561
California, 34, 45, 64, 66, 68–69, 73–74, 80–81, 144–145, 162, 190, 193, 255, 266, 310
Canada, 44, 59, 65, 79, 81–82, 101, 175, 177, 190, 192, 216, 221, 250, 255, 292, 294, 346, 375, 383, 409, 415, 417, 421, 427, 435, 451, 469, 480, 497–498, 502
carbonate, 58, 70, 72, 83–89, 91–92, 94–95, 97–98, 100–102, 104, 111, 122, 124, 188–189, 191, 220–221, 246, 252, 256, 278, 303–304, 313, 315, 319, 328, 331, 333, 335, 342, 345, 350, 352–355, 373, 380, 384–388, 390, 394–395, 397, 402, 404, 406, 410, 413, 416–417, 544, 553
Caribbean, 70, 75, 96, 100–101, 107, 122, 224, 231, 270, 276, 328, 450–451, 498, 552
Carolina, North / South, 30, 33, 45–46, 154, 193–195, 209, 212–215, 217, 230–231, 247, 250, 265, 300, 305–306, 309–310, 327–328, 361, 365, 371, 411, 416, 498, 502, 513
Cascadia, 4, 24, 42, 44–46, 65, 80, 215, 306–309, 435, 497–498
Cayman Islands, 101, 190, 278, 329
charcoal, 56–58, 184–185, 189, 306, 373, 378–380, 384, 503, 539
Chile, 4, 9, 24, 30, 42–43, 45–46, 78, 81, 193, 215, 234, 242, 247, 282, 294, 383
China, 103, 124, 146, 160, 228, 231, 233, 239, 248, 278, 311, 330, 376, 379–380, 384–385, 451, 515–516, 553
chronohorizon, 227–229, 373, 375, 377, 379, 381, 383, 385, 507
chronomarker, 376, 378–381
compaction, 7, 17, 23–24, 47, 55, 60, 64, 141, 147, 243, 313, 323, 326–329, 366, 369, 414, 430, 448, 450, 452–455, 457–461, 463, 465–469, 505, 515–516, 520, 523–524, 527, 535–536, 540, 551–553
confidence, 9, 13, 16, 19–20, 23, 59, 62, 114, 118, 163, 168–169, 201, 206, 214, 242, 255, 351, 359, 366, 370–371, 432, 479, 494–495, 500–501, 540, 546
Connecticut, 24, 48, 193–195, 197–198, 204, 209, 214, 231, 326–327, 329, 377, 468, 477–478, 480, 491, 553
contaminant / contamination, 53, 90, 101, 112, 184, 188, 196, 206, 303, 326, 350–352, 369, 385, 388, 390, 394, 405, 431, 500, 502
coral / coralline, 9, 17, 35, 37, 59, 66–71, 75, 80–82, 96, 99–100, 102, 104–134, 136–145, 151, 175–176, 249, 251, 262, 268, 270–272, 275–280, 312, 331, 357, 359, 386–403, 411, 430, 432, 435, 538, 548, 552–553
coseismic, 6, 9–10, 15, 24–25, 35, 39, 42–43, 45, 70–72, 74–75, 78, 81–82, 123, 126–127, 129, 138–139, 145, 152, 189, 242, 247, 269–270, 278, 306–308, 319–320, 325, 473, 481–483, 493, 498

databank / database, 9, 14–16, 19, 31, 179, 224, 230, 240, 244, 352–354, 359–360, 384, 393, 395, 401, 434, 443–445, 451, 525, 528, 536–553, 561–562
datum, 8–9, 12–14, 35, 37, 41, 61, 66, 76, 158–159, 166–167, 171–180, 268–269, 299, 302, 393, 409, 444, 448, 451, 458, 472, 476–479, 491, 524, 527, 529–530, 537–538, 541, 544–545, 548, 552, 557, 559, 563, 565–566, 570–572, 574
decade / decadal, 6, 8, 37, 40–41, 47, 58–59, 66, 70, 79, 101, 104, 106–107, 127, 129, 138–139, 150, 159, 161, 163, 189, 234, 242, 265, 275, 320, 323, 326, 356–358, 365, 370–371, 374, 386, 404, 409, 415, 424–425, 429, 454, 467, 500, 536, 559, 566, 568–570, 572
decompact / decompaction, 322–323, 453, 457–459, 463, 466–467, 540
decomposition, 20, 55–57, 59, 185, 188, 301–302, 305, 307, 309–310, 313, 318, 322, 325, 439, 458, 460, 482
deglacial, 24, 81, 99, 115–118, 120, 123–124, 254, 430, 436–437
deglaciation, 24, 106–107, 115, 120–121, 124, 201, 254, 360, 402, 421, 424, 430–431, 436
delta / deltaic, 13, 39–40, 48, 64, 143, 153, 188, 216, 220, 230, 239–240, 248, 278, 282, 296, 300–301, 306, 310, 313–314, 321–323, 326, 328–330, 332–333, 345–346, 355, 358–359, 370–371, 388, 411, 417, 454, 469, 514–516, 518, 528–535, 539, 545, 553
Denmark, 64, 278, 411–412, 415–417
diatom, 20, 22, 34, 45, 56, 58, 97, 189–190, 205, 217–218, 230, 233–236, 239–248, 251, 254, 284, 293, 299, 306, 308, 312, 315, 317, 320–321, 327, 329, 448, 472–479, 481, 483, 488–492, 494–495, 497–499, 530, 540
dissolution, 74, 84, 86–88, 90, 92, 99, 101–102, 124, 151, 200, 251, 325, 336, 390, 394
dunes, 30, 39–40, 44, 46, 76, 80–81, 332, 341, 405, 407–412, 415–417, 515, 519–522, 524, 534–535

Earth rheology / rheology, 81, 423–424, 428, 430, 434–437; see also GIA / Glacial isostatic adjustment
earthquake, 4–5, 9, 15, 18, 22, 24–25, 30, 35, 37–39, 42–46, 54, 63–65, 67, 70–71, 74–75, 77–79, 81–82, 129, 132, 141, 143–145, 172, 179, 190, 233, 242, 246–248, 270, 279, 294, 306–309, 319–320, 330, 383, 473, 497–498, 513, 568–569; see also seismic
earthquake deformation cycle, 15, 24, 39, 42, 383, 498
England, 17, 24–25, 44, 48, 64, 86–87, 97, 101, 175, 187–188, 190, 193, 197, 209, 212, 214–215, 217, 230–231, 247, 279, 283, 303, 311, 327–328, 345, 377, 383–384, 415, 454, 468–469, 498, 552–553
ENSO, 37, 126, 128, 139, 143, 566–567
error, 4, 11, 13–20, 23, 35, 41, 44, 53, 72, 127, 134, 138–139, 141, 143, 148–150, 158–161, 167–169, 172, 175–176, 179, 183, 199, 202, 206, 228, 240–241, 251, 264, 272, 292, 299, 303–306, 308, 316, 326, 338, 350–355, 362, 364, 367–368, 370, 389, 395, 400, 407, 409, 412, 414, 427, 442, 444–445, 452–454, 464, 466–467, 471–472, 474–475, 478, 481–482, 484–486, 488, 490–491, 494, 501–503, 505, 507, 511–513, 519, 524, 527, 530, 536–537, 540–551, 564–565, 568, 571
estimate, 8–9, 13–14, 16–18, 20–21, 23, 34–35, 37, 42, 44–46, 55–57, 59, 88–89, 91, 104–105, 107, 112, 114–115, 123, 131, 134, 138–139, 142, 148–149, 152, 158, 161, 163, 166–169, 176, 185, 201, 229–231, 240–242, 244, 249, 265, 270, 284, 286, 291, 294, 296, 300, 305–306, 313, 323, 326, 328, 362, 364, 366, 368, 370, 376, 379, 381, 383–384, 387, 392–394, 399, 401, 410–411, 415, 417, 425, 427, 430–433, 435, 445, 450, 454, 458, 462–464, 467–468, 471–473, 475, 477, 483–485, 488, 490–491, 493–494, 499–511, 513, 527, 529–530, 536–537, 539–540, 545, 549, 562, 569, 571
estuary / estuarine, 1, 3, 13–15, 19–21, 37, 39, 44, 48–49, 63–65, 80, 100, 172, 178, 186, 188–189, 215–217, 219, 230, 233, 236, 239, 242–243, 247, 249, 251–256, 261–262, 265–266, 271, 275, 278, 281, 293–294, 296–297, 300–301, 305, 307–311, 320, 325–330, 332, 336, 353–355, 360, 365, 371–372, 380, 383–384, 408, 411, 414, 416–417, 444, 447, 451, 475–476, 514–516, 518, 522–525, 528–530, 534–535, 538, 540–541, 543–544, 552–553, 570, 572
Europe, 32, 49, 57, 63, 146, 155, 160, 176, 184–187, 211, 228, 234, 244, 249, 251, 265, 300, 326, 356, 370–371, 374–380, 384–385, 435, 451, 468, 487, 552–553, 558
eustasy / eustatic, 6–7, 17–19, 23–24, 67–68, 70, 80, 84, 105, 117, 121, 154, 269, 360, 400–401, 412, 422–424, 427, 430–437, 448, 515–516, 531–532, 534, 552–553
experiment, 1, 23, 144, 216, 231, 285, 291, 311, 345, 361, 364, 370, 383, 428, 435, 468, 507

facies, 49, 62–64, 93–101, 107–109, 116, 247, 310, 313–314, 319, 326–327, 341, 343, 345–346, 386, 391, 394, 397, 409–410, 412–413, 515–518, 525, 527–528, 530–531, 534–535, 539–540
Florida, 45–46, 69, 81, 83, 86, 90, 94, 96, 99–101, 103, 190, 227–232, 252, 256, 266, 273, 278, 328, 349, 386, 411, 416–417, 549–553, 560
foraminifera, 24, 33–34, 45, 58, 81, 84, 94, 97–98, 102–105, 117–118, 190–208, 211, 213–218, 251, 254–256, 281–284, 287–288, 291–294, 297–299, 305–310, 312, 320–321, 327–328, 354–355, 357, 412, 437, 448, 472–483, 488, 490–495, 497–498, 505, 513, 518, 530, 540, 551, 571
France, 46, 90, 102, 146, 148, 152–155, 175, 215, 279, 376, 383, 515

geochemical, 53, 63–64, 91–92, 98–99, 101–102, 104, 122–123, 190, 254, 278, 295, 297, 299, 301, 303, 305, 307–311, 326, 329, 365, 370, 372–373, 376, 381–384, 387, 390, 394, 402, 407, 416, 551
geochronology, 100, 121, 144, 231, 254, 278, 361, 364, 368–372, 384, 415–417, 513, 552
geodesy, 37, 39, 46, 70, 115, 157–162, 165, 167, 170, 174–175, 179, 444, 451, 537–538, 541, 544, 548, 550, 557, 569, 571, 574
geoid, 6–8, 81, 142, 157–158, 166–167, 172, 176, 532
Germany, 81, 190, 230, 256, 268, 339, 376, 523, 528, 553
GIA / Glacial isostatic adjustment, 6, 8–9, 16, 18–23, 67–69, 75, 78–79, 96, 115, 117–118, 167, 169, 400, 421–425, 428, 430–435, 445–447, 536, 540, 544–545, 568–569
GIS, 29–30, 60, 81, 525, 534, 540
glacioeustatic, 67, 69, 75, 83, 85, 96, 123; see also eustasy / eustatic
GPS, 8–9, 13, 41–42, 45, 60–61, 133, 142, 157–170, 172, 175–177, 179, 273, 277, 550, 562, 564, 569–573
Greenland, 30, 45, 111, 119, 177, 360, 364, 374, 383, 430–436, 441, 499, 513
groundwater, 55, 83–87, 89, 91–92, 94, 96–102, 186–187, 391, 455, 460, 468, 516, 519–520, 525, 527, 529–530, 534, 569

Hong Kong, 233, 374, 383
Honolulu, 157, 168, 564
hypothesis, 2–7, 14–15, 18–22, 24, 29, 35, 44, 101, 120, 359, 384, 402, 552, 573

Iceland, 376–377, 381, 383–385, 427, 436, 502–503, 505–507, 509–510
India, 146, 154, 156, 293, 329, 376
Indian Ocean, 25, 35, 44, 74, 100, 110, 115, 121–122, 124–126, 128, 138–139, 145, 156, 216, 247, 278, 385, 402, 414, 568, 573
indicative meaning and/or indicative range, 9–14, 16–17, 22, 47, 49–50, 66, 69, 74–76, 83, 85, 90–98, 112–113, 127, 131, 142, 147, 152, 171, 174, 176, 186–189, 220, 230, 240–242, 262, 282, 299, 302, 305–306, 319–320, 323, 335, 342, 353, 367, 448–449, 453, 456, 464, 472–473, 486, 491, 514–516, 519, 527, 529–530, 532–533, 537–538, 540, 545–546, 548
Indonesia, 24, 45, 71, 81, 119, 130, 132, 137–138, 140, 145, 193, 215, 230, 242, 247

infaunal, 199–200, 205, 215–216, 262, 282
intertidal, 7, 9, 12–13, 24, 39–40, 42, 45, 74, 99, 102, 127, 145, 148–151, 171, 183, 186, 188, 191–192, 194, 196, 199, 201, 203, 205–208, 213–215, 217, 227, 229, 231, 233, 236, 248–249, 255, 268–269, 272, 274, 276–277, 281, 288–289, 293, 295–297, 299–301, 305, 307, 312, 316, 319–321, 323–324, 327–328, 332, 361, 368, 370, 374, 377, 381, 384, 404, 408, 412–413, 415, 417, 443–444, 452–455, 457–458, 461, 463–464, 467–468, 471–472, 474, 476–481, 483, 486, 488, 490, 492–493, 498, 515, 518, 520, 522–523, 525–528, 538, 553
Ireland, 24, 81, 191, 276, 383–384, 500
isostasy, 6–8, 15, 18, 20, 24–25, 30, 37–38, 67–69, 74, 79–82, 105, 114–115, 122, 124, 167, 236, 247, 254, 266, 269, 310, 396, 400–401, 411, 421–423, 425, 427, 429, 431, 433, 435–437, 445, 451, 504, 532, 536, 551–553, 567–568, 573
isotope, 25, 53, 69, 78, 80, 87–88, 90, 92, 99, 101–105, 111–114, 121–124, 138, 144, 186, 189, 191, 217, 250, 255, 271, 277, 279, 295, 297, 299–305, 307–311, 324–329, 335, 345, 350–353, 355–356, 362, 365, 369, 371–372, 374–378, 383–385, 387–390, 393–395, 397, 399–402, 409, 411, 430, 437, 500, 503, 513, 541–544
Israel, 149, 154, 156, 266, 417
Italy, 90, 92, 99, 146–150, 152, 154–155, 255–256, 266–268, 272–273, 277–280, 370, 376, 415, 516, 519
ITRF, International Terrestrial Reference Frame, 158, 161, 165, 167, 175, 179

Japan, 29, 32, 34, 42, 44–45, 49, 52, 54, 63, 102, 104, 109–110, 120, 122–124, 193, 233–235, 247, 252, 256–257, 279, 294, 310, 329, 357, 503, 516, 568–569

karst, 83–88, 93–94, 99–103, 151, 274, 329, 384
Kiritimati, 128, 136, 142

lagoon, 39–40, 52–53, 62, 64, 69–70, 86, 108–109, 117, 123–125, 140, 251–252, 256, 265, 270–271, 275, 278, 325, 328–329, 332–333, 343, 345–346, 355, 370, 412–413, 416, 515, 518, 529–530, 572
LiDAR, 32, 34–36, 41–42, 45, 528
lithofacies, 47–49, 55–58, 60–62, 100, 110, 332–333
lithology, 49–50, 55–60, 62, 72, 85, 93, 97, 117–118, 199, 201, 222, 260, 299, 373, 376, 454–455, 457–458, 460–461, 463–464, 482, 516, 525
luminescence, 404–417, 519, 541

macrofossil, 56–58, 183–190, 306, 312, 352, 355–357, 360, 530, 538, 543, 546
macrotidal, 203, 319–320, 325, 328, 383, 416, 538
Maldives, 4, 24, 165
Mallorca, 88, 90–91, 95, 100, 102–103, 278, 416
mangrove, 84, 94–96, 102, 185, 188–190, 192–193, 200, 205, 214–217, 224, 228–231, 241–242, 247–248, 252, 256, 281, 293, 296, 308–309, 311, 319, 322, 325, 328–330, 473, 518, 553
marshes, 4, 15, 34, 54, 63, 94, 190, 193, 195, 197–198, 203–204, 211, 214, 216, 220, 228, 247, 249, 255, 281–282, 285, 289–290, 292–293, 308–310, 320, 326–329, 383, 385, 412–413, 468, 472–474, 490, 497, 513–514, 527, 529–530, 545–546, 551
Maryland, 63, 256, 471, 552
Massachusetts, 120, 165, 193, 195, 197–198, 204, 214, 265, 327, 371, 409, 468, 552
Mediterranean, 74, 80–81, 88, 99–100, 102–103, 146–147, 150–156, 213, 249, 254–255, 270–271, 274–280, 282, 309, 329, 416, 435, 552
megathrust, 22, 35, 39, 42, 44–46, 70, 78, 81, 144–145
mesotidal, 142, 413, 538
microatoll, 66, 70–71, 82, 125–145, 175–177, 180, 270, 276, 280
microbial, 260, 288, 293, 296, 300–302, 308, 379, 453, 537, 552
microfossil, 13, 16, 20, 22, 34, 48, 53, 64, 84, 94, 96, 99, 196, 205, 214–215, 220, 231, 240, 244, 248–249, 251, 256, 286–287, 294, 297, 302, 306–308, 320, 330, 430, 471–473, 476, 479–481, 483, 492–493, 497–498, 501, 546
microtidal, 142, 268, 306, 319, 468, 572
minerals, 58, 88, 92, 101, 113, 314–316, 321, 325, 335–336, 362, 380, 387, 402, 404, 406
minerogenic, 17, 284, 319, 322, 325, 373, 376, 382, 384, 454–455
Mississippi, 39, 46, 48, 64, 230, 306, 310, 329, 355, 358–359, 371, 375, 377, 450, 454, 469, 516, 553
molluscs, 56, 74, 111, 151, 258–268, 273–274, 276, 279, 312, 354–355, 386, 404, 518, 521, 530

Netherlands, 24, 52, 63–65, 144–145, 175, 189, 191, 214, 227, 232, 248, 309–310, 326, 328–329, 384, 409, 514–517, 521, 523, 525–528, 534–536, 552–553
New Jersey, 190, 231, 258, 295, 297–298, 306, 309, 326–327, 449–450, 480, 490–491, 498, 552
New Zealand, 24, 30, 39, 46, 63, 74, 78, 80–82, 193, 215, 294, 381, 383–385, 513, 516
Newfoundland, 192–195, 209
Norway, 24, 79, 292–293, 470
notches, 9, 39, 66, 69, 74–75, 80–81, 87, 89, 151, 153, 269, 272–274, 276, 294
Nova Scotia, 73, 193, 216, 255, 291–292, 502

Oregon, 24, 30–34, 37, 39, 41, 44–46, 54, 64–65, 80, 190, 193, 215, 235, 247, 255, 306, 308–309, 438, 474–475, 477–478, 480–482, 490–491, 493–494, 498, 536, 551
ostracoda, 58, 84, 97–99, 199, 214, 218, 249–257, 312, 314, 473

Pacific / Pacific Ocean, 31, 41, 44–45, 71, 74, 77, 80–81, 100, 107–110, 115, 119–124, 126, 128, 139, 145, 175, 180, 214, 216, 247, 249, 275, 278–279, 282, 306, 378, 383, 414, 427, 438, 441, 445, 451, 480, 567–568, 573

palaeobotany, 190, 230–232
palaeoclimatology, 25, 99, 101, 121, 123–124, 154–156, 214–217, 230, 247, 255–256, 278–280, 294, 308, 327–328, 384, 402, 415–417, 497, 552
palaeoecology, 25, 63, 99, 101, 121, 123–124, 154, 156, 189–190, 214–217, 230, 247, 255–256, 265–266, 278–280, 294, 308, 326–328, 384, 402, 415–417, 497, 552
palaeogeography, 25, 99, 101, 121, 123–124, 154–156, 214–217, 230, 247, 255–256, 265–266, 278–280, 294, 308, 327–328, 384, 402, 415–417, 497, 552
palaeohydrology, 63, 294, 326–327
paleoceanography, 120, 122, 294, 553
paleoenvironment, 48, 63, 94, 101, 120, 154, 188, 211, 215–216, 230, 232, 255, 258, 266, 281, 286–287, 295, 297, 302, 305, 307, 313, 341, 343, 370, 379, 470–472, 474, 480–481, 484–485, 487, 491, 494, 515–516, 518, 527–528, 540, 545
paleotidal, 196, 319, 325, 439, 445–447, 449, 451, 471, 525, 527, 532, 537, 544–546, 551
Papua New Guinea, PNG, 105, 107–108, 112–113, 115
peat, 12–13, 17, 19–20, 24, 39–40, 48, 53–60, 63–64, 94–96, 102, 184–190, 200, 216, 220, 228–230, 281–282, 284–285, 291–292, 294, 298, 300, 306–307, 311–315, 317, 319–320, 322–324, 326, 328–329, 343, 352–353, 355, 360, 368, 384–385, 411, 452–454, 457, 461, 467–469, 473–474, 479, 495, 502, 505, 515, 518–520, 522, 524–531, 535, 539–540, 543, 551–553
Pennsylvania, 230–231, 295, 551
Peru, 77, 81–82, 375
pollen, 24, 94, 103, 183, 185, 188–190, 205, 215, 218–221, 223–232, 284–285, 294, 305, 308, 310, 312, 329, 368, 370, 373, 378–380, 382, 384, 470, 473, 487, 497–500, 503, 507–508, 530, 540
Porites, 70, 107–110, 116, 118, 120, 125, 127, 134, 140, 142, 144–145, 262, 276
Portugal, 411, 415, 417, 525, 530–531, 535

radiocarbon / radiocarbon dating, 8, 15–16, 20, 48, 97, 104, 110–111, 115, 118, 121, 124, 139, 150, 183–184, 264–265, 271, 273–274, 277–278, 312, 349, 351–353, 355–360, 402, 407, 409–412, 414, 417, 500–505, 507, 510–513, 519, 527, 530, 551–553
reference water level, 9–14, 18, 22, 35, 66, 76, 129, 142, 171–175, 179, 186, 319, 448–449, 537, 545–546
regression, 44, 47, 236, 244, 265, 366, 397–398, 423, 464–465, 484–485, 489, 498–499, 502, 504–510, 512, 517, 532–533, 540, 552, 567
regressive / regressive contact, 20, 48, 76, 80, 187, 236, 239, 255, 291, 319, 324, 328, 331, 452–453, 456–457, 553
replicate, 1, 17, 23, 134, 205, 277, 406–407, 414, 461, 463, 473, 496
safety, 49–50, 58, 64, 84, 285
saline / salinity, 10, 20, 69, 84–89, 96–97, 99–100, 103, 150, 185–186, 188–189, 191–192, 196–199, 204, 210, 213–214, 228–229, 233–234, 236, 242, 244–246, 248–252, 256, 284, 288, 291, 299–300, 302, 306, 308, 312, 320, 327, 336, 353–354, 445, 449, 474–475, 481, 484, 572
saltmarsh, 33–34, 48, 79, 86, 177, 184, 186–187, 191–195, 197–200, 203–206, 208, 214, 231, 233, 236, 240–241, 247–248, 277, 281–289, 291–293, 296–300, 305, 307, 309, 312, 319–323, 325–326, 328–329, 364, 373–380, 412, 448–449, 454, 458–459, 471, 473–479, 481–483, 492–493, 500–503, 505, 507, 509, 511, 513–514, 516–518, 527–528, 530, 537–538, 541, 546, 569–572; see also tidal marsh
satellite, 1, 31–32, 35, 39–40, 44, 46, 129, 157–158, 160–166, 169–170, 172, 176, 179, 438, 451, 550, 562, 570
scale, 7, 9–10, 13–15, 17–18, 20–23, 31, 33–34, 47–52, 56–57, 59–60, 62, 66, 71, 86, 111, 114, 116, 124, 127, 132, 143, 147, 152, 156, 159–160, 166–167, 169, 174, 179, 185–186, 189, 195, 197, 199, 201–202, 209, 212, 217, 220, 225, 234–235, 248, 251–253, 259, 262–263, 277–278, 287, 292, 295, 307, 313, 323, 325–326, 332, 334, 339–341, 345–346, 354, 356–361, 365, 371, 378, 390, 393–394, 396, 413, 415–416, 421, 425, 427–428, 431, 439, 441–447, 452–453, 457, 462–463, 480–481, 492, 499, 514–516, 518, 521–522, 527–528, 532, 534, 540, 543, 545, 551, 557, 567–568
Scotland, 14–15, 24–25, 176, 180, 231, 236, 241, 247–248, 276, 278, 294, 306, 309–310, 320, 328–329, 377, 384–385, 421–422, 436, 469, 474–475, 477–478, 488–491, 493–494, 552, 573
SCPs, spheroidal carbonaceous particles, 373, 380, 382, 384–385
sea-level tendency, 8, 14–16, 20, 22, 38, 47, 70, 125, 417, 457
seismic, 5, 10, 24–25, 39, 42, 44–46, 82, 117, 127, 139, 155, 188, 242, 247, 279, 282, 306, 423, 427–428, 516, 519, 531–533; see also earthquake
sensitivity, 107, 110–113, 118, 169, 191, 201, 219, 253, 262, 307, 319, 345, 362, 376, 406, 421, 427, 429–430, 432, 445, 476, 490, 504, 552, 565
shore platform, 30, 42, 66–67, 69, 72–75, 78, 80–82, 269–273
sinkhole, 83–87, 89, 91, 93–95, 97–99, 101, 103, 346
SLI, SLIP, sea-level index point, 8, 83, 91–92, 96, 99, 202, 319–320, 323, 325, 452–455, 457, 464, 466–467, 500–501, 503, 505–510, 514–519, 523–525, 527–532, 536–538, 540–541, 544–546, 551
Spain, 100, 103, 120, 253, 266, 279, 345, 370, 374, 513, 516, 573
speleothem, 83–86, 88–94, 96, 98–103, 111, 123, 271, 273–275, 277–278, 386–388, 390, 393–394, 401
stalactite, 88, 90–91, 99, 278
stalagmite, 86, 88, 90–92, 99, 387, 391, 394

storm surge, storm, storminess, 4, 9, 23, 42, 76–80, 84, 98, 127–128, 141, 151, 189, 200, 242, 258–259, 262–266, 272, 289, 329, 332–333, 346, 371, 411–412, 416, 450, 469, 515, 518–522, 527, 534, 552–553, 557–558, 561, 566–567, 574
subduction, subduction zone, 4–5, 24, 35–38, 42, 44–46, 54, 78–80, 82, 145, 242, 247, 306, 432, 435, 498, 513
submergence, 4–6, 35, 37, 48–49, 66, 71, 75, 81–82, 84, 88–89, 92, 94, 97, 99, 101–102, 111, 115, 117, 123, 147–148, 151, 154, 178, 188–190, 231, 236, 249, 252, 274, 277–278, 320, 328–329, 386–387, 397, 401, 518, 553, 572
subsidence, 18, 21–22, 24–25, 35, 37–38, 42–47, 81–82, 115, 126, 136, 139, 155, 161, 167–168, 215, 242, 247, 282, 306–309, 319–320, 323, 328–329, 424, 436, 468–469, 473, 481–482, 493, 497–498, 515, 527, 534, 553, 569
substrate, 48, 107, 117, 130, 132, 141, 186, 196, 199, 205, 233, 243, 249, 252, 268–269, 273–275, 302, 320, 322, 457, 460, 519, 524, 538–540
subtidal, 148–149, 151–152, 178, 199–200, 213–217, 249, 268–269, 301, 305, 412, 471, 518, 520, 527
Sumatra, 35, 37, 46, 70–71, 81–82, 128–130, 132, 136, 139–142, 145, 306, 308
Sunda / Sunda Shelf, 45, 70, 81, 115, 119, 122, 144–145, 306, 433
supratidal, 148, 151, 249, 268–269, 289, 291, 301, 305, 320, 477–478, 515, 518, 520, 522, 527–528, 530
Sweden, 24, 67, 189–190, 359, 421–422, 568–569
SWLI, standardised water level index, 195, 204, 298–299, 465, 472, 475, 477–483, 488–491, 493–497

Tahiti, 96, 99, 115–123
taphonomy, 190, 200, 205, 207, 214, 216, 227, 247, 251, 258–266, 311
Tasmania, 24, 379, 383, 502, 507–510, 513, 565
taxonomy, 120, 183, 193, 207–208, 216, 221, 230, 243–246, 249–250, 256, 260, 282, 286–287, 294, 392, 472
tectonic, 7, 18, 23–25, 30, 35–39, 45, 68–70, 72, 75, 78–81, 88, 92, 104, 113–114, 117, 120–122, 126, 138–139, 143, 149, 152, 154, 164–165, 167–169, 254–255, 266, 278–279, 306, 402, 409, 412, 425, 430–431, 448, 532
temperature, 69, 71, 104–107, 110–111, 120–121, 123, 128, 142, 150, 152, 162–163, 191, 196, 199, 215–217, 233–234, 246, 249, 251, 297, 301–302, 304, 306, 308–309, 313–318, 320, 352, 380, 382, 402, 427–428, 432–433, 437, 445, 449, 451, 470, 559, 561, 567–568
tendency, 8, 14–16, 20, 22, 38, 47, 70, 184, 261, 302, 342, 374, 395, 417, 457, 490
tephra / tephrochronology, 58, 380–385, 500, 503, 505–506
terrace, 36, 39, 46, 66–70, 72, 78, 80–82, 88, 105, 107–108, 114–115, 120–124, 258, 276, 386, 389, 398, 403, 411, 530–531
testate amoebae, 84, 97, 102–103, 281–289, 291–294, 320, 498
tidal; see intertidal; macrotidal; mesotidal; microtidal; paleotidal; subtidal; supratidal
tidal amplification, 178, 441, 447, 449–450, 531, 552
tidal marsh, 15, 49, 54, 229, 247, 251–252, 294, 308, 310, 328, 408, 412, 470, 473, 497, 518, 520, 530–531, 551; see also saltmarsh
tide gauge, 1, 9, 12–13, 33–39, 41, 45, 129, 157–170, 174–179, 291–292, 431, 467, 472, 476, 495, 544–550, 557–574
transfer function, 13, 16, 20, 191, 201, 205, 214, 216–217, 227–228, 238, 240–244, 247–248, 252, 256, 284, 286, 288–294, 298–299, 306, 320–327, 412, 470–499, 513
transgression, 44, 47, 90, 101, 155, 189–190, 236, 244, 254, 265, 328–329, 415, 423, 479, 515, 519–520, 528–529, 532–533
transgressive / transgressive contact, 20, 33, 48, 79–80, 236, 239, 255–256, 258–259, 262, 264–265, 291, 319, 323–324, 328–329, 453, 455–457, 515, 517, 519, 524–525
Tröels-Smith scheme, 50, 55–58, 60–64, 313, 329
tropics, 71–72, 101–102, 120, 138, 145, 175, 179, 200, 205, 247–248, 252, 268, 275, 277–279, 281, 295, 306, 310, 354, 411, 416–417, 476, 480, 498, 567–568
tsunami, 9, 29, 31, 34–35, 39, 42–46, 65, 172, 180, 190, 242–243, 246–247, 258–259, 262, 264, 266, 306, 310, 345–346, 482–483, 493, 498, 557, 561, 564, 572–573

UK, 1, 3, 14, 19, 24–25, 59, 64, 82–83, 146, 170–171, 175–177, 180, 183, 193, 196, 207–209, 212–215, 231, 236, 240, 247–248, 254, 281–282, 288, 290, 293–295, 305–306, 309, 311–312, 327–329, 346, 370, 373, 375–377, 379–380, 383–385, 404, 412, 415, 417, 436, 438, 450, 452, 464, 469, 474, 498–500, 515, 536, 541, 552–553, 557, 559–560, 573
uncertainties, 8, 13, 15, 17, 23, 44, 70, 76, 78, 85, 88, 99, 105, 111–112, 117, 131, 134, 149–150, 199–200, 206, 215, 243, 266, 277, 305–306, 350–353, 373, 376, 382, 387, 396, 400, 407–408, 413, 427–428, 430–432, 434, 445, 448, 457, 467, 471, 482, 487–488, 491, 494, 500–510, 519, 529–530, 536, 541, 543, 548, 571
unconformity, 47, 58, 95, 98, 135, 401
underestimate, 18–19, 23, 30, 318, 326, 339, 345, 430, 452–453, 459, 505, 509–510, 544, 568
uplift, 7, 10, 15, 17–18, 20–21, 30, 35–37, 39, 42, 44–47, 66–72, 74–75, 78–82, 87–88, 101–102, 105, 107–108, 113–115, 120, 123–124, 126, 128–130, 132, 138, 143, 147, 149, 152, 154–156, 167, 242, 247, 272, 278–280, 282, 386, 412, 421, 425, 437, 498, 513, 568
Uranium, 101–102, 104–105, 111–113, 118, 120–121, 123–124, 138, 370, 386–391, 393, 395, 397, 399, 401–404, 406, 415
USA, 1, 25, 29–35, 37, 41, 44–47, 54, 59–60, 63–64, 66, 73, 79, 83, 100, 103, 157, 165, 174–175, 190, 194–195, 197, 208, 214–216, 218, 224, 228, 230–232, 234–235, 243, 246–247, 249–250, 253–255, 258, 265, 268, 273, 279, 290, 294–295, 305–306, 309–310, 327–329, 346, 349, 353–356, 359–361, 365, 375–377, 385–386, 409, 411, 416–417, 438, 447, 449–451, 454, 468, 470, 474, 477–478, 480–482, 490–491, 493–494, 497–499, 513, 515–516, 536–538, 541, 544–545, 549–553, 557, 559–561, 573

validation, 14–15, 31, 325, 345, 370, 450, 459–460, 464, 467, 470, 472, 478–480, 482–483, 487–488, 490–491, 494–496, 561, 569, 572
Vanuatu, 70, 82, 115, 120, 145, 402
vegetation, 17, 20, 24, 35–36, 39, 94, 183, 185–187, 189–190, 192, 200–201, 205, 218–219, 226–228, 230–231, 252, 268, 288, 294–295, 297–298, 300, 302–303, 305–308, 311–312, 319, 321, 327, 329, 359, 379, 458–459, 464, 474, 518, 541, 547, 549, 551
volcanic / volcano, 38–39, 115, 117, 122, 168, 233, 373, 377, 380–385, 503

Wales, 82, 293
Washington, 24, 37, 39, 44–46, 51, 63–64, 80, 102, 189–190, 214, 243, 246–247, 357, 371, 402, 449, 451
wetlands, 24, 49, 59, 64, 156, 188, 216, 220, 224–225, 230–232, 236, 247–248, 281, 294–297, 299, 301–303, 305–309, 311–312, 320, 323, 325–328, 361, 364, 468–469, 519–520, 528–530

XRD, X-ray diffraction, 112, 118, 390, 393
XRF, X-ray flourescence, 376, 381, 385

zonation, 108–110, 147, 150–151, 185, 187–188, 192, 194, 199, 202–204, 206, 214, 216, 228, 268–269, 279–280, 282–284, 286, 289, 291–293, 320–321, 326, 455, 498, 546

www.ingramcontent.com/pod-product-compliance
Ingram Content Group UK Ltd.
Pitfield, Milton Keynes, MK11 3LW, UK
UKHW051029041026
15470UKWH00015B/1791

9 781118 452585